Manufacturing Processes & Materials

Manufacturing Processes & Materials

Fourth Edition

George F. Schrader
Ahmad K. Elshennawy

Society of Manufacturing Engineers
Dearborn, Michigan

Based on *Manufacturing Processes and Materials for Engineers* previously published by Prentice-Hall, Inc.

98765432

Library of Congress Catalog Card Number: 00-131494
International Standard Book Number: 0-87263-517-1

Additional copies may be obtained by contacting:

Society of Manufacturing Engineers
Customer Service
One SME Drive, P.O. Box 930
Dearborn, Michigan 48121
1-800-733-4763
www.sme.org

SME staff who participated in producing this book:

Millicent Treloar, Senior Editor
Rosemary Csizmadia, Production Supervisor
Frances Kania, Production Assistant
Kathye Quirk, Production Assistant/Cover Design

Cover photo courtesy of Precision Balancing & Analyzing Co., Machine Tool Spindle Repairs & New Spindles

Printed in the United States of America

About the Authors

Co-authors George F. Schrader and Ahmad K. Elshennawy have the wealth of practical experience and technical knowledge of manufacturing processes necessary for the compilation of a comprehensive text on the subject.

Dr. Schrader, currently Professor of Engineering, Emeritus, at the University of Central Florida in Orlando, has an educational background in mechanical engineering, applied mathematics, and industrial engineering. Since 1945 he has taught a variety of courses and coordinated laboratory exercises on the subject of manufacturing at several major universities in the U.S. In addition to his academic work, he has served as a consultant to many manufacturing industries and worked with a number of technical societies on industry-related activities.

Dr. Elshennawy, currently an Associate Professor in the Department of Industrial Engineering and Management Systems at the University of Central Florida, has an educational background in production engineering and industrial engineering. Since 1978 he has taught many courses on manufacturing and manufacturing-related subjects at Alexandria University in Egypt, Pennsylvania State University, and the University of Central Florida. Dr. Elshennawy's greatest expertise is in the area of precision measurement and manufacturing automation, subject areas in which he acquired considerable experience while working at the National Institute of Standards and Technology (NIST) in Gaithersburg, Maryland.

Table of Contents

4 IRON AND STEEL

5 NONFERROUS METALS AND ALLOYS

6 ENHANCING MATERIAL PROPERTIES

11 HOT AND COLD WORKING OF METALS

12 METAL SHEARING AND FORMING

13 WELDING PROCESSES

14 OTHER CUTTING AND JOINING PROCESSES

Preface

PURPOSE OF THIS TEXT

Manufacturing involves a complex system of people, machines, materials, and money organized to produce a product. There are a number of components to every manufacturing organization, each of which requires people with different education, training, and experience with different levels of skills. The technical departments within such an organization, for example product design, production engineering, manufacturing engineering, industrial engineering, tool engineering, quality assurance, and the production function itself, all require technical personnel with an appropriate degree of knowledge of the manufacturing process. This text is dedicated to providing the reader with an understanding of the basic processes and equipment used in manufacturing so that he or she might work more productively within those technical areas of manufacturing.

Since the scope of manufacturing is extremely broad, a single textbook cannot expect to address the whole spectrum of machines and processes that might be applicable to such a diverse field. Instead, different textbooks tend to limit their scope to those areas of manufacturing wherein the authors' interest and proficiency are greatest. In this text, the scope of coverage is more or less limited to the basic machines and processes used in the forming, machining, and fabricating of products and parts made of metallic and nonmetallic materials. It should be noted that the price/cost figures cited within this book are in 1999 dollars.

TEXT BACKGROUND

Much of the coverage of the basic manufacturing processes stems from the work of Professor Lawrence E. Doyle, Professor of Mechanical Engineering at the University of Illinois-Champaign/Urbana. Professor Doyle, with the assistance of contributing authors C. A. Keyser, J. L. Leach, J. L. Morris, G. F. Schrader, and M. B. Singer, prepared three successive editions of *Manufacturing Processes and Materials for Engineers* (Prentice-Hall, Inc., 1961, 1969, and 1985). As popular textbooks on manufacturing, these writings by Professor Doyle and his colleagues have contributed immensely to the education of literally thousands of students of the manufacturing industries for nearly half a century.

In addition to the background provided by Professor Doyle and his colleagues, recognition must be given to Dr. Vimal H. Desai, Associate Professor of Engineering at the University of Central Florida, for his consultative input concerning the formulation and treatment of engineering materials in this text.

This new edition is an extensive revision reflecting the major changes in the manufacturing industry that have taken place since 1985.

SCOPE OF COVERAGE

The basic processes of manufacturing have not changed significantly since the industrial revolution. For example, metals are still cast in sand molds, formed metal parts are still stamped on punch presses, cylindrical parts are turned on lathe-like turning machines, and surfaces are ground with abrasive wheels and stones. However, the supporting technologies, such as machines, cutting tools, controls, and measuring instruments for these processes, have made tremendous advances. This has permitted manufacturing companies to improve the efficiency and effectiveness of operations and the quality and reliability of the products produced.

This edition focuses on the basic machines and tools applicable to the job shop, toolroom, or small volume manufacturing facility. At the same time, it will expose the reader to some of the more advanced equipment used in larger volume production environments.

USE AND APPLICATION

Manufacturing Processes and Materials has been designed for use at several levels of the informal and formal educational process. It can be used as an introductory text for in-plant training of manufacturing personnel. Or, at the other extreme, it can be used as an advanced text at the college or university level where it will provide a comprehensive manufacturing educational background for technical students in a variety of disciplines. Because of the breadth of coverage, it is recommended for a two-semester or two-quarter sequence in conjunction with a manufacturing laboratory. In addition, the text will be useful as a reference for technical students and manufacturing personnel.

ORGANIZATION OF THE TEXT

Chapter 1 introduces the reader to traditional manufacturing. It is a must read for students who have not been exposed to a manufacturing environment or who may not have any knowledge or appreciation for the complexities of that environment. Chapter 2 describes many of the challenges that manufacturing establishments must face if they expect to remain competitive in a global environment.

The next five chapters are concerned with engineering materials, their physical properties, testing, treatment, and suitability for use in manufacturing. These chapters should be required reading for students with little or no preparation in these subject areas.

The chapters concerned with the machines, tools, and processes of manufacturing are arranged in accordance with the traditional hierarchy for conversion of raw materials into a finished product via a variety of casting, forming, joining, and machining processes.

Chapter 15 introduces the concept of quality assurance in the manufacturing environment. A number of the basic statistical process control (SPC) approaches and techniques are presented to introduce the reader to methods used to recognize and control excessive variability of manufacturing processes. Prior exposure to the basic concepts of probability and statistics would help the reader to better understand the SPC techniques presented in this chapter. Chapter 16 follows with a rather extensive treatment of measuring and gaging instruments used for assessing conformance to quality specifications.

Chapter 29 introduces the reader to many of the systems concepts currently used in manufacturing practice. In addition, Chapter 31 covers computer numerical control (CNC). These chapters are a must for those students who expect to take more advanced courses on manufacturing systems and controls. It will also familiarize a variety of technical manufacturing personnel with practices they will encounter in the most progressive manufacturing environments.

Production planners and manufacturing engineers will agree that the manufacturing planning process is filled with choices. With the current emphasis on just-in-time and lean/agile manufacturing, making the right choice the first time is critical to the competitive status of companies. Thus, it is important that personnel involved in planning are knowledgeable about the alternative processes available, the capabilities of those processes, and the economic advantage of one process over another. For example, as explained in Chapter 21 on milling, there are probably 40 or 50 different operations that can be performed on the versatile milling machine and its newer likeness, the machining center. These operations range from drilling a hole to cutting a keyway. Each of these operations can be done on any one of a dozen or more types of machines ranging from a simple column-and-knee type, manually operated mill to a very sophisticated and expensive multi-axis, multi-spindle CNC machining center. In addition, a variety of types and sizes of cutting tools are available to do each operation. Sometimes the choices are clear and simple, and at other times, complex.

Chapter 19 introduces the reader to the planning process and to a number of economic methods for comparing alternatives. In addition, many of the other chapters include materials on process planning and economic analysis with reference to a particular set of processes or machines. The importance of planning in any manufacturing environment must be emphasized if the results are to be cost-effective, on-time, and on-quality.

Acknowledgements

The authors wish to acknowledge the recommendations and suggestions contributed by reviewers of the manuscript for this text. Those reviews often provided a valuable perspective of subject areas in the text that had not been fully developed by the authors.

In addition, an expression of gratitude is extended to the many machine tool manufacturers, dealers, trade associations, and technical societies who contributed photographs, technical data, and other information for use in this publication. Without their cooperation and assistance, it would have been impossible to assemble the depth and breadth of illustrative detail provided in this book.

Last, but certainly not least, the authors are deeply grateful and appreciative of the tremendous assistance provided the Book Publishing staff at the Society of Manufacturing Engineers. Textbook authors could not hope for a more enthusiastic, knowledgeable, and cooperative editor than Millicent Treloar. Rosemary Csizmadia has done an exemplary job of copy editing and managing the myriad production details. Thanks are due to Kathye Quirk for working miracles with the illustrations, designing the cover, and for the clean, easy-to-follow layout of the text.

Manufacturing Foundations

1

A tool is but the extension of a man's hand and a machine is but a complex tool.— Henry Ward Beecher

1.1 MANUFACTURING

Manufacture means to make goods and wares by industrial processes. The derivation of the word *manufacture* reflects its original meaning: to make by hand. Today, however, manufacturing is done largely by machinery and, as the technology of manufacturing advances, less and less manual labor is involved in the making of a product. In fact, most manufacturing firms in the U. S. strive to minimize the labor cost component of their products to remain competitive. Thus, machinery, vis-a-vis technology, has and will probably continue to replace the human labor element in manufacturing much the same as it has done in the U. S. agricultural industry. In a contemporary sense then, manufacturing involves the assembling of a system of people, money, materials, and machinery for the purpose of building a product.

1.2 HAND TOOLS TO MACHINE TOOLS

1.2.1 Early Hand Tools

Tools of one kind or another have enabled mankind to survive and contribute to societal development for over a million years. If the original meaning of manufacture, "to make by hand" is applicable, then manufacturing in some form has existed over that time. Early prehistoric mankind learned to retain certain skeletal remains of animals, such as horns, tusks and jaw bones, and fashion them into hand tools for use in hunting and preparation of food. Later on, as the evolution of "tool making" progressed, an even greater variety of tools were made from stone and wood. During this period, flint stone was recognized as a very hard material and became a common substance for use in fashioning spears, axes, arrowheads, and even crude saws and drills. The *Bronze Age*, beginning about 6,500 years ago, ushered in the use of metal as a primary element in the construction of hand tools. For the most part, these tools were still relatively primitive, with the bronze metal being used primarily to replace the stone axe heads, spear heads, and hammer heads that were popular during the Stone Age. However, the Bronze Age did see some very slight transformations of hand tools to what might be called semimachine status. For example, the bow drill, which used a bow string to rotate a bronze drill, provided some mechanical advantage to the rotational process.

1.2.2 The Iron Age

The *Iron Age*, beginning about 3,400 years ago, gave birth to a broad spectrum of new hand tools for many different trades and a refinement of the tools from previous periods. Early in this period, hand tools were hammered out of meteoritic iron removed from meteorites that were embedded in the earth. However, the use of large quantities of iron and steel for tools and other implements did not take place until after the invention of the blast furnace in Europe at around 1340 A.D.

The installation of an operating blast furnace in the U. S. in 1621 facilitated increased production of a large variety of hand tools, semimachines, horse-drawn vehicles, agricultural implements, and so on. The machines and vehicles during that period were powered or driven or propelled by water, animal, or human energy. A variety of devices were employed, such as water wheels, treadmills, windlasses, horse-drawn whims, and the like. In addition, many creative devices were used to obtain a significant amount of mechanical advantage. For example, the development of a fitted horse collar to replace the traditionally used yoke made it possible for draught animals to increase their pulling power nearly fourfold. Many machines were operated by foot treadles, and in the early 1700s, a simple windlass was used to pull a rifling broach through the barrel of a rifle. Finally, in 1775, John Wilkinson developed a water wheel-powered horizontal boring mill in Bersham, England that permitted James Watt and Matthew Boulton to bore a true hole in the cylinder of their steam engine. Thus, the age of the engine-powered industrial revolution was born.

1.2.3 Industrial Revolution

With power available to drive them, hand tools were rapidly converted into machine tools, and thus the *industrial revolution* began in Europe and the United States. The boring machine developed by John Wilkinson in 1775 led to the development of the first engine lathe in 1794 by Henry Maudsley. A few years later, he added a lead screw and change gears to that lathe, thus giving birth to the screw cutting lathe. The need for further versatility in machine tools then inspired the invention of the planer in 1817 by Richard Roberts of Manchester, England and the horizontal milling machine in 1818 by Eli Whitney of New Haven, Connecticut. Those three machines, the lathe, planer, and mill, not only provided a basis for producing a large variety of products, but also enabled the entrepreneurs of that era to build additional similar machines that could be used to produce other products. During the late 1700s and early 1800s, most manufacturing was performed in family workshops and small factories. The availability of power to drive machine tools was, to a great extent, a controlling factor in the movement and expansion of the industrial revolution. As is evident from the timetable in Table 1-1, the steam engine was the most significant source of power for the machines of production for more than 50 years. In the early periods, a centralized engine was used to drive line-shafts which, in turn, provided power to many individual machines. Later on, as steam engines became more compact and efficient, smaller engines were placed in strategic positions around a factory to drive machine groups.

Probably one of the most significant developments occurring during the early stages of the industrial revolution was the introduction of the concept of *interchangeable manufacture*. (Interchangeable manufacture means that the parts for one particular product will fit any other product of that same model.) This idea apparently manifested itself almost simultaneously in Europe and the United States in the late 1700s via the use of templates or patterns, often referred to as filing jigs. Eli Whitney was one of the early pioneers to take advantage of this concept in the building of musket parts for the U. S. military in about 1798. Although the concept of interchangeable manufacture is usually credited to Eli Whitney, it should be pointed out that the accomplishment of this process through the use of filing jigs was mostly a manual operation, not a machine process. The credit for machine-produced interchangeable manufacture should probably go to Elisha Root, who was the chief engineer for the Colt Armory in Hartford, Connecticut. In about 1835, Root and Samuel Colt engineered the machine production of over 300,000 units of different models of the Colt revolver to a significant degree of precision. This accomplishment is often heralded as a milestone in the development of the concept of interchangeable manufacture and *mass production* in the U. S.

Table 1-1. Manufacturing process and machine tool design timetable

2000
- Coated cutting tools (1974)
- Numerically controlled jig boring machine (1974)
- Wire electric discharge machining (1969)
- Flexible manufacturing (1969)
- Silicon chip (1969)
- Programmable logic controller (1968)
- Industrial robots (1963)
- NC machining center (1958)
- Transistor (1957)
- Synthetic diamond (1955)
- Numerically controlled vertical milling machine (1953)
- Stored program digital computer (1951)

1950
- Electronic digital computer (1946)
- Electrical discharge machining (1943)
- Tungsten-carbide cutting tool (1926)
- Stainless steel (1913)

1900
- Generating-type gear shaper (1899)
- High-speed cutting tools (1898)
- Aluminum oxide (1893)
- Silicon carbide abrasive (1891)
- Gear hobbing machine (1887)
- Band saw blades (1885)
- Hydraulic forging press (1885)
- Electric motors (1885)
- Surface grinder (1880)
- Board drop hammer (1880)

1900 (cont.)
- Automatic turret lathe (1873)
- Four-stroke gas engine (1873)
- Universal grinding machine (1868)
- Dynamo electric generator (1867)
- Open-hearth steelmaking (1866)
- Tool steel cutting tools (1865)
- Water-cooled gas engine (1860)
- Turret lathe (1855)
- Milling-type gear cutter (1855)
- Two-stroke gas engine (1855)

1850
- Drill press (1840)
- Gravity drop hammer (1839)
- Mass production (1835)
- Gas engine (1833)
- Precision measuring screw (1830)
- Gage blocks (1830)
- Reproducing lathe (1820)
- Horizontal milling machine (1818)
- Planer (1817)
- Thread-cutting lathe (1800)
- Electroplating (1800)

1800
- Interchangeable manufacture (1798)
- Engine lathe (1794)
- Double-acting steam engine (1787)
- Steam-powered coining press (1786)
- Horizontal boring mill (1775)
- Atmospheric steam engine (1775)

Another significant milestone in the industrialization process was the development of precision measuring devices in about 1830 by Joseph Whitworth. As a protégé of Henry Maudsley, Whitworth pioneered early screw thread designs and then incorporated that work into the development of the micrometer screw. The ability to measure was, of course, a fundamental prerequisite to a successful interchangeable manufacturing process.

The spectrum of manufacturing capability was further enhanced in about 1840 by the development of a drill press with power feed by John Nasmyth, also a student of Henry Maudsley. About 15 years later, mass production capability in the U. S. was greatly improved by the introduction of the turret lathe by Elisha Root and Samuel Colt. Forty years or so later, the development of the surface grinding machine and the metal saw blade completed the stable of machine tools available to the early manufacturer. Thus, during the late 1800s and early 1900s, these basic machine tools: the boring mill, lathe and turret lathe, milling machine, broach, planer, shaper, surface grinder, and saw, served as the workhorses for the ever expanding industrial capacity in Europe and the United States.

1.2.4 Automation

As indicated in Table 1-1, a large proportion of the basic machine tools used in discrete parts manufacture were introduced prior to 1900. These machines and the engine power required to drive them were key elements in the industrial revolution. In the early days of that period, the machines were essentially manually operated with the quality and quantity of product output being almost totally dependent on the skill and ingenuity of the craftsmen who operated them. Recognizing the difficulties inherent in a skill-dependent production system, the machine tool builders gradually improved the operational features of their machines to lessen the level of skill required to operate them. In essence, they were gradually *automating* their operation, while at the same time improving precision, reliability, and speed.

Although not recognized as such, one of the pioneering efforts in the automation movement was made by an Englishman, Thomas Blanchard, who developed a reproducing lathe for wood turning in about 1820. Blanchard's lathe was used to turn and form the intricate shape of a wooden rifle butt. Replacing manual carving by woodworking craftsmen, Blanchard's early design of a reproducing lathe was able to produce two rifle butts in an hour. Later improvements enabled him to increase production to as many as a dozen an hour.

The conversion to automatic machine tool operation on metal products was spearheaded in 1873 by an American, Christopher Spencer, one of the founders of the Hartford Machine Screw Company. Spencer's so-called "automat" was essentially a turret lathe equipped with a camshaft and a set cams that moved levers which, in turn, changed the turret position and fed the tools forward. As the forerunner of the automatic screw machine, Spencer's machine was extremely well received by industry and used extensively for producing screws, nuts, and other small parts in large quantities. In a sense, Spencer's automat was reprogrammable by simply changing to a different set of cams.

The evolution of machine tool automation continued during the early 1900s largely through technical improvements to the concepts introduced by Spencer's automat. Electrical, pneumatic, and hydraulic servomotors were added to effect tool and workpiece position changes but, for the most part, these were still automated by various types of cams to carry out a specified program of cutting operations. The introduction of high-speed steel cutting tool materials in 1898 by two Americans, Frederick W. Taylor and Mansel White, permitted the use of higher cutting speeds on these automatic machines, thus increasing production rates. Since higher cutting speeds increased the rate of metal removal, increased horsepower for spindle motors was required. In addition, higher cutting forces required machines of greater strength and rigidity. Similarly, modifications to machine tool design were required by the introduction of tungsten carbide and other hard metal-cutting tool materials (Chapter 17) in about 1926.

Although technical improvements on the automatic and semi-automatic machines of production during the early 1900s were significant, they were, to some extent, lacking in the high degree of flexibility and precision required in the highly competitive and ever-changing world marketplace that evolved after World War ll. This weakness was mitigated to a great extent by the introduction of numerical control (NC) technology (Chapter 31) in 1952 by the Massachusetts Institute of Technology and its adaptation for mass-produced milling machines by Giddings & Lewis in 1955. Numerical control technology was followed by the development of the programmable logic controller (PLC) in 1968. With the development of microcomputer technology in the late 1970s and early 1980s, most NC controllers have been built around that technology. Thus, modern machine tools are referred to as computer numerically controlled (CNC) machines. Computer numerical control of the machines of production provides the basis for accomplishing a multiplicity of operations and operational flexibility in manufacturing that was not possible with its predecessor machines.

Another element of the manufacturing automation scenario, the industrial robot (Chapter 29), was developed in the U. S. and first appeared in the marketplace in 1963. Generally, a robot consisted of an extended arm with a gripping mechanism, a power unit, and a control unit. In theory, the robot was designed to emulate the action of the human arm and hand in reaching for, grasping, and transferring a part from one location to another. Thus, the early robots, with limited degrees of freedom, were designated as "pick and place" devices to be used to load and/or unload parts into or from machines. Now, programmable robots with many degrees of freedom and precise movements are used in a variety of manufacturing situations to complement the automation process.

The ultimate concept and scenario in manufacturing for many manufacturing engineers and executives is to achieve a completely automated manufacturing system (automated materials handling, machining, and assembly) to permit the operation of a "hands off/lights out" factory. Although feasible for some types of manufacturing situations, this concept has yet to be demonstrated on a large scale. Needless to say, progress in automated manufacturing has been spectacular since Spencer's "automat" in 1873, and it can be predicted that further progress will be made in the next century.

1.3 TYPES OF PRODUCTS

For statistical purposes, the U. S. Department of Commerce groups manufacturing establishments into 20 commodity categories according to the two-digit standard industrial classification (SIC) code shown in Table 1-2. These 20 major groups include establishments that are primarily engaged in the mechanical or chemical transformation of materials or substances into new products. These establishments are usually referred to as plants, factories, mills, or shops and they characteristically use power-driven machines and material handling equipment applicable to the type of manufacturing involved.

The 20 major groups of manufacturing establishments given in Table 1-2 are further subclassified into 139 three-digit industry groups, and then into 459 four-digit industries. The four-digit industries are then further subclassified into five-digit product classes and a seven-digit product line classification scheme. For example, the hierarchy of "Miscellaneous Fabricated Metal Products" (SIC 349) stemming from the major classification number 34, "Fabricated Metal Products," is given in Table 1-3.

Every 5 years for years ending in 2 or 7, the U. S. Bureau of Census conducts a census of manufacturing establishments to obtain information on that industry sector. This information, which is available through the U. S. Government Printing Office, is very useful to the government in determining national economic conditions and to the individual manufacturing establishment for comparative purposes. In 1992, all manufacturing establishments employing one or more persons at any time during the census year were included in the census. During that year, the manufacturing universe included approximately 380,000 establishments, of which 237,000 were selected for inclusion in the census report. (The North American Industry Classification System [NAICS] is replacing the SIC system. The Census Bureau's reports will be converted to NAICS beginning with those published in 2000. An outgrowth of SIC, NAICS is intended to recognize new industries, provide better international comparability, and permit more consistency in grouping industries.)

To illustrate the economic contribution of just one industry class, the 1992 census lists employment in the industry class 3491 (industrial valves) as 51,400, which was up some 12% from 1987. The report also indicated that the total value of shipments from the 493 firms included in that class was $6.8 billion and that the average hourly wage of production workers was $12.51. These and other statistics in the 1992 report exemplify the tremendous contribution the manufacturing sector makes to the economic health and well-being of this country.

1.4 ORGANIZATION FOR MANUFACTURING

1.4.1 Types of Manufacturing Systems

In general, the design of a manufacturing organization is dependent to a great extent on the type of manufacturing system involved. As indicated in Table 1-2, manufacturing establishments are classified into 20 product categories. Some of these categories represent process-type manufacturing systems, while others represent discrete parts or fabricating systems. Process types of establishments generally manufacture

Table 1-2. Standard industrial classification (SIC) of manufacturing establishments

Primary SIC Code	Product Group	All Establishments		All Employees Number (1,000)	Production Workers Number (1,000)	Value Added by Manufacture (million $)	Value of Shipments (million $)
		Total Number	With 20 Employees or More				
	All industries	381,870	124,927	18,253.80	11,654.10	1,428,707.40	3,006,275.20
20	Food and kindred products	20,792	9,325	1,504.80	1,100.00	156,843.40	403,836.00
21	Tobacco products	114	79	38.40	27.10	27,167.10	35,136.70
22	Textile mill products	5,887	3,203	614.80	527.00	29,862.10	70,694.20
23	Apparel and other textile products	23,048	8,564	985.60	824.40	36,357.00	71,617.00
24	Lumber and wood products	35,834	6,909	658.20	541.70	33,352.40	81,797.60
25	Furniture and fixtures	11,630	3,758	473.20	373.60	22,820.80	43,688.40
26	Paper and allied products	6,435	4,253	626.20	478.50	59,922.70	132,954.40
27	Printing and publishing	65,466	12,760	1,505.50	790.20	113,244.30	167,284.10
28	Chemicals and allied products	12,042	5,059	850.30	479.10	165,134.80	305,761.00
29	Petroleum and coal products	2,125	686	114.40	73.60	23,797.20	149,960.80
30	Rubber and miscellaneous plastic products	15,823	7,967	906.90	696.50	58,477.00	113,543.90
31	Leather and leather products	2,035	756	100.80	82.60	4,516.70	9,676.50
32	Stone, clay, and glass products	16,285	4,850	469.90	357.00	34,557.80	62,479.10
33	Primary metal industries	6,568	3,568	663.00	508.10	51,816.40	138,333.20
34	Fabricated metal products	36,357	13,522	1,369.80	999.30	83,870.80	167,015.00
35	Industrial machinery and equipment	34,973	14,103	1,742.10	1,087.70	132,143.60	258,273.10
36	Electronic and other electric equipment	16,952	7,677	1,444.30	913.50	121,949.60	217,905.70
37	Transportation equipment	11,259	4,289	1,646.40	1,079.00	161,058.40	401,213.90
38	Instruments and related products	11,346	4,282	910.10	460.90	89,805.80	135,479.20
39	Miscellaneous manufacturing industries	17,056	3,475	365.30	254.40	22,009.70	39,625.60

(Census of Manufactures 1992)

Table 1-3. Hierarchy of SIC 34

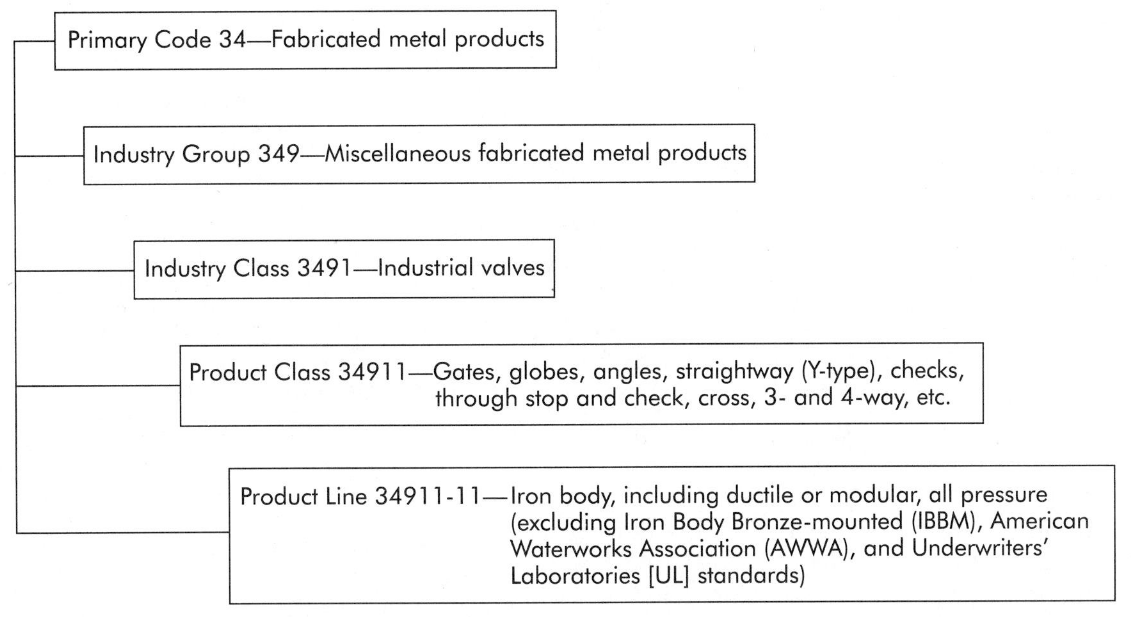
Primary Code 34—Fabricated metal products

Industry Group 349—Miscellaneous fabricated metal products

Industry Class 3491—Industrial valves

Product Class 34911—Gates, globes, angles, straightway (Y-type), checks, through stop and check, cross, 3- and 4-way, etc.

Product Line 34911-11—Iron body, including ductile or modular, all pressure (excluding Iron Body Bronze-mounted (IBBM), American Waterworks Association (AWWA), and Underwriters' Laboratories [UL] standards)

a product by means of a continuous series of operations, usually involving the conversion of a raw material. Food, chemical, and petroleum products are often produced by processes that are generally particular to each of the raw materials being converted. Thus, they are usually referred to as process industries even though discrete products, such as bottles of milk, bags of fertilizer, or containers of motor oil are the end products.

This text is primarily concerned with the discrete parts or fabricating types of manufacturing systems that make discrete items of product, such as nails, screws, wheels, tires, and paper clips, or assembled products, such as autos, televisions, and computers. A variety of manufacturing systems are employed to manufacture such products, including job shops, flow shops, project shops, continuous processes, linked-cells, and computer-integrated systems, all of which are described in Chapter 29.

1.4.2 Small Organizations

The four major ingredients of a manufacturing organization, *people, money, materials* and *machines,* must be brought together in an organized fashion to maximize their combined effectiveness and productivity. It is particularly important in a manufacturing environment that the structure and operating characteristics of the organization support, rather than impede, the process of building a quality product for a reasonable price. This is essential in a highly competitive marketplace.

The structure of a manufacturing organization depends on a number of factors, including the size of the establishment, the type of manufacturing system, and the complexity of operations involved. A simple *line organization,* as depicted in Figure 1-1, is often used when a company starts up with a small number of employees. A line organization, as the name implies, consists of a vertical line of organizational components, all representing personnel who are directly involved in producing a product or supervising those who are producing a product. This form of organization is often used in small family-owned and -operated firms in which a family member serves as president and general manager of a small number of employees. In this case, the general manager/owner handles all or most of the functions incident to the operation of the business, personnel matters, finance and bookkeeping, sales and marketing, as well as

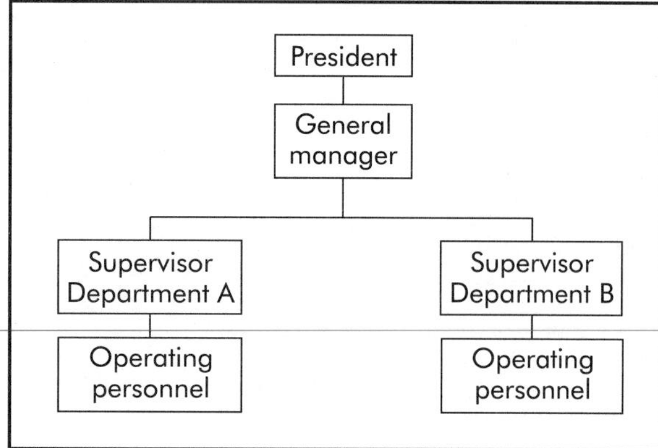

Figure 1-1. A simple line organization.

managing the production function. Quite often the line supervisors will assist the general manager in taking care of many of the technical details, such as production planning, tool design, and inspection. In many cases, certain business functions, such as payroll, accounting, and tax preparation, may be contracted out to service organizations who specialize in that kind of support service to small business establishments. In addition, many small shops may even subcontract a number of technical activities, such as product design, tool and die design, and fabrication.

According to Table 1-2, only about one-third of the manufacturing establishments counted in the 1992 census had 20 or more employees. Thus, some 256,943 establishments, representing two-thirds of the total number of manufacturing plants in the U. S., have less than 20 employees. Most of these could very well be operated via a simple line organization.

As an organization grows and the owner/manager and line supervisors find that they do not have the time or skills to handle many of the business and technical details of a manufacturing organization, specialists may be employed to take care of those activities. As this occurs, the line organization evolves into a *line and staff* type of structure. As shown in Figure 1-2, three staff specialists, a bookkeeper, chief engineer, and sales manager, have been added to the organization chart of Figure 1-1. These staff specialists provide support to the line personnel, but, in most cases, they do not have direct authority over the line operations. They report to the president and any line-related recommendations that they generate are transmitted from the president to the general manager, and on down the line. This conforms to the "one boss" principle of management that is necessary to prevent conflicting demands on operating personnel.

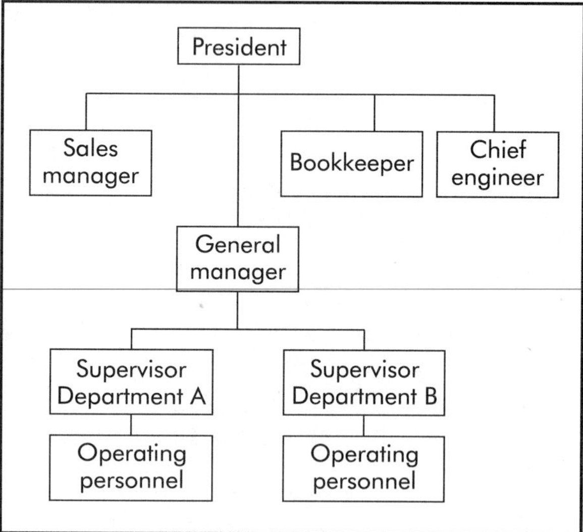

Figure 1-2. Small line and staff organization.

1.4.3 Large Organizations

The structure of a manufacturing organization generally continues to expand as the size of the organization increases. In other words, if the number of line employees is increased, then it can be expected that some increase in the number and size of the staff groups will be required. The increase in staff should not be in constant proportion to the size of the line as some economies of scale should be expected. For example, as a manufacturing firm expands its line of products or product models, it is likely that the sales or marketing staff will be increased. In time, that entity will become large enough to be a department, as will other staff groups in the organization.

The grouping of staff functions or departments and their designations varies with different organizations. Attempts have been made to classify activities as either service, advisory, coordinative, or control. However, some staff departments function within more than one and sometimes all of these categories.

The organization chart of Figure 1-3 illustrates an expanded line and staff hierarchy with seven staff groups and a superintendent of plant operations all reporting to a general manager. Some of the staff groups provide direct support to the line operations, while others have a less direct relationship. In most cases, however, the staff functions do not have direct authority over the line operations, and any recommendations that these groups make must, in theory, go through and be approved by the general manager. In practice, though, some staff groups routinely transmit information, schedules, design changes, unit costs, guidelines, etc., to different line elements for response.

Although all of the staff groups shown in Figure 1-3 interact to some degree with line operations, the three staff groups, Product Design and Test, Manufacturing Engineering, and Quality Assurance usually have a closer association. As implied by the title, the Product Design and Test group is responsible for the engineering design of new products or new models of existing products, design changes, maintaining design standards for products and components thereof, and developing and conducting feasibility and functionality tests on prototypes for these products. In modern manufacturing organizations, this group is often referred to as the Research and Development (R & D) group and it is responsible for long-range planning and research for new product development. In addition, the R & D group often conducts research on new materials for use in existing product lines and on new applications for products. The R & D group usually works closely with the Sales and Marketing staff to identify new products and to determine if product modifications may be necessary to maintain and possibly expand the firm's customer base.

The purpose of the Quality Assurance group is to provide the necessary surveillance and control of the manufacturing system to assure that product quality is consistent with customer requirements. The organization and functions of that group are described in Chapter 15.

1.4.4 Manufacturing Engineering

The planning, tooling, coordination, and control of manufacturing processes are critical to the operation of an effective and efficient

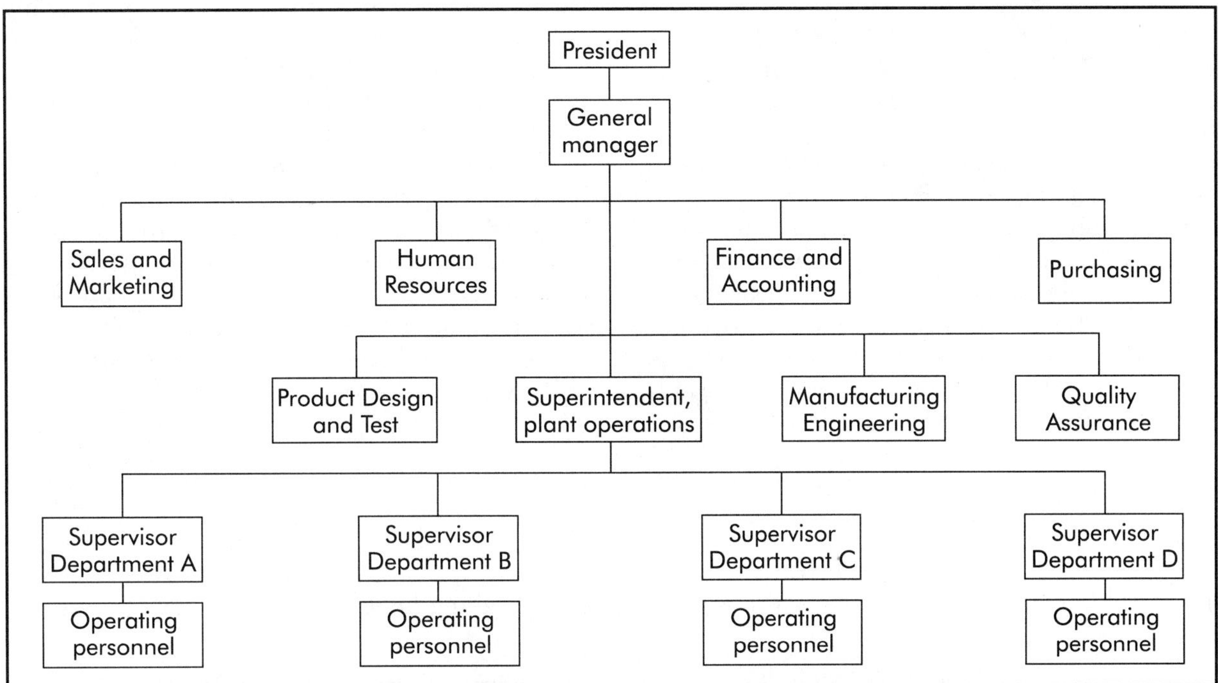

Figure 1-3. Line and staff organization for a medium-size manufacturing establishment.

manufacturing system. In fact, some manufacturing executives contend that a large proportion of the problems encountered are systems problems and are not necessarily the result of faulty machines or processes. In many large manufacturing organizations, the task of providing systems support and service to the manufacturing group is centralized in one comprehensive staff group referred to as either Manufacturing Engineering, Production Engineering, or Industrial Engineering. In other firms, many of the activities or elements of manufacturing engineering are decentralized and assigned to other staff groups or set up as stand-alone entities.

Regardless of how it is organized, the Manufacturing Engineering group or department is a staff service organization whose main role is to provide support to the manufacturing operations on production plans, processes and tools to be used, information and instructions on methods and procedures, labor standards, and assistance in solving problems. In addition, Manufacturing Engineering must work closely with Product Design and Quality Assurance to facilitate the infusion of new products and new quality standards into the manufacturing operation's product mix. It is particularly important that the manufacturing engineering group be involved in nearly every step of the product design process to assure the *manufacturability* of new products. "Manufacturability" infers that a product be designed in such a way that it can be produced in a cost-effective manner.

Depending on the extent of support activities required, the manufacturing engineering function is usually divided into several specialty areas, a number of which are shown in Figure 1-4. The Fabrication Processes group is responsible for developing production plans for the various processes involved in the manufacture of a product and its component parts. Thus, if one of those parts has to be cast, machined, and then cleaned and painted, the Fabrication Processes group will work up a set of plans encompassing the five subspecialties under that heading. Similarly, the Assembly Processes group develops plans and procedures for the various activities involved in the assembly of products and their components.

The other specialty areas shown in Figure 1-4 have definite line operations support responsibilities. The Tool Control group is responsible for providing the tools, dies, jigs, fixtures and other pieces of equipment required for both the fabrication and assembly operations. The Facilities Maintenance group provides the various utilities required to operate the manufacturing equipment, maintains the equipment, and also maintains the plant environment.

The Industrial Engineering group plays a major coordinative role in the manufacturing process through its activities in establishing work standards, setting up and balancing production schedules, and providing timely and accurate information on the status of many elements of the manufacturing system. The function of this group, often called the Manufacturing Systems group, is to assure that the manufacturing system works; that it functions smoothly and builds products on-time, on-cost, and on-quality. The coordinative role played by this group becomes increasingly important as a manufacturing organization moves from the more traditional mass-production type of operation to a more agile and flexible mass-customization type of manufacturing system. In the mass-customization environment, it is particularly important that a centralized and constantly updated computerized manufacturing information system be available to serve as the eyes and ears of the manufacturing operations.

1.5 QUESTIONS

1. Define the term "manufacturing."
2. What type of metal was used to replace stone for making hand tools during the seventh century?
3. Who is credited for developing the first machine tool and when did this occur?
4. How was the first machine tool powered?

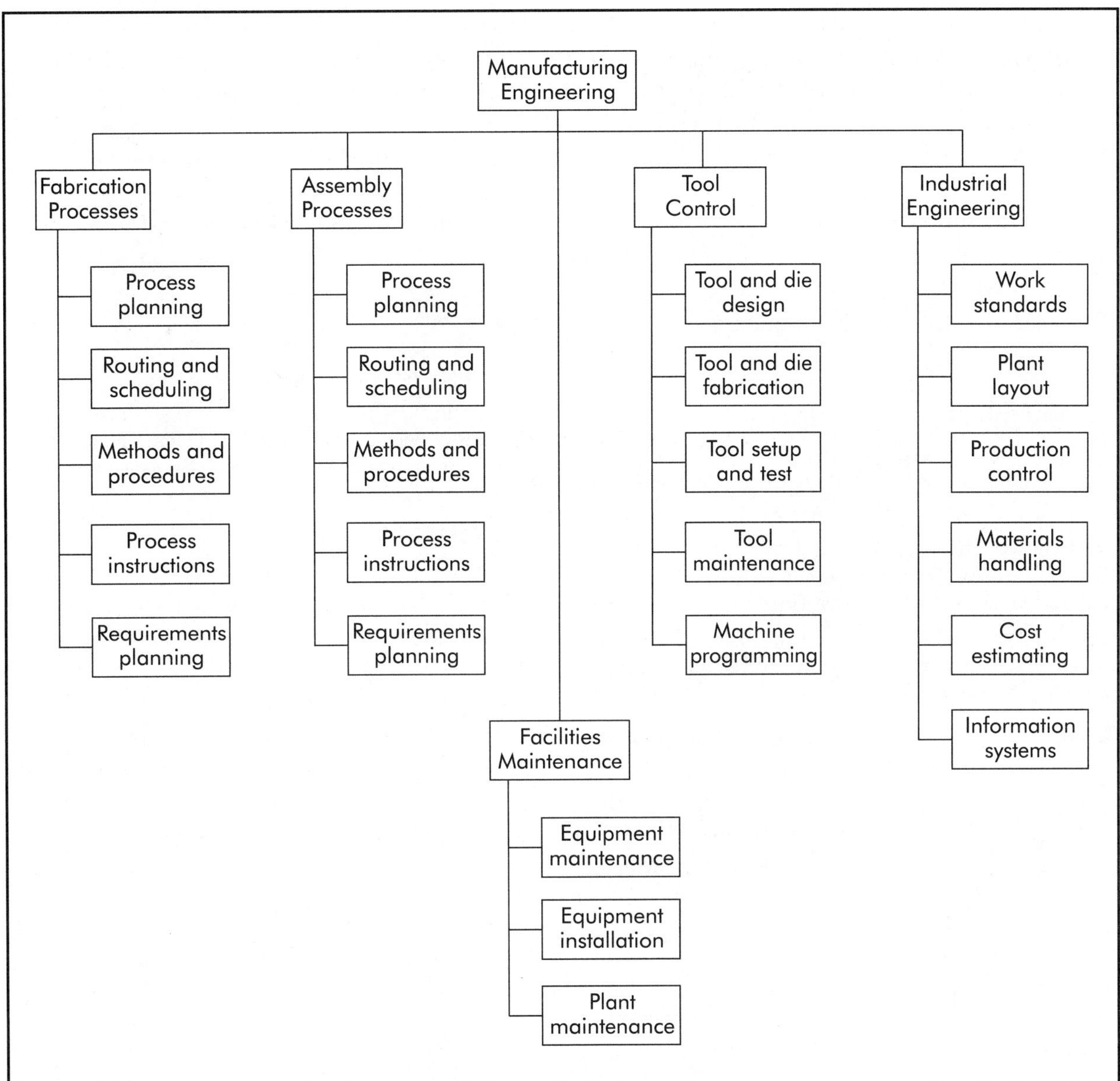

Figure 1-4. Typical areas of specialization for the Manufacturing Engineering group.

5. About when did the industrial revolution begin?
6. Define the term "interchangeable manufacture."
7. Who is credited with the early pioneering work on precision measuring devices and when was it done?
8. What are some of the advantages of automating the operation of machine tools?
9. When was high-speed steel introduced as a cutting tool material and who was responsible for its development?
10. What is the difference between numerical control (NC) and computer numerical control (CNC)?

11. Explain the hierarchy of the SIC classification for fabricated metal products.
12. Of the 20 major product groups listed in Table 1-2, which one had the largest number of establishments in 1992, which one had the greatest number of employees, and which one added the most value by manufacture?
13. What is the difference between a process-type manufacturing system and a discrete parts system?
14. Explain the difference between a line organization and a line and staff organization.
15. How are the line functions and the staff functions in a line and staff type of organization differentiated?
16. Define the term "manufacturability."

1.6 REFERENCES

Adam, E.E. and Ebert, R.J. 1989. *Production and Operations Management*. Englewood Cliffs, NJ: Prentice-Hall, Inc.

Amrine, H.T., Ritchey, J.A., and Moodie, C.L. 1987. *Manufacturing Organization and Management*. Englewood Cliffs, NJ: Prentice-Hall, Inc.

Census of Manufactures, Preliminary Report, Summary Series. 1992. MC92-SUM-1(P), Washington, D. C.: Bureau of the Census, U.S. Government Printing Office.

Cleland, D. 1996. *Strategic Management of Teams*. New York: John Wiley & Sons, Inc.

Dauch, R. 1993. *Passion for Manufacturing*. Dearborn, MI: Society of Manufacturing Engineers (SME).

Dickinson, H.W. 1936. *Matthew Boulton*. Cambridge, MA: University Press.

Holt, L.T.C. 1967. *A Short History of Machine Tools*. Cambridge, MA: MIT Press.

Industry and Product Classification Manual. 1992. EC92-R-3. Washington, D.C.: U.S. Department of Commerce, Bureau of the Census, U.S. Government Printing Office.

Smith, M.R. 1983. *Managing the Plant*. Englewood Cliffs, NJ: Prentice-Hall, Inc.

Standard Industrial Classification (SIC) Manual. 1987. Washington D.C.: U.S. Government Printing Office.

Starr, M.K. 1989. *Managing Production and Operations*. Englewood Cliffs, NJ: Prentice-Hall, Inc.

Strandh, S. 1979. *A History of the Machine*. New York, NY: A & W Publishers, Inc.

Termini, M.J. 1996. *The New Manufacturing Engineer*. Dearborn, MI: Society of Manufacturing Engineers (SME).

Woodbury, R.G. 1972. *Studies in the History of Machine Tools*. Cambridge, MA: MIT Press.

The Competitive Challenge in Manufacturing

2

The best way to predict the future is to create it. — Peter Drucker

2.1 IMPORTANCE OF MANUFACTURING AS AN ECONOMIC ACTIVITY

Manufacturing constitutes the economic backbone of an industrialized nation. In general, the economic health of a country is based on the level of manufacturing activity, and the standard of living is often reflected in that level of activity. Some national leaders contend that a nation's economy can only be as strong as its manufacturing base and any nation that does not prosper in its economic sector cannot continue to invest adequately in itself. History books have documented the fact that few nations have prospered without a strong manufacturing and agricultural sector. One important reason for a nation to maintain a healthy manufacturing base is that it provides meaningful employment for many thousands of people with a whole host of skill levels. For the most part, the compensation received by manufacturing employees is better, or at least as good, as that of many other sectors. In addition, most jobs in manufacturing require some level of skill, and as industry adopts increasing levels of technology, workers will, of necessity, improve their skills comparatively. With such a highly qualified labor force in manufacturing, a nation is therefore able to compete in the world marketplace.

In addition to the employment base that it provides, the manufacturing sector, unlike other less technology-based sectors, invests heavily in research and development (R & D). In 1993, for example, U. S. industry-funded, private sector research and development exceeded $99 billion. This investment, plus the tremendous advances in technology that it propelled, provided this nation with the basis for economic leadership among the industrialized nations of the world.

2.2 STATE OF THE INDUSTRY

The manufacturing sector in the U. S. has undergone a series of dramatic changes during the past three decades. During the 1970s and 1980s, a number of components of that sector, particularly the automotive and steel producers, began to lose ground in the face of intense competition from emerging industrial nations. Weaknesses were experienced in both market share and profitability by even the most formidable blue chip firms. During that time, some economists claimed that the deindustrialization of the U. S. was occurring. As evidence of this, some economic reports would reference manufacturing as a wasteland of obsolete factories, failing rust-belt industries, and declining exports. In addition, they cited the decline in the percentage of workers in the manufacturing labor force from 34% in 1950, 26% in 1970, 22% in 1980, 16.8% in 1992, to 16% in 1995. Further, it was predicted that the labor force in manufacturing would follow the same trend as it did in agriculture, and that by the year 2005, the percentage of the total labor force in manufacturing would shrink to 9–13%.

While the preceding statistics are disturbing, they do not mean that manufacturing in the U. S. is disappearing or that it is no longer a viable and valuable component of the economy. To the contrary, all it means is that employment in manufacturing has not grown at the same rate as it has in other economic sectors such as service, retail trade, and government. In fact, as shown in Table 2-1, the number of manufacturing companies and the number of establishments actually increased by 11% and 6% respectively from 1977–1992. Even the number of establishments with 20 or more employees increased by over 5% during that 15-year period. On the other hand, as the number of establishments increased, the total work force in manufacturing during that period shrank by nearly 7% and the employment of production workers declined by nearly 15%. However, the significant economic point to all of these changes is reflected in the last two columns of Table 2-1. From 1977–1992, the dollar value of manufacturing shipments increased by about 121% and the value added by manufacturing rose by 144%.

The term *value added* in manufacturing refers to the increments of value that are added to a product at each step in the manufacturing process. In other words, as raw materials are transformed into usable products, their value is increased somewhat in proportion to the complexity and number of steps involved in that transformation process. Thus, it is important to note from Table 2-1 that the value of shipments and value added by manufacturing were both increasing while the labor force involved was decreasing. Even though some of the increases occurring in the statistics cited in the last two columns of Table 2-1 can be attributed to inflation, it must be concluded also that some of that increase was due to improvements in the efficiency of operation of the manufacturing establishments involved. In manufacturing, this is referred to as *productivity improvement*.

2.3 LABOR PRODUCTIVITY

Labor productivity in manufacturing is generally defined as the ratio of units of output over labor hours of input. That is,

$$\text{Labor productivity} = \frac{\text{units of output}}{\text{labor hours input}} \qquad (2\text{-}1)$$

Since *efficiency* is also usually defined as output over input, then labor productivity is, in reality, a measure of labor efficiency. For many years, labor costs constituted a large percentage of the total cost of a manufactured product. In addition, manufacturing labor was very expensive because of the skill factor involved. Because of this, most manufacturers made every effort to constrain or reduce the amount of manufacturing labor going into their products. In most cases, this meant finding ways and means of replacing human labor with some kind of labor-saving device or with machine tool components that required less skill on the part of the operator. In essence, this meant replacing the labor hour content of a manufacturing operation with a technological improvement. Thus productivity improvements in manufacturing tend to be a reflection of technological advances made in machine tools, the configuration and operation of the manufacturing system, the design of the product to simplify manufacture, and the support systems for manufacturing operations.

The U. S. Department of Labor has been maintaining records on labor productivity for many years on most economic sectors. These statistics are used by various organizations for comparative purposes and, in particular, to determine how U. S. industries stack up against overseas competitors. During the 1970s, labor productivity in U. S. factories increased at an annual rate of only about 1–1.5% as compared to other industrialized nations with annual increases of 3–4%. During the late 1980s and early 1990s, however, manufacturing staged quite a comeback with annual labor productivity increases in the range of 3.5–4.0%. These increases put U. S. factories on par or better with their competitors around the world.

As indicated previously, labor productivity is a measure of efficiency in the use of labor to achieve a certain level of output. In the past,

Table 2-1. Manufacturing establishments and employment, 1977–1992

Census Year	Companies	All Establishments	Establishments with 20 or More Employees	All Employees (1,000)	Production Workers (1,000)	Value of Shipments (millions $)	Value Added by Manufacturing (millions $)
1992	333,203	381,870	124,927	18,253.3	11,654.1	3,006,275	1,428,707
1987	310,341	368,902	126,248	18,949.2	12,279.0	2,475,786	1,165,682
1982	298,429	358,061	123,163	19,094.1	12,400.6	1,960,205	824,117
1977	300,393	359,928	118,699	19,590.1	13,691.0	1,358,526	585,165

(Census of Manufactures 1992)

an improvement in labor productivity was associated with a reduction in the unit cost of manufacturing a product, particularly for labor-intensive manufacturing industries. Conversely, declines in productivity usually contributed to an increase in the unit cost of product. However, for some types of industries, an improvement in labor productivity may or may not mean that the overall cost of producing a product will be materially reduced because of the extremely high capital costs involved in accomplishing that improvement. Labor costs in the U. S. for some manufacturing industries currently tend to be only a small percentage of total operating costs, while material and overhead costs generally constitute large percentages. Since finance charges for new equipment are included in those overhead costs, the cost of technological change might be greater than the savings in labor costs. Thus, it is important that cost evaluations be made on proposed changes in manufacturing systems before such investments are made.

While being a good measure of efficiency, labor productivity does not necessarily provide a direct measure of *effectiveness*. Effectiveness in manufacturing relates to a whole host of factors which, when executed properly, result in a successful manufacturing enterprise. A variety of models have been tested, but as yet, a completely satisfactory assembly of all the factors that contribute to productivity as a measure of effectiveness has not been achieved.

2.4 INTERNATIONAL COMPETITIVENESS

The primary mechanisms for international trade have been recognized by economists for many years. Basically, the opportunity for trade exists between two countries when each of those countries specialize to some advantage in the production of certain products. For example, country X might be able to produce computer chips with better advantage than country Y, while country Y can manufacture chemical products better than country X. The production differentials or advantages possessed by each country are often focused on price, but may include considerations of product quality, service capability, delivery time, or in some cases, even international politics. If consumers in each country recognize the value of the products involved and satisfactory tariff agreements can be reached, then merchants in those countries are likely to do business with each other. For the most part, trade between two countries is market-driven because customers are sensitive to the product advantages (price, quality, etc.) offered by each country. Thus, each country has a *comparative advantage* that helps to sustain its trade relationship.

For many years, the export of products was not a strong suit for U. S. manufacturers. During the early 1970s, export sales were typically only about 7% of factory sales as stateside industries concentrated their major marketing efforts on stateside customers with not too much competition from overseas producers. Since that time, however, foreign competitors have achieved considerable success in marketing their products in the U. S., and U. S. manufacturers have had to respond with aggressive efforts to compete in international trade. These efforts, for the most part, have been quite successful, particularly among a number of large capital goods producers such as aircraft, earthmoving equipment, and automotive manufacturers wherein exports are growing as much as 15% per year. In the field of microprocessor production, some U. S. firms have reported exports at 40% of sales in recent years.

A significant proportion of U. S. manufacturers have managed to continue their stateside production operations and export internationally quite successfully. Others have established production facilities at various locations overseas and are marketing globally from those locations. While both of these approaches result in increasing revenues for American firms, stateside operations remain the favorite among government leaders because this approach offers more opportunity for the employment of American workers.

International competitiveness is often defined as the ability of a country to proportionally generate more wealth than its competitors in the world marketplace. Table 2-2 illustrates the strong global position that the U. S. held among the major industrialized (G-7) countries in 1994. These statistics represent a model of a competitive index for those seven nations. The first statistic cited, *gross domestic product* (GDP), represents the estimated total cost of goods and services produced in the U. S. To some economists, GDP is believed to be a measure of the economic welfare of a nation. If so, then the $4,700 GDP per capita advantage held by the U. S. over six of its world competitors indicates that this country must be retaining its competitive advantage in both internal and international markets.

2.4.1 Balance of Trade

If the values of the shipments between two countries are equal, then it is said that they have an equal *balance of trade.* If the values of shipments from country X are greater than those from country Y, then it is said that country Y has a trade deficit with country X. The U.S. Department of Commerce maintains an accounting of trade activities with foreign competitors so that appropriate trade policies can be developed. This accounting is considered to be the broadest gage of trade performance as it measures trade in goods and services as well as investment flows between countries and foreign aid.

Unfortunately, the U. S. balance of trade record since the early 1980s has not been good. In 1995, the U. S. foreign trade deficit increased to $152.92 billion, the worst performance since a record high deficit of $166.3 billion was set in 1987. Thus, in spite of the fact that U. S. manufacturers exported a record $387 billion worth of goods and services in 1995 and exports continue to grow at a rate of 8.6% per year, the U. S. continues to be a debtor nation as far as exporting is concerned.

2.4.2 Trade Agreements

Although it is generally agreed that the United States cannot expect a one-for-one export/import swap with some countries, there is a great deal of optimism among industrial and government leaders on improvement potential over the long range expected from various trade agreements with foreign nations and groups of nations.

For many years, most industrialized nations have initiated trade agreements with other countries on an individual basis. For the most part, these trade agreements have given favored nations a trade advantage by lowering tariffs on imports. Other countries not included in those agreements were at a disadvantage because of high tariff situations that served as trade barriers. World leaders soon recognized the irregularities and trade war situations that this sort of practice created. They began to develop cooperative agreements between groups of nations to bring about an environment wherein trading activities could be carried out between every nation of the free world with a minimum of conflict and disagreement over tariffs.

Table 2-2. Economic comparisons between the U. S. and other G-7 countries

Economic Indicator	United States	Average of Other G-7 Countries
GDP per capita, 1992	$23,200	$18,500
Job creation, 1987–93	+8.4%	+3.9%
Growth of unit labor costs, 1985–93	+1.1%	+6.6%
Unemployment, 1994	6.1%	8.8%
Manufacturing productivity, 1985–93	2.9%	2.8%

The group of seven (G-7) major industrialized countries include Canada, France, Germany, Italy, Japan, the United Kingdom, and the United States.

(U. S. Department of Commerce Statistics 1994)

One of the early free-trade agreements was established between 12 countries in Europe. Originally referred to as the European Community (EC) and later changed to European Union (EU), this agreement linked France, Germany, Italy, Luxembourg, the Netherlands, Belgium, Denmark, Greece, Ireland, Portugal, Spain, and the United Kingdom together to establish guidelines and standards for trade among those nations. This agreement was expanded later on to include Austria, Finland, Iceland, Liechtenstein, Norway, Sweden, and Switzerland.

More recently, the worldwide expansion of the General Agreement on Tariffs and Trade (GATT) to include some 123 countries and the U. S., offers considerable promise for liberalizing trade barriers and reducing tariffs among participating countries. Along with tariff reductions, GATT establishes intellectual property agreements among the participating nations. Thus, for the first time, a worldwide agreement is in place to protect copyrights and patents from international acts of piracy. In addition, GATT established the World Trade Organization (WTO) as a kind of supreme court designed to settle disputes on trade issues between participating nations.

In an effort to improve trade relationships with its immediate neighbors, the U. S. Congress in 1993 approved the North American Free Trade Agreement (NAFTA) with Canada and Mexico. Unlike GATT, which hopes only to reduce or at least equalize tariffs and trade barriers among participating countries, NAFTA expects to eventually eliminate these and promote a completely free-trade environment.

Certainly, it is optimistic to expect that these trade agreements will solve all of the trade differences between nations. However, it can be expected that they will at least help to build bridges of mutual understanding between the industrial leaders of those countries in support of reasonably competitive manufacturing practices.

2.5 MANUFACTURING INNOVATIONS

The pressures of international competition have served as a catalyst for change in American manufacturing systems and for change in the way products are marketed. During the past 10–15 years, new manufacturing technologies, automation schemes, and system innovations have been implemented to improve machine and operator efficiency, and increase productivity to reduce the cost of manufacturing. In addition, manufacturing establishments have been and are still in the process of reorganizing and restructuring to meet the challenge of introducing quality products to global markets at a competitive price with timely delivery.

2.5.1 Machine Tools

American manufacturing industries are meeting the challenges of global competition through continued development of machine tools with more advanced digital controls, higher speeds, better accuracy, and greater flexibility. For example, numerical control (NC) (Chapter 31) has been used by U. S. manufacturers for over 40 years to control the operation of a wide variety of machine tools. This technology has undergone an amazing amount of transformation and is now referred to as computer numerical control (CNC). Now, CNC users continue to pressure control system builders to develop systems that are faster, have a full range of graphics capability, completely and accurately track the cutter path, are able to support a wide variety of sensors, handle knowledge-based software, accommodate shop-floor programming, and with a host of other features that were not even imagined 40 years ago.

Numerical control of machine tools is rapidly enabling manufacturers to close the loop on manufacturing processes by reducing operator involvement. Prior to the introduction of numerical control in 1952, a machine operator was responsible for manually operating the equipment and controlling the process. Now, on many machine tools, the operator has been replaced by a digital controller that guides the machine through a sequence of operations in accordance with a previously prepared program. More recent developments have incorporated probes, sensors, and adaptive machining processes that represent the eyes and ears of an operator. Currently, highly sophisticated knowledge-based systems models are being developed and tested, which will support a completely automatic intelligent machining workstation. These systems will close the manufacturing processes loop even further by encompassing many part programming and manufacturing engineering (knowledge-based) functions.

In addition to numerical control, innovations in machine tool design and construction have been achieved in such areas as:

1. Flexibility: As pressure continues to build for short production runs and just-in-time deliveries (Chapter 29), many machine tools are built to be more flexible and do a greater variety of jobs. For example, nearly 30% of all automatic lathes are now equipped to perform some milling operations. This trend will probably continue to the extent that most machines will be able to accommodate a variety of operations that were formerly done as secondary operations on other machines. Thus, the single-purpose machine tool has evolved into a multipurpose machine called a machining center (Chapter 30).
2. High-speed machining: Manufacturers understand that higher cutting speeds and feeds can increase production and improve quality as long as the machines and cutting tools can accommodate those conditions. Thus, milling machine spindle speeds as high as 10,000 rpm are appearing, particularly for use in machining aluminum in the aerospace industry. Some manufacturing researchers see future spindle speeds going as high as 250,000 rpm for special applications on nonferrous and nonmetallic materials.
3. Rapid movement of machine elements: To be productive, machine tools with high spindle speeds must be designed to accommodate high feed rates, rapid travel rates for advancing and withdrawing tool holder elements, and rapid indexing and positioning rates for turret-type tool elements. On the early milling machines, for example, movements of the table were accomplished manually by an operator who turned a hand crank or wheel that rotated a precision feed screw. A given number of turns of the hand wheel would position the table a specified amount, depending on the lead of the screw. Later on, the feed screw was connected to motor drives by gears and power feeds became available. Movement of the table was accomplished by a feed or clamp nut that engaged the feed screw. Unfortunately, clearance was required between the screw and the feed nut to allow the screw to turn, contributing to (backlash) errors in the positioning and repositioning of the table. Also, the occurrence of wear between these two elements contributed to additional positioning error. In more recent years, many machines have been equipped with ball screw-type feeding and positioning drives with the feed nut replaced by a recirculating ball carrier. Increased positioning speed and improved accuracies are available with ball screws, but they have their limits because of the mass involved. It is expected that in the future, linear motors will be developed to replace many applications of the ball screw, particularly for high-speed acceleration/deceleration and precise positioning of machine elements.
4. Automation technologies: Tremendous advances have been made during the past two decades in the automation of machining processes. One of the major trends includes multi-axis and multifunction machining wherein both static and rotating tools perform simultaneous machining operations along separate axes. These features, plus automatic quick-change tooling, automatic tool changing, tool storage, and modular workholding devices have greatly enhanced machine tool performance and productivity. In addition, a number of machine tool builders have developed very sophisticated systems for in-process and post-process measurement and gaging with feedback control for tool compensation. And finally, many machine tools are being equipped with tool condition sensors to monitor cutting performance and provide protection against damage that might be caused by tool failure.

2.5.2 Manufacturing Systems

Along with advances in manufacturing equipment, many significant changes have been made in the way manufacturing systems are configured and operated. One of the more important changes has been in the restructuring of production areas from a departmental style of operation into what is commonly referred to as *cellular production* (Chapter 29). In cellular production, manufacturing cells are set up to include all the equipment required to produce a certain type of product. Any order for that type of product is handled within that cell, and the component parts never leave the cell until they are completed. Parts are no longer subjected to time-consuming moves from one processing department to another. Through the cellular production scheme, less in-process inventory is required and the tracking of parts in-process is greatly simplified.

Another even more drastic transformation of many manufacturing systems has been brought about by the rapidly changing patterns of customer demands in a global economy. This has caused a significant number of industries to shift from the traditional mass-production type of manufacturing system to a more flexible and agile system often referred to as *mass customization*. However, this does not mean that mass production operations are totally disappearing from the spectrum of manufacturing operations in the U. S. Many standard product items are and will continue to be in sufficient demand to support the volume considerations for mass production. But, for those product categories wherein configuration details and specifications can be modified to meet different customer requirements, mass customization becomes necessary, particularly when order quantities are small.

The trend toward mass customization in manufacturing has also inspired the adoption of a number of other concepts and practices that contribute to the flexibility and agility of manufacturing operations. *Flexibility* or *agility* in manufacturing may be defined as the ability to produce a range of different products or component parts in a minimum period of time and with a minimum amount of changes to the manufacturing equipment. This usually involves the use of machining centers (Chapter 30) where tool and workholder equipment changes and part programming can be accomplished easily and rapidly.

Another requirement of flexible or agile manufacturing is the ability to accommodate a *family-of-parts*. Often referred to as *group technology* (Chapter 29), this practice requires the manufacturer to categorize product items with similar features into groups. For example, V-6, V-8 and V-10 engine blocks might constitute a product group since they are of similar configuration except for size and number of cylinders. In the past, traditional transfer-type machining stations dedicated to only one type of engine, for example a V-6, would take possibly months to modify and retool for one of the other engine types. Sometimes the cost for such changes would amount to as much as 80% of the original cost of the equipment. Now, new design concepts are filling the gap between machining centers and transfer machines to provide the flexibility to accommodate family-of-parts operations with minimal changeover time and costs.

A number of other concepts and practices have been incorporated within the framework of flexible manufacturing systems to make them more responsive to ever-changing customer requirements. One of these, the *just-in-time* (JIT) concept (Chapter 29), works on the *pull system* of production. In essence, this means that products or parts are not made to stock, but to order, and the concept applies to customer orders as well as the processing of component parts. For in-plant operations, machine operators initiate a request for replacement parts from a previous operation only when their supply runs low. This triggers similar requests throughout the plant. Since products and component parts are not made to stock, a delay in filling an order at any station could hold up production throughout the plant upstream. Understanding the consequences of a delay at any station in the system, JIT users must make every effort to minimize setup and material handling times, and also maintain control of quality to prevent delays that could bring production to a standstill.

There are many benefits resulting from the use of a well-managed JIT program in a discrete parts manufacturing environment, including the reduction of work-in-process and inventory, and improvements in product cycle time, quality, and cost.

A number of other systems innovations are available to manufacturing organizations which, if properly applied, can contribute to a firm's competitive status. These include artificial intelligence (AI), computer-integrated manufacturing (CIM), material requirements planning (MRP), total quality management (TQM), and a number of others. They are all good concepts and each has a potential for improving certain aspects of the manufacturing system. However, it must be understood that they, either individually or collectively, cannot solve all manufacturing problems. Further, it must be understood that global competition will require continuous improvements to manufacturing systems and this can only be achieved by the adoption of a consistent manufacturing strategy geared to such improvements.

2.6 QUESTIONS

1. Why is a strong manufacturing sector important to the economy of the U. S.?
2. What are the current trends in manufacturing employment in the U. S.?
3. What is meant by the term "value added" in manufacturing?
4. What is the trend in the value of shipments and value added by manufacturing establishments in the U. S.?
5. How is "labor productivity" defined?
6. What is the difference between efficiency and effectiveness in a manufacturing environment?
7. How do manufacturing establishments in the U. S. compare with other industrialized nations as far as labor productivity is concerned?
8. What is meant by a "comparative advantage" in international trade?
9. How is "gross domestic product" defined?
10. What is the primary purpose of the General Agreement on Tariffs and Trade (GATT) and the North American Free Trade Agreement (NAFTA)?
11. When was the use of numerical control (NC) of machine tools introduced in the U. S.?
12. What are some of the advantages of numerical control over manual control of machine tools?
13. What is the difference between a single-purpose machine tool and a machining center?
14. How are the movements of the tool-holding and workholding elements on machine tools accomplished and controlled?
15. Name some of the advances made in the automation of machine tools during the past two decades.
16. What is a cellular production system?
17. What advantages does a cellular production system have over a departmentalized system?
18. What is the difference between mass production and mass customization in manufacturing?

2.7 REFERENCES

Albert, M. 1992. "Getting Started with CNC." *Modern Machine Shop*, July.

Beard, T., ed. 1993. "Japan—Looking at all the Angles." *Modern Machine Shop*, January.

Brown, C.R. 1991. "Thoughts on the Future of Metal Cutting and Manufacturing in America." Raleigh, NC: Kennametal, Inc.

Brownstein, V. 1994. "The U. S. is Set to be the Winner from Worldwide Expansion." *Fortune*, November 28.

Census of Manufactures. 1992. Washington, DC: Bureau of the Census, U. S. Government Printing Office.

Henin, G.E. 1994. "CIM Perspectives." *Modern Machine Shop*, April.

Keremedjiev, G. 1995. "China: A Plan Awakening in Metal Forming." *Metalforming*, June.

Mason, F., ed. 1995. "High Volume Learns to Flex." *Manufacturing Engineering*, April.

Noaker, P.M., ed. 1994. "The Search for Agile Manufacturing." *Manufacturing Engineering*, November.

Owen, J.V., ed., and Sprow, E.E. 1994. "The Challenge of Change." *Manufacturing Engineering*, March.

Patterson, M.C. and Harmel, R.M. 1992. "The Revolution Occurring in American Manufacturing." *Manufacturing Methods*, January/February.

Stewart, T.A. 1992. "Brace for Japan's Hot New Strategy." *Fortune*, September 21.

——. 1996. "Craftsmanship and Modern Technology Working Side-by-side." *EDM Today*, March/April.

——. 1995. "European Directives will Affect North American Stampers." *Metalforming,* November.

U.S. Department of Commerce Statistics. 1994. Washington, DC: U. S. Government Printing Office.

Material Properties and Testing

3

Stainless steel, 1913
—Henry Brearley, Sheffield, England

Metals have a common set of properties that make them among the most useful of engineering materials. Not all metals have the same properties or properties to the same degree. Most are solid at room temperatures; mercury is an exception. Actually, the melting points of various metals range to over 3,316° C (6,000° F). Metals are relatively heavy, but densities (mass per unit volume) vary over a wide range. Among the more common metals, aluminum has a density of 0.27 g/m^3 (.096 lb/in.^{3}), and tungsten 1.88 g/m^3 (.678 lb/in.^{3}). Polished metal surfaces show a high luster, but most oxidize and corrode rapidly.

Strength, hardness, wear resistance, shock resistance, and electrical and thermal conductivity are important metallic properties. Most metals are elastic to a limit; they deform in proportion to stress and return to their original state when the stress is released. At higher stresses, they plastically deform. Some metals will accept a great deal of plastic deformation before failure, and others very little.

3.1 METAL STRUCTURES

A metal may exist as a plasma, gas, liquid, or crystalline solid. Plasmas and gases exist only at high energy levels. The liquid state results from free energy that causes the atoms to move at random; their movements are limited only by the container. At no time do the atoms take fixed positions in relation to each other in a liquid.

3.1.1 Unit Cells

The atoms of a metal assume nearly fixed positions relative to each other in the solid state. A solid metal is usually composed of a multitude of crystals. Within any one crystal, the atomic arrangement is repeated by adjacent atoms many times. An imaginary line can be drawn through a string of atoms arranged side-by-side. In fact, such lines can be drawn in three coordinate directions and form a lattice work called the *space lattice* of the crystal. The space lattice is made up of a small, repeating, three-dimensional, geometric pattern having the same symmetry as the crystal and is called a *unit cell*. The whole crystal is built up of unit cells stacked together like building blocks.

Crystals are formed out of the atoms of a liquid metal when it freezes. When the free-energy level (heat content) at any point in a liquid falls to the freezing point, atoms join together into unit cells. This may occur at many points at the same time. Unit cells that start at different points do not have the same orientation and form different crystals. All unit cells within any one crystal have the same orientation. A crystal grows by taking on atoms to form more unit cells during freezing until it meets other crystals. The crystals are called *grains*, and the orientation changes from one grain to another at the grain boundary (Figure 3-1).

There are a number of shapes and sizes of unit cells. The three most important in metals are illustrated in Figure 3-2. The *face-centered-cubic* (FCC) unit cell (Figure 3-2[A]) has an atom at each corner of a cube and an atom in the center of each face of the cube. Note that in the lattice, each atom at the corner of one cube is at the same time in the face of a different cube, and so on. The *body-centered-cubic* (BCC) cell (Figure 3-2[B]) has an atom at each corner of a cube and an atom in the geometric center of the cube. Note that the latter atom is also at the corner of another cube. The *hexagonal-close-packed* (HCP) array (Figure 3-2[C]) has a honeycomb shape. The top and bottom of a cell are parallel hexagons, and halfway between is a triangle with an apex pointing to every other side of the honeycomb walls. Each apex is halfway between the longitudinal centerline and a side. An atom is located at every corner of the cell.

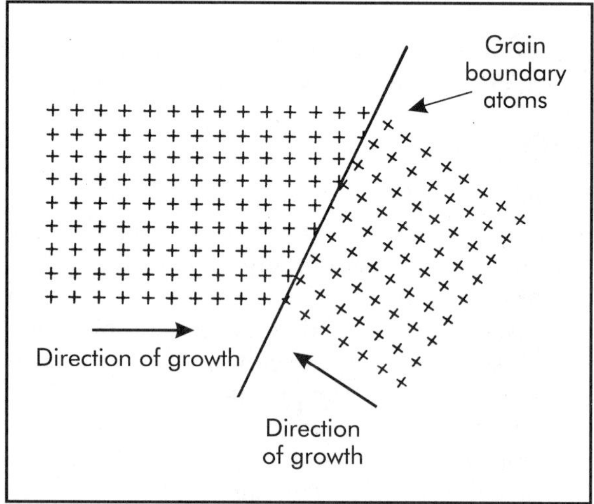

Figure 3-1. Schematic depicting the nature of a grain boundary.

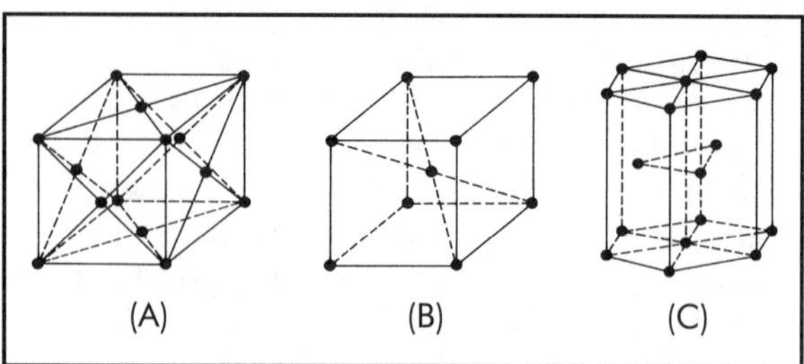

Figure 3-2. Common unit cell structures: (A) face-centered-cubic (FCC), also called cubic-close-packed; (B) body-centered-cubic (BCC); (C) hexagonal-close-packed (HCP).

3.1.2 Changes in Crystal Structure

Normally, any solid metal has a definite cell shape and size at a certain energy state, but in some metals the shape as well as the size changes from one energy state to another. The energy state is usually changed by adding or taking away heat. Such a process is called *heat treatment*. A space lattice changes to whatever shape is most stable at each energy level. Such a change is called an *allotropic transformation*. A space lattice is usually stable over a wide range of energy levels, and a metal may have to be heated to high temperatures or cooled to low temperatures to make its space lattice change.

An important variation of the BCC structure occurs from the distortion of the space lattice. The atoms of a pure metal are all of the same size and are regularly arranged in positions where the uniform forces from one atom to another are in equilibrium. An atom of foreign material trapped in a cell is of a different size, exerts different forces, and distorts the shape of the cell. Under these conditions, the cell is no longer cubic but becomes a body-centered-tetragon with one coordinate axis a little longer than the other two. This does not happen to every unit cell, and BCC and tetragonal unit cells exist side-by-side in the lattice.

3.1.3 Crystalline Structure and Physical Properties

The type of space lattice and the degree of perfection of the space lattice have much to do with the physical properties of a metal. The face-centered-cubic space lattice is in general more ductile and malleable than the body-centered type. The body-centered type is usually the harder and stronger of the two. The close-packed-hexagonal type lacks ductility and accepts little cold-working without failure. There are exceptions to these rules.

The crystals of a metal change shape when subjected to stresses and heat. Imperfections in the space lattice help determine the strength of a metal. If a stress is imposed on a crystal, some or all of the atoms are moved from their equilibrium positions, and the crystal is deformed. If the atoms are not moved out of the regions of influence to their neighbors, they return to their original positions after the stress is removed. The deformation is said to be *elastic*. If enough stress is applied to permanently deform the lattice, the atoms do not return to their original positions and the deformation is said to be *plastic*. If this occurs below what is called the *recrystallization temperature*, the metal is said to be *cold worked* (Chapter 11). Cold working distorts, elongates, and fragments the grains. At or above the recrystallization temperature, the atoms become mobile enough to form new strain-free grains that nucleate from points of high strains in the old grains. The crystals grow until they meet each other. The number of new grains formed depends upon the number of nuclei, which in turn depends upon the amount of cold working. The more the metal is strained, the smaller the grains after recrystallization. Each metal has its own recrystallization temperature.

It has been found that the actual stress required to deform a crystal is only a small fraction of what is theoretically necessary to displace all the atoms involved at the same time. Thus it is obvious that all the atoms do not move at the same time, but in sequences. Experiments have indicated that these atomic movements emanate from and are affected by imperfections in the crystals.

There are several kinds of crystalline imperfections. An atom may be missing from the place where it should be, and this is called a *vacancy*. Or, a whole plane of extra atoms may appear in a lattice to form an *edge dislocation* as depicted in Figure 3-3. Part of the planes of a lattice may be offset in a *screw dislocation* represented by Figure 3-4. And, space lattice mismatching occurs between crystals at grain boundaries (Figure 3-1). Large or small atoms distort the lattice and small interstitial atoms may bulge the lattice. There are probably other kinds of imperfections not yet recognized; investigations are far from complete. The number and distributions of the imperfections have a great effect on the properties of a metal.

Plastic distortion takes place when one part of a crystal slides on another. It appears that slip occurs between atomic planes in the lattice that are spaced farthest apart and have the highest atomic population. These are called the *glide planes* and are usually not the planes that bound the regular geometric shapes of the cells. As has been pointed out, one plane does not slip over another all at once, but glides in a series of movements. This is illustrated by Figure 3-5. As a shear stress is applied between two planes, a dislocation is strained until it is moved to the next cross plane, and so on. With a myriad of atoms in even a small crystal, there exists a large number of dislocations. Also, some types of dislocations, called *sources*, regenerate and create new

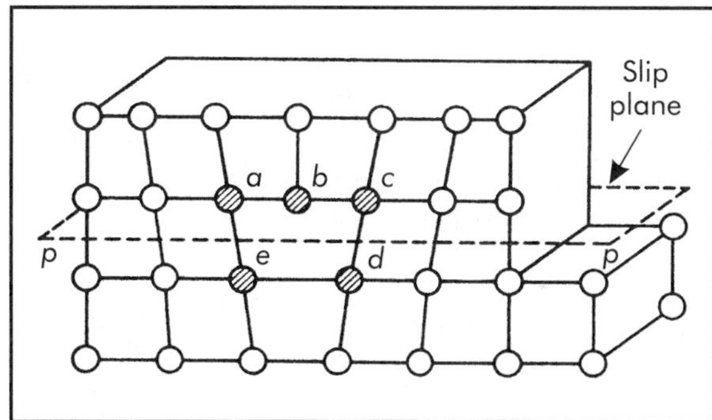

Figure 3-3. Schematic depicting an edge dislocation.

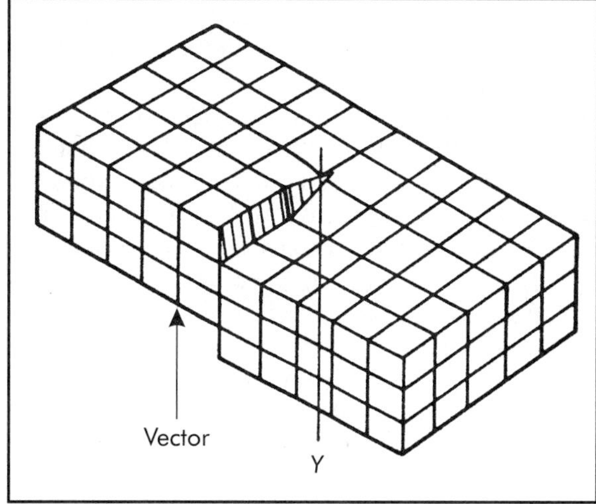

Figure 3-4. Schematic depicting a screw dislocation.

dislocations. So as stress is continued or raised, more and more dislocations are moved to cause more plastic deformation. Dislocations travel through a crystal on many planes until they reach grain boundaries or imperfections in the lattice that stop them. Other dislocations from behind interact with those stalled ahead, and movement becomes more difficult. The cold-worked metal is said to *work-harden* or *strain-harden* because a higher stress is necessary to move the entangled and crowded dislocations. As dislocations are piled up under higher and higher stress, they are forced to combine into small cracks that ultimately grow to fractures in the metal.

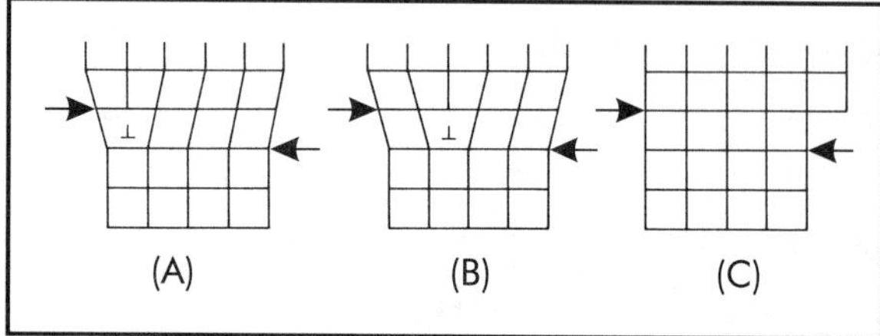

Figure 3-5. How a dislocation travels across a lattice under stress, causing displacement of the lattice.

As just indicated, anything that interferes with the flow of dislocations across a grain makes that grain harder. The interference may be a distortion of the space lattice or the presence of a foreign material. The first is the mechanism in what is called *solid solution hardening*, and the second *dispersion hardening*. As will be shown, heat treatment is a process often used to control these conditions and, thus, the physical properties of metals. Alloying of metals is another way of controlling lattice conditions. Metals like gold, silver, zinc, tin, and copper are often used in nearly pure states, but most metals are alloyed with others for best utility. The space lattice of an alloy is commonly distorted because atoms of different metals are of different sizes and exert different atomic forces. Added atoms may or may not replace atoms of the parent metal in a space lattice but, in any case, they do cause lattice distortions. More aspects of alloying will be discussed later in this chapter in connection with equilibrium diagrams.

The sizes and diverse orientations of grains in a metal largely affect its properties. A fine-grained metal is likely to have a better distribution of grains oriented to respond to stresses in any direction than a coarse-grained metal. Of more importance, fine grains present more grain boundaries to inhibit the propagation of dislocations. For these reasons, a fine-grained metal as a rule has a greater yield strength (the level of stress required to start plastic deformation), ultimate strength (the level of stress at failure), hardness, fatigue strength, and resistance to impact.

Grain orientation in a piece of metal becomes more uniformly directed when the metal is cold worked. The grains in a metal cooled slowly from high temperature have random orientation so that the path of plastic slip has to change direction from one grain to the next. Those grains favorably oriented to the applied stress are deformed most, but if enough stress is applied, all grains deform to some extent. The slip planes glide over one another, and the corresponding parts of a crystal turn with respect to each other and have a tendency to reach orientation in the direction of the stresses. The larger the applied stresses, the more numerous become the grains that are strongly oriented in the directions of the stresses. Under subsequent stresses, the metal shows *directional properties*; that is, it yields more readily if stressed in some directions than in others.

A secondary mode of crystal deformation is called *twinning*, a limited and ordered movement of a large block of atoms in a definite section of a crystal. Twinning can account for only small strains and does not occur in some materials. However, it has some importance in that it can reorient atomic planes more favorably for slip.

3.1.4 Fracture

A piece of metal breaks in one or both of two general ways after deformation by sufficient stress. One mode of fracture is termed *ductile*, and the faces of the break may be described as gray, fibrous, and silky. The parting surfaces are wiped across each other by shear stresses, and the break occurs after a large amount of deformation. Some classify this as mainly fracture through the grains (transgranular). The other kind is *brittle* fracture, called *cleavage*, where the material is actually pulled apart across atomic planes within the crystals or along the grain boundaries. The metal may first plastically deform to some extent until the forces holding it together are overcome. Then it snaps sharply in two, leaving a rough, granular, and rather bright fractured surface.

The relationship between ductile and brittle failure is one of degree. Nodular cast iron fails in a ductile manner as compared to gray cast iron, but is considered brittle in comparison with steel. Frequently, the two types of fracture exist in the same rupture because of progressive hardening as failure takes place. The outside of a break may shear in a ductile manner while the center may fail as a brittle section. Many steels fail by ductile fracture at high temperatures and cleavage at low temperatures.

3.2 FUNDAMENTALS OF METAL ALLOYS

A metal melts if heated to a high enough temperature. If heat is added continuously, the temperature of a piece of metal rises with time as indicated in Figure 3-6. When melting starts, the temperature does not rise (as shown by the plateau of the diagram) until melting is complete if the liquid is kept well mixed. This is because the liquid exists at a higher state of energy than the solid. Heat energy added during melting is used to cause a change of state rather than to increase temperature. This energy is called the *latent heat of fusion*. The process is reversible, and the same heat is given off when the metal cools and solidifies.

Another kind of change of state involves a relocation of the atoms in a solid metal. This causes a change in the space lattice and is an allotropic transformation as described in the preceding section. For each allotropic metal, the space lattice changes at a specific temperature as heat is added or taken away. The heat given off or absorbed is called the *latent heat of transformation*. If heat is withdrawn rapidly, there may be a small dip in temperature as indicated in Figure 3-7.

3.2.1 Metallic Solid Solutions and Compounds

An alloy consists of two or more metals, or at least one metal and a nonmetal, mixed intimately by fusion or diffusion. *Diffusion* is the movement

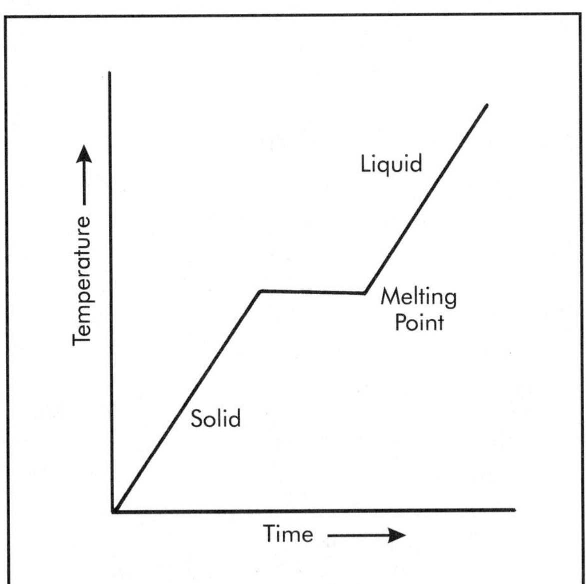

Figure 3-6. Typical time-temperature relationship of a metal being heated.

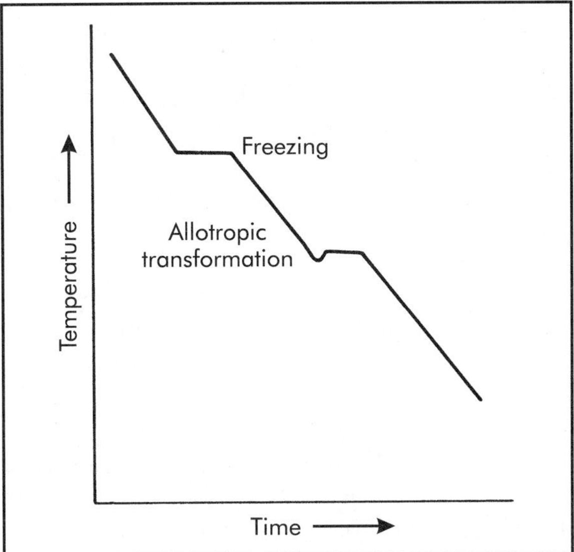

Figure 3-7. Time-temperature relationship of a metal that solidifies and passes through an allotropic transformation as it is cooled from the liquid state.

of atoms of one material among the atoms of another material. This is a well-known action in liquids, such as when sugar or salt dissolves in water. There the atoms or molecules of the solute move around in the liquid solvent. A similar action can occur in the solid state. A material is said to be dissolved in metal in the solid state when the atoms of the solute move about among the atoms of the solvent when the proper stimulant is applied. A certain resistance exists to the movement of the solute atoms in a solid, and it must be overcome before diffusion can occur. The energy that just overcomes such resistance is called the *activation energy*. More energy only increases the rate of diffusion.

One form of solid solution has each atom of the solute replacing an atom of the solvent in its space lattice. This is a *substitutional solid solution*; conditions favorable to it are:

1. The atoms of the solute and solvent differ in diameter by no more than 15%.
2. The space lattices of the solvent and solute are similar.
3. The substances are near each other in the electromotive series; otherwise, a chemical or intermetallic compound may form.

This is not to say that a substitutional solid solution cannot form if the foregoing conditions are not strictly met. It may only mean that the amount of solution formed may be limited.

A second form of solid solution occurs when the atoms of the solute take position interstitially in the space lattice of the solvent. Conditions favorable to such a solution are:

1. The diameter of the solute atom is no larger than 59% of the diameter of the solvent atom.
2. The solvent metal is polyvalent.
3. The substances are within proximity to one another in the electromotive series.

Again, limited or no solubility may result if conditions differ from these ideals.

As in liquids, more solute can be held in a solid solution at higher temperatures. Solute is likewise precipitated out on cooling. Each temperature has its own saturation amount. When a solute precipitates from a solid solvent, it often forms a chemical or intermetallic compound with the solvent. Chemical compounds like Fe_3C and Cr_4C and intermetallic compounds, such as $CuAl_2$ and Mg_2Si, have definite lattice structures and are hard and brittle as a rule.

An alloy of a particular composition contains one or more phases. A *phase* is defined as a physically homogeneous portion of matter. It cannot be subdivided by mechanical means or resolved into smaller parts by an ordinary optical microscope. As examples, molten iron is a phase and so is solid (pure) copper.

3.2.2 How Alloys Melt

The melting temperature of an alloy depends upon its composition. Consider an alloy of two or more pure metals. There may be one proportion of the constituents that has a lower melting point than any other. In some cases, this is lower than the melting temperature of any of the pure metals in the alloy. This composition is known as the *eutectic composition*, and its melting temperature is the *eutectic temperature*.

Figure 3-8 shows the behavior of three alloys as they are heated. Each is a different proportion of the same two pure metals. There is an arrest at T_1 for each; this is the eutectic temperature at which melting begins. The temperature T_2 at which melting is complete, is different for each alloy. For each case, there is a phase in excess of what is needed for the eutectic composition. This excess phase is what is being dissolved between temperatures T_1 and T_2.

3.2.3 Equilibrium Diagrams

A *constitutional equilibrium diagram* is constructed by plotting points T_1 and T_2 (like those in Figure 3-8) on coordinates of temperature and composition for a number of alloys of the same two or more metals. Lines are then drawn through corresponding points to show changes in phases.

There are three standard types of equilibrium diagrams. Only simple examples of each will be shown and discussed to illustrate principles. The same principles apply to actual alloys and more complicated diagrams.

A *type I equilibrium diagram* as shown in Figure 3-9 depicts alloys of metals soluble in the liquid state but insoluble in the solid state. An explanation of this diagram can be given for the

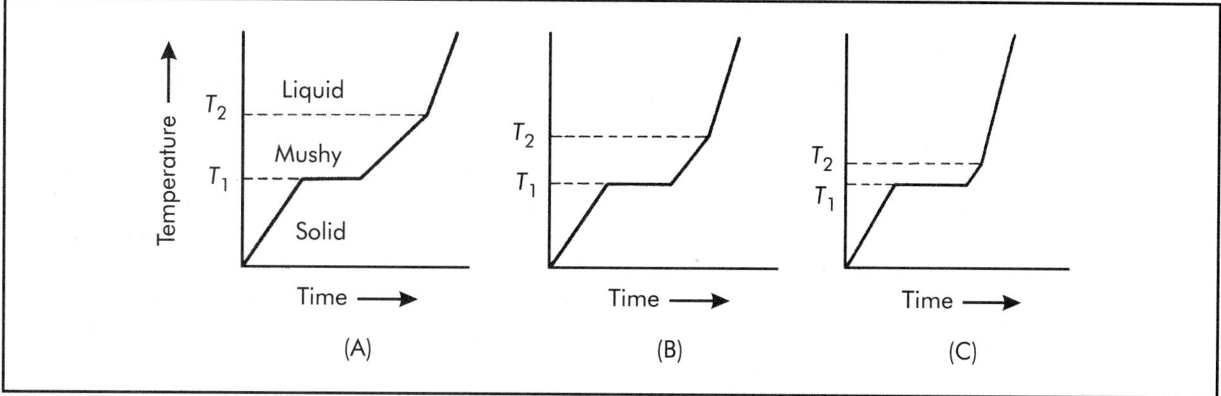

Figure 3-8. Time-temperature curves for three alloys of two pure metals.

composition of 75% A and 25% B. If these proportions of metals A and B are put in a crucible and heated to any temperature, nothing changes at first, except that the materials get hot. At the eutectic temperature (T_1) enough of metal A and metal B combine and melt to form a liquid of eutectic composition. Melting continues without temperature rise until all the metal B is consumed in making the eutectic composition. Some metal A is left over and remains solid. As the temperature is increased, more and more metal A goes into the liquid solution until T_2 is reached. At this temperature, all the metal A is dissolved in the liquid. Above T_2 there is no further metallurgical change for this composition.

The process just described is reversible. If the composition under discussion is cooled from the liquid state, some metal A begins to solidify at temperature T_2. More and more metal A freezes from temperature T_2 to T_1, and the composition of the liquid approaches that of the eutectic. Then at T_1 the rest freezes. A eutectic can occur only if there is some lack of solubility in the solid state.

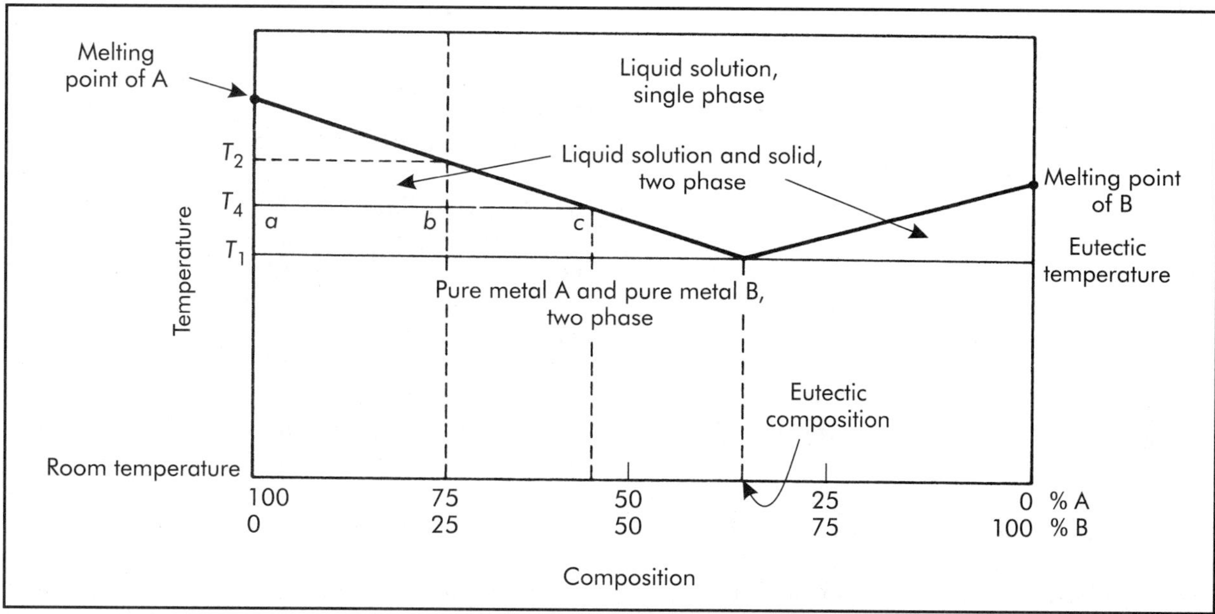

Figure 3-9. Type I equilibrium diagram of metals A and B, which are completely soluble in the liquid state and totally insoluble in the solid state.

It must be emphasized that the ideal conditions of the diagrams are realized only if heating or cooling is done slowly to approach equilibrium conditions.

A device called the *inverse lever rule* is an aid to obtaining information from equilibrium diagrams. It discloses the relative amount and analysis of each phase present in a two-phase area. As an example, consider the 75% A-25% B composition of Figure 3-9 at temperature T_4. A horizontal line is drawn to represent the temperature from a on the 100% A ordinate to c on the boundary between the liquid and mushy regions. This is the *lever*. It intersects the 75% A-25% B composition ordinate at b, which is the *fulcrum*. The amount of each phase present is signified by the inverse end in proportion to the total length of the lever. Specifically, in this example,

$$\frac{ab}{ac} \times 100 = \%\text{ of liquid phase} \qquad (3\text{-}1)$$

(liquid is approximately 58% A + 42% B)

$$= \frac{25}{42} \times 100 = 59.5\%$$

$$\frac{bc}{ac} \times 100 = \%\text{ of solid phase (pure A)} \qquad (3\text{-}2)$$

$$= \frac{17}{42} \times 100 = 40.5\%$$

Because they are used as ratios, numbers representing ab, bc, and ac can be taken directly from the composition scale without regard to units.

There are definite limits to the use of the lever rule. To gain information about phases, the lever may be used only in a two-phase area and must not cross a phase boundary. Subject to these rules, the lever may be used on any equilibrium diagram.

A *type II equilibrium diagram* shows alloys of base metals soluble in both the liquid and solid states. An illustration is given in Figure 3-10 for a composition of 83% metal C and 17% metal D. No metallurgical change takes place as the solid solution is heated from room temperature until temperature T_1 is reached. The composition is the same and entirely liquid above T_2. Gradual melting occurs between T_1 and T_2. In this region, there is a continual change in the composition of both liquid and solid phases as the temperature is raised or lowered. The compositions and proportion of solid and liquid phases at any temperature in the range can be ascertained by means of the lever rule. For example, consider the cited composition at temperature T_3. A horizontal line at temperature T_3 intersects the liquidus at a, the composition line at b, and the solidus at c. Point a' represents the composition of liquid, and c' the composition of solid in the crucible at temperature T_3. As for proportions, from Equations (3-1) and (3-2):

$$\frac{ab}{ac} \times 100 = \%\text{ of solid phase}$$

(composition is 62% C – 38% D)

$$= \frac{10}{31} \times 100 = 32.3\%$$

and

$$\frac{bc}{ac} \times 100 = \%\text{ of liquid phase}$$

(composition is 93% C – 7% D)

$$= \frac{21}{31} \times 100 = 67.7\%$$

A *type III equilibrium diagram* is for metals completely soluble in the liquid state, but only partly soluble in the solid state. An illustration is given in Figure 3-11. This diagram is a composite of type I and type II diagrams. An alloy of 60% G and 40% H acts like a type I alloy. At temperature T_1 and above, it is a liquid solution. At T_2 it is composed of a liquid solution and solid solution. The proportion and compositions of the constituents can be found from the lever *a-b-c*. At T_3, the lever *d-e-f* indicates a mixture of α solid solution and β solid solution. At T_4, the lever *g-h-i* shows the proportions of α solid solution and pure metal H at that temperature. The diagram designates that β solid solution does not exist at a temperature as low as T_4.

The alloy of 90% G and 10% H in Figure 3-11 is comparable to a type II alloy. Only a liquid solution exists at temperatures above the liquidus. This alloy is made up of a mixture of

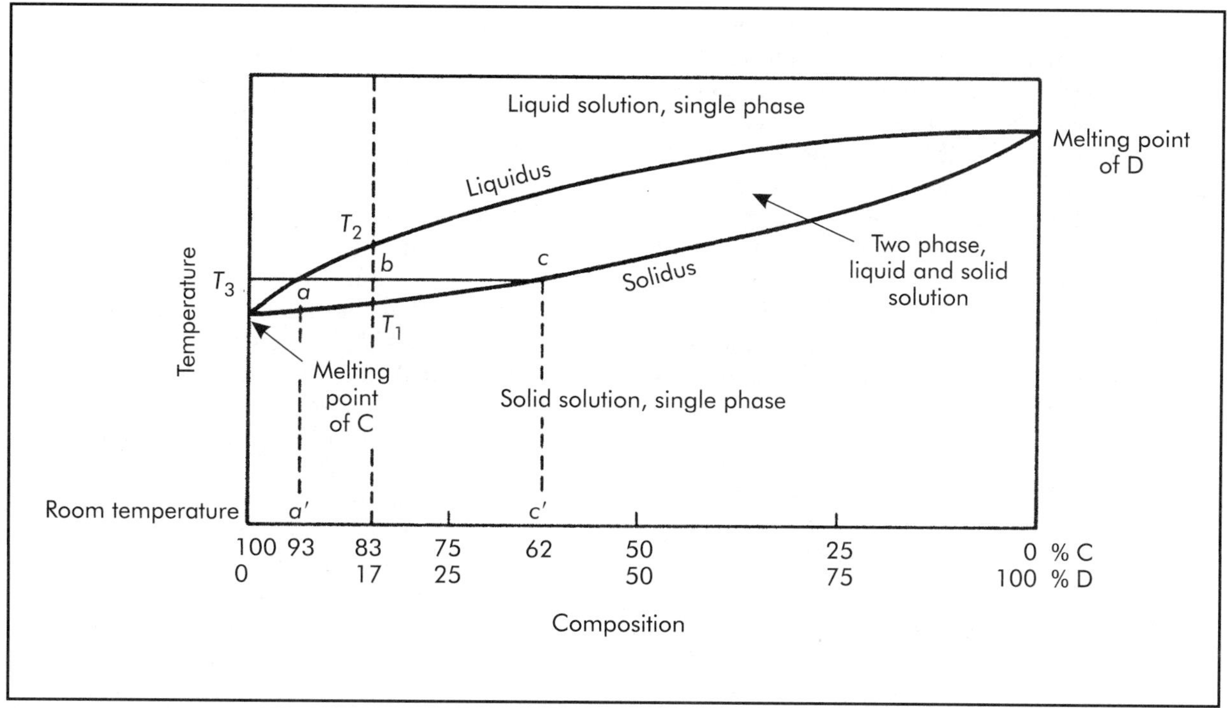

Figure 3-10. Type II equilibrium diagram of metals C and D, which are completely soluble in both liquid and solid states.

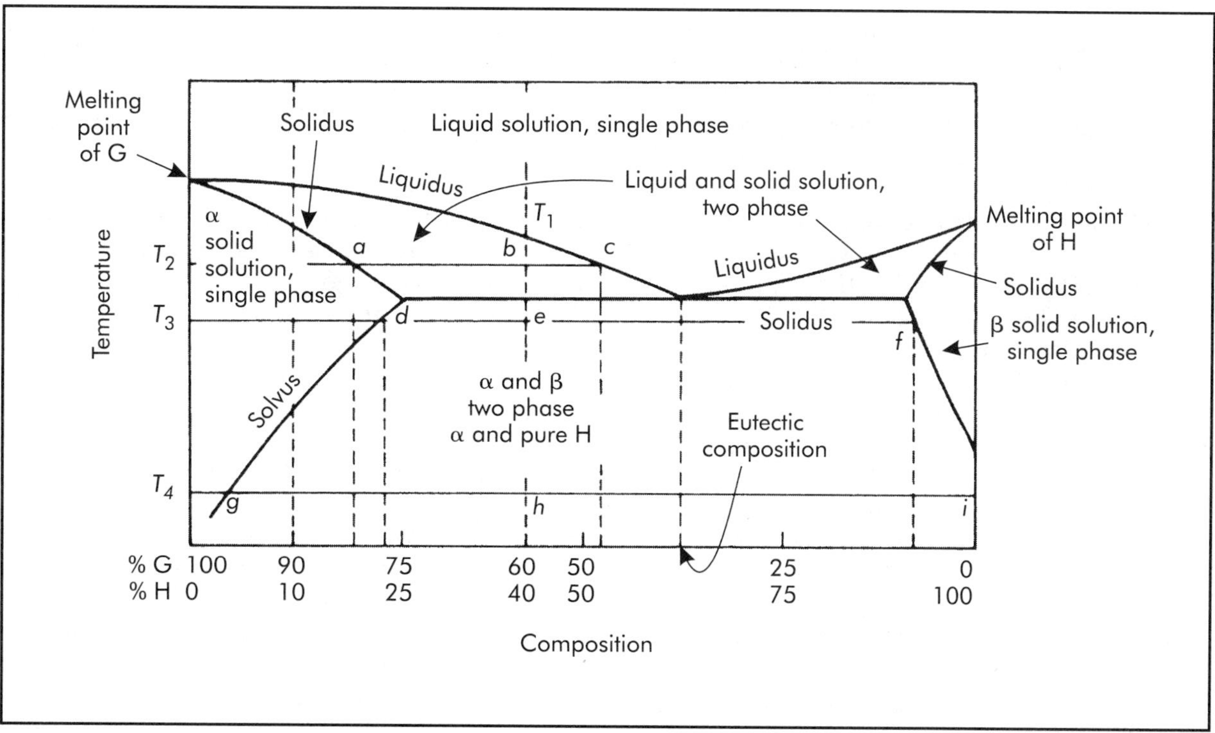

Figure 3-11. Type III equilibrium diagram of metals G and H, which are soluble in the liquid state and partially soluble in the solid state.

liquid solution and α solid solution at temperatures between the liquidus and solidus lines. The proportions and compositions of each in this region can be ascertained by use of the inverse lever rule. Only the α solid solution exists at temperatures between the solidus and solvus lines labeled in Figure 3-11, and its composition is fixed at the basic composition of the alloy. The lever rule is not applicable in any single-phase region, and thus not in the α region. Below the solvus at T_4, both the α solid solution and pure metal H exist. This is a two-phase area, and the lever rule applies.

3.2.4 Alloys in the Solid State

Alloys exist in several forms at room temperature after being cooled from the liquid state.

1. An alloy may separate into its original pure metals if they are not soluble in each other at room temperature (type I). Each grain consists of only one of the pure metals, but grains of each metal may exist side-by-side. In another alloy, the grains of one metal may be uniformly distributed through the other metal, or one metal may tend to segregate in groups of grains in the other metal.
2. An alloy may exist as a solid solution of one metal dissolved in another (type II). The solid solution may be saturated or supersaturated. The latter results from rapid (nonequilibrium) cooling in some cases and is not a stable condition but sometimes may continue to exist over a long period of time.

   Controlled decomposition of supersaturated solid solutions is the basis for precipitation hardening of metals described in Chapter 6. Solid solutions are relatively ductile, soft, and malleable, and have good shock resistance. They are desirable for cold-rolling and deep-drawing operations. Some of the popular brass and bronze alloys are solid solutions. Martensite is a supersaturated solid solution and is not ductile (exception).
3. An alloy or phase may exist as a chemical or intermetallic compound. Some are stable at low but not at high temperatures and thus form only on cooling. They commonly appear well below the solidifying temperature of an alloy.

   Intermetallic compounds are harder than the elements from which they are derived and thus are usually brittle with low tensile strength and little shock resistance. For hardness and wearing qualities, dispersions of compounds are beneficial for tools and solid bearings.
4. An alloy may exist as a mechanical mixture of two or more of the forms already described. For instance, grains of pure metal may exist side-by-side with grains of solid solution. As another example, single grains may be composed of chemical or intermetallic compounds in a matrix of pure metal or solid solution. Eutectic alloys with low melting points are desirable for solders. Such alloys may contain hard particles in a lamellar aggregate that interfere with dislocation movements in the primary phase and impart hardness to the alloy. This accounts for the wearability of some soft bearing materials of eutectic compositions.

3.2.5 Grain Growth

The way grains form from the atoms of a liquid metal was explained early in this chapter when crystalline structure was described. Molten metal solidifies when first in contact with the cooler container walls. Heat must flow from inside through the outer layer, and the inside cools at a slower rate.

The shape, size, and composition of the grains in a piece of metal depend partly upon the way it has solidified. Solidification follows a pattern as indicated in Figure 3-12. Metal nuclei form at some points on the surface, and other atoms join to them to build crystalline lattices. As illustrated, nuclei start at such points as *a* or *b* and other atoms attach to them. The solid grows in a stem perpendicular to the surface. Branches grow perpendicular to the original stem. As it solidifies, metal gives up its latent heat to its surroundings, and further solidifying is stopped for the moment. When this heat is dissipated, solidifying starts again, building on the already solid metal. The effect from a curved wall is an elongated grain with a tree-like skeletal form as indicated in Figure 3-12(B). From a flat surface, the skeleton has a ladder-like or network shape, as in Figure 3-12(C). These structures are called *dendrites*.

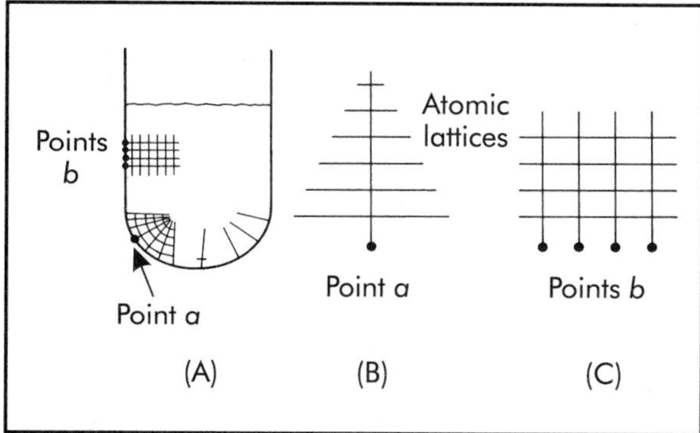

Figure 3-12. Schematic diagrams of the growth of dendritic grains.

A crystal grows from a nucleus on solidifying until it reaches neighboring grains. Thus, the size of grains produced depends upon the number of nuclei formed. Factors that produce more nuclei are a rough container surface and a fast cooling rate.

The use of the inverse lever rule in the liquid-solid area of an equilibrium diagram shows that the composition of the solid changes as the temperature drops, if the solid formed is a solid solution. This means that the first metal to solidify at the surface has one composition. The next to form on the skeleton has a different composition, and so on. So, there is a definite segregation of the various compositions formed on freezing. This is said to be *dendritic segregation*. Obviously, there can be no segregation if a pure metal solidifies from the melt.

3.3 METALLURGY OF IRON AND STEEL

Carbon is the basis for the wide range of properties obtainable in iron and steel. It forms different compositions with iron when combined in different ways and amounts. Thus carbon is the primary means for making iron or steel soft and ductile, tough or hard. Most other alloying elements, in effect, modify or enhance the benefits of carbon.

The fundamental effects of carbon on iron are shown by the iron and iron carbide equilibrium diagram shown in Figure 3-13. This is a type III diagram. The carbon commonly appears as iron carbide, but not always; for example, graphite in cast iron. Iron carbide consists of 6.67% carbon and 93.33% iron by weight and has the average formula Fe_3C. It is the hardest constituent in carbon iron and carbon steel and is quite brittle and white in color. It is called *cementite*.

3.3.1 Iron and Iron Carbide Solid Solutions

The line for zero carbon in Figure 3-13 shows that pure iron or *ferrite* solidifies at about 1,538° C (2,800° F). Over a short range of high temperature, it has a body-centered-cubic structure called *delta iron*. When cooled to 1,400° C (2,552° F), the structure changes to face-centered-cubic *gamma iron*. This is the main part of *austenite*, which may dissolve up to 2.11% carbon as iron carbide. Below 910° C (1,670° F), ferrite transforms to body-centered-cubic alpha iron. Thus, iron is allotropic. It also changes from a nonmagnetic material at high temperatures to a magnetic one somewhat below 799° C (1,470° F). Alpha iron dissolves only a small amount of carbide. Ferrite can also hold such elements as nickel, silicon, phosphorus, and sulfur in solution in amounts depending on temperature.

3.3.2 Pearlite

Steel containing 0.77% carbon is an important iron alloy. It starts to solidify when the molten solution is cooled to about 1,482° C (2,700° F) and is completely solid at about 1,249° C (2,280° F). No change occurs in the austenite until the low temperature of 727° C (1,341° F) is reached. This is at minimum in a solid solution and comparable to a eutectic in a liquid solution, and is called a *eutectoid*. At this point, the gamma turns to alpha iron, and the iron carbide is forced out of solution if cooling is slow. The eutectoid alloy increases in volume on transformation, and the resulting formation consists of a series of plates of iron carbide interspersed with plates of ferrite in each grain. This lamellar structure is known as *pearlite* and is illustrated under high magnification in Figure 3-14. What is shown is coarse pearlite produced by slow cooling. Faster cooling rates cause closer spacing of the plates in what are known as medium and then fine pearlite. Hardness increases from coarse to fine pearlite.

The interlayers of the two phases in pearlite reinforce each other. Ferrite has a tensile strength of about 293 MPa (42,500 psi) and elongation of

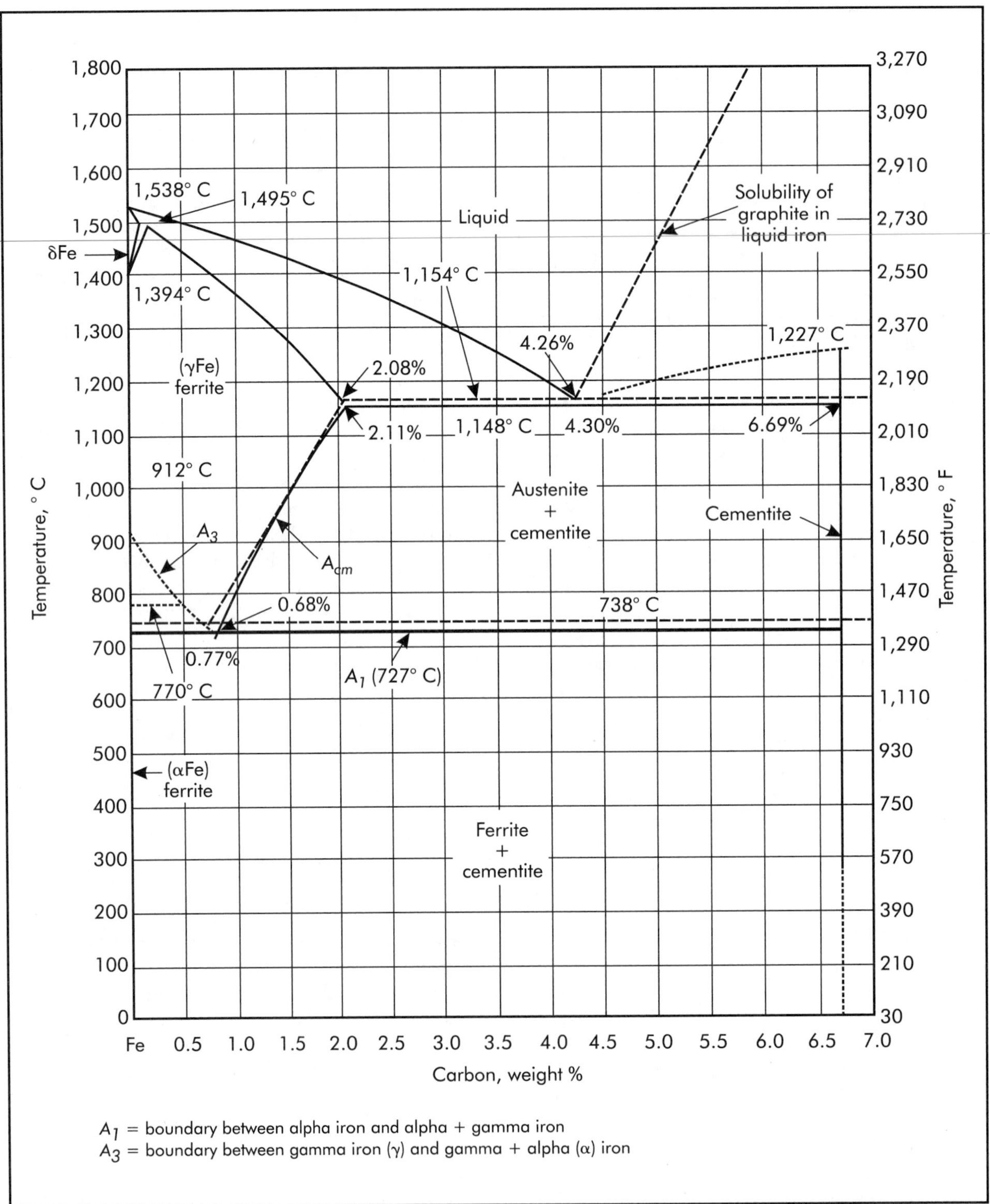

Figure 3-13. Expanded iron-carbon phase diagram showing both the eutectoid and eutectic regions. Dotted lines represent iron-graphite equilibrium conditions and solid lines represent iron-cementite equilibrium conditions. The solid lines at the eutectic are important to white cast irons and the dotted lines are important to gray cast irons (Metals Handbook Desk Edition 1998).

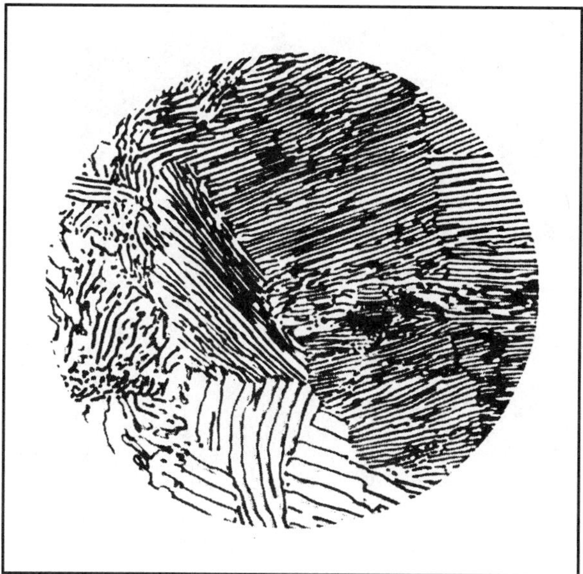

Figure 3-14. Microstructure of pearlite (etched and magnified 1,000×).

40%, and cementite a strength of 35 MPa (5,000 psi) and negligible elongation. The pearlite combination has a tensile strength of 827–862 MPa (120,000–125,000 psi) and elongation of 10–15%.

3.3.3 Hypoeutectoid and Hypereutectoid Steels

Steel containing less than 0.77% carbon is called *hypoeutectoid* and with more than 0.77% carbon, *hypereuctectoid*. The first may be exemplified by a steel containing 0.4% carbon on the diagram of Figure 3-13. This composition starts to solidify at around 1,495° C (2,723° F). First delta iron is precipitated. Then the delta iron changes to gamma iron, and solidification is complete in austenite at 1,394° C (2,541° F). No further change occurs with falling temperature until a little above 799° C (1,470° F) is reached at the A_3 line. Then ferrite precipitates as the temperature is lowered to 727° C (1,341° F), at which point the remaining austenite transforms to pearlite. The resulting structure consists of about 55% grains of ferrite interspersed with 45% grains of pearlite. A hypoeutectoid microstructure is depicted in Figure 3-15. The magnification is not sufficient to reveal the laminations in the pearlite grains. The proportion of ferrite to pearlite depends on the carbon content.

A hypereutectoid alloy may be exemplified by a steel containing 1.4% carbon. Solidification commences at about 1,449° C (2,640° F) with austenite separating out until solidification is complete at about 1,160° C (2,120° F), and the structure becomes all austenite. As indicated in Figure 3-13, no change occurs in the austenite until the temperature falls to about 1,038° C (1,900° F). Iron carbide is rejected below that point; at 899° C (1,650° F), for instance, the austenite contains only 1.1% carbon, and at 799° C (1,470° F), a little over 0.77% carbon. The excess iron carbide is rejected by the gamma iron. At 727° C (1,341° F), the remainder of the austenite is transformed to pearlite on slow cooling with the rejected iron carbide interspersed in the structure and in the grain boundaries. The lever rule (Equations [3-1] and [3-2]) applied to the region between the eutectoid at 0.77% carbon and cementite of 6.67% carbon shows the proportions of the constituents of the 1.4% carbon steel to be

$$\frac{6.67-1.4}{5.77}\times 100 = 91.3\%\ \text{pearlite}$$

and

$$\frac{1.4-0.77}{5.77}\times 100 = 10.9\%\ \text{cementite}$$

Cementite outside the pearlite exists in hypereutectoid steel in proportion to the amount of carbon beyond the eutectoid point. Iron containing more than 2.11% carbon is usually not considered to be steel.

3.3.4 Martensite

The structures considered up to now have been those formed by cooling at a slow rate. Now, assume the hypoeutectoid steel (Figure 3-13) is heated from room temperature to above 727° C (1,341° F). As the temperature is raised above that point, austenite is formed and dissolves all the carbon. The excess ferrite is dissolved between the A_1 and A_3 lines. At a temperature above the A_3 line, the metal is all austenite and contains all carbon, in an interstitial solution. If the alloy is quenched in water from above the A_3 temperature, it is cooled so quickly that the transformation of gamma to alpha iron does not have time

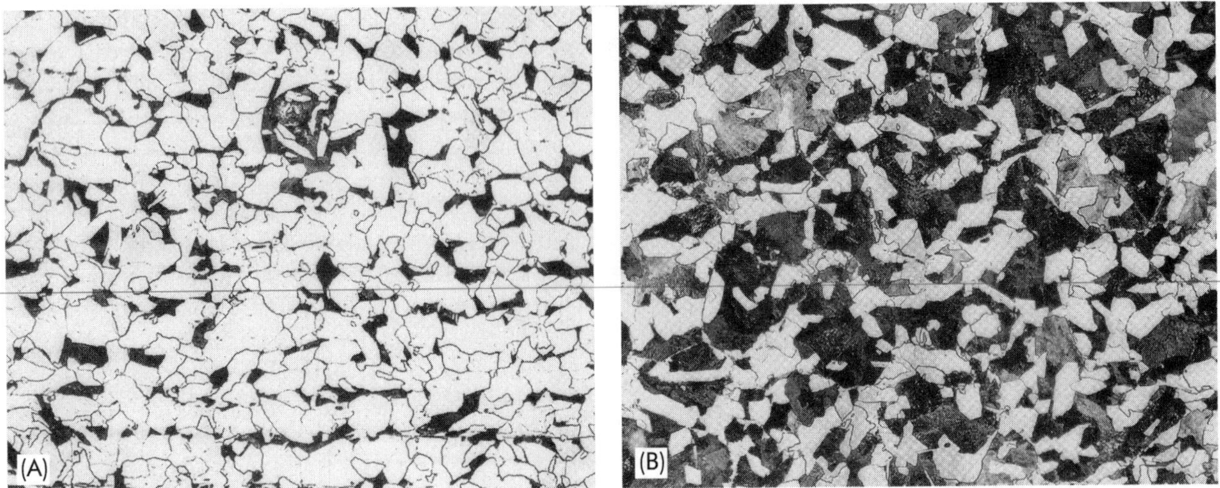

Figure 3-15. Microstructure of typical ferrite-pearlite structural steels at two different carbon contents: (A) 0.10% carbon; (B) 0.25% carbon, 2% nital + 4% picral etch, magnified 200× (Metals Handbook Desk Edition 1998).

to occur at 727° C (1,341° F). Instead, the change is suppressed to some low temperature, say around 204° C (400° F), and there is not enough energy available to cause diffusion of the carbon atoms. They remain in and distort the lattice. This makes a hardened steel called *martensite*. Its needle-like or acicular microstructure is depicted in Figure 3-16. The more highly strained layers in the lattices are attacked more readily by the etchant and appear as dark lines. Martensite is formed and hardening occurs when any steel (hypoeutectoid or hypereutectoid) is quenched fast enough from above the critical temperature.

3.3.5 Other Structures of Steel

Other steel structures may be produced by various heat treatments as described in Chapter 6. If instead of being quenched from the austenite region, a steel is cooled slowly and held for a period of time at around 704° C (1,300° F), the iron carbide will disperse in small spheroidized particles instead of lamellar plates in the ferrite. This is called *spheroidite*. On the other hand, if a steel is not fully quenched, but cooled quickly to and held at a temperature above 232° C (450° F) for a period of time, it is transformed to *bainite*. This is an intermediate structure between fine pearlite and martensite and appears to be a mechanical mixture of ferrite and minute carbide particles.

3.3.6 Practical Aspects of Carbon in Steel

Steels are selected with specific carbon contents to suit certain purposes. Easily formed sheet steel for cans and automobile fenders and body panels contains 0.1% carbon or less. Tensile strength is about 345 MPa (50,000 psi) and elongation is 35% or more. Structural steel, like that in I-beams and channels for buildings, has around 0.2% carbon with tensile strength of about 414 MPa (60,000 psi) and elongation of 30% or so. Medium carbon steels, widely used for machine parts like gears and axles, contain 0.30–0.45% carbon, have up to 689 MPa (100,000 psi) tensile strength, and 20–25% elongation. Steels with carbon contents over 0.30% have the added advantage that they can be hardened, particularly on wearing surfaces, to the extent indicated in Figure 6-4. A carbon content of 0.70% may be chosen for wear resistance without brittleness, as for railroad rails, with a tensile strength of about 827 MPa (120,000 psi) and elongation of 10%. Higher-carbon steels are used largely for tools and dies. Other alloying elements enhance heat treatment and properties of steel as described in Chapters 4 and 6.

3.3.7 Grain Size of Steel

The grain size of a steel nominally refers to the size of the austenitic grains before the steel

Figure 3-16. (A) Microstructure of typical lath martensite, 4% picral + HCl (hydrochloric acid) magnified 200×. (B) Microstructure of typical plate martensite, 4% picral + HCl, magnified 1,000× (Metals Handbook Desk Edition 1998).

is cooled to room temperature. Grain size is important because it influences many of the physical properties of steel. Occasionally, the size of the ferrite grains is important, as in deep-drawing sheet steel, where it is definitely specified as the "ferrite grain size." Actually, the austenitic grain size is not altered much by the rate of cooling to room temperature. Coarse austenitic grains raise hardenability, normalize tensile and creep strength, and improve rough machinability. Fine grains increase impact toughness, improve machining finishes, and mitigate quenching cracks, distortion in quenching, and surface decarburization.

The grain size of a steel depends upon its method of manufacture, heat treatment, the amount of hot and cold working, and alloying elements. A deoxidizer added to a molten steel to eliminate dissolved gases and reduce FeO leaves minute oxide particles in the metal. These act as nuclei and, if numerous, promote a fine-grained steel. Steel recrystallizes when heated above the A_1 line (Figure 3-13). The higher the temperature and the longer the time in the austenitic region, the more the grains grow in size, although the actual amount of growth in any case depends upon the composition of the steel. A highly worked steel starts with many small grains from numerous nuclei when recrystallized. As a rule, alloying elements like vanadium that form carbides or fine oxides tend to increase resistance to grain coarsening. This is thought to occur due to the fact that the carbides and oxides resist solution in the austenite and help fix grain boundaries.

3.3.8 Solidification of Cast Iron

The changes that occur when cast iron cools will be described for a composition of 3% carbon. Solidification from the liquid occurs at about 1,329° C (2,425° F), and austenite, with a lower carbon content, separates from the melt. The remaining liquid is enriched with carbon and the last of it reaches the eutectic composition of 4.3% carbon when the temperature has fallen to the solidus at 1,148° C (2,098° F). At that point, the austenite contains 2.11% carbon and the last of the liquid solidifies as a eutectic mixture (of austenite and primary carbide) known as *ledeburite*. On further cooling at a rate to prevent breakdown of the carbide, secondary carbide is rejected from the austenite in the mixture. At 727° C (1,341° F), the austenite contains 0.77% carbon and is transformed to pearlite. This includes the austenite in the eutectic mixture (ledeburite), and the resulting mixture of primary cementite and pearlite is called *transformed ledeburite*.

With slow enough cooling or with alloys that reduce the stability of iron carbide, graphite is formed instead of iron carbide. That is, graphite is deposited when the eutectic solidifies and is rejected by the austenite on further cooling. At transformation temperature, the austenite may be changed to pearlite or partially or wholly to graphite and ferrite. Graphite commonly exists in cast iron in flakes as shown in Figure 8-34(A), but under certain conditions may also appear as nodules. The iron graphite diagram is slightly different from the iron carbide diagram.

Differences in cooling rates, compositions, alloying elements, and subsequent heat treatment produce a variety of cast irons. The main ones are described in Chapter 8.

3.4 TESTING OF ENGINEERING MATERIALS

Testing is performed on materials, components, and assemblies. It consists of measuring fundamental properties or responses to particular influences such as load, temperature, and corrosive substances. Inspection is closely related to testing and is also applied to materials, components, and assemblies. Inspection is concerned with the geometry of objects, detection of internal defects, and examinations for performance, finish, color, and general appearance. This section is devoted to testing and to those inspection methods (often loosely called test methods) aimed at detection of internal defects. Other kinds of inspection are discussed in Chapter 15.

Tests and inspection methods may or may not be destructive of the object being examined, and so, testing is commonly subdivided into two major areas: destructive and nondestructive testing. Tests are also classified as physical, chemical, or mechanical tests. Physical tests include measurement of such quantities as specific gravity, and electric, magnetic, thermal, and optical properties. These are usually performed in scientific laboratories, rather than in engineering laboratories, and are not mentioned further here. Chemical tests, by which chemical properties are determined, are generally in the realm of the scientist rather than the engineer and, with the exception of corrosion testing, are not discussed in this text. Mechanical tests are most often performed in engineering laboratories. These include measurements of properties such as hardness, strength, and toughness, which, along with inspection for internal flaws, are the object of this section. These tests require special equipment and techniques.

3.4.1 Tension Test

The tension test is one of the most widely used of the mechanical tests. There are many variations of this test, specified in detail by the American Society for Testing and Materials (ASTM), to accommodate the widely differing character of materials, such as metals, elastomers, plastics, and glass.

The common round tension test specimen is shown in Figure 3-17. Its diameter is either 12.5 ±0.25 mm (SI), or .490 ±.010 in. (English), with respective gage lengths of 50.80 ±0.13 mm (2.000 ±.005 in.). Additional dimensional details are given in Figure 3-17.

The specimen is gripped in a machine that applies loads along the axis of the specimen. If necessary, an extensometer for measuring changes in

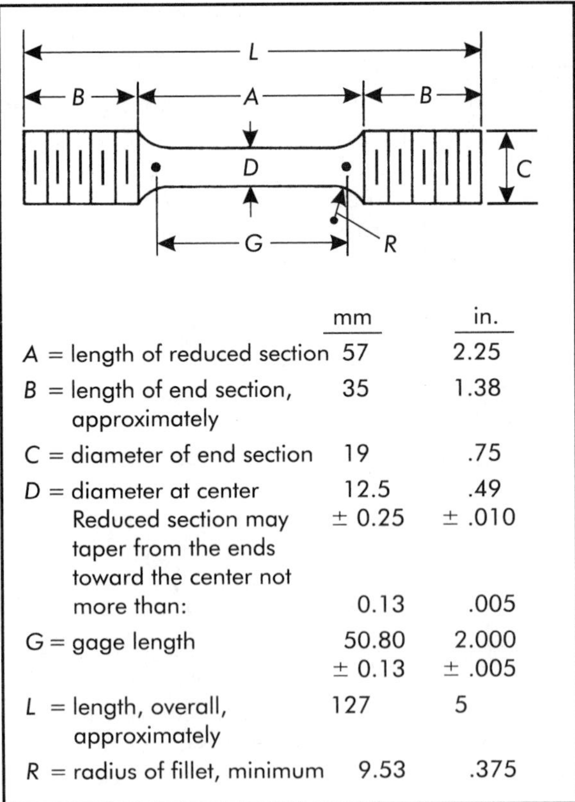

	mm	in.
A = length of reduced section	57	2.25
B = length of end section, approximately	35	1.38
C = diameter of end section	19	.75
D = diameter at center	12.5 ± 0.25	.49 ± .010
Reduced section may taper from the ends toward the center not more than:	0.13	.005
G = gage length	50.80 ± 0.13	2.000 ± .005
L = length, overall, approximately	127	5
R = radius of fillet, minimum	9.53	.375

Figure 3-17. Standard round tension test specimen (ASTM 1973).

length can be attached to the specimen. As the specimen is slowly elongated, simultaneous measurements of applied load and length are either noted manually or recorded automatically. It is convenient to convert the loads to *unit loads* or *stresses*, and the changes in length to *unit changes in length*, or *strains*. Stresses are calculated by dividing the load by the cross-sectional area over which the load acts. Strains are found by dividing the change in length by the gage length, that is, by the original length of the specimen. Stress-strain relationships determined in this way can be applied to specimens and structural members whose dimensions differ from those of the test specimen. The test is completed when the specimen finally breaks.

Typical load and length data for a mild steel, and their conversion to stress and strain are shown in Table 3-1. A plot of stresses and strains obtained from these data is shown in Figure 3-18. It should be noted that, for most ductile metallic materials, stress and strain are initially proportional to one another. Response of this kind is known as *elastic action*. The constant of proportionality between elastic strain e and elastic stress S is known as the *modulus of elasticity* or *Young's modulus*, and is denoted by the symbol E. Young's modulus is indicative of the property called *stiffness*; small values of E indicate flexible materials and large values of E reflect stiffness and rigidity. Comparing a spring of steel with one of brass, each having the same dimensions, shows the steel spring to be stiffer than the brass spring. This is as it should be, for brass has a modulus of elasticity of about 110 GPa (16×10^6 psi), while steel has about 206 GPa (30×10^6 psi). The modulus of elasticity must be taken into consideration in the forming and metal-cutting operations discussed in Chapters 12 and 17.

The property of *springback* is a function of the modulus of elasticity and refers to the extent to which metal springs back when an elastic deforming load is removed. If steel and brass wires of the same diameter are to be coiled to make springs of the same diameter, the brass will have to be bent under load to a coil of smaller diameter than the steel, because brass has a lower modulus of elasticity and so has more springback. (This assumes both the brass and steel have the same yield strength.) In metal cutting, the modulus of elasticity of the workpiece affects its rigidity, and the modulus of elasticity of the cutting tool and the toolholder affect their rigidity. These are major factors in the achievement of dimensional control.

As the stress and strain increase in a tensile test, eventually a point is reached at which stress and strain are no longer directly proportional to one another. The lowest stress at which stress and strain are no longer proportional is known as the *proportional limit*. Beyond the proportional limit, strain increases faster than stress, and the material is permanently deformed. Permanent deformation or set is known as *inelastic action*, and the first stress at which this occurs is known as the *elastic limit*. Strains below the elastic limit are known as *elastic strains;* elastically strained objects recover their original dimensions upon removal of a load. In mild steels, with increasing strain, a point is reached at which strain increases without further increase in stress. This stress is known as the *yield point*. The yield point is more readily determined than the proportional limit or the elastic limit. It is taken as a practical measure of the limit of elastic action, that is, the stress above which the material is permanently deformed. The practical limit of elastic action is extremely important in manufacturing operations such as rolling, drawing and deep drawing, and spinning, for it shows what stresses must be exceeded if permanent deformation is to be achieved (see Chapters 11 and 12).

Determination of the limit of elastic action in metals and alloys that do not have a yield point depends upon the measurement of a limiting value of stress required to produce an arbitrarily selected amount of permanent strain. The selected strain is often 0.2% or 0.002 mm/mm (.002 in./in.) strain. The corresponding stress is called the *yield strength*. Yield strengths, like yield points, are more easily determined than elastic limits. Both, of course, indicate practical limits of elastic action. Yield strength measurements for mild steels give essentially the same values as yield points.

If an extensometer is not available for making the strain measurements needed for yield point and yield strength determinations, approximate values can be obtained by coating the

Table 3-1. Stress-strain data for mild steel

D_o = initial diameter = 12.75 mm (.502 in.)
D_f = final diameter = 7.47 mm (.294 in.)
L_o = initial length = 50.80 mm (2.000 in.)
A_o = initial area = 127.7 mm² (.198 in.²)
A_f = final area = 43.9 mm² (.068 in.²)
L_f = final length = 68.6 mm (2.700 in.)

Load kN (lbf)	Length mm (in.)	Diameter mm (in.)	Stress MPa (psi)	Strain
2.6 (590)	50.805 (2.0002)		20.7 (3,000)	0.0001
5.0 (1,130)	50.810 (2.0004)		39.3 (5,700)	0.0002
7.6 (1,700)	50.815 (2.0006)		59.3 (8,600)	0.0003
10.6 (2,380)	50.820 (2.0008)		82.7 (12,000)	0.0004
12.8 (2,870)	50.825 (2.0010)		100.0 (14,500)	0.0005
15.6 (3,510)	50.831 (2.0012)		122.0 (17,700)	0.0006
18.5 (4,160)	50.836 (2.0014)		144.8 (21,000)	0.0007
20.7 (4,650)	50.841 (2.0016)		162.0 (23,500)	0.0008
23.5 (5,290)	50.846 (2.0018)		184.1 (26,700)	0.0009
26.4 (5,940)	50.851 (2.0020)		206.8 (30,000)	0.0011
29.9 (6,730)	50.861 (2.0024)		234.4 (34,000)	0.0012
31.7 (7,120)	50.876 (2.0030)		248.2 (36,000)	0.0015
32.7 (7,340)	50.902 (2.0040)		255.8 (37,100)	0.0020
32.3 (7,250)	50.927 (2.0050)		252.4 (36,600)	0.0025
31.6 (7,110)	50.952 (2.0060)		247.5 (35,900)	0.0030
32.6 (7,330)	50.978 (2.0070)		255.1 (37,000)	0.0035
34.3 (7,720)	51.003 (2.0080)		268.9 (39,000)	0.0040
37.2 (8,370)	53.290 (2.0980)	11.9 (.469)	291.7 (42,300)	0.049
42.3 (9,500)	— —	— —	331.0 (48,000)	—
37.2 (8,370)	65.024 (2.5600)	9.86 (.388)	291.7 (42,300)	0.28
26.1 (5,870)	68.580 (2.7000)	7.47 (.294)	204.8 (29,700)	0.35

surface of the specimen with a brittle lacquer that will flake off at the yield load. If an unmachined, hot-worked, scale-coated rod is tested, yielding causes the scale to pop off, and the corresponding load can be taken as the yield load.

Beyond the limit of elastic action, the tensile specimen at first elongates quite uniformly along its length. Eventually, the elongation tends to become concentrated in one region, and since the volume of the material does not change, this results in development of a constriction. This is called *necking down*. The change in length divided by the original gage length is expressed as

$$E = \frac{L_f - L_o}{L_o} \times 100 \quad (3\text{-}3)$$

where

E = elongation, %
L_f = final length, mm (in.)
L_o = initial length, mm (in.)

The percentage of elongation is a factor in the ability of the material to deform inelastically. This ability is called *ductility*. The change in area divided by the original cross section of the specimen is expressed as

$$R_A = \frac{A_o - A_f}{A_o} \times 100 \quad (3\text{-}4)$$

where

R_A = reduction of area, %
A_o = original area of cross section, mm² (in.²)
A_f = final area of cross section, mm² (in.²)

This is also a factor in the ability of the material to deform inelastically. Lack of ductility is known as *brittleness*. Ductility and brittleness are very important in forming operations, for they tend to determine just how severely the material can be deformed without tearing or rupturing. Annealing (see Chapter 6) tends to restore ductil-

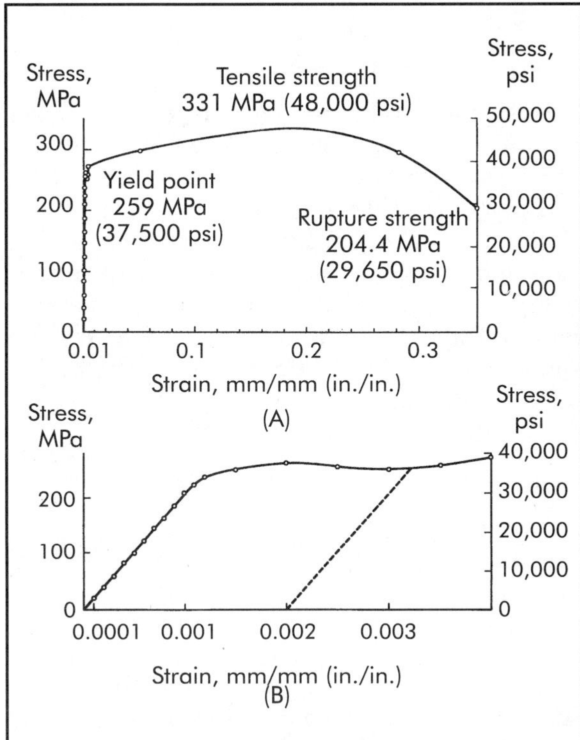

Figure 3-18. (A) Stress-strain curve for a mild steel plotted from data given in Table 3-1; (B) a portion of the same curve shown in part (A) with the strain axis expanded to permit determination of the yield strength and modulus of elasticity.

ity to cold-deformed metallic materials so that forming operations can be continued (see Chapters 11 and 12). Ductility is related also to metal cutting (Chapters 17 and 18). Generally, excessive ductility is associated with machined parts that develop rough surface finishes and high rates of tool wear.

The tensile stress-strain relationship just described, and for which a curve is plotted in Figure 3-18, is known as the *apparent stress-strain* or the *engineering stress-strain relationship*. The true, or actual, stress at any given instant during the test is greater than the apparent, or engineering, stress because the actual area over which the load is distributed is smaller than the initial area. Thus,

$$\sigma = F/A_a \tag{3-5}$$

$$S = F/A_o \tag{3-6}$$

$$\sigma > S \tag{3-7}$$

where

σ = true stress, MPa (psi)
F = applied load, kN (lbf)
S = engineering stress, MPa (psi)
A_a = actual area over which load is distributed, mm^2 ($in.^2$)
A_o = initial area over which load is distributed, mm^2 ($in.^2$)

The difference between the actual and initial areas is negligibly small until the specimen begins to neck down. Likewise, increments of true strain are defined as

$$dL/L \tag{3-8}$$

where

dL = incremental change in length caused by an incremental increase in load, mm (in.)
L = actual gage length at any moment, mm (in.)

The total true strain at any load is found from

$$\in = \int_{L_o}^{L_l} \frac{dL}{L} = ln\frac{L_l}{L_o} \tag{3-9}$$

where

$\in$ = total true strain, MPa (psi)
l = load, kN (lbf)
L_o = initial length, mm (in.)
ln = natural (hyperbolic) logarithm
L_l = gage length under load l, mm (in.)

Since the volume remains constant during straining, then

$$V_o = A_oL_o = V_l = A_lL_l \tag{3-10}$$

where

V_o = volume before load is applied, mm^3 ($in.^3$)
A_o = initial area over which load is applied, mm^2 ($in.^2$)
L_o = gage length before load is applied, mm (in.)
V_l = volume under load l, mm^3 ($in.^3$)
A_l = initial area under load l, mm^2 ($in.^2$)

So,

$$\in = ln\frac{A_o}{A_l} \tag{3-11}$$

Since

$$e = \frac{L_l - L_o}{L_o} = \frac{L_l}{L_o} - \frac{L_o}{L_o} \tag{3-12}$$

where

e = apparent strain

then

$$e+1=\frac{L_l}{L_o}$$

and

$$\epsilon = ln\ (e+1) \tag{3-13}$$

True stresses and true strains have been calculated from the data of Table 3-1 and shown in Table 3-2. Beyond the point of necking down, true strains should not be calculated from Equation (3-13), which uses apparent strain. Errors result from such a calculation because the strain is not uniformly distributed over the length of the specimen. This source of error is avoided if Equation (3-11) is used. However, the true stress calculation is still not exact because stress intensification occurs as a result of the notch that accompanies necking down. Therefore, the real stresses are even greater than calculated true stresses. These latter errors do not, however, seriously detract from the value of true stress-true strain relationships calculated using the preceding equations. For refined results needed in critical work, tension-testing, machine-computer systems are available on which the loads, displacement, and specimen responses are continuously measured and analyzed. The conditions of the test are altered as necessary and interpreted to yield a true stress-strain relationship.

A plot of true stress versus true strain in the *elastic* range yields a straight line identical with the engineering stress-strain curve, since

$$\sigma = S \text{ and } \epsilon = e \tag{3-14}$$

where

σ = true stress, MPa (psi)
S = engineering stress, MPa (psi)
ϵ = total true strain, mm/mm (in./in.)
e = apparent strain, mm/mm (in./in.)

See Figure 3-19 and compare with Figure 3-18. Similarly, a log-log plot of *elastic* true stress-true strain yields a straight line since

$$\sigma = E\epsilon \tag{3-15}$$

and

$$\log \sigma = \log E + \log \epsilon$$

where

E = modus of elasticity

This is shown in Figure 3-4. The value of E can be obtained by extrapolation of the plot to a strain $\epsilon = 1$, for which $\log \epsilon = 0$ and $\log \sigma = \log E$.

A plot of true stress versus true strain in the *plastic* range on linear coordinates deviates from the engineering stress-strain plot, as can be seen in Figure 3-19. A plot of the true stress-strain data of Table 3-2 on a log-log plot is shown in Figure 3-20. In the plastic region, a straight line is obtained, indicating that

$$\sigma = k\epsilon^m \tag{3-16}$$

or

$$\log \sigma = \log k + m \log \epsilon \tag{3-17}$$

where

m, k = constants

The value of k can be determined by substituting the value for $\log \sigma$ at a strain $\epsilon = 1$, for which $\log \epsilon = 0$. That is, k is the true stress at a true strain = 1. Knowing k and substituting corresponding values of $\log \sigma$ and $\log \epsilon$ in Equation (3-17) yields a value for m. The constant m is a property of the metal or alloy and is known as the *work-hardening coefficient*.

The *engineering tensile strength*, or apparent tensile strength (also called ultimate strength), is the figure given in most reference books. It is obtained from

$$S=\frac{F_m}{A_o} \tag{3-18}$$

where

S = engineering stress, MPa (psi)
F_m = maximum or ultimate load sustained by the specimen, kN (lbf)
A_o = initial area over which load is applied, mm^2 ($in.^2$)

The *true tensile strength* is obtained from

$$\sigma=\frac{F_R}{A_f} \tag{3-19}$$

where

σ = true stress, MPa (psi)
F_R = breaking or rupture load, kN (lbf)
A_f = final area, mm^2 ($in.^2$)

Table 3-2. True stresses and true strains as determined from Table 3-1

Load kN (lbf)	Diameter mm (in.)	Area, A_l mm² (in.²)	True Stress, σ MPa (psi)	True Strain, ϵ A_l/A_l	$ln\ (A_0/A_l)$
37.2 (8,370)	11.90 (.469)	111.5 (.1728)	333.7 (48,400)	1.14	0.131
42.3 (9,500)	— —	— —	— —	—	—
37.2 (8,370)	9.86 (.388)	76.3 (.1182)	488.1 (70,800)	1.68	0.519
26.1 (5,870)	7.47 (.294)	43.8 (.0679)	596.4 (86,500)	2.92	1.072

Note: Apparent and true values are practically identical for loads up to 34.3 kN (7,720 lb) and are not tabulated again. Refer to Table 3-1.

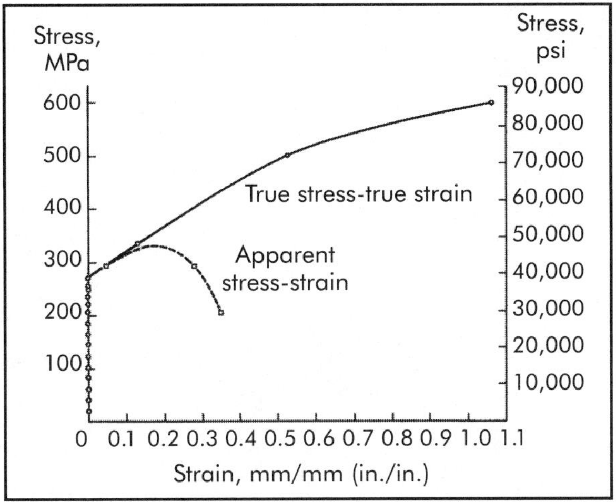

Figure 3-19. Comparison of the true stress-true strain curve with the apparent stress-apparent strain curve (both plotted on a linear scale). The data from which these curves were constructed appears in Tables 3-1 and 3-2.

A comparison of Figure 3-18 with Figure 3-19 shows that the true tensile strength is always greater than the engineering tensile strength. The true stress-true strain relationship is important for it shows that metallic materials continue to get stronger as the amount of cold deformation continues, up to the point of rupture. It is for this reason that as drawing and cold work proceed (see Chapters 11 and 12), the stresses involved become larger. The extent to which a metal or alloy is strengthened by cold work is indicated by the work-hardening coefficient m. Large values of m show that strength and hardness increase more for a given amount of cold work than when values of m are small. The value of m is indicated by the slope of the log-log plot of the true stress-true strain curve in the plastic region.

Work-hardened metals affect metal cutting (see Chapters 17 and 18) because the chip becomes more severely deformed before it finally breaks away from the workpiece. The great difficulty of machining austenitic stainless steels can be explained on the basis of the large value of the work-hardening coefficient of these alloys. Since austenitic steels (Chapter 4) are quite ductile, a great deal of work hardening occurs before the point of rupture is reached. This adversely affects power consumption and tool wear.

The area under the stress-strain curve indicates the *toughness*, that is, energy which can be absorbed by the material up to the point of rupture. Although the engineering stress-strain curve is often used for this computation, a more realistic result is obtained from the true stress-true strain curve. Toughness is expressed as energy absorbed in the material being deformed, expressed in J/cm³ (in.-lb/in.³). This is obtained from the stress-strain curve by multiplying the ordinate in MPa by the abscissa in mm/mm = m/m (psi by in./in.).

3.4.2 Hardness Testing

Although hardness testing does not directly give as much detailed information as tensile testing, it is so fast and convenient that it is more widely used. *Hardness* is usually defined as resistance of a material to penetration. In the most generally accepted tests, an indenter is pressed into the surface of the material by a slowly applied known load, and the extent of the resulting impression is mechanically or optically measured. A large impression for a given load and indenter indicates a soft material, and the opposite is true for a small impression.

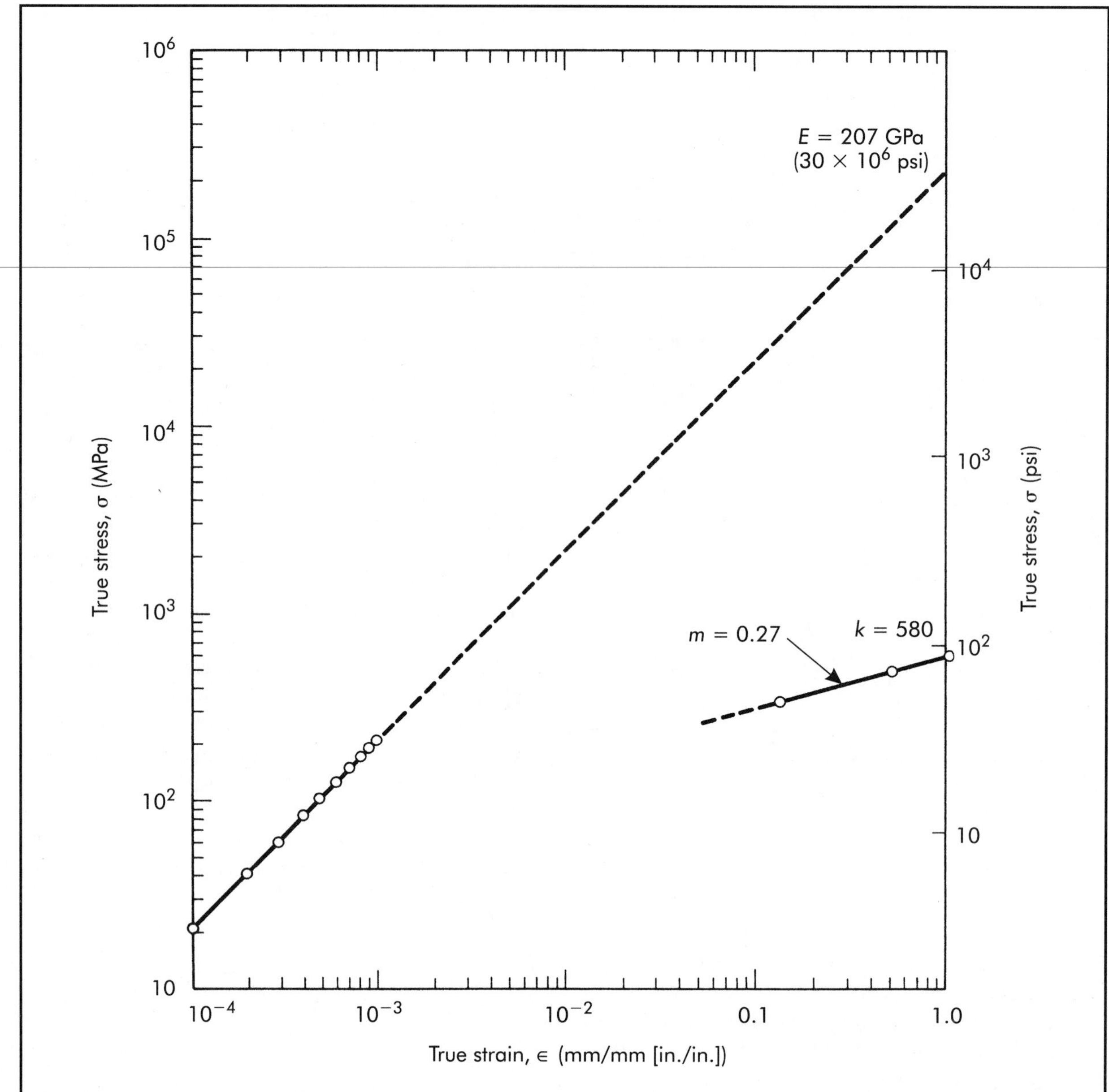

Figure 3-20. True stress-true strain curve for a mild steel, plotted on a log-log scale. The data from which this curve was constructed appears in Tables 3-1 and 3-2.

Hardness is primarily a function of the elastic limit (that is, yield strength) of the material, and to a lesser extent a function of the work-hardening coefficient. The modulus of elasticity also exerts a slight effect on hardness. Since for a given metallic material the elastic limit depends upon such factors as the previous history of the material, hardness measurements are useful in determining if processing techniques are meeting their objectives. Among the processing techniques that can be successfully monitored are heat treating (see Chapter 6), casting and foundry processes (Chapters 8 and 9), forming processes (Chapters 11 and 12), and welding and joining processes (Chapter 13). Since the elastic limit, tensile strength, ductility, and toughness bear a fixed relationship to one another for a given material having a given history, it is possible to deduce these mechanical properties from hardness readings on a given material whose history is known. The most common hardness tests can be classed as macro- or microhardness tests. Macrohardness tests scrutinize a fairly large area of the surface and their impressions are visible to the naked eye. Microhardness test impressions are very small, requiring a microscope to see them.

The Brinell hardness tester forces a hardened steel or carbide ball with a diameter of 10 mm (.39 in.) into a metal by means of a fixed load. A 29,420 N (6,614 lbf) load is used for testing ferrous alloys and alloys of similar hardness. When brass and soft alloys are tested, a 4,903 N (1,102 lbf) load is used. The time of loading is specified between 10 and 30 seconds, depending upon the alloy being examined. After the load is removed, the diameter of the impression is measured to the nearest 0.01 mm (.00039 in.) by a microscope or a laser scanner for automatic reading. The hardness, which is actually the load divided by the area of the impression, is read directly from tables for which hardness has been calculated for various diameter impressions.

The Brinell penetration is so large that it results in an averaging effect and is not as sensitive to surface roughness, light scale, or dirt, as are the Rockwell and microhardness tests. However, specimens must be thicker, successive impressions must be made further apart, and the test is destructive if applied to small components or specimens. Representative Brinell hardness numbers (Bhns) are 425 for white cast iron (very hard), 160 for gray cast iron, and 105 for wrought iron (very soft).

The Rockwell hardness test is probably the most widely used method of hardness testing. Rockwell testers use much smaller penetrators and loads than the Brinell tester. Four sizes of hard balls from 1.6–12.7 mm (.063–.5 in.) in diameter are available, as well as a cone-shaped diamond. For testing metallic materials, the 1.6 mm (.063 in.) ball and the diamond penetrator are most commonly used. The penetrator chuck is mechanically connected to a dial indicator that responds to vertical motion of the penetrator (Figure 3-21). Since the penetrators are small, the surface of the specimen should be ground smooth and clean. The specimen is placed on the anvil of the machine and the penetrator is seated by means of a 98-N (22-lbf) minor load. The dial indicator is zeroed and then a major load of 588, 981, or 1,471 N (132, 221, or 331 lbf) is applied, forcing the penetrator into the specimen. Upon removal of the major load, the indented specimen recovers slightly, and the final depth of penetration is registered directly on the dial indicator as a hardness number. Various combinations of penetrator and major load are used and designated by a series of letters. The two most common scales are the R_B scale and the R_C scale, respectively standing for the 1.6 mm (.063 in.) ball with the 981-N (221-lbf) major load, and the diamond penetrator with the 1,471-N (331-lbf) major load. In general, very hard materials are tested with the diamond penetrator. Mild steel might have a R_B reading of 90; hardened alloy steel might have a R_C of 55. These are stated as 90 R_B and 55 R_C.

The impressions made by the Rockwell tester are much smaller than those of the Brinell penetrator. Hence the Rockwell test should not be used on nonhomogeneous alloys, such as cast iron, because it is subject to errors from small

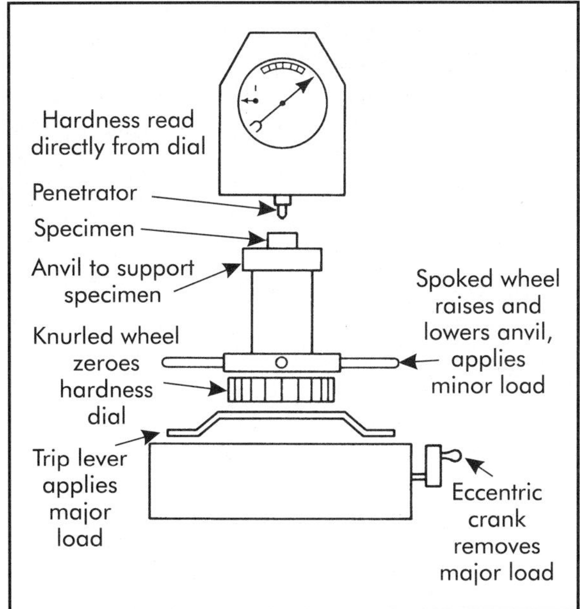

Figure 3-21. Schematic of Rockwell hardness tester. Weights, not shown, hang from rear of machine.

soft or hard spots, and from small voids. On the other hand, since the impressions are so small, the test is considered nondestructive in many applications. In addition, thinner specimens can be tested by the Rockwell than by the Brinell test. The Rockwell test can be used to test materials over a greater range of hardness because of the many combinations of penetrators and loads available. It can be used on plastics as well as metallic materials.

A series of superficial Rockwell scales are available for testing the hardness of very thin materials or case-hardened steels (see Chapter 6). The superficial Rockwell tester uses a 29-N (6.6-lbf) minor load and major loads of 147, 294, or 441 N (33, 66, or 99 lbf).

The most frequently used microhardness testers are the Vickers and Tukon testers. (The Vickers is not always considered to be a microhardness tester.) Both use a diamond penetrator. The Vickers diamond penetrator produces a square-shaped impression; the Knoop penetrator used with the Tukon tester produces a diamond-shaped impression (see Figure 3-22). Impressions formed by both machines are measured by means of microscopes, and the hardness number is obtained from tables listing the load and diagonal measurement of the impression. The Tukon tester produces the smaller impression of the two, and is a true microhardness tester. It can be used to measure the hardness of alloy components and to measure the changes in hardness at minute increments below the surface of

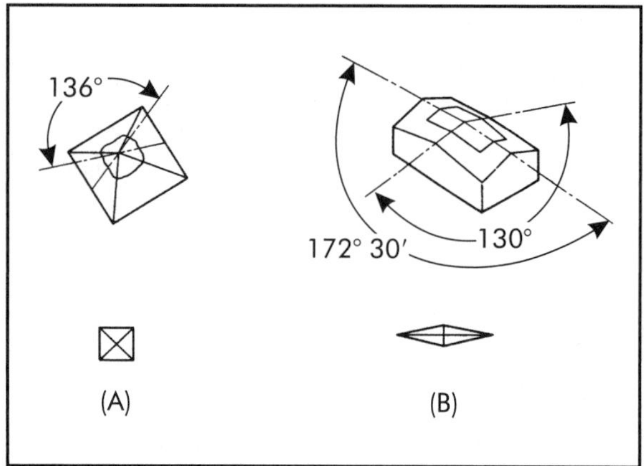

Figure 3-22. (A) Vickers diamond penetrator and impression; (B) Knoop diamond penetrator and impression.

case-hardened steel. The Tukon tester has been applied to studying metal-cutting operations to measure the hardness over small areas of chip and workpiece surface. Both tests require careful polishing and etching of the specimen surface before measurements are made.

A less used hardness tester is the scleroscope. A diamond-tipped hammer is dropped 254 mm (10 in.) on the workpiece surface, and the height of its rebound is read on a graduated scale to obtain a hardness number. This device mars the surface the least and can be easily moved and applied to huge pieces.

3.4.3 Notched-bar Impact Testing

Contrary to what might be expected from the name, impact testing does not provide a means of studying response of materials to high-velocity loading. The results obtained would not greatly differ if the loads were slowly applied, as in the tensile test. Notched-bar impact testing does provide a quick way of loading and measuring the toughness of a notched bar, that is, its ability to absorb energy. The results are not directly comparable to the results obtained by integrating the area under the tensile stress-strain curve because of the variable response of materials to the effects of notching. However, the test results are useful in comparing, for a given composition, the effects of prior history on toughness. Of particular interest in this regard is the effect of heat treatment on the properties of steel (see Chapter 6). Overheating steel impairs toughness. This can be readily detected by a notched-bar impact test and by examining the appearance of the fractured surface.

Another important application of notched-bar impact testing is to study the effects of tempering cycles on hardened steels. Certain cycles are harmful to the toughness of some alloy steels, and this test is useful in detecting these. Still another application is in the determination of embrittling temperatures, particularly for ferrous alloys. Many steel compositions tend to lose toughness as subfreezing temperatures are approached. The notched-bar impact test provides a means of determining the embrittling range for different compositions.

The most common kinds of impact tests use notched specimens loaded as beams. The beams may be simply loaded (Charpy test) or loaded as cantilevers (Izod test). The notch is usually a V-notch cut to specifications with a special milling cutter. Other types of notch have been used but they are not popular (see Figure 3-23).

The specimen is held in a rigid vise or support, and struck a blow by a pendulum traveling at a specified speed; for example, 5.3 m/sec (17.5 ft/sec). The energy input is a function of the height of fall and the weight of the pendulum. The energy remaining after fracture is determined from the height of rise of the pendulum and its weight. The difference between the energy input and the energy remaining represents the energy absorbed by the specimen. Modern machines are equipped with scales and pendulum-actuated pointers that yield direct readings of energy absorption (see Figure 3-23).

The function of the notch is to insure that the specimen will break as a result of the impact load to which it is subjected. Without the notch, many alloys would simply bend without breaking, and it would be impossible to determine their ability to absorb energy.

The appearance of the surface of the fractured impact specimen can be correlated with the energy absorbed by a given composition of specimen. Ductile, tough compositions produce fractures with a fine, silky appearance, and evidence of deformation caused by bending before fracture. Brittle compositions produce a surface that appears coarsely crystalline and shows no evidence of deformation by bending. There are, of course, degrees of brittleness and toughness, and the proportion of surface that is coarse versus silky shows the variations in properties. The transition from ductile to brittle fracture as the temperature is lowered, when such a transition occurs, can be followed by observation of fracture appearance.

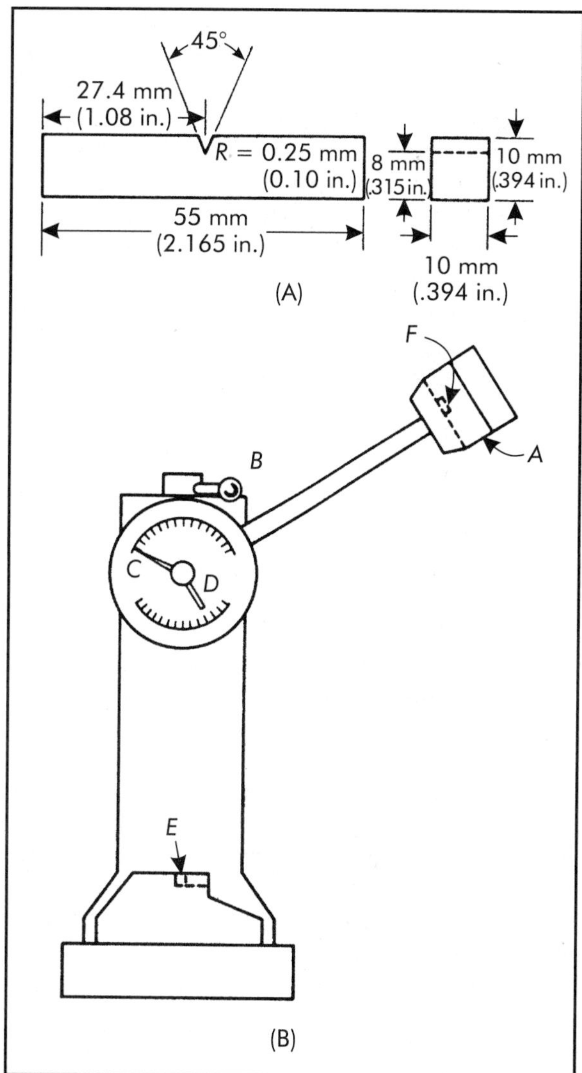

Figure 3-23. (A) Charpy impact specimen; (B) typical impact testing machine. A, pendulum; B, release and brake lever; C, pointer and scale to indicate energy absorbed; D, drive arm that pushes pointer around scale; E, anvil on which Charpy specimen rests; F, striker head.

3.4.4 Bend Tests

Multipurpose bend tests enable determination of maximum fiber stresses, known as *flexure strength*, and modulus of elasticity in materials that are brittle or very hard, which do not lend themselves to tensile testing. Plastics, glass, cast iron, and concrete are commonly tested as flexed beams. Bending tests are also useful for ductility determinations in sheet and barstock, and for testing the soundness of welds. These tests often depend upon achieving a certain angle of bend, and tend to be qualitative rather than quantitative.

3.4.5 High-temperature Tests

The kinds of tests described thus far can be conducted at high temperatures. The techniques are similar to those previously described. However, a furnace (often a vacuum furnace) to heat the specimens before or during the test is provided.

High-temperature variations of tensile, hardness, and impact testing are considered *short-time tests*. Some special considerations regarding high-temperature testing should be mentioned. Among these are the effects of speed of heating and time at temperature of the test piece before and during testing. Since the structure and properties of metallic materials may be significantly altered by high temperatures as discussed in this chapter and Chapter 6, these variables must be controlled. In hot-hardness testing, the effect of the high temperature and atmosphere on the penetrator must be kept in mind. Diamond penetrators will oxidize in the air, and diamond will dissolve in hot steel. Evacuated surroundings or special atmospheres are used to protect against oxidation and sapphire indenters are used to prevent dissolution.

In addition to the short-time, high-temperature tests, there are *long-time high-temperature tests* of importance. *Creep testing* aims at relating time, temperature, stress, and strain. Tests may run from a period of days to many years. For each different stress-temperature combination an individual specimen is required. Since each specimen requires its own loading device and strain measuring and recording equipment, the facilities and space needed are considerable. Some efficiencies can be realized by using multi-specimen furnaces.

The progress of an ideal creep test is shown in Figure 3-24 for a fixed stress and temperature. Upon initially loading the specimen, instantaneous strain results, composed of elastic and plastic components. Following this, the rate of plastic strain declines, during what is known as the *primary stage of creep*, until a steady state is reached. While the rate of creep is constant, the specimen is said to be undergoing *secondary creep*. Finally, a stage is reached during which small internal cracks form and grow. These act to intensify stress and increase the rate of strain until rupture occurs. The three stages are not always well defined, and the amount of creep occurring in the third or final stage is often variable and unpredictable. Extrapolations to predict behavior beyond the time limits of the test are occasionally made, based on the assumption that the steady state will continue to the time limit for which the extrapolation is made.

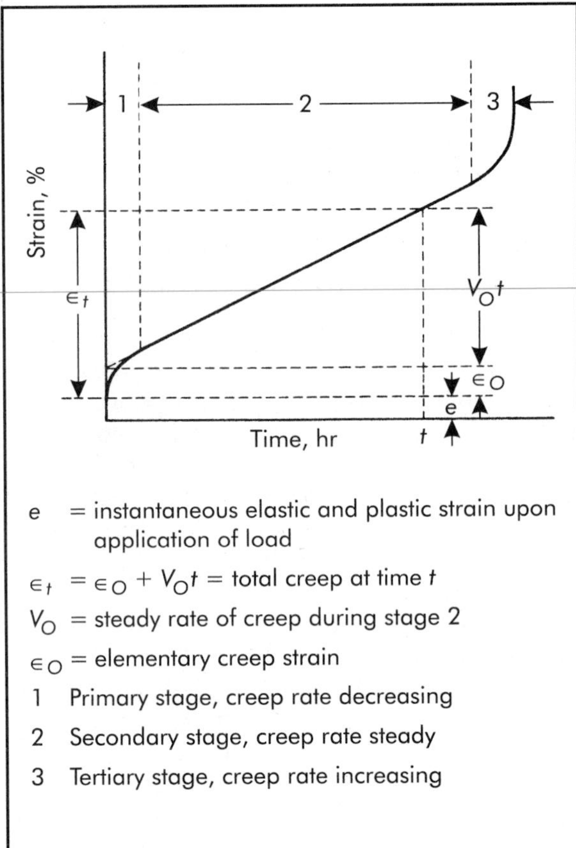

Figure 3-24. Idealized creep curve.

Stress-rupture testing is also used for long-time, high-temperature research. In this type of test, the stress, strain rate, and time are such that the amount of strain at rupture is not considered important. Thus, only stress and temperature are controlled, and time to rupture is recorded. Strains are not measured. Data for stress-rupture behavior can be obtained from relatively short-time tests mathematically manipulated to make predictions of long-time behavior under a variety of stress-temperature combinations.

Creep and stress-rupture tests do not provide information relevant to processing methods. However, the high-temperature properties are very important in design applications, such as gas and steam turbines, high-pressure steam lines, rocket engines, pressurized nuclear power sources, and chemical-processing equipment that operates at high temperatures and pressures.

3.4.6 Fatigue Testing

Probably more metallic components fail from fatigue than any other mechanical cause. For this reason, fatigue testing and determination of fatigue properties are extremely important.

Fatigue failure occurs as a result of repeated application of small loads individually incapable of producing detectable plastic deformation. Eventually, these repeated loads cause a macrocrack to open and spread across the piece. Stress intensification occurs and ultimately a sudden, brittle fracture results. Ferrous metals and alloys have a limiting value of repeated stress that can be applied and reversed for an indefinitely large number of cycles without causing failure. This stress is known as the *fatigue limit*. As illustrated in Figure 3-25, the fatigue limit is not reached even beyond 10^8 cycles. For these materials, fatigue strengths are usually given as stresses for which failure can be expected to occur at 10^7, 10^8, or some other specified number of cycles of loading. The results of fatigue tests are usually plotted on semilog plots as shown in Figure 3-25.

Fatigue tests are most often performed on a specimen having a round cross section, loaded at two points as a rotating simple beam, and supported at its ends (see Figure 3-26). The top surface of such a specimen is always in compression and the bottom surface is always in tension. The maximum stress always occurs at the surface, halfway along the length of the specimen, where the cross section is a minimum. For each complete rotation of the specimen, a point in the surface originally at the top center goes alternately from a maximum in compression to a maximum in tension and then back to the same maximum in compression.

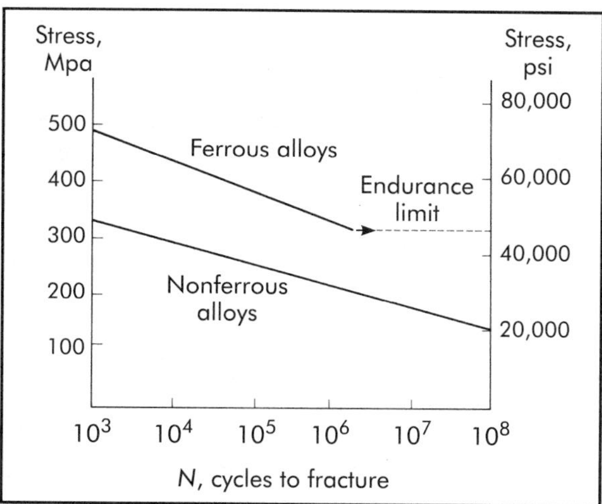

Figure 3-25. Idealized fatigue curves for ferrous and nonferrous alloys. For ferrous alloys, an endurance limit, below which stress failure does not occur, is reached near 10^6 to 10^7 cycles. For nonferrous alloys, a fatigue limit is not reached, even beyond 10^8 cycles.

Fatigue failure starts at the point of highest stress. This point may be determined by the shape of a part; for instance, by stress concentrations in a groove. It can also be found in tool marks or scratches on the surface finish, in internal voids such as shrinkage cracks and cooling cracks in castings and weldments (see Chapters 8, 9, and 13), in defects introduced during mechanical working (see Chapter 12), and in defects and stresses introduced during electroplating (see Chapters 26 and 27). It must be remembered that surface and internal defects are stress raisers, and the point of highest actual stress may occur at these rather than at the minimum cross section of highest nominal stress. Thus processing methods are extremely important as they affect fatigue behavior.

Another important characteristic of fatigue behavior, no doubt related to the distribution of defects in metals, is the wide spread in results. Fatigue test data should be subjected to statistical analysis, and it should be remembered that predictions based on tests reflect the probabilities of failure after a certain number of cycles at a particular stress. The fact that, in service, the stress pattern is rarely a regular repeated cycle, but rather a variable cycle, is also a complicating feature.

Fatigue testing of rubber and plastics is of two types: static fatigue testing is analogous to creep testing of metals; and dynamic fatigue testing corresponds to fatigue testing of metals.

3.4.7 Fracture Toughness Tests

Fracture toughness tests are an extension of notched impact tests and fatigue and static tensile tests of notched specimens. The technique gives results useful in predicting the ability of a material to arrest crack propagation and thus prevent eventual catastrophic failure at stresses below the yield strength. Fracture toughness tests

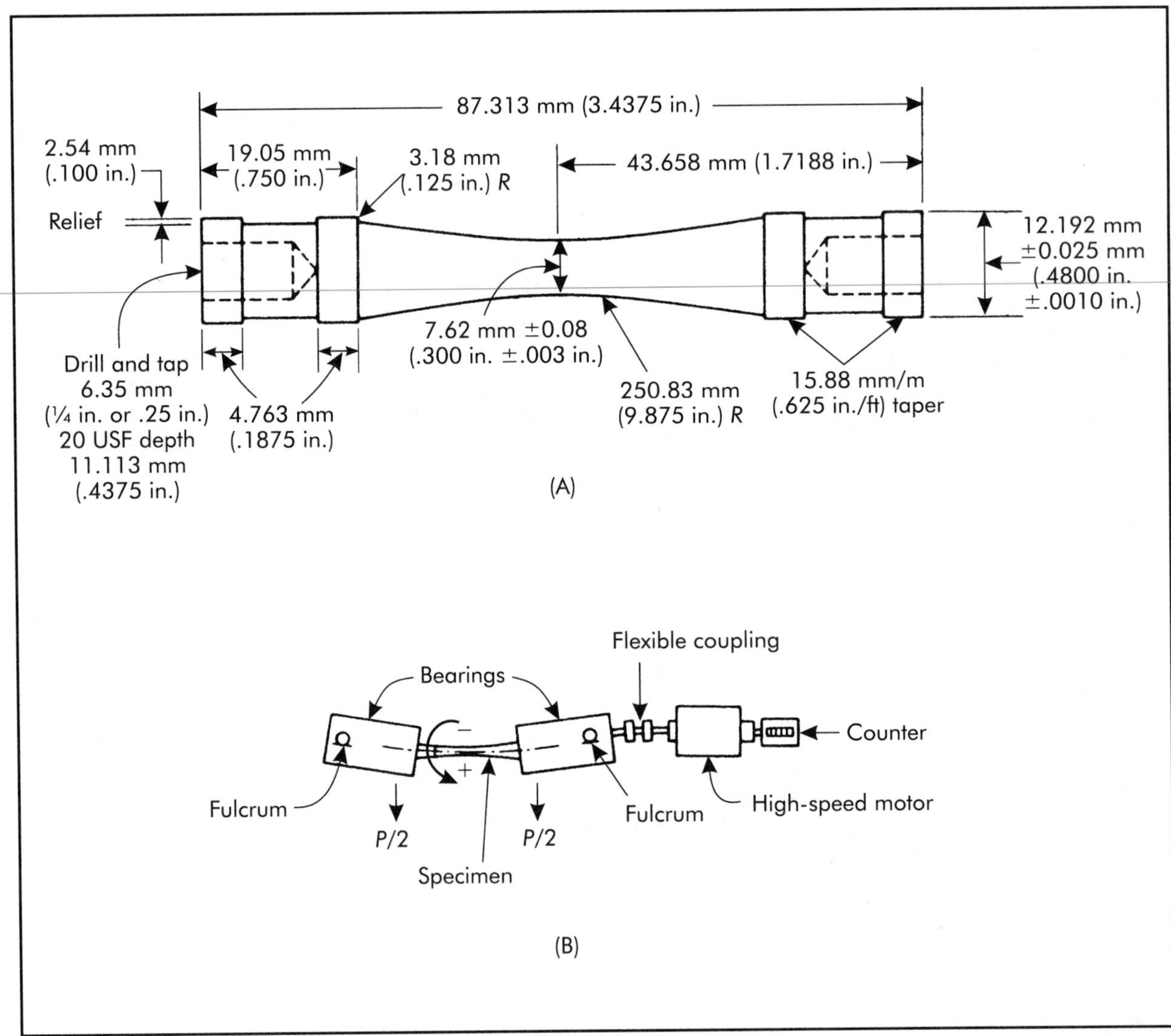

Figure 3-26. (A) Fatigue test specimen; (B) R. R. Moore fatigue-testing machine. The distance between the fulcrums is 305 mm (12 in.); between the points of loading and the nearest fulcrums is 102 mm (4 in.); and between the points of loading is 102 mm (4 in.). In a half-revolution of the specimen, the top fibers go from compression to an equal and opposite tension, and the bottom fibers go from tension to an equal and opposite compression.

are useful because they take into account the presence of cracks too small to be detected by nondestructive tests, but which are assumed to be present in all components. The tendency for these cracks to grow depends upon the material and its processing history. Studying fracture toughness as a function of processing variables, for instance, tempering temperatures or welding techniques (see Chapters 6 and 13), enables matching of materials and processing for the best performance. In particular, fracture toughness testing has been applied to such components as pressure vessels, rocket motor cases, and high-performance aircraft structural parts.

Fracture toughness tests are carried out on tensile specimens in which small fatigue cracks of controlled and measurable size have been initiated. The results are then mathematically manipulated to obtain a parameter known as the *critical stress intensity factor* or *fracture toughness*. This factor can be applied to design and to predictions of failure based on the assumption that defects of a certain maximum size are present. The end product is then tested by nondestructive means to make sure it has no flaws larger than those considered in the design.

3.4.8 Nondestructive Testing

Most nondestructive test (NDT) methods are intended to detect internal flaws likely to cause fatigue or static load failure. These tests are performed on work in process so that defective parts can be rejected early, thus saving additional expenditure of effort. The tests are also performed on used components during disassembly of equipment that has been in service in an attempt to detect the start of fatigue cracks. The most common NDT methods are briefly described here.

X-ray, gamma ray, and *neutron radiography* act upon sensitive photographic film or paper. Neutrons first strike a transfer screen to produce a radiation image after they have passed through the material. Radiation is passed, absorbed, and scattered by opaque objects. The thicker and more dense the object, the smaller the proportion of incident radiation passed. Thus, if an object has an internal defect, more radiation will be passed through the defective area than through the sound region, and the defect will show up as a dark area on the film. Unfortunately, scattering tends to obscure the defect, so that small defects may not be detected. Plane defects, such as the surfaces produced by a crack, are not detected by X-rays, but three-dimensional defects, such as gas porosity in a casting, are quite readily discerned. Sand and slag, being much less dense than metal, are also easily found by X-raying castings.

Although X-ray equipment is available in many sizes and capacities and gives the best results for most applications, other methods are superior in some cases. Gamma-ray sources can be small and convenient to use, for example, to inspect from inside a pipe. The short wavelengths of gamma rays are effective for thick material sections. The absorption of neutrons is, figuratively, the reverse of X-rays or gamma rays. For instance, neutrons are more readily absorbed by hydrogen than by lead. Thus neutron radiography can show up conditions not revealed by the other methods. In all events, care must be taken to avoid the hazards of radiation to human beings.

Magnetic methods for detection of defects are based on the principle that flaws or discontinuities distort a magnetic field induced in a ferrous material. There are several methods for detecting the irregularities. The most common is *magnetic-particle testing* or *Magnaflux™ testing*, which involves dusting the magnetized piece with fine magnetic particles or coating it with a liquid suspension of such particles. The particles are held at leaks in the magnetic field at the edges of cracks. Fluorescent magnetic particles and ultraviolet light serve in a refinement of the process to enhance the appearance of the crack. Magnetic particle testing detects only cracks on or near the surface.

Some electromagnetic methods act by inducing eddy currents into a specimen and detecting the forms of, or irregularities in, the induced currents or in magnetic fields in ferrous materials. In one method, the response is compared to that from a standard piece. These techniques may be applied to detecting flaws (even deep internal ones); separating mixed alloy stock; gaging size, shape, plating, or insulation thickness; or the depth of case hardening of steel.

Ultrasonic testing can be applied to a wide variety of materials and shapes, including metals, plastics, ceramics, glass, laminates, castings, weldments, and forgings. An ultrasonic wave is transmitted through a substance and its action is interpreted in three ways to show defects. In the *pulse-echo technique*, waves are reflected back from the far surface and from the internal surfaces of flaws. An oscilloscope trace of the reflections (spikes) indicates the position and sizes of the defects. The sound is passed through the material and picked up on the other side in the *transmission method*. The energy lost in transit because of reflections indicates the sizes and positions of defects. The *resonance technique* applies frequencies over a range. Energy drawn by the transducer increases at the resonant frequency of the specimen. If this frequency is different from that of a standard part of the same size, a defect is indicated. Ultrasonic testing is useful for revealing cracks, voids, and defects far below as well as near the surface. Skill and care is required for its application and interpretation.

Visual inspection methods have been improved by the use of dyes and fluorescent penetrants. The parts are immersed in one of these fluids and then wiped dry of any excess. The absorbed fluid is then drawn from cracks that come to the surface when the part is dusted with an absorbant powder. Dye penetrants can be viewed under ordinary light, while fluorescent oils are best seen under ultraviolet light. The method is not restricted to ferrous alloys, but is restricted to detection of thin cracks that intersect the surface.

3.4.9 Corrosion Testing

The two major places where corrosion testing is performed are (1) in the laboratory, and (2) in the field. Laboratory testing is more formalized and standardized than field testing.

Laboratory corrosion tests include three major types of tests. The first is the total immersion test for stainless steel and nonferrous metals and alloys. The second is the alternate immersion test, which requires cyclic immersion and withdrawal of specimens. Finally, there is the salt spray test, which is carried out in a closed chamber containing a foggy atmosphere of moist salt air. The temperature and composition of the corrosive substances must be carefully controlled in all three tests.

Fatigue tests also may be conducted under corrosive conditions by placing a wick, immersed in the corrosive substance, in contact with the specimen. In still another version of this test, the specimen may be immersed in a corrosive substance contained in a leakproof box. Tests of this type are known as corrosion-fatigue tests, and are intended to show the accelerating effect of corrosion on fatigue failure.

Stress corrosion is a phenomenon in which corrosion, combined with static tensile stresses at the surface of a metallic object, produces cracks. A common way of testing for stress corrosion is to immerse an elastically bent, flat, bar-type specimen in a corrosive substance. The stresses can be calculated from the dimensions of the specimen and the extent of the bend. A tendency toward stress corrosion may result from residual tensile stresses introduced by cold-forming operations (see Chapter 11).

Field corrosion testing aims at determination of corrosion resistance under environmental conditions expected in actual service. Thus, specimens are mounted on insulated racks and exposed to the atmosphere. Tests are commonly conducted in marine and industrial atmospheres since these represent very severe conditions. Rack-mounted specimens are commonly immersed in seawater. Specimens are also buried in soil to study long-time resistance. Since soil composition varies greatly, tests from one site may not be useful in predicting behavior at another site. Plant corrosion testing consists of placing small specimens in actual pieces of equipment in the plant.

The progress of corrosion tests is followed by periodic examination of the specimens for weight loss, pitting, depth of pits, length and depth of cracks, or mm (in.) of penetration (for uniform attack), as appropriate, for the composition tested and the corrosive medium.

The results of corrosion testing are useful as an aid in the choice of metals and alloys for particular applications, and as a supplement to published data on the chemical activity of metallic materials.

3.5 QUESTIONS

1. Indicate the four states in which metals may exist.
2. What happens to the atoms of a metal in a liquid state?
3. Name and describe the three most important types of unit cells in metals.
4. What is meant by the term "allotropic transformation"?
5. Describe the behavior of the atoms of a metal under elastic and plastic deformation.
6. What happens to the crystals or grains of a metal during cold working?
7. What is the recrystallization temperature and how does it affect the cold working of metal?

Load (lbf)	Length (in.)	Load (lb)	Length (in.)	Load (lb)	Length (in.)	Load (lb)	Length (in.)
320	2.0002	1,600	2.0010	2,760	2.0018	3,220	2.0060
590	2.0004	1,880	2.0012	2,920	2.0020	7,220	2.3000
920	2.0006	2,270	2.0014	3,020	2.0030	8,960	3.1000
1310	2.0008	2,560	2.0016	3,080	2.0040	6,210	3.3200

Ultimate load = 9,960 lb; D_f = final diameter = .2525 in.; L_f = final length = 3.32 in.

Calculate stresses and strains, percent elongation, and percent reduction in area. Plot a stress-strain curve on a scale that will permit accurate determination of the modulus of elasticity and yield strength. Plot a second stress-strain curve showing behavior up to the point of rupture.

7. Given a standard tension test specimen made of brass, D_0 = 12.50 mm and L_0 = 50.00 mm. Load-length readings were made as follows:

Load (N)	Length (mm)	Load (N)	Length (mm)	Load (N)	Length (mm)	Load (N)	Length (mm)
1,350	50.005	6,760	50.025	11,660	50.045	13,610	50.150
2,495	50.010	7,940	50.030	12,340	50.050	30,510	57.500
3,890	50.015	9,590	50.035	12,760	50.075	37,860	77.500
5,540	50.020	10,820	50.040	13,020	50.100	26,240	83.000

Ultimate load = 42,090 N, D_f = 6.25 mm, L_f = 83.00 mm.

Calculate stresses and strains, percent elongation, and percent reduction in area. Plot a stress-strain curve on a scale that will permit accurate determination of the modulus of elasticity and yield strength. Plot a second stress-strain curve showing behavior up to the point of rupture.

8. In addition to the data presented in Problem 6, diameter readings were made as follows:

Load (lbf)	Diameter (in.)
320-3,220	.505
7,220	.466
8,960	.370
6,210	.2525

Calculate true stress-true strain data and plot elastic and plastic true stress-true strain curves. From these, determine the value of the modulus of elasticity E, the work-hardening coefficient m, and the constant k, in $\sigma = k\epsilon^m$.

9. In addition to the data presented in Problem 7, diameter readings were made as follows:

Load (N)	Diameter (mm)
1,350-13,610	12.50
30,510	11.53
36,860	9.16
26,240	6.25

Calculate true-stress strain data and plot elastic and plastic true stress-true strain curves. From these, determine the value of the modulus of elasticity E, the work-hardening coefficient m, and the constant k, in $\sigma = k\epsilon^m$.

10. Notched-bar impact test results were obtained on a heat of steel as follows:

Energy absorbed, J (ft-lbf)	Temperature, °C (° F)	Energy absorbed, J (ft-lbf)	Temperature, °C (° F)
171, 161, 151 (126, 119, 111)	93 (200)	61, 60, 38 (45, 44, 28)	4 (40)
168, 159 (124, 117)	82 (180)	43, 27, 27 (32, 20, 20)	−12 (10)
160, 153, 152 (118, 113, 112)	71 (160)	33, 23, 16 (24, 17, 12)	−18 (0)
155, 153, 146 (114, 113, 108)	60 (140)	22, 14, 10 (16, 10, 7)	−29 (−20)
156, 152, 146, 141 (115, 112, 108, 104)	49 (120)	15, 12, 8 (11, 9, 6)	−40 (−40)
146, 126, 121, 106 (108, 93, 89, 78)	38 (100)	15, 12, 12 (11, 9, 9)	−51 (−60)
110, 81, 69 (81, 60, 51)	21 (70)	20, 11, 10 (15, 8, 7)	−62 (−80)

Plot the results and determine the approximate critical temperature range over which the steel changes from ductile to brittle. For what purpose can the notched-bar impact test be used?

11. Describe and explain the applications of hardness testing to acceptance and processing of materials.

12. What variables are usually related in long-time, high-temperature tensile tests? In short-time high-temperature tests?

3.7 REFERENCES

ASTM A370-73. 1973. Revised at various intervals. Philadelphia, PA: American Society for Testing and Materials.

Boyer, R., Collings, E., and Welsch, G. 1994. *Materials Properties Handbook: Titanium Alloys*. Materials Park, OH: ASM International.

Brady, George and Clauser, Henry. 1995. *Materials Handbook*, 13th Ed. New York: McGraw-Hill.

Hardness Testing. 1987. Materials Park, OH: ASM International.

Keyser, C.A. 1985. *Materials Science and Engineering*, 4th Ed. Columbus, OH: Charles E. Merrill.

Metals Handbook, 9th Ed. 1980. Materials Park, OH: ASM International.

Metals Handbook Desk Edition, 2nd Ed. 1998. Materials Park, OH: ASM International.

Smith, William F. 1996. *Principles of Materials Science and Engineering*, 3rd Ed. New York: McGraw-Hill.

Iron and Steel

4

The Bessemer Process, 1855
—Sir Henry Bessemer, London, England

4.1 IRON, STEEL, AND POWER

Pig iron, which contains impurities amounting to 7% by weight, is made in a blast furnace by reduction of iron ore. It is further refined to produce cast iron or steel. The availability of raw materials, particularly fuels, is the major factor in determining the location of blast furnaces and steel mills. The equivalent of about 2 tons of fuel is required for the manufacture of 1 ton of hot-rolled steel product. The dilemma of undeveloped countries, which lack adequate fuel sources, is that it is politically desirable but economically impractical to build steel mills. The same circumstances exist to suppress the agitation for steel mills in New England.

The ability to produce steel is basic to the development of economic, political, and military power. A study of steel production statistics over the last century shows this to be true. The westward expansion of the U. S. would have been impossible without the production of the tons of steel needed for the creation of an efficient land transportation system. Exploitation of the West was limited until the middle of the 19th century, when economical methods were devised for production of steel on a large scale. Increasingly, modern industrial societies require efficient methods of converting fuel to energy, and for this, machinery made of steel is essential.

4.2 IRON MAKING

The raw materials for making 1 ton of pig iron are approximately 2 tons of ore (or other source of iron), almost 1 ton of coke, nearly .5 ton of limestone, and about 3.5 tons of air. Thus, the blast furnace produces about 1 ton of principal product for every 7 tons of material that enters it. Although the composition of pig iron is variable, a typical analysis might be: carbon 4%, silicon 1.5%, manganese 1%, sulfur 0.04%, phosphorus 0.4%, and the balance iron. The exact composition will depend upon the composition of the iron source, the flux, and the coke, and the operating conditions of the blast furnace.

The major sources of iron in the blast furnace charge or *burden* are ore or agglomerates. Minor sources are scrap, roll scale from hot-rolling operations, and slag. The principal ores used are hematite (Fe_2O_3) and limonite ($Fe_2O_3 \cdot xH_2O$). Magnetite (Fe_3O_4) and other ores are less commonly used. Agglomerates are made from naturally fine-particle ores, flue dust, and ore concentrates. They are available as sinter, which resembles medium-sized clinkers, or pellets, which are rounded particles less than about 25.4 mm (1 in.) in diameter. Sinter is made by mixing the iron-bearing fines with high-ash coke and fine particles of limestone flux, and carefully burning the mixture. Pellets are made from very fine particles of iron-ore concentrate, fuel, binder (bentonite clay), and water. The mixture is placed in a rotating drum where it is formed into small pellets. These are then fired to achieve the strength necessary to prevent them from crumbling when placed in the blast furnace. If the solids charged into the blast furnace crumble, they will become so compacted that the air blast is restricted, and effectiveness of the furnace declines. Most ores and sinter contain 50–65% iron. Pellets, often made from magnetic fines obtained by processing low-grade taconite rock, contain about 60–67% iron. Other ingredients of the iron-source materials are alumina (Al_2O_3), silica (SiO_2), limestone ($CaCO_3$), carbonaceous materials, and chemically combined water.

Limestone ($CaCO_3$) and/or dolomite ($CaMg[CO_3]_2$) are the principal blast furnace fluxes. The function of the *flux* is to react with the principal impurities, such as alumina (Al_2O_3) and silica (SiO_2), forming a low-melting slag. The slag is lighter than the molten iron and floats on the latter in the bottom of the furnace. The composition of the slag is adjusted to assist in lowering the sulfur content of the iron. At temperatures encountered in the blast furnace, calcining takes place: $CaCO_3 \rightarrow CaO + CO_2$. The basic lime (CaO) then reacts with the acidic impurities (Al_2O_3 and SiO_2) to form slag.

The metallurgical coke used in blast furnaces must be strong enough to prevent crumbling and blocking of air passages. Since most of the sulfur in the iron originates in the coke, the sulfur content of the coke should be less than 1.5%, and usually is in the range of 0.4–1.2%. Ash and phosphorus contents also should be low. Finally, although the coke should be free from fines, the pieces should not be so large that optimum burning rates are not reached.

The coke serves two functions: (1) it provides heat, necessary for the attainment of desirable chemical equilibria and adequate rates of reaction; and (2) it provides the reducing gas carbon monoxide (CO), largely responsible for reduction of iron oxide. Additional sources of heat include the hot-air blast of about 1,093° C (2,000° F) and gaseous fuel injected with the air. A typical energy balance for a furnace with an 8.5-m (28-ft) diameter hearth producing about 2,840 metric tons (2,800 tons) of pig iron per day is shown in Table 4-1.

Coke is produced by the destructive distillation, in the absence of air, of special grades of bituminous coal. In addition to metallurgical coke, a large quantity of important by-products is obtained. They consist largely of gas, compounds of ammonia, coal tars, and light oil containing benzene, toluene, naphtha, etc. Thus the steel industry, through coking, is intimately connected to the production of fertilizer, and a host of organic chemical products such as dyes, plastics, solvents, drugs, and explosives. Typical yields from coal are shown in Table 4-2.

In addition to solids that are charged into the furnace at the top, air is blown in at the bottom of the blast furnace. This air is at a temperature of between 760–1,093° C (1,400–2,000° F) and is under pressure. The air is enriched with gaseous fuels and moisture is added in controlled amounts. These additions to the air blast enable control of flame temperature and produce the extra quantities of reducing gases needed to keep up with the high-temperature operation of modern furnaces. Coke alone in the burden will not produce enough reducing gases for maximum efficiency.

Table 4-1. Energy balance for a blast furnace

	MJ/kg of Pig Iron	Millions Btu/2,000 lb of Pig Iron
Energy input		
Coke	14.9	12.80
Hot-air blast	2.0	1.68
Gaseous fuel	1.1	0.96
Total input	18.0	15.44
Energy consumed: reduction of iron and other oxides, heat lost with hot metal and hot clay	10.6	9.07
Energy output: heat carried away by recoverable top gas	7.4	6.37

(McGannon 1971)

Table 4-2. Yields from coking

	Per 1,016 kg (metric ton) of Coal	Per 2,000 lb (ton) of Coal
Blast furnace coke	544–635 kg	1,200–1,400 lb
Coke breeze (fines)	45–91 kg	100–200 lb
Coke-oven gas	269–326 m^3	9,500–11,500 ft^3
Tar	30–45 L	8–12 gal
Ammonium sulfate	9–11 kg	20–25 lb
Ammonia liquor	57–133 L	15–35 gal
Light oil	10–15 L	2.5–4 gal

(McGannon 1971)

4.3 THE BLAST FURNACE AND ITS CHEMISTRY

The reactions taking place in the blast furnace can be substantially represented as follows:

$$2C + O_2 \rightarrow 2CO \quad \Delta H = -221.1 \text{ kJ } (-209.6 \text{ Btu}) \qquad (4\text{-}1)$$

$$C + H_2O \rightarrow H_2 + CO \quad \Delta H = 131.3 \text{ kJ } (124.5 \text{ Btu}) \qquad (4\text{-}2)$$

$$Fe_2O_3 + 3CO \rightarrow 2Fe + 3CO_2 \quad \Delta H = -25.5 \text{ kJ } (-24.2 \text{ Btu}) \qquad (4\text{-}3)$$

$$CO_2 + C \rightarrow 2CO \quad \Delta H = 172.6 \text{ kJ } (163.6 \text{ Btu}) \qquad (4\text{-}4)$$

$$Fe_2O_3 + 3H_2 \rightarrow 2Fe + 3H_2O \quad \Delta H = 97.9 \text{ kJ } (92.8 \text{ Btu}) \qquad (4\text{-}5)$$

$$H_2O + C \rightarrow CO + H_2 \quad \Delta H = 131.3 \text{ kJ } (124.5 \text{ Btu}) \qquad (4\text{-}6)$$

$$CaCO_3 \rightarrow CaO + CO_2 \quad \Delta H = 179.3 \text{ kJ } (169.9 \text{ Btu}) \qquad (4\text{-}7)$$

$$XCaO + YSiO_2 + ZAl_2O_3 \rightarrow XCaO{\cdot}YSiO_2{\cdot}ZAl_2O_3 \quad \text{flux reaction}$$

A minus sign with a ΔH value means that heat is liberated by the reaction.

Essentially, the *blast furnace* is an irregularly shaped cylinder into which solids are fed at the top and enriched air is blown in at the bottom. Liquid slag and metal are tapped from the bottom of the furnace and flue gases and dust escape from the top of the furnace. The furnaces operate continuously for 5–7 years and then must be rebuilt. The dimensions and material balance for a typical blast furnace are shown in Figure 4-1. As the solids in the charge progress downward in the furnace, they are heated and expand. At the same time, the gases become cooler as they approach the top of the furnace. In the bosh (see Figure 4-1), the iron and slag melt and drip from the still solid portion of the charge, thus reducing the solid bulk. The reduction in diameter of the bosh, along with a central pillar of unburned coke, supports the charge. Gas volume increases in the bosh as a result of heating and because of oxidation of coke (see Equations 4-1, 4-2, 4-4, and 4-6). The initial increase in gas volume is also accommodated by the shape of the bosh.

Special refractories and brick are used in furnaces. At the top of the furnace a super duty, hard-fired brick is used, which is resistant to the abrasion of the charge dumped into the furnace. High-temperature resistance is not a factor here. Brick in the inwall zone (see Figure 4-1) must withstand moderately high temperatures and moderate abrasion from the charge as it moves downward. The brick in the hearth and bosh must withstand very high temperatures as well as erosion and slag attack. Carbon brick is occasionally used in the hearth. The refractory brick in the inwall and bosh are cooled by hollow copper plates fitted between the courses of brick. Water is circulated through these plates as well as through the tuyere (air-blast nozzle) linings. A furnace producing about 2,840 metric tons (2,800 tons) of pig iron per day uses over 37,854 m^3 (10,000,000 gal) of recirculating water for cooling purposes.

The flue gases leaving the top of the furnace are passed through a dust catcher to remove solids and are then burned in stoves used for preheating the air blast. The dust is agglomerated and used in the burden.

4.4 STEELMAKING

Steel is made by refining either pig iron or scrap steel, or by refining a combination of both materials. Pig iron supplies, by far, most of the steel produced, but scrap steel plays a vital role in steel production. For steel produced from pig iron, the basic problem is to oxidize the impurities present, which are then removed either as a

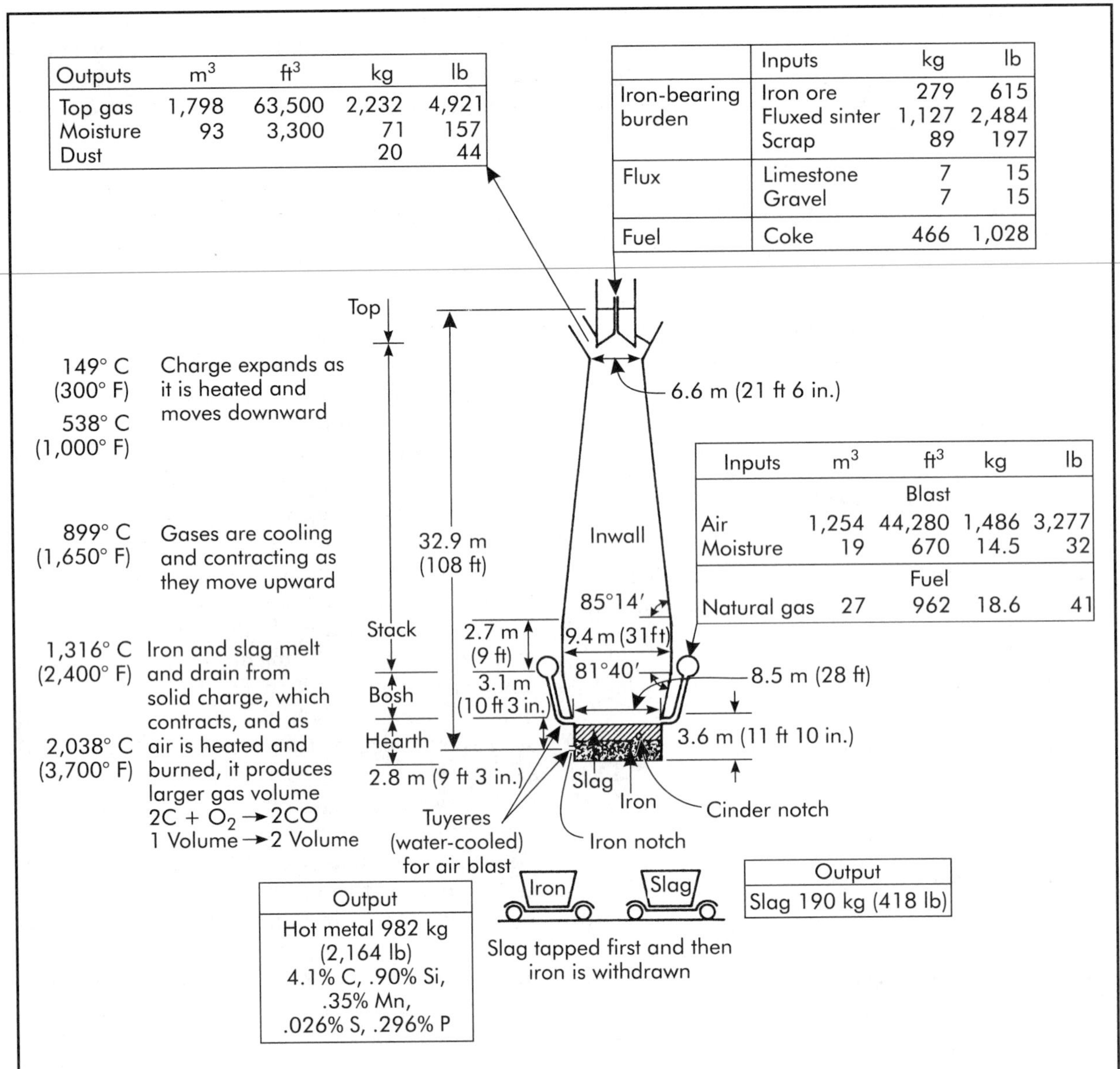

Figure 4-1. Typical blast furnace, showing inputs, outputs, dimensions, and temperatures developed in the furnace. The abbreviations m³ and ft³ refer, respectively, to standard cubic meters and standard cubic feet, measured at 16° C (60° F) and 762 mm (30 in.) of mercury (McGannon 1971).

gas (in the case of the major impurity, carbon) or in the slag (in the case of silicon, phosphorus, and sulfur). Oxygen is supplied either by an air blast (as in most of the older methods), as pure oxygen (in the most modern methods), or by means of oxides (such as iron ore or rusty scrap).

In the last two decades there has been a revolution in the methods used for producing steel. For instance, 80% of production came from basic open hearths in 1963. By 1984, this figure dropped to 7%. More efficient processes have been adopted. Basic oxygen converters currently account for 63% of production, while 30% is made in electric furnaces. The latter are especially suited for small, nonintegrated specialty mills known as mini-mills.

4.4.1 The Basic Oxygen Process

The basic oxygen process is a derivation of the Bessemer process, the first method by which large-scale tonnages of steel were produced. The *Bessemer process* relied on blowing air from holes in the bottom of the converter through the molten pig iron charge. Oxidation of impurities supplied not only enough heat to keep the charge molten, but also enough to maintain favorable chemical equilibria. In the basic oxygen process, air is replaced by pure oxygen, introduced through a lance whose end is just above the surface of the molten metal.

The most widely used basic oxygen method is known as the *L-D process*, the name for which derives from the towns of Linz and Donawitz in Austria, where it was first used. A typical furnace is shown in Figure 4-2. The end of the water-cooled oxygen lance is suspended about 0.9 m (3 ft) above the surface of the charge and supplies oxygen at a pressure of 965–1,241 kPa (140–180 psi). The furnace, a cylindrical vessel about 9.1 m (30 ft) high with an inside diameter of about 5.5 m (18 ft), has a mouth about 2.7 m (9 ft) in diameter. It is tilted to receive its charge, first scrap and then molten pig iron, after which it is brought to a vertical position under a water-cooled hood. The oxygen lance is lowered and the blow is started. The oxygen quickly produces iron oxide in the melt and this, in turn, oxidizes carbon, causing vigorous agitation of the melt as carbon monoxide and carbon dioxide are evolved. Fluxing agents, such as lime and fluorspar, are dropped from a hopper through a chute after the oxygen blow has commenced. The lance is removed after the impurities have been oxidized. Then the furnace is tilted, first to one side to tap the steel through a taphole, and then to the other side to pour off the slag. Final adjustment of the composition is made by ladle additions. The time of blowing varies according to composition and heat size, but about 25 min is typical. Gases and slag particles carried along with the gas stream are scrubbed before being exhausted to the atmosphere.

Basic oxygen furnaces range from 45–320 metric tons (50–350 tons) capacity. They use from 12–30% scrap in the charge, which is below the minimum acceptable for basic open hearths. (Both the basic open hearth and electric furnace can use up to 100% scrap in the charge.) But the basic oxygen furnace can produce steel at the rate of about 406 metric tons (400 tons) per hour per furnace, whereas the basic open hearth could only produce at the rate of about 61 metric tons (60 tons) per hour per furnace. Since each heat from a basic oxygen furnace requires only 30–45 min (compared to up to 10 hr per basic open-hearth heat), quick and precise methods of checking composition and computing requirements for adjustment of the charge are essential.

The basic oxygen furnace can be used for the same composition of pig iron as the basic open hearth. High-phosphorus (up to 2%) pig iron can be refined if a double-slagging method is used. In this process, the first slag to form is removed and replaced with a fresh slag to remove additional phosphorus.

The same grades of steel are produced in the basic oxygen process as in the basic open-hearth process and, in addition, some grades of stainless and higher-alloy steel can be produced that could not be made in the basic open hearth. The quality of product is as good or better than that produced in the open hearth. Very fluid slags yield a low phosphorus content. Nitrogen content is also low because pure oxygen rather than air is used for the blow. Surprisingly, oxygen content is also low and this reduces requirements for aluminum deoxidizing additions. Sulfur is low because of the avoidance of sulfur-bearing fuels as a heat source. Manganese economies also have been realized because of better control of the process.

About half of the basic oxygen steel produced is used for sheets, plates, and structural steel of welding quality. The other half is used for rimming steel (explained later), from which deep-drawn parts are made, such as automobile body parts.

4.4.2 The Electric-furnace Process

Most electric-furnace steel produced in the U. S. is made in furnaces heated by an arc formed between electrodes and the metal of the charge. This is the type of furnace whose operation is described here. Some steel is made in electric furnaces in which the heat is radiated to the bath

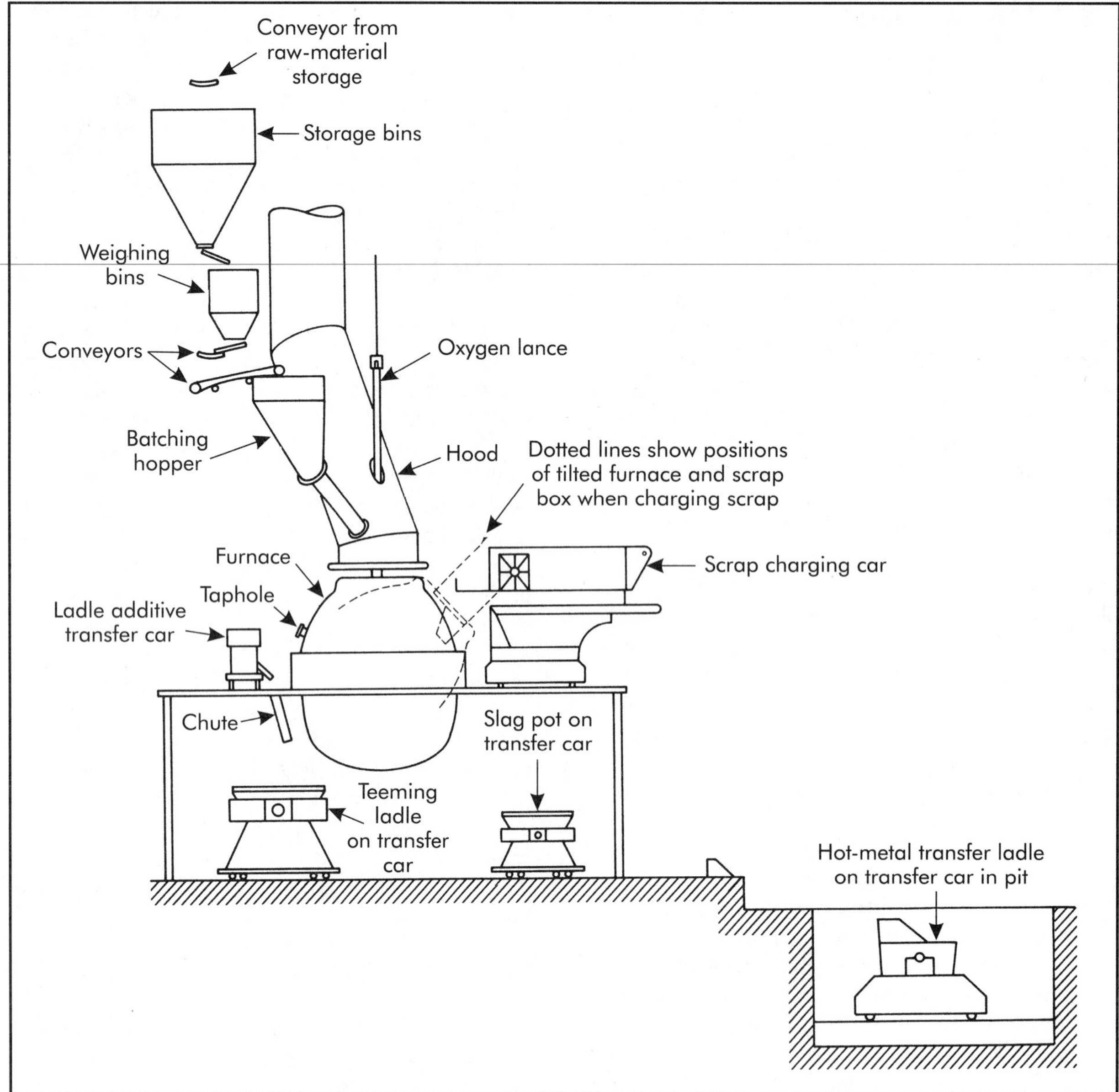

Figure 4-2. Schematic elevation showing the relative locations of various operating units of the basic oxygen process (McGannon 1971).

from an arc formed between two electrodes positioned over the charge. A small amount of steel is also made by induction-heated furnaces. Both acid and basic linings are used, depending upon the nature of the charge and the product desired. In general, the highest-quality steels are produced in the electric furnace. Since hydrocarbon fuels are not needed for heat, this source of contamination is eliminated. Better than 30% of all steel made in the U. S. is produced in electric furnaces.

The procedure followed in making electric-furnace steel depends upon the product being made. If the steel is to contain an appreciable percentage of easily oxidized alloying elements, such as chromium, tungsten, and molybdenum, two slag covers are used during a heat. An oxidizing slag promotes oxidizing and fluxing of carbon, phosphorus, and silicon. The oxidizing slag is then removed and replaced by a reducing slag in which CaO and CaC_2 are important ingredients. This slag blanket helps remove sulfur and oxide impurities and affords protection against oxidation of alloy element additions. In making steel for ordinary castings, the second slag is not needed, since the easily oxidized elements found in stainless and tool steels are not present.

An *electric arc furnace* is shown in Figure 4-3. Electrodes, which can be raised and lowered, project through the refractory-lined top. The top can be swung aside when the electrodes are raised so that the charge can be dropped into the furnace. The electrodes are made from petroleum coke, bonded with pitch or tar and shaped to approximate size. The rough electrodes are heated to 2,204° C (4,000° F), which converts the coke to graphite. Electrodes up to 76 cm (30 in.) in diameter and 213 cm (84 in.) long are available. Furnace capacities vary from 1–406 metric tons, (1–400 tons), but furnaces in the range of 15–152 metric tons (15–150 tons) are most common.

A typical, but by no means universal procedure for operating an electric furnace is as follows. After a heat is tapped, the furnace is inspected for damage and repaired as needed. Selected scrap is charged through the top of the furnace by means of a drop-bottom bucket. If ore is included in the charge, it is added with the solid scrap or after partial or complete melting. Modern practice tends toward the use of oxygen to lower carbon content, rather than the use of ore. Some nonoxidizable alloys may be added prior to meltdown. The electrodes are then lowered, the power is turned on, and an arc is struck. As melting proceeds, the electrodes burn down through the metal charge and a pool of molten metal forms in the hearth of the furnace.

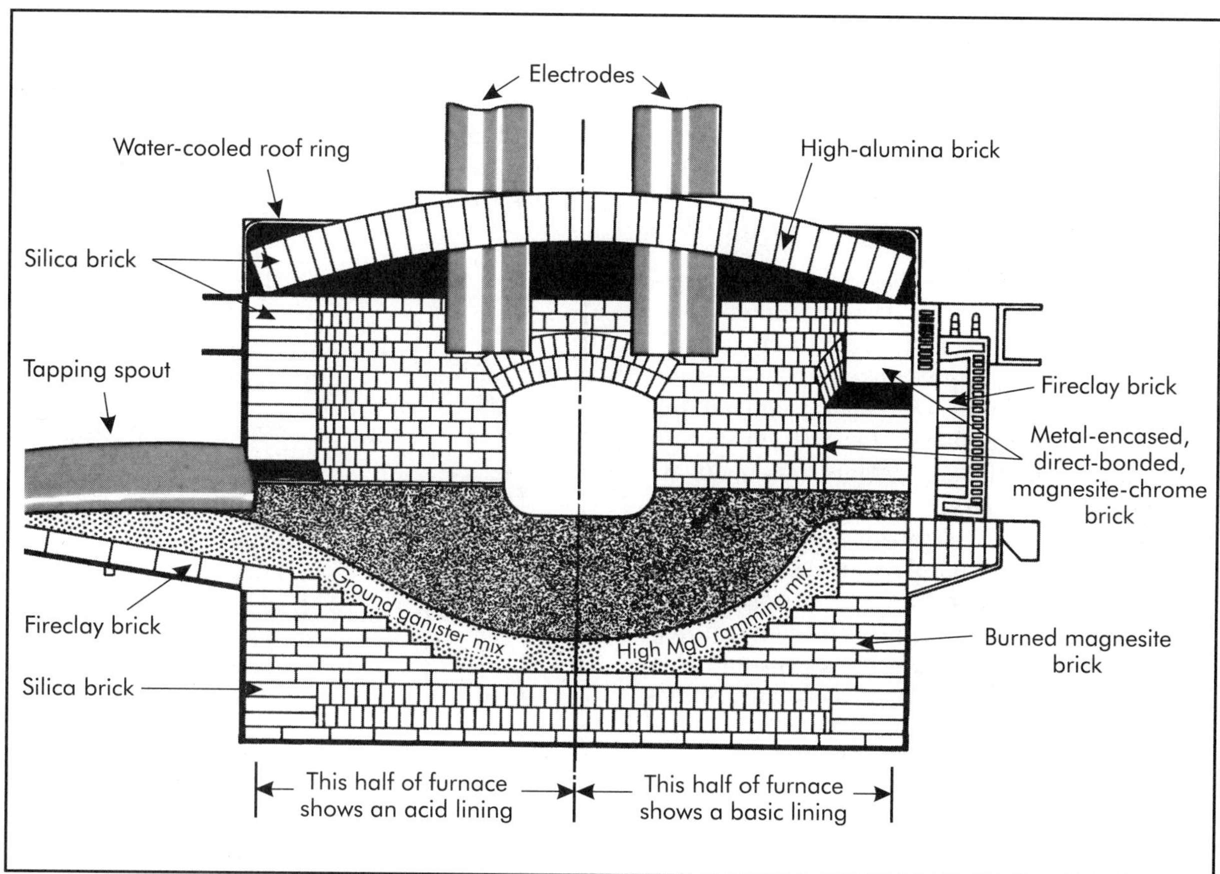

Figure 4-3. Schematic cross-section of a Heroult electric furnace (McGannon 1971).

A slag forms from oxidized impurities and reaction with lime or the furnace lining. After oxidation is complete, this slag is drawn off and replaced by a new slag cover in which the principal ingredients are lime, silica, magnesia, and calcium carbide. As soon as the final analysis of slag and bath have been adjusted to proper levels, necessary alloy additions are made and the furnace is tapped. When the furnace is tilted for tapping, the molten steel remains covered and protected by the slag until the furnace is empty. The time elapsed from charging to tapping is dependent upon the size of the furnace and nature of the product, but about 4 hr is typical.

Although the cost of making steel in an electric furnace is generally higher than for steel made in basic oxygen furnaces, the quality of the product achieved is better. Steel of the lowest possible phosphorus and sulfur contents and having the fewest nonmetallic inclusions is made by the electric-furnace process, and only by this process can high-alloy steels, such as some stainless grades and some tool and die steels, be made. The higher cost of the process is justified where small-scale, intermittent steel production requirements would not support a basic oxygen, blast-furnace installation.

4.5 FINISHING AND INGOT TEEMING

After hot metal in the basic oxygen furnace has been brought to the desired composition specifications, it is tapped into a refractory-lined ladle and separated from its oxidizing slag cover. It is now possible, if desired, to add deoxidizing agents such as ferrosilicon, ferrotitanium, and ferroaluminum. If deoxidation had been attempted before separation from the slag cover, phosphorus impurities in the slag would have been reduced and redissolved in the metal.

When steel is made in the electric furnace, the oxidizing slag containing phosphorus impurities is usually removed partway through the refining process. It is replaced by a fresh reducing slag so that return of the phosphorus to the molten metal is not a problem. The steel is then tapped into a ladle.

Pouring the steel from a ladle into ingot molds is known as *teeming*. If deoxidation is not carried out, the oxygen dissolved in the molten steel reacts to form increasing amounts of carbon monoxide and solid iron oxide as the temperature of the steel in the mold drops. The carbon monoxide is evolved as a gas, whose bubbles are trapped in the solidifying mass. The iron oxides form nonmetallic inclusions that are harmful to the mechanical properties of the finished steel. The control of the amount of carbon monoxide evolved is very important and leads to three kinds of steel: rimmed, semikilled, and killed.

Rimmed steel is essentially not deoxidized, although a small amount of aluminum may be added to the steel in the mold to prevent excessive carbon monoxide evolution. The bubbles of carbon monoxide are trapped below the first surface layers to solidify near the top of the ingot. The evolution of gas compensates for shrinkage that normally accompanies solidification and eliminates pipe formation (see Chapter 8). This results in larger yields per ingot, for when pipes occur (see Figure 4-4), the piped end must be cropped off and remelted as heavy scrap.

The gas bubbles in rimmed ingots are formed by carbon monoxide and are far enough below the rim, or outer skin, of sound steel that they do not become oxidized upon subsequent heating and hot rolling of the ingot. Therefore, they are welded shut. The surface of the rimmed steel ingots is nearly pure iron, and after hot and cold

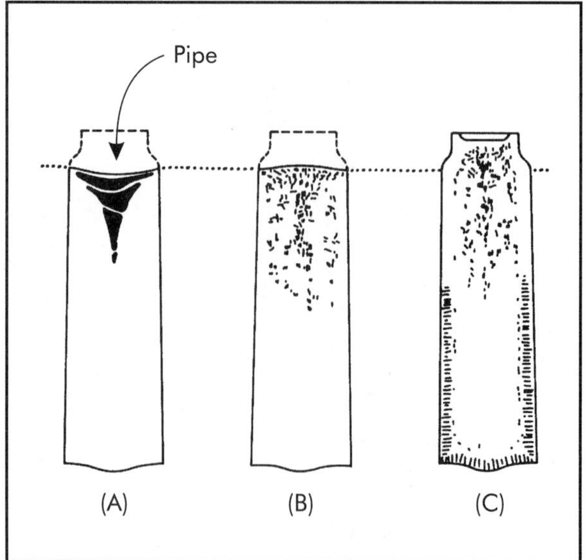

Figure 4-4. Typical ingot sections showing (A) killed steel, (B) semikilled steel, (C) rimmed steel (McGannon 1971).

rolling, these ingots are well suited for deep-drawing sheet (see Chapter 12), strip, and tin plate. Rimmed steels are low-carbon grades.

Semikilled steel is made by addition of appreciable quantities of aluminum and ferrosilicon in the ladle, but not enough for complete deoxidation (see Figure 4-4). A compromise between good yield and quality, the quality of semikilled steel is suitable for structural steels and heavy plates. Semikilled steels are of intermediate carbon content.

Killed steel is made by following the treatment for semikilled steel, except that final quantities of aluminum are added in the mold to deoxidize the steel completely. Since there is no evolution of gas from the freezing mass, the metal lies quietly in the mold and hence is called *killed* steel. Large pipes form in killed steel ingots and large cropping losses result. Killed steel ingots are the soundest since they are free from gas bubbles and have fewer entrapped inclusions. The cleanliness of this steel results in part from the opportunity of the oxide inclusions to rise to the surface as the steel solidifies quietly in the mold. Killed steels are considered high-carbon grade.

The choice of deoxidizing agent permits a variation in the quality of the semikilled and killed steels. If ferrosilicon is used as the deoxidant, the grain size of the steel is relatively coarse. If aluminum or ferrotitanium is used, Al_2O_3 or TiO_2 forms as a result of oxidation and remains trapped in the molten and solidified ingot. In low-carbon steel, this results in a fine grain size, particularly useful for deep-drawing applications. In medium- or high-carbon steel, a fine-grained structure results, which has superior toughness when quenched and tempered (see Chapter 6).

4.6 SPECIAL TECHNIQUES IN STEEL REFINING

In the quest for steels of higher quality, particularly those lower in gas content (dissolved hydrogen, oxygen, and nitrogen) and with fewer nonmetallic inclusions (oxides), various vacuum techniques have been developed. These are discussed in Chapter 8. Continuous casting, described in Chapter 9, accounts for more than 30% of the steel made in the U. S.

The *electroslag remelting* (ESR) process is basically a method by which an electrode of a given steel composition is remelted and recast under a slag cover to produce a more refined ingot of essentially the same makeup as the starting composition. A refining slag is placed on the base plate of a water-cooled mold and the electrode to be refined is lowered. At first, arcing melts the slag, but eventually the electrical resistance of the molten slag, into which the end of the electrode is immersed, produces enough heat to melt the electrode itself. The water-cooled mold solidifies the refined metal, which rises slowly as more refined metal drips from the electrode, and the length of the new ingot grows. Ingots of very high-quality tool steel, bearing steel, high-speed steel, and ultraservice steel are produced in this way. They vary in size up to 165 cm (65 in.) in diameter and 4.9 m (16 ft) long. The ESR process, shown schematically in Figure 4-5, is a competitor of *vacuum arc remelting* (VAR), which is also known as *consumable electrode melting*. In this process, no slag cover is used. Heat is supplied by an arc between the electrode being remelted and the new ingot, and the water-cooled mold is evacuated. Somewhat larger ingots are made by VAR than ESR, but the quality of the steel produced by both is similar. Both processes also are used for refining nonferrous superalloys.

4.7 ALUMINUM

One of the most abundant metals on the surface of the earth is aluminum, occurring as an oxide in most clay. However, it is not now economically feasible to concentrate this oxide to the point where it can serve as a source of aluminum. The ore used as a source of aluminum is *bauxite*. Bauxite is essentially hydrated aluminum oxide, $Al_2O_3{\cdot}3H_2O$, but it also contains silica, SiO_2, titania, TiO_2, and iron oxide, Fe_2O_3. After crushing the ore, impurities are removed by treatment with hot sodium hydroxide:

$$\underset{\text{insoluble in } H_2O}{2NaOH + Al_2O_3{\cdot}3H_2O} \underset{\text{cold}}{\overset{\text{hot}}{\rightleftharpoons}} \underset{\text{soluble in } H_2O}{2NaAlO_2 + 4H_2O}$$

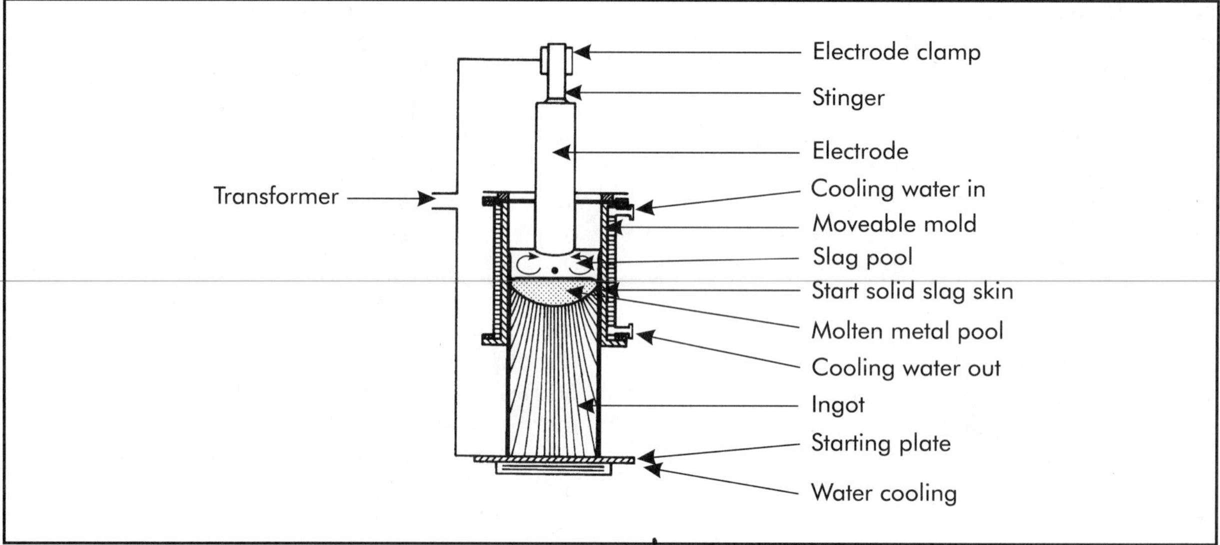

Figure 4-5. Features of a typical electroslag remelting furnace (McGannon 1971).

Titania and iron oxide do not form soluble compounds with sodium hydroxide, although some of the silica may dissolve to form a complex sodium aluminum silicate. If the solution is filtered while still hot, the TiO_2, Fe_2O_3, and most of the SiO_2 are removed. The concentrated hot sodium aluminate is cooled, seeded with $Al_2O_3{\cdot}3H_2O$ crystals, and precipitation of $Al_2O_3{\cdot}3H_2O$ occurs. The complex sodium aluminum silicate remains in solution and can be filtered away. The pure, hydrated aluminum oxide is heated to form pure alumina:

$$Al_2O_3{\cdot}3H_2O \xrightarrow[980^\circ\ C\ (1{,}796^\circ\ F)]{982^\circ\ C\ (1{,}800^\circ\ F)} Al_2O_3 + 3H_2O$$

Aluminum oxide is too stable for reduction by carbon monoxide in a blast furnace, and is instead reduced in an electrolytic cell. The purified alumina is dissolved in molten *cryolite*, Na_3AlF_6, to which small amounts (under 10%) of aluminum fluoride and calcium fluoride have been added. The voltage across the cell attracts positively charged Al^{3+} to a negatively charged carbon cathode lining. Molten aluminum with a melting point of 649° C (1,200° F) collects on the bottom of the cell, from which it is periodically tapped. Negatively charged oxygen, O^{2-}, is attracted to carbon anodes, where it reacts to form CO. From time to time the alumina is replenished. Carbon anodes are consumed at the rate of about 600 kg/metric ton (1,300 lb/ton) of aluminum produced.

4.8 COPPER

The sources of copper include native copper, sulfides, oxides, and silicates. In the U. S., the principal ore contains about 1% copper as a sulfide, and the method described for extracting copper is based on this ore.

Since the copper content of the native ore is so low, it is necessary to effect a concentration. The ore is first crushed in gyratory or jaw crushers and then screened. After ball milling and further screening, it is finally ground wet and mixed with a coating of oil. The oil wets the sulfide particles but not the particles of gangue, which consist of SiO_2, Al_2O_3, $CaCO_3$, and other elements. The finely ground, oil-coated ore is placed in a tank of water and agitated. The oil-coated sulfide particles pick up small air bubbles and form a froth rich in copper and iron sulfides, while the gangue settles out. The froth overflows and is filtered to produce a solid concentrate containing about 40% copper.

Roasting of the concentrate dries and eliminates some sulfur by reactions such as $CuS \rightarrow Cu_2S + S$. This sulfur is oxidized to SO_2. Some of the iron and copper sulfides may be converted to oxides, while arsenic, antimony, selenium, and zinc sulfides are converted to oxides and volatized. Gold and silver, which are valuable by-products, remain with the copper and iron. The product consists mainly of Cu_2S, FeS, and impurities, of which SiO_2 is the major one. Modern roasting furnaces consist of two hearths at the top of a cylindrical furnace. The concentrate is fed into these, where it is dried, preheated, and partially roasted. Rotating rakes work the concentrate toward holes in the hearths and finally the concentrated ore falls through the central part of the furnace where calcination to Cu_2S occurs and some oxidation takes place. The final reactions occur in other hearths at the bottom of the furnace.

Smelting of the roasted ore often takes place in *reverberatory furnaces*. These are furnaces in which a shallow pool of impure metal and slag is heated by a flame playing over the surface of the charge. A mixture of roasted concentrate and a limestone flux is charged through the top of the furnace at one end. Powdered coal, oil, or gas are introduced at the charging end of the furnace. The charge melts rapidly and separates into two layers. The light slag layer contains most of the gangue, some iron (as iron silicate), and about 4% copper. The slag is sent back through the reverberatory furnace to recover the copper. The heavy layer, called *matte,* consists essentially of sulfides of iron and copper, plus gold and silver. A slag taphole at the end of the furnace allows continuous removal of slag. The matte is tapped periodically through tapholes at the bottom of the furnace. The smelting operation removes some of the iron and most of the other impurities (except gold and silver).

The matte (essentially Cu_2S and FeS) is often refined in a converter similar to the L-D basic oxygen converter, except that air nozzles are located in the sides of the converter and air is blown across the bath. Heat is supplied by oxidation of sulfur to sulfur dioxide. The copper sulfide is converted to impure metallic copper and sinks to the bottom of the bath, and the iron sulfide is converted to FeO, reacting with a silica flux or the silica lining to form an iron silicate slag. The converter is tipped to discharge the slag, and the impure copper, containing precious metals and small amounts of Ni, Bi, Se, Te, and S, is cast into pigs for further refining. This impure copper is called *blister copper* because it solidifies with the evolution of gas and has a porous, blistered surface.

Blister copper is placed into a reverberatory furnace operated with an excess of oxygen blown into the charge. This oxidizes sulfur and the oxidizable impurities, which leave as a gas or form a siliceous slag. At the time of slag removal, the bath contains up to about 6% Cu_2O. This is reduced by *poling,* which consists of stirring the bath with green wood poles. These evolve hydrocarbons, reducing the Cu_2O back to Cu. Poling is stopped at an oxygen content of about 0.04%. This amount of oxygen is necessary to prevent impurities from reverting into the copper from the slag. This copper is known as *tough-pitch copper* and is nearly pure copper.

Tough-pitch copper is often electrolytically refined to improve its conductivity, or to recover valuable quantities of precious metals, or both. The electrolytically deposited copper is then remelted and cast into slabs, billets, or bars. This grade of copper cannot contain impurities of more than 0.01%, excluding silver.

The stages of roasting, smelting, and refining copper as described have been practiced in the U. S. for over a century. In recent years, demands for conservation of energy and the environment have forced changes and improvements in the processes. One improvement has been *continuous copper smelting*. A typical installation features three stationary furnaces with the molten products flowing by gravity from a smelting furnace to a slag-cleaning furnace to a converting furnace. Stages in the newer processes are not necessarily identical to the older ones. The most undesirable emission from copper smelting and refining is SO_2 gas. In some installations, electric heating has been adopted to avoid diluting the SO_2 gas with other combustion products, so that the gas can be utilized in acid manufacture. In some cases, sulfide concentrations and oxygen are injected into the furnace and burned to add heat and further concentrate the SO_2 gas. Some smelters have found it necessary to add scrubbers to comply with state and federal regulations for safe emissions.

4.9 MISCELLANEOUS METALS

There is great similarity between the extractive metallurgy of iron, aluminum, and copper, and the extractive metallurgy of other metals. It is not practical to detail here the methods used for all metals. For the reader who is interested in additional information, attention is called to the references at the end of this chapter. This section outlines the general scheme followed for the production of magnesium, zinc, lead, tin, titanium, and tungsten.

4.9.1 Magnesium

The Dow process was the first commercial method and is still the most important means for production of magnesium. Hydrated $MgCl_2$ is obtained from natural brines by recrystallization. It is dehydrated and mixed with NaCl and KCl to form a mixture that will melt at about 704° C (1,300° F). Current is passed through the melt in an electrolytic cell. Magnesium forms at the cathode and floats to the surface, from which it is periodically ladled. Chlorine forms at the anode and is collected as a valuable by-product. Magnesium chloride for electrolysis is also obtained from seawater and magnesite ores.

4.9.2 Zinc

The ores of zinc usually must be concentrated by flotation methods. Concentrated sulfide ores are converted to the oxide or sulfate by heating in air. Zinc oxide ores, carbonate ores, and the converted zinc sulfate or zinc oxide can be heated with coal. The latter acts not only as a source of heat but provides the medium for reduction to metallic zinc. Since the retorts operate above the boiling point of zinc, the metal passes off as a vapor and is condensed as a liquid near 482° C (900° F). Further purification can be accomplished by redistillation. A small amount of zinc is produced by electrolysis of zinc sulfate solutions. Environmental demands have closed down many zinc smelters in recent years. As these demands become more stringent, it is considered likely that the electrolytic process will become the only one feasible in the U. S.

4.9.3 Lead

In the U. S., the principal ore is galena or PbS. This is concentrated by a flotation process and is then roasted to convert it to PbO. The lead oxide is charged into a blast furnace with coke and a flux of lime and iron oxide. Molten lead collects in the bottom of the furnace from which it is tapped at intervals. Impurities are also tapped periodically and treated to recover valuable quantities of other metals. Further refining of the lead is often necessary. This step consists principally of heating the lead in a reverberatory furnace in the presence of air. When so heated, most of the impurities form oxides that pass into the exhaust gases, or into a skin that can be skimmed.

Environmental demands are dictating drastic changes and improvements in the lead refining industry. Measures must be taken to clean stack gases. In one new process, galena concentrates are first oxidized and then reduced in the molten state in a single sealed reactor. The process gases contain 20% SO_2 and are suitable for acid manufacture.

4.9.4 Tin

Although no significant amount of tin is produced in the U. S., this country is the largest consumer of the metal. The principal ore contains the mineral cassiterite, SnO_2, which is often reduced by reaction with coal in a reverberatory furnace. High-purity tin is obtained by electrolytic refining.

4.9.5 Titanium

The sources of titanium are ores containing the mineral ilmenite ($FeO \cdot TiO_2$) or rutile (TiO_2). High-carbon ferrotitanium is made in an arc furnace by reducing ilmenite with carbon. Low-carbon ferrotitanium is made by reduction with aluminum. These alloys are used to deoxidize steels and to make some alloy steels. For the manufacture of titanium used in titanium-base alloys, rutile is reduced to TiC in an arc furnace and then converted to $TiCl_4$ by roasting in a chlorine atmosphere. This is then reduced with magnesium at about 760 °C (1,400° F) to form a mixture of titanium, magnesium, and magnesium chloride. The latter impurities are removed by distillation and leaching, and the sponge titanium is vacuum arc melted to form a compact mass.

4.9.6 Tungsten

Tungsten has such a high melting point (about 3,399° C or 6,150° F) that it is not easily converted to the liquid state. Tungsten is mined as an ore containing the mineral wolframite, $FeWO_4$, or scheelite, $CaWO_4$. This is converted to tungstic acid, H_2WO_4, which is reduced to tungsten powder by hot hydrogen. The powder is compressed to form a briquet and sintered in a hydrogen atmosphere at about 1,093° C (2,000° F) using the techniques of powder metallurgy described in Chapter 10. Following sintering, the dense, solid bar is mechanically worked to improve its properties and convert it to wire filaments. Most tungsten is used as an alloying element in steels. For this purpose, the ore concentrate is smelted with coal, flux, and iron chips to form carbonized ferrotungsten.

4.10 STEEL

Strictly speaking, *iron* refers to one of the chemical elements. As used by engineers, iron includes commercially pure iron, such as ingot iron, or highly refined iron, such as electrolytic iron and carbonyl iron. These pure irons are used for a few magnetic applications, but otherwise are relatively unimportant. Engineers also apply the term *iron* to the ferrous alloys known as cast iron. These contain fairly large amounts of carbon, usually over 2.5%, and other alloying elements. They are discussed in detail in Chapter 8. Steels are ferrous-base alloys containing significant amounts of one or more alloying elements either intentionally added or retained during the refining process.

4.11 EFFECTS OF ALLOYING ELEMENTS IN FERROUS ALLOYS

The properties of *all* alloys are determined by the kinds and amounts of phases of which they are composed, by the properties of the phases, and by the way in which these phases are distributed among one another. (See Chapter 3 for a discussion of phases.) Ferrous alloys consist of two or more phases known as ferrite, austenite, carbides, and graphite. The alloying elements in ferrous alloys affect the stability of these phases, the relative amounts of the phases, and how the phases are distributed or dispersed throughout one another. The alloying elements also affect the properties of the phases in which the elements exist. Thus, the alloying elements achieve control over the properties of ferrous alloys.

Carbon is probably the most important alloying element found in ferrous alloys. In slowly cooled alloys containing small amounts of carbon (usually much less than 0.5% carbon), the structure consists of soft, ductile ferrite and hard, brittle carbide phases. Most of the carbon exists in the iron carbide. As the carbon content increases, the amount of iron carbide increases, and these alloys become harder, stronger, and less ductile. When the amount of carbon in slowly cooled alloys increases beyond 2%, as in the cast irons, there is an increasing tendency for the carbide phase to decompose into iron and graphite. This reaction is known as *graphitization* and can be represented by

$$\underset{\text{iron carbide}}{Fe_3C} \rightarrow 3Fe + \underset{\text{graphite}}{C}$$

Although increasing the carbon content increases the tendency toward graphitization, carbon content alone does not control the tendency. Since graphite is much softer than iron carbide, cast irons in which graphitization has occurred are softer than those in which it has not occurred. The form and distribution of the graphite phase determine whether these cast irons are ductile or brittle. This is described in more detail in Chapter 8.

The control of graphitization is important in the cast irons and is achieved by a combination of the adjustment of: (1) cooling rate and (2) alloy content. Slow cooling of the melt favors graphitization; rapid cooling impairs graphitization. Cooling rates can be increased locally by insertion of pieces of metal, known as *chills,* in sand molds (see Chapter 8). These produce areas consisting of hard carbide in castings which elsewhere have a soft, graphitic structure. The alloying elements in ferrous alloys can be classed as carbide formers or as graphitizers. The principal strong graphitizers include silicon, nickel, and, as already mentioned, carbon itself. The strong carbide formers tend to stabilize carbides and resist graphitization. The strongest of these include titanium, vanadium, molybdenum, tungsten, and chromium. The other common alloying elements have weak graphitizing or weak carbide-forming tendencies, unimportant for most purposes. Cast iron compositions are adjusted, or balanced, to yield either a hard carbide structure, a soft graphitic structure, or an in-between structure in which graphitization is partial but not complete. The strong carbide formers play an important role in quenched and tempered steels as will be discussed later.

In steels hardened by quenching and tempering, carbon performs at least two functions. All alloying elements (except cobalt), when in solution in austenite at the time of quenching, tend to shift the T-T-T curve (see Chapter 6) to the right, and thereby increase hardenability. Carbon is no exception to this generalization, although its contribution to hardenability is not great. It also must be kept in mind that carbides *not* dissolved in austenite when quenched will detract from hardenability. In alloy steels of higher carbon content it is difficult to achieve complete solution. In addition to its minor contribution to hardenability, carbon is largely responsible for improving the hardness of steel. This is its most important function.

The most important function of the alloying elements in steels to be hardened by quenching is to increase hardenability, that is, to retard the direct transformation of austenite to a coarse dispersion of ferrite and carbide. The more effective the alloying elements are in suppressing this transformation, the slower the cooling rate required to attain martensite. Thus, thermal shock and cracking tendencies are minimized and, with a given cooling rate, heavier sections of alloy steels can be transformed to martensite than for carbon steels. In the amounts used, as the percentage of an alloying element is raised, hardenability increases, although the effectiveness and cost of the elements vary. The greatest improvement in hardenability per unit cost of alloying element used is achieved by manganese, followed probably by molybdenum and chromium. Boron is extremely valuable in improving hardenability, but loses its effectiveness when more than 0.004% is present. Manganese is used in amounts up to 1.75%, chromium up to 1.55%, and molybdenum up to about 0.25% for the purpose of improving hardenability. Phosphorus is a low-cost element that improves hardenability, but its use is limited by virtue of its embrittling tendencies.

In steels hardened by quenching, a second and important function of alloying elements is to control properties upon tempering. The strong carbide formers are useful because they produce a steel that resists softening when tempered. Titanium, vanadium, molybdenum, tungsten, and chromium are known to impart this quality to steel. In fact, in some tool steels, very large amounts of the strong carbide formers are added because they produce a secondary hardening effect after the initial softening that accompanies tempering. The strong, hard, persistent carbides found in tempered steels containing these elements improve wear resistance, abrasion resistance, resistance to creep, and toughness. They permit greater stress relief in steel tempered to a given hardness, because higher tempering temperatures or longer tempering times can be used without adverse effects.

In slowly cooled steels, the most important effects of alloying elements other than carbon are: (1) their tendency to strengthen ferrite *when dissolved in it,* and (2) except for cobalt, their tendency to increase hardenability.

The tendency to strengthen ferrite is important in slowly cooled steels because ferrite is the matrix phase in these steels. An important principle of alloying is that the properties of the alloy tend to be governed by the quality of the matrix. Thus, if the alloy additions strengthen the ferrite, they tend to strengthen the alloy. The alloying elements most commonly used as ferrite strengtheners are phosphorus and silicon. Beryllium and titanium are also potent ferrite strengtheners. However, beryllium is far too expensive, and titanium is removed from solution in the ferrite by reaction with always-present carbon to form titanium carbide, and hence cannot strengthen the ferrite. Cobalt is known as a ferrite strengthener at high temperatures. It is used in steels that operate at or near red heat, where hot hardness is required. These steels, however, are quenched and tempered, rather than slowly cooled.

The tendency to increase hardenability in slowly cooled steels is important because it effectively reduces the amount of ferrite in the structure, replacing it with harder, stronger pearlite. The pearlite is also finer, and therefore stronger, than would be the case if the alloying elements were absent. The result of these two effects is a mild steel of improved strength in which there has been a negligible loss of ductility.

Corrosion resistance in stainless steels is obtained by the use of chromium as an alloying element in amounts exceeding 12%. The chromium forms a tenacious oxide film that prevents rusting. In American Iron and Steel Institute (AISI) alloy steels (described later), which usually contain less than a total alloy content of 5%, rust resistance is improved by small amounts of phosphorus, copper, and chromium.

Magnetic characteristics can be improved in steels used as temporary or soft magnets for alternating current applications through the use of alloy additions. Pure iron has desirable magnetic qualities but suffers from high eddy-current losses. These are reduced by alloy additions that have a minimal effect on hardening and a maximum effect on increasing the electrical resistance of iron. At the same time, the alloy additions should be compatible with fabricating requirements. Silicon is by far the best alloy addition in these respects. The silicon steels used for soft magnets often contain 3.5% silicon, but they may contain up to 5% silicon.

Machinability (see Chapter 17) of steel can be improved by several alloy additions. Sulfur improves machinability, but cannot be used in steel unless there is sufficient manganese to combine with it to form manganese sulfide. If manganese is not present, sulfur forms a low-melting iron sulfide network around the austenite grain boundaries. Steel with this structure is very weak and brittle at hot-working temperatures and suffers from what is called *hot shortness*. Manganese sulfide exists as an isolated phase, rather than a network, and acts as an internal lubricant and chip breaker. Internal lubricants reduce friction and wear between tool and workpiece and between tool and chip. Lead, in finely dispersed form, is another element that greatly improves machinability. It is added to ingots in the form of lead shot and is distributed throughout the solidified steel in the free state since it is insoluble in steel. It improves machinability in the same way as manganese sulfide, acting as a chip breaker and an internal lubricant. Lead is used in carbon steels and AISI steels and does not adversely affect, in the amounts used, the mechanical properties of either unhardened or hardened steels. Selenium is used for the same purpose in some stainless steel compositions.

The toughness of hardened steels is a function of the grain size of the austenite at the time of quenching. Alloying elements that form insoluble particles in the steel tend to minimize austenite grain growth during heating for hardening. Thus, aluminum-killed steels (see Section 4.5) are finer grained and tougher than silicon-killed steels, because the insoluble aluminum oxide particles resulting from deoxidation remain suspended throughout the solidified steel. The strong carbide formers also result in a fine-grained steel. The carbides of these elements are very persistent and difficult to dissolve in austenite and thus restrict grain growth. Vanadium and molybdenum are particularly noteworthy in this regard. Elements that promote fine-grained steel structures are known as *grain refiners*.

4.12 CARBON STEELS

Carbon steels are those in which carbon is the alloying element that essentially controls the properties of the alloys, and in which the amount of manganese cannot exceed 1.65% and the copper and silicon contents must each be less than 0.60%. The carbon steels can be subdivided into those containing between 0.08–0.35% carbon, those containing between 0.35–0.50% carbon, and those containing more than 0.50% carbon. These are known respectively as low-carbon, medium-carbon, and high-carbon steels. The price of steel depends on whether the steel is in bars, sheets, etc.

Low-carbon steel is relatively soft and ductile and cannot be hardened appreciably by heat treatment. It represents the largest tonnage of all steel produced. It is used for tin plate, automobile body sheet, fencing wire, light and heavy structural members (for example, auto frames and I-beams), and for hot- and cold-finished bars used for machined parts. Cold finishing improves surface finish, mechanical properties, and machinability of these compositions.

Medium-carbon steel is used for high-strength steel castings and forgings, such as railroad axles, crankshafts, gears, turbine bucket wheels, and steering arms. Medium-carbon steel can be hardened by heat treatment, but it cannot be through-hardened in sections greater than about 13 mm (.5 in.) thick.

High-carbon steel is used for forgings, such as wrenches and railroad wheels, and for hot-rolled products, such as railroad rails and concrete-reinforcing bars. High-strength wire products, such as piano wire and suspension bridge cable, are also made from high-carbon steel. High-carbon steel tools are among the most useful general-purpose tools for applications such as blanking dies, sledges, chisels, and razors.

4.13 ALLOY STEELS

Alloy steels contain appreciable quantities of alloying elements in addition to carbon. They include: (1) low-alloy, high-strength structural steels; (2) quenched and tempered low-carbon constructional alloy steels; (3) American Iron and Steel Institute-Society of Automotive Engineers (AISI-SAE) alloy steels; (4) alloy tool steels; (5) stainless steels; (6) heat-resisting steels; and (7) magnet steels.

Low-alloy, high-strength structural (HSLA) steels contain insufficient carbon and alloying elements to be hardened effectively by quenching to martensite. This is advantageous because it enables them to be welded without becoming brittle (Chapter 13). At the same time, the alloying elements they contain alter the microstructure so that it resembles a higher-carbon steel cooled at moderately fast (air-blast quench) rates. In addition, these steels contain slightly more phosphorus and silicon than the carbon steels, thereby strengthening the ferrite network. These changes in microstructure raise the yield strength about 40–50% above that of carbon structural steel. Lighter sections, of lower cost, can therefore be rated to carry a given load. Since rust and corrosion become increasingly important as section sizes become smaller, particularly in unpainted applications, HSLA steels can be made corrosion resistant (as much as eight times more than carbon steel). This is done by adding proper proportions of phosphorus, copper, silicon, chromium, and molybdenum. In that role, they are sometimes known as *weathering steels.* HSLA steels are typically used for railroad cars to save weight, and for bridge and building structural members to avoid painting.

A class of low-alloy but high-strength steels is that of the *dual-phase steels.* These have a microstructure of islands of high-carbon martensite-austenite (M-A) in a matrix of much softer ferrite. They are strengthened considerably when worked and offer weight reduction in parts because they provide high strength with less material. Thus, dual-phase steels have become important sheet metal materials in the automobile industry.

Quenched and tempered low-carbon constructional alloy steels are also known as *low-carbon martensites.* They are similar to low-alloy, high-strength structural steels, except that their alloy content permits quenching to bainite or martensite (see Chapter 6). The low-carbon martensite of these steels retains toughness to –46° C (–50° F), and these compositions produce welded joints fully as strong as the unwelded base metal. The low-carbon martensites tend to have slightly higher alloy contents than the low-alloy, high-strength structural steels. In addition, they may contain alloying elements, such as B, V, and Mo, all of which contribute to hardenabil-ity. And, the presence of V, Mo, and Ti forms persistent carbides that resist softening upon tempering. These steels are used in the form of plate for the construction of welded pressure vessels and as structural members for large steel structures, mining equipment, and earthmovers.

AISI-SAE alloy steels are steels whose compositions have been standardized by the American Iron and Steel Institute and the Society of Automotive Engineers. A numbering system using as many as five digits designates the composition of the alloy. The nominal carbon content is given in hundredths of a percent by the last two numbers when four digits are used and by the last three numbers when five digits are used. Thus, 4140 and 52100 indicate respective carbon contents of 0.40 and 1%. The first two numbers show which alloying elements are present (see Table 4-3).

With the exception of the low-carbon, plain-carbon steels, the AISI-SAE steels are always used in the heat-treated condition. Although the low-carbon, plain-carbon steels are sometimes

Table 4-3. Alloy steel carbon content designation

AISI-SAE Designation	Type
10XX	Plain-carbon steel
11XX	Resulfurized steels for free machining
12XX	Rephosphorized and resulfurized steels
13XX	Manganese steels
2XXX	Nickel steels
3XXX	Nickel-chromium steels
4XXX, 8XXX, 43XX, and 98XX	Steels containing molybdenum alone or in combination with nickel and/or chromium
5XXX or 5XXXX	Chromium steels
6XXX	Chromium and vanadium steels
92XX	Manganese and silicon steels

used in the unhardened condition, they are also sometimes surface-hardened by carburizing (see Chapter 6). The low-carbon AISI-SAE alloy steels are used for carburizing if somewhat better core properties and a greater depth of hardening is required. The steels containing more than 0.25% carbon are used in the quenched and tempered condition. Higher carbon contents, up to about 0.60% carbon, provide tempered martensite of greater hardness. Higher alloy contents serve to increase the depth to which hard martensite will form beneath the surface. In other words, the maximum hardness achievable is a primary function of carbon content, and the proportion of the section hardened is a primary function of alloy content, particularly manganese, molybdenum, chromium, and nickel.

AISI-SAE steels are used for applications such as carburized or through-hardened gears, steering mechanism parts, transmissions, shafting, and ordnance parts. The 52000 series are used mainly for ball and roller bearings.

Alloy tool steels represent a small, but extremely important percentage of total steel production, since they are essential to the processing of all other steel and engineering materials. If a tool has a simple shape, does not need to be hardened too deeply (for example, less than 6.4 mm [.25 in.]), and is to be used near room temperatures, it can be made of carbon tool steel. But when the shape becomes complex, and if hardenability, toughness, wear resistance, hot-working, and other requirements become critical, alloy tool steels are needed. The AISI-SAE designations list 13 different classifications of alloy tool steels. The major groups are listed in Table 4-4. The names of the various types of tool steel indicate some of their special properties. All tool steels should be hard, tough, and wear resistant, although the relative importance of these properties varies from application to application.

The *cold-work steels* include the water-, oil-, and air-hardening tool steels (W, O, A steels), all of which have a fairly high carbon content (0.60–2.25%) and varying degrees of hardenability. As indicated by their names, they are controlled by the amount and kind of alloying elements they contain. The shock-resisting tool steels (S steels), which can be considered special cold-work tool steels, have a lower carbon content (0.50%) to improve toughness. It is characteristic of the tool steels that toughness increases with decreasing carbon content, and the shock-resisting tool steels have measurable ductility (needed for toughness, see Chapter 3) even at 60 R_C. The high-carbon, high-chromium tool steels (D steels) have large amounts of chromium (12%)

Table 4-4. AISE-SAE alloy tool steel designations

AISI-SAE Designation	Type
	Cold-work steels
W2-W7	Water-hardening tool steel
O1-O7	Oil-hardening tool steel
A2-A7	Air-hardening tool steel
S1-S5	Shock-resisting tool steel
D1-D7	High-carbon, high-chromium tool steel
F1-F3	Carbon-tungsten tool steel
L1-L7	Low-alloy, special-purpose tool steel
P1-P20, PPT	Low-carbon mold steels
	Hot-work steels
H11-H43	Hot-work tool steels
	High-speed steels
T1-T15	Tungsten high-speed tool steels
M1-M36	Molybdenum high-speed tool steels

and other carbide formers. These alloy additions produce an air-hardening composition with excellent wear resistance, useful for blanking dies, thread-rolling dies, and brick molds. Since the strong carbide formers have such a great affinity for carbon, the carbon content is raised in these steels to assure that there is enough uncombined carbon remaining in the austenite to yield a martensite matrix of adequate hardness upon hardening. The carbon-tungsten tool steels (F steels) are similar to the oil- and water-hardening steels, but they have extra amounts of tungsten, resulting in improved wear resistance because of tungsten carbide particles. The low-alloy, special-purpose tool steels (L steels) are similar to the W steels, but have higher amounts of strong carbide formers for improved wear resistance. The low-carbon mold steels (P steels) have the lowest carbon content of all the tool steels. They are carburized for improved wear resistance after machining or pressing them to shape.

The *hot-work steels* (H steels) contain fairly large amounts of the strong carbide formers, which resist softening at operating temperatures. Chromium, tungsten, molybdenum, and vanadium are found in these compositions. Their carbon contents are below 0.65% carbon so that they exhibit moderately good toughness at high-strength levels. They are used for forging dies, extrusion dies, and die-casting dies. One composition has been found useful for structural members in supersonic aircraft, where it resists softening when subjected to temperatures of 538° C (1,000° F) or more for long times.

The *high-speed steels* contain either tungsten (T steels) or molybdenum, usually with tungsten (M steels) as the principal carbide formers. Both T and M steels contain chromium and vanadium. The high-carbon content is necessary to satisfy the carbide-forming tendencies and produce excellent wear resistance and hardness at red heat. At the same time, the carbon content is not so high that toughness is lacking. Since molybdenum is a cheaper alloying element than tungsten and is about twice as effective, the T steels have been almost entirely replaced by M steels. There is no significant difference between the performance of the two major classes of high-speed steels. Cobalt is used in some compositions because it dissolves in and imparts to the ferrite matrix a specially high strength at red heat. The high-speed steels are used for taps, reamers, milling cutters, broaches, and heavy-duty, high-temperature aircraft bearings. High-speed steel cutting tools are discussed in Chapter 17.

Magnet steels can be divided into two broad classes: (1) permanent magnets and (2) temporary magnets. The best permanent magnet materials are not steels, although steel is sometimes used for this application. Permanent magnets are physically hard, and any steel that can be hardened by heat treatment can serve as a permanent magnet. The temporary magnet steels are vastly more important than the steels used for permanent magnets. The temporary magnets are mechanically soft materials, readily magnetized and demagnetized by alternating current. The magnetic performance of these steels is impaired by any composition change or treatment that hardens them. This includes even the cold work associated with blanking and stamping, as well as that associated with stacking and clamping laminates.

Hardness tends to make the steels more "permanent," that is, more difficult to magnetize and demagnetize in an alternating current field, resulting in power losses known as *hysteresis losses*. Another source of power loss is by induced eddy currents. This kind of power loss is minimized by using insulated laminates in an assembly and increasing the electrical resistance of the laminate material. Two fundamental changes accompanying solid-solution formation are: (1) an increase in electrical resistance and (2) an increase in hardness. Therefore, alloying to increase electrical resistance and reduce eddy-current losses is accompanied by an increase in hardness and an increase in hysteresis losses. Silicon is the alloying element that most effectively increases resistance with a minimum increase in hardness. The sheet steel used for temporary magnets is a silicon-bearing steel, of very low carbon content. The presence of carbon tends to harden the steel and stabilize austenite (see the discussion of stainless steels). Since hardness is undesirable for reasons already mentioned, and austenite is nonferromagnetic, low-carbon contents are essential for the highest-quality temporary magnet steels.

Stainless steels rely primarily upon the presence of chromium for the achievement of stainless qualities. In general, the higher the chromium content the more corrosion resistant the steel. There are three common classes of stainless steel: (1) austenitic, (2) ferritic, and (3) martensitic. The names of these classes reflect the microstructure of which the steel is normally composed. The alloying elements in steel can be classed as austenite stabilizers and ferrite stabilizers. The *austenite stabilizers* of importance are carbon, nickel, nitrogen, and manganese. These elements enhance the retention of austenite as steel is cooled. When 12% or more manganese is present, or when 20% or more nickel is present, it is impossible to cool steel slowly enough to allow austenite to transform to ferrite. Even with much lower nickel and manganese contents the transformation is very sluggish and austenite is stable at room temperature. The *ferrite stabilizers* of importance are chromium and the strong carbide formers. The ferrite stabilizers tend to prevent transformation of steel to austenite upon heating. Whether a steel is austenitic, ferritic, or martensitic depends upon the balance between the amounts of austenite and ferrite stabilizers present, and the heating-cooling cycle to which the steel has been subjected.

The *austenitic stainless steels* are produced and used in the greatest tonnage. Although they all contain nickel, occasionally manganese and nitrogen are used as nickel substitutes. These three elements are responsible for the austenitic structure. The austenitic stainless steels contain, as do all stainless steels, chromium, necessary for corrosion resistance. For chromium to be effective in imparting corrosion resistance, it must be in solid solution in the austenite. These steels lose corrosion resistance if the chromium exists in a second phase, such as chromium carbide. Chromium tends to precipitate as a carbide from austenite at the grain boundaries when these austenitic stainless steels are cooled from a temperature near 816° C (1,500° F). This depletes the grain boundary region of the chromium necessary for corrosion resistance, and renders the grain boundaries susceptible to a form of attack known as *intergranular corrosion*. Corrosion resistance can be restored by heating the steel above 816° C (1,500° F), followed by quenching to prevent formation of chromium carbide, thus retaining the chromium in solution in austenite.

Welding (see Chapter 13) results in cooling rates near the weld that cause sensitization of austenitic stainless steel to intergranular attack. If the structure does not lend itself to subsequent heating and quenching for restoration of corrosion resistance, special grades of austenitic stainless steels should be used. The simplest modification is a low-carbon grade, in which the maximum permissible carbon content is below 0.03%. There will then be insufficient precipitation of chromium to be harmful. Carbon is also rendered harmless by adding alloying elements that have a greater carbide-forming tendency than chromium, thus leaving the chromium in the uncombined state. Elements that will achieve this include titanium, columbium, and molybdenum. Austenitic stainless steels containing these elements are known as *stabilized grades*.

The *ferritic stainless steels* contain chromium, no nickel, and tolerate only small amounts of austenite-stabilizing carbon. If the carbon content is increased, the chromium content must be increased to maintain balance and a ferritic structure. In this balanced condition, these steels can be heated to the melting point without transforming to austenite. Thus, it is impossible to harden them by quenching and tempering.

The *martensitic stainless steels* contain balanced amounts of chromium (ferrite stabilizer) and carbon and nickel (austenite stabilizers), so that upon heating the steel becomes austenitic, but upon cooling tends to revert to ferrite. These compositions can be heated to the austenitic range of temperatures and will transform to martensite upon cooling at suitable rates. The carbon content is sufficient to produce a martensitic hardness adequate for cutlery and surgical instruments.

Precipitation-hardenable stainless steels have either austenitic, martensitic, or semi-austenitic structures, achieved by adjustment of the amounts of austenite and ferrite stabilizers, principally chromium and nickel. Lowering the chromium/nickel ratio tends to stabilize the austenitic condition; raising it promotes transformation to martensite. Hardening is accomplished by precipitation of titanium or copper from martensitic types, precipitation of aluminum from semi-austenitic types, and precipitation of carbide from austenitic types. Precipitation may result from simple heating and aging cycles, perhaps following a subzero treatment or cold-working treatment to transform an austenitic structure to a martensitic one. The precipitation-hardenable steels were developed for applications, such as aircraft structural members, where the size and shape of the structure prevent hardening by cold work or conventional quenching and tempering.

The austenitic stainless steels cost about 10 times as much as ordinary steel and are used where corrosion resistance, high-temperature strength, and oxidation resistance are critical. Textile machinery, chemical equipment, food-processing equipment, and architectural trim are examples of austenitic stainless steel applications. The ferritic stainless steels are used in applications where lower strength and corrosion resistance can be tolerated, as in automobile body trim. The martensitic stainless steels cost about seven or eight times as much as carbon steel. They are used where wear and corrosion resistance are prime considerations, as in cutlery, razor blades, and instruments.

AISI-type numbers have been assigned to about 40 stainless and heat-resisting steel compositions. These are three-digit numbers (for example, 2XX and 3XX). The second and third digit of a number designates a specific composition. Specifications for standard types are given in reference books and handbooks.

Maraging steels develop martensite upon cooling from the austenitizing temperature, but the martensite formed in these steels is, unlike the martensite of AISI alloy steels, ductile and tough. The ductility and toughness of this martensite result from its low carbon content, which is below 0.03%. In the martensitic condition these steels can be cold-worked and hardened by precipitation at temperatures below the austenitizing temperature, for example, 482° C (900° F). Hardening is believed to result from precipitation of compounds such as Ni_3Mo and Ni_3Ti. The hardened maraging steels have yield strengths up to 2 GPa (300,000 psi) and Charpy V-notch impact strengths well over 14 J (10 ft-lbf). The impact strength for maraging steels with about 1.4 GPa (200,000 psi) yield strength is in the range of 68–81 J (50–60 ft-lbf). These steels are particularly useful in the manufacture of large structures having critical strength requirements, such as space-vehicle cases, hydrofoil struts, and extrusion press rams.

4.14 QUESTIONS

1. What raw materials are required for the production of pig iron?
2. Distinguish between ore and agglomerate.
3. What functions are performed in the blast furnace by (a) limestone? (b) ore? (c) air?
4. (a) Find ΔH for $C + O_2 \rightarrow CO_2$. Is heat absorbed or liberated by the reaction?

   (b) Prove that ΔH for $Fe_2O_3 + 3CO \rightarrow 2Fe + 3CO_2$ is 26 kJ (24.6 Btu) using Equations 4-4, 4-5, and 4-6.
5. (a) What are the principal impurities in pig iron that are largely removed when it is converted to steel?

   (b) How are the impurities referred to in part (a) removed?
6. (a) What is the source of heat in the basic oxygen furnace?

   (b) What processes has the basic oxygen process almost entirely replaced?
7. Why are two slag covers needed for stainless steels and tool steels produced in the electric furnace?
8. Describe the difference between rimmed, semikilled, and killed steel.

9. (a) What are the reasons for using vacuum-melting techniques in steelmaking?
   (b) What is the difference between vacuum melting and vacuum degassing?
10. (a) How are the principal impurities removed from bauxite to make pure alumina? (b) How is aluminum obtained from pure alumina?
11. Describe (a) concentration, (b) roasting, (c) smelting, (d) converting, and (e) refining as applied to the extraction of copper from its ores.
12. Of the metals whose refining is discussed in this chapter, list those whose primary refining is by reduction with coal, coke, hydrocarbons, or hydrogen, and those whose primary refining is by electrolytic action.
13. In what ways do alloying elements affect the control of the properties of ferrous alloys?
14. In what forms does carbon exist in ferrous alloys? How does the form in which carbon exists affect the properties of the alloys?
15. What two functions are performed by carbon in steel?
16. How does an increase in carbon content affect the mechanical properties of ferrous alloys not hardened by quenching?
17. Is ferrite strengthening a more important function of alloying elements in slowly cooled steels or in quenched steels? Why?
18. How can graphitization be promoted in ferrous alloys?
19. What is the most important function of alloying elements in ferrous alloys hardened by quenching?
20. What changes in properties and microstructure accompany the tempering of martensite?
21. How does hardenability differ from hardness?
22. What is secondary hardening? In what types of ferrous alloys does it occur?
23. Why is silicon used as an alloying element in magnet alloys?
24. Explain the effects of lead and sulfur on the machinability of steels.
25. What is the effect of alloying elements on the toughness of hardened steels?
26. Why is low-carbon steel not usually hardened by heat treatment?
27. What are the principal uses for (a) low-carbon, (b) medium-carbon, and (c) high-carbon steels?
28. What functions do the alloying elements perform in the low-alloy, high-strength structural steels?
29. How do the composition and properties of the low-carbon constructional steels differ from those of the low-alloy, high-strength structural steels?
30. In a quenched and tempered piece of steel, upon what does (a) the maximum hardness depend, and (b) the depth of hardening depend?
31. For what purposes are AISI-SAE steels commonly used?
32. In tool steels, what is the relationship between (a) carbon content and properties, and (b) carbide formers and properties?
33. Correlate the mechanical and magnetic properties of the magnet alloys.
34. What is the effect of the presence of austenite in magnet steels?
35. What determines whether a stainless steel is austenitic, ferritic, or martensitic?
36. What sensitizes stainless steel to intergranular attack and what steps can be taken to overcome this difficulty?

37. What are common applications for (a) austenitic, (b) martensitic, (c) ferritic, and (d) precipitation-hardenable stainless steels?
38. How is hardness achieved in the maraging steels? How does the martensite of maraging steels differ from the martensite of AISI-SAE steels?

4.15 REFERENCES

Brady, G.S. and Clauser, H.R. 1985. *Materials Handbook,* 12th Ed. New York: McGraw-Hill.

Harvey, P.D., ed. 1982. *Engineering Properties of Steel.* Metals Park, OH: American Society for Metals.

Langford, Jr., W.T., Samways, N.L., Craven, R.F., and McGannon, H.E. 1985. *The Making, Shaping and Treating of Steel,* 10th Ed. Pittsburgh, PA: United States Steel Company.

McGannon, H.E., ed. 1971. *The Making, Shaping, and Treating of Steel,* 9th Ed. Pittsburgh, PA: United States Steel Company.

Metals Handbook, 9th Ed. Vol. 1: *Properties and Selection: Irons and Steels.* 1978. Metals Park, OH: American Society for Metals.

Metals Handbook, 9th Ed. Vol. 3: *Properties and Selection: Stainless Steels, Tool Materials and Special Purpose Metals.* 1980. Metals Park, OH: American Society for Metals.

Schlecten, A.W. 1980. "Revolutionary Changes in Extractive Metallurgy." *ASM News, Steel Products Manual,* April. New York: American Iron and Steel Institute.

Tool and Manufacturing Engineers Handbook, 4th Ed. Vol. 3, *Materials, Finishing, and Coating.* 1985. Dearborn, MI: Society of Manufacturing Engineers (SME).

Nonferrous Metals and Alloys

5

Titanium, 1896
—Henry Moissan

Although pure iron is not of great importance, the same cannot be said of many pure nonferrous metals. The nonferrous metals are used in pure form because of such superior properties as electrical and thermal conductivity, corrosion resistance, high melting temperature, and special electrical, optical, and chemical properties. As in the case of ferrous alloys, alloying the nonferrous metals usually results in improved mechanical properties.

In this chapter, the effects of alloying on the properties of pure metals are first discussed. The following metals and their alloys are then considered: the light metals, that is, aluminum, magnesium, copper, zinc, titanium, nickel, and finally, white metals, refractory metals, and precious metals.

5.1 EFFECTS OF ALLOYING ON PROPERTIES

When any pure metal is alloyed with other metals or nonmetals, some properties are significantly impaired, others are significantly improved, and still others are not greatly altered.

The strongly metallic elements suffer a decrease in electrical and thermal conductivities as a result of alloying. The semiconductor germanium, not strongly metallic, shows an increase in conductivity as a result of alloying (or doping). Aluminum, copper, and silver, often selected for uses requiring excellent thermal or electrical conductivity, suffer significant decreases in these properties as a result of alloying. For instance, the electrical conductivity of standard copper (taken as 100%) drops to about 93% as a result of only 0.02% aluminum, even though both pure metals are excellent conductors.

The specific gravities of pure metals are raised, lowered, or unchanged by alloying, depending upon the character and amounts of the alloying elements used. When the additions represent about 10% or less of the total weight of the alloy, the changes in specific gravity are, for most purposes, insignificant. For instance, pure aluminum has a specific gravity of 2.70, while the 7075 alloy (5.5% Zn, 2.5% Mg, 1.5% Cu, 0.3% Cr) has a specific gravity of 2.80.

The moduli of elasticity of pure metals are not greatly affected by alloying with small amounts of other elements. For instance, annealed nickel has a modulus of elasticity of about 206 GPa (30×10^9 psi), and annealed Monel® metal, which consists of 34% other elements with the balance nickel, has a modulus of elasticity of about 179 GPa (26×10^9 psi). It can be reasonably assumed that for most purposes, the modulus of elasticity and shear modulus of the alloy have about the same values as the base metal.

Yield strength, tensile strength, fatigue strength, and high-temperature resistance are generally improved by judicious alloying. Because of this, alloys are more commonly used than pure metals where mechanical behavior is the primary criterion of performance. In addition, alloying sometimes produces compositions whose mechanical properties can be further enhanced by heat treating. Heat treating is applied to pure metals only to remove the effects of cold work; pure metals cannot be hardened by heat treatment.

Ductility is often reduced by alloying, but an important exception to this is the initial increase in ductility that accompanies the alloying of copper with zinc and other metals.

5.2 ALUMINUM

Aluminum is the most abundant metallic element, representing about 8% of the earth's crust. The primary source of aluminum is bauxite, which is processed into hydrated aluminum oxide by the Bayer process. From aluminum oxide, 99.5–99.9% pure aluminum is extracted by the Hall process, which uses electrolytic cells where the oxide is reduced to pure aluminum.

Pure aluminum is known for its excellent electrical and thermal conductivity, corrosion resistance, nontoxicity, light reflectivity, low specific gravity, softness, and ductility. The electrical conductivity of EC- (electrical conductor) grade aluminum is 61% of the conductivity of standard copper, based on equal cross sections. If equal weights of copper and aluminum conductors of a given length are compared, it will be found that aluminum conducts 201% as much current as copper. Since the price of aluminum is generally much lower than copper, and since the specific gravity of aluminum is only 2.7 compared to 8.9 for copper, the advantages of aluminum as an electric conductor are obvious. However, an important limitation of aluminum in this regard is the difficulty of soldering or joining it. This has been overcome by chemically coating it with tin, followed by plating with other metals if necessary. Aluminum can be joined by welding (see Chapter 13). The good thermal conductivity of aluminum has led to its application as radiator fin material in baseboard heating and air-conditioning units. Excellent light reflectivity and corrosion resistance account for its use as a sheet metal light reflector and a coating for high-grade optical reflectors. Its softness and ductility, coupled with its corrosion resistance and nontoxic nature, have resulted in its use as a foil and packaging material.

In 1994, the transportation industry became the largest consumer of aluminum, surpassing the container and packaging industry.

5.2.1 Aluminum Alloys

Aluminum alloys can be wrought (rolled, gorged, extruded, or drawn) or cast by any of the casting methods, including die casting. Major factors in determining the suitability of an alloy for die casting are its melting temperature and the corrosive nature of the molten alloy with respect to the dies and die-casting machine. Aluminum alloys rank in this regard as superior to copper alloys but inferior to zinc-base die castings.

Aluminum alloys, in general, are considered easily machined, though in some cases good surface finish is difficult to obtain. This can be improved by the use of cutting fluids (see Chapter 17). Another problem connected with machining aluminum arises from its relatively large coefficient of thermal expansion and low modulus of elasticity. These properties are likely to make dimensional control difficult, either because of inadequate cooling or excessive deflection of members as a result of cutting forces.

Joining of aluminum alloys (see Chapter 13) is accomplished successfully by a variety of welding and brazing methods. Inert-gas, shielded-arc welding is popular. Soldering is difficult and not generally recommended unless the alloy has been tin coated.

Aluminum alloys are readily mechanically worked, either cold or at elevated temperatures, since they are relatively soft and have good ductility.

Although aluminum alloys can be electroplated, they are more difficult to electroplate than ferrous, copper, and zinc alloys.

Aluminum can be hardened by solid-solution hardening, cold working, or precipitation hardening. In fact, the precipitation hardening of aluminum by copper was the first process investigated, and these alloys were the first ones commercially exploited. Among the elements added to aluminum for precipitation hardening are copper, manganese, nickel, and silicon. Silicon is used mainly to improve castability. Zinc is used as a solid-solution hardener, and magnesium is added to improve corrosion resistance.

A range of aluminum alloys with varying mechanical properties are available with strengths in the vicinity of 600 MPa (8,700 psi).

In addition to mechanical properties, several other factors should be kept in mind. One of the most important of these is the specific gravity of aluminum alloys. Dividing the strength values by the specific gravity yields a number known as the *specific strength.* Comparison of specific strengths shows that most, but not all, aluminum alloys are superior to most steel compositions. Specific strengths have an important bearing on payloads and dead weight. When limitations are placed on section size and bulk, steel may be favored over aluminum alloys. When structural rigidity is required, the additional bulk of aluminum alloys is advantageous. Considerations, such as the need to develop optimum properties by heat treating, and the need to maintain these after joining, are also important in deciding whether steel or aluminum alloys should be used for a specific application.

Besides specific gravity and specific strength, the cost of aluminum must be considered. On the basis of equal masses, aluminum alloy structural shapes cost about eight times more than steel. Cost per kilogram (pound) and specific gravity considerations usually favor aluminum alloys over copper alloys and magnesium alloys, but not over steel.

The principal use of aluminum alloys is in the transportation industry, such as for aircraft and aerospace applications, where low density and high-strength properties are important considerations. The automobile industry has increasingly used aluminum alloys because of their low density, corrosion resistance, flexibility in designs, and strength. Followed by transportation, the container and packaging industries are the second biggest users of aluminum alloys, such as for cans and foils. These industries are followed by building and other types of construction, consumer durables, etc.

Wrought aluminum and its alloys are designated by a four-digit number for identification. The first digit indicates the major alloying element. The second digit indicates a modification to the alloy or its impurity limit. The third and fourth digits indicate aluminum purity (for 1XXX series) or identify the aluminum alloy.

The cast aluminum alloy is also designated by four digits where the last digit is separated by a decimal point and indicates the product form, such as casting (indicated by 0) or ingot (indicated by 1). The first digit indicates the alloy group. The second and the third digits indicate aluminum purity (for 1XXX series) or identify the aluminum alloy. However, aluminum cast alloys are commonly represented by just three digits.

The significance of the first digit in aluminum alloy classification for wrought and cast alloys is summarized in Table 5-1.

Following the four-digit classification, the aluminum alloys are further described by temper designations. The temper designation is separated from the alloy designation by a hyphen. The basic temper designation is followed by additional digits to indicate a strain-hardened or heat-treated subdivision. These designations are summarized in Table 5-2.

5.3 MAGNESIUM

At about 2%, magnesium is the third most abundant metallic element in the earth's crust. Its worldwide production was 320,000 million tons in 1995. With a density of 1.74 g/cm^3 (.063 lb/in.^{3}), it is the lightest of all the commonly used structural metals. Metallic magnesium is produced by several extractive processes. The most widely used method is electrochemical conversion of magnesium chloride ($MgCl_2$) into magnesium and chlorine gas. Magnesium metal is also produced by thermal reduction of mineral rocks containing magnesium. Magnesium comes off as vapor at high temperature due to chemical reaction and is then condensed into crystals.

Tensile strength of commercially pure magnesium is only about 186 MPa (2,700 psi) and hence it is usually alloyed with other metals. Magnesium used as an alloying element with aluminum represents the most dominant market for primary magnesium, accounting for 52% in 1995. Primary magnesium for die castings is the second largest market at 21% in 1995. The third largest market for primary magnesium is

Table 5-1. Significance of first digit in aluminum alloy classification

First Digit of Classification	Wrought Aluminum Alloys (Major Alloying Element)	Cast Aluminum Alloys (Major Alloy Group)
1	(≥99%)	aluminum (≥99%)
2	copper	copper
3	manganese	silicon with copper and/or magnesium
4	silicon	silicon
5	magnesium	magnesium
6	magnesium and silicon	unused series
7	zinc	zinc
8	other element	tin
9	unused series	other element

Table 5-2. Aluminum alloy designations

Basic Temper Designations	Definition
F	As fabricated
O	Annealed and recrystallized
H	Strain-hardened by cold working
T	Heat treated
Strain-hardened Subdivisions	**Definition**
H1	Strain-hardened only; a second digit (2–8) indicates the degree of strain hardening from quarter to fully hardened, respectively
H2	Strain-hardened and partially annealed; the second digit indicates the degree of partial annealing (2, 4, 6, 8) of cold-worked materials
H3	Strain-hardened and stabilized; the second digit indicates temper (2, 4, 6, 8)
Heat Treat Subdivisions	**Definition**
T1	Cooled from hot-working temperature and aged at room temperature
T2	Cooled from hot-working temperature, cold-worked, and aged at room temperature
T3	Solution-treated, cold-worked, and aged at room temperature
T4	Solution-treated and aged at room temperature
T5	Cooled from hot-working temperature and aged at higher temperature
T6	Solution-treated and aged at higher temperature
T7	Solution-treated and stabilized by overaging
T8	Solution-treated, cold-worked, and aged at higher temperature
T9	Solution-treated, aged at higher temperature, and cold-worked
T10	Cooled from hot-working temperature, cold-worked, and aged at higher temperature

as a desulfurizing agent for iron and steel, accounting for about 12% in 1995.

5.3.1 Magnesium Alloys

Magnesium alloys may be cast (die, sand, permanent mold, or investment), forged, extruded, or flat rolled, but are usually die cast because of the low cost, high quality, and accuracy afforded for producing intricate shapes.

In general, magnesium alloys are weaker, more brittle, and have poorer high-temperature properties than aluminum alloys. The principal alloying additions are aluminum and zinc. Aluminum forms a precipitation-hardenable alloy, and zinc is a solid-solution hardener and improves corrosion resistance. Corrosion resistance is good only in the absence of moisture.

Among the most machinable of the metallic materials, magnesium alloys have the ideal combination of properties for good machinability, namely softness and brittleness. Their brittleness makes cold mechanical working difficult. However, they are readily hot worked and die cast.

Magnesium alloys are used mainly when weight savings is a special requirement such as in aircraft and missiles, portable power tools, material handling equipment, bicycles, roller skates, golf clubs, tennis racquets, baseball bats, luggage, and ladders. They are also used in computer housings and printing and textile machinery. Recently, the market for magnesium alloys has opened up in the automotive industry with dozens of applications for parts, such as steering wheels, valve covers, and seat frames. Light weight, a high productivity rate in die casting, high thermal conductivity, high damping capacity, creep resistance, and good machinability are offered by magnesium alloys.

The tensile strength of magnesium alloys range from about 220–240 MPa (3,190–3,480 psi) for Mg-Al and Mg-Al-Zn cast alloys up to 275 MPa (4,000 psi) for Mg-Zn-Zr cast alloys. The amount of cold working tolerated by the hexagonal, close-packed, crystal structure of Mg is restricted to mild deformations. Most magnesium alloys are, therefore, hot worked. The tensile strength of wrought magnesium alloy may reach 317 MPa (4,600 psi) (ZK60A-T6).

Poor galvanic corrosion resistance, poor cold-working capacity, and relatively high cost compared to aluminum are the major disadvantages of magnesium alloys.

Magnesium alloys are usually designated by two capital letters followed by numbers. The first and second letter indicate alloying elements in the highest and second highest concentration, respectively. The first number following the letters stands for the weight percent of the first letter element and the second number for the second letter element. These numbers may be followed by letters A, B, C etc., which indicate a modification to the alloy, usually in the form of its impurity level. Finally, this syntax is followed by the heat treatment designation using the same symbols used for aluminum alloys. The letters used for magnesium alloying elements are outlined in Table 5-3.

Table 5-3. Magnesium alloying elements

A = aluminum	P = lead
E = rare earths	Q = silver
H = thorium	S = silicon
K = zirconium	T = tin
M = manganese	W = yttrium
Z = zinc	

5.4 COPPER

Pure unalloyed copper is used for electric wire, bars, and buses, and for tubing and pipe because it has outstanding electric and heat conductivity, corrosion resistance, and solderability.

5.4.1 Copper Alloys

Copper can be hardened and strengthened by cold working and solid-solution alloying with zinc, tin, aluminum, silicon, manganese, and nickel. It forms a precipitation-hardenable alloy when small amounts of beryllium are present. The ductility of many of the solid-solution alloys is greater than that of copper.

The room-temperature mechanical properties of copper alloys are intermediate between those of aluminum alloys and steel. The elevated-temperature properties, though superior to those of the aluminum alloys, are not outstanding. The copper alloys are known for attractive appearance and corrosion resistance, accounting for its large-scale use in hardware, particularly marine hardware. Wide sheets and drawn-ingot copper rods are 12–14 times more expensive as equal masses of carbon steel.

Copper-zinc alloys are called *brasses*. Alpha-phase brass with up to 36% zinc includes ductile, easily worked compositions, such as cartridge brass for deep drawing. The beta phase appears with a larger proportion of zinc. It is more brittle and less easily cold worked but more corrosion resistant and machinable. Colors vary by composition from red to yellow. Brass shapes in large quantities cost from 8–14 times more than the same mass of carbon steel.

Most bronzes are alloys of copper with up to 12% tin. They are noted for toughness, wear and corrosion resistance, and good strength. Typical applications are for bearings and worm gears.

Fabrication of copper alloys is readily accomplished by casting, cold working, machining, powder metallurgy, and joining by soldering. The alloys take a variety of finishes and are easily electroplated. Die casting, because of the relatively high melting temperatures involved, is appreciably more difficult than for aluminum, zinc, and magnesium alloys. The higher melting temperatures result in greater thermal shock to the dies and shorten die life. Hot working can be carried out, but some of the alloys are hot short.

The copper-base alloys include some of the most machinable metal compositions. Particularly noteworthy are the leaded, free-machining brasses, in which lead acts as an internal lubricant and chip breaker.

5.5 ZINC

Industrial pure zinc with a tensile strength of 110 MPa (1,600 psi) is the fourth most used metal after iron, aluminum, and copper. Zinc ore (typically ZnS) is converted to ZnO_2 first by heating in air and then reducing to zinc electrolytically.

The primary use of pure zinc is as a coating for iron and steel wires or sheets for corrosion protection, called *galvanizing*. Zinc has remarkable atmospheric corrosion resistance properties and yet is anodic to iron and steel. Hence, in the event of a scratch in the coating, it would undergo sacrificial corrosion, galvanically protecting iron or steel. Corrosion protection is found to improve with the addition of aluminum to zinc in galvanized coatings. The growth rate of Galfan® alloy (5% Al) and Galvalume® alloy (55% Al) far exceeds the growth rate of galvanized products. The other major use of zinc is in the form of alloy die casting. Zinc has a low melting point of about 420° C (788° F) and zinc alloys permit a faster rate of die casting than other competing alloys (aluminum and copper). The third major use of zinc is as an alloying element for brass (an alloy of Cu and Zn).

Recently, a new barrier application of zinc was developed to protect umbilical control and supply lines on offshore oil production heads at depths to 1,646 m (5,400 ft). The umbilical lines are up to 96 km (60 mi) long and connect the deep-water oil heads with shallow-water platforms in the Gulf of Mexico. The zinc coating on low-carbon steel umbilical lines is 0.8 mm (.03 in.) thick, applied through extrusion technology (where molten zinc is a feed material) and is expected to provide protection for 30–40 years.

5.5.1 Zinc Alloys

Zinc as a structural material is used with alloying addition of aluminum to improve strength. Conventional die-casting alloys are based on 4% Al addition. Magnesium is added in small amounts to improve intergranular corrosion resistance. Addition of small amounts of copper improves stability and strength. Another class of zinc-aluminum (ZA) casting alloys uses nonconventional compositions with additions of Al up to 27%. Small amounts of copper and magnesium are added for optimum strength, stability, and castability. The strength of the alloys is increased from about 283–407 MPa (41,000–59,000 psi). Recently, ternary zinc-copper-aluminum alloys have been developed (ACuZinc) that are suitable for high load and elevated temperature applications. These alloys are primarily composed of zinc with copper (5–6%) and aluminum (2.8–3.3%). In addition to the die castability of ZA alloys (hypereutectic alloys developed in the 1990s) and Zamak® alloys (hypoeutectic alloys developed in the 1930s), the ACuZinc alloys offer the advantage of higher strength and greater creep and wear resistance.

The use of wrought zinc alloys is somewhat restricted because rolled zinc alloys exhibit deformation anisotropy due to zinc's hexagonal close-packed (HCP) crystal structure, and because zinc has a tendency to creep at room temperature. The superplastic Zn-22% Al alloys are an important family of wrought zinc alloys. These alloys, with small amounts of copper and/or magnesium, can be easily proned into intricate shapes since they can undergo large amounts of uniform plastic deformation at moderately high temperatures. They can be later heat treated to increase strength.

5.6 TITANIUM

Essentially no high-purity titanium is produced. Commercially pure titanium, sometimes called *unalloyed titanium*, contains up to about 1% of various alloying elements, mainly oxygen, iron, carbon, nitrogen, and hydrogen. This grade of titanium is useful mainly because of its corrosion resistance and specific gravity of 4.5. It is as strong or perhaps slightly stronger than most copper-base and aluminum-base alloys, and low-carbon steels, but is weaker than alloyed titanium. It is, however, the most ductile and the least difficult titanium composition to fabricate. Commercially pure titanium (at about 100 times the cost of carbon steel and about five or six times the cost of stainless steel) is used because of its corrosion resistance for chemical piping, valves, tanks, and prosthetic devices. In the aerospace industry, it is used mainly for applications where high-temperature rather than specific strength requirements are of greater importance. These applications include fire walls, tail pipes, and jet-engine compressor cases.

5.6.1 Titanium Alloys

The corrosion resistance and mechanical properties of titanium alloys compare favorably with those of austenitic stainless steel. In general, the titanium alloys are used as substitutes for stainless steel, particularly the austenitic grades, in applications where the lower specific gravity of titanium alloys justifies the much higher cost. About 90% of titanium alloy production is used in aerospace applications. Titanium alloys occur as three different structures: (1) alpha, which has a hexagonal-close-packed structure and cannot be hardened by heat treatment; (2) beta, which has a body-centered-cubic structure and can be age hardened; and (3) alpha-beta mixtures, which can be hardened by heat treatment. Better than 60% of titanium alloy production is in the form of mixed alpha-beta structures.

Aluminum is a major alloying addition for titanium alloys. It functions as an alpha stabilizer, dissolves in the alpha, and strengthens the alpha while in solution. For this reason, the presence of aluminum is important in alpha or alpha-beta alloys to be used at temperatures above 371° C (700° F). Aluminum-free alloys lose strength rapidly when service temperatures rise above 371° C (700° F). Most of the other alloying elements used in titanium alloys are beta stabilizers. The beta phase, containing dissolved elements such as vanadium and chromium, tends to be stronger than the alpha phase. The mixed alpha-beta alloys contain aluminum and beta stabilizers. They can be quenched to a metastable structure containing more than equilibrium amounts of beta. Upon aging, alpha precipitates and the alloys are strengthened.

Fabrication of titanium alloys is much more difficult than for aluminum-base and copper-base alloys. Machinability is about the same as for stainless steels, that is, difficult. Arc welding can be done only if shielding is provided by helium or argon, since titanium has a great affinity for the oxygen and nitrogen in air. Both oxygen and nitrogen have strong embrittling tendencies. Titanium alloys can be mechanically formed, but frequent annealing is required, and galling and seizing of dies are a problem.

Titanium alloys are used for applications such as aircraft gas-turbine compressor blades, forged airframe fittings, missile fuel tanks, structural parts operating for short times at up to 593° C (1,100° F), and autoclaves and process equipment operating at up to 482° C (900° F). Low specific gravity, good strength even at high temperatures, and corrosion resistance justify the use of these materials that cost, on a weight basis, about five times more than ordinary stainless steels.

5.7 NICKEL AND ITS ALLOYS

Almost 60% of all the nickel produced is used as an alloying element in steel and iron, particularly the austenitic stainless steels. Nickel plating consumes the second largest quantity of nickel, followed by the high-nickel alloys that account for about 15% of total nickel production. These include the *Monel*® alloys, which are essentially nickel-copper alloys. The Monels are quite similar to the stainless steels in corrosion resistance, appearance, and properties. They are difficult to machine, but welding does not sensitize them to corrosion, as happens with the austenitic stainless steels. *Nichrome*® is a nickel-base alloy containing chromium or chromium and iron. It is known for high electrical resistance and oxidation resistance at red heat and is therefore used for electrical-resistant heater elements. *Inconel*® is a nickel-iron-chromium alloy known for high-temperature oxidation resistance. Nickel powder is mixed with carbide powders, then compressed and sintered to make carbide cutting tools as described in Chapter 10. Pure nickel is also used as a catalyst for chemical reactions. Nickel ingots cost about 16 times more than an equal mass of carbon steel, and electrolytic cathodes and alloys about 20 times more.

5.8 THE WHITE METALS

The *white metals* are low-melting alloys in which lead, tin, or antimony predominate. Zinc-, aluminum-, and magnesium-base castings are sometimes erroneously called white-metal castings. The three most important classes of white metals are: (1) fusible alloys, (2) type metals, and (3) bearing alloys. Of these, the most important, from an engineering standpoint, are the bearing alloys. These are known as *babbitts.*

True babbitts are tin-base alloys strengthened by the presence of antimony and copper, and occasionally containing lead. Lower-cost, lead-base babbitts are also used in place of the true babbitts. The microstructures of all these alloys are similar. They consist of hard, primary cube-shaped particles in a soft matrix. If primary particles separate by gravity segregation and sink to the bottom of the casting, a structure of inferior wear resistance results. Segregation is prevented by the use of copper that solidifies first as an interlocking network of spiney crystals. This produces a structure in which the hard, cube-shaped particles are uniformly distributed in the matrix. The babbitt-bearing alloys are quite readily cast in the shop.

5.9 REFRACTORY METALS

Tungsten has the highest melting temperature of the metals, (about 3,399° C [6,150° F]). It is extremely strong, approaching a tensile strength of 3.5 GPa (500,000 psi) in wire of 0.1 mm (.004 in.) diameter. Pure tungsten is used for the filament wires in incandescent light bulbs. Tungsten is used as an alloying addition to steel and in the form of tungsten carbide in carbide tools. It is also used as the electrode in inert-gas-shielded-arc welding.

Tantalum and *molybdenum* are also refractory metals, melting respectively at 2,996° C (5,425° F) and 2,621° C (4,750° F). Both are used in electronic tubes, as alloying elements in steel, and as carbides in cemented carbides. Tantalum is a *getter*, that is, it reacts with all but inert gases at temperatures above 316° C (600° F) in electronic tubes. It is also used for surgical implants. Molybdenum is useful as an electrical contact material and as filament supports in light bulbs.

Beryllium is an expensive light metal (specific gravity 1.82); it costs 1,200 times more than carbon steel. It has the highest specific strength of any metal and is used in applications where weight savings is important enough to justify its cost. A few critical rocket motor parts are made from beryllium. Beryllium foil is used for the windows in x-ray and counter tubes, since it has a very low absorption coefficient for short-wavelength radiant energy. It is used in atomic energy installations as a neutron reflector and moderator.

Germanium is a semiconductor and is used in the manufacture of rectifiers, transistors, and similar devices. Silicon, however, performs similar functions at a lower cost.

Zirconium is an efficient getter for electronic tubes and, because of its low neutron absorption coefficient and strength (comparable to low-carbon steel), it is used in atomic piles.

5.10 PRECIOUS METALS

The most important precious metals are platinum, gold, and silver. *Platinum* is essential to the chemical industry as a catalyst material and to a lesser extent as a highly corrosion-resistant metal from which reaction vessels can be made. *Gold* is used as a lining material for vessels and as a plated coating where corrosion resistance is of extreme importance. *Silver* is the best electrical conductor. It is also a basic material used in the photographic industry in the form of photosensitive silver salts. Like the other precious metals, it has good corrosion resistance.

5.11 QUESTIONS

1. What effect does alloying have on the following properties of pure metals? (a) modulus of elasticity; (b) thermal conductivity; (c) electrical conductivity; (d) specific gravity; (e) high-temperature resistance; (f) yield strength; (g) fatigue strength.
2. List the outstanding properties of aluminum and its typical applications.
3. What are some of the factors that might be considered in deciding whether a retractable landing gear member for a huge transport plane should be made of an aluminum alloy, H-11 steel, or a maraging steel?
4. Compare the fabricating properties of aluminum, titanium, copper, magnesium, and zinc alloys.
5. What is the function of aluminum used as an alloying element in titanium-base alloys?
6. State some typical applications of unalloyed titanium and titanium alloys. What other alloy do titanium alloys most resemble?
7. Compare the mechanical properties of aluminum-, titanium-, copper-, and magnesium-base alloys.
8. Which is preferred for die-cast automobile hardware: aluminum-, magnesium-, or zinc-base alloys? Why? If the cost of these and copper-base alloy ingots were all the same per unit volume, which would be preferred for making automobile hardware die castings? For aircraft die castings?
9. List an outstanding property and use for each of the following: tantalum, beryllium, molybdenum, zirconium, and germanium.
10. What property is possessed by all the precious metals?

5.12 REFERENCES

Aluminum Standards and Data. 1982. Washington, D.C.: The Aluminum Association.

Betterridge, W. 1984. *Nickel and Its Alloys.* New York: John Wiley.

Donachie, Jr., M., ed. 1982. *Titanium and Titanium Alloys: Source Book.* Metals Park, OH: American Society for Metals.

"Material Selector." *Materials Engineering.* Cleveland, OH: Penton/IPC.

Metals Handbook, 9th Ed. 1990. Vol. 2. *Properties and Selection: Nonferrous Alloys and Pure Metals.* Metals Park, OH: American Society for Metals.

Society of Manufacturing Engineers (SME). 1997. *Cutting Tool Materials* video. *Fundamental Manufacturing Processes* series. Dearborn, MI: SME.

Tool and Manufacturing Engineers Handbook, 4th Ed. Vol. 3. *Materials, Finishing and Coating.* 1985. Dearborn, MI: Society of Manufacturing Engineers (SME).

West, E. G. 1982. *Copper and Its Alloys.* New York: John Wiley.

Enhancing Material Properties

6

Metallographic photomicrography, 1880
—A. Martens, Koniglich Technische, Versuchsanstalt

6.1 HEAT TREATMENT PRINCIPLES

The heat treatment process changes the strength, hardness, ductility, and other properties of metals. It is effective only with certain alloys because it depends upon one element being soluble in another in the solid state in different amounts under different circumstances. The basic metallurgy of heat treatment was presented in Chapter 3.

Hardening (or *strengthening*) is done by heating an alloy to a high enough temperature, depending on the material, and cooling it rapidly. A solid solution of the alloying elements is formed at the high temperature. This becomes supersaturated upon cooling, and desired hardness is obtained by controlling the decomposition of the constituents. Under proper conditions, the solute is dispersed in fine particles in the crystal lattice and serves to block dislocation movements when stress is applied. The added resistance to stress makes the metal act stronger and harder.

The amount of hardening that takes place in an alloy depends upon the size, shape, and distribution of the particles and the amount of coherence between the particles and the matrix. The size, shape, and distribution of the particles result largely from the dispersion of the solute in the material, which depends in turn upon the temperature and time of heating. There is an optimum particle size in each case that gives the best results. Large particles are imposing obstacles but are far apart, and dislocations pass between them. Particles too small do not greatly hinder the movements of dislocations. The degree of coherence between the particles and the primary phase has the most influence on hardening. Hardening is enhanced if the boundary between the particles and matrix is coherent. This is called *coherency hardening* and is effective because each particle distorts the space lattice of the phase around itself and thereby extends its influence in blocking the movement of dislocations. Zones of influence may overlap. *Aggregate hardening* occurs with particles that have incoherent boundaries with the primary phase. Examples are plates (such as in pearlite in steel) or globular particles. Dislocations are impeded only partially by such particles and pass readily between them.

Alloys hardened by heat treatment can be divided into two major classes. The first is *nonallotropic* and is hardened by age or precipitation treatment. The second is *allotropic* (which means the crystal structure is different at higher than at lower temperatures) and can be hardened by suppressing the decomposition of the structure on cooling. The first class contains mostly nonferrous and only a few ferrous alloys;

the second class consists of steels and irons. Some metals, such as certain stainless steels, are heat treated by a combination of the two methods.

The reverse of hardening, which is softening of metals, is done by heating alone or by heating and slow cooling. The effects are to gather together and coarsen the dispersed particles, control the grain size, and improve ductility and impact resistance. Common processes of this sort are annealing, normalizing, and tempering. Other purposes of heat treating are to relieve stresses, modify electrical and magnetic properties, increase heat and corrosion resistance, and change the chemical composition of metals (as by carburizing steel).

The processes of hardening and softening metals will be explained in detail in this chapter, including the techniques and equipment for the processes.

6.2 HEAT TREATMENT OF NONALLOTROPIC ALLOYS

An alloy that does not change in lattice structure when heated can be hardened under certain conditions. If the alloy is more soluble in the primary phase at higher rather than at lower temperatures, it can be hardened. Such an alloy is raised to a temperature that causes the greatest amount of solution without melting any phases or causing excessive grain growth. The alloy is then quenched fast enough to prevent immediate separation of the solute. Thus, it retains the solid solution even though solubility is much less at the lower temperatures. Next, particles of solute are precipitated in the alloy in what is generally called *precipitation hardening*. In some cases this occurs at room temperature over a period of time and is called *natural aging* or *age hardening*. Other alloys must be heated to bring about or hasten precipitation, in what is called *artificial aging*.

Aluminum-copper alloys are common precipitation-hardened alloys. Aluminum and copper combine chemically to form copper aluminide, $CuAl_2$, containing about 54% copper by weight. Copper is soluble in aluminum in the α phase shown in Figure 6-1(A); some consider the copper aluminide to be in solution. The proportion increases from almost nothing at low temperatures to a maximum of about 5.5% copper at 548° C (1,018° F). At lower temperatures some portion is precipitated out as $CuAl_2$.

Containing about 4.5% copper, 2024 aluminum is an important commercial alloy. As a first step in hardening, it is given a solution treatment by being heated to 491–499° C (915–930° F). At this range of temperature, most of the copper is in solution in the α phase. For economy, the material is held at solution temperature no longer than necessary to achieve homogeneity. A typical time is 1 hr. It is important that the temperature be kept below that of a eutectic melting temperature (502° C [935° F] in this case) that occurs for an alloy of 33% copper in aluminum. If that temperature is reached, the material is permanently damaged by melting in the grain boundaries.

Solution-treated aluminum is quenched in clean cold water immediately after soaking. The more rapid the quenching rate, the stronger the product and the more it resists corrosion after precipitation. After quenching, 2024 aluminum ages to practically maximum strength in 24 hr at 20° C (68° F). It can be aged to full strength in 2 hr at 175° C (≈350° F). Another aluminum alloy (6061) takes 30 days at room temperature to become fully aged. Most aluminum alloys are nearly as formable right after quenching as in the annealed state; some more so. Thus, they are often bent or straightened at that time. Aging may be retarded by refrigeration to keep material in a workable condition longer, as in the case of rivets.

An advantage of precipitation hardening is that its costs are relatively low. Most alloying elements are cheap, temperatures are not high, and procedures are simple. Because temperatures are not high, distortion, scaling, and cracking are minimized. Up to precipitation temperature, most hardened alloys retain strength better than alloys not treated.

Precipitation-hardened alloys may be reheated and cooled slowly to soften. Aluminum alloys may be given a *stress relief anneal* or a *full anneal*. In the first instance, the effects of strain hardening may be removed in most cases by heating to about 343° C (650° F). For a full anneal, a coarse and widely spaced precipitate results from soaking for 2 hr at 416–441° C (780–825° F) followed by slow cooling.

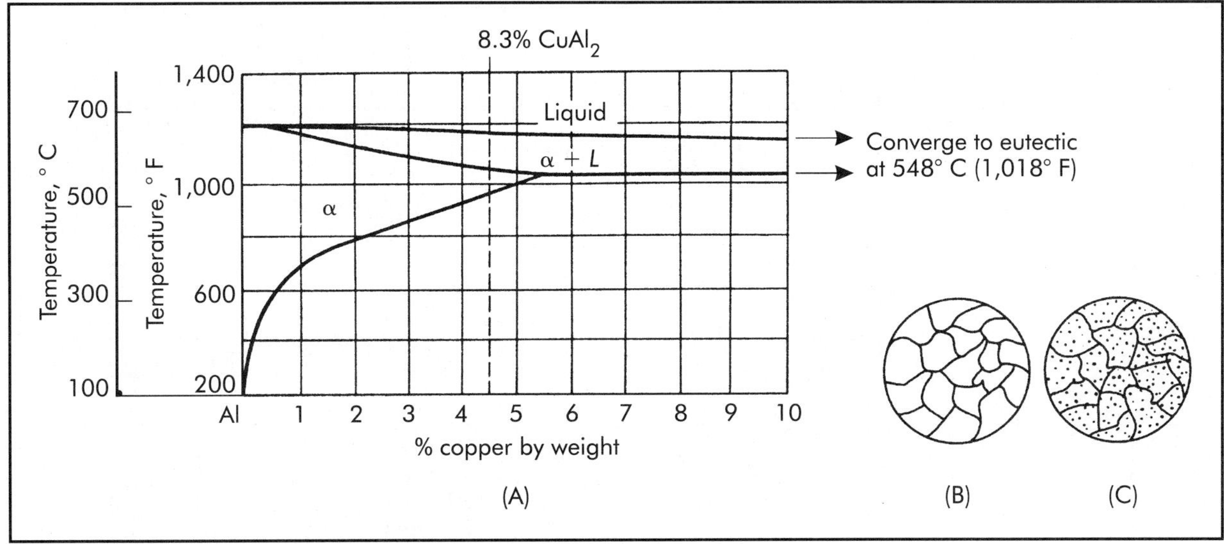

Figure 6-1. (A) Aluminum-rich end of the aluminum-copper equilibrium diagram; (B) sketch of the magnified structure of a 4.5% Cu-Al alloy heated to 491° C (915° F) for solution treatment; (C) the same alloy after quenching and precipitation of particles to induce hardening.

6.3 HEAT TREATMENT PROCESSES FOR STEELS

Major properties of steel, such as strength, hardness, durability, and toughness, influence the metal to withstand scratching and resist wear. These properties largely depend on the extent of heat treatment. Heat treatment modifies and improves the microstructures, and thus produces several mechanical properties that are important for different applications. These improved properties add to the value of the metal by improving its performance when used in the production of different components, such as gears, cams, tools, and dies.

This section presents some of the common heat treatment processes for steels, along with the different methods and techniques of heating and altering surface chemistry.

6.3.1 Steel Hardening

Steel is hardenable because carbon is more soluble in the face-centered-cubic structure at high temperatures (austenite) than in the body-centered structure (ferrite) at low temperatures. These regions are shown in Figure 6-2, which is the steel section of the iron-carbon diagram in Figure 3-13. If steel is heated to the austenite region and held there until its carbon is dissolved, and is then rapidly cooled by quenching, the carbon is not given a chance to escape and is trapped as dispersed atoms or fine particles in a strained low-temperature lattice. This sets up a distorted structure (martensite) that is quite hard and strong, but brittle.

The changes that occur when steel is cooled from the austenitic range may be depicted by *transformation diagrams* or *S-curves* like those shown in Figure 6-3. These are schematics for one type of steel; each analysis of steel has its own S-curve, and many have been published. These show what takes place in nonequilibrium cooling in contrast to the iron-iron carbide diagram for equilibrium conditions. Any cooling rate can be designated by a line like *AB* or *AD* (see Figure 6-3) on a diagram. Figure 6-3(A) shows an *isothermal transformation curve* for steel cooled to below the critical temperature and held there for a period of time while transformation takes place. No change occurs in the area to the left of the S-curve. For example, if the steel is cooled at a rate denoted by *AB* and then held at constant temperature until time *C*, it is transformed from an austenitic to a coarse pearlitic structure. If another sample is cooled rapidly from *A* to *D*, to the left of the

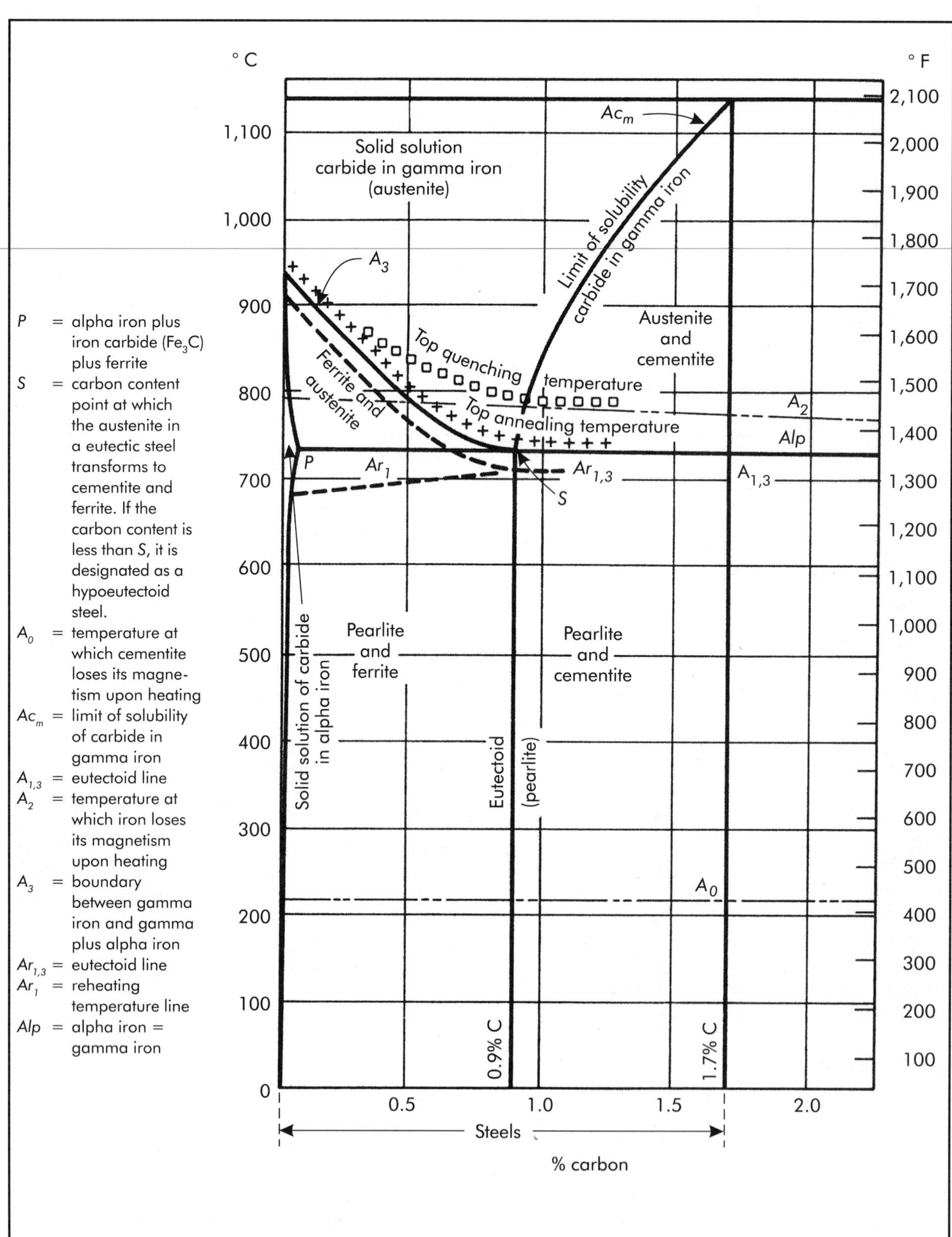

Figure 6-2. Iron-iron carbide diagram for steel.

nose of the S-curve, no transformation occurs. Then if the sample is held at temperature D until time E, the structure is transformed to bainite.

A *continuous cooling transformation curve* like Figure 6-3(B) is a modified S-curve. It shows the changes that occur when austenite is transformed over a range of temperatures rather than at one temperature. The isothermal diagram is drawn for comparison. The continuous cooling curve is represented by the boundaries of the crosshatched areas. It is seen that the transformation begins later and at lower temperatures when cooling is continuous. The formation of bainite may be disregarded for continuous cooling of carbon steels and some alloy steels. Thus the bainite region is omitted from Figure 6-3(B). This is not the case for many alloy steels.

Examples of the changes that occur when steel is cooled at various rates are given by the lines in Figure 6-3(B). Steel cooled at the rate depicted by AF is transformed to medium-coarse pearlite. Any steel cooled rapidly along a line to the left of the nose of the curve, such as line AG, is kept austenitic until it reaches the M_s temperature. There it starts to transform to martensite, and transformation is complete when the M_f temperature is reached. The line that just passes the nose of the curve represents the *critical cooling rate*. If cooling takes place along line AH in Figure 6-3(B), the transformation to fine pearlite may be only partially complete by the time the boundary designated (3) is reached. That is where transformation stops. The remaining austenite is then changed to martensite below the M_s temperature. The

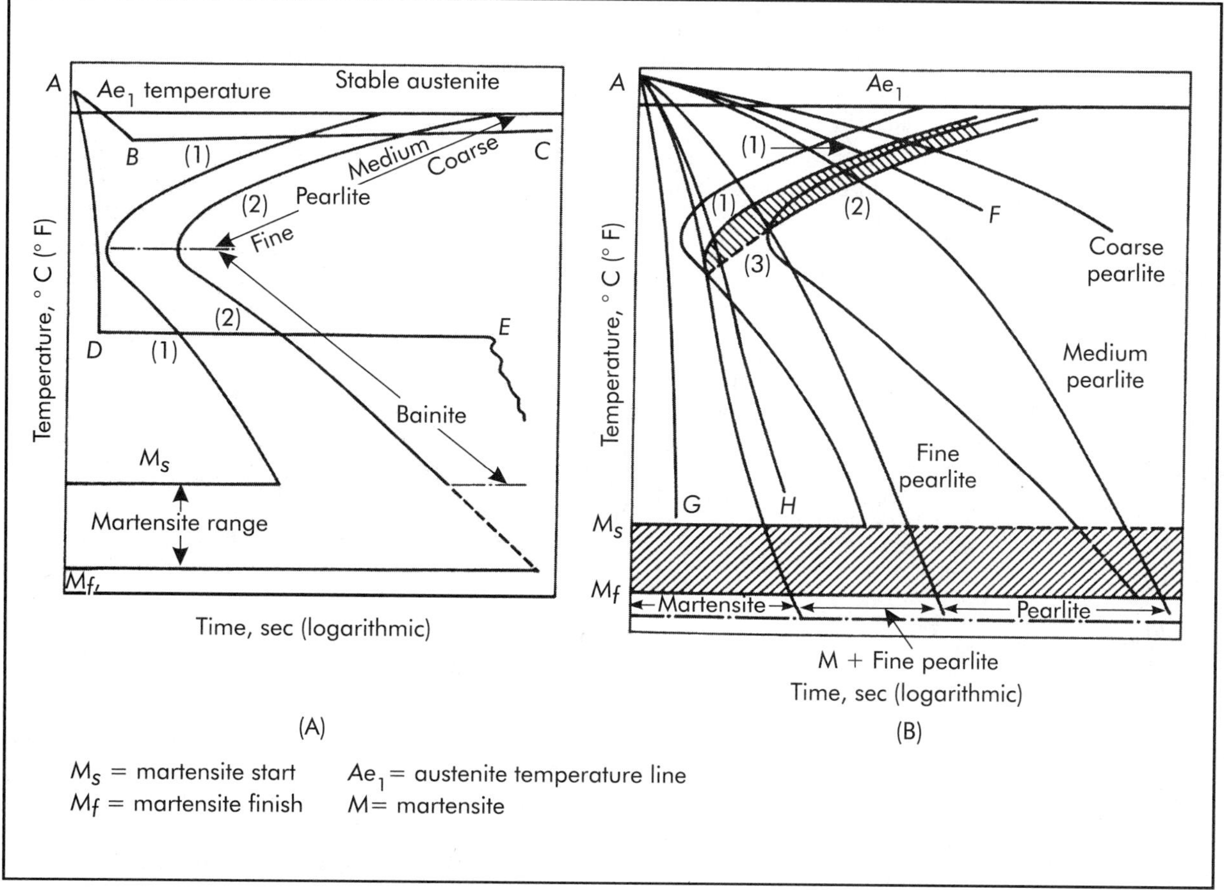

Figure 6-3. (A) Schematic transformation diagram or S-curve for a eutectoid carbon steel; (B) schematic CC curve, which is an S-curve for a eutectoid carbon steel modified for continuous cooling. Symbols: (1) austenite to pearlite transformation begins; (2) transformation complete; (3) decomposition of austenite stops.

result is a mixed martensitic and fine pearlitic structure.

The lower the temperature of transformation for a given steel, the harder and stronger the product. Thus medium pearlite is harder than coarse pearlite, and fine pearlite is harder than medium pearlite. The finer pearlite spacing offers more resistance to the flow of dislocations. For the same reason, bainite is harder than pearlite, and martensite is the hardest. Bainite and martensite contain minute and widely dispersed particles that present even more resistance.

The maximum hardness attainable in quenched steel depends on the amount of carbon it contains. As indicated by Figure 6-4, appreciable hardening does not occur with less than 0.30% carbon, and there is almost no increase for more than 0.60% carbon.

Steel must be heated above the A_3 line of Figure 6-2 and held there to dissolve the desired amount of carbon for hardening. Usually, no more than 28–56° C (50–100° F) into the austenite region is enough. Temperatures should not be higher nor soaking time longer than necessary to avoid excessive grain coarsening and burning of the steel. A rough rule is to allow 1 hr of heat time at final temperature per 25.4 mm (1 in.) of thickness in the heaviest section of a steel workpiece.

Quenching

Heat may be removed from hot metal by immersion in brine, water, oil, molten salts, or lead, or by exposure to air or gases, or by contact with solid metallic masses. Water and oil are the most common media for full quenching. Relative quenching rates of the common medias are indicated in Figure 6-5. Larger pieces are cooled more slowly with more differences between inside and outside cooling rates than small pieces. The severity of water quenching cracks some parts; oil quenching is less harmful, and air quenching is even better. Oil and air quenching require alloys of higher hardenability to make steel as hard as is possible by water quenching. Synthetic, oil-free, water-base fluids have been developed in recent years that yield quenching results that fall between that found with water and oil.

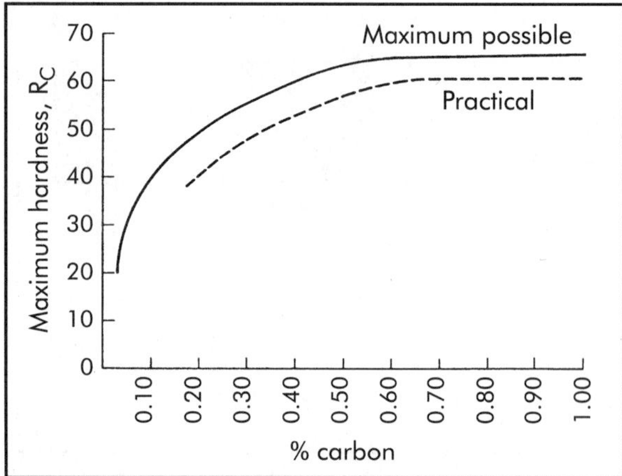

Figure 6-4. Relationship between the hardness and carbon content of quenched carbon steel.

Quenching sets up stresses that warp workpieces, and precautions are necessary to avoid distortion. Slender shafts, thin walls, and thin and thick adjacent sections are particularly vulnerable. A long slender shaft is ordinarily suspended from one end when plunged into a quenching tank. Production parts, such as gears with thin webs, may be *die-quenched*. This means that the piece is clamped firmly in a die in a press while lowered into the quenching medium. The die is made to contact and thus chill selected areas of the part. Coolant is admitted at different rates to various sections to regulate cooling and thus the warpage throughout the part.

Hot workpieces must be moved quickly and safely from the heating device to the quenching medium. This may be done by hand tongs, one piece at a time, for job work. For repetitive production, work may be transported by conveyor, and large pieces by cranes or cars. Small pieces are normally handled in wire baskets or on racks. The coolant is ordinarily agitated or swirled vigorously to achieve uniform cooling and may be circulated through cooling coils.

Tempering

When steels are hardened by heat treatment, brittleness occurs and an undesirable increase in residual stresses may result. Annealing is a heat treatment process that is used to reduce brittleness and residual stresses, while improving ductility and toughness. The process of annealing is performed by heating the steel to some temperature, which varies depending on the composition, and then cooling at a specified rate.

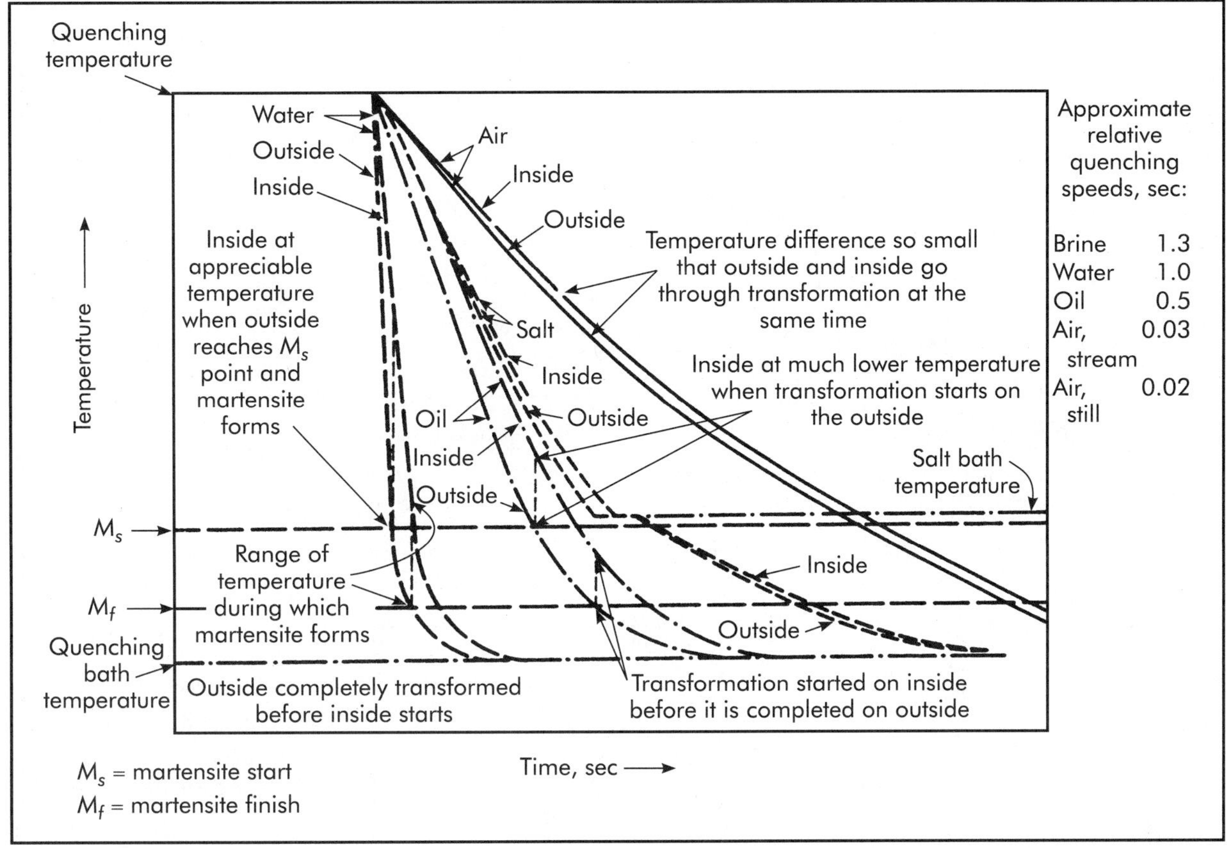

Figure 6-5. Time-temperature cooling curves for several quenching media.

Steel Hardening Methods

Direct and full quenching are the oldest methods of hardening steel and are still common practice. Quenching is economical and yields the highest immediate hardness. The essential structures produced are martensite and retained austenite. Their proportions depend upon carbon and alloy content, austenitizing temperature, quenching medium, and part geometry. One practice is to freeze the steel to −71° C (−95° F) or lower, to transform the retained austenite. Few parts are left in the as-quenched state because the *fresh martensite* is quite brittle. When the fresh martensite is heated to below the critical temperature, it becomes softer and more ductile, and internal stresses are relieved. Little benefit is obtained at temperatures below 149° C (300° F). The initial change is to *tempered martensite*, in the range of 149–177° C (300–350° F) with only slight changes in properties. The second stage, at about 177–371° C (350–700° F) depending upon the steel, is characterized by transformation of the retained austenite to bainite. Carbon from the martensite appears to become combined into finely dispersed particles of cementite. In the third stage, from about 288–704° C (550–1,300° F), the cementite agglomerates and coalesces. The structure becomes an aggregate of ferrite with cementite in quite fine spheres, referred to as *tempered martensite* and *tempered bainite*. The structures may become more or less uniformly spheroidized from prolonged heating at the upper end of the range. Reheating after quenching is called *tempering*, but some give the name of *drawing* to reheating below 316° C (600° F). A typical direct hardening and tempering cycle is depicted on the schematic T-T-T diagram in Figure 6-6(A).

The best combination of strength, hardness, ductility, and toughness for most applications may be obtained by quenching steel to martensite and then tempering as desired. This pro-

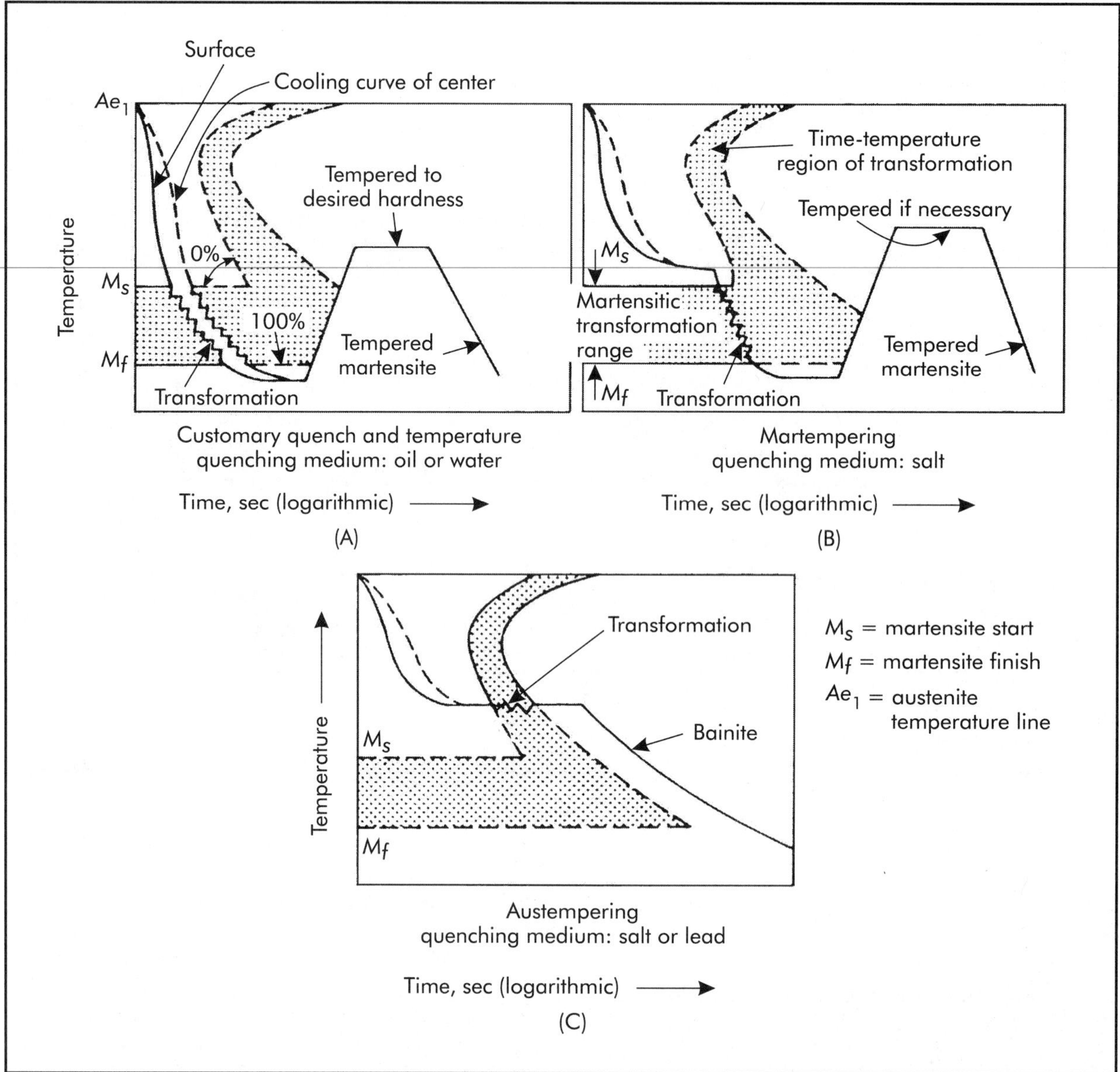

Figure 6-6. Heat-treating practices.

cessing should be done without delay because martensite can crack, even overnight. Tempering softens, but both time and temperature determine the hardness obtained when steel is tempered, as shown in Figure 6-7. Each analysis of steel has its own set of curves. In this case, the same hardness results from heating for 5 hr at 204° C (400° F) or 8 sec at 371° C (700° F). Recommended practice is to use the lowest possible temperature that gives the required results in a reasonable length of time. At higher temperatures, strength usually decreases but may increase at low tempering temperatures as residual stresses are relieved. The higher the tempering temperature, the more the quenching stresses are relieved, but the most benefit is obtained at 260–316° C (500–600° F). As hardness decreases, ductility increases, but toughness does not improve uniformly. After an initial increase as the temperature is raised, impact toughness drops off for most steels before it begins to rise again. This low toughness commonly obtained from tempering at 204–371° C (400–700° F) is called *blue brittleness* or *blue heat phenomenon* because it occurs at temperatures that leave a blue oxide film on the steel. It is ascribed to the precipitation of oxides and nitrides. If impact toughness is desired, tempering must be done at a higher temperature, and a lower hardness must be accepted. The impact toughness of some alloy steels is impaired if they are cooled slowly after tempering in the range of 450–600° C (≈850–1,100° F). This is called *temper brittleness*. It may be avoided by quenching from tempering temperature. *Toughening* is the name given to tempering at 538–704° C (1,000–1,300° F) when high hardness is not needed. Usually, this gives maximum impact toughness.

Steel is heat treated in other ways besides full quenching and tempering. Much modern practice utilizes interrupted quenching methods that hold the steel for a time at relatively high temperatures to equalize temperatures inside and out. This helps to harden the work uniformly throughout and to eliminate cracking and warpage. The leading methods, martempering and austempering, are described in the following paragraphs.

Martempering or *marquenching* starts with quenching austenitized steel in a molten salt bath as indicated in Figure 6-6(B). A piece is held in the bath just above the M_s temperature (where martensite starts to form) until its temperature is uniform throughout. Then the piece is cooled in air through the zone of martensite formation. This is generally followed by an ordinary tempering treatment as desired. Martempering is limited to carbon steel sections less than about 6.4 mm (.25 in.) thick and is better adapted to alloy steel. That is because the quenching rate of the salt bath is relatively slow, as shown in Figure 6-5, and its results do not carry carbon steel readily past the nose of the S-curve.

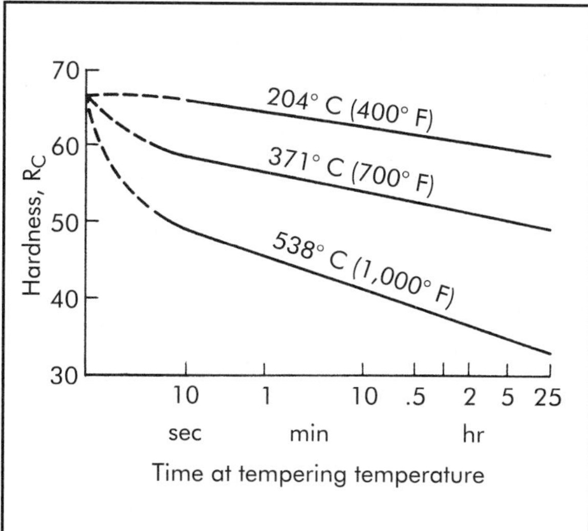

Figure 6-7. How tempering time and temperature affect hardness of 0.80% carbon steel.

Steel is quenched in a heated bath at 149–427° C (300–800° F) for *austempering* as depicted in Figure 6-6(C). The work is then held in the bath a sufficient time for the austenite to transform to bainite isothermally. Then the piece may be cooled at any rate; no subsequent tempering is needed. Hardness of 45–60 R_C may be obtained depending upon carbon content and transformation temperature. A practical range is 50–55 R_C, with the advantage of more toughness than is generally obtained by other methods. Application to carbon steel is limited because of the slow cooling rate of the high-temperature bath.

Hardenability of Steel

Several steels of different compositions may be hardened by quenching in exactly the same way. However, they will be found to differ in both intensity and depth of hardness. *Hardenability* refers to the degree and depth of hardness obtained in a heat treatment. Any austenite that is transformed to pearlite is lost to the formation of martensite, and hardening is decreased. Greater hardenability means that more austenite is transformed to martensite and its derivatives. The factors related to the suppression of pearlite and thus to hardenability are:

1. All alloying elements that dissolve in austenite (including carbon to 0.9% but not cobalt) push the nose of the S-curve to the right (Figure 6-3) and make it easier to quench the insides as well as the outsides of parts past the pearlite zone. A comparison of the hardenability of an unalloyed and an alloyed steel is shown in Figure 6-8.
2. A homogeneous austenite structure increases hardenability by holding the S-curve uniform.

3. Coarse austenitic grains push the S-curve to the right and increase hardenability. They may be caused by heating the steel too high or too long before quenching. Coarse grains are not desirable because they reduce the toughness of the hardened steel.
4. Undissolved carbides and nonmetallic inclusions in the austenite decrease hardenability because they provide nuclei for fine pearlite formation.

The *Jominy end-quench test* holds all factors constant except composition to measure the hardenability of steel. A bar 25.4 mm (1 in.) in diameter × 76.2 or 101.6 mm (3 or 4 in.) long is properly austenitized and quenched on the end in a standardized way as illustrated in Figure 6-9. Heat is substantially removed from the quenched end surface and is withdrawn at different rates along any one bar, but in the same way along any bars of steel tested. The result is a gradient of hardness along the bar that depends only on the composition of the material. After the piece has cooled to room temperature, two flats are ground lengthwise on diametrically opposite sides. Rockwell hardness readings are taken along the bar at 1.588-mm (.0625-in.) intervals to 25.4 mm (1 in.) and from there at 6.4 mm (.25 in.) intervals to 51 mm (2 in.). They are then plotted in a manner like that of Figure 6-8. The hardness illustrated for a plain carbon steel (1040) drops off rapidly a short distance from the end. An alloy steel shows a much smaller rate of decline or none, as the curve for 4340 steel illustrates.

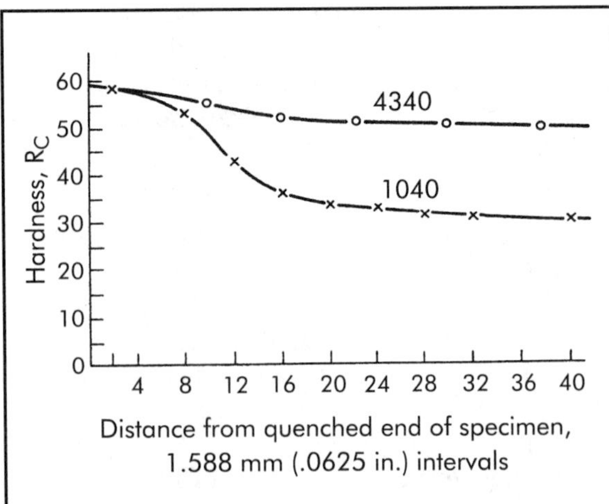

Figure 6-8. Typical Jominy curves for a plain carbon (1040) steel and an alloyed (4340) steel.

Empirical equations have been developed for calculating the Jominy curves of steels from their carbon and alloy contents. Further calculations, derived from the steel composition and method of heat treatment, can give the hardness of points within round bars and other sections.

6.3.2 Annealing of Steel

Annealing in its broadest sense means heating a metal to where a change occurs and then cooling it slowly. A main reason for annealing is to soften steel, but it also serves to relieve stresses, drive off gases, and alter ductility, toughness, electrical or magnetic properties, or to refine the grains. The end purpose may be to prepare the steel for further heat treatment or mechanical working or to meet the specifications for the finished product. The role of annealing in heat treatment is illustrated by slow cooling through the upper regions of the S-curve, such as along the path *AF* of Figure 6-3(B). There are several ways that annealing is

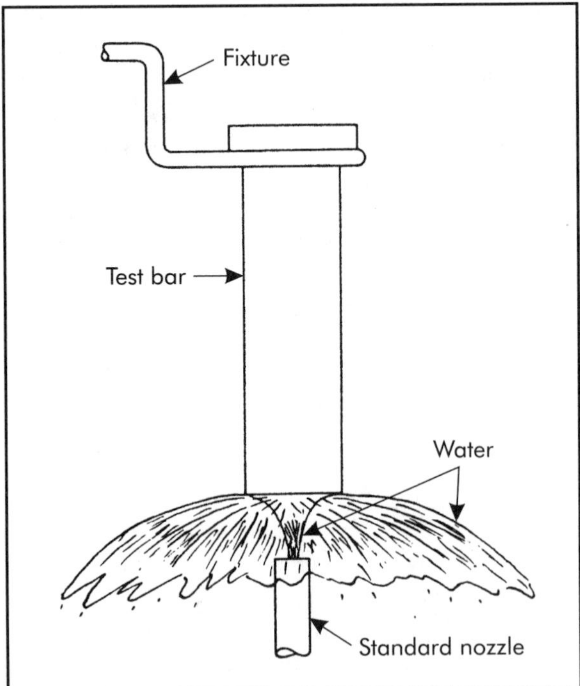

Figure 6-9. Jominy end-quench.

done. Many paths are possible and each gives somewhat different results.

Full annealing consists of heating an iron-base alloy from 28–56° C (50–100° F) above the critical temperature, holding it there for uniform heating, and cooling it at a controlled slow rate to room temperature. The work may be held in a heavily insulated furnace with the heat cut off or it may be buried in an insulating material such as ash or asbestos.

Normalizing consists of heating about 56° C (100° F) above the transformation temperature and cooling in still air. The purpose of normalizing a steel is to obtain a homogeneous structure. It usually also imparts moderate hardness and strength. Normalizing is commonly done to restore wrought steel to its near-equilibrium state after cold or hot working or overheating. Steel castings are normalized to modify grain structure and relieve stresses. Thin sections can be cooled rapidly and appreciably hardened.

Process or *commercial annealing* consists of holding iron-base alloys at a temperature a little below the critical temperature for 2–4 hr and cooling for desired results. Some softening occurs, but the main benefit is stress relief. The operation is sometimes called *stress relieving*. The advantage of the process is that warpage of thin sections and surface corrosion and scaling are slight because temperatures can be kept low. The process has wide application in preparing steel sheets and wire for drawing or redrawing, for stress relief of weldments and castings, and to remove the embrittling effects of heavy machining and flame cutting (Chapter 13). Large work may be heated locally by a torch; smaller pieces in a furnace.

Cycle annealing is done by cooling austenitized steel at a rate to reach a desired zone on the S-curve. The metal is held at the chosen temperature until transformation is complete. Then the work may be cooled in any way practicable, by quenching or in air, because no more transformation occurs. The main advantage is a short cycle time, 4–8 hours as compared to 5–30 hours for conventional annealing. The end structure may be pearlite or spheroidite (or a mixture of both) depending upon selection of temperature and time. *Spheroidizing* is the name given to the process when the carbon is collected into coarse round carbide particles, especially in high-carbon steels. This is a desirable structure to machine because the hard particles in a soft ferrite matrix are readily pushed aside by a cutting tool.

Both wrought and cast steels are annealed in essentially the same ways and for the same reasons. There is one important difference, however. A steel casting solidifies with a coarse dendritic structure and considerable segregation. This structure is not broken up and homogenized as is done by working a wrought steel. The cast structure must be refined by heat treatment. Both full annealing and normalizing are applicable. The first gives maximum softness and ductility; the second provides finer subdivision of the dendritic grains, more homogeneity, and higher strength.

Some work must be annealed two or more times to correct faults (such as gross lack of uniformity) or get desired results. A typical *double anneal* consists of heating 100–150° C (≈200–300° F) above the A_3 line for thorough diffusion, then air cooling below the critical temperature to inhibit ferrite separation, followed by regular annealing at 28–56° C (50–100° F) above the line to refine the grains and finally slow cooling.

6.4 SURFACE-HARDENING OF STEEL

A principal reason for hardening steel is to retard wear on bearing and rubbing surfaces, but hard steel is brittle and not fatigue- and shock-resistant. Therefore, for high strength along with durability, it is desirable to harden selected outer surfaces of many machine parts against wear and leave their cores soft and ductile for shock resistance. Heat treatment can be done at lowest cost when applied only to surfaces where needed. Medium- and high-carbon steels can be surface-hardened by induction- and flame-hardening. Electron beam and laser techniques, as well as electric arc methods, are modified for surface-hardening. They are primarily welding processes and are described in Chapter 13. All these processes heat only selected surfaces to austenitizing temperatures. The steel is then quenched to harden the surfaces. Surfaces of low-carbon steels may be enriched with carbon by carburizing or case-hardening and then can be hardened.

6.4.1 Induction-hardening

Induction heating is done by passing a high-frequency alternating current through a water-cooled coil or inductor around the workpiece or over a surface. The cyclic magnetic field that is generated induces alternating currents that heat the workpiece, as indicated in Figure 6-10. The depth of current penetration is

$$\delta = 1.98\sqrt{\rho \div \mu f} \qquad (6\text{-}1)$$

where

δ = depth of current penetration
ρ = resistivity
μ = magnetic permeability (μ = 1 for nonmagnetic materials)
f = frequency in hertz

The lower the frequency, the deeper the penetration, and vice versa. Actually, the current is not uniform for its full depth but drops off exponentially from the surface and heats the metal accordingly. Magnetic hysteresis adds to the effect for magnetic materials. Steel is less magnetic at higher temperatures and escapes overheating. Induction heating is done to melt metals (Chapter 8), for brazing and soldering (Chapter 13), and to heat stock for forging (Chapter 11), as well as for hardening steel and iron.

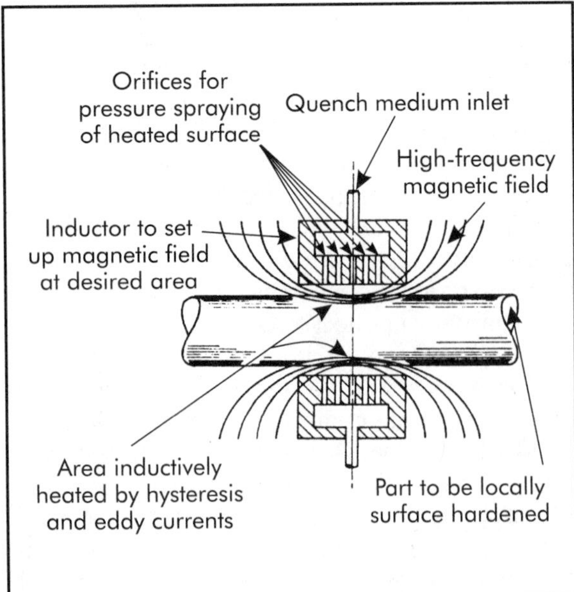

Figure 6-10. Scheme of induction-hardening.

A piece may be hardened after induction heating by dropping it into a quenching medium or flooding it with a stream of coolant. A small area on a large piece may be heated quickly and then effectively quenched by the mass of the piece drawing off the heat. The depth of hardening depends upon how deeply the steel is austenitized. That depends upon the power input, time, heat lost, and the frequency of induction-hardening. Since steel is nonmagnetic at austenitic temperatures, it is sometimes economical to heat with low frequencies at low temperatures followed by higher frequencies at higher temperatures. Motor-generator sets are the usual sources for frequencies up to 10 kHz and power to 9 J (2.5 kWh). An installation that includes a 4 J (1.1-kWh), 10-kHz motor-generator unit, heat station with inductor, and quenching fluid system costs more than $100,000. Spark gap circuits are used for intermediate frequencies, and electronic (vacuum tube) generators for high frequencies (into the MHz range).

6.4.2 Flame-hardening

The surface of a workpiece may be locally or progressively heated by a gas flame as depicted in Figure 6-11. An oxyacetylene or oxy-MAPP flame, as described for welding in Chapter 13, is preferred for flame-hardening because heating is most rapid. The torch head is commonly given a shape to match the contour of the work surface. Hardening results when the austenitized surface is quenched by the spray (usually water) that follows the flame. Equipment to harden small pieces (up to 152 mm [6 in.] in diameter) costs only a few thousand dollars.

6.4.3 Comparison of Methods

In general, any plain-carbon or alloy steel having 0.40% carbon or more may be induction- or flame-hardened, but the high hardenability of some alloys adds problems. The most usual range is from 0.40–0.60% carbon, and such steels are hardened to 40–63 R_C. Examples of work hardened by these methods are gears, tool drivers, wrist pins, crankshaft bearing journals, cylinder liners, rail ends, machine tool ways, and pump shafts. Cast iron and steel are hardened by both methods, and nodular iron and pearlitic malleable iron (described in Chapter 8) are induction-hard-

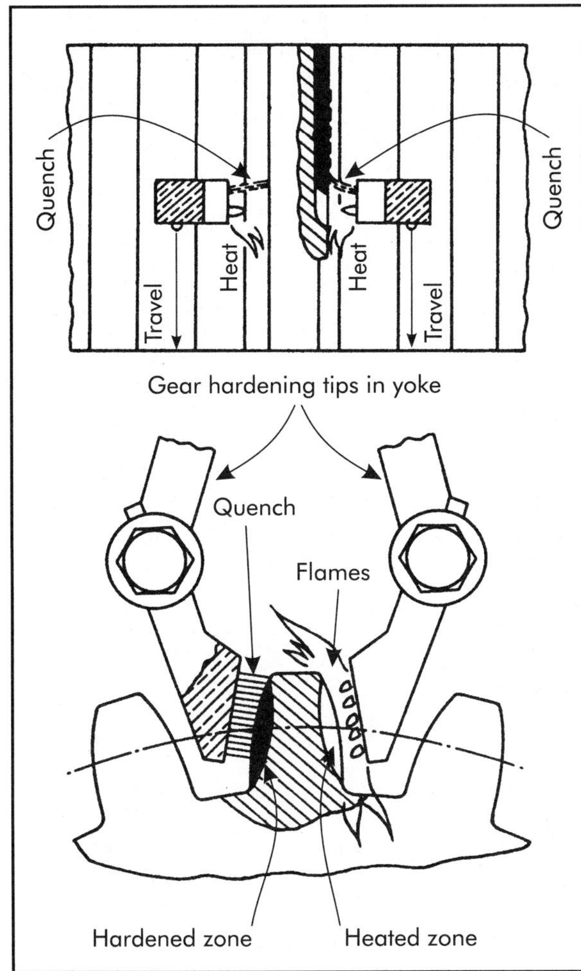

Figure 6-11. Flame-hardening of gear teeth.

ened. Case depths from less than 0.5 to over 6.4 mm (.02 to over .25 in.) are common.

Each method has certain advantages. Flame-hardening is sometimes more adaptable to surface configuration. For instance, flame heads can be adjusted to heat corners uniformly with adjacent areas. Induction-hardening is better for a shallow case with a narrow transition band, and flame-hardening for a deep case. Induction-hardening is fast and readily automated and therefore advantageous for large quantity production. Flame-hardening is more suitable for small quantities because it is versatile and equipment cost is less.

The selection of method for surface-hardening depends upon cost in most cases. Electron beam and laser hardening methods are costly and are not used extensively. Flame-hardening may be chosen for parts too large for practical furnace heating and immersion quenching. Flame- or induction-hardening may be economical where only some of a part needs to be heated and hardened. On the other hand, a complex part that must be heat treated throughout and can be handled in batches is probably best heat treated in a furnace.

6.4.4 Carburizing to Case-harden

For a ductile and shock-resistant core and a hard surface to resist wear and abrasion, a part may be machined a little oversize from low-carbon plain or alloy steel. It is then carburized to increase the carbon content of the skin and heat treated to bring out the best properties of case and core. The piece is finally ground to size. A cross section of a carburized and hardened surface is shown in Figure 6-12. Typical parts that are case-hardened are gears, splines, wrist pins, bearing balls, universal joint spiders, and valve tappets.

Steels for carburizing should have from 0.10–0.20% carbon. When a steel is heated to its austenitizing temperature, it absorbs carbon in the presence of CO gas. Theoretically, as much as 1.7% carbon may dissolve in gamma iron, but in usual carburizing practice, the carbon content seldom exceeds 1.2%. The carbon content decreases from the surface through the case to the core. The *effective case depth* is defined in

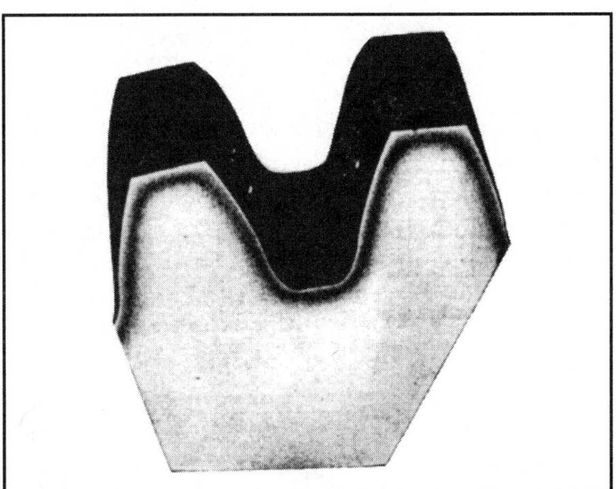

Figure 6-12. Carburized and hardened case of gear teeth.

several ways: as the depth to the point of 0.4% carbon, where the structure is 50% martensite, or the hardness is 50 R_C in the hardened part.

The amount of carbon absorbed and the depth of the case depend upon the temperature, time of exposure, carbon potential of the carburizing medium, and the composition of the original steel. Carburizing temperatures may range from 788–1,093° C (1,450–2,000° F), but usually lie between 899–949° C (1,650–1,740° F). The higher the temperature, the more rapid the carbon absorption, but also the more furnace deterioration and undesirable grain growth. Case depth increases with the square root of time at any given temperature. Typical conventional practice calls for 12 hr for soaking and carburizing to produce a case 1.5 mm (.06 in.) deep at 899° C (1,650° F). A case may be made as thick as desired, depending on length of exposure, but seldom is over 3.18 mm (.125 in.). Temperature and time in the austenitic region promote grain growth. Steels with less tendency to coarsen are preferred for carburizing. Where grain growth is excessive, subsequent heat treatment, such as double annealing, may be performed to break down grain size.

There are several methods of carburizing steel. *Gas carburizing* is favored for efficient production and consists of heating the work in a furnace with a highly carburizing atmosphere containing largely hydrocarbon gases or vapors. *Pack carburizing* is done by completely surrounding the workpiece with a carbonaceous material in a closed container. The CO gas for carburizing is obtained by heating the packing material. Much mass besides the work must be heated, and that makes the job slower, harder to control, and heat inefficient. On the other hand, pack carburizing has advantages for small lots and large pieces, particularly because it can be done in almost any furnace. The packing may also serve as insulation to cool the work slowly after carburizing. Carbonaceous material is heated in the furnace but not in contact with the workpiece in *retort carburizing*. *Liquid carburizing* can be done by immersing the workpiece in a cyanide bath, as in cyaniding described later. The salt bath composition for liquid carburizing yields a case rich in carbon in contrast to the cyanide bath that yields a cyanide case high in nitrogen.

Most carburized parts are hardened by quenching directly from carburizing temperature or after a slight cooling to just above the critical temperature. Some parts are cooled to room temperature and reheated once or twice for quenching. This gives better quality through less distortion, better carbon diffusion, grain refinement, and core toughness. Cooling and reheating may be convenient for parts requiring special handling for quenching, for example, a part held rigidly in a die while quenching to control distortion.

Selective carburizing or *hardening* is often required to leave some surfaces soft after heat treatment. One way is to initially copper plate surfaces that are to be left soft. The carbon does not penetrate the copper. Another way is to cool the piece and machine material from the surface to be soft, which removes the case before hardening. Also, selected surfaces may be softened by torch heating and slow cooling after the entire part has been heat treated.

6.4.5 Cyaniding

Cyaniding or *liquid carbonitriding* imparts a file-hard and wear-resistant case to steel by immersing it in a molten cyanide salt bath for a time and then quenching it. A cyanide case is seldom over 0.3 mm (.01 in.) thick, and carburized cases are usually thicker. However, cyaniding requires lower temperatures, usually below 871° C (1,600° F), and less time (30–60 min) than carburizing.

Extreme care must be exercised with cyaniding. The cyanides are fatally poisonous if taken internally and highly toxic when placed in contact with scratches or wounds. When cyanides are brought in contact with acids, fatally poisonous fumes are evolved. Appreciable amounts of cyanide salt are transferred to the quenching liquid and must be neutralized before waste is discharged to the sewer. Sludge accumulates in salt pots and poses a disposal problem.

Carbonitriding, also known as *dry-* or *gas-cyaniding, nicarbing*, and *nicarburizing*, adds both carbon and nitrogen to a shallow case on steel surfaces. It is done in much the same way and with the same equipment as gas carburizing in a carbon- and nitrogen-rich atmosphere.

The advantage is that the nitrogen-enriched case can be quenched in oil rather than in water, which is required for unalloyed steels after plain carburizing. The oil quench causes less distortion and cracking.

6.4.6 Nitriding

Steel is gas-nitrided in a furnace at 510–566° C (950–1,050° F) with an atmosphere, commonly ammonia, which permeates the surface with nascent nitrogen. The basic process takes a long time; for instance, with Society of Automotive Engineers (SAE) 7140 steel at 524° C (975° F), the case depth reaches 0.5 mm (.02 in.) at 50 hr and 1 mm (.04 in.) at 200 hr. Liquid nitriding also is done at 510–566° C (950–1,050° F) in a bath of molten salts. Quenching is not needed because the case consists of inherently hard metallic nitrides. For efficient results, nitridable steels alloyed with aluminum, chromium, vanadium, and molybdenum are used to form stable nitrides.

Various nitriding modifications are practiced to speed up the process. The *Nitemper*[SM] process operates at about 577° C (1,070° F) with an atmosphere of half endothermic gas and half ammonia. After being heated for only about 1.5 hr, the work can be quenched in oil for maximum fatigue strength. Nitempering utilizes a gas mixture that is explosive and must be treated carefully, however, it gives fast results. The case is much thinner than that from regular nitriding but contains an extremely hard, complex iron-carbon-nitrogen compound. *Chapmanizing* or *liquid pressure nitriding* entails passing anhydrous ammonia through the nitriding salt bath under pressure while the work is being treated. *Pressure nitriding* is done with the work in a sealed retort holding an ammonia atmosphere under pressure. *Glow discharge nitriding* or *ionitriding* is done with the work as the cathode in an anodic retort. An electric current heats the work and produces a glow discharge that ionizes the nitrogen atmosphere.

Nitriding is more expensive than other hard-case processes but offers several advantages. It is performed below the critical temperature without detriment to the strength and other properties of the steel core. The case is quite hard (70 R_C and over) and notably wear, fatigue, and corrosion resistant (except for stainless steel) and stays hard at temperatures up to about 427° C (800° F). Nitriding is applied, for example, to high reliability gears, bushings for conveyor rollers that handle abrasive alkaline materials, antifriction bearings, and gun parts.

6.4.7 Laser Beam Hardening

Laser beam hardening is used for hardening surfaces of small and large components. Laser beam surface hardening has the ability to harden and reach areas that are not practically accessible by other techniques. Other advantages of laser beam surface hardening are close control of power input and less distortion. The drawback of this technique is the high initial investment required and the relatively small depth of the hardened layer. The source of energy required for laser beam surface hardening is a continuous-wave CO_2 laser.

6.4.8 Electron Beam Hardening

Electron beams can be used for surface hardening of steels with high efficiency. The electron beam hardening technique utilizes high-velocity electrons to strike the surface of the part and generate heat. This technique is also practical to harden hard-to-locate surfaces of small and large parts. High initial investment is a governing factor and the depth of the hardened surface layer is normally small, up to 2.5 mm (.10 in.).

6.5 HEAT TREATMENT OF NONFERROUS METALS

When steels are heat-treated, they experience phase transformations. For this reason, and other fundamental differences in hardening and strengthening mechanisms, nonferrous alloys cannot be heat-treated using the techniques employed for ferrous alloys.

Hardening of most alloys, such as aluminum alloys and copper alloys, is accomplished by precipitation hardening.

6.6 HEAT-TREATING FURNACES

Usually, a heat-treating furnace consists of a box-like structure with a steel shell and an access door, a refractory lining, and temperature controls and indicators. Some of these features are lacking or are different in some cases. Furnaces may be classified by the ways the work is handled or by the means of heating. For work handling, the basic types are batch furnaces and continuous furnaces. As for fuel, oil and gas predominated at one time, but electricity has gained in popularity. Close process control is easier with electricity. With combustion heating, the work must be contained in a separate chamber if it is to be kept from the gases.

Heat-treating furnaces use large quantities of energy. Improvements in furnace design in recent years have been directed toward more efficient processing, improved insulation, more effective combustion control, and recovery of heat from fuel gases that would otherwise be wasted. An example of heat recycling has two facets. Usable sensible heat exhausted by high-temperature furnaces is passed through heat exchangers to furnish about 30% of the needs of the tempering furnaces. Also, radiant heat is collected by steel walls enclosing the furnaces and delivered by circulating air to heat the plant in winter. Each type of furnace has several varieties. The main forms will be described in the following sections. Two major kinds of batch furnaces are hearth furnaces and bath or pot furnaces.

6.6.1 Hearth Furnaces

The *direct fuel-fired furnace* (Figure 6-13) burns fuel in the space occupied by the charge. It is low in cost and suitable for all ordinary temperature ranges. This furnace is used for rough heating, such as forging, but may serve for heat treating, particularly at lower temperatures.

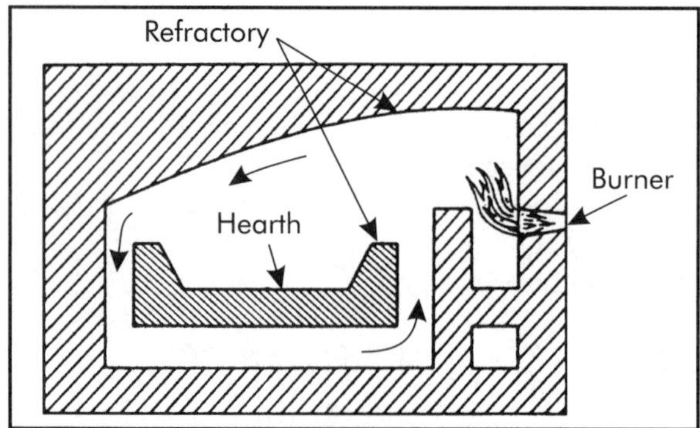

Figure 6-13. Direct fuel-fired furnace.

The *indirect-fired furnace* has a heating chamber and a muffle that separates the combustion space from the work space. The upper temperature limit for this furnace is approximately 1,093° C (2,000° F). Reduced scaling and contamination from fuels are the advantages of its use.

A *recirculation furnace* is indirect-fired but the gases of combustion are circulated through the work space from the combustion area. The hot gases are channeled so the heating is uniform. Application temperatures are mostly below 704° C (1,300° F) and they are commonly used in ovens for tempering, toughening, and stress relieving.

Muffle and *retort furnaces* are indirect-fired. The work is in a protective or carburizing atmosphere and separated from the gases of combustion. The work space of a muffle furnace is surrounded by refractory material sufficiently tight to keep out contaminants. Semimuffle furnaces have a small amount of baffling in the work space to prevent direct impingement of the combustion flame upon the workpiece. A retort furnace takes a heat-resistant retort that is loaded with work outside the furnace, capped and sealed, and then placed in the furnace to be heated. A *radiant-tube furnace* has a tightly encased, refractory-lined chamber that forces hot gases through radiant tubes arranged around the work space.

Electric furnaces look much like other kinds. A common type, such as that shown in Figure 6-14, has heating elements enclosed in preformed panels in the side walls and top and bottom of the hardening chamber. The heating unit is segregated in some furnaces, and the atmosphere is blown over into the work chamber. Electric furnaces are more costly to buy and operate than gas furnaces, but have the advantages of cleanliness, convenience, controllability, and few environmental restrictions.

In any furnace, heat is convected better and distributed more thoroughly by circulating the atmosphere. The more rapid the circulation, the more the heat is transferred. Pressure in the furnace is commonly kept above that outside to prevent air from entering when the door is opened. Some furnaces have a flame curtain at the door to burn any oxygen that may enter.

Figure 6-14. Electric, box-type, heat-treat furnace with separate hardening and drawing chambers. (Courtesy McEnglevan Industrial Furnace Company, Inc.)

A *fluidized-bed furnace* has the work immersed in a bath of inert (for example, aluminum oxide) particles in an insulated container. A heated gas mixture flows upward through the beds and holds the particles in suspension. One make operates at temperatures up to 1,010° C (1,850° F). It has a capacity of 356 mm (14 in.) diameter × 457 mm (18 in.) deep.

Batch-type furnaces have a number of different shapes. A *box furnace* is shown in Figure 6-14 with a hardening chamber on top and a drawing chamber below. Instrumentation for both chambers is provided by microprocessor-based digital controllers located on the front of the furnace. A *pit furnace* has a vertical work chamber with an opening at the top to receive the charge. A *bell furnace* has a box-like cover over the hearth, and the cover is lifted off to remove and place a charge. An *elevator furnace* has a vertical work chamber with an opening at the bottom for charging.

A *car-bottom furnace,* like the one in Figure 6-15, has a moveable hearth like a flat car that is rolled out for unloading and loading. Commonly, the load is stacked on heat-resistant alloy or refractory piers and spacers to facilitate circulation of gases and heat the materials uniformly. In the same family is the *tip-roof furnace,* where the box is raised and tipped to allow work to be inserted. Such a furnace with space of 1.2 × 1.8 × 7.3 m (4 × 6 × 24 ft) is capable of taking loads up to about 9 metric tons (10 tons), includes a hydraulic manipulator and adjacent quench tanks, and costs from $500,000–$750,000.

Vacuum furnaces are necessary for some space-age materials and electronic components, and where surface finishes must be preserved on ordinary work. They are made to be tightly sealed and have auxiliary equipment to draw and hold a vacuum like the vacuum melting furnaces of Chapter 8. There are both batch- and continuous-type models. Heat treating is commonly done by radiant heating in a vacuum.

Figure 6-15. Car-bottom furnace with hydraulic-driven car and chart recorder. (Courtesy McEnglevan Industrial Furnace Company, Inc.)

6.6.2 Rotary Furnaces

Rotary furnaces are used for both batch and continuous production. One kind is the *rotary-hearth furnace* that is built in a wide range of sizes to heat a few hundred kilograms to 50 Mg (≈50 tons) or more per hr. It has a round shell and horizontal hearth that turns slowly. Material is charged through a door and may be taken out of the same door or another one. The speed of the hearth is set so that the heating cycle is completed when the hearth makes one turn.

A *rotary-retort furnace* consists of a revolving retort with a horizontal axis inside a heating chamber. Small parts are loaded at one end, tumbled, and pushed along by ribs or vanes, and emerge from the other end of the retort. Sizes are available to take loads from about 45–680 kg (100–1,500 lb). Workpieces must be able to withstand a certain amount of rough treatment and cannot be precisely metered out of the furnace for uniform quenching.

6.6.3 Continuous Furnaces

A *continuous furnace* is the type most used for in-line, large-quantity production. It typically has a horizontal work chamber and a mechanical means of conveying the work from one end to the other. Heat is generated by electricity or burning fuel. Many furnaces have different temperatures that are precisely controlled in several zones for multistage heat treating.

Some authorities call only those furnaces "continuous" in which the workpieces all move at the same rate from one end to the other. The name of *cycling* or *semicontinuous furnace* is given when the work is held for a different preset time in each section of the furnace. In either case, small pieces may be carried through in baskets or on trays. Common forms of semi- and fully continuous furnaces are described in the following paragraphs.

A motor-driven *belt* or *chain-conveyor furnace* transports the work on an endless hearth of heat-resistant chain or links. An example is given in Figure 6-16.

In a *roller-hearth furnace*, the work rides on driven rollers, one set after another through the length of the furnace. This type is suitable for uniformly sized parts on trays.

The work sits on rollers or skids that are not driven in a *pusher furnace*. Instead, the workpieces or trays are pushed, one against the other, by mechanical, pneumatic, or hydraulic means in a timed cycle.

A *screw-conveyor furnace* has a coarse-pitch powered screw extending through the chamber.

The hearth and work are gradually accelerated forward in a *shuffle-* or *shaker-hearth* or *reciprocating furnace*. When the hearth is suddenly stopped, the work slides forward. The motion is not a destructive vibration but is repeated only from time to time.

The work is shifted along by beams that rise as they move forward and fall as they move backward in a *walking-beam furnace*.

6.6.4 Furnace Atmospheres

Metals are subjected to high temperatures for appreciable periods of time during heat treatment. This damages metallic surfaces if they are not protected. Oxygen, carbon dioxide, and water vapor are the most injurious gases, among others, and they are removed in the preparation of furnace atmospheres. Oxygen rusts, corrodes, and scales most metal surfaces and removes carbon from steel; thus it degrades surface finish, impairs size, and keeps surfaces from being hardened. Carbon dioxide scales or decarburizes steel in combination with carbon monoxide in some ratios, but is a neutral atmosphere and suitable for bright annealing in other ratios. Water vapor oxidizes iron and steel and is the main agent that turns steel surfaces blue when they are cooled. It combines with carbon in steel to form carbon monoxide and hydrogen.

Some gases that are harmful under certain circumstances serve as protectors when properly utilized. Nitrogen in the molecular state is satisfactory as a furnace atmosphere for bright annealing of low-carbon steel but does decarburize surfaces in the presence of even a trace of moisture. Atomic nitrogen may form hard surface iron nitrides. Hydrogen is absorbed by steel at certain temperatures and causes embrittlement. Dry hydrogen reduces iron oxide and does not cause scale but does decarburize steel under certain conditions. Hydrogen forms undesirable water vapor in reactions with carbon dioxide or

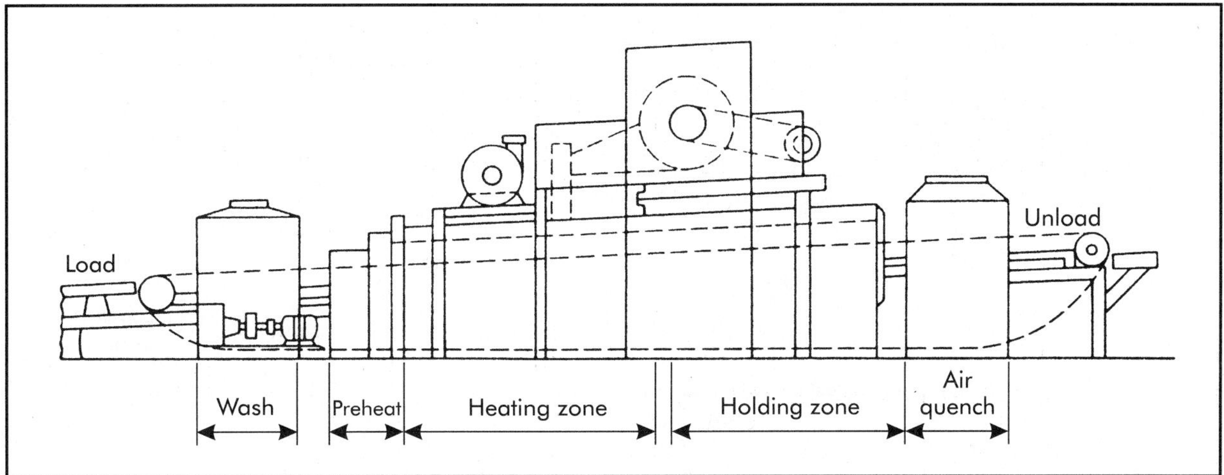

Figure 6-16. Chain-conveyor furnace for washing, drawing, and cooling hardened gears in large quantities.

oxygen. Hydrocarbons present in a furnace tend to decompose at heat-treating temperatures and to liberate hydrogen and deposit soot.

An atmosphere is generated and induced into a heat-treating furnace to keep from changing the chemistry of and prevent discoloring of the work surface. About a dozen kinds of atmospheres are widely used. The American Gas Association has divided them into six classes, as described in Table 6-1, according to how they are made or what they contain. Each of the six classes may have variants that are designated by numbers added to the second and third digits to the right. For instance, 302 indicates a rich endothermic atmosphere, and 501 a lean exothermic-endothermic mixture.

6.6.5 Molten Baths for Heat Treating

Metals are often heated or cooled by being immersed in molten salts or lead. Heating is uniform and rapid, temperature can be closely controlled, and the workpiece is shielded from the air. Molten baths are widely used for interrupted quenching, as in martempering and austempering, and for surface treatments like cyaniding.

A salt bath protects metal. When a cold piece of metal is placed in fused salts for the purpose of heating, the salt immediately contacting the piece freezes and clings tightly to the workpiece, thus forming a salt encasement that serves to prevent rapid surface heating and thermal shock. When the temperature reaches equilibrium, the frozen shell disappears, and further heating occurs by conduction. An additional advantage of this method is that the molten salt in direct contact with the workpiece more efficiently transfers heat (four to seven times faster) than would a gaseous atmosphere. Totally immersed, the materials being heated have no contact with the atmosphere and, therefore, compositional changes are avoided (where the salt bath is neutral). At the end of the heating cycle and when the part is removed, a film of molten salt adheres to the part and protects it against atmospheric attack while it is being moved to another bath or to the quenching tank.

Common salts are sodium and potassium chlorides, nitrates, and cyanides. They are mixed in various proportions and with other salts to obtain different melting points for application temperatures ranging from 163–1,399° C (325–2,550° F). Formulas are given in reference books and handbooks.

6.6.6 Bath Furnaces

Bath furnaces may be gas- or oil-fired or electrically heated. They are mostly of small and medium size because it is not economical to keep a large mass of salt heated for large parts, particularly for intermittent use. *Gas- and oil-fired salt-bath furnaces* have low initial and operating costs and are versatile. They can be restarted easily, and pots can be interchanged in one furnace to use a variety of salts. The bath may be

Table 6-1. Classes of atmospheres for heat-treating furnaces

Class	Base	Composition	Uses	Relative Cost ($/Unit Volume)	Note
100	Exothermic	70% or more N_2; remainder CO_2, CO, H_2	Bright annealing	1	Generated by controlled combustion of hydrocarbon; lowest cost
200	Prepared nitrogen	97% N_2	Neutral atmosphere	1.3	Often needs additives; not dependent on natural gas supply
300	Endothermic	40% N_2, 20% CO, 40% H_2	Carburizing, neutral hardening, sintering	1.5	Explosive; not for stainless steel
400	Charcoal	65% N_2, 35% CO	Carburizing, sintering	6	Low equipment cost; for small-scale operations
500	Exothermic-endothermic	70% N_2, 30% CO	General purpose	2	Can be modified to serve in place of most other classes
600	Ammonia	80% N_2, 20% H_2, typical	Bright annealing; sintering, neutral heating	9	Good quality atmosphere suitable for stainless steel

heated externally, as in Figure 6-17, or by immersed radiant tubes, which help keep temperature uniform in the bath.

There are several kinds of *electrically heated salt-bath furnaces*. All are surrounded by insulated casing. Heating may be done by resistance elements around the pot in the externally heated-type depicted in Figure 6-18. Well-insulated heating elements are put directly in the bath in the immersion-heating element-type of furnace. Temperatures are usually limited to 593° C (1,100° F) for satisfactory resistor life. Higher temperatures can be held by passing electricity through the bath between electrodes. The *immersed-electrode salt-bath furnace* has electrodes immersed in a metal pot. The *submerged-electrode furnace* has water-cooled electrodes extending through the sides into a ceramic brick

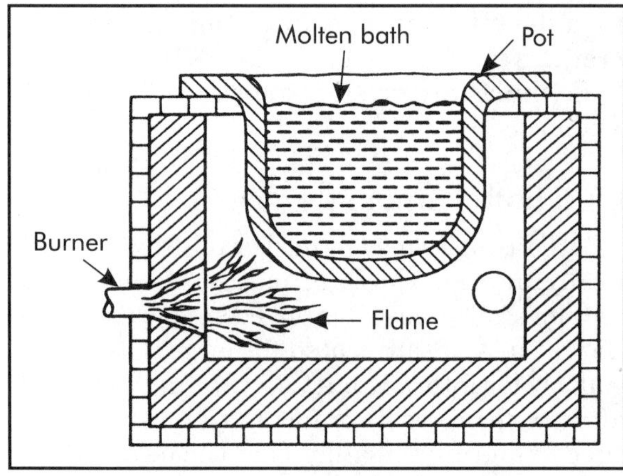

Figure 6-17. Externally heated gas- or oil-fired salt-bath furnace.

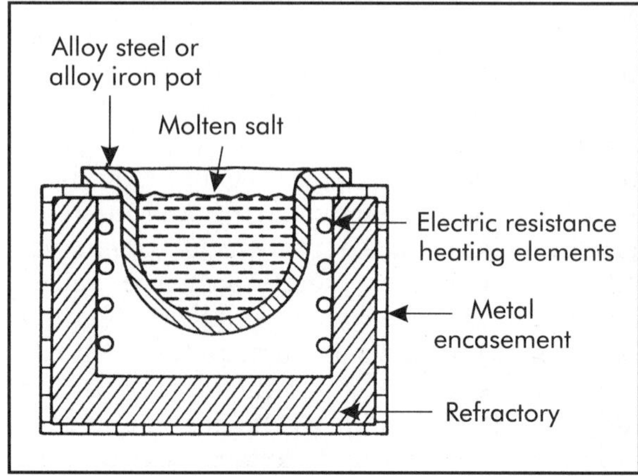

Figure 6-18. Electrically resistance-heated salt-bath furnace.

pot. The molten salt penetrates the refractory material until it reaches a zone cool enough to freeze and thus seals the pot. Electrode furnaces use alternating current transformed to low voltages (5–15 V) because direct current decomposes the liquid salt. Temperatures are easy to control within 3° C (5° F). Electrode furnaces occupy a minimum of floor space but are not easy to restart because the frozen salt is not a conductor. They are best for large-quantity continuous production.

6.7 DESIGN CONSIDERATIONS FOR HEAT TREATMENT

Reference books and handbooks specify the properties of materials that may be produced by heat treatment, as well as specifications for time and temperature, correct hardenability, etc., for each case.

It is important to recognize that most steel mill barstock and forgings have a decarburized skin that must be removed before heat treatment. It will not harden, cannot be carburized properly and, especially in nitriding, leaves quite a brittle layer that peels off easily.

Pieces that are long, or large in area and thin, unsymmetrical, or have holes, deep keyways, and grooves are difficult to heat treat. Troubles arise in nonuniformity, warping, and cracking.

Parts with threads, fine splines, sharp edges, and other thin sections may become faulty from carburizing if their thin sections harden throughout and support of the tough core is lost.

Hardened surfaces are expensive to machine. The bulk of metal removal and forming should take place before hardening. Finishing is usually necessary after hardening to correct unavoidable distortion and must be done by expensive abrasive methods, so it should be confined to essential surfaces and the removal of a minimum amount of stock.

6.8 COST CONSIDERATIONS

The selection of equipment for heat-treatment operations is dependent on many factors, including required strength, furnace parameters, material, temperature ranges, and other aspects of the operation.

The cost range for heat-treating equipment varies significantly. Furnaces may cost as little as $50,000 up to over $750,000. For example, a fluidized bed furnace operating at temperatures up to 1,010° C (1,850° F) costs over $50,000. A car-bottom furnace with up to 9 metric tons (10 tons) capacity costs over $500,000. Electric furnaces are more costly, but they are cleaner, more convenient, and provide better controllability.

6.9 PROCESS AUTOMATION

Current trends in heat-treating technology indicate a great potential for automating the processes and other aspects of heat treatment. Studies and research in heat-treatment techniques are continuing to provide better ways of process control, material property control, and defect control. Better methods for temperature control in heat-treating furnaces through the use of sensors and computer interfacing are emerging. Computers are increasingly being used to monitor and control process parameters and thus provide improved process performance.

6.10 QUESTIONS

1. What is the basic scheme for hardening metals and why is it effective?
2. What are coherency-hardening and aggregate-hardening and what are their relative merits?
3. What is the necessary condition for hardening a nonallotropic alloy? How is the condition utilized?
4. How is 2024 aluminum, as an example of a nonallotropic alloy, solution-treated and age- or precipitation-hardened?
5. What is the basic mechanism for the hardening of steel?
6. What is a T-T-T diagram or S-curve and what does it show?

7. Describe three principal ways of hardening steel after heating to austenitic temperature. What results and advantages are obtained from each way?
8. Describe the hardenability of steel and what factors influence it.
9. How is hardenability measured?
10. Describe full annealing, normalizing, process or commercial annealing, and cycle annealing. What does each process do?
11. Describe the processes for surface-hardening of steel. What are their applications and merits?
12. What means are found in heat-treating furnaces to hold protective or carbon-rich atmospheres around the workpieces?
13. What devices are utilized for continuous handling of the work in production heat-treating furnaces?
14. What gases and vapors are detrimental in a heat-treating atmosphere? What gases are mostly utilized for a protective atmosphere?
15. Describe bath furnaces for heat treating and what advantages they offer.

6.11 REFERENCES

Brooks, C.R. 1995. "Principles of the Heat Treatment of Plain Carbon and Low-alloy Steels." Materials Park, OH: American Society for Metals.

Case Hardening of Steel. 1987. Materials Park, OH: ASM International.

Handbook of Quenchants and Quenching Technology. 1992. Materials Park, OH: American Society for Metals.

Huntress, E.A. 1980. "Nitrogen: All Purpose Atmosphere." *American Machinist,* April.

Krause, G. 1980. *Principles of Heat Treatment of Steel.* Materials Park, OH: American Society for Metals.

Metals Handbook, 10th Ed. 1991.Vol. 4: *Heat Treating.* Materials Park, OH: American Society for Metals.

Reynoldson, R.W. 1993. "Heat Treatment in Fluidized Bed Furnaces." Materials Park, OH: American Society for Metals.

Society of Manufacturing Engineers (SME). 1999. *Heat Treating* video. *Fundamental Manufacturing Processes* series. Dearborn, MI: SME.

——. 1997. *Cutting Tool Materials* video. *Fundamental Manufacturing Processes* series. Dearborn, MI: SME.

Thelning, K.F. 1984. *Steel and Its Heat Treatment,* 2nd Ed. Stoneham, MA: Butterworths.

Vaccari, J.A. 1981. "Fundamentalsof Heat Treating, Special Report 737." *American Machinist,* September.

Nonmetallic Materials

7

Polyvinyl Chloride (PVC), 1930
—W. L. Semon, B. F. Goodrich Co., USA

7.1 NONMETALLIC MATERIAL FAMILIES

Engineering materials, in general, may be classified as metals and nonmetallic materials. Nonmetallic materials, in turn, are classified into plastics, ceramics, and composites. This chapter presents the fundamentals of nonmetallic materials, including plastic materials, plastics processing, design of plastic parts, and ceramics.

Plastics are mostly products of this century. Natural shellac and bitumen have been known for ages. Celluloid was discovered by Hyatt in 1868 as a substitute for ivory in billiard balls. The modern plastics industry started with the development of Bakelite® by Baekeland in 1909. Since then, growth has been rapid, now with more than 30 chemically distinct families of plastics, hundreds of compounds, and thousands of products.

Plastics have been increasingly accepted because they offer unique combinations, a wide variety of properties, and are particularly fitted for many modern developments. Plastics are light in weight; most weigh less than magnesium. Outside of especially light-foamed plastics, the average specific gravity is about 1.4. This has suited them for trim and accessories in commercial aircraft and has helped lighten many hand implements. Plastics are good electrical insulators at low and high frequencies and have important uses in the electrical industry. They are good heat insulators too. Some weather well and are highly resistant to corrosion and chemicals.

The strengths of plastics cover a broad spectrum. Most materials are in the low end of the range, but some have quite high strengths. Their main structural advantage is a high strength-to-weight ratio. The tensile strength of some plastics is only 7 MPa (1,000 psi) or so; most range from 35–138 MPa (≈5,000–20,000 psi), but some reinforced plastics have strengths to 3 GPa (440,000 psi). As an example of weight advantage, an epoxy reinforced with woven glass fabric, with a tensile strength of 586 MPa (85,000 psi) and a specific gravity of 1.85, has a strength-to-weight ratio of almost twice that of heat-treated alloy steel and only a little less than heat-treated titanium alloy.

Most mechanical properties of plastics are inferior to those of metals. Plastics are subject to some dimensional instability, with appreciable creep and cold flow at all temperatures, and some are swollen by moisture. Thermal expansion is large compared to metals. Top operating temperatures range from 66° C (150° F) to around 316° C (600° F) for plastics, but most metals serve well above that range. At best, the rigidity and fatigue strength of plastics are well below those of metals.

Manufacture of plastics is economical because most products can be entirely finished by molding and forming without secondary operations.

Molding and forming operations are aided by the pliability and heat sensitivity of plastics; for instance, a plastic part can be extricated more easily from a mold than a metal part. Intricate plastic parts can be molded, and it is not unusual for one plastic molding to take the place of several formed metallic parts.

Probably of most significance, plastics offer a greater range of appeal to the eye than any other class of materials. Molding usually leaves a good surface finish, but a higher luster can be added by a simple solvent treatment or buffing. All colors are available, but not in all plastics. Color is impregnated into an object and is more lasting than paint. Some plastics offer true transparency without the brittleness of glass. Plastic materials are commonly used for protective and decorative coatings.

Rubber and synthetic elastomers have some of the same properties as plastics and will be discussed in this chapter.

7.2 PLASTIC MATERIALS

The term *plastic* in its original sense applies to a material that can be made to flow so that it can be molded or modeled. This is true of metals, clay, and other materials, but the name has come to specifically designate a group of organic solids that can readily be made to flow by heat or pressure, or both, into valuable commercial shapes.

A plastic may be in one or more of three forms in the raw material state. First are powders, flakes, or granules for molding plastic pieces. Second are liquids for castings, impregnated laminates, adhesives, paints, and mixed molding compounds. Third are the filaments, films, sheets, rods, and tubes to be fabricated into finished articles, such as by weaving into cloth, cutting and joining into wrappers, or machining.

7.2.1 Resins and Polymers

Many kinds of materials are called plastics. Most of them are based on some form of synthetic organic resin that is a compound containing carbon as the central element. Other elements, such as hydrogen, oxygen, nitrogen, and chlorine, are linked to the carbon atoms to form the molecules. The properties of a plastic material depend upon the atoms it contains, the way these atoms are arranged in molecules, and the ways the molecules are arrayed and related in the mass.

An example of an organic compound is methane, one of the simplest. Its formula is written in Figure 7-1. Methane is a gas and not a plastic. It is the first of a series of organic chain compounds that are said to be *saturated* because all the valence bonds of the carbon atoms are satisfied. The second compound of the series is ethane; its formula is also given in Figure 7-1. A compound in which the valence bonds of the carbon atoms are not satisfied is said to be *unsaturated*. The structures of two simple unsaturated compounds, ethylene and acetylene, are also indicated in Figure 7-1. The double and triple lines in the formulas designate the lack of saturation.

With proper pressure, temperature, and catalysts, a number of ethylene molecules with unsaturated bonds can be joined together into a long molecule as indicated in Figure 7-2. A substance composed of basic molecules is a *monomer*. The formation of larger molecules from smaller ones is called *polymerization*. The substance thus formed is a *polymer*. As the molecules grow larger, a gas like ethylene becomes a liquid and ultimately a solid. Polymerization leads to an increase in the boiling point of a liquid and the melting point of a solid.

Polymers are formed in two ways. The first is *linear addition*, as exemplified in Figure 7-2, and occurs when the final mass of each molecule is the sum of the masses of the original monomers. Polymers may contain thousands of monomers. Some molecules are chains of one kind of monomer, like the polyethylene illustrated; others contain two or more different kinds. When different monomers are combined, the process is called *copolymerization*, and the product is a *copolymer*. The second way of forming polymers is by *linear condensation*, also exemplified in Figure 7-2. There the components react, and a by-product, usually water, is cast off.

Various hydrogen atoms in a chain may be replaced by other elements, such as chlorine or fluorine, which retard combustion, for instance, or by radicals, such as methyl CH_3. These affixed atoms or radicals denoted by the letter *R*

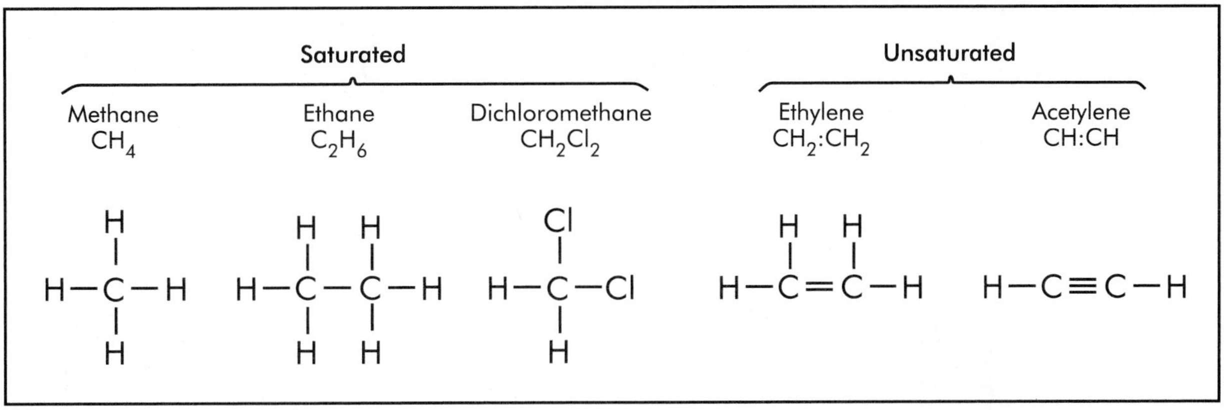

Saturated | Unsaturated

Methane CH_4 | Ethane C_2H_6 | Dichloromethane CH_2Cl_2 | Ethylene $CH_2{:}CH_2$ | Acetylene CH:CH

Figure 7-1. Designation of some simple organic compounds.

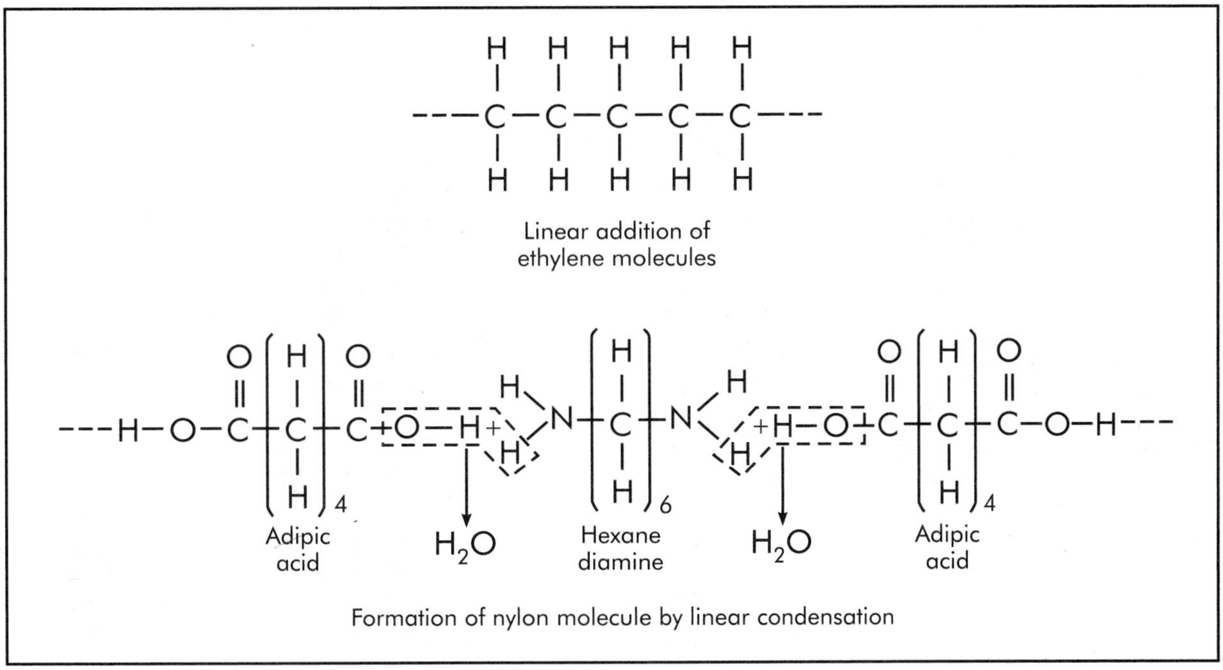

Linear addition of ethylene molecules

Adipic acid | H_2O | Hexane diamine | H_2O | Adipic acid

Formation of nylon molecule by linear condensation

Figure 7-2. Two examples of polymerization.

in Figure 7-3 may be randomly oriented on a chain (*atactic*), all on one side (*isotactic*), or alternately on one side and then the other (*syndiotactic*). Such chains can be more closely packed and have higher melting points in the order of atactic $<$ syndiotactic $<$ isotactic.

Thermoplastic Materials

One major class of plastics, called the *thermoplastic* materials, is composed of materials having chemically separate long molecules. They are held together by secondary bonds, which are associative attractive forces between the molecules that are weaker than the primary bonds between the atoms within each molecule. When the molecules are activated and separated by heat, the secondary bonds are weakened, and the material softens and ultimately melts.

Most commercial thermoplastics are either *amorphous* or *crystalline*. In the amorphous or glassy form, the molecules are intertwined in no apparent order. Examples are polystyrene, poly-

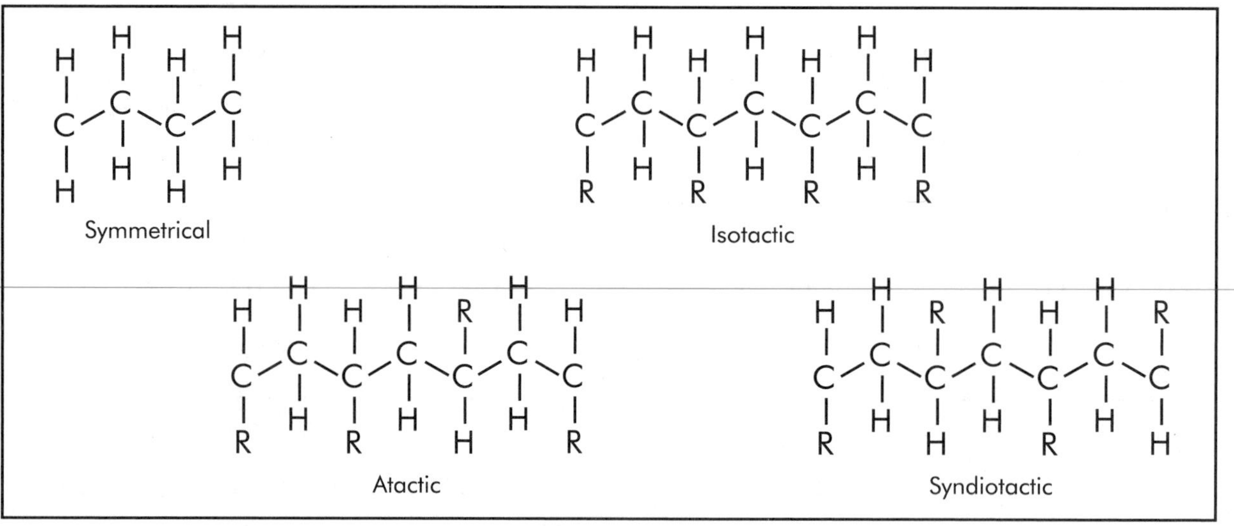

Figure 7-3. Molecular structure of plastics.

methyl methacrylate, and polycarbonate. In the crystalline form, the long molecules are folded in regular array in small thin *crystallites* gathered into bundles in larger aggregates called *spherulites*, which compose the gross internal structure like the grains in metal. The crystallites and spherulites are surrounded by amorphous material, but appear to be interconnected by multitudes of links of small fibers of crystalline material. Examples are polyethylene, isostatic polypropylene, acetal, and nylon. A third and less common form is that of rubber-modified plastics, like acrylonitrile-butadiene-styrene (ABS), in which the rubber exists as small round particles well dispersed in a rigid glassy matrix. This toughens a hard and brittle material.

The nature of the molecules, as well as the way they are arranged, determines the properties of a thermoplastic material. The forces between individual particles depend directly upon the masses and inversely upon a power of the distance between them. Thus, long and heavy molecules have strong associative forces and make up strong and tough plastics. Relatively straight molecular chains can be closely packed and form stiff plastics; materials with coiled or branched molecules are more flexible. Substitutional radicals make molecules bipolar and set up a magnetic attraction between molecules. Isostatic molecules, in particular, are strongly bipolar and can be closely packed; they tend to form materials that soften at higher temperatures. Amorphous plastics have fewer associative bonds and soften at lower temperatures. The more crystalline materials have more numerous links and can be made to withstand the highest temperatures that organic substances can endure. Clear plastics are generally amorphous, because crystallites scatter light and make a material translucent.

A thermoplastic subclass is that of the *ionomers*, which contain strong metallic polar groups distributed along nonpolar chains (for example, polyethylene). The polar groups tend to cluster together as much as they can. They inhibit crystalline formation and thus yield almost a clear plastic. Loss of crystalline strength is more than made up for by the strong polar forces, and the ionomers are tough and highly elastic.

Thermosetting Materials

A second major class of plastics, called *thermosetting* materials, is composed of materials having long molecules linked to each other in three dimensions by primary or valence bonds. They are not broken by heat until the compound is decomposed. An example of the formation of a thermosetting plastic is given by the reaction between phenol and formaldehyde to produce a phenolic resin as indicated in Figure 7-4. Under proper conditions of temperature and pressure, some of the formaldehyde molecules form links in linear chains, and others form links between the chains of phenol molecules. If the product is heated to an excessive temperature, it chars and decomposes, but short of that it does not soften once formed.

The thermosetting plastics are generally stronger, harder, more resistant to heat and solvents, and lower in material cost. The secondary bonds of the thermoplastic materials allow deformation and flow under stress. This tends to redistribute stress concentrations under load and makes the material tough. However, the thermosetting plastics are brittle. Because of heat susceptibility, thermoplastics can be molded easily but are found mostly in packaging and consumer goods for uses at room temperatures. Even so, the distinctions between the two classes have been fading away as thermoplastics have been developed to be serviceable at 427° C (800° F), well above the top working temperatures of most thermosetting materials. Thermoplastic *superpolymers* have been developed in recent years with service temperatures above 204° C (400° F), in the realm of the thermosets. They have exceptional resistance to solvents, oils, corrosive substances, and even burning. Notable among these are the fluoroplastics, polyimides,

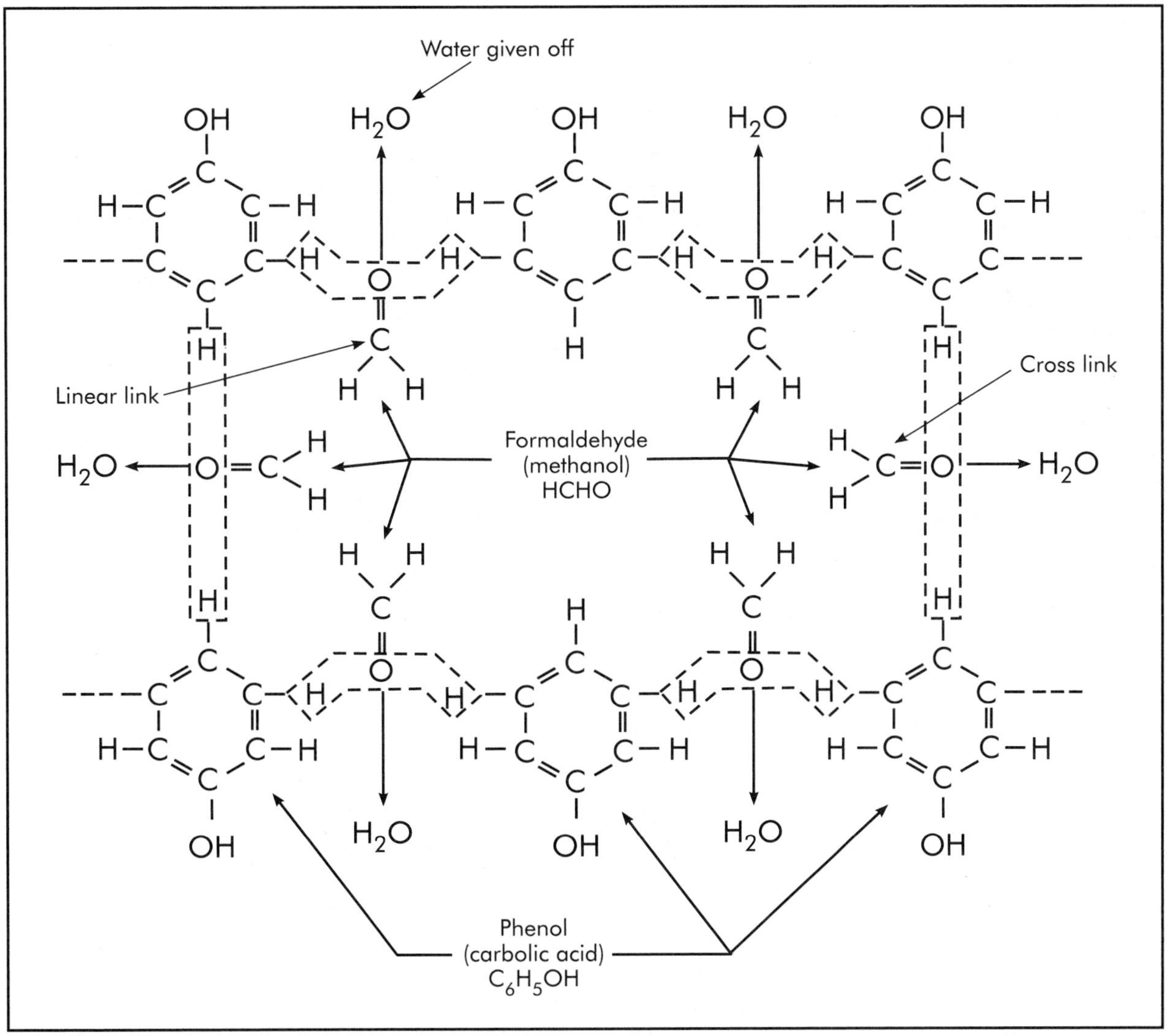

Figure 7-4. Polymerization of a thermosetting phenolic resin.

and polysulfones of Table 7-1. Their usage is limited because their costs are high and they usually are difficult to process.

Processes

The forms and properties of polymers depend largely upon how they are prepared. The lengths of the chains, side chains, how the molecules are arrayed, and other features, vary with the proportions of ingredients, catalysts, control of temperature and pressure, and commercial processes used. The four general commercial processes are the *bulk*, *emulsion*, *solution*, and *suspension methods*.

A polymer is formed in a mass of the monomers by the bulk method. Other substances are not needed, and impurities are avoided, but high temperatures harmful to the product may be hard to control. The monomer is dispersed or dissolved in another medium, often water, in the emulsion, solution, or suspension methods. Such may be necessary to control the monomer or obtain the product in a certain form, like fine particles rather than lumps.

New materials and methods are appearing in the plastics industry. The realm of plastics is growing beyond the hydrocarbon compounds into that of inorganic polymers. The silicones that are based upon chains of silicon rather than carbon are well established and will be described later. Other series are emerging, built upon such elements as phosphorus and boron. In general, organics cannot stand heat; inorganics cannot stand strain.

In processing, irradiation is used to produce polymers without heat or catalyst, induce crosslinks even in thermoplastics, and bring about reactions between polymers and other materials to change properties.

7.2.2 Additives

In most cases, the plastic resin is mixed with other substances to make the final product. The additives may be *fillers, pigments* or *dyes, stabilizers, plasticizers*, or *lubricants*.

Fillers are added to most thermosetting plastics and many thermoplastics. As much as 80% inert material may be put in molded parts. Most common earthy materials, such as calcium carbonate at \$0.11–\$0.44/kg (\$0.05–\$0.20/lb), depending upon quality, provide bulk alone. Wood filler for around \$0.22/kg (\$0.10/lb) gives bulk with some strength and good moldability, but tends to absorb moisture. Macerated cloth and fibers at higher prices add more strength. Carbon or silver additives make plastics conductive. Other fillers are used for particular purposes.

Reinforced plastics or *composites* have fibers added to a resin matrix mainly for superior strength. Glass-reinforced plastic (GRP) is the most common. The *aramids* comprise another class of high-strength fibers; Kevlar® is a common trade name.

Properties depend upon the forms of the materials and processing, but strengths may be increased with glass fibers by a factor of two or more. Some typical prices are \$1.76/kg (\$0.80/lb) for filament or roving, \$2.20/kg (\$1.00/lb) for mat, and \$6.61–\$11.02/kg (\$3.00–\$5.00/lb) for glass cloth. Asbestos costs more, but gives high strength plus heat resistance and dimensional stability. Advanced composites contain boron or graphite fibers or filaments at upward of \$22/kg (\$10/lb). However, they provide specific moduli, specific strengths, and fatigue strengths superior to GRP and, in some cases, surpass the best metals. As yet their applications are limited by cost.

Normally, neither the resins nor fillers have attractive colors by themselves. Pigments are added to impart color by their presence; dyes are added to color the resins or fillers.

A solid or liquid plasticizer is intended to make the product more flexible or less brittle. It may also help the flow of material in the mold.

Various lubricant or slip agents may be added to aid the removal of parts from molds or to make film surfaces slippery and prevent them from sticking together.

7.2.3 Plastic Products

A main purpose of the discussion so far has been to point out the many factors that affect the makeup of a plastic material. Thus, a manufacturer can vary the properties of a product greatly, and there is a large variety of different plastics available. Phenolic resins are examples. The two main phenolic resins are nominally phenol-formaldehyde and phenol-furfural compounds. Each maker of these resins has its own

Table 7-1. Thermoplastic plastics families

Class Name	Some Trade Names	Tensile Strength MPa (psi)	Maximum Service Temperature ° C (° F)	Average Relative Cost $/kg ($/lb)	Important Properties	Fabrication Processes (Raw Materials)	Typical Uses
Acrylonitrile-butadiene-styrene (ABS)	Abson®, Cycolac®, Marbon, Seilon	28–55 (4,000–8,000)	121 (250)	1.43 (0.65)	Hard, rigid, and tough; wide range of properties; weather degradable	Extrusion, molding, cold forming, calendering (resin and additives)	Auto trim, impellers, cases, piping, helmets, knobs, grilles, housings
Acetals: homopolymers copolymers	Delrin®, Celcon®	55–69 (8,000–10,000)	82–104 (180–220)	2.43 (1.10)	Strong and rigid with good moisture, heat, and chemical resistance; resists most solvents but not strong mineral acids	Extrusion, molding, forming, machining (resins, some with fiberglass and other fillers, and additives)	Gears, sprockets, casters, leaf springs, bearings, levers, fans, piping, valves
Acrylics: ethyl and methyl-methacrylate	Acrylite®, Lucite® Perpex, Plexiglas®	41–69 (6,000–10,000)	60–110 (140–230)	1.54 (0.70) molding grade	Moderate strength, soft, low heat resistance in most grades; good optically, clear to colored; good electrical resistance	Extrusion, molding, casting, machining, (molding compounds and cast sheets)	Lenses, signs nameplates, decorations, display novelties, dials, glazing, bottles, models
Cellulosics: cellulose acetate (butyrate), cellulose nitrate cellulose propionate, ethyl cellulose	Ethocel®, Lumarith®, Tenite®	10–59 (1,500–8,500)	49–93 (120–200)	2.21 (1.00)	Tough, easy to process; good transparency and surface gloss; many colors; moderate resistance to heat and etchants	Extrusion, molding, thermoforming, coating, machining, (molding compounds, films, sheets, rods, powders)	Knobs, handles, appliance housings and trim, glazing, packaging, billiard balls, pipe, steering wheels
Fluoroplastics[a], fluorocarbons[b], fluoropolymers[c]	Fluorthene, Halar®, Polyfluoron Teflon®, Tefzel®	17–45 (2,500 –6,500)	177–288 (350–550)	18.74 (8.50)	Outstanding inertness and chemical, electrical, temperature, and weather resistance; low friction; low strength but some reinforceable; tough at low temperatures	Extrusion, molding, coating, forming, dispersion casting, machining (granules, pellets, powders, and dispersions, plus fillers, films, sheets)	Nonstick, high-temperature coatings, bearings, seals, piping, electrical insulation, enamels, release surfaces, ablative shields

[a] tetrafluoroethylene (TFE), fluorinated ethylene-propylene (FEP), and perfluoroalkoxy (PFA)
[b] ethylene-tetrafluoroethylene (ETFE) and ethylene-chlorotrifluoroethylene (ECTFE)
[c] chlorotrifluoroethylene (CTFE), etc.

Table 7-1. Thermoplastic plastics families *(continued)*

Class Name	Some Trade Names	Tensile Strength MPa (psi)	Maximum Service Temperature ° C (° F)	Average Relative Cost $/kg ($/lb)	Important Properties	Fabrication Processes (Raw Materials)	Typical Uses
Ionomers	Iotek®	14–35 (2,000–5,000)	71 (160)	1.87 (0.85)	Light, tough, transparent, and flexible; not stiff and some creep	Extrusion, molding, thermoforming (resins and stabilizers)	Films, toys, containers, trays, wire insulation
Phenoxies: phenylene oxide base	Bakelite®, Noryl®	48–117 (7,000–17,000)	77 (170)	3.31 (1.50)	Good ductility, stability, and low-temperature properties	Extrusion, molding, thermoforming; foamable and some platable (various grades of resins, some reinforced)	Water-flow parts, electronic devices, auto trim, appliance housings and parts
Polyamides	Nylon®, Ultramid®, Versalon®	55–207 (8,000–30,000)	121–149 (250–300)	4.30 (1.95)	High strength, rigidity, and impact, temperature, electrical, and chemical resistant; absorb water; solvent softened	Extrusion, molding, sintering, forming, casting, coating, machining (solid and liquid resins with fillers and reinforcements)	Cloth, bristles, sutures, tubing, bearings, cams, gears, gaskets, insulation
Polycarbonates	Lexan®, Merlon®	62–72 (9,000–10,500)	121 (250)	2.76 (1.25)	High strength, ductility, rigidity, and electrical resistance down to –171° C (–275° F); transparent	Extrusion, molding, foaming, machining (resins and sheets)	Safety sheets, signs, lenses, covers, sight gages, globes and lighting aids, armor, bearing balls
Polyesters: polyterephthalates	Celanex®, Tenite®, Valox®	55–121 (8,000–17,500)	110 (230)	2.21 (1.00)	Good chemical, water, abrasion, and electrical resistance, tough	Extrusion and molding (resins; some reinforced)	Pumps, meters, gears, cams, rollers, electromechanical components
Polyethylenes	Alathone, Ethylux, Polythene, Spectex® Nomafoam® Foama-Spheres®	4–48 (500–7,000)	93 (200)	0.88 (0.40)	Tough to –98° C (–145° F); good chemical, moisture, and electrical resistance; low friction; most used plastic; many grades; flexible to rigid	Worked by all processes (resin with additives)	Housings, piping, ducts, bottles, pails, tanks, insulation, housewares, toys, coatings, films, packaging

Table 7-1. Thermoplastic plastics families *(continued)*

Class Name	Some Trade Names	Tensile Strength MPa (psi)	Maximum Service Temperature °C (°F)	Average Relative Cost $/kg ($/lb)	Important Properties	Fabrication Processes (Raw Materials)	Typical Uses
Polyimides	Gemon, Kapton®, Vespel®	69–172 (10,000–25,000)	316 (600)	26.46 (12.00)	Strong, rigid, and stable, with excellent heat, abrasion, creep, and radiation resistance; flexible at –268° C (–450° F)	Molding and sintering (powders, coatings, films, solid forms)	Valves, electrical insulation, hundreds of parts in every jet engine
Polypropylenes	Escon®, Propylux®, Tenite®	35–59 (5,000–8,500)	121 (250)	0.86 (0.39)	Chemical, moisture, and electrical resistance; special grades for impact strength and high- or low-temperature service	Extrusion, molding, laminating, coating, sintering (resin with additives)	Electrical equipment, hinges, piping, packaging, luggage, auto trim
Polystyrenes	Cerex®, Loralin, Lustron®, Styron®	14–55 (2,000–8,000)	60–79 (140–175)	0.79 (0.36)	Good electrical and stain resistance	Extrusion, molding, thermoforming, foaming (resin with additives)	Piping, dials, toys, battery boxes, dental plates, dinnerware, auto and appliance parts, lenses
Polysulfones	Udel®	69 (10,000)	149–260 (300–500)	7.94 (3.60)	Rigid and ductile to –96° C (–140° F); stable, and electrical resistance	Extrusion, molding, thermoforming (resin with additives)	Auto and electrical parts, housings, piping, cable insulation
Vinyls: polyvinyl chloride, acetate, etc.	Chemaco, Elvanol®, Saran®, Vinylite	7–48 (1,000–7,000)	60–104 (140–220)	0.88 (0.40)	Very flexible to rigid; good flame, electrical, chemical, oil, abrasion, and weather resistance in various grades; colorable and attractive; easy to process	Extrusion, molding, calendering, coating, casting, foaming (resin with additives, dispersions, sheets, films)	Floor and wall covering, upholstery, rainwear, house siding, tubing, toys, insulation, safety glass

formula. For instance, the phenol may be carbolic acid or any one of a number of related substances. In addition, a variety of fillers in various amounts may be added to the resin. The result is that phenolics are available with a wide range of properties; tensile strength is only one. It may be as low as 35 MPa (5,000 psi) in the ordinary run of molded parts, but can be made as high as 83 MPa (12,000 psi) where needed. Exceptional properties, of course, cost more, and the attempt of each supplier is to meet its customers' needs at the lowest cost.

Full specifications of the properties of plastics are tabulated in reference texts and handbooks. Only the chief features of major plastics will be described in this text.

7.2.4 Thermosetting Plastics

Thermosetting plastics are polymerized when molded or formed. A primary mixture is subjected to heat, pressure, or a catalyst, singly or in combination, over a time to enforce the links within and between the large molecules. The mixture first softens and can be forced into the shape desired, but then hardens permanently.

In this chapter, prices are given for large quantities of plastic materials as a basis for comparison. Their purpose is to indicate relative costs. They can be expected to change with economic and other conditions and quantities. Material cost is only part of the cost of the finished product. Processing cost is also a substantial item and is to a large extent based on time. In the case of thermosetting plastics, polymerization takes time, from a fraction of a minute in some molding operations to hours for some castings.

The six principal thermosetting plastic families are described briefly in Table 7-2. The phenolics are the oldest and most used because they serve many purposes at low cost. Other materials offer advantages for particular applications.

7.2.5 Thermoplastics

Thermoplastics are available as resins and compounds for molding as sheets, rods, tubes, and other forms, for fabricating in textile fabrics, as liquids for paints and adhesives, and as film and foil for packaging. A common way to classify the important commercial plastics is to divide them into families as shown in Table 7-1, but each family has many variations. For instance, one supplier offers over 400 different types of polyethylene. Compounds between different families are common.

Since thermoplastics are available in many grades, varieties, and combinations, the designer using plastic material must select the combination to suit requirements at lowest cost. The desired properties may be physical, mechanical, serviceable, or visual. Some materials have unique qualities. For instance, cellulosics offer high surface gloss; vinyls offer fluid and wear resistance with unusual flexibility. The polystyrenes are used more than any other for molded articles because of their good physical properties and low cost.

Low processing cost is a main advantage of the most used thermoplastics. Studies have indicated a ratio of two for total cost to material cost for injection molded plastics. In comparison, for zinc die castings ratios are reported to be from three to five. Processing cost is also important in the choice of a particular plastic. For example, polyethylene costs more in some grades, but it can be injection-molded faster and more economically than styrenes in some cases.

7.2.6 Elastomers (Rubbers)

Elastomers are polymers that in a primary state are tacky and flow readily at room temperature. To make them useful, their molecules are cross-linked at widely separated points (as a quasi-thermoset) into a network. Their molecules tend to curl up in random fashion, but when stretched, must act in a concerted manner. This property results from vulcanization in natural rubber. Originally, *vulcanization* was done by heating with sulfur, but today selenium, tellurium, and organic sulfur compounds also serve as vulcanizing agents. Other additives to rubber are substances to accelerate vulcanization, activators for the accelerators, antioxidation agents, plasticizers, reinforcing agents, stiffeners, fillers, and pigments or coloring agents to meet specific service conditions.

The main synthetic materials with rubbery properties of commercial importance are described in Table 7-3. Still others are available but less used. Some synthetic rubbers, like isoprene,

Table 7-2. Major thermosetting plastics

Class Name	Some Trade Names	Tensile Strength MPa (psi)	Maximum Service Temperature ° C (° F)	Average Relative Cost $/kg ($/lb)	Important Properties	Fabrication Processes (Raw Materials)	Typical Uses
Alkyds	Durez®, Plenco®, Plaston	21–62 (3,000–9,000)	149 (300)	1.76 (0.80)	Good electrical insulation, dimensional stability, and impact resistance	Molding (powders, liquids, soft sheets, ropes, logs, or slugs)	Electrical equipment
Allylics	Acme®, Dapon®, Plaskon®	28–55 (4,000–8,000)	177 (350)	8.60 (3.90)	High moisture and chemical resistance, stability, and dielectric strength	Molding and extrusion, lamination (powders, liquids, prepregs)	Electronic gears, lenses, laminates
Aminos: urea- and melamine-formaldehyde	Bakelite®, Beetle®, Melmac®	35–69 (5,000–10,000)	77–99 (170–210)	1.54 (0.70)	Colorful, hard; resist scratches, detergents, and many liquids	Molding and laminating (powders, granules, liquids, foams)	Tableware, distributor caps, countertops, appliance housings
Epoxies	Durez®, Hysol®, Polymeric®	35–207 (5,000–30,000)	260 (500)	2.09 (0.95)	Good electrical and mechanical properties, stable, resist heat and chemicals, strong adhesive	Casting, extrusion, molding, and potting (powders, liquids, foams)	Adhesives, tanks and enclosures, tools, dies
Phenolics: phenol-formaldehyde and -furfural	Bakelite®, Genal, Textolite	35–69 (5,000–10,000)	149–260 (300–500)	1.10 (0.50)	Rigid, stable, good electrical and chemical resistance, limited colors	Molding and casting (powders, pellets, solutions, impregnations)	Electrical gears, appliance parts, laminated panels, grinding wheel bonds
Polyesters	Dacron®, Mylar®	7–345 (1,000–50,000)	66–149 (150–300)	1.10 (0.50)	Make tough reinforcements; resist most solvents, acids, and bases	Molding, casting, laminating (powders, liquids, sheets, rods, and tubes)	Auto body parts, decorations, boats, luggage

Table 7-3. Major elastomers

Names	Properties*	Average Relative Cost \$/kg (\$/lb)	Features and Particular Uses
Natural rubber, natural polyisoprene (NR)	A:R, B:21 (3,000), C:7.5–8.5	1.32 (0.60)	Excellent physical properties; good resistance to cutting, gouging, and abrasion; low heat, ozone, and oil resistance
Isoprene, synthetic polyisoprene (IR)	A:R, B:17 (2,500), C:3.0–8.0	1.54 (0.70)	Same as natural rubber but requires less mastication; for auto tires, power belts, hoses, gaskets, seals, rollers
Government rubber-styrene or Buna S, styrene-butadiene (SBR)	A:R, B:1.7 (250), C:4.0–6.0	0.99 (0.45)	Good physical properties when reinforced; excellent abrasion and water resistance; not oil, ozone, or weather resistant
Butyl isobutylene isoprene (IIR)	A:R, B:19 (2,700), C:7.5–9.0	1.54 (0.70)	Excellent weather and heat resistance; low gas permeability; good chemical, ozone, and age resistance; fair strength and resilience; for tire inner tubes, steam hoses, and insulation
Chlorobutyl-, chloroisobutylene-isoprene (IIR modified)	A:T, B:19 (2,700), C:7.5–9.0	1.10 (0.50)	Properties similar to butyl with service temperatures to 204° C (400° F) and good oil resistance when blended; for inner tubes and curing bladders
Polybutadiene, *cis*-4 (BR)	A:R, B:4 (600), C:4.0–10.0	1.21 (0.55)	Overall properties like rubber and SBR but better abrasion and weather resistance, low temperature service and resilience; usually used in blends
Ethylene propylene (EPM) (terpolymer [EPDM])	A:R, B:7 (1,000) maximum, C:Poor	1.76 (0.80)	Good mechanical properties when reinforced; exceptional sunlight, oxygen, and ozone resistance; good electrical and temperature properties; for insulation, footwear, weather stripping
Neoprene, chloroprene (CR)	A:S, B:24 (3,500), C:8.0–9.0	2.31 (1.05)	Excellent ozone, heat, weather, and flame resistance, and mechanical properties; good oil and chemical resistance; for oil hoses, tank linings, and insulation
Buna N, nitrile, acrylonitrile-butadiene (NBR)	A:S, B:5 (700), C:4.5–7.0	2.54 (1.15)	Excellent oil and good chemical resistance; fair mechanical and poor low-temperature properties; for carburetor, gas tank, and pump parts, gaskets, printing rolls
Hypalon (HYP), chlorosulfonated polyethylene (CSM)	A:S, B:28 (4,000) maximum, C:maximum 6.0	2.43 (1.10)	Excellent ozone, weather and acid resistance, and color stability; fair oil and low-temperature resistance, serviceable to 121°C (250° F); for chemical and petroleum hoses, connectors, shoes, and flooring

Table 7-3. Major elastomers *(continued)*

Names	Properties*	Average Relative Cost $/kg ($/lb)	Features and Particular Uses
Urethane, polyester U (AU), polyether U (EU)	A:S, B:35 (5,000) and up, C:5.4–7.5	5.18 (2.35)	Exceptional abrasion, cut, and tear resistance; high strength, modulus, and hardness; good oxygen, ozone, and sunlight resistance, especially for vibration dampening and sound deadening; low moisture and heat resistance
Silicone rubbers (MQ), (PMQ), etc.	A:T, B:7 (1,000), C:1.0–5.0	10.47 (4.75)	Temperature −84 to 316° C (−120 to 600° F); high oxygen, ozone, and radiation resistance; high compression set; low strength, wear, and oil resistance; for insulation, seals, gaskets
Viton, fluorocarbon elastomers (FKM)	A:T, B:14 (2,000), and up	38.91 (17.65)	Temperature −40 to 316° C (−40 to 600° F); outstanding oil and chemical resistance, especially at high temperatures; good mechanical properties; for aircraft and industrial equipment
Acrylic rubbers, polyacrylate (ACM)	A:T, B:2 (300), C:4.5–7.5	3.53 (1.60)	Excellent ozone and oil resistance; poor water resistance; for seals, gaskets, hoses, and O-rings
Thermoplastic elastomers, thermolastic	A:S	3.97 (1.80)	Good elastic and mechanical properties; flexibility at low temperatures; processing is fast and at low cost; number of kinds with wide ranges of properties

* Code for designation of properties: A = service: R = no resistance to oils, S = specific resistance to oils, T = prolonged exposure to abnormal temperatures and compounded oils; B = pure gum relative mean tensile strength—MPa (psi); C = pure gum elongation at rupture, 100%.

are chemically and physically much like natural rubber and compete on a price basis. Others have properties that rubber lacks, like oil or temperature resistance, and serve where needed, but at higher cost.

Properties

The best known feature of elastomers is that they can be stretched to at least twice their original length. They usually do not conform to Hooke's law, as indicated by Figure 7-5, and return only approximately to original length. The loss in size is called *tension set* or *compression set,* according to the mode of loading. This is an indicator of how an elastomer will act in service. *Modulus* is the amount of load required to stretch a test piece to a given elongation. The modulus and the full strength vary with the hardness of an elastomer, as typified in Figure 7-5. Rubber can respond to flexion repeatedly; this property is needed for gaskets, printing rolls, upholstery, and hoses. The ability to give back energy upon release is called *resilience.* The high resilience of natural and some other rubbers is desirable in articles like golf balls. An elastomer of low resilience would be used for items such as bowling alley backstops. Another feature is impermeability to liquids and gases, in different degrees for different elastomers. This is desirable for rainwear, protective coatings of all kinds, and insulation.

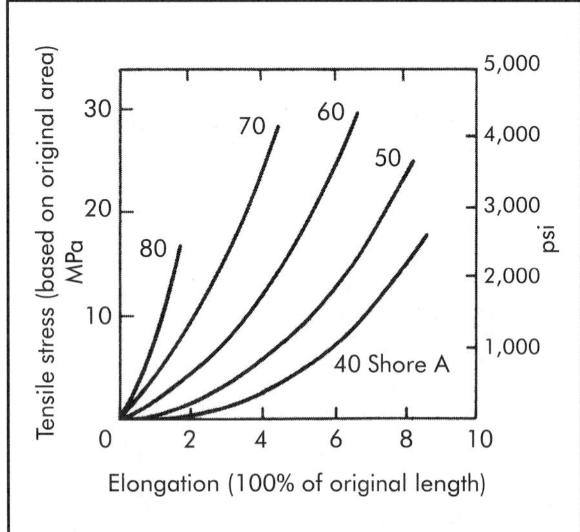

Figure 7-5. Typical stress-strain curves for an elastomer of various grades of hardness designated by Shore durometer numbers.

Reinforcing agents, such as carbon black, improve the strengths of most elastomers. Even so, the strength of rubbers is far below that of metals and many plastics. For high strength, rubber is coated on fabrics, cords, glass fibers, or wire—largely as flexible insulation against friction. Examples are automotive tires and conveyor belts.

The standard to measure the hardness of rubber is the Shore durometer, a simple and widely accepted pocket-size device. It is pressed against the rubber surface to apply an indenter and obtain a hardness reading. There are two scales: Shore A is based on a blunt-end indenter, and Shore D on a needle-point indenter. Readings are usually rounded to 10-point increments. Soft and medium rubbers are gaged on the A scale from 20 (softest) to 100 (hardest). The D scale is set to 100 on a glass surface; 50 on D is roughly the same as 100 on A.

Commonly, the properties of an elastomer can be varied by compounding and vulcanizing. Hardness is varied in these ways. Fillers and reinforcing agents can produce hardness of 50–90 Shore A in vulcanized rubbers. Natural and similar rubbers and neoprene are made harder with more sulfur. The hardest elastomers consist of a mixture of natural, styrene-butadiene, or acrylonitrile rubbers, with 32% sulfur. Other elastomers, like urethane, are made hard by varying their basic structures. The highest strength of a typical elastomer occurs at a 60 Shore A hardness as shown in Figure 7-5. Material in this range generally has the best balance of properties (strength, wear resistance, durability) and is commonly chosen for severe service, such as for tire treads.

Standard systems are in use for specifying properties of elastomers and matching them with service requirements. The most common is that of the American Society for Testing Materials (ASTM) and the Society of Automotive Engineers (SAE). By this standard, the properties of each elastomer are described by a line of numbers and letters. Other systems define the specifications for the aircraft industry, the military specifications of the armed services, and the Rubber Manufacturers' Association (RMA) standards.

7.2.7 Silicones

The *silicones* are compounds built around a basic silicon-oxygen unit in a manner similar to the way the organic compounds are formed from their basic carbon and hydrogen atoms. The silicone compounds can be polymerized in chains and rings to produce a large number of compounds. They are less active chemically than similar organic compounds and more resistant to heat. They stand temperatures in various applications to 427° C (800° F). Some are molded thermosetting plastics, others are varnishes for high-temperature electrical insulation, and still others are in the forms of high-temperature greases and oils. They are found in such products as waxes and polishes, paints, cosmetics, antifoaming agents, and dielectric fluids. Some properties of silicone elastomers are given in Table 7-3.

7.2.8 Adhesives

Adhesives have served to join weak materials like paper, plastics, and wood for many years. Discoveries of new plastics and elastomers have brought forth stronger adhesives, suitable for more materials and applications. This has led to the growing use of adhesives for joining metals to metals and other materials. Certain metal honeycomb structures, for instance, would not be possible without adhesives. Adhesive joints can be made strong enough for many purposes, do not require overheating the work, save weight, are relatively cheap, and can serve as sealers and insulators and add resistance to corrosion, vibration, and fatigue. Screws and rivets form projections on a surface and act as stress raisers; adhesive joints do not have these faults.

An adhesive joint must be designed properly. An adhesive does not act alone but with the materials it unites. No adhesive can bond anything to everything and stand up under all conditions. Whatever the joint, an adhesive must be selected to suit the materials to be joined and the temperatures, atmospheres, and other conditions under which it must hold. Adhesive joints must be properly sized. An adhesive bond is about one-tenth as strong as brazing on the same area. Thus, a welded, brazed, bolted, or riveted joint of relatively small area cannot be directly replaced by an adhesive. An even larger than proportional area is needed because strength is not uniform over an adhesive interface. A joint should be designed for straight shear or tension uniform over the whole area to avoid bending and peeling and stress concentrations.

Casein and natural glues are widely used, but thermosetting plastics, thermoplastics, and elastomers are becoming more important in critical industrial applications. Many formulations are available. Epoxides are popular because they do not shrink much on setting, are stable under a wide range of conditions, and adhere to many materials. Thermoplastics are not used for heavy, long-time loading because of creep. Acrylic adhesives offer a matchless combination of clarity and long life. Most contact adhesives are based on neoprene elastomers in a vehicle. Some adhesives are mixtures of different polymers to obtain properties for particular applications. Examples are epoxy phenolics or epoxy nylons that are among adhesives suitable for temperatures to below −240° C (−400° F). Silicones and polyimides make adhesives serviceable for short exposures to 482° C (900° F). Brittle ceramic adhesives based on glass frit stand up to about 816° C (1,500° F). Adhesives for particular purposes cost upward of several dollars per liter (quart). That usually means a fraction of a cent per square centimeter (square inch)—a small part of the total cost of the product.

Adhesives harden and develop cohesive strength by chemical curing, drying, or freezing. Chemical reactions to polymerize thermosetting plastics may be brought about by heat or a catalyst. Many plastics and elastomers may be dissolved in water or an organic solvent that is evaporated or diffused for drying. This method wets surfaces well and needs little or no heat. It is popular for bonding fabric to rigid surfaces. Thermoplastics may be heated for application and then cooled to set.

Adhesive joints are made in a number of ways. In any case, the surfaces to be joined must be clean. A cleaned surface primed with a thin film of adhesive solvent can be stored a long time until final adhesive application. Liquids or pastes are applied by brush, trowel, dip, flow, gun, or roller. Films cut to size are placed on the surface. Powders may be sifted over an area. A bonding layer 76–127 μm (3–5 mils) thick is best for most adhesives. Fixtures are commonly used to hold and press parts together while the adhesive dries, is heated, or cures. Heating is commonly done in ovens.

7.3 PLASTICS PROCESSING

There are two main steps in the manufacture of plastic products. The first is a chemical process to create the resin. The second is to mix and shape the material into the finished article or product.

Plastic objects are formed by compression, transfer, and injection molding. Other processes are casting, extrusion and pultrusion, laminating, filament winding, sheet forming, joining, foaming, and machining. Some of these and still others are used for rubber. A reason for the variety of processes is that different materials must be worked in different ways. Also, each method is advantageous for certain kinds of products. The principles of operation and merits of the processes will be discussed.

7.3.1 Compression Molding

In *compression molding*, a proper amount of material is squeezed by a *punch*, also called a *force,* into a cavity of a mold. The plastic is heated in most cases to between 121–260° C (250–500° F), softens, and flows to fill the space between the force and mold. The mold is kept closed for enough time to permit the formed piece to harden. This is done in a press capable of exerting 14–55 MPa (2,000–8,000 psi) over the area of the work projected on a plane normal to the ram movement, depending on the design of the part and the material.

Compression molding is mostly for thermosetting plastics that have to be cured by heat in the mold. Other methods are faster for large-quantity production of thermoplastics. Loose molding compound may be fed into a mold, but a cold-pressed tablet or rough shape, called a *preform,* may be prepared for more rapid production. For efficient heat transfer, parts should be simple with walls uniform and preferably not over 3.2 mm (.125 in.) thick. Even so, it may take several minutes to heat and cure a charge. This time may be reduced by as much as 50% by preheating the charge. To speed the process as much as possible, molding presses are usually semi- or fully automatic.

The three basic types of compression molds for plastics are shown in Figure 7-6. The force fits snugly in the *positive-type mold.* The full pressure of the force is exerted to make the material fill out the mold. The amount of charge must be controlled closely to produce a part of accurate size.

The force is a close fit in a *semipositive-type mold* only within the last 1 mm (.04 in.) of travel. Full pressure is exerted at the final closing of the mold, but excess material can escape, and the charge does not have to be controlled so closely. This type is considered best for large-quantity production of quality parts.

The force does not fit closely but closes a *flash-* or *overflow-type mold* by bearing on a narrow flash ridge or cutoff area. The amount of material does not need to be controlled closely, and the excess is squeezed out around the cavity in a thin flash. Some material is wasted and all pieces must be trimmed. Full pressure is not impressed on the workpiece. A mold of this kind is usually the cheapest to make.

Elaborate molds are used for certain purposes. *Multiple-cavity* or *gang molds* are economical for large-quantity production of bottle caps, for instance. A *subcavity gang mold* has a common loading chamber for a number of cavities. A *split-cavity mold* can be opened to remove a piece with undercuts.

Compression molding is mostly accomplished by heat, but some *cold molding* is done, particularly for refractory-type compounds because it is fast. The material is pressed to shape in the mold and then it is baked in an oven until cured. This method does not control size as well nor give as good a surface finish as hot molding.

Akin to the molding of thermosets is the *forging* of thermoplastics. A heated preform is placed in a die which is then closed to apply pressure to the material and make it fill the cavity. Injection molding is usually preferable for most thermoplastic parts, but forging can be done at lower pressures and with cheaper tooling. It has been

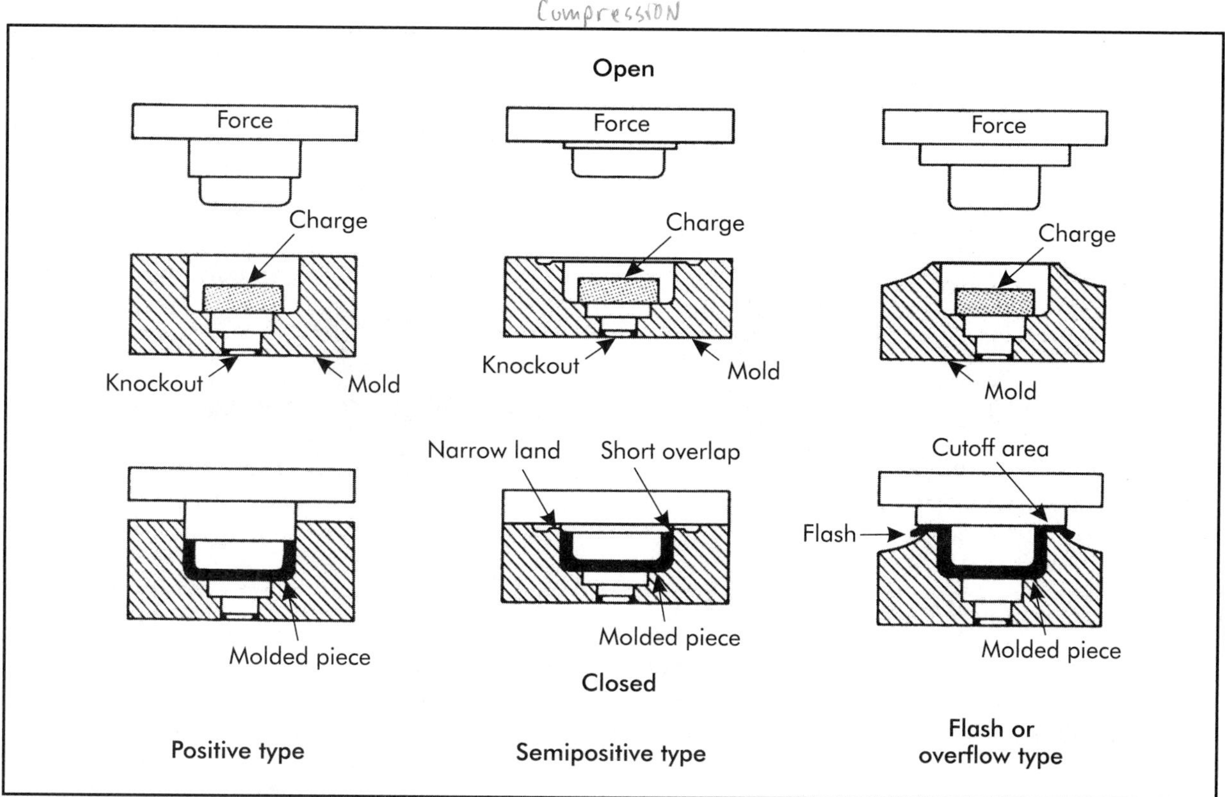

Figure 7-6. Three types of compression molds for plastics.

found economical for parts with sections thicker than 6.4 mm (.25 in.), small quantities, and materials that soften at temperatures too high for practical injection.

7.3.2 Transfer Molding

In *transfer molding*, also called *extrusion* or *gate molding*, the material is heated and compressed in a chamber and forced through a sprue, runner, and orifice into the mold cavity (Figure 7-7). The mold is costly, but closer tolerances and more uniform density can be held. Time is generally shorter for thick sections, and thick and thin sections and inserts can be molded with less trouble than in other molds. The reason is that the material enters the mold under pressures of 41–83 MPa (6,000–12,000 psi) and acts like a fluid.

7.3.3 Injection Molding

Three methods of injection molding plastic material are illustrated in Figure 7-8. The oldest is the single-stage plunger method shown in Figure 7-8(A). When the plunger is drawn back, raw material falls from the hopper into the chamber. The plunger is driven forward to force the material through the heating cylinder where it is softened and squirted under pressure into the mold. In a two-stage system, the material is plasticized in one cylinder, and a definite amount is transferred by a plunger or screw into a shot chamber from which a plunger injects it into the mold (Figure 7-8[B]). The single-stage reciprocating screw system has become more popular because it prepares the material more thoroughly for the mold and is generally faster (Figure 7-8[C]). As the screw turns, it is pushed backward and crams the charge from the hopper into the heating cylinder. When enough material has been prepared, the screw stops turning and is driven forward as a plunger to ram the charge into the die.

An injection molding machine heats to soften, molds, and cools to harden a thermoplastic material. Operating temperature is generally between 149–371° C (300–700° F) with

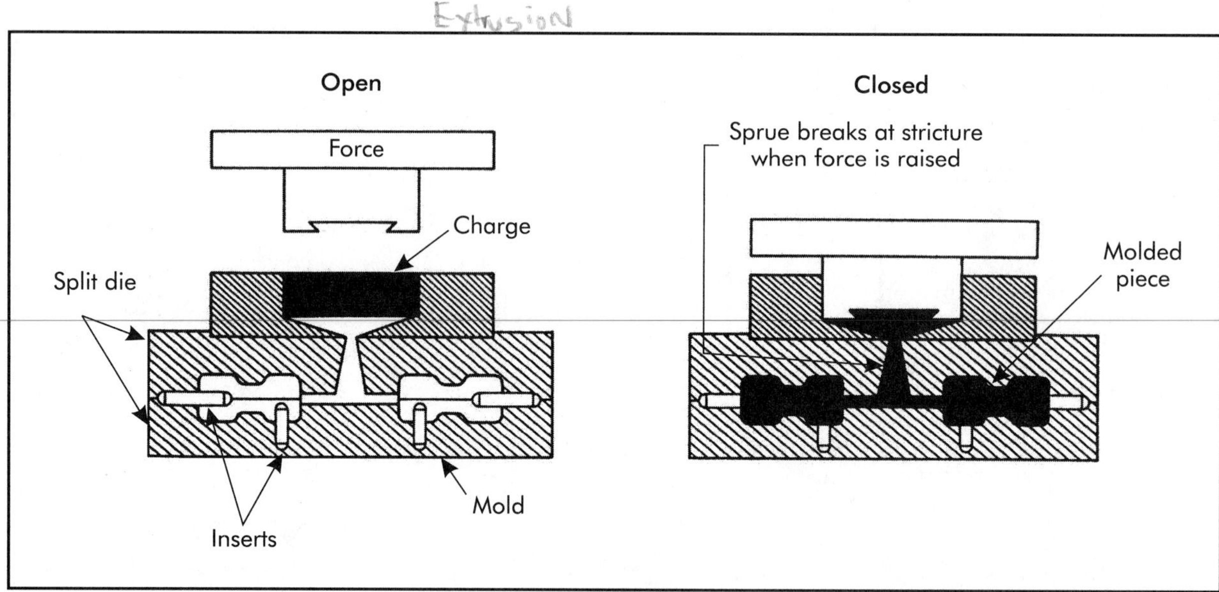

Figure 7-7. Transfer mold.

full pressure usually between 35–345 MPa (5,000–50,000 psi). The mold is water cooled. The molded piece and sprue are withdrawn from the injection side and ejected from the other side when the mold is opened. The mold is then closed and clamped to start another cycle. Thermosetting plastics can be injection molded but have to be polymerized and molded before they set in the machine. This may be done in a reciprocating screw machine where one charge at a time is brought to curing temperature. By another method, sometimes called *jet molding,* preforms are charged one at a time into a single-stage plunger machine.

Machines are available for molding sandwich parts. One cylinder and plunger injects a measured amount of skin material into the die, and then a second cylinder squirts the filler inside the mass. Then a final spurt from the first cylinder clears the core material from the sprue. The aim is to produce composites with optimum properties. Either case or core may be foamed.

In *liquid reaction molding* (LRM), two highly reactive liquids are injected in a mold where they combine to form the plastic product. Pressures are low, and large parts, like auto bumpers and body panels, can be molded without excessive die forces.

The capacity of an injection molding machine is designated by the maximum amount of material it injects efficiently at each shot and the force it can exert to lock the die. The force that opens the die is equal to the molding pressure times the projected area of the workpiece. A medium-size, reciprocating-screw injection molding machine rated at 113 g (4 oz) and 667 kN (75 tons) has a die area of 305 × 508 mm (12 × 20 in.), an injection pressure of up to 155 MPa (22,500 psi), a weight of 2,835 kg (6,250 lb), and a price of over $30,000.

Injection molding is a low-cost way of making thermoplastic parts in large quantities. With fully automated equipment, three to six shots a minute are common for moderate-size work. Intricate parts can be made to close tolerance with no need for secondary operations. Scrap loss should be less than 10%. However, molds cost from about $2,000 for a single-cavity die to over $50,000 for complex parts in multiple cavities. As few as 500 pieces of one part have been economically injection molded, but as many as 10,000 are required in some cases for economical operation. Injection molding of thermosets is on the average 25% faster than compression molding but requires careful attention.

7.3.4 Casting

Liquid resins are cast in molds made of relatively soft materials, such as rubber or plaster. The molds may be formed around a model eas-

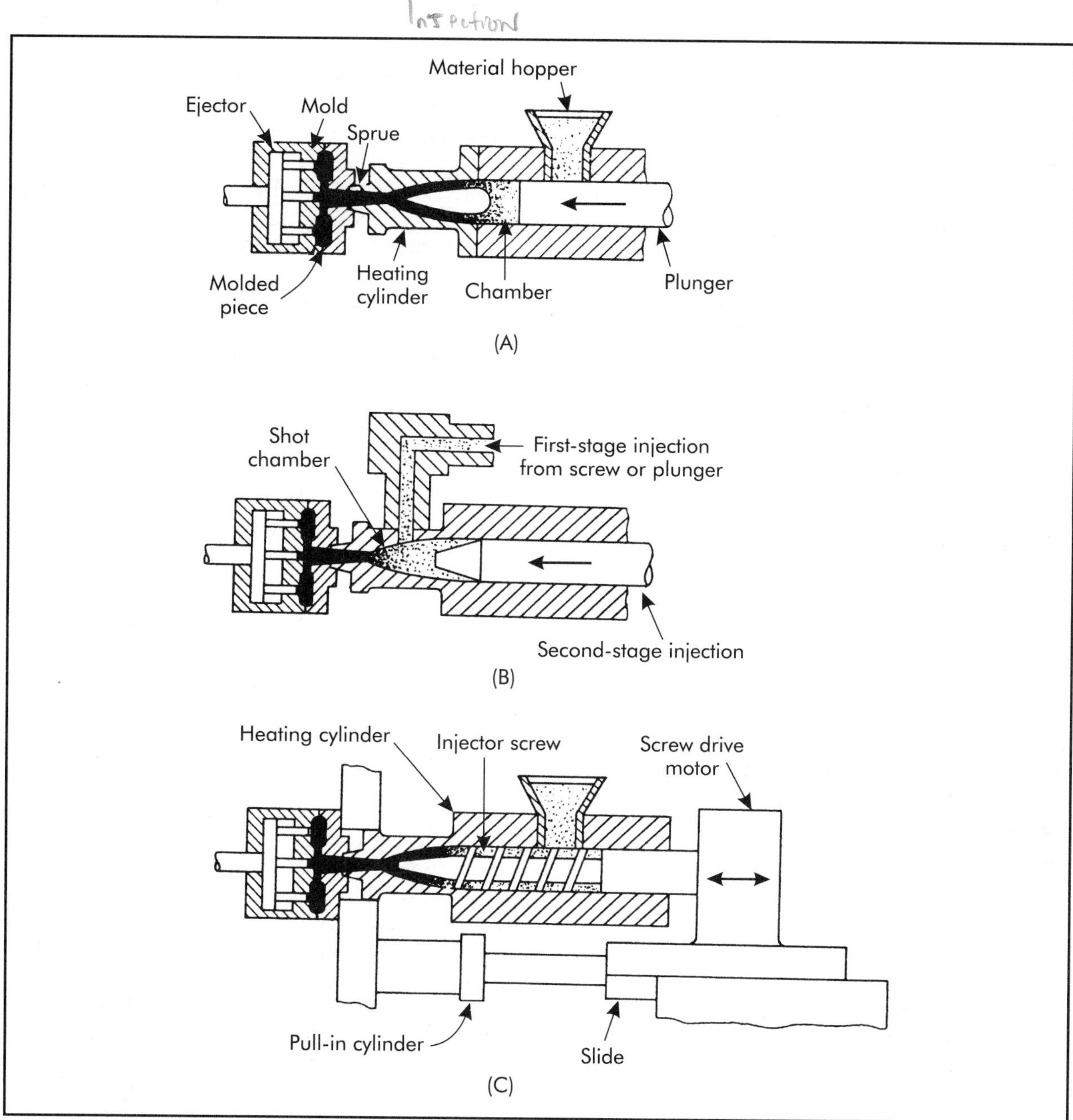

Figure 7-8. Injection molding systems: (A) conventional single-stage plunger type; (B) two-stage plunger or screw-plasticizer types; (C) single-stage reciprocating screw type.

ily shaped from materials such as wood, plaster, or metal. A catalyst is added to polymerize the resin, which is commonly heated in an oven for hours or days at about 66–93° C (150–200° F) to harden. Equipment cost is low, but the process is slow. Ornaments, prototypes, dies, and encapsulated electrical parts are examples of cast plastic products.

7.3.5 Extrusion and Pultrusion

Thermoplastics and elastomers are extruded into successive strips of uniform sections, such as rods, tubes, or angles, by the intermittent action of a plunger in a cylinder. A continuous method is depicted in Figure 7-9(A). The material drops from a hopper into a heated cylinder and then is pushed through the die by a screw.

This is faster and cheaper than molding. Insulated cable and wire commonly has its covering extruded around it as it is pulled through an extrusion die.

Strong and stiff reinforced thermosets are pultruded as shown in Figure 7-9(B). Reinforcing material (for example, glass fiber or paper) is pulled through a resin bath, forming and squeezing rolls, and a long heated die in which the plastic is cured and hardened. A typical pultrusion system can only process about 16 kg/hr (35 lb/hr) as compared to an average of 91 kg/hr (200 lb/hr) by extrusion because of the time needed to cure the thermoset.

7.3.6 Foams

Cellular plastic foams are made by chemically or mechanically expanding resins. Their structure provides thermal insulation, buoyancy, cushioning, and light weight. Strength-to-weight ratios can be two to five times those of conventional structural metals. Cells may be opened or closed and parts may be molded with smooth skins. Plastic foams may be flexible or rigid; dense or open. They may be foamed in place, sprayed on a surface, and then expanded (as on tanks), molded, extruded, or cast, and are available in stock shapes.

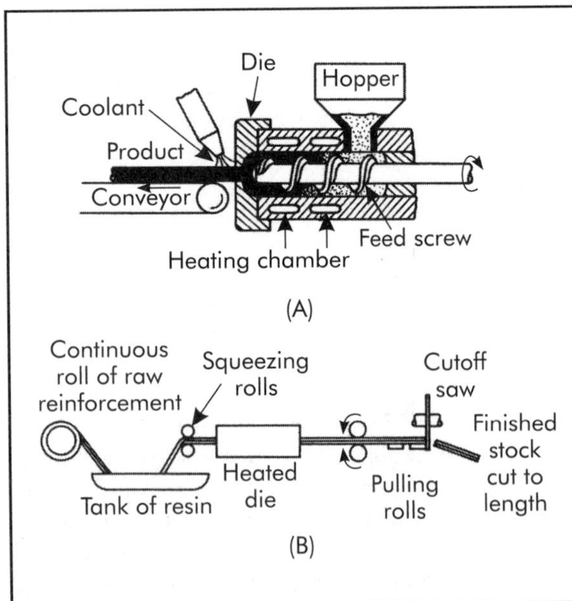

Figure 7-9. (A) Extrusion of plastic material; (B) pultrusion.

Plastics of all classes may be foamed. Depending upon the material, foaming is done by: (1) injecting a gas under pressure into a soft plastic mass; (2) adding a low-boiling-point solvent or chemicals and heating to release vapor or gas; (3) mechanical aeration or frothing by mixing, stirring, etc.; (4) including chemicals with the components of a plastic to form gas when mixed; and (5) mixing fine metal particles and gas with a plastic mass under pressure to cause bubbles to form on the metal particles when the pressure is released.

7.3.7 Laminates and Reinforced Plastic Molding

Many thin-walled objects, parts, and structures are made by mixing and laminating plastic resins or films with reinforcing materials. Two kinds of products are: (1) *high-pressure laminates*, and (2) *low-pressure* or *reinforced plastic moldings*.

High-pressure laminates are available commercially as sheets, rods, and tubes in standard sizes and are fabricated in special shapes. Several trade names are Formica®, Micarta®, and Lamicoid. Reinforcement materials include paper, cotton or glass cloth, asbestos, and nylon. The resins ordinarily used for impregnation are phenolics, melamines, silicones, and epoxies. The material is cut to size, arranged in layers, compressed at over 7 MPa (1,000 psi), and heated to around 149° C (300° F) to harden the resin (Figure 7-10). Operations on flat and curved sheets are depicted in Figures 7-10(A) and (B). Plastic films are laminated to metals in sheets or rolled strips.

Reinforced plastic moldings include such products as storage bins, loudspeaker horns, machinery housings, truck cabs, and aircraft panels. The most common reinforcing material is glass as fabric or fibers, but others are asbestos, boron, cotton, and nylon fibers. Mostly thermosetting, but some thermoplastic resins are used.

The simplest way of preparing reinforced plastic shapes is by *contact laminating* or *molding*. Layers of reinforcing material are placed by hand over a low-cost form or mold and resin is brushed or sprayed on each layer. Commercially available sheet-molding compound (SMC) containing glass fibers also may be spread over the form.

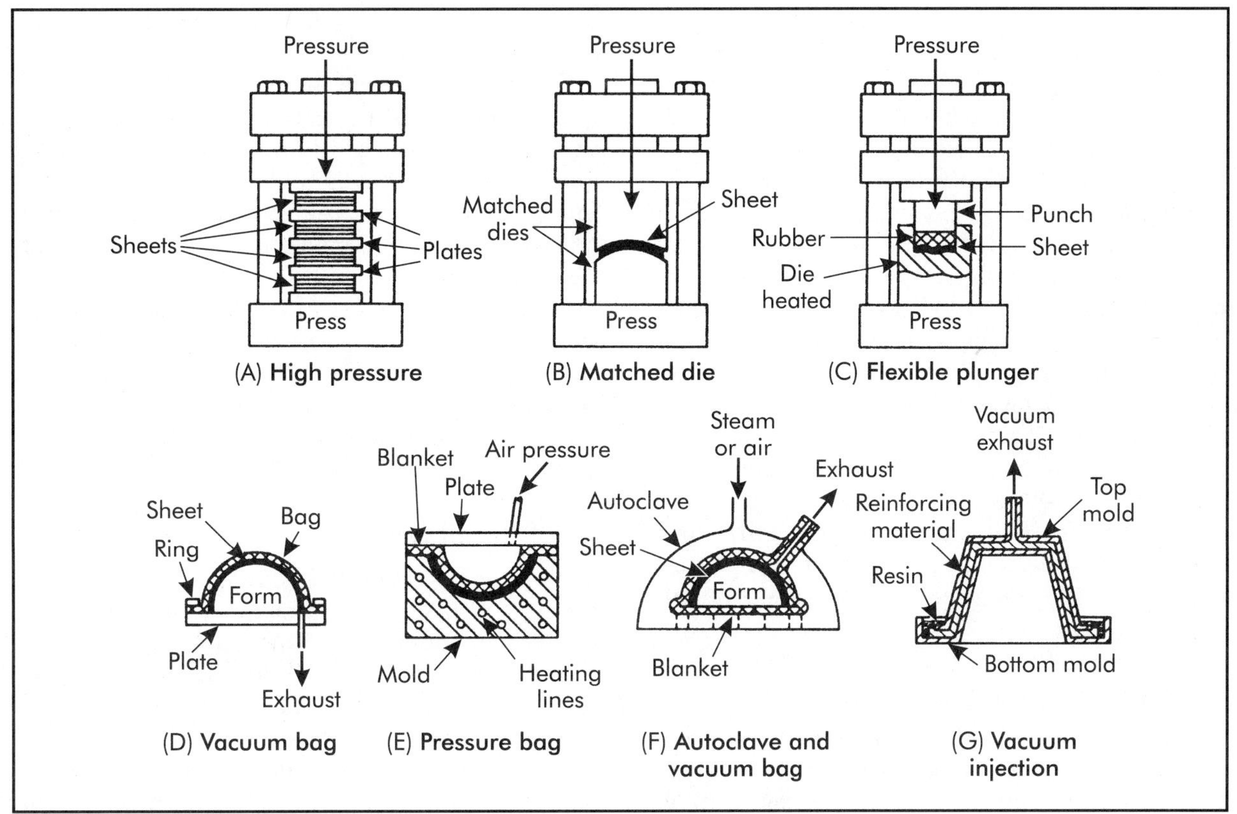

Figure 7-10. Methods of molding laminated and reinforced plastics.

For quantities of parts, molding may be done by *preforming* or *premixing*. To preform, chopped glass roving, plastic resin, and fillers are blown together into a die, mold, form, or screen by a spray gun. Good strength and appearance result because the material is evenly distributed. Structures as large as 4.9-m (16-ft) boats and swimming pools are fabricated in this way because the only limitation is the size of the form. In contrast to preforming, the resin, fibers, and fillers may be premixed and spread over a form and cured.

After reinforced plastic moldings have been formed, the resin may be catalyzed to harden without heat or pressure, or it may be baked, or the molding may be finished in a vacuum bag or matching die operation. Reinforced plastic moldings may be cured under pressure in *vacuum bag* or *pressure bag molding*, as depicted in Figures 7-10(D) and (E). A workpiece also may be placed between matched dies in a bag. Pressure is applied by exhausting the air from the bag and is limited to atmospheric pressure. Up to about 2.1 MPa (300 psi) may be applied by a pressure bag. Either arrangement may be put in an autoclave with hot air or steam to increase pressure and temperature as indicated in Figure 7-10(F). These operations are quite slow; some take hours. Tooling cost is low; one estimate is \$269–\$431/m^2 (\$25–\$40/ft^2) of workpiece area. The methods were developed and are widely used in the aircraft industry for radomes, wing sections, missile noses, and sonar domes. They are not considered economical for more than 1,200–1,500 parts of one type.

In the *vacuum injection process*, illustrated in Figure 7-10(G), reinforcing material is clamped between two matching nonporous molds. The resin is put in a trough around the bottom and is drawn up by a vacuum through the material to the top mold. The saturated molding is held in position until it sets.

Reinforced plastic moldings may be finished by rigid or flexible punches and dies in presses as indicated in Figures 7-10(B) and (C). Rigid dies are usually heated by steam or electricity to about 121° C (250° F). Pressures may be up to 7 MPa (1,000 psi) but usually are much lower. Steel or cast iron dies are fastest and last longest, but cheaper kirksite dies serve for moderate quantities. The cycle time in a press may be several minutes but, in some cases, has been reduced to a fraction of a minute, depending upon the material, workpiece size, and the amount of automation. This is still longer than the few seconds required to form a part from sheet metal. However, sheet metal dies must be more durable and are costly. As an example, tooling for a plastic sports car body is reported to have cost $500,000 as compared to an estimated $4.5 million for a sheet metal body. Parts are made as plastic moldings in quantities of approximately 100,000 pieces and from sheet metal in larger quantities.

7.3.8 Forming Plastic Sheets

Thermoplastic sheets are softened by heating and *formed* or *thermoformed* into a large variety of thin-walled articles, such as display packages, bowls and trays, refrigerator door liners, lighting fixtures, safety helmets, and luggage. The material is held in the desired shape until it cools and becomes rigid. Sheets may be formed by pressing between molds or mechanical bending as shown in Figure 7-11(A). Mechanical pulling or stretching is done in some cases. Another group of methods is based upon blowing or drawing the sheets by air pressure or vacuum. In this way, certain shapes may be free-formed as exemplified in Figure 7-11(B). This avoids marring the surface, an advantage in products such as clear windshield canopies for aircraft. Plastic sheets are also blown or drawn into shapes against molds by *drape-vacuum forming* and *snap-back forming* as illustrated by Figures 7-11(C) and (D).

A major advantage of sheet forming is that the sheets may be predecorated, coated, or preprinted for effects such as wood grain, lettered, or metallic finishes. The formed parts may have thin walls or large areas that are difficult to mold in other ways. Sheet forming is not hard on molds, which may be made of plastic, wood, plaster, masonite, soft metals, or cast iron. They seldom cost more than a few hundred dollars and can be made quickly.

Thermoforming is done manually for small quantities (such as prototypes) and is a slow process. The process is automated for large-quantity production. One example is that of one-piece, double-cavity refrigerator liners turned out on a four-station machine at the rate of about 35 parts/hr.

Some plastic material in sheets can be cold-formed rapidly by metalworking methods with suitable modifications. One example is a small container, press-drawn at the rate of 1,500 parts/min. When injection molded in a 64-cavity die, only 384 parts/min were turned out.

7.3.9 Shell Molding

Powder molding of hollow parts is done by fusing a thin layer of powdered thermoplastic resin in a heated mold. This may be done by exposing the mold over a fluidized bed with particles charged at up to 80,000 V and thus strongly attracted to the mold. Unfused resin is dumped out of the mold. By another method, called *rotational molding*, the powder is placed in a mold that is heated while being rotated around two axes (at 90°) at the same time. This distributes the material uniformly and fully closed shells may be produced. The mold is cooled and the workpiece is removed. Another method, called *slush* or *dip molding,* employs a liquid dispersion, such as a vinyl plastisol, instead of a powder. Powder molding and the like produce thin-walled parts like toys, tanks, and other containers. Walls are stress-free and crack resistant because there is no molding pressure and no flow of material. Wall thickness is uniformly controllable within about ±10%. Operation time is more, but mold cost is less than for injection molding. The break-even point is reported at somewhat less than 10,000 parts.

Blow molding is performed in four steps as illustrated in Figure 7-12. First, a heated length of thermoplastic tube (called a *parison*) is placed on an air nozzle between the halves of the open mold. The parison is extruded in place on some automatic machines. Second, the mold closes and pinches shut the open end of the parison. Then air is blown in to expand the parison to the walls of the mold, which are usually cooled to set the

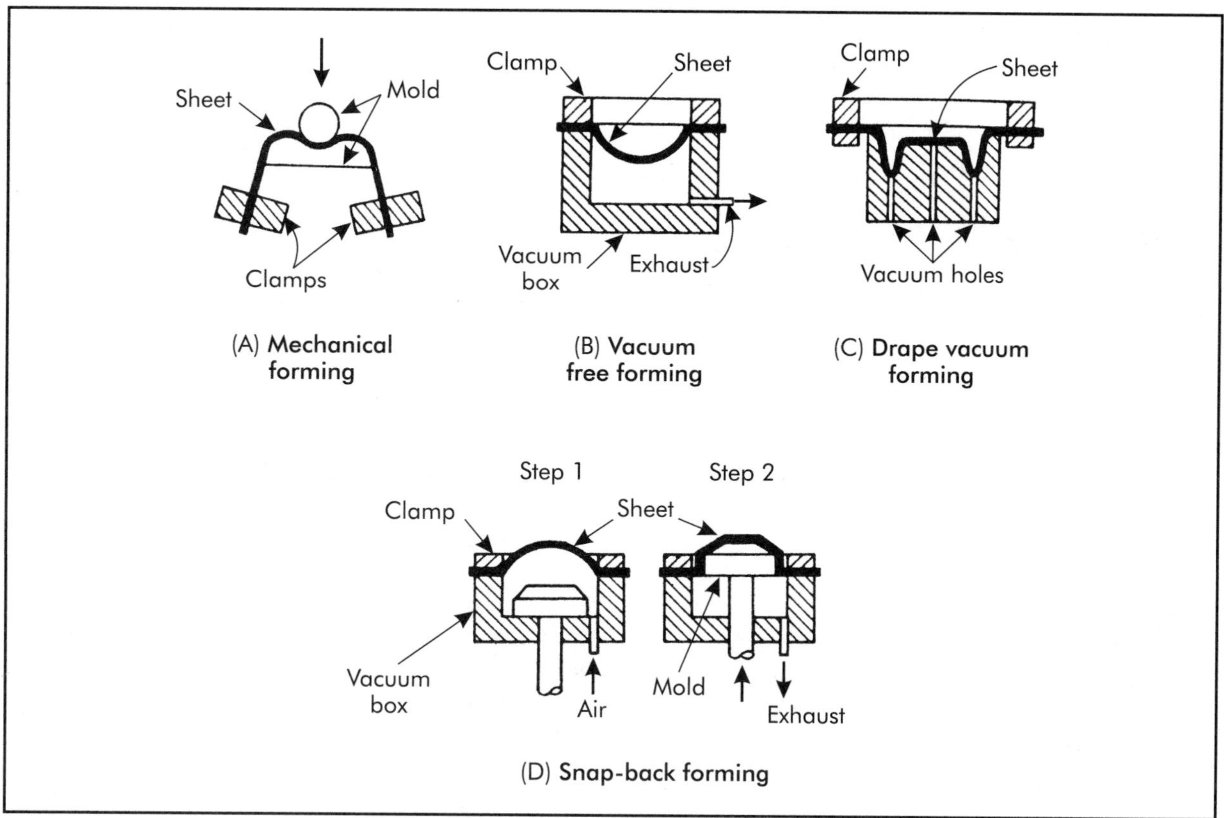

Figure 7-11. Typical methods of forming plastic sheets.

plastic. Last, the mold is opened, and the finished product is ejected. The blowup ratio may be as high as 6:1 but is commonly 3:1.

Among the best known products produced by blow molding are rigid plastic detergent bottles and flexible squeeze bottles. Most thermoplastics can be blow molded, and objects as large as 189-L (50-gal) drum liners have been produced. A major drawback is that walls cannot be held uniformly because different parts are stretched by different amounts. Also, walls cannot be thickened for ribs or bosses. Blown surfaces do not have a high gloss because pressure in the mold is not over about 700 kPa (100 psi). The process is readily automated and fast. Dies cost one-half to one-fifth the cost of those for injection molding because they do not need features such as cores, runners, or gates.

7.3.10 Joining Plastics

The four basic methods of joining plastics are: (1) cementing with solvent; (2) bonding with an adhesive; (3) thermal and ultrasonic welding; and (4) mechanical fastening.

Only certain soluble thermoplastic materials can be cemented with a solvent. The strength of the joint is comparable to that of the parent material. All plastics may be adhesive bonded to metals and other materials as well as to each other. A large number of solvents and adhesives are available and the properties of the joint depend upon the material selected.

All except highly inflammable thermoplastics can be softened or melted for welding. Softened surfaces may be pressed together or staked, by mashing a projection, to interlock one part with another. Melted joints may have filler plastic added and coalesce on cooling and hardening. Common welding means are: (1) hot gas or air; (2) a heated tool; (3) induction heating; and (4) friction (ultrasonic or other rubbing).

Plastic parts may be joined with other parts by means such as screws, bolts, nuts, rivets, swaging or peening, and press or shrink fitting.

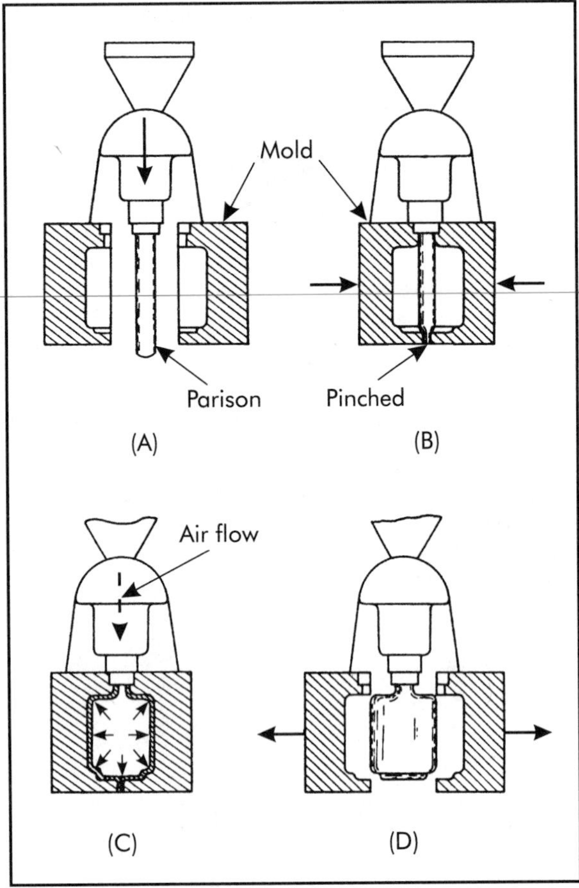

Figure 7-12. Steps in blow molding.

In designing serviceable joints, allowance must be made for the weakness and localized stress susceptibility of plastics.

7.3.11 Machining Plastics

Molding and forming are the usual ways of making plastic parts, but these methods require costly molds and forms. Machining from standard shapes may be cheaper for small and moderate quantities; the break-even point has been reported as high as 15,000–60,000 pieces with automated machining. Some parts can be made most economically by slicing them off of extruded shapes. Hard-to-mold forms, such as threads and dimensions with small tolerances, often can be most cheaply obtained by machining. Some cutting to remove excess material in flash, runners, sprues, or flanges is done on almost all molded and formed plastic parts.

Most of the techniques and equipment described in this text for machining metal are also applicable for plastics. However, there are some principles of machining that apply to plastics alone because the properties of plastics are different from those of metals. Because there are many plastics, there are many specifications for cutting plastics. This is not the place to list them, but specifications for tool shapes, speeds, feeds, etc. for particular applications are given in reference books and handbooks. What is discussed in this text are the important principles that explain why and how plastics must be treated differently from metals when cut.

The properties of plastics that determine machinability are: (1) they do not readily conduct but are easily affected by heat; (2) some contain abrasive fillers; (3) they are mostly soft and yielding; and (4) some are quite brittle though soft.

Thermoplastics soften, lose shape, become gummy, and clog cutters; thermosetting plastics deteriorate and char at high temperatures. The heat generated is not conducted away rapidly by a plastic when it is cut, so good results depend upon practices conducive to cool cutting. This calls for tools with keen cutting edges and smooth polished faces. For sawing, a band saw is preferable to a circular saw because the teeth on a long band have more time to cool. Air, water, and oil coolants are commonly used for cutting plastics.

Plastics with abrasive or high-strength fillers (composites) wear tools rapidly and require cemented carbide, sintered oxide, or even diamond tools. Care is required to avoid breakout of the reinforcing fibers of the composites.

Cutting tools withstand high speeds with most plastics because the materials are soft. As for tool shape, a relatively obtuse cutting angle is usually necessary to keep the tool from digging into the plastic, just as for brass and copper. In fact, a common practical rule is to set up to cut a plastic as though cutting brass or copper if specific information is not available. Then make adjustments during the operation to reach optimum conditions.

Machine tools can work to as close tolerances in cutting plastics as with any material, but in most cases, a small tolerance is futile because the plastic will not retain its size. Many plastics are yielding and change appreciably in size under cut; some absorb moisture and swell. In some cases, plastic pieces are chilled to make them rigid for cutting.

Some plastics are brittle and chip readily when cut. For these, tools must be sharp and cuts light. Shearing is commonly performed with sharp dinking dies or rule cutters. In some cases, plastics are heated before being cut.

7.3.12 Rubber Processing

Raw rubber is composed of long molecules of high molecular weight and is, therefore, quite springy and resistant to being worked. First, these molecules must be broken up so the rubber can be molded and formed. For this purpose, crude rubber may be extruded in a *plasticator*, masticated by revolving beaters in a *Banbury mixer*, or plasticized between rolls in a *rubber mill*. The last may be performed alone or after one or both of the other treatments.

In the rubber mill, the rubber is squeezed between two rolls turning toward each other on the entering side. One revolves up to one-third faster than the other and induces a severe shearing as well as compressive action in the rubber. The sheet coming out of the rolls is commonly fed back for a time into the entering mixture for thorough blending. In the mixer and mill, the rubber is impregnated with the substances added to it for the final product. Much the same processes are used for some plastics; one example is the calendering of decorative vinyl sheets and films. Most synthetic elastomers do not respond to mastication or working and must be synthesized to an amenable state.

The second step is to mold or form the rubber as desired and vulcanize it. The sulfur in vulcanizing forms cross-links between the rubber molecules. This makes the rubber elastic, nonsticky, strong, and more resistant to heat and solvents. The structure of most of the synthetic elastomers is set up when they are polymerized, so vulcanization is not necessary. Actual forming and molding are done in a number of ways common to plastics. Rubbers are extruded, compression and injection molded, and formed. Rubbers and thermoplastics are calendered between rolls into films (0.25 mm [10 mils] and under) or sheets. Also the materials may be pressed or wiped into the voids of fabrics. Some rubber and thermoplastic products are made on forms by dipping into, spraying, or electrodepositing latex or gel.

Thermoplastic elastomers can be molded when heated but act as cross-linked elastomers on cooling. Such a material has short sections of stiff monomers in the long elastomeric chains. These sections become mobile at high temperatures.

7.4 DESIGN OF MOLDED PLASTIC PARTS

A molded plastic part should be designed to serve its purpose at the lowest possible cost. Many objects are made from plastics for appearance, and art should have a place in their design. Still, the part must be serviceable and economical to make.

Many of the same rules and their reasons given for casting and molding of metal apply to the molding of plastics. Taper is essential to remove parts from molds; 3° is considered standard for plastics. Thick sections should be avoided because they take more material, cool slowly, and retard molding. Instead, ribs, beads, and flanges should be used to add strength where needed. Transition should be gradual between thick and thin sections to promote uniform cooling and avoid stress. Adequate radii and fillets should be provided to eliminate sharp edges and corners wherever possible. This makes parts stronger and more durable, cuts mold cost, and helps the material to flow properly in the mold. One rule is that a radius or fillet should be at least 25% of the wall thickness and never less than 0.8 mm (.03 in.). Mold parting lines should be placed to assure low mold cost, simple flash removal, and easy part ejection.

The molding of plastics is different from metals in some ways. Thick and thin sections under one surface should be avoided because the thicker section will shrink more and causes noticeable heat sinks or dimples. If these cannot be avoided over ribs or bosses, a pattern may have to be put on the surface to disguise the defects. Plastic walls must not be too thin or weak; generally not less than 1.5–2.3 mm (.06–.09 in.). Holes should be cored wherever possible to avoid machining. A through hole is better than a blind hole because its core pin can be supported at both ends. Reference texts and handbooks give many pointers on the design of holes and openings, threads, knurling, lettering, inserts, and undercuts.

The surface finish and color specified for a plastic part influence all costs. First they enter into the selection of the material. The specified surface finish dictates the degree of finish required and thus the cost of the mold. Combinations of materials and colors can be obtained by various mold design techniques and operating procedures. Most plastic parts have flash, runners, or sprues that must be removed, usually by secondary operations. Much thought is given to designing parts so that this finishing can be easily done without detracting from the appearance of the product. A typical solution is to mold a bead around a piece at the parting line. Flash can be trimmed from the bead, which serves as a guide, without marring large smooth areas of the piece. Where extra effects are required, masking, buffing, painting, and vapor metal coating operations may become necessary.

Practical tolerances for molded plastic parts depend largely upon size. Dimensions of 25.4 mm (1 in.) or less on small parts can be held to within ±51 μm (±.002 in.). For larger dimensions, not over 203 mm (8 in.), realistic tolerances may be ±10–20 μm/cm (±.001–.002 in./in.) across die parting lines. In any event, no tolerances closer than really needed should be specified.

The design for production of plastic products is not a simple matter. Even the most experienced plastics engineers commonly take the precaution of first building a single-cavity experimental mold to perfect a process before putting a part into production with a multiple-cavity mold. Many viewpoints help, and the designer of plastic products can obtain valuable aid by consulting the supplier of the plastic material and the molder.

7.5 CERAMICS

Some engineering applications require the use of materials with some properties that are not present in traditional materials. These properties may include high temperature strength, high wear resistance, low thermal conductivity, high hardness, and other desirable thermal and electrical properties. Ceramics have such general properties that make them useful in many industrial products. The list of engineering and industrial products that utilize ceramics is exhaustive. A partial list of these products includes:

- glass products, such as light bulbs, bottles, and window glass;
- cutting tool materials, such as tungsten carbide and aluminum oxide;
- fiberglass and fiber optics used for insulation applications and communication networks;
- abrasives, such as aluminum oxide and silicon carbide;
- clay products, such as bricks and tiles;
- pottery, china, and porcelain products; and
- construction materials, such as cement and concrete.

In this chapter, ceramics are generally divided into two main categories. These are *commercial ceramics*, also called *traditional ceramics,* and *fine ceramics*, also called *engineering* or *new ceramics.* Examples of commercial ceramics include pottery products, tiles, and abrasives. Fine ceramics are used for many automotive, aerospace, and other manufacturing applications as cutting tool materials and fiberglass and ceramic insulators.

7.5.1 Structure

Ceramics have a crystal-like structure that is generally more complex than that of other materials. Ceramic components are characterized by strong bonding, namely *covalent* and *ionic bonding*. This results in high hardness and significantly higher thermal resistance than most metallic components. Ceramics may take the form of a single crystal or a polycrystalline structure. A major factor influencing the mechanical and physical properties of a polycrystalline structure is the grain size. Finer grain sizes normally result in higher strength and toughness. *Glass* is one kind of ceramic material that assumes an amorphous structure, rather than a crystalline form. The properties of glass may be varied by adding other ceramic materials to the glassy or amorphous structure, such as aluminum oxide or magnesium.

Ceramic materials are generally classified by the raw materials of which they are composed:

oxide, carbides, nitrides, cermets, and *glass*. The principal raw materials for ceramics are *clay* and *silica*. The oldest is clay, which consists of aluminum silicate bonded by weak ionic layers of silicon and aluminum. Silica, such as quartz, is the principal component of glass and other ceramic materials.

Oxide ceramics have substantially improved toughness and strength characteristics in comparison to other ceramics. *Aluminum oxide*, also known as *alumina*, is the most common form of oxide ceramic. It has high strength. The properties of alumina can be controlled, since its manufacture is totally synthetic. Because of these controlled physical and mechanical properties, it is most suitable for cutting tools, abrasive wheels, and other thermal insulators. *Zirconium oxide*, also called *zirconia*, is another oxide ceramic with better wear and corrosion properties and low thermal conductivity. Zirconia is better suited for the manufacture of engine components, such as valves and electronic insulators.

Carbide ceramics include *tungsten carbides, titanium carbides*, and *silicon carbides*. Tungsten and titanium carbides are mostly used in manufacturing applications as cutting tool materials and die components. Silicon carbides are mainly used in the manufacture of grinding wheels as abrasives. They have higher resistance to wear and corrosion, a low friction coefficient, and maintain strength at elevated temperatures.

Nitrides are an important class of ceramics that are hard and brittle. Compared to carbides, they are less resistant to heat at elevated temperatures. Important nitrides include *silicon nitrides, boron nitrides,* and *titanium nitrides*. Silicon nitrides provide high thermal conductivity, low thermal expansion, and high resistance to thermal shock and creep at high temperatures. Silicon nitrides are mostly used in applications with high temperature exposure, such as gas turbines and engine components.

Boron nitride is an extremely hard ceramic material that is mostly used for cutting tools and abrasives. Compared to diamond, the hardest known substance, boron nitride is not as hard, but constitutes the second hardest substance. While boron nitrides are generally used for cutting steels, diamond is mostly applied in nonsteel machining applications.

Cermets are ceramic-based materials in which ceramics are bonded with a metallic structure. Cermets can be regarded as a composite material combining different structures of ceramic materials with metals, produced by several powder metallurgy techniques. Therefore, they provide many desirable characteristics, such as toughness, good thermal and corrosion resistance at high temperatures, and high structural stability. Applications for cermets include cutting tools, dies for powder metallurgy and sheet metal fabrication, and other applications requiring hardness, wear resistance, and high temperature.

7.5.2 Clay Products

Because of their excellent characteristics, ceramics are widely used in different applications requiring brittleness, high hardness and thermal resistance at elevated temperatures, low thermal expansion, low electrical and thermal conductivity, high wear and corrosion resistance, and other physical and mechanical properties. Clay products include a wide variety of ceramic-based products that are commonly used in many commercial and industrial applications. Perhaps the oldest applications of ceramics are pottery, stoneware, china, and similar products. Other clay products include building brick, roof tiles, and clay piping, formed primarily by shaping and/or molding low-cost clays and firing at low temperatures. Another clay-based ceramic product is porcelain, providing good electrical and magnetic properties.

7.5.3 Refractory Materials

Widely used in furnaces, material-melting crucibles, and similar industrial applications, refractory materials provide desirable characteristics, such as high heat resistance, thermal insulation, high strength, and high resistance to chemical interaction with metals. Aluminum oxide, or alumina, combined with other oxides, is a widely used refractory material possessing high temperature resistance. It is used for applications such as thermal and electrical insulation.

7.5.4 Glass

Glass is a nonmetallic compound that is cooled off at a very high rate not allowing time for crystals to form. Silica is the main ingredient in glass, adding to its unique properties, such as resistance to corrosion and chemical effects. Properties of glass, however, can be modified by the addition of oxides, thus changing its composition. The amount and types of additives, also called modifiers, depend on the application. Compared to metals and plastics, glasses have a lower coefficient of thermal expansion. In addition, glasses have low thermal conductivity and high resistance to electric shock and chemical reaction. Glasses are widely used as optics because of their optical properties, such as refraction, absorption, and transmission.

Glasses are used in many products, including soda-lime glass, window glass, drinking glasses, light bulbs, glass fibers, laboratory glasses, and optical glass.

A combination of glass and some other oxides produces a glass ceramic, providing more strength and the combined properties of all elements. Glass ceramics are used for heat exchangers, cookware, and other electric and electronic applications.

7.5.5 Cermets

Cermets may be considered composite materials. They combine ceramics (various oxides) and a metallic bond. This provides the desirable characteristics of ceramics, such as high temperature resistance, together with metallic properties, such as toughness and ductility. Depending on the type and amount of oxides, cermets have many applications that include cutting tools, mechanical sealing, and dies for metal forming and sheet metal operations.

7.5.6 Mechanical and Electrical Applications

Porcelain is one of the ceramic materials that is being used successfully in the electrical and electronic industries. It has high electrical resistance and magnetic properties that make it suitable for certain applications.

Other mechanical, electrical, and electronic applications of ceramics include:

- microelectronic chips;
- transmission components;
- fiber-optic communication lines;
- electrical insulation;
- spark plugs;
- electronic components;
- gas turbines;
- heat exchanger components;
- automotive industry applications, such as valves, rocker arms, and cylinder liners; and
- heat engines.

7.5.7 Ceramic Cutting Tools

Because of high strength and toughness, resistance to elevated temperature, as well as wear and corrosion resistance, ceramics have many applications as cutting tools. They are used in the production of cutting tool materials and abrasives for grinding and other machining operations. Ceramics are also used for the dies for sheet metal operations, wire drawing, and dies for powder metallurgy.

7.6 QUESTIONS

1. What advantages do plastics offer as a class?
2. Describe the differences in chemical composition and physical properties of thermoplastic and thermosetting resins.
3. What kinds of substances are added to plastic resins and for what purposes?
4. Discuss the relative merits of the major thermosetting plastics.
5. Discuss the relative merits of the major thermoplastic materials.
6. How is rubber obtained and made serviceable?
7. Discuss the relative merits of the principal synthetic elastomers.
8. How do the silicones differ from organic materials and what makes them useful?

9. How are adhesives used in metal fabrication?
10. Describe compression molding of plastics and the three basic types of molds used.
11. How is injection molding performed and what are its advantages and disadvantages?
12. How are plastics extruded?
13. What are "plastic foams" and how are they made?
14. What are "high-pressure laminates" and how are they made?
15. Describe the principal ways of making reinforced plastic moldings and their relative advantages.
16. What is "filament winding" and what are its advantages?
17. How are parts made from plastic sheets and when is it economical to do so?
18. Describe "blow molding" and its advantages and limitations.
19. What are the common methods for joining plastics?
20. What are the properties of plastics that affect machinability? What are some of the resulting practices?
21. How are rubber products made?
22. What are the major considerations in the design of plastic parts?
23. What are "ceramics"? Differentiate between the properties of ceramics and those of metals.
24. List some of the different physical characteristics of ceramics.
25. What is meant by "atomic bonding" in ceramics?
26. What are "refractory materials"? List some of their applications.
27. What is the main ingredient in glass?
28. Describe the major mechanical and electrical applications of ceramics.
29. List some of the applications of ceramics as cutting tools.
30. What are "cermets"?
31. How can ceramics be made tougher?

7.7 PROBLEMS

1. A case for an electrical instrument can be made from either an amino urea compound by compression molding or from cellulose acetate butyrate by injection molding. The volume of the article is 148 cm^3 (9 $in.^3$). The urea compound costs 0.12 cent/cm^3 (1.97 cent/$in.^3$). It takes 90 sec to mold one piece in a mold that costs $1,500. The cellulose compound costs 0.18 cent/cm^3 (2.95 cents/$in.^3$). A piece can be molded from it in 15 sec, but the injection mold costs $4,500. Time is worth $12/hr. When should each material and process be selected?
2. A rectangular plaque 76 × 127 mm (3 × 5 in.) is to be compression molded from a phenolic molding compound. A mean pressure is required. What force in kN (tons) should the press be able to exert to process four pieces in a gang mold?
3. A nylon gear may be injection molded or machined from barstock. A die for injection molding costs $4,800, material costs $0.0656/piece, and operation costs $0.0372/piece. The stock for machining costs $0.263/piece and operation time costs $1.14/piece. What is the cost for each method for the following numbers of pieces? (a) 1,000; (b) 3,000; (c) 6,000.
4. An acetal flanged bushing made in different ways is subject to the following costs. Assume that only one lot is to be made. For what quantities should each method be used?

Factors	Injection Mold	Conventional Machining	Automatic Machining
Die or special tools	$5,400	–	$90
Setup	–	–	$22
Material/piece, cent	0.033	0.204	0.1734
Labor/piece, cent	0.030	1.043	0.036

7.8 REFERENCES

Chanda, M. and Roy, S. 1990. *Plastics Technology Handbook.* New York: Marcel Dekker.

Harper, C. 1992. *Handbook of Plastics, Elastomers, and Composites,* 2nd Ed. New York: McGraw-Hill.

Musikant, S. 1991. *What Every Engineer Should Know About Ceramics.* New York: Marcel Dekker.

——. 1980. "Fillers: A Bigger Bargain for Improving Resins." *Modern Plastics,* April: 84.

Plastics–Military Standardization Handbook. 1975. MIL-HDBK-700 A. Washington, DC: U.S. Government Printing Office.

"Reinforced Plastics: the Composites Lead the Way." 1981. *Manufacturing Engineering,* July: 56.

Rubin, L.L., ed. 1990. *Handbook of Plastic Materials and Technology.* New York: John Wiley.

Schwartz, M., ed. 1992. *Handbook of Structural Ceramics.* New York: McGraw-Hill.

——. 1985. *Engineering Applications of Ceramic Materials.* Metals Park, OH: American Society for Metals.

Society of Manufacturing Engineers (SME). 1998. *Plastic Injection Molding* video. *Fundamental Manufacturing Processes* series. Dearborn, MI: SME.

——. 1998. *Plastic Blow Molding* video. *Fundamental Manufacturing Processes* series. Dearborn, MI: SME.

Vaccari, J.A. 1981. "Stamping Plastics Almost Like Metal." *American Machinist,* June: 131.

——. 1981. "Winding Plastics for New Jobs." *American Machinist,* May: 125.

Wood, A.S. 1980. "For Really Better Parts: New SCM Technologies." *Modern Plastics,* January: 56.

Metal Casting Expendable Molds

8

Sand Casting, 1540
—Biringuccio's Pirotechnia, Egypt

Chapters 8 and 9 are devoted to the foundry processes and include the common ways of producing castings. Although these two chapters do not cover the casting of metal in detail, they illustrate the more important principles of the field.

Founding, or *casting*, is the process of forming objects by pouring liquid or viscous material into a prepared mold or form. A *casting* is an object formed by allowing the material to solidify. A *foundry* is a collection of the necessary materials and equipment to produce a casting. Practically all metal is initially cast. The ingot from which a wrought metal is produced is first cast in an ingot mold. A *mold* is the container that has the cavity (or cavities) of the shape to be cast. Liquids may be poured; some liquid and all viscous plastic materials are forced under pressure into molds.

Founding is one of the oldest industries in the metalworking field and dates back to approximately 4000 B.C. Since this early age, many methods have been employed to cast various materials. In this chapter, sand casting and its principles receive first attention because they are most used; over 90% of all castings are sand castings. Sand casting is best suited for iron and steel at their high melting temperatures but the process is also predominantly used for aluminum, brass, bronze, and magnesium. In Chapter 9 other processes of commercial importance are covered, specifically those for nonferrous metals using permanent molds.

The elements necessary for the production of sound castings are considered throughout this chapter. These include molding materials, molding equipment, tools, patterns, and melting equipment. These basic ingredients must be combined in an orderly sequence to produce a sound casting.

8.1 SAND CASTING PRINCIPLES

Castings have specific important engineering properties; these may be metallurgical, physical, or economic. Castings are often cheaper than forgings or weldments, depending on the quantity, type of material, and cost of patterns as compared to the cost of dies for forging and jigs and fixtures for weldments. Where this is the case, they are the logical choices for engineering structures and parts.

Properly designed and produced castings do not have directional properties. No laminated or segregated structure exists as it does when metal is worked after solidification. This means strength, for instance, is the same in all directions, and this characteristic is especially desirable for items such as gears, piston rings, and engine cylinder liners. The ability of molten metal to flow into thin sections of complicated design is a very desirable characteristic. Cast iron is unique in that it has good dampening characteristics, which are desirable in bases for machine tools, engine frames, and other applications where minimal vibration is important.

8.1.1 The Behavior of Cast Metal

When molten metal is poured into a mold, the casting begins to cool inwardly from all bounding surfaces because the heat can flow only outwardly through the mold. The metal on the surface is more or less chilled because at first the mold is relatively cool. If the chilling is severe, the surface may be appreciably hardened. Under usual conditions, a fine, close-grained structure occurs near the surface, and coarser grains occur toward the center where cooling is slower. If a section is thick, enough metal may be withdrawn by contraction from the center before it cools to leave a void or cavity as indicated in Figure 8-1(A). Such a defect in the casting may be avoided by providing a supplementary mass of metal, called a *riser*, adjacent to the casting as shown in Figure 8-1(B). The purpose of the riser is to feed liquid metal by gravity into the body of the casting to keep it full. The riser is cut off after the casting has cooled.

Thin sections cool more rapidly than thick ones. One result is that thin sections benefit more from a "quench effect" and are likely to be stronger and finer grained. Figure 8-2 shows this effect on a Class 40 gray cast iron. At the other extreme, if a section is too thin, metal flowing through the narrow passage may be frozen before it has a chance to fill in the wall completely. The practical lower limit of section thickness depends upon the design of the casting and the fluidity of the metal. Iron can be conveniently handled and cast appreciably above its melting temperature. It is commonly cast in sections as thin as 3.18 mm (.125 in.). Steel melts at a much higher temperature than iron (Chapter 3), and a minimum thickness of 4.78 mm (.188 in.) is recommended. Phosphorus increases the fluidity but weakens iron, so less phosphorus and higher pouring temperatures are sometimes preferred. Aluminum may be cast with walls as thin as 3.18–4.78 mm (.125–.188 in.).

Sections of different thicknesses cool at different rates. This leads to difficulty if a casting is not designed with uniform sections throughout. Walls that shrink at different rates pull at each other and set up residual stresses. The situation can be eased by providing gradual tapers or changes in thickness where sections of different sizes must meet.

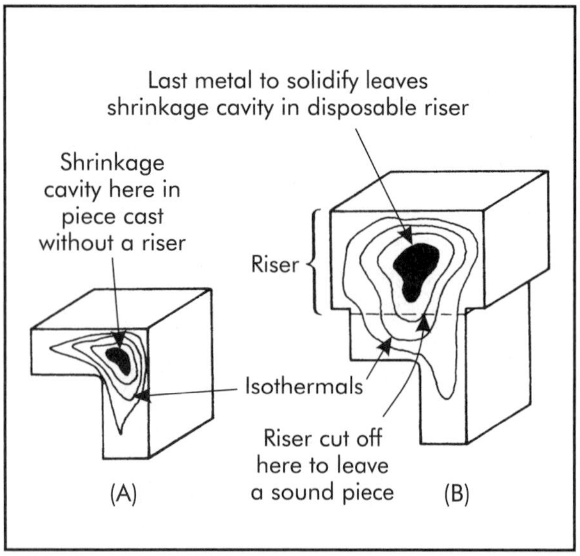

Last metal to solidify leaves shrinkage cavity in disposable riser
Shrinkage cavity here in piece cast without a riser
Riser
Isothermals
Riser cut off here to leave a sound piece
(A)
(B)

Figure 8-1. Example to illustrate the purpose of a riser on a casting.

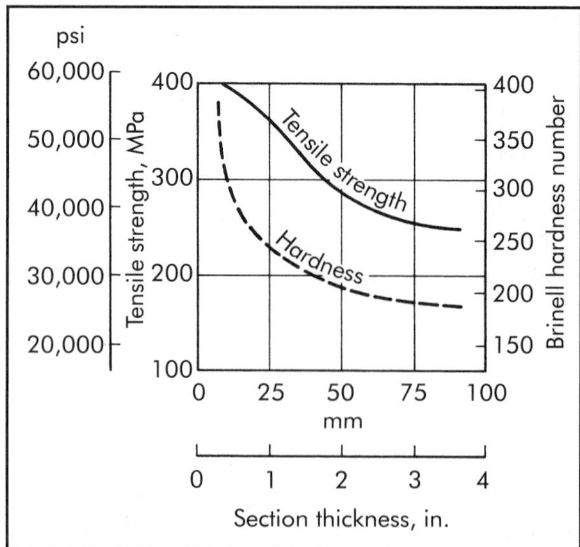

Figure 8-2. Strength and hardness of different thicknesses of a class 40 gray cast iron.

A thick section may be located where it cannot be readily fed by a riser, particularly where it must be fed through thin sections that solidify first, thus developing voids or tears. Metal concentrations or *hot spots* of this kind are indicated in Figure 8-3. The correction is in the design of the casting to make the sections uniform. A remedy may be to chill the metal at the hot spot to make it freeze before the adjoining sections, which can be fed more directly as they cool.

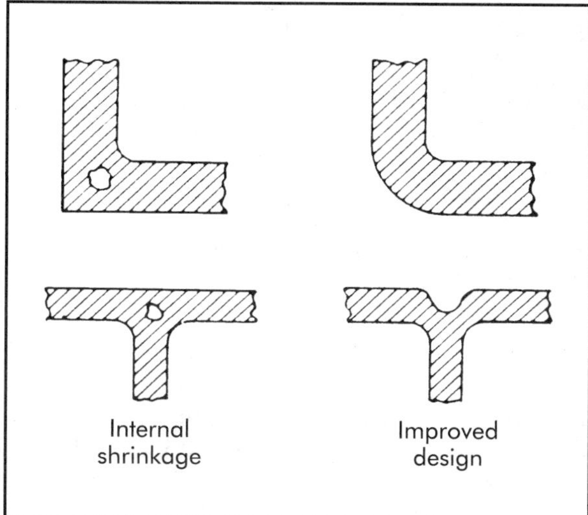

Figure 8-3. Examples of internal shrinkage in castings.

8.1.2 The Mold and Its Components

Good castings cannot be produced without good molds. Because of the importance of the mold, casting processes are often described by the material and method employed for the mold. Thus sand castings may be made in: (1) green sand molds, (2) dry sand molds, (3) core sand molds, (4) loam molds, (5) shell molds, and (6) cement-bonded molds. The major methods of making these molds are called: (1) bench molding, (2) machine molding, (3) floor molding, and (4) pit molding.

In producing a sand mold, the molder's skill is of great value. A molder must know how to prepare a mold with the following characteristics:

1. The mold must be strong enough to hold the weight of the metal.
2. The mold must resist the erosive action of the rapidly flowing metal during pouring.
3. The mold must generate a minimum amount of gas when filled with molten metal. Gases contaminate the metal and can disrupt the mold.
4. The mold must be constructed so that any gases formed can pass through the body of the mold itself, rather than penetrate the metal.
5. The mold must be refractory enough to withstand the high temperature of the metal and strip away cleanly from the casting after cooling.
6. The core must collapse enough to permit the casting to contract after solidification.

A *flask* is a wood or metal frame in which a mold is made. It must be strong and rigid to not distort when it is handled or when sand is rammed into it. It must also resist the pressure of the molten metal during casting. Pins and fittings align the sections of a flask. They wear in service and must be watched to avoid mismatched or shifted molds.

A flask is made of two principal parts, the *cope* (top section) and the *drag* (bottom section). When more than two sections of a flask are necessary to increase the depth of the cope and/or the drag, intermediate flask sections known as *cheeks* are used.

Figure 8-4 is a diagram of a typical mold and its principal parts. The features and functions of the parts are more fully explained later in this chapter.

Gates, Risers, and Chills

Gates, risers, and chills are closely related. The function of a mold's gating system is to deliver the liquid metal to the mold cavity. The riser's function is to store and supply liquid metal to compensate for solidification shrinkage in heavy sections. The function of the chill is to cause certain sections of a casting to solidify before others, often to properly distribute the supply of metal from the risers. A good gating system may be nullified by a poor riser concept. Improper use of chills can cause the scrapping of well-gated and properly fed castings.

Gates. An example of a gating system in a mold is shown in Figure 8-5. The gating system must: (1) introduce the molten metal into the mold with as little turbulence as possible,

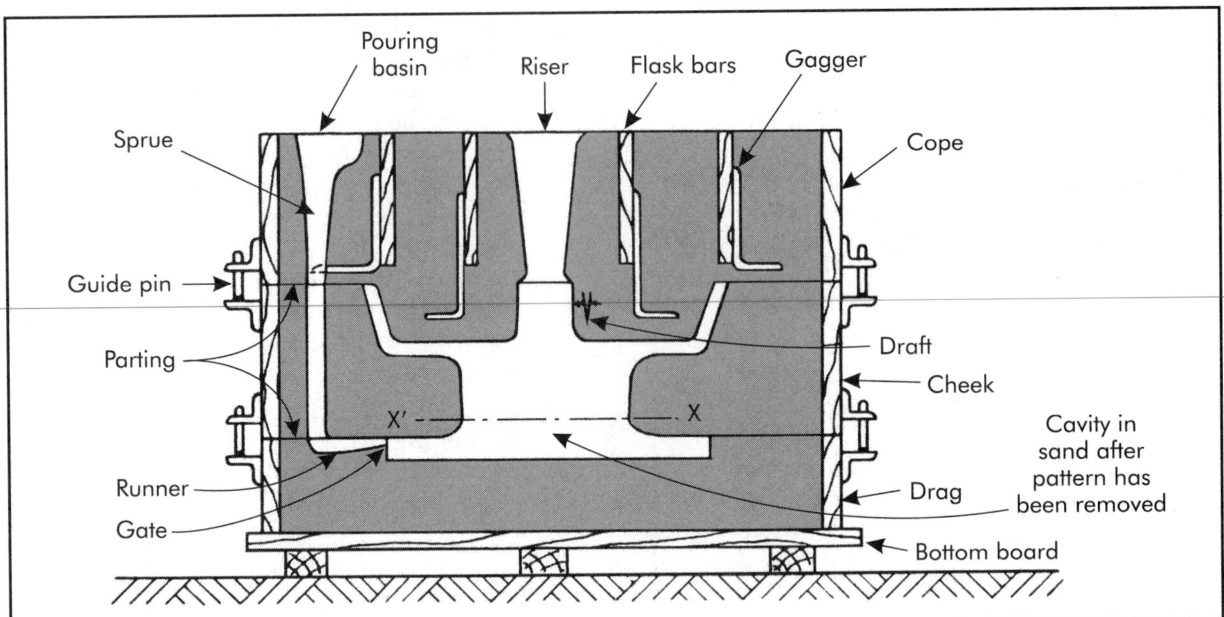

Figure 8-4. Cross-sectional view of a three-part sand mold with the parts labeled. The line X'-X indicates parting in the pattern.

(2) regulate the entry rate of the metal, (3) permit complete filling of the mold cavity, and (4) promote a temperature gradient within the casting to help the metal solidify with the least conflict between sections.

For good gating, the sprue should be tapered with the larger end receiving the metal to act as a reservoir. Generally speaking, a round sprue is preferred for diameters up to about 20 mm (≈ .75 in.); larger sprues are often rectangular. There is less turbulence in a rectangular sprue, but a circular sprue has a minimum surface exposed to cooling and offers the lowest resistance to flow.

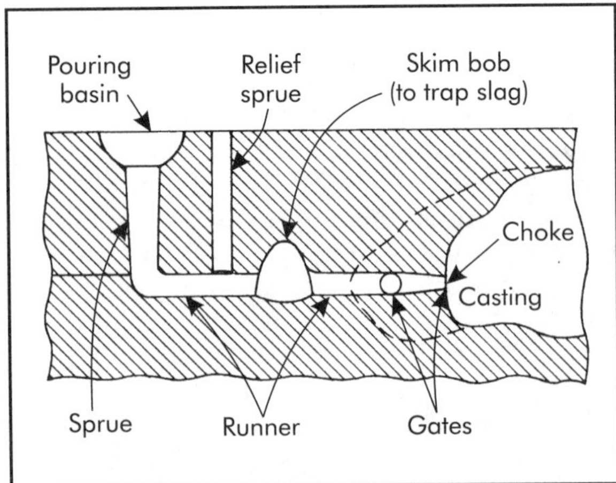

Figure 8-5. Example of a gating system.

Gating systems with sudden changes in direction cause slower filling of the mold cavity, are easily eroded, and cause turbulence in the liquid metal resulting in gas pickup. In particular, right angle turns should be avoided.

A definite relationship must exist between the sizes of the sprue, runners, and in-gates to realize the best conditions for filling the mold. The cross section of a runner should be reduced in area as each gate is passed, as indicated in Figure 8-5. This helps keep the runner full throughout its entire length and promotes uniform flow through all of the gates. As another consideration, the rate at which metal can flow into the mold should not exceed the ability of the sprue to keep the entire gating system full of liquid at all times.

The gating system should be formed as part of the pattern whenever possible. This allows the sand to be rammed harder and helps prevent erosion and washing as the metal flows into the mold.

Several in-gates rather than one help distribute the metal and fill the mold quickly, reducing the likelihood of overheating spots in the mold.

In-gates should be placed in positions where they will direct the metal into the mold along natural channels. If metal is directed against the mold surface or cores, burning is likely, and loose sand may be washed into the casting. The opening of an in-gate into a mold should have as small an area as possible, except in cases where the gates are through side risers. An in-gate may be reduced in area or "choked" where it enters a mold cavity. This holds back slag and foreign material. However, the area at the entrance should not be so small as to cause a shower effect. This may give rise to turbulence resulting in excessive oxidation of the metal.

Types. The three main types of gates are: (1) parting, (2) top, and (3) bottom, as illustrated in Figure 8-6.

The *parting gate* between cope and drag is the easiest and fastest for the molder to make. Its chief disadvantage is that the metal drops into the drag cavity and may cause erosion or washing of the mold. In the case of nonferrous metals, this drop aggravates the dross and entraps air in the metal, which makes for inferior castings.

Top gates are used for gray iron castings of simple designs, but not for nonferrous alloys since they have a tendency to form excessive dross when agitated. An advantage of top gating is that it is conducive to a favorable temperature gradient; a big disadvantage is that of mold erosion.

A *bottom gate* offers smooth flow with a minimum of mold and core erosion. Its main disadvantage is that it creates an unfavorable temperature gradient. The metal is introduced into the bottom of the mold cavity and rises quietly and evenly. It cools as it rises, and the result is a condition of cold metal and cold mold near the riser and hot metal and hot mold near the gate. The riser should contain the hottest metal in the hottest part of the mold so it can feed metal into the mold until all the casting has solidified.

Wherever possible, gating through side risers should be done. Gating directly into the casting results in hot spots because all the metal enters the casting through the gate and the sand around the gate becomes hot. Cooling in that area is retarded. Unless risers are provided to feed those localities with molten metal, shrinkage cavities or defects result.

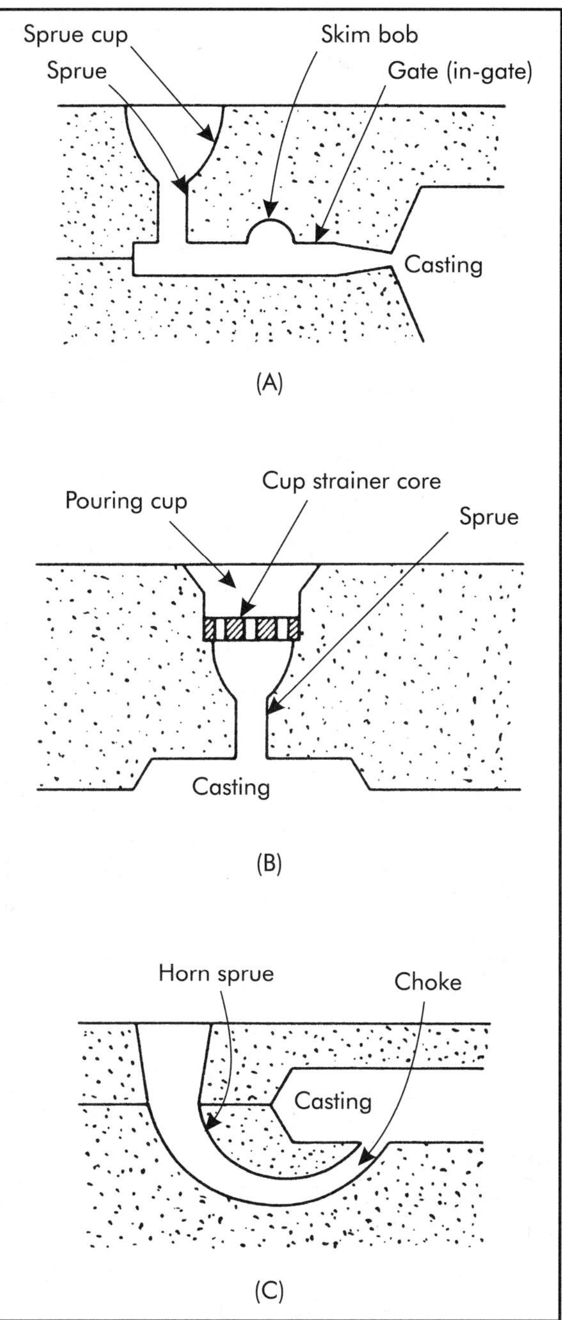

Figure 8-6. Three main types of gates for molds: (A) parting gate; (B) top gate; (C) bottom gate.

Many types of special gates besides those described are used in foundries.

Risers. In addition to acting as a reservoir, a riser mitigates the hydraulic ram effect of metal entering the mold and vents the mold. It must be the last to solidify, and to serve efficiently must conform to the following principles:

1. The volume of a riser must be large enough to supply all the metal needed.
2. The gating system must be designed to establish a temperature gradient toward the riser.
3. The area of the connection of the riser to the casting must be large enough so it does not freeze too soon. However, the connection must not be so large that the solid riser is difficult to remove from the casting.

The shape of a riser is an important consideration. Experience has shown that the most effective height of a riser is one and one half times its diameter to produce maximum feeding with a minimum amount of metal. As the area over volume ratio of a molten mass decreases, less chance is offered for heat to escape, and the solidification rate decreases. Table 8-1 provides the solidification times for various cast shapes. A sphere stays molten longest, but is not ideal as a riser; a cylinder is next best and is practical.

Chills. *Chills* are metal shapes inserted in molds to speed up solidification of the metal. Examples are given in Figure 8-7. Two types are external and internal chills. An internal chill becomes part of the casting and should be made of the same metal as the casting. An external chill should make enough contact and be large enough not to fuse with the casting. The shape, size, and use of a chill must be proportioned with care to avoid too-rapid cooling, which may cause cracks and defects in the casting.

Vents

Vents are small holes made by perforating the sand just short of the pattern in the mold with a wire or vent strip. The function of a vent is to permit gases to escape from the mold cavity and prevent them from becoming trapped in the metal or from raising back pressure to oppose the inflow of metal. Vents should serve all high points of the mold and be open to the top. Many small vents are better than a few large ones.

Table 8-1. Solidification time for various cast shapes

Shape	Sphere	Cylinder	Cube	Thick Plate	Thin Plate
Solidification time (min)	7.2	4.7	3.6	2.7	1.9

Note: All shapes have a volume of 1.85 dm^3 (113 $in.^3$) and mass of 14.5 kg (32 lb).

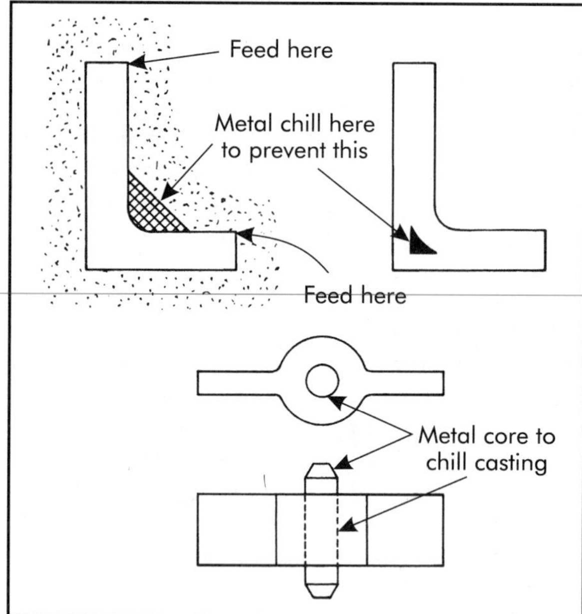

Figure 8-7. Typical forms of chills for castings.

8.2 MAKING MOLDS

8.2.1 Hand Tools for Molding

Some of the basic tools used for making a sand mold are shown on the molder's bench in Figure 8-8. A large sand *riddle* is shown on the shelf under the molders bench. A riddle is used for sifting the sand over the surface of the pattern when starting a mold. The size of the riddle is determined by the number of meshes per inch. Castings with fine surface details require fine sand and a fine riddle.

8.2.2 Mold-making Steps

A flask is selected larger than the mold cavity it is to contain to allow for risers and the gating system. There also must be enough mold mass over and under the cavity to prevent any breakout of the metal during pouring. Many castings

Figure 8-8. A molder's bench showing two molding flasks, a sand riddle, and an assortment of hand tools commonly used for making sand molds. (Courtesy McEnglevan Industrial Furnace Company, Inc.)

are lost, or require extra cleaning, and many injuries to personnel are caused by undersized flasks. A mold cavity may be carved in the sand, but this requires considerable skill for all but the simplest shapes and is seldom done in practice. Normal procedure is to first make an image of the piece to cast and then form the mold around it. This is called the *pattern*. Patterns and pattern making will be discussed more fully later.

The major steps involved in making a green sand mold are illustrated in Figure 8-9. In this example, a mold is being prepared for casting the metal pipe elbow shown at the top of the figure.

The first step in the sand mold-making process is to prepare a pattern of the part to be cast. The pattern is then cut in half and the halves are mounted on metal plates as shown in Figure 8-9(A). Care must be exercised in the mounting of the two symmetrical halves of the pattern to assure that they form matching cavities in the cope and drag sections of the mold. Quite often, the halves of a pattern are mounted on opposite sides of a metal plate to form a *match plate* pattern.

The next step in the mold building process is to prepare cores that form the open or hollow sections of a casting. A core for the metal elbow casting is illustrated in Figure 8-9(B). Note that the core is split in the middle to permit one-half of the core to be placed in the cope section of the mold and the other half in the drag section. Cores and core making are discussed later in this chapter.

Preparation of the cope section of the mold is shown in Figure 8-9(C). In this step, the cope section of a metal flask is placed upside down on top of the cope pattern plate. Note that in addition to the cope pattern, other mold elements, a sprue and two risers, have been placed on the cope pattern plate. The cope section of the flask is then filled with molding sand to form the cope section of the mold. In filling the cope section of the mold, *facing sand* is riddled to a depth of about 25.4 mm (1 in.) over the pattern and the other elements on the pattern plate. Riddling is necessary for good reproduction of the surface detail of the pattern. The riddled sand is then tucked into all pockets and sharp corners around the pattern either by hand ramming or by jolting the flask assembly.

Backing sand is then put into the flask to cover the facing sand to a depth of 76.2–101.6 mm (3–4 in.). It is then compacted by means of hand ramming or pressure applied to a *squeeze board*. The mold must be compacted uniformly to obtain a smooth, easily cleaned casting surface and to avoid metal penetration into the sand, swelling of the mold, breakouts, or other casting defects.

After compaction, the excess sand is struck off by means of a straight edge called a *strike bar*. After leveling the sand, a bottom board is placed and clamped onto the bottom of the flask and the flask is turned over. Then, the pattern plate, pattern, sprue, and risers are removed, and the cope section of the core is placed in position within the pattern cavity.

As illustrated in Figure 8-9(D), the drag section of the mold is prepared in much the same way as the cope section. When the two halves of

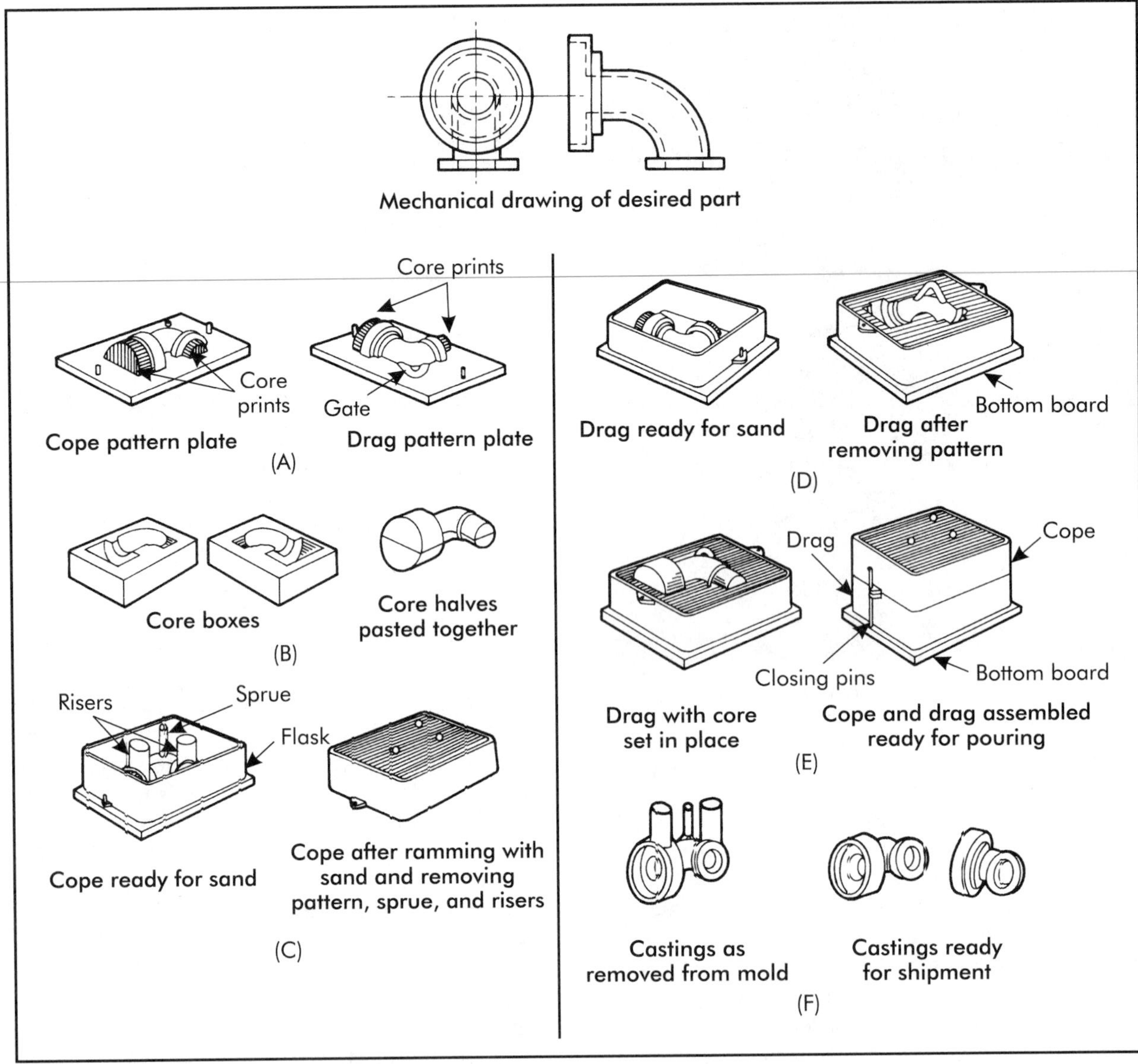

Figure 8-9. Typical steps involved in making a casting from a green sand mold. (Courtesy Steel Founders' Society of America)

the mold are completed, parting material is dusted from a bag over the mold joint or parting surface of the cope and drag sections. The parting material prevents the sand in the cope from sticking to the sand in the drag. Parting material for large molds is usually fine silica sand and for medium and small molds is finely ground powders, such as talc or silica flour. Then, the patterns are removed, the cores are carefully placed in position, and the cope section is placed on top of the drag as shown in Figure 8-9(E). Pins at each end of the flask serve to locate the cope and drag sections so that the cavities in each section are lined up. After the mold is closed, it is clamped and readied for pouring. Weights are often set on top of some molds to keep them from coming apart due to the hydrostatic pressure of the liquid metal. Molten metal is then poured through the sprue into the mold cavities.

After solidifying and cooling, the mold is broken open and the casting and molding sand are separated. The raw casting appears as shown in

Figure 8-9(F) with the sprue and two risers in place. The raw casting is then cleaned, often by sand blasting, and the sprue and risers are then cut off, usually with a band saw. Depending on the quality requirements for appearance, the new castings may be wire brushed or sanded with flexible abrasive strips to remove saw marks and parting line projections. After that, they are ready for secondary machining, if required, or shipment to another location.

Several molding procedures depend upon the size of the casting. *Bench molding* is used for molds small enough to be made on a workbench such as the one shown in Figure 8-8. *Floor molding* involves molds too large for a bench. The molds are made in flasks on the floor of the foundry or on machines standing on the floor. Molds too large to be made entirely in flasks are constructed in pits below the foundry floor. This is called *pit molding*.

8.2.3 Molding Machines

At one time, all molding was done by hand, but today's labor costs and competition make machine molding mandatory in industry. Molding machines offer higher production rates and better quality castings at lower costs in addition to requiring less heavy labor.

Molding machines serve in two general capacities: (1) to pack sand firmly and uniformly into the mold, and (2) to manipulate the flasks, mold, and pattern. Properly controlled and applied machine ramming is more uniform, dependable, and produces more and better molds than hand ramming. Manipulation is done in various degrees on different machines and may include turning over parts or all of the mold and lifting the patterns and flasks. Molding machines are available in a number of makes, models, and sizes and perform the foregoing functions in various combinations and ways. Generally, they fall into one of three classes and will be illustrated here by a typical form for each class: the jolt-squeeze, jolt-rollover, and sand-slinger molding machines.

Jolt-squeeze Machine

A jolt squeezer, shown in Figure 8-10, basically consists of a table actuated by two pistons in air cylinders, one inside the other. The mold on the table is jolted by the action of the inner piston, which raises the table repeatedly and drops it down sharply on a bumper pad. Jolting packs the sand in the lower parts of the flask but not at the top. The larger cylinder pushes the table upward to squeeze the sand in the mold against the squeeze head at the top. Squeezing compresses the top layers of sand in the mold but sometimes does not penetrate effectively to all areas of a pattern. In the case shown, the cope is squeezed so as not to damage the finished drag by jolting. Some machines do simple jolting; others squeezing alone. For high production, one jolt squeezer may be set up for the drag portion of a mold and another for the cope. A vibrator may be attached to a machine to loosen the pattern, allowing it to be to removed easily without damaging the mold.

Jolt-rollover and Pattern-draw Machine

The jolt-rollover and pattern-draw machine is designed to mold cope or drag. A flask is set over a pattern on a table and it is filled with sand and jolted. Excess sand is struck off, and a bottom board is clamped to the flask. The machine raises the mold and rolls it over onto a table or conveyor. The flask is freed from the machine. The pattern is vibrated, raised from the mold, and returned to loading position. Similar machines squeeze as well as jolt.

Sand-slinger Machine

The sand slinger achieves a consistent packing and ramming effect by hurling sand into the mold at a high velocity. Sand from a hopper is fed by a belt to a high-speed impeller in the head. A common arrangement is to suspend the slinger with counterweights and move it about to direct the stream of sand advantageously into a mold. Mold hardness can be controlled by the operator by changing the speed of the impeller and movement of the impeller head.

Sand slingers can rapidly deliver large quantities of sand and are especially beneficial for ramming big molds. Their only function is to pack the sand into molds. They are often operated together with pattern-drawing equipment.

8.3 CORES

A *core* is a body of material, usually sand, used to produce a cavity in or on a casting. An ex-

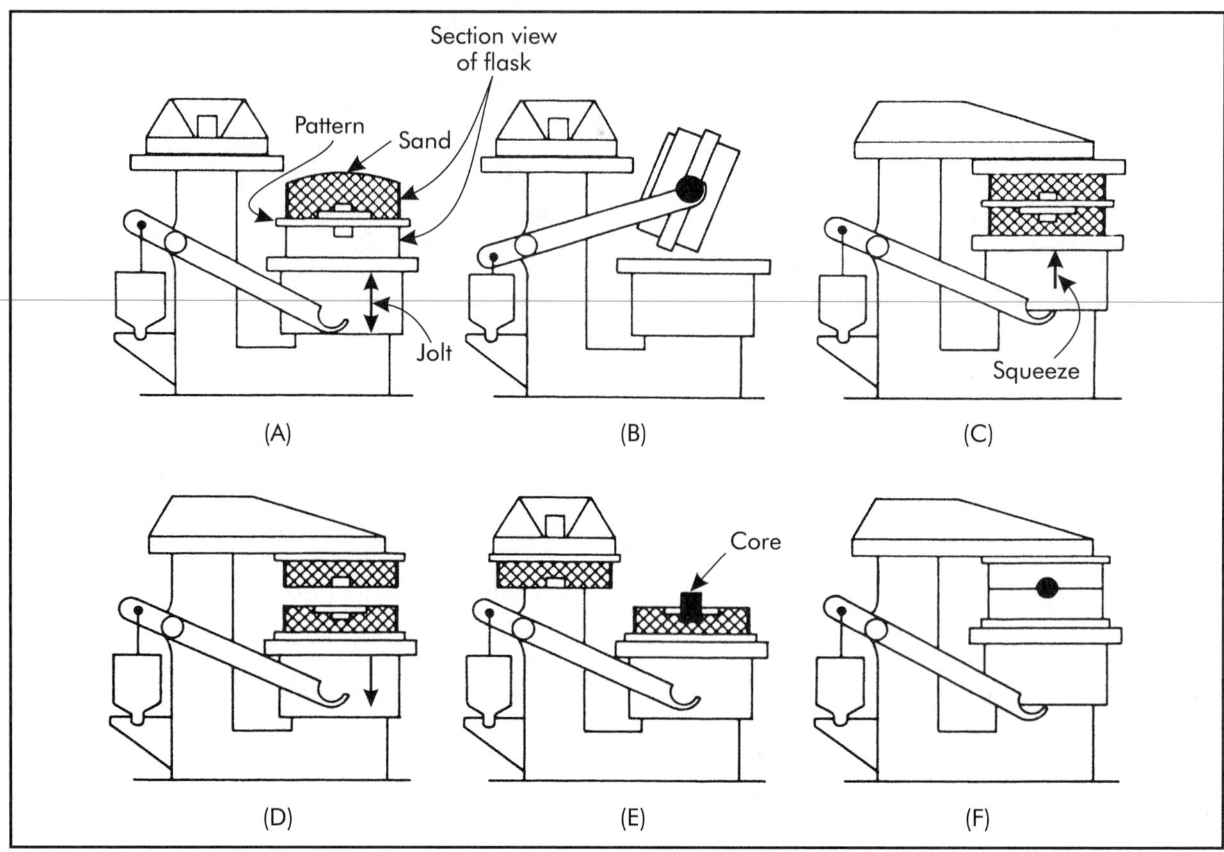

Figure 8-10. Steps in making a mold on a jolt-squeeze machine: (A) jolting the sand on the pattern in the inverted drag; (B) rolling over the flask to fill the cope; (C) squeezing to pack the sand in the cope; (D) raising the cope to remove the pattern; (E) placing a core; (F) replacing the cope on the completed mold.

ample of a core in a casting is one forming the water jacket in a water-cooled engine block; an example of a core on a casting is the one forming the air space between the cooling fins of an air-cooled cylinder.

A number of properties are essential for good cores. They are mentioned briefly here and will be discussed in more detail later. A core must have: (1) *permeability* (that is, the ability to allow steam and gases to escape), (2) *refractory* (that is, the ability to withstand high temperature), (3) green strength so that it can be formed, (4) dry strength so that it will not wash away or change size when surrounded by molten metal, (5) *collapsibility* (that is, the ability to decrease in size as the casting cools and shrinks), (6) *friability* or the ability to crumble and be easily removed from the casting, and (7) a minimal tendency to generate gas.

8.3.1 Core Making

The tools used for the production of cores are much the same as for making a mold, with the addition of the core box, core driers, and special material and equipment for venting. The core receives its shape from the core box. Driers are special forms or racks used to support complicated cores during baking. Since core driers are quite expensive, they are used only when large numbers of cores are required. Complicated cores often can be made in parts on flat plates and then assembled with paste.

Core making is much like molding except that the core sand is placed in a core box. It can be blown, rammed or packed by hand, or jolted into the box. The excess sand is struck off, and a drier plate is placed over the box. The core box is then inverted, vibrated or rapped, and drawn off the core. The core is then put in a core oven and baked. Cores must be made strong enough to withstand the handling they must endure.

Where the core does not have natural vents, supplementary vents must be provided. For a core made by pasting parts together, grooves may be cut in the faces to be joined before they are assembled. The vents are continued through to the core prints (Figure 8-11) so that molten metal will not reach and plug them. If this method cannot be used, wax venting can be done by placing strips of vent wax in the sand before it is baked. If the core is large and rather complicated, the center of the core may be filled with coke, gravel, or other porous material that will give good permeability, crushability, and friability. The surfaces of some cores are given a refractory coating.

Several different kinds of cores can be found in the foundry today. *Baked sand* or *dry sand cores* have a binder that must be cured with heat. They are being made less frequently because of energy costs. *Green sand cores* are made of molding sand at the time the mold is made and are relatively cheap and popular. Where these cannot serve adequately, *cold cure cores* are used. They contain a two-part binder that is self-curing or a one-part binder that is cured by passing a gas through the core. There are many such core binders, mostly proprietary, and new ones are introduced each year. They save energy.

Shifting

If a core does not stay in just the right place in its mold, the walls of the cavity it produces will not be of proper thickness. Shifting of cores is a major cause of defective castings. A core may be of such a shape that it needs internal support. For that purpose, heavy wire or steel rods may be embedded in it. *Chaplets* (Figure 8-12) serve to support cores that tend to sag or sink in inadequate core print seats. A chaplet is usually made of the same metal as, and becomes part of, the casting.

A core is subjected to an appreciable buoyant force when immersed in the liquid metal poured into its mold. An anchor, like the one in Figure 8-12, prevents the core from rising. Chaplets also serve this purpose.

A core immersed in heavy metal is buoyed up by a force proportional to the difference between its mass and the mass of the metal displaced. As an example, a sand core with a volume of 1 dm^3 has a mass of 1.6 kg. Molten iron has a mass of 7.2 kg/dm^3. A force of $(7.2 - 1.6) \times 9.806 = 55$ kN must be resisted to keep the core submerged. The gravitational constant is 9.806. In other units, a sand core with a volume of 1 ft^3 weighs about 100 lb, and molten iron 450 lb/ft^3. The force acting to raise that core is 350 lb.

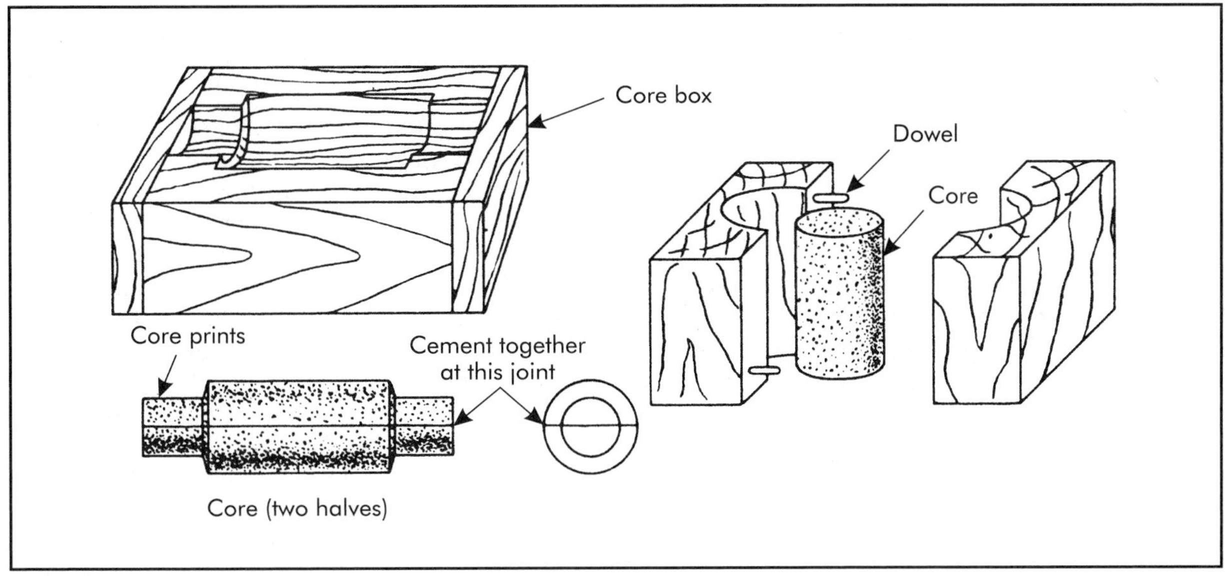

Figure 8-11. Typical core boxes.

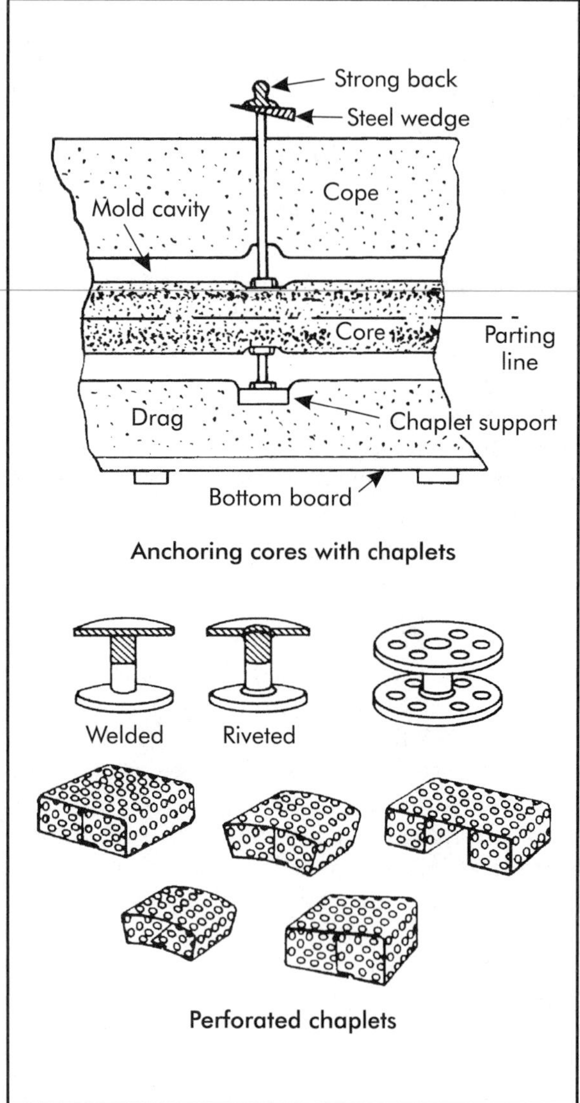

Figure 8-12. Application of a chaplet and anchor to support a core and a few examples of chaplets.

Machines

Cores of regular shapes and sections may be extruded in a machine like that illustrated in Figure 8-13 and cut to length. A central vent hole is left by a wire extending from the center of the screw.

Large cores are made like molds in jolt rollover, sand slinger, and other machines. Small and medium-size, irregularly shaped cores may be made by hand, but for quantities are produced on a *core blowing machine*. This machine blows

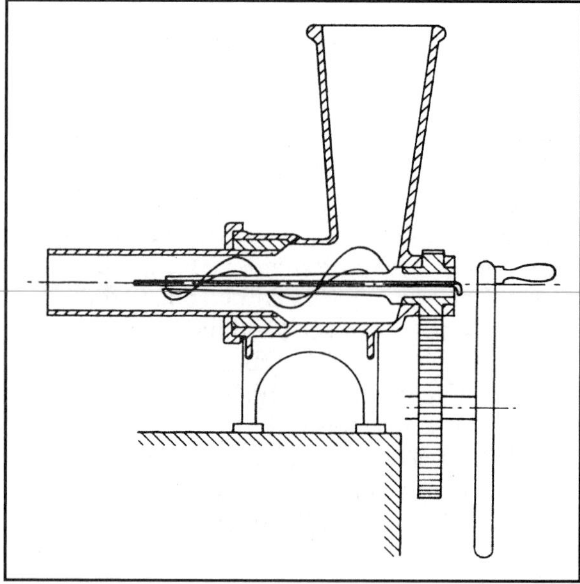

Figure 8-13. Core extrusion machine.

sand by compressed air through a core plate with holes arranged to pack the sand evenly and firmly in the core box. Each core box must be designed properly to release the air but retain the sand in a uniformly dense compact form.

Baking

Drying alone is sufficient for many cores, but others (not green sand or cold cured) are bonded by oils and must be baked for ultimate hardness and strength. The purpose of baking is to drive off moisture, oxidize the oil, and polymerize the binder.

A uniform temperature and controlled heating are necessary for baking an oil-bonded sand core. With linseed oil, a major binder, the temperature is raised at a moderate rate, held at about 204° C (400° F) for about 1 hour, and then is allowed to fall slowly to room conditions. If the same core is baked quickly at 260° C (500° F), it will be overbaked on the surface and underbaked at the center.

The size of a core affects baking. If care is not taken, the outer surface of the core will bake first and attain maximum strength. Then while the inside is curing, the outside will overbake and lose strength. To avoid this, the center of a large core can be made of a porous material, such as coke or cinders. This allows oxygen to reach the center so that oxidation and polymerization can take place.

8.4 PATTERNS

A *pattern* is a form used to prepare and produce a mold cavity. It has been said that a poor casting may be produced from a good pattern, but a good casting will not be made from a poor pattern.

The designer of a casting must look forward to the pattern to assure economical production. The design should be as simple as possible to make the pattern easy to draw from the sand. Unnecessary cores should be avoided.

8.4.1 Types

Many molds are made from *loose patterns*. Such a pattern has essentially the shape of the casting with forms for sprues, risers, etc., attached. This is the cheapest pattern to make but the most time consuming to use. A loose pattern may be made in one or more pieces. For instance, a two-piece pattern is normally *split* into cope and drag parts to facilitate molding. For a part difficult to mold, some *loose pieces* may be removable to allow the pattern to collapse for withdrawal from the sand that otherwise would not be possible.

In an emergency, an original casting or an assembly of the pieces from a broken casting may serve as a loose pattern. Of course, the part needs to be built up to allow for shrinkage.

Patterns fastened permanently to a board or *match plate* are known as *mounted patterns*. A main advantage is that a mounted pattern is easier to use and store than a loose pattern. Another advantage is that the gating system can be mounted on a match plate, and thus the time required to cut the gating system in the mold can be eliminated. Mounted patterns cost more than loose patterns, but when many castings are to be made from a pattern, the time saved in operation warrants the cost of mounting it. Patterns for a number of pieces may be mounted on one match plate.

A *core box* is essentially a type of pattern into which sand is rammed or packed to form a core as illustrated in Figure 8-11.

Symmetrical molds and cores, particularly in large sizes, are sometimes shaped by means of *sweeps* as illustrated in Figure 8-14. The sweep is a flat board with an outline of the cross section of the part to be made. It is revolved around a central axis to clear away excess sand inside the mold.

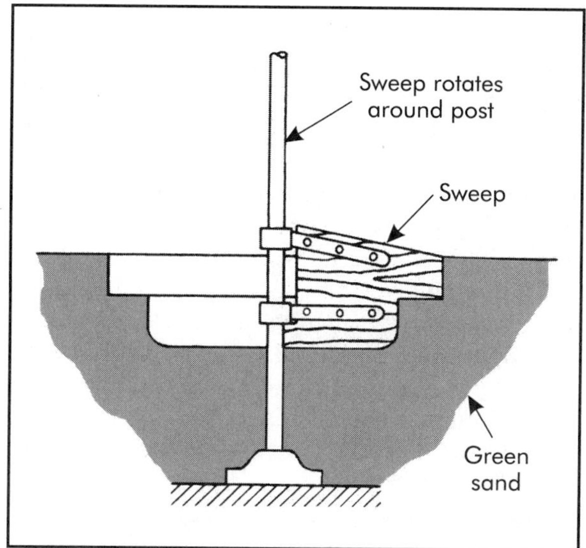

Figure 8-14. Sweep pattern.

8.4.2 Material

Wood is the most common material for patterns. It is easy to work with and readily available. Properly selected and kiln-dried mahogany, walnut, white pine, and sugar pine are often used. Sugar pine is most often used because it is easily worked and is generally free from warping and cracking. Moisture in the wood should be about 5–6% to avoid warping, shrinking, or expanding of the finished pattern.

Metal patterns may be loose or mounted. If usage warrants a metal pattern, then the pattern probably should be mounted on a plate and include the gating system. Metal is used when a large number of castings are desired from a pattern or when conditions are too severe for wooden patterns. Metal patterns wear well. Another advantage is freedom from warping while in storage. Commonly, a metal pattern is cast from a master pattern and can be readily replaced if damaged or worn.

Patterns are also made of plaster and plastics. Plaster patterns are easy to make; they can be cast where original molds are available. However, plaster is brittle and not suitable for molding large numbers of castings. Plaster molds are described in Chapter 9. Some conventional patterns are made of abrasion-resistant plastics with cost and durability performance between that of wood and metal. Certain plastics are used for making emergency patterns quickly or to salvage worn or broken patterns.

Several of the processes involved in *rapid prototyping* (Chapter 29) can be used to make a variety of patterns. One of those, the *stereolithography* process, may be used to construct plastic patterns with either simple or complex shapes. This process uses an ultraviolet laser to cure successive layers of a photopolymer (liquid plastic) to form the plastic pattern. The two-dimensional form of each layer of plastic is controlled by means of computer-generated vectors from a computer-aided design (CAD) representation of the pattern. The equipment necessary to perform these operations is known as a stereolithography apparatus (SLA).

The *evaporative casting process* (ECP), also called *full mold* (FM), *lost foam*, or *lost pattern casting*, utilizes a foamed polystyrene pattern embedded in the sand that is vaporized as the molten metal fills the mold. Each pattern is consumed for one casting. Originally, the process was confined to producing one or a few castings of a kind. The patterns were carved from polystyrene board stock. Developments have led to the production of automobile cylinder blocks, crankshafts, water pumps, and the like in large quantities. Typically, for such work, polystyrene beads mixed with pentane are heated and expanded in a die on a molding press. Sections can be joined to make a complex pattern, and risers, runners, and sprues are added. The surface of a pattern is painted with a ceramic slurry to help prevent burning the sand. There are no parting lines and no fins on the casting, and no need for draft, which may save several hundred kilograms (pounds) on a large casting. Fillets do not have to be put in the mold unless required for strength of the part. Cores and binders in the sand can be eliminated.

8.4.3 Layout

The *parting line* represents the surface that divides a pattern into the parts that form the cavities of the cope (top) and drag (bottom) of the mold. If at all possible, the parting line should be straight, which means that a simple plane divides the pattern into cope and drag sections. A straight parting line is particularly desirable for a loose piece pattern to enable the sections to lie flat on the molding board. An example is given in Figure 8-15. A straight parting line is not necessary for a match plate, but often makes the pattern easier to fabricate.

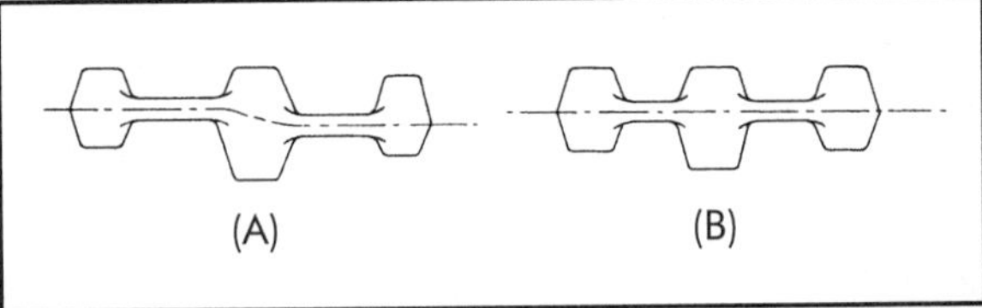

Figure 8-15. Parting line design (A) is not straight, and the piece is harder to cast than (B).

Some means are necessary to support and position cores in molds. These are in the forms of extensions, pads, and bosses on the cores and are called *core prints* (Figure 8-11). A core print must be large enough to support the core. The core weight is carried by the drag, and buoyancy is resisted by the cope.

8.4.4 Shrinkage Allowance

As metal solidifies and cools, it shrinks and contracts in size. To compensate for this, a pattern is made larger than the finished casting by an amount called a *shrinkage allowance*. Although contraction is volumetric, the correction for it is usually expressed linearly. Dimensions are not shown oversize on a part or pattern drawing to allow for shrinkage. The pattern maker measures to the finished dimensions with *shrink rules*. Such a rule has a scale that is longer than standard by a definite proportion such as 5, 10, or 15 mm/m (≈.063, .125, or .188 in./ft). Shrinkage is different for different metals, different shapes of castings of the same metal, and different molding and casting methods. As an example, light- and medium-steel castings of simple design with no cores require an allowance of 20 mm/m (≈.25 in./ft), and a rule with such a scale is used to make their patterns. In comparison, an allowance of 15 mm/m (≈.188 in./ft) is adequate for pipes and valves of the same metal because their molds and cores offer considerable resistance to contraction. Typical shrinkage allowances are shown in Table 8-2. A master pattern from which metal patterns are cast must have double the shrinkage allowance.

8.4.5 Other Allowances

Machining allowance is the amount by which dimensions on a casting are made oversize to provide excess stock for machining. Dimensions on a pattern drawing include machining allowance. The amount of metal left for machining must be no more than necessary, but enough to assure that cutters can get an ample cut beneath and completely remove the hard scale and skin on the surface of the casting. What is sufficient depends upon the kind of metal, the shape of the casting, and the methods of casting, cleaning, and machining. Typical machining allowances are shown in Table 8-3.

Distortion allowance may be added to dimensions of certain objects such as large flat plates and U-shaped castings expected to warp on cooling. The purpose of this allowance is to displace the pattern in such a way that the casting will be of the proper shape and size after it distorts in process.

Table 8-3. Typical machining allowances

	Cast Iron mm (in.)	Cast Steel mm (in.)	Brass, Bronze, and Aluminum mm (in.)
On outside surfaces	2.39 (.094)	3.18 (.125)	1.60 (.063)
On inside diameters	3.18 (.125)	4.78 (.188)	2.39 (.094)

Note: For finishing surfaces of sand castings with dimensions up to about 305 mm (≈12 in.). Allowance is added to the total dimension.

8.4.6 Draft

Draft is the taper or slant placed on the sides of a pattern, outward from the parting line as depicted in Figure 8-16. This allows the pattern to be removed (drawn) from the mold without damaging the sand surface. Draft may be expressed in mm/m (in./ft) on a side or in degrees. The amount needed in each case may depend upon the shape of the casting, the type of pattern, and the process. For example, draft of 10.42 mm/m (.125 in./ft) may be required for a certain wooden pattern. For the same casting, a metal pattern mounted on a molding machine may need only 5.25 mm/m (.063 in./ft).

8.4.7 Fillets

A *fillet* is a rounded filling along the convergence of two surfaces of a pattern as indicated in Figure 8-17. The rounded corner on the casting is also called a fillet. Fillets may be carved in wood patterns but are usually made more inexpensively of wax, plastic, wood coving, or leather. They vary in size from 3.2–25.4 mm (.125–1 in.) radius depending on the size, shape, and material of the casting. Fillets obviate sharp angles and corners and thus strengthen metal patterns

Table 8-2. Typical pattern shrinkage allowances

Shrinkage Allowance	Gray Cast Iron	Cast Steel	Aluminum	Brass	Bronze	Magnesium
mm/m	10	20	13	15	10–20	14
(in./ft)	(≈.125)	(≈.25)	(.156)	(≈.188)	(≈.125–.25)	(≈.172)

Note: For simple uncored castings with dimensions not over 610 mm (24 in.).

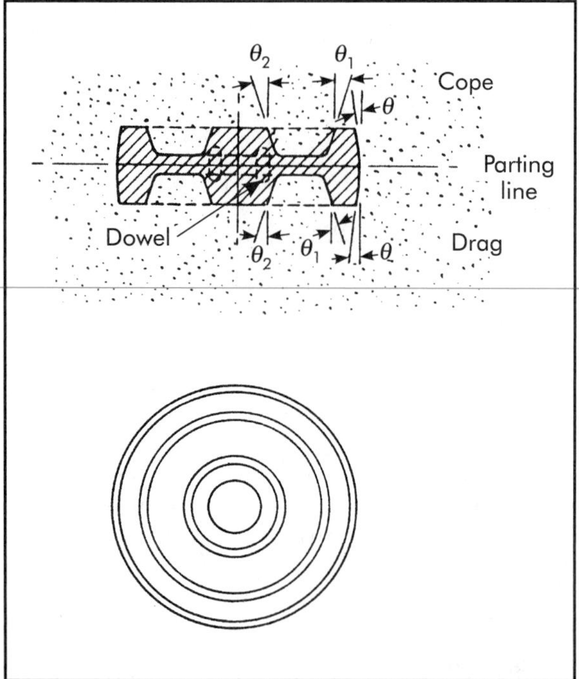

Figure 8-16. Draft on both pieces of a two-piece pattern.

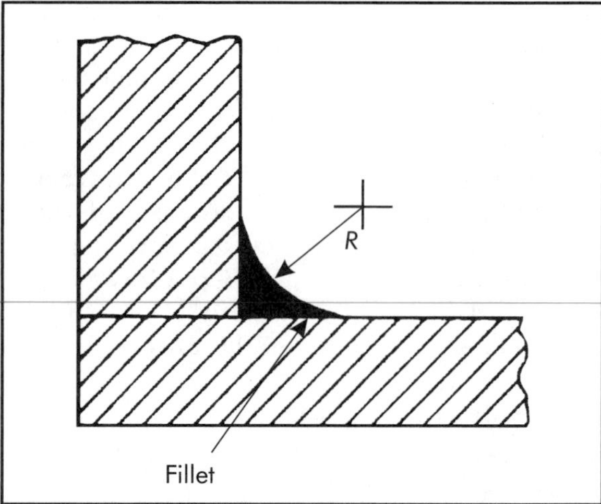

Figure 8-17. Example of a fillet on a pattern.

and castings. They provide for easier removal of the pattern from the sand, a cleaner mold, freer flow of metal through the mold, less washing of the sand in the mold, and fewer shrinking strains and hot tears between sections as the casting cools.

8.4.8 Locating Pads

Bosses or pads are commonly added to castings to provide definite and controlled spots for location in machining operations. These *locating* or *foundry pads* are gaged and may be cleaned up by filing to a relationship with the major outline of the workpiece. Care must be taken that such pads do not create sections that are too heavy or hot spots in a casting.

8.4.9 Color Coding

All surfaces of a wood pattern are coated with shellac to keep out moisture. Important parts of a pattern may be colored for identification. Some foundries adhere strictly to a color code; others do not. A widely accepted color code is one adopted by the American Foundrymen's Society (Figure 8-18).

8.5 SANDS AND OTHER MOLD INGREDIENTS

The primary function of any molding material is to maintain the shape of the mold cavity until the molten metal solidifies. Silica sand is the most widely used molding material, particularly for metals that melt at high temperatures. It serves well because it is readily available, low in cost, easily formed into complicated shapes, and able to withstand the molten metal.

The three major parts of a molding sand are: (1) the sand grains, which have the necessary refractory properties to withstand the intense heat of the molten metal; (2) a bonding material, which may be natural or added clay, cereal, etc., to hold the grains together; and (3) water to coalesce the grains and binder into a plastic molding material.

8.5.1 Molding Sand

Natural sands contain only the binder mined with them and are used as received with water added. They have the advantages of maintaining moisture content for a long time, having a wide working range for moisture, and permitting easy patching and finishing of molds. Sometimes it is desirable to change the properties of a natural sand by adding bentonite (a clay binder). Such a sand is known as a *semisynthetic sand*.

Synthetic sands are formulated from various ingredients. The base may be a natural sand with

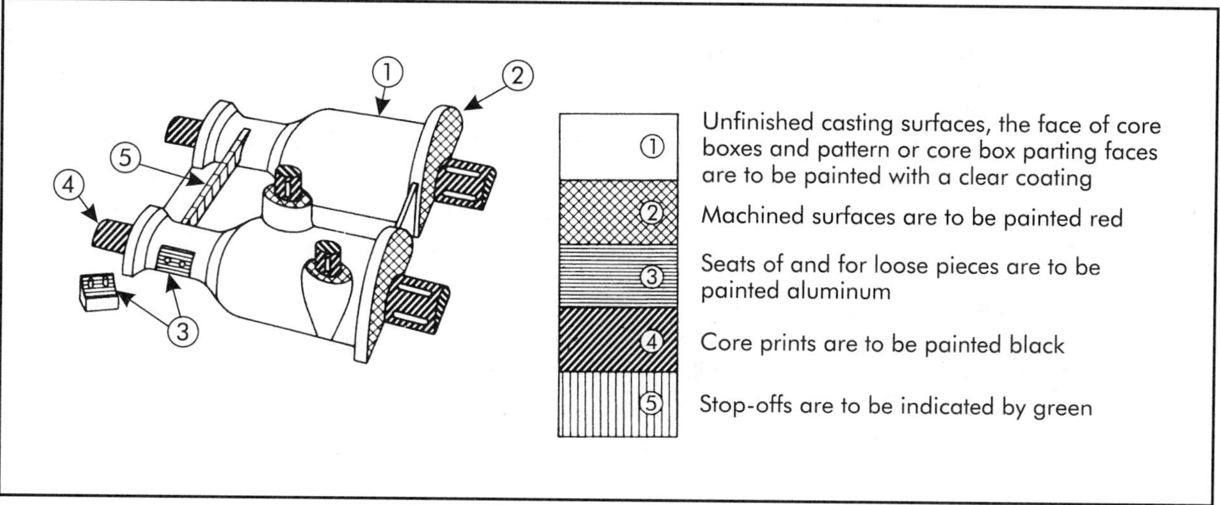

Figure 8-18. Color coding.

some clay content or a washed sand with all clay removed. A binder, such as bentonite, and water are added. The sands have these advantages over natural sands: (1) more uniform grain size; (2) higher refractory; (3) moldability with less moisture; (4) require less binder; (5) easier control of properties; and (6) require less storage space since one kind of sand may be used for different kinds of castings.

Loam sand is high in clay, as much as 50% or so, and dries hard. *Loam molding* is done by making the mold, usually for a large casting, of brick cemented and lined with loam sand and then dried.

Properties

The way sand performs in a mold to produce good castings can be tested by and largely depends upon its green permeability, green strength, and dry strength. These properties are determined chiefly by grain fineness, grain shape, and clay and moisture content of the sand. Other properties of less influence are heat strength, sintering point, deformation, and collapsibility.

Green permeability. The porosity from the openings between grains allows *permeability* of the sand. This gives passage for air, gases, and steam to escape when the molten metal is poured in the mold. Permeability is measured in a common test by passing a definite amount of air through a standard test specimen under specified conditions. The result is expressed by

$$P = 501.2/pt \qquad (8\text{-}1)$$

where

P = permeability
p = pressure, grams/cm^2
t = time, min

Thus the permeability number is larger as the sand becomes more porous.

Grain fineness is measured by passing sand through standard sieves, each with a certain number of openings per linear mm (in.). Commercial sands are made up of grains of various sizes. Grain size of a sand is designated by a number that indicates the average size as well as the proportions of smaller or larger grains in the mixture. Size is determined by a procedure described in handbooks.

Finer grains impart a smoother finish to a casting. However, permeability decreases as the grains and thus the voids between grains become smaller. The same condition results from a large proportion of fine grains in a mixture. The best compromise must be reached. For large castings that require coarse sand for high permeability, the surface of the mold cavity may have a thin layer of fine-grained facing material.

There are two distinct *shapes of sand grains*: angular and rounded, with many degrees of roundness and angularity between the two extremes. Sharp angular grains cannot pack together as closely and consequently have higher permeability than rounded grains. This is demonstrated by the graph in Figure 8-19.

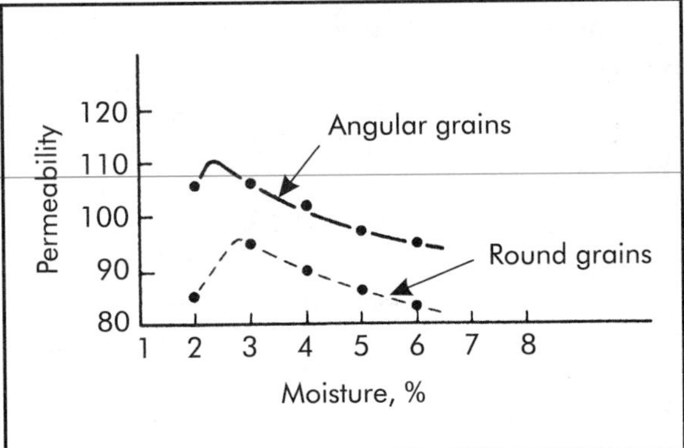

Figure 8-19. Relationship of molding sand permeability to moisture content and grain shape.

Both the type and the amount of binder have a decided effect on the permeability of sand. An illustration of the permeabilities imparted by two common types of clay is given in Figure 8-20. Over a wide range of moisture content, bentonite was found to give more permeability than fireclay. Permeability may decrease with an increase in clay content, as depicted by the 2% moisture curve for bentonite in Figure 8-21. The upper curve of 4% moisture content indicates fairly constant permeability over a wide range of bentonite content. In general, clay content is optimum when present to the extent of coating the sand particles completely without filling the spaces between the grains.

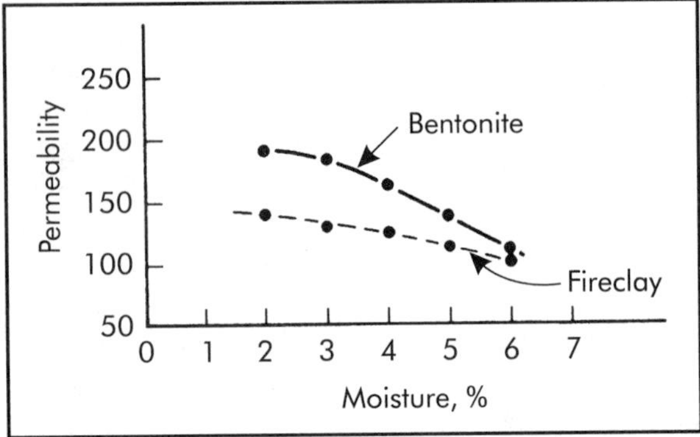

Figure 8-20. Permeability of two kinds of clay for molding sand with various amounts of moisture.

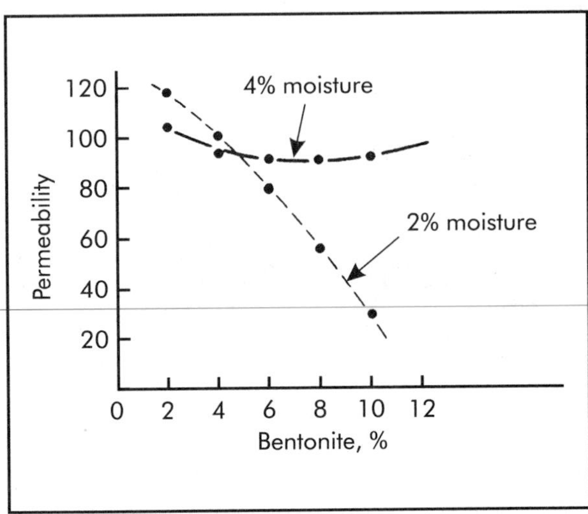

Figure 8-21. How the permeability of a molding sand varies with its bentonite content.

With a low *moisture content*, fine clay particles clog the spaces between the grains, and permeability is low. More moisture softens and agglutinates the clay around the grains for optimum conditions. An excess of moisture fills the voids and decreases permeability. Peaks of maximum permeability are seen in Figure 8-19. The optimum moisture content is not the same for all molding sands, although it generally lies between 2 and 8%.

Green strength. *Green strength* is a characteristic of a sand ready for molding, and, if the metal is poured immediately, represents the ability of the sand to hold to the shape of the mold. Green strength may be expressed in kPa (psi) required to rupture a standard specimen.

The finer the sand grains, the larger the surface area of a given bulk, and the larger the amount of binder needed to cover the area. The contacts and bonds between grains are more numerous, and thus green strength is higher with finer grains. Figure 8-22 shows that as the grain size becomes larger, green strength decreases under normal conditions.

Round grains pack together more closely than angular grains and as a result are bonded together with a higher green strength than angular grains. A comparison is given for two types of grains in Figure 8-23.

Some binders provide a higher green strength than others. Comparison of bentonite and fireclay at different moisture levels is given in Figure 8-24. Green strength increases in proportion to the amount of binder in a molding sand, but, as pointed out earlier, too much binder is detrimental to permeability, and a compromise must be accepted.

The effect of moisture on green strength is similar to the effect on permeability. Green strength increases with the first additions of moisture, reaches a maximum strength, and then starts to decrease as depicted in Figure 8-25. Also shown is that an excess of moisture has a weakening effect, even nullifying the influence of grain size.

In addition to other factors, mulling or mixing practice affects green strength as will be explained in the section on the preparation of sand.

Dry strength. *Dry strength* is a characteristic of sand that has been dried or baked. In general, dry strength varies with grain fineness, grain shape, and moisture content in the same way as green strength. However, different binders can affect dry strength and green strength differently. For example, in contrast to western bentonite, southern bentonite produces a high green strength and a low dry strength, which is conducive to easy shake out of castings.

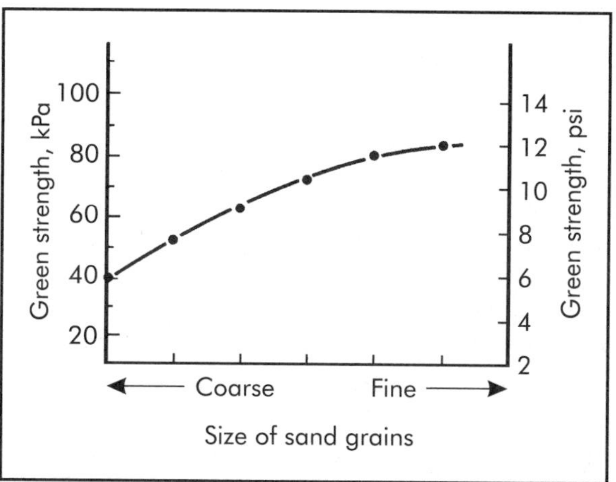

Figure 8-22. Green strength of a molding sand in relation to the sizes of its grains for one set of conditions.

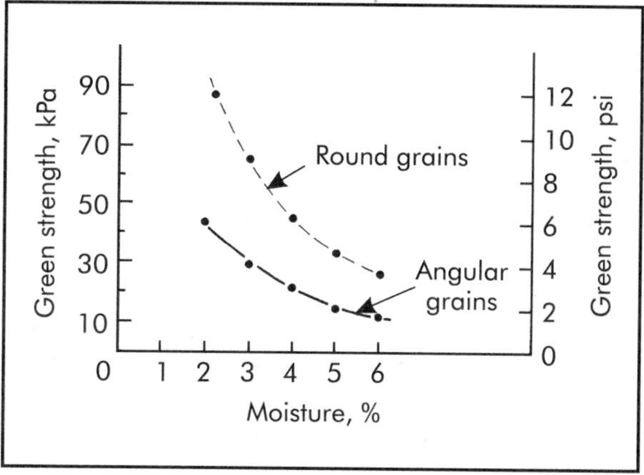

Figure 8-24. Comparison of the green strengths of two samples of molding sand with different clay binders at different moisture levels.

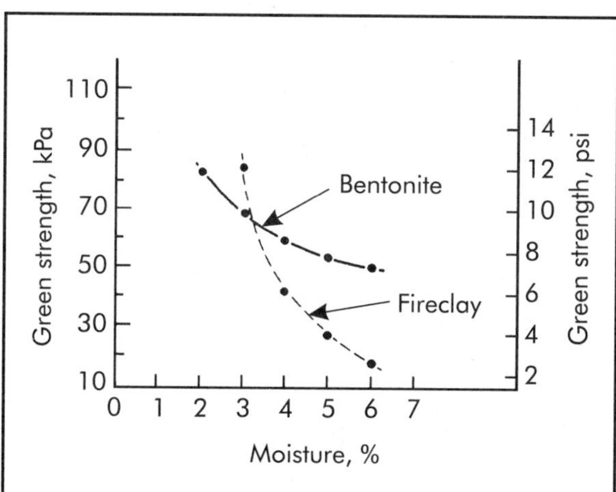

Figure 8-23. Relationship of green strength of a molding sand to moisture content and grain shape.

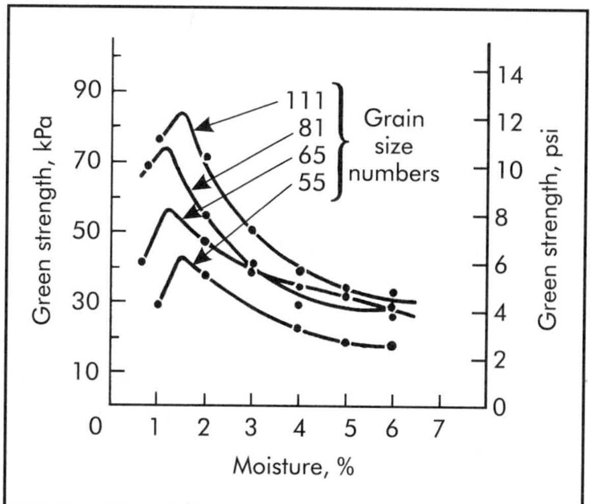

Figure 8-25. How the green strength of molding sand varies for several sizes of grains.

Control

Many substances are added to molding sands to impart or change certain properties. Many new ones become available each year, particularly substances to minimize heating and save energy. Space is not available here to list many additives, and only the basic ones are cited. Cereal (such as wheat and corn flour), dextrine, rosin, and similar substances are often added to augment or modify clay binders. It is important to realize that the action of each additive is somewhat different. Corn flour improves green strength slightly but dry strength markedly. Wheat flour improves collapsibility. Dextrine binders are a form of sugar and produce a much higher dry strength than cereal binders but decrease green strength. A rosin-bonded core has a hard surface when baked but absorbs moisture on standing and should be used as soon as possible.

The binder is burned out and the proportion of fine particles is increased when foundry sand is exposed to the heat of molten metal. As a result, green strength and permeability decrease as the sand is reused. By measuring properties periodically with sand-testing equipment, a sand technician can make appropriate additions to restore the sand before it deteriorates to the point where it must be discarded. Uniform properties may be maintained from day-to-day by continuous checks and additions of small amounts of binder.

Preparation

Under a microscope, sand grains are seen to exist in clusters of small particles adhering to large particles. This is thought to be the result of previous use. Clusters are less refractory and harder to ram than individual particles. So they need to be broken up, particularly to prepare sand for cores and for facing patterns to make mold linings. Breaking down of the clustered particles is desirable for all sand and is usually done by stirring and kneading it mechanically. A popular machine for that purpose is a *muller* that kneads, shears, slices through, and stirs the sand in a heavy iron pot by means of several revolving rollers and knives. The time needed for mulling depends upon the type of muller and the type and amount of binder. After the optimum length of mixing time in each case, there is no further increase in green strength as shown by the example in Figure 8-26. Mulling sand distributes the binder over the grains. Thus less binder is required for satisfactory green strength and permeability is higher than with hand-mixed sand.

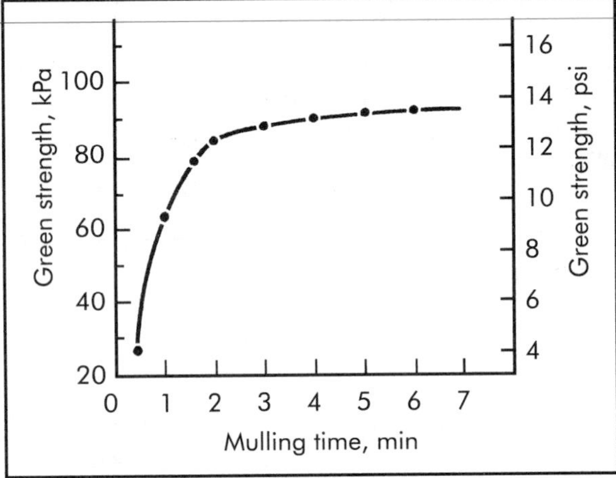

Figure 8-26. Effect of mulling time on green strength of a molding sand.

Sand is *aerated* to make it fluffy so it flows readily around and takes up the details of the pattern. This is done to some extent in mulling. One device for aeration is a *sand cutter* that hurls the sand from a rapidly moving belt against bars or springs to break the grains apart.

Sands are mixed to many formulas for various metals and purposes. For instance, for steel castings, three sands may be prepared: one for the green facing, another for the green backing, and still another for facing to be surface dried in the mold by a torch. Formulas for foundry sands are given in reference texts and handbooks.

8.5.2 Core Sand

Core sand must have much the same properties of permeability, cohesiveness, and refractory as molding sand, but must also possess collapsibility and friability. *Collapsibility* means that the core gives way easily as the casting cools and shrinks to avoid inducing hot tears and cracks in the metal. *Friability* means that the core crumbles and falls apart when it must be removed from the casting. These properties are imparted by the type and amount of binder, additives to the binder, and by the curing of the cores. For example, if cracks occur in the cored area of a casting, a weaker binder or less binder may be used, or a small amount of wood flour may be added to the binder.

Core sand mixes are started with clean, dry silica sand. Among various binders used for core sand are corn flour, dextrine, fish oil, raw linseed oil, and commercial core oils. Linseed oil is an ingredient in most proprietary core oils. A long list could be made of binders, each imparting a particular property to the core. As the ingredients are added, the sand is thoroughly mixed in a muller but not excessively, which would cause stickiness.

Oils and other core binders that must be baked have been largely replaced by other substances that require less energy. Popular among these is sodium silicate (water glass) that is mixed with the sand and hardened in the core in a few seconds when exposed to carbon dioxide (CO_2) gas. The core can then be used in the mold immediately. A major disadvantage of sodium silicate and some other no-bake binders is that they must be mixed with the sand only a short time before use because they harden rather soon on exposure to air. Other no-bake binders with longer shelf lives are available and widely used. Among them are plastics materials that can be catalyzed in the core for quick hardening. One in particular is hardened by exposure to sulfur dioxide (SO_2) for a few seconds. New cold-cure substances are being brought out each year and are too numerous to be mentioned here.

8.6 MELTING METALS IN THE FOUNDRY

The reduction and melting of iron and steel in blast furnaces, basic oxygen converters, and electric furnaces were discussed in Chapter 4. Most steel is cast from furnaces of the types already described, but over half of gray iron and malleable iron base for castings is melted in the cupola. Electric furnaces are being used more for melting iron. They cost more to operate than cupolas, but do not have the pollution control problems and expenses.

8.6.1 The Cupola

The cupola is simple and economical for melting pig and scrap iron. It is essentially a vertical steel-shrouded and refractory-lined furnace. The construction of a typical cupola is shown in Figure 8-27. Cupolas are made in many sizes, commonly about 1–2 m (≈ 4–7 ft) in outside diameter and 9 to over 12 m (≈30 to over 40 ft) high. These sizes may turn out 5–10 metric tons (≈ 5–10 tons) of melted metal per hour.

A cupola must be prepared and heated carefully to avoid damage. The refractory lining is repaired or replaced as needed, and the bottom doors are propped shut. A layer of sand sloping toward the tap hole is rammed over the bottom. Excelsior, rags, and wood are placed on the sand to protect it from the initial charge of fuel, which may be a meter or more (several feet) thick, and to ignite the fuel. Materials are charged through a door 4–8 m (≈15–25 ft) above the bottom.

After the initial charge has become hot, alternate layers of fuel and metal with flux are added. The fuel may be a good grade of low sulfur coke, anthracite coal, or carbon briquettes. The flux to help form slag to remove impurities and retard oxidation of metal is usually limestone, but sometimes soda ash, fluorspar, and proprietary substances are added. The proportion of metal to fuel by weight ranges in practice from about 6:1 to 12:1. A ratio of 10:1, which is common, means that 100 kg of coke are needed to melt 1 metric ton (≈200 lb of coke for a ton) of iron. Fuel cost is lower with a lower proportion of coke; the melting and output rates are higher with more coke.

Once the charge and cupola have had a chance to become heated uniformly in an hour or two, the forced draft is started. Air is blown through the wind box and tuyeres around the hearth into the furnace. About 850 m^3 of air are required to melt 1 metric ton of iron (about 30,000 ft^3/ton). Then, the metal begins to melt. The *tap hole* is plugged with a bott of clay. The molten metal seeps down through the coke and collects at the bottom. Tapping is done by breaking the clay plug in the tap hole. Flow may be permitted at intervals or may be continuous during the melt. For the latter case, the melting rate must be in balance with the capacity of the tap hole. The slag floats on top of the

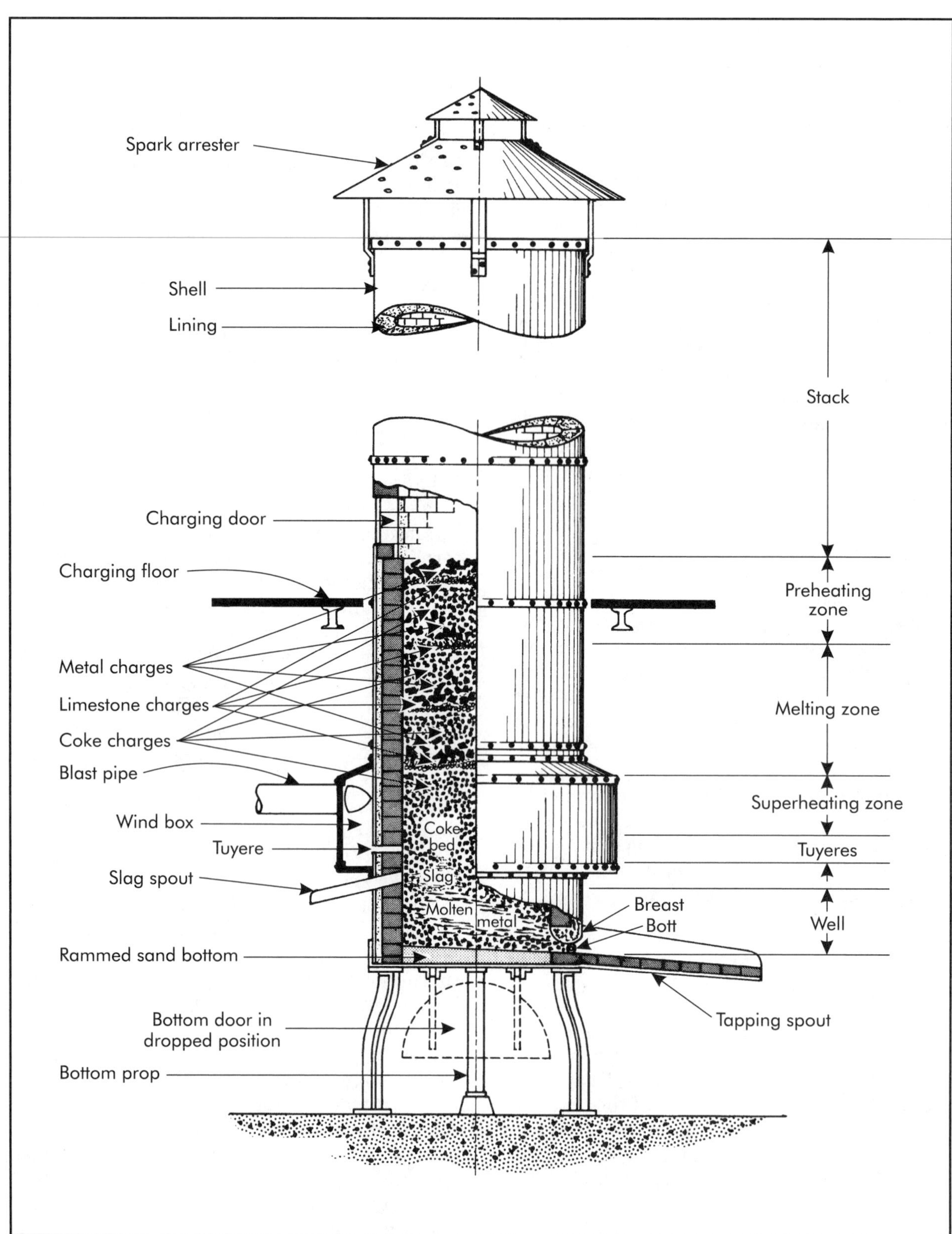

Figure 8-27. External and internal views of a classic cupola.

metal. If the molten metal is allowed to accumulate in the hearth, the slag flows off through an opening higher than the tap hole, called the *cinder notch*, in the back of the cupola. Otherwise, the slag may be skimmed off the metal as it flows out of the cupola. When sufficient metal has been melted, the bottom of the cupola is dropped to spill the remaining contents onto the ground to cool.

Calculations

The cupola does little to refine the metal, and the composition of its product depends largely upon what is put into it. The proportions of the metals charged into the cupola must be calculated carefully to assure a uniform and predictable product. These calculations are based upon knowing the amounts of carbon, silicon, manganese, phosphorus, and sulfur in the pig and scrap iron, and the nature of the reactions that take place in the cupola.

The following example shows how the calculations are made for 1,500 kg and alternately for 3,000 lb of iron fed into the cupola. The typical raw materials available in the storage yard of the foundry are listed in Table 8-4 with their analyses. In other foundries, still other materials, such as machinery steel scrap and iron of various analyses, also may be stocked depending upon the sources of supply and the desired product. In this example, on the basis of current practices and results in the foundry, it is decided to compose the charge of 10% No. 1 pig iron, 20% No. 2 pig iron, 30% new scrap iron, and 40% returns from previous melts. The amount of each element that may be expected in the product can now be ascertained on the basis of the reactions in the cupola.

1. The amount of carbon in the iron remains substantially unchanged during the process. Some carbon is oxidized, but about the same amount is picked up from the fuel. The amount contributed by each ingredient is:

   For 1,500 kg of iron:
   No. 1 pig iron
   $1,500 \times 0.10 \times 0.035 = 5.25$ kg
   No. 2 pig iron
   $1,500 \times 0.20 \times 0.035 = 10.5$ kg
   New scrap iron
   $1,500 \times 0.30 \times 0.034 = 15.3$ kg
   Returns
   $1,500 \times 0.40 \times 0.033 = 19.8$ kg
   Total 50.85 kg

   For 3,000 lb of iron:
   No. 1 pig iron
   $3,000 \times .10 \times .035 = 10.5$ lb
   No. 2 pig iron
   $3,000 \times .20 \times .035 = 21.0$ lb
   New scrap iron
   $3,000 \times .30 \times .034 = 30.6$ lb
   Returns
   $3,000 \times .40 \times .033 = 39.6$ lb
   Total 101.7 lb

   So, the final percentage of carbon equals

   $$\frac{50.85}{1,500} \times 100 = 3.39\% = \frac{101.7}{3,000} \times 100 \quad (8\text{-}2)$$

2. The silicon content can be expected to be reduced by 10% from oxidation. The amount of silicon in the charge is:

   For 1,500 kg of iron:
   No. 1 pig iron
   $1,500 \times 0.10 \times 0.025 = 3.75$ kg
   No. 2 pig iron
   $1,500 \times 0.20 \times 0.030 = 9.0$ kg
   New scrap iron
   $1,500 \times 0.30 \times 0.023 = 10.35$ kg
   Returns
   $1,500 \times 0.40 \times 0.025 = 15.0$ kg
   Total 38.1 kg

Table 8-4. Compositions of some typical metals for cupola melting (%)

	Carbon	Silicon	Manganese	Phosphorus	Sulfur
No. 1 pig iron	3.5	2.50	0.72	0.18	0.016
No. 2 pig iron	3.5	3.00	0.63	0.12	0.018
Cast-iron scrap	3.4	2.30	0.50	0.20	0.030
Returns (risers, defective castings, etc.)	3.3	2.50	0.65	0.17	0.035

For 3,000 lb of iron:
No. 1 pig iron
3,000 × .10 × .025 = 7.5 lb
No. 2 pig iron
3,000 × .20 × .030 = 18.0 lb
New scrap iron
3,000 × .30 × .023 = 20.7 lb
Returns
3,000 × .40 × .025 = 30.0 lb
Total 76.2 lb

The final percentage of silicon equals

$$\frac{38.1 \times 0.9}{1{,}500} \times 100 = 2.29\% = \frac{76.2 \times 0.9}{3{,}000} \times 100 \quad (8\text{-}3)$$

3. The manganese content is expected to be reduced by 20% from oxidation. In the same way as for the other elements, the manganese content is estimated to be 9.12 kg for a 1,500-kg charge and 18.24 lb for a 3,000-lb charge, and the final percentage of manganese equals

$$\frac{9.12 \times 0.8}{1{,}500} \times 100 = 0.49\% = \frac{18.24 \times 0.8}{3{,}000} \times 100 \quad (8\text{-}4)$$

4. Phosphorus losses in the cupola are negligible. In the same way as for the other elements, the amount of phosphorus in the charge is calculated to be 2.55 kg for 1,500 kg of iron and 5.10 lb for 3,000 lb of iron. From this, the final percentage of phosphorus equals

$$\frac{2.55}{1{,}500} \times 100 = 0.17\% = \frac{5.10}{3{,}000} \times 100 \quad (8\text{-}5)$$

5. The iron loses almost no sulfur in melting, but picks up about 4% of the sulfur in the coke. The quantity of sulfur in the metal charge is calculated to be 0.423 kg for 1,500 kg of iron and 0.846 lb for 3,000 lb of iron. The iron to coke ratio is to be 8:1, and the sulfur content of the coke is known to be 0.50%. Thus, the quantity of sulfur in the coke is (1,500/8) 0.005 = 0.9375 kg, and of that, 4% or 0.0375 kg is added to the iron for a 1,500-kg charge. Alternatively, the sulfur in the coke is (3,000/8) 0.005 = 1.875 lb, and .075 lb is added to a 3,000-lb charge. The final sulfur content is estimated to be

$$\frac{(0.423 + 0.0375) \times 100}{1{,}500} = 0.03\%$$

or (8-6)

$$\frac{(.846 + .075) \times 100}{3{,}000} = 0.03\%$$

In summary, the composition of the iron from the cupola is estimated to be 3.39% carbon, 2.29% silicon, 0.49% manganese, 0.17% phosphorus, and 0.03% sulfur.

If more or less of any of the elements is wanted, the results can be changed by specifying raw materials in different amounts and of different kinds and by adding ferroalloys of elements to the metal in the ladle after melting. Ferroalloys are less expensive than the pure elements. For instance, if in the example 2.50% silicon is desired, about 2 kg (≈4 lb) of 50% ferrosilicon can be added to each 500 kg (1,102 lb) of molten iron. Sometimes elements are added by bubbling gas carriers through the metal in the ladle.

8.6.2 Melting of Nonferrous Metals

Some nonferrous metal melting is done in almost all types of furnaces. More is done in induction furnaces for reasons of convenience, ease of operation, and fewer environmental problems, but oil- and gas-fired crucible furnaces have advantages. Although the fuel cost is about the same for both types of furnaces, an electric furnace may cost twice as much initially as an oil- or gas-fired one. Modern electric furnaces have complicated controls and are costly to service, whereas all a fossil-fuel furnace needs is relining from time to time. On the other hand, oil- and gas-fired furnaces create heat, fumes, and noise problems. A stack is needed, and a baghouse may be required depending upon local environmental regulations. It is also difficult to find people to work around hot fumes, particularly in hot weather.

Two types of crucible furnaces are stationary and tilting furnaces. The *stationary type of crucible furnace* requires that the crucible be lifted in and out for pouring. A typical furnace of this type is shown in Figure 8-28. When a stationary furnace is sunk into the floor or deck of a foundry, it is known as a *pit-type furnace*.

Figure 8-28. Stationary, high-speed, gas-fired crucible-type furnace for melting nonferrous metals. (Courtesy McEnglevan Industrial Furnace Company, Inc.)

A *tilting type of crucible furnace*, shown in Figure 8-29, requires a crucible with a suitable lip for pouring metal when the furnace is tilted.

Most nonferrous metals and alloys oxidize, absorb gases and other substances, and form dross readily when heated. Various practices are followed for each kind of metal to preserve purity and obtain good castings. Space does not permit descriptions of all practices. Full information may be found in treatises on the subject. The melting of aluminum will be treated in some detail here to point out the nature of the problems and principles.

Aluminum and its alloys have a marked tendency to absorb hydrogen when heated. This gas is released on cooling and causes detrimental pinholes and porosity in castings. Exposure to hydrogen-forming media, such as water vapor, must be avoided. Clean and dry melting stock and crucible are important; a slight excess of air in the furnace atmosphere is desirable.

Molten aluminum reacts readily with oxygen to form a film on the surface. Fortunately, the dross serves as a good shield against hydrogen and further oxidation if not broken. Excessive dross may become trapped in the metal, particularly if the metal is agitated, and appears as defects in the final casting. Both the amount of oxidation and the tendency to absorb hydrogen increase with temperature and time. Excessive

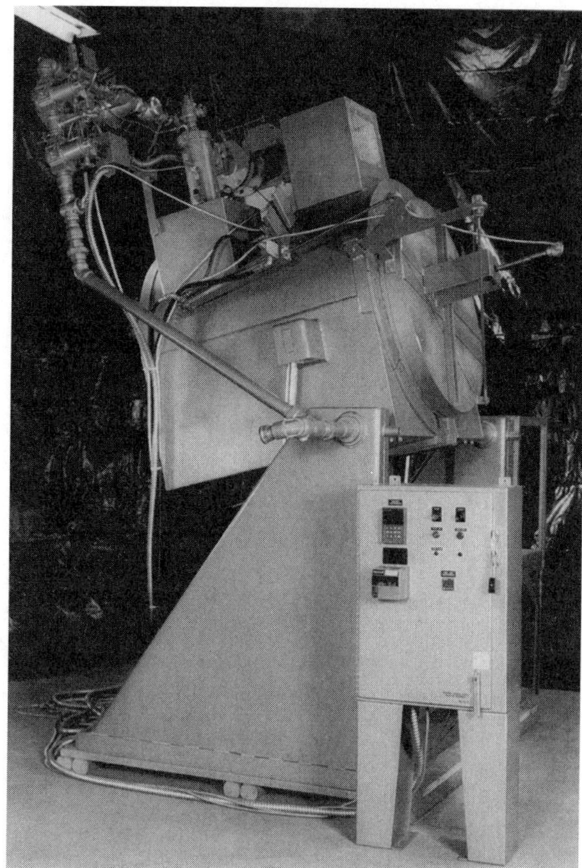

Figure 8-29. Hydraulic, nose-pour type crucible furnace for melting nonferrous metals. (Courtesy McEnglevan Industrial Furnace Company, Inc.)

temperature also causes coarse grains in castings. These conditions dictate the melting procedure for best results. Aluminum should not be heated more than about 56° C (100° F) above the necessary pouring temperature. Temperature should be checked with an immersion pyrometer. The melting time should be short with as little agitation as possible.

Aluminum does not ordinarily require as much flux for protection as some other metals because of its oxide film. However, at times the gas or oxide in the metal must be reduced. Chlorine or nitrogen may be bubbled through the molten metal. Solid fluxes, containing aluminum or zinc chloride, may be added. Various proprietary fluxes are commonly used to help dry the surface dross and facilitate skimming it from the metal.

Vacuum Melting

It has become necessary to develop ways of melting and pouring some metals and alloys in the absence of air to make and keep them pure and clean. This is done in a number of ways. A typical installation is depicted in Figure 8-30, which represents true vacuum melting and pouring. Operating temperatures with this equipment range up to 1,649° C (3,000° F). Some such furnaces operate at less than one millionth of normal atmospheric pressure, and most under 10 microns (≃133 microbars of Hg at 0° C). The metal in the crucible is melted by induction, and the induced field stirs the liquid constantly and aids in the release of gases. This is called *vacuum induction melting* (VIM). Energy for melting is supplied by electron beams (Chapter 27) in some installations. The metal is poured by tilting the crucible; in some cases, the whole furnace tilts.

Arrangements such as those already described eliminate the two contaminating media, air and slag, from the casting process. The third source of impurities, the crucible, is neutralized by the *consumable electrode* method. This utilizes an arc in a vacuum, between an initial charge and an electrode of the metal melted, inside a water-cooled copper crucible. As metal is melted off the electrode by the arc, it is quickly solidified, especially where in contact with the cooled crucible. Diffusion between crucible and melt is minimized, and crystalline structure is improved. This process is called *vacuum arc remelting* (VAR) and is commonly a second step after VIM for high-purity alloys.

Vacuum arc direct electrode remelting (VADER) is a process that remelts like VAR but is faster and produces smaller grains in the metal. Two ingots to be remelted are held horizontally in line in a

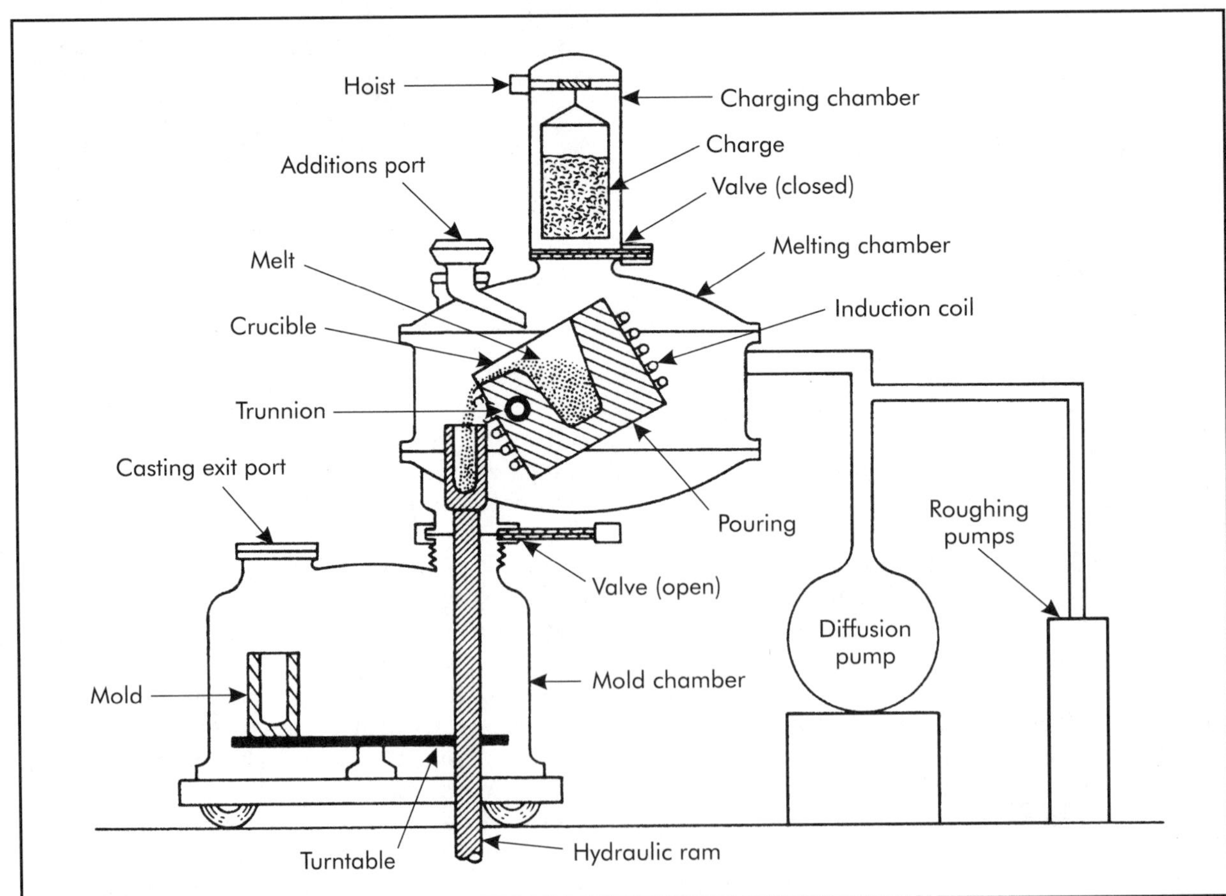

Figure 8-30. Schematic drawing of a vacuum melting furnace.

vacuum over a mold. An arc is struck between the ingots and drops of molten metal fall into the mold. All the energy is used for melting; none is required to maintain a molten pool. Melting rates at least three times that for VAR are reported.

A process called *degassing* entails pouring only in a vacuum. The metal is melted by conventional means in air. The vacuum need not be extremely high to draw off substantial proportions of hydrogen and other gases. Four common methods are depicted in Figure 8-31.

Stream degassing is done with a ladle of molten steel positioned over an evacuated chamber containing another ladle or ingot mold. When the nozzle is opened, steel pours into the chamber, giving out gas as it falls. Only one ladle is needed for *ladle degassing*; it is filled with molten steel and placed in the vacuum chamber which is then evacuated. During the operation, continuous stirring, by induction or helium injected at the bottom, constantly brings untreated steel to the surface for degassing. In the *D-H* and *R-H* processes, increments of molten steel are degassed in small chambers. The *D-H* chamber (or the ladle) surges up and down. On the first stroke, steel enters the chamber and is degassed, and on the reverse, it sinks back into the ladle. In contrast, the *R-H* process is continuous. Molten steel, aided by the injection of gas in the up leg, flows up into the chamber, is degassed, and then drops back down the down leg.

Uranium, titanium, and alloys with reactive elements can be effectively cast only in a vacuum. Vacuum melting improves the properties of metals that can be melted by ordinary means, such as steel and aluminum. Raw materials of good grade are, of course, necessary for a superior product. Metals melted and cast in a vacuum can be kept pure because they are not exposed to new contamination and can be further purified because gases are drawn from them. Oxidizers, such as manganese and silicon in steel, are not needed, and no slag is formed. In vacuum melting, precise amounts of pure carbon or hydrogen may be added to reduce oxides. As gases are evolved, they are extracted by the pumps, and the reactions are driven to completion. The addition of highly reactive elements like zirconium, titanium, and aluminum to an alloy that requires them may be delayed until after the gas content of the melt has been depleted.

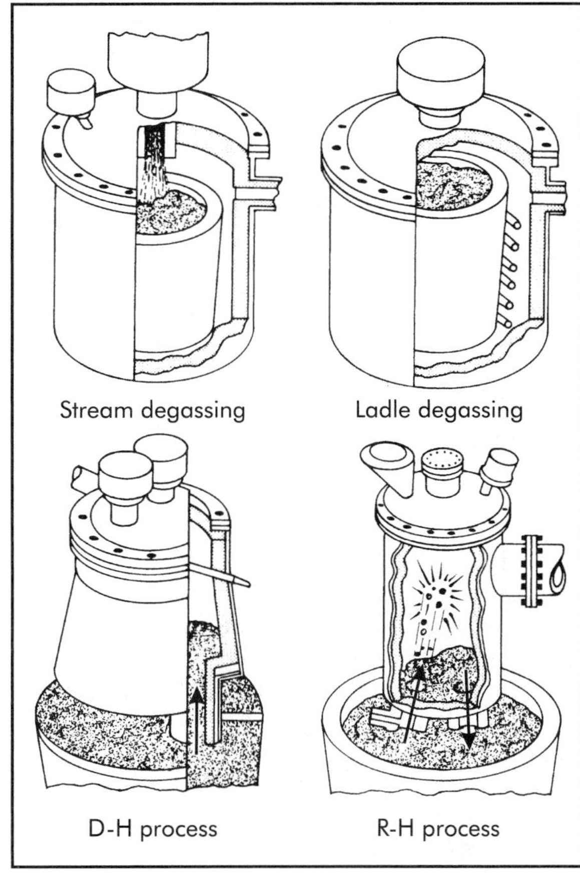

Figure 8-31. Techniques of vacuum degassing. (Courtesy Metal Progress)

Vacuum-melted steel may contain 3–20 parts of oxygen per million parts of steel. This is about one-twentieth of the proportion in commercial air-melted steel; other impurities are reduced to about the same extent. Pure and clean metals are substantially stronger, more ductile, and more resistant to fatigue and corrosion. For instance, the Charpy impact strength at room temperature has been reported to be 298 J (220 ft-lbf) for vacuum-melted 430 stainless steel compared to 11 J (8 ft-lbf) for air-melted. The benefits obtained have been found particularly desirable for high-temperature alloys, such as nickel and cobalt alloys. Thus vacuum melting is especially in demand for metals to make products such as turbine blades, buckets, bearings, castings, after-burner parts, highest-quality tool steels, and all parts that must serve at high temperatures with high strength.

Vacuum melting is inherently more costly than air melting. Although liquid melts up to 14 metric tons (15 tons) are reported for VIM, most installations are of much smaller capacity because of high equipment costs. Melts of tens of metric tons (tons) are produced by VAR, and as much as several hundred metric tons (tons) by degassing. An 1,100-kg (≈2,500-lb) capacity production unit of the type illustrated in Figure 8-30 may cost several million dollars with all equipment and accessories.

8.7 POURING AND CLEANING CASTINGS

8.7.1 Pouring Methods

Common practice for pouring is to run the molten metal from the cupola or furnace into a large receiving ladle. From that, metal is distributed to smaller pouring ladles. These range in size from ones that can be handled by one man to huge crane ladles holding hundreds of tons. The bottoms of the ladles and the sides of large ladles are lined with fire brick. The bottom and sides of a ladle are daubed with an inner coating of fire sand and clay, which is hardened by baking.

Liquid metal may be delivered to a mold in one or more ladles. It is important to have sufficient metal to fill the mold, gates, and risers completely. Pouring must be done continuously and at a uniform rate until the mold is full to keep the slag from settling. The metal temperature must be high enough for the fluid to pour easily and rapidly. Slag is always present on molten iron to some degree and it must be kept out of the mold to avoid weak slag pockets in the casting.

Foundry practices differ in some respects for small-quantity versus large-quantity production. A small quantity may be several pieces up to a few thousand castings of a kind. Low-cost patterns are used with universal handling equipment for a variety of work. Today it is common to have a computerized system guiding the flow of work even with small batches.

It is estimated that over 90% of foundries are automated for large-quantity output. An example of such a system is given in Figure 8-32. In common production practice, a match plate containing the drag half of a pattern is rigged to a molding machine on which drag halves of a mold are produced continuously. These are placed on an endless conveyor. Cores are set in position as each drag moves along. Cope halves are molded on another machine and placed on the drag halves. The closed molds pass alongside ladles travelling at the same speed on a parallel conveyor and are filled. After the molds cool enough, they are dumped onto a grating to strip out the castings. The flasks return to the molding machines, and the cycle is repeated.

In a modern high-production process called *no-flask molding*, sand molds are squeezed to shape in a chamber and pushed out of the machine into a line of molds. The molds are set with parting lines vertical and are kept closed and supported by being pressed together back to back in the line. Metal is poured into the molds at about the middle of the line. As each new mold is pushed into place, the line is moved ahead, and a new mold is shoved off onto a conveyor and stripped. This system is depicted in Figure 8-33(A).

In the *vacuum casting* process, a heated sheet of plastic is fitted over the pattern and placed in a flask. The flask is vibrated while being filled with a fine, dry, binder-free sand. The top of the flask is covered with another sheet of plastic. A vacuum of about 381 mm (15 in.) of mercury is drawn through the sand during pouring and cooling. There is no ramming, jolting, or squeezing, and no wet sand. The mold falls apart when the vacuum is released. The production rate may not be as high as with conventional molding, but smaller tolerance and better surface finish are obtained for some castings.

8.7.2 Cleaning Castings

After castings have solidified and cooled, they are removed from the sand and cleaned.

A casting may be separated from the sand in various ways. A rudimentary way for a small casting is to dump the mold assembly upside down, remove the bottom board and flask, and then pull the casting from the sand with a hook bar. After being removed from the sand, the casting is rapped with a hammer to dislodge any clinging material. Essentially the same method is done mechanically to castings made in large quantities. The flasks are emptied on a vibrating metal

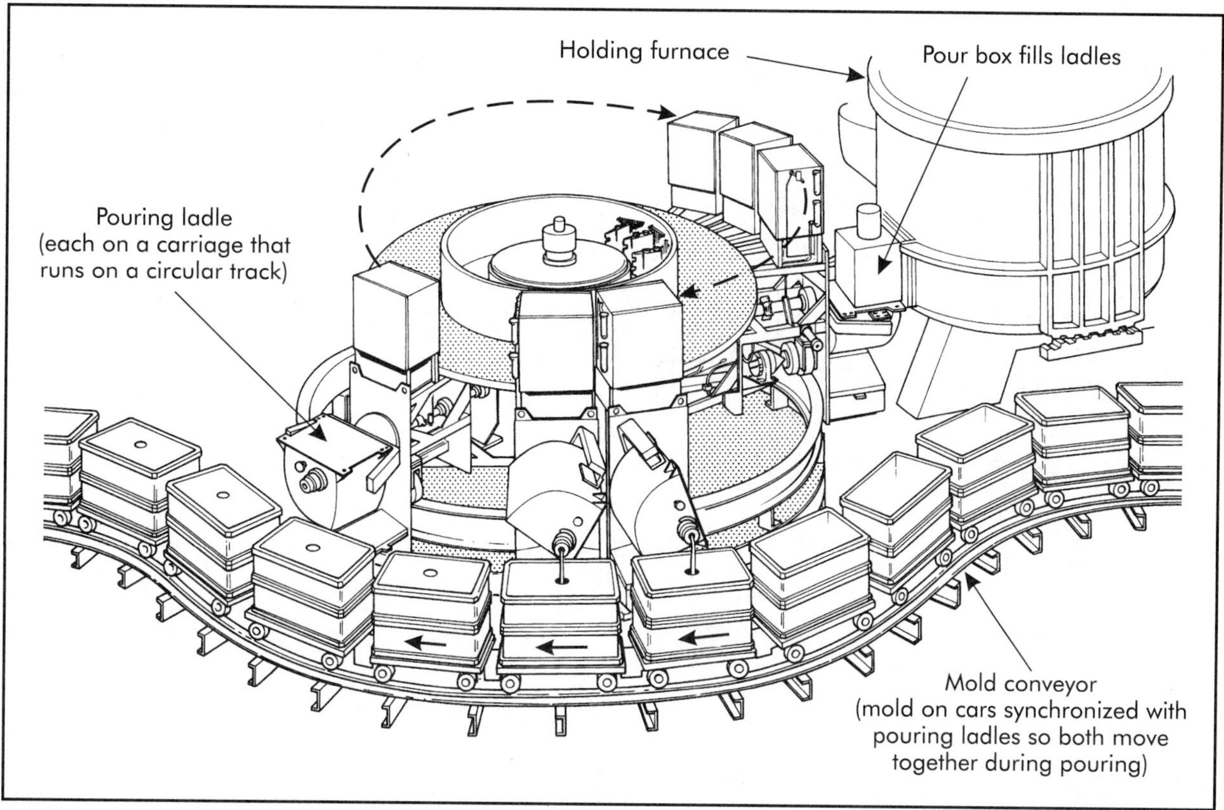

Figure 8-32. Sketch of an automated molding line capable of pouring 110 metric tons (120 tons) of iron/hr.

grating. The sand falls through, and the castings are pushed or jolted along to fall off or are stripped at the end onto a conveyor. The sand is reprocessed for future use.

Gates are often broken off gray iron castings by hammers. They are removed from steel and nonferrous castings by sawing or flame cutting. Fins, projections, and other excesses of metal may be ground or chipped away. Clean surfaces are obtained by shot or sand blasting or wire brushing. Small castings may be tumbled with scraps of metal. These processes are described in Chapter 25.

8.8 SHELL MOLD CASTING

Shell mold casting, developed during World War II, makes use of fine sand with a phenolic resin binder for a molding material. This mixture is deposited on a metal pattern heated to about 232° C (450° F); a layer around the pattern coalesces, and after a fraction of a minute, the excess is dumped off the pattern. The thickness of the shell depends upon the time the bulk of the material stays on the pattern. Thickness is usually from 4.78–9.53 mm (.188–.375 in.), as required for the job. The resin is cured, and the shell is hardened by baking at 316–427° C (600–800° F) for a fraction of a minute. The shell is stripped from the pattern, which is then cleaned and sprayed with a silicone parting compound to prepare it for the next shell.

One method of shell molding is to blow the molding material onto the pattern. The material can be deposited uniformly, forced into intricate forms, and controlled in amount. A simpler method consists of placing the heated pattern face down on top of a box partially filled with molding material. The box is turned upside down for the required length of time and then reinverted. The pattern with the green shell is then taken off the box. The *dumping method* does not always produce as accurate shells as the *blowing method,* but it is more readily adapted to automatic operation.

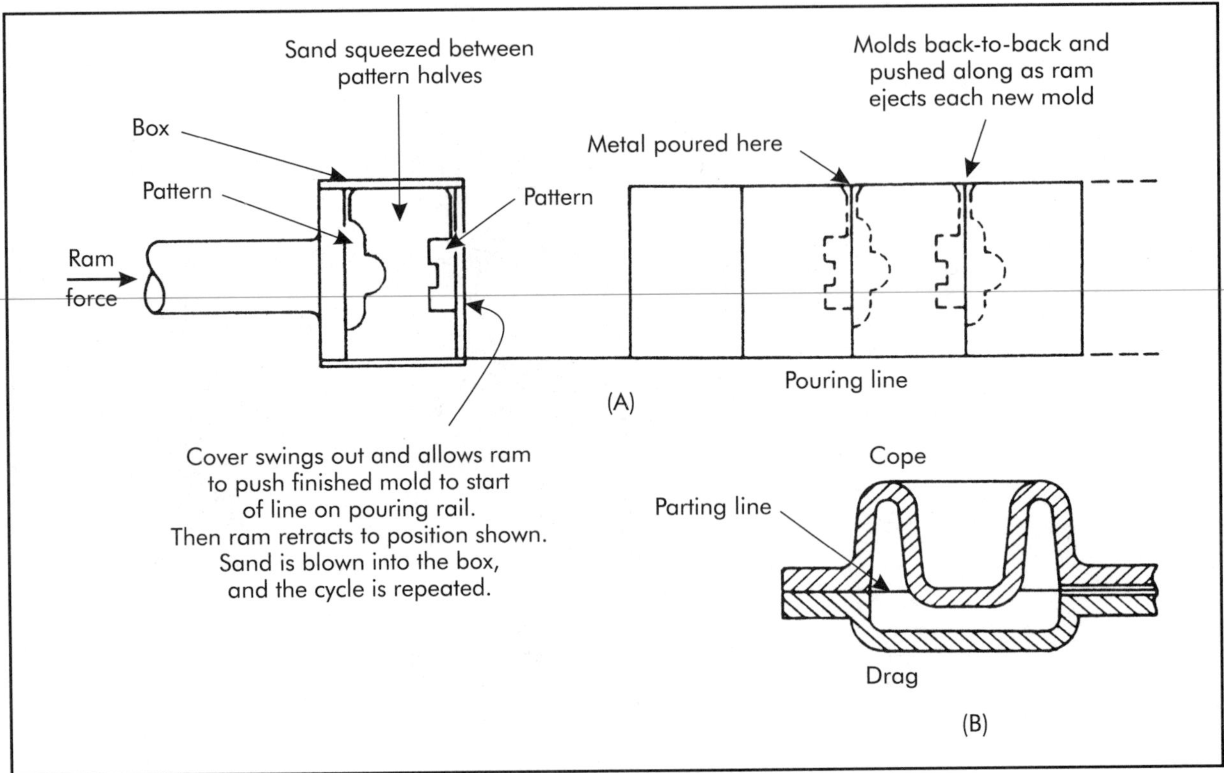

Figure 8-33. Schematics of two modern production sand casting processes: (A) no-flask molding; (B) section of a shell mold.

Cope and drag shells are made for a mold and joined together as indicated in Figure 8-33(B). They are often glued together in production. The thickness of the adhesive film may vary from 0.3–0.5 mm (.01–.02 in.), which is not desirable for precision castings. Molds pasted together may be embedded in sand or shot if reinforcement is necessary to counteract the hydrostatic pressure of the metal. Molds for a precision casting are clamped together and, in production, are held in metal backup fixtures, especially large and heavy castings. The phenolic resin binder in the shell is mostly burned away, particularly in cores, by the hot metal. The remaining sand is easily stripped from the surface and cleaned from the cavities of the casting by jarring, shaking, and tumbling.

8.8.1 Advantages

Shell molding is more expensive than green sand molding in most cases. This is because the molding material costs four to five times as much as sand alone. Some of this cost, but usually not all, is saved because of certain shell molding advantages. The savings arise because much less sand needs to be worked and handled; the molds are light, easy to handle, and can be made when convenient and stored; gases escape readily through the thin shells and fewer castings are scrapped because of blowholes or pockets; and the process is readily adaptable to mechanization.

The cost of shell molding is justified for the many jobs that it can do better than green sand molding. Complex shapes and intricate parts that cannot be cast in green sand can be made by shell molding. Usually, a better finish is obtained and closer tolerances are held by shell molding. Less stock needs to be left for machining castings made to closer tolerances, and often no machining at all is needed. A thin shell does not have the mass chill effect of a thick sand mold. Thus, shell mold castings suffer less from varying cross sections and tend to have softer skins than castings from green sand molds.

8.9 METALLURGY OF CASTINGS

Steel castings have low carbon content and possess essentially the properties as those attributed to steel in Chapter 4. With more carbon, the product is cast iron, with or without appreciable quantities of alloys. Cast iron in its common forms will be discussed in this section.

8.9.1 Cast Iron

Cast iron has several commercial forms called gray irons, white cast irons, malleable irons, and nodular cast irons. Pig iron, described in Chapter 4, is also a form of cast iron. Cast irons usually contain from 2.0 to about 4.5% carbon. Typically their constituents are those represented in that range of the iron-carbon diagram of Figure 3-13. The various forms of cast iron represent various combinations of iron and carbon and their compounds, depending upon how the material is cooled and the presence of silicon and other alloys.

Molten cast iron contains much but not all of the carbon in solution as iron carbide. When the iron is cast and cools, its ability to hold the carbide in solution decreases. At moderate cooling rates common in practice, much of the iron carbide decomposes, and the carbon is precipitated out as graphite flakes. As the metal becomes solid, iron carbide, called *cementite*, is restrained from further breakdown and ends up in a pearlitic structure. Pearlite was defined in Chapter 3. The result is a pearlite matrix with graphite flakes dispersed throughout. A microscopic view of such a structure is shown in Figure 8-34A. The edges of the graphite flakes appear as heavy black lines. The flakes have no strength of their own, break up the continuity of the iron, and make the cast iron weak and brittle. The free graphite imparts the nominal coloring to *gray iron*, which usually contains 3–4% carbon.

Other forms of cast iron than the gray iron just described result when the metal is cooled at slower or faster rates. Upon slow cooling and particularly with a high silicon and carbon content, considerable ferrite, as well as pearlite, is formed in the matrix because more of the carbon is released. The graphite becomes coarser. The ferrite is softer and not as wearable as pearlite. The product is sometimes called *open* or *soft cast iron*.

When the cooling rate is fast, the iron carbide has less chance to decompose before the metal solidifies. Thus more cementite is deposited in the mass than can be accommodated in the pearlite, which must have a definite composition.

The presence of cementite makes cast iron hard. When only part of the available carbon appears as dispersed cementite, the product is called *mottled iron* from its appearance.

White cast iron has carbon largely as cementite and consequently is extremely hard and brittle. The absence of free carbon is evident in the photomicrograph of Figure 8-34B. This condition is caused by very rapid cooling or by additives that stabilize the carbide.

It may be desirable to harden certain areas of a casting for wearability. This is achieved by cooling the areas rapidly by means of metal chills in the mold wall, and the product is called *chilled cast iron* in *chilled castings*.

Malleable iron has free carbon in tiny lumps, rather than flakes, in a ferrite matrix, sometimes with pearlite. This makes it ductile, shock resistant, and easily machinable. The structure is obtained by first producing essentially a white iron casting, reheating the casting above the transformation range, and soaking and slowly cooling it. This separates the graphite from the cementite while the metal is solid and prevents the growth of flakes like those that arise when molten metal hardens.

Ductile, nodular, or *spheroidal graphite iron* has its graphite in tiny spherulites. This form is obtained by adding (just before pouring) minute amounts of magnesium, cerium, or other elements to an iron relatively high in carbon and silicon content. Ductile iron must have a sulfur content below 0.01%.

Elements other than iron and carbon are normally present in cast iron as contamination or additional additives. The properties of cast irons can be varied widely by adding suitable kinds and amounts of alloying elements. The control of graphitization is an important reason for using alloying elements. Other purposes may be to increase strength or resistance to corrosion,

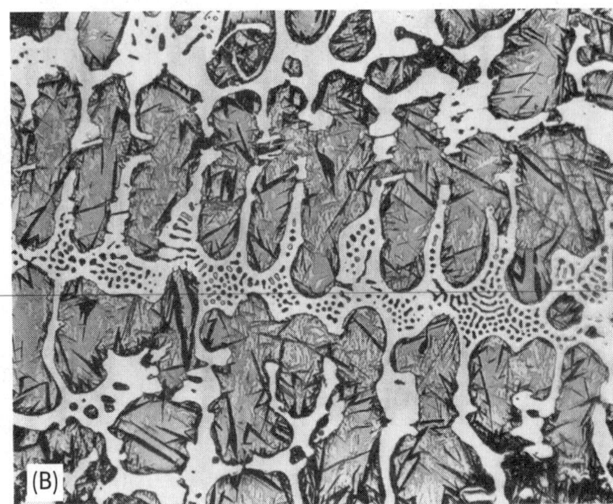

Figure 8-34. (A) Microstructure of a gray cast iron with a ferrite-pearlite matrix. Note the graphite flakes dispersed throughout the matrix, 4% picral etch, magnified 320×. (B) Microstructure of a white cast iron. The white constituent is cementite and the darker is martensite with some retained austenite, 4% picral etch, magnified 250× (Metals Handbook 1998).

heat, or wear. The principal alloying elements for cast iron and their purposes are discussed in the following section.

Alloying Elements

The most important effect of *silicon* is to promote the decomposition of cementite into ferrite and carbon. The effect increases with the amount of silicon added. Enough should be used for gray iron to decompose all the massive cementite but not the pearlite. Requirements generally range from 0.50–3%. A practice of recent years to save energy is to add an excess of silicon to the down sprues in molds to inhibit the formation of hard carbides and avoid the necessity of annealing the castings.

Sulfur exists to some extent in all irons. It comes from pig iron, scrap, and the coke consumed in cupola melting. Sulfur markedly restrains the decomposition of cementite and combines with iron to form iron sulfide, which has a tendency to weaken the grain structure.

Manganese mitigates the effects of sulfur by forming manganese sulfide, which segregates as harmless inclusions. Most irons from American ores contain only 0.06–0.12% sulfur, so 0.50–0.80% manganese is sufficient. In such amounts, manganese has little effect on the properties of cast iron other than inhibiting the action of the sulfur. In excess, manganese would tend to combine into carbides and make the iron harder.

Phosphorus content in most American irons is from 0.1–0.90%. Below about 0.50%, phosphorus has little effect upon the properties of iron. In larger amounts, it tends to form brittle iron phosphide, which may add some hardness and wearability, but weakens the iron and appreciably lowers its impact strength. Phosphorus increases the fluidity of molten iron and improves castability in thin sections. Graphite formation is promoted by the increased fluidity.

Nickel acts as a graphitizer but is only about half as effective as silicon. Small amounts help refine the sizes of grains and graphite flakes. Most additions are from 0.25–2.0%. Larger amounts, from 14–38%, of nickel are found in austenitic gray irons that resist heat and corrosion and have low expansivity.

Chromium acts as a carbide stabilizer in cast iron and itself forms carbides that are more stable than iron carbide. Thus it intensifies chilling of cast iron, increases strength, hardness, and wear resistance, and is conducive to a fine-grained structure. Most additions of chromium range from 0.15–0.90% with or without other alloying elements. Free carbides that appear with 1% or more chromium make castings hard to machine.

With 3% chromium, white cast iron is formed. Special irons that resist corrosion and high temperatures contain as much as 35% chromium.

Molybdenum in amounts from 0.25–1.5% is added alone or with other elements to improve tensile strength, hardness, and shock resistance of castings. It is a mild carbide former but also forms a solid solution with ferrite. The presence of molybdenum in cast iron produces fine and highly dispersed particles of graphite and good structural uniformity. This improves toughness, fatigue strength, machinability, hardenability, and high-temperature strength.

Copper, usually in amounts from 0.25–2.5%, promotes formation of graphite and is a mild strengthener because it helps break up massive cementite and carbide concentrations.

Vanadium is a powerful carbide former and thus stabilizes cementite and restrains graphitization. Its effect in amounts from 0.10–0.50% is to increase strength, hardness, and machinability.

Grades

A wide variation in the properties of castings is obtainable from the selection of materials put into the melt, control of the rate of cooling, and subsequent heat treatment. A summary of properties is given in Table 8-5.

Gray cast iron produces the cheapest castings and should be considered first when its properties suffice. A classic example of an application that requires a bare minimum of strength is a window weight. For it, cast iron serves adequately at the least cost. Common applications for gray cast iron are guards and frames for machinery, motor frames, motor blocks and cylinder heads, bearing housings, valve housings, fire hydrants, pulley sheaves, and miscellaneous hardware.

Each class of ferrous castings covers a range of properties. As an example, commercial gray irons are graded in the ASTM specification A48 by tensile strength. The range is from class 20 with a minimum tensile strength of 137,891 kPa (20,000 psi) to class 60 with a minimum tensile strength of 413,674 kPa (60,000 psi).

Strength is not always the major criterion for selection of a material. For instance, gray cast irons of the weaker grades have superior qualities for applications such as resistance to heat checking in clutch plates and brake drums, resistance to heat shock in ingot and pig molds, and dampening of vibrations in machine tool members.

White cast iron is brittle, but in chilled castings imparts wear-resistant surfaces to products such as plowshares, rock crushers, and mining equipment. Malleable iron, ductile iron, and steel castings provide various degrees of strength and shock resistance for machinery of all kinds. The tensile properties of ductile iron are specified by a three-number symbol. The first number refers to minimum tensile strength in ksi (1,000 psi), the second to minimum yield strength in ksi, and the third to minimum percentage of elongation in a 50.8-mm (2-in.) gage length. For example, a highly ductile grade is 60-40-18, which is generally annealed for a ferritic matrix. At the other extreme is a high-strength grade, 120-90-02, heat treated to a high but machinable hardness. This represents a remarkable range of properties obtainable in one material.

8.9.2 Nonferrous Cast Alloys

Aluminum, magnesium, brasses, and bronzes are important nonferrous alloys that are cast in sand. Their main properties were described in Chapter 5. Except for melting and pouring at lower temperatures, the techniques for sand casting nonferrous metals are much the same as for iron founding. However, nonferrous alloys are also commonly cast by other processes and will be discussed more fully in Chapter 9.

8.10 DESIGN OF CASTINGS

An engineer must learn how to design castings that do their jobs adequately and can be made economically. The treatment of the casting process in this chapter has explained the principles that must be observed for good casting design. These principles are summarized as rules for the design of castings in Figure 8-35. In essence, the rules call for walls and sections that can be kept filled upon cooling, are uniform, do not change abruptly in thickness but blend gradually one into another, and are no more complex than necessary. Sharp corners and angles must be avoided.

Table 8-5. Typical properties of ferrous castings

Kind of Metal	Ultimate Tensile Strength[a], MPa (psi)	Yield Strength[b], MPa (psi)	Hardness (Bhn)	Elongation (%)	Relative Sand Casting Cost[c]
Gray cast iron	138–552 (20,000–80,000)	—[d]	150–300	0–3	1
White cast iron	138–621 (20,000–90,000)	—[d]	300–600	—	1.2
Malleable iron	345–827 (50,000–120,000)	207–621 (30,000–90,000)	110–285	3–25	1.5
Ductile and nodular iron	414–1,103 (60,000–160,000)	310–931 (45,000–135,000)	190–425	3–25	1.8
Cast steel (plain carbon)	345–690 (50,000–100,000)	172–345 (25,000–50,000)	110–210	18–27	2
Cast steel (low alloy)	621–1,379 (90,000–200,000)	414–1,379 (60,000–200,000)	170–370	9–20	5

[a] Compressive strength of gray cast iron about 552–1,310 MPa (80,000–190,000 psi), of white cast iron over 1,379 MPa (200,000 psi), and of the others about the same as the tensile strength.

[b] Resistance to dynamic loading nil to low for gray and white cast iron, medium for malleable and ductile irons, and high for steel.

[c] Relative costs can be expected to vary with size and shapes of castings, quality, grade, supplier, market conditions, etc. Typical cost for a 25-kg (55-lb) gray iron casting is about $0.75/kg (≈$0.35/lb).

[d] Essentially the same as the tensile strength.

8.10.1 Draft

A pattern can be drawn satisfactorily from the sand and a good casting made only if its sides are tapered away from the parting line. In addition, projections such as bosses on the sides of a casting away from the parting line, as indicated in Figure 8-36, cannot be drawn out of the sand and should be avoided. Cores must be provided for them at extra expense.

8.10.2 Tolerances

Tolerances of ±1.588 mm (±.0625 in.) for dimensions up to about 300 mm (≈12 in.) are standard commercial practice for sand castings. Tolerances as close as ±0.795 mm (±.0313 in.) are held by some foundries, but at extra cost that should not be incurred unless necessary. Shell mold castings are made to tolerances of ±0.51 mm (±.020 in.) for steel and nonferrous alloys, ±0.38 mm (±.015 in.) for cast iron, and as small as ±0.08 mm (±.003 in.) in some cases at extra cost.

The smallest tolerances can be held only on dimensions that lie entirely in one part of a mold. Several times as much tolerance must be given to dimensions that extend from one part of a mold to another. This is because a dimension across a parting line is subject to variations from closing the mold, and cores may shift.

8.11 QUESTIONS

1. List the three characteristics of special interest that may be realized from properly designed sand castings.
2. List the five types of molds that may be used for sand casting.
3. What are the principal characteristics of a mold?
4. Make a sketch of a typical sand mold and name its principal parts.
5. Explain the cooling process of liquid metal in a sand mold.

Most metals and alloys shrink when they solidify. Therefore, all members of the parts should be designed to increase in dimension progressively to one or more suitable locations where feeder heads can be placed to offset liquid shrinkage. All of the rules set forth here have been proven in practice.

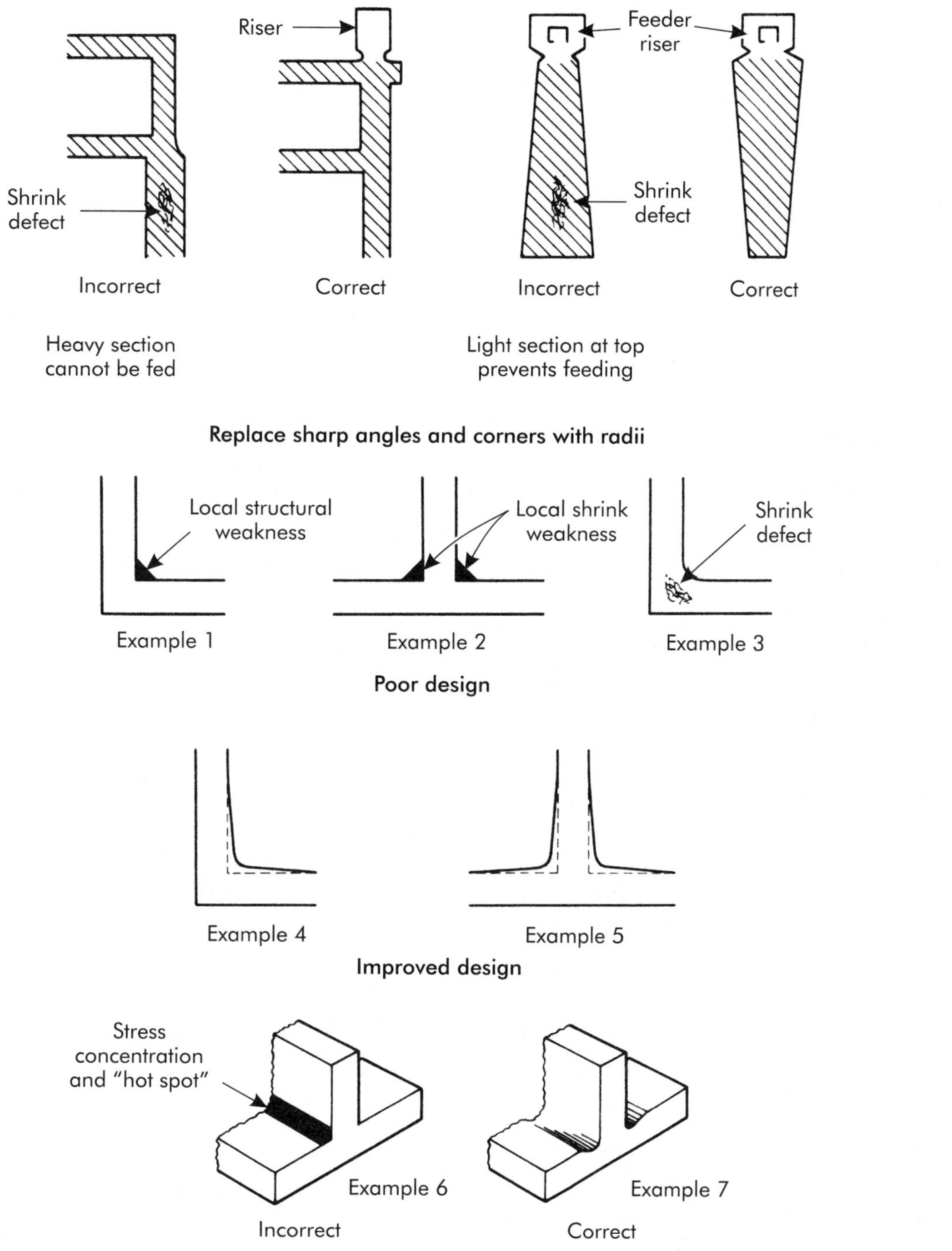

Figure 8-35. Rules for the design of castings.

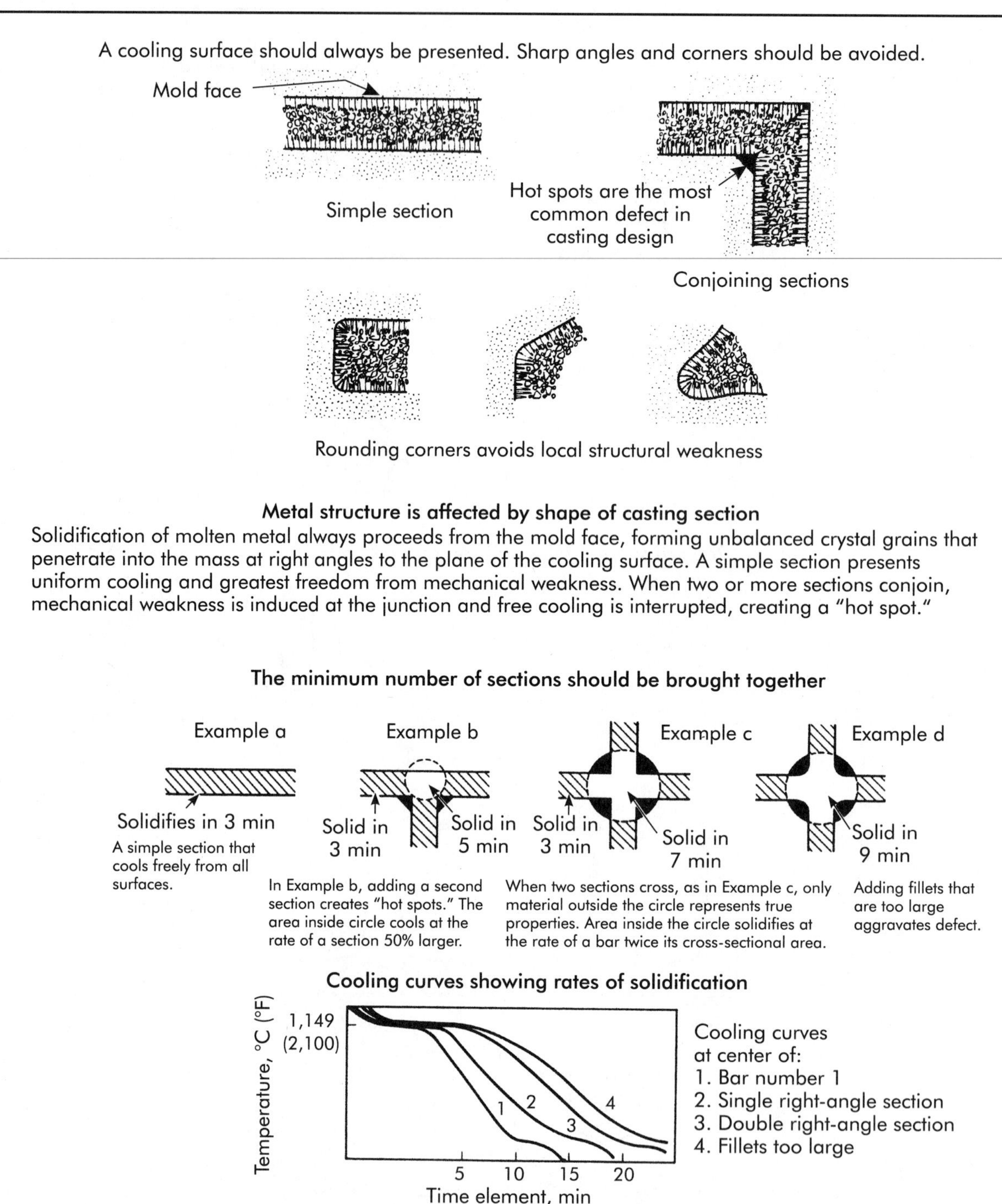

To portray the serious casting problems involved in the joining of an excessive number of members, cooling curves are made by inserting thermocouples at the adjoining sections. The results of these measurements are plotted as shown in the graph above. A well-designed casting brings the minimum number of sections together and avoids acute angles.

Figure 8-35. (cont.)

All sections should be designed as nearly uniform in thickness as possible

Cylinder with lugs

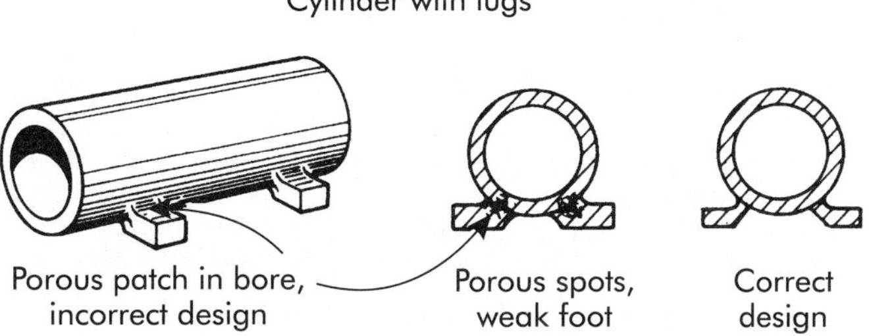

Design on left caused defects shown. Correct design shown on right. All sections should be designed as uniform in thickness as possible. Failing this, all heavy sections should be accessible for feeding.

Dimensions of inner wall should be correctly proportioned

Inner sections of castings, resulting from complex cores, cool much slower than outer sections and cause variations in strength properties. A good rule is to reduce inner sections to 9/10ths of the thickness of the outer wall. Rapid section changes and sharp angles should be avoided. Wherever complex cores must be used, the design for the section should be uniform to avoid locally heavy masses of metal.

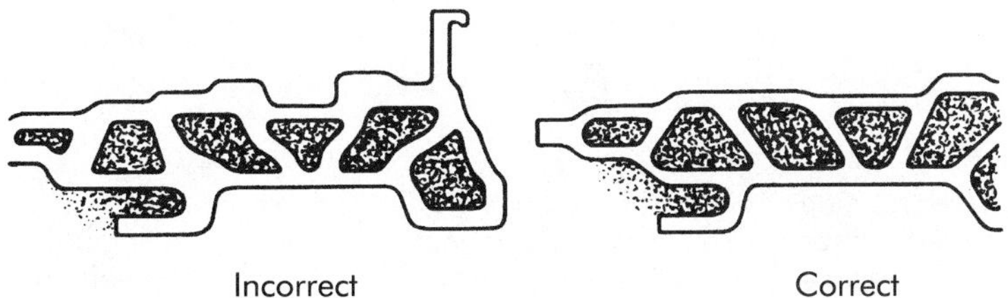

Cylinders and bushings

The inside diameter of cylinders should exceed the wall thickness of the casting.

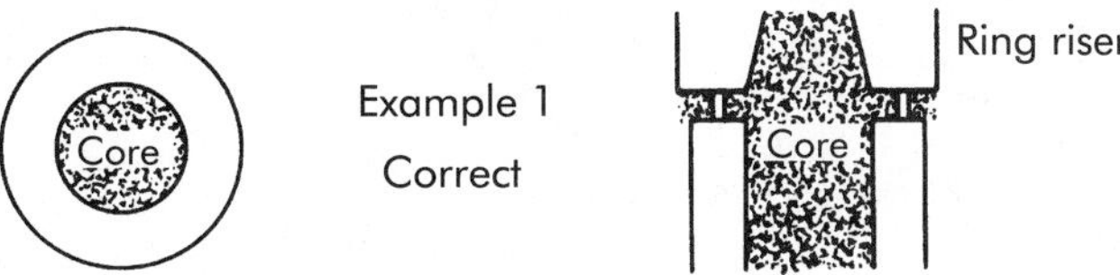

When the inside diameter of the cylinder is less than the wall thickness of the casting, as shown in Example 2, it is better to cast solid. Holes can be produced by cheaper and safer methods than by coring.

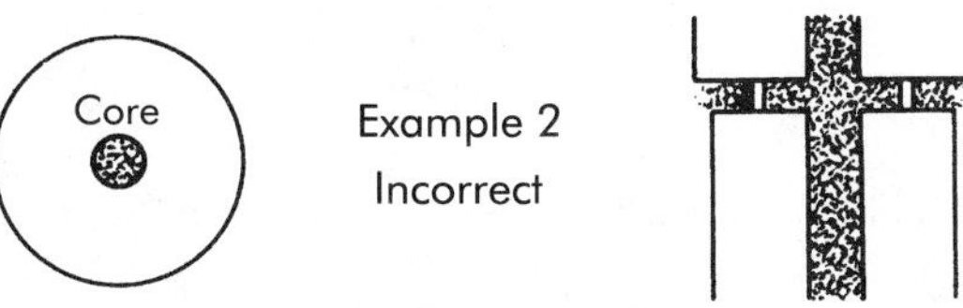

Figure 8-35. (cont.)

Ribs and brackets should be designed for maximum effectiveness

Ribs have two functions: (1) to increase stiffness, and (2) to reduce weight.
If they have too shallow a depth, or are too widely spaced, they are ineffectual.

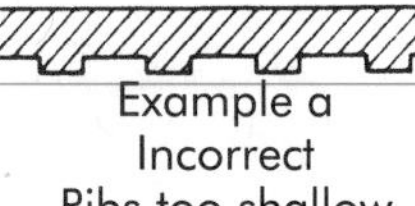

Example a
Incorrect
Ribs too shallow

Example b
Incorrect
Too widely spaced

Thickness of ribs should approximate 80% of casting thickness

Correct rib depth and spacing is a matter of engineering design.

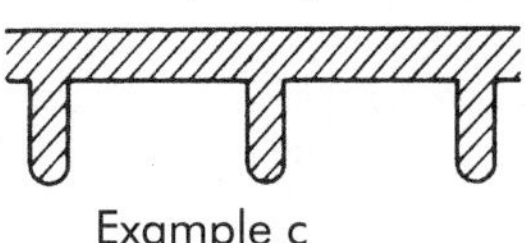

Example c
Correct

Example d
Incorrect
Thin ribs should be avoided when joined to a heavy section. Otherwise, they will lead to high stresses and cracking.

Example e
Incorrect
As far as possible, junction between ribs and main casting should prevent any local accumulation of metal.

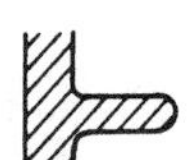

Example f
Correct
Ribs should solidify before the casting section they adjoin.

Thickness of ribs should equal 80% of casting thickness.
They should be rounded at edges and correctly filleted.

T- and H-shaped ribbed designs have the advantage of uniform metal sections and hence uniform cooling.

Design preference in average design is for ribs to have a greater depth than thickness. In general, ribs in compression offer a greater factor of safety than ribs in tension. However, castings with thin ribs or webs in compression may require design changes to provide necessary stiffening and avoid buckling.

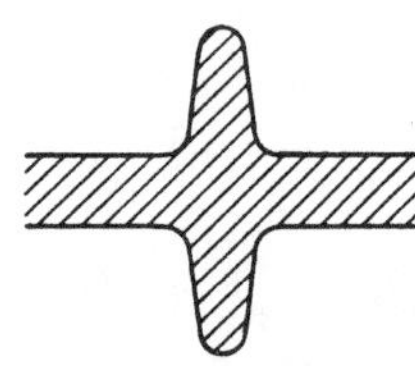

Incorrect
Ribbing on both sides of casting is undesirable from a foundry viewpoint. It increases casting defectives and cost.

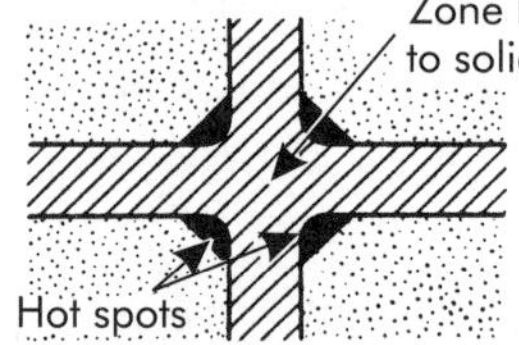

These corners of sand reach the temperature of molten iron and hold this temperature long after casting is solid.

Incorrect
Cross ribbing creates hot spots, renders feeding solid difficult, and causes local weakness and porosity.

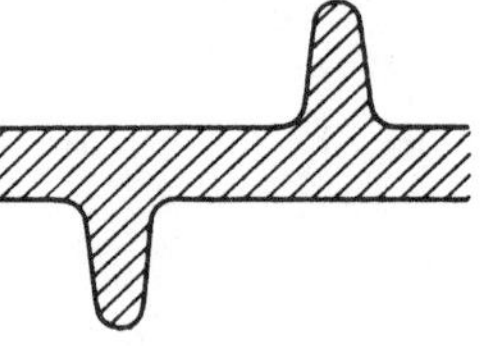

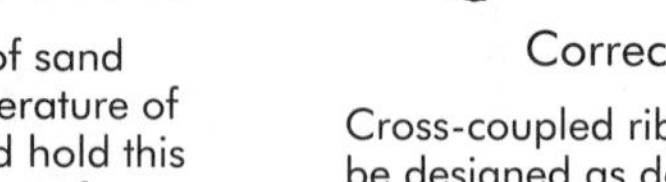

Correct
Cross-coupled ribs should be designed as double-T forms.

Complex ribbing should be avoided. It simplifies the molding procedure, assures more uniform solidification conditions, and eliminates hot spots. Casting stresses and stress distribution favor omission of ribbing if the casting wall itself can be made of ample strength and stiffness.

Figure 8-35. (cont.)

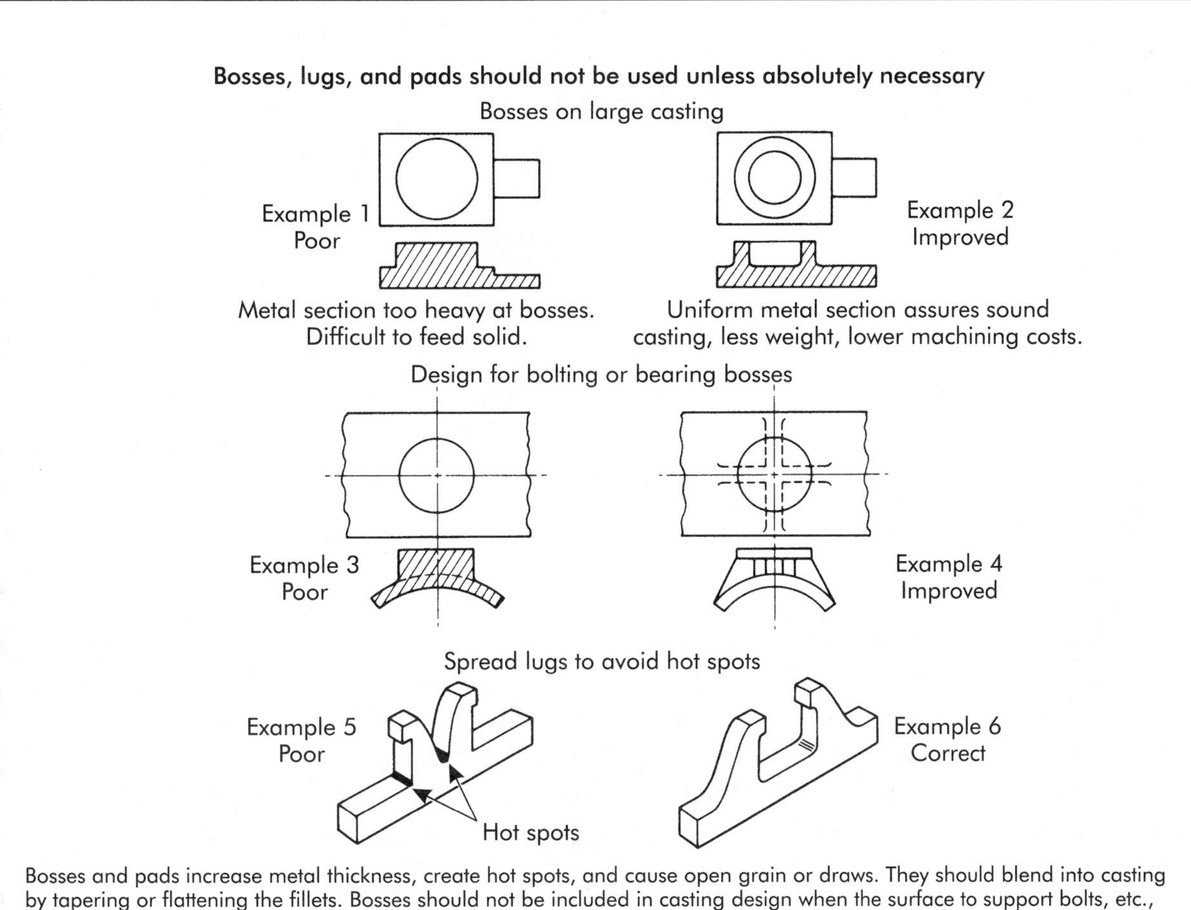

Bosses and pads increase metal thickness, create hot spots, and cause open grain or draws. They should blend into casting by tapering or flattening the fillets. Bosses should not be included in casting design when the surface to support bolts, etc., may be obtained by milling or countersinking.

A continuous rib, instead of a series of bosses, permits shifting hole location.

Preferably, thickness of bosses and pads should be less than the thickness of the casting section they adjoin, but thick enough to permit machining without touching the casting wall. Where the casting section is light and does not permit use of this rule, then the following minimum recommended heights can serve as a guide.

Approximate casting length, m (ft)	Height of boss, mm (in.)
Up to 0.5 (1.5)	6.4 (.25)
0.5–1.8 (1.5–6)	19.1 (.75)
Over 1.8 (6)	25.4 (1.0)

Example c
Incorrect

Example d
Correct

A continuous rib instead of a series of bosses permits shifting hole location.

Figure 8-35. (cont.)

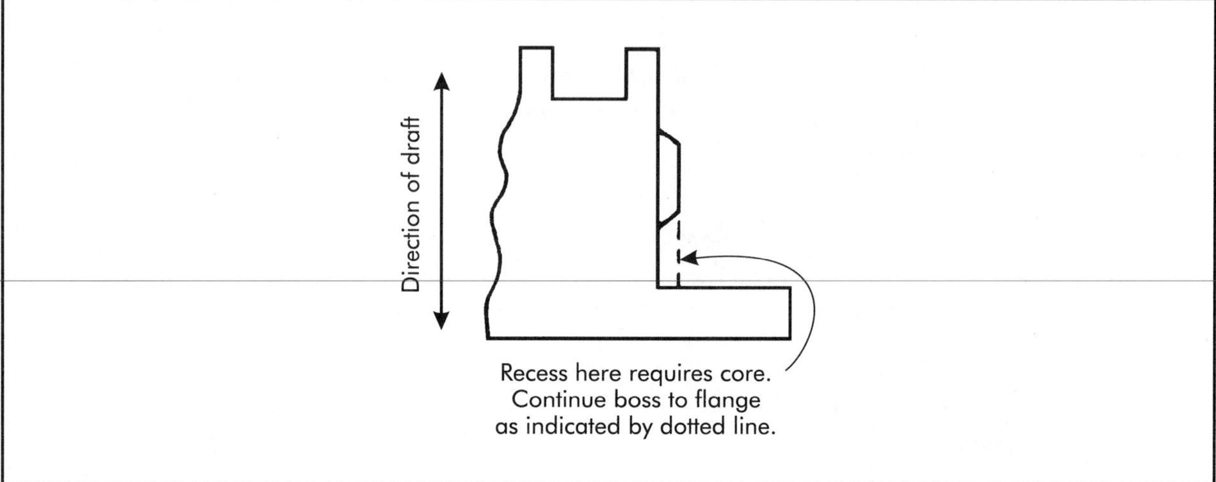

Figure 8-36. A projection increases the cost of making a casting.

6. Why is it necessary to place limits on thin sections of castings?
7. What causes "hot spots" in castings?
8. What is the purpose of a riser in the sand casting of a metal?
9. What is the purpose of a chill in the sand casting process?
10. The gating system carries out which functions in a sand mold?
11. List and describe the actions of the three types of gates employed in sand molding.
12. What are some of the hand tools used by molders?
13. Describe how a typical sand mold is made.
14. Describe the principles of operation of the three general types of molding machines.
15. What are the seven essential properties of a core?
16. What are "chaplets" and "anchors"?
17. Why and how are cores baked?
18. What are the two main types of patterns and how do they compare?
19. List the types of materials used in pattern making and cite some of the advantages of each.
20. How does the stereolithographic process form the shape of a pattern?
21. What is the purpose of a core print in a sand mold?
22. What is the purpose of a shrinkage allowance for patterns used in sand casting?
23. Explain the necessity of providing draft on a pattern.
24. Why are patterns colored?
25. Why is sand widely used for molding?
26. Describe the three major parts of a molding sand.
27. What are the differences between and relative merits of natural and synthetic molding sands?
28. What are the "green permeability," "green strength," and "dry strength" of a molding sand, and how do they determine how the sand performs in a mold?
29. What factors affect the green permeability, green strength, and dry strength of a sand, and how?
30. How many times may the same batch of molding sand be reused in the sand casting process?

31. How does core sand differ from molding sand?
32. What is a typical metal-to-fuel charge ratio for a cupola?
33. Is the cupola used to refine metals?
34. How is the composition of iron controlled in the operation of a cupola melting process?
35. What types of furnaces are used to melt nonferrous metals?
36. What are the difficulties in melting and pouring aluminum and how are they overcome?
37. Why is vacuum induction melting used for some metals and alloys?
38. How is shell molding done?
39. What are the advantages and disadvantages of shell molding?
40. What are the constituents of cast iron and how do they vary in gray, white, malleable, ductile, and nodular cast irons?
41. What are the effects on cast iron of (a) sulfur? (b) silicon? (c) phosphorus? (d) manganese? (e) nickel? (f) chromium? (g) molybdenum? (h) copper? (i) vanadium?
42. How do gray iron, malleable iron, and cast steel compare as to properties, costs, and applications?

8.12 PROBLEMS

1. A sand core in a mold has a volume of 1.6 dm^3 (100 $in.^3$). What is the buoyant force on the core if the metal poured in the mold is (a) brass or bronze? (b) aluminum? (c) steel? (d) cast iron?
2. In addition to the items listed in Table 8-4, a foundry has in its storage yard ample supplies of steel scrap that averages 0.305% carbon, 0.05% silicon, 0.05% manganese, 0.05% phosphorus, and 0.05% sulfur. Also available is No. 1 machinery scrap that contains 3.5% carbon, 1.9% silicon, 0.6% manganese, 0.405% phosphorus, and 0.085% sulfur. The iron-to-coke ratio is 8:1, with 0.5% sulfur in the coke, of which 5% is picked up by the iron. Calculate the composition of the cast iron resulting from the following charges:

   (a) 60% No. 1 machinery scrap, 25% steel scrap, and 15% No. 1 pig iron in a 1,000 kg charge.

   (b) 20% No. 1 machinery scrap, 5% steel scrap, 15% No. 1 pig iron, 10% No. 2 pig iron, 40% cast iron scrap, and 10% returns in a 10,000 lb charge.

8.13 REFERENCES

Boesch, W.J., et al., 1982. "Progress in Vacuum Melting: From VIM to VADER." *Metal Progress*, October: 49.

Flinn, R.A. 1983. *Fundamentals of Casting.* Reading, MA: Addison-Wesley.

Foundry Technology Sourcebook. 1982. Materials Park, OH: American Society for Metals and American Foundrymen's Society.

Heine, R.W., Loper, C.R., and Rosenthal, P.C. 1982. *Principles of Metal Casting.* New York: McGraw-Hill.

Jacobs, P.F. 1992. *Rapid Prototyping & Manufacturing: Fundamentals of Stereolithography.* Dearborn, MI: Society of Manufacturing Engineers (SME).

Metals Handbook, Desk Edition, 2nd Ed. 1998. Materials Park, OH: ASM International.

Mikelonis, P.J. 1982. *Foundry Technology: A Source Book.* Metals Park, OH: American Society for Metals.

Society of Manufacturing Engineers (SME). 1999. *Die Casting* video. *Fundamental Manufacturing Processes* series. Dearborn, MI: SME.

——. 1999. *Casting* video. *Fundamental Manufacturing Processes* series. Dearborn, MI: SME.

Steel Castings Handbook, 6th Ed. 1995. Barrington, IL: Steel Founders' Society of America.

Metal Casting Reusable Molds

9

Continuous Casting, 1946
—S. Junghans, Germany

Casting in sand molds as described in Chapter 8 gives adequate results at lowest cost in many cases. Other casting processes that produce more uniformly, more precisely, or at lower costs will be described in this chapter. One group utilizes metal molds that are not destroyed in service as sand molds are, but can be used repeatedly. Where metal molds are not suitable, precision casting is done in plaster and ceramic molds. Another group of processes casts metal continuously rather than in individual pieces.

9.1 METAL MOLD CASTING PROCESSES

Two processes that use metal molds are permanent mold casting and die casting. Although applied to other kinds of molds, centrifugal casting is often done with metal molds and is also included in this section.

For practical purposes, casting in metal molds is confined to metals with low to moderate melting temperatures. Some casting of iron and steel is done in refractory metal molds.

9.1.1 Permanent Mold Casting

When fluid metal is poured into metal molds and subjected only to hydrostatic pressure, the process is called *permanent mold casting*. The mold is clamped together during the operation and separates into two or more pieces to release the solidified casting. Metals commonly cast in this way are lead, zinc, aluminum, and magnesium alloys, certain bronzes, and cast iron. Typical products produced are refrigerator compressor cylinder blocks, heads, and connecting rods; flat iron sole plates; washing machine cast-iron gear blanks; automotive pistons and cylinder heads; kitchenware; and aluminum typewriter parts. Castings may weigh hundreds of kilograms (pounds), but most are under 25 kg (55 lb).

Most permanent molds are made of a closely grained alloy cast iron, such as Meehanite®, which is resistant to heat and repeated changes in temperature. Sometimes bronze molds are used for lead, tin, and zinc, and wrought alloy steel molds are used for bronzes. Cores are usually made of alloy steel but may be made of sand or plaster for severe service. Molds and cores are washed with an adhesive refractory slurry, basically graphite, clay, or whiting. This helps keep the castings from sticking, promotes easy ejection, and prolongs die life. Mold life may run from 3,000–10,000 iron castings to as many as 100,000 pieces of a softer metal.

When sand cores are used, the process is called *semipermanent mold casting*. Sand cores are cheap and easily removed from irregular holes, but the structure, accuracy, and surface finish of the cored openings are only as good as those of sand castings.

The basic steps involved in the permanent mold casting process are illustrated in Figure 9-1. Permanent mold casting may be done manually, but is usually performed by machine if large production volumes are required. A machine transfers the molds through several stations: ejection of castings, cleaning and coating the mold, placing of cores, locking, pouring, cooling, and unlocking. Some or all of the functions may be automatically performed as needed. In some variations, called *piercing* and *squeeze casting*, a punch or die is driven into the mold to compress and forge the metal while it congeals.

9.1.2 Low-pressure Casting

In low-pressure casting, molten metal is forced by gas pressure upward through a stalk to fill a mold as depicted in Figure 9-2. The metal cools inwardly in the mold to the stalk and freezes while the pressure is held. Then the pressure is released and the still molten metal in the stalk returns to the pot. The process is used mostly to cast aluminum in plaster, cast iron, and steel molds, but has been applied to other metals to a small extent.

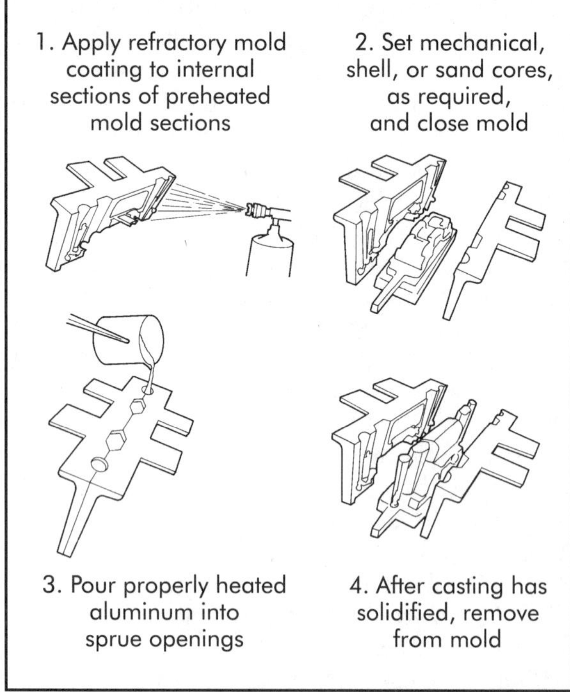

Figure 9-1. Basic steps in the permanent-mold aluminum casting process. (Courtesy Aluminum Casting Company, Division of REO Industries, Inc.)

In regard to the sizes and features of castings, low-pressure casting is at a stage between hydrostatic casting and high-pressure die casting. It gives moderately thin sections and intermediate accuracy, surface finish, density, and detail. Low-pressure castings are usually the strongest. Equipment, die costs, and production rates are in the middle range. For example, operation time for a 3-kg (6-1b) aluminum casting is reported to be 1 min, and for a 30 kg (65 1b) casting 3 min. The process is considered economical for from 500–50,000 castings per year.

Vacuum casting is a form of low-pressure casting. In this process, the metal is pushed upward from the pot by atmospheric pressure as a vacuum is drawn through the mold. Since air is drawn from the casting, porosity is low, finish is good, and the walls are more pressure tight. The castings do not blister as some others do when organic coatings are applied later. Tensile strength and hardness are optimum. Because of the lack of an air film in the die, chilling is rapid. Thus it is reported that walls can be cast thinner than, and production rates are comparable to, those of die casting.

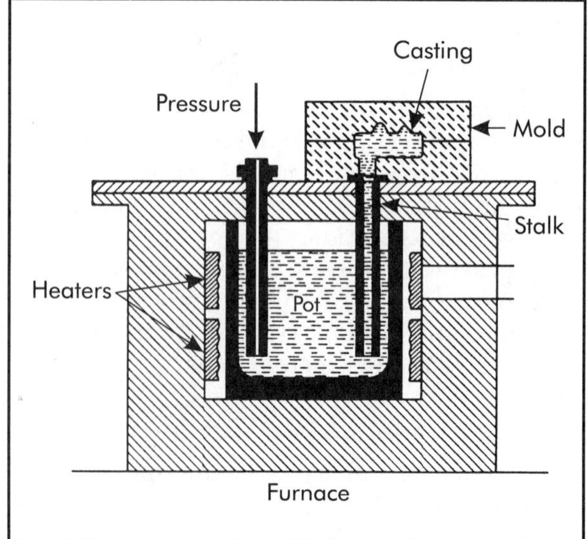

Figure 9-2. Low-pressure casting.

9.1.3 Slush Casting

For the slush casting process, molten metal is poured into a metal mold. After the skin has frozen, the mold is turned upside down or slung to remove the still liquid metal. The thin shell left behind is called a *slush casting*. Toys and ornaments are made in this way from zinc, lead, or tin alloys.

A slush casting is usually bright and well suited for plating, but is weaker and takes longer to make than a die casting. Die costs are relatively low for slush castings, which is an advantage for small quantity production.

9.1.4 Die Casting

Molten metal is forced under considerable pressure into a steel mold or die in the die-casting process. The action that takes place is depicted in a cross section of the die shown in Figure 9-3. The molten metal is shot through a runner and gate to fill the die. Vents and overflow wells are provided for air to escape. The metal is pressed into all the crevices of the die, and the pressure is held while the metal freezes to ensure density. Because of turbulence and air entrapment, porosity is often a problem, especially with thick sections. In the *pore-free process,* the die is filled with oxygen that reacts with the molten metal to produce insignificant particles of oxides rather than voids. Many dies are water cooled to hasten freezing. After the metal has solidified, the die is pulled open, and the part is ejected by pins actuated by an ejection mechanism. The product is called a *die casting*.

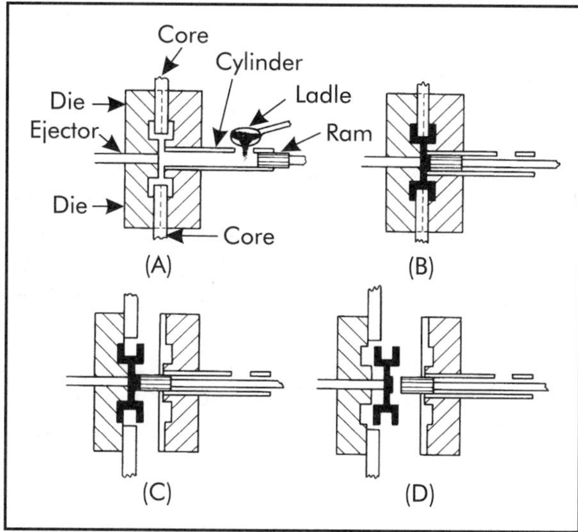

Figure 9-3. Cold chamber die-casting machine sequence of operations: (A) cylinder is filled with molten metal; (B) ram forces metal into die; (C) die is opened; and (D) casting is ejected.

Metals and alloys that are die cast include zinc, aluminum, magnesium, copper, lead, tin, and even iron and steel in a few cases. All metals available and capable of doing the job should be considered for a die casting, but the product's end usage must be taken into consideration. For example, zinc or magnesium may suffer from a corrosive atmosphere in service and would be eliminated for use under such conditions. In another case, lead, tin, or magnesium may not have sufficient rigidity for the product. Factors to be considered include environmental resistance, strength, weight, endurance, hardness, and ductility. Finally, a metal should be chosen that gives the lowest total cost on the basis of material, setup, operating cycle, and tooling. Pertinent specifications for the most popular die-casting metals in the order of usage are given in Table 9-1.

Zinc alloys serve well for many applications because they have good strength, ductility, resistance to shock at normal temperatures, and a low cost. Zinc is cast at low temperatures, so die, setup, and production (cycle) costs are low. Die life is long. Zinc alloys cost more by volume than aluminum or magnesium, but they often can be die cast with less metal in thinner sections, less draft, and faster and cheaper. Other metals are used for serviceable properties not furnished by zinc. Notably, zinc alloys have low creep strength (protracted service above 100° C [212° F] is not warranted), become quite brittle at lower temperatures, and severely corrode from steam and water. Yield and tensile strengths decrease somewhat upon prolonged aging.

In die castings, aluminum offers a high strength to density ratio, stability, service at a wide range of operating temperatures, and good resistance to corrosion. A number of alloys are available with a wide choice of properties. They stand up well in contact with food and fruit acids.

Magnesium alloys are light and fairly corrosion resistant but are badly attacked by humid tropical climates and sea water. The molten metal

Table 9-1. Properties of the most common die-casting metals

Property	Alloys of Zinc	Aluminum	Magnesium	Brass, Leaded Yellow
Tensile strength, Pa (psi)	2.8–3.3 (41,000–48,000)	2.0–3.2 (30,000–46,000)	2.3 (34,000)	2.8–3.1 (40,000–45,000)
Yield strength, Pa (psi) 0.5% offset	—	1.1–1.9 (16,000–27,000)	1.5 (22,000)	0.9–1.4 (14,000–20,000)
Elongation, %	7–10	1–9	3	15–25
Brinell hardness number	82–91	—	60	50–75 (500 kg [1,102 lb])
Density, gram/m³ (lb/in.³)	6.6 (.24)	2.66 (.096)	1.80 (.065)	8.3 (.30)
Relative strength to density (strength to weight ratio)	1.9	4.0	5.2	1.4
Minimum wall thickness, mm (in.)	0.5–2.0 (.02–.08)	0.8–2.5 (.03–.10)	0.8–2.5 (.03–.10)	1.5–3.0 (.06–.12)
Approximate relative unit volume cost	180	45	60	400
Relative maximum die life, number of pieces	10^6	10^5	10^5	10^4
Average die temperature, °C (°F)	218 (425)	288 (550)	260 (500)	500 (≈950)
Average casting temperature, °C (°F)	404 (760)	660 (1,220)	760 (1,400)	1,093 (2,000)

must be protected by a nonoxidizing atmosphere, and that adds cost to die casting.

Brass alloys have the most strength and wear and corrosion resistance but are not often die cast because their high melting point is damaging to the dies. Thus the cost of brass die castings is relatively high. They are used only when die castings of other metals are not suitable and where enough can be saved in production and machining costs over sand or plaster castings to make up for the die costs.

Die casting is mainly found in the high-production industries. Intricate parts can be produced, and inserts, such as fasteners, easily incorporated. Output is fast (100–800 pieces/hr), but usually hundreds of thousands of pieces must be made to pay for dies costing thousands of dollars. Typically, an automobile may have from 20–70 kg (≈40–150 lb) of die-cast parts: speedometer, windshield wiper motor, horn parts, carburetor body, grill work, and decorations. Also, die-cast parts are found in household appliances, business machines, bathroom hardware, outboard motors, clocks, jewelry, and tools.

9.1.5 Die-Casting Dies

Dies must be massive and strong to withstand the large loads imposed on them by die casting. A die is normally made in two parts. One part, the *front cover*, is mounted on a stationary platen and receives the molten metal from an injection nozzle on the machine. The other part is the *ejector* and is carried on a movable platen toward and away from the front cover to close and open the die. The two portions are aligned by the machine and by their own dowel pins. The two portions meet at the parting line and are locked together by the machine-locking mechanism when closed.

A die is always made so that the casting shrinks as it cools onto projections and core pins attached to the ejector portion. Thus, when the die is opened, the part clings to the movable portion, is drawn away from the front cover, and then can be ejected into the opening between the die halves. There it falls clear or is picked up by tongs.

Dies for complicated parts with undercuts, recesses, and angular holes are equipped with slides and movable core pins. Such construction is expensive but frequently makes possible the production of parts that could not otherwise be die cast. Movable cores and slides are arranged to be withdrawn before the die is opened.

Dies are of single-cavity, multiple-cavity, and combination types as illustrated in Figure 9-4. A *single-cavity die* turns out only one casting for each cycle of an operation. For large-quantity production of small and moderate size pieces, a number of cavities may be sunk in a single die and gated from a common sprue so that several castings are made at once. This is a *multiple-cavity die*. If the cavities are of two or more different shapes and produce many different parts at one time, the die is called a *combination die*.

The steel for a die depends mainly upon the material cast, temperature of the operation, and quantity of pieces produced. For die casting zinc alloys at 399–427° C (750–800° F), ordinary low-alloy steels, moderately heat treated, are satisfactory. Hot-work tool steels (high in nickel, chrome, and/or tungsten) are recommended for high temperatures and for producing a large quantity of castings.

Over the years, die-casting die design has been largely an art, and dies often have had to be modified to make them work properly. However, intensive research is providing a growing body of knowledge and literature to enable engineers to design dies and control the process more effectively. Advances in computer technology, programmable controllers (Chapter 31), and many types of instrumentation have been applied steadily to make die casting more efficient.

9.1.6 Die-casting Machines

Two basic types of die-casting machines are the hot chamber and the cold chamber machines. The hot chamber machines may be plunger or air injection operated.

Figure 9-5(A) depicts the method for operation of a typical gooseneck, plunger-operated, *hot-chamber, die-casting machine* for zinc and other low-melting-point alloys. The gooseneck contains a cylinder and curved passageways immersed in a pot of molten metal. When the plunger is retracted, the gooseneck fills with metal. When the die is closed, an air cylinder

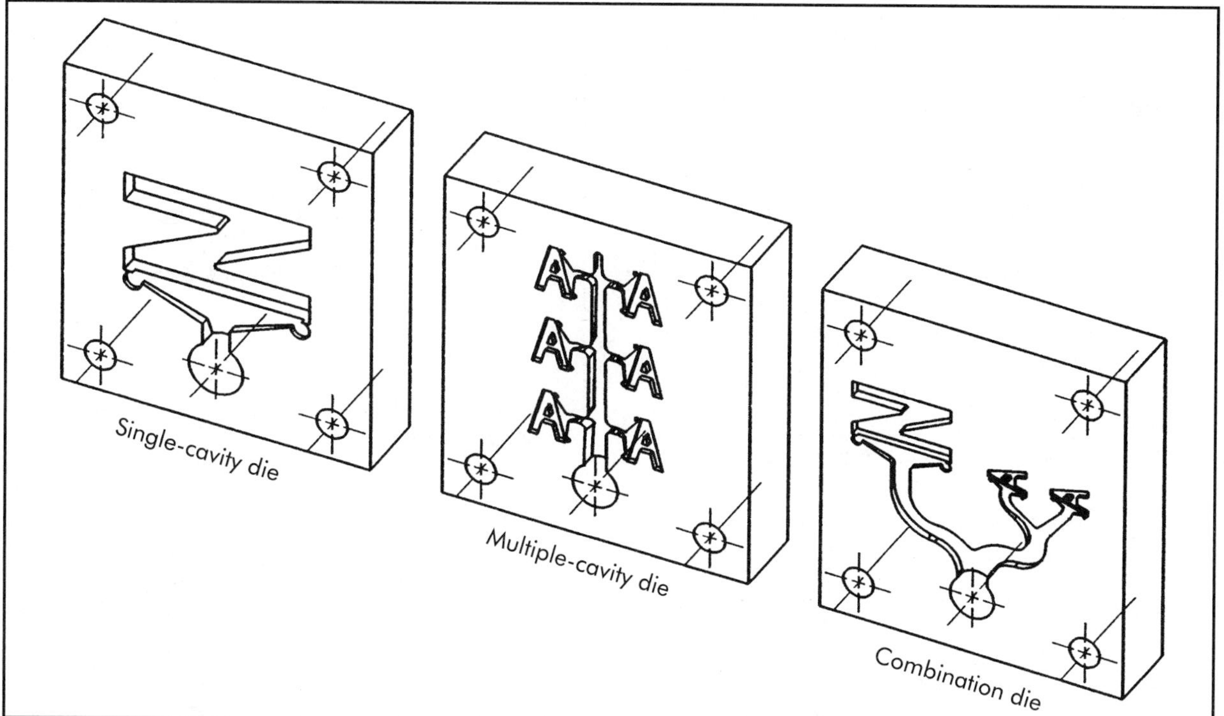

Figure 9-4. Die-casting cavity types. (Courtesy North American Die-Casting Association)

depresses the plunger and forces the metal through the gooseneck, nozzle, and die passages into the die cavity. Machines like this are more or less automatic and fast and produce from 100–800 parts/hr. Zinc alloy can be injected satisfactorily at 10 MPa (≈1,500 psi) and more than 14 MPa (2,000 psi) is rarely found necessary.

Molten aluminum attacks iron parts, such as the plunger of a die-casting machine, and in turn becomes contaminated. An *air-injection die-casting machine*, as shown in Figure 9-5(B), has a gooseneck that is lowered into the molten aluminum to receive a charge when needed. It is then raised and connects to an air line, which supplies the pressure to inject the metal into the die. Some hot chamber machines for aluminum are made with refractory components exposed to the molten metal.

In *cold-chamber die casting*, the metal needed for each shot is ladled from a separate furnace or pot into a cold chamber machine as indicated in Figure 9-5(C). A plunger is driven by an air or hydraulic cylinder to force the charge into the die. This method avoids having high-melting-point alloys in continuous contact with the working parts of the machine. Pressures are from 20–55 MPa (3,000–8,000 psi) for aluminum up to 241 MPa (35,000 psi) for copper alloys. Average production rates are reported up to 175 parts/hr for aluminum alloys, 250–300 for magnesium alloys, and 90 for copper-based alloys.

Cold-chamber die-casting systems have been developed in recent years to produce complex porosity-free aluminum parts but are adaptable to other metals. A plunger pushes the metal through a large gate in the bottom of the die so that it rises slowly to fill the cavities. The filling rate is 0.3–1 sec compared to 0.03 sec for conventional die casting. The casting is cooled from its extremity to the gate. When a skin is frozen, a small inner plunger in the charging cylinder is pushed forward to raise the pressure in the still molten metal to about 35–100 MPa (≈5,000–15,000 psi). Cooling is efficient, so total operation time is not excessive.

High pressure, depending on the metal, is necessary for dense and true die castings, but pressure alone is not enough. Selection and control of the material, design of the die, and timing and conducting of the operation must be done expertly for best results. An operation is usually performed semiautomatically for safety and uniform results. The operator presses a button to start a cycle. The die is closed and metal is injected. The die is opened after a preset time. Functions are interlocked to avoid spoiling the part, damaging the machine, or spilling metal.

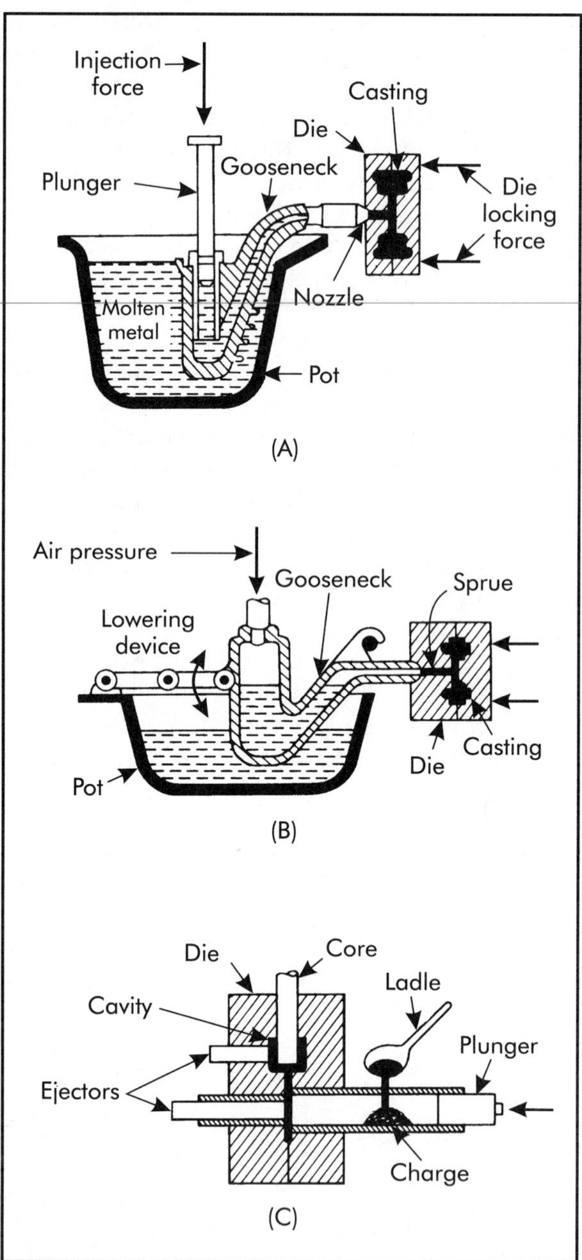

Figure 9-5. (A) Plunger hot-chamber die casting; (B) air-injection die casting; (C) cold-chamber die casting.

One way of rating a machine is by the locking force (in tons) that keeps the dies closed. The cross-sectional area of the die cavity times the injection pressure must be less than the locking force. Another rating specifies the dimensions of the space for the die.

A cold-chamber die-casting machine for precision die-casting of aluminum, brass, and zinc-aluminum alloy parts is shown in Figure 9-6. This particular machine will exert approximately 578 kN (65 tons) of clamping force with solid-state, microprocessor-based, electronic control of the operating parameters. It will cast aluminum parts up to about 0.9 kg (2 lb) in weight and 25.4 × 20.3 cm (10 × 8 in.) in breadth. A die-casting machine similar to the one shown in Figure 9-6 costs about $115,000, excluding the cost of the die. A larger machine with about 1,646 kN (185 tons) of clamping force costs about $140,000. A 1,646-kN (185-ton) hot-chamber machine for die casting zinc parts costs about $130,000.

9.1.7 Finishing Die Castings

Liquid metal forced into a die cavity under high pressure intrudes into parting face joints, around ejector pins, into overflow wells, and through any cracks. Clearances in dies are held as small as practicable, and the metal in the cracks solidifies into thin and fragile extrusions called *flash*. The flash, as well as sprues, runners, etc., must be removed to finish a casting.

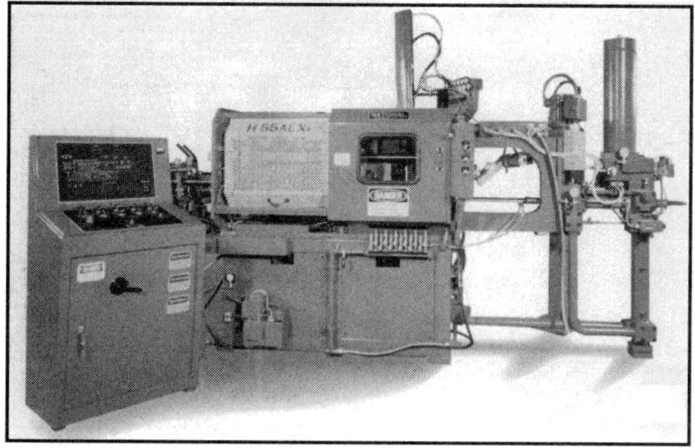

Figure 9-6. Cold-chamber die-casting machine designed for precision die-casting of aluminum, brass, and zinc-aluminum alloys. (Courtesy National Diecasting Machinery)

For small production runs, the excess metal may be broken off by hand and the edges cleaned by processes such as filing or buffing. For larger quantities, the flash can be removed quickly by a trimming die in a press. Elaborate machines (some called die-casting machining centers) are available that do die casting, trimming, and even drilling, tapping, etc., at successive stations with automatic handling between operations.

Dimensions can be held closely enough by die casting for most purposes, and only a minimum of machining is usually necessary. Some surfaces, like threads in small holes, may be more easily machined than cast. Machining of die castings is no different from machining other objects of the same materials and is governed by the principles presented in the chapters in this book on metal machining.

9.1.8 Centrifugal Casting

Centrifugal casting is done by pouring molten metal into a revolving mold. Centrifugal force creates pressure far in excess of gravity to cram the metal into the mold. For instance, an aluminum alloy about 102 mm (4 in.) in diameter is spun at about 2,600 rpm and subjected to a pressure of approximately 248 kPa (36 psi). Larger diameters require higher pressures. This is better than feeding a static casting with a head of 9 m (30 ft).

Centrifugal casting produces good quality, accurate castings, and saves material. The castings are dense and have a fine-grained structure with uniform and high physical properties. They are less subject to directional variations than static castings. Metal flows readily into thin sections, and castings come out with fine outside surface detail. Gases and dross are squeezed out of the heavier metal and the impurities float on the inside surface of the casting from which they can be later cut. Gates and risers are not needed to supply a pressure head and may be almost eliminated. This has been found to mean a savings of 40% or more in metal poured.

All the common metals may be centrifugally cast in either refractory or metal molds. One method is to introduce a ceramic slurry into the rotating flask and centrifuge it into a compact lining before the metal is poured. Rotation about a vertical axis is fast and easy to do. The hole inside the fluid metal rotated in that way becomes shaped like a paraboloid and is small at the bottom. For a straight hole, a long piece is rotated about a horizontal axis. As an example of accuracy, 4.9-m (16-ft) long stainless steel tubes for a proton accelerator are centrifugally cast with a maximum deviation on the inside surface of 0.25 mm (.01 in.) over the entire tube length.

A variation of the centrifugal casting process, called *spin casting*, is used to cast a variety of small parts such as the ones shown in Figure 9-7. The spin-casting process may be used as an alternative to pressure die-casting and other casting techniques for manufacturing and/or prototyping either functional or decorative parts. The process is used to produce high-quality parts in the standard commercial-grade pressure die-casting zinc alloys, high-strength zinc-aluminum alloys, lead and tin-based alloys, and (in limited quantities) produces small-size aluminum parts. In addition, spin casting may be used to cast thermoset plastics and wax patterns for investment castings. The process uses special rubber molds that permit setups to be made in a few hours for development or prototype parts. Spin casting is a precision process capable of meeting typical production tolerances in the ±0.05–0.08 mm/cm (±.005–.008 in./in.) range.

Figure 9-8 shows the basic components of a commercial-grade spin-casting machine with a silicone rubber mold. Cavities in the silicone rubber mold halves may be cut or molded by hand since the uncured silicone rubber is soft and hand-moldable (somewhat like clay). Cores and pullout sections can be incorporated in the mold. After preparation, the mold is closed with the part model or pattern in place. The closed mold is then placed in a vulcanizing frame under pressure and the frame is placed in an electrically heated vulcanizing press for curing. The cured mold is easily flexed to release the patterns (and later, parts) from the mold cavities. Normally, molds are designed to produce several parts per cycle with the mold cavities arranged in a symmetrical fashion around the sprue.

After curing and pattern removal, the necessary gates, runners, and air vents are cut into the rubber mold with a sharp knife or scalpel. The mold is closed again and clamped in the spin-casting machine. Instrumented controls are then set to maintain the appropriate spin speed, clamping pressure, and cycle time. After the spin cycle is started, liquid metal or plastic is poured

Figure 9-7. Example spin-cast parts. (Courtesy Tekcast Industries, Inc.)

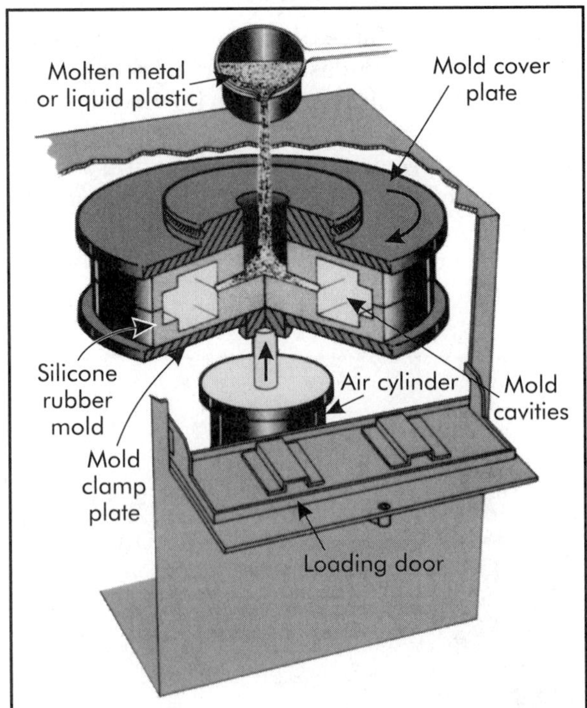

Figure 9-8. Diagram showing the basic parts of a spin-casting machine for centrifugal casting of molten metal, liquid plastic, or wax parts. (Courtesy Tekcast Industries, Inc.)

into the machine sprue. Pressure caused by centrifugal force pushes the liquid through the mold's runner system, completely filling every section, corner, detail, and surface in each mold cavity.

A medium-size spin-casting system capable of handling parts up to 229 mm (length) × 152 mm (width) × 76 mm (height) (9 × 6 × 3 in.) costs about $18,000. Each system includes a precision hydraulic vulcanizer, an electronically controlled spin-casting machine, a temperature-controlled melting furnace, and all the accessories necessary to begin production. The silicone rubber molds may be cycled 50–60 times per hour when casting metal or from 10–15 cycles per hour when casting plastics. Thus, a six-part cavity mold could produce up to 360 metal or 90 plastic parts per hour.

9.1.9 Comparison of Metal Mold-casting Methods

Almost any metal or shape or size of piece, except intricate pieces, can be sand cast, but some cannot be cast in metal molds. Some compositions are too weak at solidifying temperatures or shrink too much and thus crack around the unyielding projections in a metal mold. Many metals melt at temperatures that are damaging to metal molds. Some parts are too complex for metal molds to be practical.

Parts that can be cast in metal molds are stronger, have better surface appearance, can be held to smaller tolerances, made with thinner sections, and require less machining than equivalent sand castings. A comparison of the major casting processes is provided in Table 9-2. In general, properties are good in parts cast in metal molds under pressure, but plaster mold and investment casting (described later) are favored for premium-quality castings.

As shown in Table 9-2, tools cost most for die casting and less for the other processes. Also, the rate of production decreases and the scrap loss increases in much the same order. Sand casting is usually cheapest for a few parts, and the other processes are economical for larger quantities. The actual costs are different for each part. Higher tool costs make die casting more expensive for small quantities, but operational savings more than offset the tool costs for large quantities.

At times, metal mold castings are even competitive with other processes where sand castings are at a disadvantage. They have been found adequate to replace some forgings at less cost and may be cheaper than some drawn parts that require multiple operations. Castings may replace whole assemblies of parts made in other ways. Metal mold castings for parts have been proven more economical than some parts machined automatically where considerable metal was lost in chips.

9.1.10 Designing Castings for Metal Molds

The principles that point the way to good castings have been pointed out in Chapter 8 and apply as well to hard mold casting.

To allow for finishing, metal mold casting sections may be thinner and less stock needs to be allowed than for sand castings. From 0.79–3.18 mm (.031–.125 in.) stock, depending on the size of the part, is recommended for permanent mold castings. Die castings require little or no machining for most applications. Stock removal should be avoided or minimized because about 0.79 mm (.031 in.) of the cast surface next to the mold wall is most chilled and thus is the densest part of a metal mold casting.

Metal mold castings must have more draft than sand castings. Inside surfaces must be given about twice as much draft as outside surfaces because they shrink on their cores as they cool.

The parting line is an important factor affecting the costs of making and operating a metal mold. The line should lie as much as possible in one plane for the lowest costs. Flash can be expected to occur along the parting line and can be readily removed without defacement if the line is located on a bead or along the edge of a flange instead of across a flat surface.

9.2 PLASTER MOLD CASTING

Disposable or semipermanent plaster molds or cores for metal molds are made from plaster of paris (gypsum) with added talc, silica flour, asbestos fiber, and other substances to control setting time and expansion. A slurry is poured over a pattern and allowed to harden. The mold is dried to remove water and prevent the formation of steam upon exposure to hot metal.

Table 9-2. Comparison of casting methods

Factor	Type of Casting Process				
	Sand	Permanent Mold	Low Pressure	Die	Centrifugal (or Centrifuge)
Metals processed	All	Nonferrous and ferrous	Nonferrous[a]	Nonferrous[a]	All
Commercial sizes, kg (lb)					
minimum	Fractional	0.2 (.40)	0.2 (.40)	Minute	0.1 (.20)
maximum	Largest	136 (300)	100 (220)	50 (110)	Over 23 metric tons (over 25 tons)
Commercial surface finish, μm (μin.)	8–15 (≈300–600)	2–25 (80–1,000)	1–4 (50–150)	0.5–3 (20–125)	0.5–8 (20–300)[b]
Tolerance, mm/25 mm or less (in./first in.)	1.5 (.06)	0.38 (.015)	0.25 (.010)	0.10 (.004)	0.25 (.010)[b]
Tensile strength, Pa (psi)[c]	1.3 (19,000)	1.6 (23,000)	1.7 (25,000)	2.0 (30,000)	1.7 (25,000)
Production rate (pieces/hr)[d]	10–15	40–60	50–80	120–150	30–50
Mold or pattern cost, $[d]	1,000	6,000	9,000	15,000	2,000
Scrap loss[e]	5	4	3	2	1

[a] Iron and steel are cast in refractory metal molds but only infrequently.

[b] In metal molds.

[c] For No. 43F aluminum alloy as an example.

[d] Production and total tool cost figures are relative for a 1.4-kg (3-lb) aluminum casting of moderate complexity.

[e] 1 lowest to 5 highest.

Plaster mold casting was not a success until ways were found to make the plaster permeable so that gases could escape from the mold. Excess water helps because it is driven off when the mold is baked for drying and leaves pores. One form of the process uses a large proportion of sand with only enough gypsum for a binder. The method reputed to give the most porosity is to add the plaster slurry to an agent that is first beaten to a foam. The mold is dried below 204° C (400° F). The walls of the bubbles break as the plaster sets. The proprietary Shaw process utilizes two slurries that gel at once when mixed and poured around a pattern. Volatile agents in the mass are ignited, and the heat leaves a microcrazed structure in the mold, which is finally dried in an oven.

Only nonferrous metals, such as aluminum, copper, magnesium, and zinc alloys, are cast in plaster molds because most ferrous alloys are poured at temperatures high enough to melt the plaster. Lead is avoided because it reacts with the plaster. Typical products produced from plaster molds are aircraft parts, plumbing fixture fittings, aluminum pistons, locks, propellers, ornaments, and tire and plastic molds.

Plaster molds have low heat conductivity and provide slow and uniform cooling. Castings can be made in intricate shapes with varying sections and extremely thin walls. Most satisfactory section thicknesses are from 1.5–6 mm (≈.063–.250 in.). Inherent slow cooling precludes chilling and is conducive to a coarse-grained structure. As a result, some (but not all) metals cast in plaster molds have less strength than if cast in sand or metal molds and as much as a 25% loss in strength for some aluminum alloys.

Plaster molding material is yielding enough not to restrain cast metal when it cools and shrinks. On the other hand, plaster may break in thin mold sections and is not favored for large castings because it is weak. Most commercial plaster mold castings weigh from a fraction of a kilogram (pound) to around 9 kg (20 lb), but some have weighed up to 1,814 kg (4,000 lb).

Plaster mold casting yields more accuracy, smoother surfaces, better detail, thinner sections, and more complex shapes than sand or coated permanent mold casting. These advantages often help eliminate costly machining. Tolerances may be held as close as 76 μm (.003 in.), but common values are ±127 μm (±.005 in.) on one side and ±254 μm (±.010 in.) across parting lines for dimensions up to 25.4 mm (1 in.). Surface finishes from 0.8–3 μm (30–125 μin.) can be expected. However, plaster mold casting is on the average about three times more costly than sand casting because patterns and core boxes cost more, and the molds and cores take more time to prepare and must be baked.

Under some conditions, plaster mold casting is preferable to metal mold casting; in many cases not. The accuracy and finish produced by the two methods are comparable. Operation costs with plaster molds are generally higher than with metal molds. On the other hand, the patterns for plaster molds may cost less than metal molds. So, the overall cost then may be lower for plaster mold casting in small quantities. Plaster mold casting is of particular advantage for the nonferrous metals that melt at higher temperatures and are hard on metal molds. Upkeep may be less enough in such cases to give the cost advantage to plaster mold casting, even for large quantities. This explains why metal mold cores, which are even more exposed to heat than the molds, are often made of plaster.

9.3 PRECISION INVESTMENT CASTING

Investment casting, also called the *lost wax process*, is an ancient process that utilizes an expendable pattern of wax or plastic material. The steps in the process are depicted in Figure 9-9. A pattern may be carved for one or two experimental pieces but, for production, it is injected in a prepared die of rubber, plaster, or wood, but generally of metal for quantities. Pieces of a pattern may be joined together by heating or an adhesive. In this way, gates, risers, sprues, etc., are added. The patterns for a number of pieces may be joined together in a cluster for economical production. Complex parts and even whole assemblies may be fabricated by joining components together in the pattern stage rather than assembling finished pieces after casting.

A version of investment casting, called the *Mercast process,* starts with a pattern of frozen mercury at −57° C (−70° F) or below in a mold. Mercury changes little in size on melting and is not as likely to crack a frail mold as is wax or plastic.

A mold is invested around the pattern in the next step. The molding material is basically a refractory silica or zircon sand in a chemical binder, such as ethyl silicate, sodium silicate, or colloidal silica, to form a slurry. Film forming, wetting, and antifoam agents are also added. One method is to dip the pattern into a series of slurries to build up a layered shell of successively coarser grains about 3.18–6.35 mm (.125–.250 in.) thick. Fine grains at the inside surface of the mold impart a smooth surface finish to the casting. The wet coating may be built up by being stuccoed with refractory grains. No flask is needed, and the thin shell can be dried in 30 min or less as compared to hours for a whole mold. Another method is to position the pattern in a flask and pour molding compound around it, usually while vibrating the flask to aid in filling. Sometimes a mold started by the first method is completed by the second; that is, a shell is formed and then it is embedded in a flask.

After the mold has cured for from minutes to hours, depending on the material, it is placed in a furnace at temperatures up to 1,093° C (2,000° F) to bake and melt or volatilize the wax or plastic pattern. Melted metal is poured into the mold and may be subjected to pressure from compressed air or centrifugal force. One technique is to draw a vacuum from beneath a mold to pull out gases and concentrate atmospheric pressure

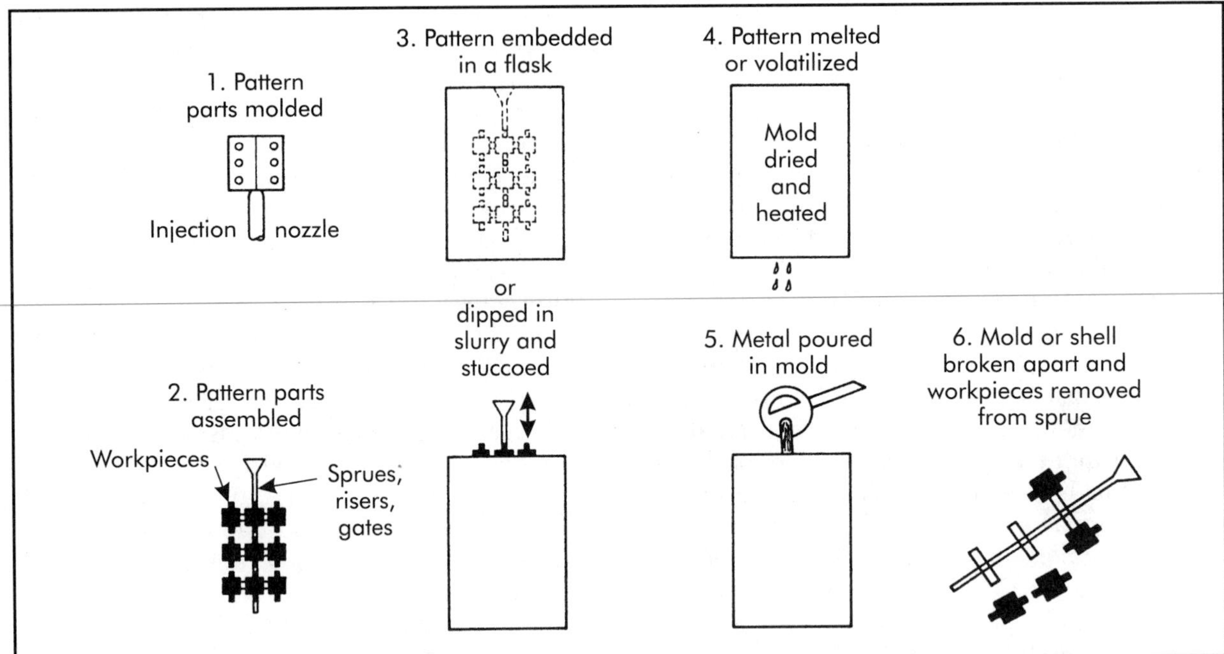

Figure 9-9. Investment casting steps.

on the metal. After the casting has cooled, the mold is broken, cores are shaken out, and sprues and irregularities are cut off or ground away.

9.3.1 Advantages and Limitations

Many common and uncommon ferrous and nonferrous alloys are investment cast—over 180 according to one authority. The most suitable metals are those with good fluidity, uniform shrinkage or freezing, and chemical inertness. This is because investment casting is usually chosen for its ability to generate small tolerances, fine details, and thin sections.

A main advantage of investment casting is that the best alloy can be selected for a part. This cannot always be done for machined parts, for instance, because extra hard or tough metals may be too expensive to cut. Most metals can be cast with equal ease, and the material cost of a small part may not affect the final cost much. As an example, a locking cam cast from soft steel (AISI 1020) at \$0.26/kg (\$0.12/lb) costs \$0.05/piece and from a hard Co-Cr-W alloy (CoJ) at \$5.31/kg (\$2.41/lb) costs only \$0.057/piece.

Tolerances as small as ±50 μm (.002 in.) are held by investment casting, but usually at extra cost. Commercial tolerances normally are at least ±50 μm/cm (±.005 in./in.). Investment casting has an advantage tolerance-wise because draft may be largely eliminated and the mold does not have a parting face, across which dimensions vary appreciably in other casting processes. Commercial surface finishes range from below 2 to 6 μm (≈60 to 220 μin.).

Investment casting is superior to other casting methods in some respects and inferior in others. It is readily automated, particularly for molding and coating the cores, and with many small pieces made and handled in clusters, cost per piece can be quite low. Processors report orders for quantities up to 1 million pieces. However, more steps are required, and investment casting is usually not competitive with sand casting, except for small tolerances and fine finishes. As a rule, shell molding is more economical for simpler parts in larger sizes, and investment casting for complex parts, particularly in small sizes. Investment casting can offer no advantage over metal mold and die casting for low-melting-point alloys, except for small quantities or complex parts, for which investment casting tooling costs less. However, it can produce results comparable to those from metal molds but with metals that melt at temperatures too high for metal molds.

Many kinds of parts are investment cast. Most weigh from a fraction of a kilogram (pound) to about 5 kg (11 lb), but some have been made as heavy as 70 kg (≈150 lb). Parts are made with sections from 0.79–1.60 mm (.031–.063 in.), but not often over about 25.4 mm (1 in.) thick. Conventionally, the process has been used for tooth fillings, surgical instruments, and jewelry. Typical products produced today include buckets, vanes, and blades for gas turbines; shuttle eyes for weaving; slides for cloth cutting; pawls and claws for movie cameras and projectors; sights, bolts, triggers, etc., for firearms; cutting tools; and waveguides for radar.

9.4 CONTINUOUS CASTING

Continuous casting consists of pouring molten metal into one end of a metal mold open at both ends, cooling rapidly, and extracting the solid product in a continuous length from the other end. This is done with copper, brass, bronze, aluminum, and, to a growing extent, cast iron and steel.

The process of continuous casting is carried out in a number of ways that differ in details. A typical process for copper-base alloys, called the Asarco® process, is depicted in Figure 9-10. Metal flows into a mold or die from below the surface of a charge in a holding furnace to keep out floating slag. It flows through a tortuous path in the furnace to protect the mold from sloshing effects during pouring. The lower part of the mold is water jacketed and freezes the metal rapidly. The solidified casting is pulled along at a controlled speed by the withdrawing rolls. The process is started with a dummy bar in the mold upon which the first metal is poured and cooled. Metals like iron and steel that conduct heat comparatively slowly freeze only skin deep while in the mold a practical length of time. They are sprayed with water and solidified throughout after leaving the mold. Some systems employ vibrating or reciprocating molds to keep the casting from sticking. Some molds are curved, and the casting veers into a horizontal position where it passes through straightening rolls. Finally, the product is cut to desired lengths by saw or torch.

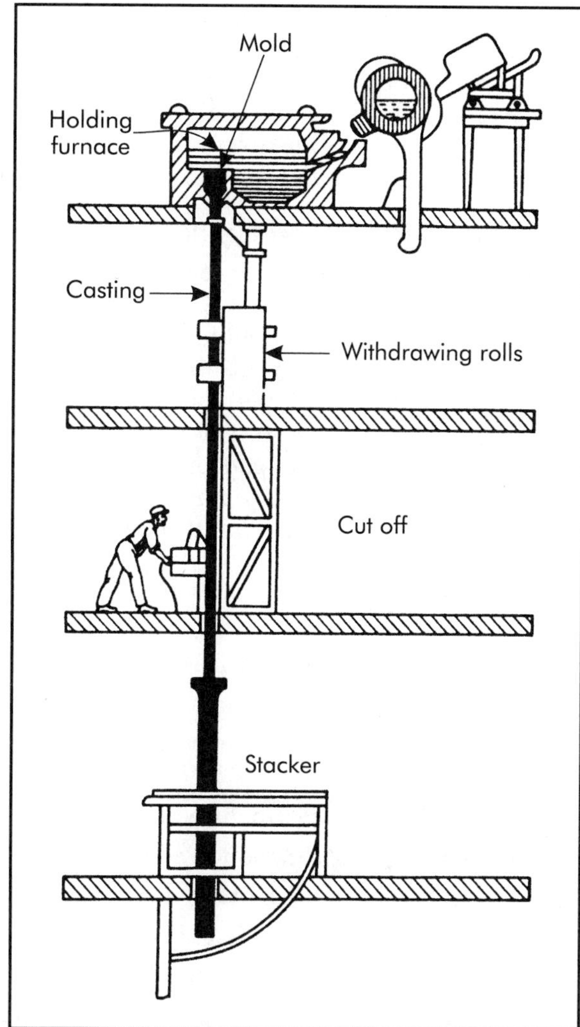

Figure 9-10. Continuous casting process.

9.4.1 Applications

Continuous casting is performed for a number of purposes. It is particularly suitable for any shape of uniform cross section: round, square, rectangular, hexagonal, fluted, scalloped, gear-toothed, and many other forms; solid or hollow. Shapes from a few millimeters to about 254 mm (10 in.) in diameter and up to 6.1 m (20 ft) long are possible. A growing use is to produce blooms, billets, and slabs for rolling structural shapes as described in Chapter 11. This is cheaper than rolling from ingots. Copper-base billets 102 mm (4 in.) square and over and steel slabs 152–254 mm (6–10 in.) thick and 0.8–2.2 m (32–88 in.) wide are produced. Continuously cast shapes may be cut to lengths and finished by light machining (bushings and pump gears, for instance). Large-diameter and long iron water pipes with flanged ends are cast in a semi-continuous manner with a separate charge of molten metal for each pipe.

9.4.2 Advantages and Limitations

Continuous casting offers several advantages. Its yield in rolled shapes is 10% or more over that from ingots. An appreciable amount of the end of each ingot must be cut off and returned to the furnace because it is porous, unsound, and full of impurities. This waste is practically eliminated by the uniformity of continuous castings. Of course, a hollow center occurs from shrinkage in a continuous casting, but it remains pure and is welded shut after four rolling passes. A continuous casting structure is dendritic as in other castings, but it is more dense and uniform than in individual castings because the whole length receives the same treatment in the same mold. Physical properties and surface finishes are comparable to those obtained in other metal mold processes.

Continuous casting is essentially automatic, and unit labor cost is low. Dies or molds are made of copper or graphite and are simple and inexpensive. Total equipment cost is high, about twice as much as for individual castings. The complete cost of a plant to continuously electrically melt and cast premium alloy steel at the rate of 1.4 Gg (3,000,000 lb) per year is several million dollars.

9.5 QUESTIONS

1. What types of metals are commonly cast in permanent molds?
2. What types of metals are commonly used for permanent molds?
3. Describe the low-pressure casting process and indicate which metal(s) are generally cast by this method.
4. What is slush casting used for?
5. How is die casting done and what are its advantages?
6. Which metals are suitable for die casting?
7. Which of the metals listed in Question 6 can be die cast at the lowest temperature?
8. Why are brass alloys generally not die cast?
9. Differentiate between single cavity-, multiple cavity-, and combination-type die-casting dies.
10. Describe the precision investment casting process and list some of its advantages.
11. What is "continuous casting" and what advantages does it offer?

9.6 PROBLEMS

1. A single gear blank, 203 mm (8 in.) in diameter, is to be die cast from zinc at 14 MPa (2,000 psi) pressure. What should be the capacity of the machine in kN (tons) of locking force?
2. When a casting is centrifuged around a minimum diameter, show that the minimum acceleration in gravitational units imposed on the casting is

$$G = a/g = 1.42 \times 10^{-5} Dn^2$$

where

G = minimum acceleration in gravitational units
a = centrifugal acceleration of the casting, ft/sec^2

g = acceleration of gravity, ft/sec^2
D = minimum diameter, in.
n = revolutions per min

Also, $G = a/g = 5.6 \times 10^{-7}\, Dn^2$ if D is in mm and g is in m/sec^2

3. Common practice with iron- and copper-base alloys is to centrifuge the castings to impose an acceleration of from 40–60 gravitational units. This yields good castings without damage to the molds. In one operation, the molds are arranged around a 350 mm (≈14 in.) diameter. At what speed should the centrifuge be rotated? Refer to the equations in Problem 2.

4. A circular disk of unit thickness of molten metal is rotated around its axis to make a centrifugal casting. Show that the pressure from the centrifugal action at radius r_2 is

$$P = Kyn^2(r_2^2 - r_1^2)$$

where

P = pressure, kPa (psi)
K = empirical constant
y = density, kg/m^3 (lb/in.^{3})
n = revolutions per min
r_1 = inside radius, mm (in.)
r_2 = outside radius, mm (in.)

For density in lb/in.^{3}, r in in., and P in psi, $K = 1.42 \times 10^{-5}$; for density in kg/m^3, r in mm, and P in kPa, $K = 5.48 \times 10^{-12}$.

5. Experience has shown that a good centrifugal casting is obtained from aluminum or magnesium if it is rotated at sufficient speed to create a pressure of 207–241 kPa (30–35 psi) on its periphery. It is assumed the casting is continuous from its center of gravity, around which it is rotated, to its periphery.
   (a) At what speed should a 508-mm (20-in.) diameter symmetrical aluminum casting be rotated?
   (b) At what speed should a 406-mm (16-in.) diameter symmetrical magnesium casting be rotated?

6. A 1-m (≈36-in.) diameter cast-iron pipe with a 25-mm (≈1-in.) wall thickness is centrifugally cast and rotated at sufficient speed to impose an acceleration of 160 g on its periphery. At what speed is it rotated? What is the peripheral pressure in the mold in kPa (psi)? See the equations of Problems 2 and 4.

7. A 1.5-kg (≈3-lb) aluminum casting for which data are given in Table 9-2 can be made satisfactorily by any of the processes listed in that table. The cost of labor and overhead is $22.50/hr. Assume a top production rate and the same setup time and material costs for all processes even though this is not always the case. What process should be selected for each of the following numbers of pieces: (a) 1; (b) 100; (c) 1,000; (d) 3,000; (e) 5,000; (f) 10,000; (g) 20,000; (h)30,000; (i) 40,000; (j) 50,000; (k) 100,000?

8. For the conditions stated in Problem 7, what range of production quantities is most economical for each of the following processes: (a) sand casting; (b) permanent mold casting; (c) low-pressure casting; (d) die casting; (e) centrifugal casting (with metal mold)?

9.7 REFERENCES

An Introduction to Die Casting. 1981. Des Plaines, IL: American Die Casting Institute.

Huntress, E.A., 1980. "Spin-casting Higher-Temp Alloys." *American Machinist*, November.

Investment Casting Handbook. 1980. Chicago, IL: Investment Casting Institute.

Kayes, A., and Street, A.C., 1982. *Die Casting Metallurgy.* London, England: Butterworths.

Society of Manufacturing Engineers (SME). 1999. *Die Casting* video. *Fundamental Manufacturing Processes* series. Dearborn, MI: SME.

——. 1999. *Casting* video. *Fundamental Manufacturing Processes* series. Dearborn, MI: SME.

Powder Metallurgy

10

Tungsten lamp filaments from Tungsten powder, 1909
—W. D. Coolidge, General Electric Company, Schenectady, NY

10.1 INTRODUCTION

Powder metallurgy is becoming a competitive process to metal casting, forging, and machining operations. It produces parts and components more economically to net shape dimensions. Gears, cams, cutting tools, bearings, bushings, and other automotive components are some examples of products made by powder metallurgy techniques.

10.2 BASIC POWDER METALLURGY PROCESS

Powder metallurgy is the manufacture of products from finely divided metals and metallic compounds. The powders are loose in some applications and pressed into pieces and parts in others.

Loose metal powders and their compounds are added to some products. Aluminum and bronze powders are mixed into paints for metallic finishes. Metal powders add strength and wear resistance to plastics. Metals in powder form in fireworks burn with colorful effects.

The most important and growing applications of powder metallurgy in industry are in the manufacture of pieces and parts. As an example, a modern automobile contains about 100 powder metal parts. Powder metallurgy has achieved this status because it excels in several respects.

Powder metallurgy is the only feasible means of fabricating some materials. The melting points of the refractory metals, such as tungsten (3,399° C [6,150° F]), tantalum (2,996° C [5,425° F]), and molybdenum (2,621° C [4,750° F]), are so high they are hard to work in appreciable quantities with available equipment. Other substances, such as zirconium (melting point 1,899° C [3,450° F]), react strongly with and are contaminated by their surroundings when melted. Powder metallurgy is a practical way of refining and fabricating such metals. It is also the only feasible way to consolidate and form the superhard tool materials, such as cemented carbides and sintered oxides.

Combinations of metals and nonmetals not obtainable economically in other ways can be made by powder metallurgy. This is of particular value to the electrical industry. Motor brushes and contact points must have proper conductivity but be resistant to wear and arcing. Brushes are made from powders of copper, graphite, and sometimes tin and lead; points require combinations like tungsten and copper or silver. Permanent magnets can be made of densely packed and finely dispersed particles of suitable substances and can be held to small tolerances. Other examples of applications are clad or duplexed parts, such as rods and slugs for nuclear reactors, bimetallic units for thermostats, and coated welding rods.

A class of materials made possible by powder metallurgy is known as *cermets*. As the name implies, they are combinations of metals and ceramics, with the strengths of the metals or alloys and the abrasion and heat resistances of the metallic compounds, which are the ceramics. One example is Al_2O_3–Cr. Cermets were developed in Germany in World War II for possible jet engine use. That usage was never realized because of their brittleness, but they have found other important applications, such as corrosion-resistant chemical apparatuses, nuclear energy equipment, pumps for severe services, and systems for handling rocket fuels.

Metal can be made porous by powder metallurgy. Practical applications are bronze, nickel, and stainless steel filter elements, more shock resistant than ceramic; and bearings, gears, and pump rotors impregnated with lubricants for long, carefree life.

Many machine and structural parts can be made most cheaply by powder metallurgy. This is true of intricate precision parts that can be produced in two quick steps by powder metallurgy instead of several costly machining operations. Examples are gears, pawls, latches, cams, valve retainers, and brackets. As another example, powder metal preforms make for simple dies and fewer strikes and thus reduce the cost of forging some parts. In some cases, higher purity, more variety, and better control of properties can be achieved by powder metallurgy than by melting and casting at fusion temperatures.

Powder metallurgy encompasses the preparation of the powders and their combination into useful articles. Basically, a powder metal is compacted into the shape desired and heated to strengthen the compact. The actual processes are many and differ to suit the materials treated and obtain the properties required in the finished product.

The powder metallurgy process is relatively standard. It consists of five basic operations as shown in Figure 10-1.

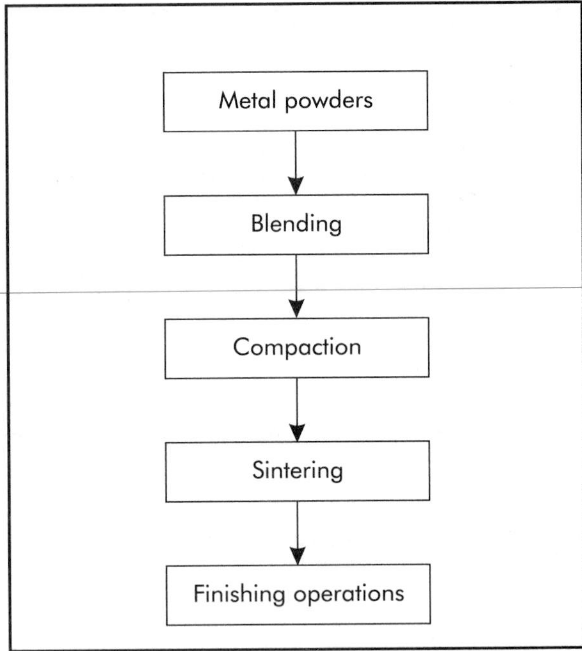

Figure 10-1. Powder metallurgy flow of operations.

10.3 METAL POWDER PRODUCTION AND BLENDING

What a metal powder will produce depends upon its composition and physical characteristics. The most used compositions are the copper-base and iron-base powders; brass, iron, and steel for structural parts, and bronze for bearings. Others of importance, though in lesser amounts, are stainless steel, aluminum, titanium, nickel, tin, tungsten, copper, zirconium, graphite, and metallic oxides and carbides.

Substantially pure metal powders are used for some parts; alloys for others. Alloys may be obtained by alloying a metal before it is powdered or by mixing together powders of the desired ingredients. The first method gives a finer and more uniform alloy. The second is easier to compound but must be sintered with care to assure that the ingredients become diffused.

The physical characteristics of a powder metal are influenced by the way it is made. The chief characteristics are particle shape, size and size distribution, purity, grain structure, density, flow rate, and compressibility. Different powders are commonly mixed together to get desired properties.

Most metal powders are obtained by reduction of refined ore, mill scale, or prepared oxides by carbon monoxide or hydrogen. Grains tend to be porous. Particle size can be made quite uniform, which contributes to uniformity in the final product. Crude reduced powder called *sponge iron* is quoted at about $1.10/kg ($0.50/lb).

Metals may be atomized in a stream of air, steam, or inert gas. Some may be melted separately and injected through an orifice into the stream; others like iron and stainless steel may be fused in an electric arc (like sprayed metal); and refractory metals in a plasma arc (Chapter 14). As an example, small droplets of titanium freeze to powder after they are flung from the end of a rapidly rotating bar heated by a plasma arc in a helium atmosphere. Atomized particles are somewhat round.

Under controlled conditions, metal powder may be electrolytically deposited. The material may have to be broken up and is milled or ground for fineness, heated to be annealed and drive off hydrogen, and sorted and blended. Electrodeposited powders are among the purest and are characteristically dendritic. Cost is about $4.01/kg ($1.82/lb) for electrolytic iron powder.

Milling or grinding in ball mills, stampers, crushers, etc., is a means of producing powders of almost any degree of fineness from friable or malleable metals. Tungsten carbide grains are pulverized in this way. Some malleable metals are milled with a lubricant into flakes, which are not suitable for molding, but are used in paints and pigments.

Nickel or iron can be made to react with carbon monoxide to form *metal carbonyls,* such as $Ni(CO)_4$. These are decomposed to metal powders of high purity, small and uniform grain size, and dense and round particles. Carbonyl powders are costly but easier to work than other forms. For instance, carbonyl iron compacts are as strong and ductile when sintered at only 649° C (1,200° F) as other iron compacts sintered at 1,093° C (2,000° F). They serve well for making continuous strips of powder and for critical electronic components, but are mostly added to electrolytic powders to add strength and ductility.

Shotting is the process of dropping molten particles from a small opening through air or an inert gas into water. This produces spherical particles, but not the smallest sizes.

Less common methods of making metal powders include ordinary machining, vapor condensation, chemical decomposition, granulation by stirring vigorously during solidification, and impacting of chips and scrap.

10.4 FABRICATION PROCESSES

The basic operations of pressing (or compacting) and sintering (or heating) may be combined in a number of ways in processes for fabricating metal powders. Common processes are depicted in Figure 10-2. The operations of pressing and sintering are varied and controlled to suit many conditions and, in some cases, other operations are added. These basic operations and the principles that govern them will be described in the following sections.

10.4.1 Pressing or Compaction

Pressing or *compaction* is a process whereby powders that are already blended are pressed into shapes using hydraulically or mechanically activated presses. Pressing is normally done in a slightly elevated temperature from room temperature. The purpose of compaction is to obtain uniform shape and size among particles and to provide enough strength for the powder to withstand further processing.

Pressure on powder metal squeezes the particles together to lock or key them in place, initiates interatomic bonds, and increases the density of the mass. The pressure applied to a compact determines its ultimate density and strength. A typical case is illustrated in Figure 10-3. Theoretically, if a powder is pressed enough, it will attain 100% of the density and strength of the parent metal, at least on being sintered. This is approached in some parts by repressing after initial pressing and presintering. High pressures, and particularly additional operations, are expensive and not warranted for parts that do not need high strength. On the other end of the density scale, little or no pressure is needed for porous parts.

Most parts are pressed cold. For more density and strength, parts may be subsequently hot pressed or forged (hammered). Hot pressing produces the most accuracy and forging the most strength, but forging costs more.

Suitable particle shape, size and size distribution, and careful selection and mixing are necessary to obtain a satisfactory pressed part. The best bonds are obtained between jagged particles, but round particles flow better into the mold and under pressure. The way the powder fills the die

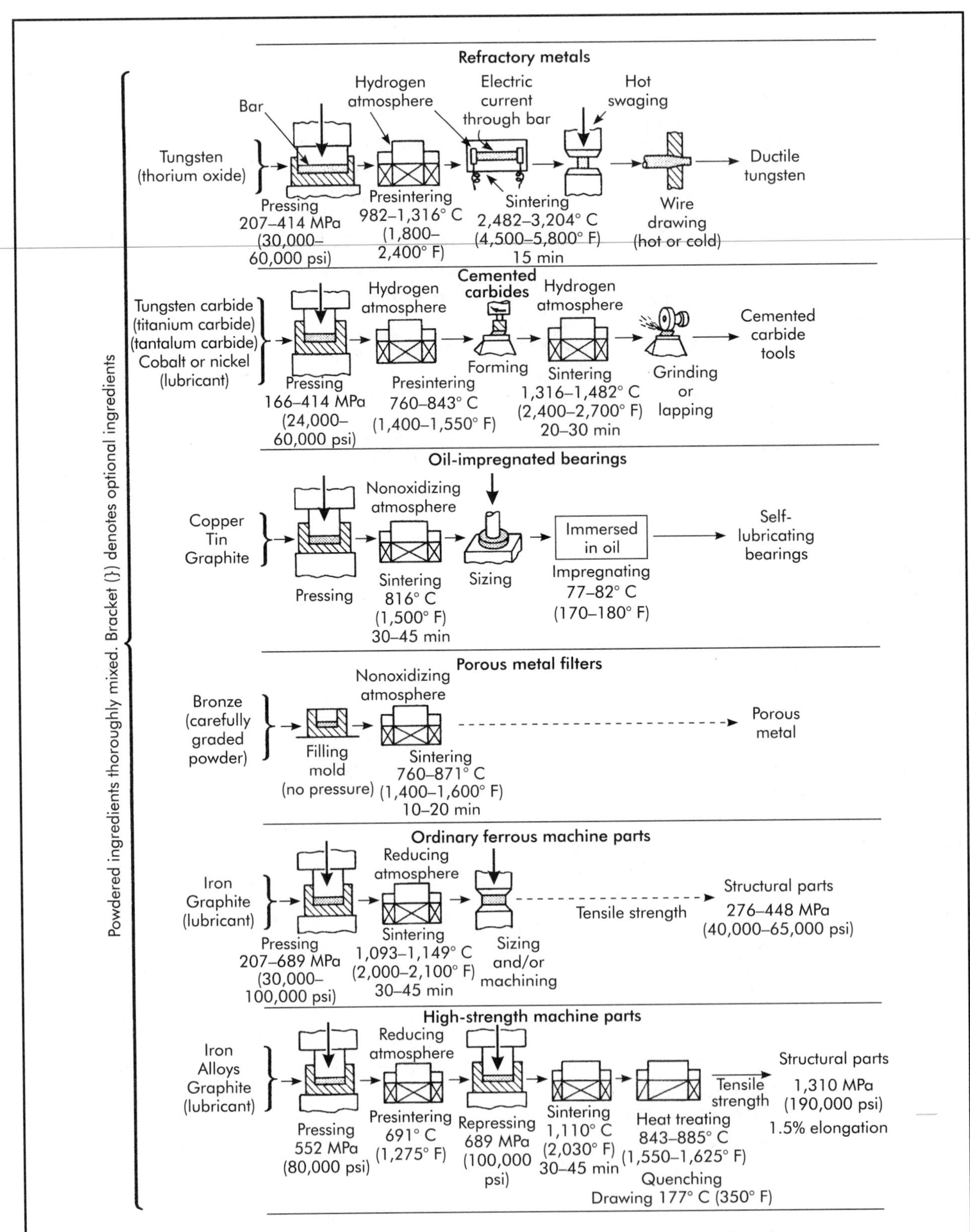

Figure 10-2. Some typical powder metallurgy processes.

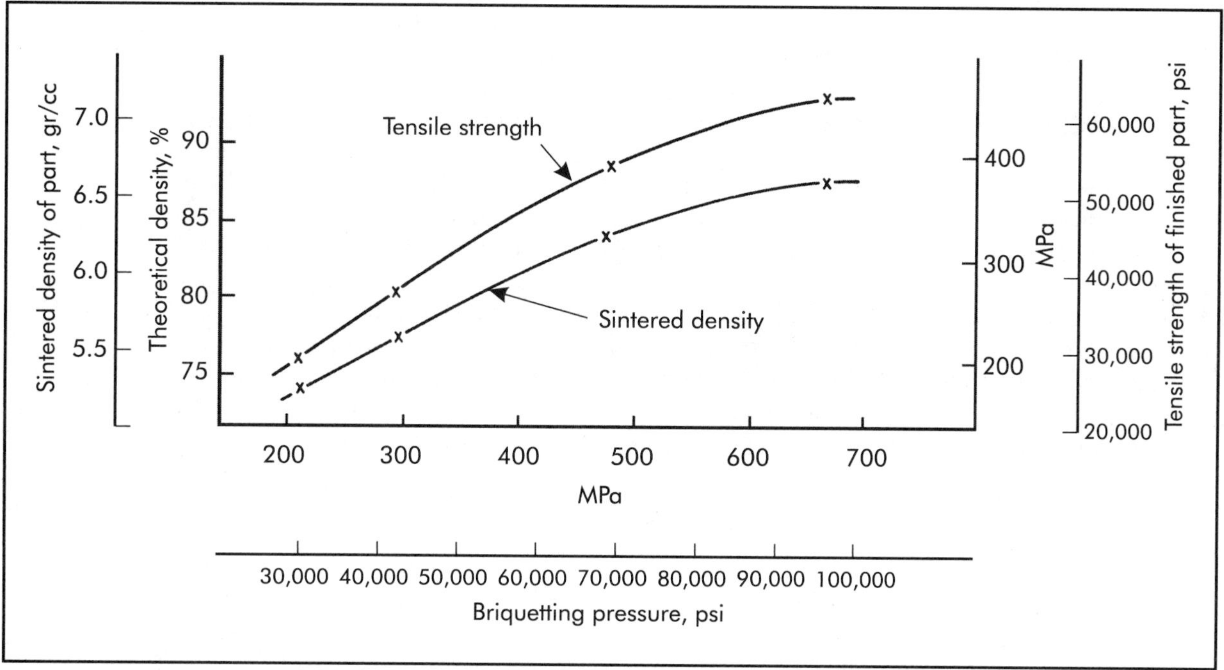

Figure 10-3. Relationships between briqueting pressure, sintered density, and tensile strength for a carbon-iron powder sintered at 1,110° C (2,030° F) for 30 min.

determines operation speed. Zinc stearate may be added to lubricate the die and the particles to minimize wear and aid in compaction. It volatizes in sintering and helps keep pores connected.

Powder metal is commonly compressed in a die cavity by one or more punches to create the shape of the part. Quality depends upon packing the material uniformly. Powder metal does not flow readily around corners and into recesses like a fluid. Friction is high between the particles and walls of the die. Thus, one punch cannot compact any but the simplest parts to a uniform density. Parts that have features, such as steps, thin walls, and flanges, must be acted upon by two or more punches to distribute the pressure uniformly through the sections. A simple example is given in Figure 10-4. The most complex parts may require as many as two upper and three or four lower punch movements, and even some side core pulling movements on the press. Some parts cannot be made with available equipment.

Production powder pressing is normally done on presses specifically designed for the purpose. Such a press is rated by the force it can deliver and the maximum depth of die chamber it can accommodate. The force rating determines the cross-sectional area of the largest part that can be subjected to a given pressure. The depth of the die chamber, called the *die fill*, determines the depth of a powder filling. This limits the length of a pressed part. The ratio of powder depth to compacted part length is from 2:1–3:1 for iron and copper and as high as 8:1 for some materials.

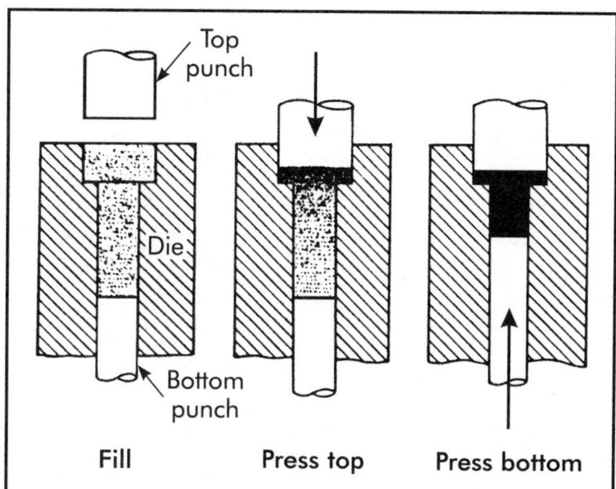

Figure 10-4. Steps in pressing a simple powder metal part.

Most presses under 1.5-N (170-tons) force capacity are mechanical for speed. A mechanical press works to uniform size, but causes differences in density with variations in conditions, such as the amount of charge. A hydraulic press may be set to exert a given pressure for uniform density or stop at a uniform size. Large hydraulic powder metal presses have ratings of 25 or more MN (several thousand tons). Presses range from single-stroke, hand-fed models for an output of a few pieces per hour (typically for large pieces in small quantities), to automatic rotary and multiple-punch machines capable of producing hundreds of parts per minute.

Other Compaction Methods

Powder metal may be *slip cast* in molds. The powder is dispersed in a liquid containing chemicals to wet the particles and help distribute and then release the mass from the mold. The mold may be porous to absorb free liquid and vibrated to densify the compact. Slip-cast parts that are later sintered have adequate strength for many purposes. Fiber aggregates are used for sound and vibration absorption and as reinforcements for plastics and metals. Mold cost is low, and the method is economical for parts that are complex or made in small quantities. For example, a stainless steel part that would have required a $6,000 hardened steel die in a 10 MN (1,200 ton) press was slip cast in a mold that cost less than $30. However, the method is slow and not justified for large quantities.

A method for heavy powders, such as tungsten carbide, is *centrifugal compacting*. The powder is twirled in a mold and packed uniformly with pressures up to 3 MPa (400 psi) on each particle. Parts must have uniform round sections. Equipment cost is not high.

Powder metal is also *injection molded*. A slurry of the powder in water or mixed with a thermoplastic material is squirted into a die. The binder is removed in sintering.

Continuous strips and rods are compacted by *rolling* copper, brass, bronze, Monel®, nickel, titanium, or stainless steel powders or fibers. In a typical process, stainless steel powder is fed from a hopper between two rolls in a horizontal plane. The material emerges as a strip with a density of 6–6.9 gram/m^3 (.22–.25 lb/in.^{3}). It is then sintered, rerolled three times, and annealed to a tensile strength of about 745 MPa (108,000 psi) and an elongation of 33%. Another process makes porous sheets for filters; a uniform layer of powder poured on a ceramic tray is sintered and then rolled to the desired density.

A means of applying pressure to obtain uniform density is to enclose powder in a shaped plastic or rubber mold and immerse it in gas or liquid in a chamber under 69–690 MPa (10,000–100,000 psi) pressure. This is *isostatic pressing*, sometimes called *hydraulic pressing* in a liquid. Complicated, asymmetrical, and large parts can be pressed more easily than in other ways. Isotropic parts are produced. Metal dies are not needed. For large work, equipment cost is about one-tenth of that for a press.

Powder metal, in metal or ceramic containers, and preforms are subjected to gas pressures as high as 345 MPa (50,000 psi) at temperatures to over 2,204° C (4,000° F) (but usually less) in *hot isostatic pressing* (HIP). This has been effective for refractory metals, ceramics, and cermets and spherical powders that do not respond to cold pressing. Close to theoretical densities can be obtained by HIP.

Stainless steel, uranium, and zirconium powders are sealed in cans and compacted by extrusion through dies while protected from contamination. The sheath may be removed to use the base metal or may be left on for further protection, as in the case of uranium reactor rods.

Long tubes can be compacted magnetically. Powder metal is poured around a mandrel inside a coaxial conductor. A current of around 1 mA is pulsed through the conductor and sets up the magnetic field that crushes the inner conductor around the compact.

High-energy rate forming (by both presses and explosives—Chapter 12) is capable of compacting powders to nearly theoretical densities with sharp definition of detail and close tolerances. As an example, a 0.33-m (13-in.) diameter and 9-kg (20-lb) iron powder dish was compacted to a 7.2-gram/m^3 (.26 lb/in.^{3}) density (7.8 gram/m^3 [.28 lb/in.^{3}] is theoretical) by 271 kJ (200,000 ft-lb) of energy on a high-energy-rate forging press like that depicted in Figure 12-18. A 62-MN (7,000-ton) hydraulic press would be needed for the same results.

10.4.2 Sintering

Sintering augments the bonds between the particles and therefore strengthens a powder metal compact. In all cases, this occurs because atoms of the particles in contact become intermingled. Generally speaking, this is brought about in one of two ways. In one way, one of the constituents of the compact melts; in the other, none melts.

An example of the first case is the sintering of cemented carbide, which is done above the melting point of the cobalt constituent. The molten cobalt fuses into a pervading matrix but also acts as a medium in which the carbide grains can grow together to form a skeleton throughout the mass. Carbide atoms appear to dissolve into the cobalt at points of high energy to build the bridges between the grains. This explains why cemented carbides are harder at high temperatures than high-speed steel that contains isolated carbides held together by a heat-susceptible ferrous matrix.

The second case is characteristic of the sintering of iron, copper, or tungsten powder. It is done usually at 60–80% of the melting temperature. The atoms at the spots of contact intermingle and migrate. The reason sintering is done below the melting point is that the process would actually be casting if the metals were melted, and powders would not be needed. Still the temperature must be high enough to excite rapid atomic mobility.

If substantially one constituent is present, as in the sintering of iron powder, a single phase is continuous from one particle to the next after sintering. In compacts of two or more different metals, alloys, or compounds, whether one melts or not, intermediate compounds or phases of the constituents are formed at the points of bonding of the particles. As sintering continues in either case, the bonded areas grow larger, and material more or less fills the voids between the particles. This is depicted in an ideal way in Figure 10-5 where particles are assumed round and metal that enters the voids is shown darkened.

When no melting takes place, atoms must migrate from the solid parent particles to form and enlarge the bonds and fill the voids. The mechanisms believed to have roles in transporting the atoms are surface diffusion, evaporation and condensation, bulk flow, and volume diffusion. All seem to take place individually, successively, or simultaneously during the sintering process, depending on the metal, powder condition, temperature, time, and atmosphere.

Diffusion and movement of atoms on the surfaces of the particles have been found to be the main activities in the early stages of sintering. *Surface tension* is the driving force to reduce the surface area, to round and smooth surface irregularities, and to segregate the unfilled spaces into fewer but larger pores. The surface area of small particles is large in relation to volume, and the influence of surface diffusion is proportionately great.

Evaporation and condensation can be expected to have more effect with a molten constituent than with a nonvolatile one. One definite effect seems to be to spheroidize angular voids.

Bulk flow is the movement of solid material under stress. Some stress may be left in the particles after pressing, but most seems to be induced by enveloping surface tension forces, which appear to cause viscous flow of the material within the particles as well as on their surfaces at sintering temperatures.

Volume diffusion originates in the random movements of atoms induced by thermal vibration. Atoms tend to move from a location of high

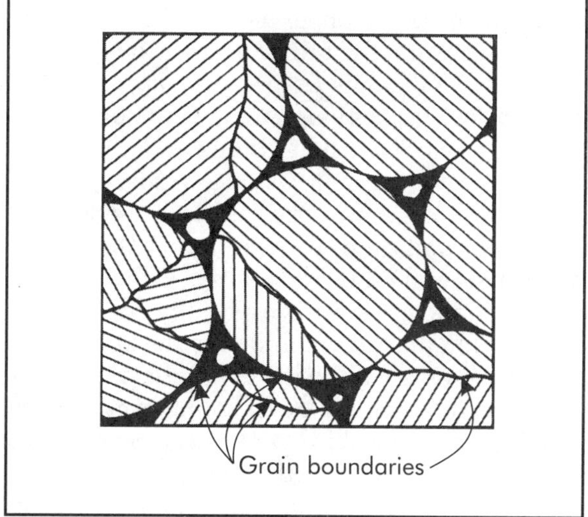

Figure 10-5. Idealized diagram of the arrangement of grains in a metal powder compact. Sections of original particles are depicted as circles.

to one of lower concentration. Within metal crystals, atoms migrate from grain boundaries to voids within the lattice structure. There is strong experimental evidence that in the later stages of sintering the major action is a movement of atoms from the grain boundaries to the voids between the particles. Below the recrystallization temperature, atoms migrate slowly. Above that temperature, the atoms move faster, but the grains grow, and the grain boundaries become fewer. It has been shown that atoms do not rapidly enter and shrink voids no longer connected by grain boundaries. The lower left-hand section of Figure 10-5 depicts the presence of grain boundaries and diminution of voids. The upper right-hand portion indicates a crystal grown to encompass several grains and the isolated pores within it that do not shrink appreciably.

As atoms leave grain boundaries, their space is filled by the grains moving together. This explains volume shrinkage often evident in sintering. After sufficient grain growth, grain boundaries become insignificant. Then atom migration, volume shrinkage, bond growth between particles, and increase of compact strength practically cease. That explains why there is an optimum time of sintering for each case. Up to that time, properties are improved, but beyond that there is little benefit.

All materials do not shrink when sintered; some grow. An example is found in copper-brass compacts in which copper atoms migrate to, and zinc atoms from, the brass. In this example, vacancies left by atoms from copper appear to be supplied by dislocations and grain boundaries, but zinc atoms leave holes in the brass. Porosity is observed to form on the brass side, and the grains do not shrink. Zinc atoms are added to the copper and the net result is growth.

Because of its relatively large surface area, a powder metal compact is especially susceptible to oxidation and attack and must be kept clean to coalesce on an atomic scale. Sintering is usually done in a neutral or reducing atmosphere and, in critical cases, in a vacuum. Sometimes reactive gases are added to speed the process and purge impurities. Solids are often mixed with powder metals to vaporize and open pores. All such measures must not contaminate the main sintering atmosphere. Precise control of atmospheres and temperature is important. Most sintering is done in semi- or fully continuous furnaces with a mesh belt, roller hearth, walking beam, or mechanical pushers and several chambers for heating and cooling in stages, such as preheat and burnoff, high heat, and cooling. Some induction sintering is done; it can be readily controlled and automated for high production rates.

Spark sintering is done by placing loose powder in a die, passing a large current through it, and applying pressure at the same time. An initial surge of current strips the oxide coating from the powder particles and leaves clean surfaces that join together. A continuing current heats the mass under pressure. This process was developed in the aerospace industry and is used for many common and unusual materials and parts. Only one step is required, and the operation is fast.

10.5 FINISHING OPERATIONS

Finishing imparts specific properties or features to metal parts. The processes include infiltration, heat treatment, impregnation, sizing, and machining.

Infiltration consists of placing a piece of copper alloy in contact with a presintered iron part and heating to 1,149° C (2,100° F). The copper melts and is drawn into the pores of the part by capillarity. The same results may be achieved in the sintering stage with copper-filled iron powder. Infiltration increases strength 70–100% and may improve machinability, but makes the product more brittle. Tolerances cannot be held as closely with infiltration as with other methods, such as double pressing. Infiltration is often more expensive than the other methods.

Iron parts may have carbon added to the original mixture or be *carburized* after sintering. They then are heated, quenched, and tempered like any other high-carbon steel product to a high degree of strength and hardness. An example is given at the bottom of Figure 10-2.

Impregnation appears in two forms. Sintered compacts, such as bearings, may be saturated with oil, about 20% by volume, by immersion, or by drawing oil through the compact by vacuum. The oil deposit is withdrawn by heat or pressure as needed in service. Powder metal parts and all kinds of castings, such as engine blocks, gear cases, pump bodies, and many more, are impregnated to seal pores and prevent leaks in service. This is done with sodium silicate, polyester resins, or anaerobic polymers.

Powder metal parts may be repressed after sintering. This is called *sizing* if done to hold dimensions, and *coining* if done to increase density. Repressing may be done in compacting presses or in conventional presses like those described in Chapter 12.

10.6 METAL COMPOSITES

A new field of metallurgy is that of adding whiskers and fibers for high strengths in metallic parts, particularly at high temperatures. The scheme is similar to that of impregnating fibers with resins for high-strength plastic parts as described in Chapter 11. Aluminum alloys can increase in strength by 345–689 MPa (50,000–100,000 psi), and increase their modulus of elasticity by 50% with a 10% addition of alumina whiskers. Nickel alloys with tensile strengths of 138–276 MPa (20,000–40,000 psi) at 927° C (1,700° F) stay at those levels at 1,093° C (2,000° F) with the addition of alumina whiskers, but drop off to one-third as much without whiskers. Composites can attain almost the stiffness of steel with as little as one-third the density, but have not been found suitable for some high-temperature applications. Costs are higher than for conventional metals but are expected to decline as the market grows. Ways of making composites include mixing fibers with molten metal, vapor- and electrodeposition of metal on fibers, powder metallurgy techniques, and eutectic solidification of the matrix and whisker constituents.

10.7 DESIGN OF POWDER METAL PARTS

Four guides to successful production of parts made from powder metal are:

1. Design parts to suit the process; modify designs if they have been made previously in other ways.
2. Design the tools to conform to the principles of powder metal operations and to compress the powder metal uniformly.
3. Standardize parts as much as possible to be able to use the minimum number of tools on most parts.
4. Strictly control all operations.

A number of rules must be observed to design parts properly. Keep parts as small as possible. Cylinders, squares, and rectangles are the easiest shapes to press, and flat pieces are best. Steps add difficulty in getting homogeneity (Figure 10-3). Sharp corners and edges, thin ridges, and deep slots should be avoided because they make tools, preforms, and finished parts weak. A part must not have side recesses that prevent it from being pushed out of the die when compacted. Walls that are too thin are difficult to fill. The length of a part should not exceed two to three times its diameter. Thin and thick sections should not be next to each other because they expand differently on heating and cause cracks. Elaborations of these rules are given in handbooks.

Tolerances as small as ±10 μm/cm (±.001 in./in.) may be held but are expensive. More practical tolerances are upward of ±20 μm/cm (±.002 in./in.) for diameters and ±30 μm/cm (±.003 in./in.) lengthwise.

Powder metallurgy is capable of producing parts with near-net-shape features. Additional machining and assembly operations are thus eliminated, adding to its features of being competitive with other processes such as forging, casting, and machining.

10.8 PROCESS AUTOMATION

Two advantageous features make the powder metallurgy process a good candidate for automation: (1) the ease of producing parts with different compositions to obtain special physical and mechanical characteristics that are difficult to achieve using other processes, and (2) high production rates of complex products are more economically achieved by powder metallurgy processing than many other competitive processes.

The use of automated equipment is very common in the powder metallurgy field. The use of computers has emerged to simulate and optimize different aspects of powder metallurgy processing, allowing better process performance under automated environments. Emerging technologies, such as rapid prototyping, discussed in Chapter 29, have found their application in powder metallurgy processing. *Selective laser sintering* is a rapid prototyping process that is based on sintering of metallic and nonmetallic powder selectively into an individual object. Ongoing research on rapid prototyping has established potential applications of other methods in powder metallurgy processing. Three dimensional printing, photochemical machining, and laminated object manufacturing (LOM) are different rapid prototyping processes that are in various stages of implementation and emergence into powder metallurgy processing.

In general, the following features affect the degree of automation possible in powder metallurgy processing:

- size and shape of parts;
- part complexity and other geometric features; and
- production rates.

Process monitoring and control can be easily done. This added feature makes powder metallurgy operations even more adaptable to automation than many others. Because of the elimination of finishing and assembly operations in many cases, waste is minimized and the need for additional machinery is avoided. The near-net-shape feature and eliminating the need for additional applications and additional processing add more justifiable reasons for automation.

10.9 QUESTIONS

1. What advantages does powder metallurgy offer?
2. How may alloys be obtained in powder metals?
3. What are the important physical characteristics of powder metals?
4. What are the principal ways of obtaining metal powders?
5. Describe the basic operations in powder metallurgy to produce ductile tungsten wire, cemented-carbide tools, self-lubricating bearings, porous metal filters, ordinary structural parts, and high-strength machine parts.
6. Why is powder metal pressed? Why is uniformity important and how is it obtained?
7. What kinds of presses are used for powder metallurgy?
8. What are the benefits of slip casting powder metals?
9. What are isostatic and hydraulic pressing and what is their main benefit?
10. How are continuous strips made of powder metal?
11. Discuss what takes place to bond powder particles together when they are sintered.
12. What four processes contribute to the movement of atoms to bring about the bonding of metal powders?
13. What factors determine the optimum temperature and length of time required to sinter a powder metal?
14. Why do some powder metal compacts shrink on sintering? Why do others grow?
15. Why must powder metals be sintered in controlled atmospheres?
16. What is "infiltration" and what does it do?
17. What is "impregnation" and what is its purpose?
18. What does coining or sizing do to powder metal parts?
19. What precautions must be taken in machining powder metal parts?

20. Describe "spark sintering" and what it will do.
21. State four guides to successful production of powder metal parts.
22. What helps and what hinders powder metallurgy in competition with other processes?

10.10 PROBLEMS

1. A bushing with 40 mm (1.58 in.) OD, 25 mm (≈1 in.) ID, and 32 mm (≈1.25 in.) length is to be made from iron and alloying materials in the manner depicted at the bottom of Figure 10-2. What force in MN (tons) and what die fill depth in mm (in.) must a press be able to provide for the first pressing?
2. The green density of the compact described in Problem 1 is 7 gram/cm^3 (.25 lb/in.^{3}). What is the approximate density of the powder metal before pressing?
3. A round piece as depicted in Figure 10-4 has a head 50 mm (≈2 in.) in diameter and 15 mm (≈.5 in.) thick. The stem has a 20 mm (≈.75 in.) diameter and is 18 mm (≈.63 in.) long. The part is to be made from brass powder. Compacting pressure must be 350 MPa (≈50,000 psi). Estimate the force and die fill the press must provide.

10.11 REFERENCES

Bradbury, S., ed. 1986. *Powder Metallurgy Equipment Manual*. Princeton, NJ: Powder Metallurgy Equipment Association.

German, R.M. 1990. *Powder Injection Molding*. Princeton, NJ: Metal Powder Industries Federation.

——. 1994. *Powder Metallurgy Science*. Princeton, NJ: Metal Powder Industries Federation.

Hausner, H.H., and Mal, M.K. 1982. *Handbook of Powder Metallurgy*. New York: Chemical Publishing Company.

Kenel, F.V. 1980. *Powder Metallurgy Principles and Practices*. Princeton, NJ: Metal Powder Industries Federation.

Metals Handbook, 9th Ed. 1984. Vol. 7: *Powder Metallurgy*. Materials Park, OH: American Society for Metals.

Powder Metallurgy Design Manual, 2nd Ed. 1995. Princeton, NJ: Metal Powder Industries Federation.

Price, P.E. 1982. "Hot Isostatic Pressing in the Aerospace Industry." *Metal Progress*, February.

Society of Manufacturing Engineers (SME). 1998. *Powder Metallurgy* video. *Fundamental Manufacturing Processes* series. Dearborn, MI: SME.

Hot and Cold Working of Metals

11

Steam-powered gravity drop hammer, 1839
—James Nasmyth, Manchester England

11.1 WROUGHT METALS

The processes described in this chapter produce what are known as the *wrought metals.* These are important engineering materials because of their strength and toughness. They are mandatory for many applications, such as critical structural members, for which cast metals do not suffice. Examples of the forms of these products are structural I-beams, channels, and angles; railroad rails; round, square, and hexagonal barstock; tubes and pipes; forgings; and extruded shapes.

Common metalworking processes included in this chapter are metal rolling, cold drawing, pipe and tube manufacture, forging, and extrusion. They all squeeze metal. Although much of their output is in final form, such as rails, most of it goes to feed secondary processes that make finished products by cutting or forming as will be described later in this book. Metals are worked by pressure in these processes for two reasons: (1) to form desired shapes, and (2) to improve physical properties. The results depend on whether the work is done hot or cold. *Hot working* is done above the recrystallization temperature. This is at or near room temperature for lead, tin, and zinc. It is above the critical temperature for steel, as depicted in Figure 11-1. Comparisons of the effects of hot and cold working are made in Table 11-1. The principles will be illustrated mostly for steel in this discussion for the sake of brevity.

11.2 HOT WORKING

The properties of a metal are different above or below its recrystallization temperature. The strength of a metal decreases as temperature rises, and its grains can be distorted more easily.

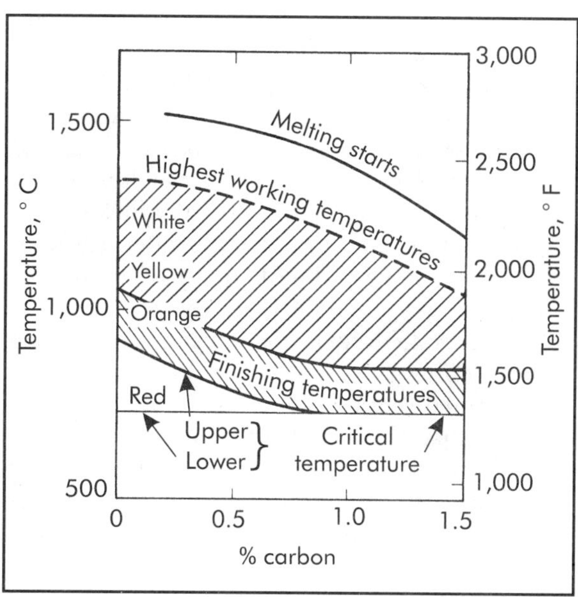

Figure 11-1. Range of rolling and forging temperatures for carbon steel.

Table 11-1. Properties of hot- and cold-worked metals

Alloy	Condition	Ultimate Tensile Strength MPa	psi	Yield Strength MPa	psi	Hardness
Aluminum 1100	"O"	110	16,000	41	6,000	28 Bhn
	H18	200	29,000	172	25,000	55 Bhn
Electrolytic tough-pitch copper	Hot-rolled	234	34,000	69	10,000	45 R_F
	Extra spring	393	57,000	365	53,000	95 R_F
Steel, SAE 1010	Hot-rolled	428	62,000	221	32,000	60 R_B
	Cold-rolled	559	81,000	345	50,000	90 R_B
Brass, yellow	Annealed	338	49,000	117	17,000	68 R_F
	Spring hard	627	91,000	428	62,000	90 R_B

If a ductile crystal is distorted by working, it does not visibly come apart, but its lattice structure is fragmented. New and smaller crystals form out of the fragments. If the temperature is dropped soon, a fine structure results. However, if the metal is held above the recrystallization temperature and below the melting point, its crystals grow larger. Small crystals tend to combine, and large ones absorb small ones. The higher the temperature, the faster the growth. The longer the time, the larger the grains become. These conditions help explain the following advantages of pressing or working hot metals.

1. True hot working does not change the hardness or ductility of the metal. Grains distorted and strained during the process soon change into new undeformed grains.
2. The metal is made tougher because the grains are reformed into smaller and more numerous crystals and its pores are closed and impurities segregated. Slag and other inclusions are squeezed into fibers with definite orientation. Chains of crystals intertwined with the filaments of impurities make the metal particularly strong in one direction. Metal is hot-worked to orient the flow lines as nearly as possible for strength in the direction of highest stress.
3. Because the metal is weaker, less force is required, the process is faster, and smaller machines can be used for a given amount of hot working as compared with cold working.
4. A metal can be pushed into extreme shapes when hot without ruptures and tears because the crystals are more pliable and continually reformed.

Hot working is done well above the critical temperature to gain most of the benefits of the process, but not at a temperature high enough to promote extreme grain coarsening. This is exemplified in Figure 11-1, which shows the range of working temperatures for carbon steels.

Hot working has several major disadvantages: it requires heat-resistant tools, which are relatively expensive; the high temperatures oxidize and form scale on the surface of the metal; and close tolerances cannot be held. Cold working is necessary to overcome these deficiencies.

11.3 COLD WORKING

Cold-worked metal is formed to shape by the application of pressure at temperatures below the critical point and, for the most part, nominally at room temperature. It is preceded by hot working, removal of scale, and cleaning of the surface, usually by pickling. *Cold working* is done mostly to hold close tolerances and produce good surface finishes, but also to enhance the physical properties of the material.

When a piece of metal is initially subjected to stress, it is strained in proportion in an elastic manner as depicted by the line *oy* in Figure 11-2. More stress causes permanent or inelastic deformation along the line *yl*. The point *y* is called the *yield point*. Stresses and strains beyond it are in the region where cold working must take place to change the shape of an object. What happens within the metal when it is cold worked is described in Chapter 3.

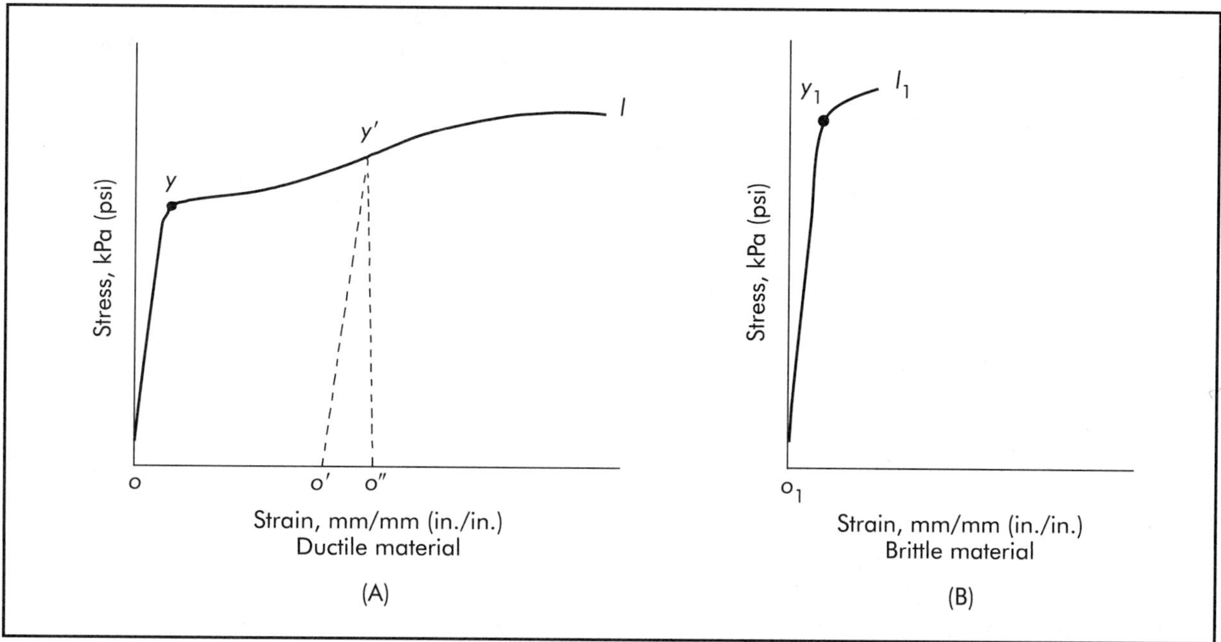

Figure 11-2. Effect of cold working a metal. Typical stress-strain relationships are depicted for room temperature.

The stress-strain diagram of Figure 11-2 typifies the tension, compression, or shear and the resulting strain in any one direction in a material. How such a curve is derived is explained in Chapter 3. The diagram in Figure 11-2(A) is characteristic of a *ductile material*, one that can stand considerable straining between its yield point *y* and point of rupture *l*. This is the kind of material amenable to cold working. Figure 11-2(B) is indicative of a *brittle material* that breaks before it is deformed appreciably, like cast iron.

Consider that cold working applies a stress to a ductile material and causes a plastic strain to point *y′* in Figure 11-2(A). When the stress is released, the strain falls back slightly to point *o′* at zero stress. This constitutes an elastic return along path *y′ o′*, which is parallel to *yo,* and accounts for the phenomenon called *springback* that always occurs when metal is cold worked. When a piece is released from the shaping tools, it returns slightly toward its original shape.

Now assume that a material has been cold worked and released to point *o′* in Figure 11-2(A). If the material is again stressed in the same way, its yield point is then *y′*, and further plastic stress takes place along the path *y′l*. In the cold-worked state, the material exhibits a new stress-strain relationship along *o′y′l*. Thus the material has a higher yield point, is harder, and has less ductility than before the original cold working. The material has become *strain hardened.* The more the material is cold worked, the closer its properties approach those of a brittle material as illustrated in Figure 11-2(B).

The natural effect of most cold-working operations is to apply much more stress in one direction than in others. Thus, the material may be strained to the equivalent of point *o′* with a yield point *y′* in one direction as in Figure 11-2(A), but to points closer to *o* and yield points closer to *y* in other directions.

Strains that occur in different or opposite directions within the same piece often react upon each other when the applicable forces and stresses are released. This keeps the material from settling back to a completely unstressed condition and causes what are called *residual stresses.* For example, a sheet may be rolled in such a way that the material is plastically compressed near its surface but not throughout its thickness. After rolling, the outer layers are kept from expanding and are held in compression by tension exerted by the inner material that resists being stretched.

Strain hardening must be relieved in some cases. It is not a desirable property in many products. If a metal is cold worked along line *oyl* to a point close to *l* in Figure 11-2(A), further cold working will lead to failure. What happens during strain hardening is that the crystals become distorted, disarrayed, and resist further change. The metal can be returned to or near its original state, depicted by point *o*, by annealing or normalizing as described in Chapter 6. Those processes heat the metal above its recrystallization temperature and reform and relax the grains.

Relatively large forces must be exerted for cold working. This means that equipment must be proportionately rugged and powerful, particularly for rapid production. Even so, many products can be finished by cold working to close limits and with good finishes at lower cost than by other means. Cold-working processes have a major and basic role in most high-production industries.

11.4 WARM WORKING

Warm working, or *warm forming* as it is often called, is performed at some intermediate temperature above room temperature, but below the recrystallization temperature of a metal. When compared to hot working, the warm working process generally:

- requires less heat energy;
- results in less scaling and decarburization; and
- contributes to improved dimensional control and surface finish.

A process known as *flashless warm forming* is being used to effect considerable scrap savings over the conventional forging process. This process uses a precision blank in a precision closed die to produce near-net-shape parts to close tolerances with practically no flash.

11.5 ROLLING

11.5.1 Principles

When metal is rolled, it passes and is squeezed between two revolving rolls in the manner indicated in Figure 11-3. The crystals are elongated in the direction of rolling, and the material emerges at a faster rate than it enters. In hot rolling, the crystals start to reform after leaving the zone of stress, but in cold rolling they retain substantially the shape given them by the action of the rolls.

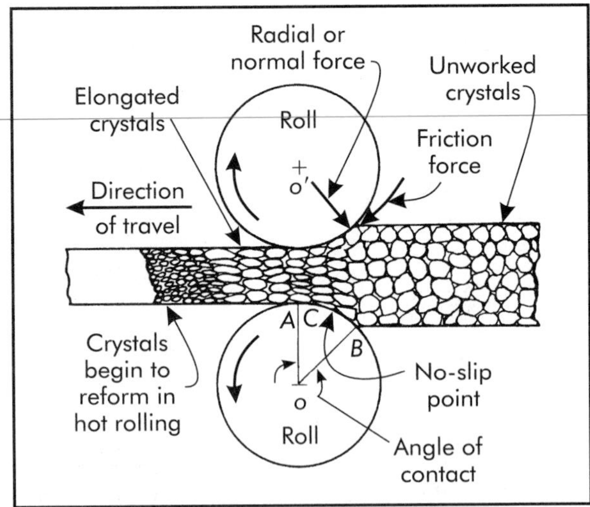

Figure 11-3. Metal rolling.

The rolls contact the metal over a length depicted by arc *AB* in Figure 11-3. At some point of contact, the surfaces of the material and roll move at the same speed. This is the no-slip point *C*, in Figure 11-3. From *C* to the exit at *A*, the metal is in effect being extruded and moves faster than the roll surface. In that zone, friction between the workpiece and rolls opposes the travel and hinders the reduction of the metal. Normal and friction forces at a point are depicted in Figure 11-3. The metal is moving slower than the rolls between points *C* and *B*, and the resultant friction force over arc *CB* draws the metal between the rolls. The position of the no-slip point *C* in arc *AB* depends upon the amount of reduction, the diameters of the rolls, and the coefficient of friction. Point *C* tends to move to *A* as the amount of reduction and the angle of contact increase. When the angle of contact (called the *angle of nip*) exceeds the angle of friction, the rolls cannot draw a fresh piece of material spontaneously into the space between them. When the angle of contact is more than twice the angle of friction between roll and work, point *C* coincides with *A*, and the metal cannot be drawn through by the rolls even if it is placed between them. This is because the horizontal component of the normal pressure of the rolls against the metal equals and nullifies the horizontal component of friction that draws the metal along.

As the metal is squeezed together between the rolls, it is elongated because it is incompressible. To accomplish this, the rolls have to apply both normal squeezing and frictional drawing pressures. Normal pressure of the rolls on the work is usually one to several times the amount of the yield stress of the metal. The pressure may rise to several hundred thousand psi in severe operations. The frictional force between the roll and work in the driving direction approaches the normal force times the coefficient of friction. This friction force times the surface speed of the rolls determines the power.

Both the frictional and normal pressures that the rolls must apply to stretch the work can be reduced appreciably if axial tension is applied to the work either fore or aft or both. For instance, it has been found with a metal having a yield strength of y psi that the maximum normal pressure of $3y$ without tension falls to y when axial tension of $.5y$ is applied fore and aft of the strip. This is commonly done in cold rolling.

The forces and power increase with the amount a piece is reduced in thickness. Thus, the strength and power capacity of the equipment and the workability of the metal determine how much a workpiece may be reduced at any one time.

11.5.2 Rolling Mills

A rolling mill is commonly designated by the number and arrangement of its rolls as indicated in Figure 11-4(A). A nonreversing two-high mill passes the work in one direction only. Stock may be passed back over the top of the rolls, but that is slow, or it may go through a series of rolls for successive reductions. The latter is faster but requires more investment in equipment. Work may be passed back and forth through a reversing two-high mill, but that takes extra time and power. The work can be passed between the bottom two rolls of a three-high mill and then raised by an elevator and passed back between the top two rolls. Large backing rolls support small working rolls in four-high and cluster mills. Small rolls are weak by themselves but are cheaper to replace as they wear, make contact with the work over less area, tend to spread the work less sideways, are subject to smaller separating forces, and require less power than large rolls. Material can be reduced more in fewer passes if it is bent back and forth as it is rolled. The variety of rolling mills shown in Figure 11-4(A) is exemplary of the alternatives engineers have in designing equipment.

Plain rolls, as depicted in Figure 11-4(B), serve for flat sections, and grooved rolls for bars and shapes. They are mostly made of cast iron or cast or forged steel. Roll design is a challenging engineering problem. The dimensions and properties of each roll must be selected for optimum conditions of hardness, wear resistance, strength, rigidity, and shock resistance. Chilled or alloyed cast iron rolls can be made hard and at low cost. However, they lack strength for severe service, and their usage is somewhat limited. Relatively new nodular-iron rolls have replaced both steel and flake-graphite iron rolls in some applications. For the most part, superior strength, rigidity, and toughness can be attained in steel rolls, particularly alloy steel, at a price. Stronger materials make smaller rolls possible with their inherent advantages. The material in a roll must not have affinity for the work material. Hot rolls are commonly rough, and even notched to bite the work, but cold rolls are highly finished to impart a good finish. A brief description of roll grinding is given in Chapter 24.

A set of rolls in their massive housing is called a *stand*. A number of stands may be arranged in a row in a continuous mill like the one in Figure 11-5. In operation, the metal runs continuously through all the stands at once. For jobbing, the stands may be side-by-side. When the work is finished in one stand, it is moved to the next, etc. These arrangements have many variations to suit specific conditions. Each mill is designed and operated for a limited range of products. For instance, a huge mill for rolling chunks of steel with a cross-sectional area of almost 0.4 m^2 (4 ft^2) is not economical for rolling small pieces of less than 0.1 m^2 (1 ft^2) in area across. The small pieces can be rolled as well on a less powerful

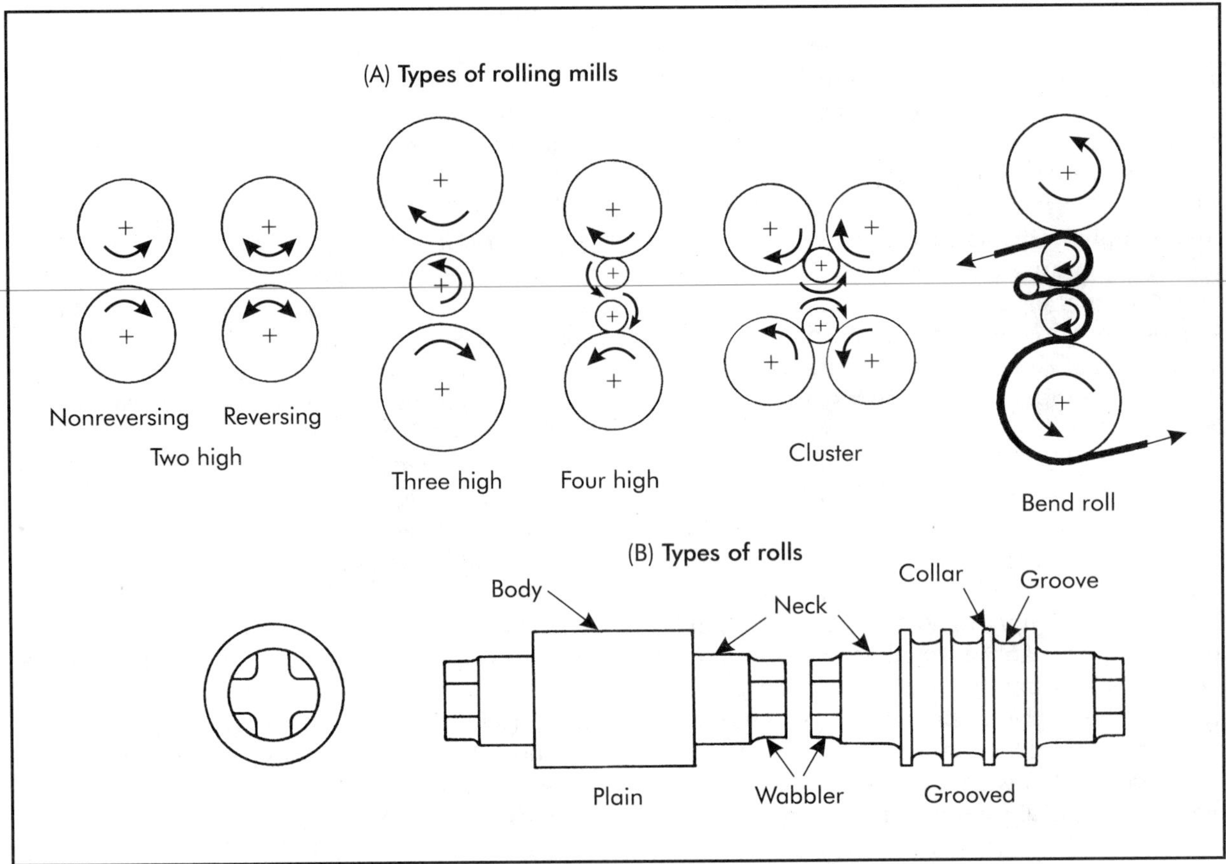

Figure 11-4. (A) Types of rolling mills; (B) types of rolls.

mill that costs only a fraction as much. On the other hand, each mill can only process pieces so large. To change rolls, particularly on a large mill, is costly. For that reason, when a mill is set up for some particular shape or shapes, it is economical to keep it on the same work.

Most rolling mills today are automated in their operation with computers controlling the setting of the rolls for each pass. The 2.8 m (110 in.), sheared-plate mill shown in Figure 11-5 is completely computerized, including automated gage control.

11.5.3 Hot-rolling Steels

After steel has been melted and refined as described in Chapter 4, traditional practice has been to cast it into a form called an *ingot.* Steel ingots are held and heated uniformly throughout (Figure 11-6) to around the highest working temperature. This is done in a *soaking pit,* which is a large furnace lined with refractory silica brick, having a neutral or reducing atmosphere, and usually loaded from the top.

Ingots are rolled into blooms, slabs, or billets as indicated in Figure 11-6. This is done rapidly before the metal cools below the working temperature. A typical cycle reduces an ingot with a square section almost 0.61 m (2 ft) on a side down to a bloom 0.15 m (6 in.) square in about 2 min, all in about 17 passes through the rolls. Most blooming mills are two-high reversible mills. The work is turned 90° between heavy reductions to work it uniformly on all sides. The ends of the blooms are sheared away to remove cavities or pipes carried over from the ingot. At the same time, the bloom may be cut to convenient lengths for later operations.

A rising trend is to produce slabs, blooms, or billets directly by continuous casting, as described in Chapter 9. This is also called *concasting* and *strand casting.* It eliminates the expensive

Figure 11-5. Plate enters the four-high finishing stand of a sheared-plate rolling mill. Steel plate is rolled to the desired thickness by being passed back and forth through the finishing stand rolls. (Courtesy Bethlehem Steel Corporation)

equipment and the steps for ingot casting and saves up to 25% in material lost in the scrap normally cropped from billets, blooms, and slabs rolled from ingots.

The great bulk of flat plates, sheets, and strips are rolled in continuous mills from slabs or directly from ingots. For premium quality products, defects are removed from the surfaces of slabs by pneumatic chipping or flame machining (Chapter 15). Structural shapes are rolled from blooms. For bars, rods, or wire, the blooms are customarily reduced to billets. If the temperature drops too low during processing, the blooms may be reheated. Blooms, billets, and slabs are semifinished shapes with rectangular sections, rounded corners, and with all dimensions over 38 mm (1.5 in.).

Because of limitations in the equipment and workability of the metal, rolling is done in progressive steps. An illustration of this is given by the 15 steps required to reduce a 102 × 102 mm (4 × 4 in.) billet to a 19-mm (.75-in.) diameter bar as indicated in Figure 11-7. From 8–10 steps are required to complete most commercial shapes, such as I-beams, channels, and rails from blooms.

11.5.4 Cold Rolling

Bars of all shapes, rods, sheets, and strips of all common metals are commonly finished by cold rolling. Foil is made of the softer metals in this way. Cold-rolled sheets and strips make up an important part of total steel production and are major raw materials for some high-production consumer goods industries, such as household appliances.

Metals are cold rolled for improved physical properties, good surface finish, textured surfaces, dimensional control, and machinability.

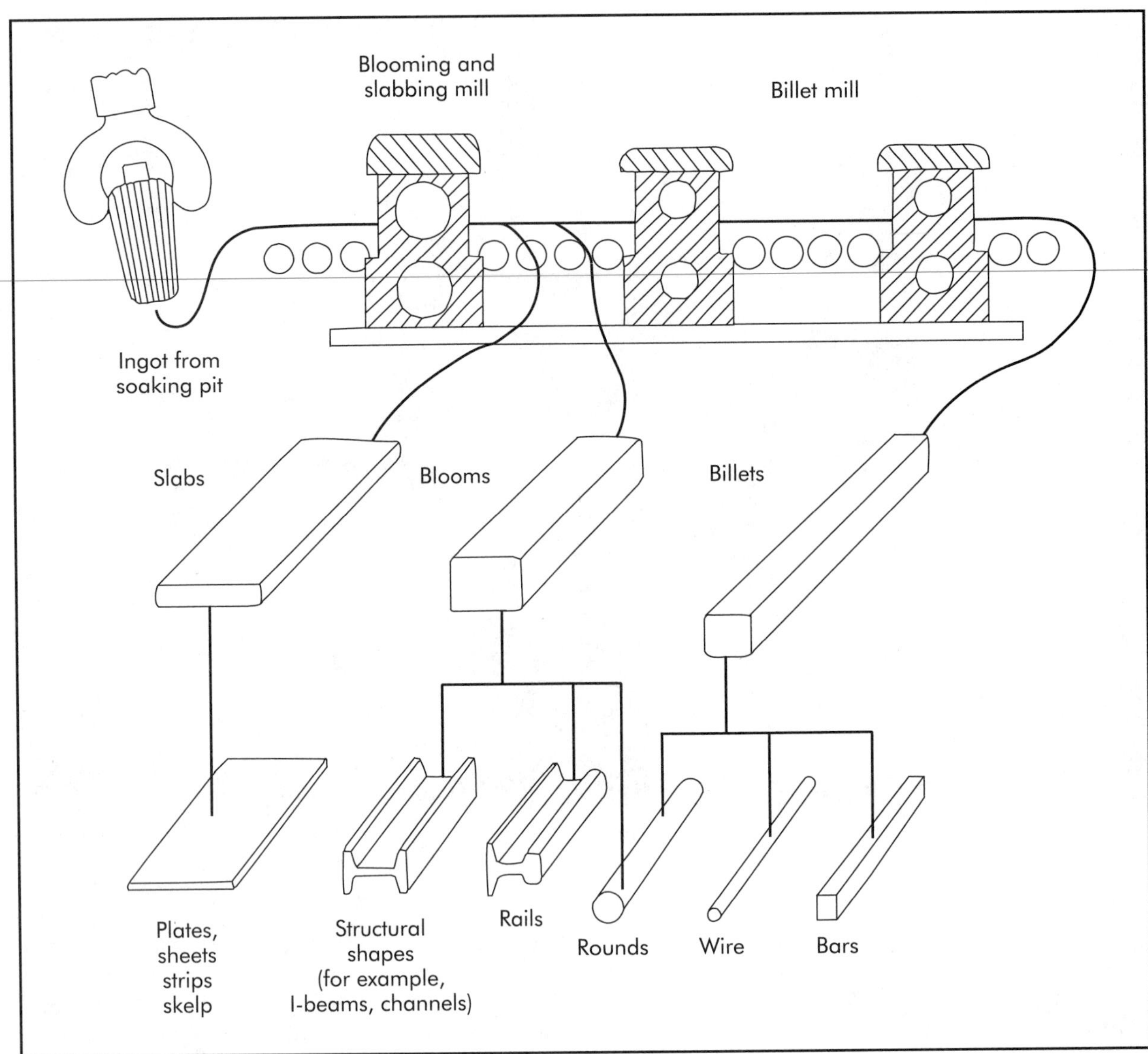

Figure 11-6. Typical steel-rolling procedure.

Sheet steel less than about 1.3 mm (.05 in.) thick is cold rolled as a matter of course because it cools too rapidly for practical hot rolling. Cold rolling is a practical means of producing the degree of hardness wanted in a material. Cold-rolled sheets and strips are classified commercially as skin rolled, quarter-hard, half-hard, and full-hard to denote amounts of cold reduction up to 50% without annealing.

If a bright finish, but not hardness, is desired, the metal may be annealed just before the final rolling pass. A large tonnage of sheet steel is precoated with an organic film, plain or figured, which becomes the ultimate finish on the final product. Various indented or raised textures or patterns are rolled into sheets of steel and other metals to increase rigidity, provide a surface that hides defects and is easier to coat, give variety in appearance, and reduce glare. Cold rolling produces uniform thicknesses and close tolerances in sheets and bars. Machinability of most steels is improved by cold working and for that reason cold-rolled or drawn stock is widely used in fast automatic machining operations.

Steel is pickled prior to cold rolling to clean the surface and remove scale. Sheets or strips

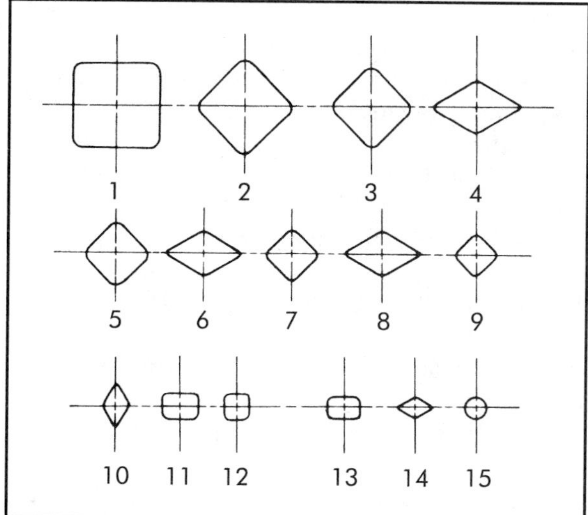

Figure 11-7. Steps taken to reduce a 102 × 102 mm (4 × 4 in.) billet to a 19-mm (.75-in.) diameter bar.

are often given a light cold rolling at first to establish a good surface finish and uniform thickness as a basis for good quality from later heavy cold rolling. Most work is done with small rolls in four-high or cluster mills. Frequently, tension is applied at either or both ends of the sheet or strip to minimize the effects of the high pressures of cold rolling.

Materials may need to be straightened at any stage of fabrication. Common practice is to pass barstock, rods, wire, sheets, or strip through a series of rollers as depicted in Figure 11-8. This is called *roller leveling*. Sheets are *stretcher leveled* or straightened by pulling them between jaws to induce a tensile stress throughout slightly in excess of the yield strength.

11.5.5 Quality and Cost

In the United States, steel shapes have been traditionally rolled in standard inch sizes, but in recent years mills also have been offering products in standard metric sizes. These conform to the preferred sizes prescribed by the American National Standards Institute (ANSI) standards listed at the end of this chapter and do not command premium prices.

Steel shapes are rolled to definite tolerances, varying with size, shape, and composition. These are tabulated in handbooks and mill catalogs. Inch sizes commonly have bilateral tolerances. As an example, thickness tolerances for cold-rolled carbon steel sheets or strips 508 mm (20 in.) wide are ±0.051, ±0.127, and ±0.152 mm (±.002, ±.005, and ±.006 in.) for thicknesses of 0.38, 1.52, and 4.57 mm (.015, .060, and .180 in.), respectively. Wider sheets are given more tolerance, and narrower ones less. Pickled hot-rolled sheets have tolerances of about 0.025 mm (±.001 in.) more for each size. A different practice is to specify a 7.6-mm (.3-in.) minus tolerance for all "metric steel plates" with a variable plus tolerance depending upon thickness and width. Alloy steel and aluminum sheet and strip tolerances are about the same as for carbon steel.

Cold-rolled sheets and strips are given surface finishes varying all the way from smooth and bright as a base for chrome plating to a rough matted finish for enameling.

Little consistency is found in tolerances from one shape to another. Hot-rolled round bars commonly have bilateral tolerances depending on size and carbon and alloy content, while cold-finished bars have a negative unilateral tolerance. As a rule, cold-drawn bars have better precision and finish than cold-rolled bars, but ground bars are the highest quality.

The values added by working steel into commercial shapes are indicated by the average market prices on one day shown in Table 11-2. These were base prices for large quantities free on board (f.o.b.) mill. Any size, composition, or requirements other than standard add to the price. The cost may be twice as much for small quantities from a distributor.

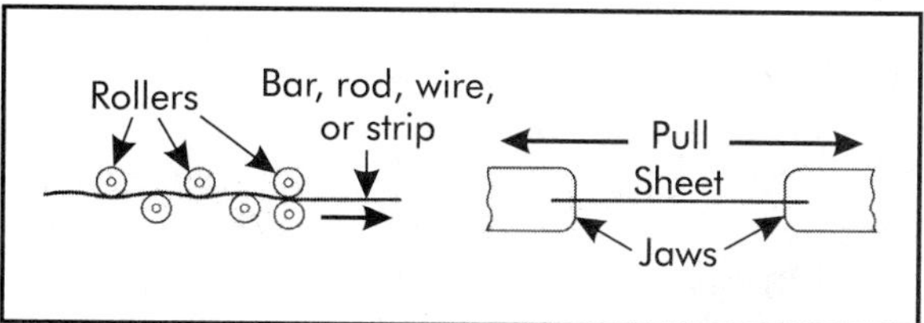

Figure 11-8. Operation of a typical straightening device.

11.6 COLD DRAWING

Round, rectangular, square, hexagonal and other shapes of bars up to about 102 mm (4 in.) across or in diameter, wire of all sizes, and tubes are commonly finished by cold drawing. Wire cannot be economically hot rolled smaller than

Table 11-2. Value added by working steel into commercial shapes

Type	Price $/metric ton	Price $/ton
Pig iron	210	213
No. 1 steel scrap	63	64
Plain-carbon steel rerolling billets	352	358
Plates	477	485
Hot-rolled strips	409	416
Cold-rolled strips	586	595
Structurals	451	458
Hot-rolled merchant bars	334	339
Cold-finished bars	615	625
Alloy steel forging billets	411	418
Plates	631	641
Hot-rolled strips	485	493
Hot-rolled bars	504	512
Cold-finished bars	673	684

about 5 mm (.2 in.) in diameter and is reduced to smaller sizes by cold drawing. Steel, aluminum, and copper and its alloys are cold drawn in large quantities.

11.6.1 Process

Hot-rolled stock is descaled, cleaned, and prepared for drawing. A common way of treating steel is to immerse it in hot sulfuric acid, then rinse, coat with lime, and bake. The leading end of a piece is tapered for insertion through the die. The action as the work is pulled through the die is illustrated in Figure 11-9. A piece is pulled through a hole of smaller size and emerges correspondingly reduced in size. Drawing pressure against a die must exceed the yield strength of the work material and commonly is as much as 0.7 to over 2 GPa (100,000 to over 300,000 psi) for steel. Steel is only able to slide through a die if coated by a lubricant that stands up under such tremendous pressures. Soap, at times with molydisulfide, is commonly applied and is affixed to the surface by the lime coating, which in turn is anchored by a soft oxide coating, called the *sull,* left by pickling. A thin copper or tin and copper plate also provides a good bearing surface on steel. It has been aptly said that what actually is drawn is a tenuous cylinder of copper or soap-lime sull, inside of which a steel core is squeezed into a new shape.

Dies must be hard and wear resistant as well as strong. They are made of chilled iron, hardened alloy steel, cemented carbide, and diamonds. The harder materials last longer but cost more.

The force to pull a wire or bar through a die is transmitted by tensile stress in the material that has just left the die. This stress increases more than proportionately with the amount the area of the wire is reduced by the draw. The stress in the material leaving the die may not exceed the yield stress. Theoretically, the yield stress is approached with a reduction in area of 50%. In practice, a reduction in area of less than 40% for each pass usually is found desirable. If the wire must be reduced more, it is passed through several dies. After a number of draws, a work-hardening material like steel becomes so brittle it must be annealed if it is to be drawn further.

Coarse low-carbon steel wire for fences, bolts, etc., may be drawn just once; hard bright wire is produced by several draws after annealing. Soft wire is annealed after being drawn. A soft bright wire is annealed and then given a final mild draw. Heat-treated steel with a carbon content of 0.3–1.2% is made into stiff and high-strength music wire, spring wire, wire for brushes, etc.

Bars are reduced from 0.41–3.18 mm (.016–.125 in.) by cold drawing from the hot-rolled size. The amount depends on the size and composition of the bar and the extent to which the physical properties must change.

11.7 MANUFACTURE OF PIPE AND TUBING

11.7.1 Butt-welded Pipe

Pipe is formed and butt welded in several ways. In one way, it is made from a strip, originally flat, called the *skelp.* Its edges are beveled enough to butt together when the skelp is rounded. The skelp is heated to welding temperature and is gripped at one end by tongs pulled by a draw chain. This pulls the skelp through a welding bell that forces it into a circular shape as depicted in Figure 11-10(A). The edges of the hot skelp are pressed and welded together.

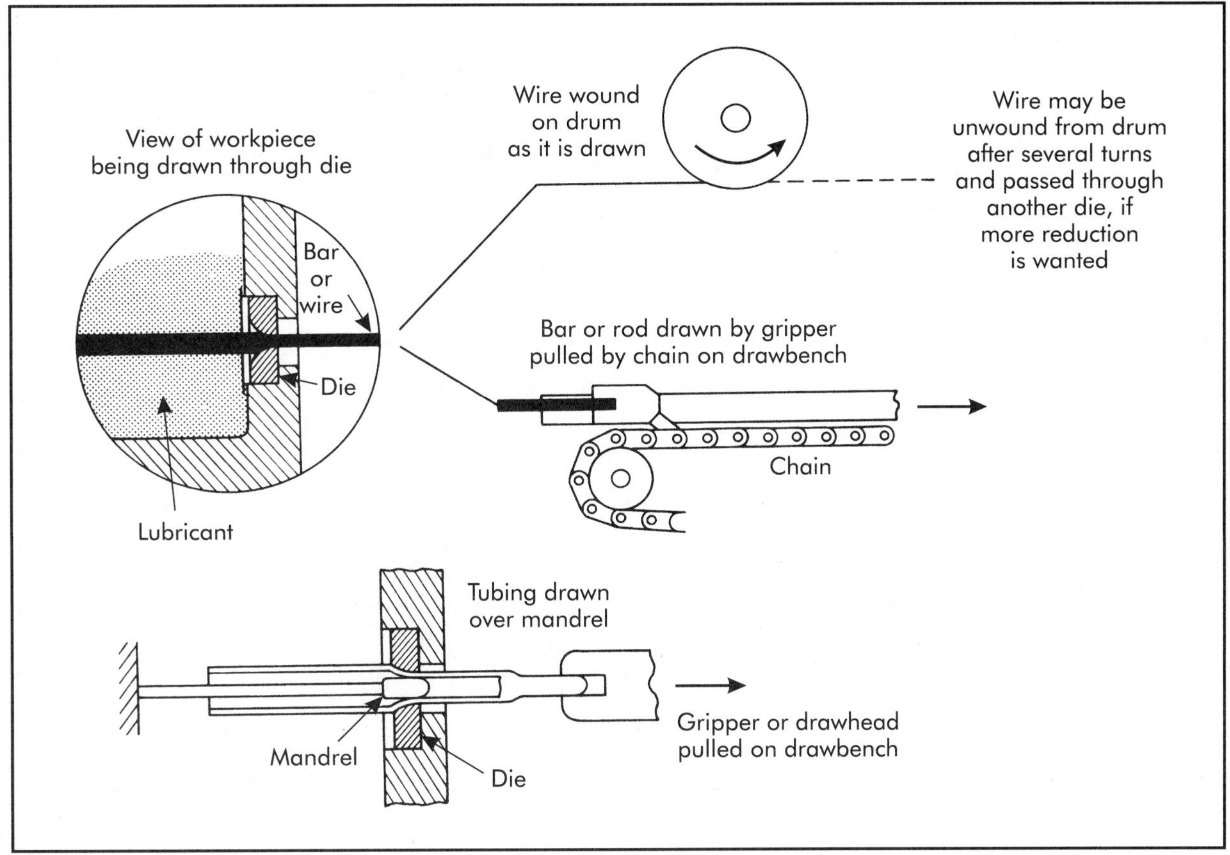

Figure 11-9. Methods of cold drawing.

For continuous butt welding of pipe, the skelp is used in coils, and the ends of the coils are welded together to make a continuous strip. Flames are directed to the edges to heat them to welding temperature as the skelp passes through a furnace. From the furnace, the skelp passes through a series of grooved rolls and is formed into pipe as illustrated in Figure 11-10(B). Pipe as large as 76 mm (3 in.) in diameter is made in this way.

Continuous steel strip is roll-formed into a circular shape to prepare it for electric-resistance butt welding. The principles of roll forming are described in Chapter 12. After being roll formed, the pipe passes between pressure rolls that hold the edges together and electrode rolls that supply current to create welding heat at the joint. This arrangement, indicated in Figure 11-10(C), is used for pipe up to 406 mm (16 in.) in diameter with a wall thickness of about 3.18–12.70 mm (.125–.500 in.). Larger pipes are commonly formed in large presses and are butt welded by the submerged-arc method.

After being formed and welded, pipe is normally passed through sizing and finishing rolls that make it round, bring it to size, and help remove scale. Continuously made pipe is cut to desired lengths. Cutters remove the extruded flash metal from both the inside and outside of the larger sizes of pipes.

11.7.2 Lap-welded Pipe

The edges are beveled as the skelp comes from the furnace to make lap-welded pipe. The skelp is then rounded in one of the ways previously described, but with the edges overlapping. It is then reheated and passed over a mandrel between two rolls as illustrated in Figure 11-10(D) to press and weld the lapped

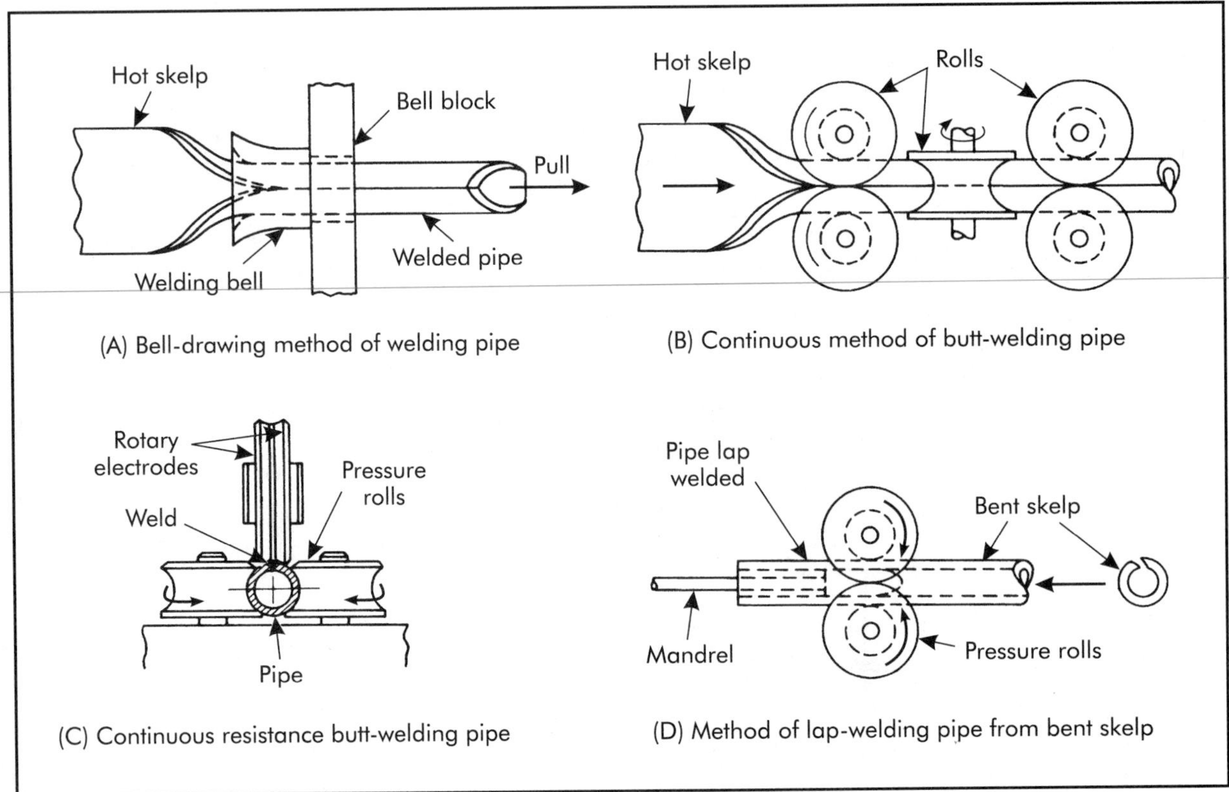

Figure 11-10. Methods of welding pipe.

edges together. Lap-welded pipe ranges in size from about 51–406 mm (2–16 in.) in diameter.

11.7.3 Seamless Tubing

Seamless steel tubing is pierced from heated billets passed between tapered rolls and over a mandrel in the manner depicted in Figure 11-11. The rolls are shaped so that their surfaces converge on the entering end to a minimum distance apart, called the *gorge*. From there the surfaces diverge to the exit end. The billet may have a small center hole drilled in the end. It is pushed and guided between the rolls, which are set to grip it in the entering taper. The rolls revolve at a surface speed of about 5.1 m/sec (1,000 ft/min) in the direction shown. Their axes are crossed, so they impart axial as well as rolling movement to the billet and force it over the mandrel. The mandrel can revolve, and the material is in effect helically rolled over the mandrel and not extruded. Shells as long as 12.2 m (40 ft) and up to 152 mm (6 in.) in diameter are produced in 10–30 sec in this way. A second piercing operation is applied for sizes up to about 356 mm (14 in.), and a third, of similar kind, for still larger sizes. The pierced shells are put through subsequent rolling and sizing operations to make finished tubes.

Seamless steel tubing is available in almost all compositions and alloys of steel and in common nonferrous metals such as aluminum, brass, and copper. It is the natural form from which to make many thin-walled round objects. Seamless

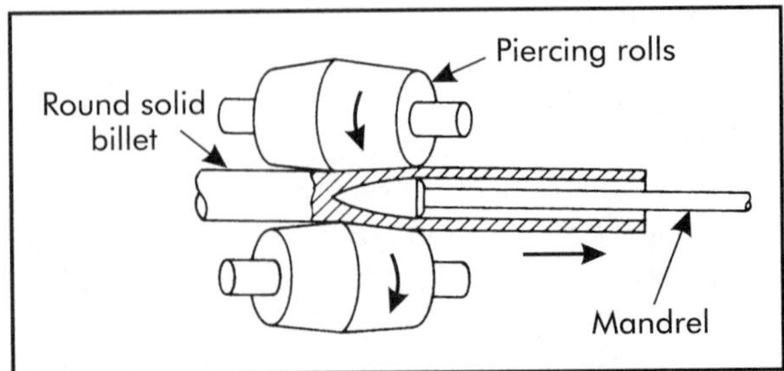

Figure 11-11. Mannesmann® process of tube piercing.

tubing is a popular and economical raw stock for machining because it saves drilling and boring of many parts.

Tubes from some hard-to-work steel alloys and nonferrous metals are not easily pierced and are produced by other methods. One way is to draw and redraw cups from hot plates in the manner described for drawing in Chapter 13. The bottom of a long cup may be cut off to make a tube. Tubes also are extruded as described later.

11.8 FORGING

Forging is the forming of metal, mostly hot, by individual and intermittent applications of pressure instead of applying continuous pressure as in rolling. The products generally are discontinuous also, treated and turned out as discrete pieces rather than as a flowing mass. The forging process may work metal by compressing its cross section and making it longer, by squeezing it lengthwise and increasing its cross section, or by squeezing it within and making it conform to the shape of a cavity.

Forging may be done in open or closed dies. Open die forgings are nominally struck between two flat surfaces, but in practice the dies are sometimes vee-shaped, half-round, or half-oval. Closed die forgings are formed in die cavities. All forging takes skill, but more is required with open than with closed dies. Faster output and smaller tolerances are obtained with a closed die. Open dies are, of course, much less costly than closed dies and more economical for a few parts. Either open or closed die forging may be done on most hammers and presses.

The high degree of skill required for forging is becoming scarce. Various systems exist in which the hammer and an automated manipulator are controlled by a computer program to execute an operation.

11.8.1 Heating the Work

It is important that a piece of material be heated uniformly throughout and to the proper temperature for forging. The proper temperature range for steel has been indicated in Figure 11-1. Heating is done in furnaces of various sizes to suit specific needs and in forms from the open forge fire to refractory-lined furnaces with precise atmospheric and temperature controls, conveyors, and rotary hearths. Automated lines for large-quantity forging commonly employ induction or electric resistance heaters.

11.8.2 Hammer Forging

A blacksmith does a simple form of open-die *hammer forging* when he strikes a hot workpiece on an anvil. This work is done mostly by machines today. The blows must be heavy to penetrate and knead the metal deeply, uniformly, and completely. Light blows affect only material near the surface. On cooling, the inner structure then differs from the outer, and the part lacks the flow lines, homogeneity, and impurity dispersion of a quality forging.

A mechanical hammer raises a heavy weight and drops it on an anvil. Various means have been employed. The *helve hammer* is used for light work, particularly to strike blows rapidly. It has a beam with a fulcrum at the middle, a heavy hammer at one end, and a revolving cam applied to the other end to raise and release the hammer repeatedly.

The most common forging hammers are steam or air operated. A single-frame, steam-forging hammer that gives access to the anvil from three directions is illustrated in Figure 11-12. A double frame is stronger and is in the form of an arch around the hammer and anvil, but the work space can be reached only from two directions. Most steam hammers today are double-acting; the hammer is driven down by the steam pressure as well as by gravity. To keep down vibration, the anvil for open die work is mounted on its own foundation separate from the frame. The effective work done by the hammer depends on the weight of the anvil. Ratios of 20:1 or more between anvil and hammer weights are common for standard hammers.

The ability of a hammer to deform metal depends on the energy it is able to deliver on impact. The energy from falling is augmented by the work derived from the steam in a double-acting hammer. Steam pressures are commonly from 517–862 kPa (75–125 psi). As an example, a hammer has a falling weight of 9 kN (2,000 lbf) and a steam cylinder bore that is 305 mm (12 in.). Mean effective steam pressure is assumed

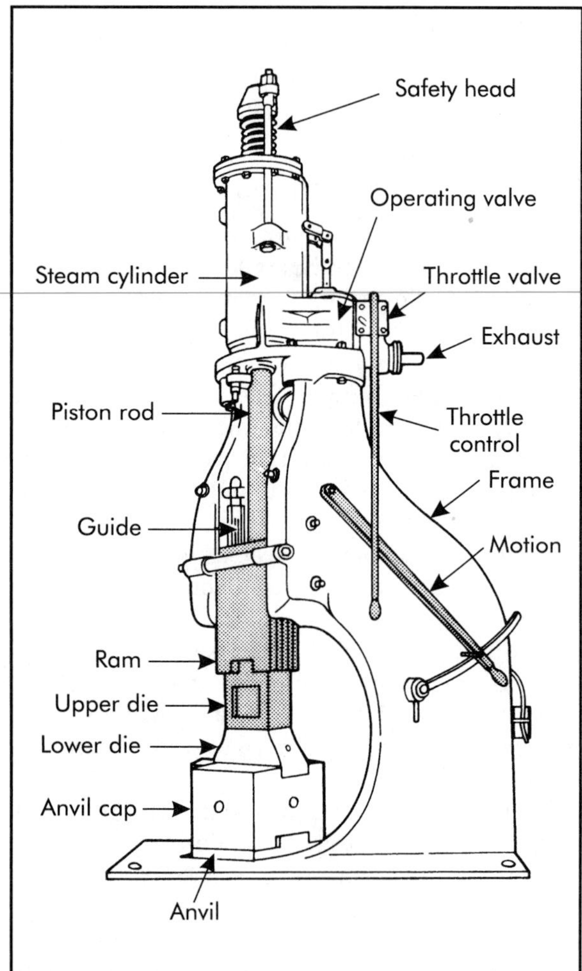

Figure 11-12. Single-frame, steam-forging hammer. (Courtesy Forging Industry Association)

to be 550 kPa (≈80 psi), and the stroke is 760 mm (≈30 in.):

$$S_f = \frac{\pi d^2}{4} \times p$$

$$S_f = \frac{\pi \times \overline{0.305}^2}{4} \times 550 = 40 \text{ kN} \tag{11-1}$$

$$\left(S_f = \frac{\pi \times 12^2}{4} \times 80 = 9{,}050 \text{ lb} \right)$$

D_f = 40 + 9 = 49 kN
(D_f = 9,050 + 2,000 = 11,050 lb)

E_b = 49 × 0.76 = 37 kJ
(E_b = 11,050 × 30 = 331,500 in.-lbf)

where

S_f = steam force, lb (kN)
d = steam cylinder bore, mm (in.)
p = mean effective steam pressure, kPa (psi)
D_f = total downward force, lb (kN)
E_b = energy in blow, kJ (in.-lbf)

If the hammer travels 3.175 mm (.125 in.) after striking the metal, the average force exerted is

$$\frac{37.2 \times 10^3}{10^{-3} \times 3.175} = 11.7 \text{ MN}$$

$$\left(\frac{331{,}500}{0.125} = 2{,}652{,}000 \text{ lb} = 1{,}326 \text{ tons} \right)$$

The amount of energy and force needed for a particular job is a matter of economics. A large hammer is expensive but can concentrate the energy to work the metal deeply and quickly. A small hammer may do the job, but works lightly and takes the time for many blows. A starting rule is that a hammer should have at least 23 kg (50 lb) of falling weight for every 6.5 cm^2 (1 in.^{2}) of cross-sectional area to be worked in the metal.

11.8.3 Drop Forging

Drop forging is the name given to the operation of forming parts hot on a drop hammer with impression or cavity dies. The products are known as *drop forgings, closed die forgings,* or *impression die forgings.* They are made from carbon and alloy steels and alloys of aluminum, copper, magnesium, nickel, and other metals.

Stock in the form of the heated end of a bar (Figure 11-13[A1]), slug, preform, or individual billet is placed in a cavity in the bottom half of a forging die on the anvil of a drop hammer. An example of a die is shown in Figure 11-13(B). The upper half is attached to the hammer or ram and falls on the stock. Generally, a finished forging cannot be formed in one blow because the directions and extent to which the metal can be forced at any one time are limited. Thus, most dies have several impressions, each a step toward the final shape. The workpiece is transferred from one impression to another between blows.

The main steps in forging a connecting rod are shown in Figure 11-13. First, the *tong hold* is formed at the right front corner of the die

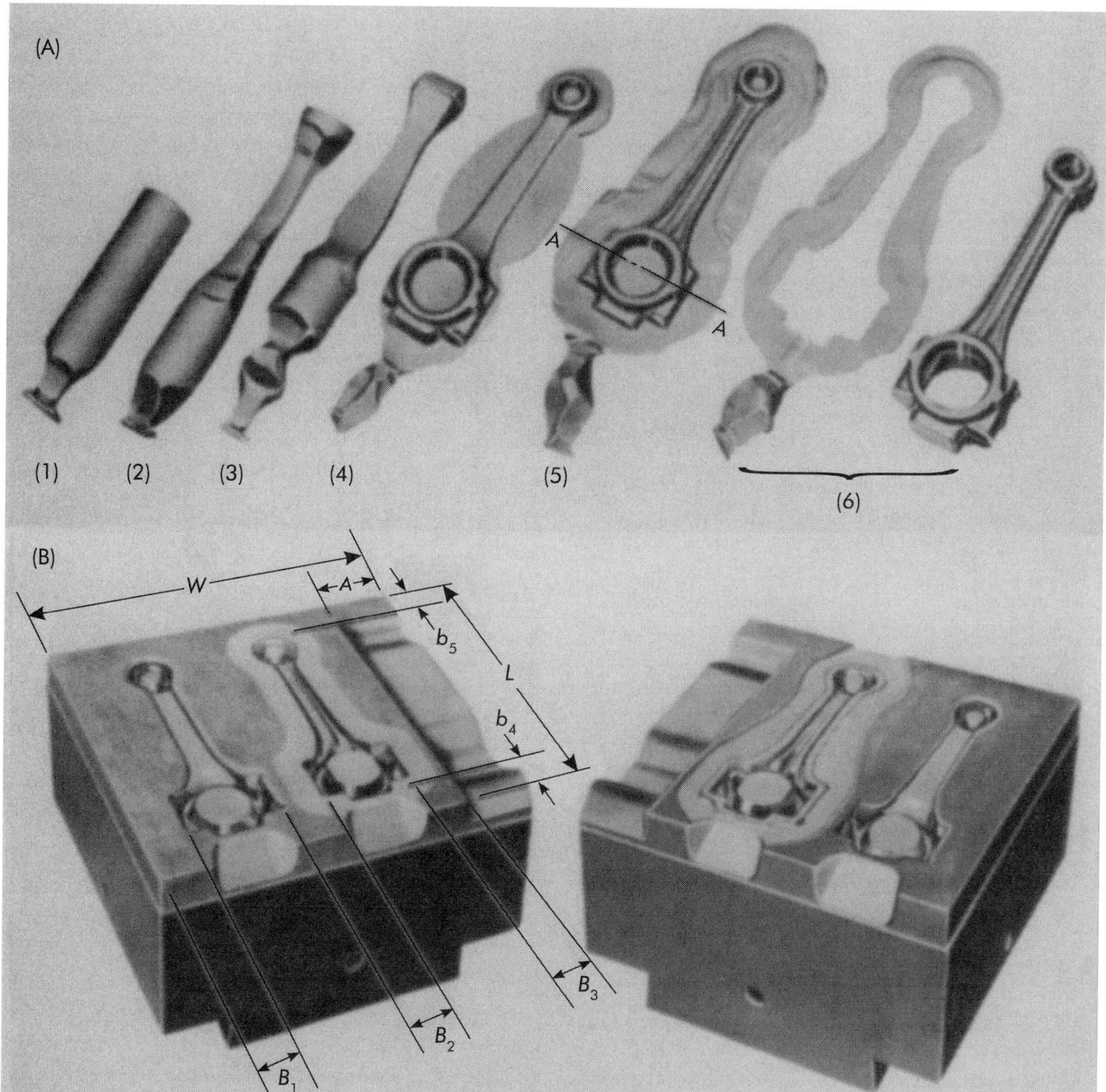

Figure 11-13. (A) Working steps in producing a connecting rod forging: (1) original bar, (2) drawn in a helve hammer; (3) result of rolling; (4) rough form after blocking; (5) finished forging as it leaves the hammer; (6) the flash and the trimmed forging. (B) Forging dies.

block. Then the stock is rolled between a number of blows and elongated at the corner and on the left side of the block. This is called *fullering*. The piece may be flattened between the surface of the die before it is *blocked* or rough shaped in the right-hand cavity. Finally, the preformed metal is pounded into the final shape of the left-hand cavity in the *finishing* operation. Excess metal is squeezed into a thin flange around the forging called the *flash*. This is sheared from the forging in a subsequent trimming operation. Actually, flash is waste metal and increases forging forces 5–10 times. Efforts have been made to eliminate it, but then filling the die requires close control of the raw stock, and that is usually more costly.

Most forging is done with unheated dies, but in some instances the dies are heated to 871° C (1,600° F). The work is chilled less and remains plastic longer. Thus it can be forged to closer tolerances and with thinner sections. Flange depths have been reported to be increased by 800%.

The die establishes the efficiency of an operation, and its design requires a high degree of skill. Some dies are cast, but more are made of heavy blocks of forged alloy-steel heat treated to less than maximum hardness, but to a toughness to resist shock. The cavities are cut or *sunk* to allow for shrinkage in the workpiece when it cools.

In addition to size, a die is designed to produce a forging with a minimum of residual stress and the most benefit from the original grain fibers of the bar. This requires enough but not too many stations, as dictated by a careful study of each case. The die is provided with draft to release the workpiece readily and with generous fillets, radii, and ribs to prolong die life. At times, locking surfaces or pins are provided to make the two halves of a die match the same way each time they come together. Multiple die impressions may be provided at each station to forge several pieces in each operation for large-quantity production.

11.8.4 Drop Hammers

A drop hammer has a guided falling hammer or ram, but differs from a forging hammer in that it has the anvil attached to the frame. This is to keep the upper and lower halves of a die aligned.

The *board drop hammer* illustrated in Figure 11-14 is suitable for small and moderate-size forgings. The ram is fastened to the lower ends of vertical hardwood boards. These pass between powered friction rolls that press from opposite sides against the boards when the ram is down. The revolving rolls raise the boards, which become latched in their uppermost position. Tripping the press releases the boards and allows the ram to drop. Maintenance and downtime cost are relatively high. One study showed that the boards had to be replaced every week, and about 14% of the work time was lost each year.

The ram is lifted by compressed air acting on a piston in an overhead cylinder on another type of drop hammer. The ram is clamped in the top position but is released and drops when the press is tripped. It takes more energy for a *power drop hammer*, but repair and downtime costs are negligible, and total operating cost is less than for a board drop hammer.

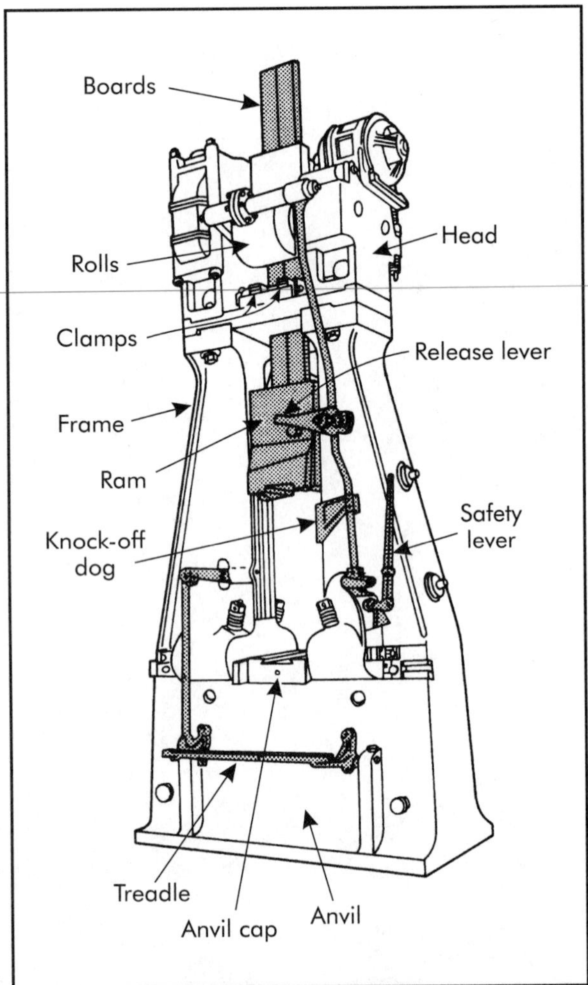

Figure 11-14. Board drop hammer. (Courtesy Forging Industry Association)

Most drop hammers are manually controlled and require considerable skill for fast and efficient operation. Some hammer systems have electronic controls that govern the number, intensity, and timing of blows in a preset cycle to produce a forging. An unskilled operator needs only to shift the stock from station to station in a routine manner.

The same factors as explained for forging hammers determine the energy and force delivered by a drop hammer. The following formula gives an estimate of the energy needed for a forging:

$$E = KA \qquad (11\text{-}2)$$

where

E = energy, J (ft-lbf)
K = value from Table 11-3
A = plan area, m^2 ($in.^2$), including flash

Flash may be 25.4 mm (1 in.) on a side for a workpiece 203 mm (8 in.) or less in diameter or width, and in proportion up to 51 mm (2 in.) for a part 610 mm (24 in.) or more in diameter or width.

A hammer is rated by the weight of its falling ram. The die adds about 25% to the falling weight. Board drop hammers are usually of 22 kN (5,000 lbf) capacity or less because larger sizes are not efficient. One line of air drop hammers ranges in size to 45 kN (10,000 lb). An 1,814-kg (4,000-lb) model with a 1.3-m (50-in.) effective stroke and 27 kJ (20,000 ft-lbf) capacity, including a 4.5 kN (1,000 lb) die, sells for about $200,000. However, the cost of a massive foundation, furnaces, and air compressor equipment must be added. A 1.36-Mg/hr (3,000-lb/hr) capacity furnace for a forge shop recently sold for $140,000, including the latest energy-saving and pollution controls. A 149 kW (200 hp) compressor that delivers 0.5 m^3/sec (1,060 ft^3/min) costs about $50,000.

Double-action air or steam hammers, called *power drop hammers*, are also used for closed die forging. A 1.36-Mg (3,000-lb) model capable of 46.6 kJ/stroke (34,400 ft-lbf/stroke) costs about $250,000 plus foundation, furnace, and air or steam equipment.

Counterblow hammers have rams driven together at the same time from the sides, or from top and bottom, to expend their combined energies on the workpiece instead of on an anvil. They may be set to strike single or repeated blows. The horizontal arrangement is easily automated in a production line because of clear space above the dies, and faster production rates can be achieved. Equipment cost is high; a model capable of delivering a 54 kJ/stroke (40,000 ft-lbf/stroke) costs about three times more than a comparable power drop hammer. Even so, in the case of a typical connecting rod forging, the output is 536 pieces/hr on a drop hammer and 972 on the two-ram counterblow hammer, with the cost per piece of the hammer about 1.5 times that of the counterblow machine.

11.8.5 Press Forging

The main feature of forging in presses is that the metal is finish formed in most cases in two or three squeezes. Preforming is usually done in other operations, such as by casting, powder metallurgy methods, rolling, and upsetting. In press forging, pressure is sustained momentarily to fully penetrate the metal and give it time to fill out the die cavity. Dies may have a minimum of or no draft, and the closest tolerances are held. By careful control of material, flash is eliminated in some operations with appreciable savings in material, energy, and trimming time. A class of work called *precision forging* is done on presses, turning out such products as gears that need almost no machining, only a little on their teeth. One type is an automatic machine that casts preforms from molten metal in one station, cools and moves them when solid but still hot to another station for press forging, and finally trims the forgings and returns the scraps to the melting pot.

Most press forging is done on upright hydraulic presses (Figure 11-15) or mechanical presses with eccentric, toggle, or screw drives like those described in Chapter 12. A press is rated on the basis of the force in tons it can deliver near the bottom of a stroke. Pressures in MPa (tons/$in.^2$) of projected area on the parting plane have been found to be 69–276 (5–20) for brass, 276 (20) for aluminum, 207–414 (15–30) for steel, and 276–552 (20–40) for titanium. These values are for

Table 11-3. Values for factor *K*

Energy (*E*)	Aluminum Alloy	Carbon Steel	Alloy Steel	Stainless Steel	Titanium Alloys
J	305–441	339–475	475–678	542–949	814–1220
(ft-lbf)	(225–325)	(250–350)	(350–500)	(400–700)	(600–900)

conventional forgings with adequate draft, fillets, etc. Precision forgings made in dies that confine the metal almost completely may require twice as much pressure. The force the press must deliver is equal to the unit pressure times the projected area.

Mechanical presses are favored over hammers for high-volume production because they are faster and require less operator skill. A press (and its tooling) may cost four to six times as much as a hammer that will do the same work, and thus the press is not justified for low production and frequent idle periods for die changes. It is reported that about two-thirds of all forgings are made on hammers.

Some unusual forging presses have multiple vertical and horizontal actions such as those shown in Figure 11-16(A).

Incremental forging makes it possible to forge large pieces that could not be formed all at one time because a press large enough to do so is not available. First, the part may be preformed in open dies as indicated on the left of Figure 11-16(B), in which the portion on the right is being forged to final shape. Finally, the remainder of the part is forged as illustrated in Figure 11-16(C). This method requires care because time lost in a number of steps may allow some portions to get too cool for forging. In some cases, the hot billet is covered with insulation to keep it hot and pliable longer.

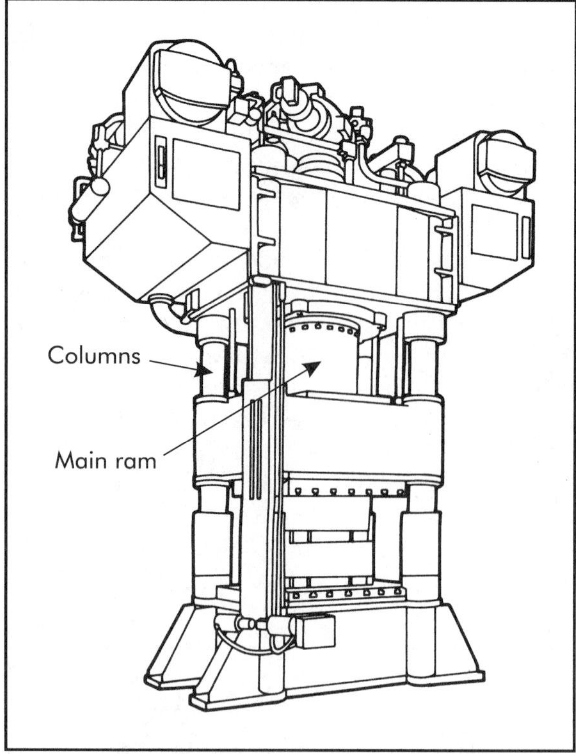

Figure 11-15. Typical four-column hydraulic press. (Courtesy Erie Press Systems)

11.8.6 High-energy-rate Forging

Although most presses do not run at high speeds, forging is done at high impact rates on some. This is different from hammer forging because the blows are not repeated. One type of machine for *high-energy-rate forging* (HERF), also called *controlled-energy flow forming* (CEFF), is illustrated in Figure 11-17.

In the ready position the ram has been lifted to the top of its stroke by the elevating pistons. The high-pressure cylinder is filled with air or gas at around 1.4 MPa (200 psi). The ram piston is pressed against a seal ring at the top. Enough high-pressure gas is admitted to the area inside the seal ring to dislodge the piston. The gas in the high-pressure cylinder then acts over the whole piston and drives the ram down at a speed of up to 20 m/sec (≈60 ft/sec) and over on the impact. Most of the energy is expended on the work and not lost in the foundation.

High-energy-rate forging does not give the metal time to cool much when in contact with the die. In fact, the temperature is raised because the heat from the work does not have time to escape. The metal can be worked more at higher temperatures and forced into intricate shapes and thin sections not possible in other ways. Thus, some parts can be made in one blow instead of in several steps by other means, and finer detail is obtained.

The process is limited mostly to symmetrical parts of small to medium size. Dies must be designed and built carefully to withstand high impact. A CEFF press with facilities to supply high-pressure air or gas costs about the same as or less than a mechanical press of comparable capacity, but operating cost is higher. Among other considerations, the time to reset the press for each cycle makes the operation rate slow, about the same as for a forging hammer. Thus,

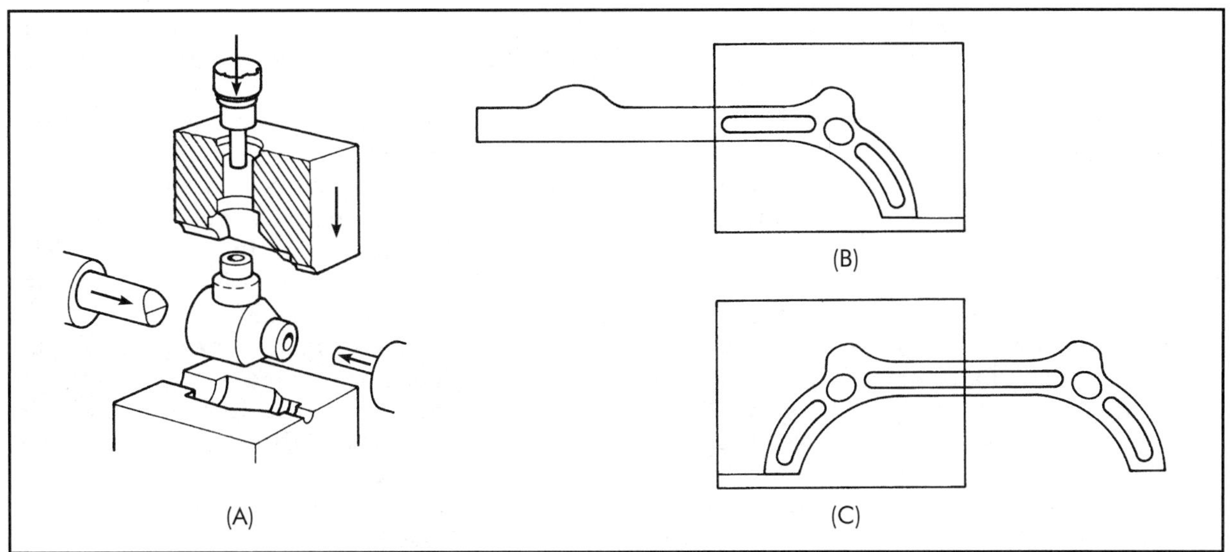

Figure 11-16. (A) Split-die forging operation. The workpiece in the center is forged in an 11,176 metric ton (11,000 ton) press (main ram rating) from a sheared length of billet stock. After the die closes, the upper piercing ram with over 25 MN (≈3,000 ton) force and the side rams with over 50 MN (≈6,000 tons) force each push the punches into the material to fill the die. (B) and (C) Steps in incremental forging.

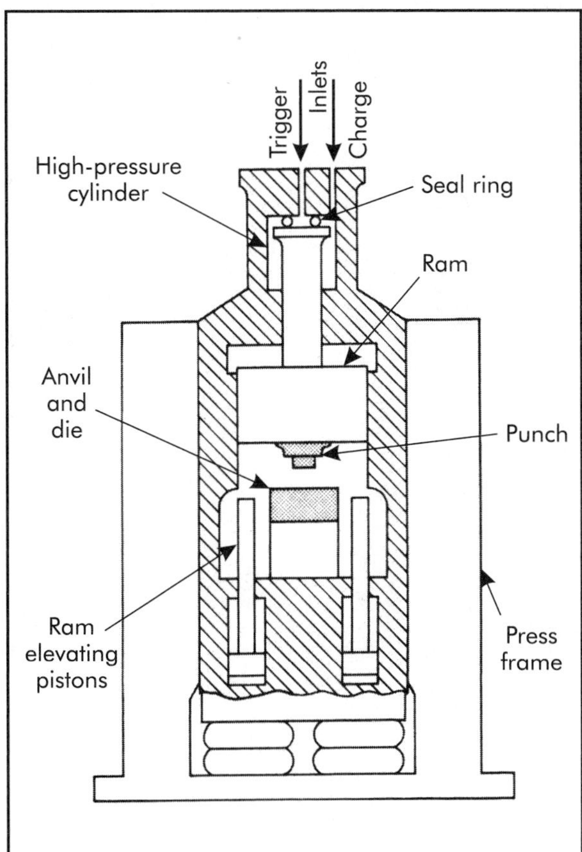

Figure 11-17. High-energy-rate forging press.

the CEFF process is not competitive unless it gives a definite benefit in workpiece quality.

11.8.7 Upset Forging

Upset forging, also called *hot heading* and *machine forging,* consists of applying lengthwise pressure to a hot bar gripped in a die to enlarge some section or sections usually on the end. Piercing also can be done. The barstock may have any uniform cross section, but is mostly round and may be steel, aluminum, copper, bronze, or other metal. Upset forgings range in weight from a few grams (ounces) to several hundred kilograms (pounds). Examples are automobile mushroom valves, gear blanks with stems, and shafts and levers with knobs or forks on their ends.

Upset forging is performed on a machine designed particularly for that purpose. A piece of hot barstock sheared to length is placed in the top cavity in one half of a die. The machine is tripped and closes the die to clamp the stock. A ram pushes the punches in a horizontal direction into the die to upset the stock. Then the punches are retracted and the die is opened. The stock is moved to the next station, and the cycle is repeated until the part is finished.

If too much stock extends from the die, it can buckle and jam between the punch and die. This can cause a serious wreck. Thus, the amount of stock that can be worked in one stage is limited. Figure 11-18 shows the basic rules found by experience. More stock can be gathered by repeated steps at more cost. Large amounts of material can be gathered by continuous upsetting on electro-hydraulic machines as indicated in Figure 11-19.

Electric upsetting is used mostly in performing operations to gather a large amount of material at one end of a round bar. The principle of operation is illustrated in Figure 11-20. A bar of circular cross section is gripped between the tools of the electrode and is pushed by the hydraulically or pneumatically operated head against the anvil plate on which the other electrode is secured (see Figure 11-20[A]). When the current is switched on, the rod section contained between the electrodes heats rapidly and the formation of the head begins (see Figure 11-20[B]). The cold bar is continuously fed between the gripping electrodes. Thus, the metal accumulates continually in the head (see Figure 11-20[C]). The anvil electrode is gradually retracted to allow enough space for the formation of the head. As soon as a sufficient quantity of metal is gathered, the machine switches off and the product can be removed by its cold end. Normally, the head is formed to final shape in a mechanical or screw-type press in the same heat. The process is suitable for preforming components like valves or steam turbine blades.

Although used for many other parts, an upset forging machine is rated by the largest bolt it can head. Some take bolts to 203 mm (8 in.) in diameter and exert 18 MN (2,000 tons) of force. A machine for bolts 13–38 mm (.5–1.5 in.) in diameter has a 11-kW (15-hp) motor and weighs 13.8 Mg (30,500 lb).

11.8.8 Forging with Rolls

Plain rolling is done on work of uniform cross section. Forging by rolling produces discrete pieces or lengths of varying cross section.

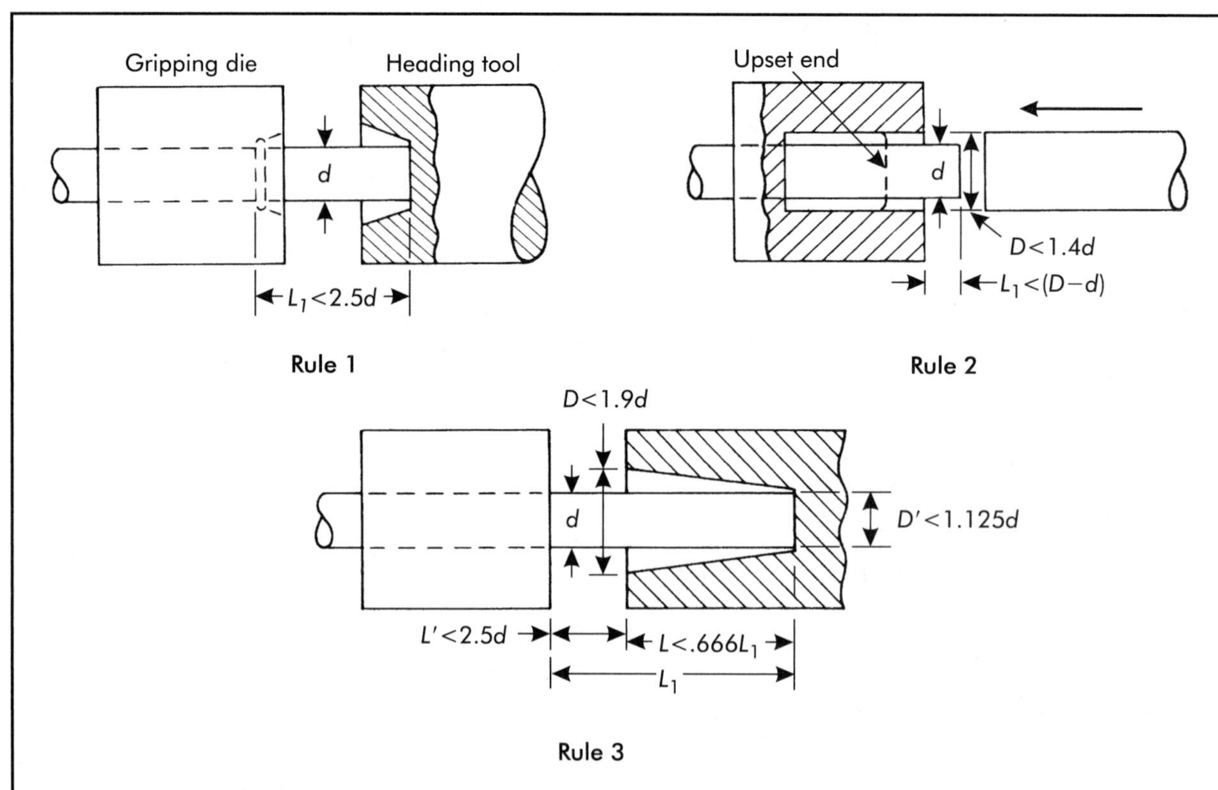

Figure 11-18. Rules for upset forging.

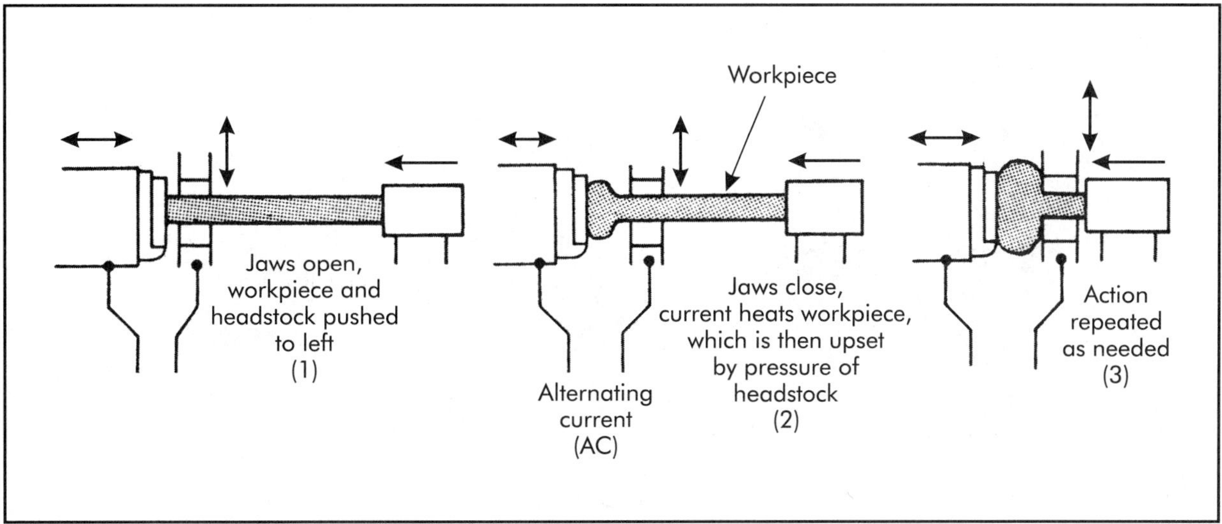

Figure 11-19. Continuous upsetting operation on an electrohydraulic machine.

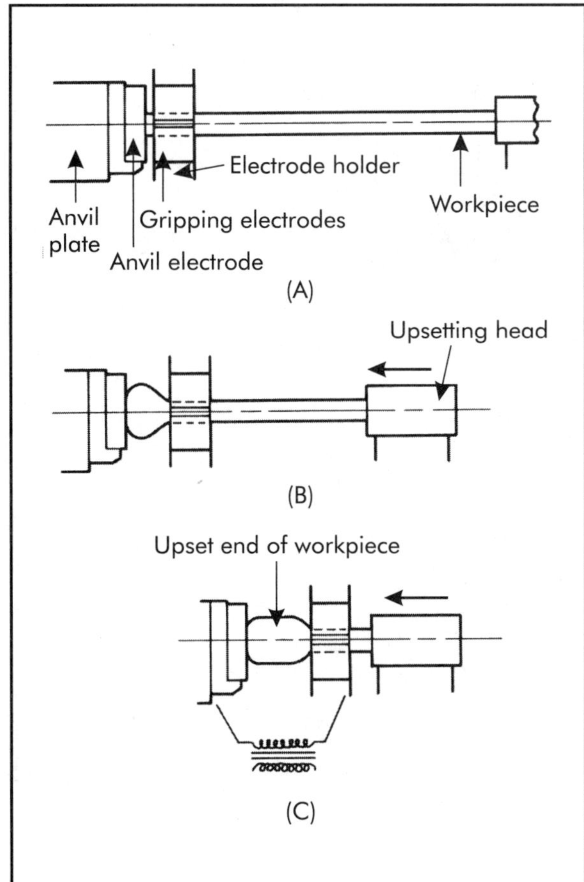

Figure 11-20. Electric upsetting is commonly used in preforming operations to gather a large amount of material at one end of a bar.

Roll forging is done with two half rolls on parallel shafts as depicted in Figure 11-21(A). These roll segments have one or more sets of grooves. A piece of stock is placed between the rolls, which then turn and squeeze the stock in one set of grooves. The stock is transferred to a second set of grooves, the rolls turn again, and so on until the piece is finished. Each set of grooved segments is made to do a specific job. By this method, barstock can be increased in length, reduced in diameter, and changed in section as desired. Because it is rapid, roll forging is of advantage in preparing some shapes for forging machines and hammers, and also for completely forging parts like levers, leaf springs, and axles.

Unidirectional or *die rolling* consists of passing stock continuously between one or more pairs of rollers with die-sunk imprints around their peripheries. One form is indicated in Figure 11-21(B). Some versions of the process are known by trade names. Parts like shafts, axles, levers, and ball blanks are produced.

Ring rolling, shown in Figure 11-21(C), starts with a small ring blank and deforms it between one or two work rolls and an idler. The ring is increased in diameter and decreased and shaped in cross section. The blank may be prepared by forging or punching. Pieces finished by this method range from small roller-bearing races to rings about 9.1 m (30 ft) in diameter.

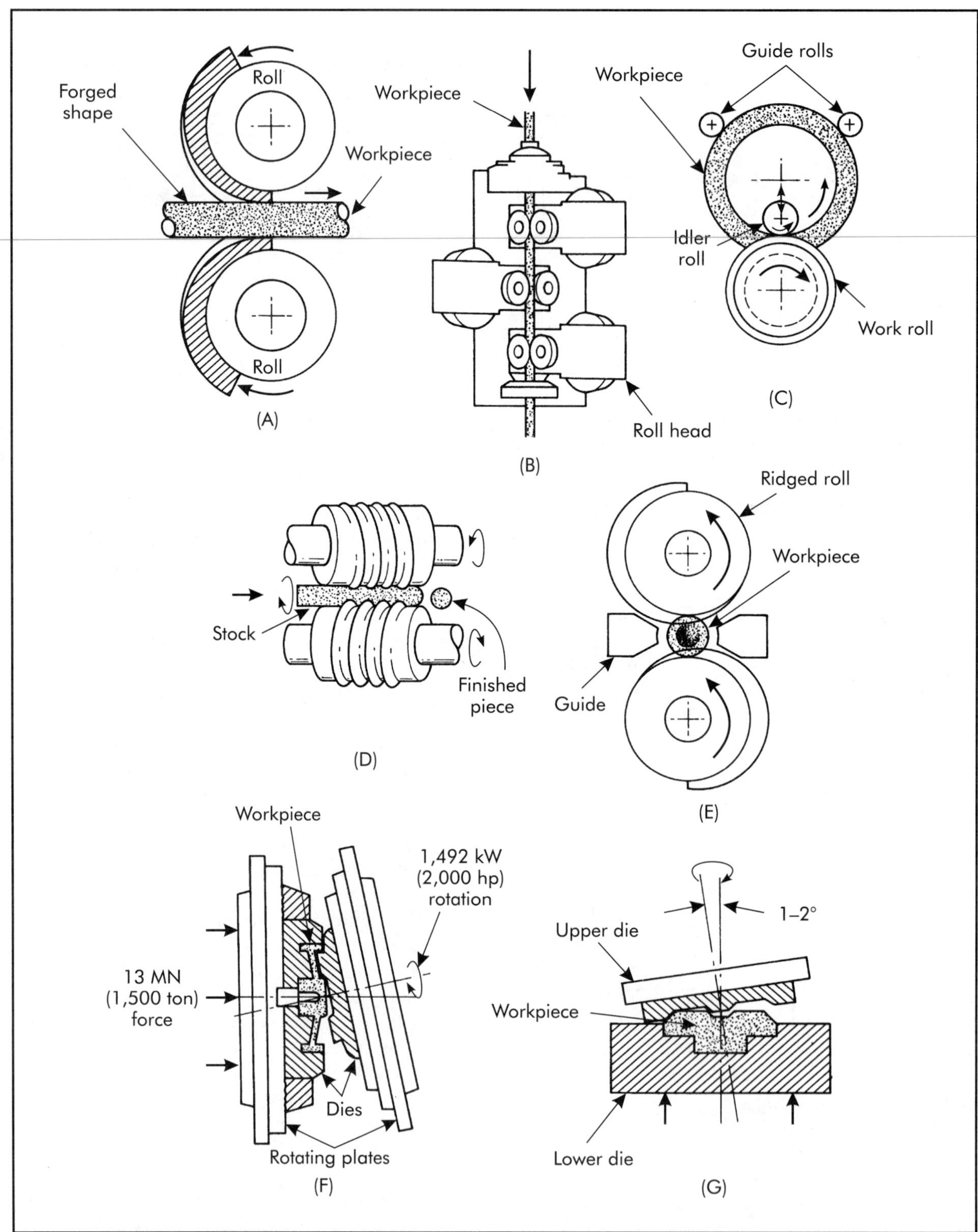

Figure 11-21. Principles of: (A) roll forging; (B) unidirectional rolling; (C) ring rolling; (D) skew rolling; (E) cross rolling; (F) circular forging; (G) orbital forging.

Skew rolling is done with two rolls on cross axes as indicated in Figure 11-21(D). Each roll has an outside helical pattern that carries the stock along and progressively shapes it as the rolls turn. Such diverse products as steel balls (up to 400/min) and railway car axles are forged by skew rolling.

Cross rolling or *wedge rolling* as shown in Figure 11-21(E) uses two synchronized rolls with a rising spiral ridge around each. Their axes are slightly crossed. A typical operation is to place a heated bar between the rolls. Then, as the rolls turn, a groove or neck is forged in the workpiece. In through-feed cross rolling, successive formed pieces can be cut off from a bar as it is fed between them. Production time is 300 small balls/min and up to 4–6 sec each for larger pieces.

Circular forging is done in a Slick mill, named after its designer, Edwin E. Slick, in the manner shown in Figure 11-21(F). A heated slug is placed between two dies on face plates that are pressed together as they turn on crossed axes. Round forgings up to 1.4 m (55 in.) in diameter are made in 55 sec.

In *orbital forging,* Figure 11-21(G), a workpiece in a lower fixed die is pressed against an upper die inclined at 1–2°. The upper die is rolled substantially in line contact around the workpiece surface and may be rocked as it goes for certain effects. The action is not fast, but is quiet and easy on dies. Available equipment forges flanged parts, disks, rings, gears, and even symmetrical parts within a 102 mm (4 in.) diameter.

11.8.9 Quality and Cost

The dimensions of a series of forgings from a die vary because of differences in the behavior of the material, temperatures, closing of the die, mismatch of the die halves, and enlargement of the cavities as they wear. A tolerance of ±0.79 mm (±.031 in.) is considered good for small carbon steel forgings and may be as large as 6.4 mm (.25 in.) in all directions for large pieces. Tolerances of 0.25 mm (.010 in.) and less have been held on precision press forgings, but at higher cost. Tolerances are larger for less workable space-age materials. Tables of commercial tolerances for various sizes and kinds of forgings are available in reference books and handbooks.

Many forgings are finished by machining to close tolerances and must have enough stock on the surfaces to be machined. The least amount of stock is about 1.5 mm (.06 in.) per surface on small forgings, and may be as much as 6.4 mm (.25 in.) or more on large ones.

Parts made from forgings also may be made as castings, cut from standard shapes, fabricated by welding pieces together, or other ways. A forging can be made stronger, more shock and fatigue resistant, and more durable than other forms. This is because it can be made of fine-grain size and fibrous structure with highest strength in the direction needed. A forging can provide required properties with less weight. Other forms are selected only when they serve as well at lower cost.

Forgings are economical in some cases because less material has to be removed from them than from barstock. An example is given by a brass compression fitting originally machined from 44.5-mm (1.75-in.) diameter hexagonal rod. Each lot of 5,000 pieces requires 717 kg (1,580 lb) of brass at $1.72/kg ($0.78/lb), a total of $1,233. A forging made from round stock costs $1.61/kg ($0.73/lb). Each lot requires 423 kg (932 lb) at a total cost of $681. In addition to the material savings of $552 or $0.11/piece, shortening of machining time raises the total savings to about $0.15/piece. A forging die costing about $2,000 is paid for by three lots.

Dies are expensive and usually rule out forging for small quantities. This is illustrated by Figure 11-22, which shows a casting to be cheaper for less than about 300 pieces. In this instance, the forging dies cost about 3.5 times more than the temporary pattern equipment and 15% more than the permanent pattern equipment. However, the scrap loss is much less for the forging.

11.9 EXTRUSION

11.9.1 Principles

When metal is extruded, it is compressed above its elastic limit in a chamber and forced to flow through and take on the shape of an opening. An everyday analogy is the dispensing of paste from a collapsible tube. Metal is extruded in a number of basic ways, as depicted in Figure

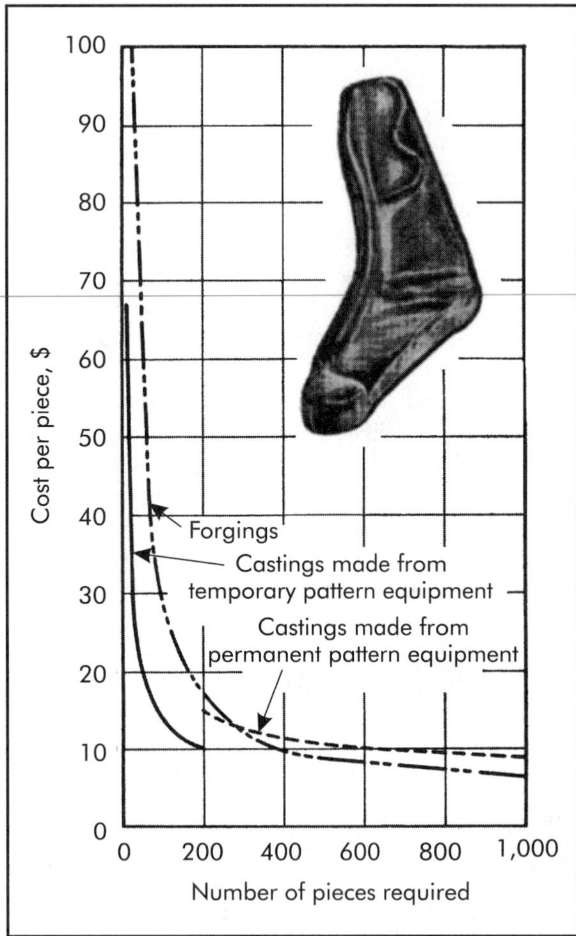

Figure 11-22. Unit cost of a 1.6 kg (3.5 lb) brace in various quantities and two methods of manufacture (Lyman 1955).

11-23. The metal is normally compressed by a ram and may be pushed forward or backward. The product may be solid or hollow. The process may be performed hot or cold. The problems and results of hot and cold extrusion are somewhat different.

11.9.2 Hot Extrusion

Forward hot extrusion of solid or hollow shapes enables the metal to be easily supported, handled, and freed from the equipment. When a piece is forward extruded, it is cut off, and the butt end is removed from the chamber.

Preservation of the equipment subjected to high temperatures is the major problem of hot extruding. Temperatures are 343–427° C (650–800° F) for magnesium, 343–482° C (650–900° F) for aluminum, 649–1,093° C (1,200–2,000° F) for copper alloys, and 1,204–1,316° C (2,200–2,400° F) for steel. Pressures range from as low as 35 MPa (5,000 psi) for magnesium to over 689 MPa (100,000 psi) for steel. Lubrication and protection of the chamber, ram, and die are necessary. Mopping these parts with an oil and graphite mixture may be sufficient at the lower temperatures. Glass, which becomes a molten lubricant, has made possible the extrusion of steel at high temperatures. Components like the ram and mandrel may be sprayed with cooling water while idle. A dummy block is used between the ram and hot metal. For extruding steel, the die is changed and allowed to cool for each piece. The best safeguard for equipment is to extrude the metal as rapidly as possible. A steel tube 152 mm (6 in.) in diameter and 15.2 m (50 ft) long is extruded in 9 sec.

Most hot extrusion is performed on horizontal hydraulic presses especially constructed for the purpose. Common sizes are rated from 2–49 MN (250–5,500 tons), but some designs in recent years have reached as high as 222 MN (25,000 tons).

Applications

Most hot extrusions are long pieces of uniform cross section, but tapered and stepped pieces are also producible. Examples of commercially extruded products are trim and molding strips of aluminum and brass, and structural shapes, rods, bars, and tubes of all forms of aluminum, steel, and titanium. A few aluminum sections are shown in Figure 11-24. Sections that can be contained within a 762-mm (30-in.) diameter circle are available for aluminum and within a 165-mm (6.5-in.) diameter circle for steel. Usual lengths are about 6.1 m (20 ft), but some are as long as 19.5 m (64 ft). Not as many shapes are available for steel as for aluminum. Examples of other kinds of hot extrusions are poppet valves, gear blanks, projector shells, piling connectors, and propeller and turbine blades.

A common tolerance for extrusions of soft metals, like aluminum, is ±13 μm (±.005 in.). Tolerances for steel extrusions range from ±0.76 mm (±.030 in.) under 25.4 mm (1 in.) to ±3.1 mm (±.12 in.) and more for the largest crosswise

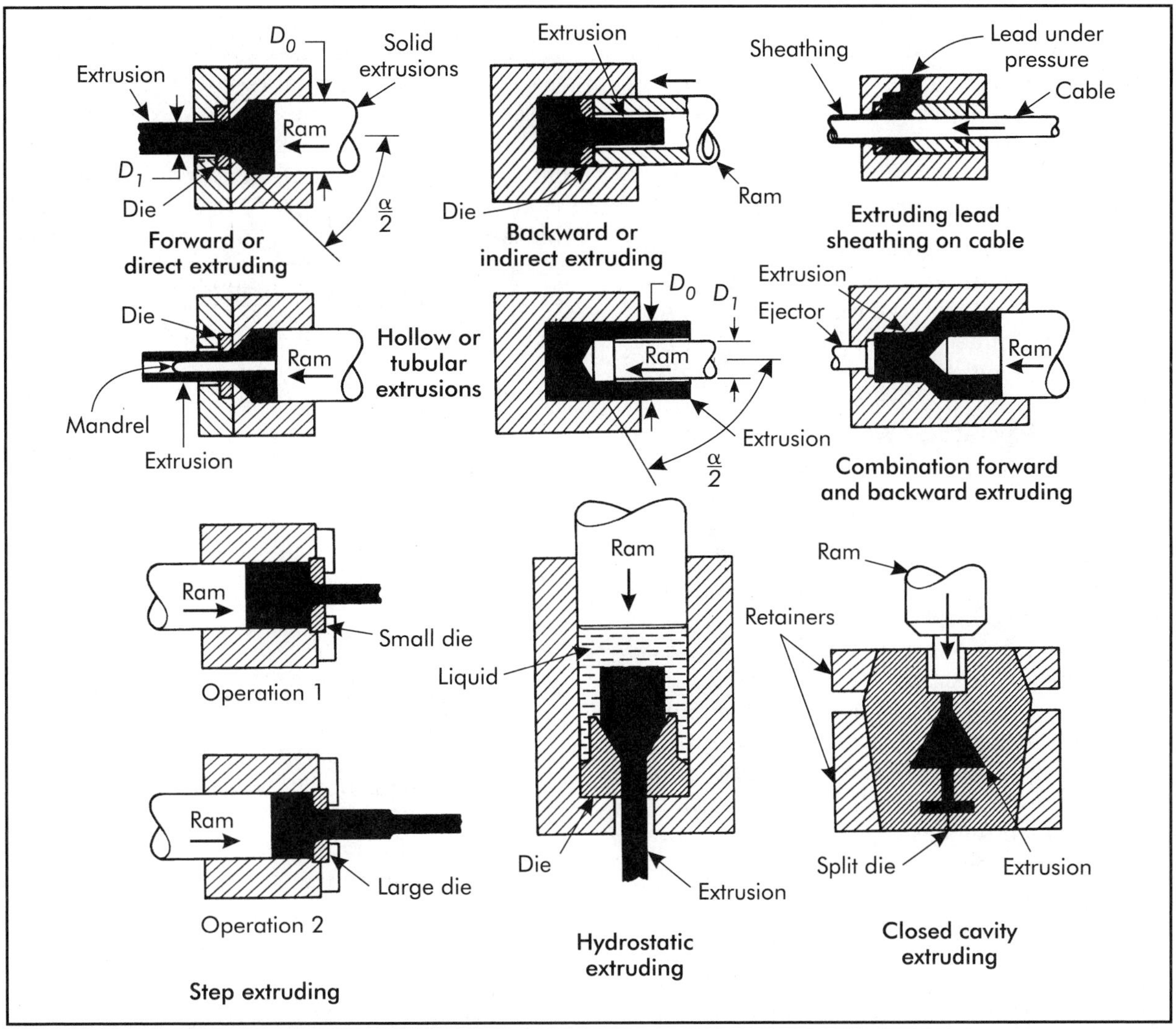

Figure 11-23. Common ways of extruding metal.

Figure 11-24. A few cross sections of extrusions.

dimensions. Surfaces are like those of hot-rolled stock.

Shapes that can be rolled are more expensive to extrude, but extrusion can produce many shapes, such as those with re-entrant angles, which cannot be rolled or produced in other ways. In some cases, extrusion offers an economical way to make large parts to replace an assembly of many individual parts and fasteners, such as a ribbed section of a bomber wing. At the other extreme is extruding to produce small parts in large quantities. A simple pump gear is an example. A long gear is extruded and then sliced into a number of individual gears. In other cases, extrusion may be the cheapest way of making parts even in small quantities. Extrusion dies are not expensive; for most shapes they average about \$500, and tools for other processes often cost much more.

11.9.3 Cold Extrusion

Cold extrusion, also called *cold forming, cold forging*, and *extrusion pressing*, is normally done at room temperature. Its forms are like those of hot extrusion illustrated in Figure 11-24. Cold extrusion is done quickly, at ram speeds of .25–1.5 m/sec (50–300 ft/min), generating heat that raises the temperature several hundred degrees, and taking less force than if done slowly. Some parts are formed in one pressing in a single die; others in two or more stages in a series of dies and sometimes in conjunction with cold heading. Presses mostly used are the same as for sheet metal forming (Chapter 12).

Several makes of continuous extruders have been developed to produce wire, rods, and various shapes from powders, ground-up scrap, or larger-diameter stock. They utilize one or more grooved wheels that take and compress the feedstock as they revolve and force it into the die through which the metal is extruded.

Warm-extrusion or *warm-forming* attempts to gain still better metal flow without changes in microstructure and surface damage resulting from high forging temperatures. It is typically done below the recrystallization temperature at 427–649° C (800–1,200° F), but is otherwise a variation of cold forming. Energy may be reduced to 35–40% below that required for cold forming, but the cost of extra induction or resistance heating equipment could nullify those savings.

An old, and until recent years, the main cold extrusion application was the manufacture of collapsible tubes from soft aluminum, lead, tin, and zinc. To perform this operation, a slug of metal is placed at the bottom of a closed cavity. A punch strikes it sharply and the metal squirts up around the punch to form the tube. The tube is blown off as the punch retracts. Production rates of 40–80 tubes/min are reported. Copper tubes are forward extruded by a similar operation. Because of the fast impulsive action, operations like these are called *impact extrusions*. Such work is done mostly on vertical mechanical and hydraulic presses.

Cold extrusion pressures range from one to three times the finished yield strength of the metal. Over 20 formulas are reported for calculating pressures, and some are quite complex. Empirical relationships developed at General Motors Institute for carbon steels have been found accurate to a few percent. For forward extrusion with dimensions defined in Figure 11-23, the pressure is

$$P = 0.45k \times [(D_0^2 - D_1^2 / D_0^2]^{0.787} \alpha^{0.375} \quad (11\text{-}3)$$

For backward extrusion, also shown in Figure 11-23,

$$P = 0.62k[(D_1^2 / D_0^2]^{0.855} \alpha^{0.355} \quad (11\text{-}4)$$

where

P = pressure, kPa (psi)
k = true stress at a true strain = 1 as explained by Equation (3-16) in Chapter 3, and is 81,050 for AISI 1008; 91,800 for 1020; 132,500 for 1035; and 142,100 for 1050 steel
D_0 = ram diameter, mm (in.)
D_1 = extrusion diameter, mm (in.)
α = extrusion throat angle, °

In spite of high pressures, the metal volume is not changed.

Adequate lubrication is mandatory for flow at the high pressures of cold extrusion. Oils, waxes, greases, some saponified, are the main lubricants. Solids like graphite, MoS_2, ZnO, and polytetrafluoroethylene (PTFE) polymer add to boundary lubrication. To extrude steel, a zinc phosphate or copper coating is necessary to help hold the lubricant to the surface.

Cold extrusion work hardens metals. For example, an aluminum alloy with a yield strength of 41 MPa (6,000 psi) and tensile strength of 110 MPa (16,000 psi) may acquire on cold extrusion a yield strength of 207 MPa (30,000 psi) and a tensile strength of 269 MPa (39,000 psi). This can be an advantage if the product can be used in the work-hardened condition with lowered ductility.

Applications

Aluminum, copper, steel, lead, magnesium, tin, titanium, zinc, and their alloys are cold extruded. Examples of products are cans; fire extinguisher cases; trimmer condensers; aircraft shear tie fittings and brackets and automotive pistons from aluminum; and projectile shells, rocket motors and heads, hydraulic and shock absorber cylinders, wrist pins, and gear blanks from steel. Stock may be solid slugs or powder metal preforms.

Advantages of cold extrusion are that it is fast, may improve physical properties and save heat treatment, it wastes little or no material, can make parts with small radii and no draft, produces to small tolerances, and saves machining. Tolerances can be as close as ±3 μm (±.001 in.), but larger tolerances are much cheaper. Walls are commonly held to ±0.3 mm (±.01 in.) tolerance when about 3.18 mm (.125 in.) thick and ±0.5 mm (±.02 in.) tolerance when thicker. Diameter tolerances normally are ±0.3 mm (±.01 in.) and length tolerances ±4.78 mm (±.188 in.). Surface finishes of 0.25 μm (10 μin.) are obtainable, but 0.5 μm (20 μin.) or rougher are more practical.

A cold-extruded part must be one with a uniform wall thickness all around and with ribs, flutes, or fins symmetrical with the part axis. The main limitation of cold extrusion is the size of equipment needed to exert the tremendous pressures required. For the most part, the process is confined to small- and medium-size pieces.

Cold extrusion is competitive with deep drawing sheet metal, described in Chapter 12, for making cups and deep shells. Extrusion has the advantage of requiring fewer steps, thus saving tooling. For example, a round can of 51 mm (2 in.) diameter and 406 mm (16 in.) deep can be extruded in one operation, but must be drawn in more than six operations. Tooling for this extrusion was estimated to cost one third as much as for drawing, even though an individual compressive die costs upward of 50% more than a drawing die. Other advantages over drawing are that shells may be extruded with thick bottoms or flanges and walls may be stepped. However, cold extrusion is not competitive with drawing of shallow cups.

Cold extrusion is competitive with casting and forging for some parts. Extrusions are usually lighter and stronger than castings. They do not need to have draft or flash to trim. Castings may be porous or brittle, but extrusions are not. Tolerances are closer and less machining is required for extruded parts. Cheaper metals can sometimes be put in extrusions because the process improves the physical properties.

An illustration of the competitive situation of cold extrusion for making an aircraft aileron nose rib is given by Figure 11-25. Little tooling investment was required to machine the part from barstock, and the cost per piece was the lowest by that method when few pieces were to be made. A forging required a little less tooling but more machining than an extrusion, which proved to be cheapest for more than 70 pieces. Break-even quantities are commonly much higher for less costly parts—over 10,000 parts/month in the high-production industries.

11.10 QUESTIONS

1. What are wrought metals?
2. What are the two primary purposes for either hot or cold working of metals?
3. Define and differentiate between the terms "critical temperature" and "recrystallization temperature" as applied to a piece of carbon steel.
4. How do the properties of a metal change when it is heated above the recrystallization temperature?
5. List several advantages and disadvantages resulting from the hot working of metals.
6. At what point on the stress-strain diagram does inelastic deformation begin?
7. Why is a metal often cold worked?

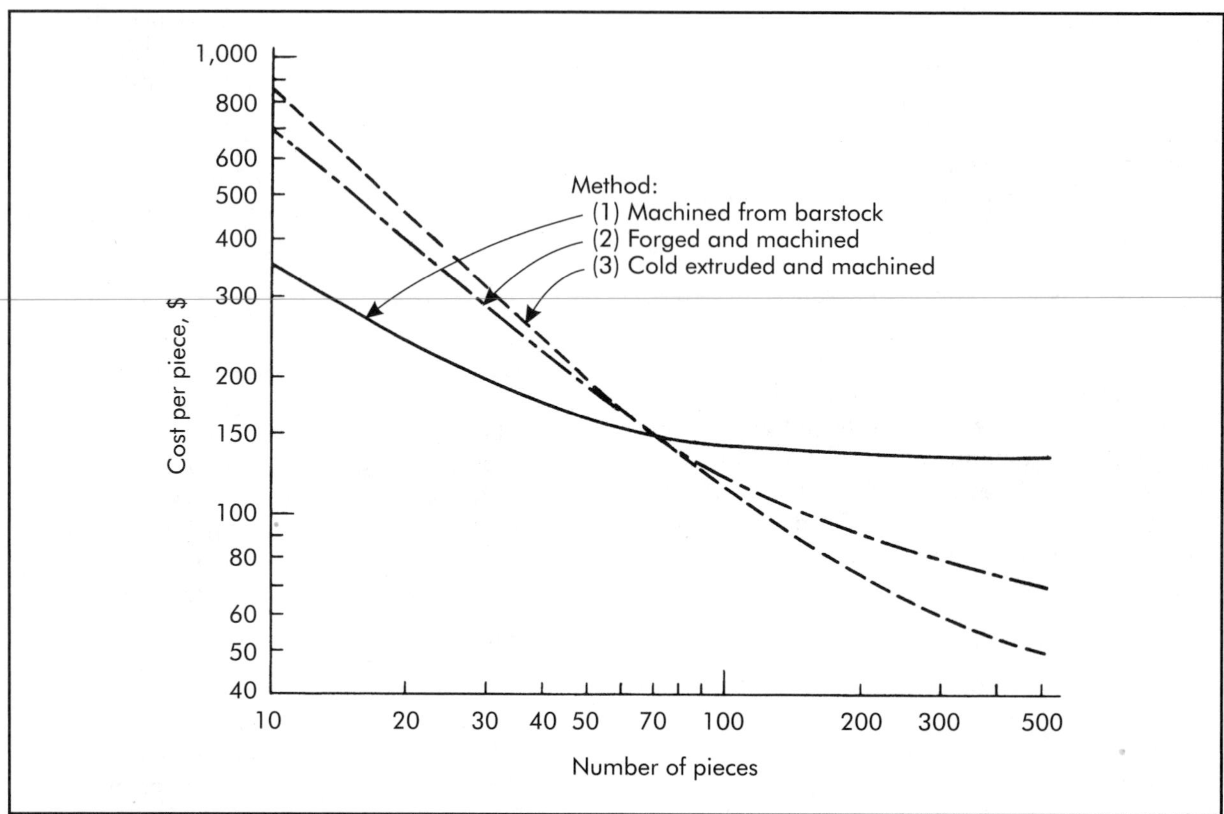

Figure 11-25. Relative costs for three methods of making an aircraft part in various quantities.

8. What are residual stresses and why do they occur in a metal?
9. Describe what occurs in metal when it is cold rolled.
10. Why are multiple passes required to roll a steel bar?
11. Describe the various hot-rolling processes used in transforming a steel ingot into different structural shapes.
12. What material property limits the amount of area reduction that can be accomplished in one pass for a cold drawing operation?
13. Describe three ways of making butt-welded pipe.
14. Describe three ways of producing seamless tubing.
15. Describe and indicate the advantages and disadvantages of the open and closed die-forging processes.
16. Differentiate between the hammer-, drop- and press-forging processes.
17. What is "upset forging" and how is it done?
18. Describe six ways of forging with rolls.
19. Under what conditions may forgings be selected for manufactured parts?
20. Discuss the "near-net-shape" qualities of the forging process as compared to casting.
21. Describe the forward and backward extrusion processes.

22. Is hot extrusion done above or below the recrystallization temperature of the material?
23. What is "impact extrusion"?
24. What is the advantage of the warm-extrusion process as compared to cold extrusion?

11.11 PROBLEMS

1. Make a sketch of the resultant forces of a set of rolls on an entering bar of metal and show the state of equilibrium that is reached when the rolls are not able to draw the metal spontaneously into the space between them. This is when the angle of contact is approximately equal to the angle of friction. Show why it is so. Depict the limiting conditions after the bar has been fully introduced between the rolls.
2. A low-carbon steel has a yield strength of 345 MPa (50,000 psi). What is the maximum force that may be exerted to cold draw a bar to 38 mm (1.5 in.) in diameter? What horsepower is required for a drawing speed of 102 mm/sec (20 ft/min)?
3. A carbon steel wire has a yield strength of 586 MPa (85,000 psi). It is to be drawn from 6.35 mm (.250 in.) in diameter to 4.37 mm (.172 in.), which is expected to stress the small size near its yield point. Calculate the percent reduction in area. Calculate the expected drawing force. What power must be applied to draw the wire at a speed of 406 mm/sec (80 ft/min)?
4. A hot steel bar 76 × 406 mm (≈3 × 16 in.) is to be reduced in one pass to 63.5 mm (2.5 in.) thick. It is assumed the 406 mm (16 in.) width does not change appreciably. Rolls are 635 mm (25 in.) in diameter. The yield stress of the hot steel is 140 MPa (≈20,000 psi), and the average pressure between the rolls and work is 1.5 times the yield stress. The coefficient of friction is 0.50. What is the force that spreads the rolls apart? How much torque must be applied to drive the rolls? What power must be delivered to the stand for a rolling speed of 127 mm/sec (25 ft/min)?
5. A 25 × 914 mm (≈1 × 36 in.) slab enters a hot-rolling mill at 2.5 m/sec (500 ft/min). It passes through seven stands and emerges as a strip 6 × 914 mm (≈.25 × 36 in.). What is the exit speed of the strip from the last set of rolls?
6. A forging can be made in 14 min by means of standard open dies and smith tools. The skilled operator's rate is $16.50/hr. A set of closed dies to make the same forging costs $3,200. The operator using these dies has a rate of $12.00/hr and can produce a forged part every 3 min. Overhead costs are the same for each method. Interest and related charges on the dies are negligible. For how many pieces is a set of closed dies justified?
7. Wire of 2.03 mm (.080 in.) diameter is to be drawn from stock that is 5.54 mm (.218 in.) in diameter. The material will stand a 60% reduction in area between annealing operations. It can be reduced 30% in area in any one draw. How many drawing operations and how many annealing operations are necessary? When must annealing be done?
8. A free-falling drop hammer weighs 680 kg (1,500 lb) and drops 0.9 m (36 in.).
   (a) With what velocity does it strike?
   (b) What is the energy of the blow?
   (c) What average force is exerted if the ram travels 1.60 mm (.063 in.) after striking the workpiece?
   (d) What diameter of cylinder for steam at an average pressure of 689 kPa (100 psi) is required to quadruple the working capacity of this hammer?
9. The actual falling mass of a steam hammer is 522 kg (1,150 lb), the cylinder is 254 mm (10 in.) in diameter, and the stroke is 711 mm (28 in.). The mean average steam pressure is 552 kPa (80 psi).

(a) Calculate the energy of the hammer blow.
(b) Calculate the average work force exerted by the hammer if it travels 3.18 mm (.125 in.) after striking the workpiece.

10. A forging has a face projected area of 129 cm^2 (20 $in.^2$). What energy must a drop hammer be able to deliver if the material is (a) alloy steel?; (b) aluminum?; (c) plain carbon steel?
11. A high-energy-rate forging machine like the one depicted in Figure 11-17 has a 229-mm (9-in.) diameter piston in a 330-mm (13-in.) diameter cylinder and a ram stroke of 508 mm (20 in.). The ram weighs 1.36 Mg (3,000 lb), and the anvil and frame weighs 13.6 Mg (30,000 lb). The gas pressure in the cylinder is 14 MPa (2,000 psi) when the press is fired and decreases adiabatically as the volume increases upon the piston leaving the cylinder. No gas can be added during the short time of the stroke. The reaction of the gas pressure raises the frame and anvil, which are freely mounted on springs. The springs are assumed completely relaxed when the anvil reaches its highest position.
    (a) What is the velocity of the ram when it strikes the workpiece?
    (b) What is the velocity of the anvil at the same time?
    (c) What energy is there in the ram when it strikes the workpiece?
    (d) How much energy is there in the anvil and frame at the same time?
    (e) From what height would a 1.36 Mg (3,000 lb) drop hammer ram have to fall to deliver the same energy as available in this press?
12. A steel cluster gear 95 mm (≈3.75 in.) on the largest diameter and 178 mm (7 in.) long can be machined from barstock of 102 mm (4 in.) diameter × 184 mm (7.25 in.) long. Machining time by this method is 22 min. An upset forging can be made from round barstock 48 mm (1.875 in.) in diameter × 229 mm (9 in.) long, and 15 min is required to machine the forging. The material costs $1.19/kg ($0.54/lb), forging costs $0.71/kg ($0.32/lb), and machining costs $15.50/hr. The forging die can be made for $1,650.
    (a) Which method should be selected for 100 pieces?
    (b) For how many pieces is the forging warranted?
13. The piece shown in Figure 11-26 is to be forward extruded at *A* and backward extruded at *B*. The material is AISI 1020 steel. Its rough size is 121 mm (4.75 in.) in diameter × 46.43 mm (1.828 in.) long.
    (a) What force in meganewtons is required for the forward extrusion? In tons?
    (b) What force in meganewtons is required for the backward extrusion? In pounds?

11.12 REFERENCES

Altan, T., Oh, S.I., and Gegel, H. 1983. *Metal Forming—Fundamentals and Applications*. Materials Park, OH: American Society for Metals (ASM) International.

ANSI B 32.1-1952 (R1988). "Preferred Thicknesses for Uncoated Thin Flat Metals (Under 6.4 mm [.250 in.])." New York: American National Standards Institute.

ANSI B 32.2-1969 (R1986). "Preferred Diameters for Round Wire (12.7 mm [.500 in.] and under)." New York: American National Standards Institute.

ANSI B 32.4M-1980 (R1986). "Preferred Metric Sizes for Round, Square, Rectangle, and Hexagon Metal Products." New York: American National Standards Institute.

Forging Handbook. 1985. Cleveland, OH: Forging Industry Association, and Materials Park, OH: ASM International.

Hosford, W., and Caddell, R. 1983. *Metal Forming: Mechanics and Metallurgy*. Englewood Cliffs, NJ: Prentice-Hall.

Laue, K., and Stenger, H. 1981. *Extrusion*. Materials Park, OH: ASM International.

Lyman, Taylor, ed. 1955. *Metals Handbook 1955 Supplement*. "Design of Closed Die Forgings." Materials Park, OH: ASM International.

Metals Handbook, 9th Ed. 1988. Vol. 14: *Forming and Forging*. Materials Park, OH: ASM International.

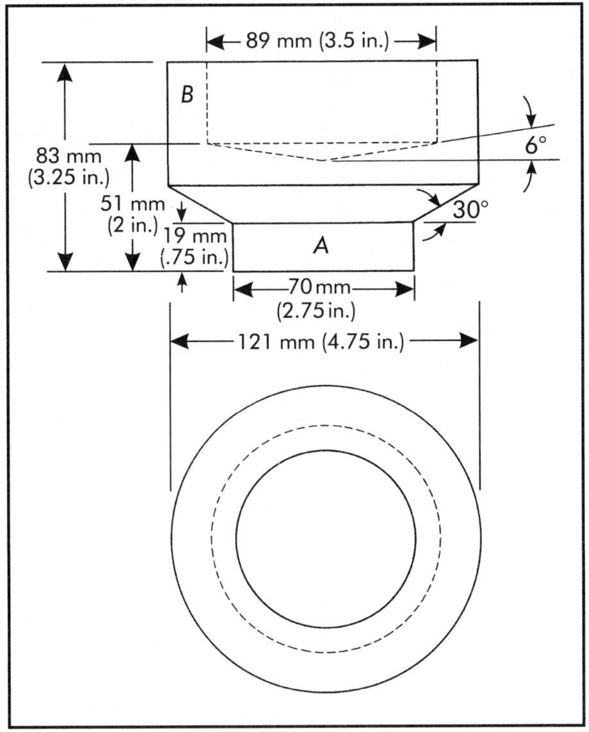

Figure 11-26. Finished size of cold-extruded workpiece.

Nickel, A.J., ed. 1994. *Roll Forming—Collected Articles and Technical Papers*. Rockford, IL: Fabricators and Manufacturers Association International.

Society of Manufacturing Engineers (SME). 1999. *Forging* video. *Fundamental Manufacturing Processes* series. Dearborn, MI: SME.

Suzuki, Y., 1982. "The Push Toward Orbital Forging." *American Machinist*, November.

Tool and Manufacturing Engineering Handbook, 4th Ed. 1984. Vol. 2: *Forming*. "Hot Forging." Dearborn, MI: Society of Manufacturing Engineers (SME).

Metal Shearing and Forming

12

Steam-powered coining press, 1786
—Matthew Boulton, Birmingham, England

12.1 INTRODUCTION

A large proportion of industrial products are manufactured by processes that shear and form standard shapes, largely sheet metal, into finished parts. Quite often, shearing and forming are all that are required to convert a flat sheet of metal into the final shape and size of a manufactured product or component. In other cases, the form produced by shearing and forming will require some secondary operations, such as drilling and tapping, or even welding, to properly prepare it for assembly into a product. In either case, the near-net-shape characteristics of the results of these processes make them an extremely cost-effective alternative for producing a product or component.

The common forms and principles of metal shearing and forming will be described in this chapter. A few examples of products produced by these processes are pots and pans, metal cabinets, door and window hardware, and automobile bodies. Other examples can be seen in almost any manufactured product because these processes are versatile, fast, and naturally adaptable to large-quantity production. Many modern stamping facilities are equipped with quick-change tooling and dies to accommodate the small lot sizes required by just-in-time manufacturing programs. These processes mostly work cold metal and produce the effects of cold working as described in Chapter 11.

As a rule, cold working and forming processes take less energy and material than the metal removal processes to produce a finished product. This is becoming more important in a world where energy and metal are growing more costly and less available.

The operations discussed in this chapter include *shearing, bending, drawing* and *stretching*, and *squeezing*. The machines and tools used for most of these operations are presses and dies. They will be discussed in the last sections.

12.2 METAL SHEARING OPERATIONS

12.2.1 Types

Operations that cut sheet metal, even barstock and other shapes, have various purposes. Common operations and the work they do are depicted in Figure 12-1. *Shearing* is a general name for most sheet metal cutting, but in a specific sense, designates a cut in a straight line completely across a strip, sheet, or bar. *Cutting off* means severing a piece from a strip with a cut along a single line. *Parting* signifies that scrap is removed between two pieces to separate them.

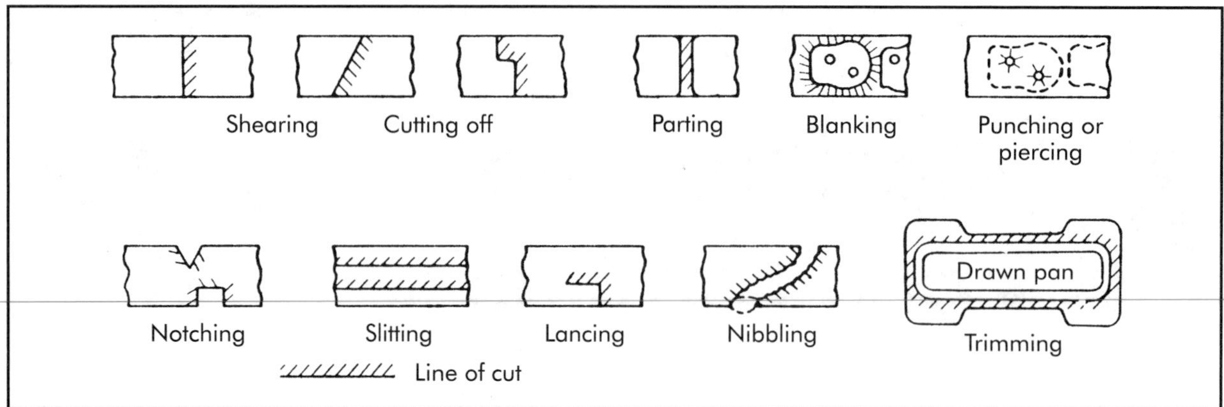

Figure 12-1. Common sheet metal cutting operations.

In *blanking,* a whole piece is cut from sheet metal. Just enough scrap is left all around the opening to assure that the punch has metal to cut along its entire edge. If the object is to cut a hole and the material removed is scrap, the operation is called *punching* or *piercing. Slotting* refers to the cutting of elongated holes. *Perforating* designates the cutting of a group of holes, by implication small and evenly spaced in a regular pattern. *Notching* removes material from the side of a sheet or strip. *Lancing* makes a cut part way across a strip. *Trimming* refers to cutting away excess metal in a flange or flash from a piece.

Coiled thin metal stock from the rolling mill may be utilized directly in dies for large parts like automobile body panels or may be cut up for use. One way is to shear the stock into sheets. Another is by *slitting,* which refers to cutting the original stock lengthwise by passing it through spaced and continuous rolls as indicated in Figure 12-2. Wider sheet metal is available, but most that is slit is rolled in widths of 1.5 m (60 in.) or less. Widths are held to within 1.5 mm (≈.063 in.) tolerance. Such strips are run through dies to make small- and medium-size pieces.

Nibbling is an operation for cutting any shape from sheet metal without special tools. It is done on a *nibbler*, which is a machine that has a small round or triangular punch that rapidly oscillates in and out of a mating die. The sheet of metal is guided so that a series of overlapping holes is punched along the path desired.

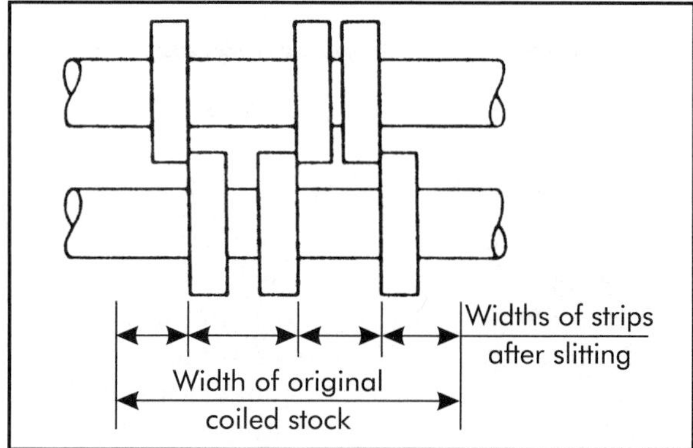

Figure 12-2. Slitting rolls.

This is a slow operation, but is economical where only a few pieces of each kind are needed because it saves costly special dies.

12.2.2 Principles

Sheet metal is sheared between a *punch* and *die block* in the manner indicated in Figure 12-3(A). The punch has the same shape all the way around as the opening in the die block, except it is smaller on each side by an amount called the *break clearance.* As the punch enters the stock, it pushes material down into the opening. Stresses in the material become highest at the edges of the punch and die, and the material starts to crack there. If the break clearance is correct, the cracks meet and the break is complete. If the clearance is too large or too small, the cracks do not meet, further work must be done to cut the metal between them, and a jagged break results. The proper amount of break clearance depends on the kind, hardness, and thickness of the material. For cold-rolled, low-carbon steels, an average clearance, C, of about 7% of the stock thickness is recommended for normal piercing and blanking operations. For a piercing operation, the punch is made to the size of the hole desired and the die hole is enlarged by an amount $2C$ to provide the proper clearance. For blanking, the die hole is made to size and the punch diameter is reduced by the total clearance $2C$.

As the punch in its downward course enters the material, the force exerted builds up as indicated in Figure 12-4. If the clearance is correct, the material breaks suddenly when the punch reaches a definite penetration, and the force vanishes as shown in Figure 12-4(A). This distance is

$$p \times t \qquad (12\text{-}1)$$

where

p = penetration, %
t = thickness of stock, mm (in.)

The maximum force is

$$F = L \times t \times S \qquad (12\text{-}2)$$

where

F = maximum force, N (lb)
L = length of cut, mm (in.)
S = shearing strength, MPa (psi)

If the piece is round,

$$F = \pi DtS$$

where

D = diameter, mm (in.)

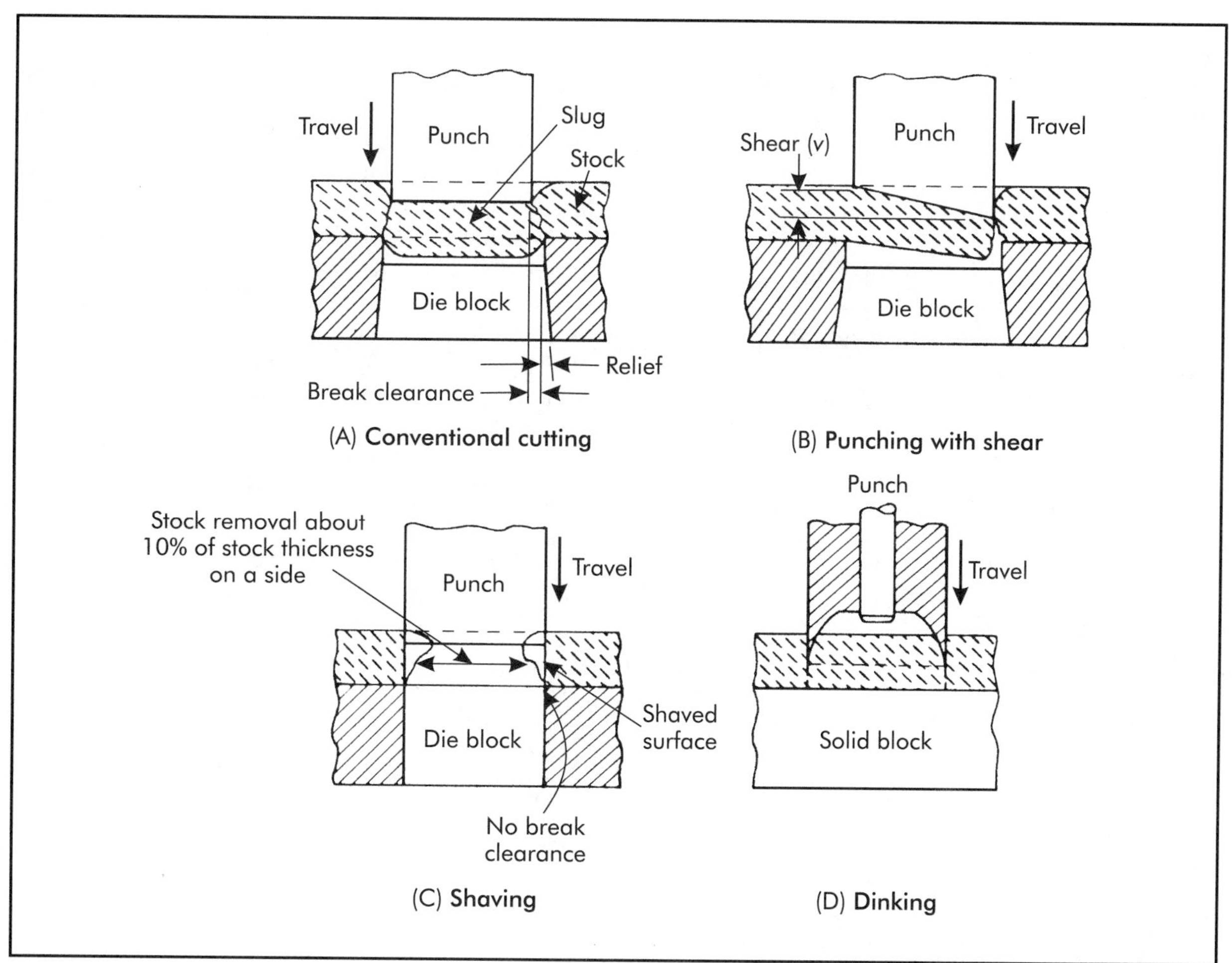

Figure 12-3. Methods of die cutting.

The shearing strength and percent penetration are properties of the material and are 55 MPa (8,000 psi) and 60% for soft aluminum, 331 MPa (48,000 psi) and 38% for 0.15% annealed carbon steel, and 490 MPa (71,000 psi) and 24% for 0.5% annealed carbon steel. Values for other metals are given in reference books and handbooks.

If the break clearance is not correct, the force does not fall off suddenly, and the curve is something like that of Figure 12-4(B). In any case, the theoretical amount of energy needed for the operation is represented by the area under the curve. In practice, the situation is not always ideal, and energy is also needed to overcome friction. A conservative estimate of energy (J) can be derived from

$$E_J = 0.00116F \times p \times t \tag{12-3}$$

where

E_J = energy, J

in the SI units specified in the preceding paragraph. Power is

$$P_W = (E_J \times N)/60 \tag{12-4}$$

where

P_W = power, W
N = number of press strokes/min

Energy (ft-lbf) is

$$E = 1.16 \times F \times p \times t/12 \tag{12-5}$$

where

E = energy, ft-lbf

and horsepower is

$$P_h = E \times N/33{,}000 \tag{12-6}$$

where

P_h = power, hp

It has been assumed so far that the end of the punch and the top of the die block lie in substantially parallel planes. If they do not, they are said to have *shear,* which can be put on either the punch or die block in a number of ways. One way of putting shear on the punch is shown in Figure 12-3(B). With shear, only part of the cut is made at any one instant, and the maximum force is much less. The energy expended to take a cut is not changed. Shear distorts the material being cut and consequently cannot be applied in many cases. An effect similar to shear can be obtained by *staggering* two or more punches that all work in one stroke. For staggering, no shear is put on the individual punches, but they are arranged so that one does not enter the material until the one before it has broken through. The most force that must be exerted during the stroke is that needed for the largest punch.

Practical tolerances in punching range from ±51 µm (±.002 in.) for the smallest parts to ±0.38 mm (±.015 in.) for dimensions over 152

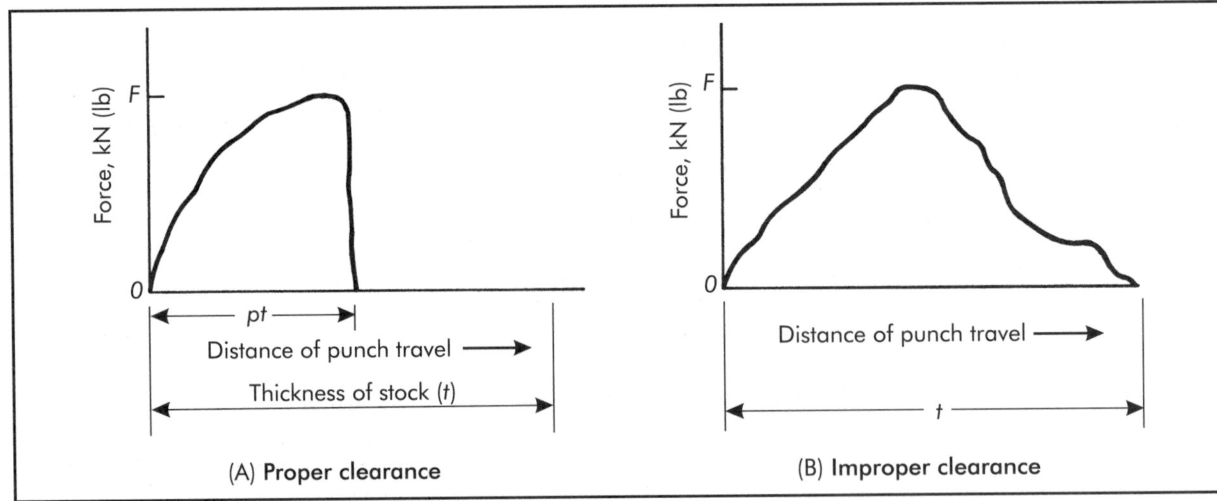

Figure 12-4. Force exerted by a punch penetrating sheet metal.

mm (6 in.). Tolerances as small as desired can be held at extra cost.

When sheet metal is sheared as shown in Figure 12-3(A), the edge is more or less jagged and not square with the large surface of the stock. Surfaces in holes may be improved by *shaving*. This involves a light cut in a second operation, as illustrated in Figure 12-3(C). Edges of blanks may be improved by milling in a costly second operation. However, a square and smooth edge may be obtained in one stroke for thick pieces by *straight-edge blanking,* also known by other names, such as *fine-edge blanking* or just *fine blanking.* Punch and die are fitted closely, typically with 5 μm (.0002 in.) clearance, and high pressure is applied by an outer punch all around the edge being sheared. A much more rigid and costly press is required, and dies cost about twice as much and last about half as long as for conventional blanking. Fine blanking can be justified only where it saves enough by eliminating machining or secondary press operations.

Paper, rubber, and other soft and fibrous materials are cut with a sharp-edged punch against a wood or soft metal block in a *dinking* operation, as illustrated in Figure 12-3(D).

12.3 BENDING

12.3.1 Punch and Die Bending

Bars, rods, wire, tubing, and structural shapes, as well as sheet metal, are bent to many shapes in dies. Several common kinds of sheet metal bends are shown in Figure 12-5. All metal bending is characterized by the condition depicted in Figure 12-6, with the metal stressed beyond the elastic limit in tension on the outside and in compression on the inside of the bend. Stretching of the metal on the outside makes the stock thinner.

In most cases, the stretching of a bend causes the neutral axis along which the stock is not strained to move to a distance of .3–.5t from the inside of the bend. An average figure of .4t is often used for calculations. The original length of the stock in the bend is estimated to be

$$L = 2 \times \pi (r + .4t) \times (\alpha/360) \qquad (12\text{-}7)$$

where

L = original length of stock, mm (in.)
r = inside radius of the bend, mm (in.)
t = original stock thickness, mm (in.)
α = angle of the bend, degrees

As has been explained, even metal that has been stressed beyond the elastic limit is prone to a certain amount of elastic recovery. If a bend is made to a certain angle, it can be expected to spring back to a slightly smaller angle when released. This *springback* is larger for smaller bend radii, thicker stock, larger bend angles, and hardened materials. Average values are 1–2° for low- and 3–4° for medium-carbon soft steels. The usual remedy for springback is to bend beyond the angle desired.

Certain limitations must be observed to avoid breaking metal when bending it. In general, soft metal can be bent 180° with a bend radius equal to the stock thickness or less. The radius must be larger and the angle less for metals of hard temper. The amount depends upon the metal and its condition. Working values are given in handbooks. A bend should be made not less than 45° and as close as possible to 90° with the grain direction of rolled sheet metal because it cracks most easily along the grain. A bend should not be closer to an edge than 1.5 times the metal thickness plus the bend radius.

The force required to make a single bend in the vee die illustrated in Figure 12-7 can be derived from a basic formula for mechanics of materials.

$$S = Mc/I \qquad (12\text{-}8)$$

where

S = stress in the outer fibers of a beam, N/mm^2 (lb/in.^{2})
M = bending moment, Nm (lb/ft)
c = thickness of beam (2c), mm (in.)
I = moment of inertia

This applies when the stress in the beam nowhere exceeds the elastic limit. So, for the case illustrated by Figure 12-7, assume $M = F_E l/4$; $c = t/2$; and $I = wt^3/12$. These values are substituted in Equation (12-8) and it is rearranged to find the elastic deformation.

$$F_E = 0.67Swt^2/l \qquad (12\text{-}9)$$

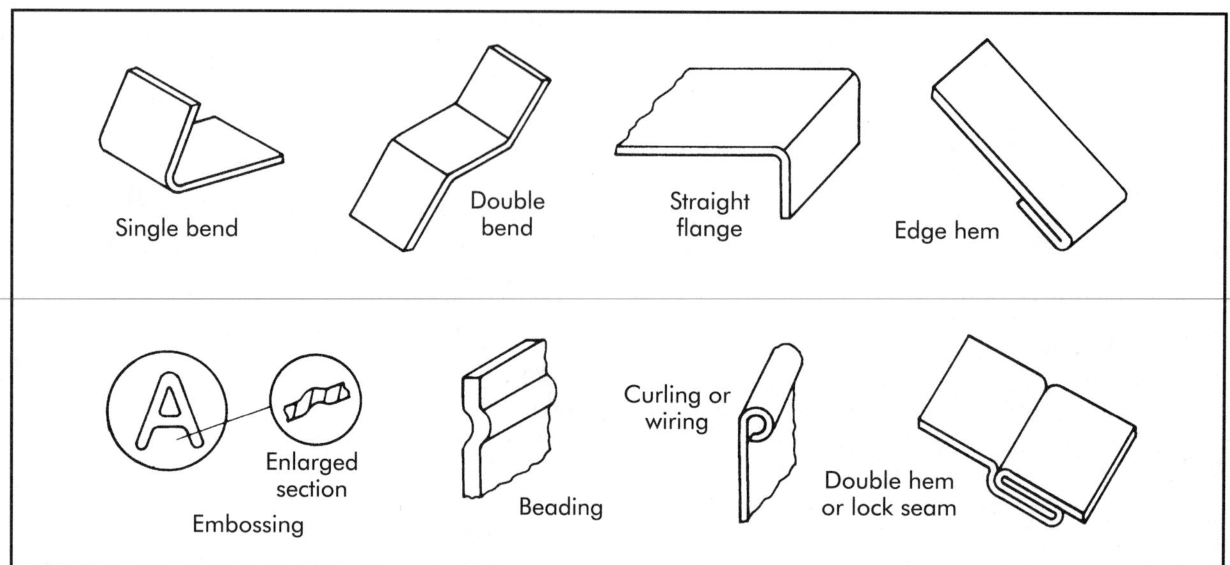

Figure 12-5. Some kinds of sheet metal bends.

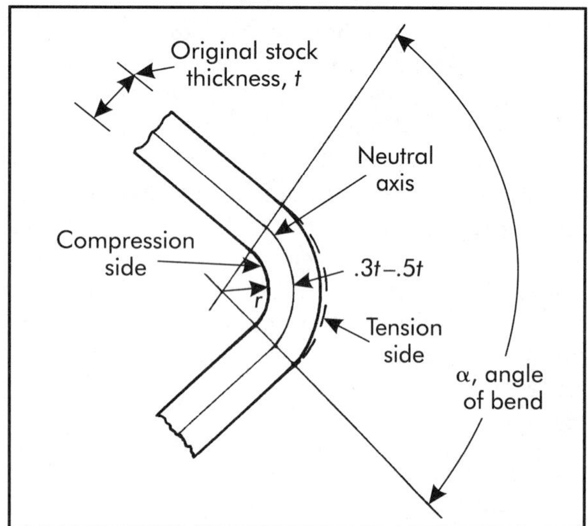

Figure 12-6. Nature of a bend in metal.

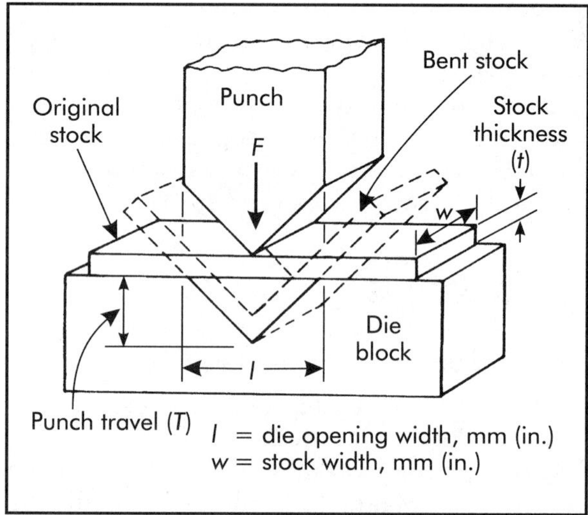

Figure 12-7. Vee-die bending operation.

where

F_E = punch force, N (lb)
w = stock width, mm (in.)
t = stock thickness, mm (in.)
l = die opening width, mm (in.)

Experiments have shown that for plastic bending the maximum force is about twice as much as indicated for elastic flexure by Equation (12-9). On this basis,

$$F = 1.33Swt^2/l$$

where

F = force, MN (lbf)
S = ultimate tensile strength of the material, MPa (psi)

For SI units, linear dimensions must be in meters to satisfy this formula. The same approach leads to variations of this formula for other types of bending operations, such as single-flange bending and for use of dies with pressure pads. Such formulas are given in texts containing a detailed treatment of bending.

12.3.2 Tube and Structural Shape Bending

Tubes, pipes, and structural shapes of all kinds are bent by methods that keep them from collapsing or distorting. Pipes and shapes may be joined at corners by fittings, welding, or brazing, but bending is cheaper and more dependable. Examples of applications are automobile exhaust and tail pipes, aircraft hydraulic lines, and structural frames. Tubes and shapes are usually supported in grooves and bent around form blocks. Bends are made in production to well within 0.5° and with the inside radius of the bend as small as the tube diameter. Common bending methods are illustrated in Figure 12-8.

The workpiece is pulled at both ends while being bent over a form block in *stretch bending* or *forming* (Figure 12-8[A]). The method is slow but almost eliminates springback. It is used to make large irregular and noncircular bends without mandrels.

Draw bending is done with the workpiece clamped against a form block, which rotates and pulls the metal around the bend (Figure 12-8[B]). The work going into the bend is supported by a pressure bar. A mandrel may be inserted in a tube to restrain flattening. Flexible ball, laminated, or cable mandrels, depicted in Figure 12-8(G), provide support around the bend length for delicate work. Draw bending is best for producing small radii and thin walls and is most versatile.

The workpiece is clamped to and wrapped around a fixed form block by a wiper shoe in *compression bending* or *forming* (Figure 12-8[C]). Flat sheet metal is commonly bent in the same way on ungrooved blocks in an operation called *wing* or *tangent bending*. The bending radius may be very small. Compression bending can make a series of bends with almost no spaces between them. A combination of stretch and compression forming is called *radial draw forming* and is advantageous for difficult curved parts.

Ram or *press bending* is done by pressing the workpiece between a moving ram block and two swinging pressure dies as indicated in Figure 12-8(D). A fixed-stroke punch press may be used, but an adjustable-stroke bending press is better. Equipment cost is a little more than for draw bending, and angles are limited to about 165°, but press bending is three to four times faster than other methods. A different press setup is required for each different bend, so the process is only suitable for quantity production, such as in furniture factories.

Roll bending of plates, bars, structural shapes, and thick-walled tubes is done with three rolls as shown in Figure 12-8(E). One roll is adjusted between the other two for the desired bend radius. Continuous coils can be made in this way. Bend radius can be changed easily, and the operation is suitable for job work, but angle control is difficult.

Roll extrusion bending is used to produce pipes over 127 mm (5 in.) outside diameter (OD) and walls to 15.88 mm (.625 in.) thick (Figure 12-8[F]). A head is rotated inside the pipe with wide thrust rollers on one side and a narrow work roller on the other. The pipe is enclosed by work rings outside of the head. The work roller is cammed in and out as the head rotates to apply pressure to extrude metal in the pipe wall on the side to make it bend. As the material is worked, the pipe is advanced past the head. This method is reported to be 10 times faster than others for large pipes.

12.3.3 Cold-roll Forming

Cold-roll forming is a continuous process wherein a flat strip of metal is passed through a series of rolls and is progressively formed into a desired uniform shape, usually a net shape, in cross section. Among its many applications are metal window and screen frame members, rain gutters for homes and buildings, picture frames, fluorescent light reflectors, metal building sections, garage door rails, steel furniture, and home appliances.

In operation, roll forming equipment feeds the metal stock, in either sheet, strip, or coiled form, through successive pairs of mating rolls that progressively shape the stock until the required form or cross section is produced. The process is essentially classified as a bending operation as the metal thickness is not altered except for some thinning at the bend radius. Practically any metallic material, ferrous or nonferrous, which is sufficiently ductile to withstand bending, can be roll formed. Often, the stock is prefinished (coated, polished, prepainted, or plated) prior to roll forming to reduce fabricating time. Stock

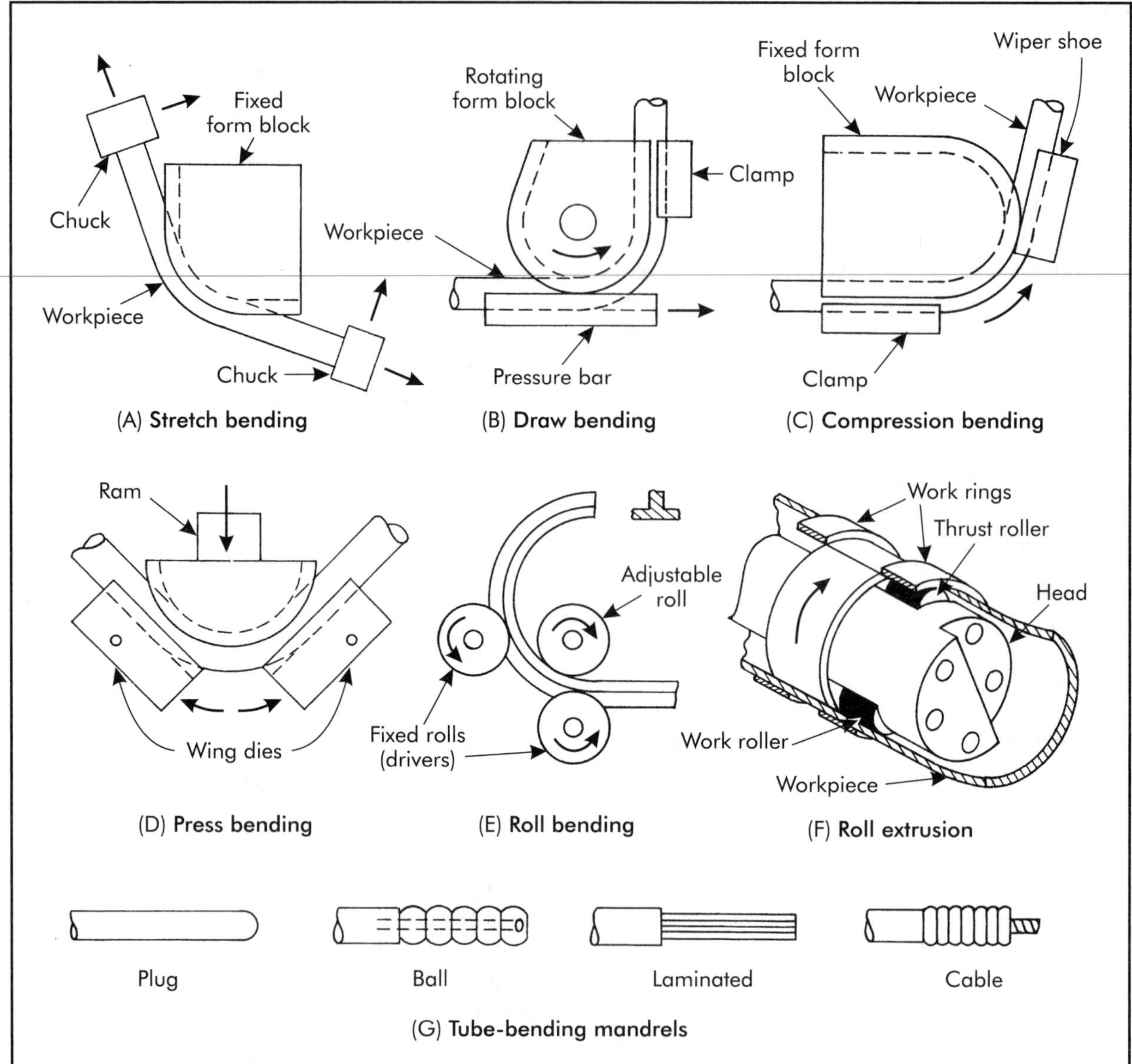

Figure 12-8. Methods for bending pipes, tubes, and structural shapes.

thicknesses from 0.13–19 mm (.005–.750 in.) and widths from 3.18–1,828.80 mm (.125–72 in.) are commonly roll formed.

The number of rolls required to form a flat strip of metal into a finished net shape is dependent on a number of factors, including the type of material being formed and its physical properties, the material thickness, the cross section to be formed, and the tolerance allowed on the finished part. The progress of the strip through a mating roll station is referred to as a *pass*, and each pass on the machine accomplishes a specified amount of bending. Care must be taken in the design of each succeeding roll to assure that the ultimate strength of the material being formed is not exceeded at any one pass. A number of empirical relationships are known to aid in the design and specification of machine and tooling parameters.

Figure 12-9(A) shows two roll-forming lines, running side-by-side, making fluorescent light parts. One line forms the housing portion of the fixture and the other forms the reflector. The fixtures are formed from either prepainted steel or anodized aluminum. Referred to as an "integrated" system, the precoated strip is fed on line from a coil reel to a precision-feed roll station, and then through a prepunch process where the required holes and tabs are punched out. The strip then goes through a section of rolls that progressively bend the stock to the required profile. After forming, the formed material is cut off to the appropriate length in a press with a flying cutoff die. A close-up of several of the top forming rolls in operation on one of the lines is shown in Figure 12-9(B). An integrated roll-forming system similar to this costs about $1,000,000.

Cold-roll forming is competitive with two major types of operations. One is the hot rolling and extrusion type described in Chapter 11. The equipment costs much less for cold-roll forming, but the raw material is more expensive. The other type of operation is the bending of shapes in discrete lengths on presses or press brakes. The equipment cost is much less, but the operation is much slower than cold-roll forming. A molding, for example, would be bent in press brake dies for quantities up to thousands of linear meters (feet). Cold-roll forming would be desirable for millions of meters (feet), but extrusion might be economical for even larger quantities.

Rotary roll forming for parts like auto and bicycle wheel rims is done by placing one side of a straight cylindrical hoop between two parallel formed rolls. The rolls rotate and close together

Figure 12-9(A). Two integrated roll-forming lines for making the housing and reflector portions of fluorescent lighting fixtures. (Courtesy Contour Roll Company)

Figure 12-9(B). Close-up photograph of several top rolls of a continuous roll-forming machine. (Courtesy Contour Roll Company)

to impress their form on the continuous ribbon as it passes around between them. The process is readily automated and capable of production rates up to 1,200 pieces/hr. In a similar manner, rings of thick sections may be squeezed and progressively formed between rolls in what is called *cold ring rolling*.

12.4 DRAWING AND STRETCHING

Drawing and stretching is used to produce thin-walled hollow or vessel-shaped parts from sheet metal. Examples of products are seamless pots, pans, tubs, cans, and covers; automobile panels, fenders, tops, and hoods; cartridge and shell cases; and parabolic reflectors. The sheet metal is stretched in at least one direction but is often compressed in other directions in these operations. The work is mostly done cold but sometimes is done hot.

12.4.1 Rigid Die Drawing

A great variety of shapes are drawn from sheet metal. The action basic to all is illustrated by the drawing of the round cup depicted in Figure 12-10. The cup is formed by being drawn from the blank shown next to it. Shaded segments of the blank and cup indicate what is done to the metal. A trapezoid in the blank is stretched in one direction by tension and compressed in another direction into a rectangle. Metal must be stressed above the elastic limit to form the cup wall, but not the bottom.

The wall of the cup may be thinned at the radius by bending and thickened elsewhere by drawing. The changes in thickness are usually negligible. If $t = t_1 \approx t_2$ and the r is ignored in Figure 12-10, then

$$D = \sqrt{d^2 + 4dh} \qquad (12\text{-}10)$$

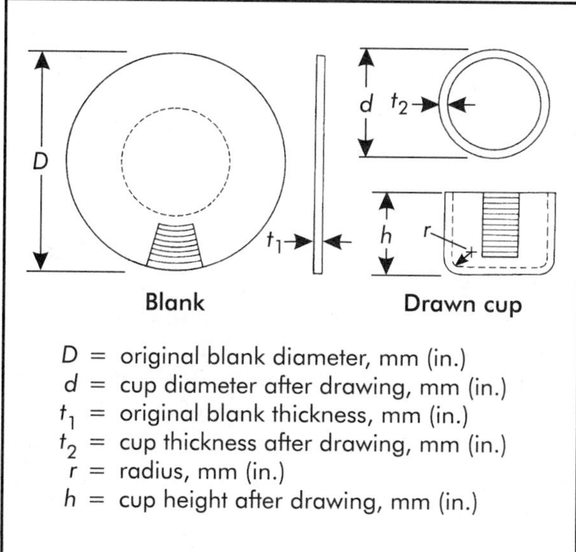

Figure 12-10. Example of a cup drawn from a round blank.

where

D = original blank diameter, mm (in.)
d = cup diameter after drawing, mm (in.)
h = cup height after drawing, mm (in.)

is obtained by equating the outside areas of the blank and cup. This is enough for practical estimates. Longer equations can be derived to take into account the radius in the cup and the changes in wall thickness.

A cup with an even edge is ideal. In most cases, the edge comes out uneven because of the anisotropy of the metal, and the cup is made higher than needed. The excess is trimmed away.

The way a cup is drawn is shown in Figure 12-11. The blank is placed on the top of a die block. The punch pushes the bottom of the cup into the hole in the block and draws the remaining metal over the edge of the hole to form the sides. The edges of punch and die must be rounded to avoid cutting or tearing the metal. The clearance between the punch and die block is a little larger than the stock thickness. As has been explained, compressive stresses are set up around the flange as it is drawn into smaller and smaller circles. If the flange is thin (less than about 2% of the cup diameter), it can be expected to buckle like any thin piece of metal compressed in its weakest direction. To avoid wrinkling, pressure is applied to the flange by a *pressure pad* or *blank holder*.

In practice, pressure is obtained from springs, rubber pads, compressed air cylinders, or an auxiliary ram on a double-action press. The force required is normally less than 40% of the drawing force. Normally the blank is lubricated to help it slide under the pressure pad and over the edge of the die.

The force applied by the punch is transmitted solely through the tensile stress in the wall of the cup. It draws the metal against friction over the edge of the hole from under the pressure pad and sets up the necessary stresses in the flange. Only as much force can be applied as the wall of the cup can support. Otherwise the cup will be torn. This limits the amount of stress that can be set up in the flange and the amount of reduction possible. Theoretically, it can be shown that a reduction of

$$(D - d)/D \quad (12\text{-}11)$$

where

D = diameter of blank, mm (in.)
d = diameter of cup, mm (in.)

or 50–70% is possible, depending upon the behavior of the material. Actually, friction from the hold-down pressure and bending forces detract from this, and in practice a reduction in diameter of only 35–47% is feasible for an initial draw.

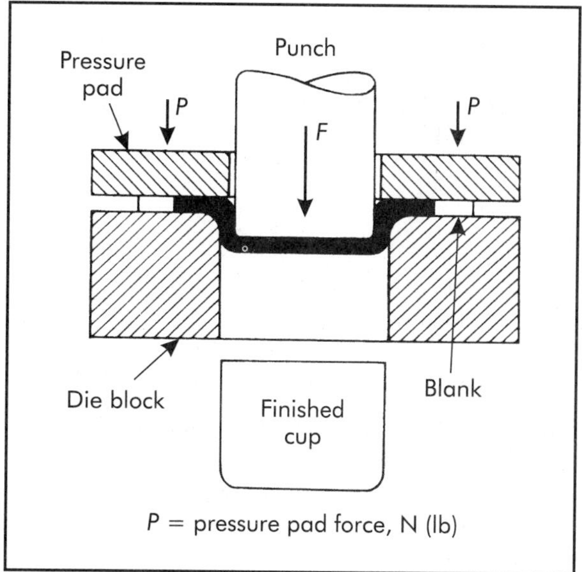

Figure 12-11. The way a cup is drawn.

Generally, thinner blanks require more hold-down pressure to prevent wrinkling and can be reduced less than thicker blanks. The results of exhaustive tests showing how much various materials may be formed in various ways are given in the references at the end of this chapter.

A drawn cup may be redrawn in another operation to form a longer cup of smaller diameter. However, metal becomes work hardened as it is drawn and redrawn and it must be annealed to prevent failure before reaching its limit. Thus, reduction limits of 30% in diameter for second draws and 20% for third and subsequent draws are often recommended if intervening annealing operations are not performed. The reduction in area during a tensile test is an indication of the total reduction in diameter to which a metal can be subjected before it must be annealed. For instance, assume a metal shows a reduction in area of 60% in tensile tests. According to Equation (12-11) indications are that a blank will be reduced such that

$$(D - d)/D = 0.6$$

This probably will have to be done in several operations in single rigid dies. Exceptional amounts of reduction (up to 14:1 length-to-diameter ratios) have been obtained by *multiple-stage drawing* in one stroke. Benefit is probably obtained from the heat of work done in consecutive steps. One method employs a telescoping punch whose parts draw the sheet metal through progressively smaller die openings.

The most force that can be applied to draw a cup is found by

$$F = \pi d t S_t \quad (12\text{-}12)$$

where

F = force, MN (lbf)
d = diameter, mm (in.)
t = wall thickness of the cup, mm (in.)
S_t = tensile strength, MPa (psi)

More exact formulas have been derived, but this one always specifies enough force. The necessary force may be appreciably less for small reductions in diameters. The energy for a draw is estimated from

$$E = CFh \quad (12\text{-}13)$$

where

E = energy, J (ft-lbf)
C = constant with an average value of 0.7
F = punch force, N (lb)
h = height of the cup, mm (in.)

Several parts drawn from sheet metal are sketched in Figure 12-12 to point out some of the considerations when drawing parts more complex than round cups. Often, two or more basic actions occur at once. In the case of the rectangular pan shown in Figure 12-12(A), the sides and ends are formed mostly by bending while the corners are drawn. Parts like those shown in Figure 12-12(B) and (C) represent control problems because they do not take the same shape as the punch until just about done. This means the punch and die cannot hold contact with enough of the metal during most of the operation to have full control. Out of control, metal tends to wrinkle or become misshapen. The part in Figure 12-12(B) may have to be formed in steps as indicated by the dashed lines before the final shape is struck. In some cases, blanks must be preformed before drawing. A flange is required on some drawn parts, like Figure 12-12(C), to hold the metal while it is stretched to its final shape. The flange may be subsequently trimmed away.

The fastest production rates are achieved by rigid die drawing, which produces from 100 or more small cups/min to four auto fenders/min with automated equipment.

12.4.2 Flexible Die Drawing and Forming

Some drawing and forming methods use either a punch or die, but not both. The workpiece is forced into the die or wrapped over the

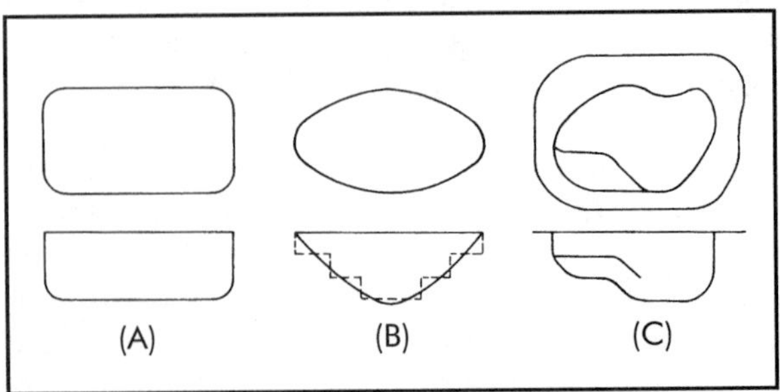

Figure 12-12. Some parts drawn from sheet metal.

punch by uniform hydraulic or rubber pad pressure. Several typical processes are illustrated in Figure 12-13.

Rubber pad forming, also known as the *Guerin process*, forms the work over an inverted punch by the action of a pad of rubber in a container attached to the ram of a standard hydraulic press as indicated in Figure 12-13(A). The operation is limited to shallow drawing, simple bending and forming, and shearing of sheet metal against sharp edges because there are no pressure pads around the punch. Typical jobs are forming flanges around flat pieces, raising ridges, forming beads to add rigidity to flat pieces, embossing, and trimming. Tolerances are not close.

In rubber forming, several punches and pieces may be placed under one pad. Loading tables that can be filled alternately are helpful to keep a press fully occupied. Lubricants are commonly used. The force applied must be sufficient to build up a pressure of 6.8–13.8 MPa (1,000–2,000 psi) over the entire bottom face of the pad, not just on the workpiece.

The *Marform process* utilizes a rubber pad to envelop the part and a blank holder or pressure pad around the punch. A blank is laid on the punch and blank holder as shown in Figure 12-13(B). The blank is drawn from between the rubber pad and blank holder as it is wrapped around the punch while the press closes. The blank holder is pushed down against hydraulic pressure. Forming pressures are mostly from 35–55 MPa (5,000–8,000 psi), but sometimes as high as 83 MPa (12,000 psi). Typical parts produced are flanged cups, spherical domes, conical and rectangular shells, and unsymmetrical shapes with embossed or recessed areas. Marforming is slower but is more suitable for deep drawing and gives better definition to shallow forms than rubber pad forming. Operation rates range from 60–240 cycles/hr.

Hydroforming employs a punch and a flexible die in the form of a diaphragm backed up by oil pressure. The piece is laid on a blank holder and over the punch as depicted in Figure 12-13(C). First, the dome is lowered until the diaphragm covers the blank and initial oil pressure is applied. Then the punch is raised, and the oil pressure is augmented to draw and form the metal to the desired shape. Hydroforming produces the same kinds of parts as Marforming with a little sharper detail, particularly in external radii.

Hydroform presses and equipment are available in 203–813 mm (8–32 in.) sizes which designate the diameter of the blank that can be drawn. Draw depths range from 127–305 mm (5–12 in.) and operating rates from 90–200 cycles/hr. The small presses are fastest.

Flexible punch forming is exemplified by the patented *Hydrodynamic process* illustrated in Figure 12-13(D). The blank is pressed between the die and a spring-loaded pressure plate. Water or oil admitted under pressure passes through an opening in the plate and pushes the work into conformity with the die cavity. The process is limited to forming shallow pieces with inclined rather than vertical sides but is able to finish in one operation some pieces that require several steps by other methods.

Thermoforming or *blow molding* (Chapter 7), as performed on plastics, are applicable to some superplastic metallic alloys, usually hot. An example is Zn 22 Al. Pressures are reported at 0.6–1.2 MPa (90–175 psi) over male molds or at atmospheric pressure drawn by a vacuum in female molds. Complex and intricately styled parts with smooth surfaces can be produced with low-cost tooling and equipment. However, forming rates are low (3 or 4 min for a medium-sized part) and conventional forming of ordinary material or die casting is preferable for large quantities. The material may be stretched and thinned appreciably over certain areas.

Applications

As a rule, the flexible die processes are not nearly as fast and cannot compete with rigid steel dies in mechanical presses for drawing or forming large quantities of pieces. Flexible dies have an advantage for quantities up to several hundred or thousands of pieces, depending on the part, because their tool costs are low and lead times short. Only one member is needed, and it generally can be made from easily machined material, such as a plastic, soft metal, or wood, because the service is not harsh. Tooling costs run from 30–80% of the cost for hard steel dies, with savings of several hundred to thousands of

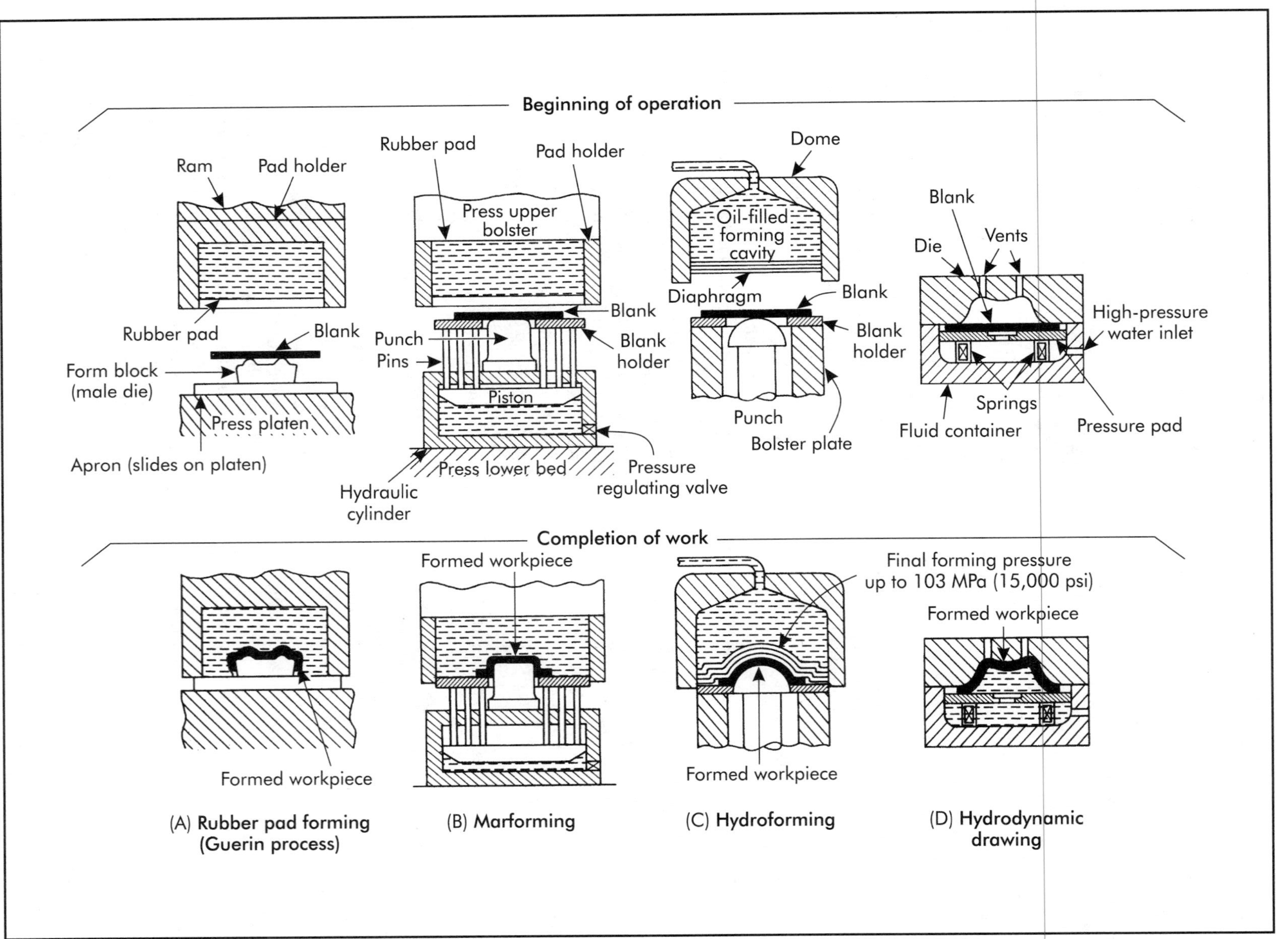

Figure 12-13. Common flexible die-forming processes.

dollars for each job. The mild forming action keeps maintenance costs low and does not mar the work material, not even prepainted sheets.

In Marforming and hydroforming, the stresses in the material being formed are low. The pressure locks the metal to the punch and prevents stress concentrations. When the metal is drawn off the blank onto the punch, the rubber does not force it to bend sharply. Because of the easier action, reductions in diameters up to and sometimes over 60% are feasible, almost twice as much as with steel dies. Many jobs can be done in fewer operations than with rigid dies, and that sometimes gives the flexible die methods the advantage for large quantities.

Equipment costs for Marforming and hydroforming are especially high and considerable work in small quantities must be available to justify the investment. A number of jobbing shops specialize in work on these forms of equipment and are resources for manufacturers who cannot individually justify the investments. Stock as thick as 6.35–9.53 mm (.250–.375 in.) is commonly worked, and even much thicker material has been formed from aluminum alloys. Tolerances of ±50 μm (.002 in.) are possible and ±127 μm (.005 in.) are practical. These are comparable to performances with the best-quality rigid dies.

12.4.3 Hydrostatic Forming

Hydrostatic pressure is uniform from all directions like that on a body immersed deep in water. Metal under hydrostatic pressure of the order of 3.5 GPa (500,000 psi) becomes extra ductile and can be ultraformed. An example of an operation in that state is hydrostatic extruding as illustrated in Figure 11-23. Cups are drawn under hydrostatic pressure with a reduction of 75% in diameter.

A commercial hydrostatic stretching operation is the bulging of a 9.53-mm (.375-in.) diameter copper tube to make a 19.20-mm (.756-in.) diameter coaxial cable connector in the manner shown in Figure 12-14. The tube is filled with and surrounded by fluid between two plungers in a sleeve. The boost pressure keeps the die parts together. The ram force creates the pressure in the tube. As the tube expands, the fluid around it cannot escape until its pressure is high enough to expand the sleeve. The resultant hydrostatic pressure makes it possible to increase the tube diameter by 100%; this is more than can be done by any other method. The production rate is 3 parts/min.

12.4.4 Metal Spinning

Parts that have circular cross sections can be made by spinning them from sheet metal (Figure 12-15). A blank of sheet metal is clamped on center against a chuck or form block. This block may be plaster, wood, or metal and is revolved on the spindle of a lathe. A rounded stick or roller is pressed against the revolving piece and moved in a series of sweeps. This displaces the metal in several steps to conform to the shape of the chuck. The pressure may be applied by hand or mechanically. The latter requires less skill and gives more uniform results.

In addition to flaring parts like reflectors, reentrant shapes required in such products as kettles and pitchers can be spun as indicated in Figure 12-15. These are spun on sectional chucks or with internal rollers in hollow chucks. Pieces may be as small as 6.4 mm (.25 in.) in diameter to as large as 6 m (20 ft) in diameter and 51–76 mm (2–3 in.) thick. Spinning is done both hot and cold. Usually, metal thickness is not substantially changed, but it can be if desired. A practical tolerance for dimensions of hand-spun articles is ±0.79 mm (±.031 in.); about half as much is quoted for mechanical spinning, and even less at extra cost. All ductile metals that can be drawn can be spun. Metal can be trimmed, curled, beaded, and burnished in the spinning operation.

Spinning is both supplementary and competitive with drawing in presses. For straightforward parts, spinning is usually slower but offers a lower tool cost and is economical for small quantities. In one case, press tools cost about $10,000 for drawing, redrawing, ironing, trimming, and beading operations to make a pan. These produced 200 pieces/hr in presses at a unit cost of $0.12/piece, not including tool cost. Tools for spinning in two operations cost $1,500; output was 18 pieces/hr; and unit cost $1.78/piece. Spinning was more economical up to about 5,100 pieces. For any amount over that, the lower unit cost of drawing more than offset the extra cost

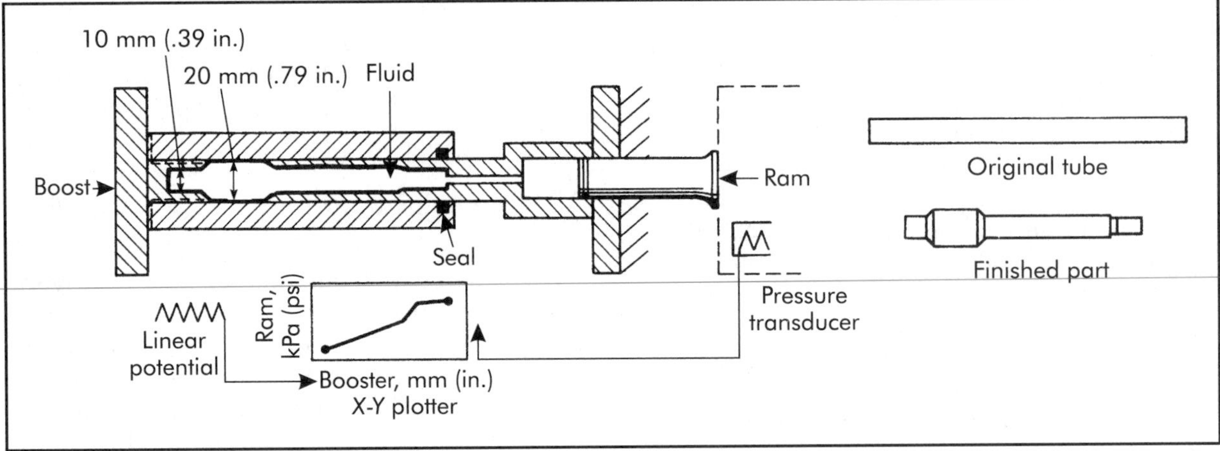

Figure 12-14. Example of a hydrostatic forming operation.

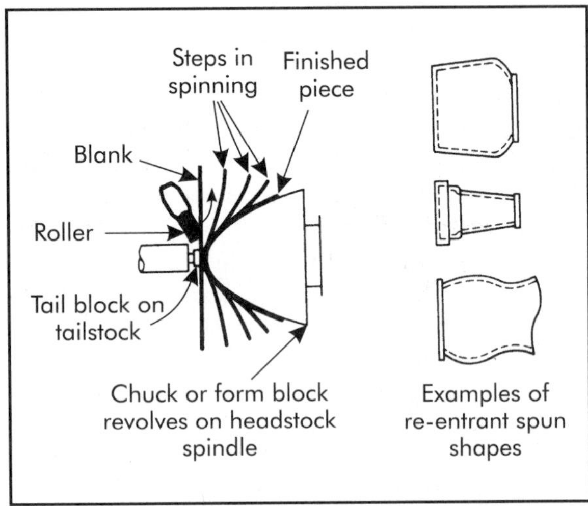

Figure 12-15. Principles of metal spinning.

of tools. In this case, the spinning tools could be obtained in 2 weeks, while the press tools would take 7 weeks. After getting started, the press output could overtake the spinning in about half a week, at about 4,000 pieces. Thus, spinning has an advantage in getting production under way in a short time and sometimes is done while preparations are made for drawing.

Spinning may be just about as fast as drawing for some parts. As an example, a conical shell required eight drawing operations versus one operation on an automatic spinning machine. Some shapes can be most economically made by a combination of spinning and drawing. Parts that are too large for available presses can be spun.

12.4.5 Roll Turning

Roll turning, also called *roll forming, hydrospinning, floturning, flow forming, power spinning,* and *power roll forming,* uses some of the techniques of spinning but is a more drastic squeezing and extruding process. Three recognized types are illustrated in Figure 12-16.

Shear forming basically extends a thick blank into a cone in such proportion that $t_2 = t_1 \sin \alpha$, as designated in Figure 12-16(A). *Tube forming* (Figure 12-16[B]) continuously extrudes a thin-walled tube along a mandrel from a short and thick-walled ring. Reductions in thickness up to 90% in low-carbon alloy steel have been accomplished. *Contour forming* (Figure 12-16[C]) acts like the other two kinds but produces parts with curved instead of straight profiles.

Roll-turning operations are also done inside of externally confined rings and cylinders to make them longer and thinner. A less penetrating operation with longer rolls that act on the surface only to improve finish is called *roller burnishing.* It is done mostly in holes, but on outer surfaces as well.

Roll turning is done on machines like lathes and vertical boring machines designed to be extra rigid for the work. Sections as thick as 51 mm (2 in.) have been reportedly worked. Reductions in thickness may be substantial with consequent cold working and changing of some properties of the metal. Parts with shapes favored by the process can be produced more rapidly than by other means. Parts that have been

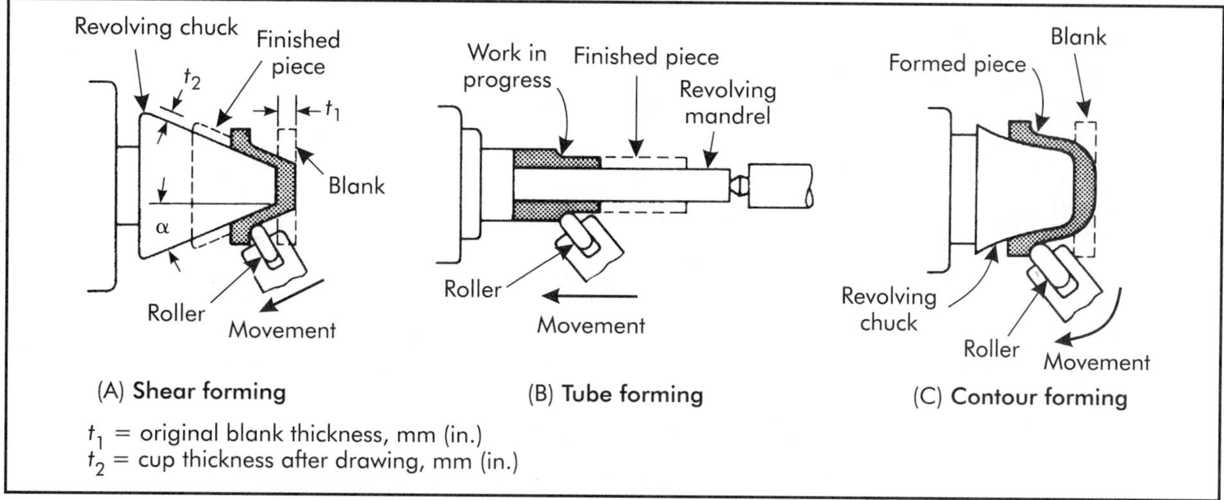

Figure 12-16. Three types of roll turning.

made in this way are television cones, cream separator bowls, missile noses, thin-walled seamless tubing, and pressure vessel components. Ductile metals may be roll turned cold, but others must be heated. Accuracy and finish are equivalent to that obtained by grinding.

If flow turning is done severely enough, it can peel away the surface. Equipment for such *flopeeling* is available to remove up to 6.4 mm (.25 in.) stock per pass.

12.4.6 Stretching and Shrinking

Several common operations in which sheet metal is formed by being stretched or shrunk are depicted in Figure 12-17. Although straight flanging is bending, the forming of curved flanges involves drawing. This may be done with solid or rubber dies.

Bulging (Figure 12-17[C]) is done by placing a cup in a die cavity of the desired shape and filling the cup with rubber or liquid. A punch is applied to create hydrostatic pressure to bulge the walls of the cup to the limits of the die cavity. The die must be split so it can be opened to remove the finished piece.

Ardforming is a process of stretch forming austenitic stainless-steel pressure vessels and rocket cases at −196° C (−320° F) in liquid nitrogen to raise yield strength above 2 GPa (300,000 psi). It also increases creep strength and toughness. The fabricated vessel is enclosed in and expanded to the walls of a die by gaseous nitrogen at high pressure.

Shallow shapes of large areas are difficult to draw from thin sheet metal. The amount the sheet must be depressed is not enough to raise stresses all over the area beyond the elastic limit to cause a permanent set. Enough localized stretching is obtained in conventional dies, such as for auto body panels, by tightly restraining the metal flange around the area drawn. Even better, *stretch forming,* in Figure 12-17(D), stretches the metal sheet uniformly to near the yield point initially, and a punch forces the work into the desired shape and readily tips the stresses into the plastic range to cause permanent set. *Stretch draw forming* (Figure 12-17[G]) is another version of the process. A sheet is gripped by rams on each side and stretched over a forming die in a press. The punch is then brought down to impress the shape in the sheet.

Stretching is commonly used to form aircraft panels. A large stretch-forming press for aluminum panels 2.5 m (100 in.) wide by 12.2 m (40 ft) long to produce fuselage panels for commercial airlines costs about $2 million. The process aids particularly in forming hard-to-work space-age materials. The metal is uniformly strengthened. Dies are less costly because less force is needed than with conventional methods. It is reported that a job usually requiring a 10.7 MN (1,200 ton) double-action press can be done on a single-action stretch-forming press rated at

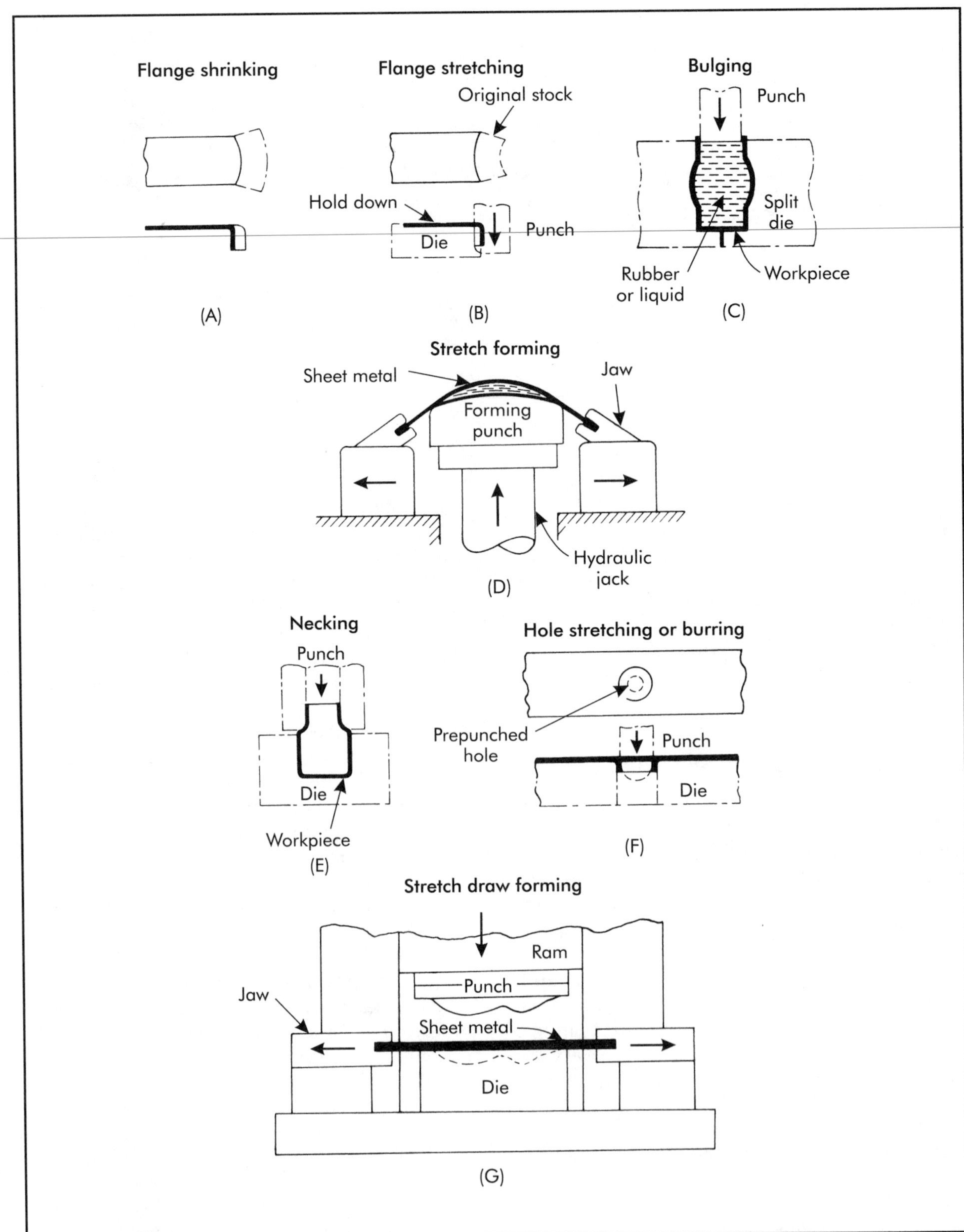

Figure 12-17. Stretching and shrinking operations.

3.6 MN (400 tons). The advantages of stretch forming have not been found sufficient to offset its relative slowness in the automobile and other high-production industries.

Expanding is an operation used to increase the size of round parts; an example is forming a bell at the end of a pipe. One type of machine called an *expander* carries a circle of shoes guided in radial slots on a face plate and driven outward by wedges to expand the work. An expander for parts up to 1.5 m (5 ft) in diameter with a wall thickness of 13 mm (.5 in.) exerts 222 kN (25 tons) of radial force and costs around $550,000. Dies for a specific job cost about $1,000 per set. It takes about 5 min to form a part of this size. Another type of machine uses a rubber bladder with oil pressure up to 35 MPa (5,000 psi) to expand and set thin-walled shells such as washing machine tubs.

Shrinking is the converse of expanding. A ring of even or varying cross section is completely surrounded by shoes tapered on their outer surfaces. Wedges act upon the shoes simultaneously to drive them inward to shrink the workpiece. The outside diameter of a wheel rim, for instance, can be accurately sized in this way.

Large sheets, such as 2.1 × 25.9 m (7 × 85 ft), wing panels, or difficult-to-form sections for aircraft are gently curved by *peen* or *ball forming*. Steel balls are slung or dropped on the work surface and induce stresses in the skin layer to deform the material. Patterns of impingement, sizes of balls, etc., can be controlled to produce shapes within close limits. Machines and tooling are simple and cheap.

12.4.7 High-energy-rate Forming

Forces applied at high velocities and over short periods of time to cut or form metal give exceptional results in some cases. High-energy-rate forming (HERF) is comprised of a number of processes. One type, called high-energy-rate forging, is illustrated in Figure 11-17. Operations mainly for forming sheet metal are *explosive forming, electrospark* or *electrohydraulic forming,* and *magnetic forming*. Examples of these operations are shown in Figure 12-18.

The relationship of high-energy-rate forming to other methods can be seen from the fact that the energy applied to form a piece of material is

$$E = Fd = WV^2/2g = MV^2/2 \qquad (12\text{-}14)$$

where

E = energy required to form a piece of material, J (ft-lbf)
F = average press force, MN (ton)
d = distance, mm (in.)
W = weight, kg (lb)
V = velocity, m/sec (ft/sec)
g = gravitational constant, $\frac{\text{m}}{\text{s}^2}\left(\frac{\text{ft}}{\text{s}^2}\right)$
M = mass = $\frac{\text{w}}{\text{g}}$, kg/m/s^2 (lb/ft/s^2)

Energy of 68 kJ (50,000 ft-lbf) may be supplied by a massive press delivering an average force of 445 kN (50 tons) over a distance of 152 mm (6 in.). The same results may be obtained from a falling drop hammer with a mass of 4.5 Mg (10,000 lb) striking the workpiece with a velocity of 5.5 m/sec (18 ft/sec). The same effect is found in explosive forming with a moving water front of 27 N (6 lbf) and an impact velocity of about 222.5 m/sec (730 ft/sec). Velocity is even much higher in some operations.

The timing of the operations just cited tells even more about them. Assume that each method is forming a 152-mm (6-in.) deep cup. The press at 0.3 m/sec (1 ft/sec) does its work in 0.5 sec and delivers energy at the rate of about 134 kW (180 hp). The drop hammer comes to rest in 0.06 sec and gives up energy at an average of about 1.2 MW (1,600 hp). The explosive operation takes place in about 0.0014 sec with an output of around 49 MW (65,000 hp). Metals are deformed at high strain rates (commonly over 100 mm/mm/sec or in./in./sec) by HERF. Yield strength is increased appreciably at high strain rates because dislocations resist rapid movement. Thus somewhat more energy usually must be applied to a HERF operation than to an ordinary one.

Many metals can be worked more fully by HERF than by other methods. Previously intractable space-age alloys are successfully formed at high velocities. Energy may be uniformly applied rather than at high spots as is done by a punch, and the metal does not thin out and fail locally.

Explosive forming is done with low and high explosives and gas mixtures. The charge may be set off in direct contact with the workpiece. In that way, pressures in the GPa range (millions of psi) are developed. However, they must be carefully controlled to avoid damage. Usually, the

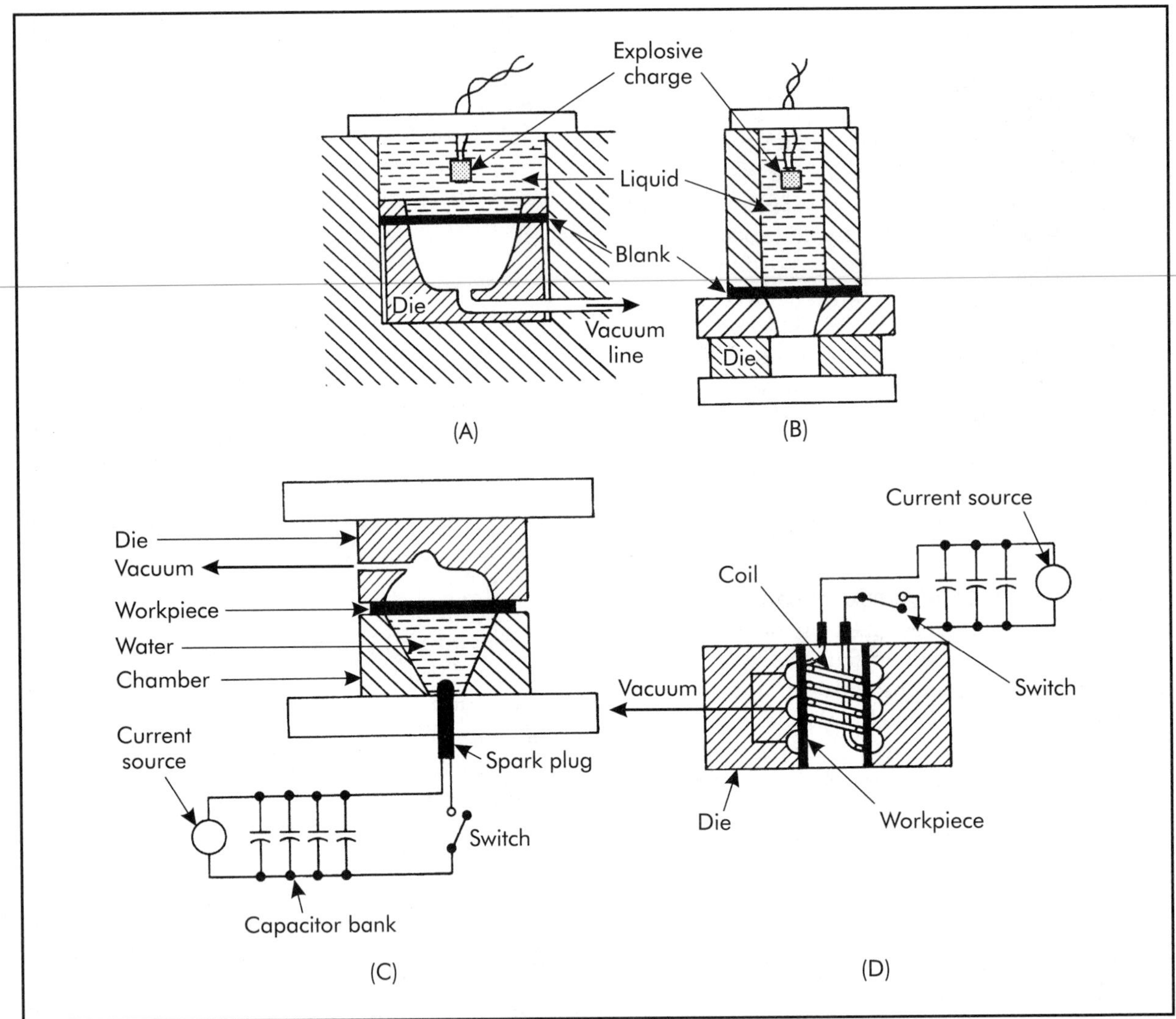

Figure 12-18. Some methods of high-energy-rate metal forming: (A) and (B) explosive; (C) electrohydraulic; (D) electromagnetic.

charge is placed at a *standoff distance* from the workpiece, as shown in Figure 12-18(A) and (B), and the shock wave is transmitted through water. In that way, pressures on the order of 345 MPa (50,000 psi) and speeds of less than 304.8 m/sec (1,000 ft/sec) are obtained. In forming, sheet metal is blown into and takes the shape of a die.

The main advantage of explosive forming is that it is a means of working difficult materials and large area or thick pieces with relatively cheap equipment (no huge presses) and a low-cost energy source. Only a punch or die is needed (not both). For a few pieces, it may be a cost-effective means to process easy-to-machine material. There is no limit to size, and tolerances equal those of other methods. The process is especially suitable for unusual and intricate shapes because pressure can be distributed over an entire surface and is not concentrated in a few spots as with a punch and die. However, explosive forming takes exceptional care, skill, isolation, and time, and cannot compete with press forming for ordinary work.

Explosive hardening is an offshoot of explosive forming. Metals are work hardened when subjected to high enough pressures. As an example, an annealed mild steel (0.17% carbon) exposed to contact explosive pressure of around 20.7 GPa (3,000,000 psi) had its yield strength increased from 241–793 MPa (35,000–115,000 psi), tensile strength from 400–862 MPa (58,000–125,000 psi), and elongation decreased from 25% to 7%. Although results are like those of cold working, the crystals are not distorted in the same way by explosive hardening. After recovery, the grains of a metal hardened by an explosion are distorted less than 5% as compared to 80% by cold rolling for the same increase in strength.

Electrohydraulic forming (EHF), also called *electrospark forming*, is done with shock waves like explosive forming, but with smaller amounts of energy. The waves may be created by one or a series of spark discharges in a liquid as indicated in Figure 12-18(C). Another way is to discharge a large current through a wire in the liquid. The wire explodes under the heavy load and sends forth a shock wave. EHF has most of the advantages of explosive forming.

EHF is not competitive with mechanical forming except for difficult materials and shapes, mostly in small and moderate quantities. The time to form a piece is only a few seconds per shot, but the number of shots depends on the energy available. As an example, a steel dome of 221 MPa (32,000 psi) yield strength, 635 mm (25 in.) in diameter, 6.6 mm (.26 in.) in thickness, and 225.43 mm (8.875 in.) in depth required 18 shots of about 70 kJ (51,629 ft-lbf) each. That amount of energy costs only a few cents, but the time is relatively long, and the total production cost is high. Equipment of up to 100 kJ (73,756 ft-lbf) capacity may cost as much as $150,000. Some machines can take parts up to about 1.5 m (5 ft) in diameter. An EHF machine can form almost any shape in its capacity range.

Electromagnetic or *magnetic pulse forming* is done in a sudden and intense magnetic field around a coil next to or inside the workpiece. Eddy currents are induced in the metal part. The wall of the piece is strongly repelled by the magnetic field, like any conductor of current, and is driven into the shape of a restraining die of dielectric material, as depicted in Figure 12-18(D). A poor conductor is formed more easily if coated with a thin layer of copper. Metal can be formed to its limit because nothing but the die need contact the workpiece, and the forces can be distributed uniformly over the part. A major use for this process is to join parts by swaging one tightly around or into grooves or recesses in another. Tubes are natural shapes for the process, but forming often is done on flat pieces. Most equipment has a capacity of less than 15 KJ (11,063 ft-lbf) and work is confined to small and moderate-size parts and areas. Operations include expanding, bulging, swaging, blanking, perforating, flanging, dimpling, and corrugating.

Electromagnetic forming is competitive for parts in moderate and large quantities. Cycle time is only a few seconds, but tooling may cost thousands of dollars. The process is easily automated. A 6-kJ (4,425-ft-lbf) capacity machine costs upward of $50,000.

12.4.8 Ultrasonic Aid to Forming

Ultrasonic radiation reduces forces and increases the amount metal can be deformed in conventional operations. It is thought that the acoustic energy is taken up mainly at the lattice defects and grain boundaries instead of uniformly throughout the crystals. This is why the effect of macrosound is even more than that of heat for the same amount of energy. As an example of *ultrasonic forming,* a copper cup was ironed (see next section) to about twice its original length between a steel die and punch. Macrosound was applied with an intensity of about 70 W/cm^2 (450 W/in.^{2}). The force was about 311 N (70 lbf) compared to 890 N (200 lbf) without macrosound.

12.5 SQUEEZING

Squeezing is a quick and widely used way of forming ductile metals. Its applications in primary metalworking in the processes of rolling, forging, wire drawing, and extrusion are described in Chapter 11. The squeezing operations of cold heading, swaging, sizing, coining, hobbing, ironing, riveting, staking, and stitching discussed in this section are considered presswork.

12.5.1 Cold Heading

Cold heading is a method of forcing metal to flow cold into enlarged sections by endwise squeezing. It is similar to upset forging, which does much the same work hot. Typical cold-headed parts are standard tacks, nails, rivets, screws, and bolts up to about 41.28 mm (1.625 in.) diameter and a large variety of machine parts, such as small gears with stems.

Cold heading is done from wire in machines specifically designed for the process. A typical series of operations is depicted in Figure 12-19. The stock is cut off at one station and transferred by mechanical fingers to the die holder, where it is struck by one or more punches as needed to give it the desired shape. Some parts are expanded in the middle in addition to or instead of the end. In the case of nails, the points are formed at cutoff. The base price for a machine that processes 6.4 mm (.25 in.) stock is about $50,000.

For common steel alloys, cold heading is considered severe if a length 2.5 times the original diameter is deformed in one stroke. The limit is usually 4.5 to 5 diameters in two strokes and 8 diameters in three strokes. Heat treatment may be necessary between strokes. Stainless steel and other metals that work harden quickly are *warm headed* by being heated to 316–538° C (600–1,000° F) to improve workability and avoid cracking.

Cold-headed parts may be heat treated, but otherwise have a bright and finished appearance. As a rule, shank diameters are held to ±76 μm (±.003 in.), head diameters to ±127 μm (±.005 in.), and lengths to ±0.7938 mm (.03125 in.). Subsequent operations, such as thread rolling, may be performed to complete the parts. The strength of cold-headed parts is improved by the cold working and the directional flow of the metal. Little or no material is wasted. The process is fast, with possible outputs of 50 pieces/min for large sizes to several hundred per minute for small sizes.

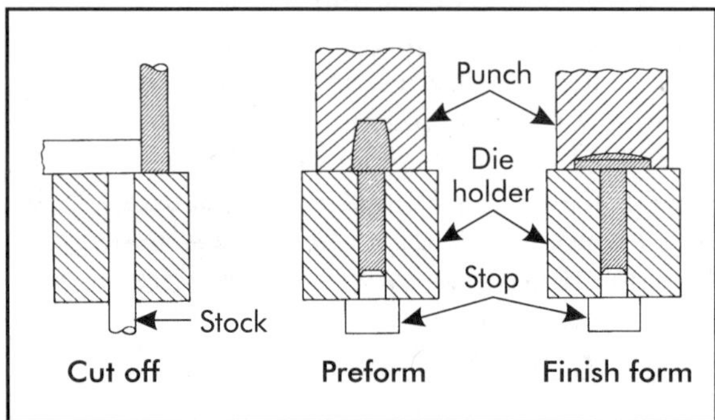

Figure 12-19. Method of cold heading a part.

12.5.2 Swaging

The term *swaging* is used loosely to designate many kinds of forming operations, often any squeezing operation in which the material is free to flow perpendicularly to the applied force. This discussion will be confined to one type of operation, *rotary swaging*, which consists of reducing the size of the diameter, usually over part of the length, of a rod, bar, or tube.

One way that rotary swaging is performed is by a pair of tapered dies as indicated in Figure 12-20(A). The dies are opened and shut rapidly. This may be done in a press while the workpiece is rotated and fed lengthwise. One type of swaging machine makes use of the mechanical device illustrated in Figure 12-20(B). The jaws are inserted in slots in a spindle, rotated, and forced together repeatedly by the rollers around the periphery, as much as several thousand times a minute. The workpiece can be fed into the jaws mechanically or by hand. All is shielded, so there is no danger. The dies only hammer in and out while the outer race and rollers revolve on some machines. Square, fluted, and other shapes of tubes may be swaged this way.

Another form of rotary swaging is depicted in Figure 12-20(C). The tubing is pushed into a bushing as either one revolves. A ring of balls in a cage may be used in place of a bushing. Tubes may be swaged on a mandrel for support or to form an internal shape like a spline.

Rotary swaging is usually done cold but may be performed on hot metal. When cold, the metal is work hardened and may have to be annealed after several passes. Reduction in diameter usually should not exceed 30% in one pass; too much causes flaking and cracking of the work. The angle of taper between the large and small diameter on a piece should not be over 10–12°. Tolerances from ±127 μm for a 25-mm diameter (±.005 in. for a 1-in. diameter) piece to as little as ±13 μm (.0005 in.) for small sizes have been held. A piece elongates as it is swaged to keep its volume unchanged. The elongation is only about 10% for tubing because the walls

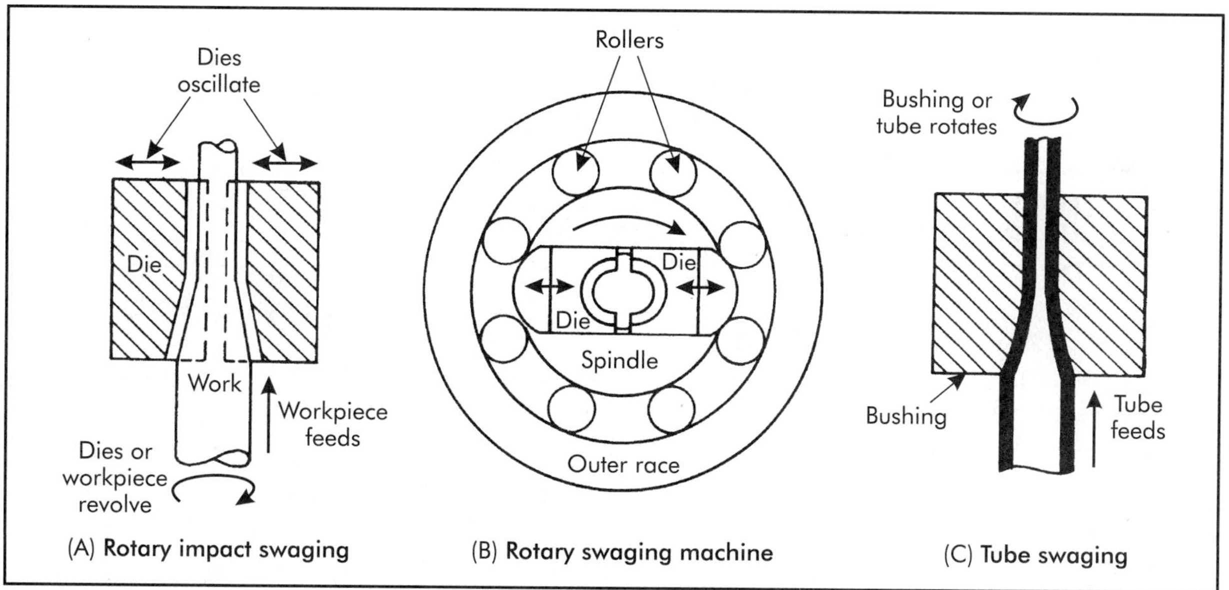

Figure 12-20. Methods of rotary swaging.

thicken. Common feed rates are 25–76 mm/sec (1–3 in./sec). The force necessary to cold swage a bar to an average diameter is found by

$$F = 1.75DLS \tag{12-15}$$

where

F = force, MN (ton)
D = average diameter, mm (in.)
L = length of contact in the swaging die, mm (in.)
S = compressive strength of the material, MPa (psi)

12.5.3 Sizing, Coining, and Hobbing

Parts of malleable iron, forged steel, powdered metals, aluminum, and other ductile nonferrous metals are commonly finished to thickness by squeezing in an operation called *sizing*. An example is given by the sizing of the bosses of a connecting rod depicted in Figure 12-21(A). Dimensions X and Y can be held between hardened blocks with a tolerance of ±25 μm (±.001 in.). Surface finish is comparable to that of milling. About 0.79 mm (.031 in.) of stock is provided for sizing. A special die is needed for almost every job, but each piece can be sized in a fraction of the time that would be required for machining. Thus, sizing is economical wherever applicable in high-production industries.

Operations like sizing have been called coining, but *coining* more truly involves the impression and raising of images or characters from a punch and die into metal. This is illustrated in Figure 12-21(B). The metal is made to flow. The designs on opposite sides of a coined piece are not necessarily related as in embossing. Hard money is probably the best known product of coining.

Hobbing or *hubbing* is a method of making molds for the plastic and die-casting industries. A punch called the *hob* or *hub* is machined from tool steel to the shape of the cavity, heat treated for hardness, and polished. It is then pressed into a blank of soft steel to form the mold; this is done slowly and carefully and sometimes in several stages between annealing operations for the blank. A retaining ring keeps the mold from spreading out of shape. A prime advantage of this method is that one hob properly applied can make a number of cavities in one mold or in a series of molds.

12.5.4 Ironing

Ironing is a name given to an operation for sizing and thinning the walls of drawn cups. As

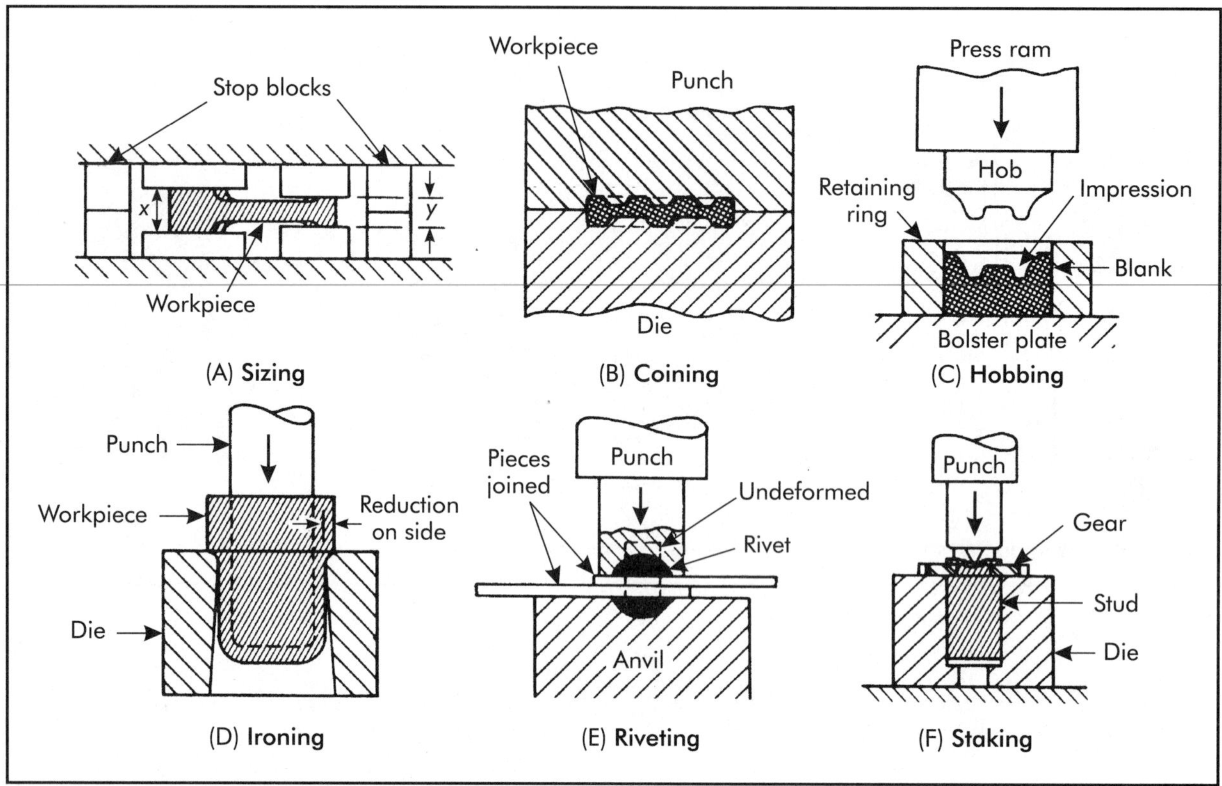

Figure 12-21. Common metal-squeezing operations.

indicated in Figure 12-21(D), the cup is squeezed between a punch and hole in a die as it is pushed through the hole. This is sometimes done at the same time as a redrawing operation. The action is similar to that of wire drawing. The maximum possible theoretical reduction in wall thickness in one pass is 50%, but more has been reported in some cases.

12.5.5 Riveting, Staking, and Stitching

A rivet is like a bolt without threads. A basic way to rivet two parts is depicted in Figure 12-21(E). A rivet is put through a hole or forced through thin material and its head is placed on an anvil. A punch with a hollowed end mashes the stem. Some rivets are hollow and their edges are pushed outward.

Staking is not done with a separate fastener, but with a projection on one of the mating pieces, which is mashed, curled over, or indented and spread by a punch with sharp edges or points as illustrated in Figure 12-21(F).

One way to join metal sheets by stitching is to pierce one or more holes through both sheets. Stock is not removed but is pushed through in tabs that are bent over to clinch the sheets together. Another way is to use staples like in paper stapling. Stitching is limited to fairly thin metal sheets but is fast. A general rule is that a staple is strong enough to hold any material through which it can be driven.

12.6 PRESSES

Presses are the machines that perform metal-forming operations. The capacity of a press depends on the following factors.

1. Dimensional size that includes:
   - (a) enough space to accommodate the tools;
   - (b) a length of stroke to drive a punch the distance required; and
   - (c) openings to get the stock, finished pieces, and scrap readily in and out of the press.
2. Strength to deliver the force required for each stroke.

3. An energy supply that can sustain the force through the working stroke.
4. Speed to deliver the required number of strokes per minute.
5. Power to maintain the energy output at the operating speed.
6. Strength and stamina to maintain alignments, hold tolerances, and produce economically for a long time.

Every press is made up of certain basic units, as indicated in Figure 12-22. These are a frame and bed, a ram or slide (or rams or slides), a drive for the ram, and a power source and transmission. Each unit can be made in a number of forms, and each form has certain advantages and disadvantages. By combining various kinds and sizes of units, a large variety of presses are constructed to suit many purposes. Each unit determines part of the physical capacity of the press. Typical units and ways of putting them together will be described and their principles explained to give a comprehensive picture of presses.

12.6.1 Frame and Bed

The lower part of a press frame on which the die is placed is called the *bed*. The bed is heavily ribbed for strength and hollow to accommodate accessories such as scrap chutes and pressure springs. A thick *bolster plate* on top of the bed gives full support to the die that is bolted to it. Important dimensions of the bed are the distances front to back and right to left available to contain a die.

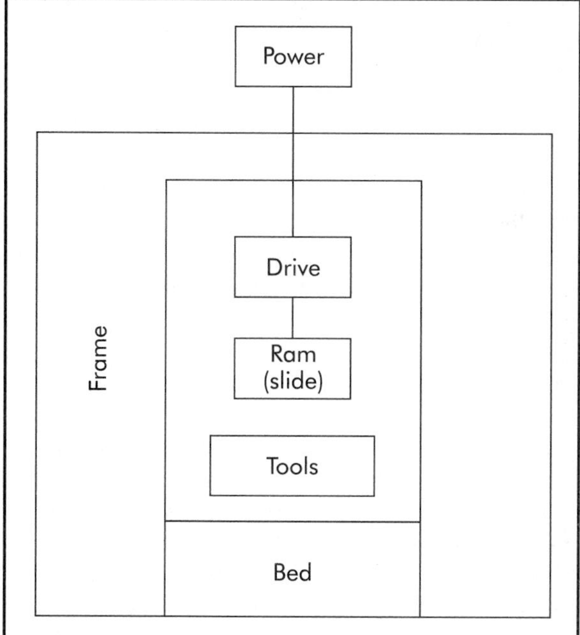

Figure 12-22. Elements of a press.

The frame contains the drive mechanism and ways to guide the reciprocating ram in a fixed path. Press frames are made of both iron and steel. They must be strong and rigid to hold true alignment between punches and dies. Common types of press frames are shown in Figure 12-23. A press is rated by the tons of force it is able to exert without undue strain. To keep deflections small, some authorities recommend choosing a press rated 50–100% higher than the force to be exerted in an operation.

Most presses up to 1.8 N (200 tons) capacity have C-type frames (Figure 12-23[A]). The C-type of frame allows full access to the die space from three sides. This speeds production because dies can be positioned and stock fed in many ways. The back also has an opening as a rule. An inherent disadvantage is that the top and bottom of a C-frame swing apart under load. This impairs alignment of the press and is detrimental to work accuracy and tool life. Of course, tie rods can be fastened to the front to reduce deflection, but they hinder access to the die area and nullify the main advantage of the press.

One of the more common types of presses used in industry is the *gap-frame,* or *open-backed stationary* (OBS) press. As illustrated in Figure 12-23, the open construction of the OBS press provides easy access to dies, ease of stock feeding, and easy ejection of finished parts. OBS gap-frame presses are applicable to blanking, forming, bending, and shallow drawing operations. They can be used for manually fed, single-stroke operations, automated transfer operations, or progressive die-coil fed operations. In addition, they are quite adaptable to short runs requiring frequent die changes.

A variation of the gap-frame press is the *open-back inclinable* (OBI) press shown in Figure 12-24. This type of press is made up of a C-frame or gap frame attached to a rocker-like base that permits the frame to be tilted backwards. This backward inclination permits either scrap or

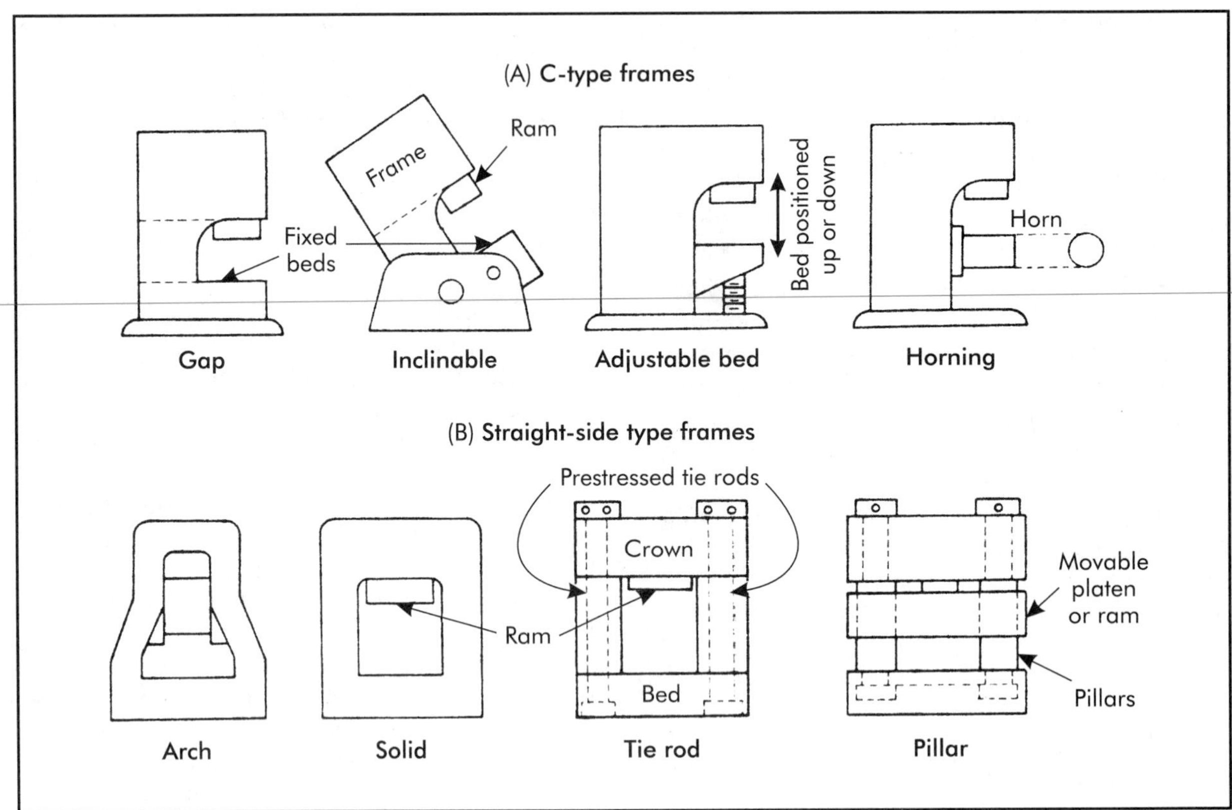

Figure 12-23. Common types of press frames.

blanked parts to be discharged by gravity through an opening in the back. The 400 kN (45 ton) press shown in Figure 12-24 is crank driven and available with either a geared or nongeared drive. With a geared drive at constant speed, it will operate at 50 strokes/min, or with a non-geared drive at 110 strokes/min. The standard length of stroke for the slide is 76.2 mm (3 in.) with the maximum force of 400 kN (45 tons) exerted at a distance of 6.4 mm (.25 in.) from the bottom of the stroke for the geared drive. The standard shut height on the bed for the machine shown is 311 mm (12.25 in.) with a slide adjustment of 76.2 mm (3 in.). The basic machine shown in Figure 12-24 costs about $36,000. The press is shown without point-of-operation guarding as would be required for an operational machine.

An *adjustable bed press* or *knee press* has the lower projection of the C-type frame in the form of an adjustable bed, table, or knee (Figure 12-23[A]). This press can be adjusted for different sizes of dies and work but loses some rigidity. A *horning press* has a large round post, called a *horn*, projecting from the front to support round pieces.

Press frames that have two or more columns like the straight-sided types in Figure 12-23(B) are more balanced and rigid than C-type frames. The design is appropriate for wide beds and long strokes. Access to the die on the bed may be somewhat restricted. Straight-sided frames are found on the medium to largest sizes of presses. The larger sizes of mechanical presses have bed, columns, and crown as separate units, but are keyed to hold alignment and bound together by tie rods. The rods are heated and shrunk in place to minimize stretching under load.

The *pillar* or *open-frame press* is usually a hydraulic press and has four pillars like the one in Figure 12-23. The pillars hold the crown to the bed but are not prestressed like tie rods. Extension under load does not matter appreciably for a hydraulic press that does not work to a

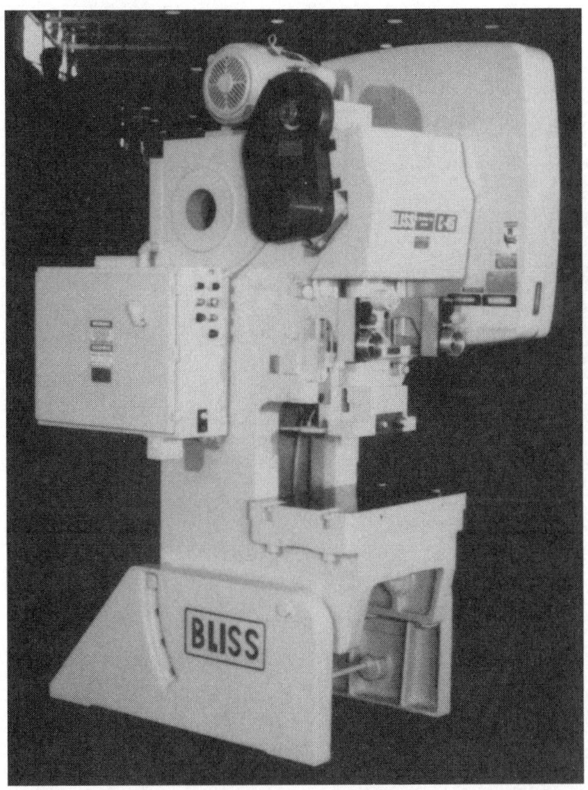

Figure 12-24. A 400-kN (45-ton) gap-frame, open-back inclinable (OBI), crank-driven press with a 0.75 × 0.46 m (29.5 × 18 in.) bed. (Courtesy E. W. Bliss Company)

fixed stop. Ordinarily, the pillars act as guides or ways for the ram.

12.6.2 Press Ram

The ram of a press drives the punch in an operation. Most mechanical presses have a fixed length of stroke. The position of the stroke can be varied by an adjusting nut and screw located just above the ram.

The alignment of the ram in its slides determines how precisely punches and dies can be set to each other in the press. A definite and often minute clearance must be held between each punch and the die opening it enters. Otherwise, cutting or forming action is not uniform all around, and product quality is not good. Also, the tools are damaged if they touch even a little and tool life is shortened. Commonly, the vertical travel of the ram is made square with the bed within 13 μm (.0005 in.). The bottom of the ram should be parallel with the top of the bed within 42 μm/m (.0005 in./ft).

Shut height is an important dimension of a press because it limits the height of the punch and die that can be used in it. *Shut height* of a mechanical press is defined as the distance from the top of the bed to the bottom of the ram when the ram is at the bottom of its stroke and the position adjustment is all the way up.

The *action* of a press refers to the number of rams. A single-action press has one ram, like those in Figures 12-24 and 12-25. A double-action press has one ram inside another. The outer ram is used to apply pressure to the flange of a piece drawn by the inner ram. A triple-action press has a double action above and a third ram that moves upward in the bed soon after the upper rams descend. This type of press may be used to combine drawing and redrawing operations.

12.6.3 Press Drives

The *drive* of a press provides the means of applying force to the ram. Two kinds of drives are mechanical and hydraulic. The mechanical devices in use today are the crank, eccentric, cam, toggle, knuckle joint, screw, and rack and pinion as depicted in Figure 12-26.

Most mechanical presses derive the movement of the ram from the throw of a crank or eccentric. These may be located on a crankshaft or eccentric shaft running from right to left or from front to back in the press, usually at the top. An eccentric provides more bearing area but has to be excessively large for long strokes. A press may have one crank, as does the one shown in Figure 12-24, so it is said to have a single point of connection or suspension. Better distribution of forces is given to large rams by having multiple points of connection such as is demonstrated in the illustration of the toggle or link-driven press of Figure 12-25.

Cam and toggle mechanisms are used to drive the outer rams of double-action presses. The purpose is to make the outer ram dwell to hold the stock during part of the stroke. A *knuckle joint press* is used for a forceful squeezing operation over a short distance, like in a sizing operation.

Mechanical presses are rated in tons exerted at or near the bottom of the stroke. This is an

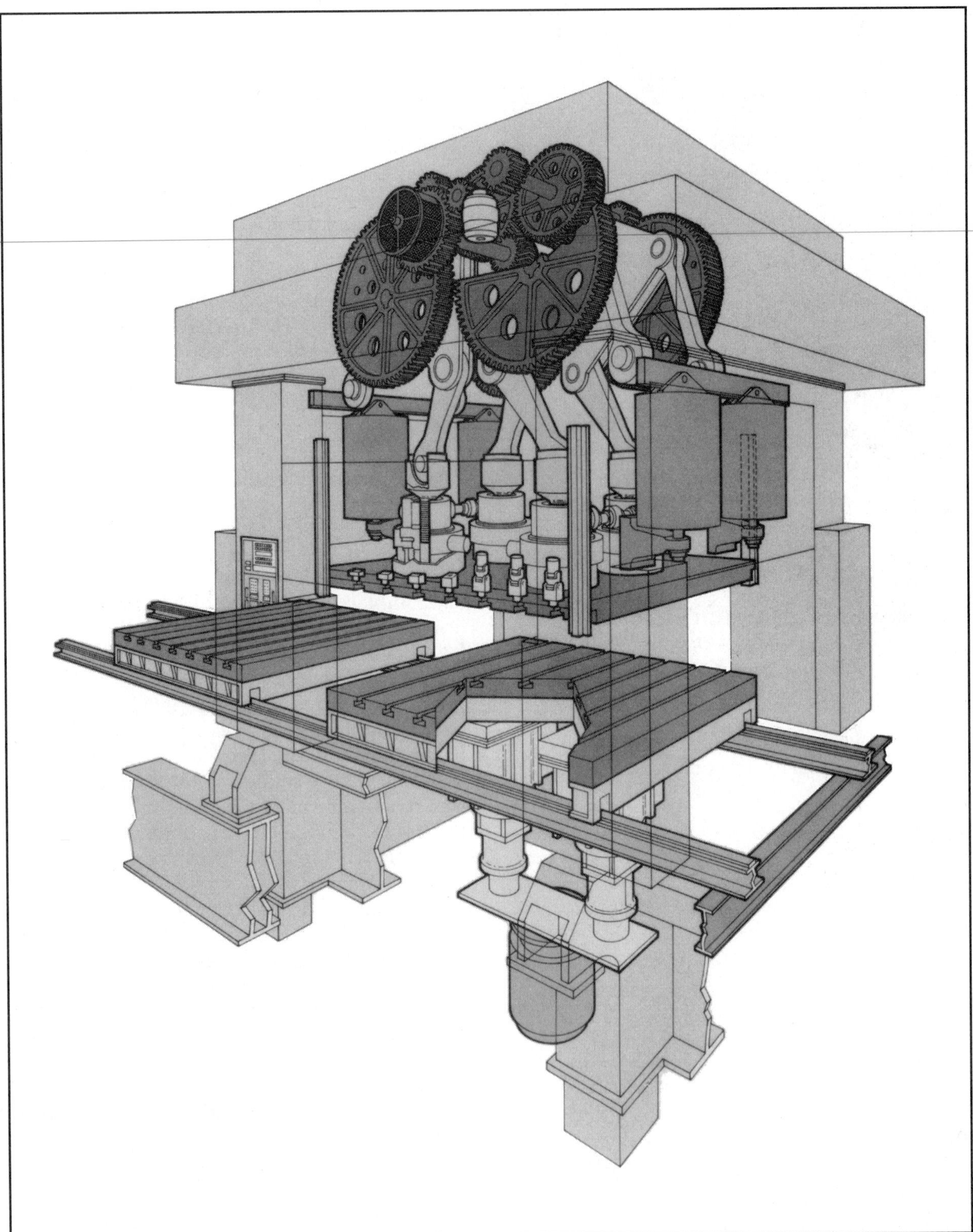

Figure 12-25. Illustration of a straight-side, link-driven press with a quick die change (QDC) moving bolster. (Courtesy Danly-Komatsu L.P.)

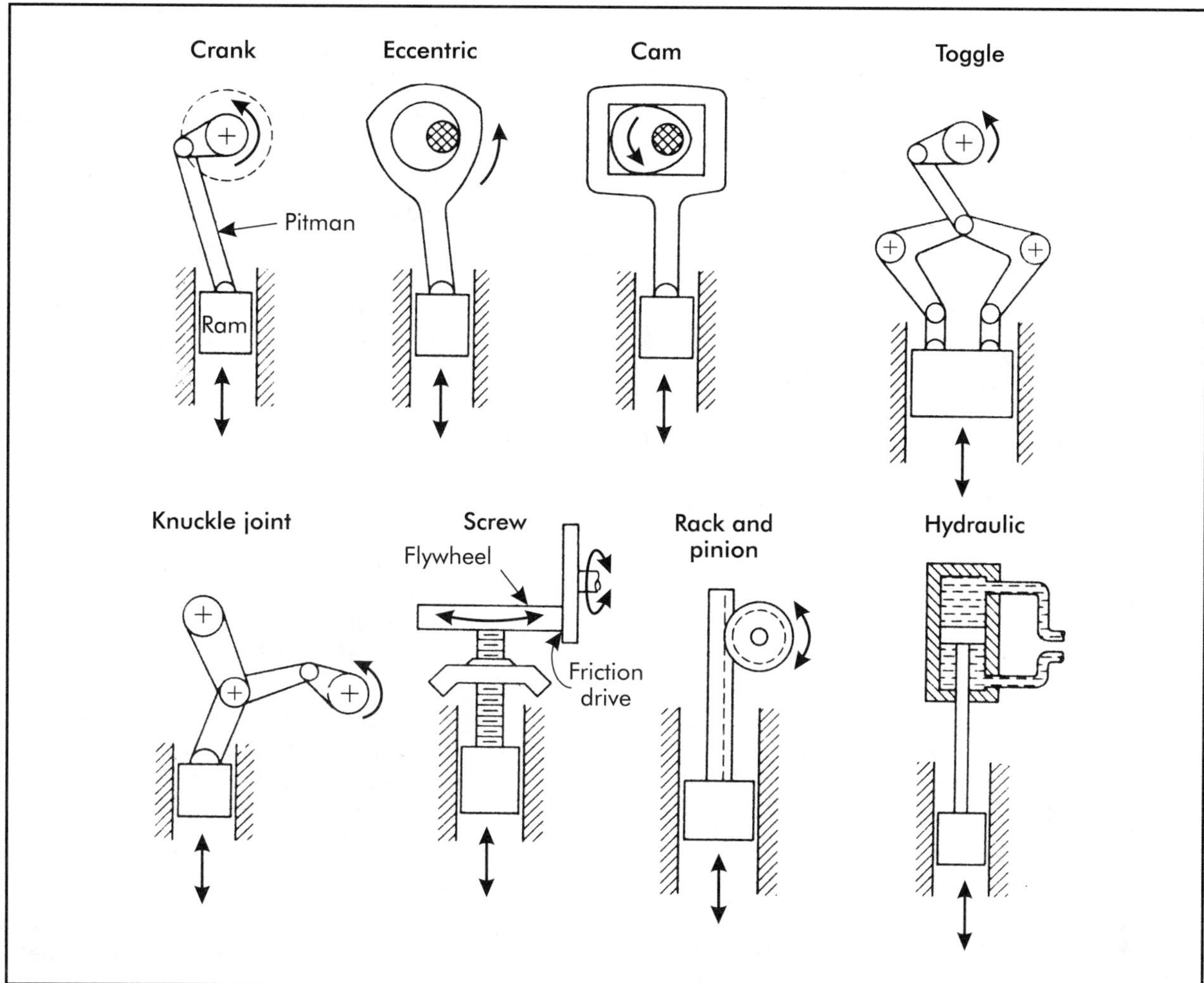

Figure 12-26. Principal kinds of press drives.

important consideration for operations, such as drawing, in which a near maximum force must be applied during a large part of the stroke. With a given rating at the bottom, the actual capacity at higher points of the stroke is less. A double-action press is rated on the basis of the capacity of the inner ram. The outer ram usually has a corresponding capacity. Press manufacturers specify these capacities in tables and charts in their catalogs.

With a crank or eccentric drive, the ram speed is highest at midstroke and zero at the bottom. A formula for ram speed is given in Figure 12-27. This may not be important for short working distances like in blanking, but it is important for work like deep drawing that starts well above the bottom. Too fast a draw makes the metal stick to the die. The proper speed depends upon the condition of the metal, lubrication, the dies, and the severity of the draw. As a rule, speeds should not be more than about 15–27 m/min (50–90 ft/min) for mild steel, 23–46 m/min (75–150 ft/min) for aluminum, and 46–61 m/min (150–200 ft/min) for brass. These limitations are met in some modern presses with two-speed drives that are fast for the approach to and return from the work and slow down for the actual drawing. Presses of this kind are making 40 strokes/min without harm to the work versus 12 strokes/min for older types.

A press should have as short a stroke as possible. In an operation where pieces must be put

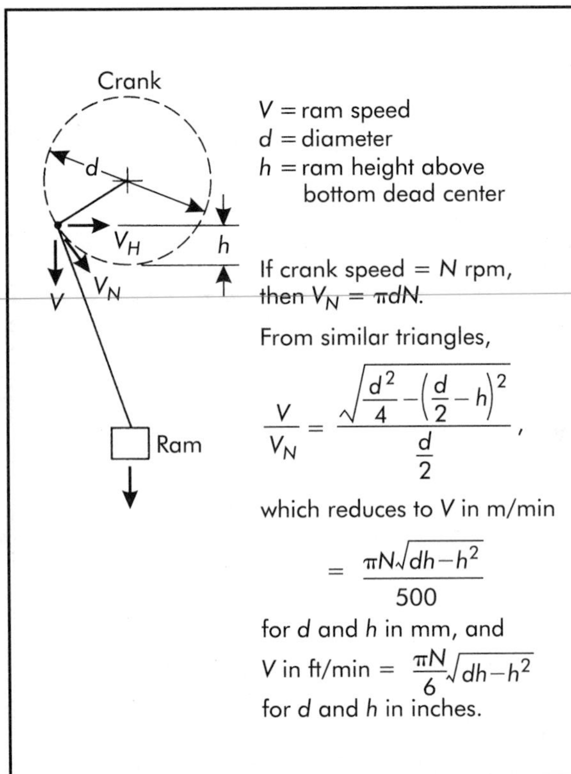

Figure 12-27. Formula for the speed of a crank-driven ram.

in or taken out of a die, the length of stroke should be at least twice the height of a piece. Where a press is used for many jobs, as most are, it should have capacity for the largest piece even though this is more than is needed for the majority.

12.6.4 Hydraulic Presses

A hydraulic press ram is actuated by oil pressure on a piston in a cylinder. Hydraulic presses are more versatile and easier to operate than mechanical presses. A hydraulic press usually has a long stroke and can deliver full-rated force over the entire length of stroke. The length of stroke and the force can be quickly changed. Force and speed can be held uniform or readily varied throughout the stroke. For example, a rapid approach and return can be combined with a slow, steady working stroke. Pressure and distances can be recorded and readily repeated when a job is rerun. A hydraulic press is safer because it will stop at a pressure setting, whereas a mechanical press must plunge through its stroke, even to destruction if it has no safety release.

The main disadvantage of hydraulic presses is that to make them as speedy as mechanical presses is costly. Short-stroke mechanical presses are running in production at speeds up to 2,000 strokes/min. Hydraulic presses of that sort have not been found generally feasible. For slow and long stroke work, like deep drawing with a 8.9-MN (1,000-ton) press and a 838-mm (33-in.) stroke at 14 strokes/min, hydraulic presses are available that operate as fast as mechanical presses and are more fully controllable, but they are more costly.

A hydraulic press must have a motor about twice as large because it cannot store energy in a flywheel during part of a stroke as does a mechanical press. As a result, most presses used for quantity production are mechanical. An exact comparison is difficult, but it has been said that a mechanical press for $100,000 is comparable for many production purposes to a hydraulic press costing $160,000. The extra cost is justified in some cases because the advantages offered by the hydraulic press are needed. On the other hand, if speed is not necessary, a hydraulic press can be built for many purposes more cheaply than a mechanical press.

12.6.5 Power Transmission

A crank or other mechanical means of driving a press ram is actuated by a flywheel through a clutch that is engaged by tripping the starting lever. If the lever is released during the stroke, the clutch is disengaged, and the crankshaft is grabbed at its uppermost position by a brake. The brake is often on the opposite end of the crankshaft from the flywheel as on the large press shown in Figure 12-24.

Many small presses have positive mechanical clutches, but some small presses and all over about 889 kN (100 ton) have friction clutches. Some large presses have electrical eddy-current clutches. A typical form of mechanical clutch has one pin, sometimes more, in a collar on the crankshaft. The pin is made to engage a hole or slot in the hub of the flywheel. Friction clutches are of the disk type and usually are air operated.

A mechanical clutch acts with a sudden shock that becomes more unendurable the larger the drive; a friction clutch is smoother. Once engaged, a positive clutch cannot be released until the end of the stroke, but a friction clutch promotes safety because it can be disengaged at any time. It can be set to slip under overload. A friction clutch costs more to make, operate (because it uses compressed air), and maintain.

The flywheel runs all the time a press is in operation and slows down to furnish the energy for each stroke. A flywheel running has a kinetic energy in joules of

$$E = MV^2/2 \qquad (12\text{-}16)$$

where

E = energy, J
M = mass, kg
V = velocity at radius of gyration, m/sec (ft/sec)

For energy in ft-lb,

$$E = WV^2/2a \qquad (12\text{-}17)$$

where

W = weight, lb
a = gravitational constant 32.2 ft/sec/sec

The more the flywheel slows down, the more energy it gives up, but more time is lost in the stroke. An economic balance must be found between the cost of the press and rapid production. Experience has indicated that this is realized if the flywheel slows down less than 10% during continuous operation. This entails a press stroke for every revolution, and as much as 20% for intermittent operation, with the operator loading each piece and then tripping the press.

According to Equations (12-16) and (12-17), for a 10% slowdown, the energy given up by the flywheel is

$$E_{10} = 0.19MV^2/2 = 0.19WV^2/2a$$

and for 20% slowdown

$$E_{20} = 0.36MV^2/2 = 0.36WV^2/2a$$

Most of the mass of an ordinary flywheel is in a heavy rim, and for practical purposes V may be taken as the speed at the mean radius of the rim. It is necessary when selecting a press to ascertain whether it will supply sufficient energy without slowing down excessively. This may have to be calculated in some cases. In others, the energy capacity is tied to the rating of the press. For instance, some manufacturers design their presses for an energy capacity in inch-tons for 15% slowdown to be approximately equal to the tonnage rating. Thus, such a 4 N (500 ton) press is able to deliver a little over 4 Nm (500 in.-ton) of energy at 15% slowdown.

Modern presses have individual motor drives. Standard motors made for constant speed applications, such as most machine tools, pumps, and generators, resist slowdown over about 5% by increasing torque and are called *stiff motors*. Such a motor is satisfactory for a press running continuously, for example, at over 40 strokes/min, but is likely to be badly overloaded on a press operated intermittently with a slowdown of as much as 20%. *Soft motors* are made for such purposes. They permit the flywheel to do most of the work and restore energy to the wheel in the relatively long period between strokes. A stiff motor would need a much higher rating to supply a major part of the work in part of one stroke. Space does not permit a discussion of motor design, but this points out some basic problems.

Various forms of power transmission between motor and crank- or eccentric-shaft are used on presses. The flywheel is usually mounted on the crankshaft of a small high-speed press and driven directly by the motor through gear teeth or belts. This is called a *direct drive* or *flywheel drive*. A large and slow press ordinarily has gear reductions between the flywheel and crankshaft. A *single-geared press* has one reduction; a *double-geared press* has two. A short crankshaft may have one drive gear on it, and that is called a *single drive*. A long crankshaft has a drive gear at both ends to reduce twist, and that is a *double drive*.

12.6.6 Applications

Most presses are mechanical because such drives are simple, durable, and fast, but certain applications require other types. Hydraulic presses provide long strokes for drawing and easy action for rubber pad forming and problematic forming and drawing. Drawing and forming have been done as a makeshift on drop hammers when presses have not been available, but this is too crude and slow for production.

Some squeezing operations, such as forging, are done using drop hammers, but mostly eccentric, knuckle joint, screw, and hydraulic presses are used. The last are finding favor because, through control of speeds, a large part of the time cycle of an operation can be devoted to the actual squeezing. This gives the metal time to flow.

Faster presses and automation of press lines have made presswork faster and faster. Presses that turn out from hundreds to tens of thousands of pieces per hour can produce large lots in short periods of time. They must, therefore, be changed over from job to job often, and the cost for setup is important. The trend is to provide means to shorten the die-change time on presses.

One system for small- and medium-size stamping presses has permanent fixtures attached to the ram and bolster plate of each press. The punches, die blocks, and other members are not mounted on die sets but on standard upper and lower die plates. These plates are precisely located from pins in the fixtures and are held by air clamps. Systems like this have made even very short runs feasible. It is reported that as many as six die setups with 10 pieces in each run have been made in 10 min.

Moveable and alternating bolster plates on small-, medium-, and large-size presses make it possible to set up a die on one plate while another is in operation in the press. Changeover is then mainly a matter of rolling one set out while the other goes into place. Quick-acting air clamps may attach the punch to the ram. Changeover is also aided by semi- or fully automatic slide positioners and feed length setters. Times for die changes on large presses have been reduced from 2–3 hr to 10–15 min.

12.6.7 Specialized Presses

Some presses have features for certain kinds of work but operate on the same principles as other presses. One of these is a *trimming press,* a crank-operated single-action press with dies to trim forgings and die castings. It has an auxiliary ram on the side of the frame for light punching, sprue cutting, etc.

Some large presses have the power transmission and drive mechanism under the bed where it is more accessible than on top. The ram on top is pulled from below. Such an *underdrive press* is mounted in a factory with the top of the bed near floor level and the mechanism extending into the basement.

A *dieing machine* is a small- to medium-size press with most of its mechanism below the die surface. The ram is above the working surface, is carried on four posts, and is pulled rather than pushed down. This helps to stabilize the ram under load, maintain alignments, and preserve the tools. The working area can be accessed from all four directions. Its low center of gravity is favorable to high-speed operation.

A *forcing press* is usually a hydraulic press capable of exerting large forces to press car wheels, gears, bushings, etc., on or off of mating parts. Some are horizontal, others vertical.

12.6.8 Press Brakes and Shears

A *press brake* is a wide press with a relatively thin bed, platen, and ram as shown on the machine in Figure 12-28(A). Generally, a press brake is used to bend a flat metal plate along a straight line. The bending process changes the shape of a single-plane workpiece into one with two or more planes. As illustrated in Figure 12-28(B), a flat plate is bent to an angle by forcing a formed punch into a corresponding die opening. The plate is bent beyond its elastic limit to the extent that it retains the angular shape imparted by the punch and die, except for a certain amount of springback. The process can be accomplished by either air bending, bottom bending, or coining, depending on the required sharpness of the bend radius.

As illustrated in Figure 12-28(B), *air bending* occurs when the metal being bent only comes in contact with the tip of the punch and edges of the die. For this operation, the punch stroke is adjusted to move down into the die opening just far enough to produce the appropriate bend angle. To accommodate springback, the included angle on the punch and die are normally less than the specified bend angle on the part. The air bending process requires the least amount of force of the three methods since only bending is involved.

Bottom bending, illustrated in Figure 12-28(C), occurs when the metal is forced to conform to the shape of the punch and die as they strike solidly and squeeze the metal at the bottom of the stroke. The force requirements for this process are usually 3–5 times that required for air bending because of the squeezing action

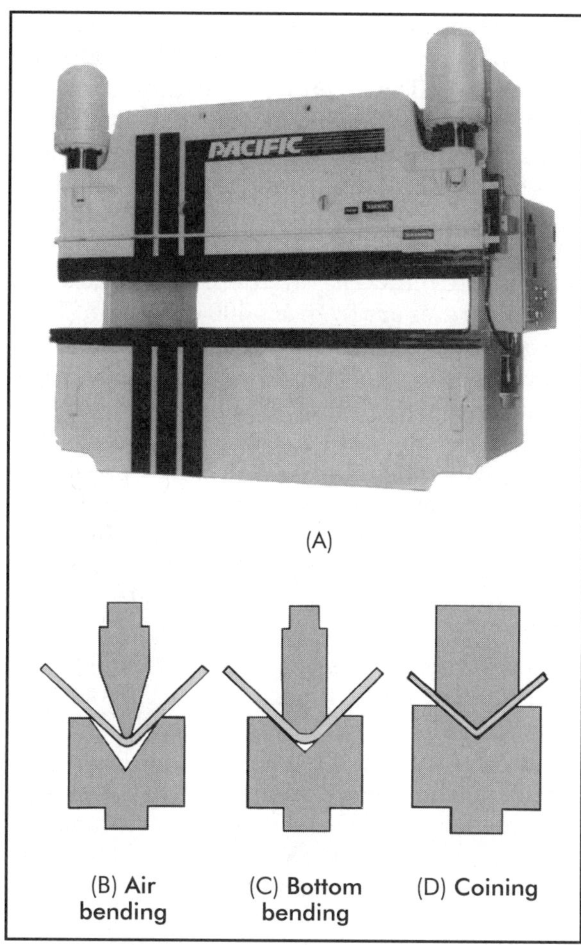

Figure 12-28. (A) 1.8-N (200-ton) capacity hydraulic press brake; (B) air bending; (C) bottom bending; (D) coining. (Courtesy Pacific Press & Shear, Inc.)

at the end of the stroke. Some springback allowance still needs to be included in the punch and die design for bottom bending, since the squeezing action is usually not sufficient to give high-strength metals a permanent set.

In press brake *coining*, the metal is squeezed to conform to the radii of the nose of the punch and the V of the die, as depicted in Figure 12-28(D). No springback is usually required for coining since the squeezing action is sufficient to set the material. Force requirements for coining may be 5–10 times that required for air bending if only extremely small bend radii are permitted.

Charts and tables indicating the tonnage requirements for air bending various materials with different die openings are available from press brake manufacturers. For the most part, the tonnage is proportional to the tensile strength of the material being bent.

Press brakes are available with several types of drives, depending on the application requirements. As with other types of presses (see Section 12.6.3), two primary types are common, mechanical or hydraulic. The main difference is that the eccentric drive mechanical press delivers maximum capacity at the bottom of the stroke while the hydraulic press is capable of delivering full tonnage anywhere within the length of the stroke.

The unit shown in Figure 12-28(A) is a 1.8-N (200-ton) capacity hydraulic-drive press brake with a 3.7 m (12 ft) ram and platen length. The length of stroke for the ram is 305 mm (12 in.) and the platen width is 203 mm (8 in.). Stroke length and ram speed change can be adjusted quickly by the operator by means of slide controls mounted on the right end of the ram. The bottom stroke setting can be adjusted to within 0.025 mm (.001 in.) by means of a micrometer-type adjusting screw. The top stroke setting is also completely adjustable to allow short stroke, high-speed operation. This press also has a ram tilt adjustment to facilitate fade-out work and to compensate for die variations. The ram may be tilted to a maximum of 25.4 mm in 3 m (1 in. in 10 ft) of ram length in either direction by simple adjustment of an automatic level control system.

The ram of the press shown in Figure 12-28(A) is powered by two hydraulic cylinders synchronized to maintain a level ram to within 25 μm (.001 in.), regardless of off-center loading. This press is equipped with a programmable logic controller (PLC) to control the sequence and timing cycle of all press operating modes. As an option, these types of presses may be equipped with a computer numerical control (CNC) system that controls up to two ram axes and five back gage axes. The basic price of a press of the type shown in Figure 12-28(A) is about $95,600. The CNC option would add another $23,000 to that cost.

A *shearing press* or *squaring shear* is like a press brake, but cuts wide sheets with a long blade or knife on its ram or slide. This upper blade is inclined at an angle to make a gradual cut through a sheet of metal across a lower blade on the bed.

A plate with hold-down fingers alongside the blades presses down on the work as it is cut.

12.6.9 Hole-punching Machines

On a conventional press, all the holes in a workpiece are punched at the same time with a specialized punch and die set for the particular job. This tooling is quite expensive, and many pieces must be produced to pay for it, but the press time per piece is very small. In contrast, on a hole-punching machine, all the holes in one piece are punched in sequence one at a time, and line shearing is done by punching a series of overlapping slots. Thus, press time per piece can be appreciable, but tooling cost is low. A punch and die module unit is required for each size and shape of hole, but these are relatively inexpensive and can often be used repeatedly for many jobs. To locate the holes accurately accounts for much of the high cost of a multi-hole die, whereas the spacing is taken care of by the movements on a hole-punching machine. Hole-punching machine operations are economical for small and moderate quantities of pieces.

Typically, on a hole-punching machine, the workpiece (a sheet of metal) is moved about on a large table and positioned as required under the punch for each hole. For this purpose, the workpiece is commonly held by mechanical fingers from a cross slide mounted on a crossbar that slides along the table. Numerical control, described in Chapter 31, has been found particularly suitable for work positioning on hole-punching machines. It is fast and does away with the need for templates that were used on older, manually operated machines. The machine illustrated in Figure 12-29(A) is computer numerically controlled (CNC), with a program to run each operation.

A different approach is offered by one press builder. The positions of holes to be punched are marked with cross lines on the workpiece. An electronic optical system aligns each hole-center mark, in turn, under the punch, again by numerical control.

Since a number of holes need to be punched in each workpiece and a different punch and die unit is required for each hole size and shape, an appreciable amount of time could be spent in changing tool units. The press illustrated in Figure 12-29(A) has 60 punch stations that can be tooled with a variety of punches. These are automatically selected and placed into position, under computer control, as needed.

Most single-station and turret presses have C-frames with a single workstation. The machine in Figure 12-29(A) carries the punches in a turret. The workpiece is automatically positioned for punching by the CNC worktable. Punching rates are from 50–500 holes/min. Turret rotation to select a tool may take place during workpiece positioning, especially with computer control, but otherwise may take 5–15 sec. A more advanced turret punch press uses a servo-controlled hydraulic ram to provide a programmable punching profile, which enhances forming and increases punching flexibility.

For typical machines of 214 kN (24 ton) capacity, the price of a manual single-station model is about $25,000; a manual turret press costs about $60,000; and a CNC turret press costs about $160,000.

A more advanced 311 kN (35 ton), 54-station turret press with CNC has an auxiliary 0.4 J (1,500-Wh) carbon dioxide CO-type laser (described in Chapter 13) to enable it to cut practically any material. Nearly any shape can be cut in a sheet guided by the CNC system. The total equipment cost is about $400,000.

12.6.10 High-production Presses

Certain gap-type and straight-side presses are known as *high-productivity presses*. They look like ordinary presses, but have features that enable them to produce faster and they cost about 50% more. Typically, their stroke is short and speed high. Ordinary presses operate up to about 200 strokes/min; above that speed dynamic balancing is necessary. Presses rated from about 222–3,559 kN (25–400 tons) are available to deliver 200–600 strokes/min. Presses with speeds over 600 strokes/min are not common but do exist.

High-productivity presses are made with exceptional rigidity and alignment to yield long die lives, especially with multiple station dies. They commonly have variable speed drives that can be set to the right speed for longest die life and best part quality. They are fitted with automatic controls and quick-acting safety controls, and

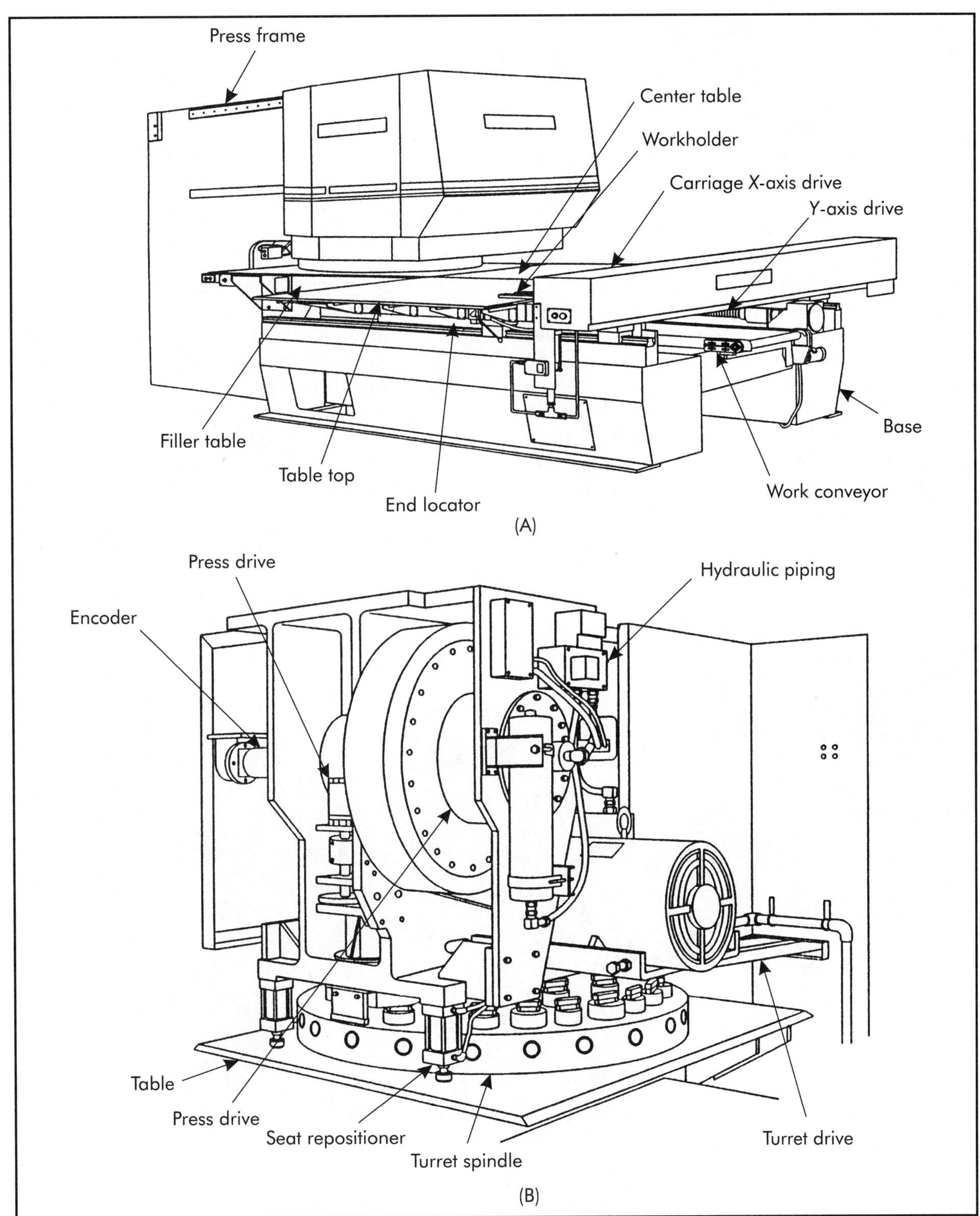

Figure 12-29. (A) A 400-kN (45-ton) turret punch press with 60 punch stations and a CNC worktable. (B) Open view of the turret punch for the machine shown in (A). (Courtesy Murata Wiedemann, Inc.)

often one can do the work of two or three standard presses.

Small sheet metal and wire objects, such as clips, lugs, brackets, eyelets, fasteners, and bottle caps, are turned out in large quantities on presses that look somewhat different from the ordinary kinds. All embody one or both of two principles. One is that there is a series of stations, each an individual press unit with one or more slides. Each station is tooled to perform one or more operations on the work, which is carried automatically from station to station. With some presses, the pieces go through individually; with others, they are kept in a strip and cut off at the end. The other principle is to provide several slides at one station. The slides converge in timed order on a workpiece to make intricate bends or forms. The Multi-Slide® press shown in Figure 12-30 utililizes both principles to form small parts from either strip or wire. The series of diagrams in Figure 12-31 illustrates typical tooling used and operations performed on a Multi-Slide press.

Figure 12-31(A) shows a completed part and the strip stock after it has been pierced in a die head and cut to length. The tooling and sequence of operations used to bend the strip stock around forming mandrels to complete the part are shown in (B) and (C). A Multi-Slide press of the size and capacity shown in Figure 12-30 costs about $125,000 without tooling.

Transfer presses are often used to perform successive press operations in several stages on one machine. Rather than accomplishing each operation separately on several individual single-stage presses, in many cases, a part can be produced on a single multiple-station transfer press. With such a machine, a part is partially formed at each station and is moved from one station to another by means of a transfer mechanism built into the press. The 11-station transfer press shown in Figure 12-32(A) is tooled to perform a

Figure 12-30. High-production Multi-Slide® automatic press with four forming slides and four die heads. This machine has a maximum length of stock feed of 203 mm (8 in.) for stock up to 38 mm (1.5 in.) width and 1.588 mm (.0625 in.) thickness. (Courtesy U. S. Baird Corporation)

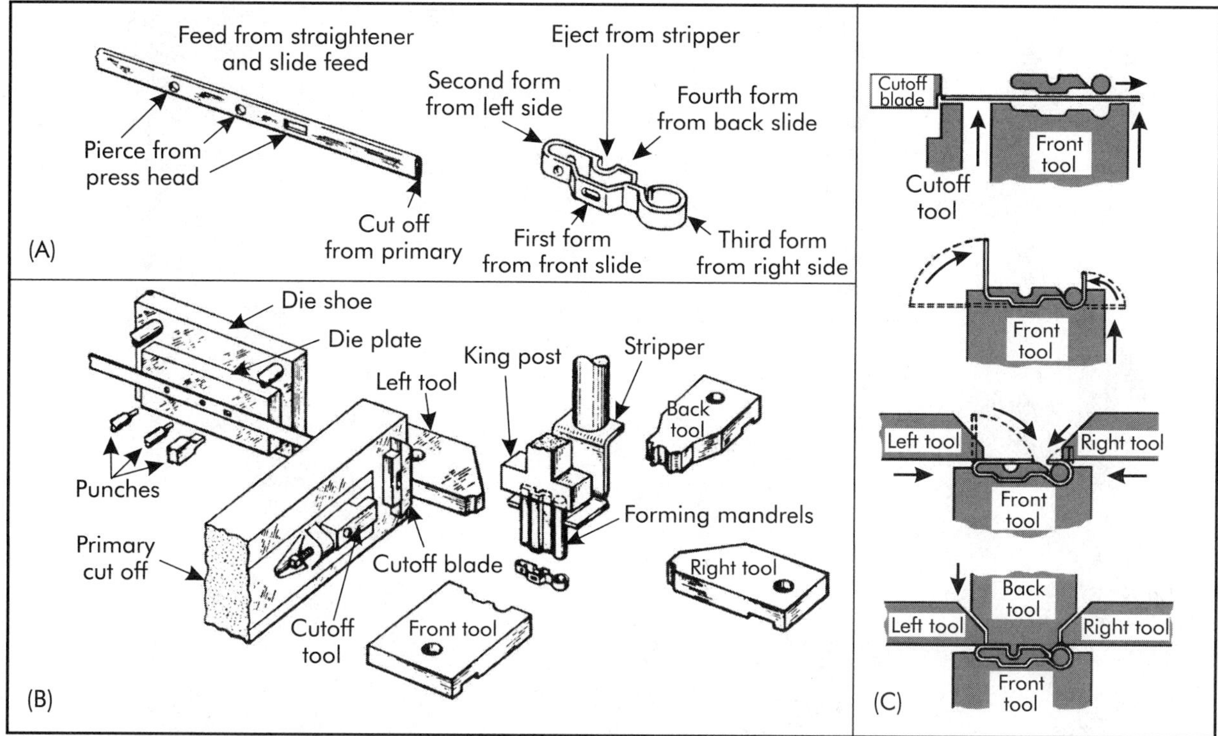

Figure 12-31. Example of forming operations used to transform coiled strip stock into a small spring clip on a high-production Multi-Slide forming press. (Courtesy U. S. Baird Corporation)

series of 11 operations on the needle bearing case shown in Figure 12-32(B). The first station blanks out a circular disk from the strip stock of 1.19-mm (.047-in.) thick steel that is fed into the press at the left front of the machine. Once all of the stations are filled, the machine will produce a finished needle bearing case on each stroke.

One of the features of the press shown in Figure 12-32(A) is the multiple-ram design, which minimizes the out-of-balance deflection that can occur in single-ram presses due to uneven loading. In addition, the 11 rams are designed to operate 7° out of phase to reduce the maximum load on the press frame. An 11-station transfer press similar to the one shown in Figure 12-32(A) costs about $750,000 and the tooling for the needle bearing case would add another $60,000. A production rate of 30–40 cases/min could be sustained on a setup of that nature.

Transfer presses produce at a very high rate and can be economically justified if several operations need to be performed on a part and production requirements are high.

12.7 PRESS TOOLS AND ACCESSORIES

12.7.1 Dies

Sheet-metal cutting and forming are done to the shape of a punch (external) or die block (internal) or with a mating punch and die block. Dies may be made of soft metal, plaster, or plastic, particularly for forming only a few pieces, but are of hardened tool steel and even cemented carbides for larger quantities.

A punch may be fastened to the ram, and a die block to the bolster plate of a press. This is common practice and saves die cost when very few pieces are to be made, but it is slow to set up and operate. Additional details are added for production to facilitate operations. The assembly of all the details is called a *die*, and the same name also refers to the die block at times.

For many pressworking operations, a die set similar to that shown in Figure 12-33(A) is purchased commercially and then die components, such as those shown in Figure 12-33(B), are in-

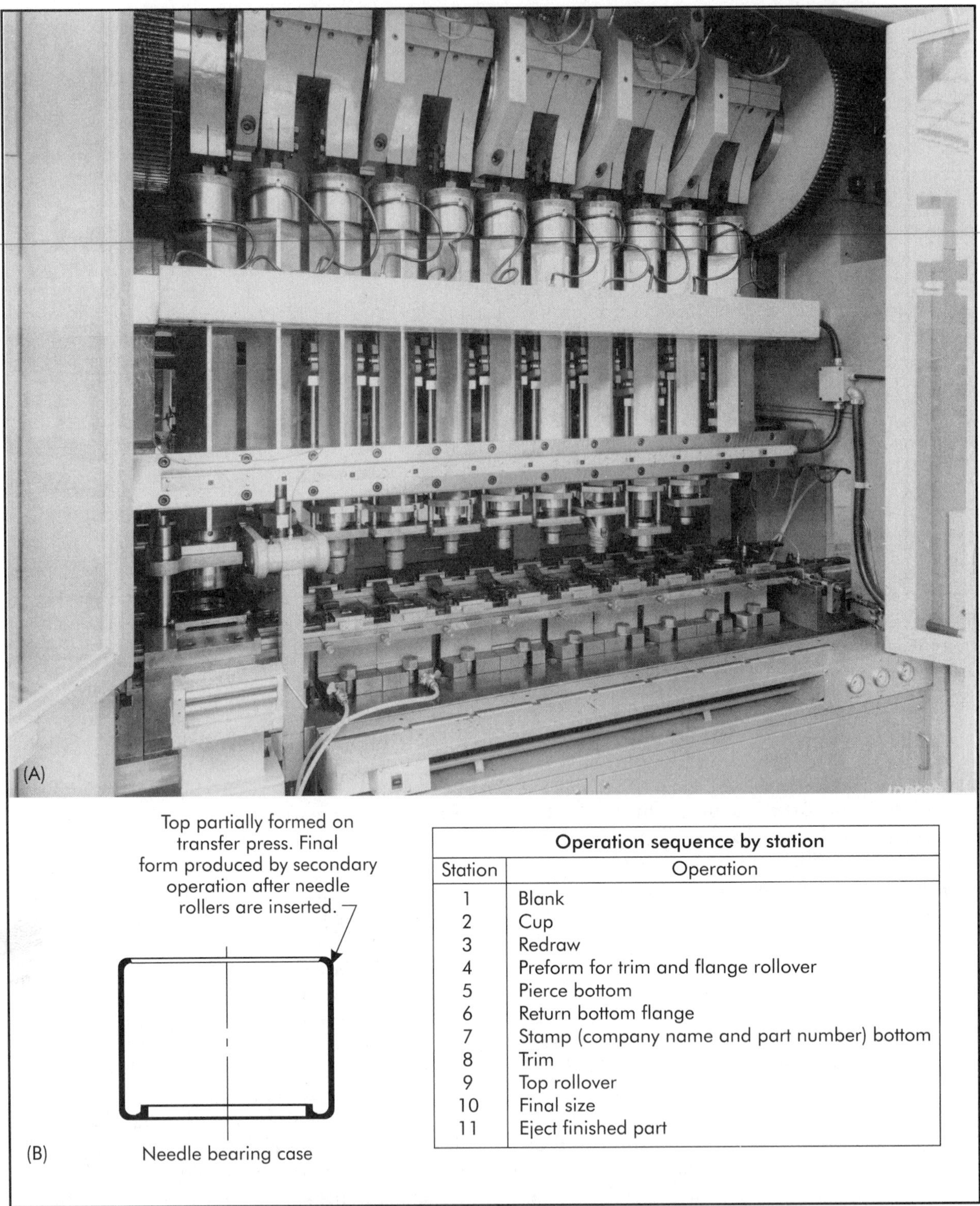

Operation sequence by station	
Station	Operation
1	Blank
2	Cup
3	Redraw
4	Preform for trim and flange rollover
5	Pierce bottom
6	Return bottom flange
7	Stamp (company name and part number) bottom
8	Trim
9	Top rollover
10	Final size
11	Eject finished part

Figure 12-32. (A) Close-up view (guards removed) of the die area of an 11-station transfer press, rated at 356 kN (40 ton) capacity for each station. The press is tooled to produce a needle bearing case from a 1.19-mm (.047-in.) thick steel strip. (B) Final part configuration and operation sequence performed at each station. (Courtesy Platarg Machinery, Inc.)

stalled as needed for a particular press operation. The die set consists of a *punch holder* on top and a *die holder* or *die shoe* on the bottom for attachment to the bolster plate of a press. Heavy *leader pins* or *guide posts* between the top and bottom of the die set maintain alignment between the punches and the die block.

When sheet metal is blanked or pierced, it grips the punch after the breakthrough. Pieces drawn or bent tend to cling to punches and stick in die openings. One variety of details added to dies has the purpose of stripping pieces from punches or knocking pieces from die openings to help remove work and scrap and speed up operations. These details are called by descriptive names such as stripper plate and knockout pad. Also, pressure plates or pads of similar construction are found in drawing dies to prevent wrinkling. Details in dies for locating workpieces may be in the form of stops, nests, or pins.

Most dies are special tools made specifically for one job. Many standard details, such as die sets, stops, and punches for common shapes, are available and widely used for economical construction. Even so, costs are high and run from several hundred dollars for even a simple die to tens of thousands of dollars for large automobile body dies. Several commercial systems are available for constructing dies from standard reusable units for blanking, piercing, and notching simple common shapes. One of these is illustrated in Figure 12-34.

Many dies are made to do single operations. Such a die may be designated as a blanking die, piercing die, drawing die, etc., according to its purpose. When the quantity of pieces to be made warrants the cost, a die is constructed to perform a number of operations on a piece. One such type is called a *progressive die*, like the one illustrated in Figure 12-35. A progressive die has a series of stations, each performing an operation on a piece during a stroke of the press. Between strokes, the pieces, usually in a strip, are transferred to their next stations, and one piece is finished for each stroke. Combination and compound dies do two or more operations in a single station. A *combination die*, like the one in Figure 12-36, does simultaneous operations of different kinds, such as cutting and drawing. A *compound die* is like a combination die, but does operations of the same kind, such as blanking and piercing.

12.7.2 Stock-feeding Devices

A variety of commercial devices are available for feeding presses. A stock feeder usually is economical for any job of more than a few thousand pieces. Two general classes are devices for feeding strips and those for feeding individual pieces.

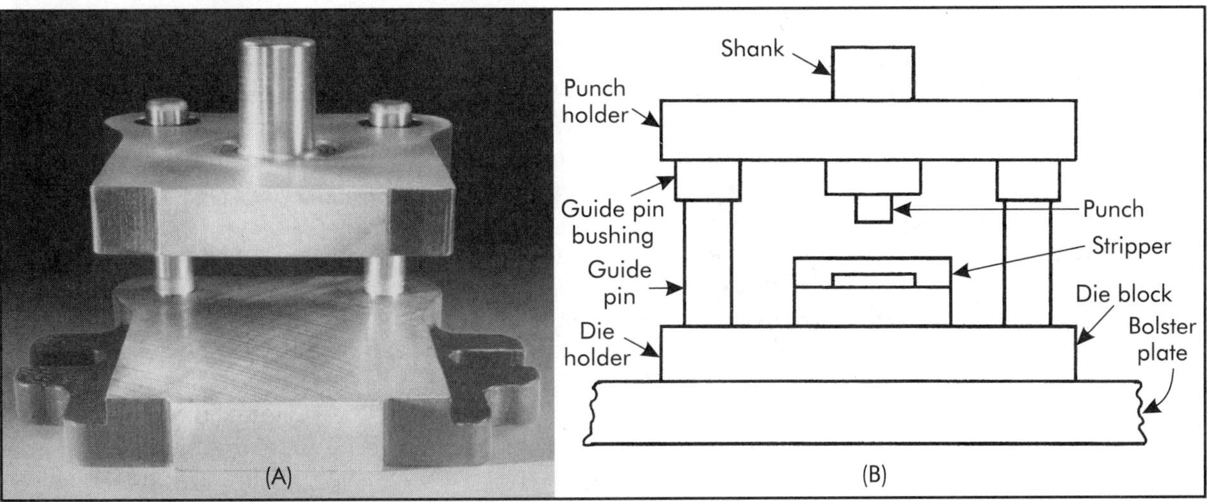

Figure 12-33. (A) Commercially available die set, and (B) simple die assembly showing common components. (Courtesy Danly Die Set, Inc.)

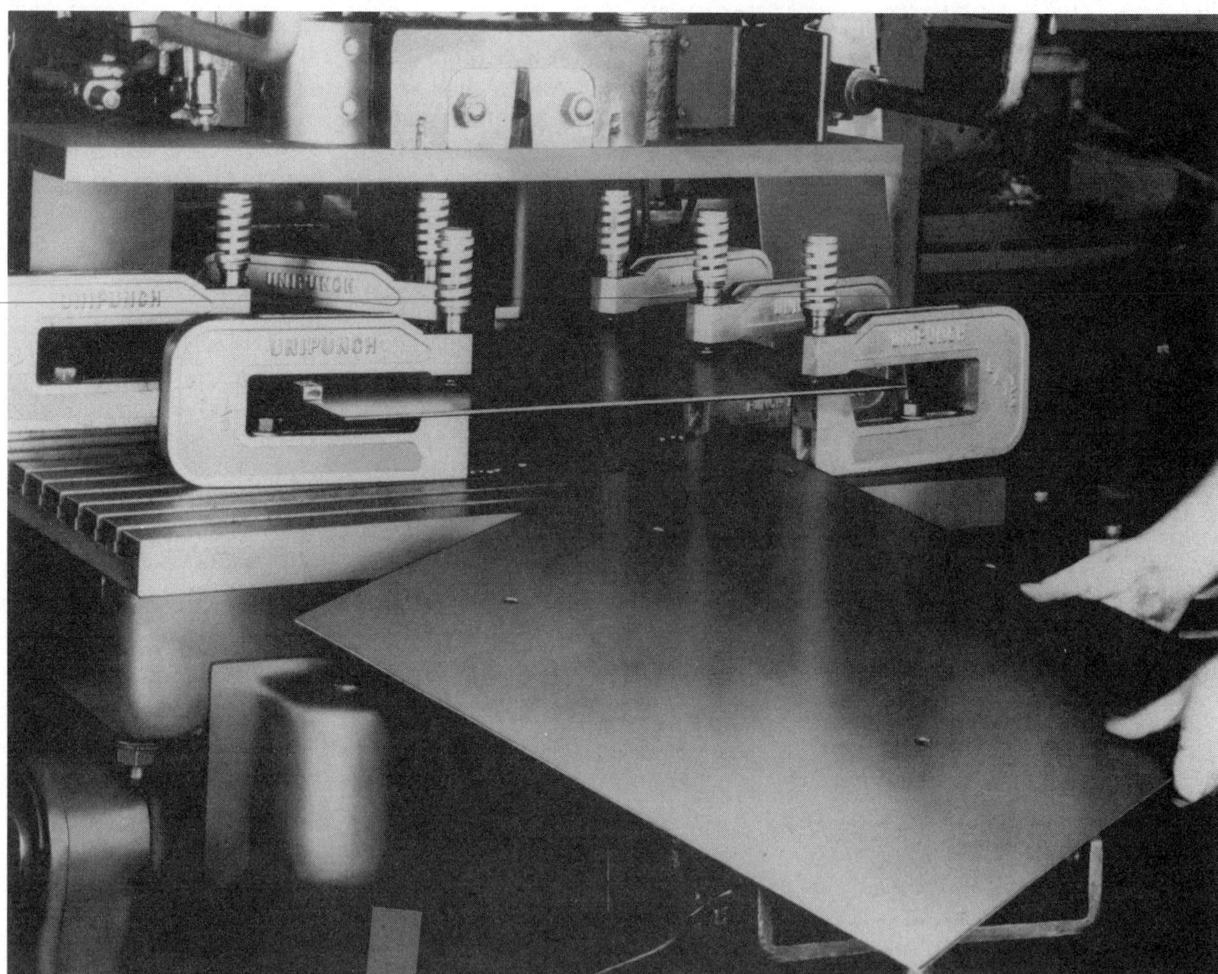

Figure 12-34. Setup of standard commercial hole-punching units on a press bed. These are quickly positioned by means of a template and then fastened in place. The punching units are actuated by the press ram when it descends. (Courtesy Unipunch Products, Inc.)

A typical attachment for feeding strips and coiled stock is the *roller feed unit* shown in Figure 12-37. The rolls are actuated through a linkage from an eccentric on the crankshaft. They release the stock when the ram descends but grip and draw the stock along as the ram rises. Adjustments can be made for the amount of feed and length and thickness of stock. Sometimes two sets of rolls are used: one to push the stock into and the other to pull it out of the die. Coiled stock is held in reels or cradles and may be passed through a straightener. Another device is the *hitch feed*, which grips the stock between blocks or shoes and pulls it along and then releases it and returns to start the next step.

A device for feeding pieces one at a time is the *dial station feed*. A round table with several equally spaced stations around it is indexed through an eccentric on the crankshaft. For each stroke, a station holding a piece is positioned under the punch. The stations are unloaded and loaded on the side away from the punch. A few among many other devices are mechanical hands, arms with suction cups, and magazines that mechanically push one piece at a time into the die.

For multiple operations, small parts are moved from one station to another in a transfer press or progressive die. Large parts, such as automobile body panels, are transferred from press to press. This is done manually for small

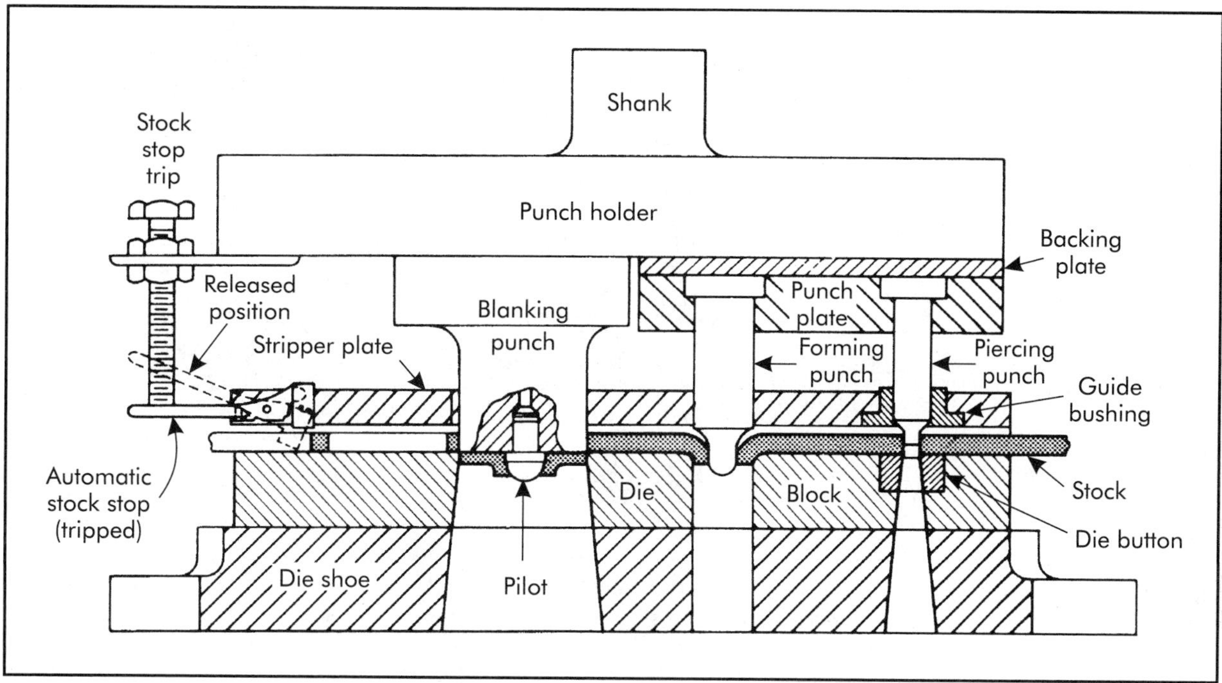

Figure 12-35. A progressive die for blanking, piercing, and forming a piece made in a strip.

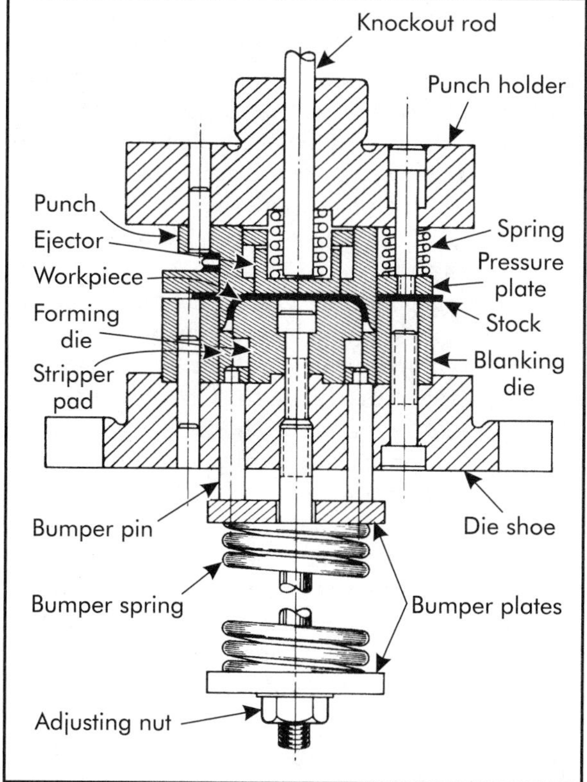

Figure 12-36. Combination blanking and drawing die.

quantities and by synchronized walking-beam devices, mechanical hands, conveyors, and robots (Chapter 30) for large quantities. Such devices can be integrated with presses into fully automated straight-line transfer machines (Chapter 30) for large quantities. A manual press line putting out 12–20 pieces/min can be made to yield 30 pieces/min by automation. A line of four presses to produce appliance panels might cost $1,000,000 alone, and fully automated as much as $1,500,000 plus another $400,000 for dies. One authority has stated that such an automated line must turn out at least 1,000,000 pieces per year to be justified.

12.7.3 Safety Devices

A noted authority once said: "When one considers that 97% of all press accidents can be traced to human failure and only 3% can be attributed to any mechanical failure, it is evident that by eliminating the human element you have gone a long way toward a solution to the problem." This places the responsibility on the manufacturing engineer to provide the means to make notoriously dangerous press operations as safe as possible.

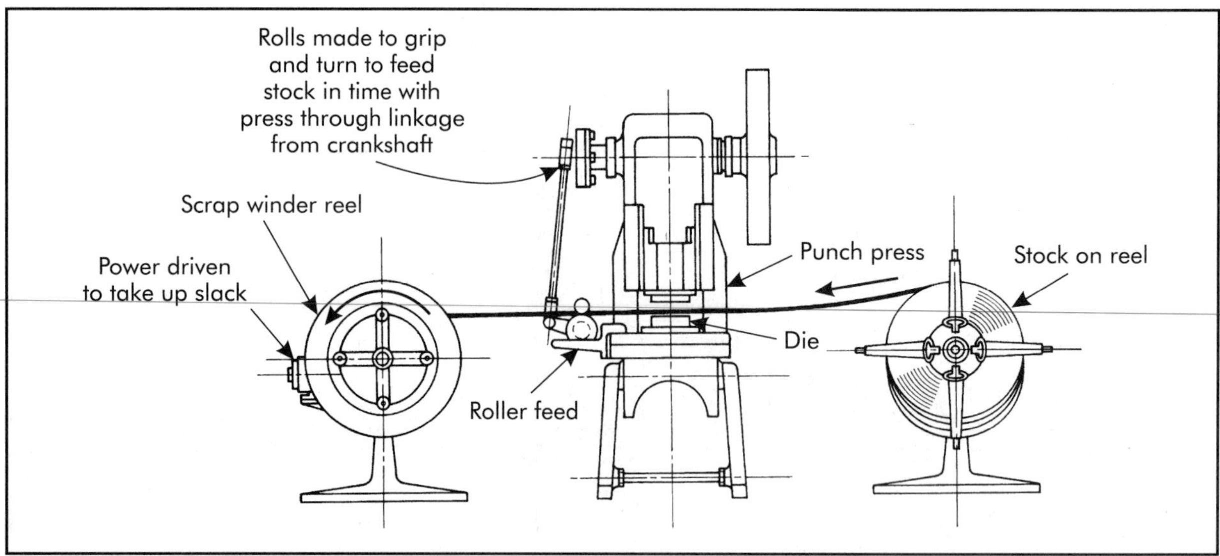

Figure 12-37. Roller feed device pulling strip stock off a reel and through a die. The scrap is wound on a power scrap winder. The parts being made are punched out of the strip as it passes through the die.

One of the best safeguards is to take the operator away from the dangerous task. This is an important benefit derived from the use of automatic feeding devices. In addition, an emphasis on the automation of the press room, including secondary operations, should contribute to a significant decline in press room accidents. However, many press operations still require an operator, and steps must be taken to insure the safety of that person. In 1970, a complete and far-reaching mandate for the safety of pressworking personnel, as well as personnel involved in other hazardous industrial operations, was given by the passage of the Federal Williams-Steiger Act (Public Law 91-596). This Act cited detailed and exacting safety standards and established the Occupational Safety and Health Administration (OHSA) to enforce those standards.

Part 1910.217 of OSHA cited in Title 29 of the Code of Federal Regulations (CFR) cites the general requirements for *point-of-operation guards* and certain performance aspects relative to safeguarding the point-of-operation of mechanical power presses. In essence, these general requirements are directed to the employer since the installation and adjustment of these safeguards is the responsibility of the employer.

Basically, two types of devices used for point-of-operation guarding are described in the OSHA standards, the *barrier guard* and the *presence-sensing device.* Essentially, barrier guards are designed to prevent the entry of hands or fingers into the point of operation of a press by reaching through, over, under, or around the guard. The presence-sensing device is designed to prevent or stop movement of the press slide if the operator's hand or other body part is within the sensing field of the device during the downstroke of the press slide.

Most barrier guards are of the fixed, adjustable/moveable or interlocked design and each must be in compliance with a number of features and characteristics cited in the OHSA standards. The moveable barrier guard shown in Figure 12-38 represents a common type of point-of-operation guard for small punch presses. In operation, the operator loads a part in the die and initiates the press cycle by means of a foot switch. This releases a transparent barrier (gate) that drops down in front of the die area prior to the downstroke of the press. Mechanical interlocks prevent the press from operating if the operator's hands or fingers impede full gate closure. On the return stroke of the press, the gate is raised by air pressure to permit the operator to remove the finished part from the die area.

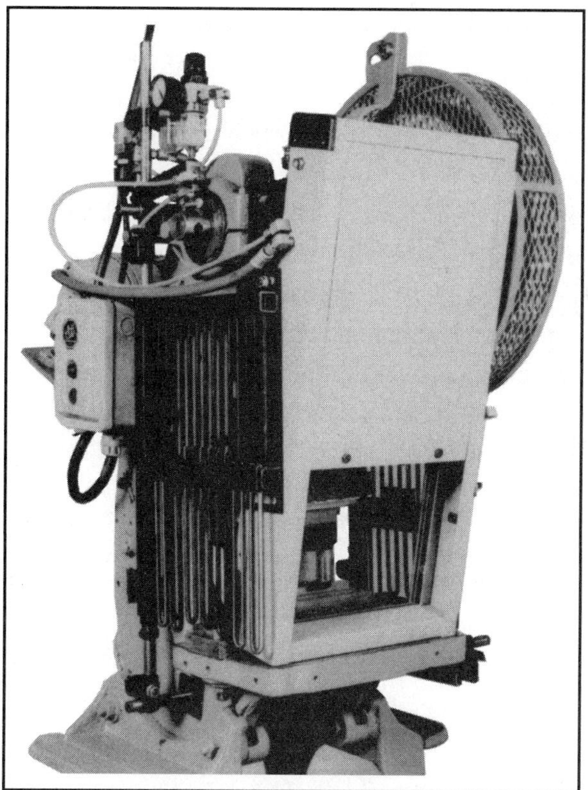

Figure 12-38. Punch press with a movable, point-of-operation guard. (Courtesy Gate Devices, Inc.)

The infrared light curtain presence-sensing device illustrated in Figure 12-39 has an infrared emitter pylon on one side of the press and a detector pylon on the other. With 19-mm (.75-in.) beam spacing, a very high scanning frequency, and microprocessor-based signal analysis capability, the device can rapidly detect a finger or hand intrusion in the point-of-operation area and stop the action of the press slide in a matter of milliseconds.

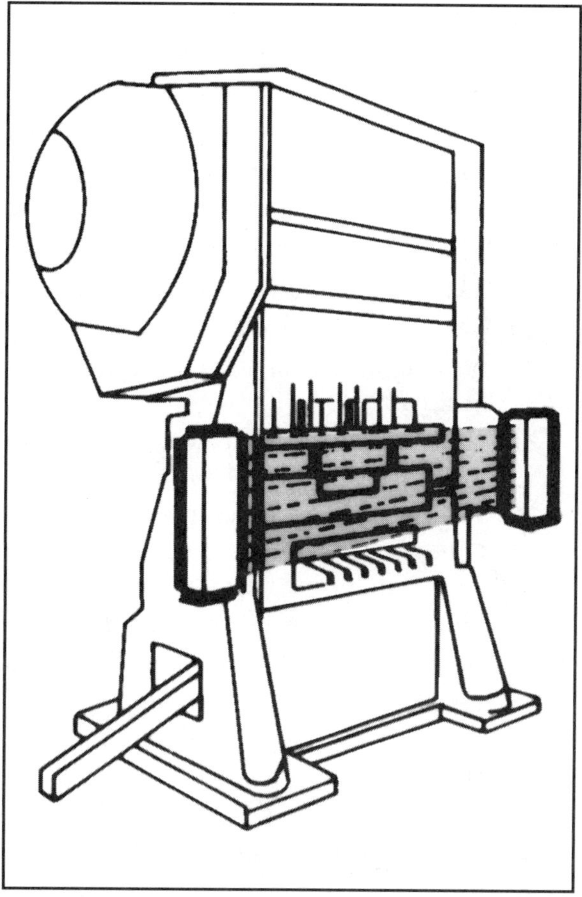

Figure 12-39. Illustration of a mechanical punch press with an infrared light curtain, presence-sensing device installed in front of the die area. (Courtesy Triad Controls, Inc.)

In addition to the point-of-operation barriers and presence-sensing devices, many presses are equipped with two-hand starting buttons that require application of both of the operators hands at a safe distance from the press.

12.8 QUESTIONS

1. What is the difference between piercing and blanking?
2. Describe what happens when sheet metal is sheared.
3. Why are punches and dies often mounted on independent die sets?
4. What is the purpose of putting a shear angle on a punch or die?
5. List and describe the 14 shearing operations discussed in Section 12.2.
6. Why is nibbling often used in sheet metal operations?

7. Why is it necessary to provide a certain amount of clearance between the punch and die opening in a piercing or blanking operation?
8. Why does springback occur in bending operations?
9. What limitations must be observed to avoid cracking sheet metal when bending it?
10. What happens to the neutral axis of a piece of metal while it is being bent?
11. Describe and compare stretch, draw, compression, press, roll extrusion, and roll bending.
12. What is "cold-roll forming" and when is it economically feasible to use this process?
13. What changes in metal thickness take place when a flat plate is drawn into a cup?
14. What is the purpose of a pressure pad or blank holder in drawing a cup?
15. What limits the amount a cup can be drawn on one operation and altogether before annealing?
16. What happens in a drawing operation when the ultimate tensile strength of the metal being drawn is exceeded?
17. What is the major difference between the flexible die-forming process and rigid die forming?
18. Compare the tolerance capability of the flexible die drawing and forming processes with those obtained in rigid die forming.
19. What is "metal spinning" and why is it used as an alternative to press drawing?
20. What is the difference between metal spinning and roll turning?
21. What is the difference between stretch forming and stretch draw forming?
22. What is "high-energy-rate forming" and how is it done? What are the main advantages of HERF?
23. List and describe the nine squeezing operations discussed in Section 12.5.
24. Why is it not feasible to cold head metals that work harden quickly?
25. What is the difference between coining and embossing?
26. Describe the rotary swaging process.
27. Describe the process of hubbing a mold cavity for use as a plastic or die-casting mold.
28. Explain the difference between riveting and staking.
29. What is the most common press frame type and what are its advantages and disadvantages?
30. What is the shut height of a press and what tooling limitations does this feature impose?
31. Describe the action of the ram or rams for a single-action press, a double-action press, and a triple-action press.
32. How do hydraulic drives compare with mechanical drives for presses?
33. In what position of the stroke is the capacity of a mechanical press rated?
34. Within what part of the stroke is the ram speed highest on a crank- or eccentric-drive press?
35. What types of work are hydraulic presses best suited for?
36. What is the difference between a knuckle joint and a toggle press and what is the purpose of each?
37. What is the difference between a press brake and a shearing press?
38. Describe a turret press and state its advantages.
39. Describe the action of a progressive die.
40. What is the difference between a combination die and a compound die?

41. Describe the action of a transfer press.
42. How does a Multi-Slide press differ from a regular punch press?

12.9 PROBLEMS

1. A 30 mm (≈1.181 in.) diameter hole is to be pierced in a 2.0-mm (.079-in.) thick sheet of AISI 1020 steel. Specify the diameters of the punch and die for this operation.
2. A round disk 127 mm (5 in.) in diameter is to be blanked without shear from 1.8-mm (.07-in.) thick stock of annealed 0.15% carbon steel. What blanking force and energy are required per stroke?
3. An irregularly shaped polygon with a perimeter of 254 mm (10 in.) is to be blanked from a 1-mm (.039-in.) thick sheet of cold-worked AISI 1040 steel with an ultimate shear strength of 517 MPa (75,000 psi) and 28% penetration. In addition, two 12.7 mm (.5 in.) square holes are to be pierced during the same stroke when the piece is blanked. Calculate the forces required for blanking and for piercing. What is the maximum force the press must exert at any one time without shear? What energy is required per stroke?
4. (a) Explain why the force in newtons (pounds) required to punch with shear is

   $$F_s = F \times p \times t/(pt + v)$$

   where

   F = force required without shear, N (lb)
   p = penetration of stock, mm (in.)
   v = amount of shear as designated in Figure 12-3(B), mm (in.)

   (b) What force is required to blank the disk described in Problem 2 with 3.05 mm (.12 in.) shear?
5. A piece of stock 3 mm (.118 in.) thick is bent to an angle $\alpha = 110°$ (Figure 12-6) with an inside radius of 7.14 mm (.281 in.). What is the original length of stock that goes into the bend?
6. How much should a low-carbon, soft steel piece be bent for a finished 100° angle bend?
7. A steel section 1.5 m (5 ft) in length has been bent in preproduction quantities on a press brake. It takes 0.8 min to bend each length, and labor and overhead on the brake are \$14.50/hr and \$3.50/hr, respectively. The part is going into production, and the purchase of a new roll-forming machine costing \$55,000 plus tooling at \$12,000 is being considered. This must be paid off in one year with an additional 25% for interest, insurance, and taxes. The new machine is expected to bend this section at a rate of 21.3 m/min (70 ft/min) with a labor charge of \$14.50/hr. What is the minimum annual order quantity required to justify the purchase of this new equipment if it is not expected to have any other use?
8. The thickness of the wall of a cup is assumed to be the same as that of the blank in Figure 12-10. Thus $t_1 = t_2 = t$.

   (a) If the bottom radius r of the cup is ignored, and the outside surface of the cup is assumed to have the same area as the blank, show that

   $$D = \sqrt{d^2 + 4dh}$$

   (b) In the case where the radius r in Figure 12-10 is large, t is small in comparison and may be neglected. Thus, the area of the outside corner at the bottom of the cup is a part of a toroid of minor radius r and may be approximated as

$A_t = \pi^2 r\ (d - 0.7r)/2$

If the outside area of the cup is assumed to have the same area as the blank, show that

$$D = \sqrt{(d-2r)^2 + 4d(h-r) + 2\pi r(d-0.7r)}$$

9. A cup 57 mm (2.25 in.) in diameter and 76 mm (3 in.) deep is to be drawn from 1.8-mm (.07-in.) thick drawing steel with a tensile strength of 331 MPa (48,000 psi). The corner radius is negligible.
   (a) What is the diameter of the blank?
   (b) What is the least number of drawing operations?
   (c) What force and energy must be applied for the first draw with a 40% reduction?
   (d) The reduction in area for this material in a tensile test is 70%. Is an annealing operation necessary?
10. A fluted cup 70 mm (2.75 in.) in diameter × 57 mm (2.25 in.) deep is to be drawn from 1.0-mm (.04-in.) thick soft aluminum. It can be drawn in one operation by Marforming at the rate of 0.40 min/piece. Marform tooling with a soft steel punch costs $600 and with a hardened steel punch $1,500. Two operations are needed with conventional steel dies, but each can produce 10 pieces/min. Soft steel dies cost $2,500 and hardened steel dies $3,000. Soft steel tools are not expected to last for more than about 5,000 pieces, but the hardened steel tools will serve for as many as will be needed. Setup time is about the same for either alternative. Manufacturing time costs $24.50 per hour. Which type of operation should be selected for each of the following quantities of pieces? (a) 1,000; (b) 5,000; (c) 10,000; (d) 25,000; (e) 100,000.
11. A bar of carbon steel with a compressive strength of 616 MPa (≈90,000 psi) and a diameter of 25 mm (≈1 in.) is to be rotary cold swaged to a diameter of 13 mm (≈.5 in.). The overall length of the finished piece is 250 mm (≈10 in.) and the length from the 25 mm (≈1 in.) diameter to the small end is 75 mm (≈3 in.). The average feed rate is 70 mm/sec (≈2.8 in./sec). The dies contact the workpiece over the 10° taper and 13 mm (≈.5 in.) beyond.
   (a) How long must the original piece of stock be?
   (b) How many passes are needed?
   (c) What is the maximum force required?
   (d) How much machining time is required for each piece?
12. For the conditions given in Problem 9, specify the kind of press needed, its tonnage requirements, stroke, speed, and power.
13. Show that the energy given up by a press flywheel (a) at 10% slowdown is $E_{10} = 0.19MV^2/2 = 0.19WV^2/2a$, (b) at 20% slowdown is $E_{20} = 0.36MV^2/2 = 0.36WV^2/2a$, as stated in this chapter.
14. A 534 kN (60-ton), open-back, inclinable press has a 550 kg (≈1,210 lb) flywheel that revolves at 90 rpm. The wheel has a 1.0 m (≈39 in.) OD and a rim thickness of 150 mm (≈6 in.). What energy is it able to furnish when it slows down (a) 10%? (b) 20%?

12.10 REFERENCES

ANSI B11.1. 1988. "Machine—Mechanical Power Presses—Safety Requirements for Construction, Care, and Use." New York: American National Standards Institute.

ANSI B11.2. 1982. "Hydraulic Power Presses—Safety Requirements for Construction, Care, and Use." New York: American National Standards Institute.

Avitzur, B. 1983. *Handbook of Metalforming Processes.* New York: Wiley Interscience.

Beard, T., ed. 1994. "The Case for the Hydraulic Punch Press." *Modern Machine Shop*, March.

Bethke, W. E. 1993. "Pressroom Automation for Secondary Operations." *Metalforming*, September.

Curry, D. T., ed. 1994. "What's New in Gap-frame Presses." *Metalforming*, December.

——. 1994. "The Science of Die Making." *Metalforming*, August.

Dobbins, D. B., ed. 1995. "Multi-slide Job Shop Handles the Hard Ones." *Metalforming*, August.

Douse, J. S. 1995. "In-die Tapping is Better than Ever." *Metalforming*, April.

Green, R. G. 1993. "Roll Forming: The Logical Choice for Linear Shapes." *Metalforming*, September.

Knight, D. 1992. "Three Ways to Eliminate Springback." *Modern Machine Shop*, March.

Kumlicka, R. 1991. "Examining Hydroforming as an Alternative to Conventional Deep Drawing." *Stamping Quarterly*, Summer.

Linders, D. "1994. Blanking Line Basics." *Metalforming*, August.

Meier, E. 1993. "Flow Forming Emerges as a Low-cost Manufacturing Alternative." *Metalforming,* July.

Mysicka, E. 1993. "Take Another Look at Fine-blanking Technology." *Metalforming*, August.

OSHA 1910.217. 1987. *Mechanical Power Presses—Requirements.* Washington, DC: Office of the Federal Register, National Archives and Records Administration.

Pacquin, J. R., and Crowley, R. E. 1987. *Die Design Fundamentals*. New York: Industrial Press.

Pietrzyk, A, and Titus, J. B. 1994. "Agile Controls Automate the Pressroom." *Metalforming*, October.

Rizzo, R. J. 1993. "Stretching Sheet Metal." *Modern Machine Shop*, August.

Schreiber, R. E. 1990. "The Power of the Punch Press." *Manufacturing Engineering*, September.

Smith, D., ed. 1990. *Die Design Handbook*, 3rd Ed. Dearborn, MI: Society of Manufacturing Engineers (SME).

Society of Manufacturing Engineers (SME). 1997. *Cutting Tool Materials* video. *Fundamental Manufacturing Processes* series. Dearborn, MI: SME.

——. 1997. *Cutting Tool Geometries* video. *Fundamental Manufacturing Processes* series. Dearborn, MI: SME.

——. 1997. *Sheet Metal Stamping Presses* video. *Fundamental Manufacturing Processes* series. Dearborn, MI: SME.

——. 1997. *Sheet Metal Shearing & Bending* video. *Fundamental Manufacturing Processes* series. Dearborn, MI: SME.

——. 1997. *Punch Presses* video. *Fundamental Manufacturing Processes* series. Dearborn, MI: SME.

——. 1997. *Sheet Metal Stamping Dies & Processes* video. *Fundamental Manufacturing Processes* series. Dearborn, MI: SME.

Tool and Manufacturing Engineers Handbook, 4th Ed. 1984. Vol. 1, *Forming*. Dearborn, MI: Society of Manufacturing Engineers (SME).

Welding Processes

13

Forge welding, circa 500 B.C.
—Glaukos, Khios Greece

13.1 INTRODUCTION

Although known and practiced in some forms before then, welding has mostly grown to its present importance in industry since World War I. It has supplanted riveting almost entirely for boilers, pressure vessels, tanks, and structural members of bridges and buildings; it is the chief means of fastening panels and members together into automobile bodies; it has taken the place of castings for a large proportion of machine, jig, and fixture bases, bodies, and frames; and it has become the means of joining at least some of the parts in most products manufactured today. A survey showed that among metalworking plants, 99% did some sort of welding.

13.2 COMMON WELDING PROCESS CLASSIFICATIONS

Welding is a means of joining metals by concentrating heat or pressure or both at the joint to cause coalescence of the adjoining areas. A good weld is as strong as the parent metal. Welding is done in a number of ways. The common processes of commercial importance are listed in Figure 13-1. In one major class of processes, metal is melted at the joint and filler may be added. *Fusion* takes place and no pressure is needed. For a *homogeneous joint*, the metal added is the same as the base metal. For a *heterogeneous joint*, the base metal is not melted. Another class of processes depends only on pressing the pieces together at the joint. The metal is usually heated locally to a plastic state, but adherence of metal can be enforced with pressure alone under favorable conditions.

Electric-arc, gas torch, energy-ray, and Thermit® welding are fusion processes wherein the filler metal is essentially the same as the parent metal in the parts being joined. Braze welding, brazing, and soldering use filler metals different from and melting at lower temperatures than the base metal, which is not melted. The main types of plastic metal joining processes are resistance, pressure, and forge welding.

Welding is done in manual, semi-automatic, or automatic operations, depending largely upon the quantity and variety of work. Automation may consist of mechanical handling and positioning of parts, feeding of filler metal and flux, or manipulation and control of the welding device. All these are handled by the operator in manual welding.

13.3 ELECTRIC-ARC WELDING

The basis for *electric-arc welding* is an electric arc between an electrode and the workpiece or between two electrodes. The arc is a sustained electrical discharge through a path of ionized particles called a *plasma*. The temperature may

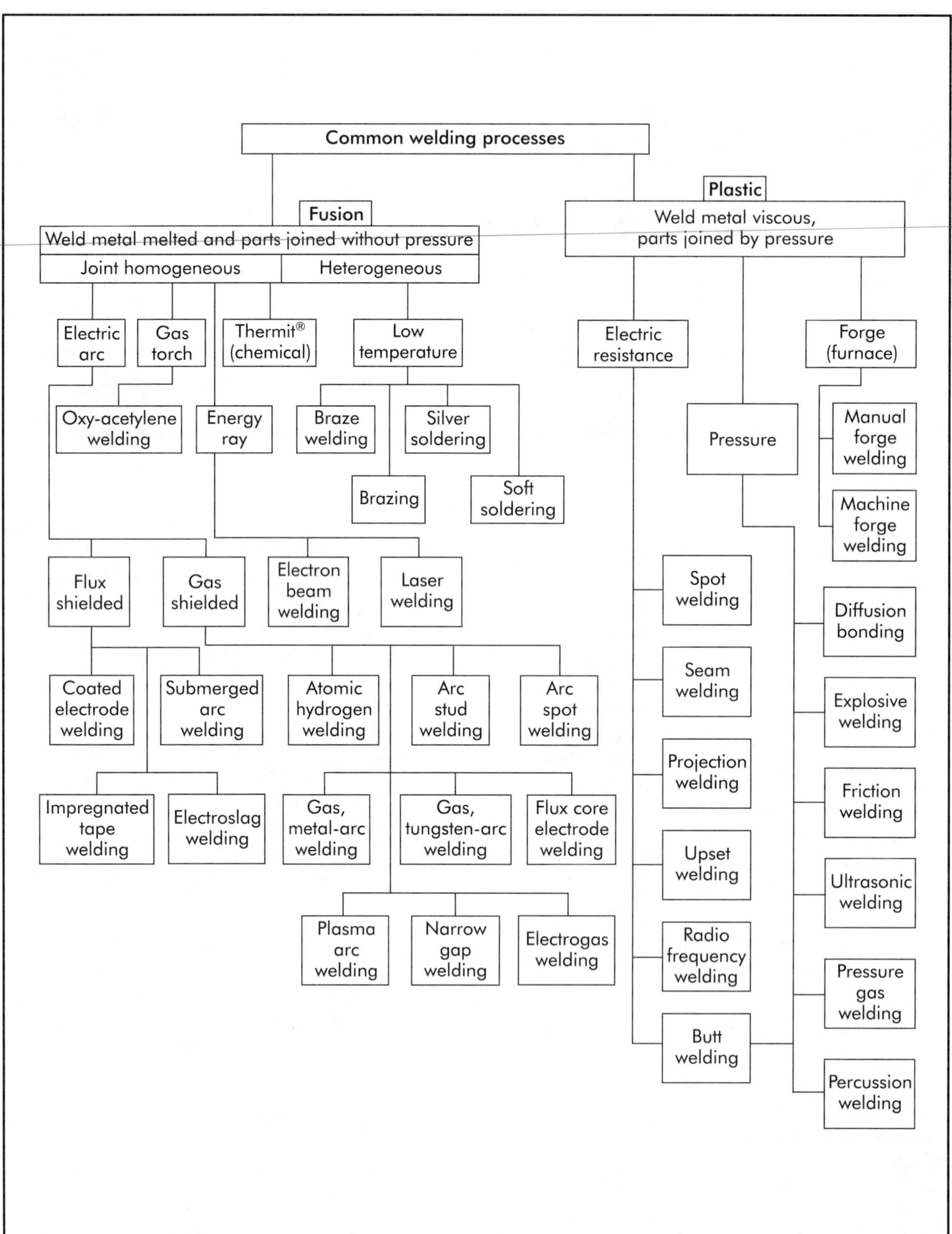

Figure 13-1. Common welding operations and their relationships.

be from about 2,760–27,760° C (5,000–50,000° F) at different parts of an arc under various conditions.

Applications are classified as to whether the electrode is *nonconsumable*, for *carbon-arc* and *tungsten-arc welding*, or *consumable*, for *metal-arc welding*. Of these, carbon-arc welding is seldom used; the others will be discussed in detail later.

The metal electrode in metal-arc welding is melted progressively by the arc and is advanced to maintain the arc length. The molten metal must be shielded from the air. Coated electrodes furnish a protective gaseous cloud around and slag that floats on top of the melt as shown in Figure 13-2. Other methods of weld protection are to pour slag-forming flux or inject inert gas around the arc and molten metal when using a bare electrode. Parent metal is melted under the arc and filler metal is added to the pool as the electrode melts.

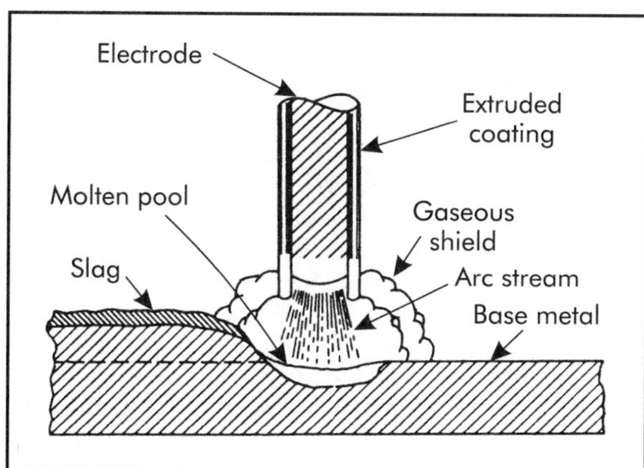

Figure 13-2. Shielded metal-arc electrode, arc stream and its shield, deposited metal, and protective slag.

A schematic representation of the *shielded metal-arc welding* operation is shown in Figure 13-3. The end of the electrode is clamped to one end of the power source and the other end is connected to the part being welded. Some features of the arc-welding operation include:

- a current value of 70–100 A with a 10 kW (13 hp) power requirement that is generally sufficient enough for the operation;
- the polarity of the electrode that may be positive or negative; and

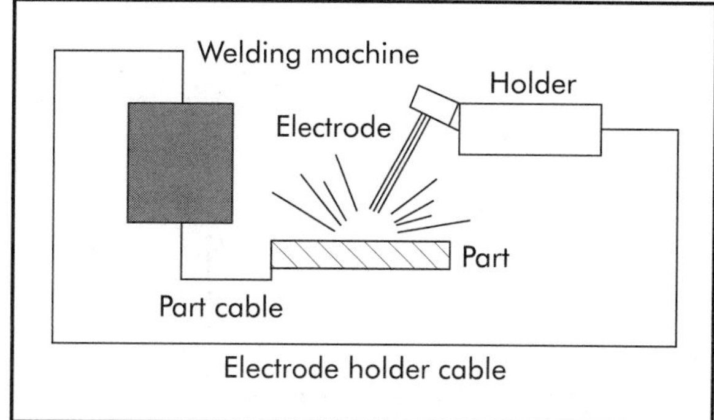

Figure 13-3. A schematic representation of the shielded metal-arc welding operation.

- the process is simple and may be performed at relatively low cost.

13.3.1 Parameters

The *rate* at which energy is delivered is expressed by

$$I^2R \qquad (13\text{-}1)$$

where

I = current, A
R = resistance, ohms

Resistance in an arc-welding operation consists of that of the electrodes, a phase between each electrode and the arc, and the resistance in the arc column. Each of these is independent of the other and may vary differently as current, voltage, shielding, and other conditions are changed. Apparent overall resistance may decrease as current is increased to around 100 A. However, a fairly constant value is assumed for currents of several hundred or more amperes commonly employed for welding. The rate at which metal is melted depends upon the energy delivered and may be expressed approximately by

$$M = k_1 I^2 \qquad (13\text{-}2)$$

where

M = metal melt rate, kg/min (lb/min)
k_i = a constant factor that takes into account average values for the heat capacity of the metal, resistance, and heat losses

The rate at which metal is melted determines the size and speed of laying a weld like the one depicted in Figure 13-4. In such a case,

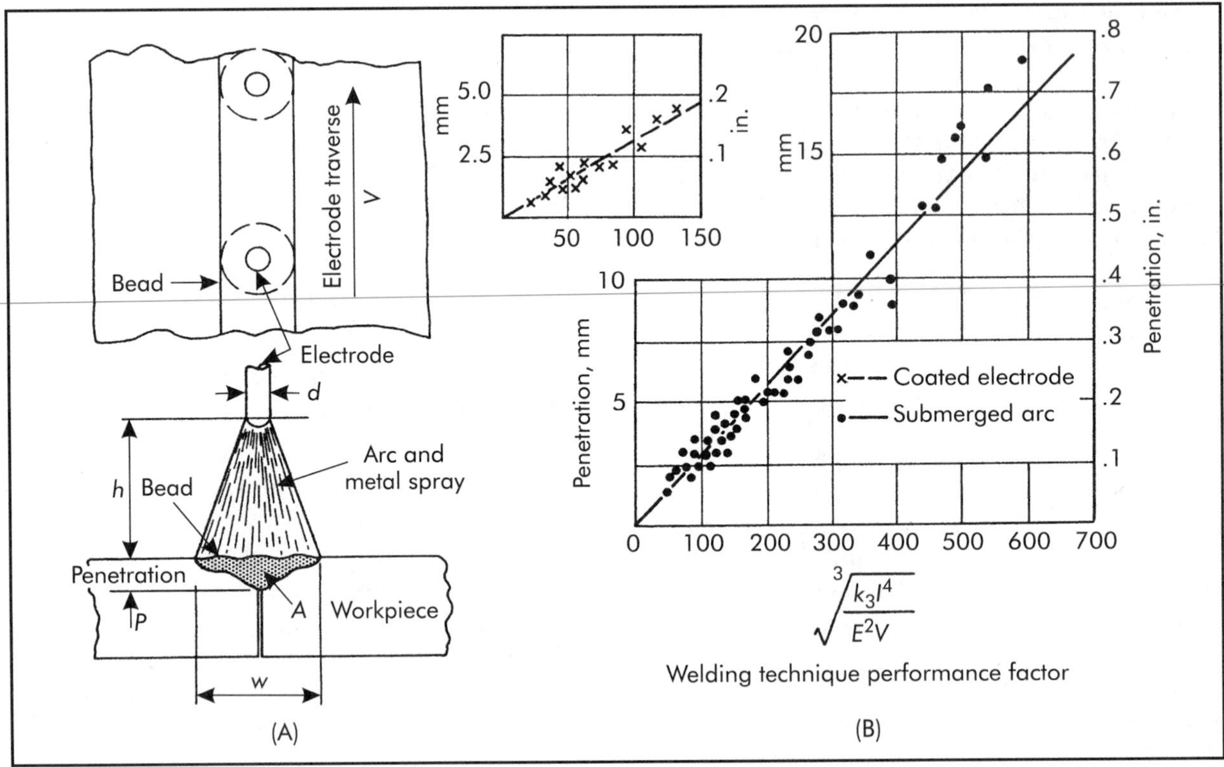

Figure 13-4. (A) An idealized model of metal-arc welding. (B) Experimental and theoretical effects of welding current, voltage, and traverse rate on penetration (Jackson 1959). Key: h = distance between electrode and weld, A = cross-sectional area of bead, w = width of bead. Ranges of welding conditions for coated-electrode process: k_3 = 0.033 for V in mm/min (.0013 for V in in./min), current (I) = 85–350 A, travel (V) = 152–457 mm/min (6–18 in./min), voltage (E) = 23–25 V, and electrode diameter (d) = 3.18–5.56 mm (.125–.219 in.). For submerged-arc process: k_3 = 0.0292 for V in mm/min (.00115 for V in in./min), current (I) = 250–1,200 A AC, travel (V) = 145–1,100 mm/min (5.7–43.3 in./min), voltage (E) = 28–45 V, and electrode diameter (d) = 3.96–6.35 mm (.156–.250 in.).

$$M = \delta AV \tag{13-3}$$

where

A = vertical cross-sectional area of the bead, including melted filler and parent metal, mm^2 (in.^{2})
V = rate of traverse of the electrode along the weld, mm/min (in./min)
δ = density, kg/mm^3 (lb/in.^{3})

The depth to which metal is deposited and melted is called the *penetration* of a joint. It determines the size and thus the strength of the bead that holds the workpiece together. The amount of penetration depends upon the cross-sectional area of the bead, the current, and the voltage. As the distance between the electrode and weld is increased, the arc flares out more and the width of bead increases. Less penetration ensues. The arc voltage increases with electrode separation; an average relationship is

$$E = 20 + 1.2h \text{ for } h \text{ in mm}$$
$$(E = 20 + 30h \text{ for } h \text{ in in.}) \tag{13-4}$$

where

E = arc voltage, V
h = distance between electrode and weld, mm (in.)

Also, as the current is increased, the more intense arc drives the penetration deeper. Experimental and analytical results indicate a fair value of

$$P = \sqrt[3]{k_2 \frac{I^2 A}{E^2}} \tag{13-5}$$

where

P = penetration, mm (in.)
k_2 = experimentally determined constant for different ranges of welding conditions: current (I), vertical cross-sectional area of the bead (A), and voltage (E)

If Equations (13-2), (13-3), and (13-5) are combined, the result is

$$P = \sqrt[3]{\frac{k_3 I^4}{E^2 V}} \qquad (13\text{-}6)$$

where

k_3 = experimentally determined constant for different ranges of welding conditions: rate of traverse (V), current (I), voltage (E), and electrode diameter (d)

The experimental results displayed in Figure 13-4 show that this is a realistic relationship of the factors that affect the depth of penetration and thus the effectiveness of a weld. Various investigators have shown that this applies to common materials and operations. The right-hand part of Equation (13-6) is called the *welding technique performance factor*. This factor can be varied most readily to control a welding operation.

Performance

Arc welding is done most efficiently by utilizing as much current as the work and equipment allow. As shown by Equations (13-2) and (13-5), a large amount of current means metal is melted faster, penetration is deeper, and fewer passes are required. Deep penetration is obtained from a short arc giving a low voltage drop. A common rule is that the arc length should not exceed the electrode wire diameter. With a large value for I and small value for E, a fast V is permissible, and even desirable, to avoid a buildup of filler metal. All this results in a higher rate of production. Limitations are imposed by the workpiece, capacity of the welding equipment, and electrode. A common example of workpiece limitation is that thin sheets cannot be welded with too intense an arc or they will be burned through and damaged.

Equipment capacity is selected upon the basis of cost and utilization. A power source costing several hundred dollars and providing 200–300 A current may be adequate for intermittent use. One worth thousands of dollars and capable of delivering 1,000 A or more is justified for continuous production of heavy work. More current can be used profitably for some operations than for others. Average ranges for some arc-welding processes are indicated in Figure 13-5.

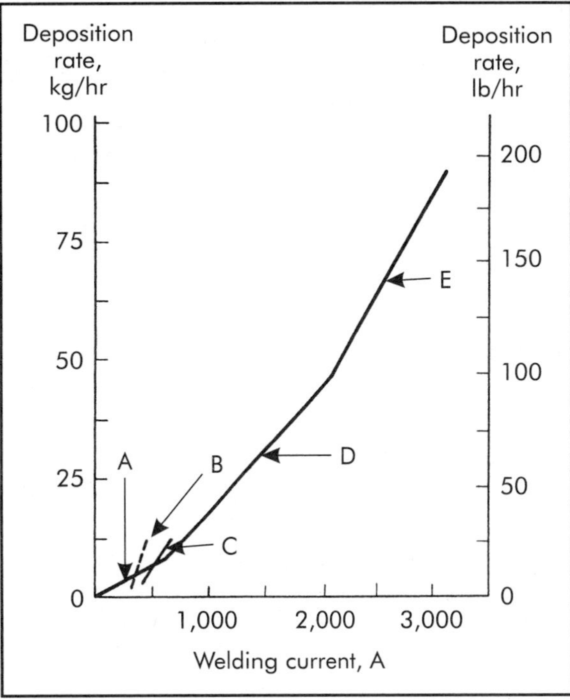

Figure 13-5. Average performance with several arc-welding processes: (A) manual shielded-arc welding with 4.8 mm (.19 in.) electrode; (B) semi-automatic gas, metal-arc welding; (C) semi-automatic submerged-arc welding; (D) fully automatic submerged-arc welding with single electrode; (E) fully automatic submerged-arc welding with multiple electrodes.

There is a proper-size electrode for each type of work, and an ideal current range for each electrode. Penetration and fusion are poor when too little current is passed through a thick electrode, and thin electrodes should be used for small currents on fine work. On the other hand, if too much current is passed through a coated electrode of a certain size, resistive heating spalls the coating. A large-diameter electrode must be used for a large amount of current. Even so, manual-arc welding with coated electrodes is not generally practiced with currents of more than about 400 A. Bare wire fed continuously as an electrode in automatic welding is not subject to the same limitations. Because a fine wire electrode is preheated by its electrical resistance, it is possible to feed fine wire at a faster rate than heavy wire in an automatic welding operation. In fact, up to 70% more metal may be deposited by a fine wire electrode as compared to a heavy wire. The submerged-arc and gas-shielded welding operations utilize bare wire electrodes, carry larger currents, and are faster than the operations with coated electrodes.

The radiation from a welding arc is harmful to the eyes. Anyone within 15.2 m (50 ft) should wear a protective shield over their face and eyes. Spatter makes protective clothing necessary. To protect others in the vicinity, welding is commonly done in booths or behind opaque screens.

Electric Current

If direct current (DC) is used for arc welding, a positive workpiece (anode) and a negative electrode (cathode) constitute *straight polarity*. Electrons flow from the negative to the positive terminal of an arc. With straight polarity, they strike the workpiece at great speeds and the material is heated faster than the electrode. This is of advantage in welding massive pieces because it puts the heat where needed. *Reverse polarity*, with opposite connections to straight polarity, limits the current that can be put through an electrode and is preferable for welding thin sections. It also has an inherent ability to scour the oxide film from the surface of aluminum, magnesium, etc.

Much welding is done with DC because it can handle all situations and jobs, usually provides a stable arc, is known to most operators, and is preferred for difficult tasks like overhead welding. Sometimes a distorted magnetic field will deflect a DC arc and degrade the work. This is called *arc blow* and it can be minimized by use of alternating current (AC). Although all work cannot be done, about 90% of all jobs can be handled by AC. It has been growing in favor because its equipment is simpler and costs only about 60% as much as for DC.

Open circuit potential is usually from 50–90 V. The electrode tip is touched to the base metal and withdrawn slightly to *strike* the arc. As the electrode melts, it must be adjusted toward the workpiece to maintain the arc. Voltage increases with the arc length. The arc is extinguished when its length exceeds the capacity of the available voltage. Metal-arc voltages range from almost zero on short circuit to a minimum of 17 V to sustain an arc, and to around 40 V for a usable long arc.

13.3.2 Applications

Arc welding has the advantages of being versatile and able to make welds under many conditions, of producing high-quality welds, of depositing metal rapidly, and of being economically competitive for many situations. As a result, it is used more than any other welding process. Examples of arc-welded structures are tanks, bridges, boilers, buildings, piping, machinery, furniture, and ships. Almost all metals can be welded by one or more of the forms of arc welding.

13.3.3 Electrodes

Practically all stick electrodes used by hand are of the shielded type, having an extruded coating over the wire. The coating may contain ingredients like SiO_2, TiO_2, FeO, MgO, Al_2O_3, and cellulose in various proportions. Over 100 formulations of electrode coatings are made. The ingredients are intended to perform all or a number of the following functions in various degrees to suit different purposes.

1. Help stabilize and direct the arc for effective penetration.
2. Provide a gaseous shield to prevent atmospheric contamination.
3. Control surface tension in the pool to influence the shape of the bead formed when the metal freezes.
4. Act as scavengers to reduce oxides.
5. Add alloying elements to the weld.
6. Form a slag to carry off impurities, protect the hot metal, and slow the cooling rate.
7. Electrically insulate the electrode.
8. Minimize splatter of weld metal.
9. Form a plasma to conduct current across the arc.

One type of electrode has a thick coating containing a substantial amount of iron powder. The coating forms a shell around and helps concentrate the arc. The iron powder adds extra metal to the weld and increases the speed of welding.

A coating with more ionizing capacity than the electrode metal is usually favored to avoid losing the arc. Such coatings on *drag* or *contact electrodes* melt more slowly than the metal so that the electrode can be positioned in touch with the workpiece without a short circuit. The proper arc length is maintained, and welding is easier.

The core of a typical electrode for welding steel is rimmed steel with about 0.1% carbon and small amounts of other elements. Some cores are killed steel and some are alloy steel, although it is cheaper to add alloying elements through the coating. Core sizes range from 1.6–8.0 mm (.063–.313 in.) in diameter.

A standard classification for electrodes, known as the AWS-ASTM (American Welding Society-American Society for Testing and Materials) classification, designates the characteristics of steel electrodes by a letter and series of numbers such as E 6010, E 7016, and E 9030. The prefix "E" indicates the filler material to be a conducting electrode. The first two, sometimes three, digits specify the minimum tensile strength in 1,000-psi units as welded. The examples stand for 60,000, 70,000, and 90,000 psi, respectively. Strengths range from 414 MPa (60,000 psi) in the low-carbon steel class to 827 MPa (120,000 psi) in the alloy steel class. The third digit indicates the welding position for which the electrode is suited: (1) all positions, (2) horizontal and flat, and (3) flat positions only. The position is determined by ingredients that control the surface tension of the molten metal. The fourth digit indicates the type of coating and conditions for which the electrode is intended, including the kind of current and polarity. In addition, the description of a low-alloy electrode carries a suffix, such as A1, B2, etc., which designates the type and amount of alloy. For example, E 8018-B2 designates steel alloyed with 0.40–0.65 Mo and 1.00–1.50 Cr. Full specifications of the properties and applications of standard electrodes are tabulated in reference books and handbooks. Some electrodes are identified by brand names.

Plain carbon electrodes cost less than $1.98/kg ($0.90/lb) in large quantities to $4.41/kg ($2.00/lb) and more at retail. Special and alloy types may cost several times as much.

13.3.4 Manual Arc Welding

The basic units for *manual metal-arc welding* (formally called *shielded metal-arc welding* [SMAW]) are: (1) a source of current called a *welder* or *welding machine*; (2) conductor cords, wires, or leads (one to the ground and one to the electrode); (3) an electrode holder; (4) an electrode; and (5) a workpiece. A complete outfit for general-purpose work can be obtained for less than $1,000.

Welding is commonly done on a metal table to which the ground lead is attached and on which the workpieces are laid or fastened. The operator manipulates the electrode. Tables or holding fixtures that tilt and turn to present joints to be welded in the most accessible and advantageous positions are called *welding positioners*. They make welding much faster. For instance, downhand welding in the most convenient position can be four times or more faster than welding in vertical or overhead positions. The positioner may be manually adjusted, and may cost only a few hundred dollars. Some are motor operated for large work and easy adjustments and may cost many thousands of dollars.

13.3.5 Automatic Arc Welding

In *automatic arc welding*, a bare wire electrode is fed continuously through motor-driven rollers with semi- or fully automatic equipment. The wire may be solid or hollow and filled with flux for benefits like those from coated electrodes; the latter is called *flux-cored arc welding*. The molten metal is surrounded by gas or flux in the protective processes described later. A semi-automatic gun or torch is moved along the weld by hand. Some have integral smoke collectors. Either the welding head or work table is mechanically traversed. Figure 13-6 shows the basic equipment for the submerged-arc welding process.

Special-purpose welding machines are common in the high-production industries. One example is a large fully automatic machine to

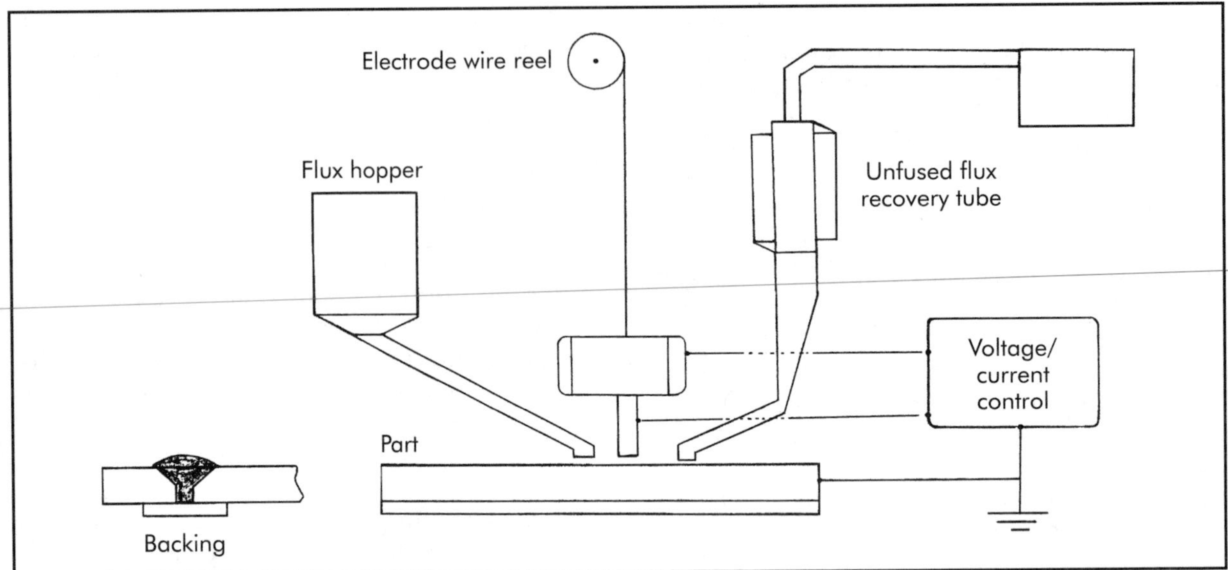

Figure 13-6. Basic equipment for the submerged-arc welding process.

produce box section-frame cross-members from stock 3.18–3.96 mm (.125–.156 in.) thick for automobiles. An operator puts two halves of a section in the first station, where they are pressed together and tacked by resistance spot welding. A walking beam transfers the section to the welding station where automatic clamps square the flanges, ensure fit-up, and position it. Four welding heads, feeding 1.98-mm (.078-in.) diameter wire and guided along the curved profiles, complete 2 m (80 in.) of welds in less than 8 sec. Special fixtures to position and hold components to be joined are necessary for mechanized welding in quantity production.

Among machines for positioning and moving work for repetitive production welding is a weld-lathe for round pieces such as tubing. An assembly to be joined is rotated and traversed between headstock and tailstock to enable one or more welding torches to move along the seams in a preset program. The price is about $30,000. Another system rotates relatively flat work around a vertical axis for automatically welding circular joints.

13.3.6 Sources of Current

An electrical generator was long the only source and is still a major source of DC for welding. It may be driven by an electric motor or by an internal combustion engine for portability. Another means is to reduce the voltage of AC through a transformer and convert it to DC through a selenium or silicon rectifier. AC is obtained directly from a transformer. Combination sources that supply both AC and DC are popular. A transformer source has no moving parts but a fan. Thus, it operates with less maintenance cost and a higher operating efficiency than a generator. One study with a 60% duty cycle and a welding current of 300 A for a 9-hr shift showed electric power cost at 3 cents/kWh to be $3.42 for a DC motor generator, $2.56 for a DC rectifier, and $2.10 for an AC transformer unit.

Arc welder sizes are designated in amperes (A) (see Figure 13-5). A 100–200-A machine is small but portable and satisfactory for light manual welding. A 300- or 400-A size is suitable for manual welding of average work like ships, structures, and piping. Automatic welding requires capacities between 800 and 3,000 A either in a single unit or a number of smaller units in parallel. Controls are provided on the best machine for varying open-circuit voltage and welding current in amperes. Some have adjustments only for current. Others have an *arc booster* that provides a momentary surge of current to give an arc a good start when it is struck.

The *duty cycle* of a welder specifies the part of a 10-min period in percent that the rated current can be drawn from the machine without overheating. For example, if a welder is rated at 300 A at a 60% duty cycle, the machine can be operated safely at 300 A welding current for 6 out of 10 min. A welder must be operated for a smaller portion of time at a higher than rated current, and the current output must be reduced if the rated duty cycle is exceeded.

A *constant-current* or *drooping power source* with characteristics typified by the solid curves of Figure 13-7 is necessary for all manual-arc welding. For any one setting of the welder, the current varies little over a wide range of voltages resulting from variations in the arc lengths. Thus, although the operator cannot hold a constant arc length by hand, he or she is given the benefit of stable melting of the electrode and easy control of the arc. The settings on the welder for the volts and amperes give the effect of a large number of curves, some steep and some flatter, to meet a variety of situations. Thus, a uniformly low current is desirable for a small electrode to weld thin sheet metal without burn-through. A flatter curve would be desirable for overhead welding. In this case, the electrode is brought close to the work to control the metal and the voltage decreases. A heavy current is beneficial to melt and drive an adequate amount of metal into the weld.

Automatic equipment may be powered by a constant current source. The electrode wire is fed by an adjustable speed motor. If the melting rate is too fast and tends to stretch the arc, a control circuit responds with an increase in voltage and speeds up the wire feed motor, and vice versa. Simpler control is possible with a *constant arc voltage* (CAV or CV) or a slightly *rising arc voltage* (RAV) source. Voltage for the conditions of the job and the rate of wire feed is selected from the controls and maintained by the source. Current is delivered from the source as needed to melt the electrode and maintain a constant arc length.

13.3.7 Gas-shielded Arc Welding

The arc and metal pool in welding may be protected by an envelope of inert or semi-inert gas. This gas may be contained in an enclosed chamber in which the welding is done, but commonly is injected in the open around the point

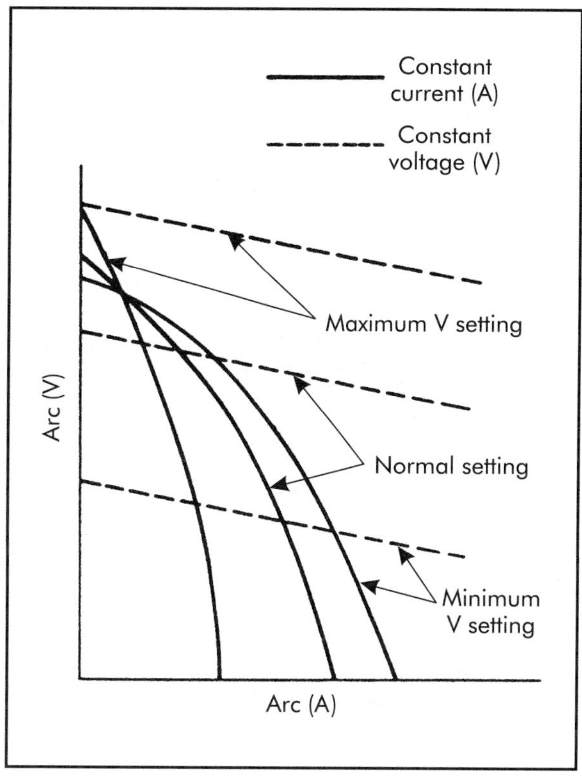

Figure 13-7. Characteristics of constant-current and constant-voltage power sources for arc welding.

of welding. Gas shielding adds cost but affords a clean and clear operation. It must be protected from strong drafts. Gas-shielded arc welding is done mostly with direct current. Thin, particularly nonferrous, sheet metal can be welded better and faster with a power supply that pulses the current at a set rate of 20–120 times per second. Equipment for *pulsed-arc welding* may be four to five times more expensive than that used with a conventional power supply.

One gas-shielded arc process is *gas tungsten-arc welding* (GTAW) also called *Tig* (tungsten inert-gas-shielded arc welding), depicted in Figure 13-8(A). It utilizes a tungsten alloy electrode that does not waste away in an inert-gas atmosphere. Filler metal is put in by a separate rod or wire as needed. Helium and argon alone, together, or mixed with other gas form the shield. The process requires care and skill, but makes clean and reliable welds. Its use is especially important in aerospace and other critical applications. GTAW is better suited for sheet metal than for thick sections. No foreign substance

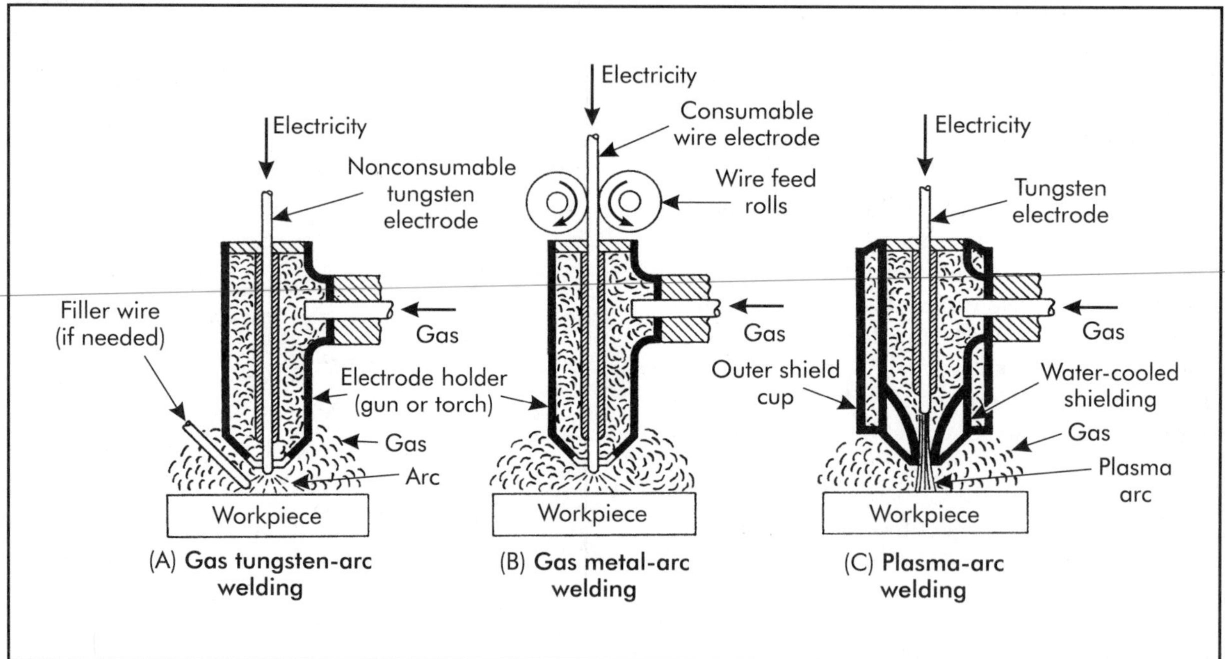

Figure 13-8. Types of gas-shielded arc welding.

need touch the weld. The arc may be initiated by a high-frequency and high-voltage pulsed discharge to obviate even touching the electrode to the workpiece.

Atomic hydrogen-arc welding is similar to gas tungsten-arc welding, but is done with an arc between two tungsten electrodes in a stream of hydrogen gas. The hydrogen gives protection from the atmosphere and enhances heat transfer from the arc to the workpiece. The process gives exceptionally clean welds, but is expensive and not used much except for deep welds, as in die blocks, and for high-temperature alloys, particularly for surfacing.

Gas metal-arc welding (GMAW), also called *Mig* (metal inert-gas-shielded arc welding), is done with a wire fed through the welding head to act as the electrode and supply filler metal as depicted in Figure 13-8(B). Inert gases are used for critical jobs, but cheaper carbon dioxide has become quite popular for a large variety of production operations. Carbon dioxide does not support an arc well. One remedy is to maintain a short arc with provisions to automatically limit the current during frequent periods of short circuiting. The carbon dioxide also dissociates into carbon monoxide and oxygen, and deoxidizers must be added to protect the weld. One form of the process utilizes a *flux-cored wire* containing arc stabilizers and deoxidizers. Another variant of the process is called *micro-wire welding* where wire is fed as small as 0.76 mm (.030 in.) in diameter at high speeds. With high current densities, up to 23,249 A/cm^2 (150,000 A/in.^{2}), the wire is well preheated and the current is concentrated at the weld. GMAW is fast, versatile, and applicable for semi-automatic welding in all positions.

Arc spot-welding is performed by Tig or Mig welding one spot at a time through two or more lapped or butted sheets or plates. An advantage is that it can be done entirely from one side without high pressure. It can weld thicker sections, and equipment costs are less than half as much as for resistance spot-welding, but the latter still is faster and more suitable for most production operations. Arc spot-welding is mostly confined to mild and stainless steel and leaves a visible lump on the surface. A way to pierce metal is to burn through a piece with a shielded electrode and blow out the molten metal. A process for quantity production utilizes a string of electrodes along a straight or curved joint. The electrodes are fired in rapid succession and produce a series of overlapping spot welds. Some joints are reported to be made faster in this way than by traversing electrodes. This is called *programmed inert-gas multi-electrode welding* (PIGME).

13.3.8 Plasma-arc Welding

Plasma is high-temperature ionized gas and occurs in any electric arc. If a stream of injected gas and the arc are made to flow through a restriction (a water-cooled copper orifice as in Figure 13-8[C]), the current density of the arc and the velocity of the gas are raised. In that way, the ionization and temperature of the gas are greatly increased. The excited particles give up large amounts of energy when they reform into atoms, some at the workpiece surface.

Most materials can be melted, many even vaporized, by the plasma-arc process and thus become subject to welding. Penetration is deep and thorough. The process can be operated at currents from less than 1 A to over 500 A and close control can be achieved; stainless steel foil as thin as 25 μm (.001 in.) is easily welded. Results are clean. The process is two to five times as fast, but equipment costs twice as much (or more) as for Tig welding.

13.3.9 Submerged-arc Welding

In *submerged-arc welding*, granular flux is poured on the weld area in advance of the moving arc. The electrode is bare wire automatically fed into the flux blanket. Flux around the arc is melted, protects the arc and weld, and is deposited as slag on top of the weld on freezing. The effect is shown in Figure 13-9. The flux may be neutral or contain alloying elements to enrich the weld. The protection affords a good quality weld and eliminates arc spatter. An operator cannot see the arc and may have trouble following a joint, but he or she does not need an eye shield. The method operates with alternating or direct current but is confined to flat or horizontal positions. Unmelted flux may be put back for reuse and slag must be removed from each pass, all adding work to the operation.

The submerged-arc method is used for semi- and fully automatic welding mostly of steel and is capable of the largest deposition rates. For example, heavy steel plates, such as for ships, are welded through from one side in one pass

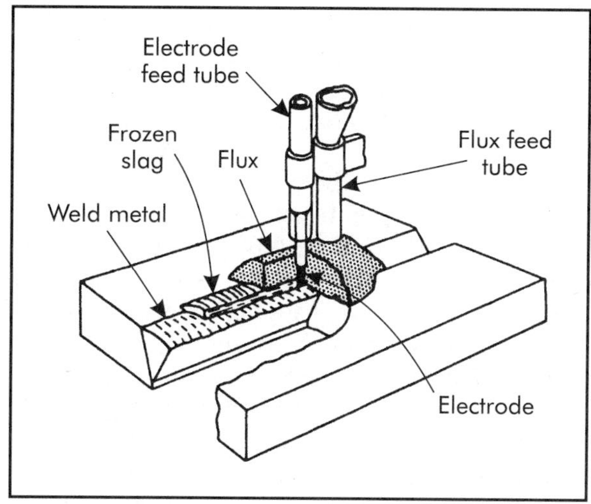

Figure 13-9. View of a submerged metal-arc weld.

with the addition of powder metal to the flux or into the gap before the flux. Two or more electrodes travel in a row with 1,500 A or more on the lead arc and 1,000 A or more on the trailing arcs for welding rates to 635 mm/min (25 in./min) on plates 25.4 mm (1 in.) thick or more.

Flux is applied to welding in other ways. *Magnetic flux welding* utilizes a flux that is attracted magnetically to and coats the electrode wire carrying a heavy current. *Impregnated-tape, metal-arc welding* is done with a machine that winds a fluxing tape around the electrode as it is fed to the arc. Also, *flux-cored, electrode-arc welding* (FCAW) with or without gas provides flux neatly with high productivity.

13.3.10 Vertical Welding

Electroslag welding (ESW), illustrated in Figure 13-10(A), is a form of vertical welding used particularly for joining steel sections more than 25.4 mm (1 in.) or so thick, such as for crusher bodies, large press frames, and pressure vessels. Welding is initiated by an arc on the bottom of the vertically positioned joint. Flux poured around the electrode is fused to slag that floats on the molten metal. After the start, there is no arc, and heat comes from the electrical resistance of the slag. Metal is added as the wire filler is fed in, melted, and solidified. This continues from the bottom to the top of the joint. The molten metal and slag are kept in the joint by dams on the sides, which may be fixed or slide upward with the progress of the operation. They usually are water cooled. More than one electrode may be fed at a time into thick joints. Simple equipment for plates as thick as 50.8 mm (2 in.) is available for below $10,000; more elaborate systems cost up to $25,000.

Electrogas welding (EGW), depicted in Figure 13-10(B), is suitable for welding sections about 12.7–38.1 mm (.5–1.5 in.) thick in one pass. The consumable wire electrode may be solid or flux cored, but most or all of the shielding is provided by an argon and carbon dioxide gas mixture injected into the gap. *Narrow gap welding* utilizes specific techniques to weld joints less than 12.7 mm (.5 in.) wide. A narrower gap requires less filler metal and time and minimizes the heat-affected zone. Vertical welding is most natural and easy for many applications. It is economical for welding sections in one pass that may require several passes by other means. In some cases it is 10–50 times faster.

Preparation is simple for vertical welding because elaborate scarf preparation and accurate fit-up are not necessary as they are for conventional arc welding. Heating and cooling are inherently slow, but there is no need for preheating or controlled cooling. However, the heat-affected zone is large, and heat treatment to obtain acceptable fatigue properties may be desirable.

13.3.11 Stud Welding

The *arc stud-welding* process utilizes a gun that automatically controls an electric arc and attaches a stud in place. The scheme of the initial setup is shown in Figure 13-11(A). A current is turned on, and the stud withdraws from the workpiece to strike an arc inside the ceramic ferrule. After a predetermined time, the current is turned off and the stud is pushed into the molten pool. After the metal solidifies, the equipment is withdrawn to leave the stud welded to the workpiece as illustrated in Figure 13-11(B). A variation of the process particularly suited for attaching small studs to thin sheets is *capacitor-discharge welding*. For this, the stud is made with a thin tip on the end as depicted in Figure 13-11(C). A heavy current discharge melts the projection and sets up an arc in the gap. The stud is then pushed against the workpiece to complete the weld.

13.4 ENERGY-RAY WELDING

In the *energy-ray welding* processes, heat is normally generated by high-velocity narrow-

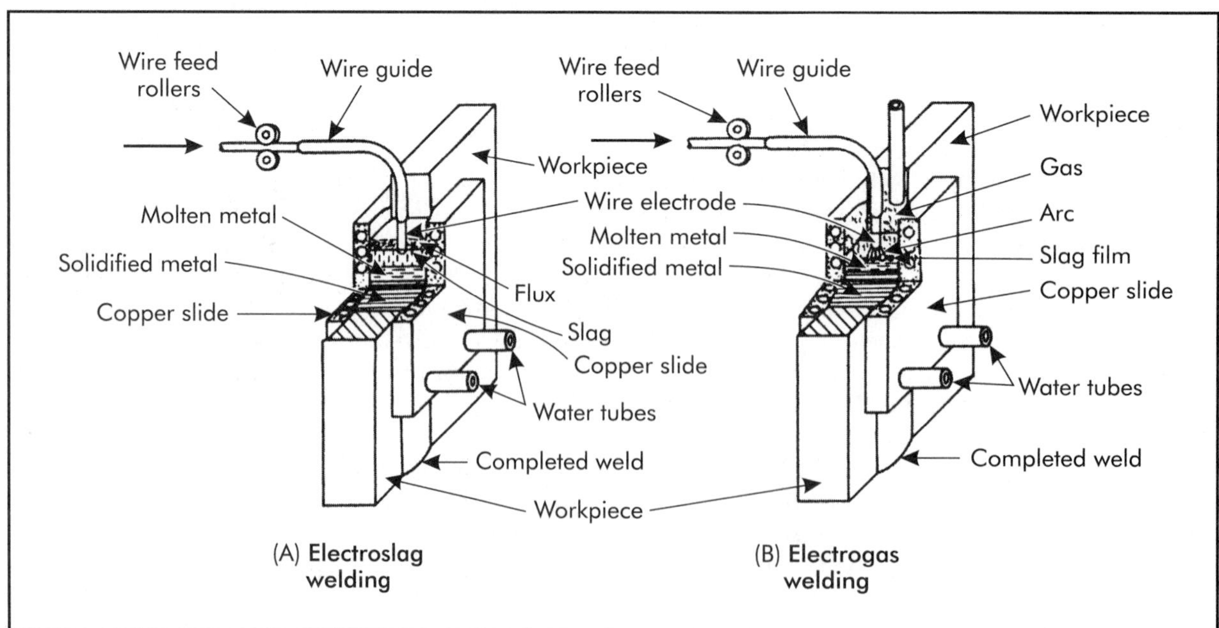

Figure 13-10. Two forms of vertical welding.

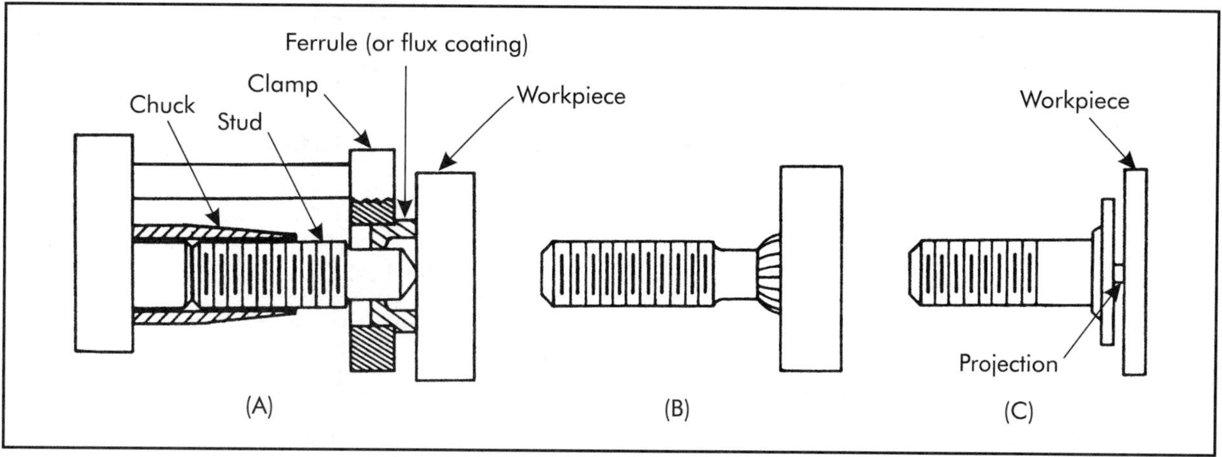

Figure 13-11. Procedures for stud welding: (A) setup for arc-stud welding; (B) arc-welded stud; (C) stud for capacitor-discharge welding.

beam electrons as in electron-beam welding, or by utilizing a high-power laser beam as in laser-beam welding. Both methods produce high-quality welds with good strength and are normally free of porosity.

13.4.1 Electron-beam Welding

Electron-beam (EB) *welding* is a fusion welding process in which the heat generated for welding is provided, like in electron-beam machining, by a highly focused, high-intensity stream of electrons hitting the work surface. Generally, all metals that are arc welded can be welded using electron-beam welding.

Energy may be supplied for welding and cutting by directing a concentrated beam of electrons to bombard the work in the manner depicted in Figure 13-12(A). The beam is created in a high vacuum. If the work is done in a vacuum of around 10^{-4} torr, no electrodes, gases, or filler metal need contaminate it, and pure welds can be made. Electron-beam welding can do the same jobs as Tig and plasma-arc welding; in some respects it is better. The beam of energy can be highly concentrated (up to 50 times stronger than in arc welding) with power densities to 3 MW/cm^2 (20 MW/in.^{2}) to melt and even vaporize ferrous and nonferrous refractory, dissimilar, and even reactive metals. Welds can be confined to shallow depth or extended to as much as 203 mm (8 in.) deep with a depth-to-width ratio as high as 100:1 (compared with 1:1 for other fusion techniques). This means that melting can be confined to very narrow limits and done fast so that the heat-affected zone is small. Welding traverse rates may be high. Grain size remains small and welds are essentially straight sided. A notable application of this process for cutting is to produce very small diameter holes, particularly in hard materials.

If welding is done in a high vacuum, considerable time may be lost if the chambers must be pumped down to accept each new piece (from 1 min or less for small chambers to 20 min or so for large ones). One machine has twin chambers in which work is done alternately. Air locks and seals have been utilized for slipping piece after piece into position or passing long pieces through a chamber. Another approach is to pass the beam through two or more chambers with the welding done in a soft vacuum (10^{-1}–10^{-3} torr) or even in the atmosphere, as illustrated in Figure 13-12(B). Total operation time may be reduced to seconds, but some of the advantages of the process are lost. An electron beam is dispersed in a gas in proportion to the density of the medium. For example, about three times as much energy as in a high vacuum is required for a given penetration after an electron beam has passed through 9.53 mm (.375 in.) of gas at atmospheric pressure. Thus the weld is wide and not as deep, and contamination occurs.

Electron-beam welding can be justified when it can produce results not obtainable by other

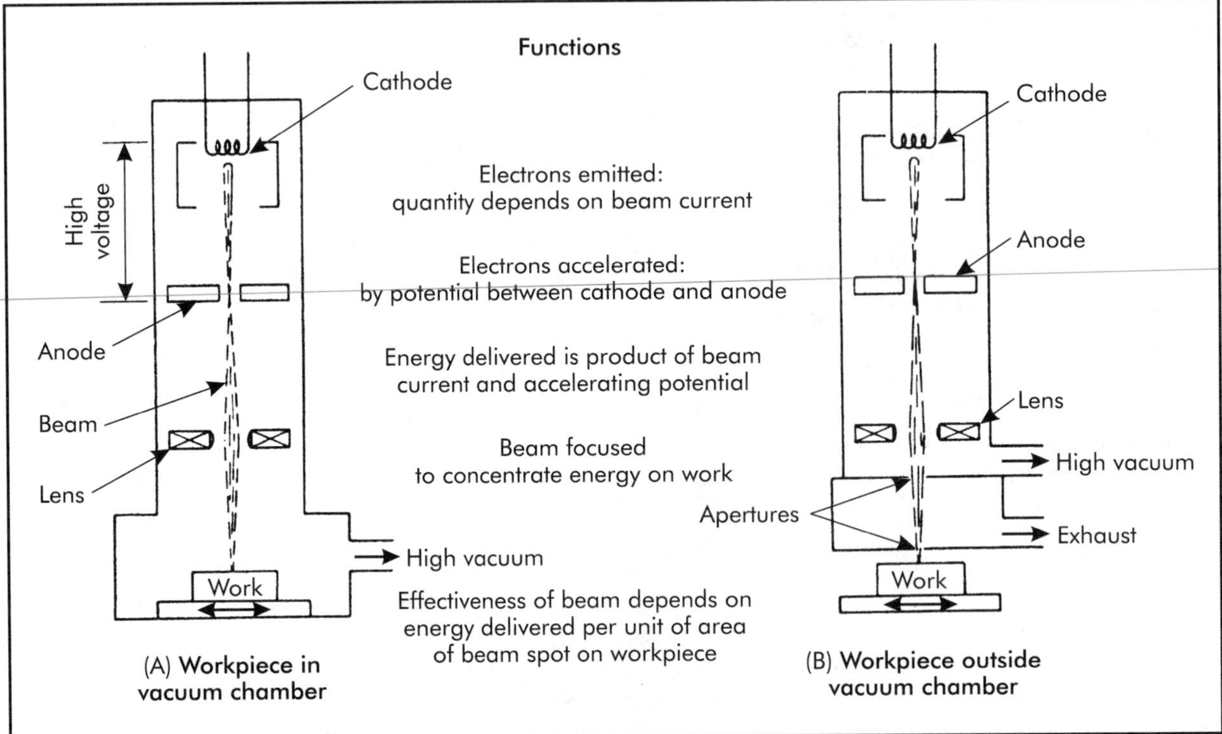

Figure 13-12. Diagram of an electron-beam welding operation.

means, that is, reduction or elimination of filler metal and contamination, less joint preparation, access to hard-to-reach places, and the least heat affect on the parent metal with little distortion. In some cases, electron-beam welding is economical because it is faster. In any case, the advantages must be enough to make up for a high first cost. Medium-size systems range in price from about \$200,000–\$400,000; the largest and most advanced costs over \$1,000,000. A leading maker offers models ranging in rated sizes from 6–35 kW for high vacuum, 10–15 kW for partial vacuum, and 12–25 kW for nonvacuum welders. Equipment cost depends to a large extent upon the rate and the amount of vacuum that must be drawn. High potentials require lead shielding to protect the operator from x-rays.

Heat treating is done by soft-focusing an electron beam and, under computer direction, making it impinge as an array of dots over the surface to control temperature. As an example, rocker arm pads are hardened in this way over an area of about 484 mm^2 (.75 in.^{2}) in 2.2 seconds at a cost of 0.365 cent with power at 5 cents/kWh. Heating is so rapid and closely controlled that the adjacent metal stays cool and serves as a quencher.

High initial equipment cost and the need to perform the process in a vacuum present some of the limitations for the electron-beam welding process.

13.4.2 Laser Welding

Laser welding is another fusion welding process where the heat of a high laser-coherent beam is focused on the joint to be welded. Process characteristics are similar to those of the electron-beam welding process. They include:

- a high quality weld;
- deep penetration; and
- a narrow heat-affected zone.

Unlike electron-beam welding, no vacuum is required to perform this process.

The Laser

A high-energy light beam capable of welding, cutting, and heat treating metal and other materials is produced by a laser (an acronym of "light amplification by stimulated emission of radiation"). A suitable medium, such as a crystal (for example, ruby) or gas mixture (mainly CO_2), is stimulated. For example, a crystal is stimulated by a bright flash of light in the manner depicted in Figure 13-13 or a gas is stimulated by an electrical discharge. Ions in the substance are raised to an unsteady energy level. Then they fall back and release an intense burst of single monochromatic light. The wavelength depends upon the medium; some are better than others for various purposes. The light is amplified by reflection from one end, and the surge of light from the other end of the laser is focused by a conventional optical system.

Solid-state laser equipment is compact but less than 2% efficient, and the heat generated limits it to a pulsed output of low capacity. One system of that kind is rated at 400 W, can deliver 29 J/pulse with a duration of 0.65 milliseconds, up to 100 pulses/sec, and costs about $40,000. With three to five axes, a part manipulator, and computer numerical control (Chapter 31), the cost is about $175,000. Gas lasers are 15% or so efficient, and gases can be circulated for effective cooling. Such systems may be pulsed but usually are lased continuously. A drawback is that CO_2 lasers do not work well on aluminum and some other materials because the wavelength is not absorbed efficiently. A machine that employs an 800-W continuous laser traversed by numerical control and applied anywhere over a 1.6 × 2-m (63 × 79-in.) work surface with facilities for oxygen-assisted cutting is around $350,000. A large 15-kW CO_2 laser unit has a price of $1,000,000. High equipment cost limits lasers to applications not quickly or readily performed by more conventional methods.

Laser welding may be done in a number of ways and has several advantages. A seam may be welded by a series of overlapping fused spots with a pulsed laser or continuously by a powerful gas laser. Gas lasers are twice or more faster than solid-state lasers. A laser beam can be precisely controlled and directed almost anywhere and concentrated as needed, even on minute spots on microelectronic devices. Welds can be made inside transparent enclosures, such as in vacuum tubes, or to join fine wires inside their insulation. Otherwise inaccessible areas can be welded by means of mirrors. The process is ideal for automated operation because the weightless beam is easy to manipulate.

Normally, a small puddle is melted and frozen in milliseconds by a laser, and the heat-affected zone is confined to 0.25 mm (.010 in.) or less. The time in the molten state is too short for appreciable chemical reaction, and little or no protection is needed from the atmosphere. The beam does not add foreign substances to the weld.

Tremendous amounts of energy can be delivered when needed by a laser. Laser beams have been concentrated to small spots of light with densities up to almost 4.6 GW/cm^2 (30 GW/in.^{2}), enough to vaporize almost any metal in an instant. It takes from 1.5–15.5 MW/cm^2 (10–100 MW/in.^{2}) for normal melting and cutting operations. The

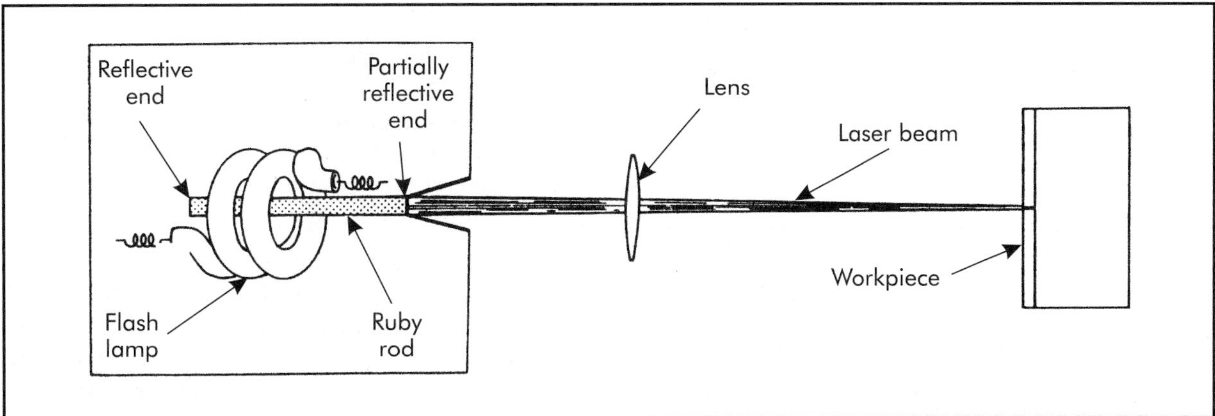

Figure 13-13. Laser system.

length of time that an intense beam impinges on any one spot is limited so that only the amount of energy needed is delivered over the area processed.

In addition to welding, lasers have many applications for surface hardening metals and cutting operations such as drilling tiny holes and slicing, trimming, and scribing of ceramics, semiconductors, films, cloth, wood, plastics, and metal sheets. A laser can do a variety of work because the light beam can be easily focused by optics and moved about as needed.

In heat treatment with lasers, the metal is heated far below the melting point. A concentrated light beam may be oscillated to produce a raster to distribute the heat over a large area and keep the temperature within the required range. Cast iron is surface treated by 5 kW of laser power traversed at 1.5 m/min (60 in./min) over a width of about 15 mm (.6 in.). Average surface energy input is about 1.3 kJ/cm^2 (8 Btu/in.^{2}).

Enough energy must be delivered to melt the metal in welding. A 5456 aluminum alloy 6.4 mm (.25 in.) thick is welded at a 2.3 m/min (90 in./min) rate with 8 kW of laser power. Average width of the weld is about 2 mm (.08 in.), and the average surface energy input is 10.4 kJ/cm^2 (64 Btu/in.^{2}).

For cutting steel with lasers, much of the heat in the kerf is obtained by injecting oxygen to burn the metal, as in flame and electric-arc cutting. Carbon steel 6.4 mm (.25 in.) thick is reportedly cut at the rate of 1.0 m/min (39.4 in./min) by an 800-W laser. Typical for cutting is a kerf of 1 mm (.04 in.). Average surface energy input is 4.6 kJ/cm^2 (30 Btu/in.^{2}). Holes from 0.3–1.3 mm (.01–.05 in.) diameter are drilled to depths up to 15 mm (.6 in.) in most alloys by pulsed lasers that generate 50-J (.047 Btu) pulses. The average surface energy input is 11.3 kJ/cm^2 (69 Btu/in.^{2}) per pulse.

Laser and electron-beam systems can do many chores alike but in effect seem to supplement each other. Both can be concentrated to melt any metal, but energy transfer is more efficient with the electron beam, and laser penetration takes more power. For laser heat treatment, the work surface needs to have a dark coating to absorb light, and that adds cost. On the other hand, a laser does not need a vacuum, the beam can travel to almost any length in air, and it is not deflected by magnetic fields.

13.5 RESISTANCE WELDING

Resistance welding is done by passing an electrical current through two pieces of metal pressed together. The pieces coalesce at the surfaces of contact because more resistance and heat are concentrated there. The heat is localized where needed, the action is rapid, no filler metal is needed, and the operation requires little skill and can easily be automated, making the process suitable for large-quantity production. All the common metals and dissimilar metals can be resistance welded, although special precautions are necessary for some. The parent metal is normally not harmed, and none is lost. Many difficult shapes and sections can be processed.

The main disadvantage of resistance welding is that equipment cost is high. Ample work must exist to justify the investment. Some jobs call for special equipment, such as fixtures, which add appreciably to the investment. A high order of skill is required to set up and maintain the apparatus.

13.5.1 Fundamentals

Resistance welding is usually done with AC from the line stepped down through a transformer and applied for a length of time controlled by a timer. A typical circuit is depicted in Figure 13-14. The heat generated in a circuit is

$$H = I^2RTK \tag{13-7}$$

where

H = heat generated in a circuit
I = current, A
R = resistance, ohms
T = duration time of the current flow, sec
K = conversion factor from kW to the unit of heat desired

The electrodes that carry the electricity to the work also press the pieces together. The current meets resistance in the metal but more so at the surfaces of contact. The most heat is generated where the resistance is highest. Thus the aim is to put the most resistance and heat at the *faying surfaces* between the workpieces to make a sure

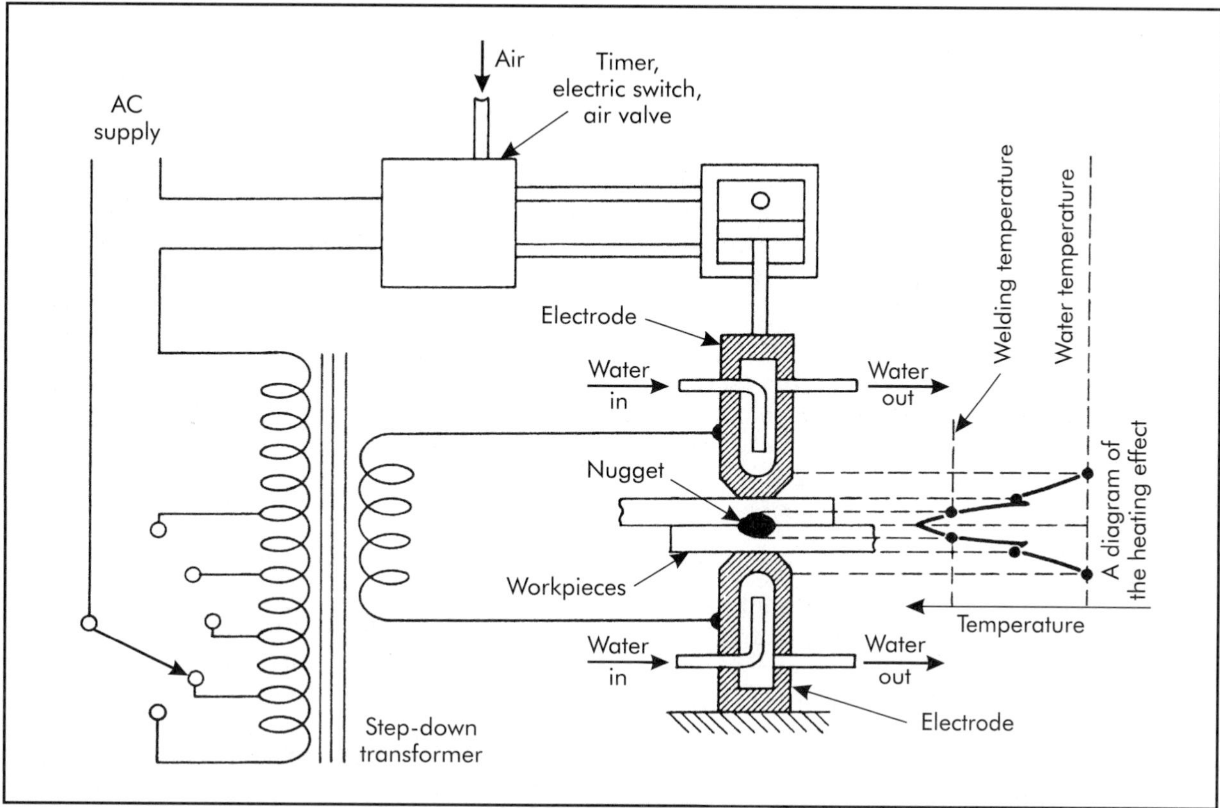

Figure 13-14. Principles of resistance welding.

weld without harming the electrodes. Under pressure, melting of the metal is not always necessary for coalescence. The heat distribution that occurs in a satisfactory operation is indicated by the diagram of Figure 13-14.

Outside the zone to be welded, the temperature should be low to conserve the electrodes. This is assured by conducting away as much heat as possible and generating as little heat as possible where it is not needed. The electrodes are normally water-cooled to carry away heat. The heat generated at any point is proportional to the resistance. Electrodes are made of copper to conduct electricity and heat well, but are alloyed and faced with other elements for good bearing strength and endurance. Clean and smooth workpiece surfaces promote low resistance and help preserve electrodes.

When the work and electrode materials have nearly the same resistance, they tend to weld together. Low-resistance metals, like copper and silver, are hard to weld. Aluminum, with a conductivity of about two-thirds that of copper, requires a high current of short duration so that the weld zone is brought to welding temperature before the heat spreads. Special copper-tungsten alloy electrodes are recommended for aluminum.

Metals of medium and high resistance, such as steel, stainless steel, Monel® metal, and silicon bronze, are easy to weld. When heat-treatable steel is welded, heat sinks rapidly into the surrounding metal mass, and the weld is more or less quenched to a hard and brittle state. This can be corrected by annealing or tempering in a separate furnace or on the welding machine. In the latter case, a second surge of metered current is put through the workpiece to reheat it to a drawing temperature before it is released.

High-frequency resistance welding is done with 400–450 kHz current commonly supplied by an oscillator. The high-frequency current readily breaks through oxide film barriers and produces a thin heat-affected zone because it travels on the surface of the material. It is applicable to joining metals from 0.1–19 mm (.004–.75 in.) thick into structural shapes, pipes, or tubes in the manner depicted in Figure 11-10(C).

The principles that have been discussed are applied in several processes to suit various purposes. These processes are spot, projection, seam, upset-butt, flash-butt, and percussion welding. The features of each will be discussed later.

13.5.2 Equipment

There is no universal machine to do all types of resistance welding, but several basic features are common to the various forms of equipment. Common to all machines are a power supply, a system of controls, a mechanical drive, and a structure.

Most resistance welders operate on AC from a single-phase transformer as indicated in Figure 13-14. This causes a low power factor and a serious unbalance in three-phase power systems. Better circuits are available, particularly on large machines, to utilize a three-phase system efficiently and reduce the load on the power source.

One capacity of a resistance welding machine is designated in kilovolt-amperes (kVA). The rating is based on a 50% duty cycle and dictates that the welder will carry that kVA at 50% duty cycle and not exceed a safe specified temperature rise. The kVA rating is the product of the welding current times the secondary voltage. Actually, a machine can do a job with a higher kVA demand than its rating, but at a lower duty cycle, or vice versa.

Any resistance welding machine must have means to adjust the amount and duration of current flow to suit various jobs and conditions. Early machines had manual or mechanical controls, as do some low-cost ones today, but these do not give accurate results and are not adequate for large currents. A simple control for the amount of current consists of taps on the transformer primary. Another is an autotransformer that varies the voltage impressed on the primary of the welding transformer. A common method that gives current adjustment in infinite steps is *phase-shift control*. This is done by altering the magnitude and wave shape of the current to the transformer primary by means of tubes or solid-state devices. By still another method, called *slope control*, the starting current is allowed to rise gradually to reach a peak in 3–25 cycles as desired.

Newer high-grade controls for switching the current on and off are electronic. They start or stop the current to the welding transformer at or just after the power factor angle in the phase relationship when the voltage is low. This eliminates transient currents that may occur when the current is switched at random in relation to its cycle. Such transients could reach several times the size of the steady current and burn the electrodes and work.

Welding currents must be switched on and off in very short lengths of time and often repeatedly. Pulsating currents with definite on and off time are often needed. A typical welding schedule might be six cycles on and four cycles off, repeated 10 times. Complex electronic controls are used to provide a large variety of pulsation schedules with times of one-half cycle (.0083 sec) or less to as long as desired with exact duplication over any time period desired.

Small resistance welders may be foot-operated to hold the pieces together. Mechanical devices, such as a power cam, air cylinders, magnetic solenoids, and hydraulic operation, are all utilized. A common need is for delivery of the force in steps. A typical condition is to apply a moderate force during passage of the welding current to assure proper contact resistance. Then, when welding temperature is attained and the current is shut off, the force is increased considerably to complete the weld and forge the metal to help refine the grain structure. On most newer welders, the mechanical schedule is automatically controlled and synchronized with the electrical program.

Many welding machines have their electrodes on the ends of two long arms or horns to accommodate the work. The leads to the electrodes constitute a loop of the secondary circuit in which a heavy current flows. A metal workpiece in the throat between the horns may add considerably to the impedance of the loop and appreciably reduce the AC that can flow at a given voltage. The loss goes into heat generated throughout the workpiece by induction. The amount of impedance increase depends upon the magnetic properties of the material and how much of the workpiece extends into the loop.

13.5.3 Spot Welding

Spot welding is the most common form of resistance welding and the simplest. It is commonly done on sheet metal up to about 3.18 mm (.125 in.) thick, but sometimes even on thicker metals. The essentials are shown in Figure 13-15. Electrodes with reduced ends are pressed against the work, the current is turned on and off, and the pressure is held or increased to forge the weld while it solidifies. This is commonly repeated in a series of spots along a joint.

If a welded spot between two sheets is sectioned, a small mass or nugget of metal is found embedded in and joining the sheets together as indicated in Figure 13-15(C). If a good weld is pulled apart, the nugget remains intact, and the sheets tear around it.

If an attempt is made to weld several thicknesses of metal, the current tends to spread out between the electrodes as indicated in Figure 13-15(D). Thus the current density is less at the inner than at the outer faying surfaces. Specific provisions must be made to assure enough heat to get good welds at the inner surfaces without burning near and at the electrodes.

Sound welds are obtained consistently by applying the proper current density, timing, and uniform electrode pressure to clean surfaces. The results of tests are given in Figure 13-16 to show the effects, in terms of weld strength, of the variables. One factor was varied at a time, and all other conditions were held constant. Welds were made with time values of 1–12 cycles at 60 Hz. No coalescence took place at less than four cycles with an 8,900-A current; that is, the peak of the temperature curve depicted in Figure 13-14 did not rise above the welding temperature. Weld incipience would occur in less time with higher currents; later with lower current. For time periods longer than the minimum for coalescence, the size and strength of the weld would increase with time, but at a descending rate. Little additional benefit can be expected from still longer time periods, and long time periods give the heat a chance to spread and harm the workpiece and electrodes.

As welding current was increased above 3,100 A with other conditions constant, the size and strength of the weld rose and then fell (Figure 13-16). The two high tests at 10,600 and 11,800 A showed considerable electrode indentation into the workpieces and squeezing of the metal out of the weld at the electrode pressure applied. Under a given set of conditions, usable current is limited.

It is commonly held that one sign of a good spot weld is little or no indentation from the electrode tip. The test results shown in Figure 13-16 indicate that there is an optimum electrode pressure for a given set of conditions. At this pressure, strength is highest and indentation low. Here are the reasons for the optimum performance. As the electrode force is increased, the resistance decreases at the faying surfaces between the metal sheets. The heat generated is proportional to the resistance. In these tests, large

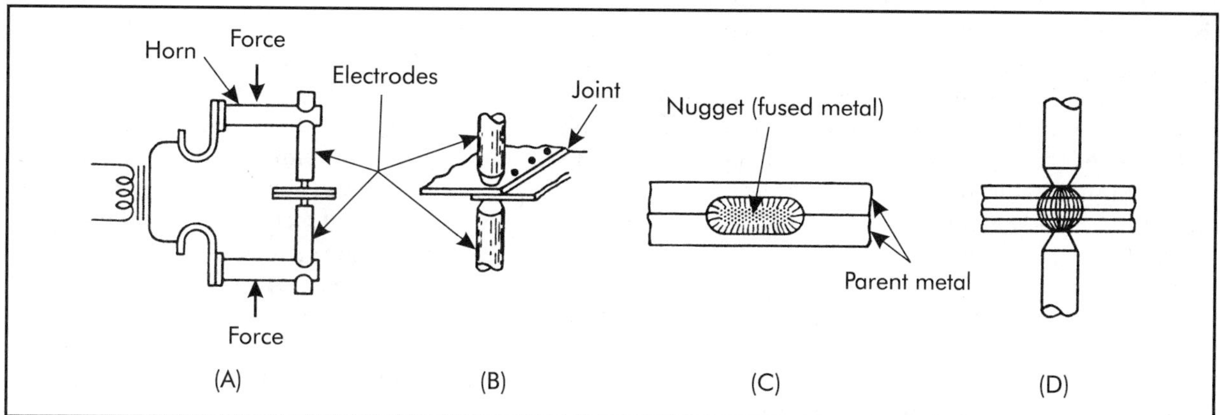

Figure 13-15. (A) Typical spot welding circuit; (B) spot-welded joint; (C) enlarged cross section through a spot weld; (D) the way the current spreads out when passed through several sheets.

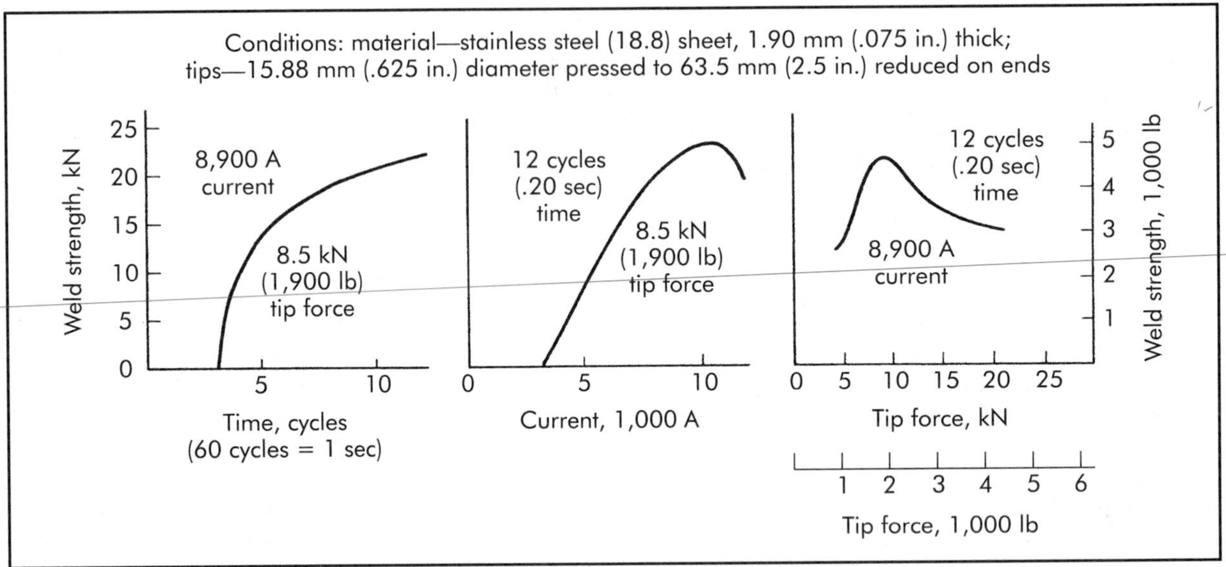

Figure 13-16. Curves showing how the strength of a spot weld varies with time, current, and tip force. (Courtesy Herbert Van Sciver)

amounts of metal melted and were squeezed out of the weld at low pressures because the resistance and heat were high. The workpiece showed large indentations below, but not above 4.5 kN (1,000 lb) tip force. A maximum strength was reached with 8.5 kN (1,900 lb) tip force. For larger forces, the resistance at the faying faces was less, the heat less, the weld smaller, and the weld strength less.

According to the test results (Figure 13-16), there is an optimum balance of current, time, and force in a resistance welding operation, and they depend on the material and its size. Too much current causes excessive indentation in a given operation. Typical values for a conventional system are indicated in Figure 13-16. However, a newer system utilizes much higher currents over time periods of a few milliseconds. It is reported to give a minimum heat effect without marring finished or painted surfaces, but the equipment costs several times more than the conventional kind.

In any case, the ideal condition is for the temperature to be low at the electrodes (Figure 13-14) and to rise sharply to a suitable peak at the interface of the workpieces. Both current and time are increased to spot weld thicker pieces because more heat is needed. Electrode size is also increased to carry the heavier current. The electrode force is increased even more. Operating specifications for various materials and sizes are given in handbooks and are usually adjusted to find the best conditions for any particular operation. Adaptive control of spot welding is described in Chapter 31.

Fatigue life of spot welds is short. Sometimes spots are coined after welding to increase fatigue life. Spot welding has not been favored for aircraft because of short fatigue life, but some combinations of spot welding with adhesive bonding have been found satisfactory.

Heat Balance

Two pieces to be resistance welded together should be heated the same amount so the weld becomes attached equally well to both. The principles will be explained for the simple case of spot welding, but the same apply to other forms.

If a thin and thick piece of the same metal are welded between electrodes as indicated in Figure 13-17, the thick one has more resistance and receives more heat. One solution is to make the electrode tip on the thinner piece of smaller diameter than the other. This raises the current density in the thin piece. Another solution is to face one electrode with a highly resistant material such as tungsten or molybdenum. The same problem exists for two metals of different con-

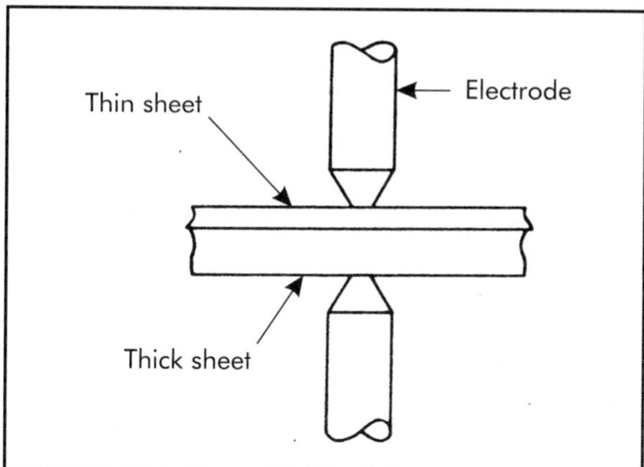

Figure 13-17. The problem of heat balance.

ductivity. The one of higher conductivity may be made thicker to realize balanced heat.

Machines

Spot welders may be classified as standard stationary machines, special multiple-electrode machines, and portable welders.

Portable spot welders are needed for large assemblies or spots that are hard to reach, for example, railroad freight cars. A *welding gun* consists of a pair of caliper arms carrying electrodes with a means for clamping them together. The transformer may be attached for low impedance or may be a separate unit. The gun is normally suspended from a spring-loaded cable on an overhead trolley. Controls are in a separate cabinet. One line of spot-welding guns ranges from manually operated models rated at about 3 kVA and weighing 14 kg (30 lb) to air-operated, water-cooled models rated from 15–85 kVA and weighing as much as 116 kg (256 lb).

A standard stationary spot-welding machine has an upper and lower horn carrying the electrodes and extending from an upright frame. Two types are the rocker-arm spot welders and the press-type spot or projection welders. The upper horn on the rocker arm type is pivoted in the frame and tilted upward to open the gap and downward to apply the electrodes. The press type has a ram on the end of the upper horn that moves the electrode straight up and down. In general, the various kinds of power sources, electrical circuits, controls, and mechanical drives described for resistance welders are found in spot welders.

Common sizes of standard stationary spot welders are 5–500 kVA. They are also rated by depth of throat, which designates the largest size of workpiece accommodated, commonly 203–914 mm (8–36 in.). Single-phase, 150-kVA spot welders are quoted at $12,000–$20,000, depending on their features. A three-phase spot welder delivers more power for the same kVA rating and is advantageous for some work, such as welding aircraft aluminum alloys and high-heat resistant, gas-turbine engine alloys. A 125-kVA, three-phase spot welder has a price of $50,000.

Special spot-welding machines with many combinations of electrodes are found on high-production jobs. Special care is taken to make multiple welds by using two or more electrodes in parallel to get equal current distribution to the welds. One system used one common welding transformer and a hydraulic cylinder for each electrode pair. The electrodes are brought in contact with the work one pair at a time. In another system, all electrodes are pressed against the work at one time, and the current from the transformer secondary is commutated to one electrode pair at a time. Some machines have a separate transformer and controls for each electrode set. Special machines are built in many sizes and shapes, each to suit a specific application. Some have been built to weld hundreds of spots in a single operation. They are rapid, obviate handling the work between welds, and are economical for large-quantity production. Some of the largest special spot welders, such as one to weld the panels and top of an automobile body together, may cost $1 million or more.

13.5.4 Projection Welding

Projection welding is done like spot welding, but the current is concentrated at the spots to be welded by projections preformed on the work as indicated in Figure 13-18. The electrodes are relatively large and are subject to a low current density, so they stand up well. The process is fast because a number of spots can be welded in one closure of the press.

Projections for welding may be made on sheet metal, cast, forged, or machined parts. A variation is called *stud welding*. A stud with its ends

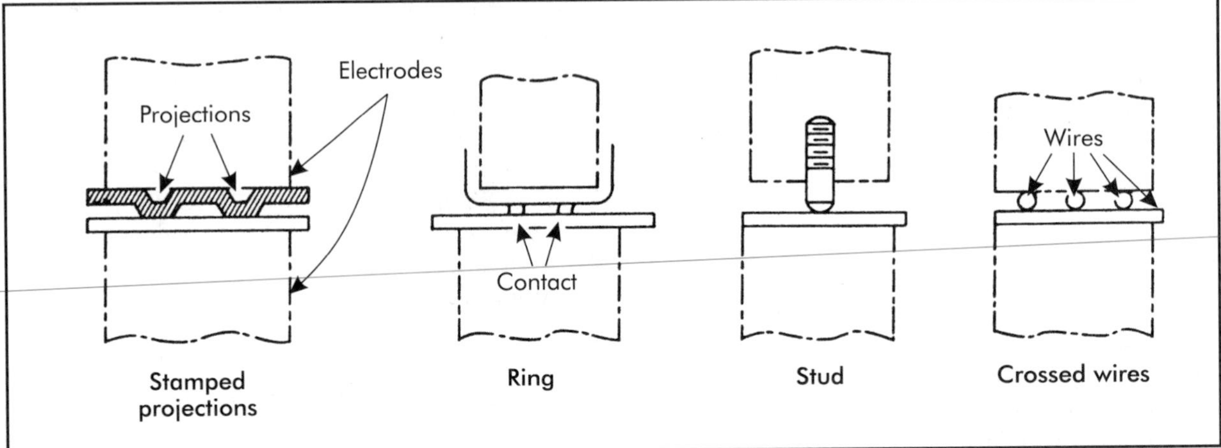

Figure 13-18. Types of projection welds.

rounded is held in one electrode and pressed against its mating part while current flows to heat the weld. The effect of projection welding is obtained with crossed wires, such as might be welded together into a grill.

A metal must have sufficient hot strength to be successfully projection welded. For this reason, aluminum is seldom welded by this process, and copper and some brasses not at all. Free-cutting steels high in phosphorus and sulfur must not be projection welded because the welds are porous and brittle. Most other steels are readily projection welded. The projections should be in the heavier of two mating parts so they will not be burned off before the weld is completed.

Spot-welding machines may be used for projection welding with the proper electrodes. Machines specifically for projection welding are like press-type spot welders and are rigidly built to hold parts in alignment during an operation. Projection welds usually require more current over shorter lengths of time and more pressure than spot welds.

13.5.5 Seam Welding

A *seam weld* is a series of spot welds either overlapping or spaced at short intervals. The latter is called a *roll spot weld* or *stitch weld*. Seam welding is done by passing the work between revolving roller electrodes as shown in Figure 13-19, or between a fixed bar and welding electrodes. A coolant is applied to conserve the electrodes and cool the work rapidly to speed the operation. Commonly, the rollers run continuously along a seam, and the current is interrupted. Another method utilizes a steady current with intermittent motion of the rollers or with notched rollers. Surfaces to be seam welded must be thoroughly cleaned, descaled, and deoxidized for satisfactory results. Pickling is a preferred method of preparing surfaces.

Both lap and butt joints may be seam welded. Continuous seams are gas and liquid tight. Seam welding is done on metal sheets and plates from 0.08–4.75 mm (.003–.187 in.) thick with standard equipment and up to 9.53 mm (.375 in.) thick with special machines. Seam welding is done mostly on low-carbon, alloy, and stainless steels, but also on many other metals including aluminum, brass, titanium, and tantalum. Typical seam-welded products are mufflers, barrels, and tanks. Specifications for proper seam proportions and designs for various materials and applications are tabulated in handbooks. A good seam is stronger than the parent metal.

A seam welding machine may be made to do either flat (horizontal) or circular work or both. Seam welders have much the same elements as spot welding machines and also are classed as rocker arm and press types. In addition, a seam welder must have a drive for turning the rollers and be able to supply heavy currents and forces. Current must be strong because much of it is lost through spots near the weld of the moment. An ordinary duty cycle for seam welding may be

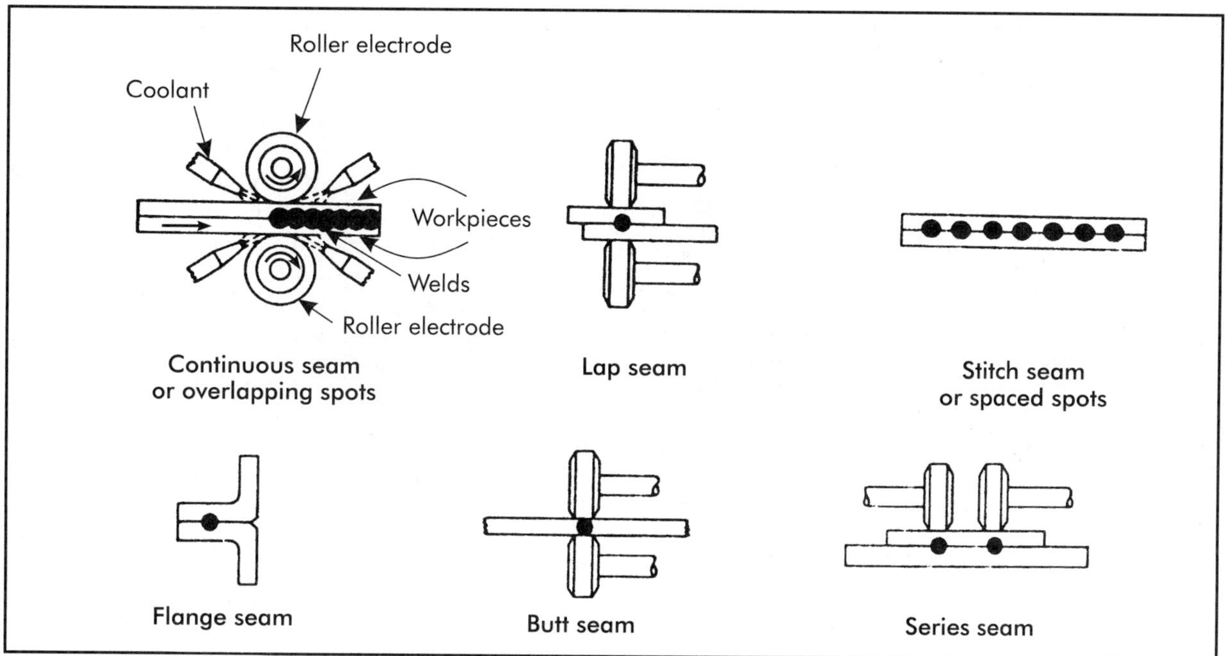

Figure 13-19. Common types of welded seams.

as high as 80%, while spot welding is seldom over 10%. A recommended pressure for low-carbon steel is 103 MPa (15,000 psi). Sturdy construction is necessary for the heavy forces. Good quality controls are necessary for precise timing. Accordingly, a fully equipped seam welder may cost 1.5 or more times as much as a spot welder with the same kVA rating and other features.

13.5.6 Upset-butt Welding

Upset-butt welding consists of pressing two pieces of metal together end-to-end and passing a current through them. Current densities range from about 310–775 A/cm^2 (2,000–5,000 A/in.^{2}). The resistance of the contiguous surfaces under light pressure heats the joint. Then the pressure is increased. This helps the metal from the two parts to coalesce and squeezes some metal out into a *flash* or *upset* as depicted in Figure 13-20. The current may be interrupted one or more times for large areas. Final pressures range from about 17–55 MPa (2,500–8,000 psi), depending on the material. Too little pressure leaves a porous and low-strength joint; too much squeezes out an excess of plastic metal and makes a joint of low impact strength.

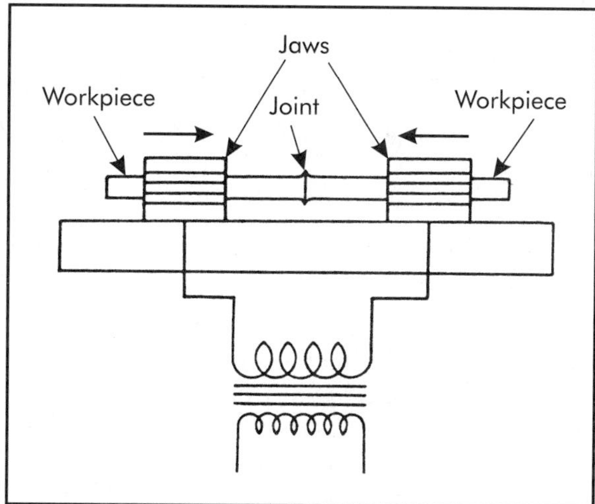

Figure 13-20. Upset-butt welding.

Metal is not melted when upset-butt welded, and there is no spatter. The upset is smooth, symmetrical, and not ragged, but the ends of the pieces usually must be machined before welding. Most metals can be upset welded. Common applications are the joining of wires and bars end-to-end, welding of a projection to a piece, and welding together the ends of a loop to make a wheel rim.

13.5.7 Flash-butt Welding

In *flash-butt welding*, two pieces to be flash welded are clamped with their ends not quite touching. A current with a density of 310–775 A/cm^2 (2,000–5,000 A/in.^{2}) is applied to a piece and arcs across the joint in a flash that melts the metal at the ends of the pieces as indicated in Figure 13-21. A pressure of about 35–138 MPa (5,000–20,000 psi) is applied suddenly to close the joint, clear out voids, and extrude impurities. On upsetting, the current density may rise as high as 7,750 A/cm^2 (50,000 A/in.^{2}). Flying particles of molten metal expelled from the joint are detrimental to equipment and personnel and present a fire hazard.

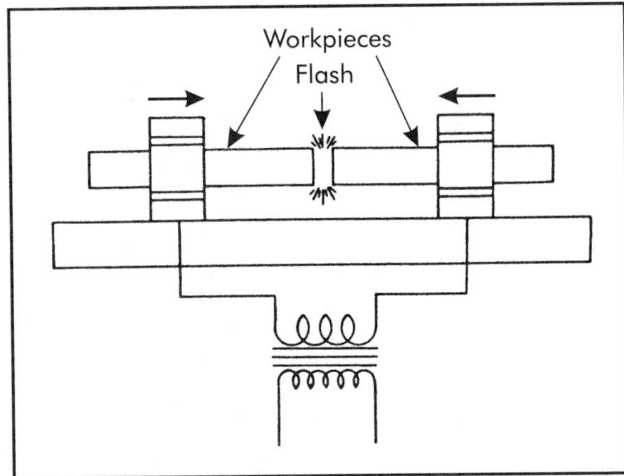

Figure 13-21. Flash-butt welding.

Most commercial metals may be flash welded. Pieces welded together in this way should each have about the same area at the joint. Common applications are the end welding of sheets, strips, and bars. An example is flash welding the end of one coil of stock to the next for uninterrupted passage of material through a continuous rolling mill. A flash weld has a small amount of sharp and ragged upset material around the joint. Because the metal is melted, a weld of full strength may be obtained even with dissimilar metals. The ends of the pieces need no preparation. Flash welding requires less power, is faster, and heats the whole piece less than upset welding.

Magnet-arc welding is a form of flash-butt welding to join steel tubes with a minimum wall thickness of 3.18 mm (.125 in.) and a maximum outside dimension of 51 mm (2 in.). A magnetic coil energized by an alternating current is placed around the joint between two workpieces held as shown in Figure 13-21. A shielding gas is injected. An arc is struck between the pieces and rotates around the joint under the influence of the magnetic field. When the ends of the pieces are heated to forging temperature, they are forced together to complete the weld. This process is reported to take less current and time, require less force, and produce less flash than any of the others for joining tubing.

Both upset- and flash-butt welding have the advantages over other joining operations of yielding reliable joints, being fast and of low cost, handling dissimilar metals and shapes not suitable for other processes, and requiring little operator skill on readily available equipment. Extruded metal generally must be removed from around the joints.

13.5.8 Percussion Welding

In *percussion welding*, two parts to be welded are clamped end-to-end but some distance apart. One is on a spring-loaded slide. When the machine is tripped, the one part is hurled into contact with the other. The parts are connected to the terminals of a high-voltage charged condenser that flashes an arc between them for about 0.001 sec before they touch. The arc fuses a layer about 76 μm (.003 in.) thick on the end of each part. The molten and softened metal is forged into a weld as the parts slam together.

Percussion welding is a fast method of joining and gives very little splatter, flash, and heat-affected zone. It can handle dissimilar metals. The process is limited to welding areas of less than 3 cm^2 (.5 in.^{2}). For connecting small wires to electrical components, percussive welding has been found to give strong and more reliable joints than crimping or soldering. However, it is more expensive for making less than millions of joints because equipment is relatively expensive.

13.6 THERMIT WELDING

13.6.1 Principles

Thermit welding is done by filling a joint with molten metal obtained by reducing its oxide by aluminum. Aluminum reduces any oxide but that of magnesium, so thermit welding is done mostly with iron, steel, and copper. Finely divided aluminum and magnetic iron oxide, in proportions of 1:3 by weight, are mixed and ignited and react as follows:

$$8Al + 3Fe_3O_4 \rightarrow 9Fe + 4Al_2O_3$$

The reaction raises the temperature to 2,500–2,800° C (≈4,500–5,000° F). The metal produced is about one-half of the original mixture by weight or one-third by volume.

13.6.2 Procedures

The first step in making a thermit weld is to prepare a mold for the metal. The pieces to be welded are positioned and fastened in place. A wax pattern of the weld is made around the joint, and a sand mold is rammed around the zone to be welded. A typical setup is illustrated in Figure 13-22. A torch is inserted into the mold to melt out the wax and heat the workpieces to cherry red.

The thermit mixture is placed in a crucible and ignited by a welding torch or by the addition of a small amount of barium peroxide and magnesium ribbon. The reaction takes about 30 sec to produce up to a ton or more of metal, which flows into the mold around the parts to be welded. The superheated weld metal fuses appreciable amounts of the parent metal, and all solidifies on cooling into a strong and homogeneous weld. Later the mold is torn down, gates and risers are removed, and the weld is chipped and cleaned.

13.6.3 Applications

Thermit welding is largely, but not entirely, applied to joining heavy sections. It is able to supply a high amount of heat rapidly to parts that have high heat capacities. Typical applications

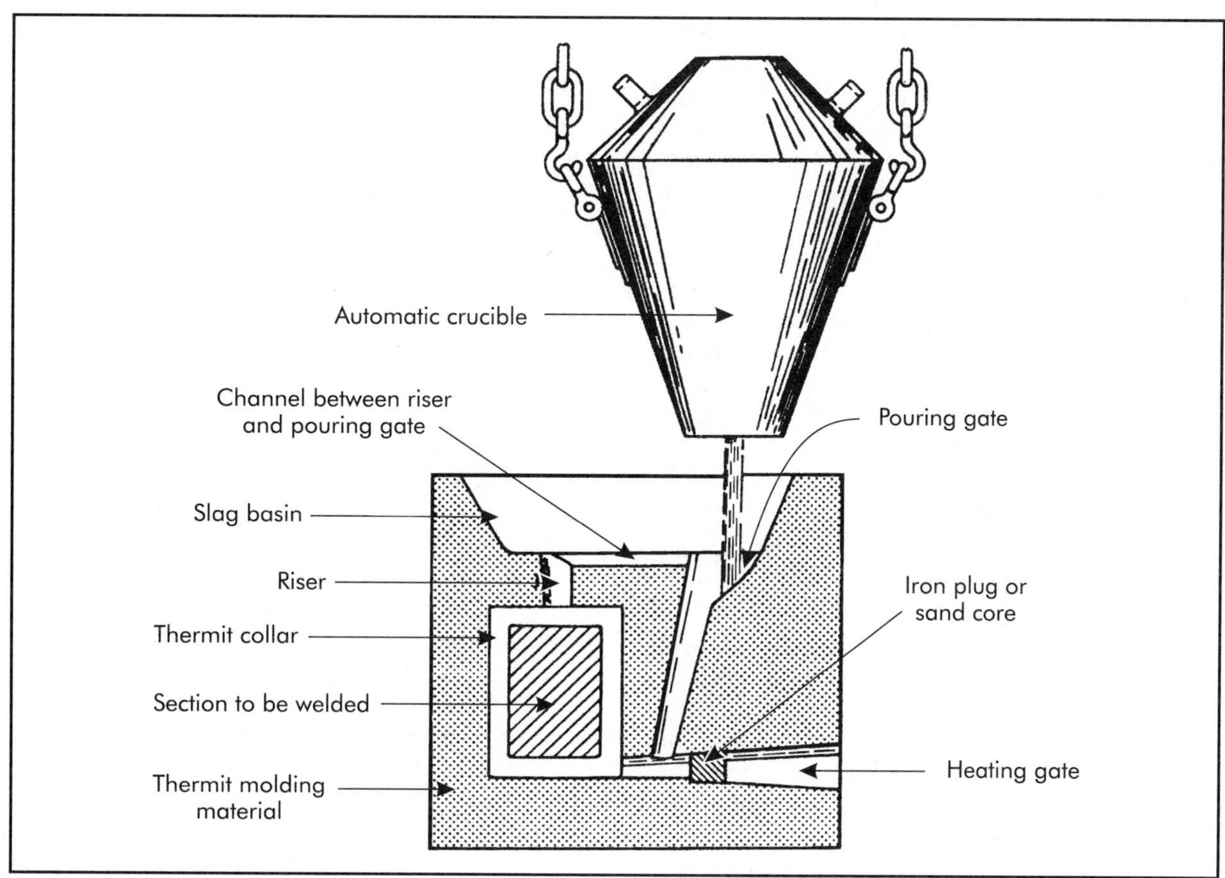

Figure 13-22. Thermit welding process.

are the joining of cables, conductors, rails, shafts, and broken machinery frames, and rebuilding large gears. Forgings and flame-cut sections may be joined in this way to make huge parts. Sometimes this is the only feasible method and often it is the fastest method of welding large pieces. The composition of the weld metal may be controlled by the addition of steel scrap or various alloys in oxide form in the original mixture. Tensile strengths from 345–758 MPa (50,000–110,000 psi) and elongations of 20–40% in 51 mm (2 in.) are commonly obtained. Thermit welds are considered better than cast steel, and test specimens can be bent flat on themselves. Complete deoxidation occurs, slag has ample egress, and air is excluded from the weld. Slow cooling relieves stresses.

13.7 GAS WELDING

Gas welding is done by burning a combustible gas with air or oxygen in a concentrated flame of high temperature. As with other welding methods, the purpose of the flame is to heat and melt the parent and filler metal of a joint.

Much gas welding has been replaced by faster electric-arc and resistance welding, but it still has important uses. Its temperatures are lower and controllable, which is necessary for delicate work, such as sheet metal and tubing. It can weld most common materials. Equipment is inexpensive, versatile, usually portable, and serves adequately in many job and general repair shops. Probably of more importance is that gas heating is the means for pre- and postheating, flame cutting, metal spraying, braze welding, brazing, and soldering.

13.7.1 Fuel Gases

Acetylene is the most important hydrocarbon in the welding industry. Newer stabilized mixtures of methylacetylene-propadiene (MAPP), have been gaining favor. Other commercial fuel gases are hydrogen, propane, butane, natural and manufactured illuminating gas, and chlorine burned with hydrogen. The properties that explain the advantage of acetylene are compared with those of propane (as an example of other kinds of gases) in Table 13-1.

Acetylene produces higher temperatures than other gases because it contains more carbon and releases heat when its components (C and H) dissociate to combine with oxygen and burn. Most other fuel gases, like propane, absorb some of the heat of combustion when their elements dissociate.

Acetylene is colorless and has a sweetish, but to many, an obnoxious odor. It is generated industrially by a controlled reaction of calcium carbide in water; the formula is $CaC_2 + 2H_2O = C_2H_2 + Ca(OH)_2$. The calcium carbide is a gray stone-like substance made by fusing limestone and coke together in an electric furnace. The gas may be generated in a central plant and compressed into cylinders for distribution or generated as needed.

The main disadvantage of acetylene is that it is dangerous if not carefully handled. By law, free acetylene is limited to pressures of 103–138 kPa (15–20 psi) because it explodes at over 172 kPa (25 psi), and sometimes below. However, it can be stored safely at about 1.4 MPa (200 psi) if dissolved in acetone. A steel tank or cylinder for storing acetylene is packed with 80% porous material such as asbestos, balsa wood, charcoal, infusorial earth, silk fiber, or kapok. The packing is saturated thoroughly with acetone, which is capable of absorbing acetylene to the extent of 25 times its volume per atmosphere of pressure. Gas is forced in to charge the cylinder and withdrawn for use through a valve on top of the cylinder. Safety fuse plugs are provided to relieve the pressure upon exposure to fire. As a safety measure, tanks are emptied in use at a rate not exceeding .20 tank/hr.

MAPP is comparable to acetylene in performance and cost as indicated in Table 13-1, but can be stabilized by additives. Thus MAPP as commercially available can be handled and subjected to shocks without danger. It can be stored at pressures of 1.4 MPa (200 psi) and over in simple tanks and used in high-pressure streams for faster operations, such as in metal spraying.

Fuel gases are burned with commercially pure oxygen for higher temperatures. For industrial purposes, oxygen is extracted by liquefaction of air and distributed in steel cylinders at about 14 MPa (2,000 psi) pressure.

Hydrogen is burned with oxygen for low flame temperatures, below 1,982° C (3,600° F), which is advantageous for thin sheets and materials

Table 13-1. Some properties of gases for welding and cutting

Property	Acetylene, C_2H_2	MAPP C_3H_4	Propane, C_3H_8
Gross heat of combustion, MJ/kg (Btu/lb)	50.0 (21,500)	49.0 (21,100)	50.4 (21,700)
Heat released, MJ/kg (Btu/lb)	8.6 (3,720)	4.7 (2,020)	
Heat absorbed, MJ/kg (Btu/lb)			2.4 (1,020)
Oxygen required, kg/kg (lb/lb)	3.1	3.2	7.3
Heat in combustion products, MJ/kg (Btu/lb)	13.56 (5,830)	11.70 (5,030)	8.02 (3,450)
Maximum flame temperature, °C (°F) (burned in O_2)	3,087 (5,589)	2,927 (5,301)	2,526 (4,579)
Cost for flame cutting*	$5.42	$5.18	$5.96

* Relative costs to cut 30.5 m (100 ft) of 25.4-m (1-in.) thick low-carbon steel plate with oxygen at $0.18/m³ ($0.50/100 ft³) (*Metals Handbook* 1982).

that must not be overheated. The flame can be made to slightly reduce for good-quality welds free of oxides. The hydrogen is stored in cylinders at up to 14 MPa (2,000 psi) pressure.

13.7.2 Oxyacetylene Gas Welding

Much *oxyacetylene gas welding* is done manually. Gas coming from either the acetylene or oxygen tank is first reduced in pressure through a *regulator*. This is a diaphragm-operated valve that can be adjusted to let only enough gas out of the tank to maintain a desired pressure on the outlet side. Gages on each regulator show the tank and the hose pressures. Basic equipment used in oxyacetylene gas welding is illustrated in Figure 13-23.

Hoses conduct the gases to the torch or blowpipe held by the operator. The torch mixes the two gases properly and emits them into the flame. The torch basically consists of regulating valves, a body, mixing head, and tip. A number of sizes of tips are available to give different sizes and intensities of flames for various purposes.

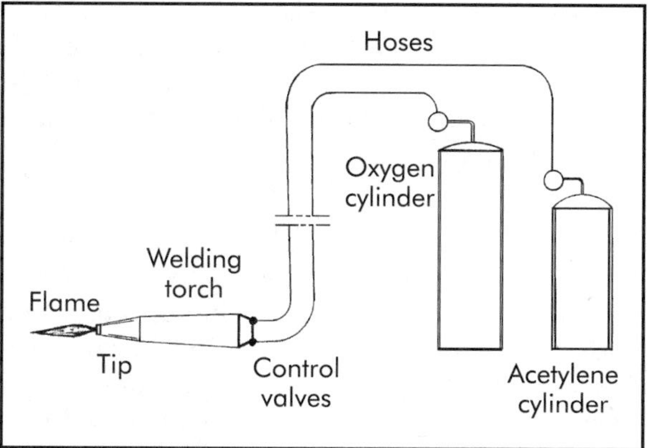

Figure 13-23. The basic equipment used in oxyacetylene gas welding.

Torches are classified as *low-pressure* or *injector*, and *medium-pressure*, also called *equal-* or *positive-pressure* types. Low-pressure acetylene torches operate at up to 35 kPa (5 psi), and medium-pressure torches in the range of 7–103 kPa (1–15 psi). In the low-pressure type, high-pressure oxygen is injected through a venturi and draws the necessary amount of acetylene along through the mixing chamber and out through the tip. Both gases pass through the medium-pressure torch at about the same pressure.

Three distinct types of flames (Figure 13-24) can be obtained from different mixtures of the gases. The *neutral flame* (Figure 13-24[A]) has no tendency to react with materials being welded. The highest temperature is at the tip of the inner cone and it is capable of melting all commercial metals and many refractories. The *carburizing flame* (Figure 13-24[B]) is distinguished by a reddish feather at the tip of the inner cone. It is capable of reducing oxides. Steel will take up carbon deposited on the surface and start melting at a lower temperature. An *oxidizing flame* (Figure 13-24[C]) assures complete combustion and the highest temperature, but has a strong tendency to oxidize metals being welded, which may be detrimental. Actually, heating is done with the inner cone, and the envelope of a flame shields and protects the weld zone from the atmosphere.

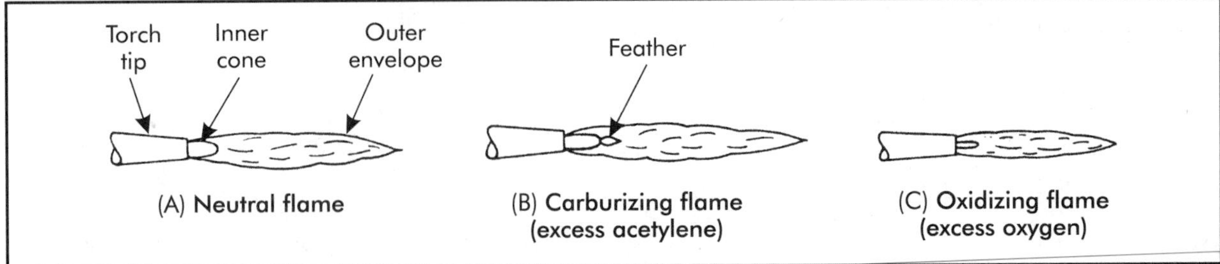

Figure 13-24. Three types of oxyacetylene flames.

A welding rod is applied to the weld. Such rods supply filler metal, help control the impact of the flame, and restore elements lost from the parent metal. Flux is also employed in the welding of such metals as cast iron, some alloy steels, and nonferrous metals to dissolve and remove impurities, control surface tension, and give protection from the atmosphere. It is usually a paste in which the rod is dipped.

The manual equipment for gas welding and flame-cutting operations costs about $300. Automatic oxyacetylene welding machines are used for some production operations, but they operate on the same principles as manual welding. They may cost several thousand dollars or more. Generally, they employ a number of small flames to obtain large amounts of heat.

Butt welding of steel can be done by oxyacetylene pressure welding. By one method called *closed-gap welding*, two pieces to be joined are pressed together, and the joint is heated by a torch or ring of torches until plastic. More pressure may then be applied, and heating is stopped to allow the joint to solidify. By a second method, the ends of two pieces are heated separately and then quickly pressed together. Applications of these methods include welding a piece of alloy steel to the end of a tool and joining pipeline sections and rails. This process is slower than electrofusion methods, but since metal is not melted, a cast structure is not formed at the joint, flash is not extruded, and cracking is minimized because the temperature gradient is not steep.

13.8 SOLID-STATE WELDING

Metals are held together by bonds between their atoms and crystals. Therefore, the atoms and crystals of two pieces must be brought into actual contact over their entire common area for a joint to be as strong as the parent material. Normally, even the smoothest surfaces touch only on a relatively few high spots when placed together. Also, metallic surfaces are covered with films of oxides, adsorbed gases, and water vapor, and are kept appreciable distances apart at the closest spots.

Fusion welding overcomes the barriers between pieces because the atoms of molten metal flow together. If cold metal pieces are pressed together hard enough, the projections can be mashed down to spread contact over wide areas, and oxides can be squeezed out. With some heat, the oxides become more fluent, the metal more plastic, and less pressure is needed for welding. This is done to a certain extent in the electrical resistance welding operation and is the basis for *forge welding*, one of the oldest forms of welding, practiced from ancient times by armor makers and blacksmiths. It consists of heating two pieces to red heat, applying a flux, such as borax, and completing the weld by hammering the metal on an anvil to press the joint together. Practice of the manual art is rare today, but some of the mechanical methods described here are counterparts. The processes described in this section are performed by application of pressure with or without heat.

13.8.1 Friction Welding

In *friction welding*, the ends of two pieces are pressed together while one is held still and the other revolved. Often the one piece is revolved in a spindle driven by a flywheel from which energy is drained, and so some people call the process *inertia welding*. Friction force between the abutting and sliding surfaces generates heat for welding and adds a forging component to that of pressure found alone in other butt-welding operations.

Common and unusual metals in similar and dissimilar pairs can be joined by friction welding for essentially round sections up to a solid 102 mm (4 in.) in diameter (much larger for tubing) on standard machines. The process is attractive for small and large quantities; operation time is short, from 0.2–2 sec. Pieces are shortened less than 0.5 mm (.02 in.) as a rule, and there is little flash and waste. Under proper control of energy and pressure, the metal is worked thoroughly, the bond is clean and usually as strong as the base metal, and results are uniform from piece to piece. The workpiece must be able to stand a large torque. The operation emits no bright light, sparks, fumes, or loud noises. A friction welding machine for rod diameters from 15.88–57.15 mm (.625–2.250 in.) or tubing from 25–89 mm (1–3.5 in.) diameter has a 30-kW (40-hp) drive motor.

13.8.2 Ultrasonic Welding

Two pieces may be bonded if pressed together and vibrated at ultrasonic frequency, one on the other, parallel to the contact interface. The ultrasonic transducer may be like the one in Figure 25-11 but acting perpendicular to the direction of pressure. The vibrations shatter surface oxides and films and can tear right through dirt and surface coatings to establish intermingling of nascent metals. The surfaces become hot and plastic, and a solid metallurgical bond is produced.

Welds may be made in spots or around rings in 1 sec or less, or along seams at up to 120 m/min (≈400 ft/min). The process is of advantage for thin wires, foils, and sheets of soft metals because no fusion occurs. There is no heat-affected zone, and dissimilar metals of similar hardness can be joined. No precleaning or filler metal is needed, and no contamination is introduced. The weld is as strong as the base metal. Even plastics of similar molecular structure are commonly joined by ultrasonic welding.

13.8.3 Explosion Welding

Surfaces may be joined by driving two pieces together with explosive force. Applications include cladding of sheets (up to 3 × 9 m [10 × 30 ft] in size) and simple forgings, lining and joining of tubes, and combining dissimilar metals into billets and various parts. Two pieces to be joined are placed a small distance apart, usually at an angle to each other, so that they collide along an advancing line when hit by the explosion. Under proper conditions, high pressures expel thin layers from impinging surfaces and crush the freshly exposed metal surfaces into intimate contact for a uniform bond. Heat is generated only incidentally and no melting occurs. Thus quite dissimilar metals may be joined without thermal interaction. The equipment is simple, but explosive shocks require some isolation.

13.8.4 Diffusion Bonding

Diffusion bonding is done by pressing pieces together while they are heated to below their melting temperatures, commonly in a vacuum or inert gas. Bonding takes place after a time because the base metals coalesce through interatomic diffusion. Surfaces to be joined must be machined flat and thoroughly cleaned to assure close contact over the entire area. Various metal interlayers (bonding aids) may be added (some are melted in the process) to reduce the temperature and pressure needed, provide closer contact, obviate oxide films, fill voids, hasten diffusion, or serve as stop-off layers in selective bonding.

Heating and pressing are done in a variety of ways for diffusion bonding. Individual assemblies may be pressed in a die with heater elements in a hydraulic press, or subjected to a dead-weight load in a furnace to apply forces in one direction. They may be processed in a heated pressure or vacuum chamber or autoclave for isostatic pressure.

In a new method, called *thermomagnetic* or *electromagnetic* forming, the pieces to be joined are placed inside an induction coil and heated by eddy currents. Then a large capacitor bank is discharged through the coil, which induces currents of 40,000–100,000 A in the parts and sets up pressure at the joint reported up to around 345 MPa (50,000 psi). A series of such pulses completes a solid-state weld in less than 10 sec, whereas older methods take minutes or hours.

Continuous strips or sheets may be heated and then rolled together in *roll bonding*. In *continuous seam diffusion bonding*, a heavy current is passed between the rolls through the work to heat the narrow zone under pressure. Roll bonding exerts high pressure under the rolls and effectively disperses surface oxides and impurities. Thus no protective atmosphere is needed for some metals.

No fusion takes place and no filler metal is needed for diffusion bonding. Thus no weight is add-ed, and the joint is as strong and temperature resistant as the base metal. No residual stress, excessive deformation, contamination, or crystal changes are introduced except by the severe rolling methods. Diffusion bonding is applicable to reactive and refractory metals, and for joining similar and many dissimilar metals, and thick and very thin pieces. The process is relatively costly and not competitive except for work difficult to join by other means, such as joining heat-sensitive or brittle materials as well as thin plate or foil. Important uses include welding metals that cannot be readily fused, such as zirconium and beryllium for nuclear reactor components and tungsten for aerospace parts, alternate layers of dissimilar metals for high strength at high temperatures, and thin honeycomb structures.

13.9 WELDING FUNDAMENTALS

Welding processes utilize different sources of energy. Thermal, chemical, electrical, and mechanical sources of energy generate a temperature that is high enough to produce a weld. The heating process brings several physical and metallurgical changes in the materials being welded. This section describes some of the fundamentals of the welding process:

- welded joints and symbols;
- metallurgy of welding;
- control of welding quality;
- welding defects;
- design for welding; and
- inspection of welded joints.

The method of heat application and other thermal properties of welded joints affect, to a large extent, the strength and toughness of welded joints. These factors govern the distribution of temperature in the welded joint. The composition and internal structure of the welded joint and other metallurgical characteristics depend on several factors, such as the amount of heat applied and the cooling method after the weld is made. Weld quality is basically controlled by these metallurgical changes in the welded joint. Design for welding is therefore an important aspect that is briefly discussed in this section.

13.9.1 Welded Joints and Welding Symbols

Common welded joints are illustrated in Figure 13-25; each has several elements. These are the type of joint, the type of weld, and the preparation for the weld. The elements can be put together in various ways. For example, a lap joint may be held by a fillet, plug, or slot weld, and a tee joint by a fillet or groove weld.

The nature of the joint depends upon the kind and size of material, the process, and the strength required. Material less than 0.25 mm (.010 in.) thick is usually lapped; thicker material is commonly butted. Butt joints are prepared for high-strength steels because they are more easily inspected and involve simpler stress patterns than lap joints. Lap joints are best for most pressure and resistance welding of sheets and for electron-beam welding where no filler metal is added. A joint may be given no particular preparation, such as a square butt joint. That joint requires no filler metal and may be as much as 80% faster to weld than one that does. However, a square butt joint depends on penetration into the edges of the piece and is not strong except for with thin material. On the other hand, a grooved joint leads to a stronger weld but requires preparation, and that adds cost. A joint is selected in each case to fulfill requirements at lowest cost.

Proportions of welded joints have been standardized. Preferred sizes, dimensions, and charts for calculating the strengths of and amounts of filler metal required for welded joints are given in reference texts and handbooks.

Position, as defined in Figure 13-25, is an important consideration for any welded joint. Gravity aids in putting down the weld metal into a flat position, which is the easiest and fastest to execute.

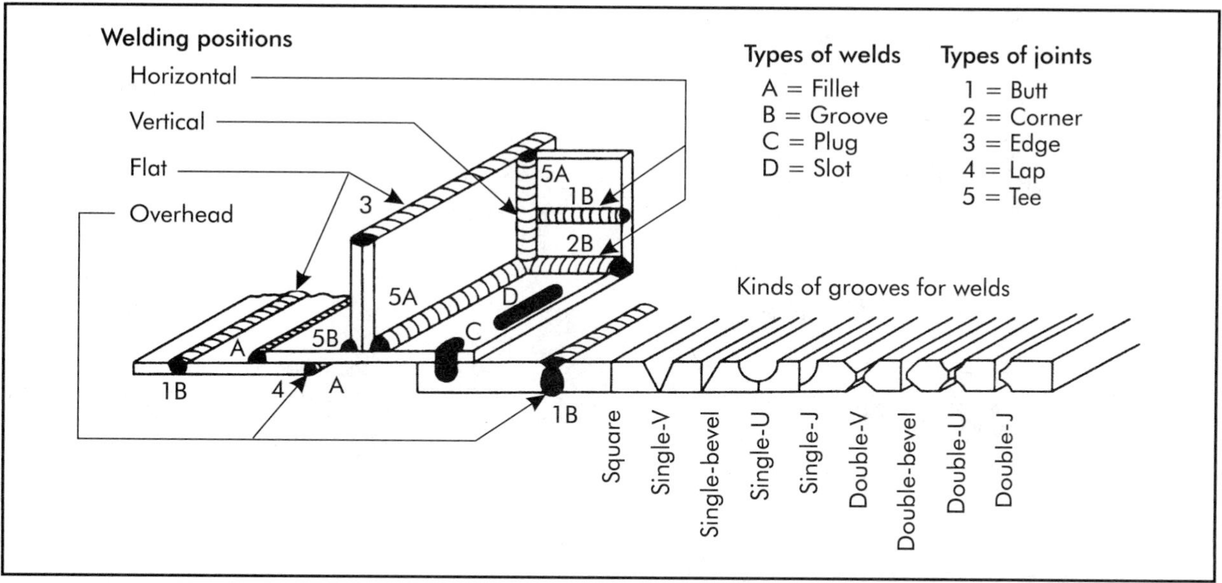

Figure 13-25. Types of welded joints.

Precise instructions for any welded joint can be given on a drawing by a system of symbols and conventions. This is a special language and is governed by definite rules, like the rules of grammar. A full list of symbols and their meanings is given in reference books. Several illustrations are presented in Figure 13-26.

13.9.2 Metallurgy of Welding

The heat of welding affects the microstructure and composition of the weld and base metal, causes expansion and contraction, and leaves stresses in the metal. A knowledge of what happens in metal when it is welded is necessary for an understanding of the welding operation, and will be explained here. For the sake of brevity, the discussion will be directed mainly to steel, but some important considerations for other metals will be included.

A cross section of a typical cooled weld is shown in Figure 13-27. The central mass, designated by *A*, represents metal that has been melted. It has the characteristic dendritic structure of a casting, which it essentially is. When the metal solidifies, it cools from the outside inward, and the crystals grow toward the center. Some segregation of constituents occurs with an alloy. The juncture of the dendrites in the center is weak, not so much in a ductile metal like steel as in a brittle one like cast iron. Slow cooling or subsequent annealing improves homogeneity and strength.

Parent metal adjacent to the molten metal in a weld is heated above the critical temperature. This is in zone *B* of Figure 13-27. Steel so heated recrystallizes to austenite around many nuclei to form small grains. The grains grow at higher temperatures and the structure becomes coarse. The metal nearest the molten pool almost reaches the melting point and becomes quite coarse. Farther from the center the structure is finer. Beyond that, in zone *C*, the parent metal is unchanged.

Coarse-grained steel hardens more readily than a finer structure of the same composition. A fine structure is tougher and stronger than a coarse one in most nonferrous metals because fine grains offer more points of resistance to slip.

Welding tends to harden high-carbon and alloy steels. The principles of hardening of steel have been explained in Chapter 6 and their role in welding will be elaborated upon here. With thick pieces, the welding zone is normally a small spot within a large mass of cooler metal. Heat flows off rapidly into the surrounding metal. When the heat source is removed, what is called a *mass quench* results. This causes hard martensite to form in hardenable steel in part or all of the zone heated above the critical temperature. Formation of martensite causes

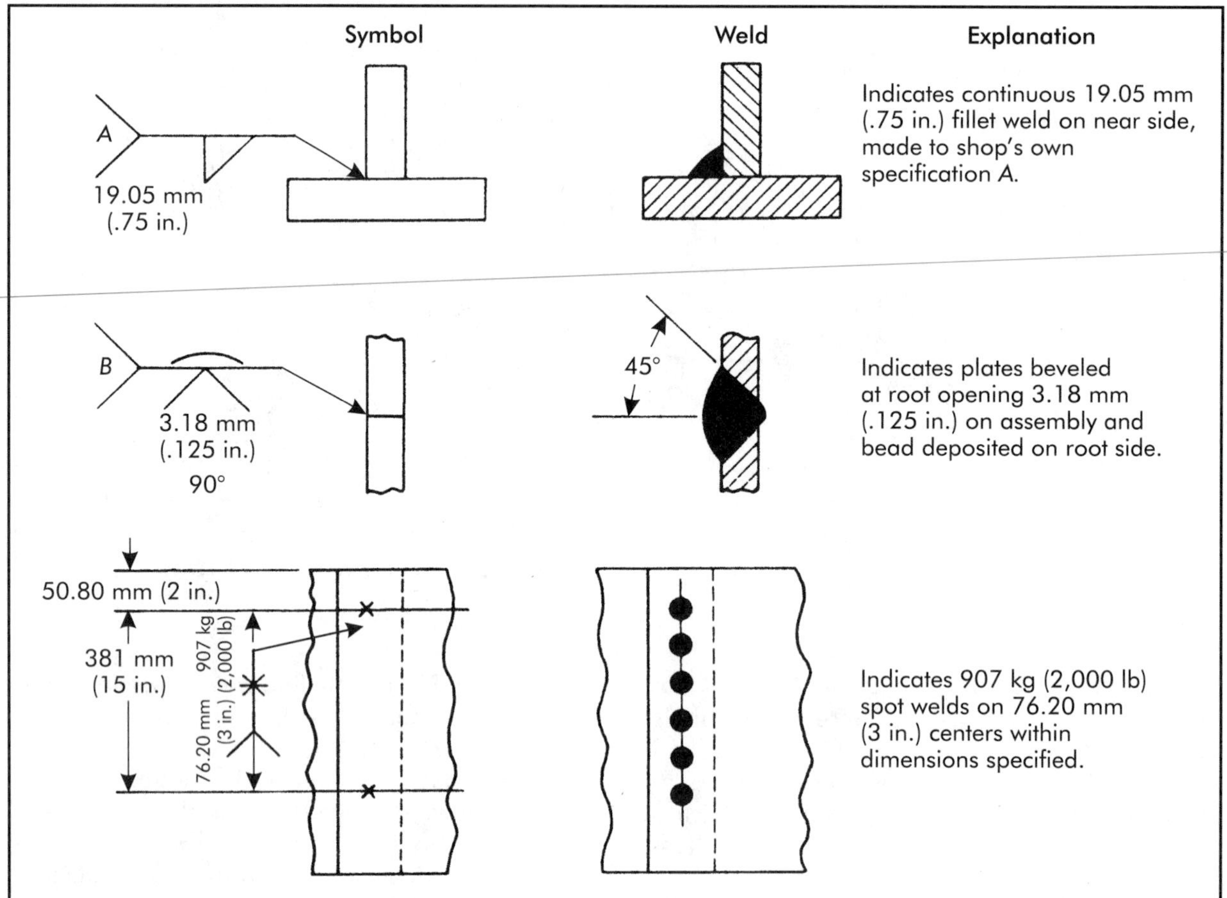

Figure 13-26. Examples of welding symbols.

cracking and must be avoided. The more hardenable the steel, the slower must be the cooling rate. The rate that heat is drawn off depends upon the mass of the surrounding metal and the difference in temperature between the weld zone and workpiece. Thin sections with little mass do not badly react to rapid cooling. A proper cooling rate for a thick section can be estimated from the time-temperature diagram for the material, like that illustrated in Figure 6-3. The workpiece temperature can be raised to retard cooling by preheating the work and by mild but pervading heat input during welding. A workpiece may be heated all over or locally; in either case, preheating adds substantially to the cost of the operation. One authority prescribes that all steel sections over 38 mm (1.5 in.) thick must be preheated. As for heat input during arc welding, it is increased by heavier currents and slower traverse rates, as explained in the discussion of Equations (13-2) and (13-5). Figure 13-5 indicates welding methods capable of large heat inputs. Care must be taken to avoid overheating and excessive penetration. Gas welding heats more slowly but more pervasively and causes less hardening than arc welding.

Chilling of welded pieces must be avoided with hardenable metals but otherwise has benefits. The metal is exposed for a shorter time to gas contamination, the heat-affected zone is smaller, the grain structure is finer, and a higher-strength weld often results.

Some aluminum and other nonferrous alloys are hardened by keying agents that are dissolved on heating. The structure of such metal may be altered in the vicinity of a weld, and subsequent heat treatment may be necessary to obtain desired properties throughout the material.

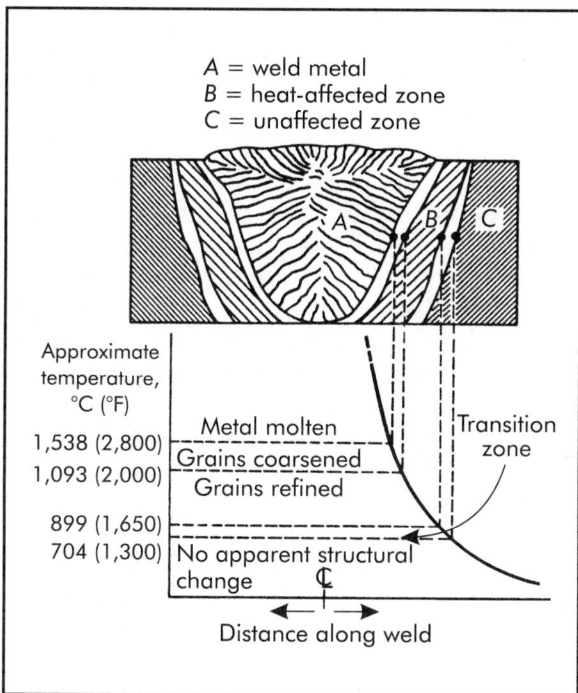

Figure 13-27. Temperatures attained and the resulting structure in a typical weld in steel.

13.9.3 Control of Weld Quality

What has been shown to happen in a weld explains the faults the weld may have and their remedies. These are discussed in the following sections.

Much can be done by design and welding practice to enable components of a weldment to yield and move slightly to alleviate stresses and warping. Effective ways are to use as thin a material and as little filler metal as possible, to preheat to minimize temperature differences between the weld and base material, and to weld from the inside or confined portion of a structure to the outside or points of most freedom. Manual welding is inherently variable; automatic welding provides better control of quality.

Weld Defects

The distortion in the welded joints caused by the stresses and heat on the part being welded, in addition to the misuse or lack of operator training, result in weld defects, some of which are listed and briefly explained in the following sections:

- cracks, voids, undercuts, and craters;
- shrinkage defects; and
- corrosion.

Cracks, voids, undercuts, and craters. A cracked weld is a defective weld and usually is rejected. Cracks may occur in the weld metal or in the parent metal, lengthwise or normal to the weld. They may occur during or after welding.

Martensite formed in welding steel is a leading cause of cracking. The martensite is brittle and does not yield but breaks when the stresses in the weld become high enough. Moreover, the martensite adds to the strains and resulting stresses. It has a lower density and occupies more volume than the softer steel from which it is formed. Measures previously described to prevent hardening and minimize stresses are effective to prevent cracking.

Some alloys weaken and crack at welding temperatures. As an example, magnesium alloys with more than about 0.15% calcium have a grain boundary constituent with a low melting point. This causes frequent cracking upon welding. The remedy is to change and control the raw material. The formation of intermediate alloys may promote cracking if dissimilar metals are welded together.

Hydrogen is a major cause of cracking in steel. It is highly soluble in hot steel but is largely expelled on cooling. If the gas cannot escape, it exerts pressure to crack the metal. This is likely to happen under the same conditions that form brittle martensite and aggravate the cracking of the martensite. Low-hydrogen electrodes and flux, cleanliness, preheating, and slow cooling to allow the hydrogen to escape help eliminate hydrogen cracking.

Tests have shown that voids up to about 7% of cross section do not materially change the tensile or impact strength and ductility of a weld. In excess of that amount, inclusions of foreign matter and blow holes or gas pockets weaken welds appreciably and act as stress risers from which cracks spread. Inclusions are usually slag, but may be scale and dirt. Grease, oil, or paint left on surfaces liberate gases when heated. Surfaces must be clean before welding to avoid contamination. Most welding is done shielded from the atmospheric gases. Much can be done with welding techniques to keep the welding pool at sufficient temperature for a long enough time to liberate gases and float out slag and other contamination. Free-cutting sulfur steels are particularly difficult to weld because of the copious release of hydrogen sulfide and sulfur dioxide gases.

An *undercut* is a groove melted in the base metal alongside a bead. A *crater* is a depression left at the end of a bead. These may be weak points in highly stressed weldments. They may be prevented by proper welding procedures.

Shrinkage. As a weld cools, it shrinks. If the members joined are restrained from moving, high stresses are induced across the weld. A concave bead has higher stresses, particularly at the surface, and is less desirable than a convex bead. An excessively deep and narrow weld also becomes highly stressed. If the stresses are high, they impair fatigue strength and cause cracking. A bead may be weakened by use of filler metal different from the base metal if there is a difference in expansion rates. This is prone to happen, for instance, if steel filler is used to weld cast iron.

Heating and cooling cause expansion and contraction in various and different places as a weld is made along a joint. The metal in the joint is hotter and tends to shrink more on cooling than the bulk of the pieces on either side. However, lengthwise shrinkage is restrained by the metal adjacent to the joint, and that induces a residual tensile stress along the joint that may exceed the yield point of the material. This pulls some parallel fibers of the base metal into light compression. If pieces being welded are restrained in a fixture or by the structure to which they are attached, stresses may be induced in any direction by contraction of the metal in the welded zone when it cools, and thus cause cracking. Even if cracking does not occur, residual stresses may impair fatigue strength significantly. This commonly causes warpage, especially when the welded material becomes unbalanced by later machining.

Annealing, called *post heating*, is a means of relieving stresses, but is not always desirable because it affects the properties of the base metal of the parts. Shot or hammer peening has been found to increase the fatigue strength of welds more than heat treatment. In that way, fatigue life may be restored to the unwelded material.

Corrosion. Welding makes metals more susceptible to corrosion in a number of ways. Actually, corrosion occurs from the attack of the air on the hot metal in the welding operation if shielding is not adequate. The intense heat of welding effaces conversion and other protective coatings from metal surfaces. Clad metals should be welded with the same composition as the surface metal, at least on top of the weld. From the standpoint of design, the inner surfaces of a lap joint are difficult to protect from corrosion. Notches and cracks that come from welding are foci for corrosion.

The heat of welding causes changes in some metals to make them more susceptible to corrosion. Welding can make stainless steel lose its corrosion resistance. The reason and remedies for this are explained in Chapter 5. When some aluminum alloys are heated by welding, parts become overaged and subject to corrosion.

Design for Welding

A product to be fabricated by welding must meet two requirements. First, it must have adequate strength, rigidity, endurance, etc. Second, it must be producible at the lowest possible cost. The first of these considerations is in the realm of mechanics of materials and is largely outside the scope of this book. For the second, the designer must understand the principles and practices of welding to utilize them most effectively. Several such points will be reviewed here to indicate how they influence welded designs.

Some of the main principles of designing for welding are illustrated in Figure 13-28. The bent enclosure welded on two ends in view (A) is easy to fit to the assembly and provides welding grooves without preparation of the edges. An inaccessible enclosure in a weldment should be completely sealed to inhibit corrosion. View (B) illustrates the rules that sections welded together should be about the same size for the least amount of heat distortion and that thin sections should be joined rather than thick ones. As indicated, provision must be made for shrinkage and machining after welding. View (C) shows a weld between thin sections. If a standard shape without a flange was available for a hub, it might be most economical even if some extra care in welding was necessary. But if a hub must be machined anyway, the flange should be added to match the spokes. View (D) illustrates how a joint should be welded if subsequent machining is to be done on one of the pieces. With the weld in the position shown, there is no need to machine it. Hard spots in the weld can damage cutting tools.

With the nature of the process in mind, a designer specifies welds in the easiest positions and provides ample accessibility to them. The principles of welding explain why high-carbon and alloy steel and cast iron are hard to weld. Low-carbon steel is used for welding wherever possible. For each weldment, thought should be given to choosing the most suitable welding process. Often several may be selected for one product for the best results. As an example, a bicycle frame is made with seven resistance welds, four spot welds, four electric-arc welds, and four oxyacetylene brazed joints, each where it serves most economically.

Welding can be a means to use material most efficiently but it must be applied properly to be efficient. A flat plate of about the proportions of Figure 13-28(E) was fixed at one end and twisted 10° at the other end under a given torque. A box section like that of Figure 13-28(F) was a little more rigid than the plate and twisted 9°. Under the same conditions, a box with diagonal braces, as in Figure 13-28(G), was twisted only 0.25°. It proved 36 times as rigid as the cross braced box with 6% less material. Many weldments are designed to replace castings for machine frames or bases, jig and fixture bodies, etc. A proven rule is that such a design should not just follow the lines of the castings, but should be a complete revision to take full advantage of the benefits of welding.

Normally, no effort is made to hold small tolerances on raw weldments. The tolerances that can be held depend upon those of the component pieces, the errors in fabrication and fitting, and the distortions from welding and heat treatment. Tolerances on weldments are commonly broad; 1.5 mm (≈.063 in.) is feasible in many cases. Surfaces that need to be held to smaller tolerances are left with stock and are machined after welding and annealing.

Inspection of Welded Joints

Welded joints may be subjected to destructive or nondestructive testing by the methods described in Chapter 3. Destructive testing may be performed on a sampling or specimen basis. Visual inspection is often the only way welds are

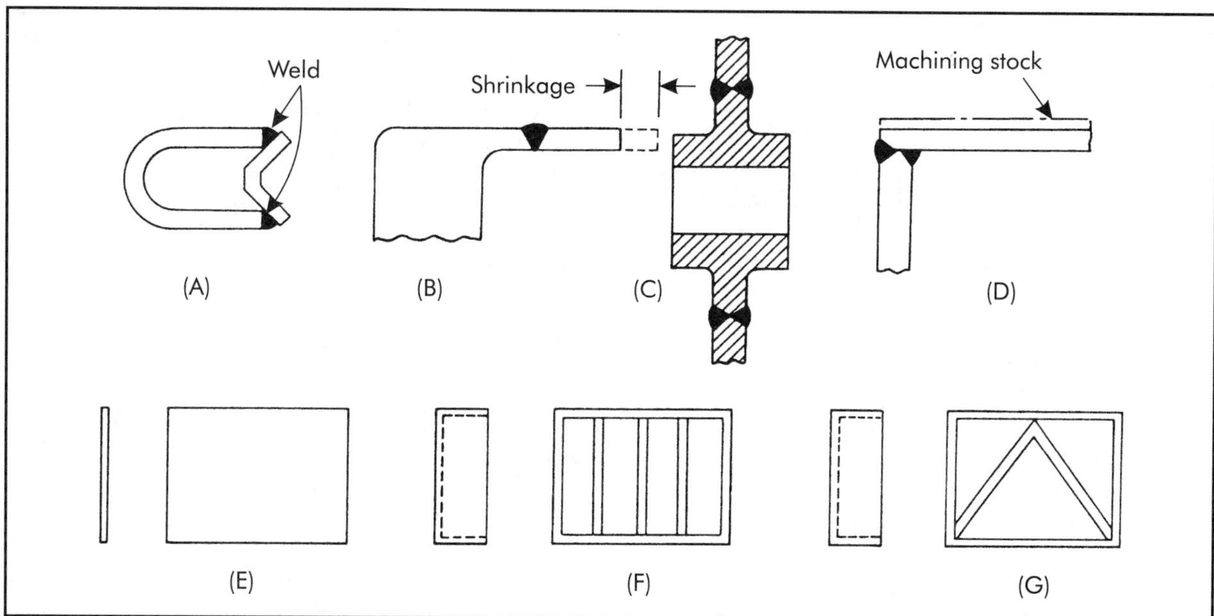

Figure 13-28. Weldment design principles.

checked. Much can be judged in this way because good workmanship shows itself in uniform, properly shaped and filled, and attractive welds. This is often the best assurance of quality.

Inspection should start on the welding floor to detect faulty methods and procedures that cause defects. Inspection is aided if weldments are designed so that welds are as accessible as possible for inspection.

Various systems have been set up to specify how welds should be inspected and what flaws are permissible, but no one code is universally recognized. An underlying idea is that the amount of effort and time put into welding a piece and inspecting the results should depend upon the severity of service to which the product is subjected. A part to be highly stressed under critical conditions should be of higher quality than one in light service.

13.10 COST COMPARISONS

Because of the many applications and processes of welding, it is necessary to provide some discussion about welding costs and cost comparisons with other competing processes. This section briefly presents some discussion related to the economics of welding, such as welding costs, a comparison of welding with casting, a comparison of welding and riveting, and a comparison of arc welding processes.

13.10.1 Welding Costs

The cost of a weldment includes a number of items. Among these may be the costs of material, machining before and after welding, forging, forming, finishing, fitting up, positioning, welding, inspecting, heat treatment, and flash removal. Most of these costs are treated elsewhere in this book, and the present discussion is confined to the actual welding costs.

The elements of cost in welding operations are those for: (1) labor and overhead; (2) electrodes, flux or gas, or rods; (3) power or fuel; and (4) equipment depreciation and maintenance. All are not pertinent to all operations, and they appear in different proportions. A leading equipment manufacturer has estimated that for manual metal-arc welding, about 80–86% of the cost is for labor and overhead, something like 8–15% for electrodes, and as little as 2% for power and equipment costs. In contrast, the costs of a projection welding operation may be mostly made up of power and equipment costs with little for labor and none for consumable electrodes.

A welder's pay rate is usually based on the position in which he or she can make a satisfactory weld. For instance, one who can do overhead welding properly receives more pay than another who can do only flat welding. Thus, the labor rate as well as the speed of doing a job depends upon the position and kind of weld.

The curves in Figure 13-29 show typical amounts of unit electrode and power consumption, welding speeds, and costs for several sizes of plain butt joints. Basic data of this sort have been compiled for many kinds of welding under various conditions and are given in charts and tables in reference books and handbooks.

Table 13-2 presents the detailed calculations for the cost of a plain butt weld in a nominal 10-mm (.375-in.) thick plate to illustrate how the total cost curve is obtained in Figure 13-29.

Notice that Table 13-2 assumes 100% efficiency. The cost of labor and overhead does not allow for the time the operator is not engaged in laying down metal on the weld. The lost time may be as much as 80% and must be determined from the performance in each shop. If 40% efficiency is assumed to apply, the cost of overhead and labor for the 10-mm-thick plate is $30/(10.9 \times 0.4) = \$6.88/m$, and the total cost is \$8.51/m of weld. For the 10 mm (.375 in.) plate, $30/(36 \times 0.40) = \$2.08/ft$, and total cost is \$2.59/ft. The unit cost is multiplied by the length of a particular weld to find its cost. The cost of electricity is relatively small and is commonly ignored.

13.10.2 Comparison of Welding and Casting

Many products, such as machine frames and other members, can be either cast or fabricated by welding. The advantage of welding is that steel structures can be made stronger, stiffer, and lighter than those of cast iron.

As a general rule, machine steel can safely be subjected to stresses about four times as high as those for cast iron. Stiffness may be illustrated

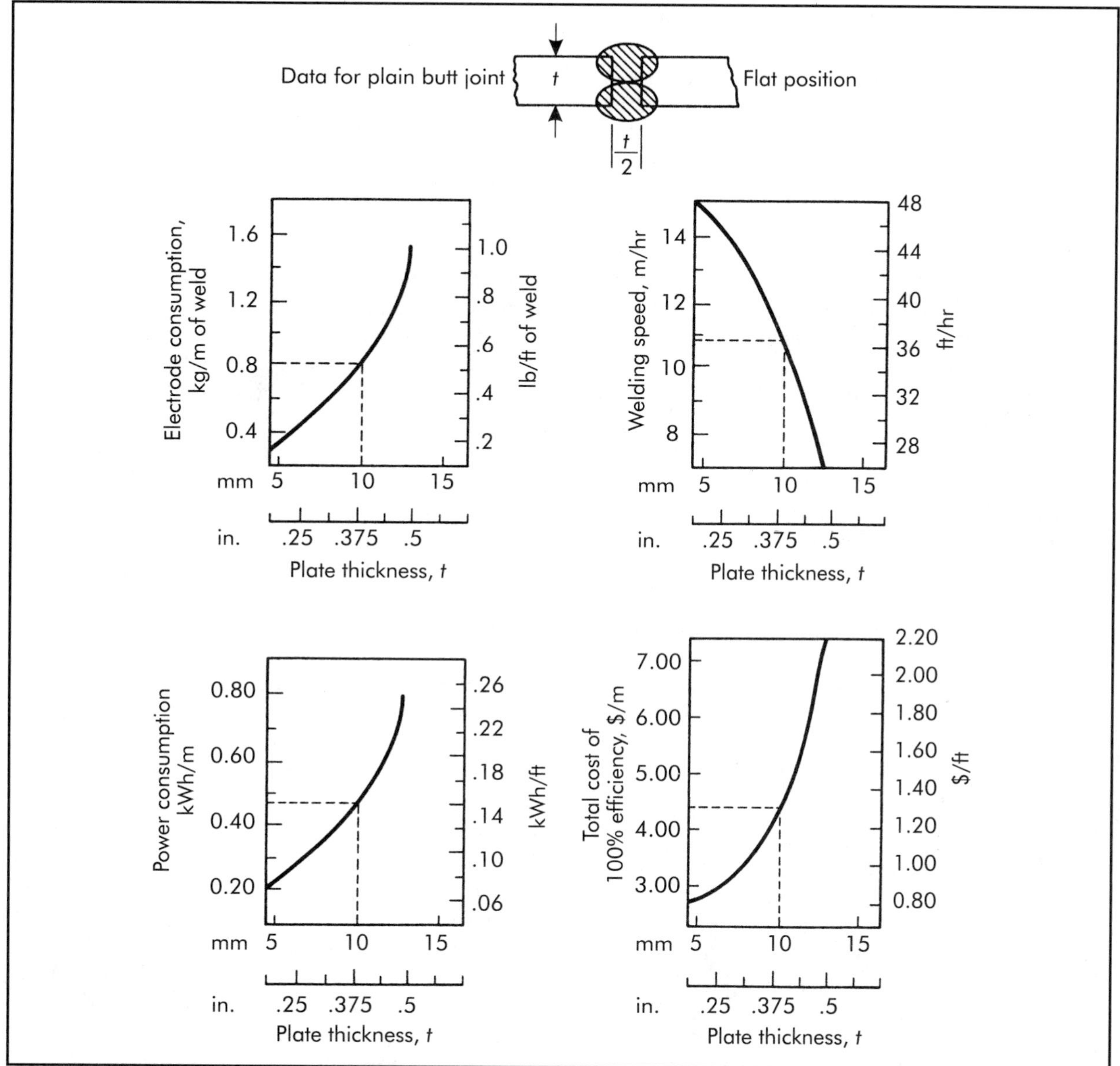

Figure 13-29. Chart of manual-arc welding performance. The total welding cost at 100% efficiency assumes power cost of \$0.06/kWh, electrode cost of \$2.00/kg (≈\$0.90/lb), labor at \$12.00/hr, and overhead of \$18.00/hr.

Table 13-2. Cost calculations for a plain butt weld

Cost	Thickness of Plate, *t*	
	10 mm	.375 in.
Labor and overhead	30/10.9 = \$2.75/m	30/36 = \$0.83/ft
Electrodes	0.8 × 2 = 1.60/m	0.55 × 0.9 = 0.50/ft
Electricity	0.48 × 0.06 = 0.03/m	0.14 × 0.06 = 0.01/ft
Total	\$4.38/m	\$1.34/ft

by the deflection of a simple beam as shown in Figure 13-30 with the fundamental equation for the amount of deflection. The deflection (*d*) is inversely proportional to the modulus of elasticity (*E*) of the material, which is 12,000,000 for cast iron and 30,000,000 for steel. For the same deflection, the width (*b*) of a steel beam needs to be only 40% of that of a cast iron beam. Approximately the same ratio applies for all kinds of bending and torsion. Good welded steel designs have about half the weight of cast iron structures that are no stronger or stiffer.

In some cases, cast iron is considered better than steel for damping vibrations. This property is of particular importance in the design of machine tools.

Steel costs about one-half to two-thirds as much per kilogram or pound as cast iron. In a well-designed steel weldment that weighs about half as much, the material cost is about one-fourth to one-third as much as for cast iron. In addition, the steel must be cut, shaped, fitted,

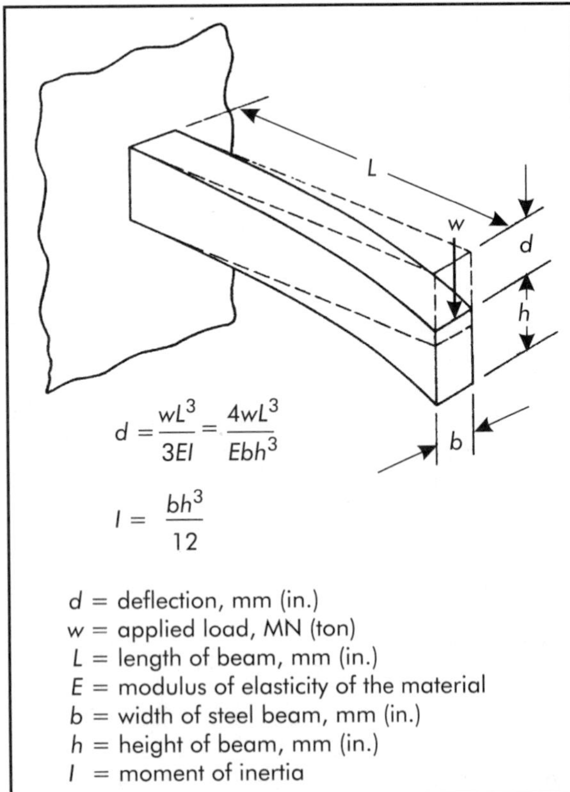

Figure 13-30. Deflection of a cantilever beam.

and welded, and the structure annealed. The casting needs only to be cleaned and chipped, but each different casting needs a pattern, which is expensive. In practically all cases, a single unit, and particularly large units, can be made more cheaply by welding than by casting. In some cases, welding is cheaper no matter how many units are made; in other cases, castings are cheaper for larger quantities. Casting is well adapted to automatic production of large quantities, especially complex pieces. For example, a number of attempts have been made to fabricate automobile engine blocks by welding, but generally none has ever displaced casting.

An example of costs for a cast versus welded base for a machine tool is illustrated in Figure 13-31. The casting weighs 360 kg (≈800 lb) and can be obtained for $1.06/kg ($0.48/lb) for one unit and at a declining rate of $0.79/kg ($0.36/lb) for 100 units. Cleaning costs $45 for each unit. A pattern costs $1,350 and serves for any number of units from 1–100. The total cost for one casting is $1,777. If a lot of 20 is made, the distributed pattern cost is $67.50/unit, the cast iron costs $1.01/kg ($0.46/lb), and each unit costs $476. A welded base for the same machine weighs 170 kg (375 lb). Steel costs $0.51/kg ($0.23/lb) and heat treatment $0.18/kg ($0.08/lb), making the total material cost $117/unit. Fabrication takes 25 hr at $18/hr, but initial preparation and setup amount to 10 hr at the same rate. By welding, one unit costs $747. The setup charge can be distributed over a lot of 20 units, and the cost per unit is $576. As seen from Figure 13-31, casting is cheaper for eight or more units; welding for seven or less.

Welding has several practical advantages. Thin and thick sections and unlike materials may be joined together, and sections may be made thin more easily in a weldment than in a casting. Holes may be prepunched in welded sections. Patterns for castings are costly to alter for design changes and to store; templates for welding are usually simpler and less bulky.

13.10.3 Comparison of Welding and Riveting

An analysis of a joint made by 19-mm (.75-in.) diameter rivets showed a cost of $0.37 for each

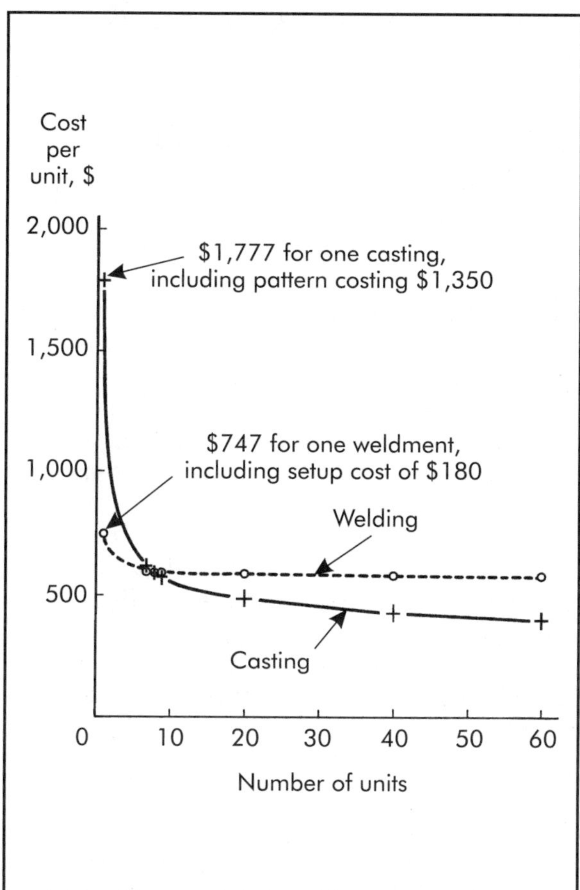

Figure 13-31. Cost versus quantity for a machine tool base made by casting or welding.

joint of 45 kN (≈10,000 lb) capacity as against $0.07 for a welded joint of equal capacity. At one time, all joints in structural steel buildings, boilers, tanks, automobile chassis, etc., were riveted. Welding has largely replaced riveting for such applications, especially where strength is important and joints permanent.

Riveting always requires lap joints. The holes and the rivets subtract from strength, and a riveted joint at best can only be about 85% as strong, whereas a good welded joint is fully as strong, as the parent metal. Joints can be made gas- and liquid-tight by welding, but must be caulked when riveted. Welded joints are easier to inspect because it is difficult to ascertain whether a rivet has been fully drawn up. The inspector must depend on the ability and integrity of the maker of manually riveted joints.

13.10.4 Comparison of Arc-Welding Processes

Manual shielded-arc welding is wholly versatile and has a low first cost. For one-of-a-kind assemblies of diverse work, the covered electrode in the hands of a skilled welder is probably the fastest and most efficient process. However, manual work is inherently slow for repetition; stick ends need to be discarded and work interrupted, and there is a definite human limit to the rate at which metal can be deposited. Semi-automatic machines that mechanize part of the welding operation cost two to four times as much as stick shielded-arc welding equipment. However, an operator with semi- or fully automatic equipment can control molten metal and the arc more readily, make use of more current, and deposit metal more uniformly and continuously. He or she is capable of welding 2–10 times faster than an operator with only a coated electrode.

A comparison is made in Figure 13-5 of performance rates reported for several arc-welding processes. The record might be somewhat different in other shops for different kinds of work. Semi-automatic, gas metal-arc welding is often faster overall than submerged-arc welding, particularly for a variety of work, because the operator can see the weld, and there is no slag to chip away and no flux to clean up. Submerged-arc welding has some advantage in that it has no loss from spatter, whereas the loss of metal may be as much as 10% for gas metal-arc welding. For fully automatic operation with complete continuous control, submerged-arc welding offers the means for utilizing heavy currents to lay down metal at the fastest rates, particularly with strip electrodes, multiple electrodes, and vertical welding techniques.

Relative costs for doing a job by several common methods are given in Table 13-3. Common for all processes are a labor cost of $12/hr, overhead cost of $16/hr, a weld deposit amount of 0.17 kg/m (.11 lb/ft) of weld, a power cost of $0.05/kWh, and a power source efficiency of 50%. For the submerged-arc process, the flux usage ratio is 1.5 kg/kg (lb/lb) of flux to metal and the flux cost is $0.88/kg ($0.40/lb). For the CO_2 gas process, the gas flow is 0.85 m^3/h (30 ft^3/h) and the gas cost is $1.41/$m^3$ ($0.04/$ft^3$).

In Table 13-3, the *duty cycle* is the percentage that arc time is of the total operation time. Other than arc or active welding time is that required to change electrodes, chip slag, handle flux, make adjustments, and for personal needs, fit-up, etc. The manual CO_2 gas-shielded process has the least lost time largely because it has no flux or slag to handle. The *deposition efficiency* is the percentage that the usable metal deposited in the weld is of the total weight of electrode used. Some of the loss is from stub ends with stick electrodes.

Table 13-3 applies to a particular situation and is presented to show how costs are compared. Under other circumstances the data might very well show the conventional metal-arc process with coated electrode or iron powder electrode to be most economical. Such a case could occur in a job shop where the costs of adapting the semi-automatic methods to a large variety of jobs would be excessive. Another case could be that of welding the framework of a high structure on which the semi-automatic equipment would be too difficult, if not impossible, to manipulate.

13.11 PROCESS AUTOMATION

Automating welding processes is an ongoing effort. The objectives are to increase weld quality, provide better control over welding variables, and minimize the need for skilled workers. Robots and other automated equipment have been used extensively for many welding applications.

Automated processes have advanced during the past decade utilizing advanced technologies, such as sensors to monitor the welding process and provide better weld quality. Current trends and ad-vances have initiated processes, such as flux-cored-arc welding and pulsed-arc, gas metal-arc welding, which are more adaptable to automation and can be used for thin metals. Because of the higher degree of automation available for some processes, such as high-energy beam welding and laser-beam welding, these processes have gained more applications in many industries.

Table 13-3. Relative unit welding costs for several common processes

	Type of Operation			
Factor of Cost	Conventional Metal-arc with Coated Electrode	Metal-arc with Powder-metal Electrode	Semi-automatic Submerged-arc	Semi-automatic Gas-shielded
Electrode diameter, mm (in.)	4.8 (.188)	4.0 (.156)	2.0 (.078)	1.2 (.047)
Travel speed, mm/min (in./min)	508 (20)	635 (25)	889 (35)	889 (35)
Duty cycle (%)	30	35	45	60
Electrode cost, $/kg ($/lb)	1.76 (0.80)	2.20 (1.00)	1.50 (0.68)	2.65 (1.20)
Deposition efficiency, %	68	70	100	92
Welding current, A	300	350	500	450
Nominal arc potential, V	30	30	35	32
Unit weld cost, $/m ($/ft)	3.51 (1.07)	2.66 (0.81)	1.67 (0.51)	1.41 (0.43)

13.12 QUESTIONS

1. What is the principle of operation of electric-arc welding?
2. What factors affect the depth of penetration in arc welding? What are their effects?
3. What is the optimum amount of current to expend in metal-arc welding?
4. What determines how large an electrode should be for manual-arc welding?
5. How do DC and AC compare for arc welding?

6. What are the functions of coatings on shielded electrodes?
7. What are the advantages of each of the several sources of current for arc welding?
8. How are arc-welder sizes designated and under what conditions?
9. What are the necessary current characteristics of a power source for manual-arc welding and why?
10. What two kinds of power sources are used for automatic welding and what are their principles of operation?
11. Explain the "Tig" and "Mig" systems of arc welding.
12. Describe plasma-arc welding. What is its main advantage?
13. What is "submerged-arc welding" and where does it serve best?
14. Describe "electroslag" and "electrogas welding" and their uses.
15. How do the main automatic arc-welding methods compare with each other?
16. What is "electron-beam welding" and when is it practicable?
17. Describe the characteristics of electrodes. How are electrodes classified?
18. Explain the meaning of "solid-state welding."
19. Name some of the applications for:
    (a) stud welding;
    (b) flash welding; and
    (c) percussion welding.
20. Explain the advantages of resistance welding over other welding processes.
21. What heat sources are used for each of the welding processes discussed in this chapter?
22. What is a "laser beam," how may it be produced, and how may it be used?
23. What are the principles of operation of "resistance welding?"
24. What are the relative and optimum conditions of time, current, and tip force in a spot-welding operation?
25. Describe and compare spot, seam, and projection welding.
26. What is "heat balance"?
27. Compare upset-butt welding, flash-butt welding, and percussion welding as to the ways they are done and their relative merits and limitations.
28. What is the principle of "Thermit® welding"? What can it do better than other processes?
29. What properties make acetylene the favorite for gas welding?
30. Describe the features of a gas welding operation.
31. What are the advantages of gas welding?
32. Describe "friction" or "inertia" welding and its uses.
33. What is "ultrasonic welding" and what are its advantages?
34. Describe "explosive welding."
35. What is "diffusion bonding" or "solid-state welding" and where does it serve to advantage?
36. What are the elements of common arc- or gas-welded joints?
37. What are welding symbols and what do they do?
38. How does welding affect the grain size and structure of a metal?
39. To what extent does steel harden in welding? Why? What effect does it have on the weldment?

40. What causes weldments to crack and what are the remedies? Explain the reasons for these events.
41. What causes inclusions and voids in a weld and how can they be prevented?
42. Why is hydrogen detrimental to welds?
43. How does welding make metals susceptible to corrosion?
44. What are the main principles of designing for welding?
45. What are the elements of welding cost and how are they estimated?
46. How do weldments compare with castings?

13.13 PROBLEMS

1. Show how the weld cost in \$/m (\$/ft) is computed from the data given in Table 13-3 and the accompanying text:
   (a) for conventional metal-arc welding with coated electrode;
   (b) for metal-arc welding with powder metal electrode;
   (c) for manual submerged-arc process; and
   (d) for the manual CO_2 gas-shielded process.
2. Arc welding to lay down 5.4 kg/h (12 lb/hr) is to be done with automatic machines capable of welding at this rate while the arc is on. Labor costs \$10.00/hr. Overhead is not thought to be much different for either case and is neglected. For the submerged-arc welding process, flux costs \$0.73/kg (\$0.33/lb) and is used at the rate of 1.5 kg/kg (lb/lb) of electrode wire consumed. The wire costs \$1.65/kg (\$0.75/lb) and the deposition efficiency is 100%. The duty cycle is 60%. For the CO_2 gas-shielded welding process, gas costs \$1.06/m^3 (\$0.03/ft^3) and is used at the rate of 0.85 m^3/hr (30 ft^3/hr) of welding time. The wire costs \$2.43/kg (\$1.10/lb) and the deposition efficiency is 90%. The duty cycle is 80%.
   (a) Which process appears more economical for this job?
   (b) Why are the deposition efficiency and the duty cycle different for the two processes?
3. Conventional welding equipment costs \$1,250 and equipment for manual CO_2 gas-shielded welding costs \$4,200. For the conditions shown in Table 13-1, how much welding must be done to justify the more expensive equipment? Interest and taxes on the investment need not be considered.
4. An end dome 25 mm (≈1 in.) thick × 610 mm (24 in.) in diameter may be welded on the end of a cylinder 610 mm (24 in.) in diameter × 889 mm (35 in.) long by electron-beam, gas tungsten-arc, or gas metal-arc welding. The following costs have been assembled for each of these methods.

Cost	Electron-beam Welding	Gas Tungsten-arc Welding	Gas Metal-arc Welding
Labor, \$/hr	7.00	6.00	6.00
Overhead, \$/hr	22.40	11.24	11.24
Welding speed, m/min (in./min)	1.0 (40)	0.1 (4)	0.25 (10)
Number of passes	1	19	10
Set up, prepare, and clean, min	47	205	205
Inspection, \$/piece	20.00	40.00	40.00
Filler metal, \$/piece	—	150.00	170.00
Gas flow, m^3/hr (ft^3/hr)	—	1.7 (60)	1.4 (50)
Power cost, \$/piece	0.48	0.64	0.48

Preparation for arc welding includes machining of a single U-groove in the joint. Labor and overhead rate for setup, preparation, and cleaning is \$15/hr. Perishable electron-beam welding elements cost \$1.00/piece and gas \$3.53/m^3 (\$0.10/ft^3). What is the cost to weld an end dome in place by each method?

5. (a) A single-phase resistance welding machine has a rating of X kVA with a 50% duty cycle. The impedance of the secondary circuit of the transformer is Z ohms, the power factor is p, and the voltage is E volts. How much heat H_1 is generated in the secondary circuit during an appreciable period of time T in minutes?
   (b) If the machine is operated to deliver Y kVA, and other conditions remain the same except that the duty cycle is changed to a value C, how much heat H_2 is generated in time period T?
   (c) Set up an expression for C in terms of X and Y for operating the machine at maximum output without overheating.
   (d) A resistance welder is rated at 50 kVA at 50% duty cycle. Secondary voltage is 5 V and does not change. A job is put on the machine to draw 15,000 A. What is the most the duty cycle may be for 15,000 A?
6. A spot-welding machine has a throat depth of 305 mm (12 in.) with 203 mm (8 in.) between horns. The secondary loop impedance is $165 \times 10^{-6}\ \Omega$. If the throat depth is changed to 762 mm (30 in.), the impedance is increased to $340 \times 10^{-6}\ \Omega$. The secondary voltage is 6 V. How much does the secondary current decrease with the change in throat depth?
7. A single-groove butt weld is to be made between two pieces of steel plate.
   (a) What kind of steel is least likely to crack from the welding? Why?
   (b) If a steel of 0.5% carbon content and medium hardenability is selected, which welding process is least likely to crack it? Why?
   (c) If arc welding is used for the steel specified in part (b), what can be done to lessen the chances of cracking?
8. Estimate the cost of a plain butt-joint weld 457 mm (18 in.) long on a 6-mm (≈.25-in.) thick plate at 70% efficiency. Power costs \$0.04/kWh, labor and overhead \$24/hr, and electrodes \$2.20/kg (\$1.00/lb).
9. It is found in a certain shop that welding speed for a square butt joint in the vertical position is 80% of that for the flat position as shown in Figure 13-29. The welder is paid \$12/hr and overhead is \$10/hr. Welding efficiency is 50%. Electrodes cost \$2.65/kg (\$1.20/lb) and electricity \$0.04/kWh. What is the cost per m (ft) of vertical square butt weld in each of the following sizes of plates? (a) 6.4 mm (.25 in.); (b) 9.53 mm (.375 in.); (c) 12.7 mm (.5 in.); (d) 5 mm (.197 in.); (e) 10 mm (.394 in.); (f) 15 mm (.591 in.).
10. A cast base for a machine tool weighs 2,222 kg (4,900 lb). The pattern costs \$1,000 and cast iron \$1.10/kg (\$0.50/lb). The cost to clean a casting is \$50. A welded design for the base weighs 1,134 kg (2,500 lb). Steel costs \$0.62/kg (\$0.28/lb). Fabrication requires 65 hr at \$28/hr for labor and overhead. Setup and preparation take 15 hr for a lot. Templates for cutting the steel cost \$100.
    (a) For how many pieces is the cost of a casting the same as the cost of a weldment?
    (b) Which process would be cheaper for 25 pieces? How much? Why might this not be the deciding factor?
11. A welding shop that needs work is willing to cut its rate for labor and overhead to \$23.00/hr. If all other conditions stated in Problem 10 are the same, what effect does the new rate have upon the choice of process?
12. A lot of 40 gear segments are to be rough-flame cut by machine and then finished by machining. The length of cut for each segment is 3.3 m (130 in.). Four segments can be cut

at one time from a template that costs \$200. Oxygen costs \$2.65/m^3 (\$7.50/100 ft^3), and acetylene \$5.83/m^3 (\$16.50/100 ft^3). The labor and overhead rate in the shop is \$15/hr. An hour is required to set up and tear down the job. What is the cost per piece for flame cutting? The material is 25 mm (≈1 in.) thick.

13.14 REFERENCES

Bolin, S. 1981. "Lasers Light the Way to Low-cost Drilling and Cutting." *Manufacturing Engineering*, December.

Cary, H.B. 1989. "Modern Welding Technology," 2nd Ed. Upper Saddle River, NJ: Prentice-Hall.

Collopy, W.F. 1982. "Powders Add Flexibility to Hardfacing Costs." *Metal Progress*, May.

"Fundamentals of Brazing, Special Report 732." 1981. *American Machinist*, April.

Gonser, T.R. 1981. "Computer Sharpens EB Hardening." *American Machinist*, November.

Jackson, Clarence E. 1959. "The Science of Arc Welding." *The Welding Journal.*

Lesnewich, A. 1982. "The Real Cost of Depositing a Pound of Weld Metal." *Metal Progress,* April.

Metals Handbook, 9th Ed. 1981. Vol. 4, *Heat Treating.* Metals Park, OH: American Society for Metals: 284.

Metals Handbook, 9th Ed. 1983. Vol. 6, *Welding, Brazing and Soldering.* Metals Park, OH: American Society for Metals.

Parmley, R.O. 1994. *Standard Handbook of Fastening and Joining*, 2nd Ed. New York: McGraw-Hill.

Principles of Industrial Welding. 1978. Cleveland, OH: The James F. Lincoln Arc-welding Foundation.

Procedure Handbook of Arc Welding Design and Practice. Cleveland, OH: Lincoln Electric Co.

Society of Manufacturing Engineers (SME). 1999. *Welding* video. *Fundamental Manufacturing Processes* series. Dearborn, MI: SME.

Source Book on Electron-beam and Laser Welding. 1980. Materials Park, OH: American Society for Metals.

Tamaschke, W. 1981. "Sheet Metal Cutting by Laser." *Commline,* May-June.

Tool and Manufacturing Engineers Handbook, 4th Ed. Volume 2, *Forming*. 1984. Dearborn, MI: Society of Manufacturing Engineers (SME).

Welding Handbook, 8th Ed. 1987. Miami, FL: American Welding Society.

Other Cutting and Joining Processes

14

Metal spraying, 1911

—Dr. M.U. Schoop, Zurich, Switzerland

14.1 INTRODUCTION

The different joining processes described in Chapter 13 have the common feature of heating the metals to be joined to some higher temperature by a thermal, chemical, electrical, or mechanical means. Such elevated temperatures cause bonding of the metals at the joint. Concerns about heating the metals and elevated temperature effects are as follows:

- some materials cannot deal with the effect of high temperatures;
- materials to be welded have very different geometric, physical, and metallurgical characteristics; and
- materials to be welded are very delicate and cannot withstand any changes in their properties.

This chapter describes some welding processes that permit lower temperatures than those normally required for other common joining processes. These processes are *brazing* and *soldering*, whose concept is based on filling the gap between the two metals to be welded with filler metals that are melted with an external source. Solidification results in a strong joint.

The chapter also describes thermal cutting processes, permanent and nonpermanent mechanical fasteners, and joining methods for plastic materials.

14.2 THERMAL CUTTING PROCESSES

Metals may be cut using mechanical means, such as sawing. Some metals also may be cut into two or more pieces with varying contours by using some source of heat that removes a narrow zone in the part to be cut. The heat sources discussed in Chapter 13 can be used for this purpose.

14.2.1 Oxygen Cutting

Oxygen cutting, also called *gas* or *flame cutting* is done by preheating a spot on ferrous metal to its ignition temperature and burning it with a stream of oxygen in the manner illustrated in Figure 14-1. The reaction that takes place at about 871° C (1,600° F) is

$$3Fe + 2O_2 \text{ —— } Fe_3O_4 + 50 \text{ MJ/mol}$$
$$(48{,}000 \text{ Btu/mol})$$

Theoretically, 1 m^3 of oxygen is required to oxidize about 430 cm^3 (1 ft^3 of O_2 for about .75 $in.^3$) of Fe. Actually, about 30–40% of the metal is melted and blown away without being oxidized, and removal of 600 cm^3 or more of iron per 1 m^3 of oxygen (1 $in.^3$ or more of Fe per 1 ft^3 of O_2) is not uncommon performance.

An oxygen jet spurts out of a hole in the center of the tip of the cutting torch depicted in Figure

14-1. Around that, gas-with-oxygen flames emerge from several holes and heat the top of the metal piece to keep it at the ignition temperature. Acetylene (or methylacetylene-propadiene [MAPP]) is widely used as fuel because of its hot and adjustable flame, among other fuels, such as hydrogen and illuminating gas. As the metal is burned and eroded away, the torch is moved steadily along the path of the cut. A uniformly wide slot, called the *kerf*, is cut by the jet of oxygen. The faster the rate of traverse, the more the bottom lags behind the top of the cut. The amount is called the *drag* and is evidenced by a series of curved lines on the sides of the kerf. Fine lines characterize a quality cut; coarse ones a fast or heavy cut. The drag must be kept small to cut a surface with a straight side around a curve.

Oxygen-cutting Equipment and Procedure

Oxygen cutting may be done by a manual torch, like gas welding. For production and more precise results, one or more torches are mounted on a cross-arm or mechanical guideway. Each torch is moved over a path determined by a tracer on a sheet metal template, an optical scanner following the lines on a drawing, or numerical control (Chapter 31). Other production features are automatic ignition and torch height control. In the automatic mode, the operation is sometimes called *thermal machining*. The work is commonly clamped on a series of disposable slats forming an open table. A production machine with two visible cutting heads is shown in Figure 14-2. This particular machine is referred to as a *multiple process machine,* since it can be used for high definition plasma, conventional plasma, or oxy-fuel thermal machining. The machine shown has a cut width of from 1.2–4.3 m (4–14 ft) and will accommodate up to eight cutting heads or stations, including an air scribe marking station.

A high degree of skill is required to get the best results from oxygen cutting. The gas and oxygen pressure, position of torch, intensity of preheat, cutting speed, and type of tip must be selected and regulated to suit the kind of material cut, work thickness, shape of the cut path, and the finish and accuracy required at the lowest possible cost for gases and time. Initial preheat tends to burn a hole in metal, so a cut is started away from the line of cut and usually at an outside edge of the plate where ignition is quick.

A principal use of flame cutting is to cut off and prepare plates and other shapes for welding. As an example, a plate may be cut off and beveled at the same time by means of two torches, one vertical and the other inclined at the bevel angle, traversed along the cut together.

Oxygen cutting usually implies cutting all the way through a piece, but a groove, cavity, hole, or even the entire top to a certain depth may be removed or cut from a piece. This is called *flame*

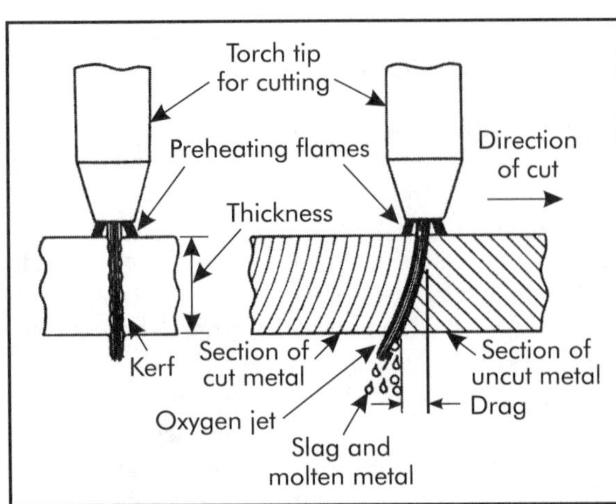

Figure 14-1. Principles of flame cutting.

Figure 14-2. Multiple-station flame/plasma cutting machine with PC-based numerical control. (Courtesy Messer MG Systems & Welding, Inc.)

machining, and specific applications of it are termed flame-planing, -milling, -turning, -drilling, and -boring, like the analogous machining operations. In flame machining, a cutting torch is inclined at an angle and adjustments are made so that the stream of oxygen does not penetrate through the piece, but curves around and returns to the top. The torch is then moved to generate the shape desired.

One variation of oxygen cutting, called *oxygen-lance cutting*, is done with a long iron pipe of small diameter with oxygen flowing through it. The end of the pipe is set aflame and furnishes preheat for the oxygen stream. This torch is able to pierce holes in all kinds of metal. One example of its use is to burn holes in a hunk of metal frozen in a ladle. Dynamite is then put in the holes to break up the chunk and remove it.

Applications of Oxygen Cutting

Most ferrous metals are oxygen cut. Thicknesses up to 1.5 m (60 in.) can be cut in the ordinary way, and virtually any thickness by the oxygen-lance method. A number of plates may be piled on top of each other and cut at one time in what is called *stack cutting*.

In addition to cutting out sections for weldments and parts, such as gears, sprockets, and cams from steel plates, oxygen cutting is used to cut off structural shapes and barstock, cut up steel scrap, cut off rivet heads, and remove butt ends and seamed, cracked, or defective sections from billets, slabs, and rounds, and for many other purposes. Rough oxygen cutting is done just to cut off pieces; precision oxygen cutting to produce fairly accurate shapes.

The metal on the sides of the kerf is heated above its critical temperature but the effect seldom penetrates more than 3.18 mm (.125 in.) below the surface. In steel of less than about 0.3% carbon and little alloy content, the result is normally confined to changes in grain size and structure near the surface. This may improve surface strength and toughness. A thin layer is likely to be hardened on high-carbon and alloy steel surfaces and cause high internal stresses and cracks. Heating before or annealing after cutting, or both, are often done to relieve this situation.

Oxygen cutting can be done at moderate cost, leaving negligible drag lines, with edges sharp, kerf uniform in thickness, and surfaces square with tolerances satisfactory for many purposes. Good oxygen-cut surfaces compare favorably with machined surfaces. Thick pieces do not warp as a rule, but plates 13 mm (.5 in.) or less thick may require careful treatment, such as staggering of cuts, to avoid excessive warpage.

Ordinarily, a tolerance of ±1.5 mm (≈±.0625 in.) is considered practical in cutting plate up to 15 mm (≈.5 in.) thick, or about ±0.8 mm (±.0313 in.) more for each additional 25.4 mm (1 in.) of thickness. Closer tolerances are held but at higher than usual commercial costs. In one case of cutting gears about 2.4 m (8 ft) in diameter in segments from 127-mm (5-in.) thick steel plate, the tooth contours were reported held to ±152 μm (±.006 in.). In precision work, cross-sectional squareness is held from 76 μm (.003 in.) for 25.4-mm (1-in.) thick sections to 762 μm (.030 in.) for 152.4-mm (6-in.) thick sections. For the most precise results, a piece may be oxygen cut oversize to remove most of the stock and finish machined to the degree of accuracy required.

Costs of Oxygen Cutting

The main elements of operating cost for oxygen cutting are: (1) labor and overhead, and (2) gas. Labor and overhead costs are based upon the time required to do a job. This usually includes time for preparation, setup, and tear down. The cutting time is calculated by dividing the length of cut by the cutting speed. Typical values for cutting speeds and rates of gas consumption, which vary with plate thickness, are given by the curves of Figure 14-3. Ranges of values apply for any one thickness depending on the skill of the operator, operating conditions, and results required. For instance, the cutting speed must be slow to hold small tolerances, resulting in increased cost. Similar tables and charts for various conditions can be found in reference texts and handbooks as a guide for cost estimating. In addition to operating cost, the cutting of a shape may call for a special template chargeable to the job.

Comparison with Other Methods

Oxygen cutting is directly competitive with shearing for cutting off plates, bars, rails, and other shapes. However, oxygen cutting is more

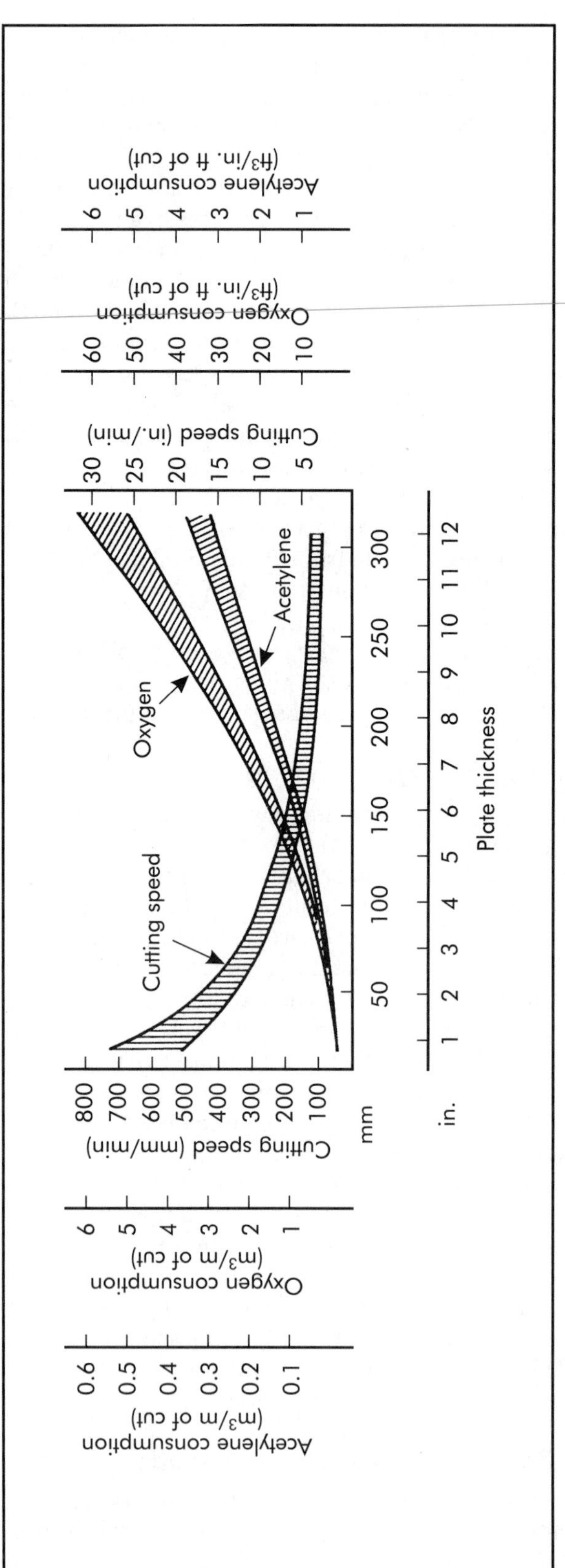

Figure 14-3. Performance data for machine oxygen cutting of mild steel not preheated.

practical for very thick materials or a wide variety of work. Big shears are needed for thick sections and do exist, but are not available nor justified in all plants. Single-torch, oxygen-cutting equipment is relatively cheap (like gas-welding equipment), adaptable to all sizes of work, and portable to work that cannot be easily moved.

Friction sawing is more economical than oxygen cutting but not as versatile. For instance, rails are cut to desired length by friction sawing when they come out of a rolling mill. Occasionally, a rail going through the mill runs afoul and becomes tangled in the mechanism. Oxygen-cutting equipment is then brought up to slice the rail into pieces to free it from the mill.

Irregular shapes may be cut from stock 12.7 mm (.5 in.) or less in thickness by nibbling small quantities and by die cutting large quantities. Intermediate quantities are sometimes more economically done by oxygen cutting, particularly by means of multiple-torch cutting with templates and stacking of workpieces. Thick irregular pieces may be sawed from plates, but oxygen cutting is more economical and commonly utilized.

Powder Cutting of High Alloys

Oxygen cutting of some high-alloy steels, such as stainless steel and many nonferrous alloys, is difficult because the alloying elements, such as chromium and nickel, oxidize along with the base metal. Many of these oxides do not melt at attainable temperatures and form an insulating coating on the work that hinders progress of the cut. Free carbon forms a like barrier in cast iron. Also, the combustion of alloys does not add enough heat to the operation.

The same kind of work done on low-carbon steel by oxygen cutting can be done successfully on stainless steel, cast iron, copper, Monel® metal, and other materials by the *powder-cutting* and *scarfing process*. In addition to the usual procedure for oxygen cutting, a fine iron-rich powder is injected into the flame and oxygen stream. This burns, and the added heat aids melting. Aluminum powder may be added to cut copper, glass, and even concrete.

A somewhat larger torch tip must be used for powder cutting to pass the extra ingredient, and the cut is rougher and wider than for straight oxygen cutting. Otherwise, the results are comparable. Stainless steel can be powder cut about as fast as low-carbon steel can be oxygen cut, but at about twice the cost. Most of the additional cost is for powder.

In *flux injection*, also called *chemical-flux cutting*, a fine stream of flux powder is injected into the oxygen stream to increase the fluidity of refractory oxides so that they can be blown easily from the kerf.

14.2.2 Electric Arc Cutting

Although most of the electric-arc welding operations can be adapted for cutting metals, only three are of commercial importance: air carbon-arc, oxygen-arc, and plasma-arc cutting.

Air carbon-arc cutting is a method of melting metal by an arc from a graphite electrode and blowing away the molten metal with a jet of compressed air. This may be done for through cutting or for gouging out grooves when the electrode and airstream are inclined to the work surface. This method is applied mostly to removing gates, risers, and defects from castings, removing weld defects, and preparing plates for welding.

Oxygen-arc cutting or *metal arc cutting* uses a flux-covered steel tube as an electrode. The coating insulates the electrode from the sides of a cut and augments metal fluidity. Metal is melted by an arc from the electrode and is burned and blown away by a stream of oxygen blown through the tube. This method was developed for underwater cutting but also can be used in air for almost any metal thicknesses. It does not give good surface finish or accuracy.

A plasma-arc torch cuts metal fast because of the high temperatures it generates. *Plasma-arc cutting* of 6.4-mm (.25-in.) thick metal at up to 5 m/min (200 in./min), is reported to be as much as 10 times faster than gas cutting. It is capable of penetrating almost any material that conducts electricity, and is used especially for those metals difficult to oxygen cut. Surface finish may be better than for gas cutting in some cases, but accuracy is generally not as good. Power demands are high, and production machines are generally limited to one, or at most two, plasma arc torches each. Arc radiation is intense. Equipment is expensive. A numerically controlled, 27 metric-ton (30-ton), hole-punching, plasma-arc, contour cutting machine has a base price of around $275,000.

14.2.3 Plasma-arc Cutting (PAC)

Plasma-arc cutting combines electrical arcing with a high-velocity flow of gas, which produces good quality cuts in metals with good conductive properties. As explained in the section on plasma-arc welding (Chapter 13), plasma is a high-temperature ionized gas that occurs in any electric arc. The plasma, which reaches extremely high temperatures of 11,093°–27,760° C (20,000°–50,000° F) is formed by ionization of a suitable gas and then passing the arc and the ionized gas through a restrictive nozzle. A high-voltage, direct-current power supply sustains an arc between the positively charged workpiece and the negatively charged arc electrode. The arc exits from the nozzle at near sonic velocity to permit the high-temperature plasma to melt and blow away the workpiece material.

One type of plasma-arc nozzle is shown in Figure 14-4. This patented design creates a virtual nozzle composed of a cool, swirling boundary layer of gas along the inner wall of the nozzle. This vortex constricts and stabilizes the plasma column, producing smooth, clean, precise cuts. Either oxygen or air is generally applied as a plasma gas at a supply pressure of approximately 827 kPa (120 psi). In addition, the nozzle incorporates a small flow of nitrogen and oxygen used as a shield gas to control both top and bottom dross formation. This permits the production of excellent edge quality without the need for secondary finishing operations.

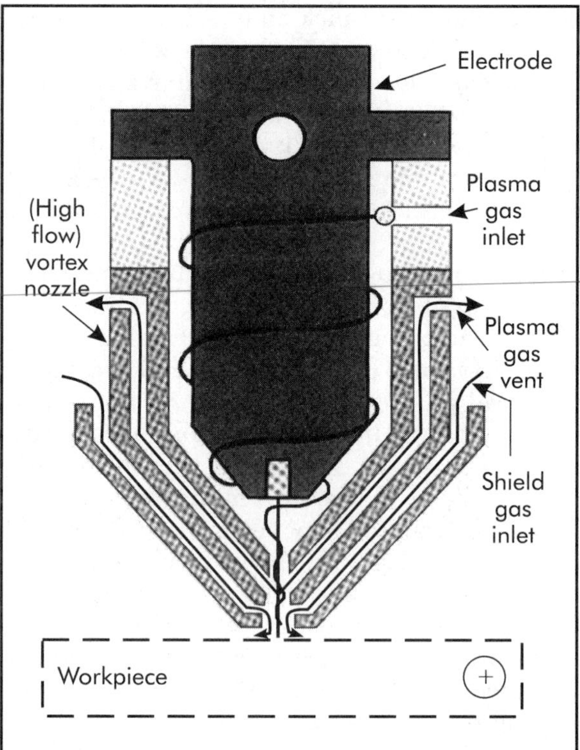

Figure 14-4. Schematic view of a patented plasma-arc nozzle. (Courtesy Hypotherm, Inc.)

Plasma-arc cutting may be performed manually with a hand-held torch or by machine-guided, torch-cutting operations. A portable unit for manual cutting operations with air as a plasma gas is shown in Figure 14-5. The hand-held torch for this unit has a patented front-end nozzle shield (stand-off device) that allows the operator to drag the torch directly on the workpiece at full output to protect the nozzle from molten spray. The portable unit has an adjustable output current of from 20–50 A. Typically, at 30 A, a 2.54-mm (.10-in.) thick mild steel plate can be cut at approximately 2 m/min (80 in./min) cutting speed. At a 50 A setting, a 12.7-mm (.5-in.) thick mild steel plate can be cut at about 0.635 m/min (25 in./min).

For high-production operations, plasma-arc torch-cutting heads are available for either robotic three-dimensional operation or conventional *X-Y* mechanized table cutting. The robotic application torch head, such as the one shown in Figure 14-6, has a pointed nozzle that permits cutting into angles and irregular shapes associated with three-dimensional parts. The conventional two-dimension cutter head has a straight head sleeve for ease of mounting on *X-Y* cutting tables. With the patented head design, precision cuts may be made in conductive materials up to 12.7 mm (.5 in.) thickness. Using oxygen as a plasma gas on mild steel, precise cuts are possible with edges approximating laser quality in terms of squareness and smoothness. High-quality cutting using air as a plasma gas is also possible on stainless steel, aluminum, and copper. Typically, a 8.9-mm (.35-in.) thick stainless steel plate may be cut at 1 m/min (39.4

Figure 14-5. Manual plasma-arc cutting with a portable unit. (Courtesy Hypotherm, Inc.)

in./min) with this type of torch at a power setting of 70 A.

Robot-mounted, plasma-arc cutting torch systems are also available for a wide range of applications. Both precision *X-Y* and robotic systems are applicable to CNC operation, and include control of the arc current and gas-flow settings.

The plasma-arc process may be used to cut just about any material that is a good electrical conductor. It may be used for such applications as beveling, shape cutting, gouging, and piercing.

14.3 METAL SPRAYING

14.3.1 Principles and Methods

Metals and nonmetals are sprayed by being melted or softened and are then swept away at high speed and atomized in a jet of air or gas. The material is fed into some spray guns as wire and into others as powder. Tiny plastic or fluid particles are hurled against a workpiece surface; they spread around and interlock with projections, embed in pits, and freeze quickly upon contact with the cool surface. Separate particles overlap and intertwine with one another to form a coherent structure.

Equipment utilizing a gas flame as depicted in Figure 14-7(A) is obtainable for $1,500–$3,000 and is satisfactory for many metals. An electric-arc spray gun that atomizes droplets from two continuously fed wire electrodes is also depicted in Figure 14-7(B). A gun of this type costs about $18,000, but is reported to operate at one-third the cost and three to five times faster than oxyacetylene equipment. As a rule, plasma arc, shown in Figure 14-7(C), is the most costly spray equipment to buy and operate, but it is an economical way to spray materials that melt above 2,204° C (4,000° F) and ceramics. One report was of a cost per pound of material deposited about one-third of that for the oxyacetylene method.

By a method called *flame plating* or *detonation-gun coating*, heated particles are shot in bursts from a gun barrel by rapidly successive oxyacetylene explosions and impacted into the workpiece surface. Excellent bonding and density are obtained without excessively heating the part. Equipment costs are quite high, particularly because elaborate isolation measures are necessary to reduce high noise levels.

A metal spray gun may be directed by hand or mounted on a machine. A typical application is to spray a round piece that turns between centers on a lathe with the gun mounted on the carriage for a uniform feed. Most sprayed metals adhere to a surface essentially through mechanical bonding, so the surface must be roughened first and free of dirt, oil, and grease. Surfaces are prepared by rough machining or grinding, knurling, blasting with sand or steel grit, inserting screws or studs, tack welding, or undercoating and painting. One method is to spray first with a high-melting-point metal that fuses and sticks to the surface.

14.3.2 Applications

All commercial metals are sprayed (in fact, the process is sometimes called *metallizing*), as

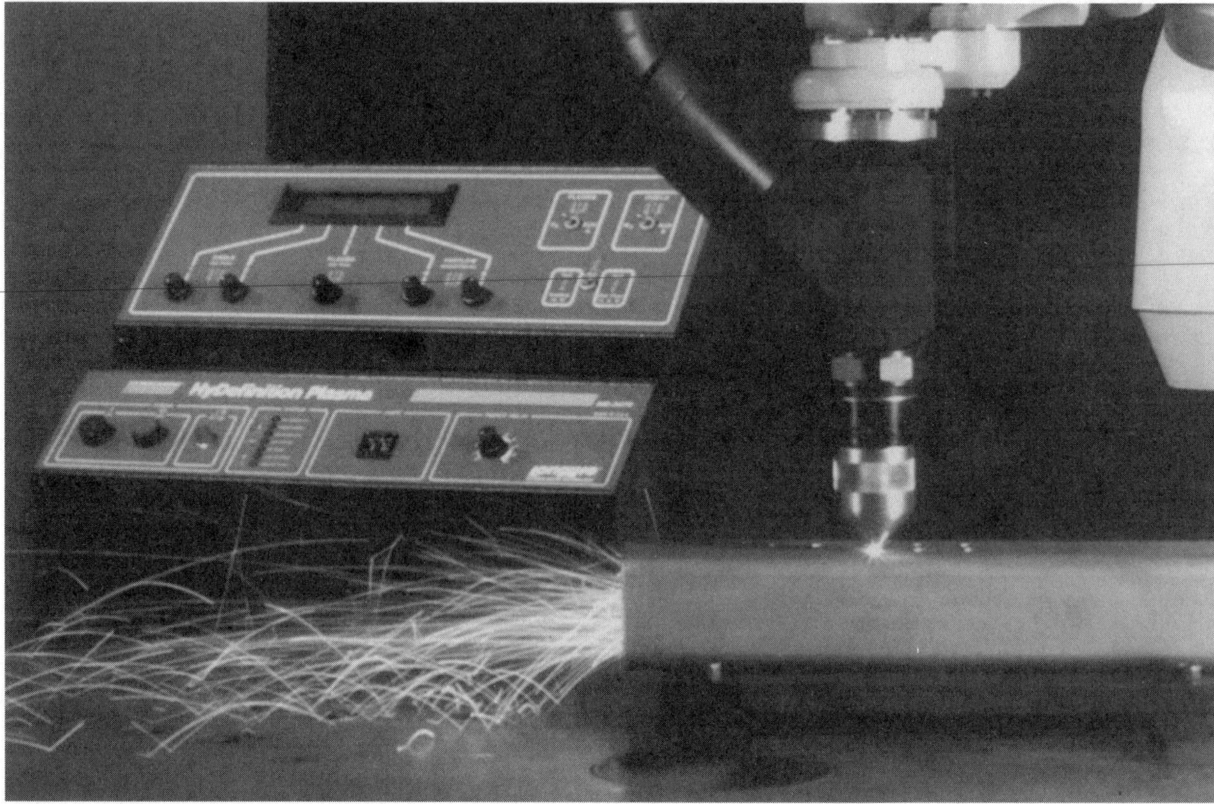

Figure 14-6. Robot-mounted, plasma-arc cutting torch for three-dimensional cutting. (Courtesy Hypotherm, Inc.)

well as nonmetals like oxides and ceramics. Spraying may be done on metallic or nonmetallic surfaces such as fabrics, ceramics, and plastics. It may be done to replace material cut away by mistake, worn away, or to confer some physical property on a surface. An example of rebuilding is a 457-mm (18-in.) diameter axle worn to 419 mm (16.5 in.) diameter. Alone, metal spraying is not a precision process, so the axle was sprayed with a 13% chromium steel at the rate of 12 kg/hr (26 lb/hr) with a 63.5 kg/journal (140 lb/journal) and reground to original size. The cost was about half that of a new part.

Other uses for spraying are to impart corrosion resistance by a coat of stainless steel, heat resistance by a coat of zirconium oxide in the combustion chamber of a jet engine, and wear resistance by a coat of aluminum oxide in an extrusion die, or tungsten carbide on a new metal-forming die. Sprayed surfaces are porous and hold lubricant well.

The appearance of cast surfaces can be improved by metal spraying. Sprayed metal can be decorative, like aluminum or bronze on cast iron. Thin-walled parts can be made in shapes or of materials that cannot be fabricated easily by other means by spraying a coating on a formed mandrel and then removing the mandrel.

Sprayed metal can be expected to oxidize somewhat, be up to 15% less dense, and be weaker than the parent material in a part. Because oxyacetylene flame spraying is done at lower temperatures and is slower than the other methods, it results in two to three times as much oxidation and less interfusion of the particles and surface. Bond strengths of 10–20 MPa (1,500–3,000 psi) are achieved in flame spraying, 20–70 MPa (3,000–10,000 psi) in electric-arc and plasma-arc spraying, and 55 to over 172 MPa (8,000 to over 25,000 psi) for detonation-gun coating. Partial welding between particles and surface appears to occur with the higher-temperature methods. Sprayed coatings serve best in compression and may be made quite hard. When hardenable steel is sprayed, it is quenched rapidly on contact. A coating may be as thin as

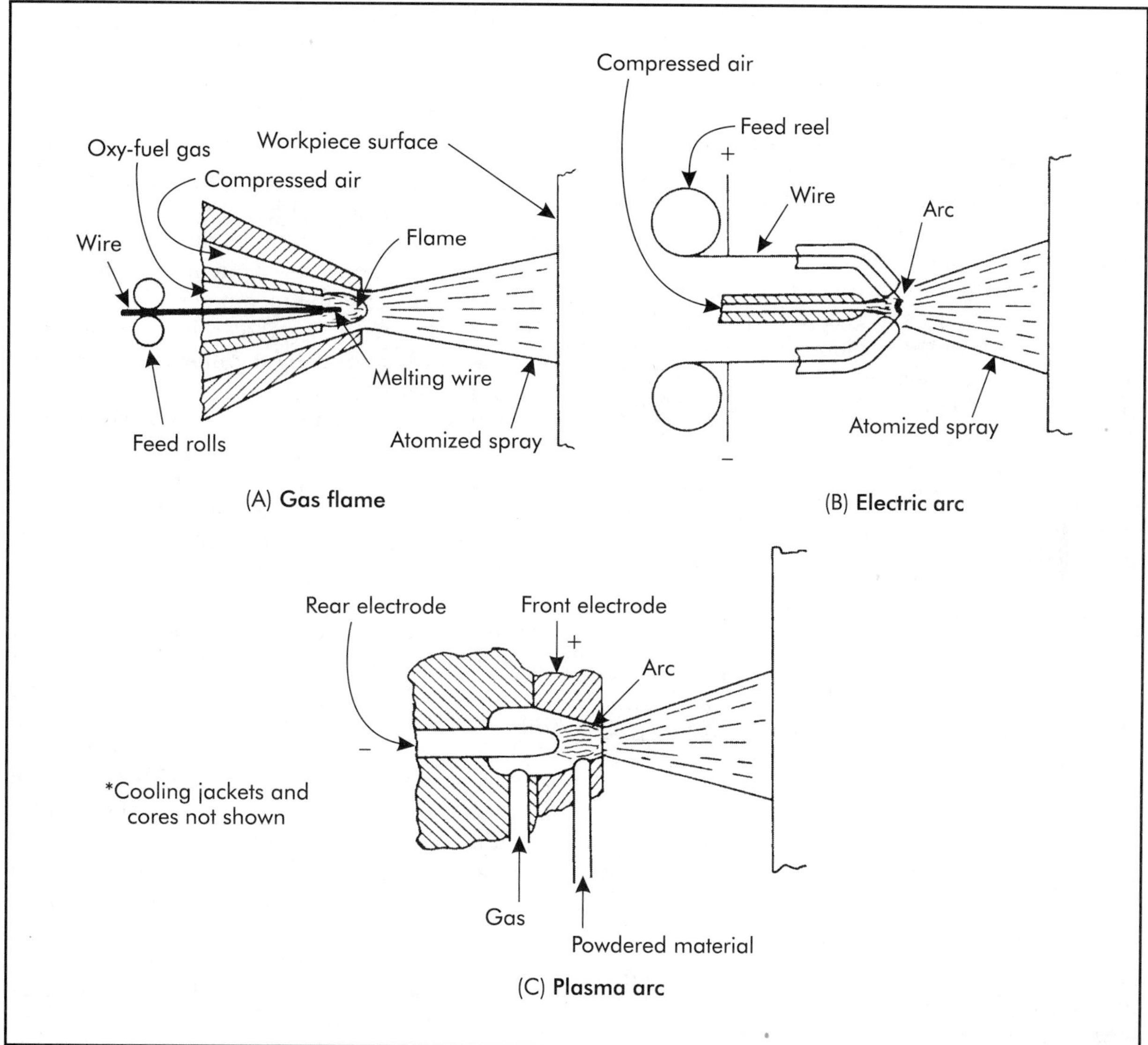

Figure 14-7. Several types of metal spray guns.

51–76 μm (.002–.003 in.) and there is no upper limit to its thickness if properly applied.

14.4 SURFACING AND HARD-FACING

14.4.1 Principles

Surfacing or hard-facing is a process of welding wear- or corrosion-resistant metal on a part to make it serviceable or to rebuild or repair it. It is commonly done on metalworking dies and tools, oil well drilling tools, parts of earth-moving and excavation equipment, rolling mill rolls, and tractor parts. As an example, it has been found that parts for heavy tractors can be rebuilt by surfacing at about 40% of the cost of replacement parts and last about 50% longer. On the other hand, new forming dies for sheet metal are commonly hard faced at critical spots to make them run longer before they have to be repaired.

All but a few hard-facing materials consist of hard particles of carbides of chromium, tungsten, molybdenum, and other metals uniformly distributed in a soft matrix. The hardness of an alloy depends on the types and amounts of carbides it contains. No one surfacing material can serve all purposes.

The many materials used may be classified in groups. For the ferrous alloys, *group A* contains those with less than 20% alloy content and *group B* those with more than 20%. The lower alloys are generally tougher and more shock-resistant but not as wear resistant as the higher alloys. The nonferrous alloys in *group C* are highly resistant to wear and corrosion. They include the cobalt-chromium-tungsten alloys, such as stellite, and the nickel-base alloys like Hastelloy®. *Group D*, known as the diamond substitutes, are bits of cemented carbides or sintered oxides welded or brazed to oil well-drilling or metal-cutting tools. *Group E* consists of metal strips, tubes, or rods as electrodes containing crushed carbide particles.

14.4.2 Procedures

Low- and high-carbon, low-alloy, stainless and manganese steels and cast iron can be hard faced with some difference in methods for each. Except for those attached as bits, surfacing metals are commercially available as electrodes or rods and are laid down as molten filler metal on the workpiece surface metal in a conventional welding operation.

All forms of gas and arc welding are utilized to suit various materials and applications. Much surfacing is manually done on diverse repair, rebuilding, and maintenance jobs. Rapid area coverage on production jobs is achieved with submerged-arc welding using wide ribbon electrodes with large amounts of current. Another production method is *bulk-process welding* in which granular metal and flux are spread on the surface and melted along with the surface by an imposed arc. This gives higher deposition rates than other methods.

A process for surfacing with a tungsten carbide electrode utilizes direct current (DC) arc discharge at the rate of 8,000 pulses/sec. Steel surfaces are impregnated in this way with a layer of tungsten carbide 51–76 μm (.002–.003 in.) thick having a hardness of 72–74 R_C to improve wear resistance of dies and cutting tools.

14.5 BRAZE WELDING, BRAZING, AND SOLDERING

14.5.1 Braze Welding

Braze welding, also called *bronze welding*, is like fusion welding, in that a filler is melted and deposited in a groove, fillet, plug, or slot between two pieces to make a joint. In this case, the filler metal is a copper alloy with a melting point below that of the base metal, but above 427° C (800° F). The filler metal is puddled into and not distributed in the joint by capillarity, as in plain brazing described later.

Metals with high melting points, such as steel, cast iron, copper, brass, and bronze, are braze welded. The base metal does not melt, but a bond is formed, sometimes stronger than the base metal alone. Studies of the bonds between copper alloys and cast iron have shown three sources of the forces that bind braze-welded joints. One is the atomic forces at the interface of metals in close contact. A second is alloying that arises from diffusion of the metals in a narrow zone at the interface. The third is intergranular penetration.

Braze welding may be done by oxyacetylene, metallic-arc, or carbon-arc welding. The filler metal is applied by a rod or electrode together with a suitable flux. Most metals require little, if any, preheat, with the exception of cast iron, in which localized heating may set up enough stresses to cause cracking.

The main advantages of braze welding result from the low temperature of the operation. Less heat is needed and a joint can be made faster than by fusion welding. The filler metal yields substantially on cooling to 260° C (500° F) or below, without weakening, and residual stresses are small. Dissimilar metals not amenable to fusion welding may be joined by braze welding.

Braze-welded joints are not satisfactory for service at over about 260° C (500° F) or for dynamic loads of 103 MPa (15,000 psi) or more.

14.5.2 Brazing

Brazing is the name given a group of welding operations in which a nonferrous filler metal melts at a temperature below that of the metal joined, but is heated above 427° C (800° F). The molten filler metal flows by capillarity between the heated but unmelted adjacent or overlapping joint members or is melted in place between those members. Examples of the kinds of joints made in this way are shown in Figure 14-8.

Two classes of filler metals for most work are copper alloys and silver alloys. An exception is aluminum that is brazed with aluminum alloys that melt just before the parent metal. There are a number of different alloys in each class. Some are considered general-purpose alloys and are satisfactory for a range of jobs. References and treatises prescribe proper alloys, fluxes, joint designs, and methods of operation for best results in specific cases.

Copper alone or alloyed with other metals is applied in brazing at 704–1,177° C (1,300–2,150° F). Some of the alloys withstand temperatures to 427° C (800° F) in service, but most are not considered serviceable above 260° C (500° F).

Brazing done with silver alloys is called *silver brazing* or *silver soldering*. Silver alone or alloyed with other metals is applied at 635–843° C (1,175–1,550° F). Some metals and parts must be joined at these lower temperatures to avoid overheating or warping. Service temperatures should usually not be over 260° C (500° F). Silver-brazed joints are about as strong as and sometimes stronger than copper-brazed joints, but the silver alloys are relatively expensive.

Aluminum alloy brazing is performed at temperatures of from 566–616° C (1,050–1,140° F). Up to 100% joint efficiency in tension is obtainable. For economical operation, parts are commonly fabricated from brazing sheet, which consists of a layer of brazing alloy bonded to one or both surfaces of a core alloy.

The bond between a brazing metal and the clean metal of the parts joined is due to some diffusion of the brazing metal into the hot base metal and to some surface alloying of the metals. A good joint is at least as strong as the filler metal. If the base metal is weaker than the filler metal, failure occurs outside the joint. If the base metal is stronger, the joint fails, but at a stress much higher than the strength of the filler metal. A good joint can only be obtained with a minute clearance between the surfaces being joined and by proper brazing techniques that assure full penetration of the joint by the filler metal. Misfits that cause voids too large to be filled naturally, as suggested in Figure 14-8, result in joints of less than full strength. Tests of stainless steel with a tensile strength of 1.1 GPa (160,000 psi)

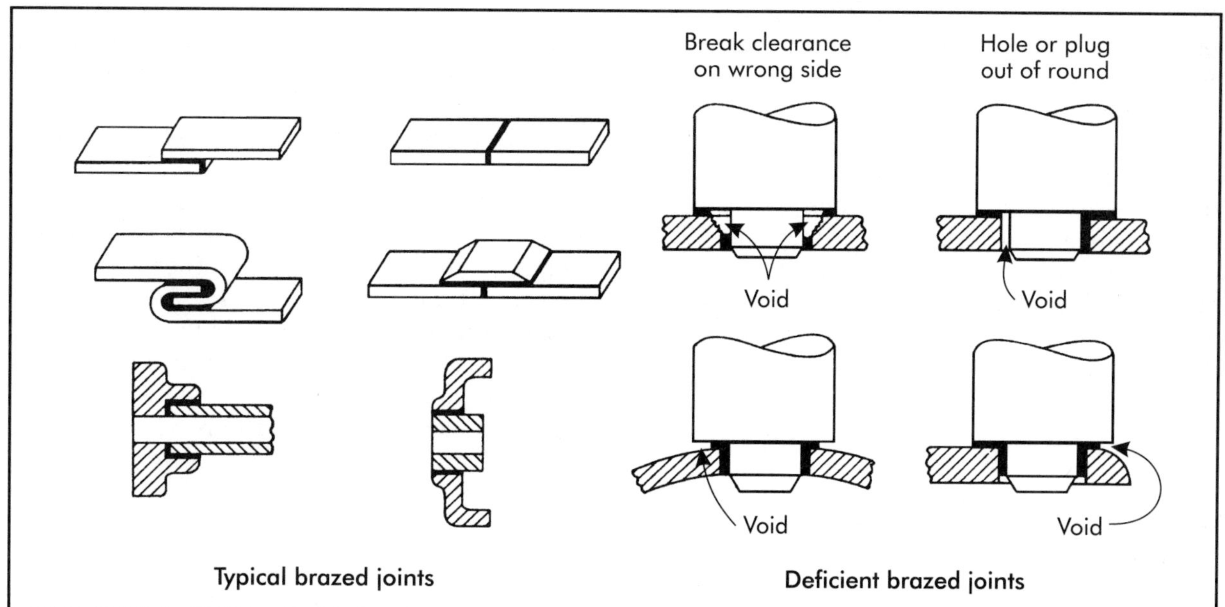

Figure 14-8. Brazed joints.

joined in a butt joint by means of a silver alloy of 483 MPa (70,000 psi) strength showed a maximum joint strength of 896 MPa (130,000 psi) with a clearance of 38 μm (.0015 in.) in the joint. The strength dropped off almost proportionately to about 276 MPa (40,000 psi) as the clearance was increased to 0.6 mm (.024 in.). In addition to strength, good brazed joints are ductile and resistant to fatigue and corrosion.

Surfaces must be thoroughly cleaned before brazing and then usually are covered with a flux to break down and prevent further formation of oxides. Borax is a widely used flux, but many proprietary brands are available. Parts should be tightly held together in process; they may be staked or crimped together or held in a fixture for production. The filler metal may be in the form of wire, preformed shapes, or alloy paste and may be applied before or during heating.

The actual heating may be done in a number of ways. Torch brazing is popular because it is convenient and adaptable. Furnace, electrical resistance, and induction heating serve in various ways in production. Parts may be dipped in molten salt or flux for preheating or final heating. With or without preheating, parts may be dipped in molten filler metal; this is particularly of advantage when brazing or soldering a number of joints simultaneously and uniformly in one assembly. Equipment costs are highest for dipping. Brazing and soldering operations are readily and commonly automated for production of from 40–20,000 assemblies/hr.

Steel, copper, brass, bronze, aluminum, and many less common metals are brazed together and to each other. Hardened steel and some aluminum alloys cannot be brazed satisfactorily. The strengths, operating temperatures, and costs of brazed joints are less than those for fusion welding but more than those for soft soldering.

14.5.3 Soldering

Soldering or *soft soldering* is the process of joining metals by means of alloys that melt between 177–371° C (350–700° F). The alloys are generally lead and tin alloys. The metals mostly joined by soldering are iron, copper, nickel, lead, tin, zinc, and many of their alloys. Aluminum can be soldered by special means.

The strength of a soldered joint depends on surface alloying and upon mechanical bonding, such as crimping, between the joined parts. The solder alone has a low unit strength, little resistance to fatigue, and is limited to service temperatures below 149° C (300° F). Tests on various soldered copper sleeve joints showed shear strengths from about 21–41 MPa (3,000–6,000 psi) at 29° C (85° F). Thin films are necessary for the strongest joints. Thicknesses of 76 μm (.003 in.) on copper and 127 μm (.005 in.) on steel are desirable.

For soldering, a flux is generally necessary to rid the surface of oxides, to promote wetting, and obtain intimate contact between the solder and base metal. Zinc chloride is most efficient, but corrosive as a flux. Rosin does not clean as well, but is noncorrosive and must be used for dependable electrical connections.

There are a number of soldering techniques, some used for particular materials. In general, the flux is applied first, the parts are heated at the joint to just above the melting point of the solder, and then the solder is touched to and flows into the joint. Any source of clean heat is satisfactory. A heated copper soldering iron or soldering copper is widely used for general-purpose work. An open flame from an alcohol, gasoline, hydrogen, or gas torch may be applied directly on the joint. A heated neutral gas, like nitrogen, may be necessary for some extremely fussy work. Induction or resistance heating or immersion of the workpieces in a solder bath are fast methods and adapted to production.

Soldering produces liquid- and gas-tight joints quickly at low cost. Temperatures are not high, equipment is simple, and the method is the most convenient and feasible means of making joints in the workshop, laboratory, or home where often the equipment for other processes is not available. Soldering provides positive and dependable electrical connections. If a soldered joint is strong and durable enough for a particular purpose, it is usually the most economical.

14.5.4 Process Automation

Brazing and soldering processes can be automated and used for large production quantities. Automated equipment speeds these processes and reduces the cost of manual labor. Generally, the cost of automated soldering equipment depends on the degree of complexity of the process and the desired features of the equipment. The cost ranges from about $100 for a simple soldering iron to over $50,000 for an automated system. Brazing equipment costs are higher than those for soldering. Those costs range from about $15,000 to over $300,000, depending on the capabilities of the equipment, the degree of automated features in the process, brazing method, and process complexity.

Brazing and soldering are slow processes and require highly skilled operators. Soldering, in general, is a slower process than brazing. For these main reasons, automation is generally an attractive alternative that can be economically justified for large production quantities.

14.6 MECHANICAL FASTENING OR JOINING

Ever since the invention of the button over 5,000 years ago, improvements have been made in the way that articles of clothing and the component parts of goods, such as appliances, automobiles, houses, writing instruments, and computers are joined or fastened together. Today, literally hundreds of methods and devices are available for fastening objects or parts together, ranging from the common straight pin to intricate mechanisms that, because of safety or security reasons, need special tooling to operate. The selection of a particular fastening method or device depends on many factors, the most important being whether or not the joining process is to be permanent or nonpermanent. If provisions must be made for repair or replacement of the components of an assembled part, then nonpermanent fasteners are required. Permanent fastening methods may be used for nonrepairable or throwaway assemblies.

The selection of a fastening or joining method or device is also dependent on a number of other factors, such as the material or materials to be joined, strength and reliability of the joint, appearance, and cost.

14.6.1 Nonpermanent Fasteners

A variety of standard nonpermanent fasteners are commercially available, the most common being the *threaded fastener*. These appear as bolts, screws, and studs and are covered by specific standards, either inch or metric. A comprehensive set of definitions that describe the characteristics of threaded fasteners is given in the American National Standard Institute/American Society of Mechanical Engineers standard ANSI/ASME B1.7M-1984 R92, "Nomenclature Definitions and Letter Symbols for Screw Threads." Bolts and screws are commonly designated by the following sequence of information: (1) nominal size or basic diameter, (2) threads per mm (in.), (3) product length, mm (in.), (4) product name, (5) material, and (6) protective finish.

As illustrated in Figure 14-9(A), a *bolt* is an externally threaded fastener that is assembled with a nut (and often a washer) to perform its intended function. A bolt is traditionally tightened or loosened by applying torque to a nut. Bolts and accompanying nuts may be obtained commercially in accordance with the American National Standards Institute U. S. and metric standards, unified standards, and International Organization for Standardization (ISO) metric standards. Standard bolts are available with hexagonal, square, round, countersunk, or T-heads. The round-headed bolts are often constructed with an underhead configuration that locks into the joint material to resist rotation when the mating nut is torqued.

A *screw*, such as is illustrated in Figure 14-9(B) is an externally threaded fastener that is usually assembled by torquing the head into a previously threaded hole. Because of their comparable thread design, it is possible to use some screws with nuts. Common classifications for screws are machine screws, cap screws, sems (screw and washer assemblies), and tapping screws. Probably the most common is the cap screw available with either hex, socket, or slotted heads. Tapping screws are commonly used in joining thin materials by cutting or forming mating threads as they are driven into mating holes.

A *stud*, such as is shown in Figure 14-9(C), is a cylindrical rod that is threaded on both ends or throughout its total length. Generally, the metal end of a stud is screwed into the component of an assembly while the other end, called the "nut" end, extends out of the component to accommodate a nut. Quite often, studs are se-

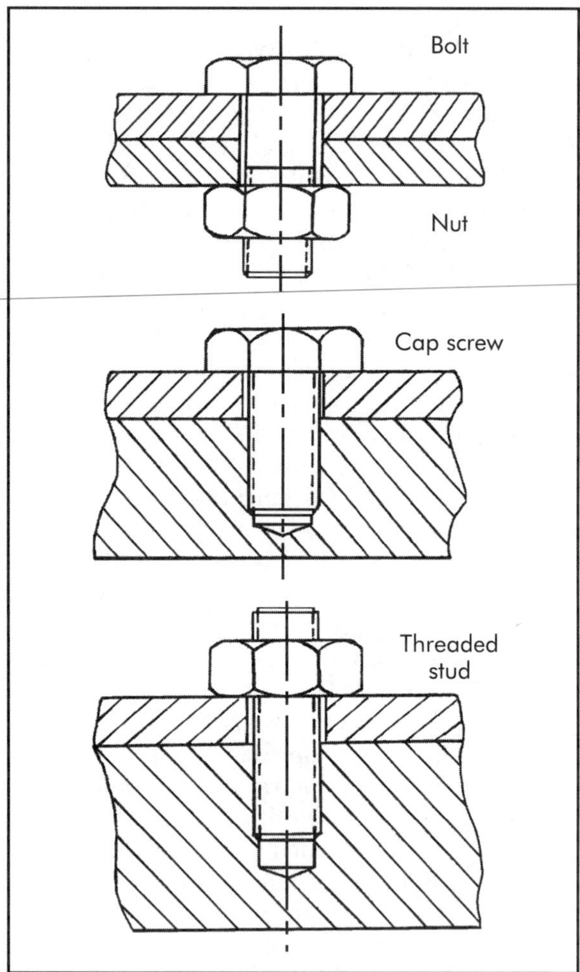

Figure 14-9. Common types of threaded fasteners: (A) hexagonal head bolt and nut, (B) hexagonal head cap screw, (C) threaded stud with hex nut.

lected in such a way that there is an interference fit between the metal end of the stud and its mating threaded hole. This provides a holding torque for assembly of the nut. It is often advantageous to use studs in some applications as pilots to facilitate the assembly of mating components.

The assemblies shown in Figures 14-9(A) and 14-9(C) are both secured by means of a nut. Thus, nuts are an internally threaded component of bolt and stud assemblies and are an important part of the nonpermanent joining process. Nuts come in a variety of shapes, including hex nuts, and square nuts, often referred to as whole nuts. Hex-slotted nuts, hex-flange nuts, hex-castle nuts, jam nuts, and acorn nuts are also available for certain purposes. In addition, single thread or spring nuts, wing nuts and other types of stamped nuts can be used for some applications.

For most joining applications, *washers* are used to provide uniform load distribution and protect against loosening of the nut resulting from vibration. Plain washers, helical spring-lock washers, and internal-external tooth-lock washers are commonly used for various applications.

In addition to threaded fasteners, other devices such as retaining rings, hinges, slip pins, quick release mechanisms, stamped nuts, and snap-in or snap-on devices are used as nonpermanent fasteners.

14.6.2 Permanent Fasteners

Many assembled parts do not require the capability of being disassembled and therefore, may be joined by permanent fastening methods or devices. Perhaps the most commonly used of this category of fasteners is the pin-type *rivet*. Generally, there are two kinds of riveted joints, the lap joint and the butt joint. For a *lap joint*, two plates overlap each other and are joined together permanently by one or more rows of rivets. In a *butt joint*, the two plates are allowed to remain in the same plane and are joined by means of a cover plate or butt strap riveted to both plates.

Solid rivets are made of various metals depending the strength needed and are made with a formed head (for example, round, button, flat, or oval countersink) and cylindrical- or pin-type body. The pin section is inserted in a prepunched or drilled hole in the two sections of metal to be joined. The protruding end of the rivet is then deformed or headed to form a second head that holds the joint together.

Staking is another joining process used to secure a plate to a post that protrudes through a hole in a plate. Used commonly in the assembly of instruments, staking is a quick method of permanently attaching plates to poles that separate the instrument parts.

A variety of processes are used to permanently assemble sheet metal parts, including crimping, edge seaming, lanced or shear-formed tabs, stapling, and wire stitching.

14.7 ADHESIVE BONDING

Adhesive bonding is used to create a permanent joint between two surfaces. Formerly limited to the joining of wood or plastic materials, adhesive bonding is now applied to metals and nonmetals. The adhesive materials used in this process cover a broad spectrum of thermoplastic resins, thermosetting resins, artificial elastomers, and some ceramic materials. Applications for adhesive bonding range from the bonding of load-bearing components using structural adhesives to relatively light-duty holding and positioning of components. Adhesive bonds also may be used to form gas- or liquid-tight seals.

Adhesives for bonding have been divided into about nine classes, all of which have different characteristics. For example, a Class 1A adhesive with multiple components can cure at room temperature and the application of heat will increase the cure rate. On the other hand, a Class 1B adhesive is heat-activated and requires the application of heat to initiate the adhesive's chemical reaction. These classes of adhesives are discussed at length in the reference selections at the end of this chapter.

The two major advantages of adhesive bonding are: (1) efficiency of production and, (2) reasonable cost. Thus, continued developments are being made in the field of adhesives to broaden their applicability and improve their toughness and resistance to various environmental factors.

14.8 QUESTIONS

1. How does an oxygen jet cut steel?
2. What is the difference between oxygen cutting and flame machining?
3. What is "oxygen lance cutting" and for what is it used?
4. How does oxygen cutting compare with metal shearing? With friction sawing? With machining?
5. Why is it difficult to oxygen cut some materials and what can be done to minimize the difficulty?
6. How is electric-arc cutting done and what are its limitations? How does it excel?
7. How is metal sprayed? Describe three kinds of equipment.
8. For what purpose are metals and other materials sprayed?
9. What types of materials can be used for spraying?
10. What is "surfacing" or "hard-facing" and why is it done?
11. What are the primary materials used to hard-face metals?
12. How do brazing and braze welding differ and how are they alike? How do they differ from soldering? What are their advantages?
13. What are the fundamental requirements of a good brazed joint?
14. What causes the bond between surfaces in a brazing operation?
15. How does the strength of a soldered joint compare with that of a brazed joint?
16. What is the primary consideration in the selection of either a permanent or nonpermanent joining process or device?
17. What is the difference between a bolt, a screw, and a stud?
18. Why are studs used in some assemblies?
19. Why are washers generally used with threaded fasteners?
20. What is the difference between riveting and staking?
21. What are the two major advantages of adhesive bonding?

14.9 REFERENCES

Adhesives in Modern Manufacturing. 1970. Dearborn, MI: Society of Manufacturing Engineers (SME).

ANSI B18.12-1962 (R1981). "Glossary of Terms for Mechanical Fasteners." New York: American National Standards Institute.

ANSI B18.2.1-1972 (R1981). "Square and Hex Bolts and Screws, Inch Series." New York: American National Standards Institute.

Brazing Handbook. 1991. New York: American Welding Society.

Collopy, W. E., 1982. "Powders Add Flexibility to Hard Facing, Cut Costs." *Metal Progress,* May.

Handbook of Adhesives, 2nd Ed. 1977. Irving Skiest, ed. New York: VanNostrand Reinhold.

Landrock, A. H. 1985. *Adhesives Technology Handbook.* Park Ridge, NJ: Noyes Publications.

Rahn, A. 1993. *The Basics of Soldering*. New York: John Wiley & Sons.

Schaffer, G. 1981. "Fundamentals of Brazing, Special Report 732." *American Machinist*, April.

Schwartz, M. M. 1990. "Ceramic Joining." Materials Park, OH: American Society for Metals.

Society of Manufacturing Engineers (SME). 1997. *Electrical Discharge Machining* video. *Fundamental Manufacturing Processes* series. Dearborn, MI: SME.

——. 1997. *Thermal and Abrasive Waterjet Cutting* video. *Fundamental Manufacturing Processes* series. Dearborn, MI: SME.

Tool and Manufacturing Engineers Handbook, 4th Ed. 1986. Vol. 4. *Quality Control and Assembly.* Dearborn, MI: Society of Manufacturing Engineers (SME).

Quality Assurance

15

Quality control charts, 1924

—Dr. Walter A. Shewhart, Bell Telephone Laboratories

15.1 INTERCHANGEABLE MANUFACTURE

Modern manufacturing systems, based on the concepts of interchangeable manufacture, require that each part of an assembly going into a final product be made to definite size, shape, and finish specification. The mass production of both consumer and producer goods relies on interchangeability, and interchangeability requires fabrication to exacting dimensions and close tolerances. Compressor pistons, for example, must be machined within limits so that a piston selected at random will fit and function properly in the compressor model for which it was designed. This interchangeable manufacturing system is, to a large degree, responsible for the high standard of living enjoyed today by the people of the United States. It makes possible the standardization of products and methods of manufacturing, and provides for ease of assembly and repair of products.

15.2 QUALITY ASSURANCE

The quality of a product may be stated in terms of a measure of its conformance to specifications and standards of workmanship. These specifications and standards should reflect the degree to which the product satisfies the requirements of a particular customer or user. The *quality assurance* function within a manufacturing organization, however it may be structured, is charged with the responsibility of maintaining product quality consistent with those requirements.

15.2.1 Organizational Approaches

A variety of organizational programs and practices are employed by different manufacturing firms to obtain a level of product quality sufficient for their needs. These programs and practices range from simple quality evaluation procedures performed by machine operators to very sophisticated and comprehensive quality management systems.

Make It Right the First Time

Some firms place the major responsibility for product quality on the product designers and machine operators who produce the product. In effect, these firms place great emphasis on the requirement that the product be appropriately designed for manufacture. Theoretically, with design for manufacture, the people in manufacturing are able to fulfill their objective of producing a product free of defects (*zero defects*) without additional costly inspection operations and personnel. This philosophy of "making it right the first time" is a state that all manufacturers would like to achieve. However, many firms are unable to achieve and maintain a satisfactory level of product quality without the addition of some components of quality surveillance and control. This is because of the complexities of many products and the many sources of variation within the raw materials and manufacturing processes.

Concurrent Engineering (CE)

The concepts and practices embodied in concurrent engineering (CE) and design for manufacturing (DFM) are becoming more important to manufacturing firms interested in producing the high-quality and low-cost products necessary for global competitiveness.

In traditional product design practices, component and part designs were handed off from one department to another in a serial fashion and the final product design was then "tossed over the wall," so to speak, to manufacturing. The manufacturing organization would then have to go through a very time-consuming process of transforming those designs in such a way as to permit the product to be manufactured. This often involved changing part drawings and tolerances, developing parts lists and assembly drawings, prescribing tooling requirements, and establishing supplier contacts. This post-design conversion process often required many hours of negotiation between various departments within the firm to resolve design, marketing, and manufacturing issues.

Concurrent engineering (CE) practice requires the earliest possible integration of a manufacturing firm's organizational components (R & D, design, marketing, and manufacturing) for the development of new products. This practice, when properly applied, changes the product development process from a serial to a parallel one and thereby shortens the development time considerably. Recent developments in just-in-time (JIT), total quality management (TQM), and CE are enabling manufacturing firms to respond to global competition by meeting customer demands for high-quality products, at low cost, in reasonable time frames.

Total Quality Management (TQM)

Since the beginning of the industrial revolution in the U. S., the word *quality*, when used by product manufacturers, has taken on several different meanings. Traditional definitions often describe a quality product as one that is well built and reliable. Other definitions refer to a quality product as one that is the best or the best for the money. More recently, such definitions have taken on a more strategic meaning by relating the quality of a product to the concept of meeting the needs of the customer.

The *total quality management* concept has expanded the strategic direction for quality management via a structured model for continuous improvement. The model builds on three fundamental principles of total quality: (1) emphasis on the customer, both internal and external; (2) emphasis on improving work processes to produce consistent, acceptable products; and (3) emphasis on the involvement of the total organization. These fundamental principles are then supported and coordinated by the organizational elements shown in Figure 15-1.

Customer emphasis implies that each person in a manufacturing organization has one or more internal customers and the expectations of those customers must be appropriately satisfied if the overall organization expects to meet the needs of external customers. Thus, every person in the

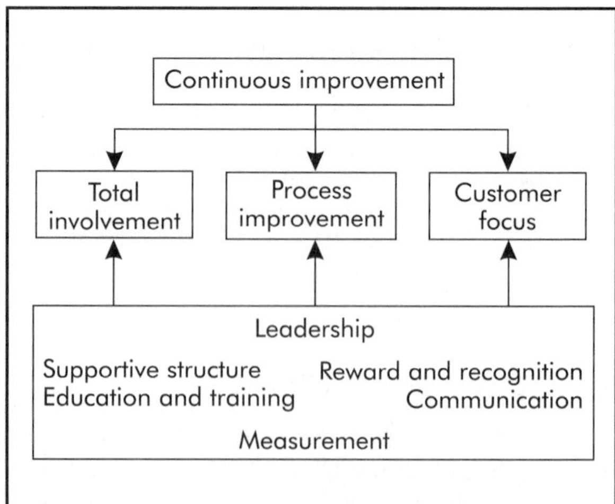

Figure 15-1. Total quality management concept model.

organization is involved and contributes to the overall effort of satisfying the needs and requirements of the customer.

TQM relies heavily on the premise that any manufacturing process can be improved. Further, any manufacturing organization that expects to be successful in international competition must consciously and continuously search for and exploit improvement. Total quality management often requires a cultural change in the attitudes and behavior of personnel who, in the past, have had a somewhat cavalier regard for product quality. TQM represents a total departure from the "let the buyer beware" (caveat emptor) philosophy prevalent in the early history of product quality development.

International Organization for Standardization (ISO) 9000 Standards

ISO 9000 is comprised of a series of standards on quality management and assurance that were promulgated by the International Organization for Standardization (ISO) in 1987. The ISO, with a worldwide membership of over 90 countries, developed the ISO 9000 quality standards in an attempt to harmonize the large number of national and international quality system standards in existence at that time. These standards have been adopted by the European Economic Community (EEC) countries. Thus, it is important that U. S. manufacturing firms acquire certification by ISO 9000 standards if they expect to do business with EEC countries or other international marketplaces. The ISO is represented in the United States by the American National Standards Institute (ANSI) and the American Society for Quality (ASQ). The U. S. version of the ISO 9000 standards is known as the ANSI/ASQ Q90-1987 series. There are five basic standards within the ISO 9000 series:

1. ISO 9000, "Quality Management and Quality Assurance Standards-Guidelines for Selection and Use." This standard essentially provides guidelines for selecting and using the next three standards in the series.
2. ISO 9001, "Quality Systems—Model for Quality Assurance in Design/Development, Production, Installation, and Servicing." This is considered the most comprehensive standard in the series. It provides a model of quality management and assurance for use when conformance to a specified requirement is to be assured by the supplier during several stages, which may include design/development, production, installation and servicing of products, or service.
3. ISO 9002, "Quality Systems—Model for Quality Assurance in Production and Installation." This standard is for use when conformance to specified requirements is to be assured by the supplier during production and installation. It is particularly pertinent to the process industries (such as, chemical, food, and pharmaceutical) where product requirements are dictated by established design or specification.
4. ISO 9003, "Quality System—Model for Quality Assurance in Final Inspection and Test." This standard is for use when conformance to specified requirements is to be assured by the supplier solely at final inspection and test. It is particularly pertinent to small shops, divisions within organizations, or equipment distributors.
5. ISO 9004, "Quality Management and Quality System Elements—Guidelines." This standard provides guidelines for designing and developing an effective quality management system that will satisfy customer needs and expectations while serving to protect the firm's interests.

ISO 9000 certification. Manufacturers in the U. S. who wish to be certified in compliance with ISO 9000 standards must be prepared for an extensive third-party audit of their quality management system. The third-party audit within the ISO 9000 certification process places considerable emphasis on quality systems. The philosophy behind this is that the quality system audit is similar in importance to a firm's product quality reputation, and just as important as an annual financial audit is to their financial reputation. ISO 9000 certification is conducted by a registrar whose competence in a particular industry environment has been accredited by an ISO member accreditation body.

15.2.2 Components of the Quality Assurance Function

In a manufacturing context, the quality of a product is normally stated in terms of the measure of the degree to which it conforms to specifications and standards of workmanship. These specifications and standards should reflect the degree to which the product satisfies the requirements of a particular customer or user. Regardless of how a particular manufacturing firm's quality efforts are organized, the quality assurance function is charged with the responsibility of maintaining product quality consistent with those requirements. Generally, this effort will involve the following basic activities:

- quality specification;
- inspection;
- quality analysis; and
- quality control.

Quality Specification

The product design engineer provides the basic specifications of product quality by means of the various dimensions, tolerances, and other requirements cited on the engineering drawings. These basic specifications are further refined and elaborated upon on the manufacturing drawings used by the methods engineering personnel to specify the manufacturing and fabrication procedures. In many cases, the quality assurance group will provide a further interpretation of those specifications as a basis for specifying inspection procedures.

Inspection

The mass production of interchangeable products is not altogether effective without some means of appraising and controlling product quality. Production operators are required, for example, to machine a given number of parts to the specifications shown on a drawing. There are often many factors that cause the parts to deviate from specifications during the various manufacturing operations. A few of these factors are variations in raw materials, deficiencies in machines and tools, poor methods, excessive production rates, and human errors.

Provisions must be made to detect errors so that the production of faulty parts can be stopped. The inspection department is usually charged with this responsibility, and its job is to interpret the specifications properly, inspect for conformance to those specifications, and then convey the information obtained to the production people, who can make any necessary corrections to the process. Inspection operations are performed on raw materials and purchased parts at the time of receiving. In-process inspection is performed on products during the various stages of manufacture, and finished products may be subjected to a "final inspection."

Quality Analysis

During the various inspection processes, quality information is recorded for review by representatives from manufacturing and quality assurance. In essence, this information provides a basis for analyzing the quality of the product as it relates to the capability of the manufacturing process and the quality specifications initially prescribed for the product. Generally, it is not possible to mass produce products that are 100% free of defects except at considerable expense. The quality analysis will determine the defect level of the product so that a decision can be made as to whether or not that level is tolerable. If a given defect level is not tolerable, decisions must be made relative to: (1) what to do with the process, and (2) what to do with the product.

Quality Control

The word "control" implies regulation, and of course, regulation implies observation and manipulation. Thus, a pilot flying an aircraft from one city to another must first set it on the proper course heading and then continue to observe the progress of the craft and manipulate the controls to maintain its flight path in the proper direction. *Quality control* in manufacture is an analogous situation. It is simply a means by which management can be assured that the quality of product manufactured is consistent with the quality-economy standards that have been established. The word "quality" does not necessarily mean "the best" when applied to manufactured products. It implies "the best for the money."

15.3 STATISTICAL METHODS

Increasing demands for better product quality often result in increased cost of inspection through surveillance of processes and product. The old philosophy of attempting to inspect quality into a product is time consuming and costly. It is, of course, more sensible to control the process and make the product correctly rather than have to sort out defective product from good product. Certain statistical techniques have been developed that provide an economical means of maintaining continual analysis and control of processes and product. These are known as *statistical process control* (SPC) or *statistical quality assurance* methods and they involve the integration of quality control in each stage of producing the product. They represent a powerful collection of tools that implement the concept of prevention, as a shift from the traditional quality by inspection/correction process. Figure 15-2 presents the concept of process control.

In the 1930s, Walter Shewhart introduced the concept of control charts. An attempt was made to use Shewhart's concepts during World War II. After the war, SPC concepts were not further implemented in American industry. It wasn't until the late 1970s that American industry realized the importance of statistical process control. In the meantime, Japan asked for help from Deming (1900-1993) and later Juran to rebuild its industrial and economic base.

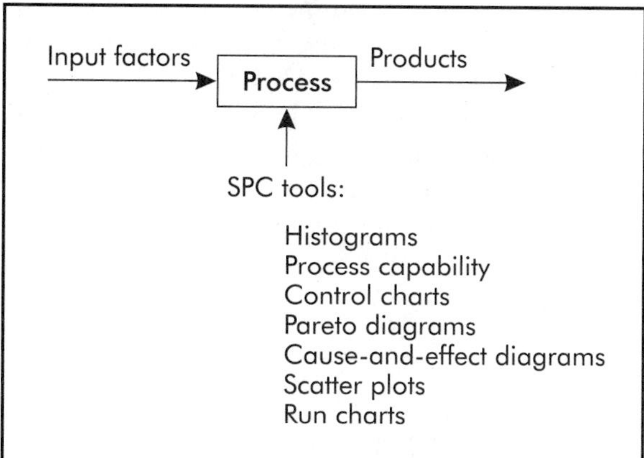

Figure 15-2. The concept of process control.

SPC uses simple statistical tools to control, monitor, and improve processes. All SPC tools are graphical and simple to use and understand. The main objective of any SPC study is to reduce variation. Any process can be simply considered a transformation mechanism of different input factors into a product or service. Since inputs exhibit variation, the result is a combined effect of all variations. This, in turn, is translated into the product. The purpose of SPC is to isolate the natural variation that can be traced or whose causes may be identified. There are two different kinds of variation that affect the quality characteristics of products, *common cause variation* and *special cause variation*.

15.3.1 Common Causes of Variation

Variations due to common causes are inherent in any process. They are inevitable and can be represented by a normal distribution. Common causes are also called *chance causes* of variation. A stable process exhibits only common causes of variation. The behavior of a stable process is predictable or consistent; the process is said to be *in statistical control*.

15.3.2 Special Causes of Variation

Special causes, also called *assignable causes*, of variation are not part of the process. They can be traced, identified, and eliminated. Control charts are designed to hunt for those causes as part of SPC efforts to improve the process. A process with the presence of special or assignable causes of variation is unpredictable or inconsistent; the process is said to be *out of statistical control*.

15.3.3 Statistical Process Control (SPC) Tools

Among the many tools for quality improvement, the following are the most commonly used tools of SPC:

- histograms;
- cause-and-effect diagrams;
- Pareto diagrams;
- control charts;
- scatter or correlation diagrams;
- run charts; and
- process flow diagrams.

Histograms are visual displays that show the variability of a process or any other quality characteristic. They can be used to illustrate the specification limits on a product in relation to the natural limits for the process used to produce it. Histograms also can be used to identify possible causes of a problem experienced by the process. It provides a powerful analytical tool to understand the process.

A *Pareto chart* is another powerful tool for statistical process control and quality improvement. It focuses the attention on those few problems that cause trouble in a process. With a Pareto chart, facts about the greatest improvement potential can be easily identified.

Cause-and-effect diagrams, also called *Ishikawa diagrams* or *fishbone diagrams*, provide a visual representation of the factors that most likely contribute to an observed problem or an effect on the process. The relationships between such factors can be clearly identified. Problems, therefore, may be identified and the root causes may be corrected.

Scatter diagrams, also called *correlation charts*, show the graphical representation of the relationship between two variables. In statistical process control, scatter diagrams are normally used to explore the relationships between process variables and may lead to identifying possible ways to increase process performance.

Control charts are graphical representations of process performance. They provide a powerful analytical tool for monitoring process variability and other changes in process mean or variability deterioration. Control charts were basically developed to hunt for special causes of variation.

Run charts depict process behavior against time. They are important in investigating changes in the process over time. Any changes in process stability or instability can be judged from a run chart.

Process flow diagrams are graphical representations of the process. They show the sequence of different operations that make up a process. Flow diagrams are important tools for documenting processes and communicating information about processes. They also can be used to identify bottlenecks in a process sequence, or to define points where data or information about process performance needs to be collected.

Most of the preceding tools are formulated on the following concepts:

1. A *population or universe* is the complete collection of objects or measurements of the type that are of interest at a particular time. The population may be finite or infinite.
2. A *sample* is a finite group or set of objects taken from a population.
3. The *average* is a point or value about which a population or a sample set of measurements tend to cluster. It is a measure of the ordinariness or central tendency of a group of measurements.
4. *Variation* is the tendency for the measurements or observations in a population or sample to scatter or disperse themselves about the average value.

In many cases, the sizes observed from the inspection of a dimension of a group of pieces have been found to be distributed as shown in Figure 15-3. If the pins represented in Figure 15-3 include all the existing pins of a particular type, then this would be referred to as a *population distribution*.

The pattern of variation shown in Figure 15-3 is typical of that obtained from data taken from many natural and artificial processes. It is referred to as a *normal distribution*, and the smooth curve formed by this distribution is called the normal curve. This, like other distributions, can be described by its *mean* (measure of central tendency), and its *standard deviation* (measure of variation or dispersion). The formulas for these measures are:

$$\overline{X} = \frac{\sum_{1}^{n} X_i}{n} \tag{15-1}$$

$$S = \sqrt{\frac{\sum_{1}^{n}(X_i - \overline{X})^2}{n-1}} \tag{15-2}$$

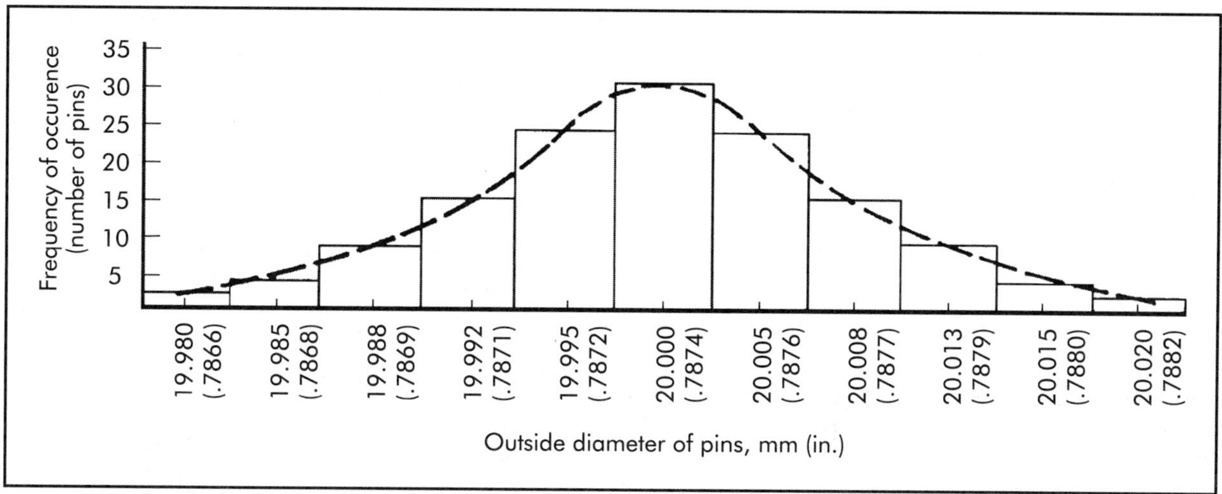

Figure 15-3. Frequency distribution of outside diameters of ground pins.

where

$\overline{X}$ = sample mean
S = sample standard deviation
n = sample size or number of data points in the sample
X_i = a value or a reading of the quality characteristic being measured

The $\overline{X}$ and S values are computed from sample data and are therefore referred to as *sample statistics*.

If the mean and standard deviation are calculated from a population distribution, they are called *population parameters*. Then μ and σ may be computed as follows:

$$\mu = \frac{1}{n}\sum_{i=1}^{n} X_i \qquad (15\text{-}3)$$

$$\sigma = \sqrt{\frac{\sum_{i=1}^{n}(X_i - \mu)^2}{n}} \qquad (15\text{-}4)$$

where

μ = population mean
σ = population standard deviation
n = population size
X_i = a value or a reading of the quality characteristic being measured

The range R is often used as a measure of variation or dispersion for small samples. The *range* is the difference between the largest and smallest observed values in a sample set of values:

$$R = X_{max} - X_{min} \qquad (15\text{-}5)$$

The greater portion of the area under the normal curve is included between the limits μ ±3 σ. The curve actually continues out to plus and minus infinity, but the area under it, beyond μ ±3 σ, is practically negligible.

The areas under intervals of the normal curve commonly used in statistical quality control practices are shown in Figure 15-4. These mean that if a population of values is normally distributed, 99.73% of the values from that population will probably appear within the limits of μ ±3 σ. Only 0.27% are expected to fall beyond those limits. Similarly, 95.46% of the values normally fall within μ ±2 σ, and 68.26% within μ ±1 σ.

If the manufacturing specification on the pins shown in Figure 15-3 was 20.0000 ±0.020 mm (.78740 ±.0008 in.) and if the limiting dimensions of that specification, 19.980–20.020 mm (.7866– .7882 in.), corresponded to μ ±3 σ, 99.73% of the pins from that population would be expected to be within specifications. Only 0.27% would be expected not to conform to specifications. Most distributions encountered in industrial inspection activities are not exactly normal, but many approach normality. As a distribution approaches normality, its properties approximate those given. Tables of areas under

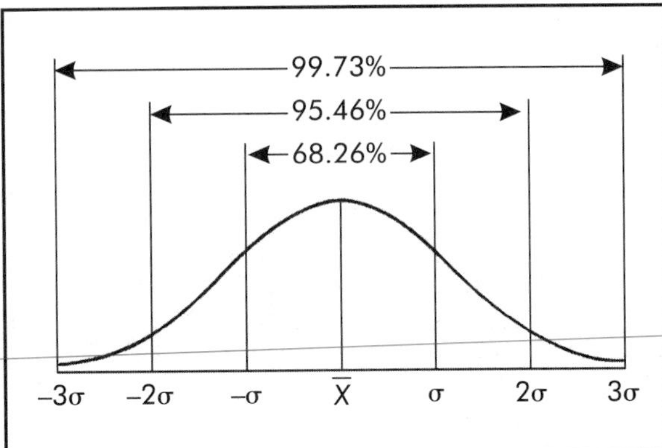

Figure 15-4. Areas under the normal curve.

the normal curve for various values of standard deviations from the mean are given in texts on statistical quality assurance.

Process Control Charts

In 1931, Walter Shewhart developed the concept for using control charts to control variability in processes. Control charts are perhaps the most powerful tools used to analyze variation in most business, manufacturing, and service processes. Control charts are line graphs that display a dynamic image of process performance and are mainly used to monitor and control the process.

Types. There are different types of control charts used for controlling and monitoring processes.

1. Control charts for variables: these are plots of specific measurements of a process characteristic, such as temperature, size, weight, or sales volume. Control charts for variables are generally most useful in controlling processes since they provide records of measurements that may be evaluated at a later stage. They are also more costly to administer because of the time required to capture and analyze data for each separate variable. Most commonly used control charts for variables are *average* and *range* ($\overline{X}$ and R) *charts* and *average* and *standard deviation* ($\overline{X}$ and S) *charts*.
2. Control charts for attributes: these plot general measurement of the total process, such as the number of errors per order, number of orders on time, or number of rejects in a batch. There are four types of attribute charts, summarized as follows:
   a. *p-charts* to control the proportion of defective units from the process; the p-chart is normally used when the samples have variable sizes;
   b. *np-charts* for the number of defectives or nonconforming units; normally, samples are of the same size;
   c. *c-charts* to control the number of defects or nonconformities; samples are usually of the same size; and
   d. *u-charts* for the number of defects or nonconformities per unit; like the np-chart, samples are of variable sizes.
3. Special control charts, such as:
   a. moving average ($M\overline{X}$) and moving range (MR) charts;
   b. individual X and MR charts; and
   c. median and range charts.

Steps for constructing $\overline{X}$ *and* R *charts*. The associated terms used in constructing $\overline{X}$ and R charts are as follows:

n = sample size
X = a value or a reading of the quality characteristic being measured
$\overline{X}$ = average of all readings in a sample, also known as the sample mean
$\overline{\overline{X}}$ = average of all the Xs or the value of the central line on the X chart
R = the range, or the difference between the largest and the smallest value in each sample
$\overline{R}$ = average of all the Rs, or the value of the central line on the R chart
UCL = upper control limit, or the upper control boundary for 99.73% of the population; it is not a specification limit
LCL = lower control limit, or the lower boundary for 99.73% of the population; it is not a specification limit

The following steps are used in constructing the charts:

1. Determine the sample size (n = 3 to 10) and the frequency of sampling.
2. Collect at least 10 time-sequenced samples of size n.

3. Calculate the average, $\overline{X}$, for each sample.
4. Calculate the range, R, for each sample.
5. Calculate the $\overline{\overline{X}}$ (the average of all the $\overline{X}$s). This is the centerline of the $\overline{X}$ chart.
6. Calculate the $\overline{R}$ (the average of all the Rs). This is the centerline of the R chart.
7. Calculate the control limits.

   For the X-bar chart:

   Centerline = $\overline{\overline{X}}$

$$UCL_{\overline{X}} = \overline{\overline{X}} + A_2\overline{R}$$
$$LCL_{\overline{X}} = \overline{\overline{X}} - A_2\overline{R} \qquad (15\text{-}6)$$

   For the R-chart:

   Centerline = $\overline{R}$

$$UCL_R = D_4\overline{R}$$
$$LCL_R = D_3\overline{R} \qquad (15\text{-}7)$$

8. Plot the data and interpret the chart for special or assignable causes.

The charts should look similar to those shown in Figure 15-5. The factors A_2, D_3, and D_4 are taken from Table 15-1.

The construction of $\overline{X}$ and S charts is done in a similar manner using the following formulas.

For the X chart:

Centerline = $\overline{\overline{X}}$

$$UCL_X = \overline{\overline{X}} + A_3\overline{S} \qquad (15\text{-}8)$$

$$LCL_X = \overline{\overline{X}} - A_3\overline{S}$$

For the S chart:

Centerline = $\overline{S}$

$$UCL_S = B_4\overline{S} \qquad (15\text{-}9)$$

$$LCL_S = B_3\overline{S}$$

The factors A_3, B_3 and B_4 are available from tables included in textbooks on quality assurance.

Figure 15-5 shows an example of $\overline{X}$ and R charts for a rough turning operation on the outside diameter of steel shafts. The first point on the $\overline{X}$ chart was determined by measuring the diameters of five shafts and then calculating the average of those measured values. Each succeeding point is also an average of the measured diameters of five shafts. Average values are plotted instead of individual values because sample averages tend to be more normally distributed than single values. The central line of the $\overline{X}$ chart represents the grand average of the subgroup averages and is calculated by using the relationship where k is the number of subgroups of five shafts each. The control limits are set at 3σ of the sample averages from the grand average.

$$\overline{\overline{X}} = \frac{\sum_{i}^{k}\overline{X}}{k} \qquad (15\text{-}10)$$

The range of values for the R chart of Figure 15-5 is obtained from the same subgroups of five samples each as are the $\overline{X}$ values. The central line $\overline{R}$ represents the average of the subgroup ranges. Control limits on range charts are calculated by multiplying the average range by the D_4 and D_3 factors for $n = 5$ in Table 15-1.

The $\overline{X}$ chart of Figure 15-5 indicates that the process average remained in a state of statistical control until the 9:00 inspection period on Thursday, at which time a point exceeded the upper control limit. In this case, the out-of-control point was found to be caused by a worn cutting tool. Replacement of the tool caused the subsequent points to fall well within the control limits. This illustrates how a major change in the production process is reflected by the chart. Points that fall outside the limits of either chart occur because of chance causes or actual changes in the process.

It must be emphasized that the control limits of a process control chart do not represent the performance limits of the process nor the specification limits of the dimension. The performance limits of the process are the limiting dimensions within which practically all the parts fall. These are the ±3σ limits of the population distribution, if it is near normal. For a very large (theoretically infinite) population,

$$\sigma = S_{\overline{X}}\sqrt{n} \qquad (15\text{-}11)$$

where

σ = population standard deviation
S = sample standard deviation

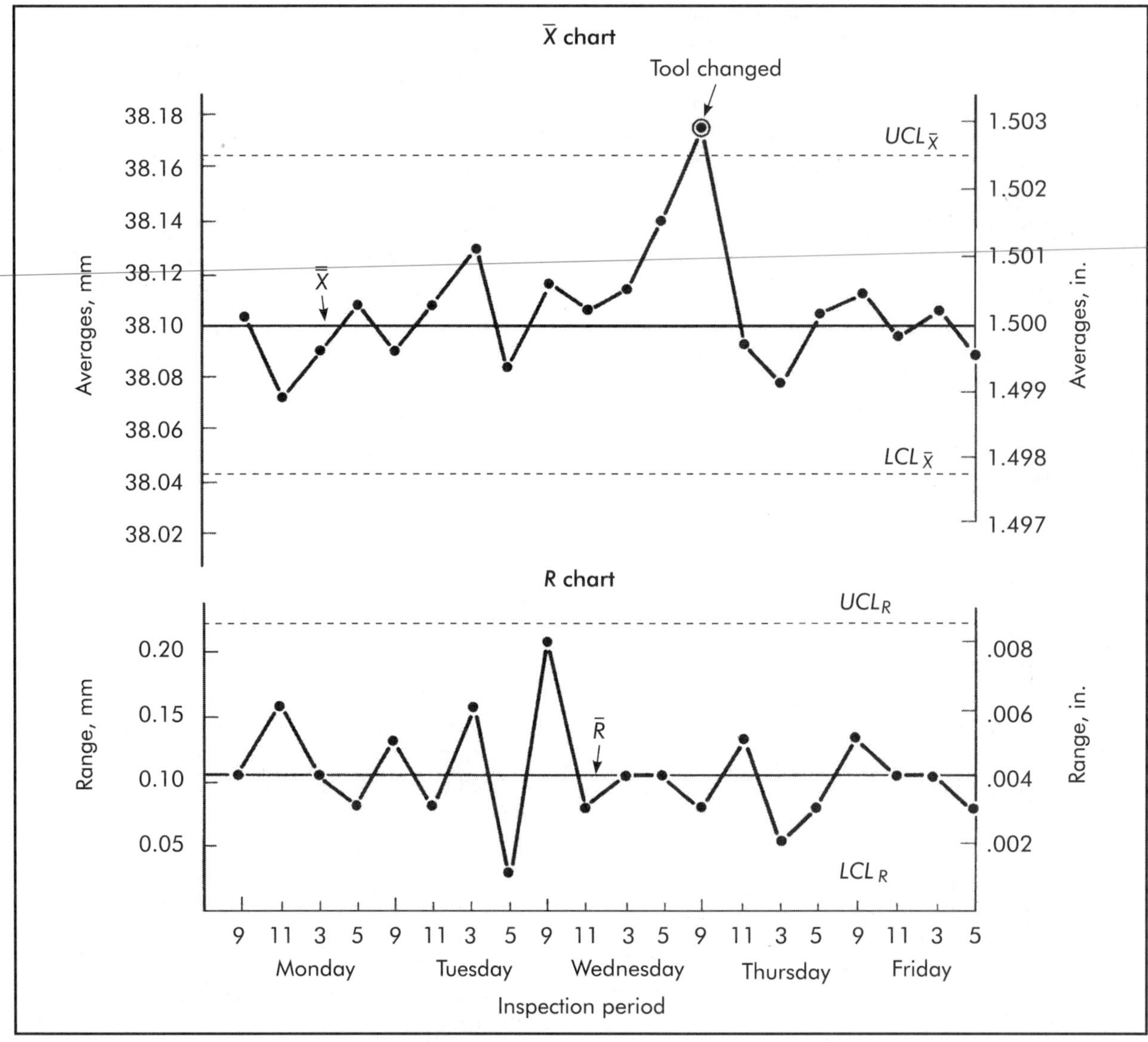

Figure 15-5. Quality control chart showing averages and ranges of samples.

Table 15-1. Factors for control charts

Sample size, n	Factor A_2	d_2	D_3	D_4
2	1.880	1.128	0	3.268
3	1.023	1.693	0	2.574
4	0.729	2.059	0	2.282
5	0.577	2.326	0	2.114
8	0.373	2.847	0.136	1.864
10	0.308	3.078	0.223	1.777

$\overline{X}$ = sample mean
n = number of items in each sample

From this relationship, it can be shown that the process represented by Figure 15-5 is producing shafts from 37.97–38.23 mm (1.495–1.505 in.). For manufacturing economy, it is normally desired that the process be designed to perform within the specification limits.

15.3.4 Process Capability

In the design and development of a manufactured product, the individual part sizes are specified in accordance with some conventional practice. Most commonly, the parts are given a nominal size, such as 30 mm (1.1811 in.), based on the functional requirements of the part. Since design engineers know that a part cannot be manufactured to an exact size, a manufacturing tolerance is then added to the dimension, for example, 30 ±0.03 mm (1.811 ±.001 in.), to accommodate the variability inherent to the manufacturing process. The so-called *5 Ms* of manufacturing, *men and women, machines, materials, methods,* and *measurements,* all contribute to this variability or error source in the manufacturing process. The magnitude of the tolerance that can be assigned to a dimension depends on certain functional aspects of the part as will be explained later in this chapter. The amount of tolerance that needs to be assigned also depends on the capability of the process used to machine or fabricate the part. There are, of course, trade-offs between these two considerations.

Process capability, the natural range of variability inherent to a particular manufacturing process, determines the degree to which a product can be built to specifications. Information on process capability is important to the product designer to aid him or her in assigning practical tolerance limits. It is also useful to production personnel to enable them to predict the percentage of product that can be expected to meet specifications and to schedule appropriate inspection procedures.

There are three primary considerations or conditions involved in the evaluation of process capability as follows.

1. The limiting dimensions of a part specification.
2. The mean value of centering a process within those limiting dimensions.
3. The spread or range of variability demonstrated by the process.

Four possible combinations of those conditions are illustrated in Figure 15-6. In Figure 15-6(A), it appears that the process variation is well within the specification limits. This could represent an ideal situation if some shifting of the process mean were anticipated. The process variation of Figure 15-6(B) coincides with the specification limits, and this could be a satisfactory relationship as long as the process stays well centered and the magnitude of the variation does not increase. The conditions in both Figures 15-6(C) and (D) would result in a percentage of nonconforming product, thus requiring some changes in either the process or the specifications. Several different actions could be taken, depending on the limitations of the process, product design limitations, and the economics of the actions.

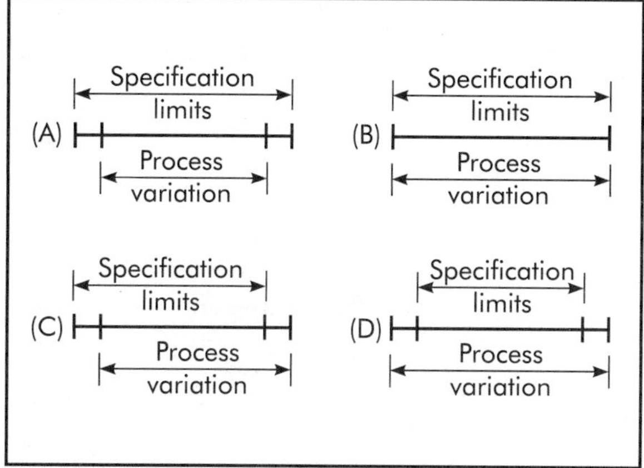

Figure 15-6. Process variability versus specification limits.

Since most contemporary quality assurance efforts are focused on process improvement, the process capability evaluation provides a basis for determining the proper courses of action.

Measurement

Process capability is determined from the total variation caused only by common causes, after all assignable causes have been removed. It represents the performance of a process that is in statistical control. When a process is in statistical control, its performance is predictable and can be represented by a probability distribution.

The proportion of production that is out of specification is a measure of the capability of the process, which may be determined using the process distribution. If the process maintains statistical control, the proportion of defective or nonconforming production remains the same. There are several ways to measure the capability of the process.

Using control charts. When the control chart indicates that the process is in statistical control and when the control limits are stable and periodically reviewed, the chart can be used to assess the capability of the process and provide information to document such capability. From the $\overline{X}$ and R control charts, $\overline{\overline{X}}$ can be used to estimate the process mean μ and $\overline{R}$ can be used to estimate the process standard deviation. That means,

$$\mu = \overline{\overline{X}} \quad (15\text{-}12)$$

and

$$\sigma = \frac{\overline{R}}{d_2} \quad (15\text{-}13)$$

where

d_2 = factor taken from Table 15-1 in accordance with the sample size used in the $\overline{X}$ and R charts

As an example, suppose that a process is in control. The X-bar and R control charts show the following results: centerline = 80, $LCL_{\overline{X}}$ = 76.54, $UCL_{\overline{X}}$ = 83.46, R = 6, and sample size, n = 5. The capability of this process can be determined by estimating the process mean and process standard deviation by means of the relationships from equations 15-12 and 15-13.

$$\mu = 80$$

$$\sigma = \frac{6}{2.326} = 2.58$$

From this, an estimate of the process capability will be 6 × 2.58 = 15.48.

Using histograms. If a histogram is available that shows measurements of a process or a quality characteristic of a process, the mean and standard deviation computed from the histogram can be used to estimate the process mean and process standard deviation. Histograms also show the relationship between the process variability and specification limits. From a histogram, $\overline{X}$ is used to estimate the process mean, and S is used to estimate the process standard deviation.

Indices

Process capability indices are simply a means of indicating the variability of a process relative to the product specification tolerance. The most commonly used indices are C_p and C_{pk}, computed as follows:

$$C_p = \frac{USL - LSL}{6\sigma} \quad (15\text{-}14)$$

$$C_{pk} = \text{the minimum of } \frac{USL - \mu}{3\sigma} \quad (15\text{-}15)$$

$$\text{or } \frac{\mu - LSL}{3\sigma}$$

where

C_p = spread of the process
USL = upper specification limit
LSL = lower specification limit
σ = process standard deviation or its estimate
C_{pk} = spreading and setting of the process
μ = process mean

As an example, consider a process whose mean (μ) = 650 and standard deviation (σ) is estimated to be 50. If the USL and LSL are 750 and 250 respectively, C_p and C_{pk} may be computed as follows:

$$C_p = \frac{750-250}{6(50)} = 1.66$$

$$C_{pk} = \text{minimum of } \frac{750-650}{3(50)} \text{ or } \frac{650-250}{3(50)}$$

$$C_{pk} = 0.66$$

If $C_p > 1$, the process spread is capable of producing parts well within specified tolerance. If $C_{pk} < 1$, the setting of the process is incorrect and needs to be adjusted.

15.3.5 Sampling Inspection

Inspection operations are costly and do not contribute directly to the value of the product. Therefore, the amount of inspection should be kept to a minimum but consistent with the quality requirements of the product. To be assured that every part produced conforms to specifications, it is usually necessary to perform *100% inspection* on them. However, if a few defective items can be tolerated among a large number of good items, then an inspection procedure known as *sampling* can be applied. In this procedure, a given number of parts are chosen at random from a group or lot of parts and are inspected for conformance to certain specifications. According to previously determined acceptance criteria, the lot of parts is judged acceptable or rejectable on the basis of the sample results.

There are quite a number of published sampling plans available based on single sampling, double sampling, or some form of sequential or group sequential sampling. A typical single-sampling plan might require the inspection of 110 random samples from a lot of 1,300 parts. If only three or less of the 110 were found to be defective, the lot would be accepted. The lot would be rejected if more than three defective parts were detected in the sample of 110.

Sampling plans are selected on the basis of the amount of risk that can be tolerated in accepting defective material and rejecting acceptable material. Information about these risks is obtained from the *operating characteristic curve* of the sampling plan. Figure 15-7 shows an operating characteristic curve for the single-sampling plan described earlier. From this curve, it is observed that with the quality of the material coming into inspection being 1% defective (point p_1 on the abscissa), the material would be accepted by the sampling plan about 97% of the time. If the material coming in were 6% defective (point p_2 on the abscissa), it would be accepted only about 11% of the time. Thus, a sampling plan should be selected that will satisfy the demands of both the producer and the consumer of the material being manufactured. Each party should be aware of the risks inherent in sampling.

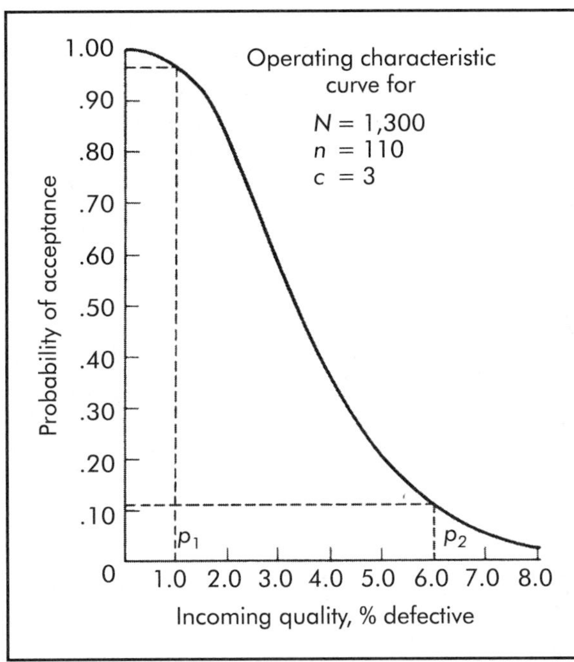

Figure 15-7. Operating characteristic curve for a single-sampling plan.

15.4 MANUFACTURING SPECIFICATIONS

Effective interchangeability relies upon complete specifications of the dimensions of a part or group of parts making up a product. Modern mass manufacturing systems are so complex that it is usually not economical for one skilled worker to machine and fabricate all of the individual parts that go into an assembly. Rather, it is necessary that the various parts be worked upon by numerous people who are responsible for only one or perhaps a few operations on one of the component parts of a product. Thus, exacting specifications relative to the dimensions of these parts must be supplied to the workers so that they may produce articles that can be assembled requiring a minimum of costly hand fitting and reworking.

15.4.1 Tolerances

Ideally, it would be desirable to manufacture parts to an exact size, but this is not physically practical nor economically feasible. To grind a cylindrical shaft to exactly 25.4 mm (1 in.) diameter would require a grinding machine of ultimate perfection—perfect spindle bearings, perfect way surfaces, perfect balance of the grinding wheel, etc. Even though such a machine could possibly be built and maintained in such a state of perfection, it would be beyond the realm of practical measurements to determine whether or not the shafts ground on the machine were exactly 25.4 mm (1 in.) in diameter. For this reason and others cited previously, it is necessary that some deviation be allowed from the exact desired size of a part. This deviation is usually referred to as a *manufacturing* or *working tolerance* and may be defined as the permissible variation in a dimension.

Types

Manufacturing tolerances may be assigned according to three different systems as shown in Figure 15-8. Part *A* is assigned a *bilateral tolerance;* part *B* has a *unilateral tolerance;* and part *C* has the size specified in terms of *limiting dimensions* or *manufacturing limits.* Notice that all three parts have the same total amount of tolerance and the same limiting dimensions. There are certain advantages to specifying tolerances according to one or the other of these systems.

A unilateral tolerance often implies that it is more critical for a certain dimension to deviate in one direction than in another. Unilateral tolerances are also more realistically applied to certain machining processes where it is common knowledge that dimensions will most likely deviate in one direction. For example, in drilling a hole with a standard-size drill, the drill is more likely to produce an oversize rather than an undersize hole. Unilateral tolerances also facilitate the specification and application of standard tools and gages. The application of unilateral tolerances permits easy revision of tolerances without affecting the allowance or clearance conditions between mating parts. Thus, this system of tolerancing usually finds the greatest acceptance in industrial application.

More recently, the trend has been to utilize the bilateral tolerance system for manufacturing dimensions. This system vividly points out the theoretically desired size and indicates the possible and probable deviations that can be expected on each side of it. Bilateral tolerances are easier to add together.

15.4.2 Clearance

One of the basic reasons for specifying precise limiting dimensions for machined objects is to be assured that component parts of an assembly will fit properly and function well under certain operating conditions. For example, a crankshaft must be free to rotate within the confines of the main bearings. Yet it must not fit so sloppily that excessive vibrations cause possible damage to the shaft or engine block. There must be a limited amount of free space between the shaft and the bore. This free space is usually referred to as *clearance*, and it may be defined as the intentional difference in the sizes of mating parts.

In many cases, it is necessary that mating parts be assembled more or less permanently; one part being pressed firmly into another. Thus, a given amount of what is termed *negative clearance* or *interference* must be assigned to this

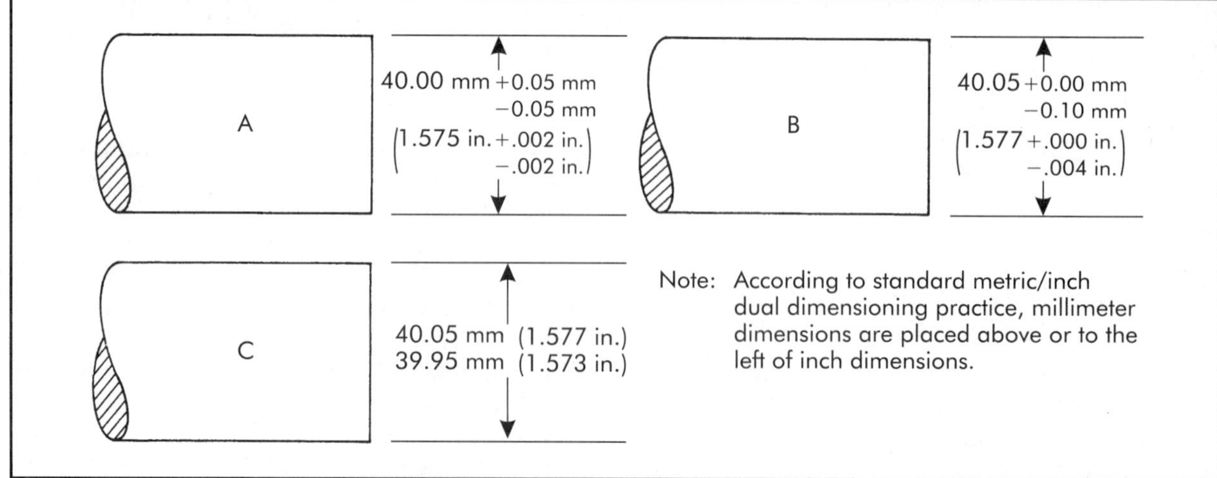

Figure 15-8. Methods of assigning tolerances.

condition to assure a tight fit. The minimum clearance or maximum interference between mating parts is called an *allowance*. Other mating parts must fit together with no perceptible freedom and yet require little or no pressure on assembly. This is usually termed a *metal-to-metal* or *transitional fit*. These three general categories of fits (clearance, transition, and interference) may be broken down further to suit specific conditions.

15.4.3 Development of Manufacturing Specifications

The tolerance that should be assigned to a dimension is that which gives an economic balance between quality and cost. For a given dimension, some parts can be expected to be further from the most desirable size than others and be able to function less well or wear not as long. If a larger tolerance is assigned, more parts can be expected to be inferior, but still within limits, and the total value of the product will be less. In general, the value of a product will decrease in some manner as tolerance value is increased. A hypothetical case is shown by the upper curve of Figure 15-9.

A large tolerance for a given dimension can be held more easily and is therefore cheaper than a small tolerance. Each machine and process has a certain capability. The closer the tolerances that must be held, the more care and skill required. If smaller tolerances are demanded that are beyond the capability of the machine and process, resort must be made to other and usually more costly processes. Thus, in a typical case depicted by the lower curve of Figure 15-9, cost increases as the tolerance decreases. The curve, of course, is different for every situation. Such a curve may represent grinding a diameter, turning, or even superfinishing to the smallest tolerances.

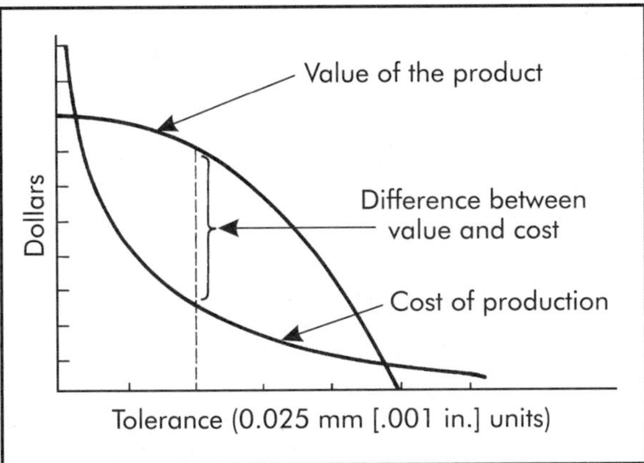

Figure 15-9. Curves of value and cost, showing location of ideal tolerance.

If for a particular situation the value and cost curves are plotted on the same scale, the result is as shown in Figure 15-9. The ideal tolerance is that which gives the largest difference between value and cost. It would be worthwhile to ascertain this ideal tolerance for large-quantity production because of the savings involved. It could be approximately set because tolerances within a range are close enough for practical purposes.

Certain dimensions of nonmating parts or products, such as the length of a handle or diameter of a handwheel, may not require very precise specifications because their functional effectiveness does not depend on whether or not their dimensions vary by as much as, for example, ±6.4 mm (±.25 in.). However, for the sake of uniformity of method, stock usage, packing, and other factors, a nominal amount of tolerance, such as ±0.41 mm (±.016 in.), is usually specified.

Assembly Methods

Basically, there are two methods of assembly for mass-produced components, random assembly and selective assembly. In *random assembly*, the mating parts must be made to a tolerance that will permit any component selected at random to fit properly with any randomly selected mating component. In *selective assembly*, the components are put into groups according to size and then assembled with mating components also segregated by size. For example, a shaft diameter may be produced to a tolerance of 0.030 mm (.0012 in.), with a nominal diameter of 25 mm (≈1 in.) and actual limits of 25.00 to 24.97 mm (≈1.0000 to .9988 in.). The bearing may likewise have a tolerance of 0.030 mm (.0012 in.) with actual limits of 25.03 to 25.00 mm (≈1.0012 to 1.0000 in.).

The clearance between parts assembled at random may be anywhere from 0–0.061 mm (0–.0024 in.). In the event that the clearance must not be over 0.030 mm (.0012 in.) without excessive losses in the product, two courses of action are open. One is to reduce the tolerance on each part, at greater cost. The other is to make and divide the shafts into two groups, one with limits from 24.985 to 25.000 mm (≈.9994 to 1.0000 in.) and the other with limits from 25.000 to 25.015 mm (≈1.0000 to 1.0006 in.), which still embraces a production tolerance of 0.030 mm (.0012 in.). So, the bearings, made to the same tolerance as before, are divided into two groups with ranges of 25.000–25.015 mm (≈1.000–1.0006 in.) and 25.015–25.030 mm (≈1.0006–1.0012 in.). Now by assembling from proper groups, clearances can be held to 0.030 mm (.0012 in.) or less. This selective assembly, in some cases, is more economical than manufacturing to small tolerances. A practice common in tool and job shop work but uncommon for production of quantities is to machine one part of a pair to a size and the mating part to a clearance within a specified range.

Tolerances on Linear Dimensions

The exact desired size of a part is usually referred to as the *basic size,* and it is the size from which variation is permitted. A *standard size* is defined by an integer or subdivision of a unit of length. Whole-number millimeters and common fractions of an inch are examples.

Tolerances are usually developed to conform to certain standardized design procedures. Two commonly used procedures are called *standard hole practice* and *standard shaft practice.* In standard hole practice, the basic size of the hole is assigned a standard size, and a positive unilateral tolerance is applied. The basic size of the mating shaft is determined by subtracting the prescribed allowance from the standard size for a clearance fit, or adding the prescribed allowance to the standard size for an interference fit. A negative unilateral tolerance is assigned to the shaft. Notice that the holes and shafts of Figure 15-10 are dimensioned according to standard hole practice. This system permits the use of standard reamers for finishing holes and standard limit plug gages for checking.

Standard shaft practice designates the maximum limiting size of commercial shafting as the standard size and applies a negative unilateral tolerance to it. The basic size of the hole is determined by adding the prescribed allowance to the standard size for a clearance fit, or subtracting the prescribed allowance from the standard size for an interference fit. The hole is assigned a positive unilateral tolerance.

Standards for Allowances and Tolerances

One very common method for determining allowances and tolerances for mating parts is through the use of formulas or tables prepared by technical organizations. For example, the American National Standards Institute (ANSI) has published standard tables prepared by the American Society of Mechanical Engineers (ASME) on preferred metric limits and fits for mating parts in which hole and shaft limits are provided for three basic types of fits (ANSI B4.2-1978).

Table 2 of the standard, a portion of which is shown in Figure 15-11, gives limits for clearance fits applicable to basic hole sizes ranging from 1–500 mm (0.0394–19.685 in.). Five subclasses of this type of fit are provided, ranging from *loose running fits,* intended for wide commercial tolerances or allowances on external members, to *locational clearance fits,* providing a snug fit for locating stationary parts, but which can be freely assembled and disassembled. Other tables are given in the standard for both transition and interference types of fits. One of the transition types, the *location transition fit,* is a compromise between clearance and interference, and is used for accurate location of mating parts. Three subclasses of interference fits are given, ranging from *locational interference fits* to *force fits. Force fits* are suitable for parts that can be highly stressed or for shrink fits where heavy pressing forces are impractical.

Hole and shaft dimensions for a free-running fit and a sliding fit are shown in Figure 15-10. Basic hole size in each case is 40 mm (1.5748 in.).

Tolerances for 100% Interchangeability

If the amount of clearance or interference required between mating parts is known, the tolerances for 100% interchangeability can be easily determined by simple addition. The shaft and bearing assembly of Figure 15-12 illustrates this procedure. For this assembly, it is assumed

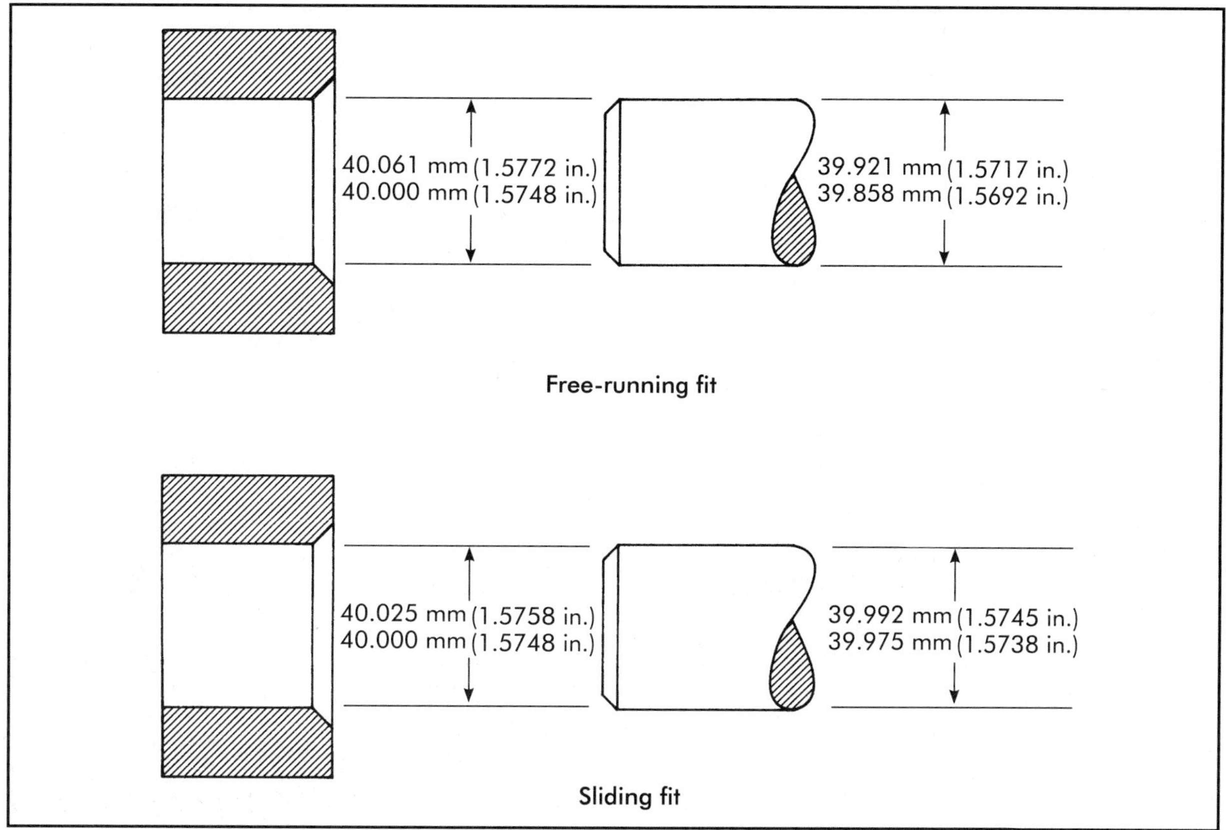

Figure 15-10. Dimensions of shaft and hole for a free-running fit and a sliding fit.

that the nominal assembly size is 50 mm (≈2 in.) with a minimum total clearance of 0.05 mm (.002 in.) and a maximum total clearance of 0.20 mm (.008 in.). Tolerance is shown bilateral from basic size, but could be specified in any way. For purpose of analysis, Figure 15-12 shows the shaft resting on the bottom of the bearing so that the maximum and minimum total clearances can be observed. Tolerances and basic sizes may be obtained by simply summing dimensions from a reference surface of the part under consideration. The tolerance on the shaft may be determined as follows: starting at *B*,

$$\overline{BD} + \overline{DE} - \overline{EB} = 0 \qquad (15\text{-}16)$$

where

$\overline{BD}$ = tolerance on the shaft, mm (in.)
$\overline{DE}$ = half of the minimum clearance, 0.05 mm/2 (.002 in./2)
$\overline{EB}$ = half of the maximum clearance, 0.20 mm/2 (.008 in./2)

Thus,

$$\overline{BD} + 0.05 \text{ mm}/2 \text{ (or .002 in./2)} - 0.20 \text{ mm}/2 \text{ (or .008 in./2)} = 0$$

or

$$\overline{BD} = 0.10 \text{ mm (or .004 in.)} - 0.025 \text{ mm (or .001 in.)}$$

$$= 0.075 \text{ mm (or .003 in.)}$$

Other dimensions may be determined by a similar process.

Determination. Methods of statistical analysis may be applied to increase tolerances and effect manufacturing economies. The statistical concepts applicable to this problem are those of the normal distribution and its parameters $\overline{X}$ and S defined by Equations (15-1) and (15-4). Theory and experience have shown that if the variations of the dimensions of the parts approximate a normal distribution, as they commonly do, and the parts are selected at random, the

Clearance Fits										
		Free running			Close running			Sliding		
Basic size		Hole H9	Shaft d9	Fit	Hole H8	Shaft f7	Fit	Hole H7	Shaft g6	Fit
40	Maximum	40.062	39.920	0.204	40.039	39.975	0.089	40.025	39.991	0.050
	Minimum	40.000	39.858	0.080	40.000	39.950	0.025	40.000	39.975	0.009
50	Maximum	50.062	49.920	0.204	50.039	49.975	0.089	50.025	49.991	0.050
	Minimum	50.000	49.858	0.080	50.000	49.950	0.025	50.000	49.975	0.009
60	Maximum	60.074	59.900	0.248	60.046	59.970	0.106	60.030	59.990	0.059
	Minimum	60.000	59.826	0.100	60.000	59.940	0.030	60.000	59.971	0.010
80	Maximum	80.074	79.900	0.248	80.046	79.970	0.106	80.030	79.990	0.059
	Minimum	80.000	79.826	0.100	80.000	79.940	0.030	80.000	79.971	0.010
100	Maximum	100.087	97.880	0.294	100.054	99.964	0.125	100.035	99.988	0.069
	Minimum	100.000	99.793	0.120	100.000	99.929	0.036	100.000	99.966	0.012
120	Maximum	120.087	119.880	0.294	120.054	119.964	0.125	120.035	119.988	0.069
	Minimum	120.000	119.793	0.120	120.000	119.929	0.036	120.000	119.966	0.012
160	Maximum	160.100	159.855	0.345	160.063	159.957	0.146	160.040	159.986	0.079
	Minimum	160.000	159.755	0.145	160.000	159.917	0.043	160.000	159.961	0.014
200	Maximum	200.115	199.830	0.400	200.072	199.950	0.168	200.046	199.985	0.090
	Minimum	200.000	199.715	0.170	200.000	199.904	0.050	200.000	199.956	0.015
250	Maximum	250.115	249.830	0.400	250.072	249.950	0.168	250.046	249.985	0.090
	Minimum	250.000	249.715	0.170	250.000	249.904	0.050	250.000	249.956	0.015
300	Maximum	300.130	299.810	0.450	300.081	299.944	0.189	300.052	299.983	0.101
	Minimum	300.000	299.680	0.190	300.000	299.892	0.056	300.000	299.951	0.017
400	Maximum	400.140	399.790	0.490	400.089	399.938	0.208	400.057	399.982	0.111
	Minimum	400.000	399.650	0.210	400.000	399.881	0.062	400.000	399.946	0.018
500	Maximum	500.155	499.770	0.540	500.097	499.932	0.228	500.063	499.980	0.123
	Minimum	500.000	499.615	0.230	500.000	499.869	0.068	500.000	499.940	0.020
Dimensions in mm = .03937 in.										

Figure 15-11. Abbreviated table of clearance fits (ASME 1978).

distribution of fits obtained is practically normal. The mean value of the clearance is

$$\overline{X}_{\text{clearance}} = \overline{X}_{\text{bearings}} - \overline{X}_{\text{shafts}} \quad (15\text{-}17)$$

and the standard deviation of the variations in the clearance is

$$\sigma_{\text{clearance}} = \sqrt{\sigma^2_{\text{bearings}} + \sigma^2_{\text{shafts}}} \quad (15\text{-}18)$$

where

$\overline{X}$ = sample mean
σ = standard deviation of the variation in size from the mean

In the case of the bearing and shaft assembly of Figure 15-12,

$$6 \times \sigma_{\text{clearance}} = 0.20 - 0.05 = 0.15 \text{ mm}$$
$$(.008 - .002 = .006 \text{ in.})$$

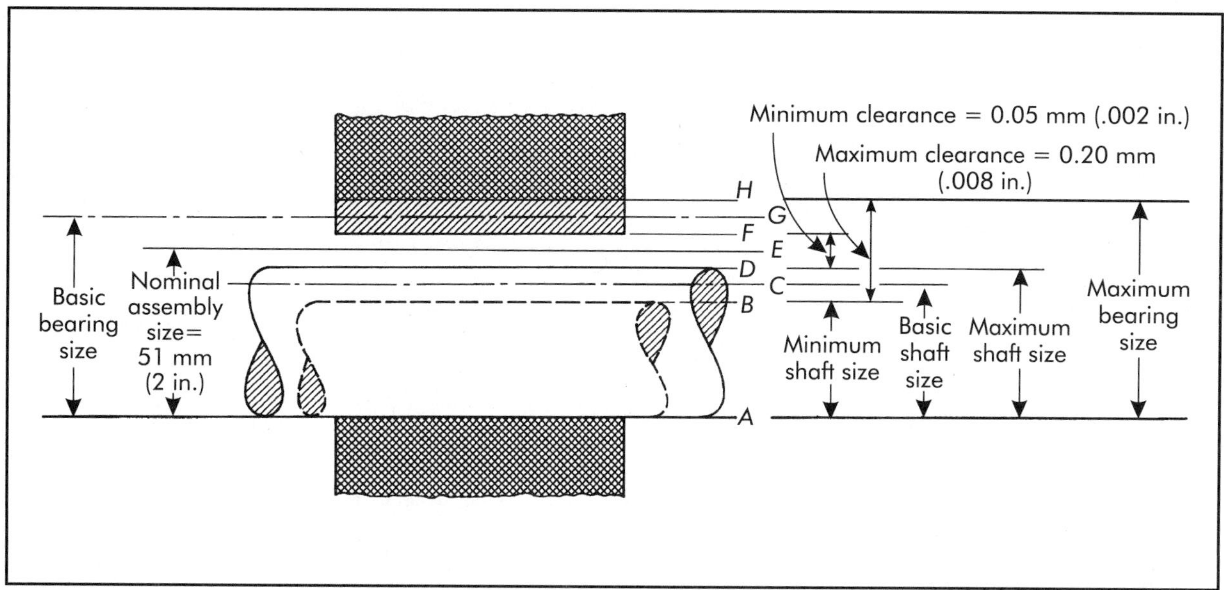

Figure 15-12. Extreme positions for a shaft and bearing.

and

$$\sigma_{clearance} = 0.025 \text{ mm } (.001 \text{ in.})$$

If not specified otherwise,

$$\sigma_{bearing} = \sigma_{shaft} = \sigma_p, \text{ and } \sqrt{\sigma_p^2 + \sigma_p^2}$$

$$= 0.025 \text{ mm } (.001 \text{ in.}) \text{ from Equation } (15\text{-}18)$$

where

σ_p = standard deviation of the variation in bearing size or shaft size

Thus, $\sigma_p = 0.018$ mm (.0007 in.), and the tolerance for the bearing or shaft is $6\sigma_p$ or 0.108 mm (≈.0042 in.). This permits 40% more tolerance on the parts than is required for 100% interchangeability. Yet if the prescribed conditions are met, fewer than three assemblies out of 1,000 can be expected to have clearances greater or less than specified.

Geometric Dimensions and Tolerances

The limit dimensions of the simple cylindrical piece at the top of Figure 15-13 define the maximum and minimum limits of a profile for the work. The form or shape of the part may vary as long as no portions of the part exceed the maximum profile limit or are inside the minimum profile limit. If a part measures its maximum material limit of size everywhere, it should be of perfect form. This is referred to as the *maximum material condition* (MMC) and is at the low limit for a hole or slot but at the high limit for parts, such as shafts, bolts, or pins.

If it is desired to provide greater control on the form than is imposed by the limit dimensions, then certain tolerances of form must be applied. In most cases, these tolerances appear in the form of notations on the drawing as is illustrated at the bottom of Figure 15-13.

Positional Tolerances

Positional tolerancing is a system of specifying the true position, size, or form of a part feature and the amount it may vary from the ideal. The advantage of the system is that it allows the one responsible for making the part to divide tolerances between position and size as he or she finds best. The principles are illustrated for two simple mating parts in Figure 15-14. The basic dimensions without tolerances are shown at the bottom and right side of each part. Beneath the size dimension for holes or posts is a box with the notations for positional tolerancing. Actually, a number of specifications are possible, but only one set is shown here as an example. The circle and cross in the first cell of the box is

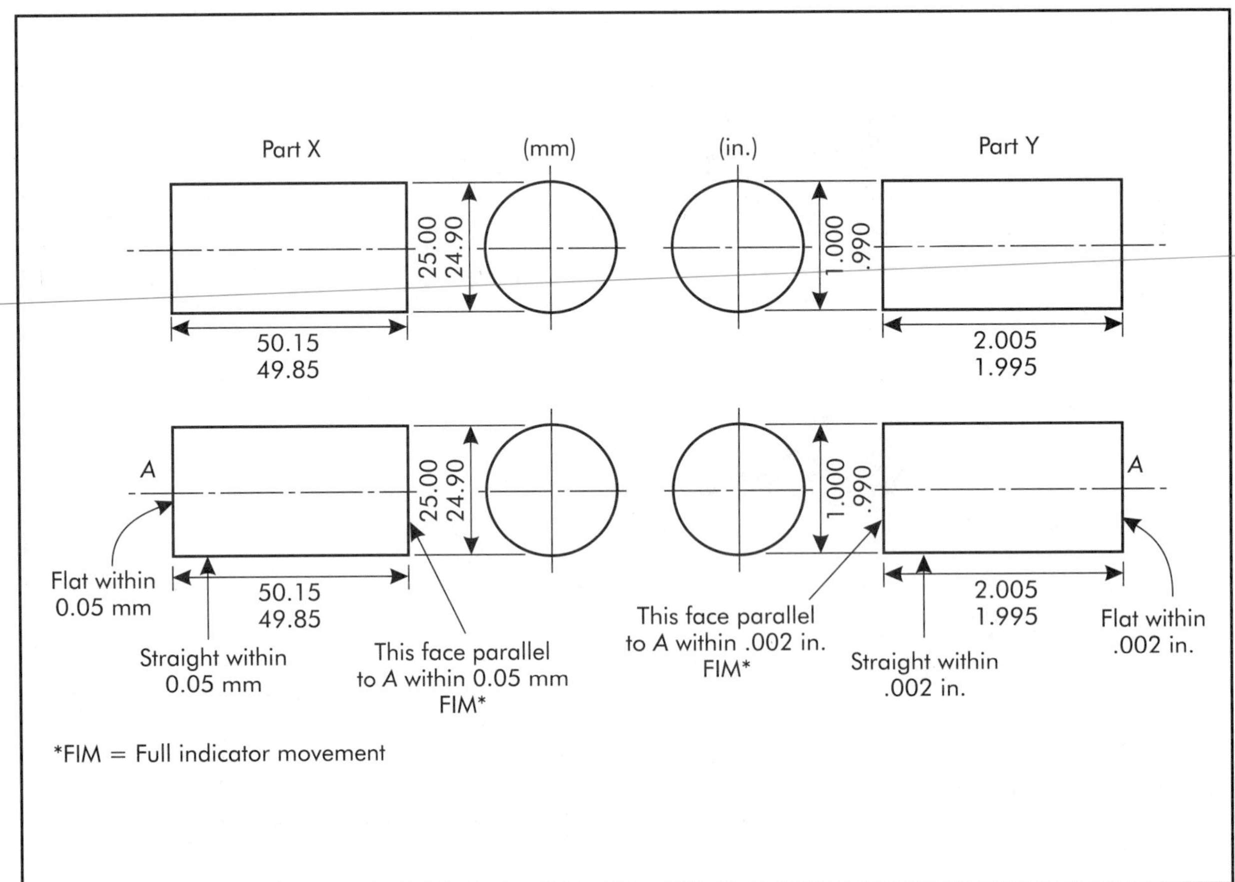

Figure 15-13. Part drawing with and without tolerances of form.

the convention that says the features are positionally toleranced.

Part I in Figure 15-14 introduces the idea of the MMC utilized in most positional tolerancing. This is designated by the letter M in a circle and means that the smallest hole (12.70 mm or .500 in.) determines the inner boundary for any hole. The "Ø0.20 mm (.008 in.)" notation in the box specifies that the axis of any minimum-size hole must not be outside a theoretical cylinder of 0.20 mm (.008 in.) diameter around the true position. A 12.50-mm (.492-in.) diameter plug in true position will fit in any 12.70-mm (.500-in.) diameter hole with its axis on the 0.20-mm (.008-in.) diameter cylinder. Any hole that passes over such a plug is acceptable, provided that its diameter is within the high and low limits specified.

The letter "A" in the specification box designates that the theoretical cylinder bounding the hole axes must be perpendicular to the datum surface carrying the "A" flag. Features usually are referred to with three coordinate datum surfaces, but for simplicity, in this case, the holes are related only to each other and surface "A," and not to the sides of the part.

Part II of Figure 15-14 introduces the idea of zero maximum material condition specified by "Ø0.000" before the MMC symbol. This means the axis of the largest-diameter post (12.50 mm [.492 in.]) must be exactly in the true position, but smaller sizes of posts may vary in position as long as they do not lie outside the boundary set by the largest. Thus, if the posts are held to a tolerance smaller than the 0.20 mm (.008 in.) specified, say to a tolerance of 0.05 mm (.002 in.), the difference (0.15 mm [.006 in.]) is then available for variations in post positions. The advantage of zero MMC is that only one limit of the feature, in this case the lower limit of the post diameter, needs to be checked along with position.

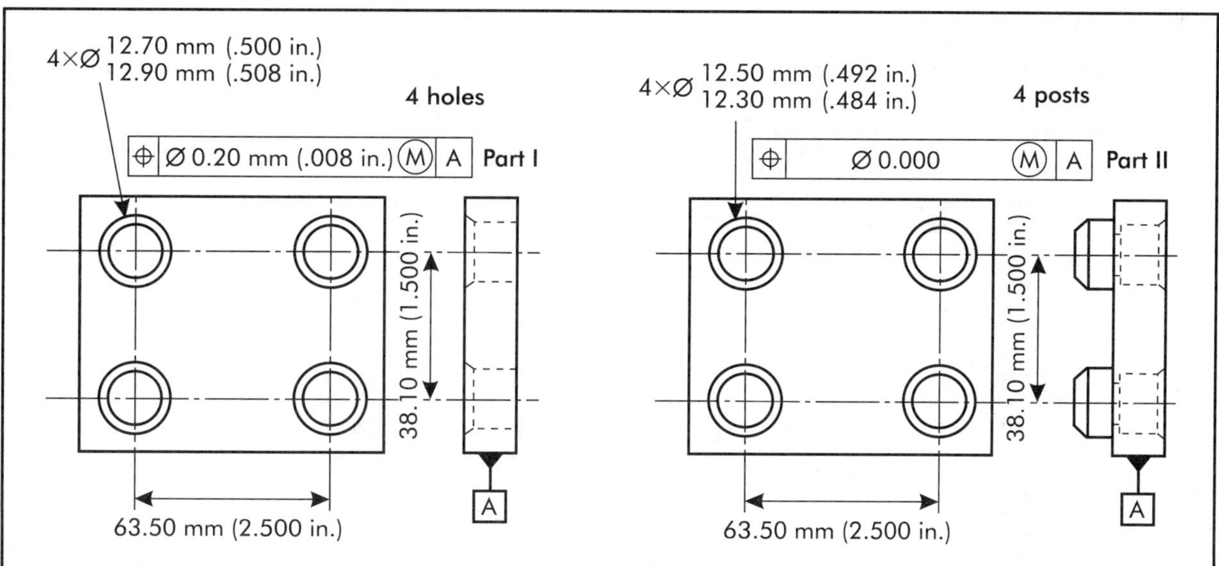

Figure 15-14. Two parts dimensioned with positional tolerances.

15.5 QUESTIONS

1. What is meant by the term "interchangeable manufacture"?
2. Explain the difference between a "serialized" and a "parallel" product design-development-manufacture process.
3. Name and describe the three fundamental principles of total quality emphasized in the total quality management system.
4. Why is it important that U. S. manufacturers be certified in compliance with ISO 9000 standards?
5. Name and describe the four basic components or activities involved in the typical quality assurance function.
6. During what stages in the manufacturing process are inspection operations typically performed?
7. Explain the difference between a sample statistic and population parameter.
8. Why are average values ($\overline{X}$) plotted instead of individual values (x) on a control chart?
9. Explain the difference between "process control" and "process capability."
10. What courses of action might be considered in the event of a situation such as is depicted in Figure 15-6(D)?
11. Explain the difference between the terms "tolerance," "allowance," and "clearance."
12. Give three different ways of specifying tolerance and cite the advantages and disadvantages of each.
13. Give three general classifications of fits between mating parts.
14. Why do manufacturing costs generally increase as manufacturing tolerances are decreased?
15. Explain the difference between "standard hole practice" and "standard shaft practice."
16. Explain the difference between "100% interchangeability" and "statistical average interchangeability."
17. Why is it sometimes necessary to include tolerances of form on some manufacturing drawings?

15.6 PROBLEMS

1. The following measurements on the outside diameter of brass bushings from an automatic turning machine were taken by process inspectors. A sample of five bushings were measured and the data was recorded every 2 hr as shown in the following table. Each unit in the table stands for 0.01 mm, with 25 = 0.

1	2	3	4	5	6	7	8	9	10	11	12	13	14	15
+1	0	0	−1	0	−2	+1	−2	+1	0	+2	−1	−1	+2	+1
0	+2	−1	−1	0	+2	−3	−1	+1	−2	+1	+2	0	−1	0
−1	0	−2	0	−1	+4	0	−1	−1	+2	0	+1	+2	0	−3
+1	0	0	0	+3	+1	0	−1	−1	0	+1	+1	−2	0	−1
−1	−3	+1	0	+2	0	−3	0	0	+5	+1	+2	+1	0	−2
16	**17**	**18**	**19**	**20**	**21**	**22**	**23**	**24**	**25**	**26**	**27**	**28**	**29**	**30**
0	+1	0	0	+5	+1	−2	+3	+1	+3	0	0	+2	0	−1
+3	−2	−1	+1	0	−1	−1	0	0	0	−2	0	−1	+1	−1
−1	−2	+2	−1	+1	0	+2	0	+1	−2	+3	+2	−1	+2	0
−1	+3	+2	−3	+1	+1	0	−1	−3	0	+2	0	0	0	0
−1	+3	−2	0	+1	+3	0	+1	−2	0	−2	−1	+4	+3	+2

   (a) Plot the individual measurements in the form of a frequency distribution similar to the one shown in Figure 15-3.
   (b) Compute the average ($\overline{X}$), range (R), and standard deviation (S) for all of the units measured.

2. For the measurements given in Problem 1:
   (a) Compute the average ($\overline{X}$) and range (R) for each of the 30 subgroups.
   (b) Compute $\overline{\overline{X}}$, $S_{\overline{X}}$, and $\overline{R}$ from the subgroup average and range values.
   (c) Construct $\overline{X}$ and R charts for the 30 subgroups of measurements. Calculate and show control limits for $\overline{X}$ and R (UCL for $R = 2.114\ \overline{R}$).
   (d) From the data given, estimate the tolerance limits for the process.

3. A hole and mating shaft are to be machined to provide a free-running fit. The basic hole size is 60 mm (2.362 in.). Using standard hole practice and unilateral tolerance:
   (a) determine the manufacturing specifications for the hole and shaft according to the standards given in Figure 15-11; and
   (b) give the maximum and minimum clearance that could occur with these specifications.

4. A hole and mating shaft are to have a nominal assembly size of 40 mm (1.575 in.). The assembly is to have a maximum total clearance of 0.025 mm (.0010 in.). Determine the specifications for the hole and the shaft:
   (a) for 100% interchangeability; and
   (b) for statistical average interchangeability.

15.7 REFERENCES

ANSI B4.2-1978 (Reaffirmed 1984). "Preferred Metric Limits and Fits." New York: American Society of Mechanical Engineers (ASME).

ANSI/ASQC E-2-1984. "Guide to Inspection Planning." Milwaukee, WI: American Society for Quality Control.

ANSI Y14.5M-1994. "Dimensioning and Tolerancing." New York: American National Standards Institute (ANSI).

ANSI/ASME Y14.5.1-1994. "Mathematical Definition of Dimensioning and Tolerancing Principles." New York: American National Standards Institute (ANSI).

Boznak, R.G. 1991. "Manufacturers Must Prepare for International Quality Initiative." *Industrial Engineering*, October.

Chapple, A. 1993. "Quality Standard Key for Work in Europe." *Engineering Times,* December.

Gunter, B. 1991. "Process Capability Studies Part 1: What is a Process Capability Study?" *Quality Progress,* February.

Lamprecht, J.L. 1991. "ISO 9000 Implementation Strategies." *Quality*, November.

Pfau, L.D. 1989. "Total Quality Management Gives Companies a Way to Enhance Position in Global Markets." *Industrial Engineering*, April, 21:4.

Woodruff, D., and Phillips, S. 1990. "A Smarter Way to Manufacture: How Concurrent Engineering Can Invigorate American Industry." *Business Week,* April 30.

Measurement and Gaging

16

When you can measure what you are speaking about you know something about it.—Lord Kelvin

The purpose of the inspection function in quality assurance is to determine, by means of a measurement or gaging process, whether or not manufactured materials and products conform to specifications. In addition, machine setup personnel and machine operators must use a variety of measurement and gaging devices in their work with the manufacturing equipment. Thus, to be globally competitive in precision-engineered products, manufacturers require a high degree of accuracy and precision on the part of the measuring and gaging instruments used to setup machines and evaluate product conformance to specifications.

Generally, the dimensions of manufactured parts have to be checked as they are being made *(in-process inspection)* and after they have been finished *(final inspection)* to assure they conform to specifications. As a general rule, *measuring* is done to determine the actual size of a dimension, while *gaging* usually only shows whether a dimension is within specified limits. Measuring and gaging may be done manually or by a variety of instrumented systems employed to automatically perform those activities.

Measuring and gaging devices may be of the *contact* or *noncontact* type. As the designation implies, the contact types actually touch the workpiece at one or more points, while the noncontact instruments use a variety of sensor elements that do not come in contact with the workpiece. The basic measuring and gaging devices discussed in the earlier sections of this chapter are generally contact instruments. The more technologically advanced noncontact instruments are introduced in the latter part of the chapter.

16.1 STANDARDS

If manufactured parts of a kind are to be interchangeable, all measurements must be based on a set of reliable standards. This is especially true when mating parts are made in different places in the nation or the world. The parts will not fit together unless each maker uses the same standard of measurement.

16.1.1 Length Standard

In 1975, the U. S. Congress passed the Metric Conversion Act, which was intended to end the United States' isolation in a metric world. This legislation called for a voluntary conversion to the international metric system, *SI* (from the French Systeme International d'Unites), and it was expected that transition to that system could be effected in an orderly manner within a 7-year period. The Metric Education Act was approved by Congress in 1978 to assist in the conversion process. Those pieces of legislation, along with the Omnibus Trade Act of 1988 and a variety of subsequent free-trade agreements among many industrialized nations of the world, have inspired U. S. manufacturers to accelerate their metric conversion programs.

The *SI* system consists of a foundation of seven base units from which all others are derived. It represents a logical framework for all measurements in science, industry, and commerce. The *meter* is the basic length standard under the metric system and it is defined as the length equal to 1,650,763.75 wavelengths in a vacuum of the radiation corresponding to the transition levels $2p_{10}$ and $5d_5$ of the krypton-86 atom.

16.1.2 End Standards

Gage blocks are the practical length standards for the manufacturing industry. They are rectangular, square, or round blocks of steel, carbide, or ceramic materials. Each has two faces flat, level, and parallel to within approximately ±0.00003 mm to +0.00015 or −0.00005 mm (±.000001 to +.000006 or −.000002 in.), depending on the length and accuracy grade. There are four basic classes of gage blocks that conform to specified federal accuracy standards (Federal Specification GGG-G-15). *Laboratory master blocks,* which conform to federal accuracy grade 0.5 (formerly AAA grade), must be accurate to about ±0.00003 mm per 25 mm of length (±.000001 in./in.) and are intended for use in checking other gage blocks or setting very precise laboratory equipment. *Federal accuracy grade 1 blocks* (formerly AA grade) are accurate to about ±0.00005 mm per 25 mm of length (±.000002 in./in.) and are used as reference blocks to calibrate measuring instruments, set gages, and for very close layout work. *Federal accuracy grade 2 and 3 blocks* are normally used as working gage blocks in manufacturing operations for establishing machine settings and layout inspection. Grade 2 blocks through 25.4 mm (1 in.) length have a length accuracy of +0.00010 mm or −0.000050 mm (+.000004 in. or −.00002 in.), while grade 3 blocks of similar lengths have a length accuracy of +0.00015 mm or −0.00005 mm (+.000006 in. or −.000002 in.).

Gage blocks may be purchased individually or in sets that can be put together to obtain various lengths. The individual blocks are *wrung* together with a film of less than 0.000008 mm (.0000003 in.) between them. A variety of gage blocks are commercially available in either metric or inch sizes. The set of rectangular steel blocks shown in Figure 16-1 consists of 81 blocks, which can be combined to obtain any length from .050 in. to over 10 in. in steps of .0001 in. These steel blocks in a Federal accuracy grade 2 are available for about $1,000, while a similar set of

Figure 16-1. Set of 81 steel gage blocks. (Courtesy Federal Products Company)

tungsten carbide blocks would cost from $2,500–$3,500.

A number of quality factors other than just length tolerance are considered in the selection of gage blocks for a particular activity. Flatness, parallelism, dimensional stability, surface finish, hardness, or resistance to abrasion may contribute to the utility and reliability of gage block applications. One of the major sources of error in gaging to submillimeter tolerances in precision-engineered products is temperature variation. When a gage block is said to have a certain length, it is understood to be the length at the standard temperature of 20° C (68° F). Thus, steel, carbide, or ceramic gage blocks used in a laboratory environment maintained at standard temperature would be expected to retain a high degree of dimensional stability. If the blocks are to be used in a work environment wherein considerable temperature variation occurs, then a gage block material would need to be selected that responds to temperature variations in about the same manner as the workpiece or machine on which they are used. Since wear resistance is related to hardness, most steel gage blocks are hardened to about 65–68 R_C, while chrome carbide and tungsten carbide blocks have hardness values in the 70–73 R_C range.

16.2 INSTRUMENTS

Metrology, the science of weights and measures, has many applications and requires a wide variety of instruments in just about every facet of science and industry. In the competitive manufacture of precision-engineered products where a high degree of quality is required, it is particularly important that the proper measuring and gaging instruments be employed to provide accurate, reliable, and cost-effective inspection results. Some industry studies have indicated that dimensional tolerances on state-of-the-art manufactured products are shrinking by a factor of three every 10 years. Thus, the selection of appropriate measuring and gaging instruments will become even more critical and demanding in the future.

16.2.1 Principal Technologies

Manufacturing metrology is concerned with the measurement or gaging of a variety of workpiece characteristics including: length, diameter, thickness, angle, taper, roughness, concentricity, and profile. Different sensing technologies may be employed to measure or gage those characteristics, depending on the requirements for accuracy and other considerations. There are basically five different technologies that may be used individually or in combination to perform these inspection functions:

1. mechanical;
2. electronic;
3. air;
4. light waves; and
5. electron beam.

In general, the mechanical and electronic types of measuring and gaging instruments have sensing devices or probes that come in contact with the workpiece and are referred to as contact instruments. Air instruments, while employing contacting elements, rely on air pressure differences to effect measurement. Thus, they are basically noncontact instruments. Although different technologies are involved in the light-wave and electron-beam instruments, they both utilize a variety of optical systems. Thus, they are often grouped together as optical noncontact instruments.

16.2.2 Selection

There are many factors to consider in the selection of a measuring or gaging instrument or system for a particular manufacturing inspection operation. In general, a reference to the *Rule of Ten* will serve as a baseline or beginning of that selection process. The Rule of Ten, often referred to as the *Gage Maker's Rule*, states that inspection measurements should be better than the tolerance of a dimension by a factor of 10, and calibration standards should be better than inspections measurements by a factor of 10. If, for example, the tolerance on a shaft diameter is ±0.025 mm (±.0010 in.), then the increment of measurement on the inspection instrument should be as small as 0.025/10 = 0.0025 mm (.00010 in.). Similarly, the increment of measurement for the calibration standard for that inspection instrument should be as small as 0.0025/10 = 0.00025 mm (.000010 in.).

Once the smallest increment of measurement for an instrument has been determined, then candidate instruments need to be evaluated in terms of the degree of satisfaction they offer relative to the following performance criteria.

1. *Accuracy*: The ability to measure the true magnitude of a dimension.
2. *Linearity*: The accuracy of the measurements of an instrument throughout its operating range.
3. *Magnification*: The amplification of the output reading of an instrument over the actual input dimension.
4. *Repeatability*: The ability of the instrument to achieve the same degree of accuracy on repeated applications (often referred to as "precision").
5. *Resolution*: The smallest increment of measurement that can be read on an instrument.
6. *Sensitivity*: The smallest increment of difference in dimension that can be detected by an instrument.
7. *Stability* or *drift*: The ability of an instrument to maintain its calibration over a period of time.

Consideration of these factors, along with cost and operating convenience, should help in selecting an appropriate measuring or gaging device for a particular inspection operation. For operating convenience, most instruments are or can be equipped with discrete digital readout devices. Most of these can be connected to microprocessors or computers for data recording and analysis.

16.2.3 Basic Linear Measuring Instruments

Measuring instruments may be direct reading or of the transfer type. An ordinary steel rule, such as the one shown in Figure 16-2 (row 5), contains a graduated scale from which the size of a dimension being measured can be determined directly. The spring caliper in row 3 contains no scale graduations. It is adjusted to fit the size of a dimension being measured and then is compared to a direct reading scale to obtain the size of the dimension.

Most of the available measuring instruments may be grouped according to certain basic principles of operation. Many simple instruments use only a graduated scale as a measurement basis, while others may have two related scales and use the vernier principle of measurement. In a number of instruments, the movement of a precision screw is related to two or three graduated scales to form a basis for measurement. Many other instruments utilize some sort of mechanical, electrical, or optical linkage between the measuring element and the graduated scale so that a small movement of the measuring element produces an enlarged indication on the scale. Air pressure or metered airflow is used in a few instruments as a means of measurement. These operating principles will be more fully explained later in the descriptions of a few of the instruments in which they are applied.

People in nearly every craft, whether they be television repairmen, automotive mechanics, or telephone installers, have a set of tools applicable to their work. Similarly, machinists, machine operators, toolmakers, etc., have a variety of basic measuring and gaging devices that they use in their daily work and that may be considered tools of their trade. Many of these are often referred to as standard or general-purpose tools, and for the most part they are relatively simple in construction. The results obtained from the use of these tools are considerably dependent upon the skill and dexterity of the person using them. For instance, the accuracy obtained in many cases depends upon the amount of pressure applied to the measuring elements. Thus the craftsman, through training and experience, acquires the sense of touch necessary to apply the tools properly. A group of basic instruments is shown in Figure 16-2. The group includes direct-reading and transfer-type linear measuring instruments, angular measuring devices, fixed and adjustable gaging devices, layout tools, and other miscellaneous metalworking accessories.

Direct-reading Instruments

Most of the basic or general-purpose linear measuring instruments are typified by the steel rule, the vernier caliper, or the micrometer caliper.

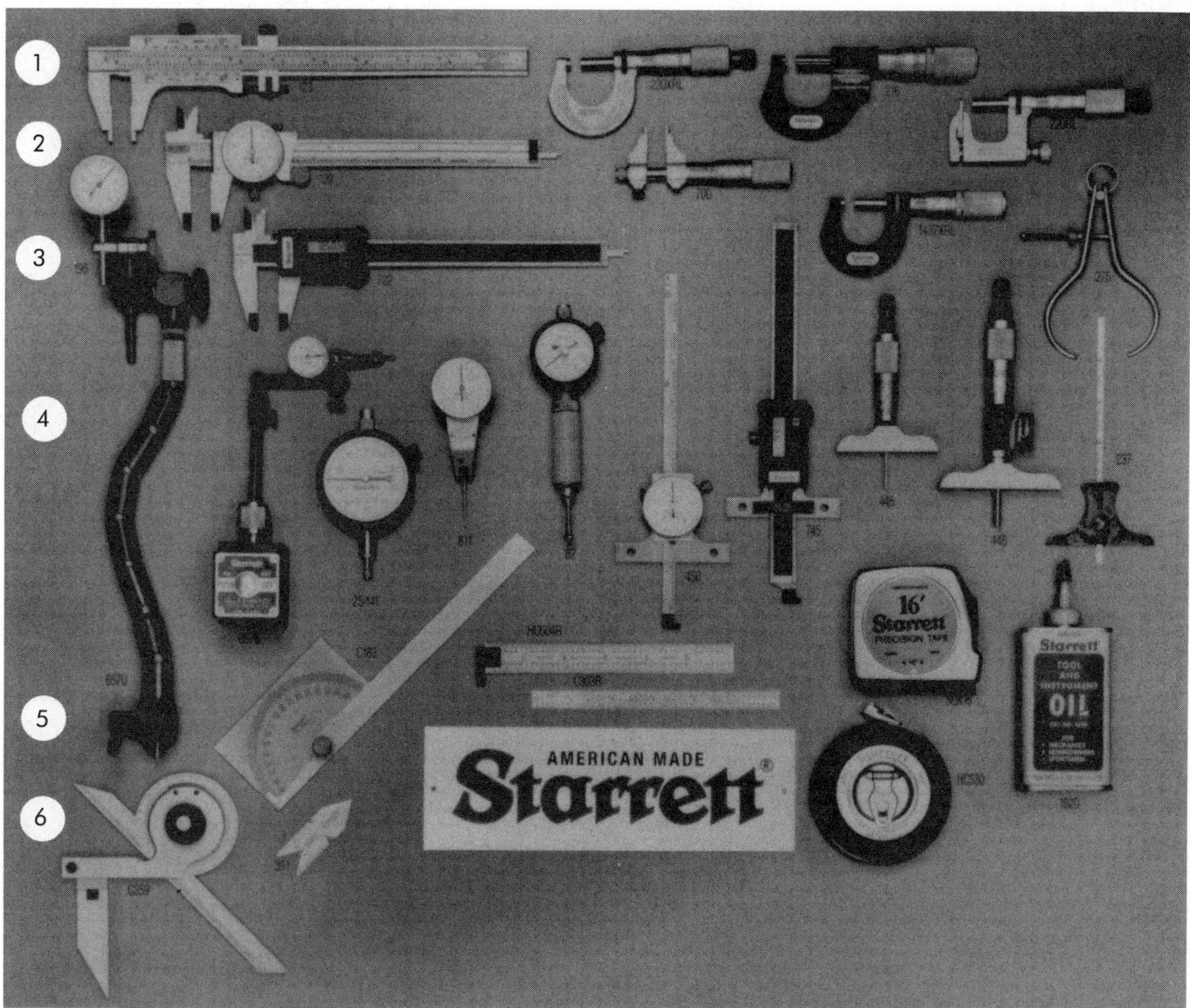

Figure 16-2. Standard measuring instruments: Row 1 (from left to right), master vernier caliper, outside micrometer, digital micrometer; row 2, dial caliper, inside micrometer caliper, mul-T-anvil outside-micrometer caliper; row 3, universal dial-test indicator, electronic digital caliper, outside micrometer, toolmaker's outside caliper; row 4, flex-o-post dial-indicator holder, magnetic base-indicator holder, dial indicator, swivel-head dial indicator, dial bore gage, dial depth gage, electronic digital-depth gage, micrometer depth gage, digital-micrometer depth gage, rule depth gage; row 5, protractor, steel rule, steel rule, precision measuring tape, tool and instrument oil; row 6, universal bevel protractor, center gage, steel measuring tape.

Steel rules are used effectively as line measuring devices, which means that the ends of a dimension being measured are aligned with the graduations of the scale from which the length is read directly. A depth rule (Figure 16-2, row 4) for measuring the depth of slots, holes, etc., is a type of steel rule. Steel rules are also incorporated in vernier calipers, as shown in Figure 16-2 (row 1), where they are adapted to end measuring operations. These are often more accurate and easier to apply than in-line measuring devices.

Verniers. The *vernier caliper* shown in Figure 16-3 typifies the type of instrument using the vernier principle of measurement. The main or beam scale on a typical metric vernier caliper is numbered in increments of 10 mm, with the smallest scale division being equivalent to 1 mm. The vernier scale slides along the edge of the main scale and is divided into 50 divisions, and

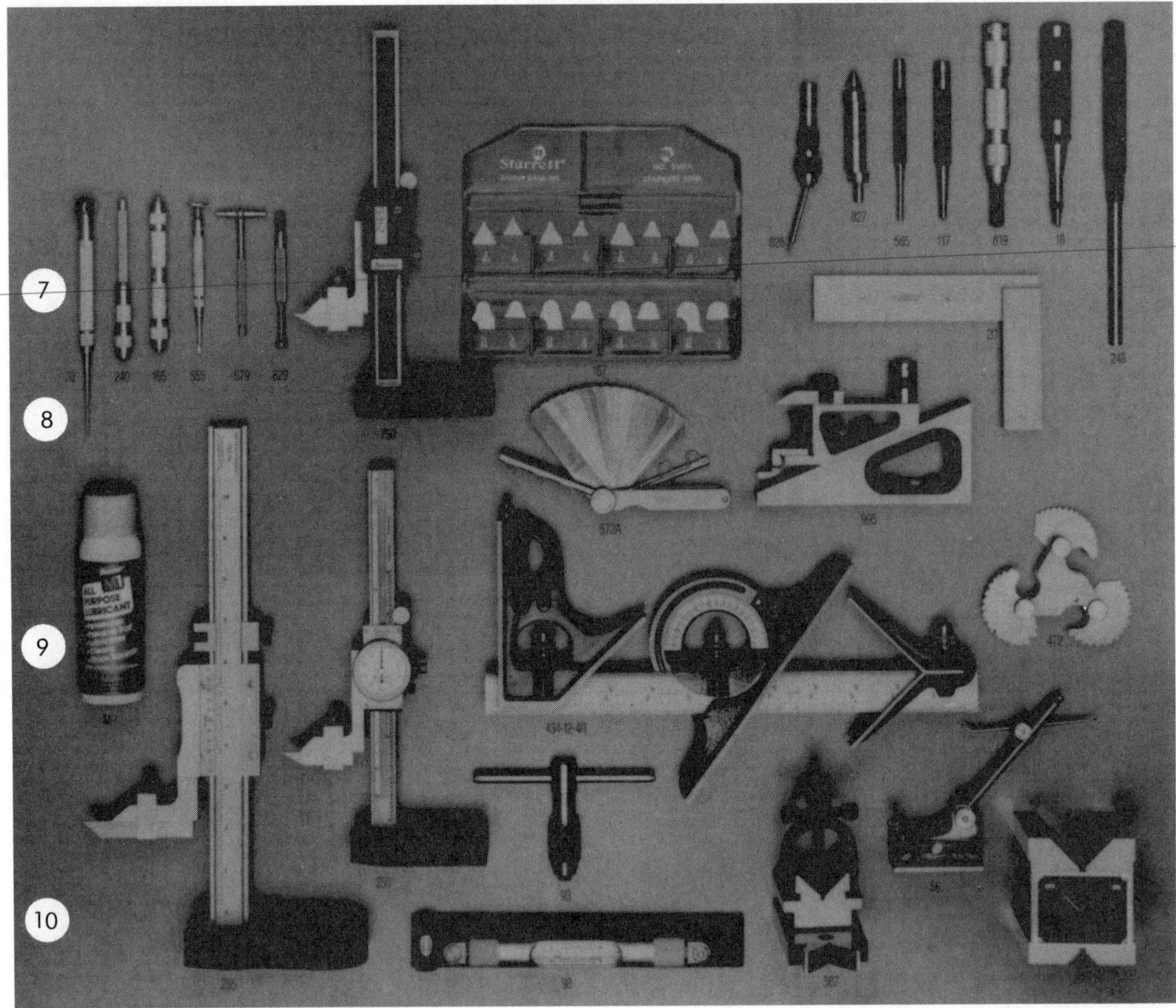

Figure 16-2 (continued). Standard measuring instruments: Row 7, pocket scriber, pin vise, double-end pin vise, jeweler's screwdriver, telescoping gage, small hole gage, electronic-digital height gage, radius gages, center finder, edge finder, drive pin punch, center punch, automatic center punch, automatic center punch; row 8, thickness gages, universal precision gage, steel square; row 9, all-purpose lubricant, vernier height gage, dial height gage, combination set, screw pitch gage; row 10, mechanic's level, T-handle tap wrench, V-block and clamp, toolmaker's surface gage, magnetic V-block. (Courtesy the L. S. Starrett Company)

these 50 divisions are the same in total length as 49 divisions on the main scale. Each division on the vernier scale is then equal to $\frac{1}{50}$ of (49 × 1) or 0.98 mm, which is 0.02 mm less than each division of the main scale. Aligning the zero lines of both scales would cause the first lines on each scale to be 0.02 mm apart, the second lines 0.04 mm apart, etc. A measurement on a vernier is designated by the positions of its zero line and the line that coincides with a line on the main scale. For example, the metric scale in Figure 16-3(A) shows a reading of 12.42 mm. The zero index of the vernier is located just beyond the line at 12 mm on the main scale, and line 21 (after 0) coincides with a line on the main scale, indicating the zero index is 0.42 mm beyond the line at 12 mm. Thus 12.00 + 0.42 = 12.42 mm.

The vernier caliper illustrated in Figure 16-3 also has an inch scale so that it can be used interchangeably for either inch or millimeter mea-

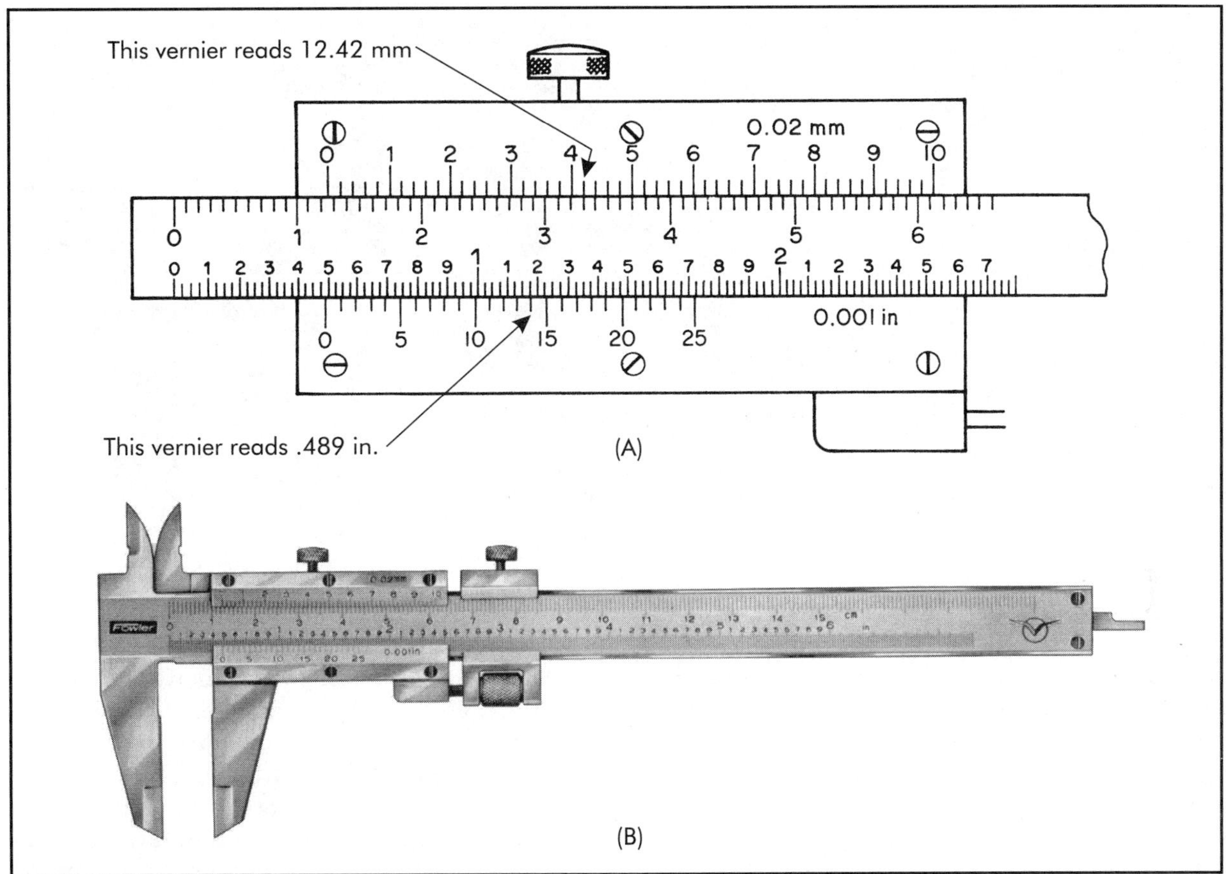

Figure 16-3. Fine-adjustment style vernier caliper. (Courtesy Fred V. Fowler Company, Inc.)

surements. The smallest division on the main scale represents .025 in. and the vernier is divided into .001 in. increments. Thus the measurement illustrated is .475 from the main scale plus .014 from the vernier scale, for a total of .489 in.

A vernier may be designed to indicate practically any increment of length (within practical limits of visual perception) according to the relationship

$$\frac{d_m(d_s+1)}{I}=1 \qquad (16\text{-}1)$$

where

d_m = number of divisions in one unit of length on the main scale

d_s = length of the secondary scale measured in divisions of the main scale

$1/I$ = smallest increment to be read with the vernier

The vernier caliper shown in Figure 16-3(B) consists of a steel rule with a pair of fixed jaws at one end and a pair of sliding jaws affixed to a vernier. Outside dimensions are measured between the lower jaws; inside dimensions over the tips of the upper jaws. It costs about $40.

The *digital reading caliper* shown in Figure 16-4 provides LCD readouts in either millimeters or inches and operates by a microprocessor-based system. The caliper has a measuring range of 0–152 mm (0–6 in.) with readings in increments of 0.013 mm (.0005 in.). The unit is capable of retaining a reading in the display when the tool is used in an area where visibility is restricted. It is powered by long-life disposable batteries and costs about $175.

The *vernier height gage* of Figure 16-2 (row 9) is similar to a vernier caliper except the fixed jaw has been replaced by a fixed base, and the sliding jaw may have a scriber attached to it for

Figure 16-4. LCD digital-reading caliper with 0–152 mm (0–6 in.) range. (Courtesy Fred V. Fowler Company, Inc.)

layout work or a dial indicator for measuring or comparing operations. A more sophisticated version of the vernier height gage is represented by the microprocessor-based digital height gage shown in Figure 16-5. This instrument can easily measure in two dimensions in either rectangular or polar coordinates. It can measure external, internal, and distance dimensions, as well as perpendicularity, flatness, straightness, centers, and diameters.

Vertical measurements are made on the gage shown in Figure 16-5 in either metric or inch units to a resolution of 0.0005 mm (.00002 in.) by an opto-electronic sensor moving over a high-accuracy glass scale. The gage head moves on a cushion of air generated by a completely self-contained pneumatic system. Dedicated function programs, along with a keypad and interactive LCD display, are designed to guide operators smoothly and efficiently through a variety of measurement operations. The gage can be equipped with an RS 232-C interface for direct data transfer to other data collection devices. Instruments of this type cost between $8,000 and $14,000, depending on size and the type of control panel required.

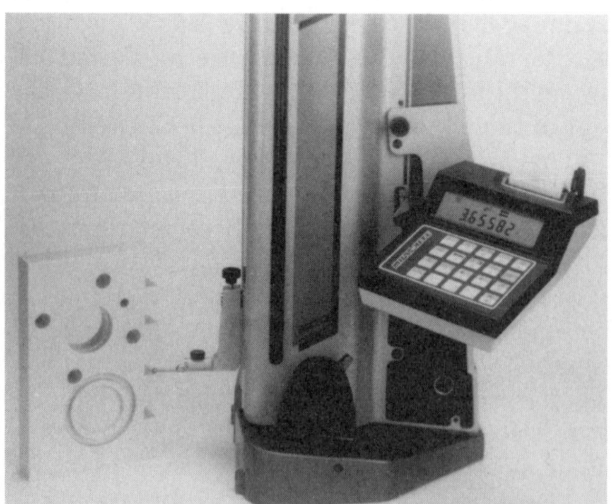

Figure 16-5. Digital-reading, single-axis height gage for two-dimensional measurements. (Courtesy Brown and Sharpe Manufacturing Company)

Micrometers. The *micrometer caliper* illustrated in Figure 16-6 is representative of the type of instrument using a precision screw as a basis for measuring. The measuring elements consist of a fixed anvil and a spindle that moves lengthwise as it is turned.

The thread on the spindle of a typical metric micrometer has a lead of $\frac{1}{2}$ or 0.5 mm, so that one complete revolution of the thimble produces a spindle movement of this amount. The graduated scale on the sleeve of the instrument has major divisions of 1.0 mm and minor divisions of 0.5 mm. Thus, one revolution of the spindle causes the beveled edge of the thimble to move through one small division on the sleeve scale. The periphery of the beveled edge of the thimble is graduated into 50 equal divisions, each space representing $\frac{1}{50}$ of a complete rotation of the thimble, or a 0.01 mm movement of the spindle.

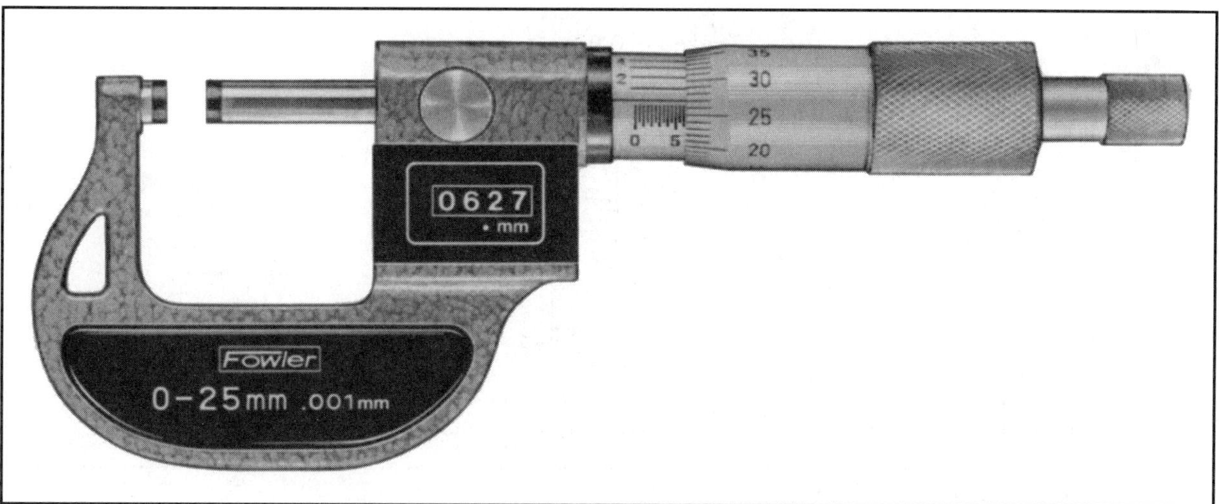

Figure 16-6. A 0–25-mm micrometer caliper. (Courtesy Fred V. Fowler Company, Inc.)

Micrometers with scales in inch units operate in a similar fashion. Typically, the spindle thread has a lead of .025 in. and the smallest division on the sleeve represents .025 in. The periphery of the beveled edge of the thimble is graduated into 25 equal divisions, each space representing $\frac{1}{25}$ of a complete rotation of the thimble or a spindle movement of .001 in.

A reading on a micrometer is made by adding the thimble division that is aligned with the longitudinal sleeve line to the largest reading exposed on the sleeve scale. For example, in Figure 16-7 the thimble has exposed the number 10, representing 10.00 mm, and one small division worth 0.50 mm. The thimble division 16 is aligned with the longitudinal sleeve line, indicating that the thimble has moved 0.16 mm beyond the last small division on the sleeve. Thus the final reading is obtained by summing the three components, $10.00 + 0.50 + 0.16 = 10.66$ mm.

A *vernier micrometer caliper*, such as that represented by the scales shown in Figure 16-8, has a vernier scale on the sleeve permitting measurements to 0.001 mm. The vernier scale shown has 10 divisions over a length equivalent to 19 divisions around the periphery of the thimble. Thus the difference in length of a division on the vernier scale and two divisions on the thimble is $0.02 - (1/10)(19 \times 0.01) = 0.001$ mm. Thus the reading illustrated in Figure 16-8 is $10.00 + 0.50 + 0.16 + 0.006 = 10.666$ mm.

Digital micrometers. Micrometers with digital readouts are also available to make readings faster and easier for inspection personnel regardless of their degree of experience. The digital micrometer shown in Figure 16-6 represents one instrument of this type for use in measuring to a resolution of 0.01 mm. The instrument shown in Figure 16-9 has a digital readout with a resolution to .0001 in. When equipped with vernier scales, the resolution may be increased to 0.001 mm (commonly .0001 in. in the case of an inch-reading device).

Micrometer caliper. The *micrometer caliper*, or *mike* as it is often called, is an end-measuring instrument for use in measuring outside dimensions. Although the mike is fairly easy to apply, the accuracy it gives depends upon the application of the proper amount of torque to the thimble. Too much torque is likely to spring the frame and cause error. Thus it is important that personnel using these instruments be trained in their use, and also that they be periodically required to check their measurements against a standard to minimize errors. The indicating micrometer of Figure 16-10 has a built-in dial indicator to provide a positive indication of measuring pressure applied. The instrument can be used like an indicating snap gage.

A standard metric micrometer is limited to a range of 25 mm (1 in. for a micrometer reading in inch units). Thus different micrometers are needed to measure a wide range of dimensions.

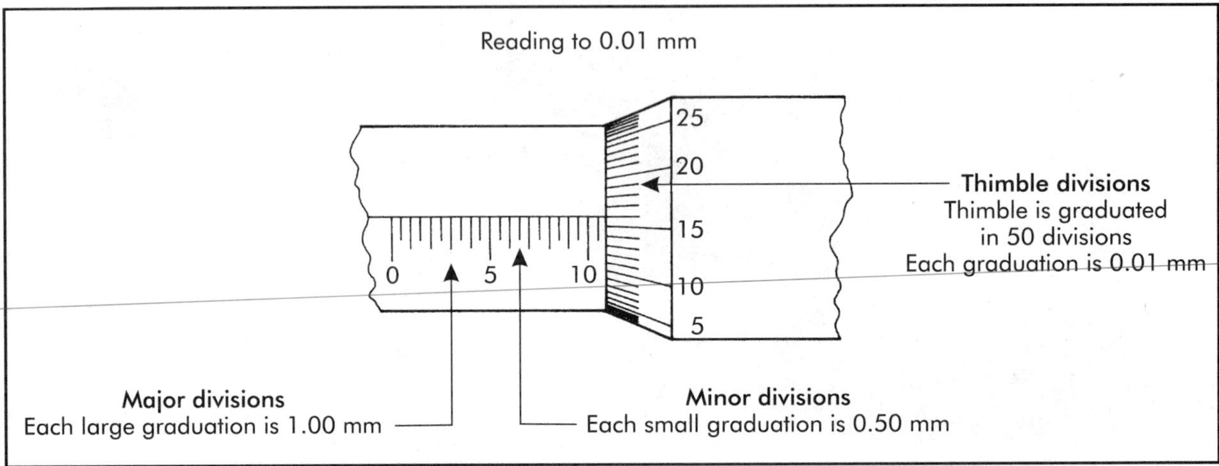

Figure 16-7. Micrometer reading of 10.66 mm.

The precision screw principle is applied directly in other measuring instruments such as the type of inside micrometer shown in Figure 16-2 (row 2), the micrometer depth gage, in Figure 16-2 (row 4), and the internal micrometer plug. It is also used as a device to provide precise calibrated linear movement to staging devices and other moving components of toolmakers' microscopes and optical projecting comparators.

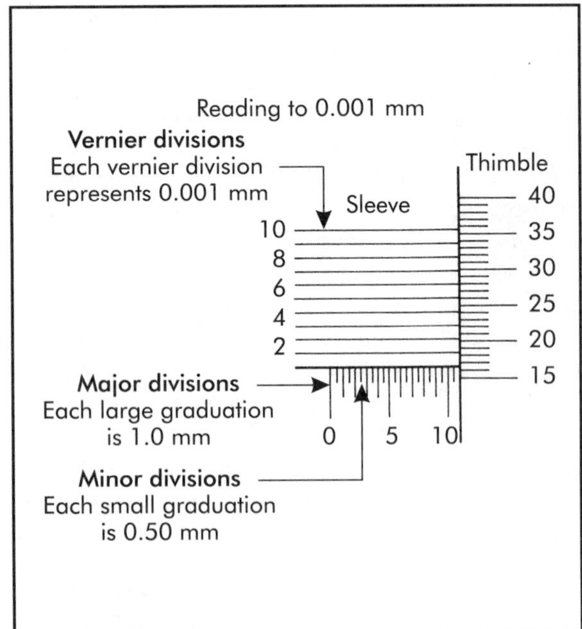

Figure 16-8. Scales of a vernier micrometer showing a reading of 10.666 mm.

Transfer-type Linear Measuring Devices

Transfer-type linear measuring devices are typified by the spring caliper, spring divider, firm joint caliper, telescoping gage, and small hole gage. Examples of each of these are shown in Figure 16-2.

The outside caliper is used as an end measure to measure or compare outside dimensions, while the inside caliper is used for inside diameters, slot and groove widths, and other internal dimensions. They are quite versatile but, due to the construction and method of application, their accuracy is somewhat limited.

Angular Measuring Devices

The unit standard of angular measurement is the degree. The measurement and inspection of angular dimensions are somewhat more difficult than linear measures and may require instruments of some complexity if a great deal of angular precision is required.

Simple tools. The *combination set* consists of a center head, protractor, and square with a 45° surface, all of which are used individually in conjunction with a steel rule. The heads are mounted on the rule and clamped in any position along its length by means of a lock screw. The parts of such a set are shown in Figure 16-2 (row 9). The center head is used to scribe bisecting diameters on the end of a cylindrical piece to locate the center of the piece. The protractor reads directly in degrees. Both the square head and the protractor may contain a small spirit

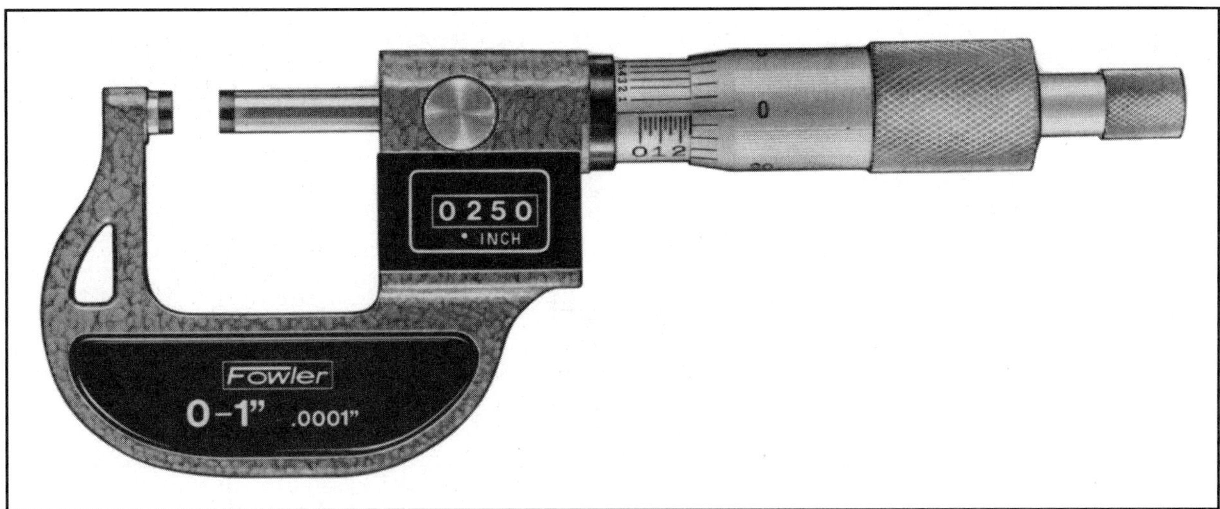

Figure 16-9. A digital micrometer. (Courtesy Fred V. Fowler Company, Inc.)

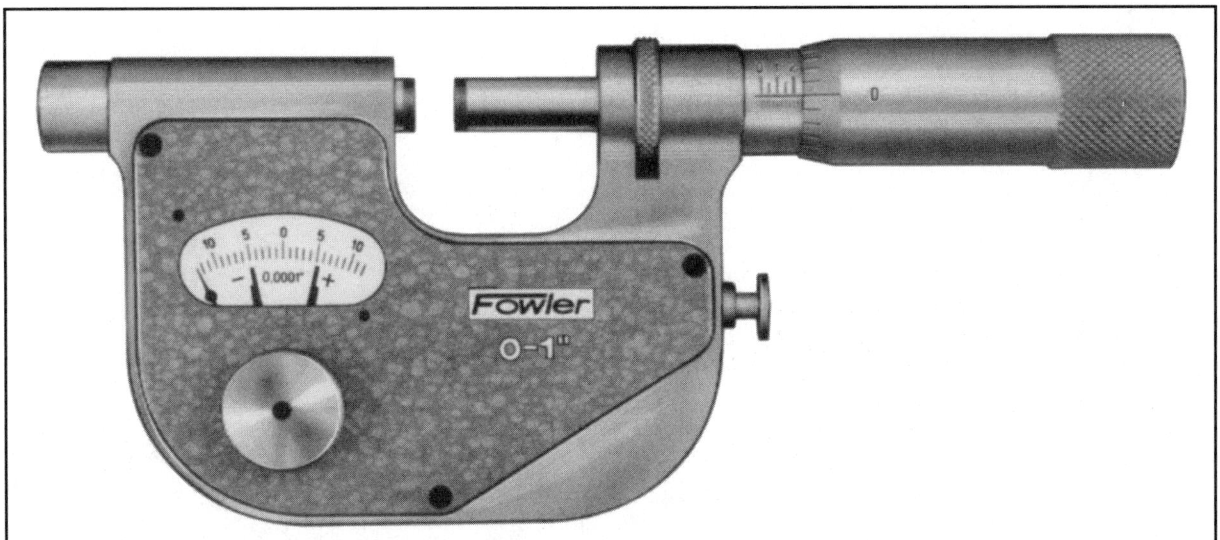

Figure 16-10. An indicating micrometer. (Courtesy Fred V. Fowler Company, Inc.)

level. A *bevel protractor* utilizes a vernier scale to show angles as small as five minutes.

Sine bar. The sine bar is a relatively simple device for precision measuring and checking of angles. It consists of an accurately ground, flat-steel straight-edge with precisely affixed round buttons, a definite distance apart, and of identical diameters.

Figure 16-11 illustrates one method of applying a sine bar in the determination of the angle α on the conical surface of the part located on the surface plate. For precise results, a sine bar must be used on true surfaces. In Figure 16-11, the center-to-center distance of the sine bar buttons is 127 mm (5 in.) and the distances A and B are determined by means of gage blocks or a vernier height gage to be 25.400 mm (1.0000 in.) and 89.794 mm (3.5352 in.), respectively. Thus the sine α equals $(89.794 - 25.400)/127.00 = 0.50704$ and from trigonometric tables the angle α is 30°28′.

Dividing heads. Mechanical and optical dividing heads are often employed for the circular measurement of angular spacing. The mechani-

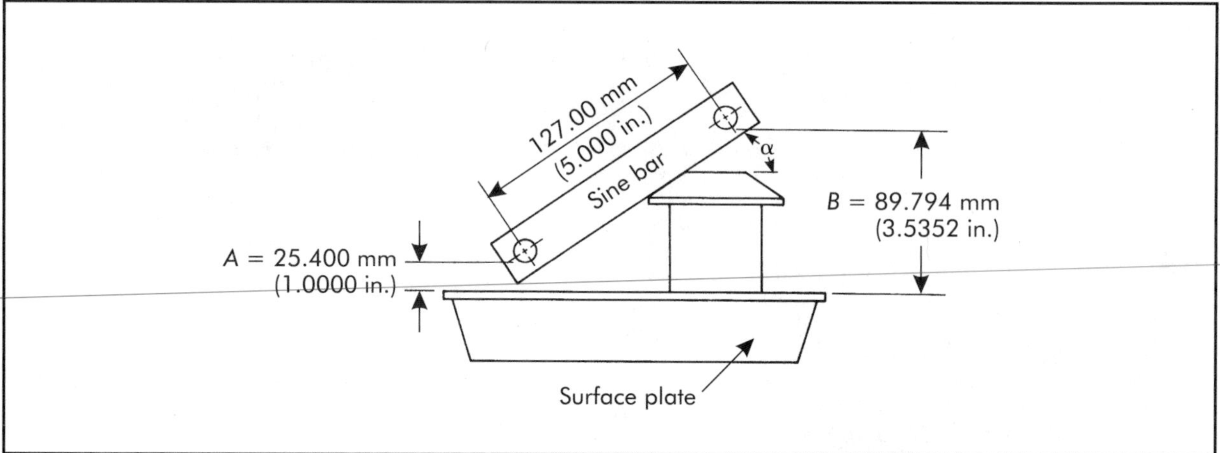

Figure 16-11. Application of a sine bar.

cal dividing head is described in Chapter 21 (see Figure 21-17). The optical dividing head performs the same function but more precisely. One make has a disk around its main spindle with graduations in fine increments inscribed around its circumference. These are viewed through a microscope.

Layout Instruments and Locating Devices

Considerable metalworking and woodworking, particularly in job shops, for pattern building, model building, and tool and die work, is done to lay out lines, circles, center locations, etc., scribed on the workpiece itself. Chalk or dye is often applied to the work surface before scribing so that the lines can be readily seen.

A *surface plate* provides a true reference plane from which measurement can be made. A cast-iron surface plate is a heavy ribbed, box-like casting that stands on three points (establishing a plane) and has a thick and well-supported flat top plate. New plates generally have an average of 18 bearing spots on an area of 6.5 cm^2 (≈1 in.^{2}) that do not vary from a true plane by more than 0.005 mm (.0002 in.). The use of natural stones for surface plates is becoming increasingly popular because of their hardness, resistance to corrosion, minimum response to temperature change, and nonmagnetic qualities. Figure 16-12 shows a granite surface plate used in inspection work. Reference surfaces also may be obtained by the use of bar parallels, angle irons, V-blocks, and toolmakers' flats.

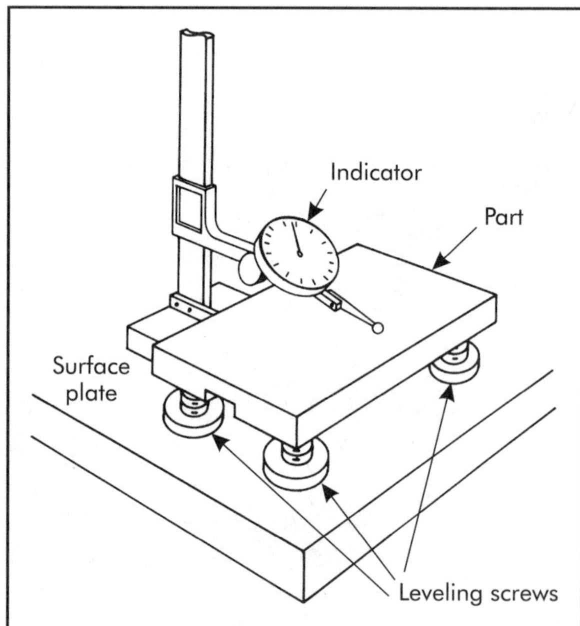

Figure 16-12. Application of a granite surface plate for checking the flatness of a part with a dial indicator and leveling screws.

A variety of hand marking tools, such as the scriber, spring divider, and center punch, are employed by the layout person. These tools are shown in Figure 16-2. The *surface gage,* Figure 16-2 (row 10), consists of a base, an adjustable spindle, and a scriber, and may be used as a layout instrument. The scriber is first adjusted to the desired height by reference to a steel rule or gage blocks and then the gage is moved to the workpiece and a line is scratched on it at the desired location. The vernier height gage may be employed in a similar manner.

Gages

Classes. In mass-manufacturing operations it is often uneconomical to attempt to obtain absolute sizes during each inspection operation. In many cases, it is only necessary to determine whether one or more dimensions of a mass-produced part are within specified limits. For this purpose, a variety of inspection instruments referred to as *gages* are employed. However, the distinction between gaging and measuring devices is not always clear as there are some instruments referred to as gages that do give definite measurements.

To promote consistency in manufacturing and inspection, gages may be classified as working, inspection, and reference or master gages. *Working gages* are used by the machine operator or shop inspector to check the dimensions of parts as they are being produced. They usually have limits based on the piece being inspected. *Inspection gages* are used by personnel to inspect purchased parts when received or manufactured parts when finished. These gages are designed and made so as not to reject any product previously accepted by a properly designed and functioning working gage. *Reference* or *master gages* are used only for checking the size or condition of other gages, and represent as exactly as possible the physical dimensions of the product.

A gage may have a single size and be referred to as a *nonlimit gage,* or it may have two sizes and be referred to as a *limit gage.* A limit gage, often called a "go" and "not go" gage, establishes the high and low limits prescribed by the tolerance on a dimension. A limit gage may be either double-end or progressive. A *double-end gage* has the "go" member at one end and the "not go" member at the other. Each end of the gage is applied to the workpiece to determine its acceptability. The "go" member must pass into or over an acceptable piece, but the "not go" member should not. A *progressive gage* has both the "go" and "not go" members at the same end so that a part may be gaged with one movement.

Some gages are fixed in size while others are adjustable over certain size ranges. *Fixed gages* are usually less expensive initially, but they have the disadvantage of not permitting adjustment to compensate for wear.

Most gages are subjected to considerable abrasion during their application and must therefore be made of materials resistant to wear. High-carbon and alloy steels have been used as gage materials for many years because of their relatively high hardenability and abrasion resistance. Further increased surface hardness and abrasion resistance may be obtained from the use of chrome plating or cemented carbides as surface material on gages. Some gages are made entirely of cemented carbides or they have cemented carbide inserts at certain wear points. Chrome plating is also used as a means of rebuilding and salvaging worn gages.

Common gages. Typical common functional gages can be classified on the basis of whether they are used to check outside dimensions, inside dimensions, or special features. Some examples of typical gages are shown in Figure 16-13. They include *ring* and *snap gages* for checking outside dimensions, *plug gages* for checking inside dimensions, and other gages for checking other geometrical shapes such as tapers, threads, and splines. Typical *plug gages,* such as the ones shown in Figure 16-13(A), consist of a hardened and accurately ground steel pin with two gage members: one is the "go" gage member and the other the "not-go" gage member (top view of Figure 16-13[A]). *Progressive plug gages* (bottom view of Figure 16-13[A]) combine both "go" and "not go" members into one. The design of the gage member and the method used to attach it to the handle depends on its size as shown in Figure 16-13(B). The gage members are usually held in the handle by a threaded collet and bushing (view 1), a taper lock where gage members have a taper shank on one end that fits into the end of the handle (view 2), or a trilock where the gage members have a hole drilled through the center and are counterbored on both ends to receive a standard socket-head screw (view 3). One way of checking a hole for out-of-roundness is to have flats ground on the side of the gage member as shown in Figure 16-13(C).

Ring gages, such as those shown in Figure 16-13(D), are used for checking the limit sizes of a round shaft. They are generally used in pairs:

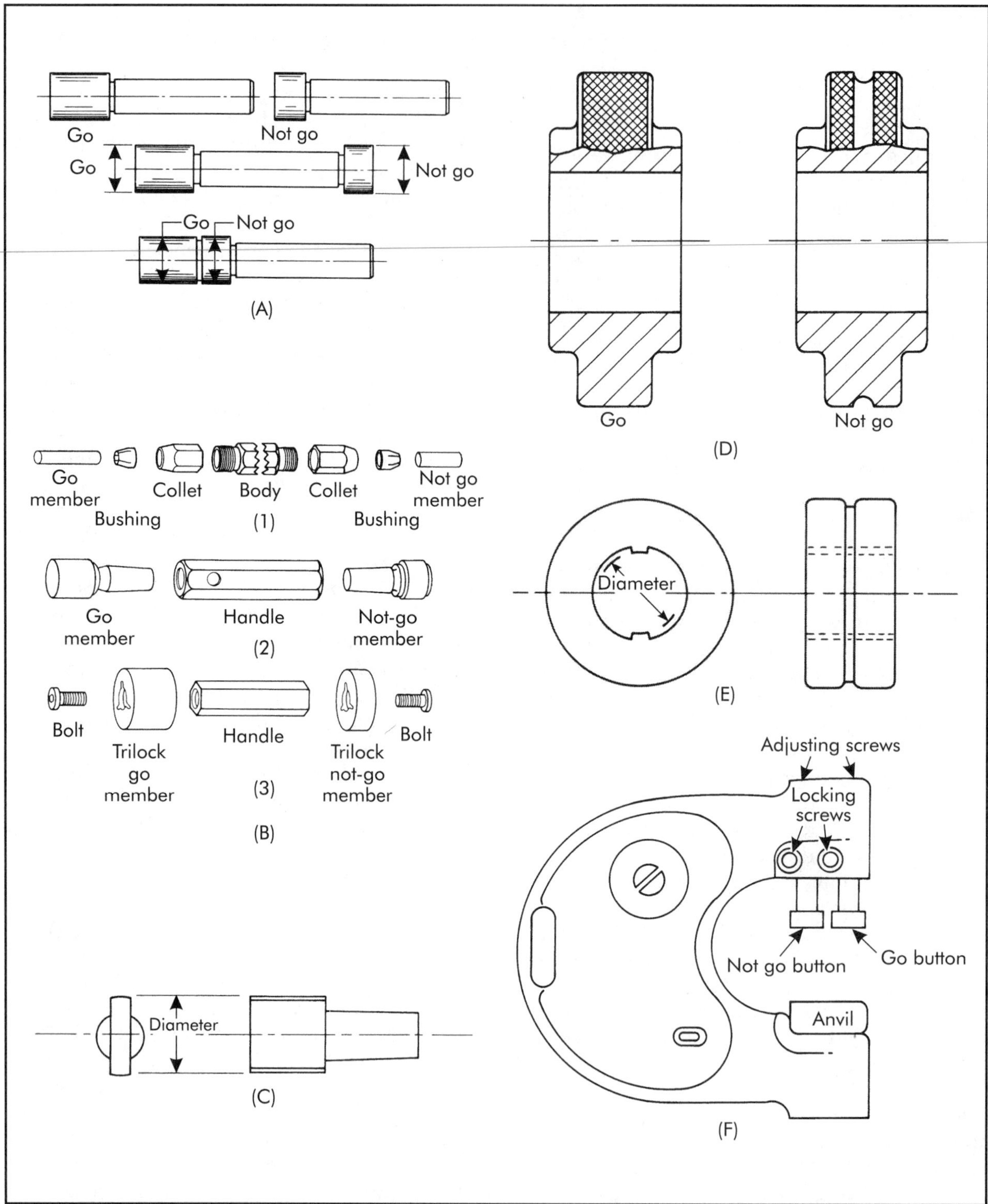

Figure 16-13. Examples of typical gages: (A) typical plug gages; (B) attachment of the gage member to the handle by: (1) threaded collet and bushing, (2) taper lock, or (3) trilock; (C) cylindrical plug gage with flats for checking out-of-roundness condition; (D) ring gages for inspecting shaft diameters; (E) ring gage for checking a shaft for an out-of-roundness condition; (F) typical snap gage.

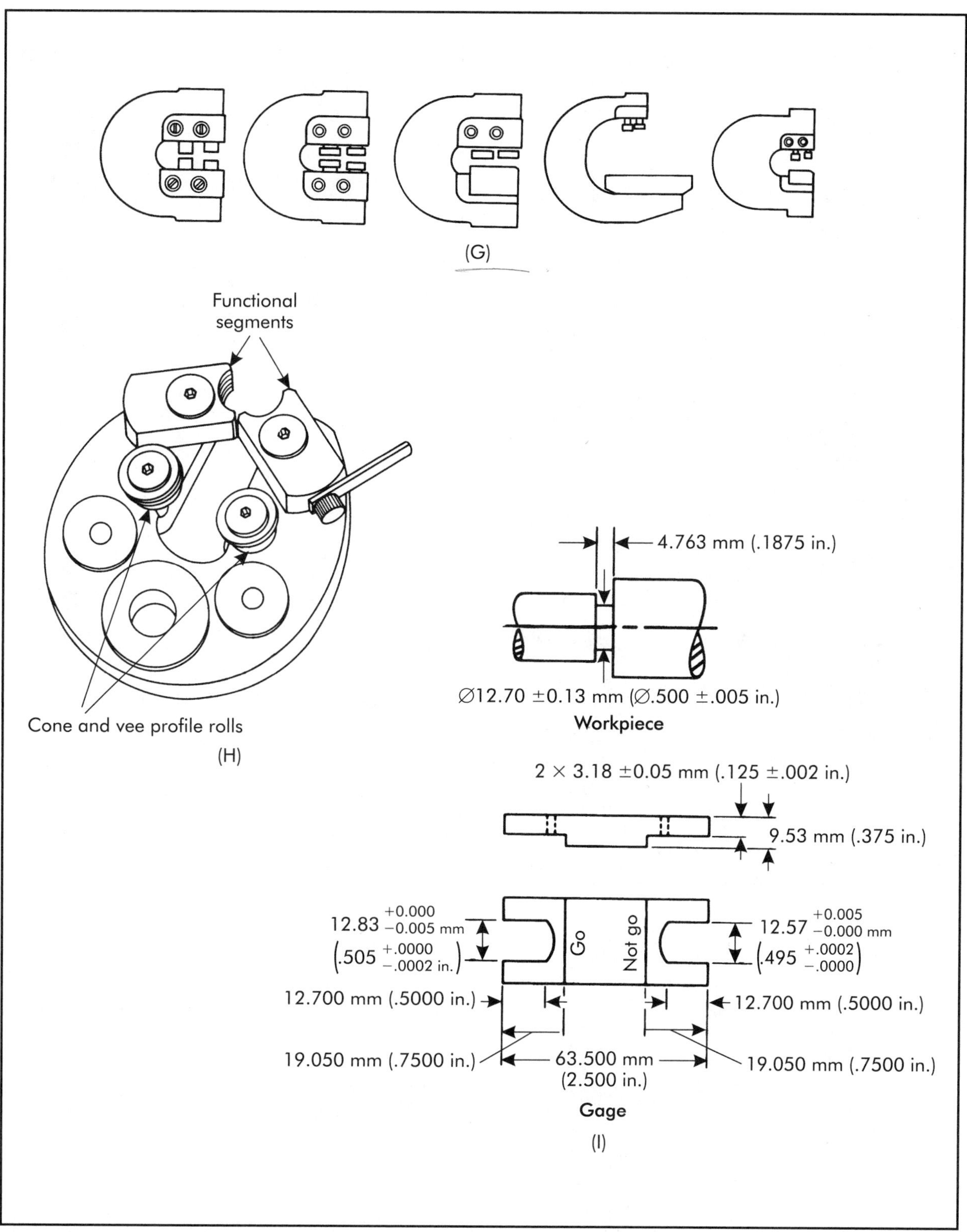

Figure 16-13 (continued). Examples of typical gages: (G) adjustable snap gages; (H) thread snap gage; (I) special snap gage.

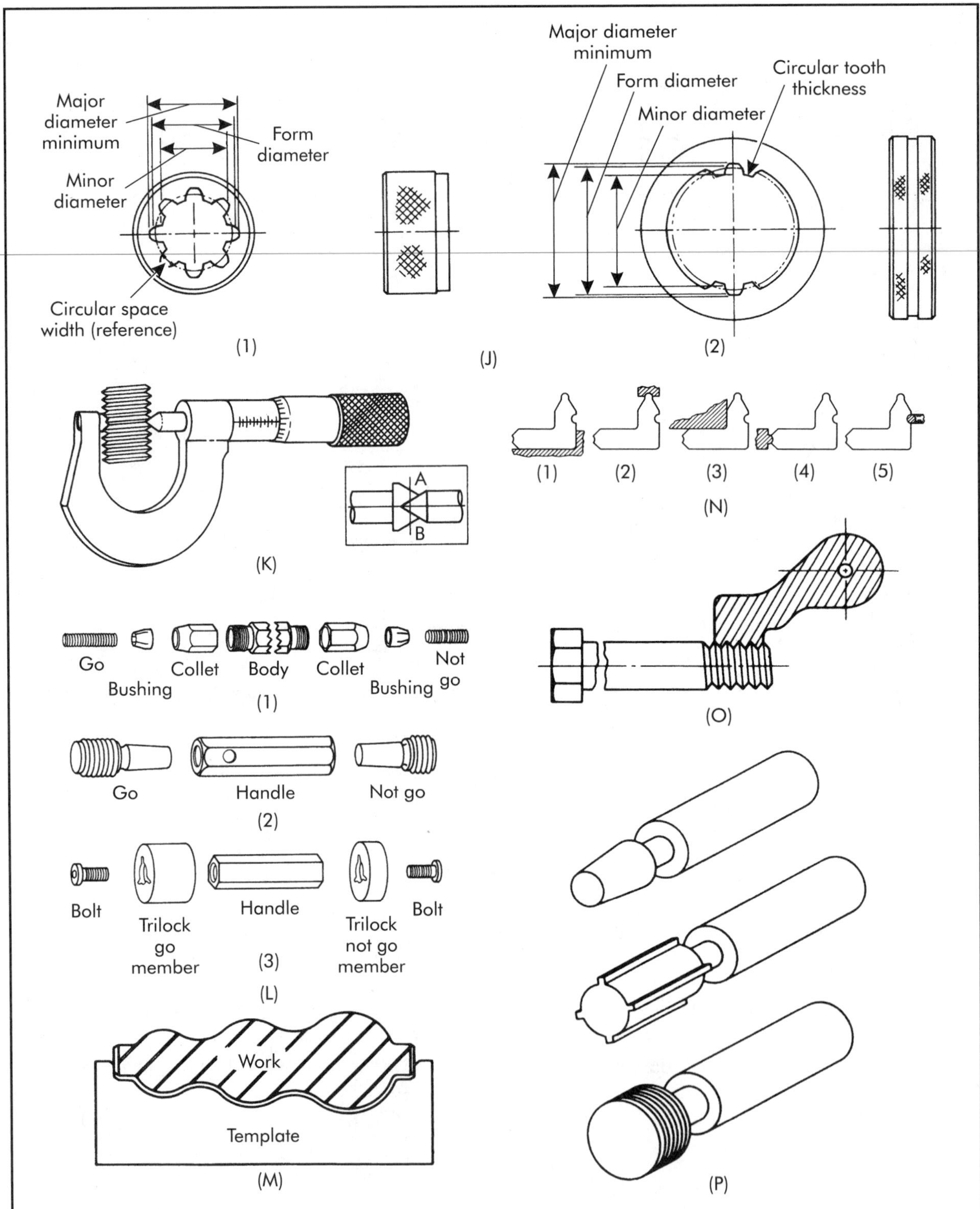

Figure 16-13 (continued). Examples of typical gages: (J) fixed-limit spline gage; (K) screw thread micrometer; (L) thread plug gages; (M) gaging the profile of a workpiece with a template; (N) radius gages; (O) screw pitch gage; (P) special plug gages for inspecting the profile or taper of holes.

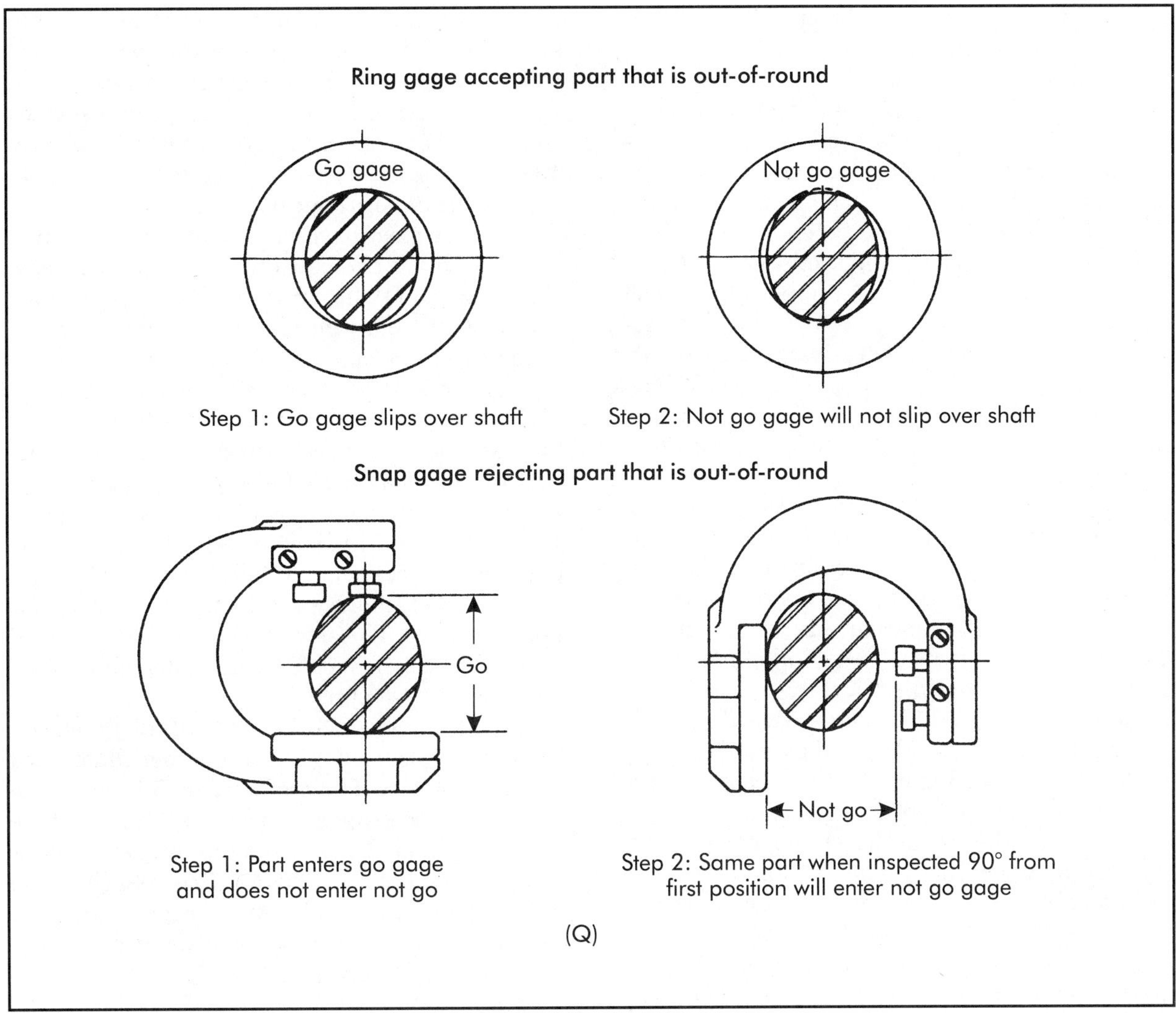

Figure 16-13 (continued). Examples of typical gages: (Q) ring and snap gage comparison.

the "go" gage for checking the upper limit of the part tolerance and "not go" gage for checking the lower limit. The "not go" ring has a groove in the outside diameter of the gage to distinguish it from the "go" ring. It is possible that a shaft is larger at its ends than the middle or it could suffer an out-of-roundness condition. This situation cannot be detected with a standard cylindrical ring gage. Such an out-of-roundness condition can be checked by a ring gage that has the inside diameter relieved such as the one shown in Figure 16-13(E).

A *snap gage* is another fixed gage with the gaging members specially arranged for measuring diameters, thicknesses, and lengths. A typical (may also be called adjustable) *external measuring snap gage* is shown in Figure 16-13(F). It consists of a C-frame with gaging members in the jaw of the frame. Figure 16-13(G) shows other types of snap gages. Threads can be checked with thread plug gages, thread ring gages, thread snap gages, or a screw thread micrometer. *Thread snap gages* have two pair of gaging elements combined in one gage. With appropriate gaging elements, these gages may be used to check the maximum and minimum material limit of external screw threads in one path. An example of a thread snap gage is shown in Figure 16-13(H). In some cases, special snap gages may be desired. The example in Figure 16-13(I) illustrates the use of a special double-end snap gage for inspecting the outside diameter of a narrow groove.

The use of a *spline gage* is a common way of inspecting splined workpieces prior to assembly. External splines are checked with internal-toothed rings, whereas internal splines are checked with external-toothed plugs. Figure 16-13(J) shows the two basic types of fixed-limit spline gages: composite and sector gages. *Composite gages* have the same number of teeth as that of the part. *Sector gages* have only two sectors of teeth 180° apart. These gages are further subdivided into "go" and "not go" gages. View 1 of Figure 16-13(J) shows a "go" composite ring gage, and a "not-go" sector ring gage is illustrated in view 2.

A *screw thread micrometer*, such as the one shown in Figure 16-13(K), has a specially designed spindle and anvil so that externally threaded parts can be measured. Screw thread micrometers are generally designed to measure threads within a narrow range of pitches. *Thread plug gages* are similar in design to cylindrical lug gages except that they are threaded. They are designed to check internal threads. Typical thread plug gages, such as those shown in Figure 16-13(L), consist of a handle and one or two thread gage members. Depending on the size of the gaging member, the member can be held in the handle using a threaded collet and bushing design (view 1), a taperlock design (view 2), or a trilock design (view 3).

To check a specified profile, *templates* may be used. They also may be used to control or gage special shapes or contours in manufactured parts. These templates are normally made from thin, easy-to-machine materials. An example of a contour template for inspecting a turned part is shown in Figure 16-13(M). To visually inspect or gage radii or fillets, special templates, such as those shown in Figure 16-13(N), may be used. The five basic uses of such templates are: inspection of an inside radius tangent to two perpendicular planes (view 1), inspection of a groove (view 2), inspection of an outside radius tangent to two perpendicular planes (view 3), inspection of a ridge segment (view 4), and inspection of roundness and diameter of a shaft (view 5).

The pitch of a screw may be checked with a *screw pitch gage*. To determine the pitch, the gage is placed on the threaded part as shown in Figure 16-13(O). A drawback of using screw pitch gages is their inability to give an adequate check on thread form for precision parts.

It is sometimes necessary to design special gages for checking special part features such as square, hexagonal, or octagonal holes. Figure 16-13(P) shows some special plug gages for checking the profile or taper of holes.

As an inspection tool, a snap gage is sometimes a better choice to use than a ring gage. Figure 16-13(Q) illustrates how a ring gage may accept an out-of roundness condition that would otherwise be rejected by a snap gage.

A *functional gage* checks the fit of a workpiece with a mating part. It normally just simulates the pertinent features of the mating part. An example of a functional gage would be a plate like part II of Figure 15-14 with four plugs of 12.50 mm (.492 in.) diameter each located in true position as nearly as possible. Any part I that would fit on that gage would pass inspection for hole positions. The sizes of the holes would need to be checked with a plug 12.70/12.90 mm (.500/.508 in.) in diameter.

A *flush pin gage* checks the limits of dimension between two surfaces in the manner illustrated in Figure 16-14. The step on pin *B* is the same size as the tolerance on the depth of the hole being checked. Thus, the step on pin *B* must straddle the top of collar *A* for the depth of the hole to be within limits. An inspector can compare the surfaces quickly and reliably by feeling them with a fingernail.

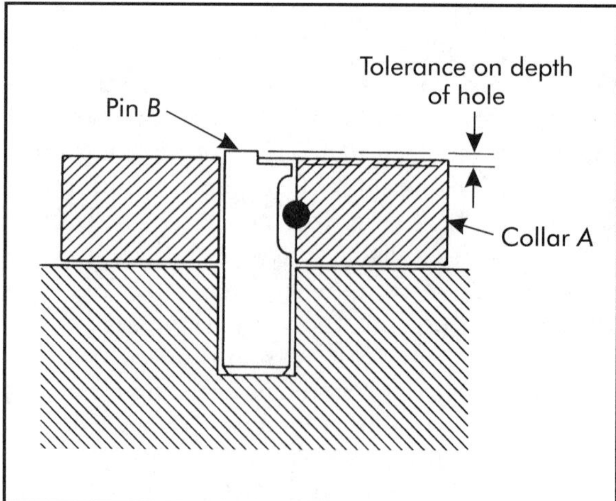

Figure 16-14. Typical flush-pin gage for gaging the depth of a hole.

Sizes. In gage making, as in any other manufacturing process, it is economically impractical to attempt to make gages to an exact size. Thus it is necessary that some tolerance be applied to gages. It is desirable, however, that some tolerance still be available for the manufacturing process. Obviously, though, the smaller the *gage tolerance*, the more the gage will cost. Along with the gage makers' tolerance, it is usually necessary to provide a *wear allowance*.

There are three methods of applying tolerances to gages, each of which affects the outcome of the inspection operation differently. These three methods are illustrated in Figure 16-15. The first is to use a unilateral gage tolerance and make the gage within the work tolerance as shown at *A*. This will result in some acceptable products being rejected. The second method is to use a bilateral gage tolerance about the limiting specifications on the part as shown at *B*. This might allow some acceptable parts to be rejected or some rejectable parts to be accepted. The third method is to use a unilateral tolerance and make the gage outside the work tolerance, as in *C*. Gages made according to this method will permit defective parts to be accepted at the start and continue to be accepted as long as the gage is in use, but it provides the most manufacturing tolerance.

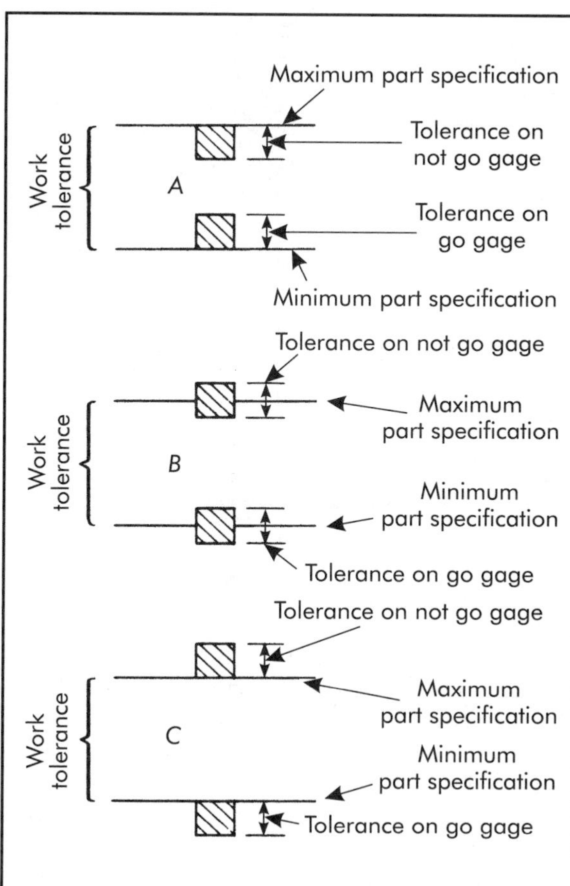

Figure 16-15. Methods of assigning gage tolerances.

There is no universally accepted policy for the amount of gage tolerance. A number of industries where part tolerances are relatively large use 20% of the part tolerance for working gages and 10% for inspection gages. For each of these gages, one-half of the amount is used for wear on the "go" member and one-half for the gage makers' tolerance on both the "go" and "not go" members. This method has been used to determine the tolerances for the plug gages shown in Figure 16-16 to check a hole with a diameter of 40.010 +0.10/−0.00 mm (1.5752 +.004/−.000 in.). The total part tolerance is 0.10 mm (.004 in.). Thus 20% of 0.10 mm (.004 in.) gives 0.020 mm (.0008 in.) for the working gage, and 10% of 0.10 mm (.004 in.) gives 0.010 mm (.0004 in.) for the inspection gage, applied unilaterally.

Indicating Gages and Comparators

Indicating gages and *comparators* magnify the amount a dimension deviates above or below a standard to which the gage is set. Most indicate in terms of actual units of measurement, but some show only whether a tolerance is within a given range. The ability to measure to 25 nanometers (nm) (.000001 in.) depends upon magnification, resolution, accuracy of the setting gages, and staging of the workpiece and instrument. Graduations on a scale should be 1.5–2.5 mm (.06–.10 in.) apart to be clear. This requires magnification of 60,000× to 100,000× for a 25 nm (.000001 in.) increment; less is needed, of course, for larger increments. Mechanical, air, electronic, and optical sensors and circuits are available for any magnification needed and will be described in the following sections. However, measurements have meaning and are repeatable only if based upon reliable standards, like gage blocks, and if the support of the workpiece and instrument is stable. An example is that of equip-

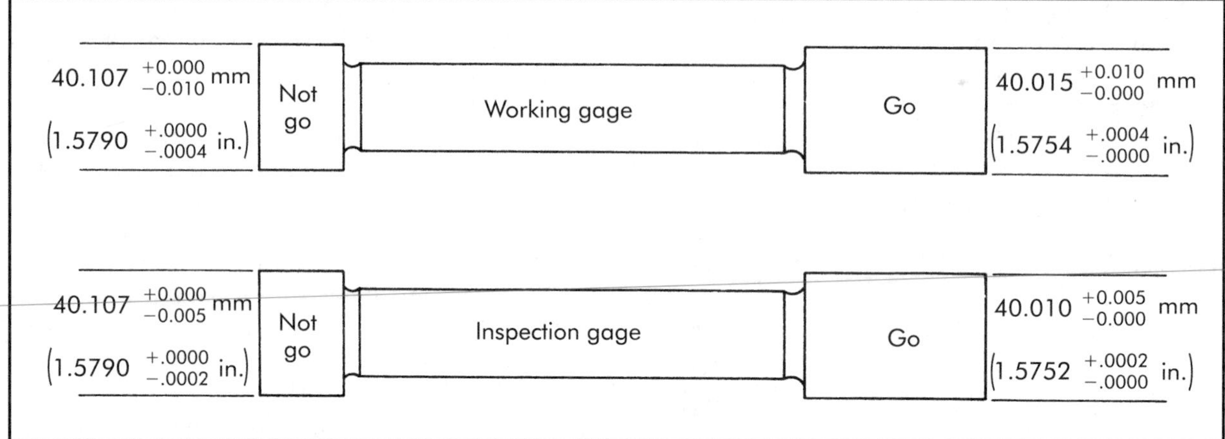

Figure 16-16. Specifications on working and inspection limit plug gages.

ment to trace the roundness of cylindrical parts. Either the probe or part must be rotated on a spindle that must run true with an error much less than the increment to be measured.

Mechanical indicating gages and comparators. Mechanical indicating gages and comparators employ a variety of devices. One type is the dial indicator depicted in Figure 16-17. Movement of stem *A* is transmitted from the rack to compound gear train *B* and *C* to pointer *D*, which moves around a dial face. Springs exert a constant force on the mechanism and return the pointer to its original position after the object being measured is removed.

Dial indicators are used for many kinds of measuring and gaging operations. One example is that of inspecting a workpiece such as the one illustrated in Figure 16-18. They also serve to check machines and tools, alignments, and cutter runout. Dial indicators are often incorporated in special gages and in measuring instruments, as exemplified by the indicating micrometer of Figure 16-10.

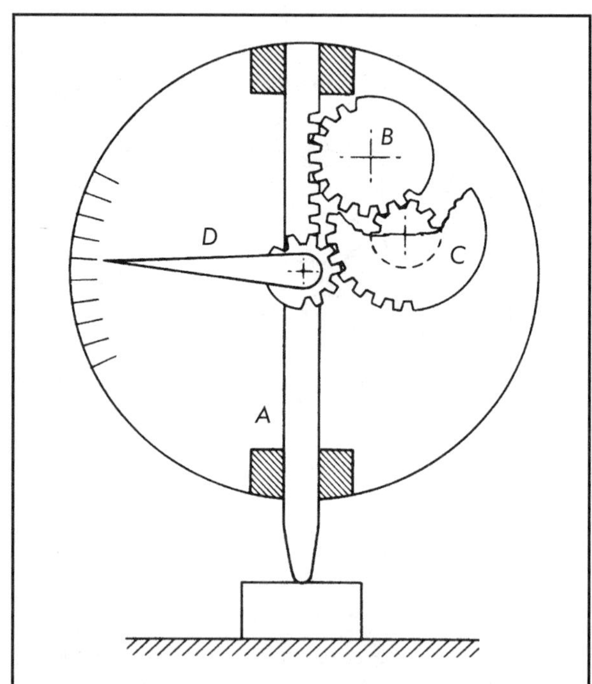

Figure 16-17. Simple dial indicator mechanism.

Electric and electronic gages. Certain gages are called *electric limit gages* because they have the added feature of a rack stem that actuates precision switches. The switches connect lights or buzzers to show limits and also may energize sorting and corrective devices.

An *electronic gage* gives a reading in proportion to the amount a stylus is displaced. It may also actuate switches electronically to control various functions. An example of an electronic gage and diagrams of the most common kinds of gage heads are shown in Figure 16-19. The *variable inductance* or *inductance-bridge transducer* has an alternating current fed into two coils connected into a bridge circuit. The reactance of each coil is changed as the position of the magnetic core is changed. This changes the output of the bridge circuit. The *variable transformer* or *linear variable displacement transformer* (LVDT)

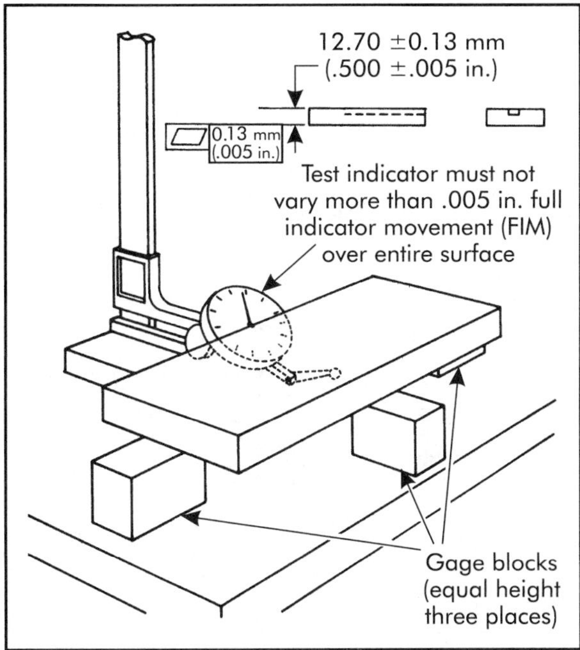

Figure 16-18. An application of dial indicators for inspecting flatness by placing the workpiece on gage blocks and checking full indicator movement (FIM).

transducer has two opposed coils into which currents are induced from a primary coil. The net output depends on the displacement of the magnetic core. The deflection of a *strain gage transducer* is sensed by the changes in length and resistance of strain gages on its surface. This is also a means for measuring forces. Displacement of a variable capacitance head changes the air gap between plates of a condenser connected in a bridge circuit. In every case, an alternating current is fed into the gage as depicted in Figure 16-19(E). The output of the gage head circuit is amplified electronically and displayed on a dial or digital readout. In some cases, the information from the gage may be recorded on tape or stored in a computer.

Depending on capacity, range, resolution, quality, accessories, etc., electronic gages are priced from a little under $1,000 to many thousands of dollars. An electronic height gage like the one shown in Figure 16-19(E) with an amplifier and digital display lists for about $1,700. A digital-reading height gage like the instrument shown in Figure 16-19(F) can be used for transferring height settings in increments of 0.0025 mm (.00010 in.) with an accuracy of 0.00127 mm (.000050 in.). The instrument costs about $2,000.

Electronic gages have several advantages: they are very sensitive (they commonly read to a few micrometers); output can be amplified as much as desired; a high-quality gage is quite stable; and they can be used as an absolute measuring device for thin pieces up to the range of the instrument. The amount of amplification can be switched easily, and three or four ranges are common for one instrument. Two or more heads may be connected to one amplifier to obtain sums or differences of dimensions, as for checking thickness, parallelism, etc.

Air gages. An *air gage* is a means of measuring, comparing, or checking dimensions by sensing the flow of air through the space between a gage head and workpiece surface. The gage head is applied to each workpiece in the same way, and the clearance between the two varies with the size of the piece. The amount the airflow is restricted depends upon the clearance. There are four basic types of air gage sensors shown in Figure 16-20. All have a controlled constant-pressure air supply.

The *back-pressure gage* responds to the increase in pressure when the airflow is reduced. It can magnify from 1,000:1 to over 5,000:1, depending on range, but is somewhat slow because of the reaction of air to changing pressure. The *differential gage* is more sensitive. Air passes through this gage in one line to the gage head and in a parallel line to the atmosphere through a setting valve. The pressure between the two lines is measured. There is no time lag in the *flow gage*, where the rate of airflow raises an indicator in a tapered tube. The dimension is read from the position of the indicating float. This gage is simple and does not have a mechanism to wear, is free from hysteresis, and can amplify to over 500,000:1 without accessories. The *venturi gage* measures the drop in pressure of the air flowing through a venturi tube. It combines the elements of the back-pressure and flow gages and is fast, but sacrifices simplicity.

A few of the many kinds of gage heads and applications are also shown in Figure 16-20. An air gage is basically a comparator and must be set to a master for dimension or to two masters

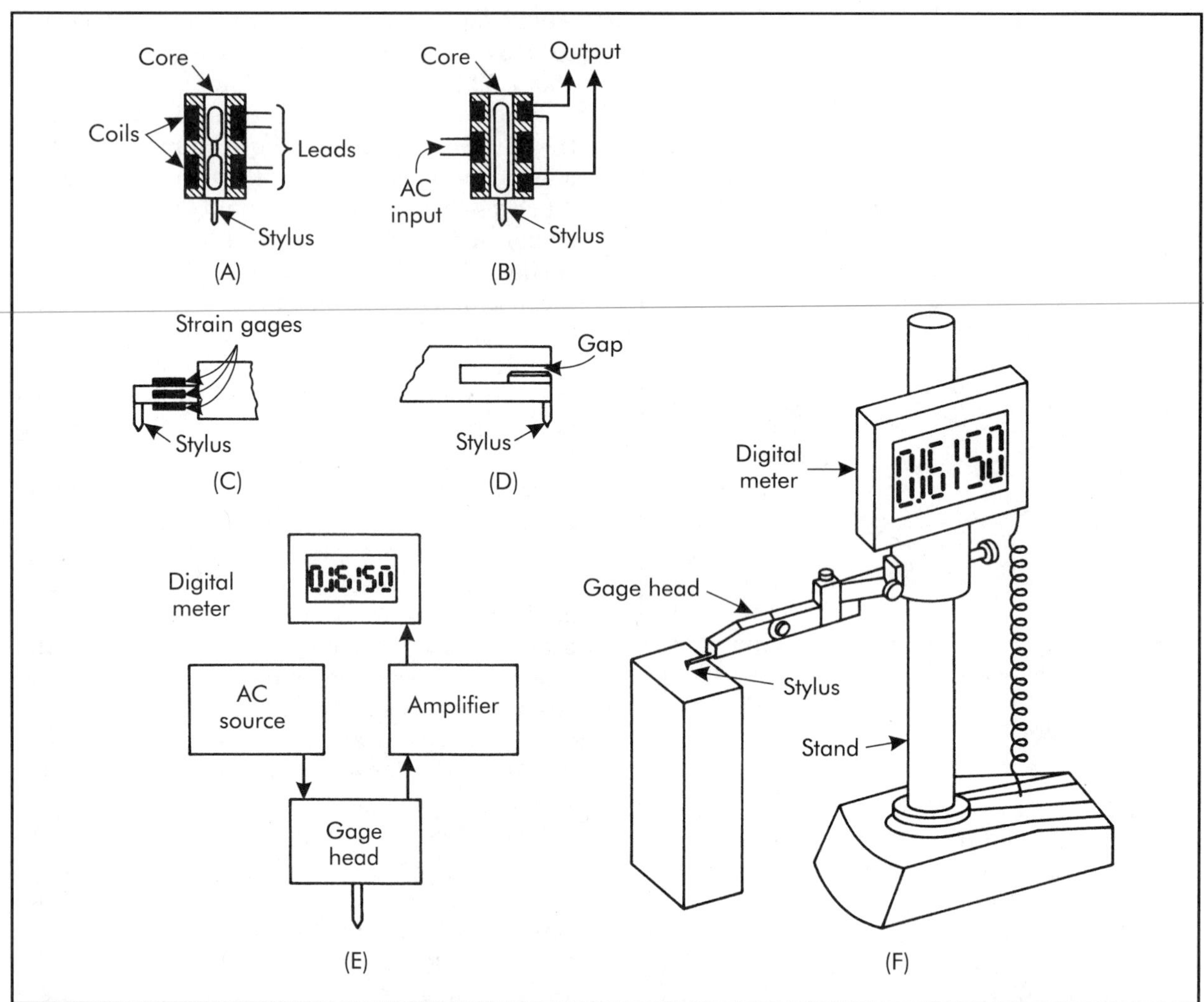

Figure 16-19. Elements of electronic gages. Types of gage heads: (A) variable inductance; (B) variable transformer; (C) strain gage; (D) variable capacitance; (E) block diagram of typical electronic gage circuit; (F) one model of electronic gage.

for limits. The common single gage head is the plug. Practically all inside and outside linear and geometric dimensions can be checked by air gaging. Air *match gaging,* depicted in Figure 16-20(I), measures the clearance between two mating parts. This provides a means of controlling an operation to machine one part to a specified fit with the other. A *multidimension gage* has a set of cartridge or contact gage heads (Figure 16-20[H]) to check several dimensions on a part at the same time. The basic gage sensor can be used for a large variety of jobs, but a different gage head and setting master are needed for almost every job and size.

A major advantage of an air gage is that the gage head does not have to tightly fit the part. A clearance of up to 0.08 mm (.003 in.) between the gage head and workpiece is permissible, even more in some cases. Thus no pressure is needed between the two to cause wear, and the gage head may have a large allowance for any wear that does occur. The flowing air helps keep surfaces clean. The lack of contact makes air gaging particularly suitable for checking against highly finished and soft surfaces. Because of its loose fit, an air gage is easy and quick to use. An inexperienced worker can measure the diameter of a hole to 25 nm (.000001 in.) in a few seconds with

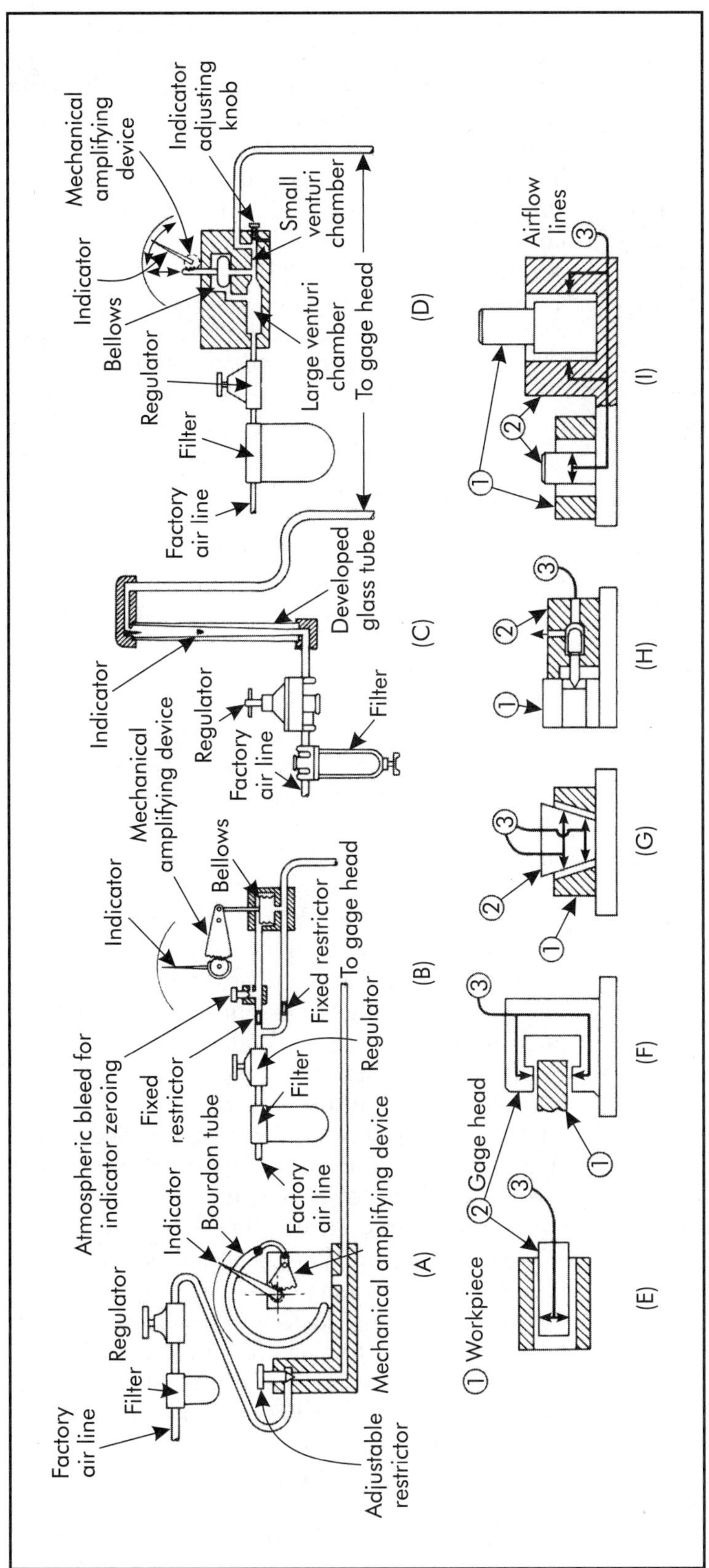

Figure 16-20. Diagrams of air gage principles. Air gage sensors: (A) back-pressure; (B) differential; (C) flow; (D) venturi. Gage heads: (E) plug; (F) thickness; (G) taper; (H) cartridge or contact; (I) matching.

an air gage; the same measurement to 25 μm (.001 in.) with a vernier caliper by a skilled inspector may take up to 1 min. The faster types of air gages are adequate for high-rate automatic gaging in production.

Optical comparators. Many industrial products and component parts are so small and of such complex configuration that they require magnification for accurate discernment. For this purpose, a number of measuring and gaging instruments using various optical systems, such as the toolmakers' microscope, the binocular microscope, and the optical projecting comparator, find wide application for the inspection of small parts and tools.

The *optical projecting comparator* projects a magnified image of the object being measured onto a screen. A workpiece is staged on a table to cast a shadow in a beam of light in *diascopic projection,* as shown in Figure 16-21. The outline of the part is magnified and displayed on a screen. In *episcopic projection,* the light rays are directed against the side of the object and then reflected back through the projection lens.

Optical projection provides a means to check complex parts quickly to small tolerances. Commonly, a translucent drawing is placed over the screen with lines drawn to scale for the contour of the part, the limits of the outline, or critical features such as angles. For instance, the outline of a part can be compared with a drawing on the screen and deviations in the whole contour can be quickly seen. A fixture or stage may be supplied for a part to mount all pieces in the same way in rapid succession. The table can be adjusted in coordinate directions by micrometer screws or servomotors to 2 μm (.0001 in.). Table positions can be determined from the micrometer readings or from digital readout devices. Thus, a part can be displaced to measure precisely how far a line is from a specified position. In addition, the screen can be rotated to a vernier scale to measure angular deviations in minutes or small fractions of a degree. Magnifications for commercial comparators range from 5× to as much as 500×. For example, at 250× magnification, 0.0020 mm (.00008 in.) on a part becomes 0.500 mm (.0197 in.) on the screen, which is readily discernible.

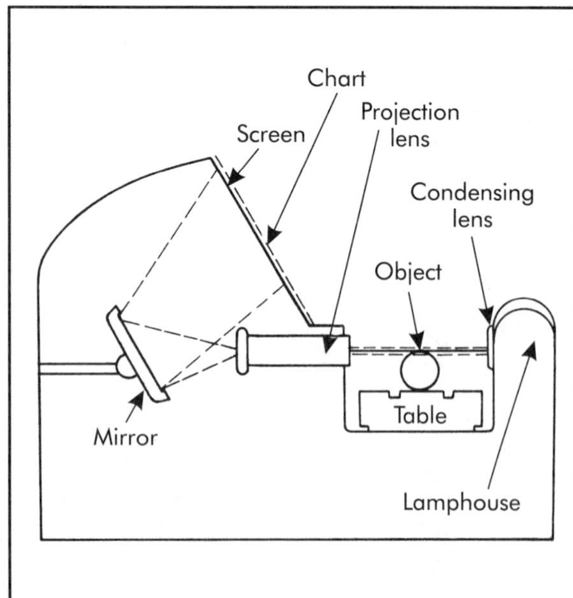

Figure 16-21. Optical comparator system.

The horizontal optical comparator shown in Figure 16-22 is table mounted and has a 356-mm (14-in.) diameter viewing screen. The designation "horizontal" means that the lens system is mounted horizontally as illustrated in Figure 16-21. Comparators are also commercially available with a vertical lens configuration to facilitate the staging of thin parts.

One of the features of the comparator shown in Figure 16-22 is a computerized digital readout (DRO) located on the top of the machine. The DRO has a two-axis digital display for establishing measurements in the *X-Y* plane. In addition, a 12-character, alphanumeric readout displays help messages, setup options, and the results of calculations. A fiber-optic edge-sensing device is also shown extending down the upper left portion of the screen. This device permits the digital readout to precisely indicate the edges of a part. A 16-key external keypad mounted on the lower base provides the option of using the dedicated keys as they are identified or redefining any or all of those keys to execute any of 20 different programs containing up to 100 keystrokes each. The keypad includes a joystick capable of *X*-, *Y*-, and *Z*-axis control.

Another feature of the comparator shown in Figure 16-22 is an electric screen protractor that reads angles directly to either a minute or 0.01°.

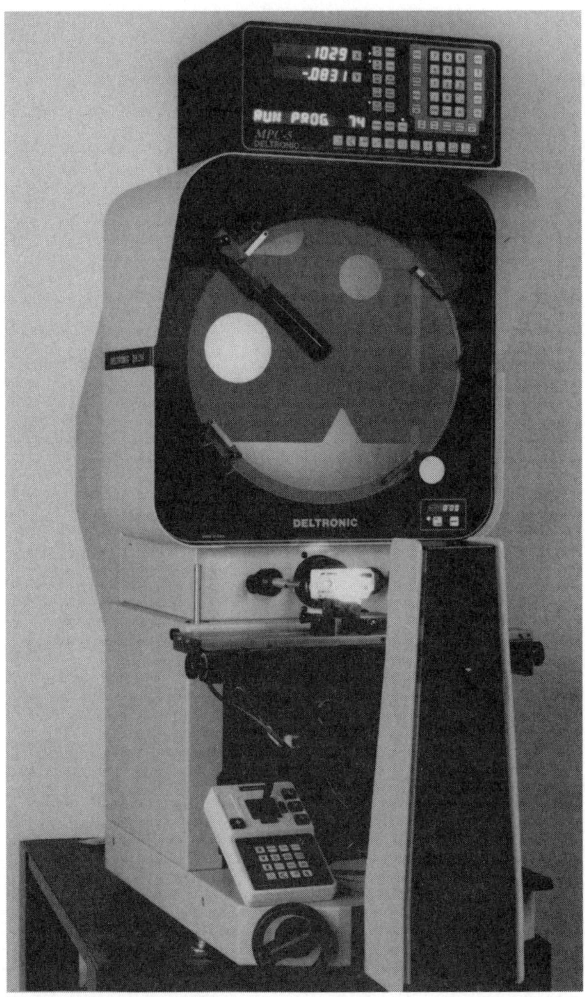

Figure 16-22. Horizontal optical comparator with a 356-mm (14-in.) viewing screen, digital readout, and edge-sensing device. (Courtesy Deltronic Corporation)

The angular setting of the protractor is displayed on an LED readout at the bottom-right of the screen. The machine has built-in provisions for either diascopic projection (contour illumination) or episcopic projection (surface illumination) via a high-intensity, tungsten-halogen light source. Lens changing is facilitated by the use of quick-change, bayonet-type lens holders. Seven different lens magnifications are available, ranging from 5× to 100×, all with an optical focusing range of 76 mm (3 in.). This comparator costs about $9,700, which includes the edge-sensing device, the computerized digital readout, and one lens. Additional lenses cost about $550 each.

16.3 COORDINATE MEASURING MACHINE (CMM)

The *coordinate measuring machine* (CMM) is a flexible measuring device capable of providing highly accurate dimensional position information along three mutually perpendicular axes. This instrument is widely used in manufacturing industries for the post-process inspection of a large variety of products and their components. It is also very effectively used to check dimensions on a variety of process tooling, including mold cavities, die assemblies, assembly fixtures, and other workholding or tool positioning devices.

16.3.1 Moving Bridge CMM

The two most common structural configurations for CMMs are the *moving bridge* and the *cantilever* type. The basic elements and configuration of a typical moving-bridge type coordinate measuring machine are shown in Figure 16-23. The base or worktable of most CMMs is constructed of granite or some other ceramic material to provide a stable work locating surface and an integral guideway foundation for the superstructure. As indicated in Figure 16-23, the two vertical columns slide along precision guideways

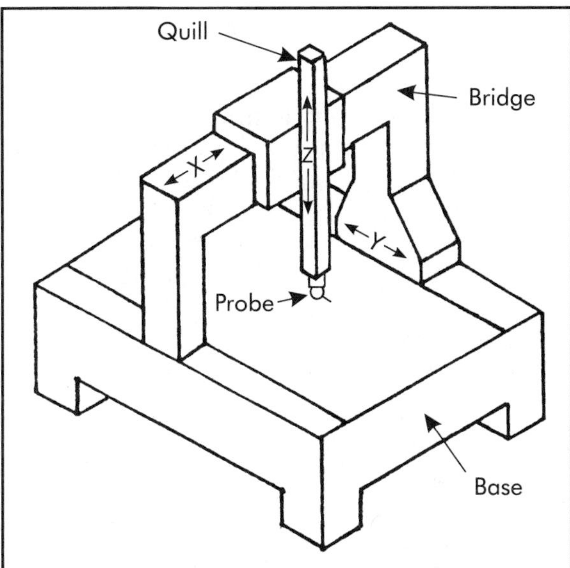

Figure 16-23. Typical moving-bridge coordinate measuring machine configuration.

on the base to provide Y-axis movement. A traveling block on the bridge gives X-axis movement to the quill, and the quill travels vertically for a Z-axis coordinate.

The moving elements along the axes are supported by air bearings to minimize sliding friction and compensate for any surface imperfections on the guideways. Movement along the axes can be accomplished manually on some machines by light hand pressure or rotation of a handwheel. Movement on more expensive machines is accomplished by axis drive motors, sometimes with joystick control. Direct computer controlled (DCC) CMMs are equipped with axis drive motors, which are program controlled to automatically move the sensor element (probe) through a sequence of positions.

To establish a reference point for coordinate measurement, the CMM and probe being used must be *datumed.* In the datuming process, the probe, or set of probes, is brought into contact with a calibrated sphere located on the worktable. The center of the sphere is then established as the origin of the X-Y-Z axes coordinate system.

The coordinate measuring machine shown in Figure 16-24 has a measuring envelope of about 0.5 × 0.5 × 0.4 m (18 × 20 × 16 in.) and is equipped with a disengagable drive that enables the operator to toggle between manual and direct computer control (DCC). It has a granite worktable and the X-beam and Y-beam are made of a extruded aluminum to provide the rigidity and stability needed for accurate measuring. The measurement system has a readout resolution of 1 μm (.00004 in.) and the DCC system can be programmed to accomplish 45–60 measurement points/min. A CMM of this size costs about $58,000 with computer support, joystick control, and an electronic touch probe. It can also be equipped with a video camera for noncontact measurement.

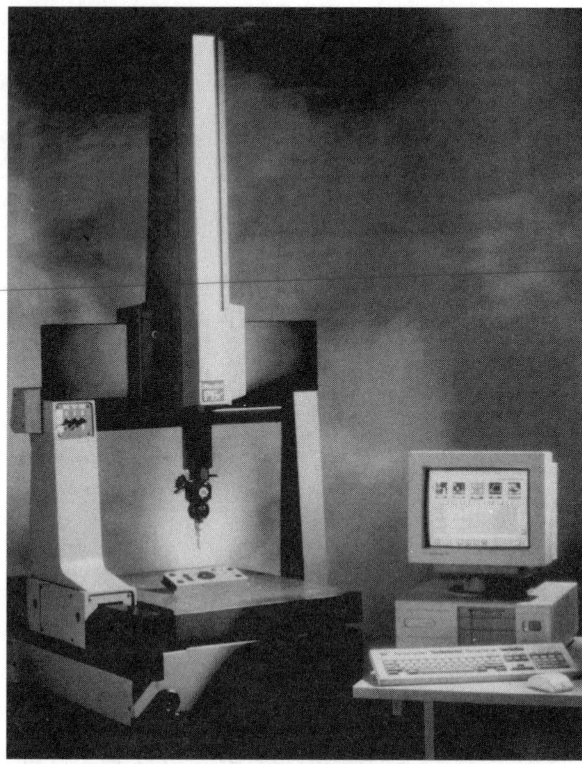

Figure 16-24. Coordinate measuring machine. (Courtesy Brown & Sharpe Manufacturing Company)

16.3.2 Contacting Probes

CMM measurements are taken by moving the small stylus (probe) attached to the end of the quill until it makes contact with the surface to be measured. The position of the probe is then observed on the axes readouts. On early CMMs, a rigid (hard) probe was used as the contacting element. The hard probe can lead to a variety of measurement errors, depending on the contact pressure applied, deflection of the stylus shank, etc. These errors are minimized by the use of a pressure-sensitive device called a *touch trigger probe.*

The touch trigger probe permits hundreds of measurements to be made with repeatabilities in the 0.25–1.0 μm (.00001–.00004 in.) range. Basically, this type of probe operates as an extremely sensitive electrical switch that detects surface contact in three dimensions. The manual indexable touch-trigger probe shown in Figure 16-25 can be used to point and probe without redatuming for each position measured. A probe of this type for use on manually operated CMMs costs about $4,000. Motorized probe heads are available at a cost of from $16,000–$19,000 for use on DCC CMMs.

16.3.3 Noncontacting Sensors

Many industrial products and components that are not easily and suitably measured with

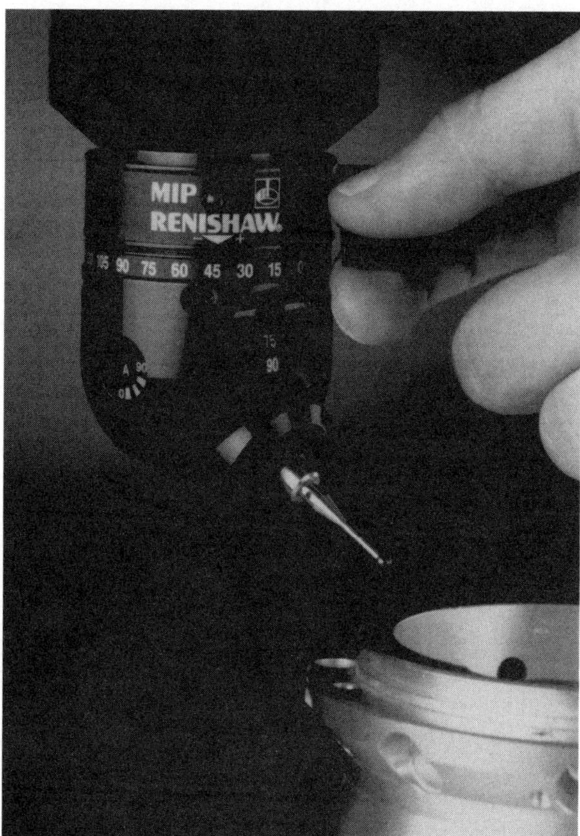

Figure 16-25. Manual indexable probe. (Courtesy Renishaw, Inc.)

surface contacting devices may require the use of noncontact sensors or probes on CMMs to obtain the necessary inspection information. This may include two-dimensional parts such as circuit boards and very thin stamped parts; extremely small or miniaturized microelectronic devices and medical instrument parts; and very delicate, thin-walled products made of plastic or other lightweight materials.

There are several noncontacting sensor systems available that have been adapted either individually or in combination with the coordinate measuring process. All of these involve optical measuring techniques and include microscopes, lasers, and video cameras. The automated, three-dimensional, video-based, coordinate measuring system appears to offer many advantages because of its ability to accomplish multiple measurement points within a single video frame. Thus, it is often possible to obtain data from several hundred coordinate points with a video measuring system in the same time it would take to obtain a single-point measurement with a conventional touch-probe system. A large number of points on a part can be imaged with CCD (charged couple device) video cameras, depending on the optical view and array or pixel size of the camera. For example, one commercially available system has a field of view range from 0.5–7.6 mm (.02–.30 in.) with a 756 × 581 pixel array and can image the field of view at 30 video frames/sec.

Video-based systems do not have the same throughput advantage for height (*Z* axis) measurements as they do for two-dimensional (*X-Y*) measurements because of the time involved for focusing. On some machines, this disadvantage is overcome by integrating laser technology with the video system.

The *multi-sensor coordinate measuring machine* (MSCMM) shown in Figure 16-26 incorporates three sensing technologies, optical, laser and touch probe, for highly accurate noncontact and/or contact inspection tasks. The machine uses two quills or measuring heads to accomplish high-speed data acquisition within a four-axis (*X*, *Y*, *Z*1, *Z*2) configuration, thus permitting noncontact or contact inspection of virtually any part in a single setup.

The sensing head on the left quill in Figure 16-26 contains an optical/laser sensor with a high-resolution CCD video and a coaxial laser. The video camera has advanced image processing capabilities via its own microprocessor, enabling subpixel resolutions of 0.05 μm (2 μin.). The coaxial laser shares the same optical path as the video image and assists in the focusing of the CCD camera. This eliminates focusing errors associated with optical systems and increases the accuracies of *Z* measurements. Single-point measurements can be obtained in less than 0.2 sec and high-speed laser scanning/digitizing can be accomplished at up to 5,000 points/sec.

The right-hand quill of Figure 16-26 contains the *Z*2-axis touch-probe sensor used to inspect features that are either out of sight of the optical/laser sensor or are better suited to be measured via the contact method.

The MSCMM of Figure 16-26 is being used to inspect a valve body with the *Z*1-axis optical/

Figure 16-26. A multi-sensor coordinate measuring machine with optical, laser, and touch probes for noncontact and contact measurements. (Courtesy WEGU, Inc.)

laser probe and a transmission case with the $Z2$-axis touch probe. The machine shown is a bench top type with a maximum measuring range of 900 mm (≈35 in.), 800 mm (≈32 in.), and 600 mm (≈24 in.) for the X, Y, and $Z1$ and $Z2$ axes respectively. Positioning of the moving elements is accomplished by backlash-free, recirculating-ball lead screws and computer-controlled DC motors. Position information is provided by high-precision glass scales. A multi-sensor coordinate measuring machine of this type with Windows®-based computer control costs about $275,000.

16.4 AUTOMATIC GAGING SYSTEMS

As industrial processes are automated, gaging must keep pace. Automated gaging is performed in two general ways. One is in-process or on-the-machine control by continuous gaging of the work. The second way is post-process or after-the-machine gaging control. Here the parts coming off the machine are passed through an automatic gage. A control unit responds to the gage to sort pieces by size and adjust or stop the machine if parts are found out of limits.

16.5 MEASURING WITH LIGHT RAYS

16.5.1 Interferometry

Light waves of any one kind are of invariable length and are the standards for ultimate measures of distance. Basically, all interferometers divide a light beam and send it along two or more paths. Then the beams are recombined and always show interference in some proportion to the differences between the lengths of the paths. One of the simplest illustrations of the phenomenon is the optical flat and a monochromatic light source of known wavelength.

The optical flat is a plane lens usually a clear fused quartz disk from about 51–254 mm (2–10 in.) in diameter and 13–25 mm (.5–1 in.) thick. The faces of a flat are accurately polished to nearly true planes; some have surfaces within 25 nm (.000001 in.) of true flatness.

Helium is commonly used in industry as a source of monochromatic or single-wavelength light because of its convenience. Although helium radiates a number of wavelengths of light, that portion which is emitted with a wavelength of 587 nm (.00002313 in.) is so much stronger than the rest that the other wavelengths are practically unnoticeable.

The principle of light-wave interference and the operation of the optical flat are illustrated in Figure 16-27(A), wherein an optical flat is shown resting at a slight angle on a workpiece surface. Energy in the form of light waves is transmitted from a monochromatic light source to the optical flat. When a ray of light reaches the bottom surface of the flat, it is divided into two rays. One ray is reflected from the bottom of the flat toward the eye of the observer, while the other continues on downward and is reflected and loses one-half wavelength on striking the top of the workpiece. If the rays are in phase when they re-form, their energies reinforce each other, and they appear bright. If they are out-of-phase, their energies cancel and they are dark.

This phenomenon produces a series of light and dark fringes or bands along the workpiece surface and the bottom of the flat, as illustrated in Figure 16-27(B). The distance between the workpiece and the bottom surface of the optical flat at any point determines which effect takes place. If the distance is equivalent to some whole number of half wavelengths of the monochromatic light, the reflected rays will be out-of-phase, thus producing dark bands. This condition exists at positions X and Z of Figure 16-27(A). If the distance is equivalent to some odd number of quarter wavelengths of the light, the reflected rays will be in phase with each other and produce light bands. The light bands would be centered between the dark bands. Thus a light band would appear at position Y in Figure 16-27(A).

Since each dark band indicates a change of one-half wavelength in distance separating the work surface and flat, measurements are made very simply by counting the number of these bands and multiplying that number by one-half the wavelength of the light source. This procedure may be illustrated by the use of Figure 16-27(B). There the diameter of a steel ball is compared with a gage block of known height. Assume a monochromatic light source with a wavelength of 0.5875 μm (23.13 μin.). From the four interference bands on the surface of the gage block, it is obvious that the difference in elevations of positions A and B on the flat is equal to $(4 \times 0.5875)/2$ or 1.175 μm ($[4 \times 23.13]/2$ or 46.26 μin.). By simple proportion, the difference in elevations between points A and C is equal to $(1.175 \times 63.5)/12.7 = 5.875$ μm ($[46.26 \times 2.5]/.5 = 231.3$ μin.). Thus the diameter of the ball is $19.05 + 0.005875 = 19.055875$ mm ($.750 + .0002313 = .7502313$ in.).

Optical flats are often used to test the flatness of surfaces. The presence of interference bands between the flat and the surface being tested is an indication that the surface is not parallel with the surface of the flat.

The way dimensions are measured by interferometry can be explained by moving the optical flat of Figure 16-27(A) in a direction perpendicular to the face of the workpiece or mirror. It is assumed that the mirror is rigidly attached to a base, and the optical flat is firmly held on a true slide. As the optical flat moves, the distance between the flat and mirror changes along the line of traverse, and the fringes appear to glide across the face of the flat or mirror. The amount of movement is measured by counting the number of fringes and fraction of a fringe that pass a mark. It is difficult to precisely superimpose a real optical flat on a mirror or the end of a piece to establish the end points of a dimension to be measured. This difficulty is overcome in sophisticated instruments by placing the flat elsewhere and by optical means reflecting its image in the position relative to the mirror in Figure 16-27(A). This creates interference bands that appear to lie on the face of and move with the workpiece or mirror. The image of the optical flat can be merged into the planes of the workpiece surfaces to establish beginning and end points of dimensions.

A simple interferometer for measuring movements of a machine tool slide to nanometers (millionths of an inch) is depicted in Figure 16-27(C). A strong light beam from a laser (described in Chapters 13 and 26) is split by a half mirror. One component becomes the reference R and is reflected solely over the fixed machine base. The other part, M, travels to a reflector on the machine slide and is directed back to merge with ray R at the second beam splitter. Their resultant is split and directed to two photodetectors. The rays pass in- and out-of-phase as the slide moves. The undulations are converted to pulses by an electronic circuit; each pulse stands for a slide movement equal to one-half the wavelength of the laser light. The signal at one photodetector leads the other according to the direction of movement.

When measurements are made to nanometers (millionths of an inch) by an interferometer, they are meaningful only if all causes of error are closely controlled. Among these are temperature, humidity, air pressure, oil films, impurities, and gravity. Consequently, a real interferometer is necessarily a highly refined and complex instrument; only its elements have been described here.

16.5.2 Optical Tooling

Telescopes and accessories to establish precisely straight, parallel, perpendicular, or angled lines are called *optical tooling*. Two of many applications are shown in Figure 16-28(A) and (B).

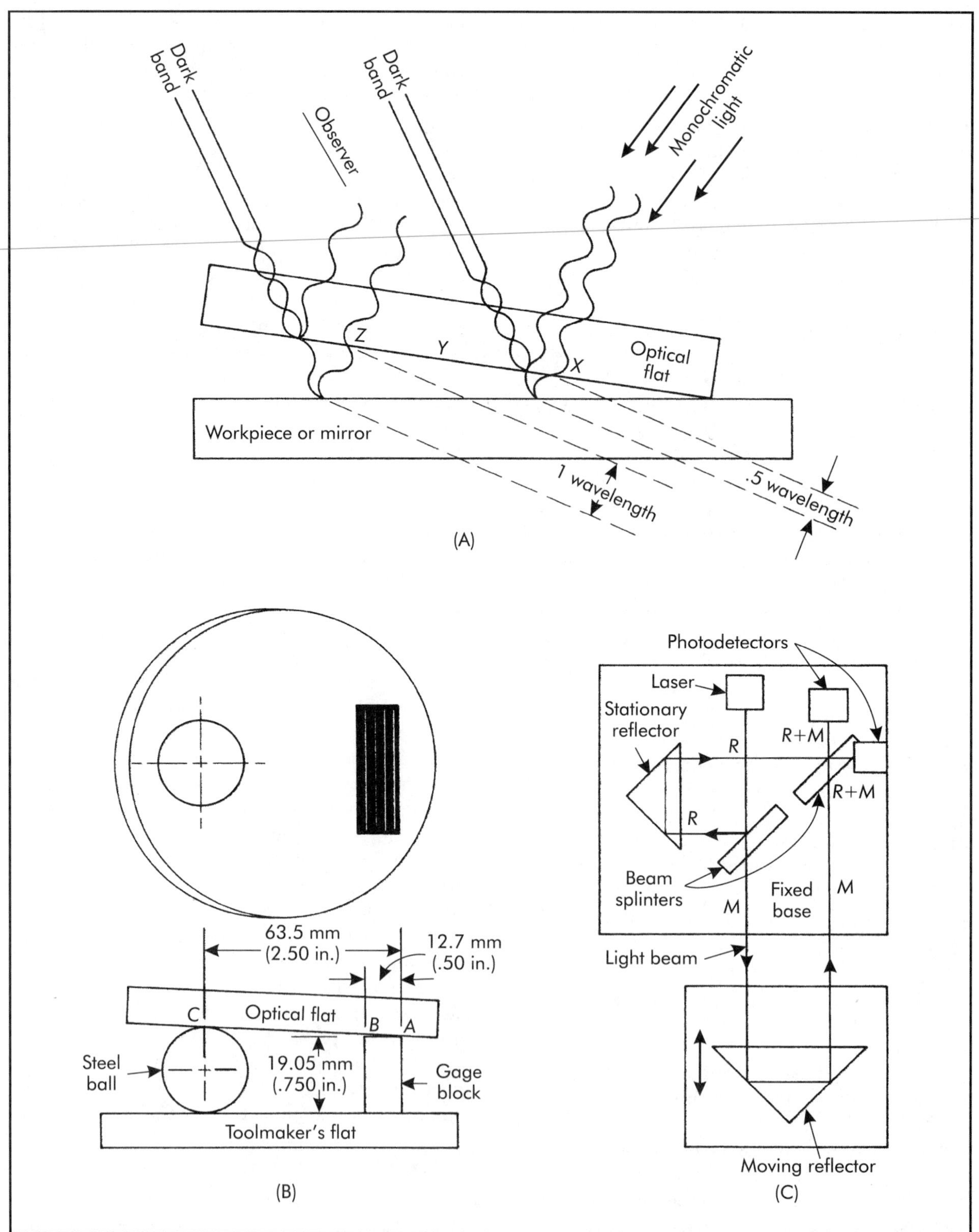

Figure 16-27. (A) Light-wave interference with an optical flat; (B) application of an optical flat; (C) diagram of an interferometer.

One is to check the straightness and truth of the ways of a machine tool bed at various places along the length. The other is to establish reference planes for measurements on a major aircraft or missile component. Such methods are especially necessary for large structures. Accuracy of 1 part in 200,000 is regularly realized; this means that a point at a distance 2.5 m (100 in.) can be located within 13 μm (.0005 in.). Common optical tooling procedures are autocollimation, autoreflection, planizing, leveling, and plumbing.

Autocollimation is done with a telescope having an internal light that projects a beam through the cross hairs to a target mirror as indicated in Figure 16-28(A). If the mirror face is truly perpendicular to the line of sight, the cross-hair image will be reflected back on it-

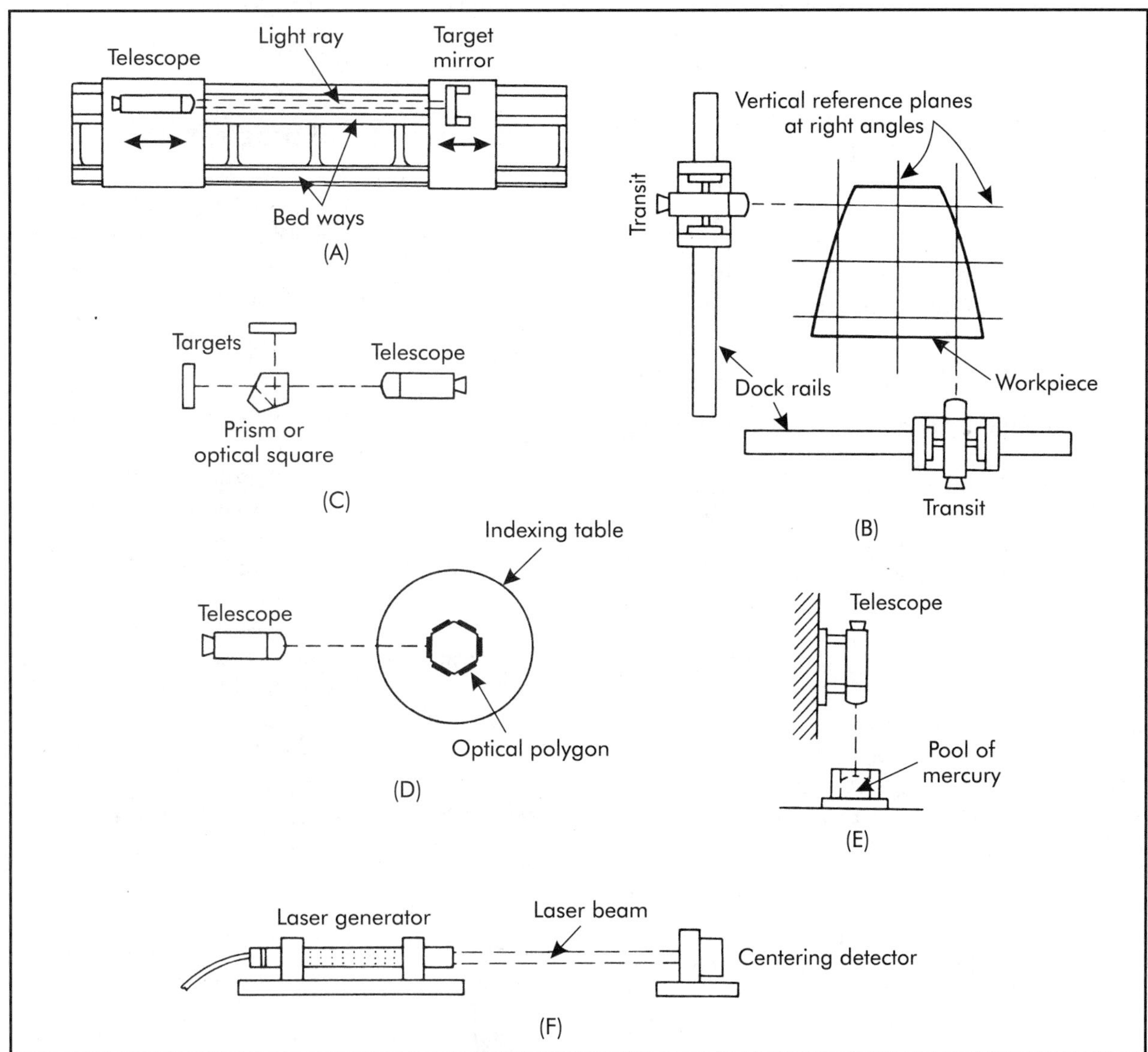

Figure 16-28. Optical tooling: (A) autocollimation, or autoreflection, for checking the straightness of the ways on a machine tool bed; (B) planizing to establish reference planes in a tooling dock; (C) establishing a line at right angles to a line of sight by means of an optical square; (D) collimating the faces of an optical polygon to check angles on an index table; (E) example of a plumbing operation to establish a vertical line; (F) use of a centering detector to establish alignment with the center of a laser beam.

self. The amount the reflected image deviates from the actual reticle image is an indication of the tilt in the target. A target may have a cross-line pattern for alignment with the line of sight. An autocollimated image is not clear for distances over 15.2 m (50 ft) and then a somewhat less accurate method must be used. This is *autoreflection,* with an optical flat containing a cross-line pattern mounted on the end of the illuminated telescope and focused to twice the distance of the target mirror. Then if the mirror is perpendicular to the line of sight, the pattern of the flat is reflected in coincidence with the cross hairs in the telescope.

Planizing is comprised of fixing planes at 90° with other planes or with a line of sight. This may be done from accurately placed rails on which transits are mounted in a tooling dock as indicated in Figure 16-28(B). A transit is a telescope mounted to swing in a plane perpendicular to a horizontal axis. Square lines also may be established with an optical square or planizing prism mounted on or in front of a telescope as depicted in Figure 16-28(C). Angles may be precisely set by autocollimating on the precisely located faces of an optical polygon as in Figure 16-28(D).

Leveling establishes a horizontal line of sight or plane. This may be done with a telescope fitted with a precision spirit level to fix a horizontal line of sight. A transit or sight level so set may be swiveled around a vertical axis to generate a horizontal plane. *Plumbing,* shown in Figure 16-28(E), consists of autocollimating a telescope from the surface of a pool of mercury to establish a vertical axis.

An advanced step in optical tooling is the use of the intense light beam of a laser (described in Chapters 13 and 26). A centering detector, shown in Figure 16-28(F), has four photocells equally spaced (top and bottom and on each side) around a point. Their output is measured and becomes equalized when the device is centered with the beam. This provides a means to obtain alignment with a straight line. Squareness may be established by passing a laser beam through an optical square.

16.6 SURFACE QUALITY

The quality of surface finish is commonly specified along with linear and geometric dimensions. This is becoming more common as product demands increase because surface quality often determines how well a part performs. Heat-exchanger tubes transfer heat better when their surfaces are slightly rough rather than highly finished. Brake drums and clutch plates work best with some degree of surface roughness. On the other hand, bearing surfaces for high-speed engines wear-in excessively and fail sooner if not highly finished, but still need certain surface textures to hold lubricants. Thus there is a need to control all surface features, not just roughness alone.

16.6.1 Surface Characteristics

The American National Standards Institute (ANSI) has provided a set of standard terms and symbols to define such basic surface characteristics as profile, roughness, waviness, flaws, and lay. A *profile* is defined as the contour of any section through a surface. *Roughness* refers to relatively finely spaced surface irregularities such as might be produced by the action of a cutting tool or grinding wheel during a machining operation. *Waviness* consists of those surface irregularities that are of greater spacing than roughness. Waviness may be caused by vibrations, machine or work deflections, warping, etc. *Flaws* are surface irregularities or imperfections that occur at infrequent intervals and at random locations. Such imperfections as scratches, ridges, holes, cracks, pits, checks, etc., are included in this category. *Lay* is defined as the direction of the predominant surface pattern. These characteristics are illustrated in Figure 16-29.

16.6.2 Surface Quality Specifications

Standard symbols to specify surface quality are included in Figure 16-29(C). Roughness is most commonly specified and is expressed in units of micrometers (μm), nanometers (nm), or microinches (μin.). According to the American National Standard ANSI/ASME B46.1-1985, the standard measure of surface roughness adopted by the United States and approximately 25 other countries around the world is the arithmetic average roughness, R_a (formerly AA or CLA). R_a represents the arithmetic average deviation of the ordinates of profile height increments of

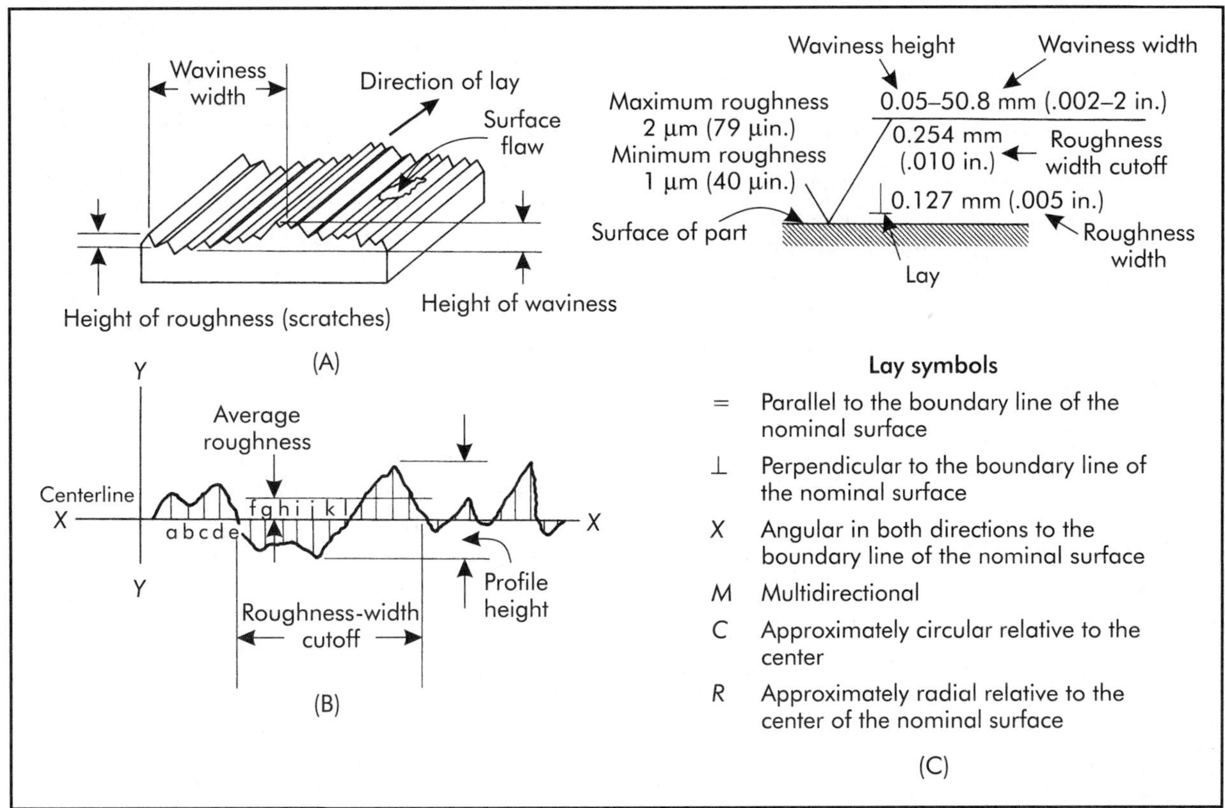

Figure 16-29. (A) Typical surface highly magnified; (B) profile of surface roughness; (C) surface quality specifications.

the surface from the centerline of that surface. Thus

$$R_a = \frac{1}{L} \int_{x=0}^{x=L} |y| \, dx \qquad (16\text{-}2)$$

where

R_a = arithmetic average deviation from the centerline
L = sampling length, mm (in.)
x = limits of integration on X axis
y = ordinate of the curve of the profile

An approximation of the average roughness may be obtained by

$$R_a = \frac{y_a + y_b + y_c + \cdots + y_n}{n} \qquad (16\text{-}3)$$

where

R_a = approximation of the average roughness
$y_a \ldots y_n$ = absolute values of the surface profile coordinates
n = number of sample measurements

The longest length along the centerline over which the measurements are made is the roughness-width cutoff or sampling length. In many cases, the maximum peak-to-valley height on a surface (R_y) is about four to five times greater than the average surface roughness as measured by R_a. This may present a problem for precision parts having small dimensional tolerances. For example, a flat surface on a part with an R_a of 0.4 µm (16 µin.) might very well have a peak-to-valley height (R_y) of 1.6 µm (64 µin.) or greater. If the tolerance on that dimension is 0.0025 mm (.0001 in.), then the 0.4 µm (16 µin.) surface finish represents nearly two-thirds of the permissible tolerance.

Prior to 1955, the root-mean-square (rms) was used to designate surface roughness. If measurements are made (plus or minus) from the centerline like the ones in Figure 16-29(B), the rms average is

$$[(\Sigma y_i^2)/n]^{1/2} \qquad (16\text{-}4)$$

where

y_i = sum of measurements taken from the centerline
n = number of measurements

Rms values are about 11% larger than R_a figures. Some surface roughness measuring instruments have a scale with numbers labeled "rms" and a scale calibrated in R_a values. On such instruments, the numbers called "rms" are root-mean-square values only when the profile is sinusoidal.

Waviness height alone may be specified, or it may be accompanied by a width specification. Thus, in Figure 16-29(C), the specification 0.05–50.8 mm (.002–2 in.) means that no waves over 0.05 mm (.002 in.) high are allowed in any 50.8 mm (2 in.) of length. If no width specification is given, it is usually implied that the waviness height specified must be held over the full length of the work. Other specifications in Figure 16-29(C) are less common.

16.6.3 Surface Finish and Cost

Each manufacturing process ordinarily produces surface finishes in a certain range, as indicated in Figure 16-30. Some may overlap with others, but each has its own surface pattern. Turning and shaping leave parallel feed lines, and face milling puts curved or crossed lines on a surface. The large number of small cutting edges acting at random in grinding give a directional pattern of small scratches that vary in length and often overlap. Honing and lapping may produce multidirectional or crisscross patterns.

Generally, to improve surface finish quality requirements implies incurring increased machining costs. A principal reason is that methods that give good surface finishes cannot rapidly remove stock. Therefore, two or more steps often are necessary to get a good finish. For example, rough and finish turning followed by rough and finish grinding operations are normal to obtain a 0.5 μm (20 μin.) R_a finish on a steel shaft where 8 mm (.313 in.) of stock must be removed. These steps are usually less costly than a number of turning operations, requiring special care, to achieve the same results. On the other hand, just two turning passes should be adequate for a 3 μm (125 μin.) R_a finish. It is important that no better finish than really needed be specified for a surface. Otherwise, cost will be excessive.

16.6.4 Measurement of Surface Finish

Waviness and roughness are measured separately. Waviness may be measured by sensitive dial indicators. A method of detecting gross waviness is to coat a surface with a high-gloss film, such as mineral oil, and then reflect in it a regular pattern, such as a wire grid. Waviness is revealed by irregularities or discontinuities in the reflected lines.

Many optical methods have been developed to evaluate surface roughness. Some are based on interferometry. One method of interference contrast makes different levels stand out from each other by lighting the surface with two out-of-phase rays. Another method projects a thin ribbon of light at 45° onto a surface. This appears in a microscope as a wavy line depicting the surface irregularities. For a method of replication, a plastic film is pressed against a surface to take its imprint. The film then may be plated with a thin silver deposit for microscopic examination or may be sectioned and magnified. These are laboratory methods and are only economical in manufacturing where other means are not feasible, such as on a surface inaccessible to a probe.

Except for extremely fine surface finishes that require laboratory measurement, most manufacturers measure surface texture at or near the workplace. A variety of instruments, called surface finish gages, are commercially available and are either hand-held or table mounted. These require only moderate skill, and roughness measurements are displayed on a dial, digital readout, chart, or a digital output for statistical process control (SPC), depending on the type of instrument used. Most of these instruments employ a diamond-tipped stylus that is moved across the surface of the part to sense the point-to-point roughness of that surface. As illustrated in Figure 16-31, there are two basic types of gages, the *skid-* or the *skidless* type. The skid-type shown in Figure 16-31(A) has a hinged probe that rides the work surface in close proximity to a fairly large skid that also contacts the work surface. The skid-type instruments usually have inductive transducers and are used predominately for

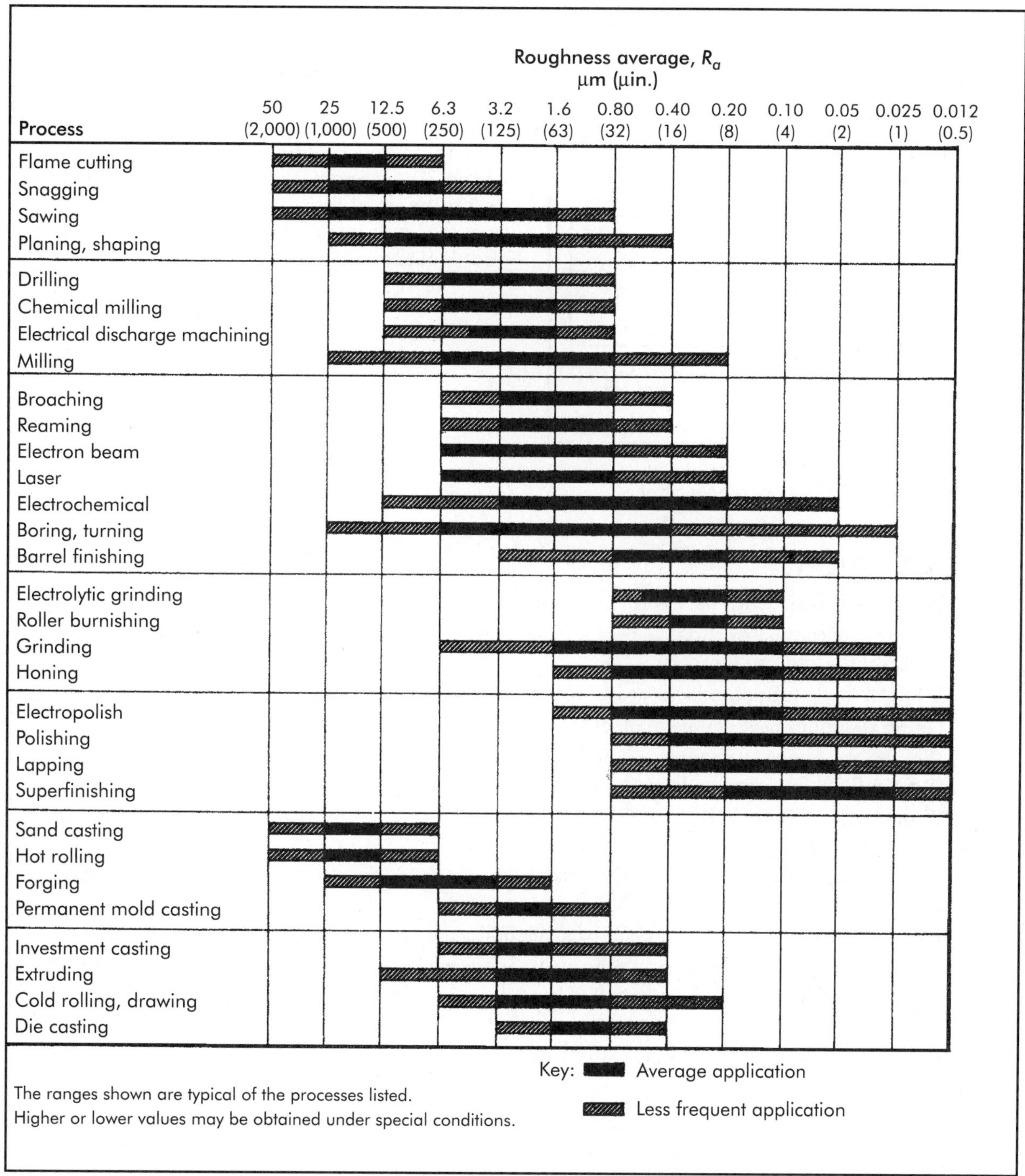

Figure 16-30. Typical ranges of surface finishes from common production processes. Higher or lower values may be obtained under various conditions (ANSI/ASME 1985).

averaging measurements of surface roughness, but not waviness. The skid filters out waviness. Most portable (hand-held) instruments are of the skid-type and they are reasonably accurate for roughness measurements in the range of 0.30–0.51 μm (12–20 μin.) R_a.

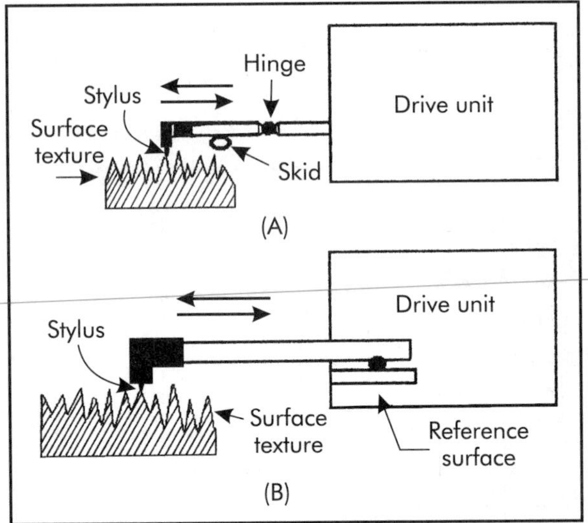

Figure 16-31. (A) Skid-type or average surface-finish measuring gage; (B) skidless or profiling gage.

The skidless type of instrument illustrated in Figure 16-31(B) has a built-in reference surface that permits the probe to sense both long- and short-wavelength variations in surface conditions. Thus, these can be used to measure both waviness and roughness, as well as surface inclination (straightness). These instruments are often referred to as "profiling" gages and they usually generate a profile chart on paper or on a computer screen.

16.7 QUESTIONS

1. What is the difference between "measuring" and "gaging?"
2. Why is it necessary at times to perform both in-process inspection and final inspection on a manufactured product?
3. What is the length standard for the international metric system, SI?
4. Name the four classes of gage blocks and state the required tolerances on each.
5. Why might noncontact types of measuring or gaging devices be preferred in some situations rather than contact types?
6. What is the "Rule of Ten" and why is it necessary to apply this rule in the design or selection of measuring instruments?
7. What is the difference between the "accuracy" and the "precision" of a measuring instrument?
8. Explain the principle of operation of a vernier scale on a measuring instrument.
9. Give one of the major sources of measurement error in the use of a micrometer caliper and indicate how this might be controlled.
10. What is a sine bar used for?
11. Name the three classes of gages and indicate how each class is used.
12. What is the difference between a "nonlimit" gage and a "limit" gage?
13. Indicate which of the three methods of applying tolerances to gages should be used if adequate customer satisfaction is the primary criterion.
14. Give three advantages that air gaging provides over other gaging processes.
15. Differentiate between diascopic and episcopic projection on optical comparators.
16. What is a "coordinate measuring machine" (CMM) and for what is it used?
17. What is accomplished by the "datuming" process on a CMM?
18. Explain the function of a touch-trigger probe in conjunction with the operation of a CMM.
19. How may CMMs be adapted for noncontact measurements?

20. Describe the principle and application of "interferometry."
21. What is "surface roughness" and how does it occur?
22. What is "waviness" and how does it occur?
23. What is the relationship between the maximum peak-to-valley height, R_y, on a surface and the average surface roughness, R_a?
24. Why is surface finish important to the functional performance of mating parts?

16.8 PROBLEMS

1. Make a sketch of the graduated scales of a vernier caliper with a reading of 78.56 mm (3.093 in.).
2. Make a sketch of the scales on a vernier micrometer caliper showing a reading of 12.735 mm (0.5014 in.).
3. Determine the angle α of the conical part in Figure 16-11 if the distance B were given as 82.55 mm (3.250 in.).
4. Specify the dimensions for the "go" and "not go" ends of a plug gage to be used in checking the holes in Part I of the mating parts shown in Figure 15-14. Use Method A of gage tolerancing as prescribed in Figure 16-15.
5. Specify the internal diameters for "go" and "not go" ring gages to be used in checking the posts on Part II shown in Figure 15-14. Use method A of gage tolerancing as prescribed in Figure 16-15.
6. Determine the specifications for the "go" and "not go" ends of a set of manufacturing and inspection plug gages to be used in checking a hole with a diameter specification of 40.00 +0.025/−0.000 mm (1.5748 +.0010/−.0000 in.)
7. An optical flat is used to check a workpiece under a monochromatic light having a wavelength of 0.5875 μm (23.13 μin.) in the manner shown in Figure 16-27. The distance between the center of one band to the center of the next is 5.08 mm (.200 in.). What is the angle between the surfaces of the optical flat and workpiece expressed in μm/m (μin./in.)?

16.9 REFERENCES

ANSI/ASME B46.1-1985. "Surface Texture." New York: American Society of Mechanical Engineers.

Ackroyd, P. 1988. "Surface Performance and Costs." *Quality*, August.

Aronson, A. B., ed. 1995. "The Growing Importance of Gaging." *Manufacturing Engineering*, April.

Auschuetz, B. 1994. "Touch Probes Touch Profits." *Modern Machine Shop*, April.

Harding, K. G. 1991. "Sensors for the '90s." *Manufacturing Engineering*, April.

McCusker, J. 1992. "You Won't Err with Air." *Modern Machine Shop*, August.

Noaker, P. M., ed. 1995. "The Well-honed Competitive Edge." *Manufacturing Engineering*, September.

——. 1991. "Scrutinizing Surface Measurement." *Manufacturing Engineering*, April.

Owen, J. V., ed. 1991. "CMMs on the Shop Floor." *Manufacturing Engineering*, April.

Rakowski, L. R. 1995. "CMMs Boost Fabricator's Part Inspection." *Metal Forming*, October.

Schuetz, G. 1994. "Start to Finish." *Modern Machine Shop*, May.

——. 1994. "Gaging and Mastering Uncertainty." *Modern Machine Shop*, June.

Society of Manufacturing Engineers (SME). 1999. *Measurement & Gaging* video. *Fundamental Manufacturing Processes* series. Dearborn, MI: SME.

Solomon, S. A. 1994. "Stamping Out Rejects with Automated Vision." *Metal Forming*, July.

Wendt, K. 1994. "When to Get in Touch with Noncontact Tool Measurement." *Modern Machine Shop*, April.

How Metals are Machined

17

High-speed steel cutting tools, 1898
—Frederick W. Taylor and Manuel White, Midvale Steel Co., Philadelphia, PA

17.1 IMPORTANCE OF METAL MACHINING

A large proportion of industrial products are fabricated from metallic materials and machining is a common method of converting those materials into the required shape and size. In some cases, a finished part is produced from a solid piece of metal by a sequence of machining operations. In other cases, metal parts or products that have been previously formed by other metal casting or working processes are subjected to finish machining operations. These are often called *secondary operations*.

The purpose of metal machining is to finish surfaces more closely to specified dimensions than can be done by other methods. Parts formed roughly by other processes, like casting and forging, normally have some or all of their surfaces refined by machining. For example, most automotive engine blocks are cast and then their cylinders, faces, and bearing surfaces are machined to size. By various machining processes, metal surfaces can be refined to practically any discernible degree of accuracy, truth, or smoothness desired. The greater the degree of refinement, the more the cost.

Metal machining is a convenient way of making one or a few pieces of almost any shape from an available piece of raw material. Large amounts of material can be cut away when necessary. But metal machining is not limited to making parts in small quantities. It can readily be adapted to fast, automatic, and accurate production. Certain metal removal processes, such as grinding, are capable of finishing very hard substances.

17.2 BASIC PROCESSES

17.2.1 Surface Form Generation

In all metal machining operations an edged tool is driven through material to remove chips from the parent body and leave geometrically true surfaces. All else that occurs merely contributes to that action. The kind of surface produced by the operation depends on the shape of the tool and path it traverses through the material. If a workpiece is rotated about an axis and a tool is traversed in a definite path relative to the axis, a surface of revolution is generated. If the tool path is parallel to the axis, the surface is a cylinder as indicated in Figure 17-1(A). This is called *straight turning* or just *turning*. An inside cylindrical surface is generated in the same way by *boring*, as depicted in Figure 17-1(B). If the tool path is straight but not parallel to the workpiece axis, a conical surface is generated. This is called *taper turning*, as in Figure 17-1(C). Both outside and inside tapers can be generated. If the tool is directed in a curved path as shown in Figure 17-1(D), a profile of varying diameter is generated by *contour turning*. In the foregoing examples, the shape of the surface generated depends more upon the path than the form of the tool. A surface of revolution also may be machined by plunging a tool into a revolving workpiece as in *contour forming* illustrated in Figure 17-1(E). The profile cut in this way corresponds to the form of the cutting edge of the tool. Straight and tapered surfaces may be formed in a similar manner.

A plane surface on the end or shoulder of a workpiece may be generated by revolving the piece and feeding a tool at a right angle to the axis as shown in Figure 17-2(A). This is called *facing*. Planes may be generated by a series of straight cuts, without turning the workpieces, as illustrated in Figure 17-2(B). If the tool is reciprocated and the workpiece is moved a crosswise increment at each stroke, the operation is called *shaping*. *Planing* is done by reciprocating the workpiece and moving the tool a little for each stroke. Formed contours can be cut by these methods by varying the depth of cut or using a formed tool as indicated in Figure 17-2(C).

Surfaces may be machined by tools having a number of edges that can cut successively through the material. Drills for opening holes are of this type. A drill may turn and be fed into the workpiece, or the piece may revolve while the drill is fed into it. Boring is often done with tools having several edges. Drilling and boring are treated in Chapters 18 and 20. Plane and contour surfaces are machined by milling cutters. A milling cutter has a number of teeth on its periphery. The cutter revolves and moves over the workpiece as illustrated in Figure 21-1.

17.2.2 Machining Parameters

The act of metal machining is in some ways like cutting a slice of bread. The knife is moved rapidly back and forth and at each stroke penetrates the bread a certain amount. When metal is cut, the workpiece surface is driven with respect to the tool, or the tool with respect to the surface, at a relatively high rate of speed. This is called the *cutting speed* or *speed*. Mostly, the tool or workpiece revolves. Almost all such machine tools are calibrated in revolutions per minute (rpm). The cutting speed is related to the rpm and thus is conveniently expressed in m/min (ft/min) or in surface feet per minute (sfpm). Speed in m/min must be converted to m/sec for calculating other quantities in SI units, such as power. Cutting speed is commonly in the range of 30–305 m/min (100–1,000 ft/min). At the same time, the tool is advanced comparatively slowly in a direction generally perpendicular to the speed. This motion is called the *feed* and is defined as the distance the tool advances into or along the workpiece each time the tool point passes a certain position in its travel over the surface. Feed is expressed in mm/rev or in a fraction or thousandths of an inch per revolution for turning; per stroke for shaping; or per tooth for milling.

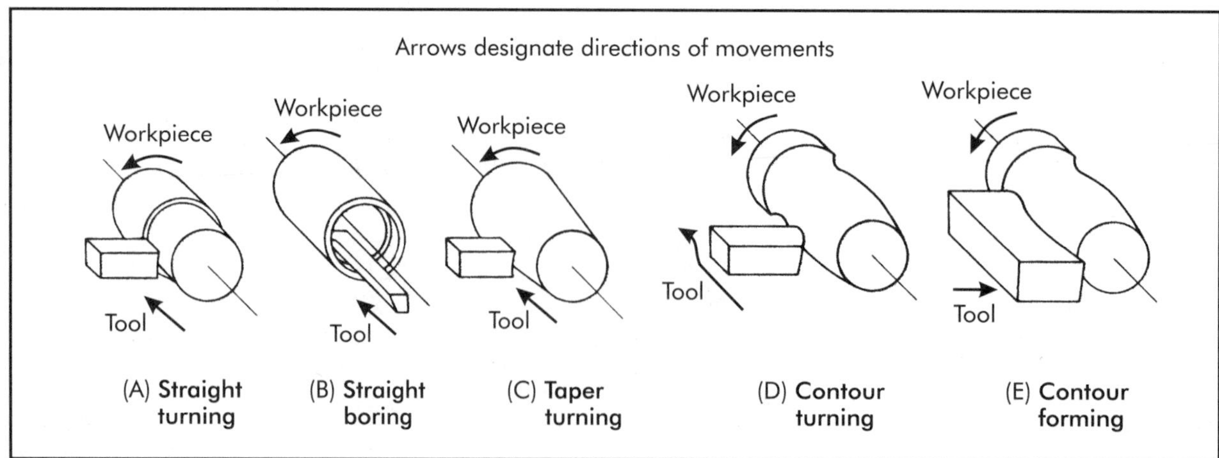

Figure 17-1. How surfaces of revolution are generated and formed.

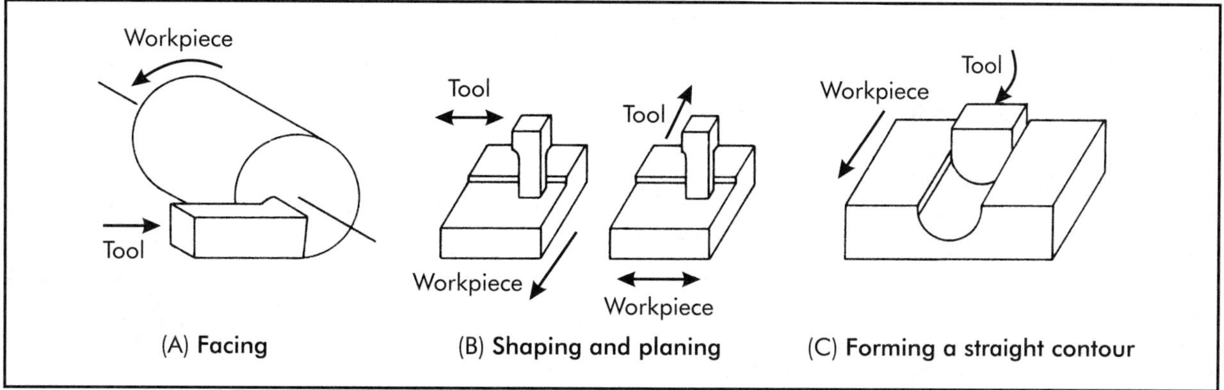

Figure 17-2. How plane surfaces are generated and formed.

The *depth of cut* is the normal distance from the surface being removed to the surface exposed by a cutting tool. It is measured in mm (in.).

In turning, the rate of travel of the surface of the workpiece is the speed; the distance the tool advances per revolution at a right angle to the speed is the feed, and half the amount the diameter is changed by the action is the depth of cut. In milling, the peripheral rate of travel of the cutter is the speed; the distance the workpiece advances between the time one tooth makes contact and the time the next tooth starts to cut is the basic feed; and the normal distance from the original surface to the surface left by the cutter is the depth of cut.

A workpiece or cutter must be revolved at the number of revolutions per minute (rpm) that will give the required surface speed. The necessary rate of rotation is found by

$$N = \frac{1000V_m}{\pi d_m} = \frac{V}{\pi d/12} = \frac{12V}{\pi d} \tag{17-1}$$

where

N = number of revolutions/min
V_m = surface speed, m/min
V = surface speed, ft/min
d_m = diameter, mm
d = diameter, in.

For practical purposes, π can be considered to be 3, and

$$N = \frac{333V_m}{d_m} = \frac{4V}{d} \tag{17-2}$$

The rate of metal removal during turning or boring is

$$Q_m = V_m \times f_m \times c_m \tag{17-3}$$

or

$$Q = 12 \times V \times f \times c$$

where

Q_m = rate of metal removal, cm^3/min
Q = rate of metal removal, $in.^3$/min
f_m = feed, mm/rev
f = width of uncut chip, in., equivalent to feed in in./rev
c_m = depth of cut, mm
c = depth of cut, in.

17.2.3 Machine Tool History

Metal may be cut by simple hand tools such as a hammer and chisel, file, saw, or stone. These are used today to remove metal in small amounts or as makeshifts. At one time, such tools were about the only means available for cutting metals. Obviously, the articles cut from metal solely by hand tools were few and quite expensive.

With the advent of the industrial revolution, the invention and development of devices like the steam engine and textile machinery called for faster and more accurate methods of cutting metals. Machines were devised to apply power and consistent precision to metal cutting. These superior tools are called *machine tools*, in contrast to hand tools, and the work done by them is called *machining*.

Machines for turning, drilling, boring, and planing came into being early. At first it was considered quite an accomplishment just to make a few articles of metal precisely; later the demand arose for varieties of products in quantities. Machining methods were applied to making firearms, clocks, the reaper, sewing machines, and a multitude of new inventions. Other machine tools, like the milling machine, turret lathe, and grinding machine, were developed to cut metal faster, reduce labor, and increase precision. To meet the demands of the present century for production in large quantities, highly specialized and automatic machine tools and machining centers have been developed.

17.3 MECHANICS OF METAL CUTTING

17.3.1 How a Tool Penetrates Metal

The most important part of a metal machining operation is the spot where the cutting tool meets the workpiece and pries away chips. An understanding of what happens in the cutting zone is necessary to appreciate what makes a good cutting tool and how it should be operated. The basic action is the same whether a single edge is being cut or several edges in a multiple tooth tool are cut at the same time or in succession.

When a tool cuts metal, it is driven by a force necessary to overcome friction and the forces that hold the metal together. The metal that the tool first meets is compressed and caused to flow up the face of the tool. The pressure against the face of the tool and the friction force opposing the metal flow build up to large amounts. Figure 17-3 is a diagram of the action in a single plane of a cutting tool forming a chip. At point *A* the material may be sheared by the advancing tool or torn by the bending of the chip to start a crack. The stress in the material ahead of the advancing tool reaches a maximum value in a band that becomes narrow at high cutting speeds and looks like a plane approximately perpendicular to the tool face. That plane is known as the *shear plane*, and one edge is depicted by the line *AB* of Figure 17-3. When the strength of the metal is exceeded, rupture or slippage occurs along the shear plane. With further movement, new material is compressed by the tool, and the cycle is repeated again and again. As it travels along, the cutting edge scrapes and helps clean up the surface.

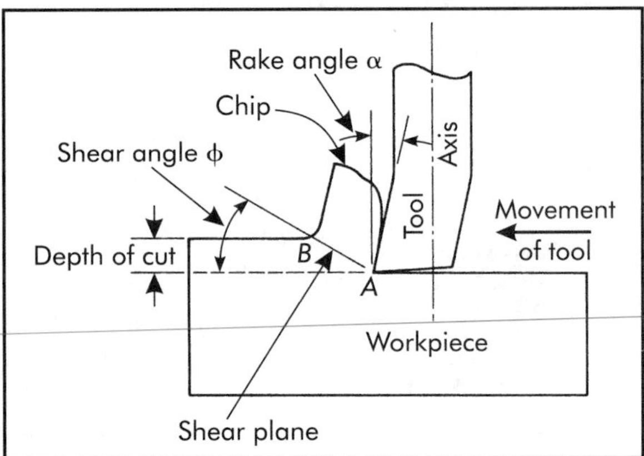

Figure 17-3. Cutting tool action.

17.3.2 Types of Chips

When a brittle material like cast iron or bronze is cut, it is broken along the shear plane. The same may happen if the material is ductile and the friction between the chip and tool is very high. The chips come off in small pieces or segments and are pushed away by the tool as illustrated in the highly magnified view of Figure 17-4. A chip formed in this way is called a *type I, discontinuous*, or *segmental chip*.

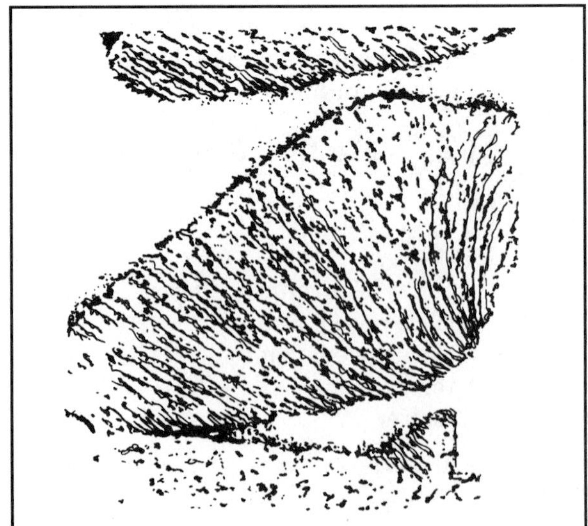

Figure 17-4. Drawing of a photomicrograph of a discontinuous chip.

A ductile material, cut optimally, is not broken up but comes off like a ribbon as shown in Figure 17-5(A). This is known as a *type II* or *continuous chip*. An evident line of demarcation separates the highly distorted crystals in the chip from the undistorted parent material. That line is the edge of the shear plane at one instant and corresponds to the line of *AB* of Figure 17-3. As the material slips along one plane, it is work hardened and resists further distortion. The stresses build up on the next plane to bring about slippage on new material, and so on.

When steel is cut, a continuous chip is usually formed, but the pressure against the tool is high, and the severe action of the chip quickly rubs the natural film from the tool face. The freshly cut chip and the newly exposed material on the face of the tool have an affinity for each other, and a layer of highly compressed material adheres to the face of the tool. Such formation is illustrated in Figure 17-6 and is called a *built-up edge*. The chip is a *type III* or *continuous chip with built-up edge*. As the cut progresses, the pile on the face of the tool becomes large and unstable. At frequent intervals, pieces topple from the pile and adhere to the work surface or pass off with the chip. The fragments of built-up edge are a main cause of roughness on a cut surface. The built-up edge pushes on ahead of the tool and to some extent protects the edge, changing the effective rake angle.

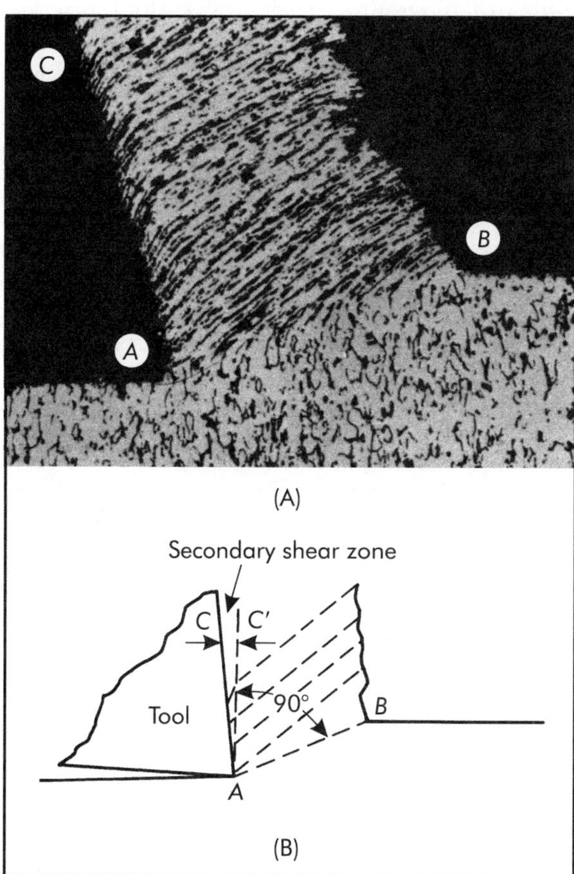

Figure 17-5. (A) Photomicrograph of partially formed continuous chip. AISI 1015 steel cut at 7.3 m/min (24 ft/min) with cutting fluid comprised of water + 1% rust inhibitor. (B) Diagram interpretation of (A).

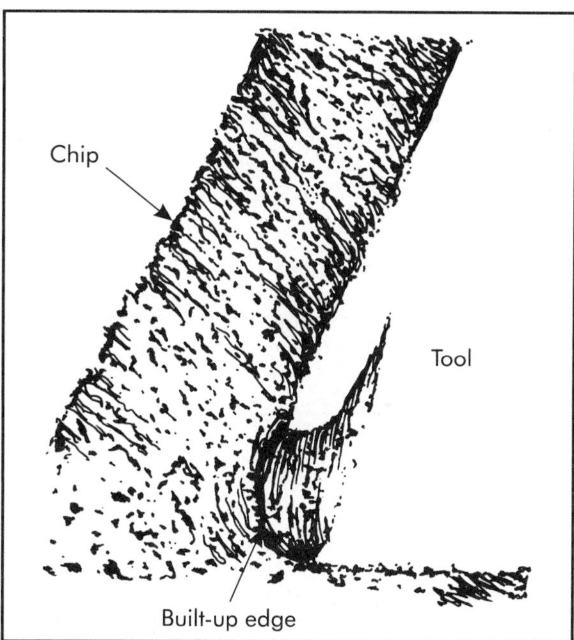

Figure 17-6. Drawing of a photomicrograph of a continuous chip and tool with a built-up edge.

17.3.3 Cutting Theory

Figure 17-7 depicts the formation of a type II chip with the single-cutting edge of the tool at right angles to the direction of movement, and the surface cut parallel to the original. This is called *orthogonal cutting*, and the principles it reveals are true for all forms of metal cutting. The tool exerts a force R on the chip with a normal component F_n and a friction component F_f that opposes the flow of the chip up the tool face. For equilibrium, the chip must be subjected to a substantially equal and opposite reaction, R', from the workpiece at the shear plane with a normal component F_N and a shearing force F_S along the shear plane. For convenience, the force R applied to the tool is resolved into a component F_C in the direction of tool movement and a normal component F_L. To illustrate the relationship, all the forces may be represented by the force R acting at the edge of the tool with the components in their respective positions as in Figure 17-8.

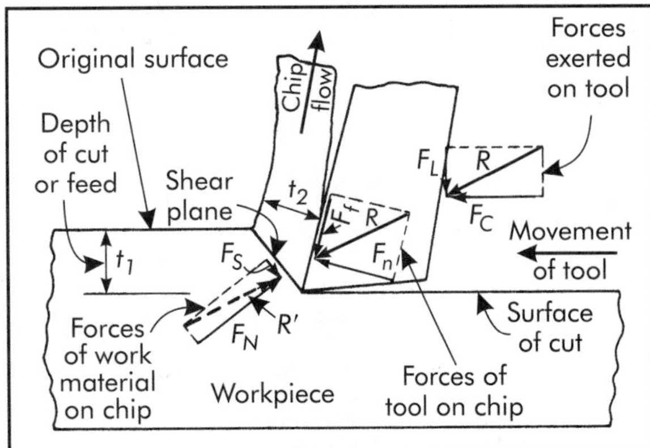

Figure 17-7. Forces exerted by a cutting tool.

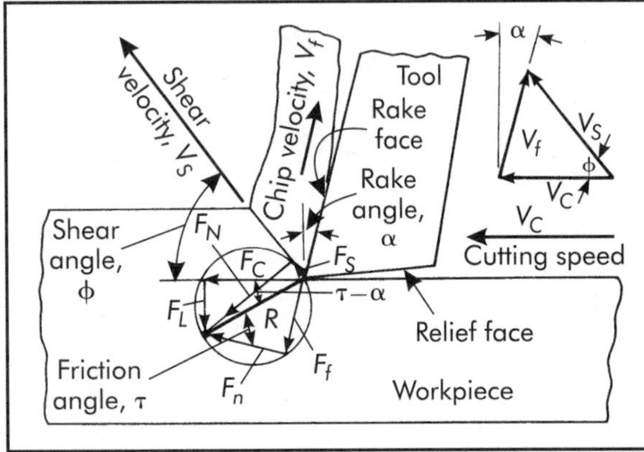

Figure 17-8. Force and velocity system of a cut.

The shear plane angle can be calculated from the change in size of the chip that is deformed along the shear plane. The precut chip thickness t_1 and rake angle α are known. The chip thickness t_2 is measured normal to the rake face of the tool along a line that makes an angle of $\phi - \alpha$ with the shear plane. The length of the shear plane is $t_2/\cos(\phi - \alpha)$, and $\sin \phi = (t_1/t_2) \cos (\phi - \alpha)$. This is reduced by trigonometry identity to

$$\tan \phi = \frac{r_t \cos \alpha}{1 - r_t \sin \alpha} \tag{17-4}$$

where

r_t = t_1/t_2 and is called the *chip thickness ratio*
α = rake angle, degrees
t_1 = precut chip thickness, mm (in.)
t_2 = chip thickness, mm (in.)

The strain that takes place in chip formation can be depicted ideally as a displacement of thin parallel plates as in Figure 17-9. Displacement occurs for one plate at a time sliding the distance *AC* along the shear plane. Shear strain is the displacement divided by the thickness of a plate. This can be expressed in terms of the angles of movement from the right triangles *ADB* and *CDB*, and the strain

$$\epsilon = \frac{\Delta s}{\Delta x} = \cot \phi + \tan (\phi - \alpha) \tag{17-5}$$

Secondary strain occurs from friction on the chip face sliding along the tool. The result is seen in the smooth face on one side of a type II chip. In contrast, the other side is rough and jagged as indicated in Figures 17-5 and 17-9.

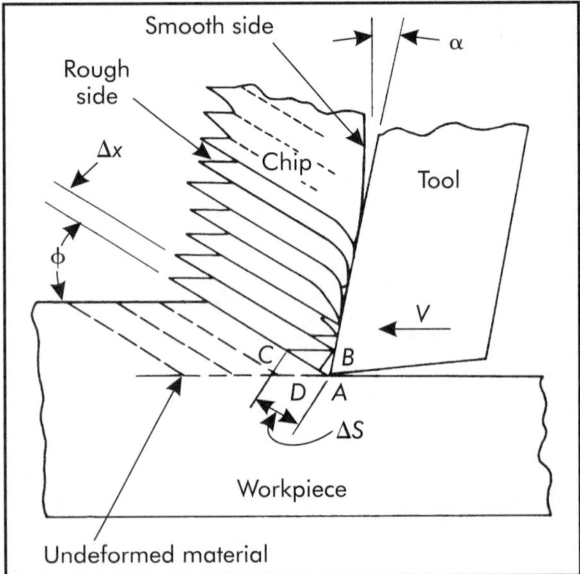

Figure 17-9. Simplified concept of chip strain and deformation.

All the components of force as arranged in Figure 17-8 are sides of right triangles with a common hypotenuse R, the diameter of the force circle, and can be readily related by simple trigonometry. The components of force, F_C and F_L, applied to the tool, can be measured by means of a dynamometer. Cutting speed is known. With α and ϕ already ascertained, all the other forces and velocities can be calculated. From the forces, velocities, and strains already identified, energy rates of shear and friction and their sum can be determined. As an additional technique, use is made of the fact that the dissimilar tool and workpiece material form a thermocouple to measure the temperature at the work-tool interface. These calculations and measurements have been performed by research investigators and are reported in detail in treatises on metal cutting. What is important here is to recognize that such calculations make it possible to express what really happens in a metal-cutting operation. This will be explained later in the discussion of how and why the power, forces, tool life, productivity, and surface finish vary as speed, feed, and other variables are changed. If it is known why certain results occur, then ways to optimize operations can be deduced.

The act of metal cutting is really quite complex and only partially understood. The direction of shear can be ascertained after a metal has been cut, but no one can predict without previous experience what the shear plane angle will be before the first trial. Merchant derived a formula for shear in the direction of the apparent maximum shearing stress and least energy. Lee and Schaefer made an analysis on the basis of the theory of ideal plasticity. Other investigators have proposed ingenious and intricate solutions. Most formulas are of the form $\phi = C_1 - C_2(\tau - \alpha)$ for the angles identified in Figure 17-8. Each theory derives certain values for the constants C_1 and C_2, but no one solution applies to all tool-work material combinations and other conditions, and no one has been able to explain fully why the shear angle is different for different materials. There may be no really unique solution because too many factors are involved: anisotropy, work hardening, variation and distribution of the coefficient of friction, thermal effects, etc. However, the fact remains that for a given work-tool combination, the shear plane angle and, thus, the shearing force, are related to the angles of friction and rake on the tool.

17.3.4 Oblique Cutting

Most cutting action takes place in three dimensions and along more than one edge. The simple two-dimensional picture already presented applies basically in normal planes but must be broadened for a full view. The next stage is *oblique cutting* with the cutting edge set at an angle θ from the Y direction in Figure 17-10. The workpiece is identified in the horizontal X-Y and vertical Z directions. The normal rake angle of the tool α_n is in a vertical plane normal to the cutting edge MN. A *velocity rake angle* α_v is in the X-Z plane (containing the cutting velocity V_C and points $ADEFO$). The chip goes off in a direction V_f in a plane $ABCO$, in which OG is perpendicular to V_C. The rake angle between OG and V_f is called the *effective rake* α_e. The chip flow in the direction V_f is at an angle β with the normal OH on the tool face.

The conditions in a normal plane through OZ and OH are like those in orthogonal cutting. The shear plane angle in this plane can be determined by Equation (17-4) with α_n substituted for α. Armarego and Brown have found that the size of the power or cutting force F_C in the direction of the cutting velocity V_C is determined by the normal rake angle. The angle of inclination, and thus the effective rake angle, have an influence upon the other components of force and the size and direction of the resultant cutting force R. The main benefit of oblique cutting is that it is a means of controlling chip flow. A curling chip takes the form of a helical spring. Stabler found that $\beta = \theta$ as a first approximation. It can be shown that under that condition

$$\sin \alpha_e = \sin^2 \theta + \cos^2 \theta \sin \alpha_n \qquad (17\text{-}6)$$

and the effective rake is always larger than the nominal rake angle of the tool.

17.4 METAL MACHINING CONDITIONS

At production rates, metal is cut under extreme conditions. The tool exerts a tremendous pressure upon the chip, normally of the order of around 1–5 GPa (145,038–725,189 psi). The

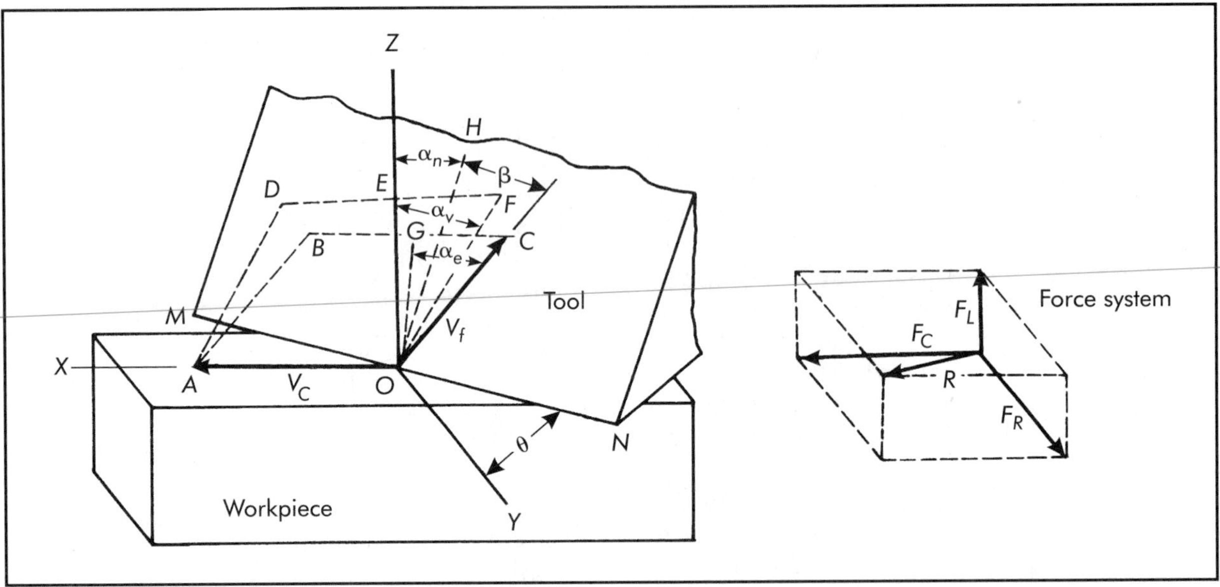

Figure 17-10. Oblique cutting.

temperature at the interface may reach 760° C (1,400° F) in cutting steel at 91–122 m/min (300–400 ft/min). The frictional resistance to the flow of the chip up the face of the tool is high. The coefficient of friction is usually over 0.5 and often over 1, as compared with a coefficient of less than 0.2 normally experienced in mechanical devices. Rubbing and temperatures near to those on the rake face occur on the relief faces behind the cutting edge. The high temperatures in metal cutting are sustained by the heat derived from the work done in sliding and shearing. About 80% or more of the heat passes off in the chips; most of the remainder goes into and largely contributes to deterioration of the tool. The forces, pressures, stresses, and temperatures determine how precisely the tool cuts and how long it lasts; these factors depend in turn upon the speed, feed, depth of cut, tool materials, and angles.

17.4.1 Effect of Cutting Speed

At low speeds, tool wear takes place largely by abrasion and interaction of surface asperities, edge deterioration under high pressures, and the formation and breakup of cold-pressure welds. Temperature becomes an important cause of tool wear at high cutting speeds. When a typical steel is cut at a speed under 30 m/min (100 ft/min), the temperature of the tool face is around 538° C (1,000° F). As the cutting speed is increased, more work is done in a unit of time in shearing the chip and sliding it over the face, and more heat is produced. The temperature increases at a declining rate; doubling the speed for cutting steel increases the temperature an average of 121° C (250° F).

Tool materials soften at higher temperatures, some sooner than others, but really serious breakdown comes from chemical activity. Above a certain *critical temperature* for each workpiece-tool combination, the two materials interfuse and weld together. A softer layer is formed on the tool surface as atoms intermingle. Chip particles welded to the surface of the tool are swept away and tear out minute chunks of adulterated tool material. This action diminishes somewhat as the work material becomes softer at still higher temperatures. The ploughing action of the workpiece and chip naturally proceeds at a higher rate at faster speeds and is even more harmful on the debased surface of the tool. Critical temperatures are reported around 621° C (1,150° F) for high-speed steel (HSS), 871° C (1,600° F) for cemented carbide, and over 1,204° C (2,200° F) for sintered oxide cutting AISI 4142 steel, as examples. Tool wear rises sharply above the critical temperature.

As speed increases and temperature rises, the chip material next to the tool face becomes weaker and shears more easily in friction. Both the coefficient of friction and the friction force decrease. Whereas the friction conditions for cutting steel at speeds below 30 m/min (100 ft/min) may be such as to cause the formation of a built-up edge, their change with increased speed leads to the diminution and eventual disappearance of the built-up edge. The surface finish produced is correspondingly improved.

The angle ø of the shear plane increases with speed. It should be recognized that the force (R) of the tool on the chip sets up stresses throughout the base of the chip and adjacent workpiece material. In a quasiplastic material, such as metal, these stresses occur in all directions with a definite intensity in each direction. The shear plane represents that direction along which the stresses have a magnitude sufficient to cause plastic deformations in the material. As the force applied to the material changes in size and direction, the pattern of the stresses changes. When the speed is increased, the friction force and friction angle (τ) decrease. This means the resultant R of Figure 17-8 must rotate clockwise. The pattern of stresses rotates correspondingly and the angle ø of the shear plane becomes larger.

As the shear angle ø increases, the distance along the shear plane from the tool point to the original surface decreases. This means the area of the shear plane is reduced. For a substantially constant shearing strength of the material, the shearing force F_S required to cause deformation of the material is smaller. As a result, the force R needed to cause formation of the chip can and does decrease, as do its components F_C and F_L. This phenomenon usually does not appear until the cutting speed rises to around 30 m/min (100 ft/min). Then the cutting force falls off rapidly at first but at a less rapid rate as the speed is increased further. Although decreases in cutting forces of as much as 50% have been found in some cases, the reduction is generally less than 15%.

The power required for a cut is

$$P = F_C V_C / 60 \tag{17-7}$$

where

P = power, kW

F_C = cutting force acting in the direction of the cutting velocity, kN

V_C = cutting velocity, m/min

In hp

$$P = F_C V_C / 33{,}000$$

where

F_C = cutting force acting in the direction of the cutting velocity, kN

V_C = cutting velocity, ft/min

If the speed is increased, the power does not increase as much because the force decreases at ordinary cutting rates. The unit power is

$$P_u = P/Q \tag{17-8}$$

where

P_u = unit power, kW/m^3/sec or KJ/m^3/sec (hp/in.^{3}/min)

P = power required for cutting, kW (hp)

Q = rate of metal removal, m^3/sec (in.^{3}/min)

Since Q increases proportionately with speed and P does not, P_u decreases slightly as speed is increased.

17.4.2 Effect of Feed

The depth of cut designated by t_1 in Figure 17-7 is equivalent to the feed in turning and most other operations, and will be referred to as "feed" in this discussion. As the feed is increased, the tool presses against more material, more work is done, and the temperature at the interface increases. However, an increase in feed causes a smaller rise in temperature than does a corresponding increase in speed. In turning steel, doubling the feed causes an average increase of around 32° C (90° F) in temperature. Thus, the rate of metal removal may be increased by increasing the feed with less temperature rise and consequently less wear on the tool than by increasing the speed.

As feed is increased with other factors held constant, temperature increases, and the coefficient of friction decreases. Friction force and related cutting force increase, but not in proportion to the increased rate of metal removal. In a typical case, the cutting force was found to increase from 667–1,779 N (150–400 lbf) when the feed was increased from 76–254 μm/rev (.003–.010 in./rev). If feed is increased and speed is kept constant, the increase in power is less and the reduction in unit power is more than if speed is increased with constant feed.

17.4.3 Effect of Rake Angle

The rake angle (α) shown in Figure 17-8 is considered positive to the right of the vertical and negative to the left. Any change in the angle changes the direction of the resultant force applied by the tool to the chip and changes the stress pattern. An increase in the rake angle increases the shear angle and decreases the forces required to deform the work material but, at the same time, decreases the mass and strength of the tool point.

17.4.4 Hot Machining

Cutting becomes easier when a workpiece is heated just in front of the tool to raise the temperature and decrease the material strength on the shear plane. The improved machinability is more than offset by the deleterious effects upon some materials, particularly refractory metals. The heating has been done by radio frequency (RF) induction currents and plasma arc or inert tungsten arc torches. Equipment is costly and not easy to apply to all operations. The process has received only limited acceptance.

17.4.5 Tool Wear

After a tool has been cutting for some time, depending upon conditions, its cutting edge breaks down. The chip rubbing over the rake face may wear a crater that grows until it undermines the edge. The relief face of Figure 17-8 rubbing over the cut surface and abraded by the chip particles may wear away just behind the edge. Too high a temperature may soften the tool, and the edge may be wiped away. The cutting edge may chip or crumble or the tool may break from high pressure, mechanical or thermal shock, or vibration. Often a combination of factors leads to tool failure. For instance, a crater on the rake face and wear land on the relief face may grow to the stage where the edge is weakened and breaks away. When the cutting edge breaks down, surface finish becomes poor and cutting forces and power consumption rise rapidly.

Deterioration of the faces of a tool may be retarded by use of a cutting fluid, a sharp and smooth tool, and the choice of work and tool materials with low natural coefficients of friction. Chipping or breaking can be avoided by selection of a tough tool material and a rigid tool, careful sharpening, uniform rather than intermittent cooling, and a rigid machine.

The length of time a tool cuts before it has to be resharpened is called the *tool life*. The amount of metal a tool has removed is sometimes used as a measure of tool life. A tool should be removed and reground before it wears to where an excessive and uneconomical amount of material must be ground from its faces to recondition it. A practical criterion for judging when a tool should be resharpened is the amount of wear on the relief face or *flank*. The criterion may be average or maximum wear because the amount of wear behind the cutting edge (the width of the land) is not always uniform across the face. There are a number of standards, but a common rule is to allow a cemented carbide cutting tool to have a wear land width of 0.4–0.8 mm (.016–.031 in.) and a high-speed steel (HSS) tool about 1.6 mm (.063 in.). After a tool has been reground a number of times, it must be reworked or replaced. Tool material and the labor for sharpening, resetting, readjusting, and reworking tools make up a definite part of the cost of any machining operation. Throwaway tools that are not reground may be run to imminent destruction.

17.4.6 Vibration and Chatter

Metal cutting is inherently cyclic. Cutting forces build up as the tool penetrates the material and it deflects, even if slightly. When rupture or shear occurs to form the chip and the forces momentarily drop, the tool springs back. Vibrations increase when the cutting forces get out of phase with the tool forces. Even in a seemingly continuous cut, fluctuations can occur because of the change in the cutting force with the relative velocity of the tool and the workpiece. The body of the tool is being driven through the material at a constant velocity. As it penetrates, the tip of the tool is sprung backwards by the increasing resistance it meets, and the relative velocity between the cutting edge and workpiece decreases accordingly. At the lower velocity the forces rise, and the tool is stressed still more. When the bending moment in the tool becomes large enough, it begins to move the tool back to its original shape. That increases the relative speed between the cutting edge and workpiece, and the cutting forces drop. The tool springs back until equilibrium is reached, the relative velocity decreases, the cutting forces start to rise, and another cycle begins. These actions are called *self-excited vibration*. In addition, *force vibrations* may be induced by nonbalanced rotating parts, by intermittent cutter impacts (as in milling), from other machines, etc. Also, cyclic variations in other factors, such as depth of cut, material properties, friction forces, and rubbing of the tool nose, affect vibration.

Strong vibration in metal cutting is called *chatter*. It may become very noisy and obnoxious, can damage tools and machines, and defaces work surfaces with patterns called *chatter marks*. The tendency to chatter depends upon the type of operation and the workpiece-tooling-machine system. The chatter proneness of a work material is measured by a factor known as its *cutting stiffness*. Some values are 1.1–1.8 GPa (159,000–258,000 psi) for B1112 steel, 3.1 GPa (453,000 psi) for 2340 steel, and 0.6–0.8 GPa (88,000–112,000 psi) for 2024-T4 aluminum, depending on operating conditions. Detailed tables are given in reference texts and handbooks. The higher the value of cutting stiffness in an operation, the more the tendency to chatter. To eliminate chatter, the dynamic stiffness of the machine tool must be at least twice as much as the cutting stiffness. Remedies to decrease chatter are to reduce speed, raise feed, and increase dampening capacity and rigidity of the tooling and machine.

17.5 METAL-CUTTING TOOLS

Metal-cutting tools may be classified as single-point and multiple-point tools. The latter are in effect, and act like, combinations of single-point tools. Thus, the factors that give single-point tools their characteristics are also basic to multiple-point tools. In this chapter, the elements of cutting tools will be discussed in detail in connection with single-point tools. Later, the features of multiple-point tools, such as drills and milling cutters, will be described along with the processes for which they are used.

The factors that determine how a cutting tool performs are: (1) the tool material, (2) the shape of the tool point, and (3) the form of the tool.

17.5.1 Tool Material

Hardness is the first requisite of a cutting-tool material because it must be able to penetrate other materials. Toughness, abrasion resistance, and thermal conductivity are other properties that often relate to the ability of a cutting-tool material to maintain a good cutting edge under various conditions. Toughness permits a tool to withstand the forces and shocks encountered in machining hard and tough materials. Abrasion resistance contributes to the maintenance of a reasonable tool life, and thermal conductivity prevents the tool from overheating during the cutting process. Cutting tools must work upon many kinds of metals and under a variety of conditions. No one cutting-tool material is best for all purposes.

Cutting-tool materials may be roughly classified into three categories: (1) the solid ferrous and nonferrous tool materials, (2) the hard insert materials and, (3) the superhard insert materials. The solid types include carbon tool steels, high-speed steels, and cast nonferrous alloys. These materials are cast to the rough shape of the type of cutting tool desired and then machined and ground to final shape and size. The hard materials include cemented carbides and sintered oxides or ceramics. They are usually designed and produced as inserts that are either brazed or clamped into softer metal tool bodies. The third type, the superhard materials, include cubic boron nitride (PCBN) and polycrystalline synthetic diamond (PCD). These are also made up as inserts or a layer of the superhard material is bonded to a carbide insert. Table 17-1 shows some typical values of hardness and thermal conductivity for the two hard materials and the two superhard materials.

Carbon Tool Steel

Carbon tool steel is the oldest kind of cutting material but is used little today. It contains from 0.90–1.2% carbon and sometimes appreciable

Table 17-1. Comparison of the physical properties of cutting tool materials

Property	Diamond	Cubic Boron Nitride	Aluminum Oxide	Tungsten Carbide
Knoop hardness, kg/mm^2 (lb/in.^{2})	8,000 (11,378,580)	4,500 (6,400,452)	2,500 (3,555,806)	2,100 (2,986,877)
Thermal conductivity cal/cm/sec (Btu in./sec/ft^2)	5.0 (4.032)	3.3 (2.661)	0.08 (.065)	0.08 (.065)

amounts of alloying elements. Its chief disadvantage is that it softens at temperatures above 204° C (400° F) and therefore is limited to very slow cutting speeds and light duty. At low temperatures carbon tool steel is hard, wear resistant, and as serviceable as more expensive materials for some applications. It is comparatively inexpensive and suitable for special tools, like odd sizes of drills, which are infrequently and lightly used and do not warrant much investment. It is easy to fabricate and simple to harden.

High-speed Steel

High-speed steel (HSS) is so named because it cuts at higher speeds than other steels. Even so, it is limited to speeds and conditions that do not give rise to temperatures above about 593° C (1,100° F). It is moderate in cost, workable, and tough, which makes it preferable in many cases to harder but more brittle and expensive materials. For instance, high-speed steel is able to withstand interrupted cuts better than harder materials. It can be more readily made into complex tools and is preferred for such tools as twist drills, reamers, taps, and form tools.

HSS is composed of alloyed elements that form hard, wear-resistant carbides (about 10–20% of the volume) dispersed in a matrix of steel. There are a number of types offering different degrees of hardness and toughness. One of the oldest is known as 18-4-1; it contains about 0.7% carbon, 18% tungsten, 4% chromium, and 1% vanadium. Molybdenum is used in some types to replace tungsten partially or wholly with comparable performance at a savings of about one-third in cost. A popular general-purpose "moly" HSS contains 8% molybdenum, 4% chromium, 1.5% tungsten, and 1% vanadium. Each type of high-speed steel is designated by a number with a letter prefix to designate the principal alloying element, for example, T1 for tungsten and M43 for molybdenum. Some alloys have 1% or more of carbon to increase hardness and some up to 12% cobalt for superior red hardness and abrasion resistance. Powder metal techniques are used to obtain HSS with exceptionally fine and well-dispersed carbides and higher alloy content. Some tools are given surface treatments (coatings) to improve performance. The more refined tools cost 50–100% more than conventional grades.

Cast Nonferrous Alloys

Cast nonferrous alloys contain no iron and cannot be softened by heat treatment for easier machining. They must be cast to shape and ground to size. One type of alloy consists mainly of chromium, tungsten, columbium, and carbon in a cobalt matrix. This material is not quite as hard or tough as HSS at room temperatures but retains hardness and resists wear up to red-heat temperatures. It serves best in the speed range of 30–61 m/min (100–200 ft/min) falling between those capable with HSS and cemented carbides. One application involves turning a large diameter with a cemented carbide tool while turning a smaller diameter on the same piece with a cast nonferrous tool so that each tool runs at its optimum speed.

Nonferrous alloys at the high end of the speed range are given outside coatings for high surface hardness. One contains 50% columbium, 30% titanium, and 20% tungsten. After the tool surface receives a nitriding treatment, it attains up to 25% more hardness than sintered oxide and 50% more than cemented carbide. Initial tool cost is high, but the alloy is quite competitive for heavy cuts at high speeds in common steels and it resists chip welding. It has not been found practical for some applications or for interrupted cuts.

Carbides

Cemented or *sintered carbides* are the most popular cutting tool materials for production operations. They are composed of hard metallic particles (about 75–95% by volume) bonded together in a metallic matrix. The materials are processed by powder metal techniques (Chapter 10). Sintering causes intergrowth of the hard particles to build a skeleton in the surrounding matrix. Carbides are, for the most part, not as strong as steel tools. Good wear resistance and cutting ability come from the high proportion of hard particles. The early commercial types of cemented carbide cutting tools, *tungsten carbide/cobalt* (WC/Co), were composed of fine particles of tungsten carbide bonded with metallic cobalt. This type of carbide was found to be very abrasive resistant and is still used today for many machining applications.

Over the years, a variety of carbide tool materials have been developed to satisfy different cutting conditions. *Tungsten-titanium carbide/cobalt* (WC/TiC/Co) grades were developed for use in cutting steel and other ferrous alloys. The TiC content of this grade helped to inhibit cratering of tool surfaces by the high cutting temperature and chemical breakdown encountered in machining high-strength steels. *Titanium carbide/molybdenum/nickel* (TiC/Mo/Ni) materials used as clamped-in-place inserts have very high indentation hardness and high resistance to cratering. With a nickel-molybdenum binder, these grades also can be used to perform a broad range of machining operations, including interrupted cutting where the tool is subjected to frequent shocks.

Cemented carbides are made in many grades to suit many purposes by varying the kinds, sizes, and proportions of the carbide particles and the amount of binder. At one extreme are the hardest but most brittle carbides with high resistance to abrasion and wear. In the other grades (with softer binders), hardness is sacrificed in various degrees for strength and shock resistance. Exceptionally small-grained cemented carbides have been developed with superior strength and wear resistance at high cost.

A variety of standards for cemented carbides have been developed over the years which have gradually evolved into a series of standards adopted by the International Organization for Standardization (ISO) (ISO standard 513), and the American National Standards Institute (ANSI) (ANSI B212). ISO 513 classifies all machining in grades of cemented carbides into three color-coded categories. Tungsten carbide grades are identified by their red color and are listed by the letter K. K-type tools are nominally prescribed for cutting nonferrous metals, grey cast iron, and nonmetallic materials. Highly alloyed carbide grades are colored blue and are called P grades. These are used primarily for machining steel and steel castings. The third grade, the M series, is yellow colored and has less alloying than the P grades. It is generally used for multipurpose tools on a variety of steels, steel castings, grey cast iron, and nonferrous metals. The letter designations for each grade are followed by a two-digit number that indicates the relative degree of hardness and toughness of a particular material. For example, P01 refers to a material that has maximum hardness and minimum toughness, while a P50 material has minimum hardness and maximum toughness.

Each of the cutting tool manufacturers have their own letter or number designations representing application codes within the ISO classification scheme. Thus, it is necessary to compare classification indices if a particular ISO letter grade is desired. In addition, most manufacturers have different letter or number designations for coated and uncoated carbide grades.

Coated cemented carbides. Cemented carbide tools are coated for some applications, giving them the advantage of composite materials. These coatings give the rake and clearance surfaces of the cutting tool improved properties over those of the body of the cemented carbide insert. For example, the benefits of TiC (also TiN and HfN) for high-speed cutting is obtained by bonding a layer of such material 5–8 μm (.0002–.0003 in.) thick on a higher-strength WC body. Chemical vapor deposition (CVD) methods are used to provide a uniform coating thickness. Generally, thin coatings are practical only for throwaway inserts since regrinding would eliminate the coating. A variety of these refractory coatings are applied to different types of hard metal or superhard metal substrate materials by either chemical vapor deposition (CVD), physical vapor deposition (PVD), or by more recently developed plasma processes. Extremely sharp cutting edges can be achieved by the PVD process by depositing ions directionally from the electrode instead of applying them evenly.

The near-ultimate in cutting tool wear resistance can now be achieved by applying coatings of boron carbide, cubic boron nitride, and diamond. Tools with these kinds of coatings are being used very successfully in the fabrication of semiconductor products and other small instrument parts.

Sintered Oxides

Sintered oxides, cemented oxides, or *ceramics* are compressed and sintered aluminum oxide powder with small to moderate additions of other metallic compounds. With metallic additives, such as TiC (among many), the materials may be called *cermets*.

Ceramic tool material is harder than most carbides, retains its hardness and strength up to about 1,093° C (2,000° F), has a low coefficient of heat conductivity, and a low coefficient of friction in cutting common metals. It performs best at speeds of 152–610 m/min (500–2,000 ft/min) but has been applied successfully up to 5,486 m/min (18,000 ft/min). This material is brittle and has a low rupture strength with little resistance to mechanical and thermal shock. As a result, for the most part sintered oxide tools have earned a poor reputation for roughing and have not stood up to interrupted cuts, but improvements are slowly being made. Some shops do not have the skill or equipment or take the care necessary to operate these tools properly. Still there have been many reports of remarkable performance when the tools have been properly applied to make cuts up to 10 mm (≈.375 in.) deep.

Oxides have made their best marks with light cuts for good finishes at high speeds and are best suited for long, steady dry cuts. They are at most advantage in machining hard and abrasive ferrous materials, of little advantage for soft materials that can be cut easily by other tool materials, and are not practical for some alloys, such as certain stainless steels and high-temperature alloys. Oxides must be used on machines that have adequate speed and power and are rigid and do not vibrate. Cutting edges must be fully supported.

Cemented carbide tools with an aluminum oxide coating 8–10 μm (.0003–.0004 in.) thick are available, often with other coatings added. They offer the high cutting speeds of the oxides and the greater strength of the carbides. They are of no advantage where solid oxide or carbide tools work well but extend the high-speed capabilities of ceramics into areas not fully served otherwise.

Superhard Crystalline Materials

Whole diamonds have had limited application for many years as cutting tools for some hard, nonferrous, and abrasive materials, and to produce fine finishes at high cutting speeds. For the most part, they were expensive and difficult to shape into usable cutting tool configurations. In addition, natural diamonds were available only from overseas sources and were therefore subject to supply and demand variations, which were not always stable.

To overcome some of the foregoing limitations, the General Electric Company set out to develop the *industrial* or *man-made diamond* in 1951 and successfully produced the first diamond ever manufactured in 1955. Industrial or synthetic diamonds are formed by subjecting graphite, in combination with a molten-metal solvent-catalyst (iron, nickel, or cobalt), to extremely high pressures and temperatures.

Polycrystalline diamond (PCD) tools are produced as either a solid compact insert or as a thin coating applied to a carbide backing. One type of commercial tool has a 0.5 mm (.020 in.) polycrystalline diamond layer on a substrate of cemented carbide about 2.5 mm (.1 in.) thick.

Because diamonds react with iron, nickel, and some other metals, they are limited to specific applications, such as for hard cemented carbides, and abrasive and difficult-to-machine nonferrous alloys, such as silicon aluminum castings, and nonmetals such as reinforced or filled plastics and high alumina ceramics.

Generally, cutting speeds can be raised as high as the machine can go, with the diamond tools lasting many times longer than other tools. As an example, one manufacturer of graphite electrodes for electrical discharge machines (EDM) typically used carbide insert-type end mills running at 3,000 rpm and 7.6 cm/min (30 in./min) feed for high-speed machining of graphite electrodes. These carbide tools required changing every other day because of excessive wear that caused dimensional variations to occur on the electrodes. Changing from a carbide tool to a PCD tool permitted it to run at 8,000 rpm and 15.2 cm/min (60 in./min) feed with tool life extended to 3–4 months. Other users have reported increased removal rates of up to 200% with PCD tools, and that such tools often last up to 100 times longer.

Polycrystalline cubic boron nitride (PCBN) is almost as hard and not as reactive as diamond. It has found widespread acceptance in machining hard steels with very little chemical breakdown or cratering. PCBN tool material is commonly used either brazed or clamped to a sintered carbide backing. PCBN, with its high hardness, good abrasion resistance, and chemical stability has the ability to machine a variety of ferrous materials as well as other high-alloy space-age materials difficult to cut with other tool materials. The main disadvantage of this tool material is its high cost compared to many of the hard metal tools.

Comparison of Materials

A comparison is made of typical cutting-tool materials in Figure 17-11. The stronger the material, the deeper the cuts, and the more shocks and abuse it can take. The speed ranges given cover most practices, but other speeds are at times used for any of the tool materials, especially with uncommon operations and work materials. Relative costs may vary with market conditions, quantities, and tool shapes. Of the sintered tool materials used, about 60–70% are straight tungsten carbides, 25% coated, 5% titanium carbides, and 2% sintered oxides.

17.5.2 Shapes and Forms

Tool Angles

The *point* is that part of a cutting tool where cutting edges are found. It is on the end of the *shank* or *body*. The surfaces on the point bear definite relationships to each other and are defined by angles. The angles of a single-point cutting tool of a type used on a lathe, shaper, or planer are sketched in Figure 17-12. The main elements that define the shape of a tool point are the back rake, side rake, end relief, side relief, end-cutting edge, side-cutting edge angles, and the nose radius. The angles are measured in degrees and the radius in inches. The shape of the tool may be described by a series of numbers specifying the values of the angles and radius in the order just listed. Thus a tool with a shape specified as 8-14-6-6-6-15-0.05 has 8° back rake, 14° side rake, 6° end relief, 6° side relief, 6° end-cutting edge, 15° side-cutting edge angles, and a 0.05-in. nose radius.

The positions of the angles on a tool are determined by the way the tool is designed to act. One kind of lathe tool, depicted by Figure 17-12, is designed to enter the material top first and the rake angles are on the top face. The sizes of the angles largely affect tool performance.

Relief angles. The purpose of a *relief angle*, as the name implies, is to enable the side of the tool to clear the work and not rub. The least amount of relief needed for this purpose depends on the kind of cut. As an example, a turning tool is fed sideways into the work, and the side relief must be greater than the helix angle of the cut.

Investigations have shown that the larger the relief angle, the lower the rate of wear on the flank. On the other hand, the cutting edge is weakened and crumbles if the relief angle is too large. Thus, for most purposes, the relief angle should be as large as possible and still give enough support to the cutting edge. The angle may be larger for tough HSS (5–12°) than for a brittle tool material like carbide or ceramic (5–8°). It may be larger for cutting soft materials or for a continuous cut than for cutting a hard material or for an interrupted cut. Often, a secondary relief angle, called a *clearance angle*, is ground on the shank below the insert of a carbide or ceramic tool. A shaper or planer tool is subjected to repeated shocks as it enters the material at the start of its stroke and is given relief angles as small as 4°. As long as a relief angle is large enough to avoid rubbing, it has no effect upon forces, power, or surface finish.

Rake angles. The *rake angle* of a tool affects the angle of shear during the formation of a chip. The larger the rake angle, the larger the shear angle, and the lower the cutting force and power. A large rake angle is conducive to a good surface finish. However, increasing the rake angle decreases the cutting angle and leaves less metal

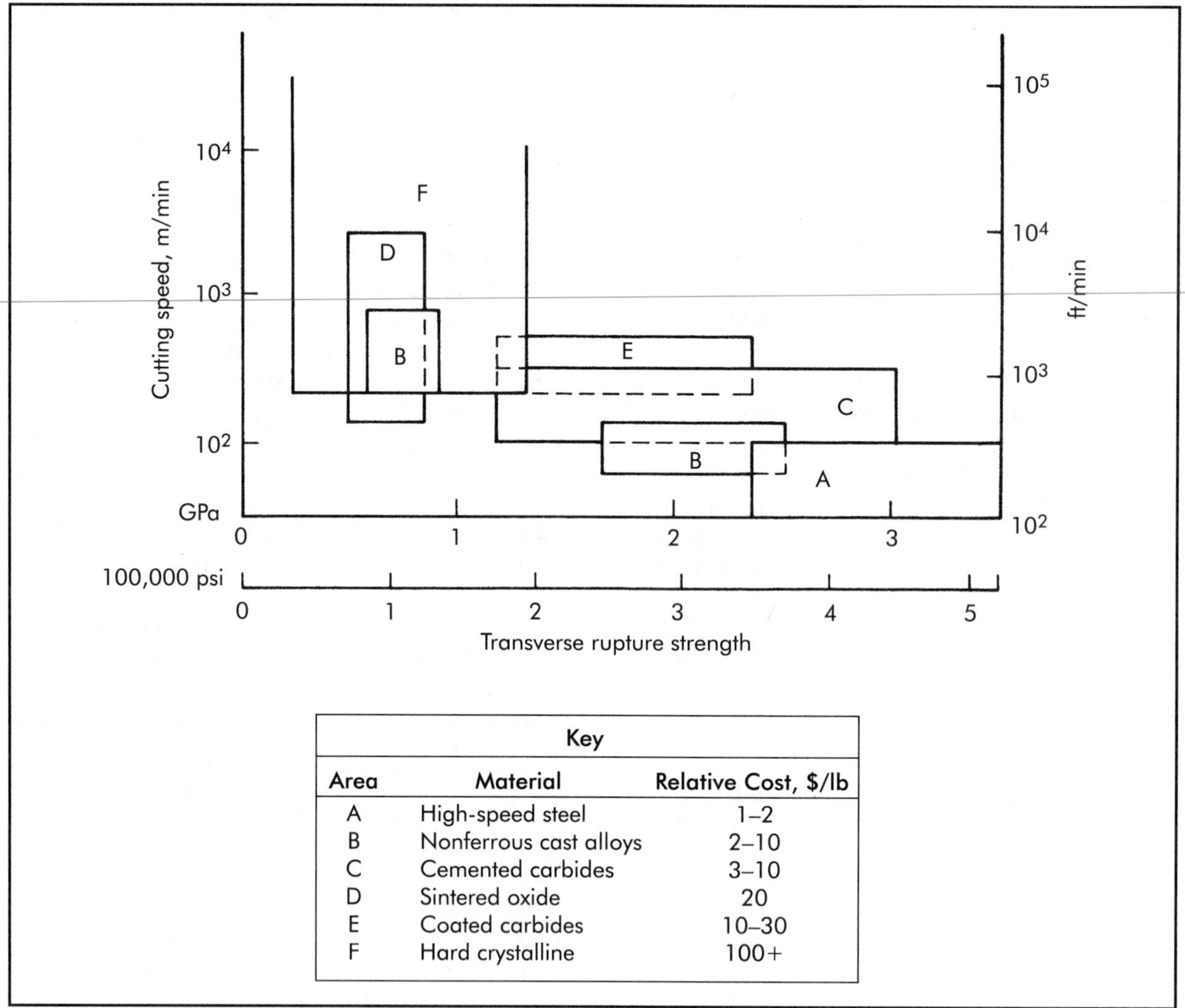

Area	Material	Relative Cost, $/lb
A	High-speed steel	1–2
B	Nonferrous cast alloys	2–10
C	Cemented carbides	3–10
D	Sintered oxide	20
E	Coated carbides	10–30
F	Hard crystalline	100+

Figure 17-11. Relative areas of strength, common cutting speeds, and cost of tool materials.

at the point of the tool to support the cutting edge and conduct away heat. Harder tool materials are generally given smaller rake angles. A practical rake angle represents a compromise between a large angle for easier cutting and a small angle for tool strength. In general, the rake angle is small for cutting hard materials and large for soft ductile materials. An exception is brass, which is cut with a small or negative rake angle to keep the tool from digging into the work.

The two conventional components of rake are back rake and side rake, designated by the rake angles of Figure 17-12. The tool shown is designed to cut on the side-cutting edge, the nose radius, and to some extent on the front-cutting edge. The chip separated by the cutting edges flows along a line *Z-Z*. The *true rake angle,* in a plane perpendicular to the base of the tool and through *Z-Z,* is comparable to the effective rake angle α_e of Figure 17-10. Back rake on a turning tool is like the angle of inclination θ and side rake like the velocity rake α_v of oblique cutting. A difference is that in turning, cutting takes place along two edges, on the front and side of the tool, instead of along one edge as in oblique cutting. In effect though, the directions of the resultant force and chip flow are established by the back and side rake angles ground on the tool. Typical rake angles are given in Table 17-2.

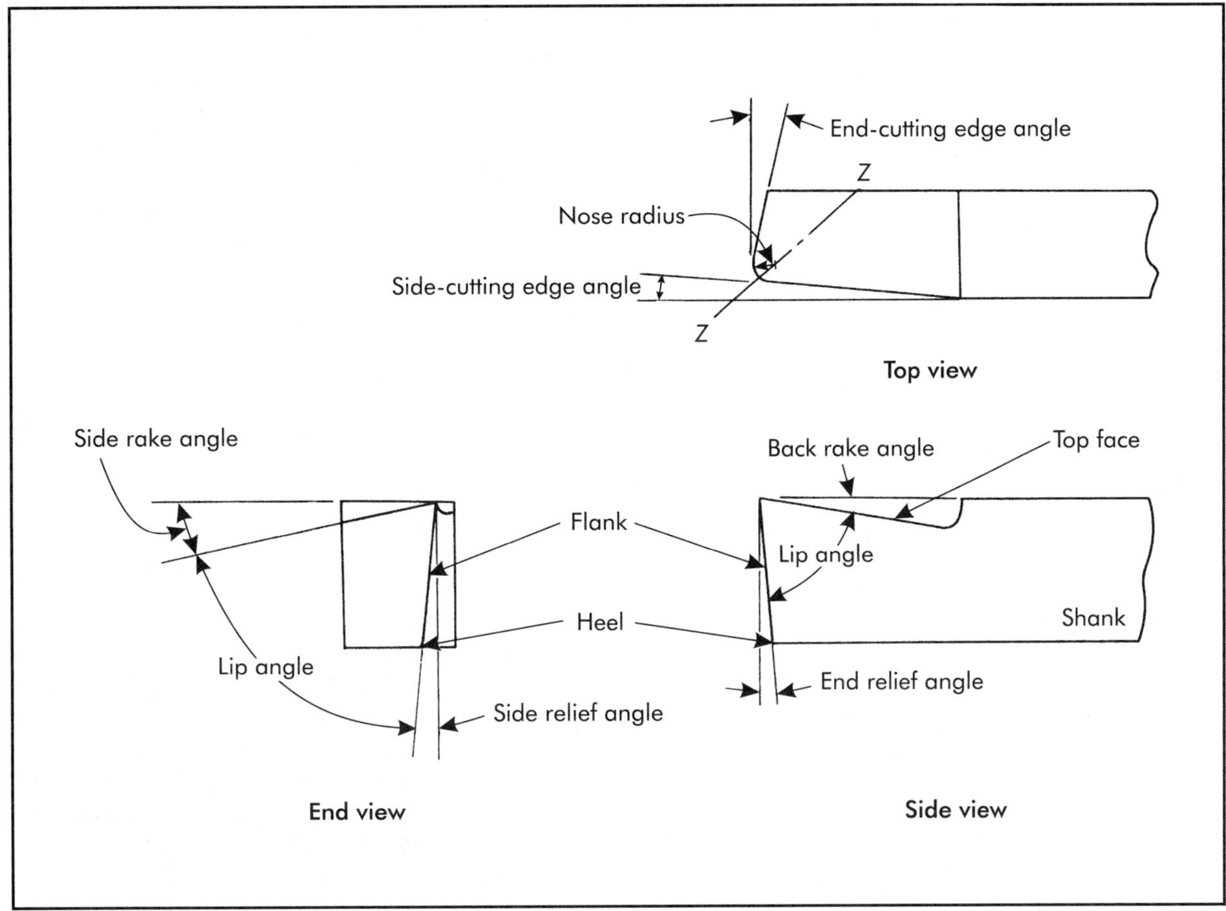

Figure 17-12. Conventional cutting tool angles. Figure 17-1(A) shows how a tool like this cuts.

Table 17-2. Rake angle recommendations

Work Material		High-speed Steel and Cast Alloys		Cemented Carbide			
				Brazed		Throwaway	
Class	Hardness (Bhn)	Back Rake (°)	Side Rake (°)	Back Rake (°)	Side Rake (°)	Back Rake (°)	Side Rake (°)
Aluminum alloys	30–150 500 kg (1,102 lb)	20	15	3	15	0	5
Cast iron	110–200	5	10	0	–6	–5	–5
	300–400	5	5	–5	–5	–5	–5
Copper alloys	40–200 500 kg (1,102 lb)	5	10	0	8	0	5
Steel	85–225	10	12	0	6	0	5
	325–425	0	10	0	6	–5	–5

(*Machining Data Handbook* 1980)

The rake angles shown in Figure 17-13 are positive. Rake angles measured counterclockwise from zero are called *negative rake angles*. They have been found to give good results on carbide and ceramic tools, particularly where the tool is subjected to shocks. A tool with negative rake receives initial impact behind the cutting edge when it starts to cut, and its edge has added material for support. A negative rake angle does increase the cutting forces at lower speeds and gives a poor finish, but carbide tools with negative rake can be run at high speeds where cutting forces drop off and surface finish improves. Tests have shown that at such operating speeds, a $-7°$ rake angle on carbide tools results in cutting forces and power requirements only 8.4% higher than with a 0° rake angle.

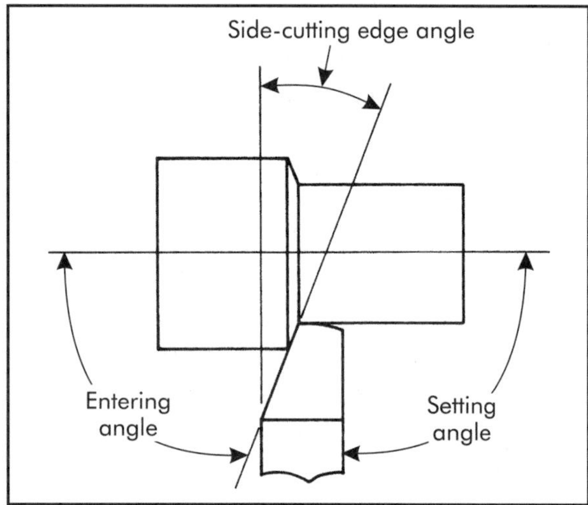

Figure 17-13. Positioning angles of a tool.

Cutting edge angles. An *end-cutting edge angle* gives clearance to the trailing end of the cutting edge and reduces the drag that tends to cause chatter. Too large an end-cutting edge angle takes away material that supports the point and conducts away heat. An angle of 8–15° has been found satisfactory in most cases for side-cutting tools, like turning and boring tools. Sometimes a flat 1.588–7.938 mm (.0625–.3125 in.) long is ground on the front edge next to the nose radius so that the edge can get in a wiping action to help produce a good finish. End-cutting tools, like cutoff and necking tools, often have no end-cutting edge angle.

A *side-cutting edge* or *lead angle* affects tool life and surface finish. It enables a tool that is fed sideways into a cut to contact the work first behind the tip. A side-cutting edge at an angle has more of its length in action for a definite depth of cut than it would without the angle, and the edge lasts longer. On the other hand, the larger the angle, the greater the component of force tending to separate the work and tool. This promotes chatter. The most satisfactory side-cutting edge angle is generally 15°, although advantage is found with angles as large as 30–45° for heavy cuts. No side-cutting edge angle is desirable when cutting castings or forgings with hard and scaly skins because then the least amount of tool edge should be exposed to the destructive action of the skin.

The actual effect of the cutting edge angles in practice depends upon the angle to which the shank of the tool is set with respect to the workpiece, as indicated in Figure 17-13. The entering angle is 180° minus the sum of the side-cutting edge and setting angles.

A *nose radius* is favorable to long tool life and a good surface finish. A sharp point on the end of a tool is highly stressed, short lived, and leaves a groove in the path of cut. As the nose radius is increased from 0°, surface finish and permissible cutting speed improvements are seen in representative tests as indicated in Figure 17-14.

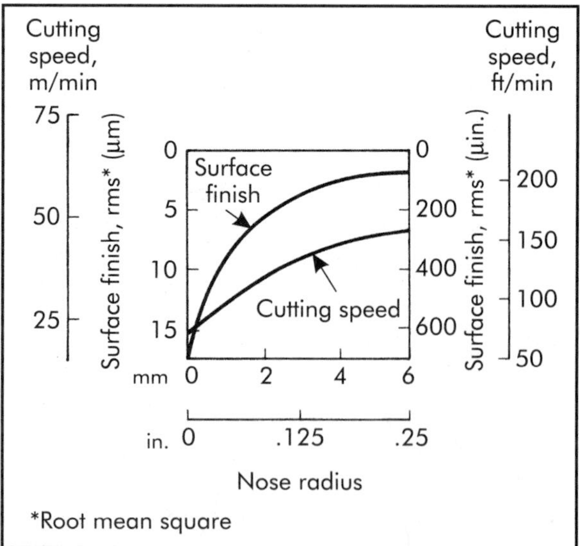

Figure 17-14. Variations of surface finish and tool life with nose radius of a turning tool.

The amount of nose radius that can be put on a tool is limited because too large a radius is conducive to chatter. Radii 0.396–6.350 mm (.0156–.2500 in.) are common but 0.795–3.175 mm (.0313–.1250 in.) are most often used. The size of a nose radius may be prescribed by the fillet that must be left in the corner at the end of a cut.

The physical condition of the cutting edge and face of a tool has a considerable effect upon performance. A tool that is ground so that the cutting edge is jagged will break down quickly. A keen cutting edge and a good finish on the tool face are helpful in minimizing the formation of a built-up edge and promote good surface finish on the workpiece. A good practice is to stone a high-speed steel tool after it is ground to whet the cutting edge and improve the finish of the face. Carbide and ceramic tools are usually finish ground with fine-grit diamond wheels. For heavy cuts, a carbide or ceramic tool for cutting steel may have its edge chamfered or "dubbed" 51–127 μm (.002–.005 in.) at 45° by means of a hand hone to remove weak irregularities along the edge.

Chip Breakers

A continuous-type chip from a long cut can be very troublesome. Such chips become tangled around the workpiece, tool, and machine members, and are dangerous to the operator because they are hot and sharp. At high speeds, chips come off fast, often out of control, and do not tend to curl. Chip breakers are features of cutting tools designed to curl and break up chips for easy disposal.

Common chip breakers are shown in Figure 17-15. The gullet type is a groove, and the stepped type an offset; both are ground or formed into the rake face of the tool. The groove weakens the cutting edge somewhat and is suitable only for moderate feeds. The mechanical type is a block of hard tool material clamped on the face of the tool and is adjustable for different feeds. The form of the land angle type is pressed into tool inserts. It does not present an obstruction to the chip and is reported to give good chip control with much smaller forces and less power than that of other types.

In recent years, considerable research has been done on the control of chip formation by installing different and sometimes intricate patterns on the rake face of insert-type cutting tools. Cutting tool manufacturers are now using very sophisticated computer simulation models to develop full three-dimensional images of metal-cutting chip formation. This permits manufacturing engineers to examine the effect of different chip-forming patterns and different cutting conditions on the performance of a cutting tool. Cutting forces, stresses, pressures, and heat generation data determined from these 3-D models help to determine the optimum dimensions for chip-forming configurations. Quite often, a difference of as little as 0.01 mm (.0005 in.) in a chip breaker groove dimension will make a significant difference in tool performance. These intricate patterns along the rake face of an insert-type tool are pressed into the carbide blank before final sintering. The improvement in chip-forming ability of these designs is extremely critical in the operation of machining centers and other automatic machining equipment. Figure 17-16 shows a variety of indexable insert shapes with different chip-forming patterns on the rake face.

Tool Shapes

The elements that have been described so far are found in all single-point cutting tools and they are put together in many ways to satisfy various requirements. Common shapes of tools used for turning, facing, shaping, planing, and boring are illustrated in Figure 17-17. These are only a few of many varieties.

Some side-cutting tools are designed to cut in one direction, others in another. These are called right-hand or left-hand tools. The hand affiliation is revealed when the tool is placed with its nose pointing at an observer. If the cutting edge is on the observer's right side, the tool is a right-hand tool, and vice-versa.

Single-point Tools

Single-point tools are formed in a number of ways to suit the various kinds of cutting materials. Small pieces of tool material, called *bits*, are ground on their ends and fastened in toolholders like those in Figure 17-18. To make a large solid tool, a piece of high-speed steel may be welded on the end of an inexpensive soft steel shank.

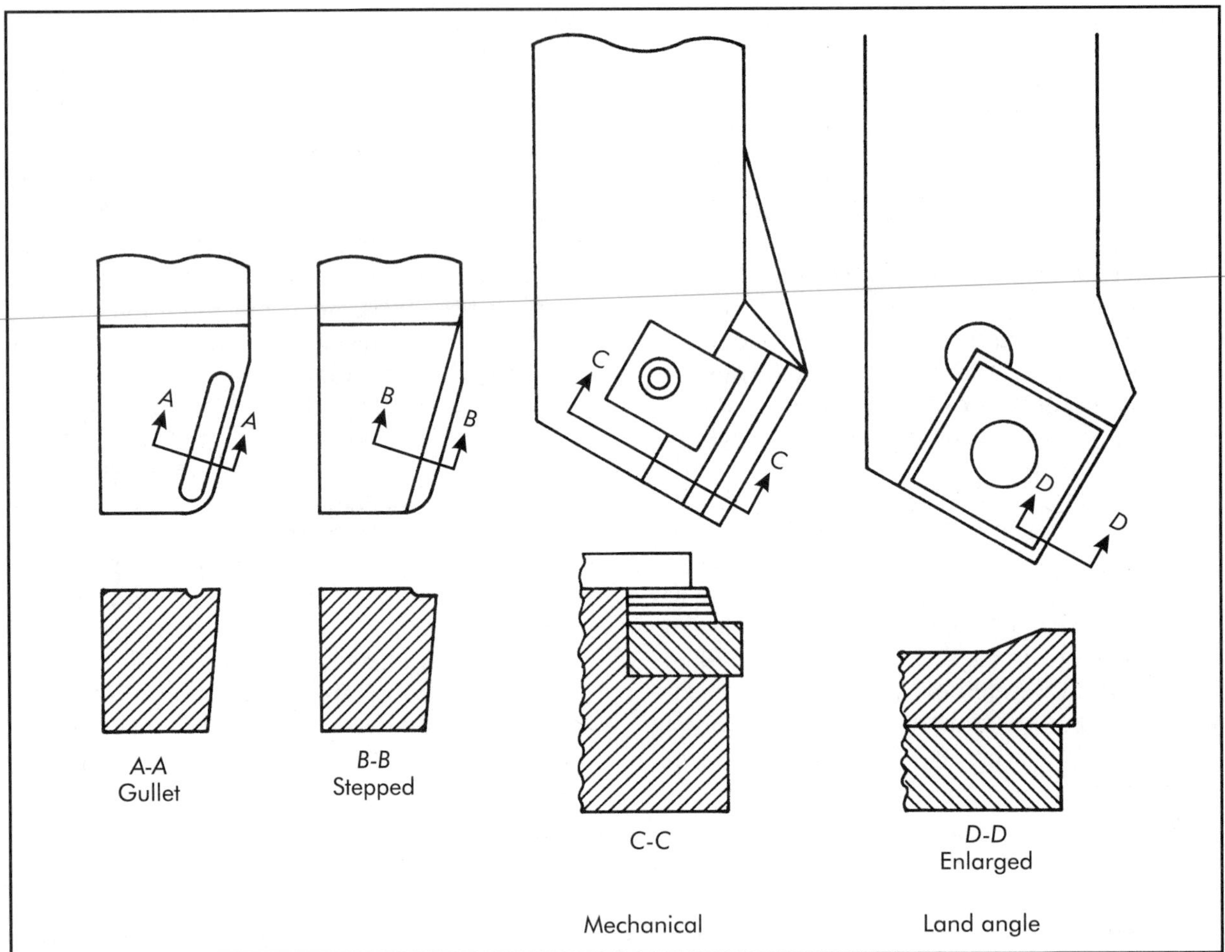

Figure 17-15. Four types of chip breakers.

Small pieces of expensive cutting materials are commonly brazed, cemented, strapped, or clamped onto heavy shanks of soft steel. Tools made in this way are said to be *tipped*. Some typical commercial tools suitable for cemented carbides, ceramics, etc., are shown in Figure 17-19. An insert may be brazed into a pocket on the end of the tool shank. Tips brazed or cemented in this way are securely held, but as a result of brazing and the different coefficients of expansion of the adjacent materials, inherent stresses may be set up that weaken or crack the tip. A sandwich consisting of a carbide chip breaker, tool bit, and support held by a mechanical clamp also may be used. The support at the bottom is made of cemented carbide because it has a high modulus of elasticity, about 7×10^7, and deflects less than half as much as steel under a load. Such an insert is positioned with a negative rake angle so both sides can be used with clearance under each edge. Thus eight corners of the insert can be indexed into position and used before the insert needs to be resharpened or discarded. In many cases, it is more economical to throw away such inserts than to sharpen them. A triangular insert is held by a central screw. Other common shapes are diamond and circular.

The tool may have a positive rake angle; its top is flat and its sides are relieved, so it cannot be turned over to obtain the use of eight edges, and only four are serviceable. Tool inserts with positive rake angles formed in both faces offer eight cutting edges. As has been explained, positive rake angles give lower cutting forces and temperatures, and are better for many operations, even though more expensive. In one case,

Figure 17-16. Indexable insert shapes with different chip-forming patterns on the rake face. (Courtesy Kennametal, Inc.)

tough Inconel® 718 was cut at 30 m/min (100 ft/min) with a cemented carbide tool having a −5° rake angle. Tool life was 1.3 minutes. The same tool material with a 15° rake angle permitted an increase in cutting speed to 56 m/min (185 ft/min) with the same tool life for an increase in productivity of 85%.

Standards for the shapes and sizes of indexable insert-type toolholders and indexable inserts are available from the American National Standards Institute (ANSI). ANSI B212.5-1986 uses a 10-position system of letters and numbers to describe the various features and characteristics of indexable insert toolholders. These

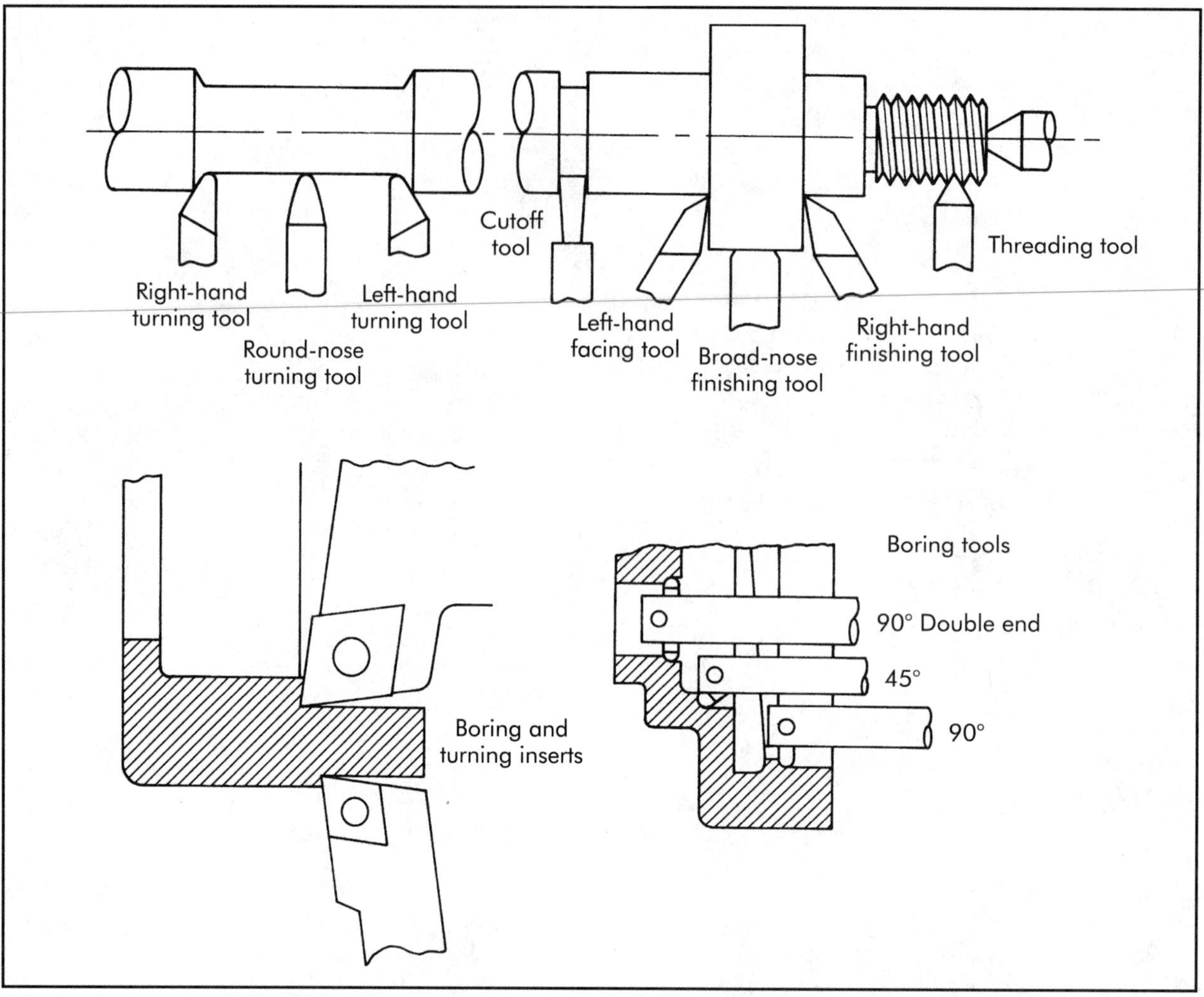

Figure 17-17. Common tool shapes.

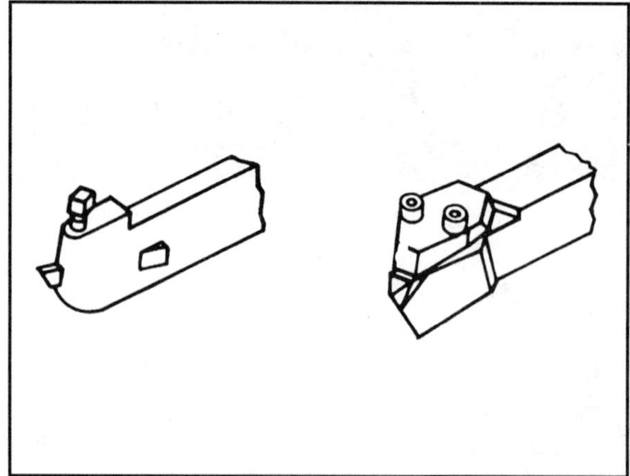

Figure 17-18. Typical holders for solid HSS single-point tool bits.

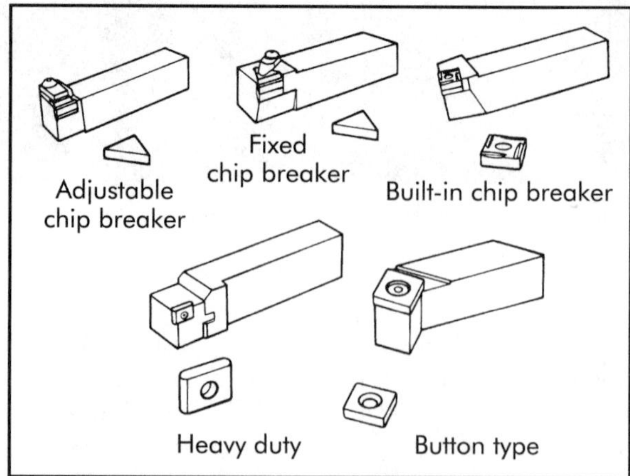

Figure 17-19. Some typical holders, indexable inserts, and chip breakers.

standards describe the method used in clamping an insert in the toolholder, the shape of the insert pocket, the shank style of the holder, hand of the tool, length and width, insert size, etc. The last position of the system refers to a tolerance on the dimensions between the insert tip and both the back end and far side of the holder to control tool positioning error during machine setup and tool change.

ANSI B212.4-1986 uses a similar 10-position system to describe the characteristics of an indexable insert. Standards for such characteristics as insert shape, relief angle, method of clamping the insert to the toolholder, size, thickness, hand, nose radius, etc., are prescribed. As in the case with toolholders, a standard is also given for the tolerance on the planer dimensions of an insert to minimize tool positioning error during setup and tool change.

Form Tools

A *form tool* has a cutting edge with a definite profile or contour and produces a desired form on the workpiece in the manner indicated in Figure 17-1(E). A form tool may be made by grinding a profile on the end of an ordinary single-point tool or bit. This is done for small and simple profiles like circular arcs. Two common types for production work are flat and circular form tools (Figure 17-20). Form tools are sharpened by grinding the rake face behind the cutting edge. Because of the angles at which a form tool must be ground and positioned, it usually must be designed with a form somewhat different from that desired on the workpiece.

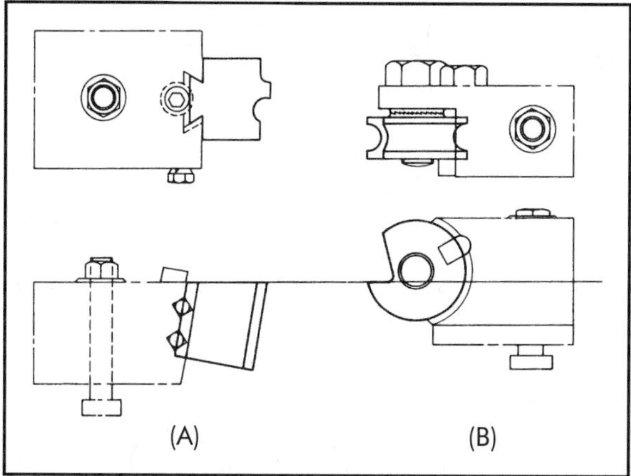

Figure 17-20. Examples of form tools: (A) flat form tool held by a dovetail; (B) circular form tool.

17.6 CUTTING FLUIDS

17.6.1 Purpose

Fluids are commonly applied to metal-cutting operations, chiefly to cool the tool and workpiece and provide lubrication. In some cases, little benefit is derived from a cutting fluid, but in most cases increases of 20–50%, and sometimes more, in cutting speed are possible with the same tool life when a cutting fluid is used as compared to dry cutting.

A cutting fluid is often called a *coolant*, and cooling is an important function at all cutting speeds. High temperatures are the chief cause of tool wear. An overheated workpiece may warp. Coolants help correct these conditions.

Cutting fluids may serve to lubricate the chip sliding over the tool face. The evidence is that capillary action tends to draw the fluid in through the natural grinding scratches on the face of the tool. Lubrication reduces friction forces acting on the tool, improving tool life and decreasing power consumption. The tendency to form a built-up edge is decreased by lubrication, and surface finish is improved. However, speed is much more effective than cutting fluid in improving finish.

The lubricating action of cutting fluids is effective only at low speeds; it is already poor at 30 m/min (100 ft/min) and totally absent at 122 m/min (400 ft/min) and above. At high speeds the fluid does not have time to reach the tool face where it can lubricate; external cooling is all that can be done.

A cutting fluid is useful in cooling and washing away the chips formed in an operation. A fluid also should be capable of lubricating exposed machine elements, preventing corrosion, and not be harmful to the operator.

A cutting fluid is usually flowed in a copious stream on top of the chip and tool. The flow must be uninterrupted with carbides because they crack easily from sudden temperature changes. Some advantage is obtained by directing the flood upward between the relief face of the tool and workpiece. Fluid may be flung away from the cut and create a mess at high speeds. Cutting fluid may be sprayed as a mist from above or below. A jet of fluid may be injected into the space between the relief face of the tool and the workpiece. Under some circumstances, it reaches the cutting edge and appreciably increases tool life. However, the method requires fairly costly equipment and extra care from the operator, and is not widely used. Flooding is as effective as any other method at low temperature and costs the least. At higher temperature, a layer of steam is trapped by flooding, thus preventing the fluid from reaching and cooling the surfaces. A mist blows away the steam layer and cools effectively. This helps only for a limited range because a thick cloud of steam is generated at still higher temperatures and keeps away even the mist or spray. In one test at 91 m/min (300 ft/min), mist cooling improved tool life by 68% over flooding and 150% over dry cutting. At 180 m/min (590 ft/min), dry cutting was as efficient as flooding, with mist cooling showing a 41% longer tool life; but dry cutting became as efficient as mist spraying at 274 m/min (900 ft/min).

17.6.2 Types

Cutting fluids may be classified as: (1) gases, (2) water solutions, (3) oils, and (4) waxes.

Gases have inferior cooling capacity and are relatively unimportant cutting fluids. Compressed air is sometimes blown on cast iron being cut. Carbon dioxide at low temperatures may be directed to cool a tool without quenching the work material. This has limited application in machining ultra high-strength alloys to keep them from hardening while being cut and from damaging the tool.

Water is the best cooling medium and the most effective fluid for high-speed cutting. However, by itself it offers little lubricating value, does not spread well over a surface to wet it because of high surface tension, and causes rust and corrosion. It is mixed with chemicals and oils to improve its use as a cutting fluid.

Chemical or *synthetic cutting fluids* primarily contain water modifiers without oils. They may be pure coolants containing just water softeners and rust inhibitors, or lubricating coolants with wetting agents and/or extreme pressure (EP) agents such as iodine, chlorine, sulfur, and phosphorus. Other chemicals used in cutting fluids include blending agents and germicides. Chemicals enhance the penetrating and cooling abilities of water even more than oils, reduce corrosion, and add lubricity approaching that of oils. However, when lubrication is of prime importance, oils are usually recommended. Chemical fluids have long useful lives, high detergency and cleanliness, and do not require degreasing after use. However, they are not usable on some machines where they can get into ways and bearings, and some substances leave abrasive residues upon evaporation, which may be harmful to slides.

Popular cutting fluids, known as *soluble* or *dispersible oils*, are emulsions of mineral oil with emulsifiers (such as soap in water) and sometimes EP agents. Mixtures vary from 5 to over 100 parts of water to 1 of oil. Other compounds are added to give various properties, such as extreme pressure resistance. The mixtures vary in character from heavy solutions with high lubricity and cushioning suitable for heavy turning, milling, broaching, and similar operations, to light solutions with little lubricity and high detergency suitable for fine-grit grinding and other high-speed operations. The cost of the soluble oil solutions is low.

Straight cutting oils are those not mixed with water. They are classified as inactive and active, according to whether or not they are mixed with chemicals.

Inactive *straight mineral oils* are not widely used but serve satisfactorily for inherently free-cutting materials. They have a specific heat about half that of water and a low degree of adhesion or oiliness, but are quite stable and do not develop disagreeable odors. They range in viscosity from kerosene used on magnesium and aluminum to light paraffins for free-cutting brass. Inactive *fatty oils* at one time were quite popular but are used little today. They are expensive. Fatty oils have better wetting and penetrating properties than mineral oils, and the two are chiefly blended together or with small amounts of chemicals to serve nearly as well as active oils, with less tendency to stain the work surface.

Sulfur and, to a lesser degree, chlorine and phosphorus, are mixed with both mineral and fatty oils to make *active cutting oil compounds* for high antiweld properties and lubrication under extreme pressures. The active ingredients combine with metallic surfaces to form tenacious but slippery films. These compounds play an important part in modern metal-cutting practice, being used extensively for heavy turning, gear cutting, broaching, etc., of tough, stringy, and unusually soft materials. Such agents are not advantageous for light cuts.

Certain *waxes* are adsorbed strongly to metallic surfaces and augment the actions of other ingredients. Cutting fluids that contain these waxes have high-pressure and high-temperature lubrication properties. It is not easy to successfully incorporate wax in a cutting fluid, and indiscriminate attempts to apply it have failed and have given wax a bad reputation as a cutting fluid ingredient. Where successful, the results with wax have been rather spectacular, but the applications are limited.

17.7 QUESTIONS

1. Why are metal parts or products machined?
2. What is meant by a "secondary operation" in machine work?
3. How does straight turning differ from taper turning?
4. Define the terms "cutting speed," "feed," and "depth of cut."
5. Show by means of a sketch the action of a single-edge cutting tool while cutting metal.
6. Name and describe three kinds of chips.
7. Show by means of a sketch the forces acting on a chip during its formation.
8. What are the conditions for orthogonal cutting?
9. What conditions of pressure, friction, and temperature occur at the cutting zone?
10. What happens to the angle ø of the shear plane as cutting speed is increased?
11. What happens to the forces in the cutting zone as the cutting speed is increased?
12. What happens to the temperature and forces as the feed is increased?
13. What effect does a change in rake angle have on the shear angle?
14. In what ways may a cutting tool wear?
15. How is "tool life" defined?
16. Name and describe the principal cutting-tool materials.
17. Name at least four physical properties desirable in a cutting-tool material.
18. Which of the cutting-tool materials discussed are the most popular for production operations?
19. Why are cemented-carbide cutting tools sometimes coated with other materials?
20. What is the difference between ceramic- and cermet-type cutting-tool materials?
22. Why are synthetic-diamond cutting tools not used for machining ferrous materials?
23. Name in conventional order and describe the angles of a cutting tool.
24. What is the purpose of a relief angle on a cutting tool?
25. What does the rake angle of a cutting tool do?
26. What is the purpose of the front-cutting edge angle? The side-cutting edge angle?
27. Specify the effect upon forces, power consumption, and tool life of each angle and the nose radius of a cutting tool.

28. What is a "chip breaker" and what purpose does it serve?
29. What is a "form tool"? Describe two kinds.
30. What are the purposes of cutting fluids?
31. What are the principal kinds of cutting fluids?
32. How are cutting fluids applied to metal-cutting operations?

17.8 PROBLEMS

1. Given the following diameters and cutting speeds, calculate the rpm setting for a turning operation on each.

	(a)	(b)	(c)	(d)
Diameter, mm (in.)	76 (3)	305 (≈12)	102 (4)	457 (18)
Cutting speed, m/min (ft/min)	33.5 (110)	23 (75)	60 (≈200)	30 (100)

2. If each of the following diameters is rotated at the rpm given, what is the surface speed in m/min (ft/min) of the bar being turned?

	(a)	(b)	(c)	(d)
Diameter, mm (in.)	102 (4)	406 (16)	254 (10)	305 (12)
Rpm	200	70	20	150

3. Calculate the rate of metal removal for each of the following sets of conditions:

	(a)	(b)	(c)
Cutting speed, m/min (ft/min)	46 (150)	18 (60)	91 (300)
Feed, mm/rev (in./rev)	0.25 (.010)	0.76 (.030)	0.46 (.018)
Depth of cut, mm (in.)	6.4 (.25)	9.53 (.375)	1.5 (.06)

4. In a cutting operation like that depicted in Figures 17-7 and 17-8, the feed t_1 is 0.127 mm (.005 in.) and the chip is found to have a thickness t_2 of 0.25 mm (.010 in.). The cutting force F_C is 1,335 N (300 lbf) and the normal force F_L is 756 N (170 lbf). The rake angle of the tool is 110°. Find:

   (a) the shear angle ø;
   (b) the size of the force R exerted by the tool on the chip;
   (c) the coefficient of friction on the face of the tool;
   (d) the magnitudes of the friction force F_f and the normal force F_n; and
   (e) the magnitudes of the shearing force F_S and the normal force F_N.

5. An orthogonal cut 3.18 mm (.125 in.) wide is made at a speed of 33.5 m/min (110 ft/min) and feed of 0.25 mm/rev (.010 in./rev) with a high-speed steel tool having an 18° rake angle. The chip thickness ratio r_t is found to be 0.58, the cutting force F_C is 1,379 N (310 lbf), and the normal force F_L is 356 N (80 lbf). Determine the

   (a) chip thickness t_2;
   (b) shear plane angle ø;
   (c) magnitude of the resultant force R;
   (d) coefficient of friction on the face of the tool, f;
   (e) magnitude of the friction force F_f and normal force F_n;

(f) magnitude of the shearing force F_S and the normal force F_N; and
(g) specific energy J/cm^3 (in.-lb/in.^{3}).

6. A workpiece is being cut at 76 m/min (250 ft/min), and the power is found to be 2 kW (2.7 hp). The feed is 0.25 mm/rev (.010 in./rev) and the depth of cut is 5.0 mm (.200 in.).
   (a) What is the cutting force in N (lbf)?
   (b) What is the unit power consumption in kW/cm^3/min (hp/in.^{3}/min)?
7. In a cutting operation like that depicted in Figures 17-7 and 17-8, the feed t_1 is 0.13 mm (.005 in.) and the depth of cut normal to the plane of the paper is 2.5 mm (.100 in.). The cutting speed is 244 m/min (800 ft/min). The cutting force F_C is found to be 1,779 N (400 lbf) and the normal force F_L is 890 N (200 lbf). The rake angle of the tool is +8°. Find:
   (a) the power required for the cut in kW (hp);
   (b) the rate of metal removal in cm^3/min (in.^{3}/min); and
   (c) the unit power in kW/cm^3/min (hp/in.^{3}/min).
8. A tool making an orthogonal cut has a −10° rake angle. The feed t_1 is 102 μm (.004 in.), the width of cut 6.4 mm (.25 in.), the speed 165 m/min (540 ft/min), and a dynamometer measures the cutting force F_C to be 1,779 N (400 lbf) and the normal force F_L to be 1,512 N (340 lbf). A high-speed photograph shows a shear plane angle of 20°.
   (a) What should the thickness of the chip be?
   (b) Compute the coefficient of friction.
   (c) Calculate the shearing and normal stresses on the shear plane.
   (d) What is the shearing strain?
   (e) What power is expended in shearing the metal and altogether in making the cut?
9. Assume an oblique cut is 6.4 mm (.25 in.) wide. The tool has a 5° normal rake α_n and is inclined at $\theta = 10°$.
   (a) What is the angle β on the tool face between the directions of the normal and effective rake angles?
   (b) Calculate the effective rake angle.
10. A cutting tool has a normal rake of 10°. How large an inclination angle is required for an effective rake angle of 15°?
11. A shaft 1.6 m (64 in.) long and 102 mm (4 in.) in diameter is turned from tough steel having a hardness of 22–26 R_C. With a high-speed steel tool, the speed is 24 m/min (78 ft/min), the feed 0.50 mm/rev (.020 in./rev), and the depth of cut 1.27 mm (.050 in.). The tool must be ground four times before it finishes one shaft and can be ground 40 times before being discarded. The cost of a tool bit is $2.50. It takes 10 min to grind and replace the tool bit. A sintered oxide tool bit with six edges that costs $3.90 is put on the job. After all six edges are dull, this bit is not sharpened but is thrown away. It cuts with a speed of 90 m/min (295 ft/min), a feed of 0.51 mm/rev (.020 in./rev), and a depth of cut of 1.27 mm (.050 in.). Two edges must be used to cut one shaft. The time to index the tool is 30 sec. The cost for labor and overhead in the shop is $26/hr. Which tool is more economical?
12. A workpiece of SAE 4340 steel (300 Bhn) is rough turned for a distance of 457 mm (18 in.) with a stock removal of 11 mm (.435 in.) on a side of a 140 mm (5.5 in.) diameter. With high-speed steel tools, the speed is 24 m/min (80 ft/min) and feed 0.30 mm/rev (.012 in./rev). The tool must be resharpened for each piece. The tool costs $2.60 and can be reground 40 times. Time for sharpening and resetting the tool is 2 min for each piece. A cemented carbide tool costs $5.48 and can be ground 10 times. It is operated at 100 m/min (≈330 ft/min) with a feed of 0.60 mm/rev (.024 in./rev). It can turn 10 pieces before having to be sharpened, but the time to sharpen and reset it is 10 min. The labor and overhead rate is $26/hr. Which tool is more economical for the job?

13. A workpiece has a hardness of 34–36 R_C and is turned on the end to a diameter of 46 mm (1.81 in.) from a diameter of 64 mm (2.5 in.) for a length of 140.5 mm (5.53 in.). A brazed carbide insert tool has been making the cut in production at 65.5 m/min (215 ft/min) with a feed of 0.28 mm/rev (.011 in./rev). Fifty pieces are obtained between tool grinds. The tool can be reground four times and costs $8.26 new. Each grind costs $2. A throwaway replaceable insert tool is tried on the job. Because the tip is free from brazing strains, this tool can be operated at 91 m/min (300 ft/min) with a 0.38 mm/rev (.015 in./rev) feed and completes 60 pieces per cutting edge. Three edges are available. Each insert costs $3.30, and the holder, which lasts for 1,000 inserts, costs $21.50. Can the use of throwaway replaceable inserts be justified on this operation where the labor and overhead rate is $15/hr?

17.9 REFERENCES

Albert, M., ed. 1994. "Square Inserts for Square Shoulders." *Modern Machine Shop*, April.

ANSI B212.12-1991. "Indexable Inserts Commonly Used for Turning Tools, Part l." New York: American National Standards Institute.

ANSI B212.12-1990. "Carbide Tips for Brazing on Turning Tools." New York: American National Standards Institute.

ANSI B212.3-1986. "Holders for Indexable Inserts, Precision." New York: American National Standards Institute.

ANSI B212.4-1986. "Cutting Tools—Indexable Inserts—Identification." New York: American National Standards Institute.

ANSI B212.5-1986. "Indexable Inserts, Metric Holders for." New York: American National Standards Institute.

ANSI/ASME B94.55M-1985. "Tool Life Testing with Single-point Turning Tools." New York: American Society of Mechanical Engineers.

ANSI B212.2-1984 (R1990). "Carbide Seats Used with Indexable Inserts for Clamp-type Holders." New York: American National Standards Institute.

ANSI B212.1-1984 (R1990). "Carbide Blanks, Brazed and Solid, Single-Point Tools." New York: American National Standards Institute.

ANSI/ISO 513-1975. "Carbides—Applications of Carbides for Machining by Chip Removal—Designation of the Main Groups of Chip Removal and Groups of Applications." New York: American National Standards Institute.

Coleman, J.R., ed. 1992. "No Myth—High Speed Machining." *Manufacturing Engineering*, October.

Koelsch, J.R., ed. 1996. "An Engineering Science—Insert Geometries Contain Less Art than Meets the Eye." *Manufacturing Engineering*, January.

——. 1995. "A Deeper Understanding of Machining." *Manufacturing Engineering*, July.

Machining Data Handbook, 3rd Ed. 1980. Vols. 1 and 2. Cincinnati, OH: Institute of Advanced Manufacturing Sciences.

Mason, F., ed. 1996. "Electronic Tool Selection." *Manufacturing Engineering*, January.

Society of Manufacturing Engineers (SME). 1997. *Cutting Tool Materials* video. *Fundamental Manufacturing Processes* series. Dearborn, MI: SME.

——. 1997. *Cutting Tool Geometries* video. *Fundamental Manufacturing Processes* series. Dearborn, MI: SME.

Sprow, E.E., ed. 1993. "Time yet for Cermets." *Manufacturing Engineering*, January.

Tool and Manufacturing Engineers Handbook, 4th Ed. 1983. Vol. 1, *Machining*. Dearborn, MI: Society of Manufacturing Engineers.

Turning, Boring, and Facing

18

Screw cutting lathe, 1800
—Henry Maudslay, London, England

18.1 TURNING OPERATIONS

Small and medium-size workpieces usually are turned around a horizontal axis. Turning operations may be divided into two classes: those done with the workpiece between centers and those done with the workpiece chucked or gripped at one end with or without support at the outer end.

18.1.1 Plain or Straight Turning

The principal kinds of operations performed on work between centers are indicated in Figure 18-1. The workpiece is driven by a dog clamped on one end. If work is to be done on both ends, the dog is clamped on each in turn, and the workpiece is reversed in position.

A workpiece held between centers deflects less under a given force than if held at only one end. Also, a workpiece runs true on good centers, and several diameters cut at the same or at different times are most likely to be concentric if turned on centers.

To be turned on centers, a workpiece must have a center hole at each end. These holes have 60° conical bearing surfaces and are cut with a combination center drill and countersink (Figure 18-2). This may be done with the workpiece chucked on a lathe or on a center drilling machine like a horizontal drill press.

18.1.2 Chuck Work

When a workpiece is chucked, the same types of operations can be done on it as those performed between centers. In addition, parting or cut off and internal operations can be done, like those illustrated in Figure 18-3. The types of tools described in Chapter 20 for drilling and related operations are used to machine holes in chuck work.

18.1.3 Taper Turning and Boring

As illustrated in Figures 18-1 and 18-3, turning produces not only straight, but also tapered outside and inside surfaces. A taper may be designated by the angle between opposite sides or by the increase in diameter along the length, in mm/m (in./ft). Certain tapers are standard. The Morse® taper, found on drilling machines and some lathes, is approximately 52.08 mm/m (.625 in./ft). Sizes, designated by whole numbers from 0 through 7 (the largest), are given in handbooks. Other tapers used on lathes are the Reed® and Jarno tapers, each of 50 mm/m (.600 in./ft).

Three methods of generating tapers by offsetting the tailstock, swiveling the compound rest, and use of a taper attachment will be described in the discussion of the equipment used. Tapers may also be form cut, but the length of

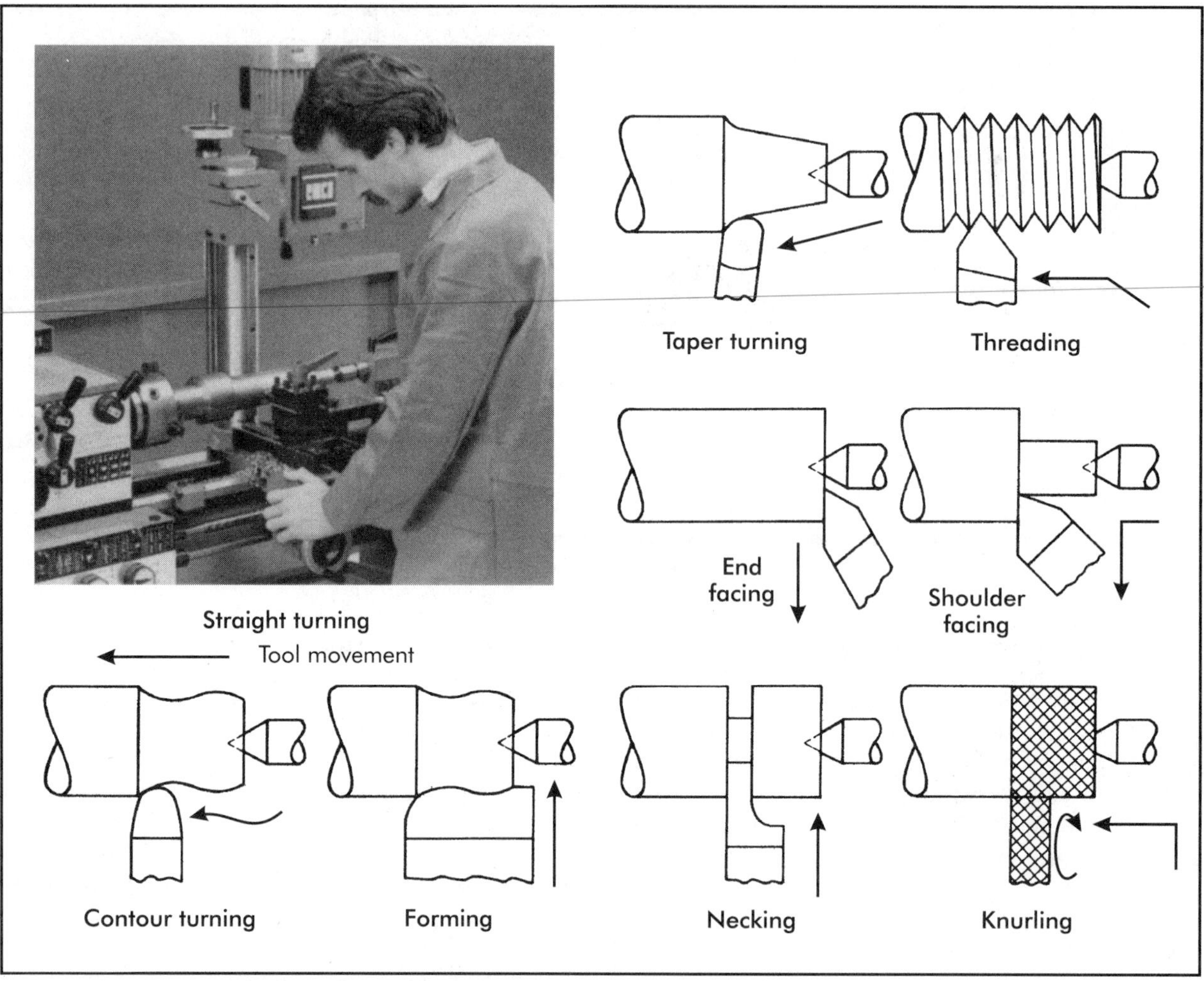

Figure 18-1. Turning operations on a lathe. (Courtesy Emco Maier Corporation)

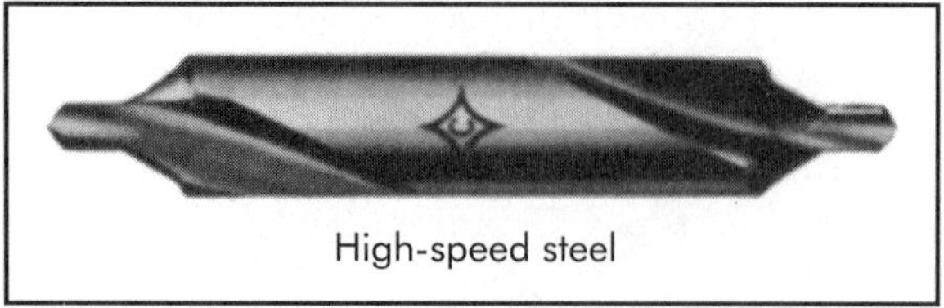

Figure 18-2. Combination center drill and countersink. (Courtesy Cleveland Twist Drill)

cut must be kept small so the force between tool and work does not become excessive.

18.2 THE LATHE

The basic machine tool on which turning operations are performed is the lathe, typified by the engine lathe in Figure 18-4. The diagram of Figure 18-5 delineates the principal parts and movements of the engine lathe.

18.2.1 Principal Parts

A lathe is built around a *bed*, made massive and rigid to resist deflection and vibration. On top of the bed at the left is the *headstock* that carries a revolving *spindle*. A lathe spindle is hollow to take long barstock, and the diameter of the hole determines the largest size barstock that can be passed through the spindle for chucking. The spindle hole has a definite taper at the front end to receive tapered center and tool shanks. The *nose* or end of the spindle has true locating surfaces for positioning and a means for

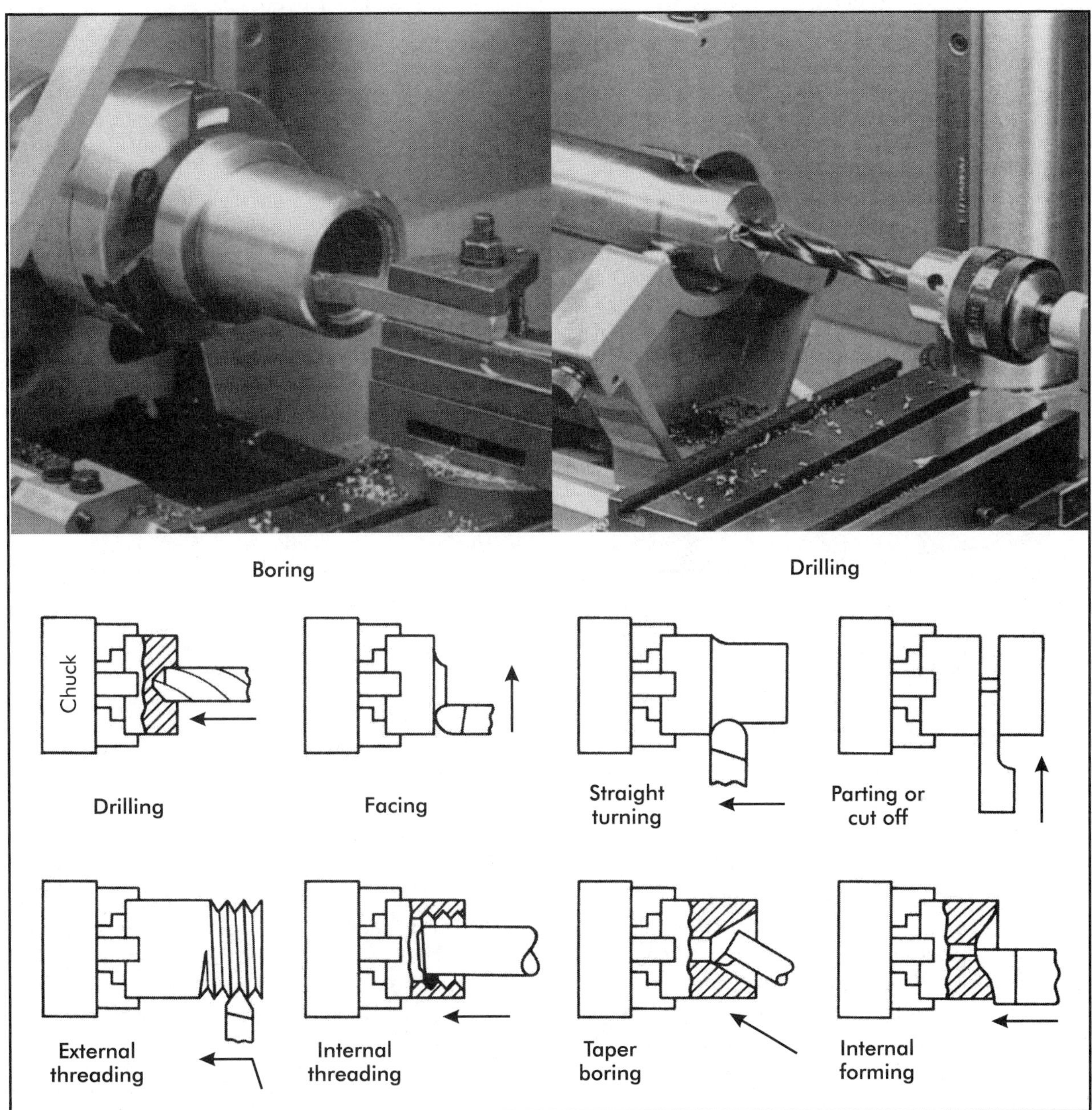

Figure 18-3. Chucking operations on a lathe. (Courtesy Emco Maier Corporation)

fastening a chuck, as shown in Figure 18-3, or a face plate or dog plate as in Figure 18-1.

The workpiece is driven by the headstock spindle. Present-day lathes have individual motor drives, usually of constant speed. Most of them have geared headstocks like the one in Figure 18-4. The drive from the motor to the spindle goes through several gear combinations shifted by means of levers or dials on the outside of the headstock to change spindle speeds. On some small lathes, the drive from the motor to the spindle is comprised of a belt and step pulleys to allow several spindle speeds. In addition, a set of reduction gears may be engaged to get a lower series of speeds.

The *tailstock* is on the other end of the bed from the headstock. Its spindle does not revolve but can be moved a few inches lengthwise and

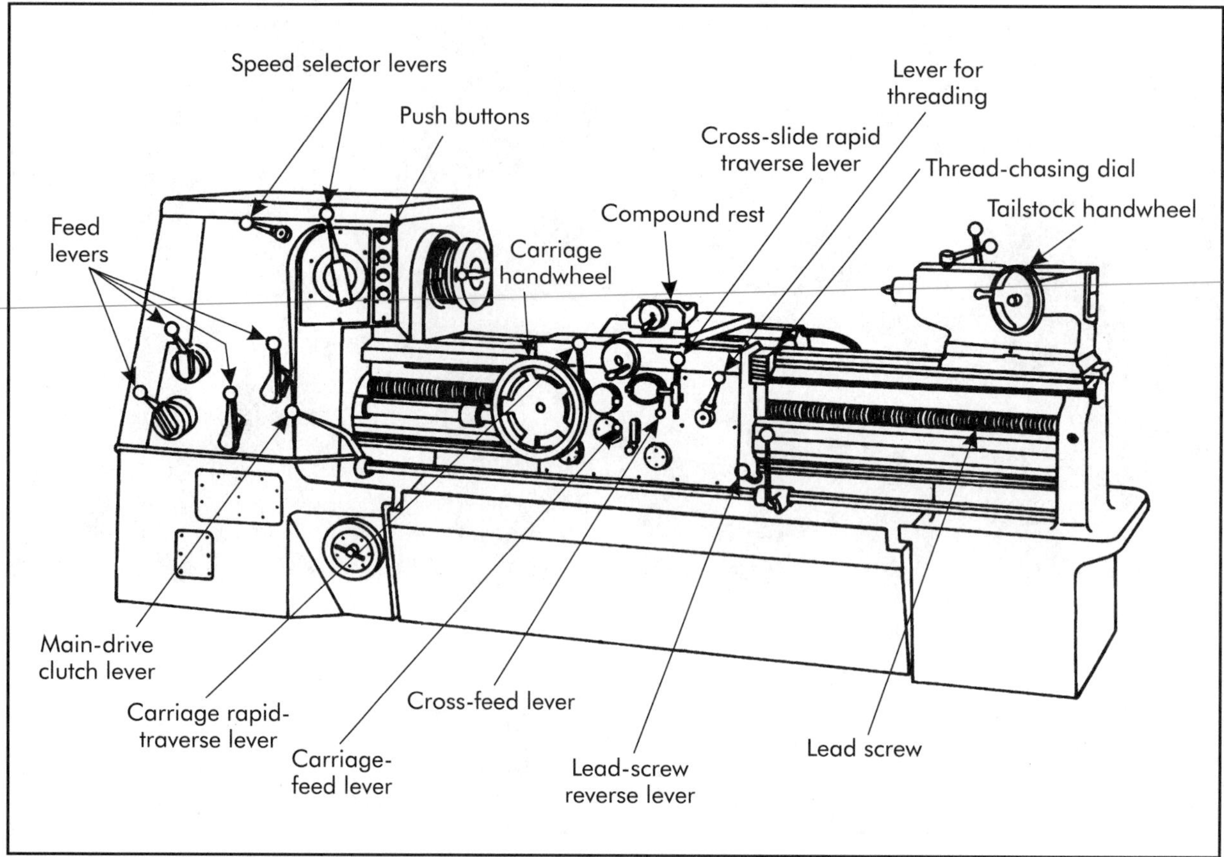

Figure 18-4. Center-type engine lathe.

clamped as desired. Drills, reamers, taps, and other end-cutting tools are held and fed to the workpiece by the tailstock spindle, which is hollow with a taper to take the shanks of centers, drill chucks, drills, reamers, etc. The whole tailstock can be moved and clamped in any position along the bed where it can best serve its purpose.

The body or top of the tailstock can be adjusted crosswise on its base. This may serve to align the tailstock with the headstock so work can be turned straight. On the other hand, the tailstock may be intentionally offset to hold a piece between centers at an angle for tapered turning.

Between the headstock and tailstock is the *carriage* or *saddle* containing several parts that serve to support, move, and control the cutting tool. The saddle slides along ways on top of the lathe bed and can be moved by means of a handwheel and crank or a power-feed mechanism as described in the next section. This provides the longitudinal (Z-axis) direction of movement shown on the apron of the lathe in Figure 18-5. On top of the saddle is a *cross slide* that is adjusted and fed at right angles to the length of the bed by means of a handwheel and crank or a feed mechanism. This accomplishes the cross-feed (X-axis) adjustment or movement of the cutting tool. An additional adjustment of the cutting tool is provided by the *compound rest* mounted on top of the cross slide. The compound rest has a graduated base and can be swiveled around a vertical axis. In this way, its slide can be set at any angle with the axis of the workpiece. The slide can be adjusted or fed by hand through a screw and nut controlled by a handwheel and crank.

A graduated dial or scale on the cross-feed screw collar enables the lathe operator to position the cutting tool along the X axis for a given depth of cut in a turning operation. Generally, the dial is graduated in increments of 25 μm (.001 in.). Thus, rotating the handwheel through one increment on the dial scale will move the cutting

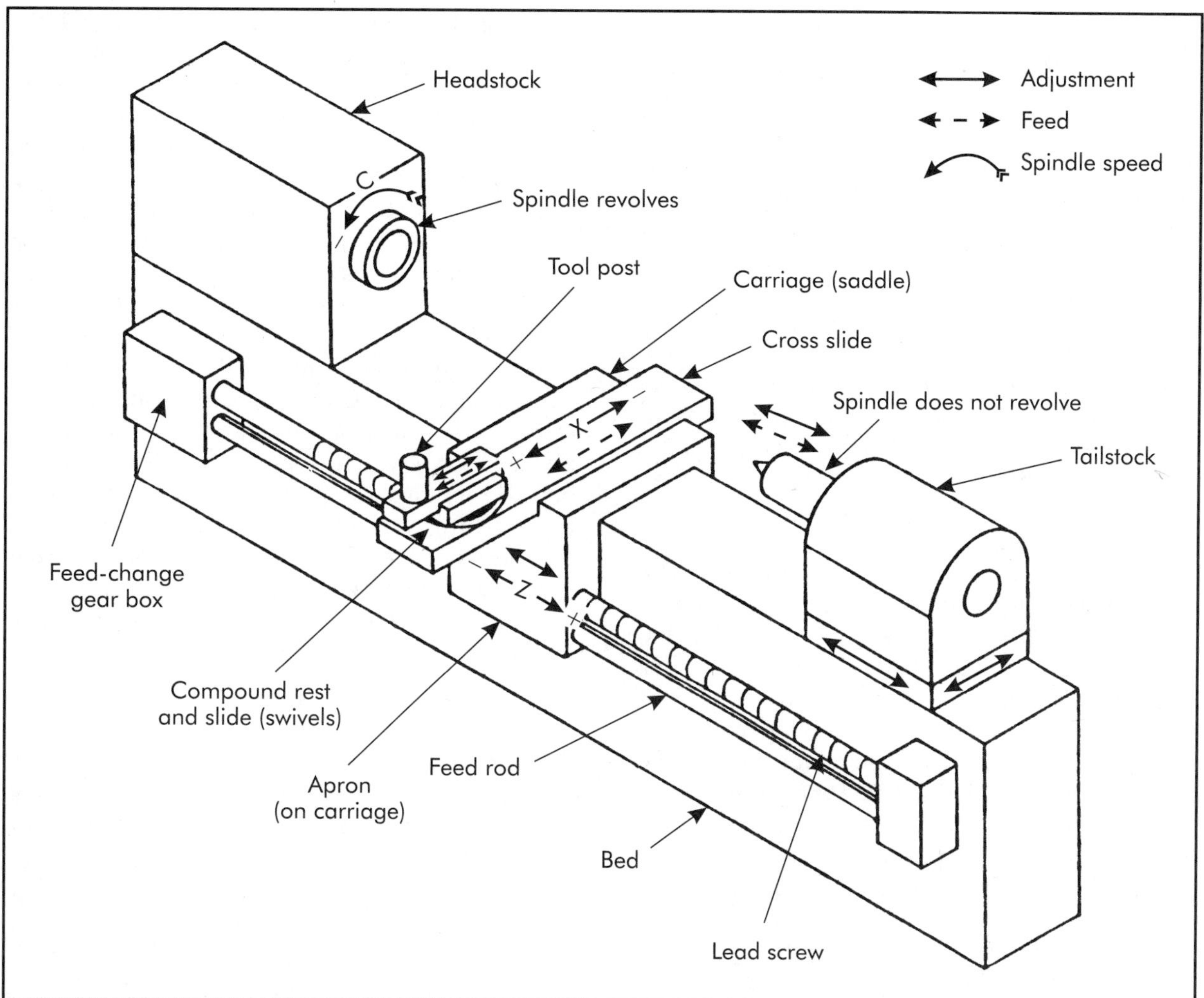

Figure 18-5. Principal parts and movements of a lathe.

tool 25 μm (.001 in.) along the X axis. A similar dial on the compound rest enables the cutting tool to be positioned accurately along the X axis, the Z axis, or at some angle to those two axes. Positioning of the cutting tool along the Z axis also can be accomplished by means of the handwheel on the lathe apron. Some lathes have a graduated dial that indicates a position along that axis, while others may use an adjustable stop collar for accurate tool location. Tool positioning speed and accuracy can be facilitated by the use of a digital readout device, as described later in this chapter.

The most common device for clamping the tool or toolholder to a lathe is a simple tool post. It is normally mounted on a compound rest on top of the cross slide. The compound rest has a graduated base and can be swiveled around a vertical axis. In this way, its slide can be set at any angle to the axis of a workpiece. The slide can be adjusted or fed by hand through a screw and nut controlled by a crank and graduated dial. The slide can travel only a few mm (in.) but is useful for feeding tools at an angle, needed for generating short and steep tapers.

The bracket that hangs from the front of the saddle is the *apron*. It contains the mechanism for moving the carriage along the bed. Turning the handwheel on the front of the apron turns a small pinion extending from the other side of the apron. The pinion is engaged with a rack attached to the bed and pulls the carriage along.

18.2.2 Power Feed

Power comes to the apron for feeding the carriage through the lead screw and feed rod along the front of the bed in Figure 18-5. The lead screw is used to move the carriage along at a precise rate for cutting a thread. It passes through a split or half-nut attached to the apron. This nut is normally spread apart but can be closed and engaged with the lead screw by throwing a lever on the front of the apron.

The feed rod is used to save the lead screw when other than threading operations are performed. It drives a train of gears in the apron. Clutches in this gear train are engaged to apply the power to either move the carriage along the bed or the cross slide across the bed.

The power-feed drive on a lathe is always taken off the headstock spindle. In that way, the feed, whether for turning or thread cutting, is always related to the speed of the spindle. The drive goes through a set of reversible gears for setting the direction of feed and through a set of change gears for setting the amount of feed. With a definite setting for the feed change gears, the carriage travels a definite distance in mm (in.) for each revolution of the headstock spindle, no matter how fast the spindle turns.

The feed change gears are in a sliding transmission gear box on the left of the bed below the headstock, as shown in Figures 18-4 and 18-5. A chart on the box tells the operator how to shift the levers to select the available feeds; changes can be made quickly. On some lathes, the changes are made by means of pick-off gears. Most lathes have a large range of feeds. For example, a general-purpose 381 mm (15 in.) lathe cuts 48 different threads from 0.1–6 mm pitch (equivalent to 4–224 threads/in.) and has 48 feed rates from 0.05–3 mm/rev (.002–.120 in./rev).

18.2.3 Types

There are many kinds of lathes, ranging from manual one-tool-at-a-time lathes like those described here, to semi-automatic, automatic, and CNC production-type lathes described later. The production-type machines are generally equipped with multiple cutting tools to eliminate the need for frequent tool changes as are required on the manual lathes when machining complex parts.

The general-purpose manually controlled lathe is known as the *engine lathe*. It has the principal parts already described and depicted in Figure 18-5. Swing capacities generally range from 229–1,270 mm (9–50 in.), and bed sizes from around 0.9–4.9 m (3–16 ft), although lengths up to 15.2 m (50 ft) and more are available. A heavy-duty 406-mm (16-in.) swing engine lathe offers 25 speeds from 14–1,800 rpm. One with 1.3 m (52 in.) of swing has a speed range of 2.3–185 rpm in 27 steps.

Other types of engine lathes are given names that describe particular features, such as *bench lathe, toolroom lathe* (exceptionally accurate), and *gap lathe* (with a removable section in the bed to increase swing). *Wheel lathes* are made for finishing the journals and turning the treads of railroad car and locomotive wheels mounted in sets. *Oil country lathes* are used to make and maintain oil-well drilling equipment and have holes from 178–406 mm (7–16 in.) through their spindles to pass long pieces.

18.2.4 Sizes

The rated size of a lathe indicates the largest diameter and length of workpiece it will handle. Thus a 356 mm (14 in.) lathe will swing a piece 356 mm (14 in.) in diameter over the bed ways. Some manufacturers designate the swing by four digits, such as 1610. The first two designate the largest diameter that can be swung over the bed ways, the last two the diameter over the cross slide (16 and 10 in., respectively, in the example). For length, some lathe makers specify the maximum distance between centers, in inches; others, the overall bed length in feet. Thus one lathe designated to take 762 mm (30 in.) between centers may be the equivalent of another specified as 1.8 m (6 ft) long.

More than the swing size needs to be considered when selecting a lathe. Complete specifications are given in manufacturers' catalogs. Lathes of the same nominal workpiece size capacity may differ considerably in ruggedness, strength, and power. A bench lathe capable of swinging pieces 311 mm (12.25 in.) in diameter × 914 mm (36 in.) long has a 0.37 kW (.5 hp) motor, weighs 227 kg (500 lb), and can be bought for $2,000. A 432 × 1,016 mm (17 × 40 in.) lathe has a 5.6 kW (7.5 hp) motor and costs about $12,000. This larger lathe may be expected to remove metal about 15 times faster than the cheaper one, but costs about six times more. Yet the extra power and rigidity may not be of any value if the larger lathe is not used most of the time or is needed only for light work that the smaller lathe can handle as well. On the other hand, power is a necessity to get the most from the better cutting materials described in Chapter 17, especially for production, where medium-size lathes with motors of 18.7 and 37.3 kW (25 and 50 hp), and strength and rigidity to suit, are not uncommon. A few are powered to over 74.6 kW (100 hp).

Large workpieces must be turned on big lathes. As an example, steel mill rolls up to 1.7 m (66 in.) in diameter × 8 m (26 ft) long are turned on a 298.4-kW (400-hp) lathe with a cut as deep as 51 mm (2 in.) and a feed of 13 mm/rev (.5 in./rev).

18.2.5 Tracer Lathe

A *tracer* or *duplicating lathe* is one with an attachment that enables the machine to turn, bore, face, and generate many kinds of contours from templates. The cutting tool is made to follow a path that duplicates the path of a stylus or tracer finger moving along the template. With two-axis control on the lathe, one direction is that of traverse and is called the *tracing axis*. That would be the lengthwise feed in a turning operation. The other axis provides the in-feed and is called the *feeding axis*.

A variety of devices are available for controlling the relationship between the template follower and the cutting tool on a tracer lathe. Those that utilize machine movement along a tracing axis are called *motor tracers*. *Slide tracer* attachments are used where both the tracing axis and the in-feed axis are controlled. These devices use hydraulic servovalves that provide positive control to facilitate the machining of steps, back tapers, and shoulders.

A tracer lathe is generally faster than a manually operated lathe when several diameters, shoulders, or faces on a part are to be machined because the time to adjust a manual machine from one surface to another is eliminated. There is, however, an investment in time and money for fabricating and setting the template on a tracer lathe. Thus, an economic evaluation of the manual lathe and tracer lathe alternatives, as well as a third alternative, the CNC turning lathe, should be made in planning for the turning of complex shapes and contours. The CNC turning lathe will be described later in this chapter and CNC programming is introduced in Chapter 31.

18.3 ACCESSORIES AND ATTACHMENTS

Lathe accessories include common workholders and toolholders, and supports such as chucks, collets, centers, drivers, rests, fixtures, and mandrels. Attachments are devices to facilitate specific operations. They include digital readout devices, stops, thread-chasing dials, taper attachments, and devices that enable milling, grinding, gear cutting, and cutter relieving on the lathe.

18.3.1 Chucks

Usually, a *universal* or *scroll chuck* has either three or six jaws engaged and moved in unison by a scroll plate located behind the face of the chuck body. Simultaneous movement of the jaws is accomplished by inserting a chuck wrench into one of the pinions located on the periphery of the chuck and rotating the wrench to either open or close the jaws. Figure 18-6(A) shows a 203-mm (8-in.) diameter universal chuck with three hard reversible jaws. The reversible jaws provide flexibility for clamping on either outside or inside surfaces.

Loss of gripping force at high speeds becomes a serious problem with ordinary lathe chucks. It is not unusual to operate lathes at over 1,000 rpm with many cutting tool materials, and the centrifugal forces on chuck jaws become appreciable. To assure their safe operation, the insurance industry has recommended rotational speed limits for chucks as outlined in Table 18-1.

An *adjustable chuck* is a universal chuck mounted on an adapter that attaches to the spindle nose of a lathe. By the use of positioning screws on the adapter, the chuck can be adjusted to a true center position. The universal chuck shown in Figure 18-6(A) can be adjusted on the adapter to run as true as 13 μm (.0005 in.) total indicator reading (TIR) repeatedly. A

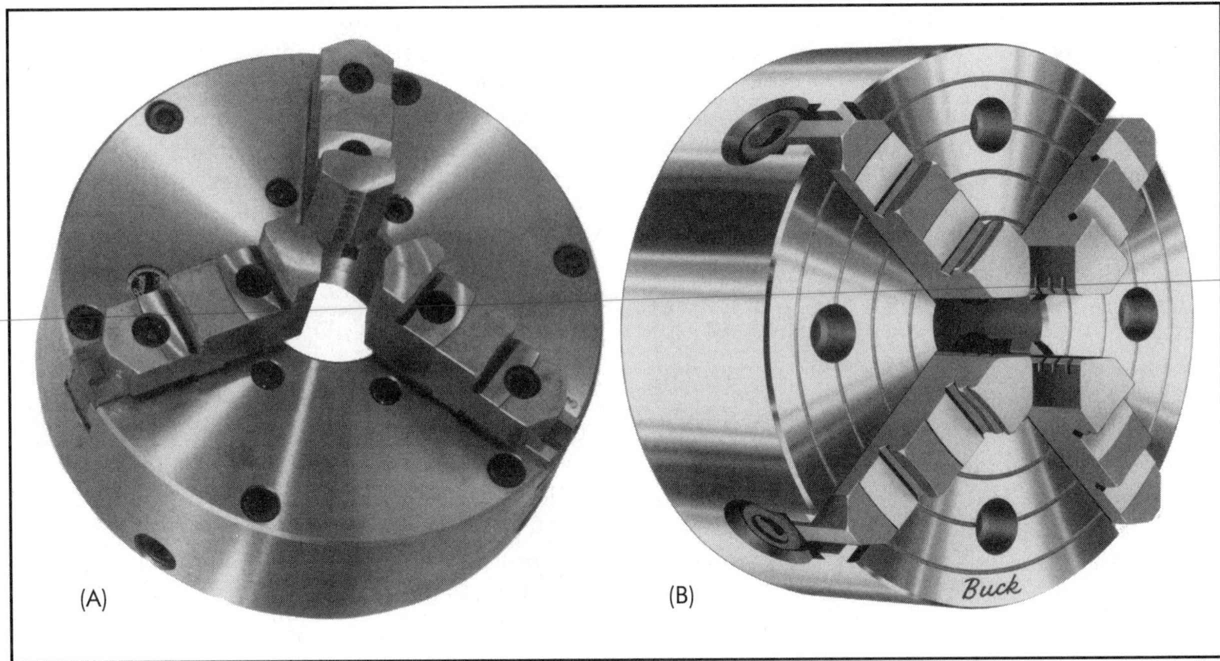

Figure 18-6. (A) A universal chuck with three hard, reversible top jaws and, (B) a four-jaw independent chuck with solid reversible one-piece jaws. (Courtesy Buck Chuck Company)

Table 18-1. Rotational speed limits for chucks

Chuck Diameter, mm (in.)	152 (6)	210 (8.25)	279 (11)	305 (12)	381 (15)	457 (18)	533 (21)	610 (24)
Maximum safe rpm	2,890	2,420	2,010	1,680	1,340	1,110	960	840

conventional universal chuck can only be expected to provide workpiece centering to 51 μm (.002 in.) TIR when new.

A four-jaw *independent chuck* has jaws that are moved separately, each by a screw. The jaws will clamp almost any shape of workpiece and can be adjusted to run as true as desired. However, the centering of a workpiece in a four-jaw independent chuck is tedious and time consuming. Figure 18-6(B) shows a 203 mm (8 in.) four-jaw independent chuck with reversible one-piece jaws. A four-jaw chuck like the one shown will cost about $700, while a similar size, universal three-jaw adjustable chuck will cost about $1,100.

A *combination chuck* has jaws that can be moved together through a scroll plate or adjusted separately.

Two-jaw chucks are adapted to hold workpieces of irregular shapes by means of slip jaws added to the permanent jaws. Each piece can be chucked in less time than with a four-jaw chuck, but sufficient production is necessary to justify the special jaws.

Air- and *hydraulic-operated chucks* are quick, grip the work strongly, and are economical for production. A cylinder is carried on the rear of and revolves with the spindle of the machine. The piston actuates a rod lengthwise through the spindle and through levers opens and closes the jaws of the chuck. Manufacturers of chucks of this kind guarantee runout of less than 13 μm (.0005 in.).

A *wrenchless chuck* is operated by a lever on a ring at the rear of the chuck body. The lever does not revolve with the chuck and can be actuated before the chuck body comes to rest. The action is fast, and wrenchless chucks are often used in production.

A *drill chuck* may be used on either the headstock or tailstock spindle of a lathe to hold straight-shank drills, reamers, taps, or small-diameter workpieces. Two types of drill chucks are available for such work: keyed or keyless. Figure 18-7 shows an illustration of a keyed type on which a geared head-driving key with a T-handle is used to open and close the jaws to grip a cylindrical drill shank or other tool. The keyed drill chuck permits the operator to adjust the grip on a tool to about any degree required to resist the cutting torque applied. A keyless drill chuck is tightened and released by grasping the chuck body in the hand. This permits more rapid tool changes and increased operator productivity. A self-tightening feature within the chuck automatically increases the gripping force proportional to the increased torque to prevent tool shank slippage.

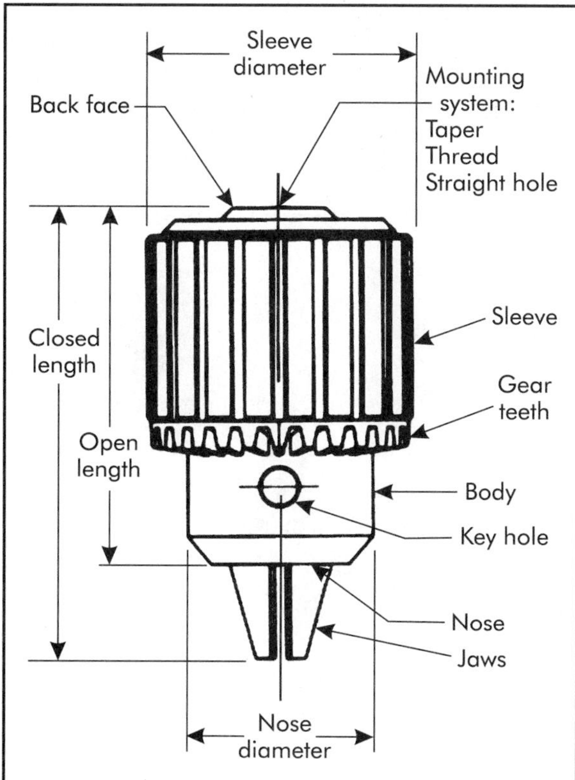

Figure 18-7. Keyed-type drill chuck. (Courtesy Jacobs Chuck Manufacturing Company)

18.3.2 Collets

A *collet* is a thin steel or brass bushing with lengthwise slots and an outside taper. When it is forced into the tapered sleeve of a collet chuck, the collet is sprung together slightly to grip a workpiece securely and accurately.

The three basic styles of collets used for metal cutting applications are stationary, push-out, and draw-in collets. They are illustrated in Figure 18-8(A). *Stationary collets* do not move longitudinally in the machine spindle, and thus variations in stock lengths are eliminated. *Push-out collets*, such as the one illustrated in Figure 18-8(B) provide accurate control of workpiece lengths when the barstock is fed up, through the open collet, to a turret-mounted stock stop as the collet closes. When this happens, the push bar or plunger pushes the tapered nose of the collet against a mating tapered bore in the spindle nose cap. The resulting action forces the collet to close and hold the barstock in place for machining. Draw-in collets are pulled into the spindle by a drawbar. The resulting action forces the tapered outside diameter of the collet head to press against the tapered inside diameter of the spindle nose cap, causing closure of the collet and a secure grip of the barstock.

18.3.3 Centers and Drivers

A lathe center has a 60° included angle taper at one end and a sticking taper at the other end to fit into either the headstock or tailstock spindle of the lathe. The solid (dead) center shown in Figure 18-9(A) has a full tungsten-carbide point for increased wear resistance in heavy turning applications. The one-piece construction of the dead center resists deflection for positive workholding. However, heat build-up between the stationary center and the rotating workpiece tends to limit cutting speeds. The rotating point on the live center shown in Figure 18-9(B) is supported by ball bearings that permit the point to rotate at the same speed as the workpiece. Live centers of this type are constructed to provide a rotational accuracy of 0.005–0.010 mm (.0002–.0004 in.) total indicator reading (TIR) and speeds up to 6,000 rpm.

A *dog plate* is shown on the spindle nose in Figure 18-10. A *face plate* is larger than a dog plate and has a number of radial slots for bolts.

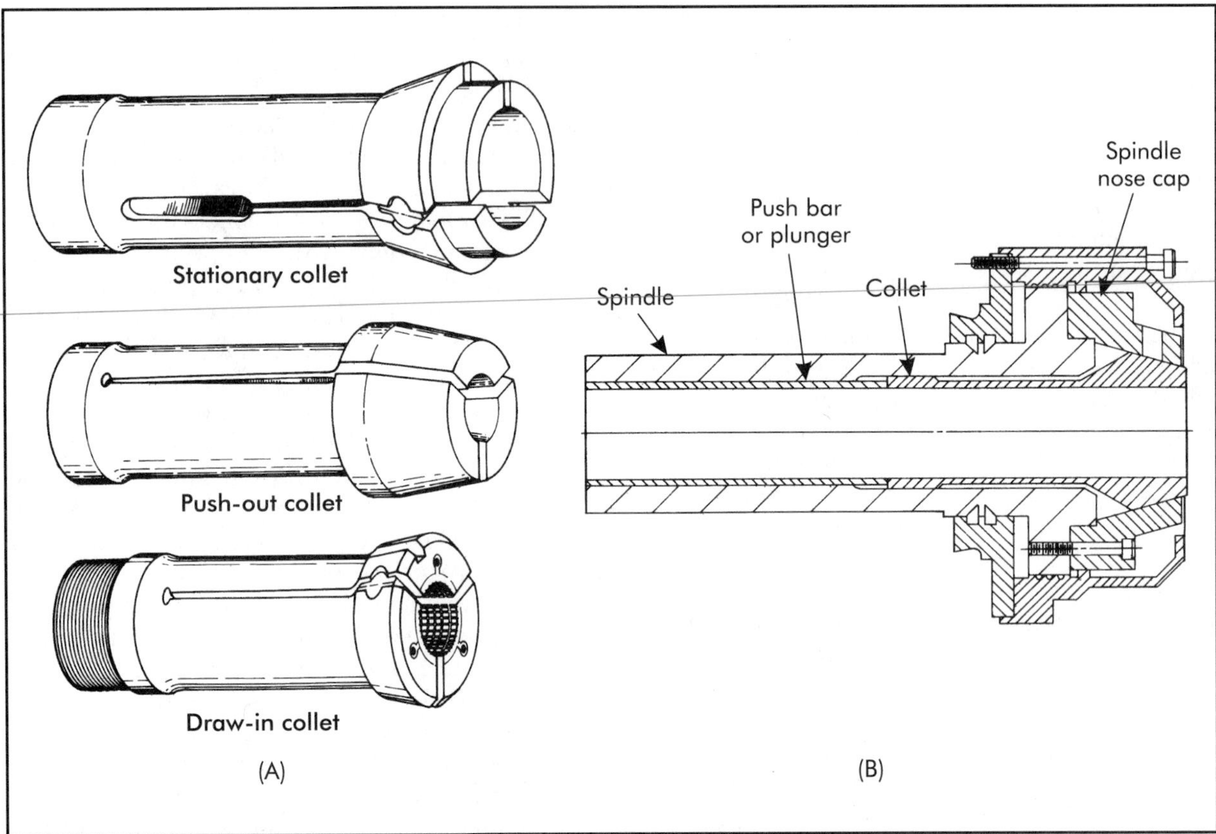

Figure 18-8. (A) Three basic styles of collets. (B) Push-out collet provides accurate length control because stock does not pull away from the stop.

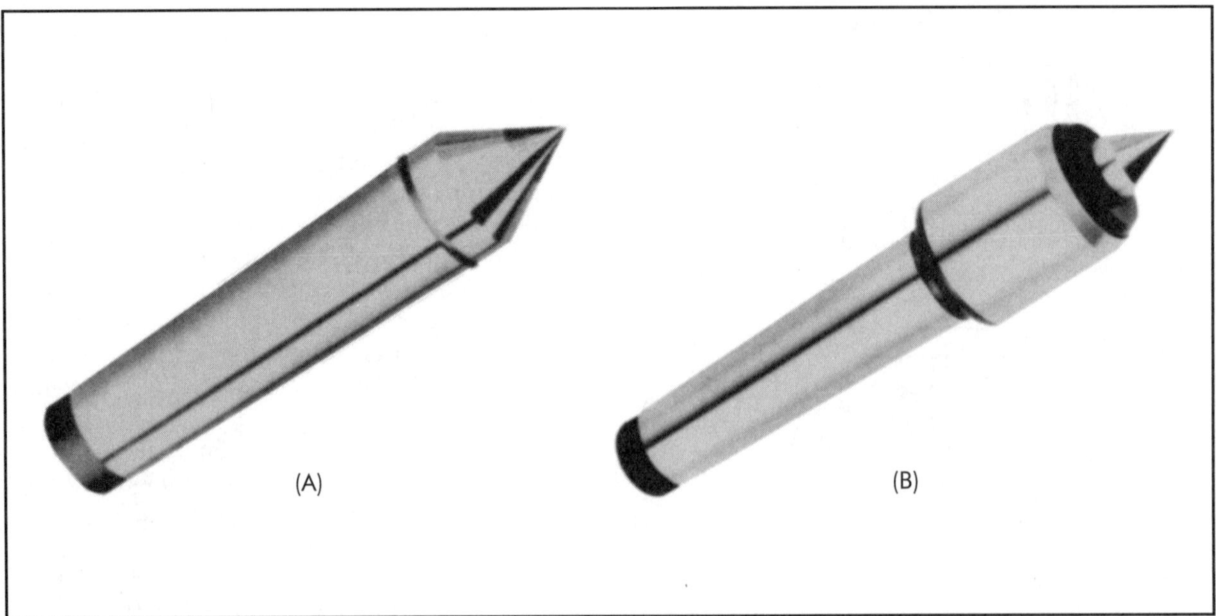

Figure 18-9. (A) Solid lathe center; (B) ball bearing live center. (Courtesy Jacobs Chuck Manufacturing Company)

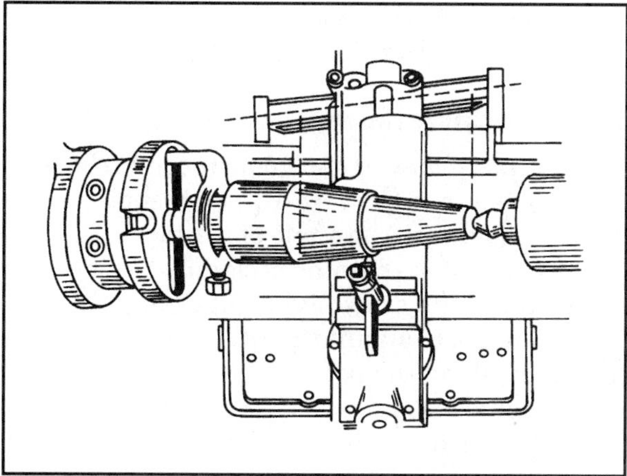

Figure 18-10. Engine lathe equipped with dog plate on spindle nose and taper attachment.

Workpieces are bolted on the front of the face plate.

A *fixture* is a special device fastened directly to the spindle nose or bolted on a face plate to hold and locate a specific piece or pieces. Fixtures are commonly used for quantity production of pieces on production, turret, and automatic lathes.

A *mandrel* locates a workpiece from a hole. Common types are depicted in Figure 18-11. A tapered mandrel is pressed into the workpiece hole.

18.3.4 Rests

A *center* or *steady rest* has three shoes that are brought up to contact and support a slender workpiece that would otherwise deflect too much under its weight or from the cutting forces. The shoes are carried in a bracket clamped on top of the bed. A *follower rest* is fastened to and moves along with the carriage with shoes that support the workpiece at the cutting tool position.

18.3.5 Digital Readout Systems

As indicated previously, the positioning of a cutting tool for a machining operation on a manually operated lathe is accomplished by rotating handwheels on the various machine elements. By reading the graduated dials or collars associated with those handwheels, the cutting tool is located in a given position in the X and Z planes on the lathe. An experienced lathe operator can perform locating maneuvers with a reasonably high degree of accuracy. With a less experienced operator, the process is slow and error prone. The *digital readout* (DRO) *device* allows the operator to determine the X and Z coordinate positions of the cutting tool at a glance, and also to reposition the tool accurately and swiftly on successive workpieces. In addition, the DRO enables accurate positioning in spite of irregularities

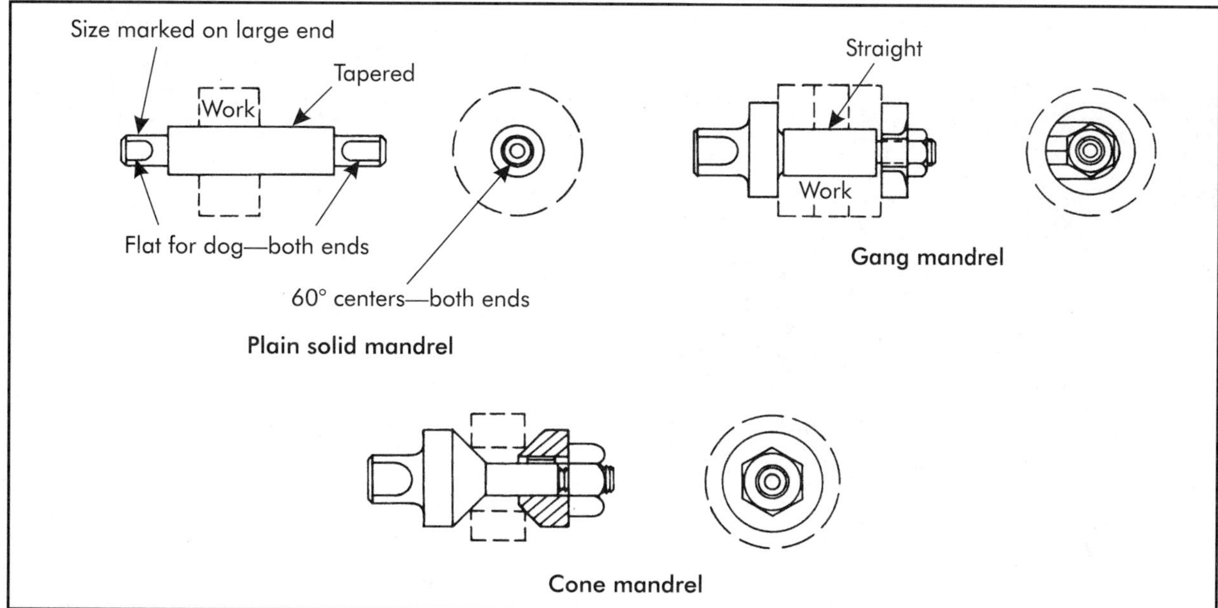

Figure 18-11. Common mandrels.

caused by a worn lead screw or other machine elements.

A digital readout system for a lathe consists of a linear encoder for each axis position and a display unit to numerically register the positions. Normally, two linear encoders are used on lathes, one to sense the position of the cross slide (X axis) and one to sense the position of the carriage or saddle along the Z axis. On larger lathes, a third encoder may be used to sense the position of the compound rest along another Z axis (Z_0 or Z_2). These three axes positions are noted on the DRO display unit of Figure 18-12 wherein a taper turning operation is being performed on a workpiece held in a chuck.

Linear encoders range from gage wheels to glass and magnetic scales. For the popular glass scale type, a light source and photodetector is moved along a glass scale. This photo-optical arrangement reads the overlapping patterns of light transmitted through the gratings on the glass. Digital readout units range from simple single-axis data displays to sophisticated multi-axis cathodic ray tube (CRT) -based units. Most standard CRT-based units feature simple positioning aids that allow operators to establish zero settings for tools, preset for datum points, switch between inch/metric measures, and convert from radius to diameter. More advanced models will perform certain calculations and/or even compile a program while machining the first piece.

Figure 18-12. Three-axis digital readout system being used to show cutting tool position on a lathe taper turning operation. (Courtesy Heidenhain Corporation)

DRO units may come equipped on new machines or be retrofitted to older machines.

18.3.6 Attachments

A *taper attachment* shown fastened to the rear of the lathe in Figure 18-10 has a slide on top that is swiveled to the angle of the taper desired. A block riding on the slide is clamped to the rear extension of the cross slide and causes the tool to cut along the angular path as the carriage is traversed along the bed.

As has been pointed out, outside tapers can be turned without an attachment by offsetting the tailstock, but the adjustments required make this method slow. Inside or outside steep tapers can be generated by swiveling the compound rest, but the length of taper must be short. Short or long and inside or outside tapers may be generated with the taper attachment, but steepness is limited. The attachment is easy to adjust, and the lathe may be easily reset for straight work.

A *relieving attachment* is a device for moving the cutting tool in and out in relation to the revolving workpiece to back off the teeth of multiple-point tools, such as milling cutters and reamers.

Stops of various kinds can be attached to the bed to position the carriage accurately and quickly for spacing grooves, facing shoulders, etc.

A variety of attachments are available for adapting the lathe to do milling, gear cutting, grinding, shaping, and fluting. Such work can be done more efficiently on other machines especially designed to do the work; the lathe is justified only as an expedient where more suitable equipment is not available.

Lathe manufacturers have developed and are able to furnish many other ingenious attachments to enable lathes to economically perform special operations at moderate and large-scale production rates. Examples of these are attachments for turning crankshafts and pieces with unusual shapes.

18.4 LATHE OPERATIONS

18.4.1 Practical Tolerances and Surface Finishes

Lengthwise dimensions (Z axis) to be machined on a lathe are normally given a tolerance of ± 0.5 mm ($\approx \pm .016$ in.) unless a micrometer dial or digital readout device is available for tool location and adjustment along that axis. In addition, some lathes are equipped with a carriage stop that can be adjusted by means of a screw to provide accurate *Z*-axis tool positioning or travel limit.

Under average conditions, reasonable tolerances for rough turning and boring range from 0.15 mm (≈.005 in.) for diameters under 15 mm (≈.5 in.) to 0.40 mm (≈.015 in.) for diameters over 50 mm (≈2 in.). For finishing operations, tolerances from 50 μm (.002 in.) for diameters below 15 mm (≈.5 in.) to 0.180 mm (≈.007 in.) for diameters above 50 mm (≈2 in.) are usually achievable. For hole finishing, a properly used fluted reamer will produce holes to within 25 μm (.001 in.) of nominal size.

According to Figure 16-30, a wide range of surface finishes from 25 μm (1,000 μin.) to as fine as 0.025 μm (1 μin.) R_a are achievable in turning and boring operations. However, conventional turning and boring operations in most production environments will only permit surface finishes in the range of 6.3 μm (250 μin.) down to 0.40 μm (16 μin.) R_a. Holes may be regularly finished to 0.8 μm (32 μin.) R_a by reaming.

Workpiece tolerances and surface finishes are, of course, dependent on the condition of the machine and the cutting tool. A machine with worn spindle bearings and ill-fitting guideways cannot be expected to yield as close a tolerance and as fine a surface finish as one that is in good condition. Similarly, a worn or poorly prepared cutting tool is likely to result in higher cutting forces and vibrations, and contribute to poor dimensional control and a rough surface finish. With the current trend of using cast or forged parts of near-net-shape configuration, machining operations are moving away from performing heavy roughing cuts to semifinishing or finishing in a single pass. This trend places more emphasis on high-speed machining with a fairly shallow depth of cut and a low feed rate. This permits good dimensional control and fine finishes without secondary lathe and/or grinding operations.

Because the simple lathe is versatile and flexible, it is not generally a fast producer. The tools can be arranged in many ways but, for the most part, the tools can be used only one at a time, and for each piece and every different cut, tools must be changed, set, and adjusted. This takes up an appreciable amount of the time to make a part. If a quantity of pieces must be made, the time soon adds up to a large amount. A skilled operator is required to operate a lathe properly, labor is expensive, and a sizable proportion of the time is spent just waiting while the machine is cutting. As a rule, it is not feasible for one person to operate more than one engine lathe at a time.

Where only a single cut is needed on a part, many pieces of that kind might well be produced on an engine lathe as economically as on any other machine tool. However, most parts require more treatment, and for them the ordinary lathe is suitable for making only one or a few pieces, except in cases where extra equipment, such as an indexing turret or tracer attachment, is added.

18.5 PRODUCTION TURNING MACHINES

Simple lathes like those described earlier are efficient for machining one or a few pieces of a kind of large variety. However, more complex machines have evolved, which are faster and require less labor and skill to operate than the engine lathe to produce duplicate parts in desired quantities. One of the basic features of some of these production machines is that once they are set up, the tools can be quickly applied to the work repeatedly and stopped precisely when they reach the end of their machining operations. Many of these machines require considerable skill and time for setup, but after they are set up, they can be operated by lesser skilled labor. Often, several automatic machines can be tended by one operator.

Until a few years ago, the hierarchy of production turning machines ranged from the manually operated turret lathes to the single-spindle automatic bar and chucking machines, on up to the extremely productive multiple-spindle automatic bar and chucking machines. The Swiss-type automatic screw machine, once used primarily for the production of small screws and other small threaded parts, had been the workhorse of the small precision turned parts industry for many years.

The early versions of these single- and multiple-spindle automatic lathes used specially designed cams and other mechanical devices to control the positioning of the spindles and the movement of the cutting tools. Thus, they were setup intensive and required the production of large batches of product to justify the setup time and expense. Many of these machines are still in use today within those manufacturing industries that require long-run production of large batches of standardized parts. However, where the flexibility of just-in-time (JIT) manufacturing and other small order quantity production programs are in effect, the cam-controlled machines have, in many cases, been replaced by newer computer numerically controlled (CNC) lathes, sometimes referred to as CNC turning centers. Since the machine elements and machining cycles on these machines are controlled by software through servomotor drives and other digitally responsive hardware, they are easier and quicker to set up and operate.

The ultimate in machine-centered automation is represented by the CNC machining centers. They perform a variety of machining operations (for example, turning, drilling, boring, threading, and milling) in sequence or simultaneously on one or more parts by means of quick-change tool magazines and other automatic machining and part-handling devices. Machining centers are introduced in Chapter 30.

Typical examples of production turning machines and their operating features are covered in the following sections. The simplest of these are the manual turret lathes that require constant operator attention. Beyond these are machines that go through every step of an operation automatically and require minimal attendance after they have been set up.

18.5.1 Turret Lathes

A *turret lathe* is a manual lathe with a hexagonal toolholding turret in place of the tailstock found on an engine lathe. Some turret lathes are designed and equipped for working on barstock and are called *bar-type machines*. The name of *screw machine* or *hand screw machine* also has been used for such machines, particularly in the smaller sizes. Other turret lathes are equipped for chuck work.

The name "turret lathe" used alone ordinarily implies a horizontal spindle machine. Vertical turret lathes are described later. The two main types of horizontal turret lathes are the ram and saddle types.

A *ram-type turret lathe* carries its turret on a ram as shown in Figure 18-13. The ram slides longitudinally on a saddle positioned and clamped on the ways of the bed. Tools in their holders are mounted on the faces of the turret, and the tools on the face toward the headstock are fed to the work when the ram is moved to the left. When the ram is withdrawn, the turret indexes, and the next face, called a *station*, is positioned to face the headstock.

A ram is lighter and can be moved more quickly than a saddle, but lacks some rigidity. Because of convenience and speed, the ram-type construction is favored for small- and medium-sized turret lathes where the ram does not have to overhang too far.

Gears are shifted manually to change speeds on some turret lathes; others have power shifts. Typical machines offer 8, 12, or 16 speeds over wide ranges.

A *reach over* or *bridge-type carriage* across the entire bed is commonly found between the saddle and headstock on a ram-type turret lathe. On the carriage is the cross slide with a quick, hand-indexed, four-station, turret toolholder on the front end and a holder for one or more tools at the rear. Heavy single-point tools are clamped in these holders and fed to the work by movement of the carriage along the bed and the cross movement of the slide.

A plain cross slide is entirely hand operated, but the universal kind is power-fed and is more common. The ram usually has power feed in addition to hand feed. Eight feeds from 0.076–0.76 mm/rev (.003–.030 in./rev) are typical.

The cross slide can be positioned accurately by means of a cross screw and micrometer dial. Individual stops for each tool station can be set to throw out the power feed and limit the longitudinal movement of the ram or carriage. In this way, the movement of any tool along the workpiece can be precisely controlled and readily repeated for as many workpieces as desired.

The hexagonal turret of a *saddle-type turret lathe* is carried directly on a saddle that slides

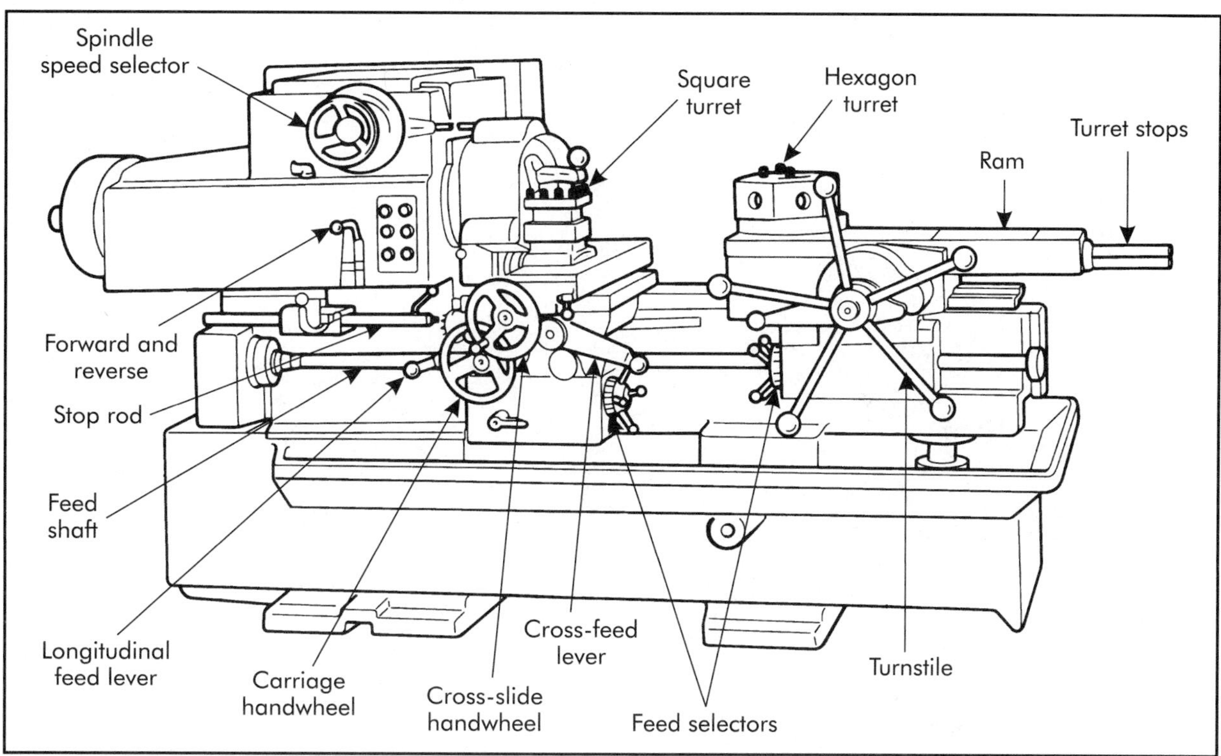

Figure 18-13. Power-fed, ram-type turret lathe.

lengthwise on the bed as depicted in Figure 18-14. This construction is favored for large turret lathes because it provides good support for the tools and the means to move the tools a long way when necessary. The saddle is moved toward the headstock by hand or power to feed the tools to the work, and is withdrawn to index the turret.

The turret is fixed in the center of the saddle on some machines. On others it may be moved crosswise. This helps reduce tool overhang for machining large diameters and is helpful for taper or contour boring and turning.

Most saddle-type turret lathes have the side-hung type of carriage, which does not extend across the entire top of the bed and allows larger pieces to be swung. This means the rear tool station on the cross slide is lost.

The cross slide and carriage usually are both power- and hand-fed. The cross slide is adjusted by a micrometer dial and screw. Stops are provided to throw out the feeds and position the tools longitudinally for any of the stations of either a square or hexagonal turret.

Sizes

The size of a turret lathe is designated by a number that indicates the diameter of workpiece that can be swung and the diameter of bar that can be passed through the hole in the spindle. Machines of different makes with the same size number may vary somewhat in capacity.

Tools and Attachments

On turret lathes, collets and collet chucks are commonly used for barstock, and hand and power chucks are used for individual pieces. Essentially the same kinds of single-point tools, form tools, and multiple-point tools, such as drills, reamers, and taps, are used on turret lathes as on engine lathes. Single-point tools are generally heavier.

Convenient devices are made specifically for holding and adjusting cutting tools singly and in groups on the hexagonal turrets. This equipment is commercially available in many standard forms and sizes and is described in manufacturers'

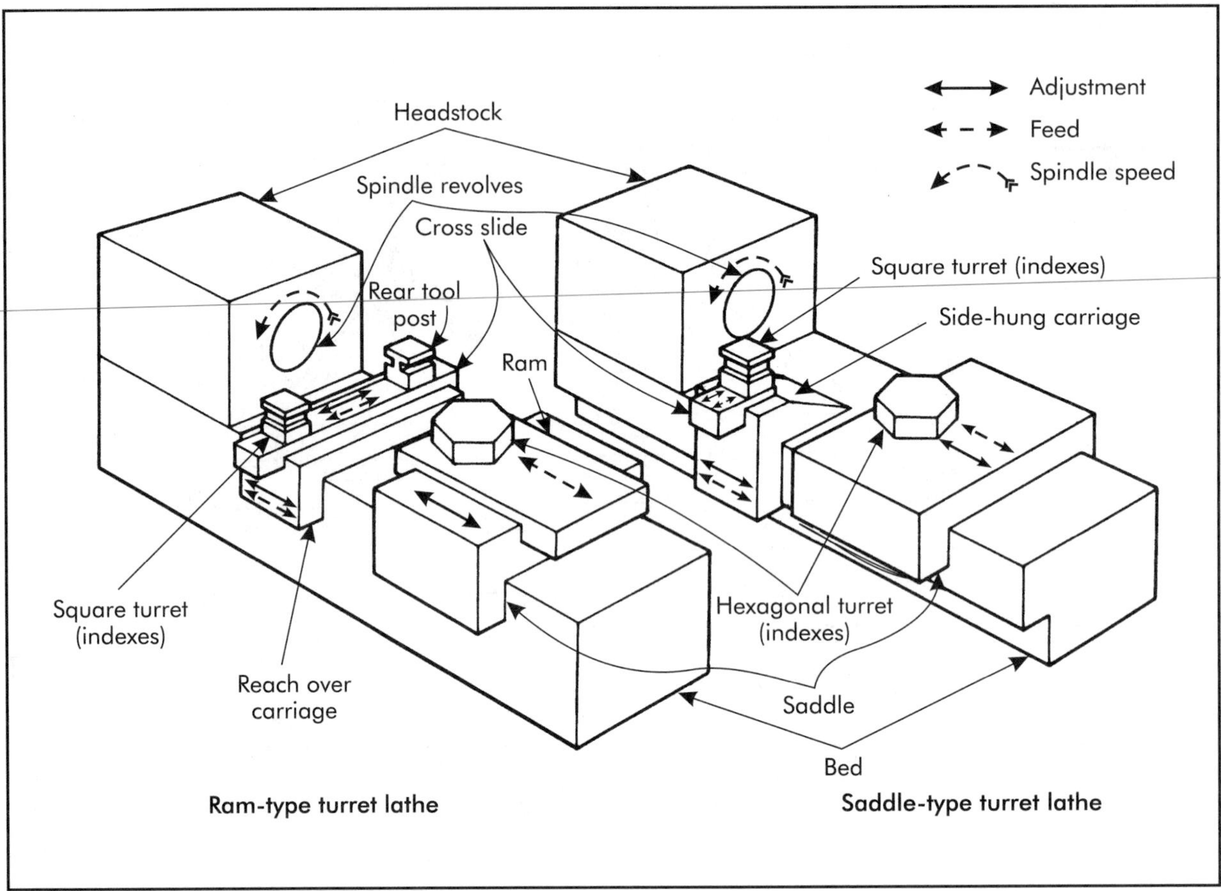

Figure 18-14. Principal parts and movements of horizontal turret lathes.

catalogs. Some equipment is intended for bar work, other for chucking, but much is suited for both applications.

Generally, barstock is supported by the tailstock center on an engine lathe, but that usually is not feasible on a turret lathe, especially when cuts are taken from the hexagonal turret. *Box tools* support overhanging bars that are being turned, faced, chamfered, or centered from the hexagonal turret. Rollers or a crotch bear against and back up the work surface opposite the cutting tool. The *bar turner* shown in Figure 18-15(A) is an example of a box tool. Other models may carry several turning bits, a facing tool, or a center drill.

The *quick-acting slide tool* in Figure 18-15(B) is designed to hold round-shank single-point tools and boring bars for fast recessing and facing cuts on both bar and chuck work. The slide that carries the cutting tools moves 12.7 mm (.50 in.) with a quarter turn of the handle. The *adjustable knee tool* in Figure 18-15(C) can be set up quickly for turning combined with drilling, boring, or centering on short pieces. The *combination stock stop and starting drill* in Figure 18-15(D) is a two-purpose tool that saves one turret face and one index. A typical set of bar equipment is illustrated on the lathe in Figure 18-13.

Chucking work often involves a greater range of diameters and more tool overhang than bar work. *Multiple turning heads* carry *cutter holders* and *boring bars* on the turret faces in line with the machine spindle. A *stationary pilot bar* may be attached to the headstock and slips into the bushings and gives extra support to the turning heads. It can be heavy and strong because it does not add to the load on the turret. *Piloted boring bars* slip into and are guided and reinforced by an internal or center pilot in the machine spindle.

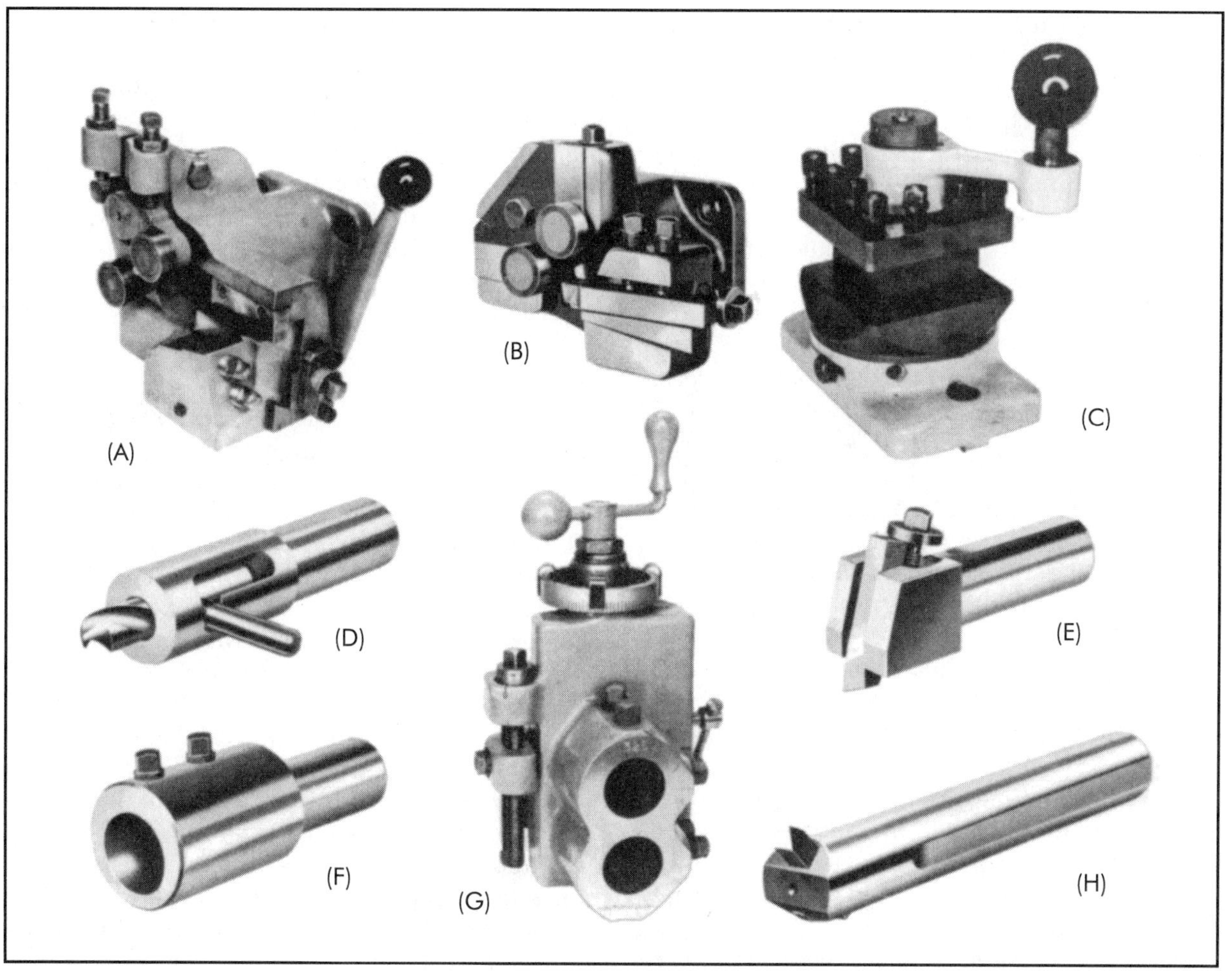

Figure 18-15. Examples of tooling for turret lathes: (A) single-cutter bar turner with roller support and quick-acting tool relief; (B) single-cutter turner with roller support; (C) indexable square turret for cross slide; (D) combination stock stop and starting drill; (E) angle cutter holder; (F) toolholder for drills, counterbores, or reamers; (G) vertical slide toolholder; (H) stub boring bar. (Courtesy Bardons and Oliver, Inc.)

Planning Operations

Time is the major item of cost in a turret lathe operation. It includes the time to set up the machine and make the actual cuts. The following principles are guides to obtaining low costs. They also apply to other kinds of operations and machine tools.

A large part of setup time is spent on mounting and adjusting the cutting tools. These costs may be minimized by using universal tooling and maintaining a permanent setup on a turret lathe. To do this, large and heavy tools that perform basic functions in most operations are permanently mounted in their logical order on the turret. All these tools are not needed for every job, but then the turret may be back or skipped-indexed. In most cases, the extra indexing time is less than would be taken to remove the tools and change them around on the turret. The lighter tools are rearranged in various combinations on the heavy toolholders for various jobs. Recommendations for proven permanent setups for various classes of work and sizes and types of machines are available from turret lathe manufacturers.

The choice of raw material affects work handling time and its selection is based upon several factors. Barstock is easily held and pieces often can be completed and cut off in one operation. A casting or forging may require that less stock be removed, but calls for the cost of a pattern or die and the founding or forging operations. As an example, a part can be cut from barstock at a cost of $40/piece. The cost of 5 pieces is $200 and for 25 pieces is $1,000. If the part is cast, a pattern must be made at a cost of $100. Molds can be made and the pieces cast for $10 each. The machining charge to finish the castings is $15 each. To make 5 pieces by casting and machining costs $225, for 25 pieces, $725. Obviously, barstock is preferable for 5 pieces, and castings for 25 pieces.

Work handling time also is affected by the choice of collets, chucks, and fixtures. For average work in small or moderate quantities, standard holding devices are best with special jaws, arbors, and simple fixtures added as justified. Special fixtures can pay for themselves on jobs where the parts are otherwise hard to hold and made in fairly large quantities.

Machine handling time consists of the time to index and position the tools and set the speeds and feeds. More time is required to advance from step-to-step in an operation and it is harder to keep up a fast pace on a large machine than on a small one because of the heavier masses that must be moved.

Parts like collars, spacers, and gear blanks can be machined in groups with turning, drilling, boring, reaming, and cutting off done on all pieces in one group at each setting. In that way, the machine movements per piece are less than if the pieces are machined individually.

Machine handling time is reduced by taking combined and multiple cuts to save indexing. Cutting time is also reduced. A *combined cut* is one where tools in both the hexagonal turret and square cross-slide turret are set up to cut at the same time. A *multiple cut* is one where two or more tools are applied at the same time from one turret station.

Internal cuts are almost always made by tools in the hexagonal turret and are planned and set up first in an operation. Provision must be made for the proper order of internal cuts. For instance, drilling should take place before boring and boring before reaming a hole.

Tolerance down to 10 mm/cm (.001 in./in.) and finishes as fine as 1.5 μm (60 μin.) R_a are practicable on turret lathes, but the more exacting the requirements, the more cost. One cut may suffice to hold a tolerance of 0.25 mm (.010 in.). But, as a rule, two cuts are needed for a tolerance between 0.15 and 0.25 mm (≈.005 and .010 in.), and three cuts for a tolerance less than 0.15 mm (≈.005 in.) with corresponding surface requirements.

Vertical Turret Lathes

The *vertical turret lathe* (VTL) is convenient for turning and boring short cylindrical workpieces over about 500 mm (≈18 in.) diameter. This is because the work can be located and clamped on the rotating table more readily than it could be hung on the end of the spindle of a horizontal turret lathe. The turret on a vertical turret lathe is carried on a ram on a cross-rail above the worktable and can be fed up or down (Y_1 axis) and crosswise (X_1 axis). There is also a side tool head, essentially a carriage and cross slide, alongside the table with a square turret that can be fed radially (X_2 axis) and parallel to the table axis (Y_2 axis). For most VTLs, the rotary table serves as a four-jaw independent chuck for supporting and clamping large cylindrical workpieces. The table also has T-slots for clamping irregularly shaped parts.

The VTL shown in Figure 18-16 has a 1,118-mm (44-in.) diameter table with 18 table speeds from 5–250 rpm. It also has 18 turret and toolhead feeds ranging from 5–1,803 mm/min (.2–71 in./min). The turret head has five-position power indexing with automatic clamping. Both the turret and side toolhead may be rapidly traversed to the cutting position via pendant control.

Larger machines similar to the one shown in Figure 18-16 are available with rotary tables up to about 3.7 m (12 ft) in diameter. These are generally configured with two vertical toolhead rams instead of turrets and are generally referred to as *vertical boring machines*. Both types are available with CNC operation. The VTL is a very versatile machine for turning, facing, boring, and grooving a variety of large castings and weld-fabricated parts for a number of product lines.

Figure 18-16. Vertical turret lathe with five-position turret and four-position square side toolhead. (Courtesy Stan-America Corporation)

18.5.2 Automatic Turning Machines

A machine that moves the work and tools at the proper rates and sequences through a cycle to perform one or more operations on one or several workpieces without the attention of an operator is commonly called an *automatic*. Strictly speaking, the machine is a semi-automatic if an operator is required to load and unload the machine and start each cycle. Often an operator can do this for several machines in a group or in a flexible cell as described in Chapter 30. Workpieces may come to a fully automatic machine on a conveyor or an operator may load a magazine or hopper at intervals. In many of the more advanced automation systems, programmable robots are used to load and unload machines. On bar-type automatic turning machines, bar-stock is automatically advanced to the cutting position and a finished part is allowed to fall into a conveyor hopper after it has been machined and cut off the bar. Although referred to as automatic turning machines, most of these types of machines perform other operations such as drilling, boring, reaming and, in some cases, thread cutting or tapping.

Many automatic turning machines are made massive, rigid, and powerful to drive cutting tools at their maximum to get the most from gangs of tools and multiple and combined tooling. Most of these machines are designed and constructed to maintain the degree of accuracy and repeatability required for the mass production of thousands of parts. Some are even equipped with sensor devices to detect worn or broken tools so that the machine can be stopped before producing many defective parts. Tool setups for automatic turning machines are planned and tools are carefully designed to get as many tools cutting at once to complete the work as fast as possible.

Automatic machines intended for large lots of finished workpieces and infrequent change-over usually are not designed for quick setup. For example, speeds and feeds are not changed by sliding levers or turning dials, as on an engine lathe. On the older automatics, pick-off gears must be removed and replaced and, on the newer machines, digital servo units must be programmed. The multiplicity of tools must be set with respect to each other as well as to the machine. In many cases, it is a good practice to preset the tools in blocks off the machines. Then the blocks go into definite positions quickly on the machine at setup time and machine downtime is reduced. Setup time may be less than an hour in favorable cases where not much has to be changed from one job to another. Other more complex change-overs may require considerably longer times.

Many manufacturers of a large number of different parts have established *group technology* programs (Chapter 29). Such programs classify parts that have similar features and require similar machining operations into groups and relate those groups to particular machines to minimize setup times. Thus, careful scheduling of parts and machines within these technology groupings can often result in considerable reductions in setup times.

Even though setup time may be extensive, an automatic machine is at an advantage when cutting because it is fast and often able to make up for the time lost in setup after perhaps a hundred or so pieces.

An automatic machine follows a definite program for work positioning, movement of tool slides or turrets, and spindle speeds to perform an operation. Programming and accomplishing the movement of these elements is done in various ways on different makes and models of automatic turning machines. A basic mechanical device for driving tool slides and other units is the cam. However, some machines require a different set of cams for each job. The shape of each cam programs one or more movements. Other machines do not require cam changes for each job and are called *camless automatics*. One type has a permanent set of cams that act through adjustable linkages. Without cams, activities are carried out through adjustable and/or programmable electro- and hydromechanical devices or digital servo units.

The designation "turning machine" is applied to a rather large spectrum of lathe-like machines in the machine tool marketplace. However, most of these machines do considerably more than just turning and boring. The emphasis in competitive manufacturing circles is *process integration*, which often calls for turning, drilling, boring, and milling to be accomplished in one setup. In addition, there is considerable economic pressure to include a variety of secondary operations within a single setup. Thus, many of today's automatic turning machines are configured and equipped to do backside operations, cross-drilling and milling, automatic loading and unloading, and even in-process measurement and gaging. Some of these operations require live tooling on one or more of the tool positions. The objective is to permit these machines to produce net-shape or near-net-shape parts in about any quantity with little or no operator attention. As these machines become more multifunctional, they are often referred to as *automatic turning centers* or *automatic turning cells* (Chapter 29).

Selected types of automatic turning machines are discussed in the following sections. They point out some of the ingenious ways in which a multiplicity of tooling can be brought into play for various first and secondary machining operations. Because of space limitations, this material is by no means exhaustive of the variety of automatic turning machines available.

Computer Numerically Controlled (CNC) Lathes

Before the 1950s, turning operations on lathes for small- or medium-volume production of a variety of parts was generally performed manually and the production rates were relatively low. Alternately, the large-volume, high-production rate turning of a single type of mass-produced part with consistent quality generally required automatic or semi-automatic lathes. As discussed earlier, these machines were mechanically controlled, usually with cams and linkages, and were quite expensive. In addition, they were setup intensive and their use was only justified when the quantity of parts to be machined was large enough to compensate for the high machine and setup costs.

Contemporary manufacturing practices, characterized by an emphasis on just-in-time and agile manufacturing concepts, requires that automatic lathes be suited to small or medium lot-size production and frequent changeover from one part or part model to another. Since the introduction of the numerically controlled (NC) milling machine in 1952 (Chapter 1) and the programmable logic controller (PLC) in 1968, machine tool manufacturers have steadily adapted their machines to the numerical control technology. With the development of microcomputer technology in the late 1970s and early 1980s, most NC controllers have been built around that technology. Thus, NC machine tools are referred to as being computer numerically controlled (CNC). In the future, it is expected that the control of preference for many machine tools will be a personal computer (PC) with a CNC motion-control function that will operate within a Windows® environment. These machines will most likely be referred to as PC-CNCs.

A typical numerically controlled lathe has the same basic elements as an engine lathe or a turret lathe in addition to the numerical control system. Typically, the numerical control system is programmed to control the movement of the various machine elements and tools involved in a particular machining sequence. Numerical control systems for lathes operate on the same principles as for other machines, as described in Chapter 31. Programming starts with a list of statements describing the steps in an operation. A simple program example is shown in Table 18-2.

These are a few lines taken from a long program. Each line, called a block, gives instructions for a step in the operation. The first line calls for tool 04 to move crosswise (toward the centerline) a distance of 1.3 units and lengthwise (toward the spindle) a distance of 0.08 units at a feed rate 0300 with spindle speed 08. The M03 function calls for the coolant. For the next step, the tool stays at the same diameter setting (no change in the *X* dimension) and is fed toward the spindle for 4.948 units. Next, the tool is moved crosswise away from the centerline a distance of 0.05 units at feed rate 0030. Then it is returned lengthwise a distance of 4.348 units at feed rate 1200. The feed and speed specifications are commonly coded. There are many program formats for NC, as explained in Chapter 31.

The NC program is fed into and interpreted by the machine control unit (MCU), which issues electrical impulses, corresponding to the numbers, to the drive units. Sensors on the activated units return signals to the MCU reporting the results. The reports are compared to the orders. If there is any discrepancy, further orders are issued until the requirements are met. Then the MCU goes on to the next step.

A major advantage of an NC lathe is fast changeover, with little lost time for setup. Thus the machine can be cutting most of the time. This is facilitated by doing programming away from the machine, presetting tools to suit the program, and utilizing universal tooling. When one or a very few pieces of a kind are to be turned, a manual engine lathe or turret lathe may be more economical; it is simpler and cheaper and requires no extra programming. But, for more than a few pieces, the NC lathe is commonly more economical. As a rule, the NC lathe is preferable for smaller numbers of pieces than can be justified on other kinds of automatic turning machines. However, for large quantities (thousands and tens of thousands of pieces or more), automatic machines that can utilize gangs or groups of tools acting optimally are usually at an advantage.

Single-spindle Automatic Bar and Chucking Lathes

Automatic machines for internal and external operations on barstock were, for many years, referred to as automatic screw machines. However, the name *automatic bar machines* is currently more appropriate because these machines are used for many more purposes than just the machining of screws. Their counterparts for individual workpieces are generally referred to as *automatic chucking machines*. Automatic bar and chucking machines may be classified as single spindle or multiple spindle, with a number of variations in each class. In fact, there is some degree of indistinction in the classification of the variety of lathe-like automatic machines. For example, the automatic lathes shown in Figures 18-17 and 18-18 could just as easily be classified as combination automatic bar and chucking

Table 18-2. Partial CNC program

011	X-013000	Z-000800	F0300	S08	T04	M03
012		Z-049480	F0115			
013	X+000500		F0030			
014		Z+043480	F1200			
015		Z+006000	F0200			

Figure 18-17. CNC gang-tool chucker and automatic bar lathe. (Courtesy Hardinge, Inc.)

machines since they are adaptable to either type of operation.

Automatic lathes have the same basic units of simple lathes: bed, headstock, tool slides, and sometimes a tailstock. In addition, an automatic lathe drives the tools through all the steps of a cycle without operator attention once the machine has been set up. They perform basically similar functions but appear in the marketplace in a variety of forms. Many have no tailstock and some work only on barstock, while others do chuck work and/or bar work. Some models have only one tool slide, and others have two or three. Some machines have level tool slide ways; others have sloping ways and are referred to as slant bed machines. Some have overhead slides for additional tools. A number of models are now equipped with turrets and are sometimes referred to as *automatic turret lathes*. Some tool slides are actuated and fed by cams and templates, some by hydraulic or air-hydraulic means, and others by digital servo drives.

Figure 18-17(A) shows a popular type of 2-axis CNC automatic lathe designed for either bar or chuck work. This machine has a horizontal cross slide with an interchangeable tool top plate to accommodate multiple toolholders (gang tools). In many cases, gang tooling is preferred because tool positioning time is minimized as compared to the indexing time for turret toolholders. The tool plate can be rapidly traversed and fed simultaneously in both *X*- and *Z*-axis directions to permit each cutting tool to perform a machining operation in some preprogrammed

Figure 18-17 (continued).

sequence. Figure 18-17(B) shows a barstock stop and three cutting tools mounted on the top tool plate for end-working operations. Since the top tool plate is interchangeable, additional plates can be set up with tooling for a range of jobs and then stored away for future runs. The machine can handle round barstock up to 27 mm (1.063 in.) and chuck work up to 102 mm (4 in.). A 32-bit CNC control unit on the machine provides a host of programming functions, including off-line and menu programming and CRT display features. A machine of the type and size shown in Figure 18-17 costs about $75,000, plus tooling costs of about $2,500.

Another variation of a CNC operated automatic lathe is shown in Figure 18-18(A). This machine, referred to as a *slant-bed lathe*, has a 12-station main turret and an optional six-station secondary turret. The main turret can be tooled for either end-working or turning operations and, as shown in Figure 18-18(B), equipped with up to six live tooling attachments for either cross-working or end-working operations, including tapping. The optional six-station secondary turret is primarily used for end-working operations, but one of the stations can serve as a tailstock for supporting lengthy parts. The turrets can be programmed to perform simultaneous machining for reduced cycle times. Capable of quick bidirectional indexing and rapid traverse and feed movements in both X- and Z-axis directions, the 12-tool turret provides excellent machining flexibility and productivity for both bar and chuck work.

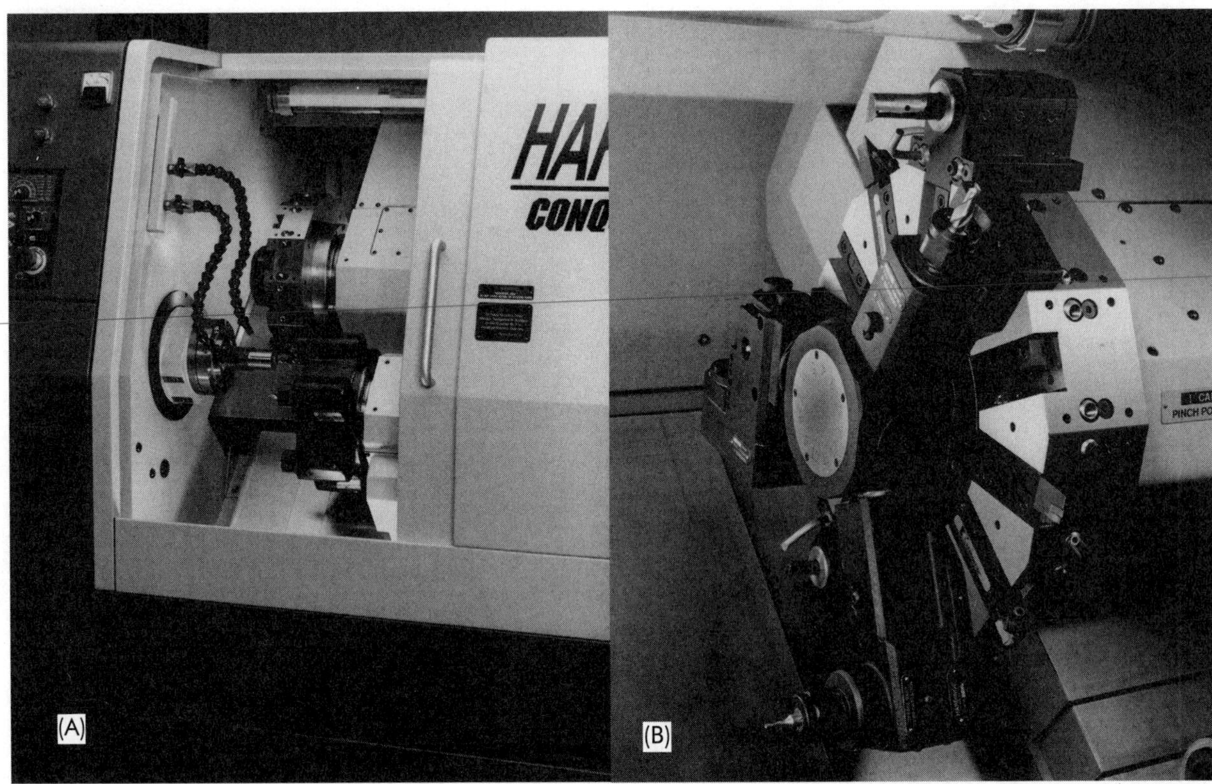

Figure 18-18. CNC slant-bed automatic lathe with a 12-station main turret and a six-station secondary turret for end-working operations. (Courtesy Hardinge, Inc.)

The machine shown in Figure 18-18(A) may be equipped with a subspindle in the place of the secondary turret. The subspindle is used to perform machining operations, including threading, on the reverse end of workpieces that have been cut off after main spindle operations. To effect transfer of the part from the main to the subspindle, the speeds of the two spindles are synchronized and the subspindle is programmed to grasp the part at the appropriate time prior to the cutoff operation. Tooling for back-working operations would then be required on the main turret. The machine shown in Figure 18-18(A) will accommodate round barstock up to 41 mm (1.625 in.) and jaw chuck work up to 203 mm (8 in.). The basic machine costs about $94,000 and tooling for an average job would cost around $3,000.

Swiss-type Automatic Lathe

Swiss-type automatic lathes have been around for generations and are known for their ability to precisely turn long, thin workpieces with length-to-diameter ratios greater than 4:1. Originally designed for the Swiss watch-making industry, these machines were, for many years, cam operated and therefore setup intensive. There are many of the cam-operated machines still in operation and they remain competitive for producing small, high-precision instrument parts, computer parts, parts for medical/dental implants, and many other precision parts in large quantities. These machines are often capable of machining small parts in less time than by any other process. However, cam-operated automatics are inefficient when producing small quantities of parts or when part dimensions must be accurate to between 0.0025–0.005 mm (.0001–.0002 in.). These types of jobs are normally best handled by CNC-operated machines.

Figure 18-19(A) shows the tooling area of a CNC Swiss-type automatic with five turning tool slides above the spindle, three end-working tool posi-

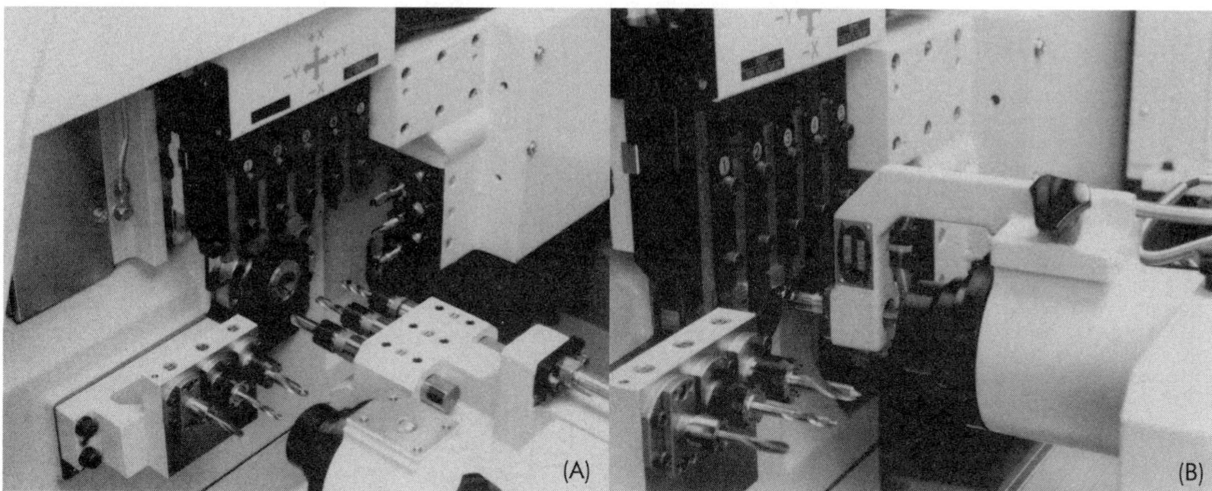

Figure 18-19. (A) Tooling area for a Swiss-type, five-axis, CNC automatic lathe showing 14 tool positions. (B) Tooling area with tool setter in place. (Courtesy Hirschmann Corporation)

tions in front of the spindle, three cross-working tool positions to the rear of the spindle, and three back-working tool positions in front of a subspindle. With this arrangement, it is possible to perform front and rear-end working operations simultaneously. Figure 18-19(B) shows the same tooling area with a tool setter in position in front of the subspindle. The tool setter automates the setting of tool diameter and tool length compensation values within the lathe's own CNC system.

The machine shown in Figure 18-19 will accommodate barstock up to 16 mm (.625 in.) in either the main spindle or subspindle. The three cross-working tools are power-driven with two-axis control to accommodate cross-drilling, milling, or tapping operations. This action, combined with a 15° main spindle indexing capability, makes the machine extremely versatile for secondary operation machining. The base price of the machine shown is about $151,300.

The diagrams of Figure 18-20 show two views of the tooling areas for the automatic lathe of Figure 18-19. Figure 18-20(A) illustrates the Z axis of movement for the main spindle and the XB and ZB axes of movement for the subspindle carrier along with the end-working tools attached to that unit. The X and Y axes of movement for the turning tool and back-working tool carrier are shown in Figure 18-20(B). The magnitude of linear travel in mm for each of those units is also given on the diagrams. With five axes of motion under computer control and 14 tooling positions, this machine is capable of a broad spectrum of machining operations on a wide variety of small parts, including those that would otherwise require secondary operations on other machines.

18.6 MACHINING TIME AND MATERIAL REMOVAL RATE

For purposes such as manufacturing planning, setup, scheduling, and costing, it is often necessary to calculate the machining time and metal removal rate for a lathe operation. For a straight turning operation similar to that illustrated in Figure 18-21, the desired workpiece rotational speed is determined from the relationship

$$N = \frac{C_S \times 1{,}000}{\pi D} \text{ or } N = \frac{C_S \times 12}{\pi D} \tag{18-1}$$

where

N = workpiece rpm
C_S = cutting speed, m/min (ft/min)
D = workpiece diameter, mm (in.)

Machining time is then obtained by dividing the distance that the cutting tool travels by the rate of travel.

$$T = \frac{L + a}{f_n \times N} \tag{18-2}$$

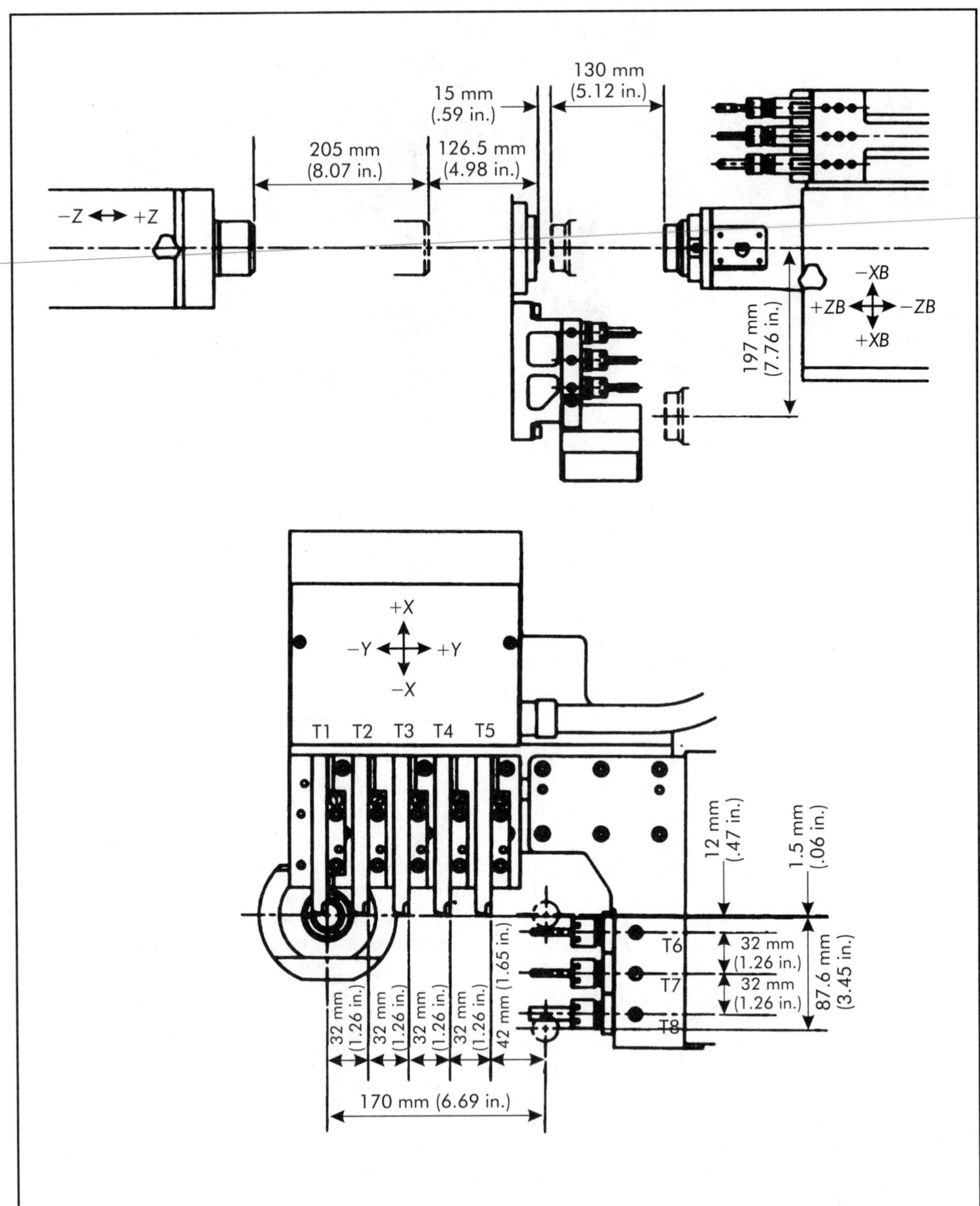

Figure 18-20. Tooling diagram for the Swiss-type, five-axis, CNC automatic lathe shown in Figure 18-19. (Courtesy Hirschmann Corporation)

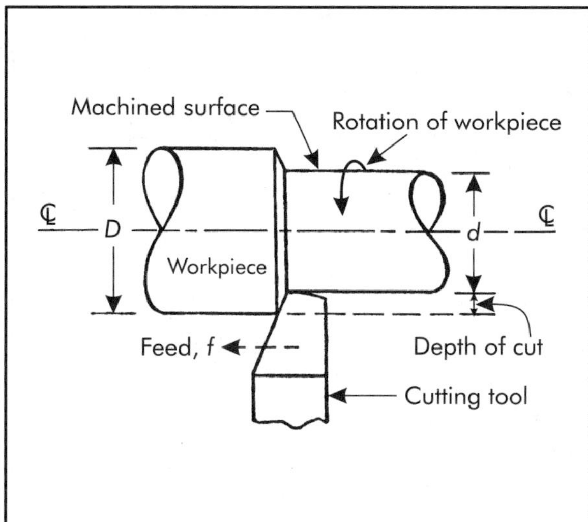

Figure 18-21. Typical straight turning operation.

where

T = machining time, min
L = length of cut, mm (in.)
a = distance allowance for tool travel to enter and clear the cut, mm (in.)
f_n = feed rate, mm/rev (in./rev)

The length of cut usually includes the length of surface to be machined plus an allowance for the distance that the tool travels to enter and clear the cut, normally about 1.6–6.4 mm (.063–.25 in.) for a turning operation. The metal removal rate for such an operation is obtained by dividing the volume of metal removed by the time required to remove it. That is,

$$Q = \frac{\pi / 4(L)(D^2 - d^2)}{T} \qquad (18\text{-}3)$$

where

Q = metal removal rate, mm^3/min ($in.^3$/min)
D, d = workpiece diameters shown in Figure 18-21, mm (in.)

If the allowance factor a used to obtain T in (18-2) is large, then that factor should be deleted from the T calculation in Equation (18-3) to obtain a true Q value.

As an example, a 305 mm (12 in.) length of 51-mm (2-in.) diameter barstock of hot-rolled AISI 1020 steel is to be rough turned in one pass between centers on a lathe to a diameter of 38 mm (1.5 in.) for a distance of 152 mm (6 in.) from one end. The steel is to be machined with an uncoated carbide turning tool at a cutting speed of 88 m/min (290 ft/min), a feed of 0.38 mm/rev (.015 in./rev), and a depth of cut of 6.35 mm (.250 in.). An allowance of 6.4 mm (.25 in.) is used for entering and clearing the cut. For this operation,

$$N = \frac{1{,}000 \times 88}{\pi 50} = 560 \text{ rpm}$$

$$T = \frac{152 + 6.5}{0.38 \times 560} = 0.745 \text{ min}$$

$$Q = \frac{\pi / 4\,(152)\,[(50)^2 - (38)^2]}{0.745}$$

$= 169{,}216\ mm^3/min$ or $169.2\ cm^3/min$ ($10.32\ in.^3/min$)

Note that the allowance factor a has been deleted from the machining time T in the Q calculation.

18.7 QUESTIONS

1. Name the four methods of generating or turning tapers on a lathe.
2. Why is a center drill used prior to drilling a hole on a lathe?
3. Name the two basic methods of holding workpieces for machining on a lathe.
4. Describe the X- and Z-axis movements of a cutting tool on a lathe.
5. How is a cutting tool positioned accurately along either the X or Z axis of a lathe?
6. What is the difference between a feed rod and a lead screw on a lathe? Why are both needed?
7. How is the size of a lathe designated?

8. What is the difference between a universal chuck and an independent jaw chuck?
9. What is a "combination lathe chuck"?
10. What is a "collet chuck" and for what is it used?
11. What is the advantage of using a live center in the tailstock spindle of a lathe?
12. What is the purpose of a digital readout device on a lathe?
13. Name the principal components of a digital readout system for a lathe.
14. Explain how a lathe taper attachment works.
15. Explain the near-net-shape concept as applied to lathe work.
16. Explain the difference between a manually operated lathe and a production turning machine as they relate to labor productivity.
17. Explain the difference between a cam and a CNC-controlled automatic lathe.
18. What is the difference between a ram-type and a saddle-type turret lathe?
19. What is the difference between a combined cut and a multiple cut on a turret lathe?
20. Describe the different ways in which the cycle of machining operations on an automatic turning machine are controlled.
21. What is the difference between cam-controlled and camless automatic turning machines?
22. What is meant by the term "process integration" with respect to manufacturing equipment?
23. What is the difference between a manually operated lathe and a computer numerically controlled lathe?
24. Explain the use of "gang tooling" on an automatic bar and chucking lathe.
25. Explain the operation of a subspindle on an automatic lathe.
26. What are Swiss-type automatic lathes used for?
27. Show by means of a sketch the five axes of movement for the Swiss-type automatic lathe shown in Figure 18-20.

18.8 PROBLEMS

1. The step-down shaft shown in Figure 18-22 is to be machined from 75-mm (≈3-in.) diameter barstock on a manually operated engine lathe. The first operation is to turn the barstock to a 65 mm (≈2.5 in.) diameter over a distance of 160 mm (≈6.25 in.) from the end in one pass. Using a tungsten carbide cutting tool, a cutting speed of 90 m/min (295 ft/min) and a

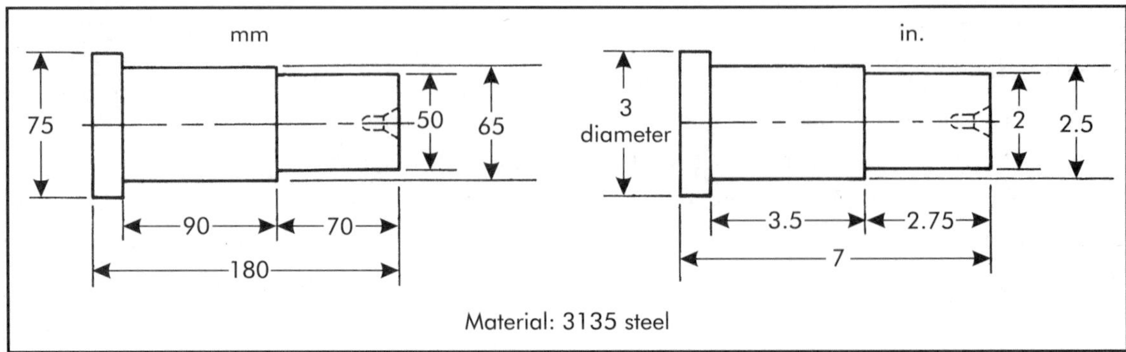

Figure 18-22. Sketches of a workpiece.

feed of 0.50 mm/rev (.020 in./rev), calculate: (a) the rpm of the workpiece; (b) the tool feed in mm/min (in./min); (c) the machining time in min; (d) the rate of metal removal in cm^3/min (in.^{3}/min).

2. The second operation to be performed on the shaft of Figure 18-22 is to turn the small-diameter section. Using the same cutting parameters as used in Problem 1, calculate the machining time for the small-diameter section.
3. A taper 150 mm (≈6 in.) long has a large diameter of 52 mm (≈2.05 in.) and a small diameter of 38.1 mm (1.50 in.). What should be the setting of the taper attachment in mm/m (in./ft) to machine this taper?
4. The large diameter of a piece measures 23.813 mm (.9375 in.); the small diameter, 11.113 mm (.4375 in.). The taper is 420 mm/m (≈5 in./ft). What is the length of the taper?
5. A taper of 26 mm/m (.313 in./ft) is to be cut on a piece 460 mm (≈18 in.) long. How much should the tailstock be set over to obtain that amount of taper?
6. A piece of aluminum is to be cut at 90 m/min (≈300 ft/min). Stock totaling 3.2 mm (.125 in.) is to be removed, leaving a finished surface of 125 mm (≈5 in.) diameter. The length of cut is 100 mm (≈4 in.) and the feed is 0.65 mm/rev (≈.025 in./rev). What is the cutting time?
7. A finish cut for a length of 250 mm (≈10 in.) on a diameter of 50 mm (≈2 in.) is to be taken in cast iron with a speed of 45 m/min (≈150 ft/min) and a feed of 0.2 mm/rev (.008 in./rev). What is the cutting time?
8. A piece 100 mm (≈4 in.) in diameter is to be cut off by a tool fed at 0.125 mm/rev (≈.005 in./rev). The lathe is set at 100 rpm. How long should the cut take?
9. A workpiece 250 mm (≈10 in.) in diameter is to be faced down to a diameter of 100 mm (≈4 in.) on the end. The lathe is equipped with an electronic device that controls the spindle speed and maintains the cutting speed at 60 m/min (≈200 ft/min) for the diameter the tool is cutting at any instant. The feed is 0.38 mm/rev (.015 in./rev). What should be the time for the cut?
10. If the spindle speed for the workpiece described in Problem 8 is set to give a speed of 60 m/min (≈200 ft/min) at 250 mm (≈10 in.) diameter and is not changed during the cut, what is the time required for the cut?

18.9 REFERENCES

ANSI B 5.16-1952 (R 1986). "Accuracy of Engine and Tool Room Lathes." New York: American National Standards Institute.

Noaker, P.M., ed. 1991. "Turning it JIT." *Manufacturing Engineering*, March.

——. 1995. "The PC's CNC Transformation." *Manufacturing Engineering*, August.

Schreiber, R.R. 1991. "The Requisite Readout." *Manufacturing Engineering,* May.

Society of Manufacturing Engineers (SME). 1995. *Turning and Lathe Basics* video. *Fundamental Manufacturing Processes* series. Dearborn, MI: SME.

——. 1996. *Introduction to Workholding* video. *Fundamental Manufacturing Processes* series. Dearborn, MI: SME.

Sprow, E.E. 1994. "Turning Machines and Systems." *Manufacturing Engineering*, August.

Process Planning and Cost Evaluation

19

Shop Management, 1903
—Frederick W. Taylor, Pittsburgh, PA

19.1 INTRODUCTION

Process planning in manufacturing involves defining the operations required to transform a product or a component part from a rough to a finished state in accordance with the specifications cited on a manufacturing drawing. In effect, process planning is the link that couples the engineering design process to the manufacturing process and makes it possible for a product to be built. In contemporary practice, process planning logically starts with the engineering design of a product to assure its manufacturability. *Manufacturability* is the ability of a product to be manufactured in a cost-effective manner.

Efficient and effective manufacturing methods do not just happen. They must be carefully planned with the many details worked out before production is initiated. Typical functions involved in the design, planning, and manufacture of a product are depicted in Figure 19-1. For a new product, a market survey is generally conducted to determine whether or not a particular type of product has a sales potential. If so, designers are given the go-ahead to start the product design. It should be noted in Figure 19-1 that there are feedback loops between many of the functions involved in the design, planning, and manufacturing processes. Feedback is necessary to ensure coordination within and between those functions. Because of the need for feedback and coordination, the time frame or lead time between product design conceptualization and getting a product into the marketplace is often extensive. However, contemporary techniques, such as computer-aided design (CAD), computer-aided process planning (CAPP), and rapid prototyping, have helped to shorten lead time.

In discrete parts manufacturing, process planning usually involves two stages: (1) preproduction or preliminary process planning and (2) detailed methods and operations planning. Cost evaluations are usually computed during each of these stages to assist manufacturing engineers in make-or-buy decisions and the evaluation of alternative processes.

19.2 PREPRODUCTION PROCESS PLANNING

Preproduction planning is usually associated with the introduction of a new product or a new model of an existing product. This phase of planning requires the simultaneous conceptualization of a product design and an applicable manufacturing plan. In the early days of manufacturing, designers worked somewhat independently of manufacturing in the design of a product. When the design was finished, the design group handed the product drawings and

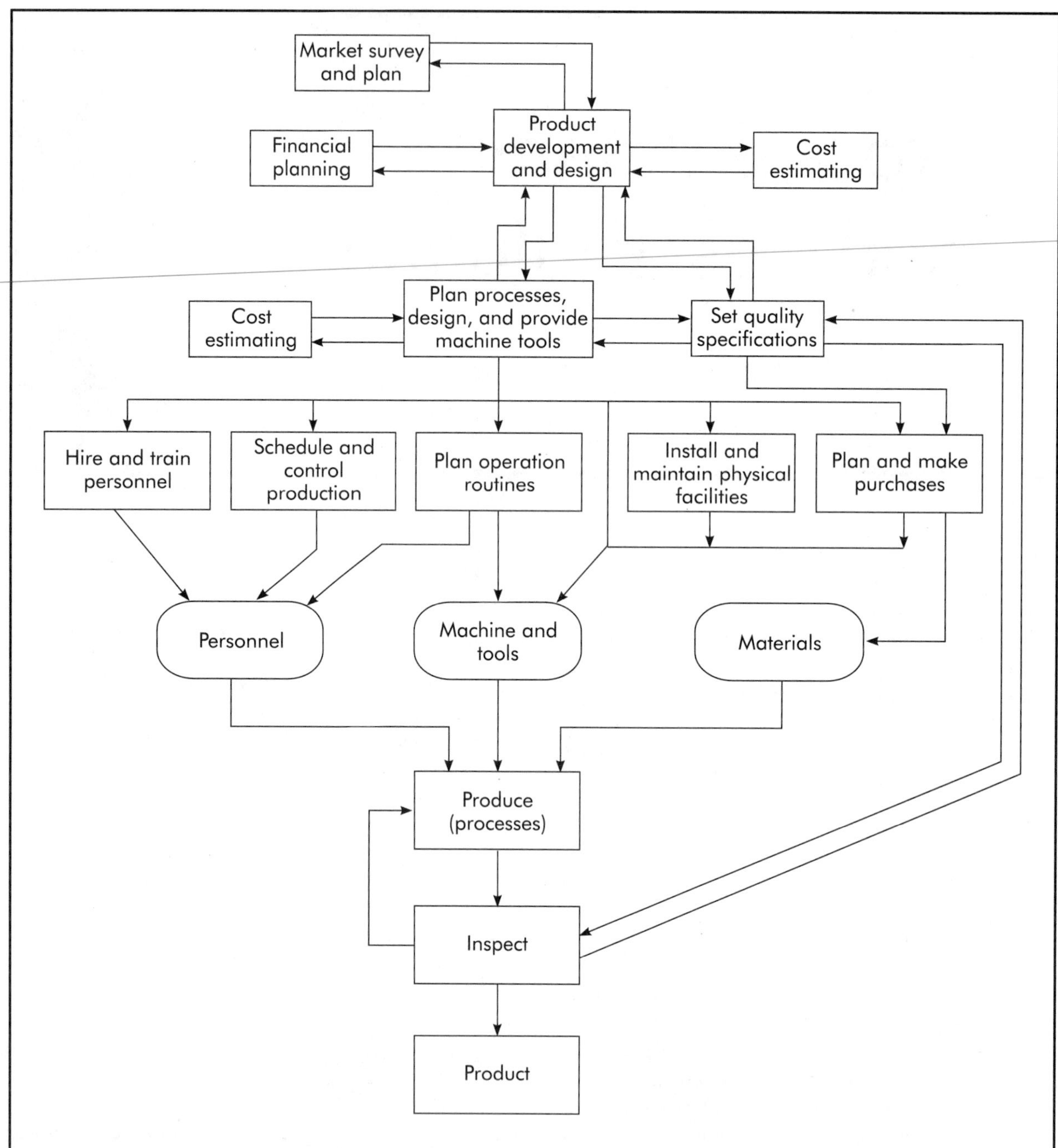

Figure 19-1. Functions that lead to efficient and effective manufacturing.

specifications over to the manufacturing group to build the product. This procedure was often referred to as *over-the wall production planning* as the product plans and specifications were literally tossed over the wall to the manufacturing planning personnel. Since manufacturing did not have an opportunity to provide input during the product design process, it was often discovered that manufacturability was seriously affected. This resulted in numerous changes in the design of the product and almost endless delays in getting a new product to market.

A product designed for ease of manufacture and assembly must involve timely input from many components of the organization, particularly manufacturing engineering, marketing, and quality assurance. Design engineers should make every effort to incorporate requirements from these groups into the initial design of the product to eliminate possible costly design changes after production has started. In contemporary product planning circles, the integration of design and manufacturing in the development of a product is often referred to as part of the concurrent engineering approach. In its broadest conceptualization, *concurrent engineering* involves the use of a systems approach to integrate the design and manufacture of a product to optimize all of the elements involved in the life cycle of a product, including costs, quality, scheduling, and final disposal.

Preproduction planning starts with the conceptualization of a manufacturing plan based on a preliminary product design concept. Through frequent consultations between the designers and manufacturing, and a succession of iterations, an initial product design is evolved that satisfies the major criteria for manufacturability. Once an initial product design is agreed upon, the manufacturing engineer can then define the elements of a manufacturing plan.

19.2.1 Design for Manufacturability

A product must be designed to satisfy a market or customer need. To be competitive in either of these situations, the product also must be designed so it can be manufactured in a cost-effective manner. Thus, a competitive advantage is achieved with the design of a quality product that can be produced by a cost-effective manufacturing process. Product design and process design are extremely critical to the success of a product.

A certain level of quality is defined in the design phase of the product and is then maintained during the manufacturing phase through a variety of control techniques (Chapter 15). To be of value to the customer, the product must: (1) possess certain capabilities and performance characteristics; (2) have appealing style and aesthetics; (3) possess durability and reliability characteristics; (4) be maintainable and carry after-sale service and support; and (5) the quality designed and built into a product must be competitive and affordable.

A product may be designed in many different ways and there are usually a number of processes used to produce it. The designer's objective must be to optimize the product design within the constraints of a given manufacturing system. That system includes the in-house processes, the capabilities of suppliers, the alternatives for materials handling, the capabilities of the available labor force, and the limitations of the distribution system. Finally, designers must design a functional product within given economic and schedule constraints. Most importantly, the product must be designed for manufacture prior to the start-up of production to avoid the costs associated with design changes after production has started.

Design for manufacturability guidelines will vary in accordance with the primary types of processes to be used. For example, a product or component part that is to be fabricated from sheet metal will require different considerations from one that is to be made from casting and machining. There are, however, some general guidelines applicable to discrete parts manufacturing that should be observed by product designers. These include:

1. Minimize the number of component parts. Since the time and cost of fabricating generally increases as the number of parts increases, it makes sense to minimize the number of components. A reduced number of parts also reduces the opportunity for error in manufacturing and assembly.
2. Design for ease of handling and orientation. Parts need to be designed to easily orient themselves for movement to and placement in processing equipment. Product configuration, symmetry, a low center of gravity, easily identifiable guiding and locating surfaces, and appropriate features for pickup and handling are important, particularly for automatic processing equipment.
3. Design for simplified assembly. Parts need to be designed so that they can be easily placed in position and either snapped together, spot welded, or bonded. The use of

screw-type fasteners (for example, bolts, washers, or nuts) for assembly should be avoided unless pressure-tight conditions are required. Usually, assembly is started with a base or body component and other parts are added, preferably by a simple stacking process. Simple nesting-type fixtures should be used to simplify assembly. A product should be designed in such a way that the assembly process is foolproof, that is, components can only be assembled the correct way and will not fit if reversed.

4. Design for achievable tolerances. The product and its components need to be designed with machine capability in mind. Design tolerances that are too tight will only lead to excessive inspection, sorting, and rework. The stacking of tolerances should be avoided on multiple part assemblies so that expensive screening and classification of parts are not required.
5. Design for use of common and standardized parts and materials. Commonality of configuration permits the use of group technology, reduces inventory, and contributes to standardization of processing, handling, and assembly. A complex and unique part design leads to increased costs and vendor difficulties. Commonality of design among parts simplifies automation and operator training.
6. Design for efficient product and component verifiability. Simple and easy-to-recognize go/no go, line match-up, or shoulder stops should be used on mechanical components. Electrical parts should be designed with built-in self-test or diagnostic capabilities.
7. Utilize modular design concepts. Modularization permits building block approaches for subassembly and assembly, and the ease of product model alteration and changeover to satisfy changing customer demand. Standardization of alternative modules can be achieved to reduce lead times.
8. Design for product robustness. Robustness in design allows for margin of error in manufacturing. Robustness compensates for a certain amount of product abuse by increasing product durability under unexpected or extreme conditions.
9. Design for serviceability and maintainability. Ease of access to parts and components should be provided in case they fail and need repair or replacement. On low-cost products, throwaway components should be considered rather than those requiring costly repairs. Modular designs should permit easy disconnection of modules.

These guidelines can help the designer develop alternative designs and evaluate the trade-offs between those alternatives. Again, it is important that the product designers frequently consult the manufacturing engineers on the manufacturability of the product as different alternatives are examined and evaluated.

In contemporary engineering design practice, a number of techniques and tools are available to assist the designer in developing and analyzing alternative designs. Most common among these are computer-aided design (CAD) programs that permit the designer to call up drawings of standard parts and components to evaluate their applicability to the product under consideration. Solids modeling, another feature of CAD programs, provides the designer with a quick means of visualizing individual parts and their relationship with other parts. In addition, solids modeling enables the designer to determine the appropriate part orientation and define the required clearances for ease of assembly.

19.2.2 Planning Methods and Procedures

The methods and processes involved in preproduction planning are varied, depending on the type of manufacturing industry involved, complexity of the product, anticipated volume of production, and expected duration of production. For some light manufacturing industries, preproduction planning is simplified and used to determine the feasibility of producing a particular item. In contrast, preproduction planning for a new model of a commercial aircraft would, of necessity, be quite extensive. With many subassemblies and components, preplanning for such a product would involve very comprehensive analyses to determine those items that would be produced in-house and those that would need to be purchased from or subcontracted to other suppliers.

19.3 PROCESS PLAN DEVELOPMENT

Process planning is often referred to by a variety of titles, including *production planning, process engineering, operations planning,* and *manufacturing activity planning*. A *process plan* summarizes and provides a record of the sequence of events required to manufacture, assemble, and/or process a product. It involves gathering together all of the details that describe the operations, the equipment to be used, and the operating parameters required to transform a part from a rough to the finished state specified on the design drawing.

As shown in Figure 19-2, the process planning activity synthesizes a series of inputs into a coherent group of outputs. These outputs provide a road map for the cost-effective production of a product. In most cases, the planning process involves the development of at least three basic sets of information: (1) routing information, (2) a process map or flow diagram, and (3) operation charts or sheets. The routing information and process map are usually developed somewhat simultaneously as the manufacturing or process engineer begins to enumerate the various steps required to finish a part or product.

19.3.1 Routing Information

In the early stages of process planning, the manufacturing or process engineer examines the design drawing(s) of a part or product and enumerates the operations or steps that must be performed to complete it. Those operations are then ordered or sequenced in accordance with good machining practice and the need for establishing and maintaining appropriate part positioning and datum reference. This delineation of an ordered sequence of operations is usually referred to as a *manufacturing routing* or a *routing chart.*

For example, there are a number of basic processing operations or steps that need to be performed on the cast aluminum-bronze spur gear shown in Figure 19-3. The manufacturing engineer will list these in the order best suited to the manufacturing equipment available and the machining practice requirements. A listing, such as is shown in Figure 19-4, can provide a certain amount of initial routing information from which to construct more comprehensive documentation as the planning process continues. It should be emphasized that this is only a preliminary or conceptual-type routing plan and is subject to change as machine availability information, tooling requirements, and other process details are considered. In addition, the manufacturing engineer may elect to change or interchange some of the steps because of advantages or limitations that may be discovered later. The step numbers in the first column of Figure 19-4 have been purposely sequenced with large numerical gaps so that such changes can be accommodated.

19.3.2 Process Mapping

Flow diagrams or flow charts (often referred to as process mapping tools) are used to trace the path and identify the workstations that a part or product must go through in a particular

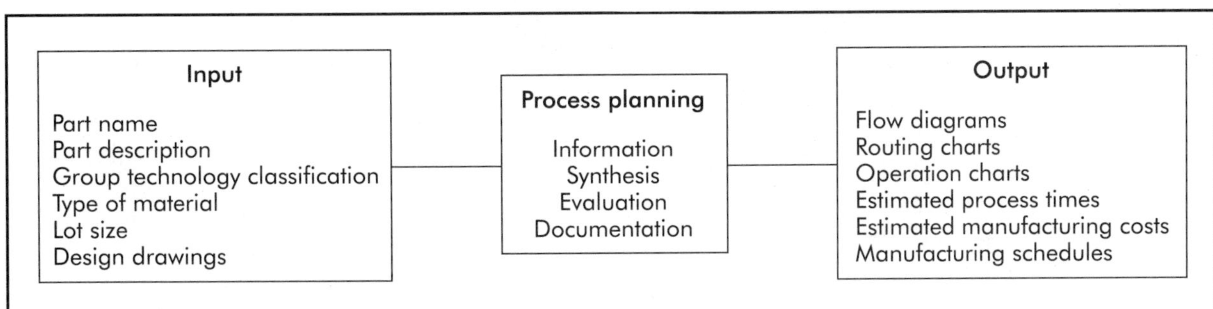

Figure 19-2. Typical process planning elements.

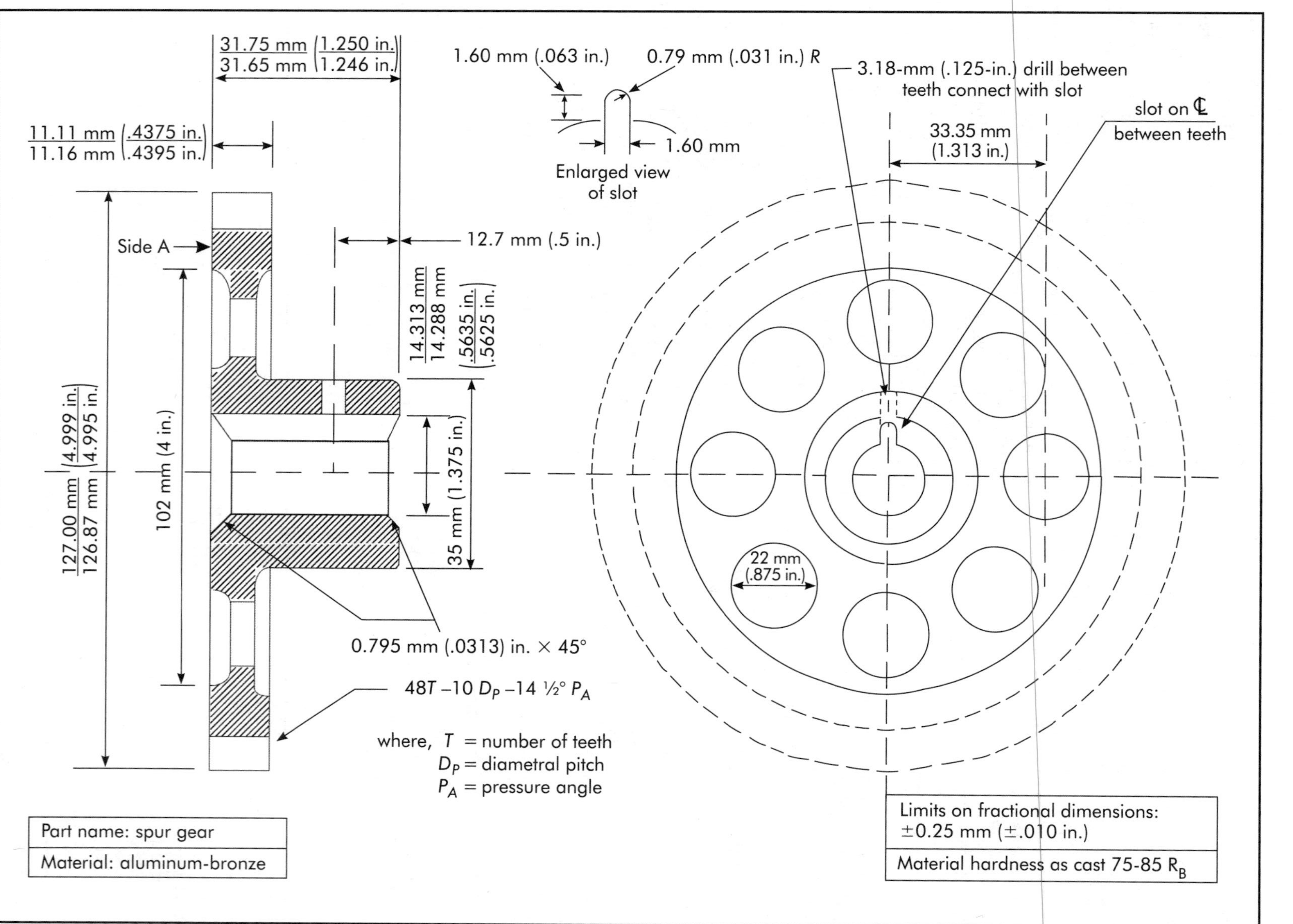

Figure 19-3. Cast aluminum-bronze spur gear.

Preliminary Routing Information		
Part Name: *Spur Gear*		Lot size: *25*
Part number: *29-60539-08C*		Material: *Aluminum-bronze*
Step number	**Description**	**Workstation**
5	Make permanent mold casting	Foundry
10	Clean casting	Foundry
15	Inspect castings	Receiving
20	Face rim and hub, drill and ream hole, chamfer hole	Lathe
25	Face back of rim and hub, chamfer hole, turn OD	Lathe
30	Broach slot	Vertical broach
35	Drill hole to slot	Drill press
40	Generate gear teeth	Gear shaper
45	Deburr	Finishing
50	Inspect	Finishing
55	Store	Storage

Figure 19-4. Preliminary routing information for the spur gear shown in Figure 19-3.

plant or shop environment to complete the operations required to finish that part. For example, it has been initially determined that the cast aluminum-bronze spur gear shown in Figure 19-3 will need to be processed at 11 different workstations. The travel path for those processing steps is illustrated in the flow diagram of Figure 19-5. In this case, it is noted that the travel path moves through different areas or departments in the plant.

The arrangement or layout of the processing equipment shown in Figure 19-5 is usually found in a *job shop*. This type of plant layout is usually associated with the processing of a variety of products or models of products in relatively small lot sizes. It is characterized by the grouping of similar types of machine tools, usually general-purpose machines, in each area or department of the plant. In the language of the plant layout specialists, this is also known as a grouping of machines by function. In other words, milling machines for a variety of surface finishing operations may be grouped together in one area, turning machines (lathes) may constitute another functional group, and so on.

One of the advantages of a job shop type of operation is that it can accommodate a large variety of different machining or process operations as well as permit frequent changeovers from one part or product to another. One fairly obvious disadvantage of this type of arrangement is that a lot of handling of the parts is required to move them from one machining department to another. This can become rather burdensome and inefficient when large numbers of parts are processed at the same time. For this reason, many manufacturing organizations have adopted other more contemporary arrangements that are more efficient than a job shop type of layout. Such arrangements are discussed in detail in Chapter 29.

19.3.3 Operation Sheets

After having developed a feasible preliminary or conceptual routing plan for a part, the process planner begins to assemble the necessary information to finalize the plan. Generally, this involves the development of an operation sheet or operation chart for each of the steps specified in the routing plan. The *operation sheet* supplies detailed information on the sequence of work to be done, tools to be used, the machining parameters, and an estimate of the time involved in

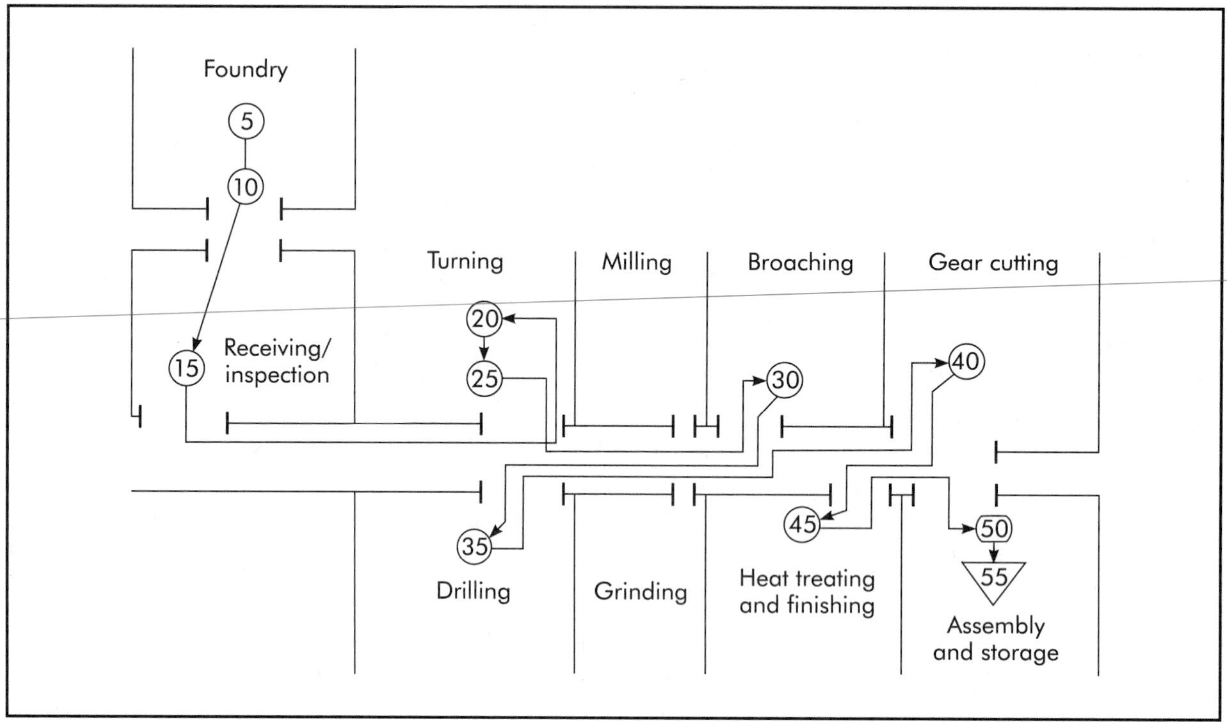

Figure 19-5. Flow diagram or flow process chart showing a preliminary routing plan for the spur gear shown in Figure 19-3.

processing a part through a particular workstation. The operation sheet also serves as a guide for setup personnel and machine operators. It establishes a standard method of accomplishing a certain segment of work so that operation time and cost estimates are reasonably accurate and consistent. A typical operation sheet is shown in Figure 19-6. The sequence of operations to be performed on the spur gear of Figure 19-3 at workstation 20 is listed on this document.

The methods and sequence of operations prescribed in Figure 19-6 are subject to possible changes as improvements in the overall process plan are developed. As with manufacturing routings, the development of an operation sheet is somewhat of an iterative process. That is, improvements and trade-offs are made as the process planner optimizes those plans prior to the start-up of production. If there are any questions concerning the feasibility of a particular operation, the process planner will generally consult with line personnel to work out the difficulties. For obvious reasons, start-up delays caused by inadequate planning can not be accommodated where just-in-time manufacturing is used.

Each of the columns on the operation sheet depicted in Figure 19-6 are numbered for easy reference. Although operation sheets used by different organizations may include other types of information, the example shown contains most of the essential elements. These include:

1. Operation numbers—the operation numbers in column 1 are all preceded by the number 20 and relate to the appropriate step number on the routing chart. The remaining two-digit numbers have gaps between each number to permit additional operations to be inserted later on, if needed.
2. Operation description—a brief description of each operation to be performed is given in column 2. Thus, operation number 20-05 instructs the lathe operator to clamp the 34.93-mm (1 3/8- or 1.375-in.) projecting hub of the spur gear casing in a three-jaw self-centering chuck. Operation 20-10 then instructs the operator to rough face the left

Operation Sheet

Part Name: Spur Gear
Routing Step Number: 20
Initial Setup Time: 18 min

Lot Size: 25
Material: Aluminum-bronze
Workstation: 356 × 762 mm (14 × 30 in.) Lathe (Variable Speed)

1	2	3	4	5		6	7	8		9
Operation Number	Operation Description	Work-holding	Cutting Tool	Cutting Speed		Depth of Cut mm (in.)	Feed mm/rev (in./rev)	Estimated Time, min		Total Estimated Time, min
				m/min (ft/min)	rpm			Machine	Non-machine	
20-05	Chuck casting on 34.93-mm (1 3/8- or 1.375-in.) hub boss	Three-jaw chuck							1.50	1.50
20-10	Rough face rim and hub boss, side A	Three-jaw chuck	Carbide insert	122 (400)	305	1.3 (.050)	.18 (.007)	1.16	1.80	2.96
20-15	Finish face rim and hub boss, side A	Three-jaw chuck	Carbide insert	152 (500)	382	0.3 (.012)	.08 (.003)	2.08	1.20	3.28
20-20	Center drill hub boss	Three-jaw chuck	HSS drill and countersink	22 (72)	2,200*		Manual	0.12	1.80	1.92
20-25	Rough drill center hole through hub	Three-jaw chuck	13.891-mm (35/64- or .5469-in.) carbide drill	76 (250)	1,740		Manual	0.10	1.50	1.60
20-30	Ream center hole	Three-jaw chuck	14.29-mm (9/16- or .5625-in.) carbide reamer	90 (294)	2,000		Manual	0.10	1.50	1.60
20-35	Chamfer center hole 0.795 mm (1/32 or .0313 in.) × 45°	Three-jaw chuck	Carbide insert	90 (294)	2,000		Manual	0.06	0.50	0.56
20-40	Chamfer rim 0.396 mm (1/64 or .0156) in. × 45°	Three-jaw chuck	Carbide insert	152 (500)	382		Manual	0.06	0.50	0.56
								Total time, min/piece		13.98

*Maximum lathe spindle rpm

Figure 19-6. Operation sheet for workstation No. 20 on routing sheet of Figure 19-4.

side (side A) of the rim and hub with a single-edge carbide cutting tool. These instructions may be clarified or expanded, if needed, in the remarks column or in a footnote at the bottom of the page. Generally, most organizations do not go into great detail in these instructions to allow machine operators some latitude in how they do a job.

3. Workholding—column 3 identifies the type of machine element or tooling accessory required to locate and hold the workpiece in place for machining. If special fixturing devices are needed for some operations, then provisions will need to be made to acquire these prior to production start-up. The design and acquisition of special tooling is another important part of the process planning activity.
4. Cutting tools—the various cutting tools required for each operation are identified in column 4. As for workholding devices, plans need to be made to have either standard or special cutting tools available prior to the start-up of production to avoid delays. In many cases, the process planner will include a standard tool identification number here to aid the operator in selecting an appropriate cutting tool.
5. Cutting speed—the linear and rotational speeds specified in column 5 are necessary components of the machining parameters prescribed for the cutting conditions and tools specified in columns 2, 3 and 4. The cutting speed for facing in operation 20-10 relates to the material being cut and the type of tool used. Typical linear cutting speeds for a number of material types and cutting tool materials are given in machine tool or cutting tool manufacturers' literature. Most cutting tool manufacturers will provide recommended cutting speeds for the various cutting tool materials and grades that they stock.

   The number of revolutions per minute (rpm) listed in column 5 is calculated using Equation 17-1, where

$$N = \frac{1{,}000\, V_m}{\pi \times d_m} \text{ or } \frac{12\, V}{\pi \times d} \tag{19-1}$$

   where

   N = rpm
   V_m (V) = surface speed, m/min (ft/min)
   d_m (d) = outside diameter (OD) of spur gear, 127 mm (5 in.)

   Thus, for a cutting speed of 122 m/min (400 ft/min), the rpm given for operation 20-10 is calculated as follows:

$$N = \frac{1{,}000 \times 122}{\pi \times 127} = 305$$

   or

$$N = \frac{12 \times 400}{\pi \times 5} = 305$$

   Since the lathe at workstation number 20 has variable spindle speeds, an rpm of 305 should be feasible. If such were not the case, then an rpm setting that is available on the machine would have to be selected.
6. Feed—as explained in Chapter 17, the surface finish obtained by machining with a single-edge tool is affected by the feed rate. That is, higher feed rates generally result in rougher surface finishes. Thus, the feed rate selected for rough cutting in column 6 is approximately twice that used for finishing cuts so that a relatively smooth surface finish is obtained on the finish cut. Care needs to be exercised in the selection of a feed rate that will not overload the cutting tool or cause a deflection of the workpiece, particularly if the workpiece clamping method could permit some deflection caused by excessive torque. For operation number 20-10 on the operation sheet of Figure 19-6, the hub boss is clamped between the three jaws of a self-centering chuck while the rim and hub are faced. Excessive forces on the cutting tool while facing the rim could cause the casting to deflect because of a rather large moment arm between the point of cutting and the location of the clamping forces. Typical feed rates are given in machine tool and cutting tool manufacturers' literature for different materials, but these will need to be modified to reflect cutting conditions.

7. Estimated operation time—the estimated time for operation 20-10 in column 8 is made up of two components: (1) machining time and (2) nonmachining time. The machining time required to rough face the rim and hub of side A is the quotient of the distance traveled by the cutting tool divided by the linear feed rate in mm/min (in./min). It is noted, however, in Figure 19-3 that there will be a gap in the cutting action as a tool is fed from the outer rim to the hub section because of the relieved web section of the gear. In addition, an adjustment in the spindle rpm may be justified as the cutting tool progresses toward the center of the hub. These are examples of the decision complications encountered by the manufacturing engineer during the development of a process plan. In the interest of simplicity, however, it was decided to machine side A with one continuous action of the cutting tool and ignore the fact that the tool will be cutting nothing during its travel across the relieved web section. Thus,

$$t = \frac{d \div 2}{f_1} = 305 \qquad (19\text{-}2)$$

where

t = machining time for side A, min
d = outside diameter (OD) of the spur gear, mm (in.)
f_1 = feed rate, mm/min (in./min)

Since the feed rate given in column 7 is in mm/rev, that rate will need to be converted to a linear feed rate in mm/min (in./min). Thus,

$$f_l = f_n \times \text{N} \qquad (19\text{-}3)$$

where

f_l = feed rate, mm/rev (in./rev)
N = revolutions per minute (rpm)

So, the machining time for rough facing side A of the spur gear is calculated as

$$f_l = 0.18 \times 305 = 54.9 \text{ mm/min}$$

and

$$t = \frac{127 \div 2}{54.9} = 1.16 \text{ min}$$

The process of estimating the nonmachining time for side A may be even more arduous than that required for estimating machining time, depending on the degree of accuracy required. Nonmachining time can be estimated from in-house predetermined elemental time data, from industry time standards, or from an actual time study of the job itself. The selection of one or the other of these approaches is, again, dependent on the degree of accuracy required and the anticipated repetition of this particular machining job. If, for example, it is not expected that this job will be repeated and the lot size is small (as it is in this case), then a rigorous, time-consuming estimate of nonmachining time would not be justified. If, on the other hand, it is anticipated that many repeat orders for this part will be received in the future, then a more substantive estimate and a follow-up time audit would be in order. Traditionally, personnel from the company's work standards group are called upon to provide technical assistance in developing rigorous time standards. For the purpose of this example, hypothetical elemental times have been assumed for the compilation of nonmachining times. These elemental times include the necessary setup activities required to perform the operation, such as mounting and clamping the carbide cutting tool in the toolholder, positioning the tool for a facing operation, and setting the proper machine speeds and feeds.

The remainder of the operations required in routing step number 20 are shown in Figure 19-6. In all, this routing step involves eight operations that are performed on side A of the spur gear casting while it is clamped in place in a three-jaw chuck.

As indicated at the bottom of Figure 19-6, the estimated time to complete routing step number 20 is approximately 14 min per piece, exclusive of any allowances for operator personal time, possible tool failures, resharpening of cutting tools, time for inspection measurements, etc. Assuming a 20% allowance for these occurrences, the time per piece is increased to 13.98 × 1.20 or 16.78 min. From this, the total time to complete routing step 20 for a lot of 25 pieces is estimated as:

$$T_{20} = 25t_p + t_s \tag{19-4}$$

where

T_{20} = total time for step 20, min
t_p = time per piece, min
t_s = initial machine setup time of 18 min

Thus,

$$T_{20} = 25(16.78) + 18 = 437.5 \text{ min or } 7.29 \text{ hr}$$

Operation sheets and time estimates will need to be developed for each of the other 10 steps listed on the routing sheet of Figure 19-4. The composite time estimate for all 11 of those steps will then form the basis for estimating the total cost of producing a lot of 25 spur gears.

19.4 ECONOMICS OF PROCESS PLANNING

The choice of a method or process for machining a part largely determines the type of machine tool needed. In the previous example for completing routing step number 20 on the spur gear of Figure 19-3, a turning machine was selected since the operations required rotation of the part. In that example, a manually operated lathe was selected because of its ease of setup for the various operations to be performed and the small lot size required. A larger lot size may have justified the selection of a turning machine with the capability of a higher rate of production, such as a turret lathe or automatic lathe. Usually, however, in the final analysis, the selection of a particular machine tool is based on a cost comparison between those machines available to do the job.

The choice of a machine tool must take into account: (1) the size and shape of the workpiece, (2) the work material, (3) the accuracy and surface quality required, (4) personal preferences, and (5) the quantity of parts and the sizes of lots required. Usually, a number of machine tools can do a job, but the one that will do the job at the lowest cost when required is the one to be chosen. Of the number of machine tools that might conceivably do a job, most can be eliminated without detailed estimates of their costs, since they are too small or too large, too weak, or in some other ways obviously deficient. The costs of the remaining few can then be ascertained to determine which tool is the best for the job.

19.5 GENERAL CONSIDERATIONS FOR MACHINE TOOL SELECTION

19.5.1 Size and Capacity

The dimensions that designate the size of a machine tool generally specify the size of the largest workpiece that can be handled. If enough pieces of one kind are to be produced to keep a machine busy practically all the time, a size just able to take the part might well be chosen. For instance, for pieces to be made from 40-mm (≈1.5-in.) diameter barstock, an automatic bar machine with a collet capacity of 40 mm (≈1.5 in.) or a little larger would be favored. On the other hand, if only a few pieces of a particular size are to be made, a larger general-purpose machine would be selected and it would be adaptable to other jobs as well. Thus pieces from 40-mm (≈1.5-in.) diameter barstock would probably be turned on lathes with 300 mm (≈12 in.) or larger swing, because lathes of that size for the run of medium-sized parts would be found in a job shop.

The particular features of a machine tool may be dictated by the workpiece size and dimensions. As an example, small- and medium-size parts are turned on horizontal lathes, but short pieces of large diameters are commonly machined on vertical lathes. Other dimensions of a machine that may have to be considered are the directions and lengths of movements of tools to assure that the surfaces to be cut can be covered, the clearances for the tools that must be used, and the provisions for the holding device. By comparison with the requirements of the job, the dimensions of a machine suitable for a particular job may be ascertained from the manufacturer's catalog.

19.5.2 Strength and Power

The rigidity and strength of a machine tool are not easy to calculate, but for reputable machine tools, are in keeping with the rated power capacity. Thus a light machine usually has a smaller motor than a heavier machine. In general, a machine tool is designed to resist the forces arising from cuts at the rated power. So the power required in an operation must be a major consideration in selecting a machine tool. The machine must have enough power, but too much is wasteful.

Ways of calculating forces and power at the cutting zone were explained in Chapter 17. A convenient way of estimating power is to multiply the rate of metal removal in cm^3/min ($in.^3$/min) by the unit power in W/cm^3/min (hp/$in.^3$/min). The unit power varies with the (1) work material, (2) type of operation, (3) rake angle of the cutting tool, (4) size of cut, and (5) speed, but mostly with the first two. Accordingly, average values of unit power may be used to estimate the power requirements of common operations.

Some loss in power can be expected in any machine tool, so the power at the motor must be more than that at the cutting zone. An efficiency of 80% is a fair estimate for average conditions.

As an example, a 16-mm (5/8- or .625-in.) diameter hole drilled in steel of 35 R_C at 550 rpm and 0.15 mm/rev (.006 in./rev) requires

$$P_D = 16^2 \times \frac{\pi}{4} \times 0.15 \times 550 \times 64 \times \frac{1.3}{10^6} = 1.4 \text{ kW} \quad (19\text{-}5)$$

or

$$P_D = \left(\tfrac{5}{8}\right)^2 \times \frac{\pi}{4} \times .006 \times 550 \times 1.4 \times 1.3 = 1.8 \text{ hp}$$

where

64 = unit power requirement, W/cm^3/min
1.4 = unit power requirement, hp/$in.^3$/min (*Machining Data Handbook* 1980)

The factor of 1.3 is to provide enough power as the drill gets dull. The motor power should be 1.4/0.8 = 1.8 kW or 1.8/0.8 = 2.3 hp.

In actual production, tools may be run until extremely dull, the operator may choose to run at higher speeds and feeds than anticipated, and the work material may vary in hardness and stock allowance. This means that if a machine tool is loaded to capacity on a job for maximum economy under ideal conditions, it may be as much as 50% or more overloaded at other times. This is a common occurrence, and good machine tools are intentionally built to withstand overloads. As an example, one leading machine tool manufacturer has recommended:

- rated capacity for continuous operation;
- 25% over the rated capacity for normal operation;
- 50% over the rated capacity for intermittent operation, 5-min maximum period, and the minimum idle time between cuts should equal one-fifth of the cutting time; and
- 75% over the rated capacity for intermittent operation, 1-min maximum period, and the minimum idle time between cuts should equal the cutting time.

Frequently, the situation is that a machine happens to be available that meets all other requirements but has a limited power rating. The problem then is to determine the speed and feed at which the job may be run. As an example, a depth of cut of 3 mm (.118 in.) is required to reduce a diameter of 40 mm (1.575 in.) to 34 mm (1.339 in.) at 60 m/min (197 ft/min) in 40 R_C steel. The lathe has a 7.5-kW (10-hp) motor and is 80% efficient. The permissible rate of metal removal is

$$Q = 7{,}500 \times \frac{0.8}{64} = 93.75 \text{ cm}^3/\text{min} = f \times 3 \times 60 \quad (19\text{-}6)$$

$$= 10 \times \frac{0.8}{1.4} = 5.71 \text{ in.}^3/\text{min} = f \times .118 \times 197 \times 12$$

where

Q = rate of metal removal
f = 0.5 mm/rev (.020 in./rev)

19.5.3 Other Considerations

Among other factors that need to be considered in machine tool selection are the accuracy and surface finishes a machine is capable of producing, the removal of hard material or large amounts of stock, the skill available and required to operate the machine, personal likes and dislikes, and availability. Most of these factors are not as basic as others, but in some cases may be decisive. For example, if fine surface finishes and small tolerances must be held, they often are obtainable only on grinding equipment. Grinding also is the only means of cutting some hard substances but is not well suited to remove large amounts of stock from any substance. And, in a situation where skill to set up automatic machines is not available, operations might have to be simplified and spread over other simpler types of machines. Often, in a shop, a machine that is available will be selected for a job rather than buying another or waiting for one that is already overloaded.

19.6 HOW COSTS ARE ESTIMATED AND COMPARED

Once the decision on the type of machine for an operation has been narrowed to one size of two or more kinds of machines, it then becomes necessary to estimate and compare the costs.

All costs that vary must be considered and for convenience may be divided into direct costs, indirect or overhead costs, and capital costs. The direct cost that varies most from machine to machine is labor, which is usually calculated by multiplying the time required for an operation by a labor rate. The time to set up and perform an operation must be estimated to find its direct cost. Other direct costs, such as power and material, do not vary much for the same job when done in one way or another and are not included as a rule. Overhead costs are commonly calculated by multiplying the operation time by an overhead rate. Such a rate is obtained by dividing the total indirect costs applicable to a production unit (such as a machine center or department) for a period of time (say a month) by the total number of hours of direct labor in the same period. Capital costs are determined by distributing the major machine and tool costs on an hourly basis or among the pieces produced.

19.6.1 Productive Time

The total productive time required to perform an operation may be divided into four parts. They are:

1. Setup time—this is the time required to prepare for the operation and may include time to get tools from the crib and process paperwork as well as arrange the tools on the machine.
2. Worker or handling time—this is the time the operator spends loading and unloading the work, manipulating the machine and tools, and making measurements during each cycle of the operation.
3. Down or lost time—this is the unavoidable time lost by the operator because of breakdowns, waiting for the tools and materials, etc.
4. Machine time—this is the time during each cycle of the operation that the machine is working or the tools are cutting.

Setup, Worker, and Downtime

Setup time is usually performed once for each lot of parts. Therefore, it should be listed separately from the other parts of the operation time. If 30 min are required for setup and only 10 pieces are made, an average of 3 min of setup time must be charged against each piece. On the other hand, if 60 pieces are made from the same setup, only .5 min is charged per piece. Thus a prorated setup time may be very misleading because it depends so much on lot size.

Both setup and worker time are estimated from previous performance on similar operations. All work on a particular type of machine tool consists of a limited number of elements. These elements may be standardized, measured, and recorded. This is the essence of time study, a large field in itself. Space does not permit a detailed treatment of this subject here. The time to perform an operation also includes time for the personal needs of the operator, time to change tools, etc.

The actual amount of downtime that will occur in a specific operation can scarcely be predicted. Some operations will run smoothly; others will be beset by troubles. The best estimate that can be made is based upon the average amount currently lost in the plant. For a comparison between two operations, the assumption is that both will be subject to the same downtime, so this part of the cost is not included as a rule.

Machine Time

The way to calculate machine or cutting time for turning operations was explained earlier in this chapter. The basic relationship for any operation is that the cutting time in minutes is equal to the distance the tool is fed in mm (in.) divided by the feed in mm/min (in./min).

The distance a tool is fed to make a cut is the sum of the distance the tool travels while cutting to full depth plus its approach distance and its overtravel. The *approach* is the distance a tool is fed from the time it touches the workpiece until it is cutting to full depth. Approach distance for a drill is the length of its point, which is about one-fourth the diameter of a standard drill. The approach of most single-point tools is negligible. *Overtravel* is the distance the tool is fed while it is not cutting. It is the space over which the tool idles before it enters and after it leaves the cut. Overtravel may be <1 to >6 mm (<.0313 to >.2500 in.).

19.6.2 How Operation Time is Estimated

The workpiece in Figure 19-7 may be made on an engine lathe or on a turret lathe with the tool layout shown. To illustrate the details of estimating operation time and as a basis for further comparison, Table 19-1 contains times compiled for both engine lathe and turret lathe operations. The speeds and feeds selected are customary ones for the operations and tools. A check of the heaviest cut, that of number 7, shows power required of 66 × 0.46 × 2.75 × 50 × 1.3/1,000 = 5.3 kW (12 × 216 × .018 × .109 × 1.1 × 1.3 = 7.3 hp) at the tool point. The standard times for the engine lathe and turret lathe were taken from industry standards for those machines.

19.6.3 Comparison of Engine Lathe and Turret Lathe

The operating differences between engine lathes and turret lathes are illustrated in Table 19-1. To produce five pieces of the part shown in Figure 19-7 takes a total of 41.70 min or 8.34 min/piece on the engine lathe as compared with 93.40 total min or 18.68 min/piece on the turret lathe. For 100 pieces, the total time is 454 min or 4.54 min/piece on the engine lathe, but only 348 total min or 3.48 min/piece on the turret lathe. If the labor and overhead rates and capital costs are nearly the same on both machines, and they usually are for two machines like these, the comparison of the operation times tells which machine is more economical for the required number of pieces. This comparison proves the principle that the engine lathe is usually economical for a few pieces, but for larger numbers, the turret lathe shows lower costs.

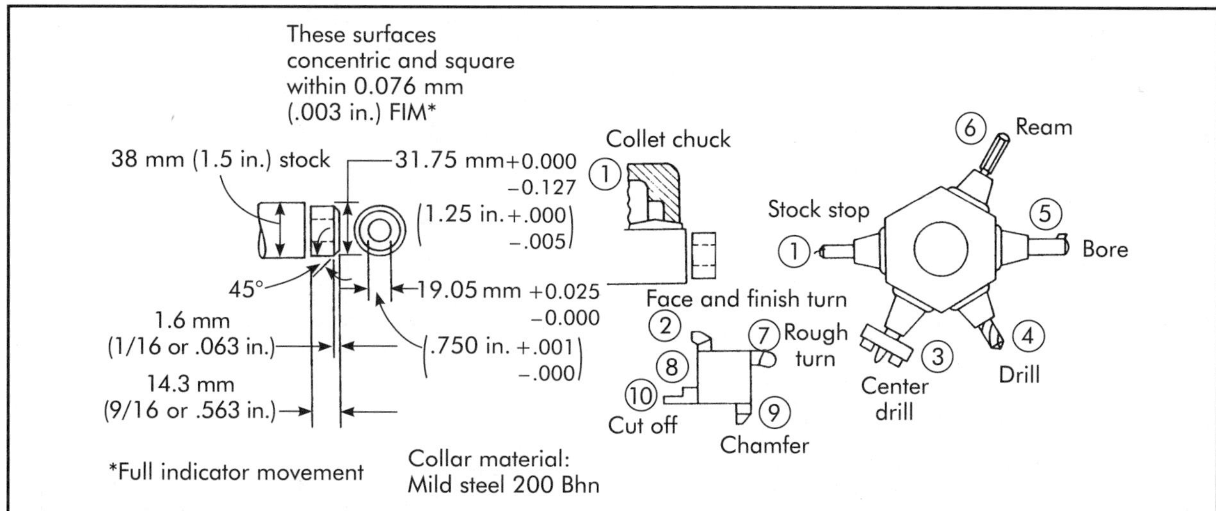

Figure 19-7. Workpiece and a turret-lathe tool layout for making it.

Table 19-1. Comparison of elemental times on 355.6-mm (14-in.) engine lathe and No. 4 turret lathe

Element			Cutting Time			Standard Time, min	
No.	Designation	m/min (ft/min)	(rpm)	[mm/rev (in./rev)]	(min)	Engine Lathe	Turret Lathe
1.	Advance and position bar					0.20	0.12
2.	Face end—set tool[a] 0.8-mm (1/32- or .0313-in.) deep cut	66 (216)	550	0.25 (.010)	0.16	0.16	0.12
3.	Center drill—set tool[b] Cut to 5 mm (≈3/16 or .188 in.) diameter		1,200	hand	0.15	0.20	0.12
4.	Drill—set tool[b] 16 mm (≈5/8 or .625 in.) diameter × 17.5-mm (≈11/16- or .688-in.) deep cut	27 (90)	550	0.18 (.007)	0.23	0.20	0.13
5.	Bore—set tool[c] 18.85/19.0 mm (.742/.747 in.) diameter × 16-mm (≈5/8- or .625-in.) deep cut	42 (138)	700	0.18 (.007)	0.14	0.36	0.13
6.	Ream—set tool[b] 19.050/19.075 mm (.750/.751 in.) diameter × 16-mm (≈5/8- or .625-in.) deep cut	12 (39)	200	0.79 (.031)	0.11	0.20	0.13
7.	Rough turn—set tool[a] 32.545 mm (1 9/32 or 1.2813 in.) diameter ×19-mm (3/4- or .75-in.) long cut	66 (216)	550	0.46 (.018)	0.09	0.26	(with No. 4)
8.	Finish turn—set tool[a] 31.750/31.623 mm (1.250/1.245 in.) diameter × 16-mm (≈5/8- or .625-in.) long cut	84 (275)	700	0.25 (.010)	0.13	0.25	(with No. 5)
9.	Chamfer—set tool[a] 1.60 mm (1/16 or .063 in.) × 45°	70 (230)	700	hand	0.10	0.15	0.13
10.	Cut off—set tool[a] Cut to 14.29 mm (9/16 or .5625 in.) thick	20 (65)	200	0.10 (.004)	0.35	0.25	0.13
11.	Break edges					0.25	0.20
12.	Check (every piece on lathe, every fifth piece on turret lathe)					0.40	0.10
				Handling time		2.88	1.31
				Applicable cutting time		1.46	1.37
				Cycle time		4.34	2.68
				Setup time		15.00	75.00
				Tear down and cleanup		5.00	5.00
				Total per lot		20.00	80.00

Notes:

On engine lathe:
[a] Place and adjust tool on cross-slide (includes 0.10 min to measure if dimensional)
[b] Place tool in tailstock and adjust
[c] Place and adjust tool on cross-slide

On turret lathe:
[a] Index square turret and position tool
[b] Index hexagonal turret, advance tool to work, and start feed
[c] Index hexagonal turret, advance tool to work, and start feed

For No. 5, bore, the standard time for the lathe of 0.36 min includes "tool adjust 0.16 min" and "measure inside diameter (ID) 0.20 min." The tool is preset to cut to size on the turret lathe, and the standard elemental time for that machine is only 0.13 min to "index hexagonal turret, advance tool to work, and engage feed." The total length of cut includes 1.590 mm (≈1/16 or .0625 in.) overtravel, and the cutting time = 17.5/(700 × 0.18) = 0.14 minute = 0.69/(700 × 0.007).

On a 7.5-kW (10-hp) turret lathe, No. 7 has a total time of 0.22 min and can be performed entirely within the cutting time of 0.23 min of No. 04. The value of 0.09 min for cutting time in No. 7 is therefore not applied to total cutting time on the turret lathe.

For No. 8, the handling time on the turret lathe of 0.13 min can be performed during the cutting time of 0.14 min for No. 5.

For No. 3, center drill, feed is by hand and machine time is not calculated. Instead, a value is obtained for "spot drill or center 0.15 min" from industry standards.

The setup time on the lathe was calculated from the following elements:

Check-in on job	1.00 min
Study blueprint	1.00
Trip to tool crib	5.00
Handle eight tools at 0.4 min	3.20
Handle four measuring instruments at 0.4 min	1.60
Install chuck	3.00
Total	14.80 min

Teardown and cleanup consists of these elements:

Sign out	1.00 min
Remove and clean eight tools at 0.3 min	2.40
Clean four measuring instruments at 0.4 min	1.60
Total	5.00 min

The setup time on the turret lathe is estimated in the same way, but is larger than for the lathe because elements must be included to mount the tools, adjust them, and set the stops to get the cuts to size.

This estimate does not include time for personal needs of the operator, changing tools, delays, etc. In a comparison between machines, the time for such items usually can be expected to be about the same for both machines and often is ignored. One way to include it is to allow a fixed amount for each hour. The productive hour may be considered to consist of only 50 minutes, for example.

19.6.4 Finding the Lowest Cost for an Operation

Commonly, an operation can be done on two or more machines. One calls for a smaller investment; another may do the job faster. Often the problem is to find which alternative promises the lowest cost to make a definite quantity of pieces. To make a sound decision, the amounts of direct costs, applicable indirect costs, and fixed or capital cost are computed for each alternative for the quantity of parts required.

Some of the fixed costs for an operation are for standard machines and tools that can be fully used for other work as well. Those costs are charged to each job in proportion to the amount of asset use on the job. Other fixed costs are for special machines and tools that have no use other than for the one job. They must be charged entirely to that job.

The costs of interest, insurance, and taxes are neglected in this presentation. This simplifies calculations and does not cause appreciable difference in relative costs. Some more refined methods of analysis take these factors into account and are recommended for advanced study.

An example of calculations for a job that can be done satisfactorily in four ways is given in Table 19-2. The single-spindle automatic machine offers the lowest total cost for the required quantity of 1,000 pieces in 10 lots. It is assumed that no more than 1,000 pieces will be made, and thus all the special tools are charged to this one job. If there were prospects of repeating the job, the special tooling cost might be prorated in some other way. The cost of the standard machines and tools is distributed on an hourly basis over the life of the equipment and 2,000 hr/yr during which the facilities are expected to be engaged with other work as well. The computed hourly rate for use of the standard equipment is added to the rates for labor and applicable overhead to make up the composite hourly rate.

19.6.5 Equal Cost Point

In the comparison of two machine tools, the equal cost or break-even point for an operation is at the quantity at which the total and unit costs are the same for both machines. The machine with the smaller fixed cost is more economical for a smaller quantity, and the other for a larger quantity. A typical example is shown in Figure 19-8.

An equal cost or break-even point analysis is useful for appraising prospects in many cases. If a preliminary study is being made, the demand for the product may not be exactly known. The equal cost point may be compared with various possible quantities of output to ascertain which machine is preferable for each quantity. Even if a definite production quantity is specified, it may be compared with the equal cost quantity to find out how much change in the quantity of output will change the advantage from one machine to the other.

For the equal cost quantity,

$$\left(\frac{Q}{P_A}+S_A\right)(L_A+O_A+D_A)+E_A$$
$$=\left(\frac{Q}{P_B}+S_B\right)(L_B+O_B+D_B)+E_B \qquad (19\text{-}7)$$

or

$$\left(\frac{Q}{P_A}+S_A\right)R_A+E_A=\left(\frac{Q}{P_B}+S_B\right)R_B+E_B$$

where

Q = the number of pieces produced at the equal cost point for two machine tools, A and B

P_A or P_B = number of pieces produced per hour on machine A or machine B

S_A or S_B = number of hours for setup and tear down on machine A or machine B

L_A or L_B = labor rate, \$/hr on machine A or machine B

O_A or O_B = overhead rate, \$/hr (not including machine depreciation) on machine A or machine B

D_A or D_B = depreciation rate on standard items, \$/hr on machine A or machine B

E_A or E_B = cost of special tools and equipment chargeable to the job, \$ on machine A or machine B

R_A or R_B = composite rate = $L + O + D$ on machine A or machine B

Table 19-2. Comparison of machine costs

	Ram-type Turret Lathe			
	With Standard Tools	With Special Tooling	Single-spindle Automatic with Proper Tooling	Multispindle Automatic with Proper Tooling
1. Cost of machine ($)	57,000.00	57,000.00	75,000.00	109,300.00
2. Cost of standard tools ($)	13,408.00	957.00	5,500.00	5,500.00
3. Total cost of standard items ($)	70,408.00	57,957.00	80,500.00	114,800.00
4. Setup time, estimated (hr)	2.5	2.5	3.5	6.0
5. Cost of special tools ($)	3,570.00	16,275.00	2,500.00	4,500.00
6. Operation time per piece, estimated (min)	6.5	4.0	3.0	1.0
7. Production per 50-min/hr (pieces/hr)	7.69	12.5	16.67	50.0
8. Annual depreciation on a 15-yr basis of standard items ($/yr)	4,693.87	3,863.80	5,366.67	7,653.33
9. Over 2,000 hr/yr—depreciation ($/hr)	2.347	1.932	2.68	3.827
10. Composite rate ($/hr)[a]	35.35	34.93	28.18	29.33
11. Total setup cost per lot ($)[b]	88.37	87.33	96.01	220.96
12. Cost per piece, without setup and special tools ($/piece)	4.595	2.80	1.69	0.587
Analysis for 1,000 pieces in 10 lots of 100 pieces per lot				
13. Direct costs ($)	4,595.00	2,280.00	1,690.00	586.70
14. Setup for 10 lots ($)	883.70	875.30	1,248.80	2,209.62
15. Special tools ($)	3,570.00	16,275.00	2,500.00	4,500.00
16. Total cost for 1,000 pieces ($)	9,048.70	19,430.30	5,438.80	7,296.32

Note: For simplicity, interest, insurance, and taxes are not included.
[a] Labor rate is $15.00/hr on ram-type turret lathe but is $7.50/hr on the automatic lathes where one person tends two machines. The overhead rate (not including depreciation) is $18.00/hr.
[b] Labor rate is $15.00/hr for setup on all machines.

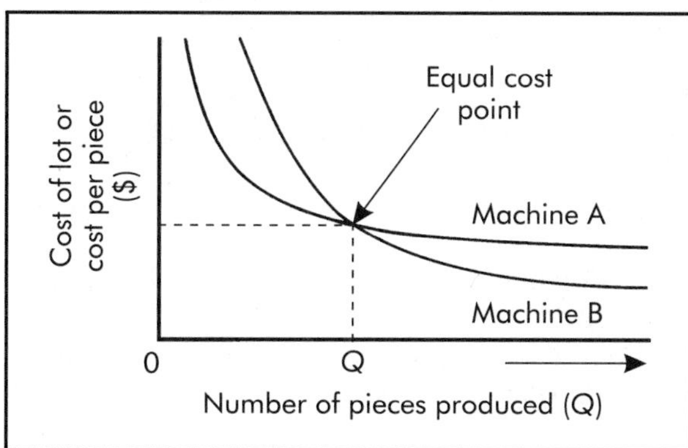

Figure 19-8. Typical equal cost situation.

The left side of the equal sign is the cost of producing Q pieces on machine A, which for the equal cost quantity must equal the cost for Q pieces on machine B, represented by the right side. The solution for the equal cost quantity is

$$Q = \frac{P_A P_B[(S_B R_B) + E_B - (S_A R_A) - E_A]}{(P_B R_A) - (P_A R_B)} \quad (19\text{-}8)$$

It must be emphasized that only the costs that actually change from one machine to the other should be included in the overhead rates O_A and O_B. For instance, whether one machine or another is used for an operation, it is not likely that there is going to be any difference in the salaries of the officers of the company, so the cost is irrelevant.

As an example, Equation (19-8) may be applied to the case described in Table 19-1, for which the rate of production is 11.52 pieces per 50-min/hr on the engine lathe and 18.69 pieces on the turret lathe. Assume the composite rate to be \$28/hr on the engine lathe and \$32/hr on the turret lathe. No special equipment is required for either machine. Then

$$Q = \frac{11.52 \times 18.69[(1.33 \times 32) - (0.33 \times 28)]}{(18.69 \times 28) - (11.52 \times 32)}$$

$Q = 46$ pieces

Thus the engine lathe is the more economical for lots of up to 46 pieces, and the turret lathe for larger lots.

The operation described by Table 19-2 also may be studied by an equal cost point analysis. In this case, the single-spindle automatic is more economical than the turret lathe with special tooling at any level of production. That is because the single-spindle automatic not only produces faster but also has a lower composite rate and tooling cost than the special tooling setup on the turret lathe. However, the single-spindle automatic requires more tooling and setup cost than the turret lathe with standard tools. Accordingly, when tooled simply, the turret lathe is more economical for a few pieces.

Using Equation 19-8, the number of pieces for the equal cost point for the turret lathe with standard tools and the single-spindle automatic can be computed.

$$Q = \frac{7.69 \times 16.7[(3.5 \times 28.18) + 2{,}500 - (2.5 \times 35.35) - 3{,}570]}{(16.67 \times 35.35) - (7.69 \times 28.18)}$$

$= 365$ pieces

So, 365 pieces will justify the extra special tool cost of the single-spindle automatic with one lot. For more pieces in a lot, the single-spindle automatic is preferred. The equal cost point for the multiple- and single-spindle automatic is

$$Q = \frac{50 \times 16.67[(6.0 \times 29.33) + 4{,}500 - (3.5 \times 28.18) - 2{,}500]}{(50 \times 28.18) - (16.67 \times 29.33)}$$

$= 1{,}882$ pieces

The multiple-spindle automatic should be chosen for more than 1,882 pieces.

19.6.6 Special Considerations

Equation 19-8 may be used to find the equal cost quantity when a special machine is involved, but care must be exercised to assign the factors their correct values. This may be illustrated by the assumption that machine B can be utilized to make the part under consideration, but in reality there is no other work that can be put on it. Another example would be when machine A is the turret lathe with standard tools, which can be kept busy when not on this job. Machine B is the single-spindle automatic and has no use other than for this job. In this case, the value of the special tools and equipment E_B is \$80,500 + \$2,500 = \$83,000. The composite rate is \$25.50 instead of \$28.18 because it includes only labor and overhead without any depreciation. The setup rate may be considered the same or even neglected because no changeover is required and setup time is small. So,

$$Q = \frac{7.69 \times 16.67[(3.5 \times 25.5) + 83{,}000 - (2.5 \times 35.35) - 3{,}570]}{(16.67 \times 35.35) - (7.69 \times 25.5)}$$

$= 25{,}897$ pieces

Another comparison is that of the standard turret lathe with special tools and the single-spindle automatic that can be used only for this job. In this case,

$$Q = \frac{12.5 \times 16.67[(3.5 \times 25.5) + 83{,}000 - (2.5 \times 33) - 74{,}232]}{(16.67 \times 33) - (12.5 \times 25.5)}$$

$= 7{,}903$ pieces

A comparison of the turret lathe with special tooling and the multiple-spindle automatic, also considered a special machine, shows an equal cost quantity

$$Q = \frac{12.5 \times 50[(6 \times 33) + 119{,}300 - (2.5 \times 34.93) - 16{,}275]}{(50 \times 34.93) - (12.5 \times 33)}$$

$= 48{,}321$ pieces

These comparisons indicate that under the circumstances the turret lathe with standard tools is the economical choice for a small quantity, the turret lathe with special tools is economical for moderate quantities, and the multiple-spindle automatic is preferable for any quantity over 48,321 pieces. The single-spindle automatic does not seem to have a place in this situation. Before, when all the machines were considered standard, the turret lathe with special tooling was not found to be economical, and the single-spindle automatic was the choice for medium lot sizes.

19.7 QUESTIONS

1. Define the term "manufacturability."
2. Why is feedback necessary between the various functions involved in the product design function?
3. What is the "concurrent engineering" approach to product development?
4. Name four product characteristics that enhance the value of a product to the customer.
5. What are the three basic information sets that form the basis for process planning?
6. What is the purpose of a flow diagram in process planning?
7. Describe the job shop type of manufacturing layout and name some of its advantages and disadvantages.
8. What types of information are provided on operation sheets?
9. Is the general-purpose machine tool with a size capacity just large enough usually selected for a workpiece? Why?
10. What should the size capacity of a special- or single-purpose machine tool be in relation to the size of part it is intended to handle?
11. What kinds of pieces are machined on vertical lathes or boring mills?
12. What is one method of estimating the power required for an operation?
13. Why is the power at the motor of a machine tool likely to be considerably higher than that estimated at the tool point?
14. Is it permissible to overload a machine tool?
15. What are some particular considerations important to machine tool selection?
16. How are labor and overhead costs usually calculated?
17. Into what parts may the time to perform an operation be divided? Describe these parts.
18. What is the basic relationship for calculating cutting time?
19. How do the engine lathe and turret lathe compare as far as quantity of production is concerned? Why?
20. What does an equal cost point show in the comparison of two machine tools?

19.8 PROBLEMS

1. Develop an operation sheet for completing the manufacturing operation specified in step number 25 of the routing sheet of Figure 19-4. Use essentially the same machine and cutting tools as were used in developing the operation sheet of Figure 19-6.
2. Calculate the cutting time for the starter pinion-gear blank shown in Figure 19-9. Assume a cutting speed of 60 m/min (≈200 ft/min) for cemented carbide tools and 18 m/min (≈60 ft/min) for high-speed steel tools (except for reaming). Barstock size is 65 mm (≈2.5 in.).
3. Estimate the total cycle time to turn out the gear blanks of Figure 19-9 on a 355.6-mm (14-in.) engine lathe with a 5.6 kW (7.5 hp) motor.
4. Estimate the total cycle time to turn out one of the gear blanks of Figure 19-9 on a No. 5 ram-type turret lathe with a 7.5 kW (10 hp) motor.
5. It is estimated that it takes 21 min to set up and tear down and 10 min for each piece shown in Figure 19-9 on a 355.6-mm (14-in.) engine lathe. Setup and tear down is estimated to be 85 min, and each piece takes 6 min on a No. 5 turret lathe. The rate for labor and overhead is $22/hr on both machines. No special tools are required.
   (a) What is the equal cost quantity for the two machines?
   (b) Calculate the cost for each machine to produce: (1) 5 pieces; (2) 10 pieces; (3) 25 pieces; (4) 50 pieces.
6. For the operation of the ram-type turret lathe with standard tools and the same machine with special tooling described in Table 19-2, calculate the following.
   (a) The equal cost quantity for the two machines.
   (b) The cost for each machine to produce: (1) 1,000 pieces in one lot; (2) 5,000 pieces total in 25 lots; (3) 10,000 pieces total in 10 lots; (4) 15,000 pieces total in 150 lots.
7. If the part shown in Figure 19-9 is made on a No. 5 ram-type turret lathe with standard tools, setup time is 85 min and the cycle time is 6 min per piece. The machine costs $40,000 and tools $5,000, depreciated over a 15-yr period and 2000 hr/yr. Labor rate is $12/hr and overhead (not including depreciation) is $10/hr. If $4,000 of special tools are applied to the job, the cost of standard tools is reduced to $2,500, and the time per piece to 5 min. Setup time is not changed.
   (a) Calculate the equal cost quantity for the two options on the basis of a 50-minute hour.
   (b) Calculate the cost for each option to produce (1) 1,000 pieces total in 10 lots; (2) 5,000 pieces total in 25 lots; (3) 10,000 pieces total in 10 lots; (4) 15,000 pieces total in 100 lots.
8. The part shown in Figure 19-9 can be made on a 63.5-mm (2.5-in.) capacity, single-spindle, automatic bar machine in 4.5 min, at the rate of 11.1 pieces in a 50-minute hour. The machine costs $60,000 and standard tools $2,700, depreciated over a 15-yr period and 2,000 hr/yr. Special tools cost $1,200. Setup time is 3 hr. The labor rate on setup is $12.50/hr, but otherwise is $6.25/hr because one operator tends more than one machine. The overhead rate except for depreciation is $14/hr. This machine can be used for other work whenever it is not busy with this job.
   (a) Calculate the cost of the operation on this machine for a quantity of (1) 500 pieces total in 10 lots; (2) 1,000 pieces total in 2 lots; (3) 1,500 pieces total in 15 lots; (4) 3,000 pieces total in 100 lots.
   (b) What is the equal cost quantity for the single-spindle automatic and the turret lathe with standard tooling for which costs are given in Problem 7?
   (c) What is the equal cost quantity for the single-spindle automatic and the turret lathe with special tooling for which costs are given in Problem 7?

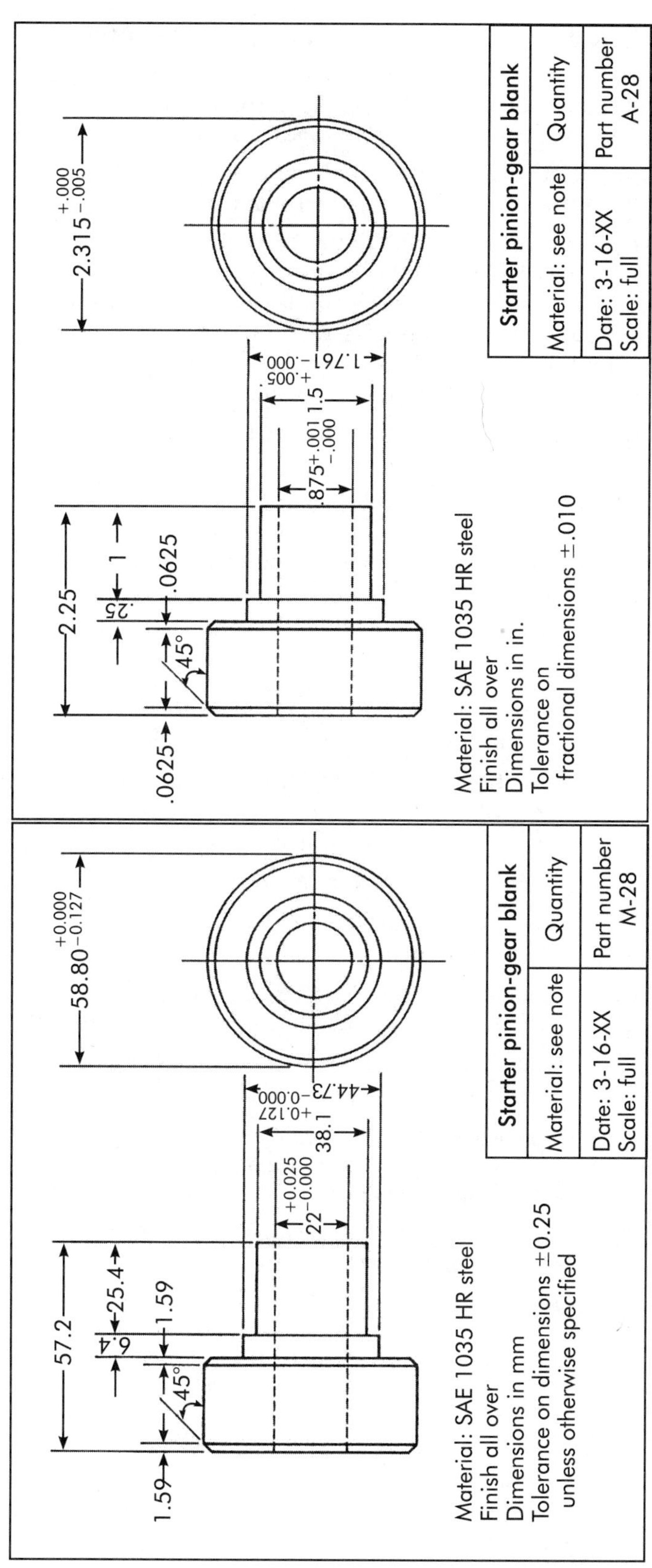

Figure 19-9. Starter pinion-gear blank.

9. If the single-spindle automatic bar machine described in Problem 8 has no use other than for this part,
   (a) What is the equal cost quantity to justify use of the single-spindle automatic in preference to the hand-operated turret lathe with the standard tooling in Problem 7?
   (b) What is the equal cost quantity to justify the use of the single-spindle automatic in preference to the hand turret lathe with the special tooling in Problem 7?
10. The part shown in Figure 19-9 can be made on a 67-mm (2.625-in.) capacity, six-spindle, automatic bar machine in 1.5 min. The machine costs $112,000 and standard tools $10,000, and it is depreciated over a 15-year period and 2,000 hr/yr. Special tools cost $7,200. Setup time is 7 hr. The labor rate for setup is $14.00/hr, but otherwise is $7.00/hr because one operator tends more than one machine. The overhead rate, except for depreciation, is $16/hr. The machine can be used for other work when it is not busy with this job.
    (a) What is the equal cost quantity for the six-spindle automatic and the turret lathe with special tooling for which costs are given in Problem 7?
    (b) What is the equal cost quantity for the six-spindle automatic and the single-spindle automatic described in Problem 8?
11. If the six-spindle automatic bar machine described in Problem 10 cannot be used for other work, what quantity must be produced on it to justify its selection?

19.9 REFERENCES

Coleman, J.R., ed. 1991. "Problem Solving with Software." *Manufacturing Engineering*, March.

Gershwin, S.B. 1994. *Manufacturing Systems Engineering*. Upper Saddle River, NJ: Prentice-Hall.

Groover, M.P. 1987. *Automation, Production Systems, and Computer Integrated Manufacturing,* 2nd Ed. Upper Saddle River, NJ: Prentice-Hall.

Machining Data Handbook, 3rd Ed. 1980. 2 Volumes. Cincinnati, OH: Institute of Advanced Manufacturing Sciences.

Mason, F., ed. 1997. "Mapping a Better Process." *Manufacturing Engineering*, April.

Noaker, P.M., ed. 1995. "Simplifying Setup." *Manufacturing Engineering,* July.

——. 1993. "Managing Manufacturing History." *Manufacturing Engineering*, November.

Ostwald, P.E. 1992. *Engineering Cost Estimates,* 3rd Ed. Upper Saddle River, NJ: Prentice-Hall.

——. 1982. "Manufacturing Cost Estimating Guide." *American Machinist*.

Owen, J.V., ed. 1991. "Justifying Manufacturing Flexibility." *Manufacturing Engineering*, March.

——. 1996. "Making Work Flow." *Manufacturing Engineering*, March.

Phillips, E.J. 1997. *Manufacturing Plant Layout*. Dearborn, MI: Society of Manufacturing Engineers (SME).

White, J.A. ed. 1987. *Production Handbook,* 4th Ed. New York: John Wiley and Sons.

Winchell, W. 1989. *Realistic Cost Estimating for Manufacturing,* 2nd Ed. Dearborn, MI: Society of Manufacturing Engineers (SME).

Wyman, T. 1994. "Total Cost of Expendable Tooling." *Modern Machine Shop*, May.

Drilling and Allied Operations

20

The drill press, 1840
—James Nasmyth, Manchester England

This chapter describes the kinds of operations involved with the opening, enlarging, and finish cutting of holes ranging from a small fraction of a millimeter (μin.) to hundreds of millimeters (in.) in diameter. In most of the operations, the tools, and not the workpieces, are revolved and fed into the material.

The common types of operations under consideration are illustrated in Figure 20-1. *Drilling* is the easiest way to cut a hole into solid metal. It is also done to enlarge holes and then may be called *core drilling* or *counter drilling*. When a hole of two or more diameters is cut by one drill, the operation is called *step drilling*.

Boring is the enlarging of a hole, sometimes with the implication of producing a more accurate hole than by drilling. Enlarging a hole to a limited depth is called *counterboring*. If the depth is shallow so the cut leaves in effect a finished face around the original hole, it is called *spot facing*. The cutting of an angular opening into the end of a hole is *countersinking*, also loosely termed *chamfering*. *Reaming* is also a hole enlarging process, but its specific purpose

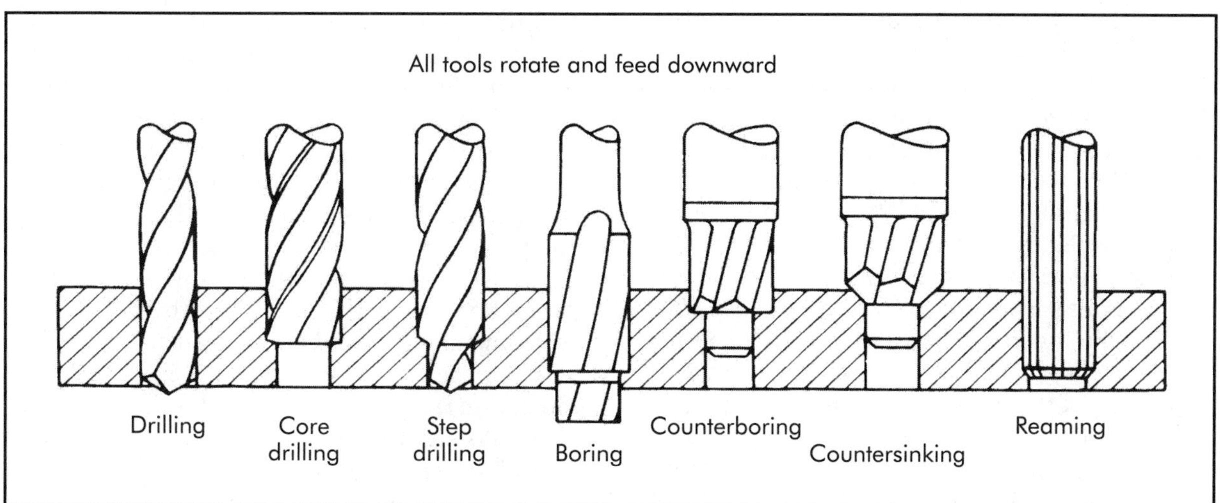

Figure 20-1. Common drilling and related operations.

is to produce a hole of accurate size and good surface finish, and stock removal is small.

Tapping is often done with other operations in a hole, but it will be discussed with other thread-cutting operations in Chapter 28.

20.1 DRILLS, BORING TOOLS, AND REAMERS

20.1.1 Common Drills

The most common form of metalworking drill is the *twist drill* with helical grooves or flutes as shown on three of the drills in Figure 20-2. Generally, these drills are furnished with two flutes as those with three or more flutes cannot be used to start holes unless the web thickness at the point is reduced. Three- and four-flute drills are generally used to enlarge holes previously drilled or cored and they are referred to as *core drills* or *core reamers*. Smaller diameter drills with straight flutes, such as shown in Figure 20-2(D) are designed to produce short chips and are best suited for use in drilling brass and other nonferrous materials.

A *multicut drill* or *step drill* makes holes of two or more diameters as in the step drilling operation of Figure 20-1. Such a drill often can be used to counterbore and countersink.

The fluted portion of a twist drill is its *body*. The *point* is the cutting end and cutting only takes place along the cutting tips of the drill point. The drill is held and driven at the *shank* on the other end. The shank may have an American National Standards Institute (ANSI) taper (Morse®) with a tang or it may be straight.

Insert drills, like the ones shown in Figure 20-3, are in wide usage, particularly in production turning machines and computer numerical control (CNC) screw machines. Many manufacturing firms claim that as much as 50% of all the machining time on their products is used to produce a variety of sizes and types of holes. Thus, a large proportion of these firms have supplanted the traditional twist drill with the replaceable insert drill. As illustrated in Figure 20-3, many of these tools are designed to include a variety of inserts to enable them to perform chamfering, spot facing, and counterboring, along with the drilling operation. A complete hole is machined in one pass, thereby eliminating several separate operations and tool changes.

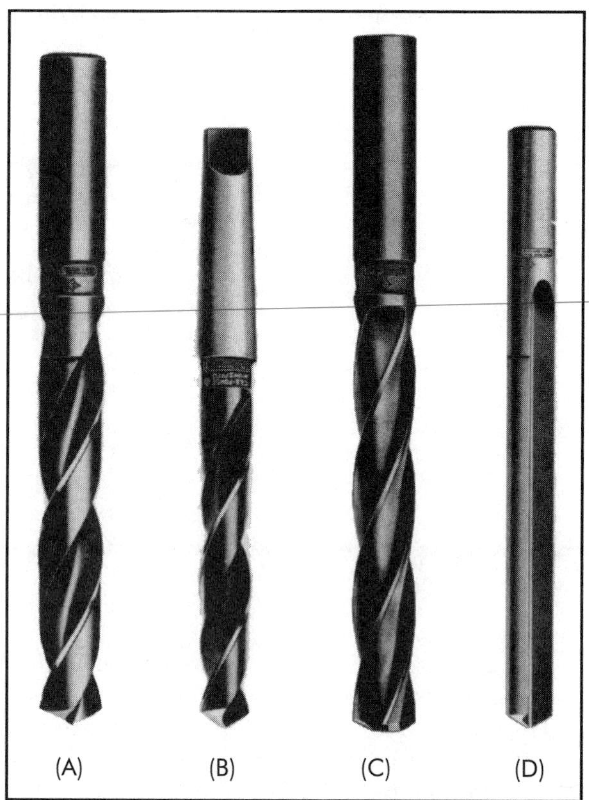

Figure 20-2. (A) High-speed steel, straight-shank, two-flute twist drill; (B) tapered shank, two-flute twist drill; (C) four-flute core drill; (D) straight-flute drill. (Courtesy Cleveland Twist Drill Co.)

The insert drills illustrated in Figure 20-3 are designed to use replaceable inserts of carbide, coated carbide, or coated high-speed steel. The helical drill-point geometry of the inserts reduces the tool thrust and increases the self-centering action of the drilling operation. As will be explained later, the conventionally ground, high-speed twist drill has a straight chisel edge with a negative rake angle that tends to plow through metal rather than creating a shearing action. Thus, the point of the drill tends to drift off center as the chisel edge becomes worn and dull.

In addition to improved accuracy in both size and centering, the tool life of the insert drill is improved significantly over that of a conventional high-speed twist drill. After switching to insert drills, one manufacturer reported an in-

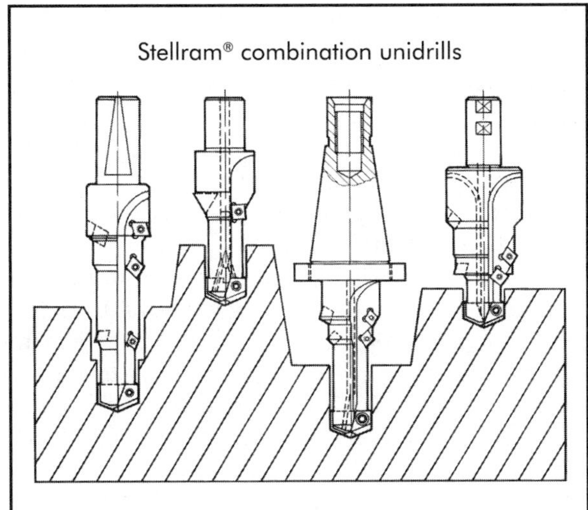

Figure 20-3. Typical configurations for multiple insert drills used to perform several hole drilling and finishing operations simultaneously. (Courtesy Giddings & Lewis, Inc.)

creased tool life 45 times longer than that of high-speed steel twist drills.

20.1.2 Drill Sizes and Materials

The size of a drill designates the nominal diameter of its body and hole it is intended to produce. Standard drills are available in *numbered, lettered,* and *fractional inch* and *millimeter* sizes.

Fractional-size drills come in 0.41 mm (1/64 or .016 in.) steps up to 44.5 mm (1.75 in.) and larger steps above that to over 75 mm (3 in.) in diameter. Numbered and letter-size drills run from 0.150–10.490 mm (.0059–.4130 in.) in diameter in between the fractional sizes, so there is only a few thousandths of an inch difference between one drill size and the next in that range. A leading maker offers carbide insert drills in high-speed steel, and straight- and taper-shank drills as outlined in American National Standard B94.11M-1979 (R1987).

The industry standard material for twist drills has been high-speed steel (HSS) for many years. However, recent developments in coating technology for carbide materials and insert geometry has made drill selection more complex, while at the same time offering opportunities for significant increases in hole-making productivity and cost savings. Along with carbon steel and HSS, twist drills are now available in high-speed cobalt and solid carbide. Both high-speed steel and carbide drills are available with titanium nitride (TiN) coating. The TiN coating increases the lubricity and hardness of the cutting edges of the drill and thereby permits the use of higher cutting speeds and feeds for a given tool life. A TiN-coated drill costs about twice as much as a noncoated HSS drill, while a solid carbide drill is about ten times as expensive as a similar size HSS drill. Thus, an evaluation of initial cost, tool life, and productivity should be made before selecting a drill material for a given production drilling operation.

Insert drills similar to those shown in Figure 20-3 are available with replaceable or indexable high-speed steel or tungsten carbide inserts, with a coating of either titanium carbide (TiC), titanium nitride (TiN), titanium carbon nitride (TiCN), or aluminum oxide (Al_2O_3). Advanced grades of insert materials, such as cermets, cubic boron nitride (CBN), or polycrystalline diamond (PCD), are also available for machining certain types of materials on machines capable of handling those tools.

The selection of speeds and feeds for drilling operations is covered later in this chapter. As in the case for most cutting tools, it is suggested that the drill manufacturer be contacted concerning recommended drilling parameters for specific production requirements.

20.1.3 Drill Angles and Edges

The body and point of a drill must have certain components for efficient performance. These are designated for the common form of twist drill in Figure 20-4. A twist drill cuts only at the point, not along the sides of the body. Both the chisel edge and the cutting lips are cutting edges.

The *chisel edge* of an ordinary drill indents as it is forced into the metal and, as it turns, partially cuts like a cutting tool with a large negative rake angle as depicted in Figure 20-5. The chisel edge does not penetrate like a sharp point and tends to make a drill start off center. Some drills are ground to reduce or eliminate the chisel edge; among a number of forms are the thinned point, crankshaft grind,

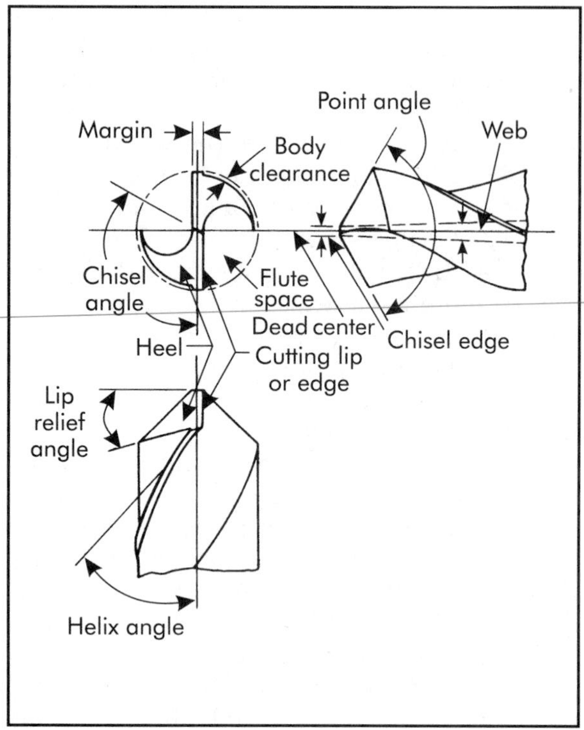

Figure 20-4. Twist drill elements.

and spiral point of Figure 20-6. Too much compensation may weaken the point, but optimum amounts have been found to reduce thrust force by one-third, increase tool life by as much as three times, and improve hole location precision as compared to the full chisel edge. Guidelines for grinding various points are given in reference texts and handbooks.

The cutting lips or edges illustrated in Figure 20-4 correspond to the cutting edges of a single-point tool. The rake angle depicted on the left of Figure 20-5 results primarily from the helix angle of the flutes. An average value for the helix angle is 30°, but angles from about 18° for hard materials to 45° for soft materials are used. The effective rake angle at the lip is larger at the periphery than toward the center of the drill. Some users grind drills at the end of the flutes to create a uniform rake angle along the lips.

The heel of the drill point must be backed off when ground to give relief behind the cutting lips, as depicted in Figure 20-4, like the relief on a single-point tool. The cutting lip relief angle is ordinarily 12–15° at the outside diameter and

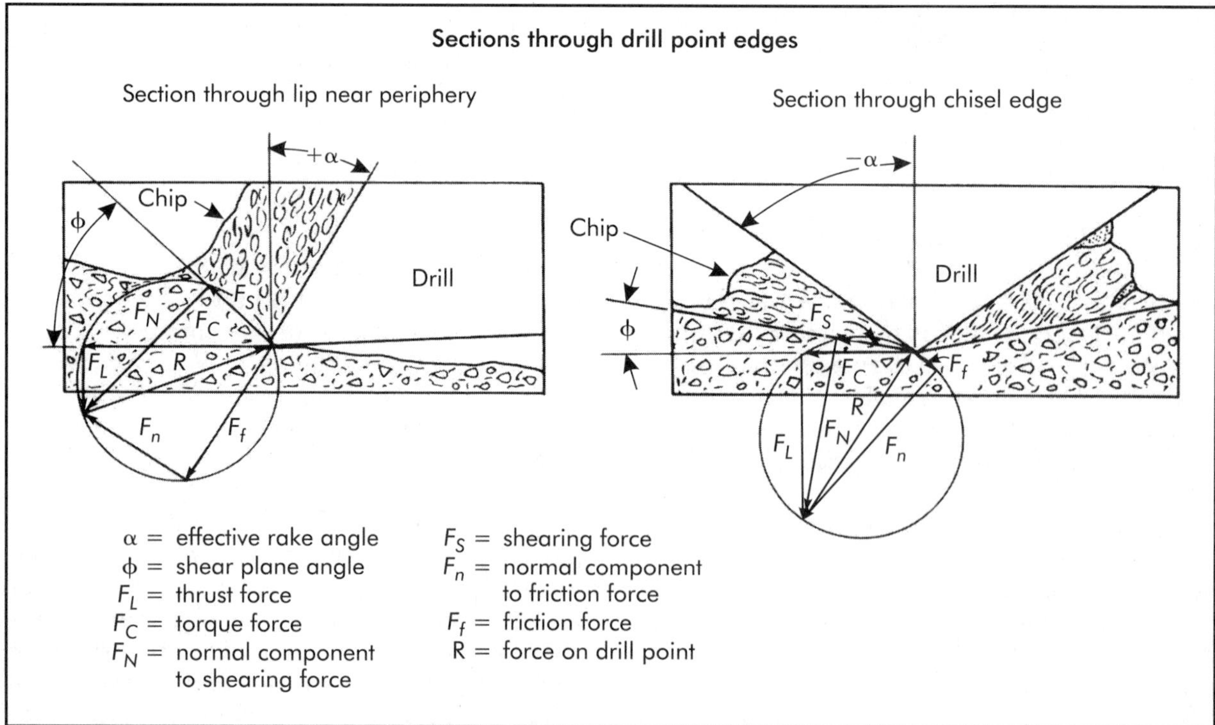

Figure 20-5. Sketches made from photomicrographs of chip formation at the two cutting zones of a twist drill. Force circles like that of Figure 17-8 are superimposed to illustrate differences in actions.

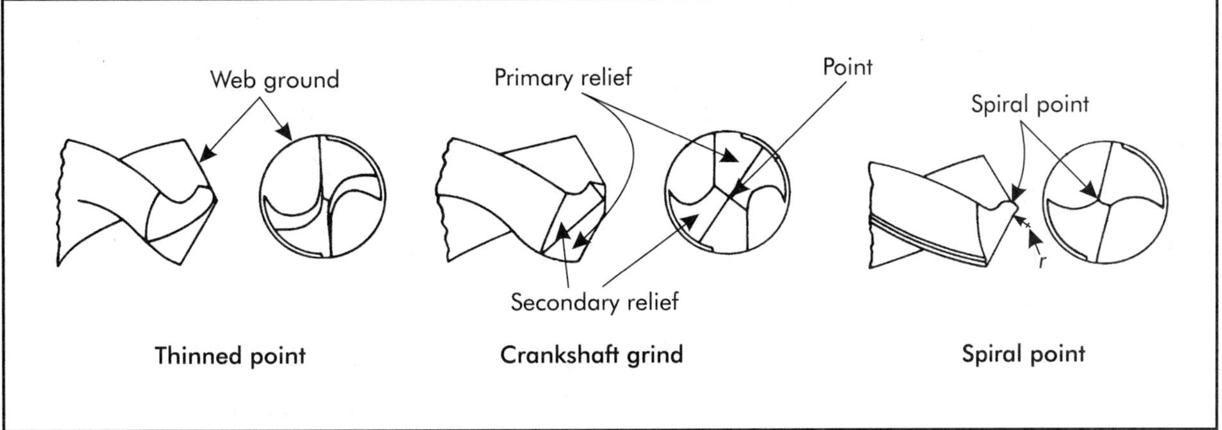

Figure 20-6. Drill forms that reduce the ill effects of the chisel edge.

decreases toward the axis of the drill. Too much relief weakens the cutting edges and shortens drill life, but the relief angle must always be larger than the lead angle of the path of the cutting edge to prevent rubbing. The lead angle (the complement of the helix angle) of the path increases for smaller diameters, and therefore the relief angle is made larger toward the center on a properly ground drill.

The point angle between the cutting edges corresponds to the side-cutting edge angle of a single-point tool. It is 118° for average work but has different values for specific purposes. For example, a point of 136° is suitable for hard manganese steel, but only 60° is needed for wood and fiber.

Chip breakers help drills, like other tools, cut stringy chip materials more efficiently. Notches, known as *chip breakers,* are sometimes ground in the cutting lips. One make of drill has a step along each flute to curl and break chips. Other drill manufactures have different styles of chip breakers.

It is particularly important that the cutting edges or lips be of equal length and lie at equal angles from the axis of the drill. Otherwise, the drill cuts erratically and has a short life. Obvious results of unequalized drill sharpening may result in oversize and misshapen holes. This is merely an exaggeration of what may be expected from ordinary errors in drill grinding and helps explain why most drills, in practice, cut oversize holes. Landberg found that drills carefully ground in the laboratory had lives up to three times longer as when manually sharpened under usual shop conditions. Common errors or imperfections in hole geometry, such as errors in shape, location, roundness, and dimension, may occur to various degrees in drilling as shown in Figure 20-7.

The *body clearance* on all but very small drills leaves a narrow margin or strip at full nominal diameter along the edge of each flute. This reduces rubbing between the drill and hole and allows cutting fluid to reach the point of the drill. Also, the body decreases a few thousandths of an inch in diameter from the point to the shank to reduce rubbing. Drill rigidity is needed for stamina and depends on the web and body thickness and flute length. Long drills have been shown to have short lives. One rule is to keep the flute length less than ten times the diameter. The longer its unsupported length, the more a drill tends to wander off center, particularly when starting a hole.

A twist drill is sharpened by grinding the heel behind the cutting edge on the point—never the outside diameter. Skilled mechanics may grind chisel-point drills off hand with passable results, but generally machine ground drills cut faster, last longer, and produce more accurate holes. Machines are definitely needed for precisely grinding particular forms, such as the spiral point. In principle, a drill grinder is a grinding stand with an attachment or fixture to hold the drill at a certain angle and swing it in a particular path with respect to the grinding wheel, or vice versa. A means is also provided to true the wheel.

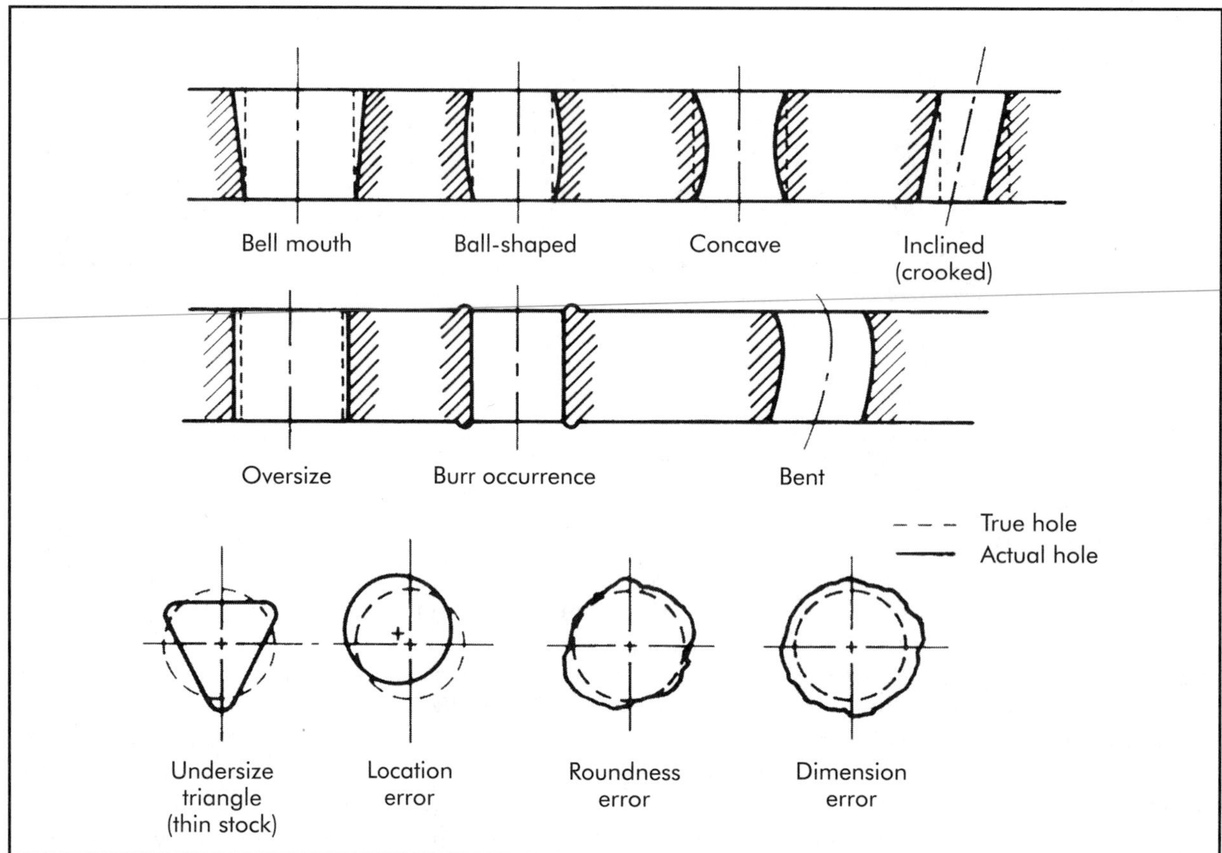

Figure 20-7. Common errors in hole geometry experienced in drilling.

20.1.4 Deep Hole Drills

Deep-hole drilling is generally accomplished by either gun drilling, ejector drilling or trepanning.

Gun drilling is most commonly used for small holes ranging from 3 mm (≈.125 in.) to around 25.4 mm (1 in.). Small hole diameters up to about 3 m (10 ft) in length are often produced by this process. A gun drill usually consists of a single-lipped carbide insert affixed to a steel drill body. The drill body is then attached to an alloy steel shank of sufficient length for the depth of hole required. The single-lip design of the cutter causes it to cut in a true circular pattern. Support or follower pads around the drill body support the drill in the hole, providing good dimensional control. Thus, a diameter tolerance of ±0.05 mm (±.002 in.) can usually be achieved in gun drilling. Solid-carbide gun drills are available with a one-piece, solid carbide tip and shank. This type of construction provides increased rigidity for gun drilling small holes in alloy steels, stainless steel, and cast iron. Solid carbide gun drills are available in sizes ranging from 0.13 mm (.005 in.) to 3.18 mm (.125 in.) with flute lengths up to 15.2 cm (6 in.).

Ejector drilling is accomplished with a proprietary tool that forces coolant through a disposable drill head tipped with cemented-carbide cutting edges. The forced coolant flow through holes in the drill head ejects the chips from the hole. Ejector drilling can be used to drill larger and longer holes than gun drilling. Diameter tolerances of +0.10 mm (+.004 in.) and surface finishes in the range of 60–70 root mean square (rms) can be achieved by this process.

Trepanning uses a tube-type cutting tool to produce a solid core or slug during the drilling operation. Thus, trepanning is often referred to as a *core-drilling* process. Typically, it is used to machine large hole diameters up to 45 mm (1.75

in.) and to depths of 15.2 m (50 ft). Diameter tolerances of around ±0.38 mm (±.015 in.) can be held by trepanning. Since this process cuts out a core, it is somewhat more efficient than the other drilling methods that have to remove all of the metal from a hole.

20.1.5 Boring Tools

Boring is essentially internal turning where a single-edge cutting tool forms internal shapes and dimensions. Thus, the guidelines discussed in Chapter 17 for single-edge cutting tools are generally applicable to single-edge boring tools. A variety of boring bar configurations are available depending on the type of machine and cutting tool insert used.

In addition to the single-edge type boring tool, a number of multiple-edge cutting tools are included in the "boring" tool vernacular. For example, the multiple-blade boring, counterboring, and countersinking tools illustrated in Figure 20-1 are often considered to be members of the boring tool family. A number of them serve dual purposes, such as the counterbore, which can also do spot facing and in that role is referred to as a *spot facer.*

To maintain concentricity, multiple-edge counterboring tools are often equipped with a pilot on one end to guide the tool in the previously drilled and reamed hole.

20.1.6 Reamers

Reamers mainly finish holes to size and are made in a number of styles. They may be hand- or machine-driven; have integral shanks or be attached to holders; be solid or have inserted blades that may be expanded or adjusted; have straight or helical flutes; have straight, tapered, or other shapes; and be designed for roughing or finishing work.

A *hand reamer* has a straight shank with a square tang for a wrench, as shown in Figure 20-8(A). It is expected to remove at most only a fraction of a millimeter of metal from a hole. Its teeth are ground with relief behind the cutting edges and taper slightly from each end to a straight portion in the middle.

Machine or *chucking reamers* are made with or without relief, with straight or spiral flutes, and solid or with inserted blades. Common forms are described in the paragraphs that follow.

A *rose chucking reamer,* like the one in Figure 20-8(B), is cylindrically ground and has no relief behind the outer edges of the teeth. It cuts on the end chamfer of the teeth. Rose reamers are used for heavy roughing cuts, particularly for clearing out cored holes, and not for especially smooth holes.

Two kinds of machine reamers with relief behind the circular margin on the outside diameter, as well as chamfer on the front of the teeth, are the *jobbers' reamer* with long flutes and the *fluted chucking reamer* with shorter flutes.

A reamer over about 19 mm (.75 in.) in diameter is often made as a shell that fits over an arbor and is called a *shell reamer* (Figure 20-9). When a shell is worn out, another one can be put on the same arbor.

An *expansion reamer* can be enlarged to remove an extra fraction of a millimeter (few thousandths of an inch) from a hole or to compensate for wear. One design is shown in Figure 20-8(C). The blades are part of or, if tipped, brazed to the body, which is hollow and has several slots running part way along the flutes. A tapered thread pin expands the body.

An *adjustable reamer* has inserted and replaceable blades locked to the body. The blade seats are tapered and the diameter of the reamer is increased by moving the blades.

As shown in Figure 20-10, the teeth of a reamer have angles corresponding to and governed by the same factors as the single-point tools described in Chapter 17. A machine reamer does most of its cutting on the end.

20.2 DRILLING MACHINES

Drilling machines are made in many forms and sizes. Portable or hand drills are well known. The drilling machines commonly used for precision metalworking are known as drill presses.

20.2.1 Vertical Drill Presses

The main features of *standard upright drill presses* are depicted in Figure 20-11. A column on a base carries a table for the workpiece and a spindle head. The table is raised or lowered manually, often by an elevating screw, and can

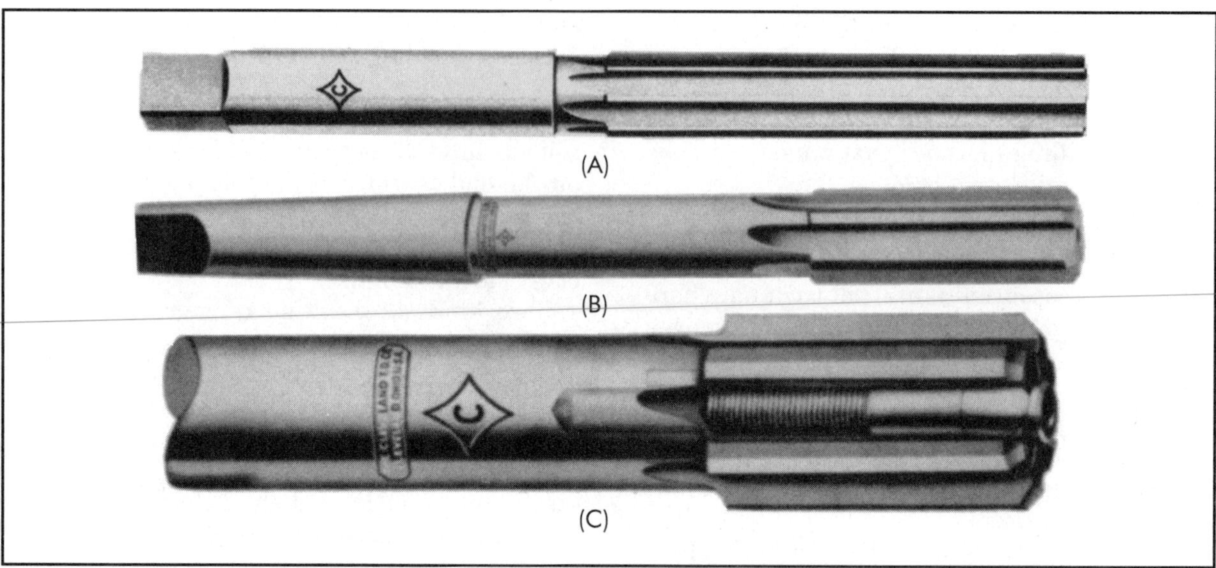

Figure 20-8. Typical HSS reamers with integral shanks: (A) straight-flute hand reamer; (B) rose chucking reamer with tapered shank; (C) straight-flute expansion chucking reamer. (Courtesy Cleveland Twist Drill Co.)

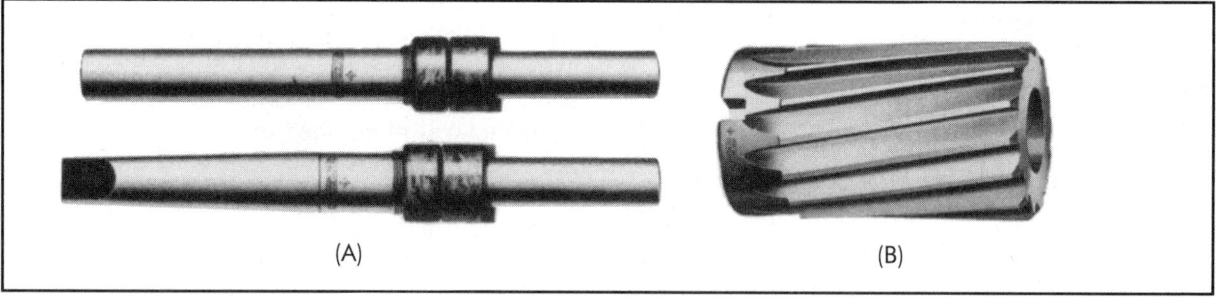

Figure 20-9. Parts of a shell reamer assembly: (A) straight and taper shank with sliding collar arbors; (B) typical HSS helical flute. (Courtesy Cleveland Twist Drill Co.)

be clamped to the column for rigidity. Some tables are round and can be swiveled. On vertical drill presses with round columns, the tables generally can be swung out from under the spindle so workpieces can be mounted on the base.

The spindle that drives the cutting tool revolves in the nonrevolving quill that is fed up or down. Some machines require hand feeding; others have power feeds. Machines of this kind may be equipped with a positive lead screw for tapping and a spindle reversing mechanism. As a rule, an adjustable stop is provided to limit the depth of travel of the quill and, with a power feed, to disengage the feed or reverse a tapping spindle at a definite depth. Most machines have a Morse taper hole in the end of the spindle, and small machines often have a drill chuck attached to the end of the spindle. The spindles of fractional horsepower drill presses are usually driven by vee belts; larger ones have gear transmissions, and some have multiple-speed motors. Numerical control is also utilized (Chapter 31).

Single-spindle, stand-alone drilling machines are available in either a round or box column design. Figure 20-12 shows a typical round-column, geared-head, precision drill press used for medium or heavy-duty continuous production. This particular machine has a 50-mm (≈2-in.) drill capacity in steel with nine spindle speeds ranging from 66–1,255 rpm. It is equipped with a power feed spindle with six different feed rates ranging from about 0.10–0.51 mm/min (.004–

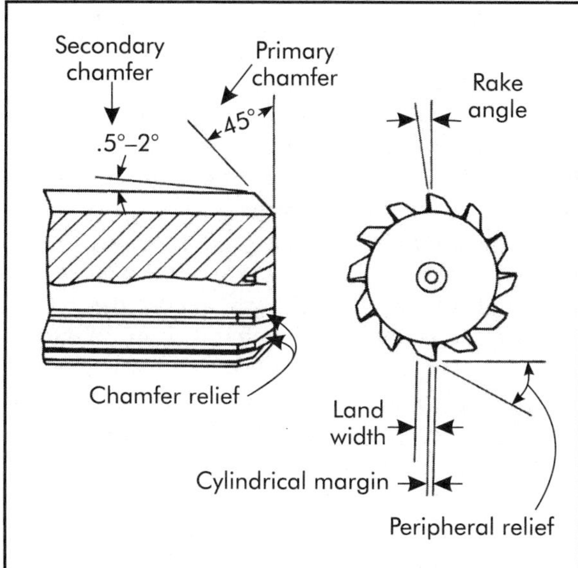

Figure 20-10. Machine reamer angles.

.020 in./min). An optional mechanism for lead screw tapping is also available. A drill press of this type and size costs about $15,000.

A *bench-type drill press* is a light machine for small work. Light bench and upright drill presses that are hand fed so the operator can feel the resistance met by the drill are called *sensitive drill presses*. They are advantageous for feeding small drills to avoid breakage.

In one application for vee bearings, very small holes must be drilled on highly sensitive presses with spindles running quite true at high speeds. The action of the drill is watched through a microscope. Holes less than 25 μm (.001 in.) in diameter have been drilled, and diameters around 127 μm (.005 in.) are common in the instrument and electronics industries.

Production drill presses are sturdily built for heavy work and simple in design, but in overall appearance look like other upright drill presses.

20.2.2 Multispindle Drill Presses

As a rule, a single-spindle drill press is not efficient when a number of pieces are to be machined and each piece requires several different holes or several cuts (such as drilling, counterboring, and tapping) on each hole. In such cases, tools, speeds, feeds, and positions must be changed repeatedly for each piece or each hole.

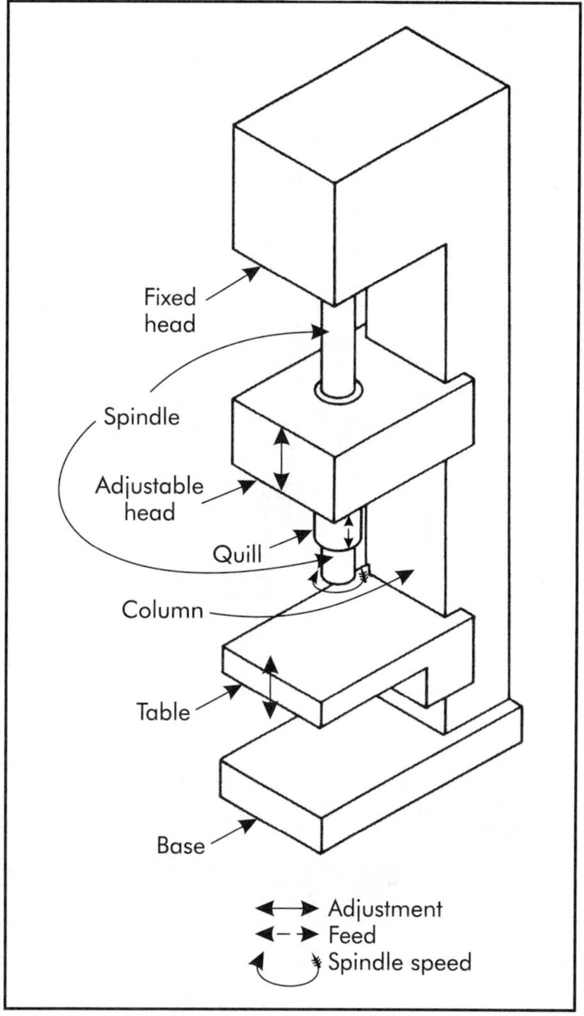

Figure 20-11. Principal parts and movements of a single-spindle upright drill press.

More efficient results can be obtained in a number of ways from multispindle drill presses, each of which has advantages for certain applications.

A *gang drill press* is the equivalent of two, three, four, or more upright or production drill presses in a row with a common base or table. For example, two or more of the drilling heads similar to the one on the machine shown in Figure 20-12 could be mounted on a common table to perform sequential drilling operations at each drill station. A gang drill can be set up so that work can be passed from spindle to spindle to undergo two or more operations. In another mode, the same operation may be performed at each spindle; the operator unloads and loads a

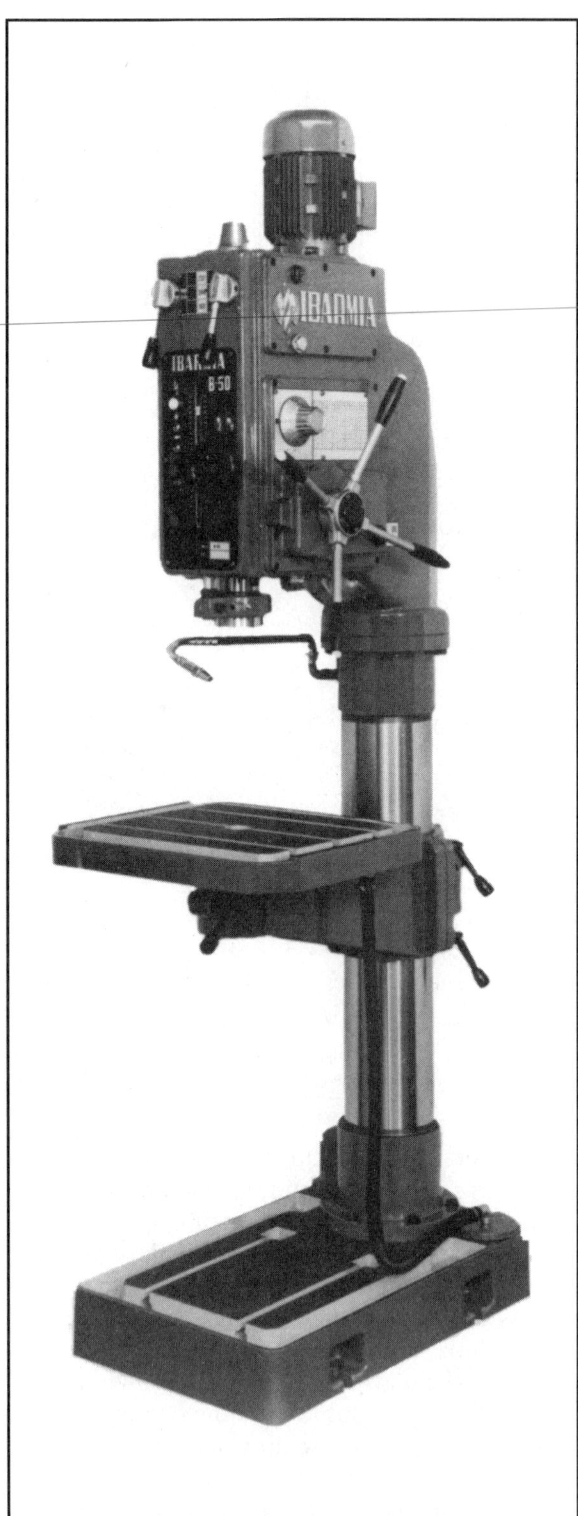

Figure 20-12. Round-column, geared-head, precision drill press with power feed. (Courtesy Reynolds Machine & Tool Corporation, Commander Multi-Drill Division)

workholding jig at each spindle in turn while the other spindles are cutting with a power feed.

Another form of production drilling machine is represented by the eight-position, CNC vertical *turret drill press* shown in Figure 20-13. Available in either vertical or horizontal design, this type of machine is capable of high-production precision machining of small ferrous or nonferrous parts requiring some combination of drilling, tapping, light milling, facing, reaming, and/or chamfering. Using a fixed-position worktable, all movements along the *X*, *Y* and *Z* axes are accomplished by a CNC-controlled traveling column. The traveling column maintains a ±0.010 mm (±.0004 in.) position accuracy and ±0.005 mm (±.0002 in.) repeatability for close, precision machining. The turret toolholders can be indexed from one position to the next position in 1 sec and the traveling column can be moved from one position to another in 2 sec or less, depending on distance traveled. With standard equipment, spindle speeds from 1–6,000 rpm are available, and optional high-speed spindle units may be installed with speeds up to 15,000 rpm. The machine's versatility and productivity can be improved by the installation of

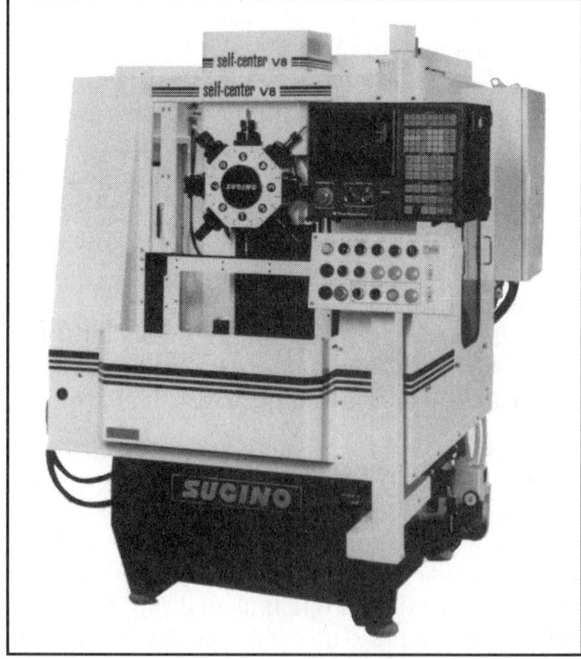

Figure 20-13. Precision CNC turret-drill press with eight tool stations. (Courtesy Sugino Corporation)

multispindle heads on the turret tool head positions. These two, three, and four spindle heads permit several operations, such as drilling and tapping, to be performed at one turret position. In addition, the fixed worktable may be replaced with an air-driven turntable that rotates 180° to facilitate loading and unloading parts while the machine is in operation. A turret drill press of the type shown in Figure 20-13 cost about $80,000.

A *multiple-spindle drill press* has a cluster of spindles on one or more heads used for the production machining of parts with a large number of hole positions. The multiple-spindle machine shown in Figure 20-14 is representative of this type of drill press. This particular machine has 100 9.525-mm (3/8- or .375-in.) hex pick-offs in the head assembly to which slip-spindle assemblies may be attached. The worktable, which will accommodate workpieces as large as 914 × 610 mm (36 × 24 in.), is fed up to the multiple-spindle head by hydraulic pressure and guided by four 76.2-mm (3-in.) diameter guide rods. The machine shown in Figure 20-14 is furnished with a fixed-position spindle plate that must be bored to accept the number and position of spindles required for a particular job. The machine drive motor is rated as having enough horsepower and thrust to drive 16 12.7-mm (1/2- or .5-in.) drills, 28 9.525-mm (3/8- or .375-in.) drills, 50 6.35-mm (1/4- or .25-in.) drills, or 100 3.18-mm (1/8- or .125-in.) drills in cast iron. Each hex pick-off is capable of drilling up to a 12.7-mm (1/2- or .5-in.) hole in mild steel. A master bushing carrier plate is located underneath the spindle plate in which drill bushings are placed to accurately guide the drills into the workpiece.

Figure 20-14. Multiple-spindle drilling machine. (Courtesy Reynolds Machine & Tool Corp., Commander Multi-Drill Division)

20.2.3 Radial Drill Presses

Radial drills are convenient for heavy workpieces that cannot be moved around easily or are too large for other kinds of drill presses. A *plain radial drill*, like the one in Figure 20-15, has a base and column that carries an arm. The arm can be raised or lowered and swung around the column. The head can be moved along the arm, and the spindle it carries can thus be positioned in a circle with a radius about as large as the length of the arm. All adjustments can be locked once the spindle is positioned. The drill spindle is fed up and down manually or by power. The maximum distance from the column to the center of the spindle on the machine shown in Figure 20-15 is 1.1 m (45 in.) and the maximum distance from the spindle to the machine base is 1.2 m (48 in.). These two dimensions essentially limit the size of a workpiece that can be accommodated by the machine. It has 12 spindle speeds, ranging from 44–1,500 rpm, and three power-spindle feed rates of 0.05 mm/rev (.002 in./rev), 0.09 mm/rev (.0036 in./rev), and 0.15 mm/rev (.006 in./rev). Sufficient motor horsepower is available to drill a 50.8-mm (2-in.) diameter hole in mild steel. A radial drill the size of the one shown in Figure 20-15 costs around $14,000.

20.2.4 Sizes of Drilling Machines

The sizes of drilling machines are designated in several ways. The most common designation

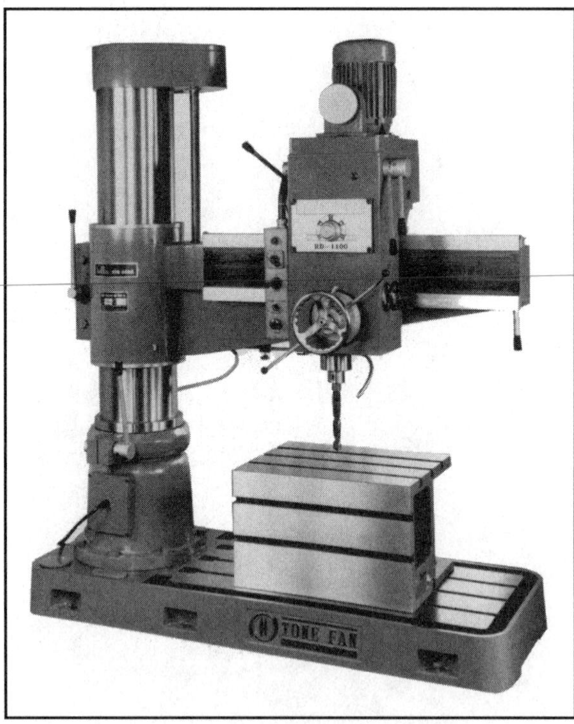

Figure 20-15. Radial drilling machine. (Courtesy Willis Machinery and Tools Co.)

for a vertical drill press is the diameter in millimeters (inches) of the largest disk or workpiece in which a hole can be drilled at the center on the press. Other designations are the diameter of the largest drill the press is designed to drive in cast iron or steel, and the Morse taper size in the spindle hole. The size of a radial drill press designates the radius of the largest disk in which a center hole can be drilled with the head at its outermost position on the arm.

20.3 DRILLING MACHINE ACCESSORIES AND ATTACHMENTS

20.3.1 Toolholders and Drivers

The essential purpose of all toolholders for drilling machines is to make the tool run true with the tapered hole in the machine spindle. Straight-shank tools may be held by a drill chuck, as on a lathe, or by a split collet. If of the right size, the tapered shank of a tool may be placed directly in the tapered hole of the machine spindle. If the tapered shank is smaller than the tapered hole, a *taper shank socket,* with inside and outside tapers, is used. A short socket is shown in Figure 20-16. The cross slot takes the tang of the tapered shank tool and allows a wedge (called a *drift*) to be used to push the tool out of the socket when the two are to be separated.

A *floating driver* is a toolholder with its two ends somewhat loosely coupled to allow a reamer or tap to follow a previously drilled or bored hole. An *adjustable extension assembly* is a form of taper socket adapter that can be adjusted for length. It is the means whereby drills of different lengths can be used together on a multiple-spindle head.

A *quick-change chuck* and *quick-change collet* allow tools to be taken off and put on the spindle of a drill press while it is running.

20.3.2 Multiple-spindle Drill Heads

A standard single-spindle drill press can be converted to a multiple-spindle machine, within limits, by attaching a multiple-spindle head to the machine spindle. Some of these heads are made for specific purposes and have their spindles in fixed positions. On others, like the one in Figure 20-17, the locations of the spindles can be changed. The six adjustable spindles on this unit are gear-driven through booted universal joints with enough strength to allow drilling up to six holes, 9.53 mm (.375 in.) in diameter in mild steel. The six drill spindles can be positioned in a variety of patterns within a 120.7-mm (4.75-in.) diameter circle. The unit can be adapted to fit over the spindle of most standard drilling machines. In addition, these drill heads can be used for multiple tapping by use of a motor reversing unit. An adjustable multiple spindle drill head like the one shown in Figure 20-17 costs about $800.

20.3.3 Workholding Devices

If a drill or other tool catches in a hole, it can twirl a workpiece that is not properly fastened, which is dangerous with all but small drills. A workpiece may be kept from turning by an obstruction placed in its way, or it may be clamped to the table by bolts and straps.

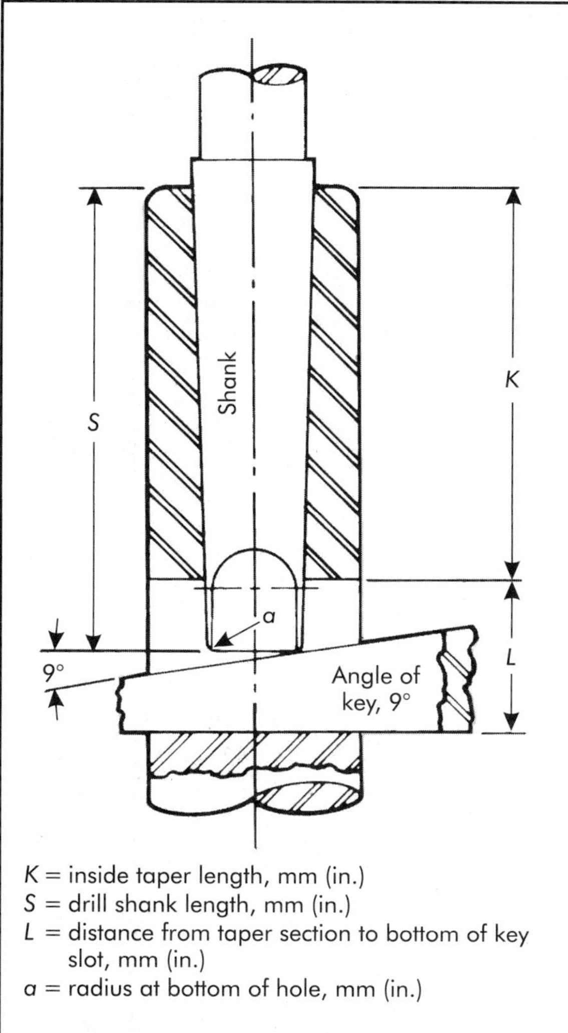

Figure 20-16. Taper shank sleeve.

Vee blocks make good casual supports and locators for round pieces. A workpiece may be clamped to an angle plate. A universal angle plate has a T-slotted surface that can be tilted and swiveled and then secured.

A variety of vises are commercially available for supporting and clamping workpieces to be drilled on a drill press. A standard drill vise has an open construction that allows a drill to pass through the workpiece. Many vises are actuated by cams or toggles and can be opened and closed rapidly. Reduced operation time can be achieved by using air or hydraulic pressure to open and close some types of vises.

Figure 20-17. Multiple-spindle drill head with adjustable arms for spindle positioning. (Courtesy Reynolds Machine & Tool Corp., Commander Multi-Drill Division)

A three-jaw chuck, like the one used on lathes, may be mounted on a base with the jaws up to hold pieces for drilling. The base often contains an indexing mechanism so that holes can be spaced and drilled at equal intervals on a circle.

20.3.4 Positioning Tables

Sometimes a table on a saddle on a base is added to a drill press or radial drill to position workpieces under the spindle with respect to two coordinate directions. Thus holes may be located as specified for the pieces. The table and saddle may be positioned by lead screws and dials, against preset stops, or by a computer numerical control system as described in Chapter 31.

20.3.5 Fixtures and Jigs

A fixture is a device that holds and locates a workpiece. Strictly speaking, conventional work-holding devices, such as chucks and vises, are fixtures, but the name is mostly given to single-purpose holding devices. Fixtures are mostly associated with lathes, shapers, planers, millers, broaches, and grinders.

A jig not only holds and locates a workpiece, but also guides the cutting tool. This is particularly desirable in drilling because the point of the drill tends to wander unless it is guided when a hole is started. Multiple-point boring tools and sometimes reamers need to be guided.

Workpiece location is particularly important in most machining operations for precision engineered products. As illustrated in Figure 20-18, a free body in space has six degrees of freedom, three represented by the coordinate axes of linear translation (*X*, *Y*, and *Z*) and three by rotation (*A*, *B*, and *C*) about those axes. Since movement may occur in a plus or minus direction along or about each of those axes, the free body is said to have freedom of movement in 12 directions. Thus, a workpiece is fully and accurately located when confined or restricted against movement in all of those directions, except those required for a machining operation.

In designing workholding devices, such as jigs and fixtures, the tool designer commonly uses the concept illustrated in Figure 20-19(A) where the workpiece is located against three mutually perpendicular planes *X*, *Y* and *Z*. This arrangement restricts nine directions of movement for the part. The remaining three directions can then

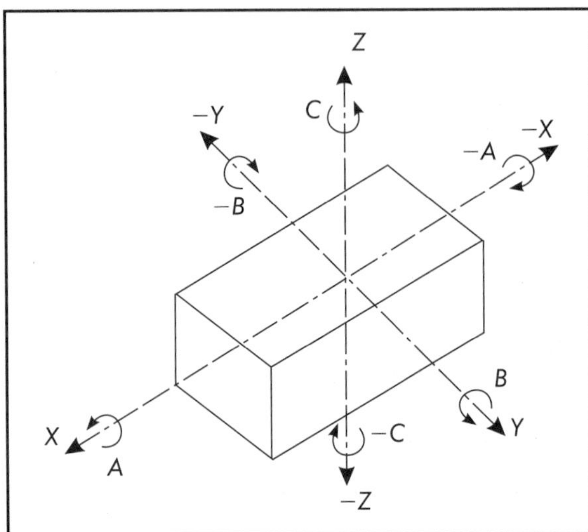

Figure 20-18. Free body with 12 directions of movement.

be restricted by the application of a clamping force over the part to hold it against those three planes. In theory, the workpiece shown in Figure 20-19(A) is precisely located if all of the contacting surfaces are perfectly flat and clean. This, however, is usually not the case in a manufacturing environment. Thus the 3-2-1 method of location illustrated in Figure 20-19(B) is generally used to accomplish the same purpose. With this method, three locating pins or buttons are used to establish the *Z* plane, two pins establish the *Y* plane, and one pin establishes the *X* plane.

The 3-2-1 location method uses the minimum number of location points needed to restrict nine directions of movement. However, full restriction may not always be required, as in the case where a workpiece must be free to rotate between centers. Thus, fewer than six locating points are supplied in some cases. A basic feature of any jig or fixture is that it provides the locating surfaces or points against which all parts of one kind can be precisely located.

The simplest kind of jig is a template that fits a part and has holes to guide one or more drills. More elaborate jigs have locating and clamping details to hold the workpiece in relation to the jig plate. The jig plate is the essence of the jig, it is made of soft steel, and has accurately located holes bored in it. Hardened bushings, inserted in the plate to withstand wear, do the actual guiding of the tools, such as drills and reamers. Slip or removable bushings are convenient where tools of different sizes are to be used in one hole.

Drill jig assemblies are commercially available in a variety of forms or they may be fabricated in the toolroom of the user firm. The commercial assemblies are usually purchased without locator pins or drill bushings and those elements are added by the user firm to suit the part configuration and drilling operations to be performed. Four commercially available jig configurations are shown in Figure 20-20. Figure 20-20(A) shows a type of *leaf* or *plate jig* with a drill bushing in place and a spring pin engaging the workpiece to hold it in position under the drill bushing. The top plate can be moved up or down by the lever on the side and locked in place by exerting some pressure on the handle. The top plate is tilted back to permit removal of the part when the drilling operation is complete. *Box jigs*,

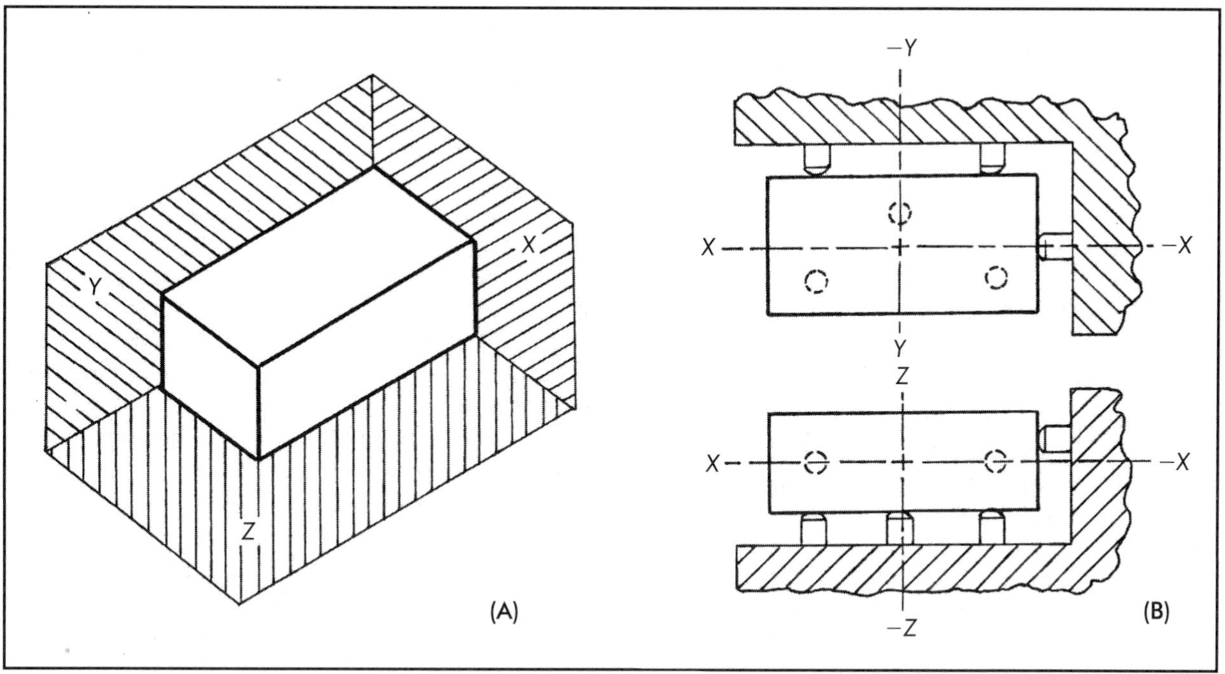

Figure 20-19. (A) Workpiece located against three mutually perpendicular planes. (B) 3-2-1 method of location restricting nine directions of movement.

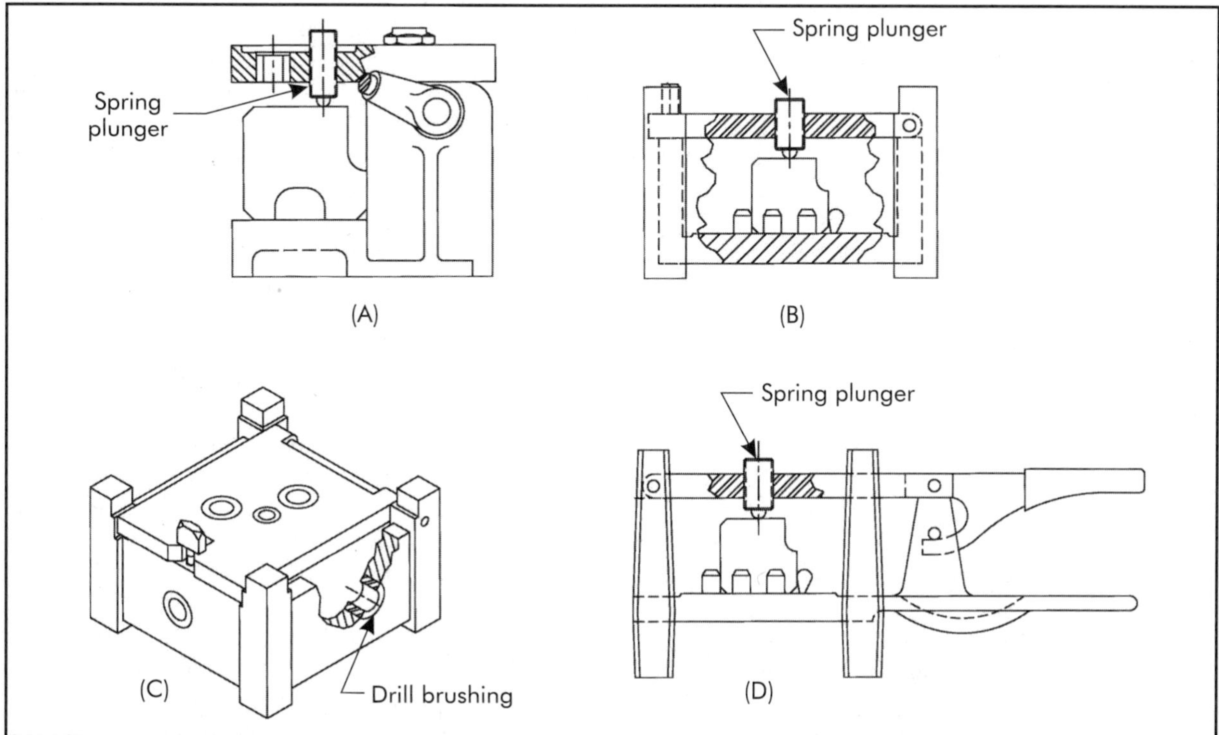

Figure 20-20. Drill jig illustrations. (A) leaf or plate jig with plate actuating lever, (B) and (C) box jigs, (D) leaf jig with cam locking device. (Courtesy Carr Lane Manufacturing Co.)

shown in Figure 20-20(B) and (C), are often used when holes are to be drilled in one or more sides of a part. The top plate is hinged so that a workpiece can be inserted and easily removed. Figure 20-20(D) shows a leaf jig with a cam locking device. An irregularly shaped part may be located in a nest of locator pins as shown in Figure 20-20(B) and (D).

20.4 BORING MACHINES

As indicated in the previous sections of this chapter, drilling operations are used to create new holes in solid materials and, because of the nature of the drilling process, a large amount of metal is often removed. This is time-consuming and costly, especially if large holes are required. To avoid this and achieve near-net-shape efficiency in manufacturing, holes in many manufactured parts are cast in place by the use of cores and other devices during the initial forming processes. In that case, all that needs to be done is to enlarge the hole slightly to true it up, finish it to size, and provide an appropriate surface finish. This is generally referred to as a *boring* operation and is usually accomplished by the use of either a single-point tool on a boring bar or an insert cutting tool similar to the ones shown in Figure 20-3.

20.4.1 Precision Production-boring Machines

Several kinds of machine tools are often used for boring operations and some are referred to as boring machines. Vertical boring machines are often really vertical lathes and do much the same kind of work as described in Chapter 18. Horizontal boring machines are often similar to the milling machines described in Chapter 21. Specialized precision boring machines are employed in many cases for high production work on automotive engine parts, pump housings, and other products where several holes need to be simultaneously finish-machined or where extreme accuracy is required.

Precision production-boring machines, using single- or multiple-insert tools, machine internal surfaces rapidly, precisely, and repetitively. They are generally capable of holding tolerances of a few micrometers (a few thousandths of an inch) and finishing surfaces to 0.2–0.4 μm (≈10–20 μin.) R_a or better. As their name implies, they are basically used for boring, but also may be tooled for facing, trimming, contouring, grooving, and chamfering. The work they do also can be done on general purpose machines, such as the lathe and mill. However, the precision boring machines are more accurate and efficient where large quantities of parts are produced because they can be operated automatically or semi-automatically.

A variety of *precision production-boring machines* are available to suit the requirements for a particular set of machining operations. These range from light-duty single- or multiple-spindle models designed primarily for light finishing cuts on small parts, to heavy-duty types for rough machining cuts on large workpieces. Figure 20-21 shows a hydraulically operated, double-end, horizontal boring, turning, and facing machine set up to perform several machining operations on two sides of a cast iron steering box. The steering box is clamped in a fixture on a worktable, which can be hydraulically fed in a longitudinal (*Z*-axis) direction. The left-hand spindle is equipped with an insert-type boring bar for boring and chamfering operations, while the right-hand spindle employs a radially fed facing head for facing operations. Machines of this type are designed for a great deal of flexibility so that they can be easily adapted to a variety of machining and production requirements. They can be equipped with as many as three spindles on each end and arranged for either hydraulic, programmable logic controller (PLC) hydraulic, or CNC operation. The hydraulically operated machine shown in Figure 20-21 costs about $400,000.

Another configuration of *horizontal production-boring machine* is shown in Figure 20-22. This machine has two chucking spindles and four tool blocks mounted on a hydraulically actuated table. The tool table can be fed in either of two directions, cross feed (*X* axis) or longitudinal (*Z* axis). For the machine shown, compressor end caps are clamped and rotated in a chucking fixture while previously micro-adjusted cutting tools in each tool block perform boring, chamfering, and facing operations on each cap. The machine is hydraulically operated.

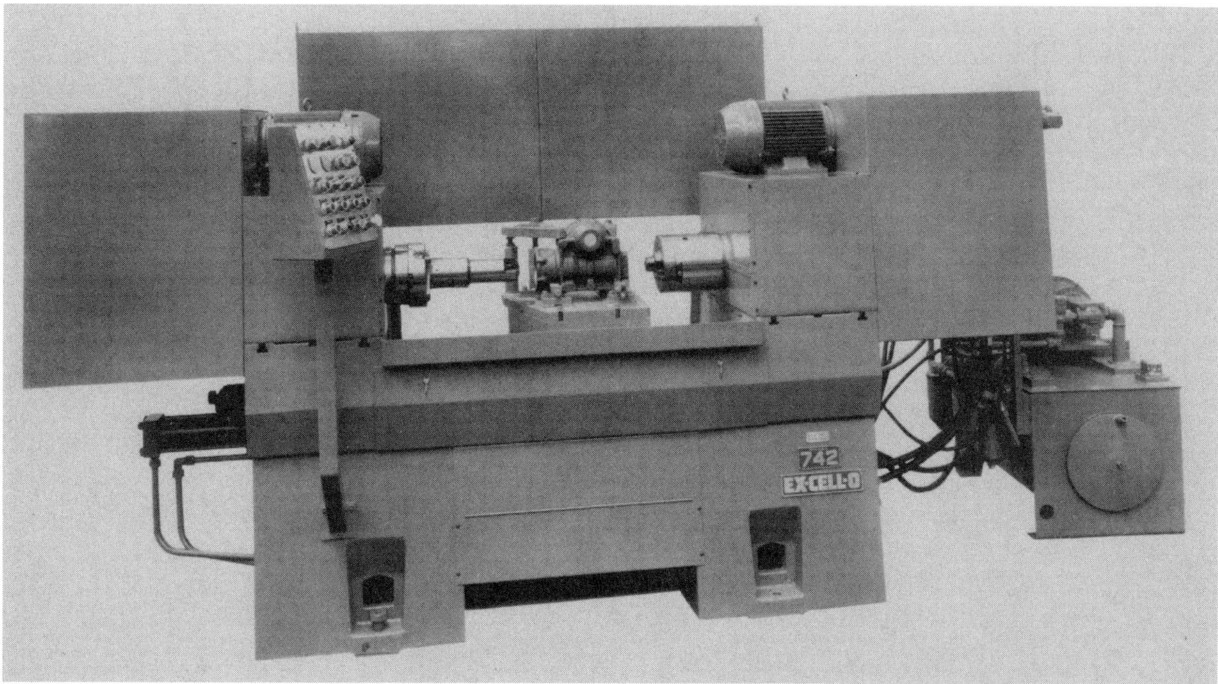

Figure 20-21. Double-end, hydraulic boring, turning, and facing machine. (Courtesy EX-CELL-O Machine Tools, Inc.)

20.4.2 Precision Jig-boring Machines

Jig boring machines are very accurate drilling and boring machines that, in some cases, resemble a vertical milling machine (Chapter 21). These machines are generally strongly built with quite rigid construction, hardened, ground and lapped ways, and precision-ground or ball-type lead screws. They are traditionally used for extremely accurate toolmaking and prototype machining in the toolroom, and on the production floor for machining exacting parts in small and moderate quantities. For many years, jig boring machines have been the workhorses of the toolmaking and die-making industry because of their ability to position a workpiece or a toolholding spindle with extreme accuracy in three coordinate directions. One type of machine has a vertical spindle capable of vertical (Z-axis) positioning and a workholding table that can be positioned in X-axis (left to right) and Y-axis (front to back) directions. On older machines, the positioning of these elements is manually accomplished by means of handwheels, and the position accuracy is established by use of micrometer handweel dials, end measuring rods, and mechanical dial indicators. Position accuracy down to 1.3 μm (≈0.00005 in.) or better often can be achieved on these machines. More contemporary jig-boring machines are generally equipped with rapid power-feed drives for positioning the various machine elements and digital readouts (DROs) for position measurement. Many of the more advanced machines are operated by CNC as described in Chapter 31.

The jig boring machine shown in Figure 20-23 has a workholding table that can be positioned in a horizontal front to back (Y-axis) plane and a cross-rail mounted spindle carrier positioned in a left to right (X-axis) plane. The spindle itself is positioned in a vertical (Z-axis) plane. The machine shown is equipped with a metric- or inch-reading DRO for X and Y axes positioning (resolution of 0.0013 mm [.00005 in.]), and a mechanical Z-axis measuring device with a resolution of 0.013 mm (.0005 in.). The worktable/spindle position can be achieved anywhere within a rectangular area of X = 650 mm (25.6 in.) and Y = 457 mm (18 in.). Position accuracy over the entire travel of the X–Y axes is 0.0025 mm (.00010 in.). A precision jig-boring machine

Figure 20-22. Double-spindle, single-end, horizontal boring, turning, and facing machine. (Courtesy EX-CELL-O Machine Tools, Inc.)

like the one shown in Figure 20-23, equipped with DRO costs about $250,000.

20.4.3 Jig Grinding Machines

A jig grinder is like a jig borer except that the spindle head of the machine carries a high-speed grinding spindle that revolves in a planetary fashion. Jig grinders are capable of finishing holes in hard materials, such as hardened steel, to a degree of accuracy equal to that of jig boring machines in soft materials and are used for finishing operations.

20.4.4 Jig Boring Operations

A jig boring machine and its accessories represent a large investment and an experienced operator is essential to get the best out of the machine. Even so, the machine is capable of producing results usually attainable on other machines only by using expensive setups and measuring equipment. The jig boring machine is capable of achieving these same results at lower cost when enough work is available to keep the machine and operator busy.

A variety of work is performed on a jig boring machine, and more time is consumed for setup, tool changes, and changeover than for actual machining. Thus, important savings in time can be realized through the use of tools and accessories that make manipulation of the machine quick and easy. This includes rapid-power feed drives for positioning, quick-change tools, DRO

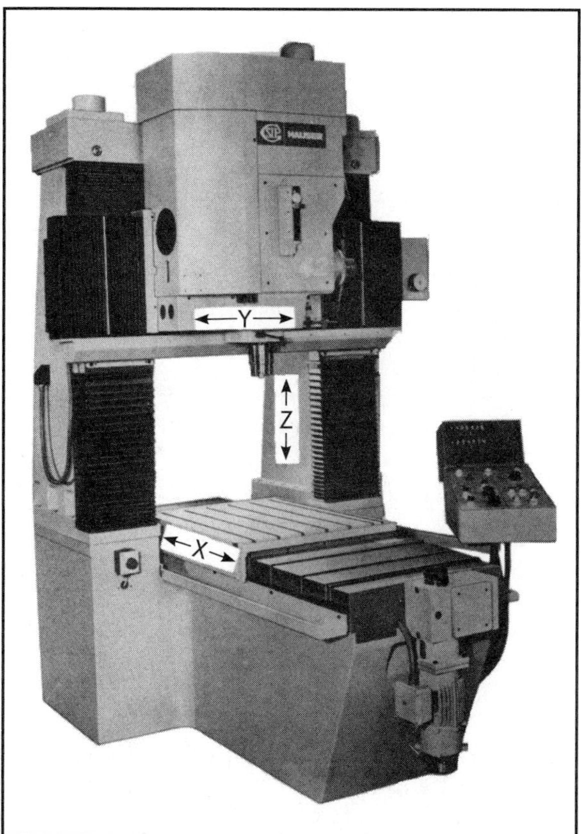

Figure 20-23. Rigid, double-column, closed-frame, jig boring machine with digital readout position measurement. (Courtesy American SIP Corporation)

devices for position measurement and, ultimately, complete CNC operation.

Generally, coordinate location is required for jig boring, and a workpiece drawing should be dimensioned according to the absolute dimensioning techniques discussed in Chapter 15. This will require the establishment of a zero datum coordinate location from which the workpiece and tool positions are subsequently referenced. Zero datum positions are generally accomplished through the use of edge finders, spindle-centering microscopes, or contacting probes (Chapter 16).

Temperature changes are important when working to small tolerances on a jig borer. These may result from the handling of measuring instruments, the actions of cutting tools, or heat from motors, pumps, and moving parts. Care must be taken to constantly control temperatures by allowing for heat dissipation and maintaining consistent room temperature.

Figure 20-24 illustrates a typical precision boring operation on the type of machine shown in Figure 20-23. In this illustration, an insert-type boring tool is used to rough and finish bore several holes located in a very precise relative position in a steel bearing plate.

Figure 20-24. Typical jig boring operation. (Courtesy American SIP Corporation)

During recent years, other machine tools have found increased use in the toolroom for certain toolmaking and die-making activities that were formerly restricted to the jig boring machine. This includes the multi-axes, CNC precision-milling machine (Chapter 21) and the CNC electrical discharge machines (EDMs) discussed in Chapter 27.

20.5 DRILLING AND BORING OPERATIONS

20.5.1 Accuracy

The accuracy of a hole involves both size and location. Hole size accuracy largely depends upon the tools and the way they are used. Under average conditions, it may be said that practical drilling tolerances range from 51 μm (.002 in.) for 4.78-mm (.188-in.) diameter holes to 0.25 mm (.010 in.) for 25–51-mm (1–2-in.) diameter holes. Boring generally gives appreciably better results. A multiblade boring tool will produce a large number of holes within 25–51 μm (.001–.002 in.) of a specified size. A fluted reamer properly used with limited stock removal (often achieved by prior boring) is a quick way to produce holes within 25 μm (.001 in.) of a specified size. Where a truly positioned and round hole with an accuracy of less than 13 μm (.0005 in.) is required, finishing by several cuts with a single-point boring tool, or grinding, lapping, or honing are the only means of assuring results. The surface finishes obtainable from drilling and related operations are indicated in Figure 16-29.

The accuracy required for the location of holes varies from 0.8 mm (.03 in.) commonly needed for clearance holes to a few thousandths of a millimeter (ten thousandths of an inch) for holes in exacting production parts, such as those for aircraft engines, and many jigs, fixtures, dies, and gages. The method selected to locate a hole or holes depends upon the accuracy required and the number of pieces to be made.

The accurate location of holes calls for three steps. The first is to establish the positions of the holes, the second is to cut the holes, and the third is to check the results. The position of holes may be established by layout, transfer, buttoning, and coordinate location. What has been said about cutting holes for size applies also to location. Of the abrasive methods, grinding alone is a way of appreciably correcting positional errors.

The *layout method* consists of drawing lines on a workpiece and center punching it for the locations of holes in the manner indicated in Figure 20-25. A drill, usually one of small diameter, is started in the center punch marks. A skilled machinist can evaluate how true a drill is cutting and make compensations to centralize the hole with the scribed circle. Even so, location to within 0.13–0.25 mm (.005–.010 in.) is all that can be expected. The method is slow and suited only for rough work in small quantities.

One form of the *transfer method* is to drill, and even ream, through the holes already in a part into a mating part. In this way, holes between two parts can be matched as precisely as in any other way. This is practical for some production assemblies and for making dies, fixtures, jigs, and gages. An extension of this method is the use of jigs, but their cost can be justified only for producing parts in quantities.

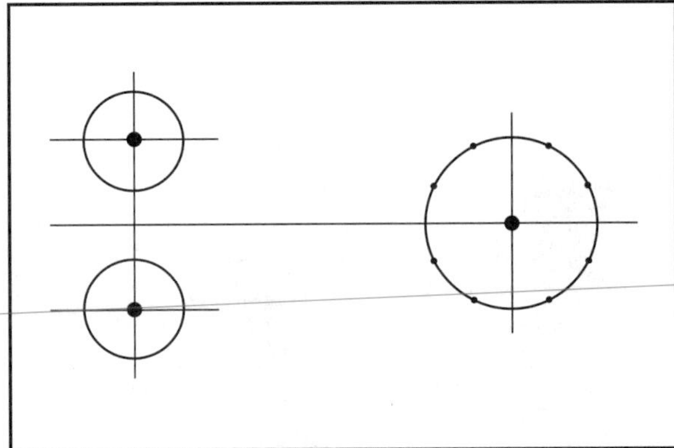

Figure 20-25. Layout for drilling three holes.

The *buttoning method* uses toolmakers' buttons. These are accurately sized hollow cylinders with squared ends. A button is clamped by a screw to the workpiece and adjusted with precision measuring tools to the desired position for a hole. The workpiece is then mounted on the face plate of a lathe with the button protruding. The workpiece is shifted until a dial indicator shows the button running true. It is secured, the button removed, and the hole is drilled and bored in the same spot. Holes may be located by this method to within 13–25 μm (.0005–.0010 in.) of true location. No expensive equipment is needed, but the method is time consuming and not practical in modern shops.

Coordinate location is accomplished by moving a workpiece from a reference point through measured distances along perpendicular axes. Movements are commonly made through cross screws or lead screws and graduated dials by means of a positioning table on a drill press or milling machine (Chapter 21), manually or numerically controlled (Chapter 31). Holes can be located to within 25–127 μm (.001–.005 in.) on such machines without extra attachments. Coordinate location is ordinarily done on jig boring machines by the refined methods already described for positioning to within 3–13 μm (.0001–.0005 in.) and closer at premium cost.

20.5.2 Speeds and Feeds

The speed at which a drill, boring tool, or reamer should be run depends on the same considerations as for other tools, as explained in Chapter 17. In general, the proper peripheral cutting speed for a drill is about the same as for a single-point tool under comparable circumstances. Table 20-1 lists cutting speeds that may be used for drilling certain materials with high-speed steel drills under normal conditions. This table indicates that lower cutting speeds are required for the harder materials, while softer materials and free-machining metals may be drilled at higher speeds. Cutting speeds for carbide and carbide-tipped drills often may be increased anywhere from 300–500% over that used for high-speed steel drills, depending on the grade of carbide material and the workpiece conditions. A variety of coatings are available for both high-speed steel and carbide drills and this normally permits a higher cutting speed to be used, sometimes as much as 20–50% higher. As is the case for other cutting tools, cutting speeds for drilling are selected to provide the appropriate tool life for a particular operation on a particular machine. The appropriate tool life is that which provides the right economic balance between production rate, tool change, and tool maintenance costs. In any case, it is recommended that the tool manufacturer be contacted for information on the appropriate cutting speeds for different types of drills used on various materials. This is particularly important for solid carbide drills, carbide insert drills, and coated drills.

The feed of a drill is the distance it advances in one revolution. Drill feeds depend to a great extent on the strength of the drill and the type of material being machined. The feed rates given in Table 20-2 are recommended for normal drilling operations with high-speed steel drills.

Hard materials, such as tool steel, high-strength alloy steel, and stainless steel, should generally be drilled at the lower feeds, while softer materials like cast iron, aluminum, and brass may be drilled at the higher feed rates. Feed rates for solid carbide drills may vary depending on the type of carbide used in the drill body itself. The use of fine-grain carbide in drill construction has increased the toughness of these tools to a comparable level with high-speed steel, which permits drilling at higher speeds and feeds.

20.5.3 Drilling Time and Cost

One of the cost components in any production drilling operation is the time required to drill a hole to a required depth or through a given thickness of material. Drilling time is based on the following relationships:

$$\text{Drill rpm} = N = \frac{C_S \times 1{,}000}{\pi d} \text{ or } \frac{C_S \times 12}{\pi d} \quad (20\text{-}1)$$

$$\text{Drill feed} = f = N \times f_n \quad (20\text{-}2)$$

$$\text{Drilling time} = T = \frac{L + l}{f} \quad (20\text{-}3)$$

$$\text{Machining cost} = C = \frac{C_h \times T}{60} \quad (20\text{-}4)$$

Table 20-1. Cutting speeds for drilling with high-speed steel drills

Material	Brinell Hardness	Cutting Speed, m/min	Cutting Speed, ft/min
Plain carbon steel	100–120	27.4–30.5	90–100
Free-machining steel	100–150	30.5–36.6	100–120
Alloy steel	120–170	22.9–25.9	75–85
High-strength steel	200–300	12.2–15.2	40–50
Cast iron	110–140	27.4–30.5	90–100
Aluminum castings		76.2–91.4	250–300
Magnesium alloys		106.7–121.9	350–400

Table 20-2. Feed rates for high-speed steel drills

Drill Diameter, D		Feed Rate per Revolution, f_n	
mm	in.	mm	in.
$D < 3.175$	$D < .125$	0.025–0.051	.001–.002
$3.175 < D < 6.35$	$.125 < D < .250$	0.051–0.127	.002–.005
$6.35 < D < 12.70$	$.250 < D < .500$	0.102–0.203	.004–.008
$12.70 < D < 25.40$	$.500 < D < 1.000$	0.178–0.356	.007–.014
$25.40 < D$	$1 < D$	0.254–0.610	.010–.024

$$\text{Metal removal rate} = Q = \left(\frac{\pi \times d^2}{4}\right)f \qquad (20\text{-}5)$$

where

N = revolutions per min
C_S = cutting speed, m/min (ft/min)
d = drill diameter, mm (in.)
f = feed rate, mm/min (in./min)
f_n = feed rate per revolution, mm/rev (in./min)
T = drilling time, min
L = drilling depth, mm (in.)
l = distance from drill point to workpiece before drilling, mm (in.)
C = machining cost, \$
C_h = cost per hr of machine time, \$/hr
Q = metal removal rate, mm^3/min (in.^{3}/min)

The use of these relationships is illustrated by the following example. A hole 19 mm (.75 in.) in diameter is drilled through a cast-iron workpiece 63.5 mm (2.5 in.) thick by a HSS drill with a surface speed of 27.4 m/min (90 ft/min) and a feed rate of 0.254 mm/rev (.01 in./rev). The depth to which the drill is fed is 63.5 mm (2.5 in.) plus an assumed overtravel of 6.35 mm (.25 in.), and an approach of 4.78 mm (.188 in.) (one-fourth of the diameter of a standard drill). Thus, the total drill travel is 74.63 mm (2.938 in.). From the relationships in Equations (20-1) through (20-5)

$$\text{Drill rpm} = N = \frac{27.4 \times 1{,}000}{\pi \times 19} = 459$$

Drill feed $= f = 459 \times 0.254 = 117$ mm/min (4.61 in./min)

$$\text{Drilling time} = T = \frac{63.5 + 11.13}{117} = 0.638 \text{ min}$$

$$\text{Metal removal rate} = Q = \frac{\pi \times 19^2}{4} \times 117$$

$$= 33{,}173 \text{ mm}^3/\text{min } (2.024 \text{ in.}^3/\text{min})$$

20.5.4 Cutting Forces and Power

The two components of force in drilling and boring operations are the thrust in the axial direction of the tool and the torque or moment. Investigations have shown that these depend upon the feed, the drill diameter, the chisel edge length, the material, and the number of teeth. Simplified formulas for two flute high-speed steel drills with a normal chisel edge length to drill diameter ratio of 0.18 have been given by Oxford as

$$T = 2Kf^{0.8}d^{0.8} + Ld^2 \text{ and } M = Kf^{0.8}d^{1.8} \qquad (20\text{-}6)$$

where

T = thrust, kN (lb)
K = 0.6035 (24,000) for 200 Bhn steel, 0.7795 (31,000) for 300 Bhn steel, 0.8549 (34,000) for 400 Bhn steel, 0.1760 (7,000) for aluminum and leaded brass, 0.1006 (4,000) for most magnesium alloys, and 0.3520 (14,000) for most brasses
f = feed rate, mm/rev (in./rev)
d = drill diameter, mm (in.)
L = 4.309×10^{-3} (625)
M = torque or moment, Nm (lb-in.)

The factors for SI units are given first and those for English units are in parentheses. Similar expressions are given in reference books and handbooks for hole-enlarging tools.

The horsepower for a drilling operation may be estimated on the basis of

$$P = MR/63{,}025 \qquad (20\text{-}7)$$

where

P = horsepower
M = torque or moment, lb-in.
R = drill speed, rpm

The power in kW is

$$P = MR/9{,}549 \qquad (20\text{-}8)$$

where

M = torque or moment, Nm

An insignificant portion of the power is related to the thrust because the feed is slow compared to the speed. Power also may be estimated by the use of unit power factors available in reference books and handbooks.

20.6 PROCESS PLANNING

20.6.1 Design for Manufacturing

In many metal parts manufacturing industries, 50% or more of all machining time is used in making and/or finishing holes. Thus, the time spent in the planning for hole-making or finishing operations is normally justifiable in terms of time and cost savings because of the many alternatives for producing and finishing those holes. As explained in Chapter 19, the design of a product often dictates the complexity or simplicity of the operations that will be required to finish it. It follows then that the planning for hole-making or finishing operations should be a major consideration during the preliminary and final stages of product design. The following are just a few basic guidelines that cover almost any part design where drilling operations may be required:

1. The drilling of large-diameter holes should be avoided wherever possible because of the amount of metal that must be removed. Large holes should be preformed by use of cores in the casting process. Also, trepanning is often an alternative choice for holes over 48 mm (1.875 in.) in diameter and lengths greater than 610 mm (24 in.).
2. Drilling should be performed on surfaces that are perpendicular to the thrust of the drill, otherwise the drill will tend to deflect.
3. Through-hole drilling is usually preferred over blind-hole drilling, and internal interrupted hole surfaces may contribute to dimensional inaccuracy and burr problems.
4. The bottoms of holes should conform to the point angle of a drill. Flat hole bottoms may require a secondary operation.
5. Manufacturing drawings should reference hole locations in such a way as to minimize positioning errors.
6. Parts with several holes to be drilled should be designed to minimize complex fixturing and relocation.
7. Part configurations and wall thicknesses should be strong enough to resist deflection during the drilling process.

20.6.2 Process Selection and Operation Sequence

The selection of a process or machine and the determination of the sequence of operations required for hole making is dependent on many factors, including the production requirements, the locational, dimensional and geometric accuracy, type of material, hole-making machines, and the tools and equipment available. Of particular importance, of course, are the production requirements, since these parameters alone will set some limits on the time and scope of process planning and the complexity of tooling that can justifiably be dedicated to the job.

As in most manufacturing situations, there are many ways and a number of different machines that can be used to produce and finish holes. Drilling machines are best for many jobs, particularly small holes, because they are simple, low cost, easy to set up and manipulate, and can be tooled at moderate cost. Drilling, boring, and reaming are usually done on other machines, such as lathes or milling machines, for particular advantages. For instance, a hole may be machined more economically and accurately on a lathe in the same setup with outside surfaces, especially when concentric or square. The need to accurately locate holes is often a prime factor; the selection of machines for that purpose was discussed in the section on accuracy. When feasible, punching is generally cheaper than any other way of producing holes. If punching is not practicable, drilling alone is most economical, but either must be followed by other operations for smaller tolerances. However, direct and tool costs increase with the number of operations, and no more precision than necessary should be specified or attempted.

Once a process or machine has been selected for making and finishing a hole, then the sequence of operations, tools, method(s) of locating and clamping the workpiece, process parameters, etc., must be prescribed. For example, the sequence of operations shown on the operation sheet of Figure 20-26 could be used to drill and counterbore the pivot block shown therein if only one or a few were to be produced. However, both the machine and the operation sequence used would most likely be changed if several thousand parts were required. The cost per piece for the method used in Figure 20-26 is likely to be high because of the number of tool changes and amount of manual labor involved. Thus, alternative approaches would probably be examined to reduce the cost per piece for large production requirements. For instance, a drilling machine with a positioning table could reduce the time and cost for locating the hole center, while the use of a simple jig might effect even greater savings. As other alternatives, the CNC turret drill press shown in Figure 20-13 and other CNC machining centers (Chapter 30) could be candidates for this job if such are available and production volume requirements were sufficient to economically justify the use of such equipment.

There are a number of methods used by different industries to evaluate economic alternatives. The relationship given by Equation (20-9) is one of the more commonly used approaches. For example, suppose 10 holes are to be drilled, bored, reamed, or tapped in a cast iron workpiece and the work takes 1 hr. A suitable jig for that part costs $500 and will be depreciated in 1 yr with a combined rate of 30% for interest, insurance, taxes, and maintenance. It costs $25 to set up a positioning table and after that the charge for its use is $2/hr. Overhead costs do not change. For how many pieces in one lot is it economical to build a jig and release the positioning table for other work?

$$Ns(1 + t) = I(A + B + D + M) + Y \quad (20\text{-}9)$$

where

The left side of the equation represents the amount saved by the more costly tooling (a jig) and the right side sums the annual cost of the extra investment and any additional setup charges. And

where

N = number of pieces to be made
s = direct savings in $/piece of one method over another
t = relevant overhead rate in $ invested in tools
I = investment in $
A = capital charges for interest
B = capital charges for insurance and taxes
D = capital charges for depreciation
M = capital charges for maintenance
Y = setup costs in $/yr of one method over another

Thus, $N(2)(1 + 0) = 500(1.30) - 25$ and, $N = 312.5$. Therefore, it would be more economical to use the positioning table for a lot size of 312 and the jig for a lot size over that amount.

20.7 QUESTIONS

1. Define "drilling," "core drilling," "counter drilling," and "step drilling."
2. Define "boring," "counterboring," "spot facing," and "countersinking."
3. What is the purpose of reaming?
4. What is the primary difference between a core drill and a metalworking twist drill?
5. What are some of the advantages of carbide drills over high-speed steel drills?
6. What types of coatings are available for drills and how do these improve drill performance?
7. What angles of a drill correspond to the rake and relief angles of a single-point tool?

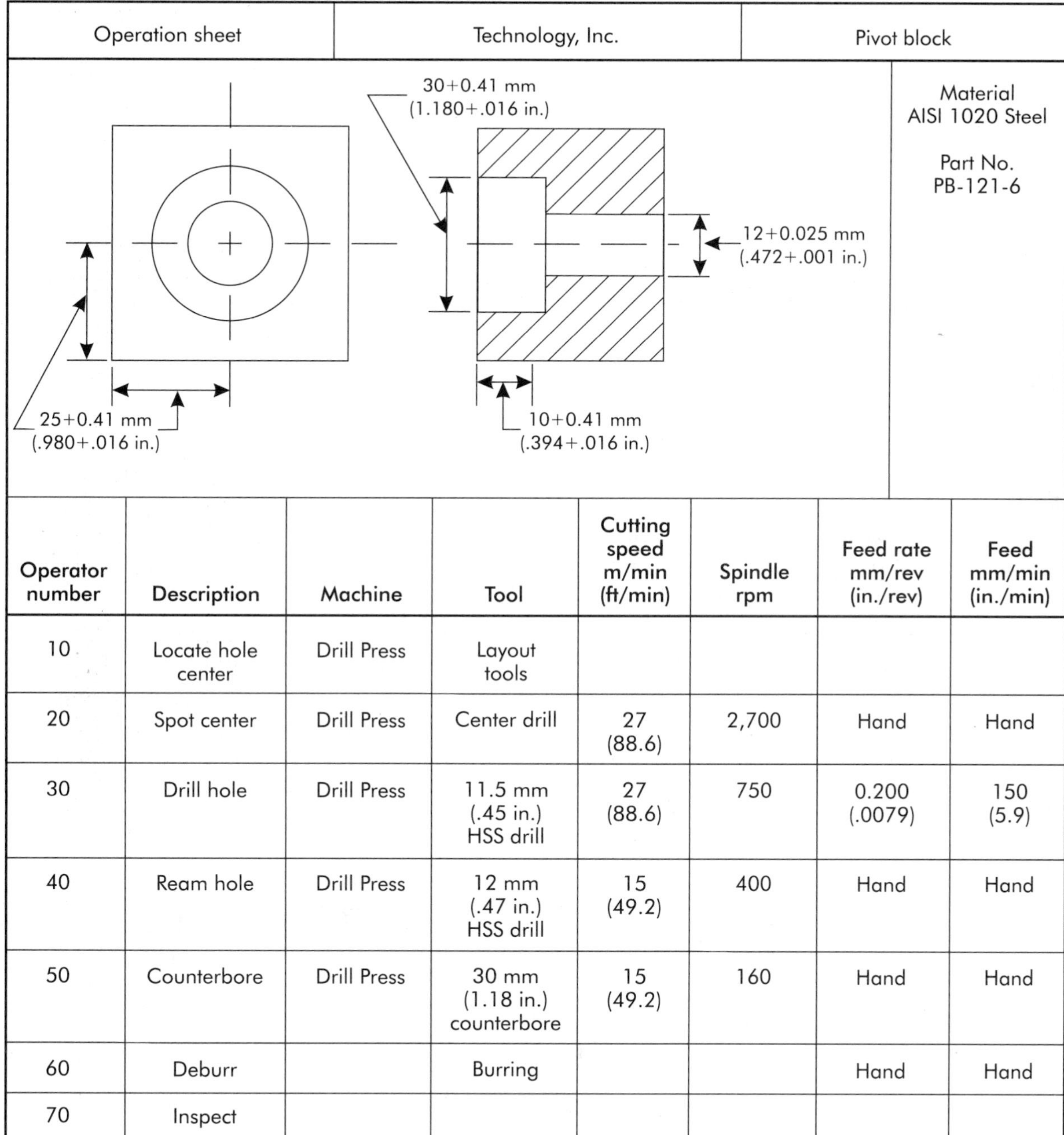

Operator number	Description	Machine	Tool	Cutting speed m/min (ft/min)	Spindle rpm	Feed rate mm/rev (in./rev)	Feed mm/min (in./min)
10	Locate hole center	Drill Press	Layout tools				
20	Spot center	Drill Press	Center drill	27 (88.6)	2,700	Hand	Hand
30	Drill hole	Drill Press	11.5 mm (.45 in.) HSS drill	27 (88.6)	750	0.200 (.0079)	150 (5.9)
40	Ream hole	Drill Press	12 mm (.47 in.) HSS drill	15 (49.2)	400	Hand	Hand
50	Counterbore	Drill Press	30 mm (1.18 in.) counterbore	15 (49.2)	160	Hand	Hand
60	Deburr		Burring			Hand	Hand
70	Inspect						

Figure 20-26. Sequence of operations for drilling and counterboring a steel pivot block.

8. What determines the proper amount of relief angle on a drill?
9. What is "trepanning" and what is it used for?
10. What methods may be used to drill deep holes?
11. Describe and designate by a sketch the principal elements of a drill press.
12. How is a drill held in place in the spindle of a drill press?

13. What is the difference between a gang drill press and a multiple-spindle drill press and how does this affect their utility?
14. Describe the motions or movements that are available on the turret drill press shown in Figure 20-13.
15. What is a radial drill press used for?
16. What is the difference between a jig and a fixture?
17. Show a sketch of a free body with its 12 directions of movement labeled.
18. Explain the 3-2-1 principle for locating a part to be machined.
19. Describe the motions or movements that are available on the boring, turning, and facing machine shown in Figure 20-21.
20. What is the purpose of a jig boring machine?
21. How is the feed of a drill specified?

20.8 PROBLEMS

1. Prepare an operations sheet similar to Figure 20-26 showing the operations required to machine all of the holes in the part shown in Figure 20-27. Specify the machine, tools, and cutting speed and feeds to be used.
2. Calculate the machining time and metal removal rate for drilling the center hole in the part shown in Figure 20-27 using the speed and feeds specified in Problem 1.
3. Calculate the machining time and metal removal rate for drilling one of the 8 mm (.313 in.) holes in Figure 20-27 using the speeds and feeds specified in Problem 1.
4. A workpiece must have eight 13 mm (.5 in.) holes finished in it. Layout time is 30 min/piece. To set up a positioning table on a drill press for a whole lot, 1 hr is required. To finish the holes requires 45 min for each piece, not counting layout or setup. The labor rate is $12.50/hr, the machine rate is $10/hr, and the use of a positioning table costs $3/hr. If the table is used, the labor cost to lay out each can be saved. Both methods give the same quality product. How large a lot justifies the use of the positioning table?
5. A part may have a hole located for drilling by layout. If a jig is provided, 30 sec is saved for each piece. The labor rate is $12.60/hr, which means a savings of $0.105/piece. The overhead rate on the labor saved is 100%. Setup time is no more with than without the jig. The combined rate for interest, insurance, taxes, and maintenance is 35%. The cost of the jig is $1,000.
   (a) How many pieces must be made in one lot to make the jig worthwhile?
   (b) How many pieces must be made on the jig in one lot each month to earn the cost of the jig in 2 years?
   (c) How many pieces must be made on the jig in one lot each year to save its cost in 2 years?
6. If one lot of 3,000 pieces is to be made under the conditions described in Problem 5, how much is justified to spend for a jig?
7. A part with seven holes can be machined on a CNC turret drill press in 15 min. The rate on the machine for labor and overhead is $28/hr. The part can be machined on a gang drill press with a special jig in 17 min per piece. The jig costs $250, and the combined rate for depreciation, interest, insurance, and taxes is 135%. The hourly rate for the machine and operator is $20. Setup time is the same for both machines. For how many pieces is it economical to make a jig?
8. A jig for machining a part with three holes costs $200 and with it the operation takes 12 min. The operation can be done without a jig on a numerically controlled (CNC) drill press

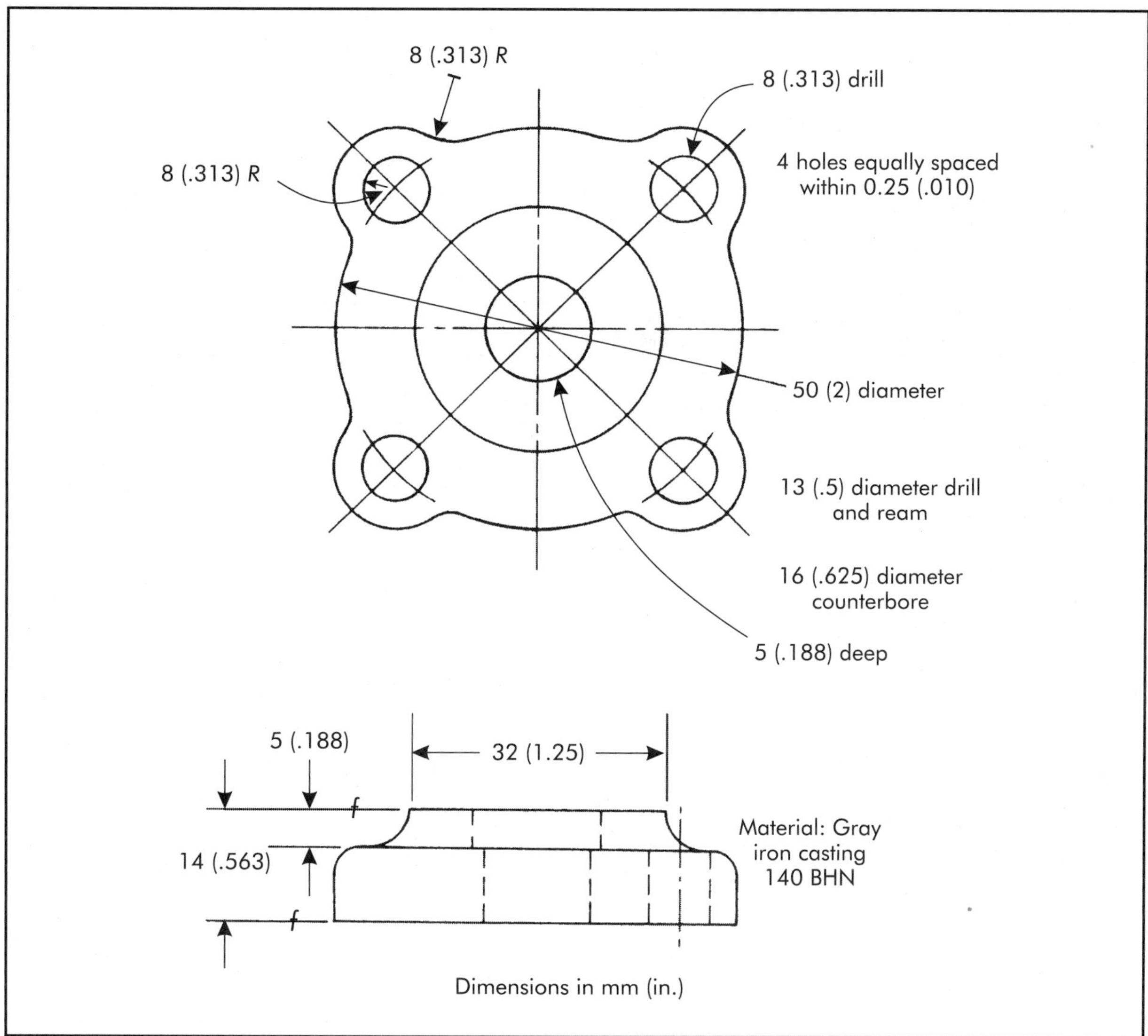

Figure 20-27. Workpiece.

in 10 min. The other conditions are the same as in Problem 7. For how many pieces is it economical to make a jig?

9. To be justified in a certain plant, a jig boring machine must show a savings of \$10,000/yr. It is estimated that on average, about one-fourth of the time now spent on locating holes by other methods can be saved by jig boring. Labor and overhead are considered worth \$30/hr. How many hours of work per year should be available for this jig boring machine to justify its purchase?

10. A jig for a job will cost \$900. The operation can be done with the jig in 4 min and on a jig borer without a jig in 12 min for each piece. Setup time is the same either way. The labor and overhead rate on the drill press with the jig is \$15/hr and on the jig boring machine \$20/hr. A composite rate for interest, insurance, taxes, and maintenance is 25%. For how many pieces is the jig justified?

20.9 REFERENCES

ANSI B11.8-1983. "Safety Requirements for the Construction, Care, and Use of Drilling, Milling, and Boring Machines." New York: American National Standards Institute.

ANSI B94.11M-1979 (R1987). "Twist Drills, Straight-shank and Taper-shank Combined Drills, and Countersinks." New York: American National Standards Institute.

ANSI B212.8-1988. "Cutting Tools—Carbide Blanks for Twist Drills, Reamers, End Mills, and Random Rod." New York: American National Standards Institute.

Coleman, J. R. 1991. "Boring In Depth." *Manufacturing Engineering*, May.

Fundamentals of Tool Design, 3rd Ed. 1991. Dearborn, MI: Society of Manufacturing Engineers (SME).

Noaker, P. M. 1991. "Holemaking's Tough Tools." *Manufacturing Engineering,* May.

——. 1992. "A New Twist on Indexables." *Manufacturing Engineering,* May.

Schreiber, R. R. 1992. "Assessing Holemaking Alternatives." *Manufacturing Engineering,* May.

Scott, N. 1994. "Hole Making has Strategic Importance." *Modern Machine Shop*, June.

Society of Manufacturing Engineers (SME). 1995. *Basic Holemaking* video. *Fundamental Manufacturing Processes* series. Dearborn, MI: SME.

Weck, M. 1984. *Handbook of Machine Tools*, 4 vols. New York: John Wiley and Sons.

Milling

21

The milling machine, 1818

—Eli Whitney, New Haven, Connecticut

21.1 EVOLUTION OF FLAT SURFACE GENERATING PROCESSES

In the early days of manufacturing, shapers and planers were used to machine flat surfaces in either horizontal, vertical, or angular planes. The shaper used a single-edge cutting tool that was reciprocated back and forth across a workpiece, but cutting was accomplished only on the forward stroke. A planer operated in a similar manner, except the workpiece was reciprocated instead of the cutting tool. These machines were reasonably easy to use and the single-edge cutting tools were easy to maintain. However, the machining process was slow because the tool was cutting in only one direction and a large amount of energy was consumed by accelerating and decelerating machine components to obtain decent cutting speeds.

A milling machine can be used to generate flat as well as curved surfaces, but it uses a rotating cutter with multiple cutting edges or teeth. With a few exceptions, the milling machine can perform any machining operation formerly done with either a shaper or planer. The milling machine is considered to be more efficient than either the shaper or planer because the multi-tooth cutting tools are cutting continuously and cutting speeds can be controlled by varying the spindle speed. In the early days, the maintenance of sharp cutting edges on solid multi-tooth milling cutters was a problem. Now, the use of inserted teeth milling cutters has minimized this aspect of milling machine maintenance.

Because of its versatility, the milling machine, particularly in its contemporary form as a CNC machine or machining center, has become a workhorse in the parts fabricating industries. This machine performs a multitude of surface generating functions and has found utility in the toolrooms, job shops, and production facilities of a large variety of manufacturing industries.

21.2 MILLING PROCESS

Flat or curved surfaces, inside or outside, of almost all shapes and sizes can be machined by milling. Common milling operations are depicted in Figure 21-1. As a rule, the workpiece is fed into or past a revolving milling cutter that has a number of teeth all taking intermittent cuts in succession. Also, the rotating cutter may be fed into the workpiece.

As indicated in Figure 21-1, the same kind of surface often can be milled in several ways. For in-stance, plane surfaces may be machined by slab milling, side milling, or face milling. The method for any specific job may be determined by the kind of milling machine used, the cutter, or the shape of the workpiece and position of the surface.

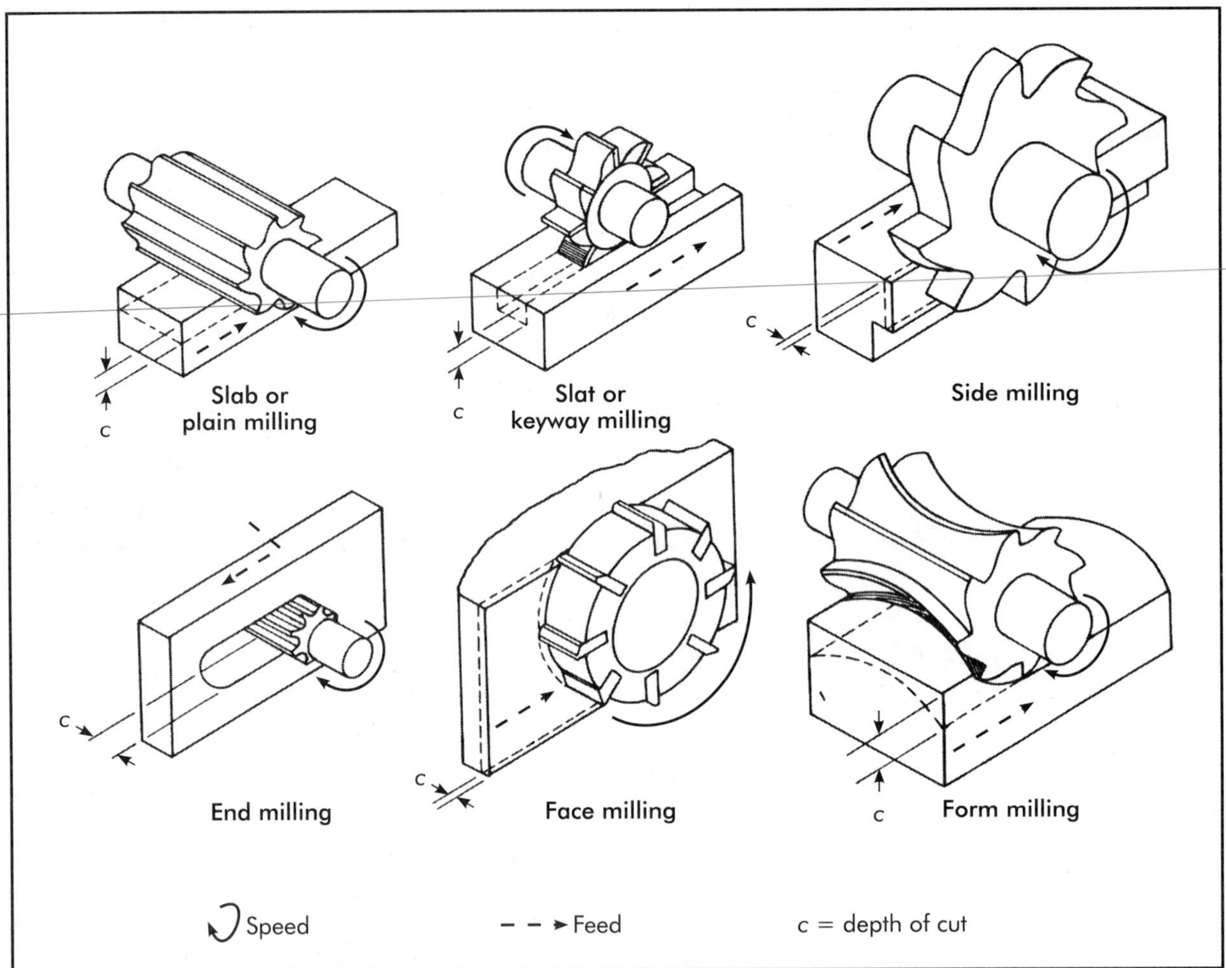

Figure 21-1. Some milling operations.

21.3 MILLING CUTTERS AND DRIVERS

21.3.1 Kinds of Cutters

Many kinds and sizes of cutters are needed for the large variety of work that can be done by milling. Many standard cutters are available, but when they are not adequate, special cutters are made. Generally, milling cutters are identified according to the type of milling operation they perform. As shown in Figure 21-1, a plain milling cutter is used to do a plain milling operation and a face milling cutter does face milling. With the exception of the face mill, all of the cutters depicted in Figure 21-1 are solid, one-piece tools, usually made of high-speed steel. The face milling cutter with inserted carbide teeth is more representative of the milling tools used in contemporary manufacturing. It is easier to maintain and has a longer tool life than its high-speed steel counterparts.

The inserted-tooth slotting cutter shown in Figure 21-2(A) is typical of the type of cutter used for production milling operations. This tool has 12 replaceable carbide inserts arranged in a staggered fashion and locked in place around the periphery of the cutter body. The bodies of large milling cutters of this type are usually made of soft steel or cast iron. Figure 21-2(B) shows an end-milling cutter with four replaceable, carbide-inserted, end-cutting teeth locked in place around the working end of the cutter. The slotting cutter shown in Figure 21-2(A) has a hole in the center with a keyway. Thus, it is generally mounted on an extension arbor or stub ar-

Figure 21-2. (A) Arbor-mounted, slotter-type milling cutter with inserted carbide teeth. (B) Integral shank, end-milling cutter with inserted teeth. (Courtesy Lovejoy Tool Company, Inc.)

bor and driven by a key. The end-milling cutter of Figure 21-2(B) has an integral shaft (a Welden shank) mounted in an adapter on the milling machine spindle.

Teeth of Cutters

The teeth of a milling cutter have cutting edges and angles related to the edges and angles of other cutting tools. In effect, each tooth acts like a single-point tool. They are evenly spaced on most milling cutters, but uneven spacing sometimes is necessary to reduce vibration and chatter. The names of the surfaces or elements and angles for a solid, plain milling cutter made of high-speed steel are shown in Figure 21-3. It should be noted that the teeth of the cutter have rake and clearance angles similar to those of a single-edge turning tool. As in single-edge tools, the magnitude of those angles is dependent on the type of machining operation and material being machined.

Cutters with inserted teeth are exemplified in Figure 21-4 for a typical face-milling cutter. The teeth cut mostly on the nose chamfer edge. Both axial and radial rake determine the true rake angle in a plane perpendicular to the cutting edge. Cutters with both rake angles negative are common for carbide and oxide inserts

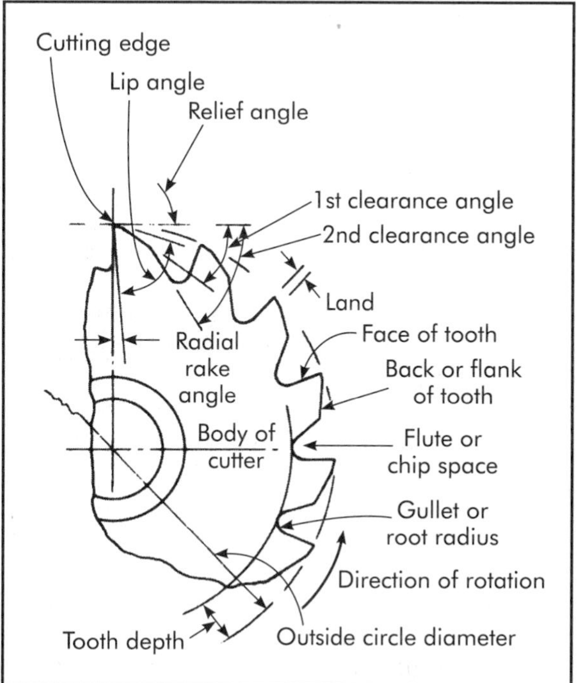

Figure 21-3. Elements and angles of the teeth of a plain milling cutter.

because they help reduce shock at the edge. However, they result in higher cutting forces that may distort weak workpieces. With both

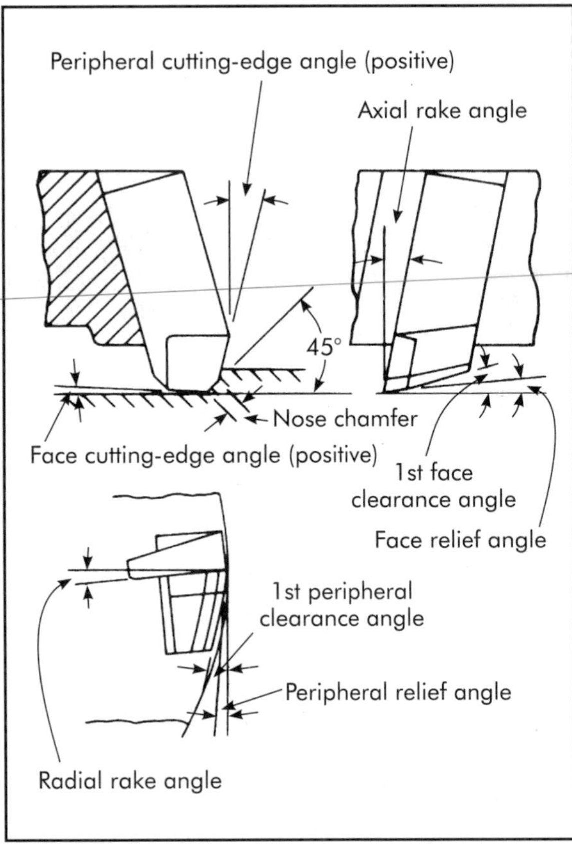

Figure 21-4. Angles of a typical face mill.

rake angles positive, forces and power may be minimized, but the cutting edges are weaker and subject to a greater wear rate. Thus, many manufacturers use negative rake angles to prolong tool life.

As in single-point tools, many different grades of carbide materials and coatings are available for use as inserted teeth on milling cutters. In addition, different insert geometries (such as square, round, triangular, diamond, and octagonal) are available with or without chip breaker provisions. Even polycrystalline diamond- (PCD) and cubic boron nitride- (CBN) tipped inserts may be obtained for use in high-precision milling operations.

21.3.2 Arbors, Collets, and Adapters

A variety of holders and drivers is needed to accommodate the many sizes and types of milling cutters. These are known as milling machine arbors, collets, and adapters.

The hole in the spindle of a modern milling machine has a taper of 88.9 mm/m (3.5 in./ft) at the cutter end. This is known as an American National Standard taper and it exists in several standard sizes.

Milling cutters are held in several ways. Face-milling cutters with an integral shank like the one shown in Figure 21-2(B) may be held in place in the machine spindle by means of a short adapter, while large face mills are located and held in place by a *stub arbor.* A cutter with a center hole such as the one in Figure 21-2(A) is held by a *milling machine arbor.* The cutter is clamped and keyed on the straight portion, and the tapered end of the arbor is held in the hole of the machine spindle. Short arbors are available for shell-end and small-face milling cutters held close to the machine spindle.

Collets and *adapters* serve to adapt taper shank cutters or drivers to the taper in the end of the hole of a milling machine spindle. An adapter usually has an outside standard taper and a smaller inside taper. Usually, collets have *sticking tapers*, such as Brown and Sharpe (12.7 mm/m [.5 in./ft]) or Morse (15.9 mm/m [.625 in./ft]), inside and outside. A quick-change adapter engages a cutter shank or driver by a quick cam action.

A *spring chuck* or *collet holder* may be mounted on a milling machine spindle to hold wire, rods, and straight-shank tools.

21.4 MILLING MACHINES

Many types of milling machines serve in various ways in industry. Some are best for general-purpose work, others are suited for repetitive manufacturing, and some are ideally arranged for special operations. Milling machines may be classified as general-purpose, production, planer-type, and specialized machines. Horizontal boring and drilling machines are included in this section because they also do milling. Small, general-purpose milling machines and boring machines are often hand operated, but the larger, more contemporary machines have power drives for operating the moving elements. The movements of production milling and boring machines are usually accomplished by power feeding and rapid traverse mechanisms. Most of the newer precision milling machines are equipped for computer numerical control (CNC) operation (Chapter 31) to provide for increased flexibility and automatic operation. It is also possible to retrofit older machines with CNC capability.

21.4.1 General-purpose Machines

The column-and-knee milling machine for general-purpose work has components and movements illustrated in Figure 21-5. This type of machine is capable not only of milling plane and curved surfaces, but also gear and thread cutting, drilling, boring, and slotting when suitably equipped. The heavy-duty horizontal column-and-knee type milling machine shown in Figure 21-6 has a hollow spindle in the column to locate and drive various types of milling cutters. The overarm extending from the top of the column carries an outer and inner support bracket to provide rigidity to cutter arbors. Arbors and adapters are either clamped on the spindle nose or held in place by a drawbar through the hollow spindle.

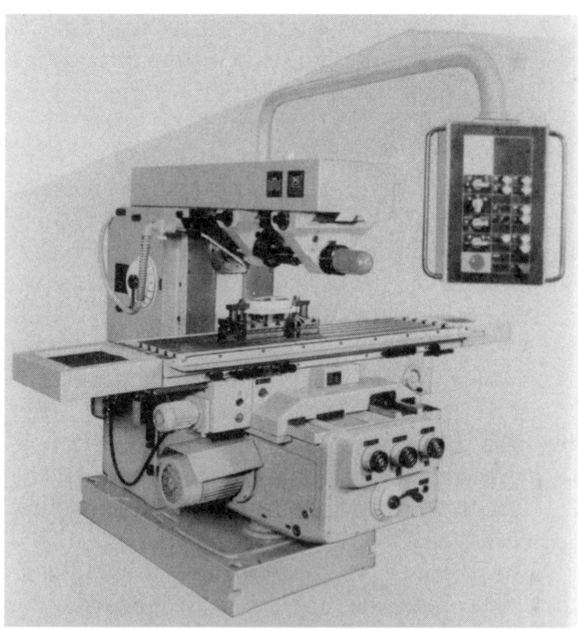

Figure 21-6. Heavy-duty horizontal column-and-knee type milling machine. (Courtesy WMW Machinery, Inc.)

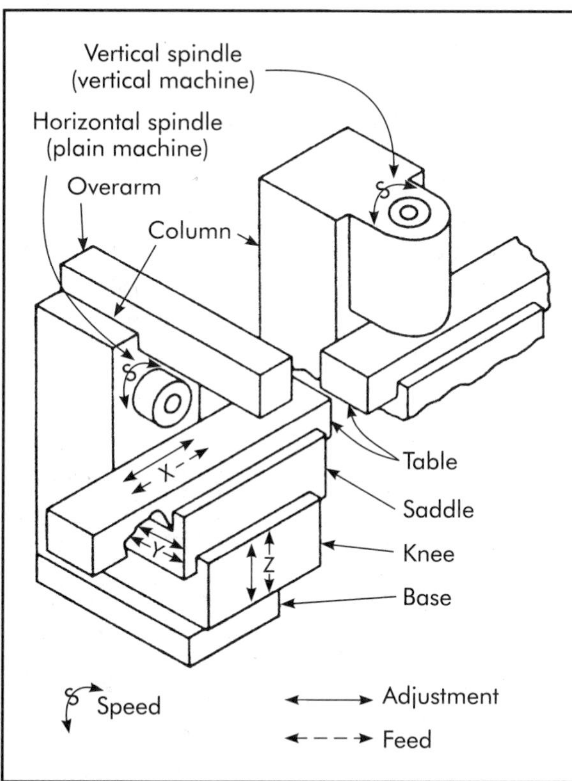

Figure 21-5. Horizontal and vertical column-and-knee type milling machines.

The knee, saddle, and table on many milling machines are manually positioned by means of handwheels or cranks attached to a transport screw, lead screw, or ball screw. For example, one revolution of a transport screw, lead screw, or ball screw under the table of the milling machine in Figure 21-5 would move that table along the *X* axis a certain distance, depending on the lead of the screw. Measurement of the amount of movement in a given coordinate direction is established by reference to calibrated micrometer dials or graduated collars behind the handwheels or cranks used to effect the movement. In most cases, the lead screw and associated micrometer dials will permit the positioning of the three elements: knee, saddle, and table, in increments of 0.025 mm (.0010 in.) or less in each of the three coordinate directions. Many contemporary milling machines are equipped with digital readout (DRO) devices that permit the machine operator to determine the position of the three elements very quickly and with less risk of error than with micrometer dials.

The knee, saddle, and table of the milling machine shown in Figure 21-6 are moved in their respective coordinate directions by mechanical drives powered by an electric motor. These movements can be activated by push buttons or switches on the pendant-mounted control box with the amount of movement indicated by graduated collars on the face of the knee. Electromagnetic clutches on the table feed drive of the machine permit the operator to position or reposition the table at a rapid traverse rate. A built-in hydraulic table rise-and-fall system causes the knee to lower about 0.7 mm (.028 in.) during rapid return of the table. This rise-and-fall action protects the workpiece surface and tool cutting edges on the table return cycle. In addition, the machine is equipped with a hydraulically operated climb-milling attachment to provide backlash compensation between the feed screw and feed nut during climb-milling operations.

A universal column-and-knee milling machine like the one shown in Figure 21-7 is like a plain type except that the table can be swiveled in a horizontal plane. This is especially helpful for milling a helix as shown in Figure 21-8. The upper part swivels on the lower and carries the table with it, to a limit of about 45°. The lead screw that feeds the table is arranged so that a geared drive can be taken from it to drive a rotary indexing unit or dividing head located on the table.

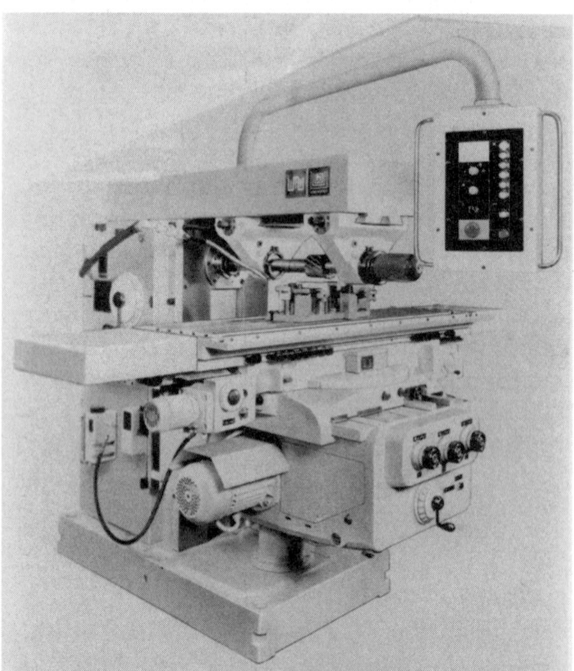

Figure 21-7. Heavy-duty universal column-and-knee type milling machine. (Courtesy WMW Machinery, Inc.)

Figure 21-8. Setup for milling a helical gear on a universal column and knee-type milling machine. (Courtesy WMW Machinery, Inc.)

A vertical column-and-knee milling machine has a vertical head and spindle as indicated in Figure 21-9, but the same feeds and adjustments as the plain machine. On some models, the head can be moved up and down, and on others it can be swiveled around a horizontal axis. The machine shown in Figure 21-9 represents a technological step between manual operation and full CNC operation (Chapter 31). The power feed movement of the table in two coordinate directions (X and Y) can be accomplished by digital

Figure 21-9. Vertical milling machine with two-axis digital control. (Courtesy Bridgeport Machines, Inc.)

control through the use of a numeric keypad. The machine operator simply enters the dimensions for each *X*- and *Y*-axis move directly from the manufacturing drawing and the screen displays and executes them. The operator can also "teach" the control system how to machine a part by moving the table manually to a specific location and striking the "Enter" key to store the table position at the end of each move. The *Z* axis on this machine is independently controlled by the operator.

Column-and-knee type milling machines can be arranged to do any kind of milling and are relatively easy to change from one job to another. However, these general-purpose milling machines have a certain amount of unavoidable looseness in the necessary joints between units and weakness from overhanging members. An operator must attend to a number of controls to govern all the available movements, adjustments, and arrangements.

21.4.2 Production Machines

Production or *manufacturing milling machines* are designed to remove metal rapidly and require a minimum of attention from the operator. They are not as easily adapted to various jobs as column-and-knee machines. For instance, adjustments are provided in three coordinate directions, but feeds are often available in only one direction and seldom in more than two directions. As a result, each workpiece must be held in a particular position to be milled. Versatility is not so important, however, because production millers are intended for long runs where setup changes are not frequent and special fixtures and cutters are often justified to hold workpieces and adapt the machine to specific jobs.

Production milling machines are made in several styles. Some small models have knees, but the most common construction is on a heavy bed that forms a base for the major units and by which the table is fully supported. They are sometimes called *bed-type milling machines*. As many units as possible are securely tied together. The construction is rigid to withstand heavy cuts. The operator is aided because the controls are simple and the action is often automatic.

Simple *hand-operated millers* are common among the smaller production machines. They are suited for light cuts, like slotting the heads of screws. As the name implies, the table is fed by hand through a handwheel or lever, but the cutter is driven by a motor.

A production milling machine with one spindle carrier is called a *plain machine*. A *duplex machine* has two opposed spindles. A rail may be placed across the columns of a duplex mill and a third spindle carrier vertically mounted on it over the table. This type of machine is often referred to as a *tri-way mill*. The characteristics of a typical medium-size duplex production-milling machine are illustrated in Figure 21-10. The only feed movement is the crosswise feed of the table between the two opposed cutter heads. The table is hydraulically fed in either direction and the length of travel is controlled by trip dogs or switches on the side of the table. In a typical cycle, the table rapidly carries the workpiece to the cutter, proceeds at a regular feed rate through the cutting area, reverses at the end of the cut, and rapidly withdraws. The spindles are started at the beginning of the cycle, stopped at the end of the cut, and withdrawn slightly to clear the workpiece as the table returns to the loading/unloading position. Once the machine has been set up, the operator starts each cycle by simply pushing a button on the control panel.

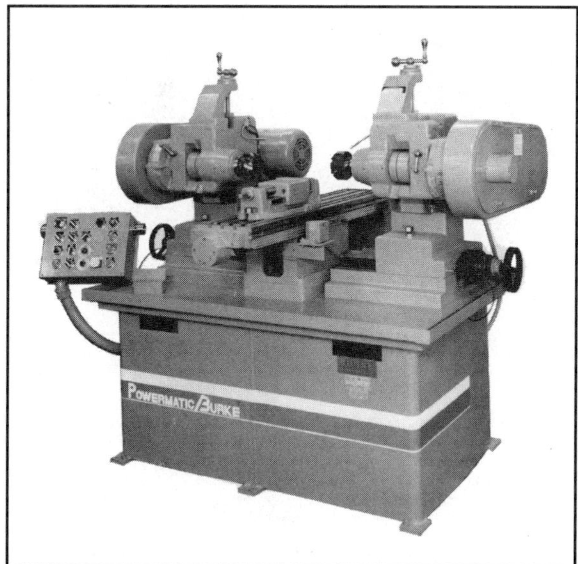

Figure 21-10. Duplex production-milling machine with hydraulic table feed. (Courtesy D. C. Morrison Company)

The spindles of the machine shown in Figure 21-10 are each carried in a quill in the spindle carrier and they can be positioned crosswise over the table by means of handwheels. The spindle carrier for each head is vertically adjustable on a column attached to the machine bed. Position adjustment for the two heads to 0.025 mm (.0010 in.) is established by reference to graduated collars or micrometer dials behind the handwheels. Parts to be machined are located and clamped in place by means of a milling machine vise or fixture. A machine similar to the one shown in Figure 21-10 costs about $30,000.

Not all production milling machines have reciprocating tables. *Continuous mills* have rotary tables, some horizontal, others vertical. A single-spindle machine with a 1.2-m (48-in.) horizontal rotary table is shown in Figure 21-11. Six air-operated fixtures are mounted on the continuously rotating table so that parts can be loaded and unloaded while the spindle continues cutting. The machine can be equipped with a skip-feed control that permits a rapid table speed as the cutter moves from one workstation to the next. The vertical slide for the spindle carrier has a micrometer adjustment that permits the cutter head to be positioned for the appropriate depth of cut. Machines of this type are of rugged construction and permit the milling of parallel part faces to exacting requirements on flatness, parallelism, and surface finish.

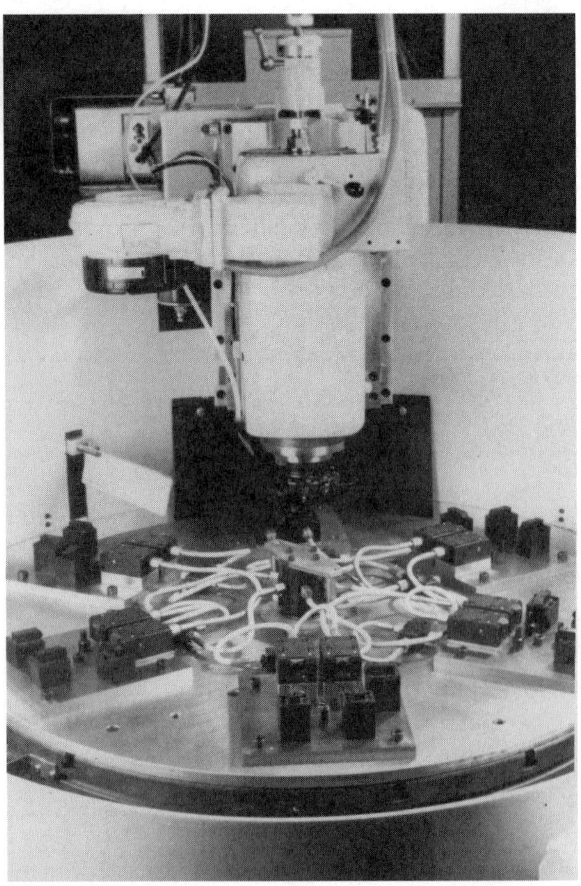

Figure 21-11. Single-spindle, continuous milling machine with horizontal rotating table. (Courtesy Onsrud Machine Corp.)

A two-spindle model of the machine shown in Figure 21-11 is available to facilitate the rough and finish milling of parts in one pass. The two spindle carriers are mounted on independent columns located at 90°, each independently controlled.

Tracer-controlled profiling or *contouring milling machines* are available for reproducing curved or irregular surfaces in moderate to large quantities. On these machines, a tracer is moved across a template surface, or the surface is moved in contact with the tracer. Through a servomechanism, the cutter spindle is made to move up and down as it traverses the workpiece the same way the tracer moves, and the cutter reproduces the surface or form covered by the tracer.

To a large extent, the tracer-controlled profiling and contouring mills are being replaced by either horizontal or vertical CNC copy and profile milling machines with three or more computer-controlled axes (Chapter 31). For example, in the aircraft industry, large vertical CNC profiling machines with three to ten axes of CNC movement are routinely used to machine aircraft wing skins, bulkheads, landing gear, and other key components. These machines, usually of a bridge-bed-rail type design, some with up to 305 m (≈1,000 ft) of bed length, are available with single or multiple spindles and speeds up to 10,000 rpm. Most machines of this class are equipped with automatic tool changing and storage mechanisms.

Computer numerically controlled (CNC) *milling machines* are becoming more prevalent in job shops, toolrooms, and production facilities.

These machines involve the same features and components as the standard machines described in this section. The CNC milling machines range from relatively inexpensive two-axes (X, Y) control horizontal or vertical machines to the more costly machines with multiple axes control, automatic tool changing and storage devices, and workpiece loading and unloading systems. When the versatility of these machines is increased to enable drilling, reaming, boring, facing, tapping, etc., along with the milling operations, they are often referred to as *machining centers*.

21.4.3 Machine Sizes

The capacity of a milling machine is determined by its table size, length of movements, and power. The more common sizes of machines have been standardized in ANSI B5.45 (see the references at the end of the chapter). Horizontal knee- and column-type millers are designated by size numbers 2, 3, 4, and 5. Minimum dimensions are specified for each size. Horizontal bed-type machines are designated by sizes 10, 14, 18, 22, and 30. Each size may have a range of dimensions. Commonly, for each dimensional size, machines are made for light, medium, and heavy duty, with corresponding strength and power. Some manufacturers use their own symbols to designate sizes.

21.4.4 Planer-type Machines

A planer-type milling machine looks like and has comparable movements to a double-housing planer, except that the table customarily feeds at a slow rate and spindle carriers for rotating cutters take the places of the planer toolheads. These machines are designed to handle large workpieces.

21.4.5 Horizontal Boring, Drilling, and Milling Machines

A horizontal boring mill is typically fitted with a backrest or end support for boring with long bars as illustrated in Figure 21-12(A). The bar is driven by the spindle in the headstock or spindle head, which can be moved up or down on ways on a column. The end support on the right end of the bar is raised or lowered by a screw in unison with the headstock. Stub boring bars are used without end support. Horizontal boring, drilling, and milling machines do milling as well as boring, usually with a face mill on the end of the spindle as shown in Figure 21-12(B).

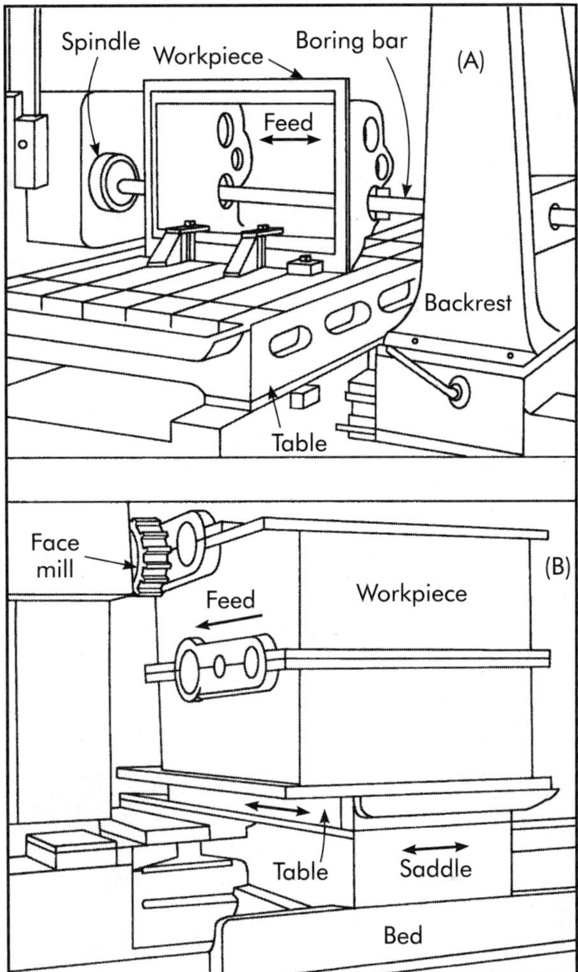

Figure 21-12. Operations on horizontal boring, drilling, and milling machines. (A) Boring a casting with a long bar supported in a backrest, (B) face milling a pad on a welded structure.

Horizontal boring, drilling, and milling machines are built to provide support and rigidity to precisely machine large castings, forgings, and weldments for products such as diesel engines, turbines, and machine tool columns. The worktable on a *table-type* machine, like the ones in Figure 21-12, is mounted and moves lengthwise and crosswise over a saddle and bed. Other types offer still more rigidity. A *floor-type boring mill* has a headstock on a column that slides on runways. Workpieces are placed on floor plates alongside the runways. A *planer-type horizontal boring mill* has a table that moves and is directly supported on a bed. The headstock on its column and the end support, or another headstock in its place, are movable on runways crosswise to and from the table.

The size of a horizontal boring, drilling, and milling machine is designated by the diameter of its spindle, customarily from 76.2–355.6 mm (3–14 in.). The table-type CNC machine shown in Figure 21-13 has a spindle diameter of 120 mm (4.75 in.) and a table surface size of 1,397 × 1,600 mm (55 × 63 in.). This machine is typically used for small-lot machining of castings, forgings, and weld fabricated parts up to 5,897 kg (13,000 lb). It is equipped with a 60-position automatic tool changer (not visible on the left), a tool shuttle and indexer visible in front, and an integral rotary table. Five-axes computer control on the machine provides control for the crosswise movement of the table (*X* axis), longitudinal table travel (*W* axis), headstock vertical travel (*Y* axis), in-and-out feed movement for the spindle (*Z* axis), and rotary motion of the table (*B* axis). Five-axes control plus the 60-position automatic tool changer gives the machine great versatility in a variety of manufacturing environments.

Figure 21-13. Table-type horizontal boring, drilling, and milling machine. (Courtesy WMO Machine Tool Corp. and O. Keller Tool Engineering Co.)

The part shown on the table of the machine in Figure 21-13 is a precision aluminum fixture for use in the assembly of automotive transmissions. An electronic probe in the machine spindle is used to obtain precision alignment of the hole in the fixture with the machine spindle. A machine of this type and size with CNC control and automatic tool-changing equipment costs about $530,000.

21.4.6 Machine Attachments

Many standard attachments are available to make milling machines easier and faster to operate, improve precision, and increase the variety of jobs that can be done. They may be divided into two general categories. One class includes attachments for positioning and holding work. The other includes attachments for positioning and driving milling cutters.

Work-positioning and Holding Devices

The movable elements of a column-and-knee type milling machine must be moved to a specific location to position the workpiece for the operation to be performed. As explained earlier, this movement is accomplished by means of handwheels and cranks or powered transport screws, and the position is established by reference to micrometer dials or graduated collars behind the handwheels. Positioning by these means is often complicated by hard-to-read dial scales and excessive backlash caused by worn mechanical elements, such as lead screw/nuts, racks, and gears. On older machines, precision machining in the range of 0.013 mm (.0005 in.) is often a matter of luck. As explained in Chapter 18, digital readout devices (DRO) sense machine slide movements directly, bypassing the backlash. Thus, the machine operator can read the exact tool position at all times on a numerical display unit. Parts can be machined according to the data given on the drawing without time-consuming calculations. These devices enable the operator to work with absolute dimensions, with several datums if necessary, in either millimeters or inches.

A digital readout system for a milling machine consists of a linear encoder for each axis to be referenced and a display unit or counter where the numerical value of the measurements is shown. The linear encoders are sensors that detect the movements on the machine slides and transforms them into electrical signals. Those signals are then fed back to the counter where the axis movement is numerically displayed. A three-axes (X, Y, Z) digital readout display unit on a milling machine is shown in Figure 21-14. This unit will display the position of a workpiece on the table with reference to datum points in the X-Y plane and the position of the milling cutter along the Z axis. When used in conjunction with an edge finder, the unit can store two datum points in a nonvolatile memory. After power interruption, the operator need only pass the table over the datum positions to re-establish its position relative to those points. The counter shown in Figure 21-14 has eight digits plus a sign for each axis and is capable of number display increments of 5 μm (.0002 in.), 2 μm (.0001 in.), or 1 μm (.00005 in.). It will also handle setup functions for "edge" and "centerline" position reference when used in conjunction with an edge finder.

Other digital readout models are available that feature conversational programming to assist the milling machine operator with various positioning tasks. Some of these units have program memories that can hold up to 99 positioning steps. More advanced models are available that support milling operation with interactive menus on the display unit screen. Many of these also provide on-screen support, such as a calculator, stopwatch, or a cutting data calculator. One model has programming capabilities that make it ideal for small-lot production on manual machine tools. Up to 20 programs and a total of 2,000 program blocks can be stored in its memory. Programs can be created either by keying them in step-by-step or generating them through actual position capture (teach-in programming). Most have preprogrammed routines for locating bolt hole circles or linear hole positions.

An inexpensive three-axes digital readout system, including linear encoders and a display unit, can be purchased for less than $3,000. These systems can be retrofitted on older machines or come pre-installed on new machines.

A very useful accessory for datum setting on a milling machine is the *edge finder*. An illustration of such a device is shown in Figure 21-15. Usually, the edge finder is used in conjunction with a digital readout unit to establish a datum. To set a datum, the operator must move the axes of the machine to some known point, such as a corner of the workpiece, from which all dimensions are to be measured. The operator then sets the display values to zero or otherwise enters

Figure 21-14. Numerical display unit featuring three-axes digital readout. (Courtesy Heidenhain Corporation)

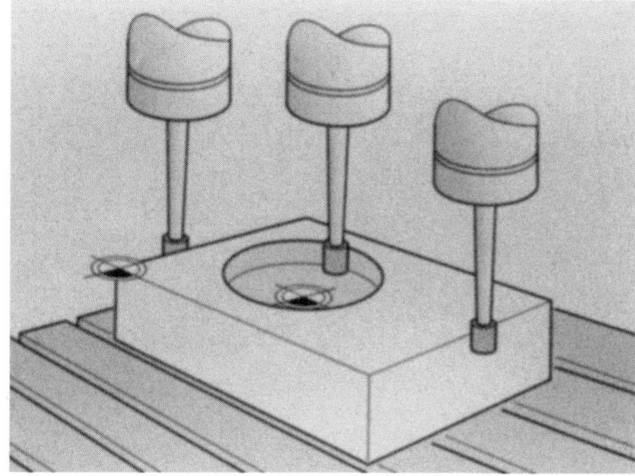

Figure 21-15. Establishing X and Y datums with an edge finder for a workpiece located on a milling machine table. (Courtesy Heidenhain Corporation)

the values of the known position to define axis positions. Generally, the cylindrical stylus is spring-mounted in the edge finder housing. When the sylus contacts the workpiece, an electrical signal is sent via a connecting cable to the display unit. Usually, the display unit automatically compensates for the cylinder radius of the stylus. As illustrated in Figure 21-15, the edge finder also may be used to establish a hole center as a datum.

A *vise* is the most common workpiece holding device found on milling machines. *Plain vises* are shown on the tables of the milling machines in Figures 21-9 and 21-10. Other kinds can be swiveled or tilted, but are not as rigid. Many vises are opened and closed by a screw, but faster acting cam and toggle devices are often used. Air or hydraulically operated vises are generally more economical for quantity production.

The vise jaws that make contact with the workpiece are held on by screws and are removable. Standard jaws for general-purpose work have flat faces. Special jaws of various shapes are often made and applied to hold specific workpiece shapes that are to be machined in quantities. Special jaws on a standard vise are often cheaper and just as effective as a special fixture.

Universal chucks like the one shown in Figure 21-16 are also employed to hold and locate cylindrical parts on a milling machine table.

Figure 21-16. Rotary indexing table with three-jaw chuck for locating and holding cylindrical workpieces. (Courtesy WMW Machinery, Inc.)

Fixtures were introduced and defined along with jigs in Chapter 20. They are widely used to hold and locate parts milled in moderate to large quantities. Fixtures are mounted on the table of the machines in Figures 21-6 and 21-7. A fixture may save time and money in an operation by holding a workpiece in a desired position so that attachments or a difficult setup are unnecessary, by providing full support to a workpiece so that heavy cuts can be taken (better pieces and less scrap are made because deflection and distortion are eliminated), and by making unloading and loading the workpieces easier and quicker. A fixture usually is good for only one operation and is justified only when it can save enough on that operation to pay for its cost. The worth of a fixture may be estimated by Equation (20-9).

A *circular milling attachment* or *rotary table* is a round table that can be turned about its axis on a base. A rotary indexing table with an indexing plate in front is shown in Figure 21-16. Some rotary tables are turned or indexed by hand; others are power-driven from the table feed drive of the machine. Cuts may be made at desired intervals around a workpiece held in the chuck on the unit shown in Figure 21-16 by turning the hand crank on the index plate a specified amount. Also, circular arcs may be milled, and other curves obtained by combining the circular motion with the linear machine feeds.

A *dividing* or *indexing head* is a mechanical device for dividing a circle accurately into equal parts. This is called *indexing*. A common indexing device for milling machines is the *universal dividing head* shown in Figure 21-17.

Direct indexing is done by means of a plate on the front of the spindle of the head. The plate has equally spaced notches or holes in one or more circles. The number of divisions available is limited to the hole circles on the plate.

Plain or *simple indexing* is done through a single train of gears. The sections of the universal dividing head in Figure 21-17 show a worm turned by gears and a crank on the side of the head. The worm is engaged with a worm wheel around the spindle. Forty turns of the crank are needed to turn the work spindle of this head one full revolution; five turns are needed on some heads. A pin in the crank handle

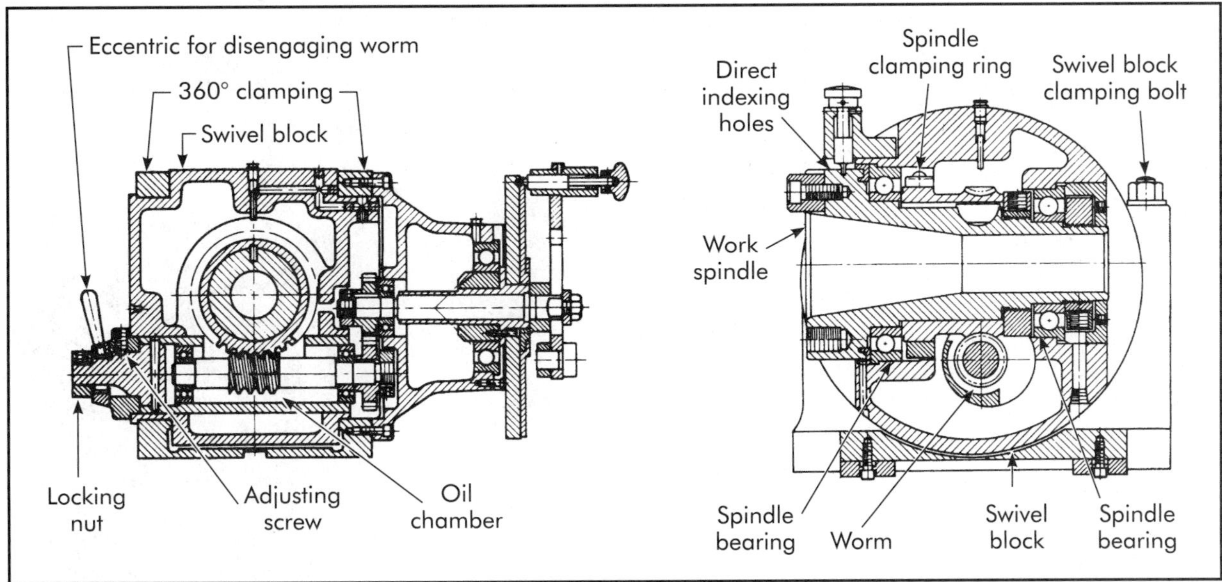

Figure 21-17. Sections through a universal dividing head to show the indexing mechanism. (Courtesy Cincinnati Milacron, Inc.)

registers in holes in a plate fixed to the side of the head. The plate and crank can be seen in Figure 21-16. The holes are arranged in circles, and a different number is equally spaced in each circle. Standard plates on one make of dividing head have holes that make it possible to index all numbers up to and including 60 and many higher numbers.

To determine the number of turns required to index the work one division,

$$C_t = R \div D \quad (21\text{-}1)$$

where

C_t = turns of the crank of the dividing head
R = ratio of the dividing head
D = number of equally spaced divisions required for one full turn of the workpiece

As an example, a dividing head with a 40:1 ratio is to be used to cut a gear having 36 teeth. The crank must be turned 40 ÷ 36 = 1-4/36 = 1-1/9 turns to index from one tooth space to the next on the gear. If the index plate has a circle with 18 holes, the crank must be advanced 18 × 1/9 = 2 spaces after each full turn.

The dividing head has been called a jewel among machine tools because of its precision. Plain indexing is commonly done on standard heads with an error on the work from setting to setting of less than 1 minute of arc. This is equivalent to 38 μm (.0015 in.) on the circumference of a circle of 305 mm (12 in.) diameter, or one part in 25,000.

Several means are available for indexing numbers not obtainable with standard plain indexing, especially large numbers. One of these is *differential indexing,* an arrangement whereby a suitable train of gears is installed between the spindle of the dividing head and a jackshaft that turns the side indexing plate. When the crank is turned and causes the spindle to turn, the plate is rotated and the actual distance the crank is moved from one hole to another depends upon the displacement of the plate. A *wide range divider* has two side plates and cranks, one that turns the spindle with a 40:1 ratio and another that furnishes an additional 100:1 ratio. This provides indexing from 2–400,000 divisions. An *astronomical dividing head attachment* has three plates and cranks arranged to divide a circle into degrees, minutes, and seconds.

A *helical milling* or *lead attachment* is mounted on a milling machine table, as in Figure 21-8, to provide a drive from the table lead screw to the dividing head to cut helical gears, worms, threads, twist drills, etc. The attachment drives the jackshaft that causes the index plate to turn. The crank, engaged with the plate, turns and causes the spindle of the dividing head to revolve. The lead of a helix cut on a milling machine is equivalent to the distance the table advances while the dividing head spindle makes one full revolution. The lead is varied by means of change gears in the driving attachment.

Linear indexing may be done on a general-purpose milling machine for jobs such as cutting rack teeth by means of the table lead screw dial or a *rack indexing attachment.* This unit consists of a gear or slotted plate connected to the lead screw through change gears and is indexed by a pin.

Cutter Driving Attachments

Attachments are commonly put on horizontal spindle milling machines to hold and drive cutters with their axes inclined or swiveled from the conventional position. This puts the cutters into position for milling helical gear teeth, racks, cams, and surfaces, grooves, or holes at all angles in workpieces that cannot be conveniently reached otherwise. A popular type is the universal high-speed milling attachment shown in Figure 21-18 that can be swiveled 360° in a plane parallel to and 45° either way in a plane normal to the machine column face. The attachment shown is driven by the machine spindle as are many others, but some have their own driving motors.

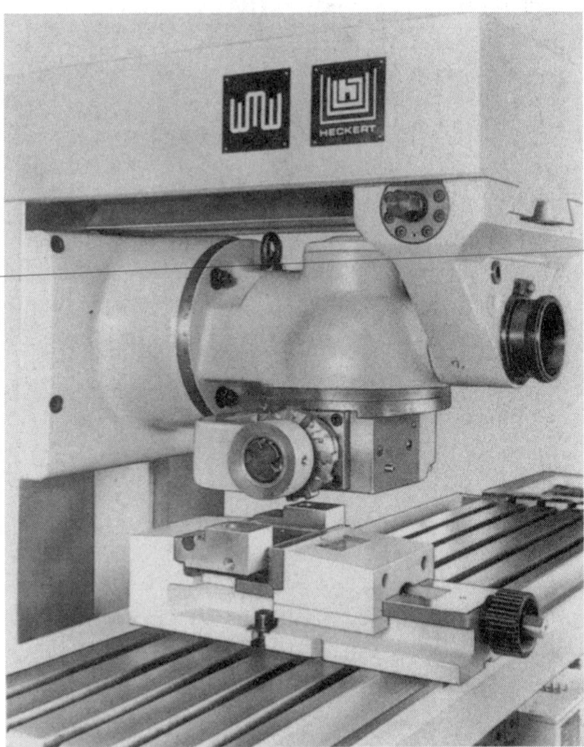

Figure 21-18. Universal milling attachment for a horizontal milling machine. (Courtesy WMW Machinery, Inc.)

21.5 PROCESS PLANNING

As in any other machining operation, planning for milling machine applications requires consideration of a number of factors, depending on the complexity of the job to be done. For example, the face milling operation being performed on a cast iron plate shown in Figure 21-19 appears to be rather uncomplicated. On the other hand, the slotting, boring, chamfering and grooving operations being done on the steel yoke illustrated in Figure 21-20 would appear to present some complications. Both of these jobs, however, require consideration of essentially the same factors. A number of those factors are discussed in the following sections.

Figure 21-19. Face- or end-milling operation on a cast-iron workpiece. (Courtesy WMW Machinery, Inc.)

21.5.1 Performance

Two styles of peripheral milling are depicted in Figure 21-21. *Conventional* or *up milling* is more common because the cutter is opposed by the feed of the work and the effect on the machine is more even. It is harder on cutters because each tooth tends to rub rather than to take a bite as it enters the cut; pronounced feed marks are left on the work surface; and forces tending to lift the work are high when the depth of cut is

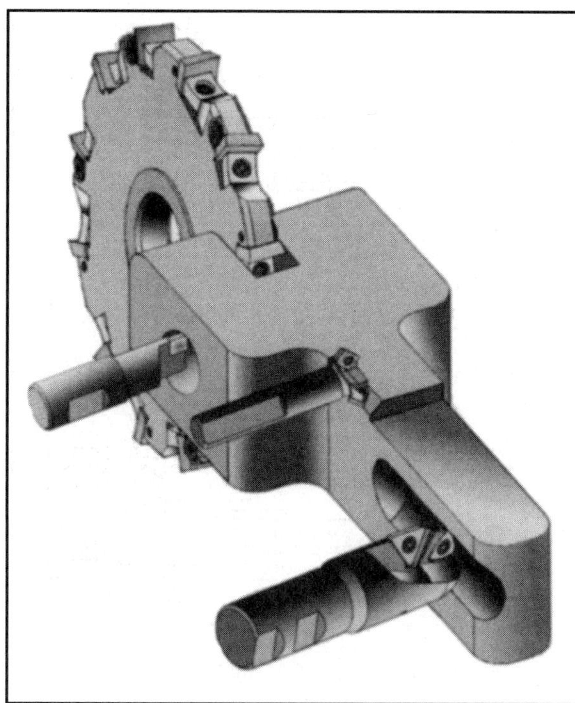
Figure 21-20. Slotting, boring, chamfering, and grooving operations being performed on a steel yoke. (Courtesy Lovejoy Tool Company, Inc.)

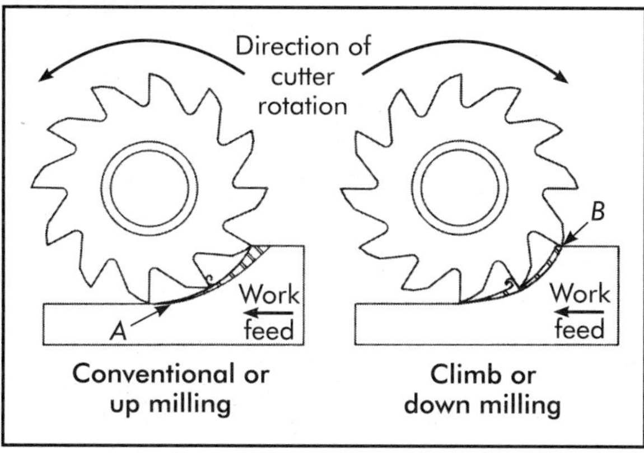

Figure 21-21. Difference between conventional and climb milling.

over about 3.18 mm (.125 in.). In contrast, each tooth enters the work with a substantial bite in *climb* or *down milling*. The cut is cooler, cutters last longer, feed marks are smaller, and the work is held down. However, the cutter tends to pull the workpiece along. Many of the older milling machines have too much backlash in the lead screw and nut, which allows the cutter to draw the workpiece ahead and take bites that are too large. Damage is likely to result unless the workpiece and fixture are strong and backlash is eliminated. On newer machines, backlash is eliminated, or at least minimized, through the use of ball screws, hydraulic feeds, or other backlash-limiting devices.

On older milling machines, tolerances of less than 25 μm (.001 in.) are difficult to maintain, even with skilled operators. However, newer machines equipped with ball screws and digital readout units are capable of consistently holding tolerances of 12.7 μm (.0005 in.) or less in many cases.

21.5.2 Economical Milling

Setup often accounts for most of the time required to mill one or a few pieces of a kind. Where a variety of work is milled, time can be saved by planning and scheduling the jobs so that similar parts are milled in succession.

It is shown in Chapter 18 that a turret lathe is tooled in different ways to get the lowest cost with different quantities of pieces. This same principle applies to milling operations as well as to most others. Some common milling arrangements are shown in Figure 21-22.

Plain or *simple milling* involves the loading and milling of one piece at a time and is the usual arrangement for one or a few pieces. Cutting time is saved in *string* or *line milling* with two or more pieces in a row because the cutter can be entering one piece as it leaves another. Efficiency is achieved by arranging for the operator to load at one station while the cut is taken at another, as in *reciprocating milling,* illustrated in Figure 21-22. There the cutter cuts upward in one direction and downward in the other. If this is undesirable, it may be eliminated by *index base milling.* An index base consists of two plates mounted one above the other on the machine table. The upper plate carries a fixture on each end and is indexed 180° when retracted from the cutters. Then one fixture is advanced to the cut while the other is unloaded and loaded. Fixtures on a rotating round table are loaded with pieces and fed continuously to the cutter or cutters in *circular* or *rotary milling.* This may be done on a continuous miller with an integral rotary table or on an auxiliary

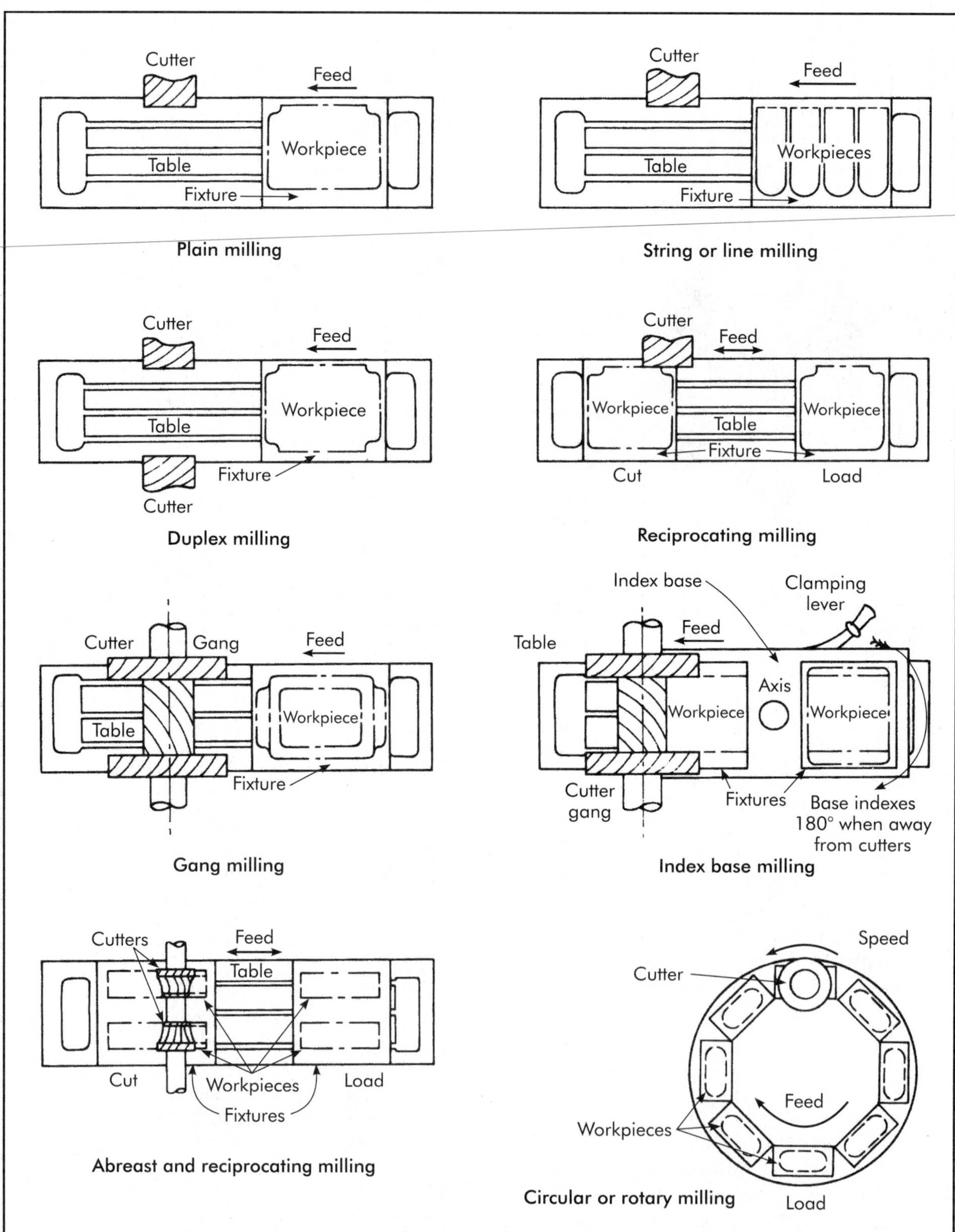

Figure 21-22. Some common forms of production milling operations.

circular table mounted on a conventional milling machine.

Cutting time is saved when two or more cutters are put to work at the same time. Common examples are given in Figure 21-22. *Duplex milling* is done on a duplex miller constructed as indicated in Figure 21-10. *Gang milling* utilizes two or more cutters on one arbor. If side milling cutters machine two sides of a workpiece at the same time, the operation is called *straddle milling*. If the purpose is to machine two or more parts side-by-side, the term is *abreast milling*.

Tooling for milling operations may be as complex as the amount of production warrants. The basic forms may be combined, as suggested by abreast and reciprocating milling in one operation in Figure 21-22. Special indexing attachments, some arranged for automatic unloading and loading of a number of pieces at a time, are examples of devices for large quantity production. In any case, each production arrangement requires a certain investment in tooling, and the principle exemplified by Equation (18-1) applies here: the amount saved per piece times the number of pieces must at least equal the tooling cost and incidental charges, including any extra setup costs.

21.5.3 Speed, Feed, and Depth of Cut

Substantially the same factors determine economical speeds for milling cutters as for single-point tools. Typical recommendations for speeds and feeds are given in Table 21-1 for roughing cuts and finishing cuts with either high-speed steel (HSS) or brazed carbide cutters. As noted at the bottom of the table, cutting speeds may be increased considerably for inserted tooth cutters and coated carbide cutters. In any case, it is recommended that the cutting tool supplier or manufacturer be contacted for cutting speed and feed recommendations for specific types and grades of milling cutters to obtain maximum cutting efficiency.

The *cutting speed* for a milling cutter is the linear velocity of a point at the periphery of the cutter. Thus, the cutter rpm is based on the cutter diameter in accordance with the relationship

$$N = \frac{C_S \times 1{,}000}{\pi d} \; or \left[\frac{C_S \times 12}{\pi d}\right] \qquad (21\text{-}2)$$

where

N = cutter rpm
C_S = cutting speed, m/min (ft/min)
d = diameter of the cutter, mm (in.)

Basic *milling feed* is the distance the workpiece advances in the time between engagements by two successive teeth. This is called *feed per tooth* in units of mm/tooth, µm/tooth, or in./tooth. However, the machine feed rate is given in mm/min or in./min and is calculated as follows:

$$f = f_t \times Z \times N \qquad (21\text{-}3)$$

where

f = machine feed, mm/min (in./min)
f_t = cutter feed/tooth, mm/tooth (in./tooth)
Z = number of teeth in the cutter
N = cutter rpm

The actual feed or maximum chip thickness in milling is considerably less than the nominal feed per tooth for shallow cuts as indicated by Figure 21-23. The two components of feed are equal only when the depth of cut is at least equal to the radius of the cutter. So, for the same tooth load a heavier feed can be taken with a shallow cut than with a deep cut.

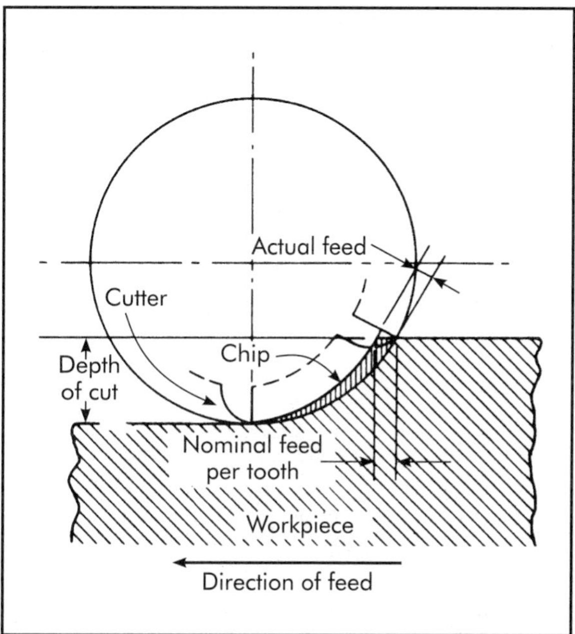

Figure 21-23. Relationship between actual and nominal milling feed.

Table 21-1. Recommended cutting speeds and feeds for milling

Material	(Bhn)	Depth of Cut (mm [in.])	Face Mill: High-speed Steel (HSS) Feed (mm/tooth [in./tooth])	Face Mill: High-speed Steel (HSS) Speed (m/min [ft/min])	Face Mill: Uncoated Cemented Carbide Feed (mm/tooth [in./tooth])	Face Mill: Uncoated Cemented Carbide Speed* (m/min [ft/min])	Plain or Slab Mill: High-speed Steel HSS Feed (mm/tooth [in./tooth])	Plain or Slab Mill: High-speed Steel HSS Speed (m/min [ft/min])
Aluminum alloys cold drawn	30–80 (500 kg)	7.62 (.300) 1.02 (.040)	0.51 (.020) 0.25 (.010)	198 (650) 366 (1,200)	0.64 (.025) 0.25 (.010)	366 (1,200) 610 (2,000)	0.41 (.016) 0.30 (.012)	259 (850) 366 (1,200)
Copper alloys cold drawn 145 to 782	50–100 R_B	7.62 (.300) 1.02 (.040)	0.46 (.018) 0.25 (.010)	122 (400) 183 (600)	0.51 (.020) 0.25 (.010)	213 (700) 396 (1,300)	0.46 (.018) 0.36 (.014)	122 (400) 191 (625)
Gray cast iron as cast Class 45 and 50	220–260	7.62 (.300) 1.02 (.040)	0.36 (.014) 0.15 (.006)	15 (50) 26 (85)	0.36 (.014) 0.18 (.007)	63 (205) 122 (400)	0.25 (.010) 0.15 (.006)	14 (45) 24 (80)
Steel hot rolled or cold drawn 1005–1025	175–225	7.62 (.300) 1.02 (.040)	0.41 (.016) 0.20 (.008)	34 (110) 58 (190)	0.41 (.016) 0.20 (.008)	95 (310) 168 (550)	0.25 (.010) 0.15 (.006)	34 (110) 56 (185)
Steel hot rolled or cold drawn 1030–1055 1525–1527	225–275	7.62 (.300) 1.02 (.040)	0.36 (.014) 0.15 (.006)	24 (80) 38 (125)	0.36 (.014) 0.18 (.007)	81 (265) 137 (450)	0.23 (.009) 0.13 (.005)	21 (70) 37 (120)
Steel Heat treated 1330–4130 5130–8630	275–325	7.62 (.300) 1.02 (.040)	0.31 (.012) 0.15 (.006)	18 (60) 31 (100)	0.25 (.010) 0.15 (.006)	72 (235) 114 (375)	0.18 (.007) 0.13 (.005)	15 (50) 27 (90)

*Notes: Speeds given are for brazed carbide teeth. For throwaway or indexable uncoated inserts, speeds may be 10–20% higher, and for coated inserts 30–50% higher. These are relatively large and strong cutters. Feeds may be less for smaller and weaker cutters, such as end mills, form cutters, and saws.

(Institute of Advanced Manufacturing Sciences 1980)

Feed per tooth should be as high as possible for fast cutting, but the heavier the feed per tooth, the greater the load on the cutter teeth, workpiece, holding device, and machine. A large face mill will withstand a greater feed per tooth than a small end mill. A light feed may have to be chosen for a fragile workpiece. The rigidity and power of a milling machine may limit the rate at which stock can be removed. A heavier feed is more possible with soft materials than with hard or tough metal. A good finish calls for a light feed.

One cut is enough for most jobs, but rough and finish cuts are required to produce the best surface finishes and hold small tolerances. The depth of a roughing cut should be as much as the cutter, machine, and work will stand. In most cases, this is the full amount of stock on a surface minus that needed for finishing (1.5 mm [≈.0625 in.] or less is normally left for finishing).

21.5.4 Estimating Time and Power

The approach of a milling cutter may be an appreciable part of the length of cut. A face mill taking a roughing cut is normally stopped when it has just cleaned the surface as indicated by Figure 21-24(A). The approach shown at the beginning of the cut can be calculated from the right triangle having sides $(d/2) - A$, $w/2$, and $d/2$.

$$A = (d/2) - \sqrt{(d^2 - w^2)/4} \qquad (21\text{-}4)$$

However, if the diameter of the cutter is only a little larger than the width of the surface, the approach is $d/2$ for practical purposes. For a finishing cut, a face mill is passed entirely over a surface so that its trailing edge can get in a full wiping action. Thus the approach for finishing is equal to d.

The approach of a cutter with its surface offset from the center of the section or surface cut is depicted in Figure 21-25. This might be a slab mill cleaning the top of a piece, a side milling cutter opening a slot, or a face mill forming a step. From the right triangle shown,

$$A = \sqrt{cd - c^2} \qquad (21\text{-}5)$$

The cutting time for a face or slab milling operation such as is illustrated in Figures 21-24 and 21-25 can be calculated as

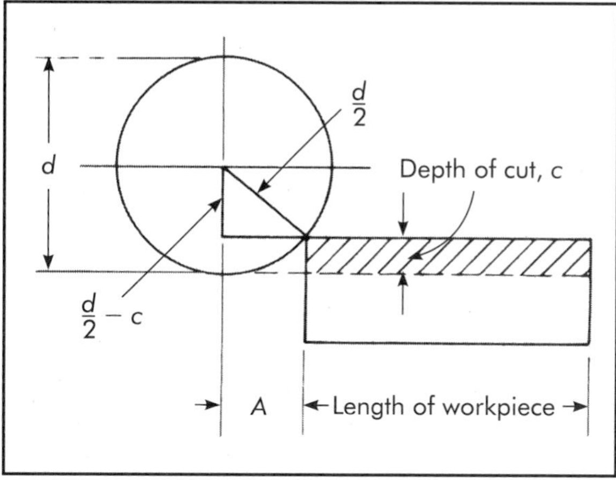

Figure 21-25. Approach of plain or slot milling cutter.

$$T = \frac{A + L + O}{f} \qquad (21\text{-}6)$$

where

T = cutting time, min
A = approach distance, mm (in.)
L = length of the workpiece, mm (in.)
O = overrun, mm (in.)
f = table feed, mm/min (in./min)

Normally, some overrun occurs at the end of a cut that needs to be added to the cutter travel distance.

Power consumption in milling is to a great extent a function of the rate of metal removal. For face or slab milling, the rate of metal removal is,

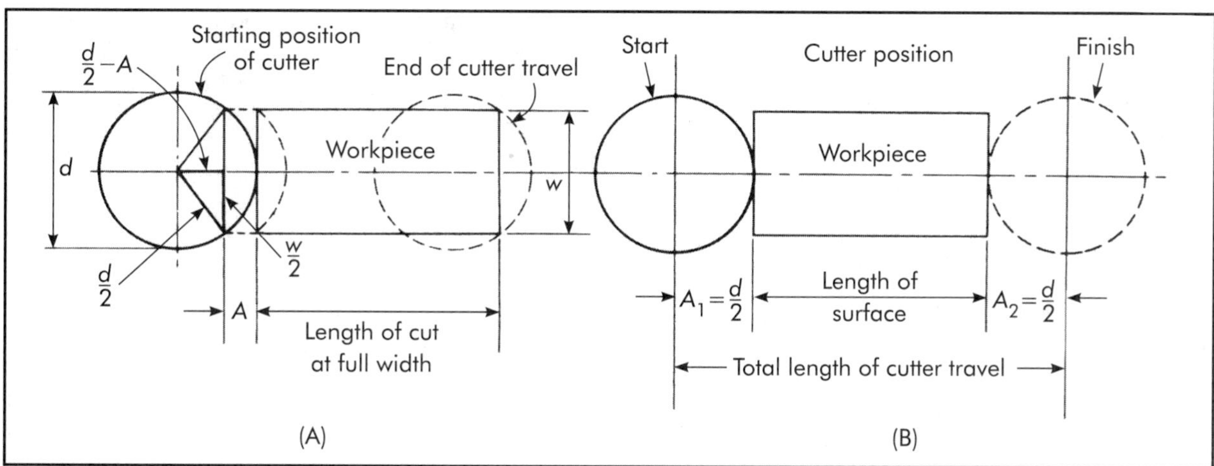

Figure 21-24. Approach of face mill for (A) roughing cut, (B) finishing cut.

$$Q = \frac{w \times c \times f}{1{,}000} \quad (w \times c \times f) \tag{21-7}$$

where

Q = rate of metal removal, cm^3/min ($in.^3/min$)
w = width of cut, mm (in.)
c = depth of cut, mm (in.)
f = table feed, cm/min (in./min)

Unit power consumption data for various materials with different types of cutters under certain cutting conditions has been developed by milling cutter manufacturers. They should be contacted for such information when precise estimates of machine consumption are required. An empirical formula proposed by one company that takes into account the effect of the rate of metal removal upon power consumption in milling is

$$P = P_u Q^{0.75} \tag{21-8}$$

where

P = power, kW (hp)
P_u = unit power, $kW/cm^3/min$ ($hp/in.^3/min$)
Q = rate of metal removal, cm^3/min ($in.^3/min$)

The unit power requirements for face milling with high-speed steel cutters range from 15 $W/cm^3/min$ (.32 $hp/in.^3/min$) for aluminum to around 118 $W/cm^3/min$ (2.6 $hp/in.^3/min$) for very hard (55–58 R_C) tool steel. Unit power requirements for inserted tooth carbide cutters may be significantly less than for high-speed steel cutters.

Using appropriate values for P_{um}, the power consumption for milling is estimated by

$$P_m \approx 2 \times 10^{-3} P_{um} Q_m^{\ 0.75} \tag{21-9}$$

where

P_m = power required, kW (hp)
P_{um} = specific power, $W/cm^3/min$ ($hp/in.^3/min$)
Q_m = rate of metal removal, cm^3/min ($in.^3/min$)

21.5.5 Comparison with Other Operations

Any surface that is accessible can be milled. This means that milling machines are to some extent competitive with all other machine tools. However, when the work is to be revolved, a milling machine can be used, but is seldom selected for the job because machines of the lathe family are inherently more efficient for such purposes.

Milling machines are capable of machining holes and locating them with a fair degree of accuracy, to tolerances of ±25 to ±50 μm (±.001 to ±.002 in.). A milling machine is economical for doing such work in small quantities without extra equipment. If the holes do not need to be located accurately, a drill press will do the job more quickly and easily. For large quantities, the milling machine is usually slower and not able to compete with the use of jigs on drilling machines or with production boring machines. Jig boring machines are necessary where holes must be located more precisely than can be done with milling machines. Really large pieces require the capacity and range of horizontal boring machines, which is beyond that of most milling machines.

Flat, straight, and many curved and irregular surfaces can be shaped, planed, or broached as well as milled. Broaching is more economical than milling in many cases for large quantities, but is at a disadvantage in other instances as explained in Chapter 22. Milling often is best for moderate quantities.

Grinding is capable of producing to closer tolerances and finer finishes and can remove harder materials than milling. Some grinding is done entirely from rough stock in what is called abrasive machining (described in Chapter 24), but many parts are milled before grinding.

21.6 QUESTIONS

1. What kinds of machining operations can be performed on a milling machine?
2. What is the difference between integral-shank and arbor-mounted milling cutters?
3. Sketch a typical face-milling cutter tooth and name its elements.
4. Name and describe the principal kinds of milling cutters.
5. Why would it be desirable to have a helix angle on a plain or slab milling cutter?
6. Name and describe the common cutter-holding devices on milling machines.

7. Name and describe the principal elements of a plain horizontal milling machine.
8. Indicate in the conventional manner the axes of movement for a universal horizontal milling machine.
9. Why would a milling cutter with inserted teeth be easier to maintain (sharpen) than a solid cutter?
10. What are the relative advantages and disadvantages of column-and-knee type and bed-type milling machines?
11. How are the knee, saddle, and table on a manually operated, horizontal column, column-and-knee type milling machine positioned?
12. How does a digital readout device improve the operation of the machine described in Question 11?
13. How are the cutters positioned horizontally and vertically on the duplex production milling machine shown in Figure 21-10?
14. Why is more than one workholding fixture provided on the rotating table of the continuous milling machine shown in Figure 21-11?
15. What is the purpose of a tool magazine on a horizontal boring, drilling, and milling machine?
16. Indicate by a line sketch the coordinate axes of movement for the horizontal boring, drilling, and milling machine shown in Figure 21-13.
17. How are workpieces held on the table of a milling machine?
18. What is "plain" or "simple indexing" and how is it done?
19. How can a helix be milled?
20. Describe the conventional and climb-milling processes and state the advantages of each.

21.7 PROBLEMS

1. If a fixture is made for a milling operation, in which pieces are now held in a vise, 30 sec can be saved for each piece made. The fixture will cost $350, and the cost of interest, insurance, taxes, and maintenance is 30% altogether. No extra setup time is required for the fixture. The hourly rate for labor and overhead in the plant is $32. For how many pieces is the fixture justified?
2. A surface 101.6 mm (4 in.) wide × 254 mm (10 in.) long is to be rough milled with a depth of cut of 8 mm (.30 in.) by a 152-mm (6-in.) diameter face milling cutter with 16 indexable, inserted carbide teeth. The material being machined is medium-hard case iron (220–260 Bhn). Estimate the cutting time and metal removal rate for this operation.
3. The estimated unit power required for milling the medium-hard cast iron used in Problem 2 is 27 W/cm^3 (.6 hp/in.^{3}). Calculate the total power required for the cutting parameters given in that problem (Institute of Advanced Manufacturing Sciences).
4. An operation takes 3 min of floor-to-floor time. The need for a vertical milling attachment can be eliminated by the use of a fixture that costs $250. The cost of the fixture must be recovered in one lot plus 25% for interest, insurance, taxes, and maintenance. The cost of this attachment is $3.50/hr, and the device can be put to work elsewhere if not needed on this operation. The hourly rate for labor and overhead is $28. Under these circumstances, how many pieces justify the cost of a fixture? What is the most that should be paid for a fixture if only 1,000 pieces are to be run?
5. A high-speed helical milling cutter (slab mill) with 12 teeth is 102 mm (4 in.) in diameter and 127 mm (5 in.) long. It is used to mill a soft-steel surface (175–225 Bhn) 76 mm (3 in.)

wide × 229 mm (9 in.) long with a depth of cut of 19 mm (.75 in.). A cutting speed of 30 m/min (100 ft/min) and feed of 0.25 mm/tooth (.010 in./tooth) are selected. What is the cutting time and how much power is required? Assume P_{um} = 50 W/cm^3/min and P_u = 1.1 hp/in.^{3}/min.

6. If the cutter described in Problem 5 takes a finishing cut 1 mm (.04 in.) deep, what time is required and what power is consumed? Assume P_{um} = 50 W/cm^3/min and P_u = 1.1 hp/in.^{3}/min.
7. A slot 15 mm (≈.5 in.) wide × 102 mm (4 in.) long through a piece of medium steel 25 mm (1 in.) thick is to be widened to 19 mm (.75 in.). Its other dimensions are to remain the same and radii are allowed at the ends. What time is required to do the operation with a 19-mm (.75-in.) diameter HSS end mill having eight teeth? The cutting speed is 24 m/min. (80 ft/min) and the feed 50 μm/tooth (.002 in./tooth). How much power is required? Assume P_{um} = 50 W/cm^3/min and P_u = 1.1 hp/in.^{3}/min.
8. The cast-iron slide block shown in Figure 21-26 is to be finish milled from a rough casting having 3 mm (.118 in.) of stock on all surfaces. Normal operational tolerances of ± 0.38 mm (.015 in.) are satisfactory. Make up an operation sheet similar to that shown in Figure 20-26 for finish machining all of the surfaces on one piece. List each of the operations to be performed and the type of milling cutter, speeds, feeds, and depth of cut to be used. Also specify how the workpiece is to be held for machining.
9. Estimate the machining time for one of the slide blocks described in Problem 8.

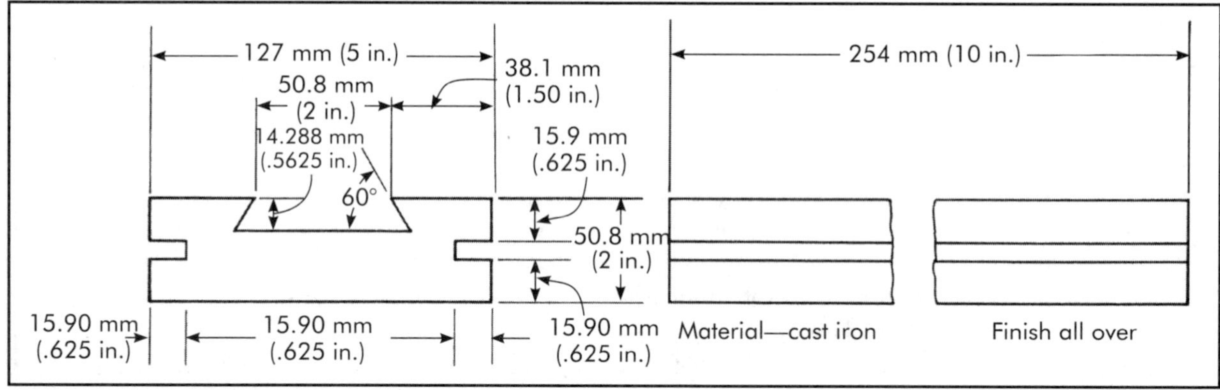

Figure 21-26. Cast-iron slide block.

21.8 REFERENCES

ANSI B5.18-1974 (R1991). "Spindle Noses and Tool Shanks for Milling Machines." New York: American National Standards Institute.

ANSI B5.45-1972 (R1991). "Milling Machines." New York: American National Standards Institute.

ANSI B5.47-1972 (R1991). "Milling Machine Arbor Assemblies." New York: American National Standards Institute.

ANSI B94.8-1967 (R1987). "Inserted Blade Milling Cutter Bodies." New York: American National Standards Institute.

ANSI B212.6-1986. "Indexable (Throwaway) Hard Metal Inserts for Milling Cutters—Dimensions—Triangular and Square Inserts." New York: American National Standards Institute.

ANSI B212.11-1988. "Cutting Tools—Indexable Insert Shank-type Milling Cutters (Inch Series)—Designation." New York: American National Standards Institute.

Drozda, T. and Wick, C., eds. 1983. *Tool and Manufacturing Engineers Handbook,* 4th Ed. Vol. 1. *Machining*. Dearborn, MI: Society of Manufacturing Engineers (SME).

Institute of Advanced Manufacturing Sciences. 1980. *Machining Data Handbook,* 3rd Ed. Cincinnati, OH: Institute of Advanced Manufacturing Sciences, Inc.

Koepfer, C. 1993. "Machine Tool Considerations Come to the Surface." *Modern Machine Shop*, Oct.

Noaker, P. M. 1994. "The Search for Agile Manufacturing." *Manufacturing Engineering*, Nov.

Seco—Guide Milling, 1993. Carboloy, Inc. Detroit, MI: A. Seco Tools Company.

Society of Manufacturing Engineers (SME). 1996. *Introduction to Workholding* video. *Fundamental Manufacturing Processes* series. Dearborn, MI: SME.

——. 1995. *Milling and Machining Center Basics* video. *Fundamental Manufacturing Processes* series. Dearborn, MI: SME.

Tofte, R. 1993. "When Ball Screws Need Reconditioning." *Modern Machine Shop*, Nov.

Broaching and Sawing

22

The band sawing machine, 1933
—Leighton Wilke, Minnesota

Broaching and sawing operations are, in general, somewhat similar in that they both use an elongated cutter with a series of cutting teeth that moves across the surface to be machined. Or, as in the case of broaching, the workpiece is often moved across the broaching tool. Broaching is usually employed to machine some form of external or internal surface on a part, while sawing is generally associated with parting or cutting off of a section or sections of a metal bar or plate.

22.1 BROACHING

In broaching, a tool with a series of teeth called a *broach* is pushed or pulled over a surface on a workpiece as depicted in Figure 22-1. Each tooth takes a thin slice from the surface. Broaching of inside surfaces is called *internal* or *hole broaching*; for most outside surfaces, it is usually referred to as *surface broaching*. Typical internal broaching operations are the sizing of holes and cutting of serrations, straight or helical splines, gun rifling, and keyway cutting.

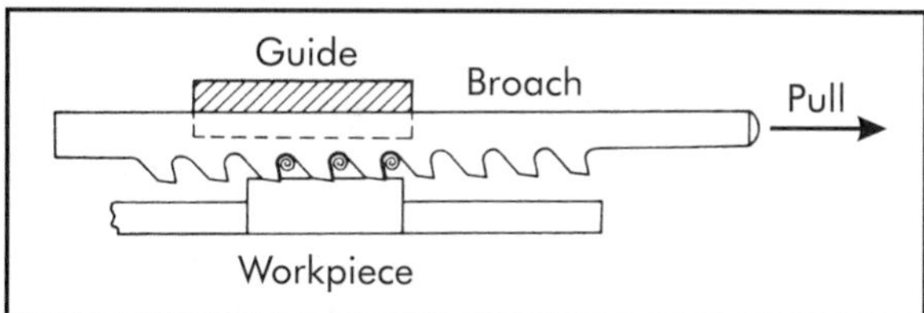

Figure 22-1. The way a broach works.

22.1.1 Types of Broaches

Almost every broach is designed to do a specific job in a certain way. Typical forms are shown in Figure 22-2. Some are intended to be pulled, others pushed, and still others to be held, either on a movable ram or in a fixed position. A broach may be made in one piece, solid, or assembled or built up from shells, replaceable sections, or inserted teeth. Replaceable sections, teeth, or shells make a broach easier to repair and, in some cases, easier to make initially.

A *burnishing broach* makes a glazed surface in a steel, cast iron, or nonferrous hole. Burnishing teeth are rounded and do not cut but compress and rub the surface metal.

A *pot broach* is a hollow cylinder with teeth inside. A common application is to cut all the teeth of an external gear at one time, while the workpiece is pushed or pulled through the broach.

Characteristics

The hole broach depicted in Figure 22-3 is gripped by a puller at the shank end. The *front pilot* centers the broach in the hole before the teeth begin to cut. The first group of teeth remove most of the stock and are called *roughing teeth*. The *finishing teeth* of a new broach are all

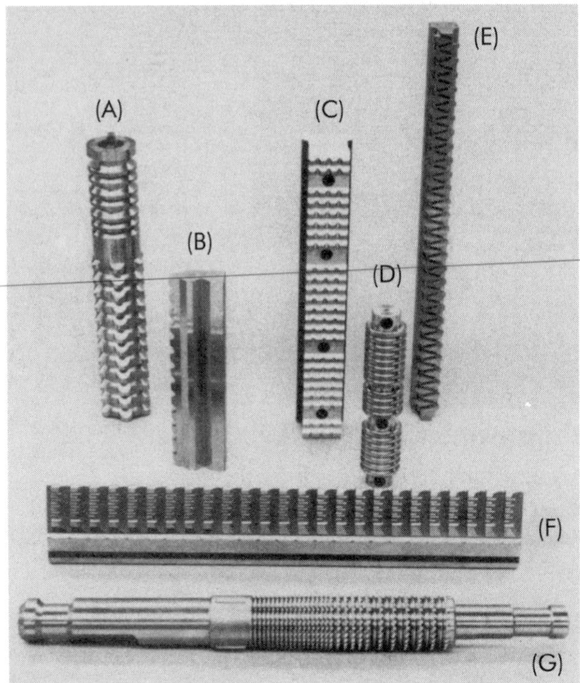

Figure 22-2. Typical broaches: (A) special form broach; (B) titanium slotting broach for compressor disks; (C) semifinish rack-gear form broach; (D) round broach used for half-round operation on connecting rods; (E) titanium dovetail roughing broach for compressor disks; (F) Inconel® pine-tree finish broach for turbine disks; (G) rebroach for hardened gears. (Courtesy Marbaix Lapointe)

substantially the same size and have the shape of the finished hole. As the first finishing teeth become worn, those behind take up the function of sizing. The *rear pilot* supports the broach after the last tooth leaves the hole. A broach handled automatically has a tail.

The enlarged section of the broach teeth in Figure 22-3 reveals features like those of other cutting tools. The *back-off angle* corresponds to the relief angle of a single-point tool, and the *hook* or *face angle* to the rake angle. A hole broach is sharpened by grinding the tooth face. Surface broaches with curved shapes may be sharpened the same way, but others are ground on the back-off angle land.

The teeth of broaches are staggered, offset, relieved, and arrayed in various ways to break up chips, reduce chip load, and enable teeth to take deep cuts under scale. One way this is done is illustrated in Figure 22-4.

Most broaches are made from high-speed steel, ground after hardening. Cemented carbide is used, especially for surface broaches, mostly for cast iron work and high production. Throw-away inserts have been found economical in some cases.

22.1.2 Pullers and Fixtures

Many surface broaches are mounted on holders bolted to the face of a ram, but some surface broaches and all pull broaches must be connected to the end of a ram by a *puller* or *puller head.* Several common pullers and broach shanks to fit them are shown in Figure 22-5.

The *threaded puller* of Figure 22-5(A) and the *key-type puller* in Figure 22-5(B) are simple and inexpensive, but relatively slow. The *automatic puller* of Figure 22-5(C) is favored for production, especially on machines with broach elevators and automatic handling equipment. When the sleeve on this puller is pushed backward by hand or by striking a stop, the puller accepts or releases the broach shanks.

Fixtures generally are justified for holding workpieces being broached because production quantities are large. Broaching fixtures are simple in comparison with those of other processes; one may even be just a plate to bear the part with a hole to pass the broach. Broaching forces generally tend to stabilize the part in the fixture but are high, and fixtures must be strong.

22.1.3 Machines

Broaching machines may be classified as: (1) push broaching machines, (2) pull broaching machines, (3) surface broaching machines, and (4) continuous broaching machines. Some manufacturers offer a basic model that can be arranged for two or more purposes, for instance for pull broaching or for surface broaching. In addition, a variety of specialized machines are available such as the rotary/turn broaching machine shown in Figure 22-6.

Broaching machines are constructed in horizontal and vertical models. A *horizontal broaching machine* has the advantage that any part of it, in particular the workstation, can be reached readily from the floor. A machine with a long stroke can be supported at many points and leveled to keep the slides straight. An ad-

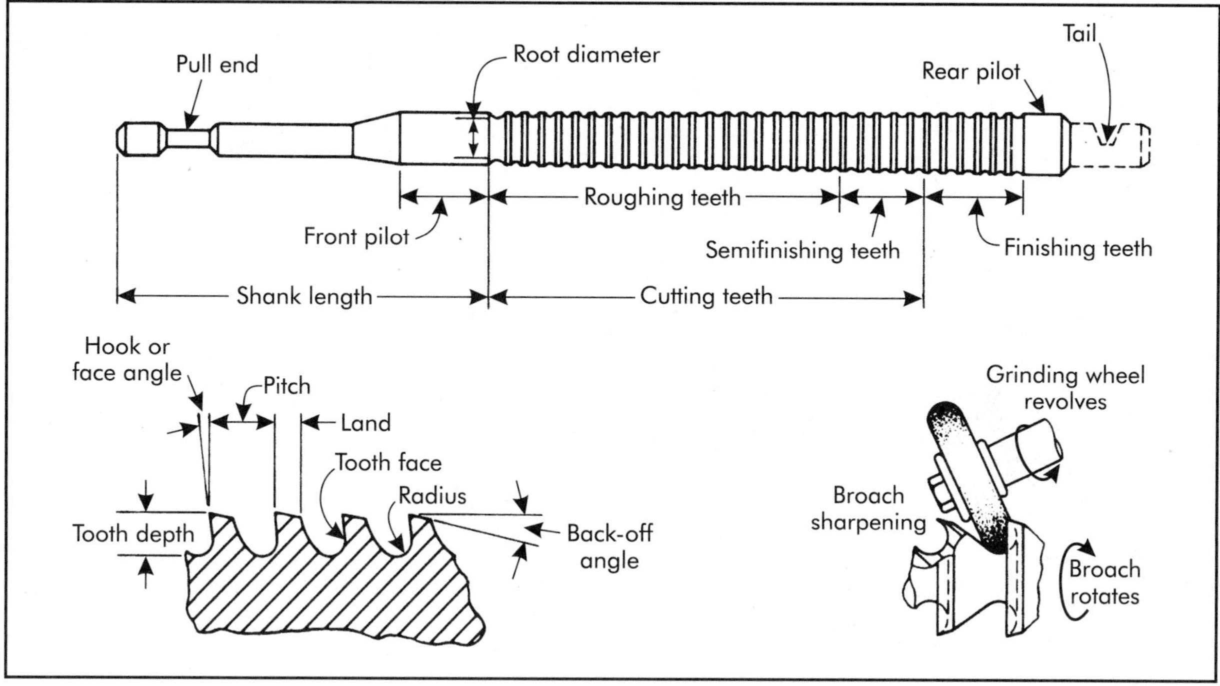

Figure 22-3. Hole broach details.

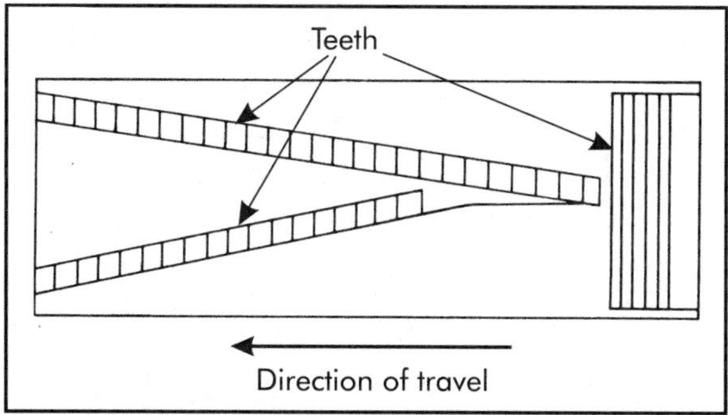

Figure 22-4. Arrangement of a progressive surface broach.

vantage of a *vertical broaching machine* is that it occupies minimum floor space, but if its stroke is long, a vertical machine must be sunk in a pit or have a platform for the operator to reach the workstation.

Internal broaches last up to twice as long on vertical machines because cutting fluid is utilized better and there is less drag of the tools.

Broaching machines have hydraulic or electromechanical drives. A hydraulic machine exerts force by means of a piston and cylinder. Such machines generally cost less, give smooth action, and are more popular. An electromechanical drive employs a screw engaged with a recirculating ball nut. This type of drive is preferred for longer-strokes and high-speed machines. It is more energy efficient and avoids hydraulic system maintenance. Most drives have infinitely variable speeds within their ranges. The size of a broaching machine is commercially designated by the force in tons that can be applied to the broach and by the length of the stroke in inches. Thus, a 10-54 machine can exert 10 tons (89 kN) of force and has a maximum stroke of 54 in. (137.2 cm).

Vertical Push Broaching Machines

Arbor presses are often used for internal push-down broaching such as hole sizing and keyway cutting. Push broaches must be short, and each cannot remove much stock. However, machines and tools of this type are rather simple and inexpensive, and operations can be set up and changed over easily. Quite often, simple manual or hydraulic arbor presses are used for very limited production broaching operations. For this type of press, the workpiece is placed on the table or bolster, sometimes in a simple fixture, under the vertical ram that pushes a short broach

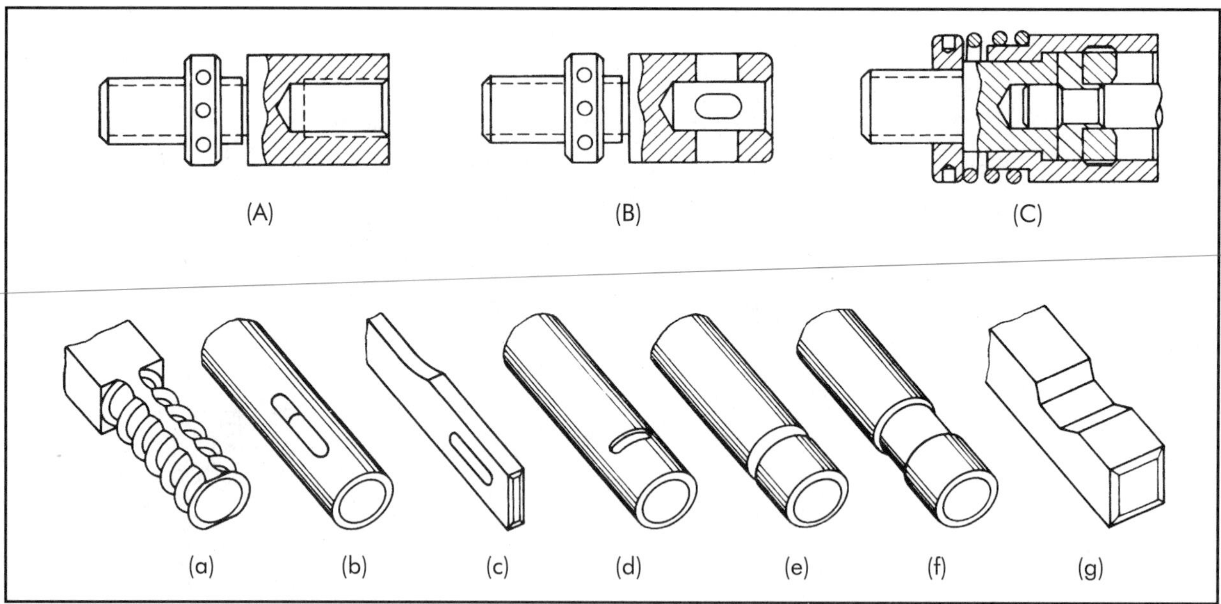

Figure 22-5. Broach pullers and shanks.

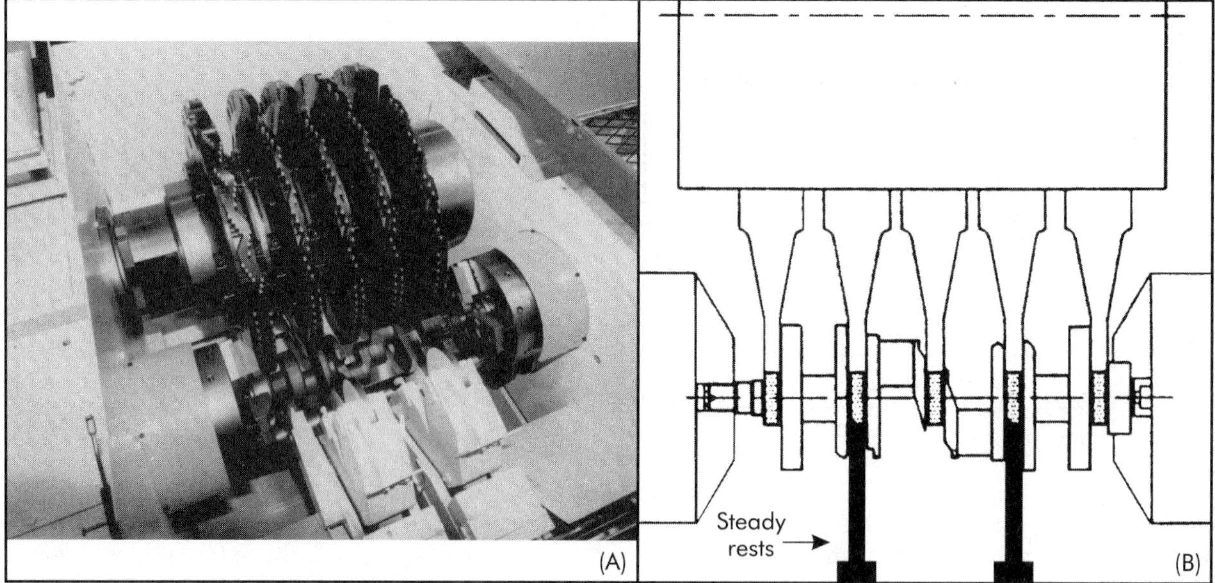

Figure 22-6. (A) Rotary/turn broaching operation on a V8 crankshaft. (B) Positions of five rotary broaches on main bearing and oil seal diameters. (Courtesy Ingersoll CM Systems, Inc.)

through a hole. Arbor presses are also readily usable for other operations such as assembling, bending, and staking.

Another form of push broaching, called *pot broaching*, is an economical method for producing precision external tooth forms. Pot broaching uses a tool with internal teeth held together in a pot. The pot is passed over a round part to produce external shapes in such forms as involute gear teeth, splines, and slots. In most cases, all of the gear teeth or other shapes are formed in one pass through the hollow pot broach. Most machines of this type push the workpiece upward as indicated in Figure 22-7 because such movement allows the chips to escape easily and the arrangement is amenable to automatic loading and unloading. However, if the workpiece is small or the stroke must be long, the push rod would be slender and subject to buckling. For such cases, pull-up pot broaching machines are available.

Pull Broaching Machines

Pull-type broaching machines are available for either internal or surface broaching. Generally, on a pull broaching operation, the shank of the broach is passed through the prepared opening in the workpiece and attached to a puller that draws it completely through the opening. Commonly, vertical broaching machines are arranged to pull broaches either downward or upward, although some do both. On a *vertical pull-down broaching machine*, the broach is fed through the workpiece from above, grasped by a puller from below the table, and then pulled down through the workpiece. This permits large and irregularly shaped workpieces to be located and held in place more easily than on a pull-up machine. Quite often, fixtures or nesting devices are used to hold the locations of broached holes in relationship with external surfaces of parts.

The broaches on *vertical pull-up broaching machines* start from below the machine and, as they are pulled upward through a part, press that part against the underside of the table. At the end of the stroke, the part is freed to fall by gravity down a chute and into a collection bin.

Figure 22-8 shows a vertical pull-down type of internal broaching machine being used to broach the bore of a truck axle frame. Broaching machines of this type are usually special-purpose machines designed for high production.

A variety of pull-type, horizontal broaching machines are also available for either internal or surface broaching. The horizontal machine shown in Figure 22-9 is used primarily for the internal broaching of keyways, involute splines, squares, rectangles, hexagons, and a variety of irregular profiles. The operation of this machine is fairly simple. A workpiece is placed in a nest or fixture on the face plate. The broach is then fed through the prepared hole in the workpiece and grasped by a puller. The hydraulically operated puller unit then pulls the broach through the hole to perform the necessary cutting operation. Again, most broaching operations are completed in one pass.

The type of machine shown in Figure 22-9 is available in sizes ranging from 56–356 kN (6.3–40 tons) and with stroke lengths from 1,245–2,489 mm (49–98 in.). Cutting speeds are variable up to 4.9 m/min (16 ft/min) to meet the required cutting conditions for different materials and material removal rates. High broach return speeds up to about 15.8 m/min (52 ft/min) ensure maximum productivity. Basic machines of this type cost between $120,000 and $250,000.

Surface Broaching Machines

Surface broaching machines are available as horizontal or vertical machines. Generally, a horizontal machine is used to broach large, heavy workpieces involving considerable metal removal. For example, automotive cylinder heads are often broached on a horizontal machine by moving the heads along a stationary broach of considerable length. In other cases, the workpiece is stationary and either single- or double-acting broach heads move across the face of the workpiece. For the double-acting type, two sets of broaches mounted on a ram permit cuts to be taken as the ram moves in either direction.

The vertical, electrical-drive broaching machine shown in Figure 22-10 is designed for high production. Parts to be broached are clamped in a fixture at the front of the machine and the surface broach is moved vertically during the cut. These machines are rated at 142 kN (16 tons) and operate with variable cutting speeds up to 18 m/min (60 ft/min). High levels of accuracy and repeatability can be achieved on this type of machine for broaching parts such as turbine blades and automotive components, including connecting rods and caps. Contour broaching can be performed when the machine is equipped with a programmable controller and a servo-driven table. This is done by control of the servo table on which the part is mounted. Radius profiles are generated on the part as it is moved along the broach. This approach eliminates the changeover time and cost associated with traditional cam-controlled systems.

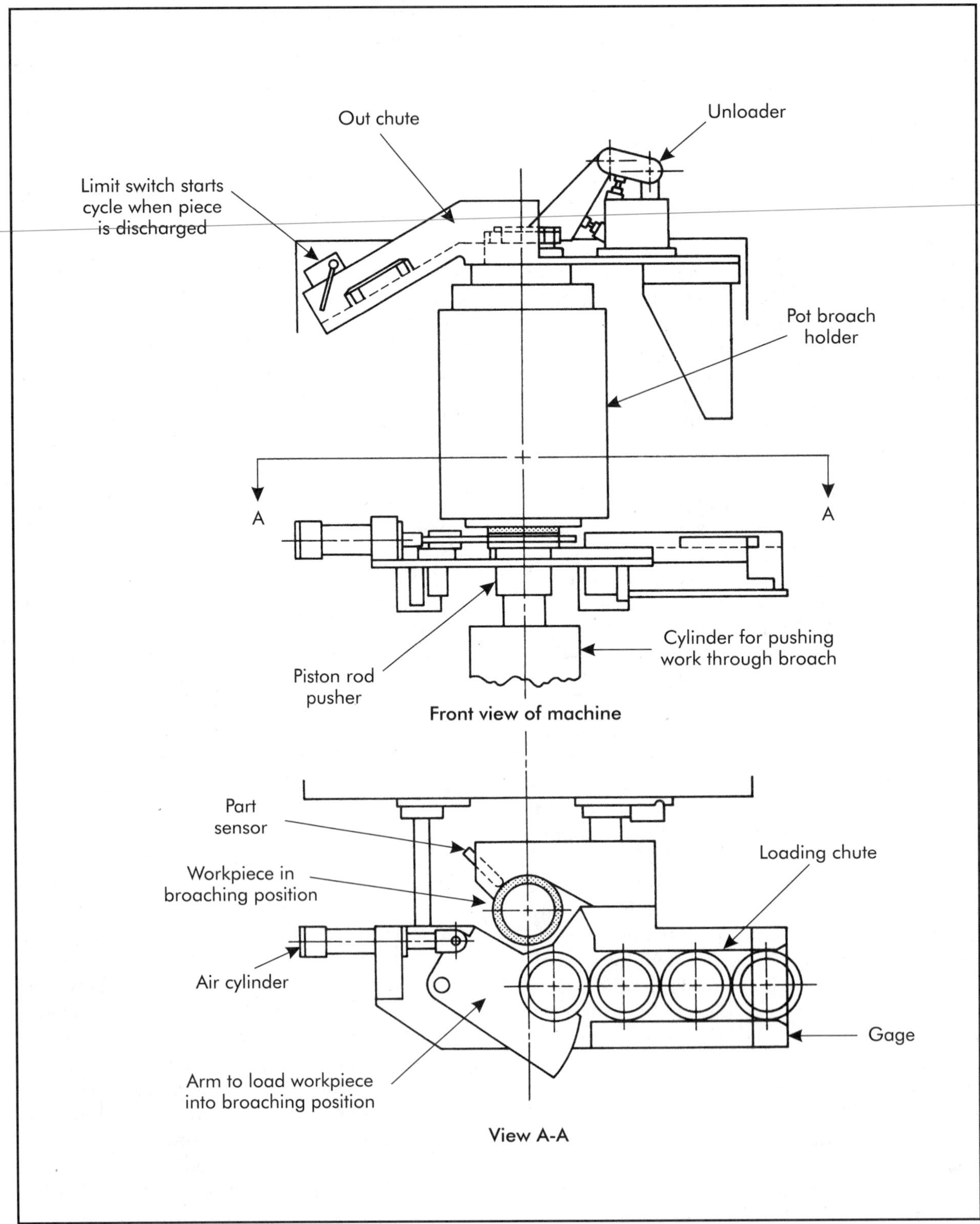

Figure 22-7. Diagram of the operating zone of a push-up pot broaching machine with a typical automatic loading and unloading system.

Figure 22-8. Vertical pull-down internal broaching machine. (Courtesy Marbaix Lapointe)

Rotary/Turn Broaching Machines

As discussed in the previous sections, most traditional broaching is done by establishing a linear relative motion between the broach and the workpiece. This often requires lengthy broaching tools and machines capable of accomplishing the rough and finish machining of some workpieces in one pass. And, the machining of external cylindrical shapes is not easily accomplished by the linear broaching process. The proprietary rotary/turn broaching system was developed in 1984 for use in turning/broaching crankshaft concentric and eccentric diameters and faces for the automotive industry. For the turn/broaching process illustrated in Figure 22-6, a series of five spiral broaches are mounted on an arbor parallel to the crankshaft axis. Each individual broach is made up of successive carbide inserts positioned progressively outward from the center of the arbor axis. The carbide inserts in this example are coated with multi-layers of aluminum oxide. Through the use of retractable quills, the crankshaft is center-mounted and chuck-driven at both ends in a rotational direction opposite to that of the cutter arbor.

Figure 22-9. Horizontal hydraulic broaching machine. (Courtesy Marbaix Lapointe)

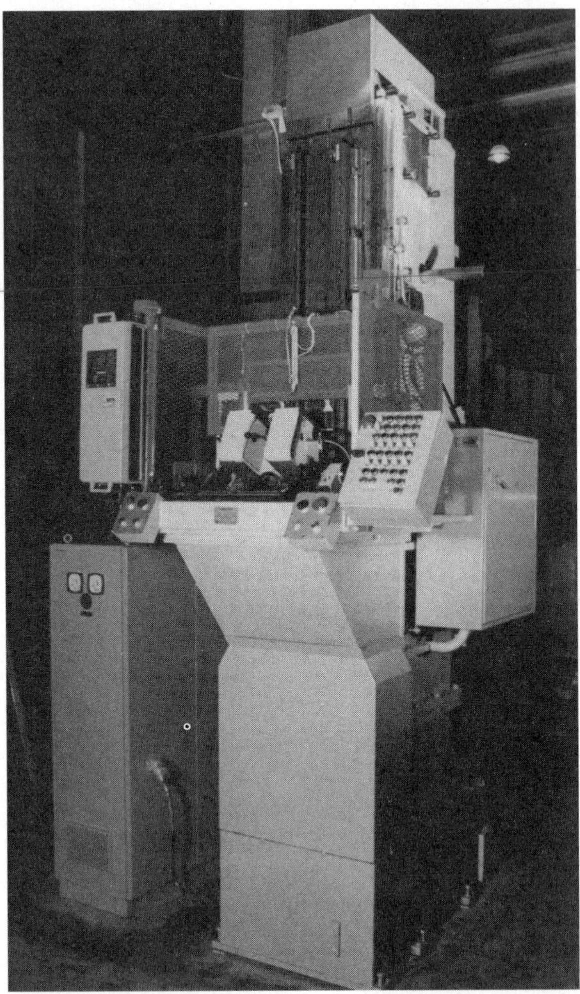

Figure 22-10. Vertical electrical-drive surface broaching machine. (Courtesy Marbaix Lapointe)

Machining of the crankshaft in Figure 22-6(A) is accomplished by in-feeding the cutter to touch the crankshaft and rotating it through one revolution to achieve the required facing and diametral machining operations. Variable arbor cutter speeds between 1 and 10 rpm and variable workpiece spindle speeds between 10 and 1,200 rpm are available to maintain a constant cutting speed for the different diameters involved. An average cutting speed for a crankshaft turn/broaching operation such this one is 137–168 m/min (450–550 ft/min).

Figure 22-6(B) shows the machining positions of the five broaching tools along the axis of a V8 crankshaft. The sequence of operations performed by one rotation of these cutters is as follows:

1. Finish machine counterweight radii, cheek faces and collar faces, and undercut fillet.
2. Semifinish broach main bearing and oil seal diameters.

Depending on customer specifications, crankshafts may require secondary operations such as the grinding and polishing of certain dimensions after broaching. However, many of the other surfaces are finish machined in the turn/broaching operation. A production rate of approximately 38 crankshafts/hr gross can be achieved by a turn/broaching operation of this type when using automatic loading and unloading equipment.

The in-feed rotary/turn broaching machine shown in Figure 22-11 utilizes a single-axis CNC control with a closed-loop AC spindle drive. It is designed to accommodate crankshafts up to 1.3 m (50 in.) in length and cut at a 0.5 m (18 in.) center distance (distance between the workpiece and the tool mounted arbor). Crankshaft locating and chucking in the machine shown is automatic. The machine can be loaded and unloaded either manually or automatically. However, the machine pictured in Figure 22-11 is not equipped with an automatic loading device. An in-feed rotary/turn broaching machine similar to the one shown costs about $900,000. A single rotary broaching cutter similar to the one shown in Figure 22-6 would cost about $125,000.

Continuous Broaching Machines

A *horizontal continuous broaching machine* is different from the machines previously de-

Figure 22-11. In-feed rotary/turn broaching machine. (Courtesy Ingersoll CM Systems, Inc.)

scribed because the broaching cutters on it remain stationary while the workpieces held in fixtures (on carriers) are forced past the cutters. This is illustrated in Figure 22-12. Parts are loaded manually or automatically at one end and may be discharged down a chute at the other end. Work can be broached as rapidly as it can be loaded.

One type of *rotary continuous broaching machine* is somewhat like a rotary miller. It has a slowly revolving round table carrying a group of fixtures in a circle. Workpieces loaded in the fixtures are carried past stationary broaches and may be ejected automatically thereafter.

22.1.4 Operations

Planning

The skill in broaching lies in how the tools are configured in setting up the equipment. Beyond that, the operation of the machine is mere routine.

The cutting time in a broaching operation is short, and the loading and unloading time takes an appreciable part of the total time. Thus labor savings and motion economy are important considerations. For production of large quantities of pieces, the machines often have several workstations or pull a number of broaches at once, broach handling is done automatically, and fixtures are made to be quick-acting to save operation time.

Surfaces that run in the same direction on a workpiece, especially if they are contiguous, can generally be broached at the same time. As many surfaces on a part as possible should be broached in one operation.

Although most broaching operations are completed in one pass, some are arranged for repeated cuts to simplify tooling and save cost where only small or moderate quantities are needed. For instance, the teeth of a gear or spline may be broached all together or indexed and cut one or a few at a time.

Free-cutting material of 25–30 R_C hardness is considered best for broaching, but an acceptable range is from 10–35 R_C, and metal as hard as 45 R_C has been broached. Material that is too hard wears broach teeth rapidly, and too soft a material is difficult to cut cleanly.

A well-designed broach should last for 10–30 or so sharpenings. A small simple broach may require less than an hour to resharpen, a large one several hours. The amount of service between grinds depends upon the length of cut and the number of times the broach is subjected to the shock of entering the cut. Probable average numbers of pieces per grind at conventional speeds have been reported by Charles O. Lofgren as 25,000 for brass and aluminum, 15,000 for bronze and SAE 1020 steel, 6,000 for alloy steel, and 800 for high-temperature alloys. Thus broaches are expected to have long lives and produce many pieces.

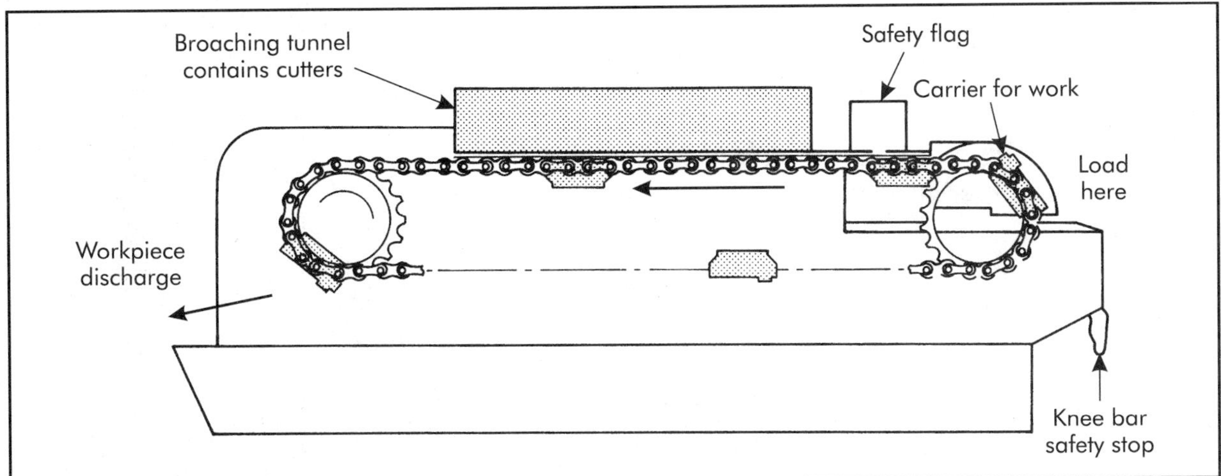

Figure 22-12. Horizontal continuous broaching machine.

Economic tool life practices dictate that broaches should have long lives because they are expensive to make and sharpen. A long life means a slow speed. For years, broaching has been done with high-speed steel tools mostly at 3–12 m/min (10–40 ft/min), but in recent years cases of up to 30 m/min (100 ft/min) and more have been reported. Carbide broaching speeds range from 30–91 m/min (100–300 ft/min). Broaching speeds are still well below conventional speeds for turning and milling.

Tolerances within 25 μm (.001 in.) can be held between surfaces and from locating surfaces with rigid workpieces in surface broaching. Less than half as much tolerance can be held to locating surfaces if the tooling is adjustable. Hole-position tolerances are quite difficult to hold in internal broaching, but 25 μm/mm (.001 in./in.) of hole diameter is feasible for size if sections and deflections are uniform along the workpiece. Smaller tolerances have been reached by trial and error grinding of broaches during tryout.

Estimating Time

The cutting time in a broaching operation is the quotient of the length of stroke divided by the cutting speed in compatible units. For example, assume a stroke of 1.2 m (48 in.) at 7 m/min (≈24 ft/min) is required. The cutting time is $1.2/7 = 0.17$ min = 10 sec (or $48/(24 \times 12) = .17$ min = 10 sec). The return stroke may be calculated in the same way from the return speed of the broach, which may be 12 m/min (40 ft/min) or more. Other amounts of time are for stopping and starting, commonly 2 sec, and unloading and loading, which depend upon the nature of workpiece and fixture.

22.1.5 Broaching Compared with Other Operations

The work done by broaching also can be done in other ways. Under certain conditions, broaching is preferable to other operations, but under some conditions broaching is either unfeasible or disadvantageous.

The main advantage of broaching is that it is fast; it commonly takes seconds to do a job that requires minutes in any other way. Little skill is needed to run a broaching machine, and automation is easily arranged. Good finish and accuracy are obtainable over the life of a broach because roughing and finishing are done by separate teeth.

The main handicap of broaching is that broaches are costly to make and sharpen. Standard broaches are available, but most broaches are made especially for and can do only one job. A standard HSS broach for a 6 mm (≈.25 in.) keyway costs about $50 and for a 25-mm (≈1-in.) internal hexagon around $400. Specials vary in price from a few hundred dollars for a fractional-inch internal spline broach to many thousands of dollars for a surface broach 1.8 m (6 ft) long or more.

Special precautions may be necessary in founding or forging to control variations in stock or extra operations to remove excess stock may have to be added before broaching to protect the broach. These add to the overall cost of manufacturing.

Some of the limitations of broaching are enough to make it impracticable for certain work. A surface cannot be broached if it has an obstruction across the path of broach travel. For instance, blind holes and pockets normally are not broached. Frail workpieces are not good subjects for broaching because they are not able to withstand the large forces imposed by the process without distortion or breaking. For example, automobile engines have been drastically redesigned to reduce weight and save fuel. The engine blocks had previously been broached, but it has been found that the thinner walls of some cannot stand the broaching forces, so more costly milling is necessary. As a rule, surfaces that run in the same general direction often can be broached at the same time, but surfaces not so related must be broached separately. For instance, a hole and a perpendicular face may be machined in one operation on a lathe or boring machine but require two passes in broaching. The lines left on a surface lie in the direction of broach travel. For instance, broaching is not capable of producing a circular pattern in a hole if such a finish is required.

When broaching can be done, it is selected because the amount it saves as compared to other operations is more than enough to pay for the cost of the tools. This may not require a very large output if the job is hard to do otherwise. The cutting of an internal spline is one case, for example, a 6 diametral-pitch spline, 51–102 mm (2–4 in.) long, in AISI 8620 steel. One piece could be cut with a single-point formed tool (costing several dollars) on a standard shaper with a dividing head, taking several hours. The job could be done with a gear shaper cutter, worth about \$200, in 4–5 min. That cutter could produce 100 pieces/grind and stand sharpening about 50 times. It could be resharpened in 15 min. The cost for a broach for the splined hole is \$2,300. It cuts a piece in 20 sec. On a vertical pull-down broaching machine it lasts for about 1,000 pieces until dull and can be resharpened 26 times. Each sharpening takes 1.75 hr. At \$25/hr in the shop, the broach is economical for any quantity of more than about 1,500 pieces. However, savings per piece are not large for some operations, and many pieces must be produced to justify broaching. To broach a surface on a cast-iron cylinder block may be only a .5 min faster than milling with cemented carbide cutters. The tooling for broaching may well cost \$30,000 more than for milling. If the labor and overhead rate is \$25/hr, the tooling is depreciated in 2 years, the annual charge for interest, insurance, property taxes, and maintenance is 25%, and setup time is ignored, an analysis based on Equation 19-7) shows $25N/(2 \times 60) = 30{,}000(0.75)$, and $N = 108{,}000$ pieces/year. This is the smallest quantity of production that justifies broaching in such a case.

22.2 SAWING

Metal is removed in sawing by the action of many small teeth. Saw teeth act in a narrow line, and a saw can sever a sizable chunk of material with a minimum of cutting. If a piece of metal is removed by milling, for instance, a large part or all of it may have to be reduced to chips. The same piece may be cut off or out by a saw acting on only a small part of the material. Thus, in many cases work can be done faster and with less power by sawing than by other methods of metal cutting, and material can be saved.

Saws have been used as hand tools since ancient times because they require little force and power to operate. Power-driven hacksaws, circular saws, and band saws are employed to cut off pieces of barstock, plates, sheets, and other shapes of metal. Versatile band saw machines have been developed for cutting out cavities or pieces of intricate shapes with a minimum waste of material and time.

22.2.1 Saw Characteristics

Three common kinds of metal-cutting saws are hacksaws, circular cold saws, and band saws. Although different in overall form, they all contain a series of cutting teeth that operate in the same basic way. The important features of a saw are its: (1) material, (2) tooth form, (3) tooth set, (4) tooth spacing, and (5) size.

Saws are made from low-alloy high-carbon steel, special alloy steels, high-speed steel, and cemented carbide. Most saws are solid; some have teeth of a hard material backed by flexible steel; and many have tipped or inserted teeth.

Saw teeth have rake and clearance angles corresponding to single-point tools as shown by the typical profiles in Figure 22-13. The teeth of most band saws for metal cutting are like those of other saws, but other forms are available for specific materials and operations. Among these are scallop-edge band saw blades for cloth, special saw blades for paper, and diamond-tooth blades for ceramic and vitreous materials.

Saw teeth are offset to the side to make the cut wider than the thickness of the back of the blade, which prevents rubbing. The width of the saw cut is called the *kerf*. Three common types of saw settings are shown in Figure 22-14. The wave tooth is suited to fine tooth spacing. Circular saws, and sometimes other kinds, have varying tooth heights and shapes such as the ones in Figure 22-13. These forms help break up the chips, distribute the load, reduce chatter, and allow some teeth to take fine cuts to leave good finishes.

Tooth spacing or pitch has an important influence upon saw performance, as explained in Figure 22-15. At least two or three teeth must bear at one time to avoid snagging of the teeth. On the other hand, enough space is needed between the teeth to take care of the chips. Straight saws are available with as many as 13 teeth/cm to less than 1 tooth/cm (from about 32 teeth/in. to 2 teeth/in.). The pitch of circular saws may run from 5–51 mm (.20–2 in.).

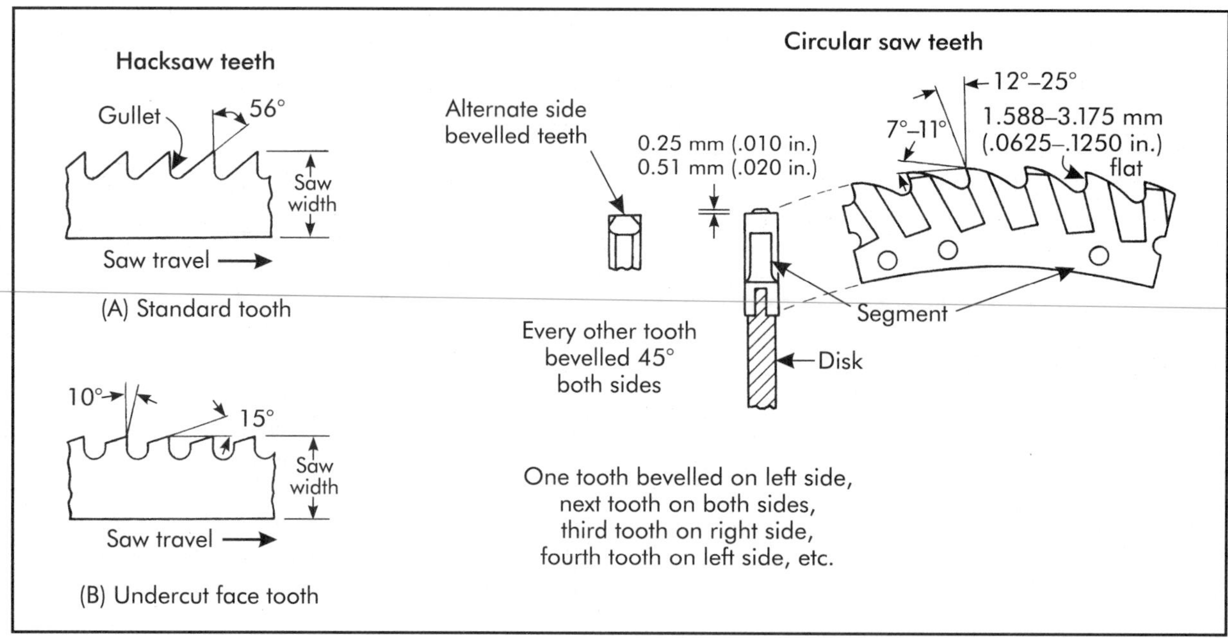

Figure 22-13. Typical saw-tooth profiles.

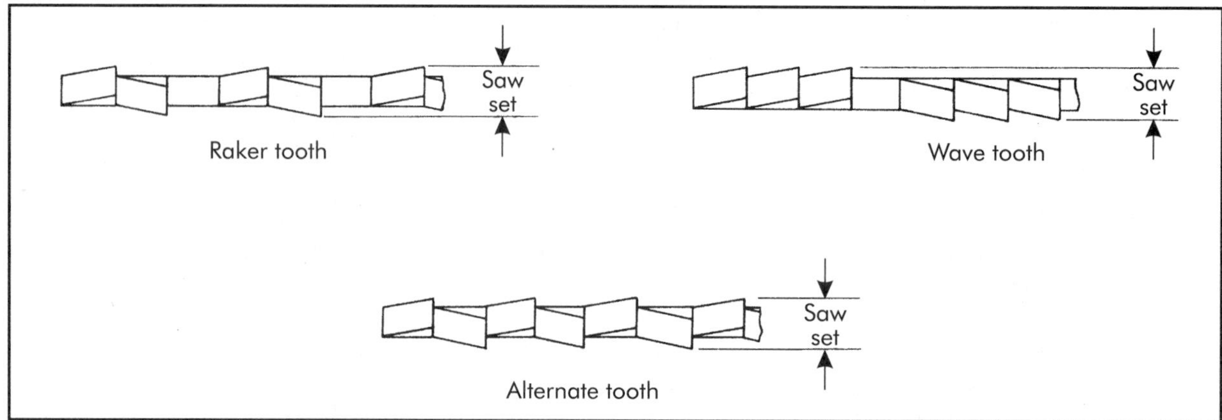

Figure 22-14. Types of saw-tooth sets.

Saws are commercially available in many sizes. Dimensions are given in suppliers' catalogs. A thin saw removes less material, but a thick blade makes a truer cut. Commonly, a circular saw cuts wider than a hacksaw, and a hacksaw cuts wider than a band saw.

Some band saws are as small as 3.18 mm (.125 in.) wide to cut around curves. Some band saws and hacksaws are as large as 25–102 mm (1–4 in.) wide for backup strength for heavy cuts. Circular saws are from about 203 mm–2 m (8–80 in.) in diameter to accommodate various depths of cuts.

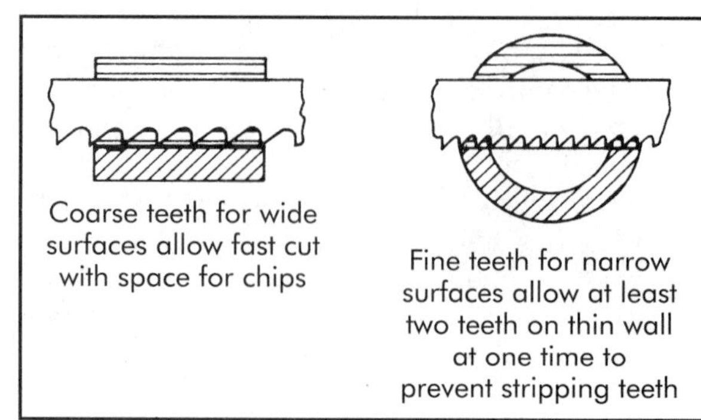

Figure 22-15. Significance of saw-tooth spacing.

22.2.2 Machines

Power Hacksawing Machines

A *power-driven hacksawing machine* drives a blade back and forth through a workpiece, pressing down on the forward cutting stroke and releasing the pressure on the return, much the same as a person does in sawing by hand. As illustrated in Figure 22-16, the saw blade is attached under tension to a C-frame guided by an overarm. In addition to tension, a saw may be fully backed up for heavy cuts. The saw is fed downward a positive preset amount on some machines; a preset uniform pressure is applied on other machines. The amount of down feed depends on the resistance the saw meets.

The machine shown in Figure 22-16 can accommodate round workpieces up to 210 mm (8.26 in.) in diameter and flat stock up to 240 mm (9.44 in.) in width. The saw is driven by a two-speed motor that can be set at either 62 or 125 strokes per min. At these stroke speeds, medium cutting speeds of 15.8 m/min (52 ft/min) or 31.7 m/min (104 ft/min) are achieved. Cutting feed or down feed for the saw blade is hydraulically controlled. A power hacksaw of the size shown in Figure 22-16 costs about $3,000.

Hacksawing operations are simple but can be adapted to production. Tolerances as small as ±0.13 mm (±.005 in.) are considered practical. Work is usually held in a vise, and the machine stops at the end of a cut. Production-type machines are arranged to feed, measure, and cut off a series of pieces automatically from one or more bars.

Because the cutting is intermittent, hacksawing is not fast, but the machines are simple and not costly, easy to change from job to job, and easy to operate and maintain. On an average job, a continuous circular or band saw may be one-third or more faster than a hacksaw, but the overall cost of hacksawing may be as low as one-half or less.

Circular Saw Machines

There are three kinds of *circular saws*. One has teeth as illustrated in Figure 22-13 for *cold sawing;* another has a smooth or nicked outer edge for *hot sawing,* and the third is a thin abrasive wheel for *abrasive sawing*.

A cold saw is commonly fed horizontally through a workpiece as depicted in Figure 22-17, particularly for heavy work. Other machines may feed the saw vertically or around a pivot with a chop stroke, shown in Figure 22-18, particularly for light work. Workpieces are usually clamped in a vise, one or more at a time. The stock is pushed against a stop and clamped, and the saw is fed by hand on manual model machines. Other models are semi- and fully automatic in operation. Some also chamfer or center the pieces that are cut off.

Cold sawing is a continuous and fast method of cutting off and leaves a smooth and accurate milled surface with few or no burrs, which may

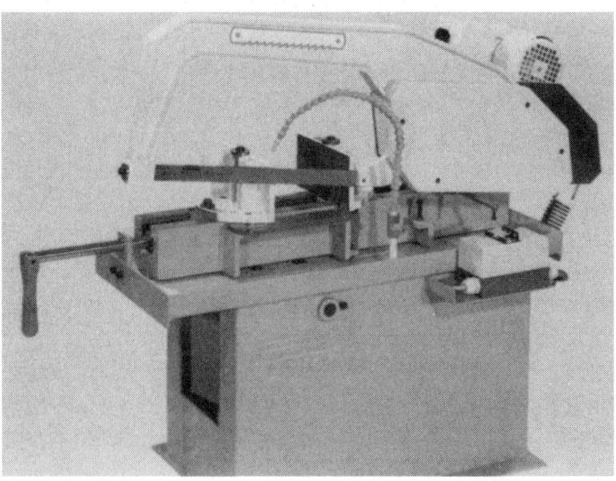

Figure 22-16. Power reciprocating saw with hydraulic down feed. (Courtesy Kasto-Racine, Inc.)

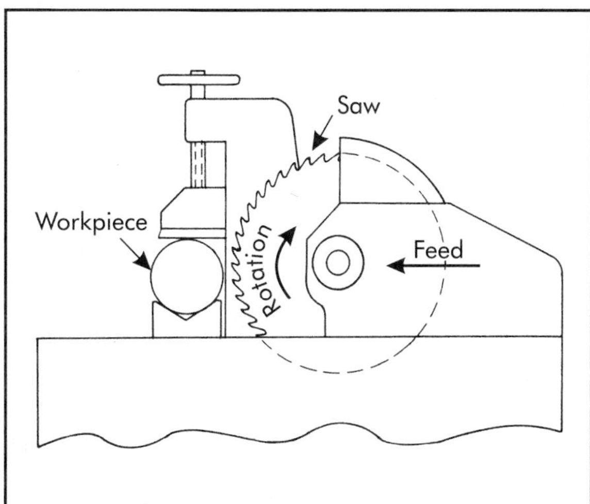

Figure 22-17. Cold circular sawing operation.

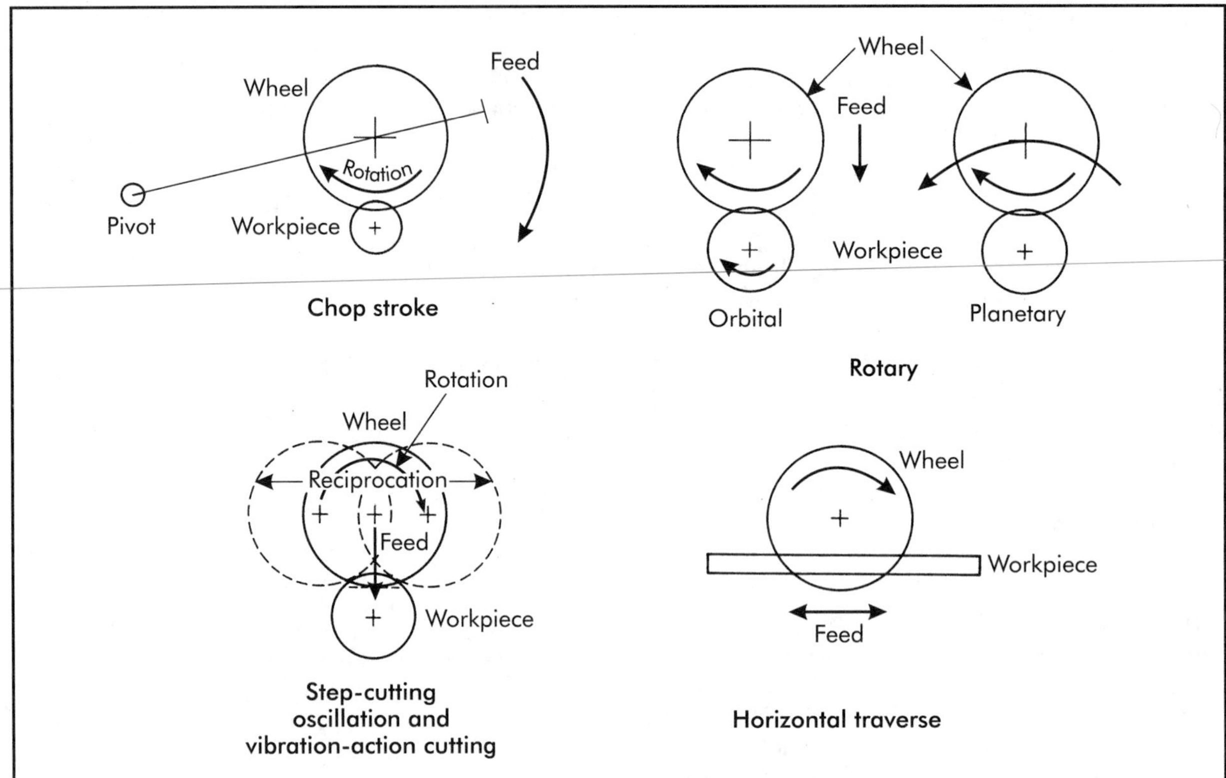

Figure 22-18. Basic types of abrasive cutoff machines.

save work in subsequent operations. A 152-mm (6-in.) diameter steel bar can be easily cut off in a minute. The typical experience of one plant is that pieces of barstock can be cut off with a tolerance of 76 μm (.003 in.) and a finish equal to or better than that obtained from a cutoff tool on a lathe.

Production cold-sawing equipment is expensive. An automatic machine with a 254-mm (10-in.) diameter capacity in steel and a 7.5 kW (10 hp) motor costs about $50,000. A segmental 711-mm (28-in.) diameter blade for the machine has a price of about $500. It may be sharpened about 35 times at about $50 for each sharpening. Light machines for small work, 51 mm (2 in.) or so in diameter, may cost a few thousand dollars, and saws less than $50.

Hot sawing is done at 3,048–7,620 m/min (10,000–25,000 ft/min). The heat of friction softens the metal in contact with the disk and the soft metal is rubbed away. Sharp saw teeth are not needed for cutting. Only a small portion of the saw is in contact at any instant, and the rest is cooled as it travels around to enter the cut again. Friction sawing is fast but leaves a heavier burr and a less accurate surface than tooth cutting does. Typical performance is 35 sec for a 254-mm (10-in.) I-beam, and 18 sec for a bar 305 mm (12 in.) in diameter.

In one form of hot sawing, a current (1,000–4,000 A at 30–55 V) is passed to maintain an arc between the disk and the workpiece. Molten metal is swept away by the high peripheral speed of the saw. The cutting rate is reported faster than by any other method. Such equipment with a 1.8-m (72-in.) diameter blade to cut 762-mm (30-in.) steel billets costs over $100,000.

Abrasive sawing is done at speeds up to 5,486 m/min (18,000 ft/min) with wheels 0.3 m (1 ft) or less in diameter for small work and up to over 1.8 (6 ft) in diameter for large billets. Common forms of abrasive sawing operations are depicted in Figure 22-18. A chop stroke is quick and popular for small work. The rotary methods are advantageous for large diameters, particularly for tubing where feed through only the wall is necessary. For step cutting and oscillation- and vibration-action cutting, the wheel is reciprocated to minimize contact area, which allows a higher pressure and feed rate. The difference between the actions is that for step cutting the wheel is reciprocated over the width of the work, for oscillation action only over about a third of the width, and for vibration action, reciprocation is quite short but rapid. A horizontal traverse action is suitable for long cuts on sheets, slabs, etc.

Abrasive sawing is advantageous for hard materials, even in thick sections. On the other extreme, cutting pressure can be kept small, and thin sections in hard or soft material can be cut easily without distortion. A thin abrasive disk removes a minimum of stock and saves expensive material, as illustrated by the results of a study presented in Table 22-1.

Band Saw Machines

A continuous saw blade or band runs over the rims of two shrouded wheels on a *band saw machine*. *Horizontal band saw machines*, like the one in Figure 22-19, are commonly used to do cutoff work. The saw is carried on a frame and is fed downward through the workpiece in a vise on the bed. The machine shown in Figure 22-19 can be fully automated to load and clamp stock in the bed vise, activate the saw for cutting, retract the saw, and unload or transfer the stock that has been cut off. It will accommodate round stock up to 660 mm (26 in.) in diameter or square stock up to a similar height and width. Cutting speeds are variable from 16–105 m/min (52–344 ft/min). The base price of a fully automated machine similar to that shown in Figure 22-19 is about $115,000.

Vertical band saw machines are also used as cutoff machines, but in addition are widely used for cutting outside and inside shapes of all kinds for tool, fixture, die and gage making, and the production of small intricate parts. For inside cutting, the band is cut, passed through a drilled hole, and butt-welded back together. The contour band-sawing machine in Figure 22-20 is a general-purpose machine that can be used to saw, file, or polish aluminum, brass, copper, mild steel, tough tool steel, stainless steel, and sheet metal. In addition, when equipped with an appropriate type of saw, it will cut nonmetals, like plastics, wood, paper, and fibrous materials. The machine shown has a 660 × 660 mm (26 × 26 in.) table that can be tilted 45° right and 10° left. It has a 330-mm (13-in.) maximum workpiece height to handle a wide variety of material sizes. A variable speed drive and two-speed transmission permits the operator to select band speeds within two ranges, 17–91 m/min (55–300 ft/min) or 293–1,585 m/min (960–5,200 ft/min). The machine shown in Figure 22-20 may be equipped with an air-operated power feed assembly that pulls the workpiece into the saw band and helps increase operator productivity.

Contour band sawing machines are adaptable to other operations. A continuous band of file segments or stones may be put in place of the band saw. Friction sawing is done on high-speed machines with bands having dull teeth. Hard crystalline materials, particularly thin sections and single crystals, are cut with wire charged with diamonds. Some machines are made especially for wire cutting.

Band sawing is steady and continuous and may be faster than hacksawing, and as fast as cold sawing within its limitations, especially through thin sections and weak workpieces. Thin band saws waste only a modicum of stock. An inherent limiting factor in band sawing is the relatively thin blade, which may diverge from a true path in a heavy cut. One manufacturer guar-

Table 22-1. Relative costs for cutting 381-mm (15-in.) diameter hard refractory alloy

	Method			
	Powder burning	Hacksaw	Cold saw	Abrasive saw
Operating and tool costs ($)	17.12	29.61	70.38	16.38
Cost of lost material ($)	66.20	11.55	62.50	16.18
Condition of surface	Uneven and oxidized	Clean	Clean and square	Straight and smoothest

Note: Estimated relative cost based on actual performance.

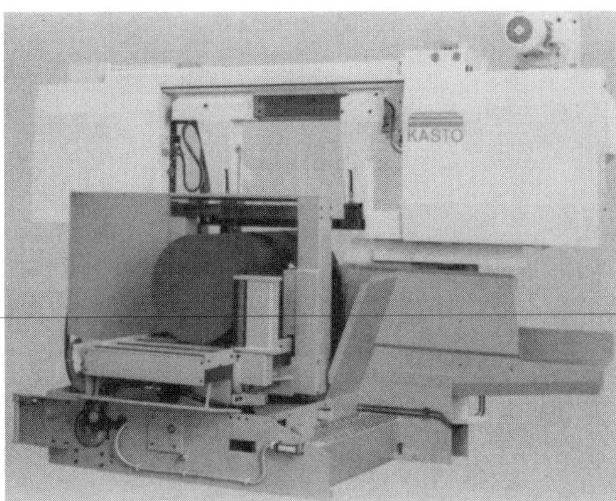

Figure 22-19. Horizontal band sawing machine. (Courtesy Kasto-Racine, Inc.)

antees a tolerance of 0.002 mm/mm (in./in.) of cutoff distance width if proper limits are observed. Contour sawing is normally done to a tolerance of ±0.25 mm (±.010 in.), but if carefully and skillfully done, ±38 μm (±.0015 in.) can be held.

22.2.3 Operations

Sawing is governed by the same considerations as other cutting operations. Saw speeds range from 15 m/min (50 ft/min) for hard and tough alloys to 914 m/min (3,000 ft/min) and more for soft materials. The speed of a hacksaw is specified in strokes/min; 60–150 represents the usual range from hard to soft materials. The rate a saw penetrates material depends upon the feeding force. Large forces tend to tear out teeth and deflect the blade from a true path. The feed rate for hacksaws and cold saws is limited by what the teeth can stand. With band saws, usually unacceptable accuracy is reached before the teeth are hurt. As an example, in sawing mild steel bars 75–125 mm (3–5 in.) in diameter, typical rates of cutting and tool cost are for a hacksaw 65 cm^2/min and \$0.0003/cm^2 (10 $in.^2$/min and \$0.002/$in.^2$), a band saw 90 cm^2/min and \$0.0005/cm^2 (14 $in.^2$/min and \$0.003 $in.^2$), and a cold saw 161 cm^2/min and \$0.0009/cm^2 (25 $in.^2$/min and \$0.006/$in.^2$). As has been shown, machine costs

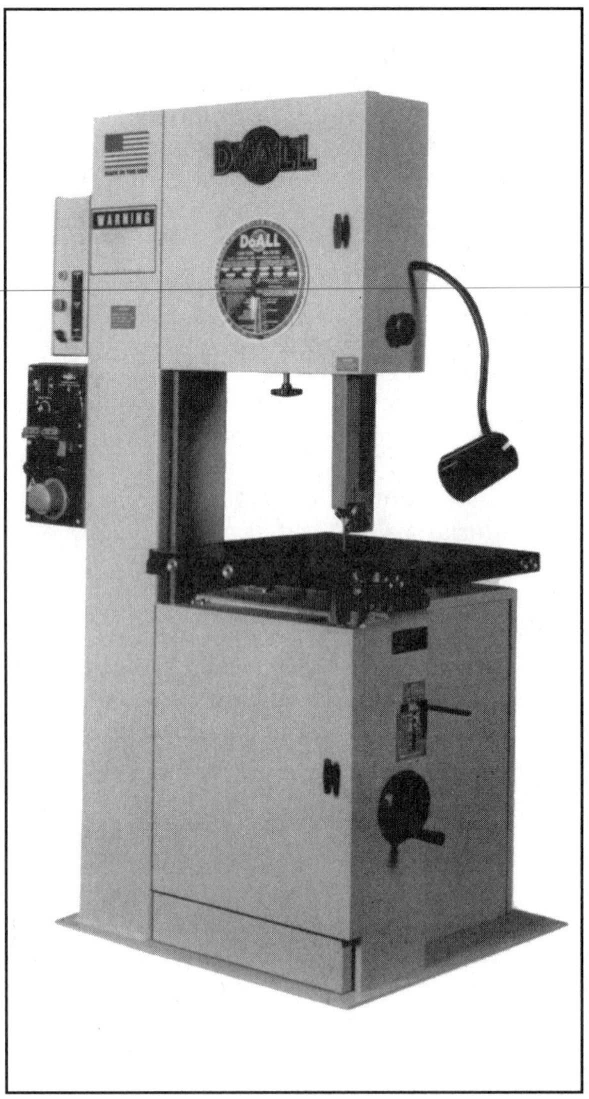

Figure 22-20. General-purpose contour band-sawing machine. (Courtesy ITS/DoAll Industrial Company)

increase in about the same order. Thus, if enough sawing can be done, a high cutting rate and low unit cost can be achieved, but at extra cost for machine and tools. Hacksawing, horizontal band sawing, and cold sawing are best for ordinary cutoff work, and other methods are advantageous for particular applications, such as extra hard materials or unusual shapes and sizes. The method that should be selected for a job depends upon the amount of sawing and the kind of work to be done.

22.3 QUESTIONS

1. Differentiate between broaching and sawing.
2. Describe a hole broach and name its principal elements.
3. What does a burnishing broach do?
4. Show by a sketch the shape of the teeth on a broach and name the details of that shape.
5. Explain the action of a pull-type hole broach and indicate how it rough cuts and finish cuts a hole to size in one pass.
6. Why are most broaches pulled rather than pushed?
7. How are broaching machines classified?
8. How is the size of a broaching machine designated?
9. Describe the action of a pot broaching operation.
10. What type of broaching operation is commonly done by pot broaching?
11. What advantage does a vertical pull-down broaching machine have over a pull-up machine?
12. What is the difference between a rotary turn-type broaching tool and a milling cutter with formed teeth?
13. How does a continuous broaching machine operate?
14. By what means is economy achieved in broaching?
15. When is broaching economical and when is it not feasible?
16. What is "tooth set" on a saw blade and why is it needed?
17. Why is tooth spacing on a saw blade important?
18. Which machine is generally faster in performing a sawing operation, a hacksaw or a band saw?
19. What is the difference between cold circular sawing and hot circular sawing?
20. When can abrasive sawing be used advantageously?
21. What is the most common use for horizontal band saws?
22. How is the sawing of internal sections accomplished with a vertical contour band saw?

22.4 PROBLEMS

1. The roughing section of a broach is to remove stock 2.78 mm (≈.109 in.) deep from a steel surface 216 mm (8.5 in.) long and 45 mm (1.75 in.) wide. Each tooth takes a cut 0.15 mm (.006 in.) deep. The pitch of the teeth is given by the relationship

$$\text{pitch (mm)} = 1.75\sqrt{\text{length of surface (mm)}}$$

$$\text{pitch (in.)} = 0.35\sqrt{\text{length of surface (in.)}}$$

(a) How long should the roughing section be?
(b) What pull must the broaching machine exert if the chip pressure is 4 GPa (≈600,000 psi) of chip area?
(c) What power is required for a broaching speed of 7.6 m/min (25 ft/min)?

2. A 1.2 m (48 in.) stroke is required for the operation described in Problem 1. The time to unload and load a fixture is .12 min and to index the machine is .04 min. What is the cycle time per piece on

   (a) a single-ram surface broach with a return speed of 13.7 m/min (45 ft/min)?
   (b) a double-ram surface broach?

3. A 24/48-pitch internal involute spline with 36 teeth and a pitch diameter of 38 mm (1.5 in.) is to be cut in steel in a hole 50 mm (≈2 in.) long. This can be done on a gear shaper in 5 min/piece. Tooling costs $650. The spline can be broached with an operation time of .5 min, and the cost of tooling is $1,450, including the broach. Setup time is negligible. Time saved is considered worth $18/hr. The rate for interest, insurance, taxes, and maintenance of tooling is 25%. For what quantity of pieces should each process be used?

4. With an assumed 80% efficiency, how much power is required at the motor when the broaching machines with ratings specified are operated at the given speeds?

   (a) an 8 × 16 broaching press at 30 ft/min;
   (b) a 10 × 54 vertical broaching machine at 40 ft/min;
   (c) a 50 × 72 horizontal broaching machine at 80 ft/min;
   (d) a 25 kN pull at 10 m/min; and
   (e) a 225 kN pull at 15 m/min.

5. A cast-iron housing has a surface 115 mm (≈4.5 in.) in width × 150 mm (≈6 in.) long from which 3 mm (.125 in.) of stock is to be removed. On a milling machine the feed is 1.2 m/min (48 in./min) and the time per piece is .30 min complete. The surface can be broached in 0.25 min/piece on a double-ram surface broaching machine. The cost for the machine, overhead, and labor is $20/hr on either machine. Setup time is the same on both. Tooling for milling costs $850, and for broaching $6,300. Tooling cost must be recovered in one run without interest. What annual rate of production justifies the use of the broaching machine?

6. In the design of a broach, common practice is to multiply the maximum cross-sectional area of chips taken at one time by a factor, C, to determine the pull required. The factor C is in GPa (psi) of chip cross-sectional area and depends mostly on the material. A broach is to be designed for a material for which C has not been determined and a value of 4.1 GPa (600,000 psi) is selected to be safe. The broach is to machine a round hole nominally 50 mm (≈2 in.) in diameter × 75 mm (≈3 in.) long. The pitch or spacing of the teeth is 11 mm (.438 in.), and each roughing tooth is 50 μm (.002 in.) larger in diameter than the one that cuts just before it.

   (a) After the broach has been put in service, the true value of the constant C is to be ascertained as a basis for designing other broaches for the same material. It is run on a hydraulic broaching machine with a cylinder pressure of 6.9 MPa (1,000 psi) when pulling 270 kN (≈30 tons) rated capacity. The pressure to move the ram under no load is negligible; a gage shows a pressure of 4.8 MPa (700 psi) in the cylinder when the broach is cutting to its full capacity. What is the actual value of the constant C?

   (b) The material from which the broach is made is considered to have a working strength of 345 MPa (50,000 psi). What should be the minimum diameter of the space between the teeth for the broach described?

7. The pitch of the teeth is 5.8 mm (.23 in.) on a 460-mm (≈18-in.) diameter circular cold saw. It is fed at a rate of 0.05 mm/tooth (.002 in./tooth) and revolves at 18 m/min (60 ft/min). How long will it take to cut through a 125-mm (≈5-in.) diameter steel bar? What is the maximum power consumption at 0.05 kW/cm^3/min (1 hp/in.^{3}/min) of stock removal? The kerf of the saw is 6 mm (.25 in.).

8. The cold sawing machine to do the operation described in Problem 7 has an overhead rate of \$5/hr. A hacksaw will cut off the same piece in 5 min and has an overhead rate of \$2/hr. Each machine stands idle for 1 min while the operator gets the stock and loads it. Which machine would you select for the job and why?
9. A cold circular saw is needed for general ferrous work that should keep it busy for the full 2,000 hr/yr in which the plant operates. It is estimated that the saw will be cutting 80% of the time and removing material at an average rate of 82 cm^3/min (5 $in.^3$/min) with a kerf 5 mm (≈.188 in.) wide. The machine costs \$15,000 and \$1,500 a year is needed to amortize its cost. To do the same work on a horizontal band saw will require about 25% more cutting time, and therefore two machines are needed. Each costs about \$8,000, and the annual cost of the two is \$1,600. Labor and overhead cost will be the same for either kind of saw. The band saw cuts up about only half as much material in making each cut, and the stock saved is considered worth \$0.28/kg (\$0.13/lb). Which sawing machine would you recommend and why?

22.5 REFERENCES

ANSI B11.10-1990. "Metal Sawing Machines, Safety Requirements for Construction, Care, and Use of." New York: American National Standards Institute.

ANSI 212.9-1986. "Carbide Blanks for Tipping Circular Saws." New York: American National Standards Institute.

Bistrick, E. J. 1983. "Broaching Internal Involute Splines to .0004 in." *Manufacturing Engineering,* April.

Conklin, D. 1993. "Broaching Goes for the Gold." *Modern Machine Shop,"* May.

Koelsch, J. R. 1995. "Better Broaching." *Manufacturing Engineering,* April.

Kusz, R. S. 1981. "Pot Broaching: High-production Gear Cutting." *Tooling and Production,* August.

Lorincz, J. A. 1992. "Broaching Fires Up Competition." *Tooling and Production,* October.

Machining Data Handbook, 3rd Ed. 1990. Vols. 1 and 2. Cincinnati, OH: Institute of Advanced Manufacturing Sciences.

Scharff, R. 1981. *Getting the Most of Your Band Saw.* Reston, VA: Reston Publishing Co.

Yera, H. J. 1991. "Proper Sharpening for Successful Broaching." *Manufacturing Engineering,* January.

Abrasives, Grinding Wheels, and Grinding Operations

23

Man-made diamond, 1955—General Electric Company, Schenectady, NY

23.1 ABRASIVES

Abrasives are hard substances used in various forms as tools for grinding and in other surface finishing operations. They are able to cut materials too hard for other tools and give better finishes and hold closer tolerances than can be obtained economically by other means on most materials.

Abrasives may be used as loose grains, in grinding wheels, in stones and sticks, and as coated abrasives. When applied most efficiently, abrasives remove metal by cutting it into chips just like other metal-cutting tools, but the chips generally are so small that they must be magnified to be seen.

The important properties of an abrasive material are: (1) hardness, (2) toughness, (3) resistance to attrition, and (4) friability. Hardness is the ability of a substance to resist penetration. An abrasive must be hard to penetrate and scratch the surface of the material on which it works. The greater the difference in hardness between an abrasive and the work material, the more efficient the abrasive.

Abrasive grains deteriorate in service by the loss of fine particles (microparticles), which flattens and dulls the edges, and by relatively large pieces of the crystal breaking away. The first action is called *attrition*, the second *fracture*. Toughness or body strength is necessary for an abrasive particle to withstand the shocks of a cut. How long it remains sharp depends on its resistance to attrition. The latter is partly related to hardness but also to the chemical affinity between the abrasive and the material it penetrates under high pressures and temperatures. An abrasive may have high attrition resistance with one material and low resistance with another of about the same hardness. After attrition takes place, an abrasive particle must be sufficiently friable to break apart in chunks to form new sharp edges.

23.1.1 Conventional Abrasives

A variety of natural abrasive materials have been used as tools of one kind or another by humans for the last two million years. Probably the most common of those was natural sandstone from which the Egyptians made a crude form of grinding wheel. Although plentiful, the natural abrasives did not prove to be satisfactory for modern high-speed production processes because of their lack of uniformity and poor physical properties as compared to synthetic abrasive materials. Two primary artificial abrasives, aluminum oxide, Al_2O_3, and silicon carbide, SiC, have served as the workhorses for industrial grinding operations since the early 1900s.

Aluminum oxide and silicon carbide have long been the mineral mainstays for the coated-abrasive industry. Coated abrasives are the products of an extremely technical process comprised of three basic elements as shown in Figure 23-1. The first element is a flexible or semirigid backing to which the abrasive grains are bonded by an adhesive. The adhesive may be glue or resin that holds the grains together and to the backing. The backing serves as the base upon which a coating of adhesive known as the "maker" coat is uniformly applied to anchor a single layer of adhesive particles. The maker coat is solidified, and a second coating of adhesive known as the "sizer" coat is applied. In coated-abrasive terminology, the maker and sizer coats are often referred to as an adhesive bond. Coated abrasives are available in sheets, strips, rolls, cones, and disks of various sizes, but belts prevail for stock removal.

A synthetic abrasive, *aluminum oxide* (Al_2O_3) is made from bauxite and other additives. The bauxite is first refined in an electric furnace and then coke and iron are added to produce a fused mass of aluminum oxide. The fused mass is then crushed to form grains of abrasive particles suitable for use in grinding wheels. In its pure form, aluminum oxide is white. However, its grinding versatility stems from its ability to chemically bond with other materials and compounds during the fusion process.

Aluminum oxide is the most popular of the conventional abrasives for a variety of grinding applications, including the grinding of most steels, ductile cast iron, and nonferrous cast alloys. Mixed with zirconium oxide or titanium oxide, it is used for high rate of metal removal snagging operations. Adding chromium oxide to the aluminum oxide produces a pink-colored abrasive wheel, which can be used to grind heat-sensitive tool steels. As indicated in Table 23-1, aluminum oxide abrasive particles are somewhat

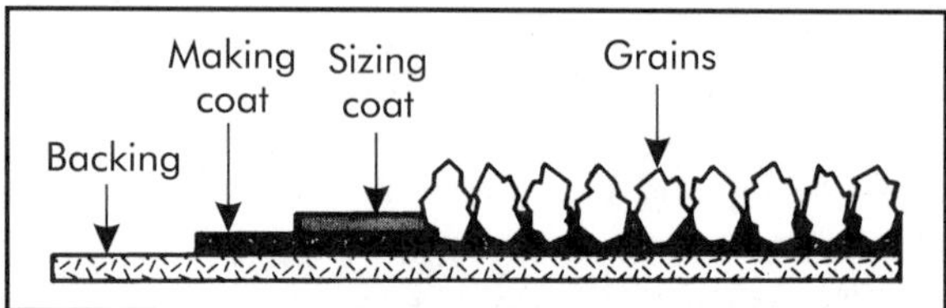

Figure 23-1. Elements of a coated abrasive.

softer than silicon carbide, but tougher. Thus, aluminum oxide wheels are often considered to be better suited for general-purpose grinding operations.

Silicon carbide (SiC), another synthetic abrasive, is produced by heating a mixture of pure sand (silicon) and fine petroleum coke (carbon) in an electric oven at around 2,204° C (4,000° F) for several days. The crystalline mass resulting from this cooking process is then crushed and graded according to the particle size desired. Silicon carbide crystals vary in color from black to green according to their purity. The pure green particles are more costly than the more common black particles. Silicon carbide grains are relatively hard and brittle, thus limiting their application to the grinding of cast iron, other low tensile strength nonferrous metals, and certain nonmetallic materials.

23.1.2 Superabrasives

Two materials, *diamonds* and *cubic boron nitride* (CBN), are generally considered to be superabrasives because of their extreme hardness. As indicated in Table 23-1, the diamond is the hardest known substance. Diamond and CBN cost much more than either of the conventional abrasives, but they perform considerably better in particular applications. For example, the cost of grinding cemented carbides and other space-age materials with a superabrasive is a fraction of what it is with conventional abrasives. Although not as hard as a diamond, CBN grinds some hard materials better because it is more inert and withstands higher temperatures.

Natural diamonds have been used commercially for many years to grind and cut hard and dense materials such as concrete, glass, granite, and marble. In more recent years, natural diamond abrasive wheels and coated products have found many applications in grinding and finishing tungsten-carbide materials.

As the demand for natural diamonds increased, it became evident in the early 1950s that a domestic source of diamonds would be required for industrial use because of the instability of the supply and the cost of imported diamonds. Thus, the General Electric Company committed significant research resources to the development of synthetic or man-made diamonds. In

Table 23-1. Properties of conventional and superabrasive materials

Material	Knoop Hardness	Compressive Strength, MPa (psi)	Thermal Conductivity W/m °C (Btu ft/hr per ft² °F)	Relative Abrasion Resistance
Aluminum oxide	2,000–3,000	3,450 (500,380)	29 (16.76)	9
Silicon carbide	2,100–3,000	1,450 (210,305)	42 (24.27)	14
Cubic boron nitride	4,000–5,000	6,520 (945,646)	1,300 (751.14)	37
Man-made diamond	7,000–8,000	8,690 (1,260,380)	2,000 (1155.60)	43

1955, they were successful in producing diamond crystals in the laboratory and, in 1957, they introduced the first man-made diamond product. These diamonds were made by subjecting a pure source of carbon (graphite), along with a proper molten-metal solvent-catalyst, to extremely high temperatures and pressures. The first man-made diamond was in the form of an elongated friable crystal with rough edges. Its properties made it suitable for use in resin and vitreous grinding wheels for grinding tungsten-carbide materials. A photograph of man-made diamond particles is shown in Figure 23-2(A).

Later on, General Electric developed other types of man-made diamond abrasive grains for use in particular applications. One type, a tough, block-shaped diamond crystal is used to grind a variety of ceramic materials, as well as cemented carbides and glass. Another type was developed for use in metal-bonded saws for cutting stone and concrete. In addition, General Electric experimented with a variety of coatings designed to reduce grinding wheel wear caused by the premature tearing away of crystals from the bonding materials. Figure 23-2(B) shows diamond grains coated with a nickel-based alloy. The metallic coating provides stronger holding power for the crystals in resin-bonded wheels.

It must be emphasized that diamond grinding wheels are effective only for grinding nonferrous materials. When used on steel, a carbon solubility potential develops wherein the steel reacts with and absorbs carbon from the diamond particles themselves. When used on steel, this chemical solubility of diamond in iron generates excessive diamond wear.

Although not as hard as diamond, cubic boron nitride (CBN) grinds some hard materials better because it is more inert and withstands higher temperatures. The synthesis of CBN is much the same as with a diamond, except that boron, a metal, and nitrogen, a nonmetal, are combined by the use of an appropriate solvent-catalyst into a tightly bonded structure similar to that of a diamond.

A variety of CBN abrasives are produced for different applications, depending on the type of bonding used, the mode of grinding, and the material removal rate required. One, the single-crystal (monocrystalline) type, has a multiplicity of cleavage planes that fracture and permit the grains to resharpen themselves as they become dull. These large fractures, called macrofractures, are particularly effective in maintaining grain sharpness when grinding steel.

Another type of CBN superabrasive grain is microcrystalline in structure and consists of an extremely large number of micron-size subcrystalline sections packed and bonded together to form a very dense structure. Cleavage occurs in much the same manner as with the monocrystalline type of grain, but at the submicron level (microfracturing). Resharpening is practically continuous, which permits a very efficient cutting action with minimal frictional heating.

As with diamond particles, CBN abrasive grains are often coated with metal to improve crystal retention in different types of bonds.

23.1.3 Grain Size

Abrasive grains are sorted into various sizes for a uniform and dependable product. This is done by passing the grains through screens in mechanical sieving machines. Grain sizes are designated by numbers as indicated in Figure 23-3. Each grain size passes through a screen of a certain size but not through smaller screens. Screen size openings in millimeters and their

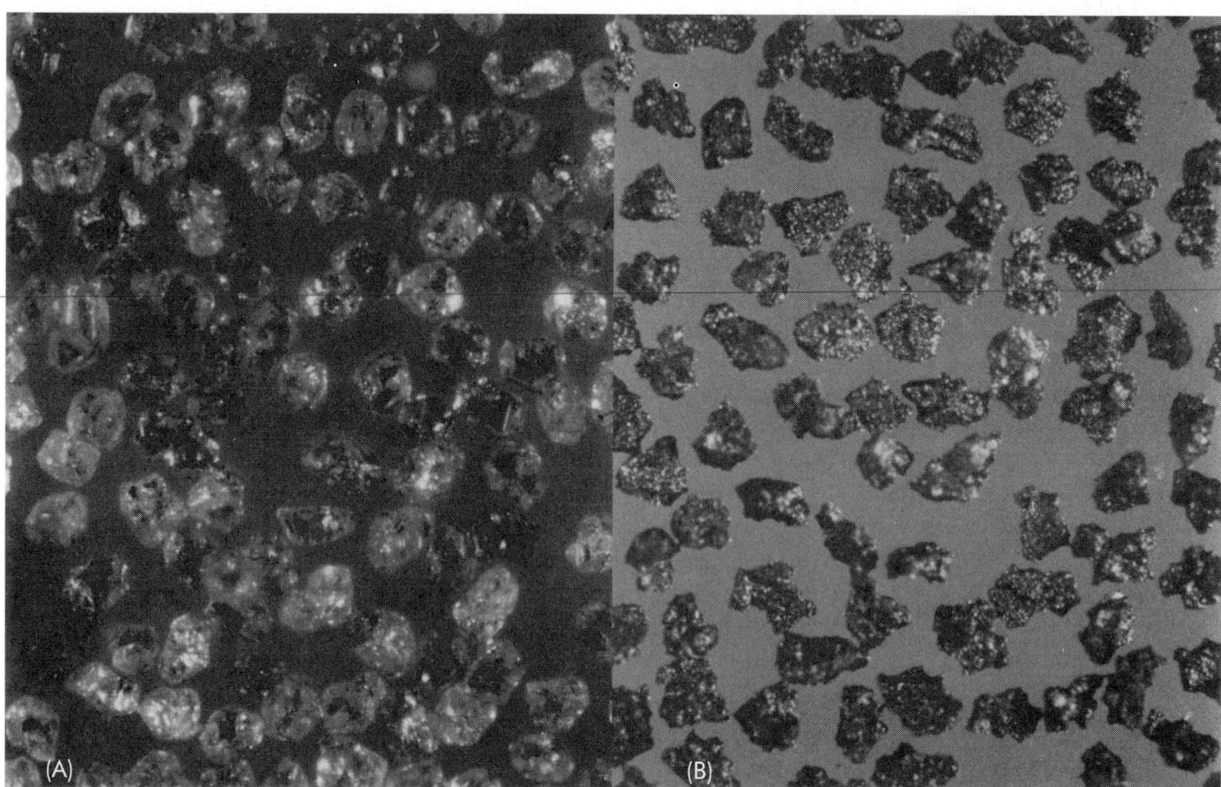

Figure 23-2. (A) Uncoated man-made diamond abrasive grains; (B) man-made diamond grains coated with a nickel-based alloy. (Courtesy G. E. Superabrasives)

relation to grain sizes are given in published tables (ANSI 1976). The finest sizes, called *flours,* are segregated by flotation methods.

23.2 GRINDING WHEELS

23.2.1 Properties

A grinding wheel (Figure 23-4) is made of abrasive grains held together by a *bond*. These grains cut like teeth when the wheel is revolved at high speed and brought to bear against a workpiece. The properties of a wheel that determine how it acts are the kind and size of abrasive, how closely the grains are packed together, and the kind and amount of bonding material.

To get the wide range of properties needed in grinding wheels, some are bonded by mixing abrasive grains with inorganic and organic materials. Inorganic bonds are vitrified, silicate, and metallic. A *vitrified bond* is a clay bond melted to a porcelain- or glass-like consistency. It can be made strong and porous for heavy grinding and is not affected by water, oils, acids, or other than extreme temperatures. Most grinding wheels have vitrified bonds. Their strength and rigidity help to control size and finish. A silicate bond is essentially water glass hardened by baking. It holds the grains more loosely than a vitrified bond and gives a cooler cut. Large wheels can be made more easily with a silicate bond. Cubic boron nitride and diamond abrasives are usually (but not always) embedded in *metallic bonds* (by electroplating or electroforming) for the utmost in strength and tenacity to hold these costly and long-wearing grains.

Wheels with organic bonds are held together by such materials as phenolic resins (resinoid), rubber, and shellac. They tend to be somewhat more flexible and resilient than inorganic bonds, and the wheels can stand more bumping and side forces. Abrasives are readily released when dull. They offer cool cutting and give lustrous finishes. Thin cutoff wheels have resinoid or rubber bonds. Some CBN and diamond wheels have resin bonds.

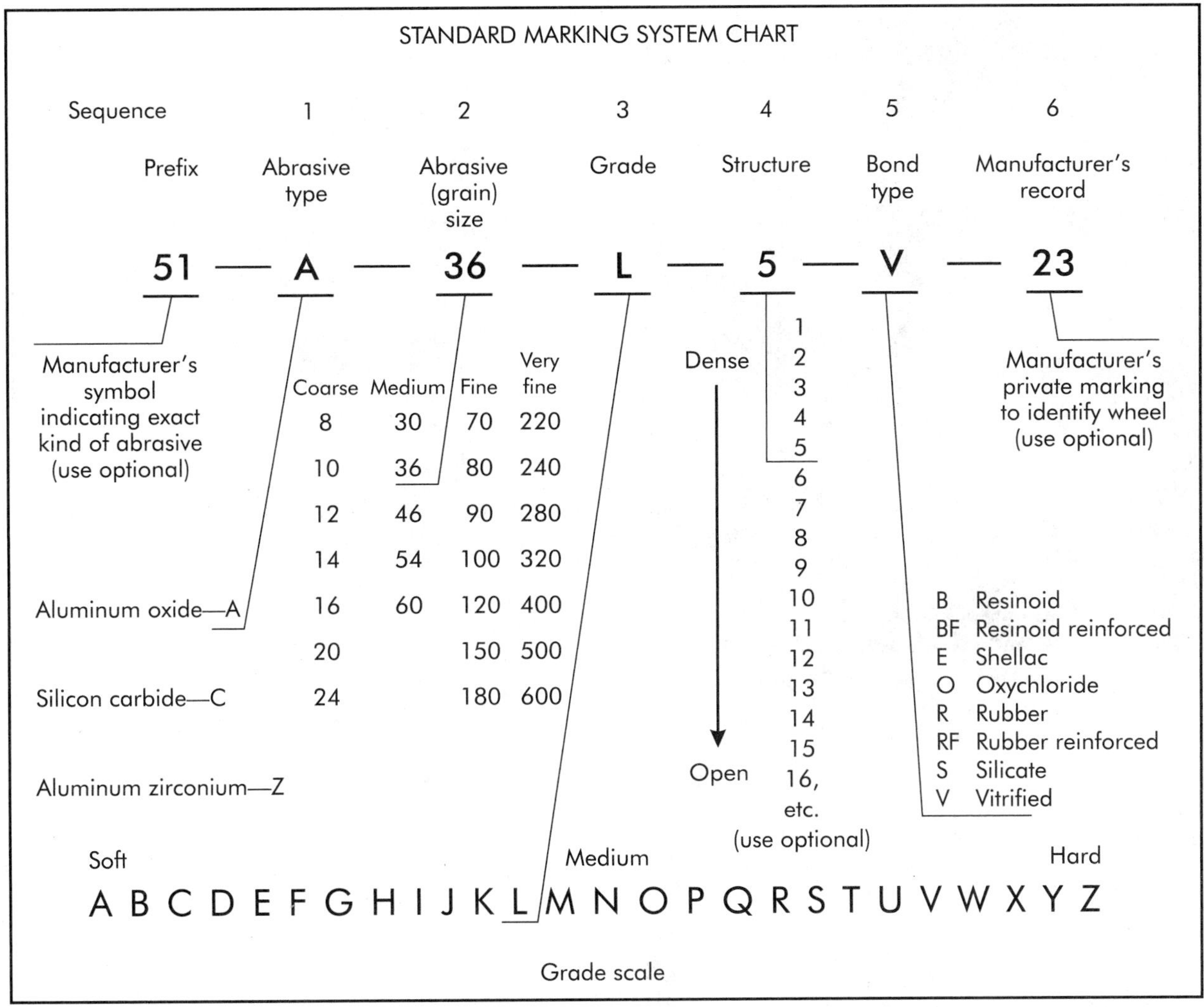

Figure 23-3. Standard marking system chart for aluminum oxide, silicon carbide, and aluminum zirconium grinding wheels. A similar chart exists for diamond and cubic boron nitride wheels (ANSI 1990).

Other grinding wheels are reinforced with such materials as steel, nylon, and fiberglass for added strength, particularly at ultra-high speeds. Organic bonds mix well with reinforcement.

The *grade* of a grinding wheel is a measure of how strongly the grains are held by the bond. The bonding material in a wheel surrounds the individual grains and links them together by connectors called *posts,* as illustrated in Figure 23-5. The sizes and strengths of the posts depend upon the kind and amount of bonding material in a wheel. The ability to hold its abrasive grains is called the *hardness* of a grinding wheel. A hard wheel holds its grains more tenaciously than a soft wheel. A wheel that is too hard for a job keeps its grains after they have become dull. A wheel that is too soft loses grains before they have done full duty. Hardness of the wheel should not be confused with hardness of the abrasive grains themselves.

Cubic boron nitride (CBN) grinding wheels are available in a variety of shapes and sizes, similar to those made of conventional abrasives. However, CBN wheels, because of the high cost of the abrasive material, are made with the abrasive material concentrated only around the rim of the wheel. Two types of bonding systems are available; impregnated-bond wheels and electroplated wheels. The *impregnated-bond wheels* have CBN abrasive grains mixed in the bonding

Figure 23-4. A selection of grinding wheels. (Courtesy Carborundum Abrasives, Inc.)

material throughout the thickness of the rim section. Such wheels are constructed with a preformed core and the abrasive rim is bonded to the core. The abrasive rim sections are usually between 1.57–6.35 mm (.062–.250 in.) in thickness, depending on the size and other characteristics of the wheel. Impregnated-bond wheels are constructed with either resin, metal, or vitrified bond materials, with the bond material designed to match the properties of the abrasive so they will both wear at the same rate.

Electroplated CBN grinding wheels are constructed with only a single layer of abrasive grains bonded to the machined surface of a metal wheel core. Bonding is accomplished in a nickel electroplating bath in such a way that an extremely strong abrasive retention is achieved. Electroplated wheels can be easily formed or contoured and are the preferred type for many form-grinding operations on gear teeth, splines, etc. In addition, electroplated wheels are usually very free-cutting, thus permitting high rates of metal removal.

The *structure* or *spacing* of a grinding wheel refers to the relationship of abrasive grains to bonding materials and of those two elements to the

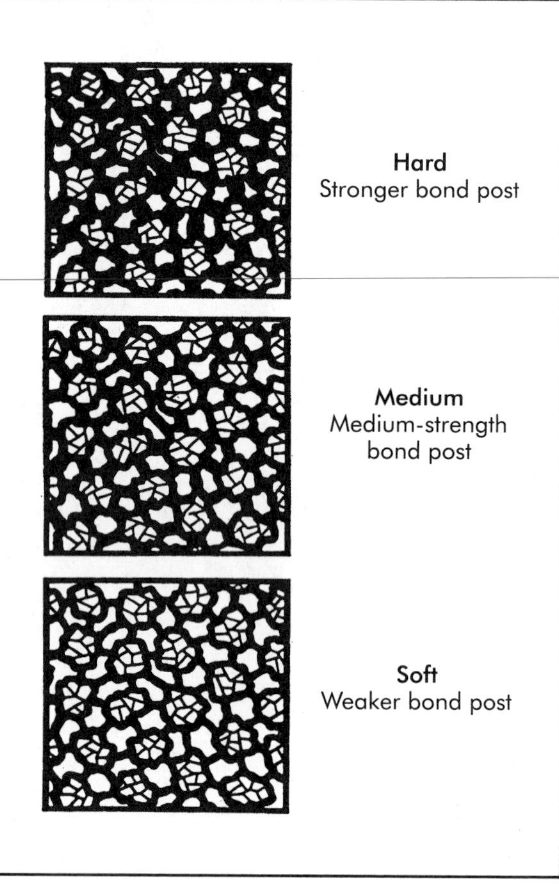

Figure 23-5. Explanation of the meaning of wheel grade. (Courtesy Carborundum Abrasives, Inc.)

voids between them. The meaning of structure is illustrated in Figure 23-6. The spaces in a grinding wheel provide room for chips to escape during a cut and for cutting fluid to be carried into a cut.

Grinding wheels are marked with symbols to designate their properties. A typical wheel marking and an explanation of its meaning were shown in Figure 23-3. Each letter or number in a certain position in the sequence designates a particular property.

All grinding wheel manufacturers use substantially the same standard wheel marking system. However, properties of wheels are determined to a large extent by the ways the wheels are made. The processes vary from one plant to another, and wheels carrying the same symbols but made by different manufacturers are not necessarily identical.

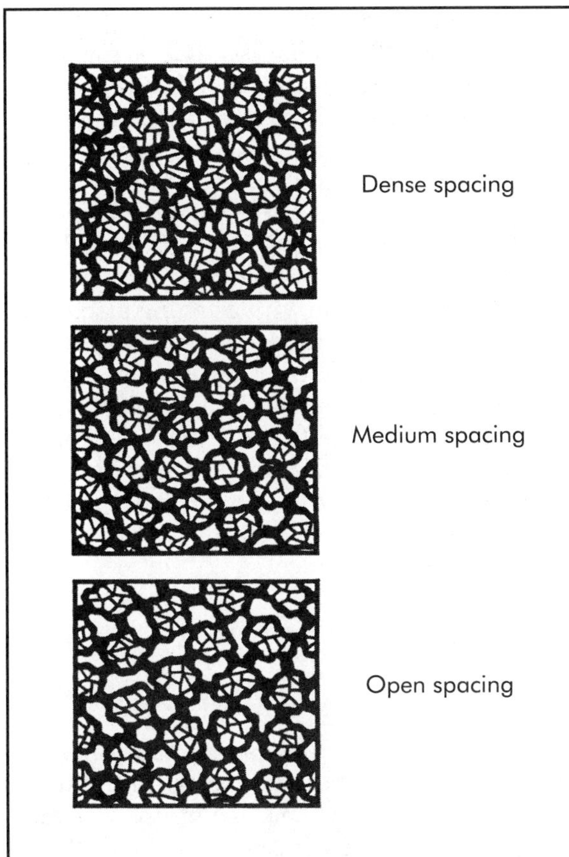

Figure 23-6. Explanation of the meaning of wheel structure. (Courtesy Carborundum Abrasives, Inc.)

Shapes and Sizes

A few typical grinding wheel shapes were shown in Figure 23-4. The nine grinding wheel shapes recognized as standard include straight cylinders with or without recesses in their sides, and others described as tapered two sides, straight cup, flaring cup, dish, and saucer. Other shapes may be obtained as specials.

The principal dimensions that designate the size of a grinding wheel are the outside diameter, width, and hole diameter. Standard wheel shapes are made in certain sizes only, but the variety is large.

Disk wheels are abrasive disks cemented or bolted to steel disks and are stronger for grinding from the side of the wheel than straight wheels alone. Side grinding wheels known as *built-up, segmental,* or *sectored wheels* are composed of bonded-abrasive blocks held in a chuck or fastened to a metal disk by wedges, steel bands, wire, or bolts. They are easier to make than solid wheels for diameters over 1 m (≈36 in.) and cut coolly because they cut intermittently. Adequately held segmental wheels can be run at higher speeds than solid wheels for peripheral grinding, such as on rough castings at surface speeds over 6 km/min (≈20,000 ft/min).

Mounted wheels and *points* are small grinding wheels, usually a few millimeters in diameter, with attached shanks. They are commonly used at high speeds on portable grinders for burring, removing excess material from dies and molds, grinding in recesses and crevices, and small holes.

23.2.2 Manufacture

Vitrified grinding wheels may be made by the puddling, tamping, or pressing processes. Pressed wheels are the most dense and puddled wheels the least dense. First clay and abrasive are machine-mixed thoroughly. In the puddling process, water is added, and the mixture is poured into molds. For pressed wheels, the dry or semidry mixture is placed in molds and squeezed in hydraulic presses. The same type of mixture is compressed less, but still firmly, in the tamping process. At this stage, the wheels are baked and dried. The puddled wheels must be trimmed to size.

Grinding wheels are vitrified by being fired for several days at high temperatures, like pottery. When hard, the wheels are trued, their arbor holes are bushed with babbitt metal or lead, and large wheels are balanced.

Other kinds of wheels are made by processes associated with the particular bonding agents. In general, the ingredients are mixed, molded, and heated as required. The finished wheels are sized, balanced, and graded.

23.3 OTHER ABRASIVE PRODUCTS

Silicon carbide and aluminum oxide abrasive grains are bonded into sticks and stones of various types and sizes. They are used for honing, touching up edges of cutting tools, and cleaning, polishing, and finishing dies, molds, and jigs.

23.3.1 Coated Abrasives

Coated abrasives are made of abrasive grains, adhesive, and backing. The adhesive may be glue or resin to hold the grains together and to the backing of paper, cloth, or plastic. For a closed coating, the abrasive grains completely cover the surface; in an open coat they are uniformly distributed over 50–70% of the surface. Coated abrasives are available in sheets, strips, rolls, cones, and disks of various sizes, but belts prevail for stock removal.

23.3.2 Polishing Wheels

Flexible wheel bodies of cloth, leather, or wood, depending upon the work, are coated with adhesive and rolled in abrasive grains of uniform sizes, coarse for roughing and fine for finishing. After the adhesive, glue or cold cement, has dried, the abrasive layer is cracked by pounding to make it resilient. The resulting polishing wheels are revolved at surface speeds around 2.3 km/min (7,500 ft/min). After its grains have become dull and worn off, a polishing wheel is stripped and recoated.

23.3.3 Abrasive Belts

Abrasive belts are primarily used for light stock removal and polishing operations, however they are being increasingly used for heavier stock removal. Abrasive-belt machines are available in a wide variety of types to suit specific applications, as shown in Figure 23-7.

23.4 GRINDING OPERATIONS

Grinding is done on surfaces of almost all conceivable shapes and materials of all kinds. It may be classified as a nonprecision or precision process, according to purpose and procedure. *Nonprecision grinding*, common forms of which are *snagging* and *off-hand grinding,* is done primarily to remove stock that cannot be taken off conveniently by other methods from castings, forgings, billets, and other rough pieces. The work is pressed hard against the wheel or vice versa. The accuracy and surface finish obtained are of secondary importance. *Precision grinding* is concerned with producing good surface finishes and accurate dimensions. The wheel or work or both are guided in precise paths.

Any grinding process is a high-energy operation and potentially very dangerous. The American National Standards Institute (ANSI) prescribes mandatory safety measures, which are the basis for most state standards and laws.

The three basic kinds of precision grinding shown in Figure 23-8 are *external cylindrical grinding, internal cylindrical grinding,* and *surface grinding*. Variations of each of these will be described in connection with grinding machines.

Grinding is able to produce accurate and fine surfaces because it works through small abrasive cutting edges, each of which takes a tiny bite. Appreciable quantities of material can be removed by grinding because a large number of cutting edges are applied at high frequency. For instance, about 39,000,000 cutting points act in 1 min when a 46-grit wheel, 457 mm (18 in.) in diameter and 51 mm (2 in.) wide, is revolved with a surface speed of 1,524 m/min (5,000 ft/min).

23.4.1 The Factors of Cost

The major costs of a grinding operation are those incurred for labor, time, overhead, and the abrasive wheel. What these amount to depends upon: (1) the selection of grinding wheel, (2) the equipment and its operation (including such items as wheel speed, work speed, feed, depth of cut, cutting fluid, and wheel dressing), and (3) the workpiece and its material. How these factors affect costs and how they should be controlled for best results will be explained in the remainder of this chapter.

Grinding Wheel Selection

A grinding wheel may be needed to remove stock rapidly, give a high finish, or hold small tolerances. Each function calls for different properties in the wheel, and no wheel can do all best. The extent to which a wheel is suited for a job is called its *grinding quality*. For a production operation, a wheel suited just for the job is chosen (quite often one for roughing and another for finishing). A compromise may need to be made if a general-purpose wheel is chosen. The following discussion is intended to provide a basic understanding of what is involved in grinding wheel selection. In practice, experts should be consulted to select wheels for important jobs.

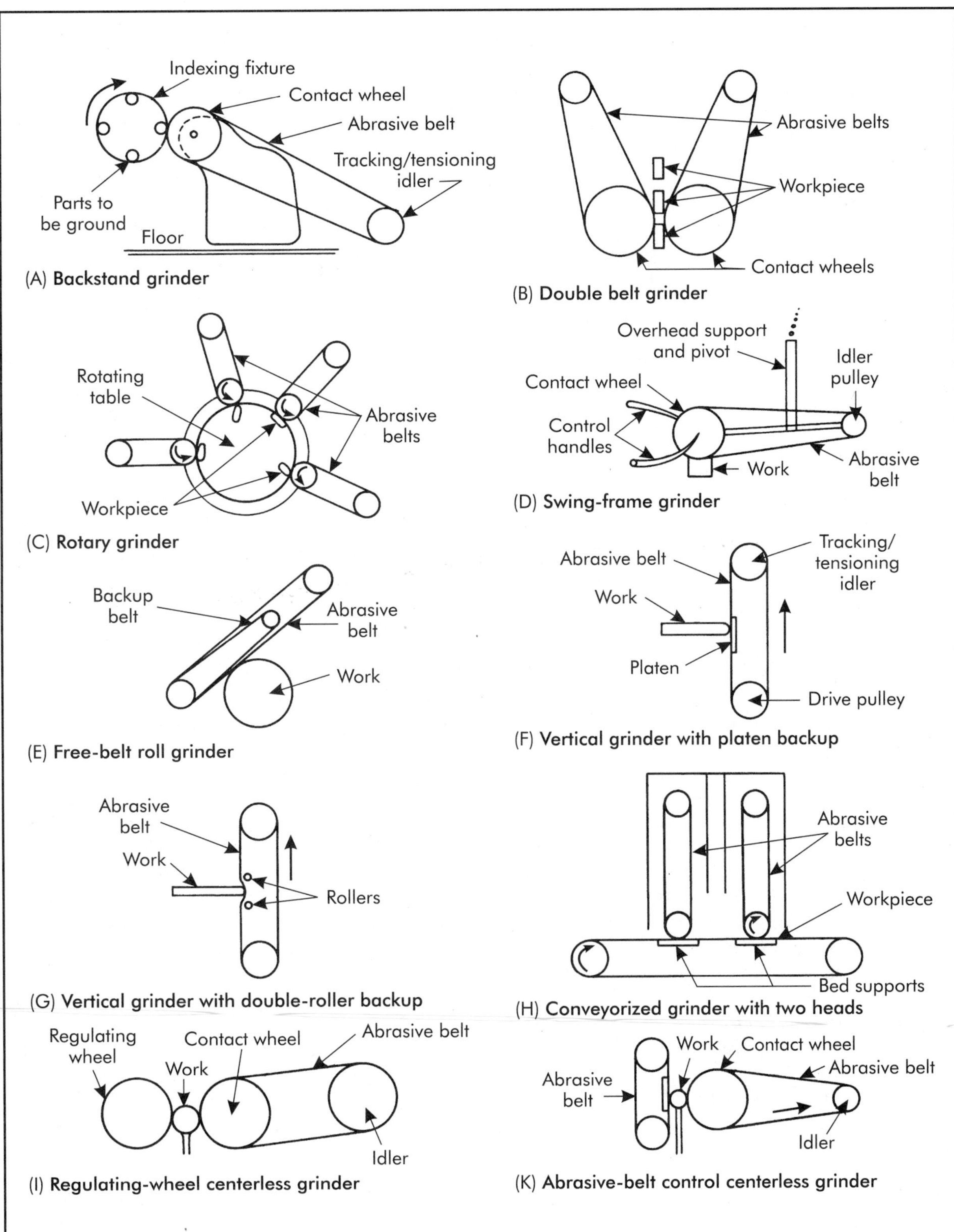

Figure 23-7. Typical applications and configurations of adhesive-belt machining.

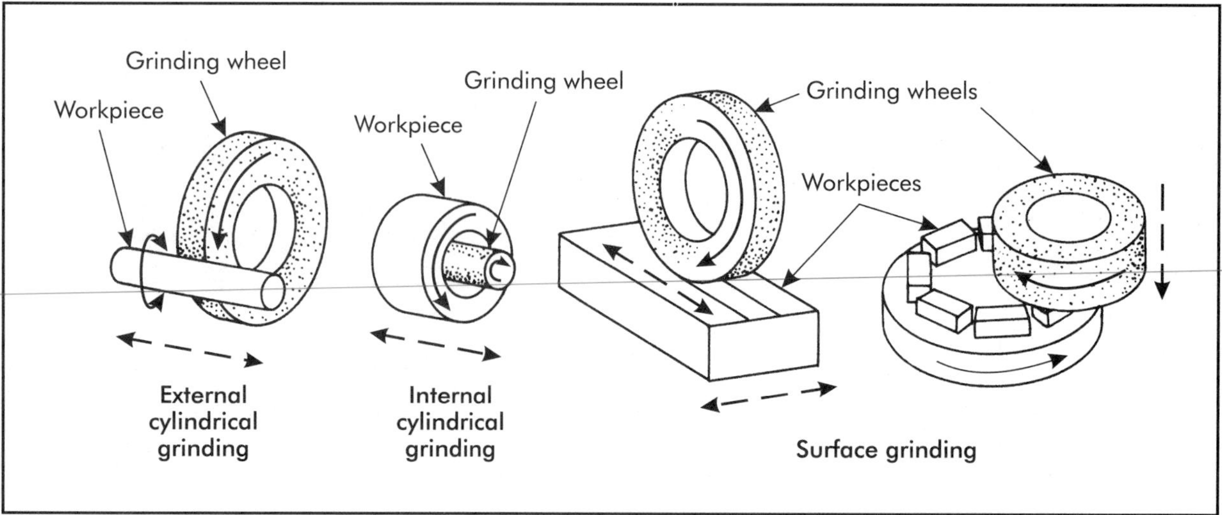

Figure 23-8. Basic precision grinding operations.

Abrasive and grain size. The resistance to fracture seems about the same for silicon carbide and aluminum oxide. Silicon carbide has a low resistance to attrition when cutting steel or malleable iron, and its grains dull quickly, so aluminum oxide is preferred for that purpose. Silicon carbide performs better as a rule for grinding cast iron, brass, copper, soft bronze, aluminum, stone, rubber, leather, and cemented carbides. The reason is that silicon carbide fractures more satisfactorily in relation to the dulling of its grains, and its cutting edges are renewed as needed. However, various kinds of aluminum oxide are made, and the crystals of some kinds are highly friable and desirable for some soft ductile materials.

If a fine finish rather than stock removal is desired, the positions of the abrasives may be reversed. In some cases, rapid dulling of the grains is desirable to produce fine finishes. Thus, silicon carbide wheels are used to get a mirror finish on hardened steel rolls, and aluminum oxide wheels for a high finish on glass.

Diamond abrasive has physical properties for grinding far superior to other abrasives, but its cost and the cost of wheels made from it are high. Diamond wheels are used for rapidly cutting and finishing gems, ceramics, stone, and cemented carbides.

Soft materials are rough-ground with coarse grains, and hard materials with fine grains. Coarse grains take big bites in and rapidly remove soft materials. Only small bites can be taken in very hard materials, and consequently small grains are advantageous because more of them can be brought to bear on the material at a given time.

Fine-grain wheels can be trued to thin sections and generally hold up better at corners and edges than coarse-grain wheels. An example is found in thread grinding, where fine-grain wheels are desirable to hold the shapes required for thread roots.

Wheel bond, grade, and spacing. Hard, tough materials rapidly dull abrasives and require soft grinding wheels that release the grains readily when they become dull. Hard wheels are suitable for soft material. However, they exert high pressure and tend to chatter, which is detrimental to good surface finish.

A common measure of wheel performance is the *efficiency ratio, grinding ratio*, or *volume ratio*. It is defined as the number of kg or cm^3 (lb or $in.^3$) of metal removed in a given time divided by the kg or cm^3 (lb or $in.^3$) of wheel loss. This may be plotted for a series of wheels of the same type but of different grades for a specific operation, as in Figure 23-9. For any one wheel, the grinding ratio decreases as in-feed is increased. This is generally the case for heavy feeds. However, the grinding ratio may rise to a peak with lighter feeds as in-feed is increased.

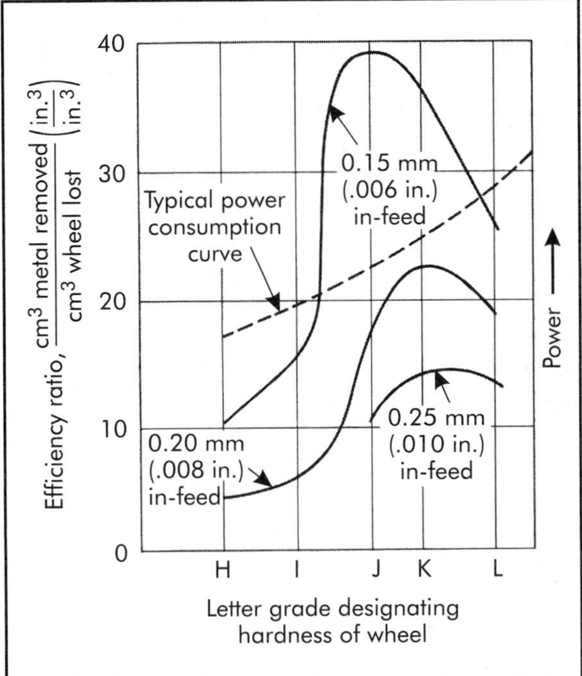

Figure 23-9. The influence of grinding wheel hardness on the efficiency ratio and power consumption in a centerless grinding operation.

For a given set of conditions in an operation, one wheel grade generally shows the highest ratio, as indicated in Figure 23-9. If wheel wear alone is to be considered, the best wheel is the one at the highest point. Wheel wear is important because of wheel cost. Also, in precision grinding, wheel wear affects wheel shape, and a wheel that wears rapidly and nonuniformly must be retrued often. This takes time, interrupts production, and is expensive. In contrast, a wheel that does not have the highest volume ratio may be faster and can show a total cost less than the wheel at the top of the curve. Power consumption and the heat generated increase with grade hardness, and that may be harmful to the work. Experts recommend that the best wheel to use for a job is the softest grade that performs satisfactorily.

Grain spacing provides the openings between grains for chips to escape during a cut. An open spacing or structure is desirable for a large area of contact and heavy roughing cuts in soft materials. It permits the grains to work at their fullest depths and provides good chip clearance space. If the face of a wheel tends to become loaded with metal, its structure is too dense for the work it is doing.

When the grains cannot bite deeply, as in grinding hard materials, chip clearance is secondary and dense spacing is desirable so that many grains can act at once. Dense spacing is also necessary for fine finishes to produce scratches close together. A wheel with dense spacing holds its shape well and contributes to dimensional control.

23.4.2 Balancing the Grinding Wheel

Grinding wheels, particularly large high-speed ones, must be balanced to produce good finishes. A wheel is mounted on the machine and trued before it is balanced. A conventional wheel mount for a straight wheel fits into the hole of the wheel and has two flanges that clamp to blotting paper rings against the sides or in the recesses of the wheel. Weights attached to one flange can be manually adjusted for static balancing, which is the most common method for large wheels. Grinding wheel mounts are available with weights that move to seek a balanced state while the wheel runs. One system for fine-balancing electronically measures vibrations and actuates a tool that abrades the side of the wheel to restore balance.

23.4.3 Dressing and Truing the Grinding Wheel

In an ideal grinding operation, a grinding wheel retains its true form and resharpens itself fully as it breaks down. The condition is only approached in most operations, and the sharpness and form of the wheel must be restored from time to time by dressing and truing as indicated in Figure 23-10. How often a wheel must be dressed or trued depends upon the type of work, the fitness of the wheel, and the skill of the operator. For internal grinding, the wheel is not uncommonly trued for each piece. For some high-production, precision, external, cylindrical grinding jobs, the wheels may run for days without being dressed.

When metal particles become embedded and fill the spaces around the surface grains, the wheel face is said to be *loaded*. Ductile materials

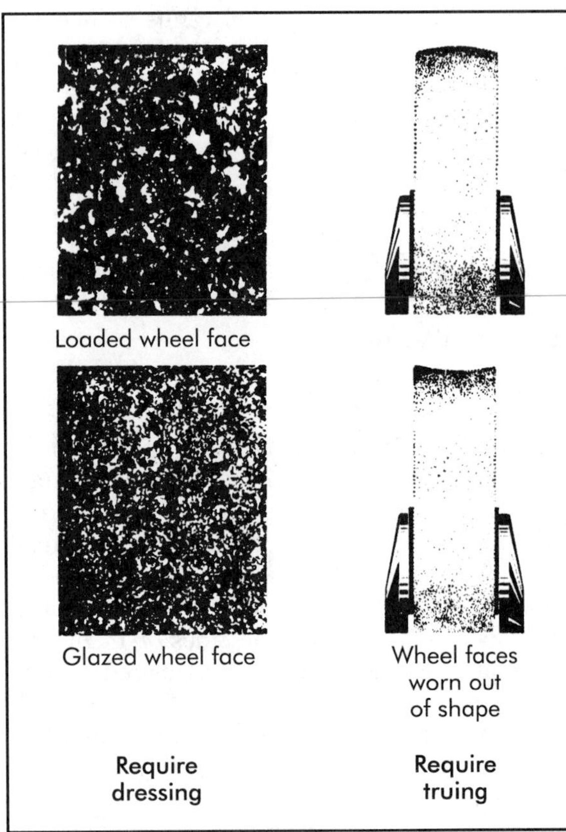

Figure 23-10. Meaning of dressing and truing. (Courtesy Carborundum Abrasives, Inc.)

and dense wheel structure particularly engender that condition. When the abrasive grains become dull and cease to cut efficiently because their edges are rubbed off, the wheel face becomes *glazed*. Dressing is done to restore the cutting action of the wheel by fracturing and tearing away the dull grains to expose fresh cutting edges or clear away the embedded material. *Truing* is done to create a true surface on a wheel. The act of truing a wheel also dresses it. Most wheels are trued and dressed at usual grinding speeds.

Steel cutters, abrasive sticks, and small abrasive wheels are relatively inexpensive and are popular tools for dressing grinding wheels. They may be pushed against and wiped across the face of the wheel by hand or attached to holders on the machine.

Diamonds unsuited for gems are used for truing wheels for precision grinding. A single large stone, a group of small ones in a setting, sintered polycrystalline compacts, or a diamond-impregnated wheel may be used. Light cuts of less than 25 μm (.001 in.) must be taken with a coolant to avoid overheating and damaging the stone. Continued use of a dull diamond makes for dull grains and a poor cutting wheel. Diamond tools are accurately fed and traversed for precision truing, in most cases by means of the normal movements of the machines. For that purpose, a standard grinder usually has an attached diamond toolholder. Truing attachments for straight surfaces, angles, radii, and almost any other form are available for addition to any grinding machine.

Crush dressing is a method of truing and dressing a grinding wheel by means of a hard roller, sometimes diamond-impregnated, pressed against the slowly revolving grinding wheel. The reverse of the form desired on the wheel is given to the roller, which displaces and crushes the surface grains and imprints the form on the wheel.

Crush dressing is often more economical than diamond truing, especially for intricate forms. Sharper crystals are left by crush dressing, and that means cleaner and cooler cutting. Diamond truing is generally more accurate and can be made to produce better surface finishes.

Reactive dressing is a routine followed in grinding operations to restore wheel cutting efficiency. When the wheel becomes dull and requires more power, the work speed is increased automatically; this causes the wheel to break down and resharpen itself because of an increase in the grain depth of cut.

23.4.4 Theory

The effects of the principal variables in a grinding operation can be seen by analyzing the grinding action depicted in Figure 23-11. The grinding wheel rotating in the direction shown with a surface speed of V m/min is assumed to have n grains/mm in a narrow band around its periphery. The average spacing from grain to grain is $1/n$. After one grain passes point A, the time until the next grain reaches the same point is $T = 1/(1000Vn)$ min. During that time the workpiece is revolving with a surface speed of v m/min, and a point on its periphery advances from A to B, a distance $AB = 1{,}000vT = v/Vn$. The chip removed in that time is represented by the area

CEAB. The maximum chip thickness t= $\overline{EB}$ largely determines what happens to the grinding wheel. The area *EAB* is small and may be closely approximated by a triangle *EAB* with the angle *EAB* = α + β. The maximum chip thickness in millimeters (in.) is

$$\overline{EB} = t = AB\sin(\alpha+\beta) = \frac{v}{Vn}\sin(\alpha+\beta) \quad (23\text{-}1)$$

where

$\overline{EB}$ = maximum chip thickness, mm (in.)
t = grain depth of cut, mm (in.)
AB = distance traveled on workpiece, mm (in.)
v = surface speed of workpiece, m/min (ft/min)
V = grinding wheel surface speed, m/min (ft/min)
n = number of grains contained on grinding wheel

The triangle O_1OA in Figure 23-11 has one side $D/2 + d/2 - C$, a second side $D/2$, and a third side $d/2$. By the law of cosines,

$$\left(\frac{D}{2}+\frac{d}{2}-C\right)^2 = \left(\frac{D}{2}\right)^2 + \left(\frac{d}{2}\right)^2 - 2\left(\frac{D}{2}\right)\left(\frac{d}{2}\right)\cos[\pi-(\alpha+\beta)] \quad (23\text{-}2)$$

where

C = depth of cut, mm

also

$$\sin(\alpha+\beta) = \sqrt{1-\cos^2(\alpha+\beta)} = \sqrt{1-\cos^2[\pi-(\alpha+\beta)]} \quad (23\text{-}3)$$

If Equations (23-2) and (23-3) are combined and solved, and terms containing C^2 are neglected because C is quite small, it is found that

$$\sin(\alpha+\beta) = 2\sqrt{C[(1/d)+(1/D)]} \quad (23\text{-}4)$$

If this result is substituted in Equation (23-1),

$$t = \frac{2v}{Vn}\sqrt{C}\sqrt{(1/D)+(1/d)} \quad (23\text{-}5)$$

This relationship is based on several approximations but is substantially the same as the exact expression that can be obtained for *t* for each type of grinding. Experience and experiments have shown that the force acting on a grain increases or decreases with the grain depth of cut *t*, although not necessarily in proportion. An increase in the grain depth of cut causes the forces to increase and the grains to fracture or break away sooner, and the wheel acts softer. Conversely, a decrease in the grain depth of cut makes the wheel act harder. Thus Equation (23-5) reveals at least qualitatively how the variables in a grinding operation should be arranged for efficient performance.

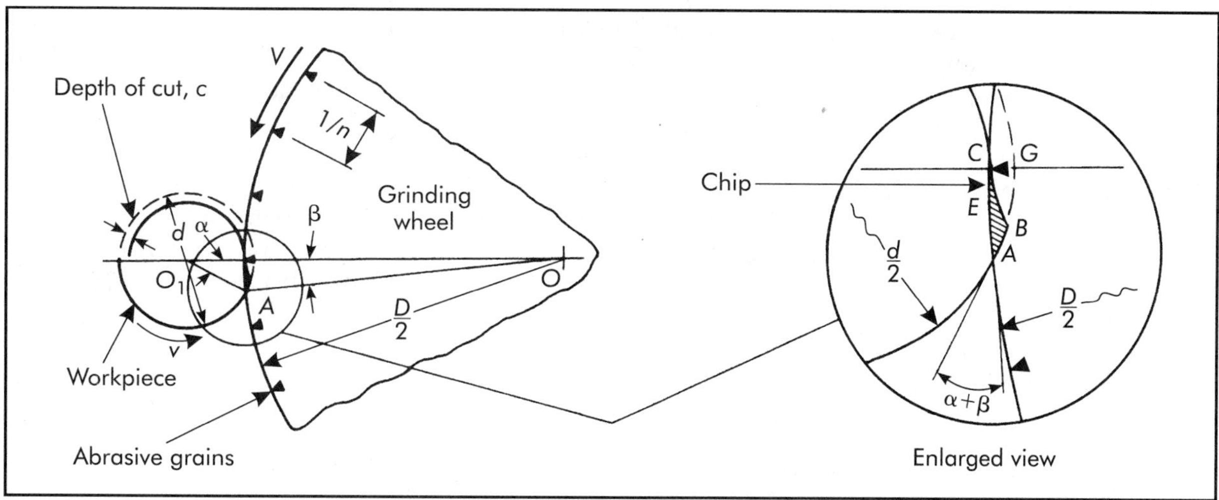

Figure 23-11. Grinding wheel action.

The grain depth of cut must be small in precision grinding to make light scratches and leave a fine finish. Even in nonprecision grinding, the grain depth of cut cannot be excessive. From Equation (23-5) it is evident that the wheel speed must be high for a small t. As wheel speed is increased and other parameters are held the same, grinding forces and wheel wear decrease, and the wheel produces a better finish over a longer time. On the other hand, production may be increased by increasing work speed and/or infeed rate along with increasing wheel speed; power rises, and more heat is generated. Low wheel speeds are desirable for some materials, particularly for heat-sensitive metals, but as a rule grinding wheels are run as fast as safety allows. Ordinary vitrified wheels are run at up to 2,591 m/min (8,500 ft/min) and resinoid, rubber, shellac, and segmental and reinforced (even vitrified) wheels up to 5,486 m/min (18,000 ft/min) and even more.

For a small grain depth of cut, the workpiece velocity (v) in Equation (23-5) should not be high and the depth of cut (C) should be shallow, especially for fine surface finishes. For external cylindrical grinding, work speeds vary from 9 m/min (30 ft/min) for hardened steel, which abrasives cannot cut deeply, to 15–31 m/min (50–100 ft/min) for average work, to as high as 61 m/min (200 ft/min). For average internal grinding, work speeds of 46–61 m/min (150–200 ft/min) are advocated. Table traverse speeds on surface grinders are generally less than 31 m/min (100 ft/min). A depth of cut of 25–102 μm (.001–.004 in.) is satisfactory for roughing when precision grinding, but less than 25 μm (.001 in.) should be used for finishing. The work speeds and feeds cited are for conventional grinding. For high-speed grinding at wheel speeds of 3,500–5,500 m/min (≈12,000–18,000 ft/min) or more, a like increase in productivity may be obtained by increasing the work speed and/or feed proportionately without changing the grain depth of cut.

It would not be reasonable to expect that exactly the best wheel would be readily available for every grinding job. Equation (23-5) reveals in which direction each variable can be changed to modify the action of the wheel. As a wheel wears to a smaller diameter and its surface speed decreases with the same number of revolutions per minute, the wheel acts softer. The surface speed of the workpiece can be increased to make the wheel act softer, or decreased for harder action. A deeper cut causes a wheel to act softer, and a lighter cut harder. Deep cuts and rapid stock removal call for hard wheels.

A small value of n in Equation (23-5) corresponds to a wheel with an open structure or with its surface coarsely dressed. Such a wheel can be expected to have a soft action. This explains the accepted practice of using a wheel for finishing that is of the same grade as for roughing but has finer grains more closely spaced, or the alternative of a softer wheel with the same grain size and spacing as for roughing. The first is a wheel that cuts finely; the second can break down readily, stays sharp, and cuts cleanly.

Wheel dressing has a pronounced affect upon wheel action. A deep dressing cut or a long lead in the path of the dressing tool (such as a single-point diamond) across the wheel produces a relatively coarse wheel face that removes stock efficiently but leaves a poorer surface finish on the workpiece than if dressed more finely.

Forces and Power

A grinding wheel exerts two components of force on a workpiece, a normal F_n and tangential F in the manner shown in Figure 23-12. The ratio $C = F/F_n$ is called the *coefficient of grinding pull* and varies from 0.1–0.3 for snagging to 0.4–0.5 for wet precision grinding. The normal force F_n determines the rate of stock removal in nonprecision grinding. As it is increased with the same wheel speed, the force F and the power increase with the stock removal. Approximately 0.9 kN (200 lb) is the heaviest load for manual application with weights. Mechanical and hydraulic means are utilized for larger forces.

An empirical formula for estimating the power needed for grinding iron and steel with a wheel running at normal grinding speed is in kilowatts,

$$P_s = K_s\sqrt{Q_s} \tag{23-6}$$

where

P_s = power, kW
K_s = value from Table 23-2
Q_s = rate of stock removal, cm^3/min

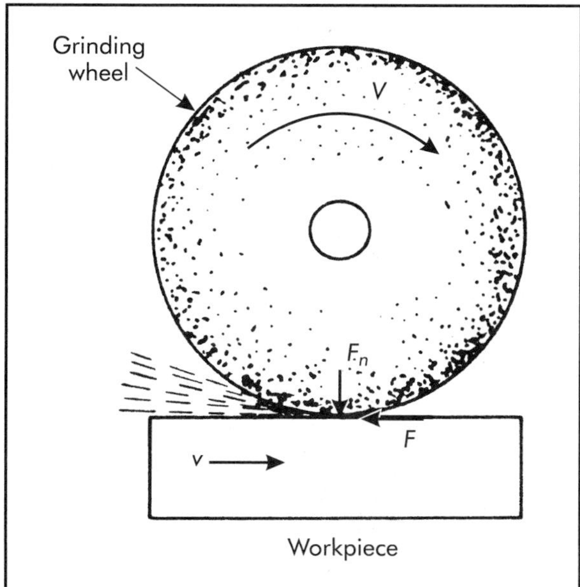

Figure 23-12. Forces exerted by a grinding wheel.

or in horsepower,

$$P = K\sqrt{Q}$$

where

P = power, hp
K = value from Table 23-2
Q = rate of stock removal, in.^{3}/min

Power is one of the least costly items in most grinding operations. However, the energy released when power is applied is important because most of it becomes heat which can crack, check, or soften the ground surface under severe conditions.

Area of Contact

An expression approximating the length of path of an abrasive grain through a workpiece is

$$l = \sqrt{CD/[1+(D/d)]} \qquad (23\text{-}7)$$

where

l = path length of an abrasive grain through a workpiece, mm (in.)
C = depth of cut, mm (in.)

This is really the length of the circular arc *CEA* in Figure 23-11. It shows the same thing as Figure 23-13, that the arc is shortest for external grinding, longer for surface grinding, and longest for internal grinding. Surface grinding with the side of a wheel gives still more contact. A large wheel has a longer arc of contact than a small wheel. A deep cut results in more contact than a shallow cut. The longer the path through the material, the longer each grain is exposed to the high temperatures and pressures of grinding, the more the grains are subject to attrition, and the harder the wheel acts. For this reason, internal grinding calls for softer wheels than external grinding under comparable conditions.

The area of contact is the product of the arc length times the width of contact. As the area of contact increases, the number of grains in contact with the work increases with a uniform grain spacing, and the force tending to separate the wheel from the work is increased. This affects dimensional accuracy and finish in precision grinding and indicates the need for a more open spacing.

Size Effect

Three stages have been discerned in the action of abrasives in grinding. When an abrasive grain first makes contact, it distorts and may displace material slightly, but if it penetrates no more deeply, in effect, it merely rubs as it continues on its course. With somewhat

Table 23-2. Typical values of *K*

	Snagging		Precision Grinding with Wheels 25 mm (≈1 in.) Wide		
	Floor-stand	Swing-frame	Surface	Internal	Cylindrical
Value of K_s	1.1	1.3	1.6	1.75	1.8
Value of K	6	7	8.5	9.5	10.0

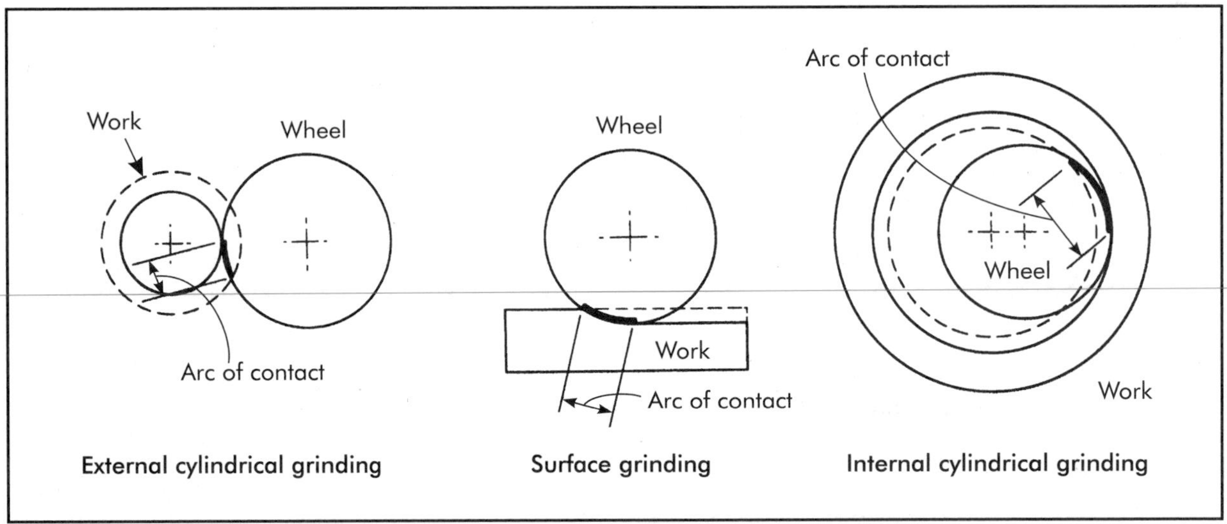

Figure 23-13. Relation between the type of operation and area of contact with the periphery of a grinding wheel.

deeper penetration, metal is plastically extruded around the grain in a ploughing action, and debris is scattered alongside the scratch. If a grain bites deeply enough to fracture metal before it, it cleaves a chip in a true cutting action. All three stages may exist with differently placed grains at one time, and a single grain may go through one, two, or all three stages as it is driven through the work material. The degree of penetration determines the forces and energy required to remove material. Ploughing predominates for chips less than about 0.8 µm (30 µin.) thick with a uniformly high specific energy of cut as evidenced by the plateau on the curve of Figure 23-14. The upturn at the top of the curve is evidence of rubbing. In grinding, the specific energy decreases as the size of cut increases as in other cutting operations.

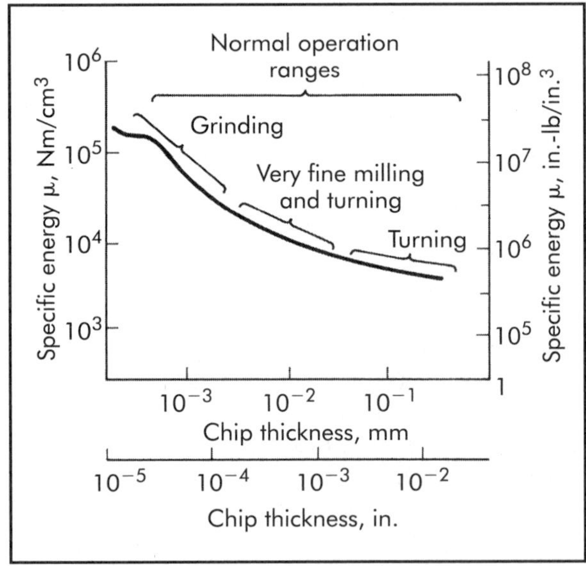

Figure 23-14. Trend of increase in specific energy as chip thickness decreases in metal cutting or grinding.

Thermal Effect

Most of the energy generated by grinding becomes heat. The bulk of the heat passes off with the chips, but some goes into the abrasive grains, and some into the workpiece, and it affects both. Indications are that the instantaneous temperature on a surface being ground and on the contact area of a grain is normally 1,093–1,649° C (2,000–3,000° F). This may cause workpiece temperatures to reach around 427° C (800° F) at 25 µm (.001 in.) depth and 316° C (600° F) at 127 µm (.005 in.) depth in normal operations.

High temperatures are among the chief causes of attrition and fracturing of abrasive grains. The high surface temperatures speed up chemical reactions between abrasive and work materials. For instance, titanium is notoriously difficult to grind because it combines with the surface layer of any common abrasive, quickly dulling the grains. In all grinding operations, sudden heating and then quenching, particularly with coolant, causes thermal shock and fracture of abrasive grains. If surface crumbling alone occurs, attrition results; if the effect is deeper, appreciable chunks are broken off and leave new sharp edges on a grain. The thermal stresses induced in a grain are dependent upon the heat injected, which Hahn has shown as

$$Q = ClA^{0.75}/V^{0.5} \qquad (23\text{-}8)$$

where

Q = thermal stresses induced in a grain (mm^2/min)
C = constant proportional to the ratio of specific heats of abrasive and work material
l = length of grain path, mm (in.)
A = contact area of the grain, mm^2 (in.^{2})
V = wheel speed, m/min (ft/min)

This explains why wheel life is better at high speeds.

The heat of grinding may cause thermal stresses in the material and change its microstructure. In some cases, such as in snagging rough castings, the effects of heat are unimportant. In other cases, such as grinding hardened steel, excessive temperature may ruin the workpiece by softening, burning, rehardening, or cracking the surface material. To correct such situations, an understanding of the factors that cause high temperatures is necessary.

Investigations have shown that surface temperature is a direct function of the depth of cut, wheel speed, and area of contact between the grains and workpiece, all of which increase the rate of energy input. Slow wheel speeds are desirable for heat-sensitive materials. An increase in work speed reduces wheel-work contact time and temperature. Surface temperature is also an inverse function of the chip thickness and thermal conductivity of the work material. A material with a high thermal conductivity value to carry away heat, like low-alloy steel, can be expected to have a lower surface temperature than one like stainless steel with a low thermal conductivity. A thick chip requires less energy for each unit of material removed, and a light cut removes little material. Thus a light cut and thick chip are favorable to low grinding temperatures. In contrast, a thin chip is desirable for good surface finish, as has been noted. Consequently, some conditions favorable to a fine surface finish are opposed to those necessary to keep temperatures down and prevent surface damage.

The Material and Workpiece

Three considerations are necessary to describe how hard or easy a material is to grind. These have been called grindability, finishability, and grinding sensitivity.

Grindability is a measure of the relative ease of removing material. The rate of wheel wear as indicated by the volume ratio is considered a satisfactory index of grindability because it generally agrees with the evaluations made in practice. Power consumption also might be considered an index because it is an indication of the amount of heat involved, but it does not correlate with the volume ratio. The grindability of hardened tool steels has been found to decrease to a great extent with an increase in the amount of hard carbide particles present in them. Sulfur in steel, particularly in tool and stainless steels, improves grindability as much as ten times. The most difficult steels to grind have been reported to cause wheels to wear 200 times faster than the easiest to grind.

Finishability is related to the relative cost of putting a fine finish on a material. In general, materials of medium grindability have been found to have the poorest finishability. As a rule, hard materials with low grindability seem to have good finishability because they dull the grains and prevent deep scratches.

Grinding sensitivity is the propensity of a material to crack or lose its surface hardness when ground. Untempered martensite in high-speed steel, excessively high carbon content in the outermost layers of carburized steels, and retained austenite in medium- and high-alloy steels are conditions that have been shown to cause hypersensitivity. Even sensitive steels can be ground if the work is done slowly and carefully enough, but that is expensive. Corrective measures during heat treatment generally can eliminate causes of ultrasensitivity and keep costs low. Where some sensitivity still exists, control of the conditions that keep grinding temperatures low is necessary.

The condition of the rough workpiece determines much of the success of a grinding operation. Too much stock requires an unnecessary amount of time and wheel wear for grinding; too little stock may mean that a piece will not clean up before finished size is reached. A case-hardened piece may be left with a soft surface if too much material is removed. Warpage and runout determine to a large extent the amount of stock that must be provided for grinding. The better the control of these factors, the more efficient a grinding operation. Typical practice is to allow stock of 0.25–0.41 mm (.010–.016 in.) for rough and 0.05–0.15 mm (.002–.006 in.) on the diameter for finish cylindrical grinding, depending on the size of the workpiece. About half as much is left on a side for surface grinding.

23.4.5 Cutting Fluids

Water solutions are widely used for grinding because they cool well. Water alone markedly increases wheel wear, but chemical additives to deter or alter reactions may materially reduce wear. Oils may lubricate chips and abrasive grains, mitigate wheel loading, and inhibit chemical reactions, thus retarding attrition. Combinations of additives to oils are beneficial; one series of tests showed that combinations of soapy fats, sulfur, and chlorine increased the grinding ratio tenfold.

Efficient use of cutting fluids presents unique problems in grinding. High wheel speed tends to eject fluid at the grinding zone, and the higher the speed, the worse the problem. In most cases, the zone is flooded copiously. Other remedies, such as high-pressure nozzles and special guards to contain the fluid, are available. All fluids are helpful in removing particles but tend to hold them in suspension, and care must be taken to filter or settle out the particles so they are not recirculated to cause scratches.

23.4.6 Economics

Experiments have shown that for conventional grinding, operating costs decrease but grinding wheel costs increase as the rate of material removal is increased. This is illustrated by the two solid line cost curves in Figure 23-15(A). Summing those two costs produces a total cost relationship represented by the broken-line curve C_t. That curve indicates that total grinding costs fall to a minimum and then rise as the rate of material removal increases. Thus, minimum grinding costs are achieved when optimal material removal rates are selected. There are, however, other factors which must be considered in the selection of grinding parameters. For example, surface finish and dimensional accuracy often suffer if material removal rates are set too high. This means that a trade-off decision must be made between the economies of the process and the accuracy of the results. Since the need for accuracy and good surface finish often dictate the need for grinding, then it probably follows that optimal material removal rates are not going to be achieved. On the other hand, accuracy and finish are not the limiting factors in many abrasive machining, roughing, and nonprecision grinding operations. The aim then is to operate at the lowest cost.

A number of empirical relationships have been developed through experimentation to define grinding cost in terms of a volume ratio R and the rate of material removal M. Generally, the volume ratio of a grinding operation decreases as the rate of material removal increases as illustrated in Figure 23-15(B). This phenomenon can be approximated by the empirical relationship

$$RM^n = k \tag{23-9}$$

where

R = volume ratio
M = rate of material removal, cm³/min (in.³/min)
n, k = experimentally determined value for each type of grinding wheel and material being ground

From this, the unit cost of grinding in dollars per unit volume of material removed is

$$C_t = C_o + C_g = \frac{L}{M} + \frac{W}{R} = \frac{L}{M} + \frac{WM^n}{k} \tag{23-10}$$

where

C_t = total cost, \$
C_o = operating cost, \$
C_g = grinding wheel cost, \$
L = cost of labor and overhead, \$/hr
W = grinding wheel cost, \$/unit volume of material removed

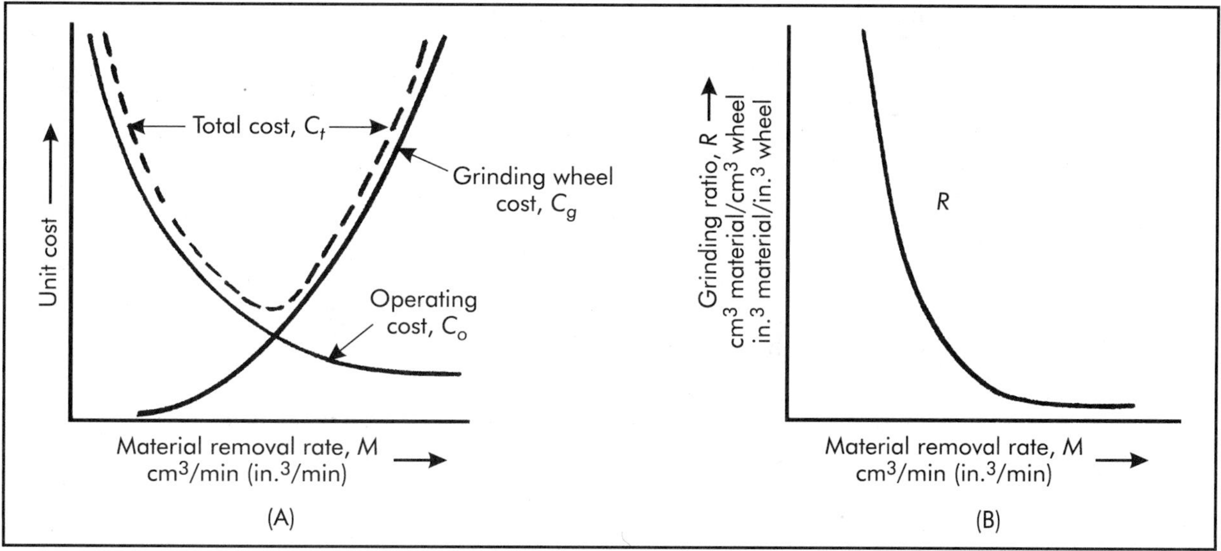

Figure 23-15. (A) Typical cost curves for grinding as a function of material removal rate. (B) Typical grinding ratio for grinding as a function of material removal rate.

If the expression is differentiated with respect to M, and the derivative is set equal to zero, the rate of material removal for minimum cost is found to be

$$M_m = \left(\frac{kL}{nW}\right)^{\frac{1}{n+1}} \tag{23-11}$$

where

M_m = rate of material removal for minimum cost, cm³/min (in.³/min)

This occurs when the machine cost equals r times the grinding wheel cost expressed as

$$\frac{L}{M} = r\frac{WM^n}{k} \tag{23-12}$$

To illustrate the use of the equations, suppose that a particular type of grinding wheel is recommended for surface grinding stainless steel plates. Experiments have shown that this wheel performs in accordance with the relationship $R_1 M_1 = k_1$, which is the same as Equation (23-9) with $n = 1$. In this relationship, R_1 is in Mg/cm³ (tons/in.³) and M_1 is in Mg/hr (tons/hr). For $n = 1$, Equation (23-12) can be rewritten as

$$\frac{L}{M_1} = \frac{LM_1}{k} = \frac{W}{R_1} \text{ for } n = 1$$

Unit wheel cost W is estimated to be \$0.02/cm³ (\$0.327/in.³), and the labor and overhead rate L = \$16/hr. Only affected overhead costs are included. Trials with wheels of two grades, X and Y, showed the results outlined in Table 23-3. A review of these experimental results reveals that the grade Y wheel removes stock at a higher rate but with greater wheel cost than grade X. However, the machine costs and total cost are less for the grade Y wheel. Thus, it would appear that the grade Y wheel would be the more economical of the two.

23.5 QUESTIONS

1. What are the four important properties of abrasive materials?
2. What are the two ways that abrasives deteriorate in service?
3. Name the two most popular conventional abrasives.
4. Describe how aluminum oxide and silicon carbide abrasives are made.

Table 23-3. Results of trials with two grades of wheels

Wheel grade	Production Rate, M_l Mg/hr (tons/hr)	Volume Ratio, R_l Mg/cm^3 (tons/in.^{3})	Machine Cost, L/M_l \$/Mg (\$/ton)	Wheel Cost, W/R_l \$/Mg (\$/ton)	Total Cost, C_t \$/Mg (\$/ton)
X	0.1823	0.00044	87.74	45.06	132.80
	(.201)	(.008)	(79.60)	(40.88)	(120.48)
Y	0.2250	0.00039	71.12	50.77	121.89
	(.248)	(.0071)	(64.52)	(46.06)	(110.58)

5. Which of the conventional abrasives is best suited for general-purpose grinding operations? Why?
6. Name and describe some of the properties of the two superabrasives.
7. Describe how man-made diamonds are produced.
8. Why are diamonds not suitable for grinding steel?
9. Why are diamond abrasives often coated with a metallic coating?
10. What is the difference between a monocrystalline and a microcrystalline type of cubic boron nitride abrasive?
11. How is the size of an abrasive grain designated?
12. Name and describe the principal bonds for grinding wheels.
13. What is meant by the grade of a grinding wheel?
14. What is meant by the structure or spacing of a grinding wheel?
15. Describe common types and shapes of grinding wheels.
16. How are grinding wheels manufactured?
17. What are "sticks," "stones," "coated abrasives," and "polishing wheels", and how are they used?
18. Describe nonprecision and precision grinding.
19. What considerations enter into the selection of a grinding wheel?
20. What considerations influence the selection of an abrasive?
21. When are coarse grains and when are fine grains preferred?
22. What considerations determine what grain spacing a grinding wheel should have?
23. How and why are grinding wheels balanced?
24. What are "trueing" and "dressing" of grinding wheels, and how are they done?
25. Explain how the grain depth of cut, surface finish, and rate of wheel wear are affected by the work speed, wheel speed, wheel depth of cut, wheel diameter, and work diameter.
26. How does the length of contact between wheel and workpiece influence a grinding operation?
27. What conditions are favorable to a low surface temperature in grinding and why?
28. How do the conditions favorable to a low surface temperature compare with those favorable to a good surface finish?
29. What is meant by "grindability," "finishability," and "grinding sensitivity"?
30. What determines the right amount of grinding stock on a workpiece?

23.6 PROBLEMS

1. Estimate the normal force that must be applied for grinding steel to:
   (a) Remove 10 kg (22 lb) of metal/hr on a swing-frame grinder with a resinoid wheel running at 2,850 m/min (9,350 ft/min).
   (b) Grind wet at the rate of 8 cm^3/min (.5 $in.^3$/min) of metal removal on an external cylindrical grinder with a vitrified wheel having a surface speed of 2,050 m/min (6,725 ft/min).
2. A 250-mm (≈10-in.) diameter soft carbon steel shaft is to be traverse ground with coolant and a wheel 50 mm (≈2 in.) wide running at 1,830 m/min (≈6,000 ft/min) on a cylindrical grinder. The motor is rated at 11 kW (15 hp) and the machine efficiency is 80%. Machine rigidity allows a work-wheel normal force of 890 N (200 lb). What is the traverse rate permitted with a depth of cut of 50 μm/pass (≈.002 in./pass)?
3. What change in each of the following variables in a grinding operation will tend to give a finer surface finish? (a) wheel speed; (b) work speed; (c) grain size; (d) depth of cut; and (e) wheel size.
4. Metal is being removed at a rate of 5 kg/hr (11 lb/hr) with a power consumption of 4 kW (5.4 hp) at the wheel in a snagging operation on a swing-frame grinder. The wheel has a surface speed of 1,830 m/min (≈6,000 ft/min) and is pushed against the work with a force of 800 N (≈180 lb). To what amount must the normal force be raised to increase the metal removal rate to 7 kg/hr (≈15 lb/hr)?
5. A motor on a floor-stand grinder is delivering 4 kW (5.4 hp) to drive a vitrified grinding wheel at 1,830 m/min (≈6,000 ft/min) to remove 16.4 cm^3/min (1 $in.^3$/min) of metal. The drive efficiency is 80%. If a resinoid wheel is put on the stand and driven at 2,740 m/min (≈9,000 ft/min) and the same normal and tangential forces are maintained at the point of contact, what power must the motor deliver at the same efficiency? If the K factor of Equation (23-6) for the resinoid wheel is 6.6 (K_s = 1.22), what rate of metal removal may be expected from that wheel?
6. Two wheels of a certain type are tried for a swing-frame grinding operation. Grade A has a life of 13.11 hr with a production rate of 0.152 Mg/hr (.168 ton/hr). Grade B has a life of 10.17 hr with a production rate of 0.173 Mg/hr (.191 ton/hr). The cost of either wheel is \$165 and the labor and overhead rate is \$18.00/hr. $RM = k$ for this type of operation. Which wheel is more economical? Is there any indication that either is the most economical available?
7. What should be the approximate properties of a grinding wheel with respect to abrasive, grain size, bond, grade, and structure for each of the following typical applications?
   (a) to commercially grind a mildsteel shaft;
   (b) for snagging cast iron; and
   (c) to sharpen a high-speed steel milling cutter.
8. Meehanite GC class 40 gray iron bars 75 × 75 × 300 mm (≈3 × 3 × 12 in.) in size were surface ground with a segmented wheel 450 mm (≈18 in.) in diameter. The relationship $RM^{3.33}$ = 40.000 was found as the grinding ratio for M in $in.^3$/min, $RM^{3.33} = 4.4 \times 10^8$ for M in cm^3/min. Abrasive cost was estimated at 1.26 cents/cm^3 (20.6 cents/$in.^3$) and labor and overhead rate was \$18.00/hr. What rate of metal removal gives the lowest cost, and what is the cost?
9. Five wheels 900 mm diameter × 100 mm wide (≈36 in. × 4 in.) tried for a precision grinding operation perform as follows:

	Wheel				
	A	B	C	D	E
Rate of production, cm^3 in 8 hr (in.^{3} in 8 hr)	14,900 (≈910)	8,620 (526)	17,370 (1,060)	11,300 (≈690)	13,600 (≈830)
Volume ratio (cm^3 stock/cm^3 wheel, in.^{3} stock/in.^{3} wheel)	4.74	3.35	2.45	6.31	5.5

Wheel cost is $0.01/cm^3 ($0.16/in.^{3}) and the overhead rate is $18.00/hr. Which wheel would you recommend for use?

23.7 REFERENCES

ANSI B74.13-1990. "Markings for Identifying Grinding Wheels and other Bonded Abrasives." New York: American National Standards Institute.

ANSI B74.3-1986. "Shapes and Sizes for Diamond or CBN Abrasive Products, Specifications for." New York: American National Standards Institute.

ANSI B74.2-1982. "Shapes and Sizes of Grinding Wheels, and Shapes, Sizes and Identification of Mounted Wheels, Specifications for." New York: American National Standards Institute.

ANSI B74.12-1976 (R1982). "Size of Abrasive Grain—Grinding Wheels, Polishing and General Industrial Uses, Specifications for." New York: American National Standards Institute.

Aronson, R. B., ed. 1995. "Where's CBN Going." *Manufacturing Engineering,* March.

Drozda, T., ed. 1982. *Manufacturing Engineering Explores Grinding Technology.* Dearborn, MI: Society of Manufacturing Engineers (SME).

Giles, M. T. 1982. "New Generation of Superabrasives." *Manufacturing Engineering*, June.

Koepfer, C., ed. 1994. "Grit, Glue—Technology Too!" *Modern Machine Shop*, December.

Lewis, K. B. and Schleicher, W.F. 1976. *The Grinding Wheel: A Textbook of Modern Grinding Practice*, 3rd Ed. Cleveland, OH: The Grinding Wheel Institute.

Owen, J.V., ed. 1991. "Dress for Success." *Manufacturing Engineering*, August.

Society of Manufacturing Engineers (SME). 1995. *Basics of Grinding* video. *Fundamental Manufacturing Processes* series. Dearborn, MI: SME.

Subramanlan, K. 1991. "Grind Your Ceramics Aggressively." *Manufacturing Engineering*, August.

Tool and Manufacturing Engineers Handbook, 4th Ed. 1983. Vol. 1. *Machining*. Dearborn, MI: Society of Manufacturing Engineers (SME).

Grinding Machines and Methods

24

Universal grinding machine, 1868
—Joseph R. Brown, Providence, RI

Grinding machines utilize grinding wheels and may be classified according to the types of operations described in Chapter 23. Thus the broad classes are precision and nonprecision grinders. The main types of precision grinding machines are external and internal cylindrical grinding machines and surface grinding machines. Certain types have been developed to do specific operations and are classified accordingly.

The basic types of manually operated grinding machines are described in the early part of this chapter. Most of these are also available with semi-automatic, fully automatic and/or CNC features, which make them more suitable for production environments. Representative types of these are briefly described in the later part of the chapter.

24.1 PRECISION GRINDERS

24.1.1 Cylindrical Center-type Grinders

Cylindrical center-type grinders are used for grinding straight and tapered round pieces, and round parts with curved lengthwise profiles, fillets, shoulders, and faces. A simple (and the original) form of cylindrical grinder is a head that revolves a grinding wheel mounted on the cross slide or compound rest of a lathe. These *tool post grinders* are made to deliver from a small fraction to about 7.5 kW (10 hp) at costs from under $1,000 to several thousand dollars. They are expedient when regular grinding machines are not available.

A workpiece is usually held between dead centers and rotated by a dog and driver on the faceplate on a *plain cylindrical center-type grinder* as depicted in Figure 24-1. The centers are held in the headstock and footstock of the machine and do not revolve to provide the most rigid work support and accuracy between centers. The headstock and footstock are carried on an upper or swivel table and can be positioned to suit the length of the workpiece. The upper table can be swiveled and clamped on a lower table to adjust for grinding straight or tapered work as desired. The amount of taper that can be ground in this way is normally less than 10° because the table bumps the wheel head if swiveled too far. The lower table slides on ways on the bed of the machine to traverse the work past the grinding wheel and can be moved by hand or power.

The grinding wheel revolves in the direction shown in Figure 24-1 on a heavy spindle—running in close fitting bearings to prevent flutter. The grinding force is directed downward for stability. The wheel head carries the wheel to and away from the work. The movement is called *infeed* and normally can be controlled to 5 μm (.0002 in.) or less on manually controlled grinders and to as little as 0.7 μm (≈25 μin.) on some automatic grinders.

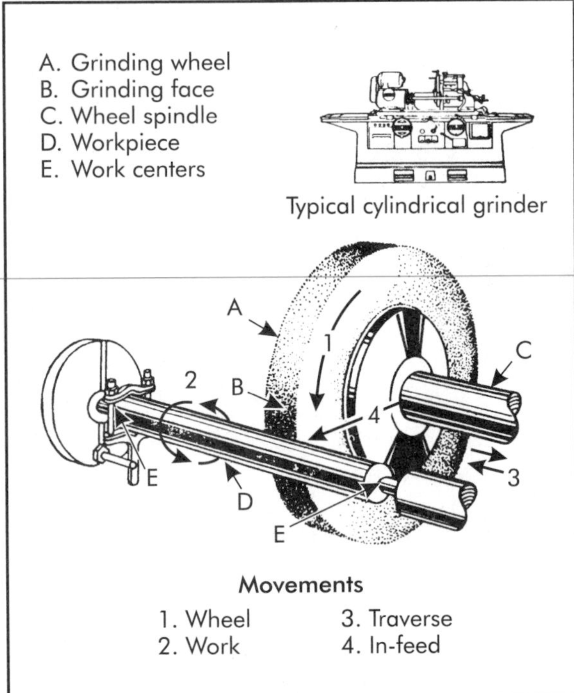

Figure 24-1. Movements of an external cylindrical grinding machine. (Courtesy Carborundum Abrasives, Inc.)

Cylindrical grinding may be done by the plunge cut or traversing methods. Similar methods are found in surface grinding. In *plunge cut grinding,* a wheel somewhat wider than the work surface is fed in as the workpiece revolves. With other factors constant, the depth of cut and rate of stock removal depend upon the rate of in-feed. A work surface wider than the wheel is *traverse ground* by traversing the revolving workpiece lengthwise with respect to the wheel, or vice versa. The rate of traverse becomes one of the conditions determining the length of grain path. The radial depth of cut is determined by the amount the wheel is fed in at each reversal of the traverse. An advance along the workpiece of one quarter to one half of the width of the wheel per revolution of the work in cylindrical grinding or per stroke in surface grinding is common. The leading portion of the wheel removes most of the stock and the remainder cleans up the surface.

The necessary steps in a precision grinding operation include bringing the wheel at a rapid rate to the workpiece (or vice versa) and then at feed rate to grind to size. The feed is continuous for a plunge cut but incremental (for each pass) for traverse grinding. The wheel is allowed to dwell momentarily at the end of the cut for cleanup and is then withdrawn at a rapid rate. The usual treatment for one piece of a kind is to rough and finish grind it completely in one operation, with the wheel trued at least once for the final finishing pass or passes. Or, a lot of pieces of the same kind may be all rough ground in one operation (leaving a small amount of stock) and then finish ground in another operation with a different feed rate, dressing, and even a different wheel.

In job shops and toolrooms, the operator may control all the steps. For example, the controls on the cylindrical grinder shown in Figure 24-2 are all conveniently located at the front of the machine. The handwheel on the left operates the table traverse along the Z axis or, that wheel may be disengaged for hydraulic operation of the table traverse. The handwheel on the right operates the in-feed of the grinding head along the X axis. Positioning of the wheel head along this axis is controlled to within 0.05 mm (.0002 in.) on the workpiece diameter by referring to a graduated collar on the handwheel. Finer adjustments to 0.0013 mm

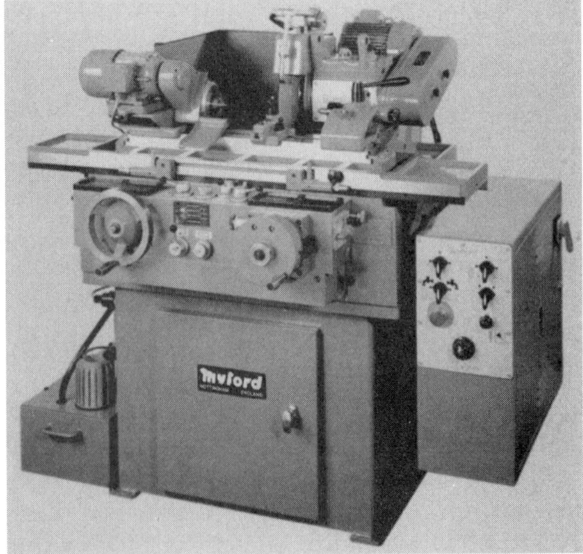

Figure 24-2. Precision hydraulic cylindrical grinder. (Courtesy Royal Master Grinders, Inc.)

(.00005 in.) on the workpiece diameter are made by means of an adjusting screw located just above the handwheel.

The grinding machine shown in Figure 24-2 has a 127 mm (5 in.) swing over the table. It is capable of grinding diameters up to 76 mm (3 in.) and workpiece lengths up to 305 mm (12 in.). To facilitate traverse grinding in toolroom and low-volume production environments, the machine has a hydraulic table traverse rate from 0–3,048 mm/min (0–120 in./min) with a variable dwell at each end of the table traverse. In addition, hydraulic rapid advance and retraction of the wheel head facilitates loading and unloading. Automatic mechanical in-feed of the wheel head is also provided for traverse grinding. The basic cost of a machine like the one in Figure 24-2 is about $45,000. The same size machine equipped with automatic controls for both traverse and plunge grinding is available at a cost of around $66,000.

Automatic cylindrical grinders may be equipped with automatic size-control devices similar to the one shown in Figure 24-3. The gage head for this unit is mounted on a slide that can be hydraulically positioned along the *X* axis for rapid measurement of workpiece diameters. With this device, wheel head in-feed changes for both traverse and plunge grinding are made without the necessity of stopping the machine for measurements. Since sizing takes place directly from the workpiece, the device also may be used to automatically compensate for grinding wheel wear. The auto-size control permits production grinding tolerances to be held to within 0.003 mm (.0001 in.).

Grinding wheels may be arranged in a number of ways. The wheel heads on some plain grinders are set at an angle, so the periphery of a wheel can grind workpiece shoulders and faces as well as diameters. One or several wheels may be mounted on a machine and trued to a certain shape or shapes for a workpiece ground in large quantities. Examples are given in Figure 24-4.

Cylindrical center-type grinders capable of swinging diameters in excess of 508 mm (≈20 in.), particularly for grinding rolls for rolling mills, are called *roll grinders*. Roll grinders commonly have cambering attachments to enable them to grind accurate and reproducible curved lengthwise profiles on rolls. Rolls are ground in this way so they become straight when deformed by high temperatures and pressures in operation.

Figure 24-3. Cylindrical grinding with automatic size control. (Courtesy Royal Master Grinders, Inc.)

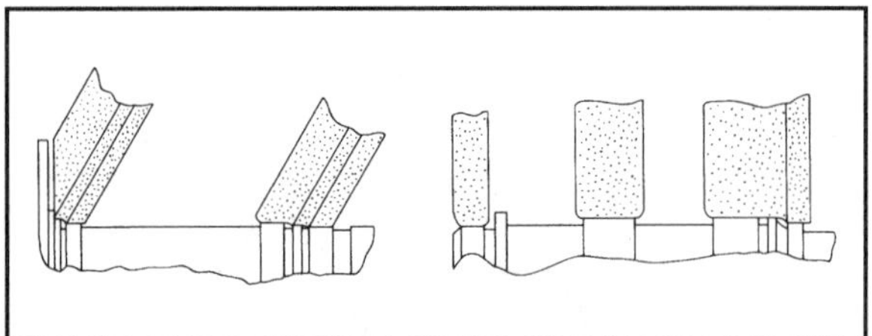

Figure 24-4. Typical multiwheel arrangements for production grinding.

A *universal cylindrical center-type grinder* has all the units and movements of a plain grinder, along with the following added capabilities.

1. Its headstock spindle may be used alive or dead (rotated or not), so that work can be held and revolved by a chuck as well as between centers.
2. Its headstock can be swiveled in a horizontal plane so that any angle, even a flat plane, can be ground on a workpiece chucked on the headstock spindle.
3. Its wheel head and slide can be swiveled and traversed at any angle in the manner indicated in Figure 24-5 so that any taper can be ground on work between centers.

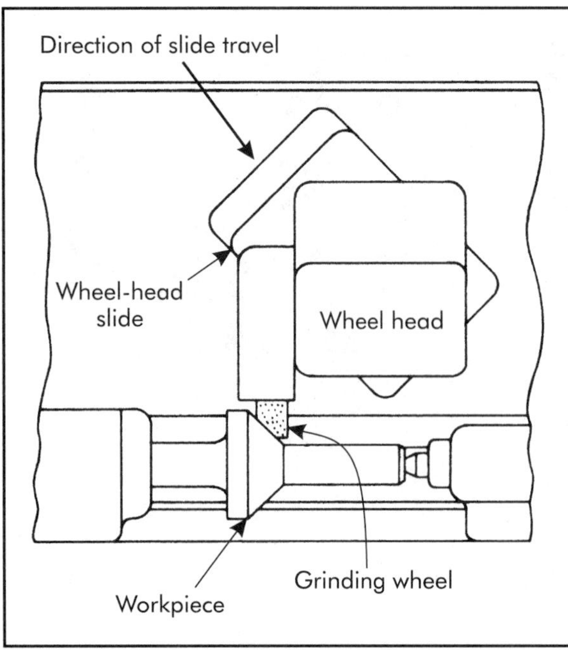

Figure 24-5. Universal center-type grinder arranged for grinding a steep taper.

Most universal grinders can be arranged for internal grinding by the addition of an auxiliary wheel head to revolve small wheels at high speeds.

Universal grinders can grind surfaces like steep tapers and holes not accommodated on plain grinders, but they sacrifice rigidity, power, and rapid output because of their flexibility. They are found in toolrooms, jobbing shops, and sometimes on production jobs where they grind shapes that are hard to grind on plane grinders.

The size of a cylindrical center-type grinder is usually designated by the diameter and length in mm (in.) of the largest workpiece the machine can nominally take between centers. Thus a 300 × 1,000 mm (≈12 × 40 in.) center-type grinder can swing a workpiece 300 mm in diameter over the table and grind it with a new wheel. A workpiece up to 1,000 mm long can be mounted between centers. Specifications of several types and sizes of center-type grinders are given in Table 24-1.

24.1.2 Chucking Grinders

Chucking grinders are designed for grinding small- and medium-diameter short parts automatically and in large quantities. Typical applications are the grinding of tapered roller bearing cone races, valve tappets, and small bevel gear shoulders and stems. The workpiece is held in a chuck, collet, or fixture.

24.1.3 Centerless Grinders

An *external cylindrical centerless grinding machine* revolves a workpiece on top of a work rest blade between two abrasive wheels as shown in Figure 24-6. The grinding wheel removes material from the workpiece. The workpiece has a greater affinity for and is driven at the same surface speed as the regulating wheel, which is normally a rubber-bonded abrasive wheel that turns at a surface speed of 15–61 m/min (50–200 ft/min).

Through-feed centerless grinding is done by passing the workpiece completely through the space between the grinding and regulating wheel, usually with guides at both ends, as indicated in Figure 24-7(A). The regulating wheel is tilted a few degrees about a horizontal axis perpendicular to its own axis. This feeds the work-

Table 24-1. Typical specifications of center-type cylindrical grinders

Type	Nominal Size, mm (in.)	Main Motor, kW (hp)	Weight, Mg (lb)	Cost
Universal, manual	320 × 991 (12.6 × 39)	3.7 (5)	3.1 (6,800)	$ 38,000
Plain, manual	445 × 2,197 (17.5 × 86.5)	11.2 (15)	6.5 (14,300)	76,000
Universal, automatic cycle	838 × 3,048 (33 × 120)	37.3 (50)	15 (33,000)	130,000
Profile and step grinder, numerical control (external and internal)	406 × 813 (16 × 32)	3.7 (5)	5.9 (13,000)	150,000

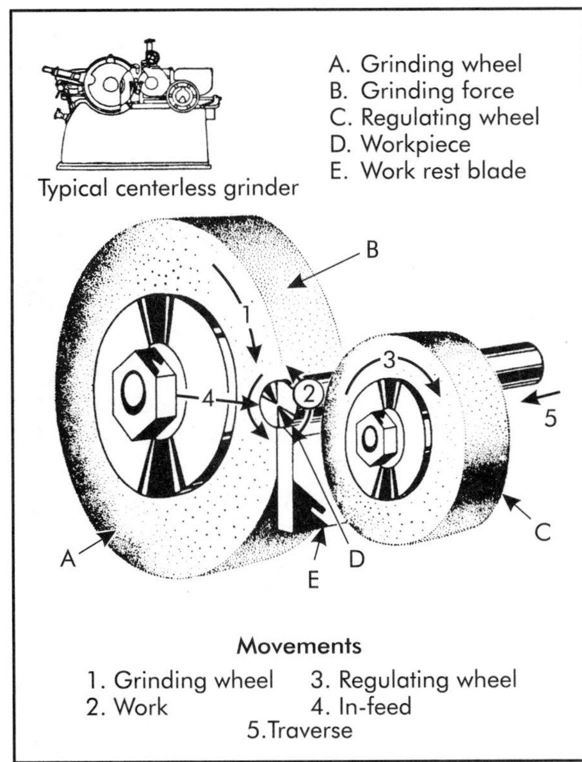

Figure 24-6. Action of an external cylindrical centerless grinding machine. (Courtesy Carborundum Abrasives, Inc.)

pieces lengthwise. A slow feed is necessary to remove relatively large amounts of stock within the power capacity of the machine and to produce accuracy and good surface finish. Most through-feed work is ground in two passes with a total stock removal of 0.25–0.38 mm (.010–.015 in.). The rate of feed depends upon the angle of inclination and speed of the regulating wheel as expressed by the relation

$$F = \pi DN \sin \alpha \qquad (24\text{-}1)$$

where

F = feed rate, mm/min (in./min)
D = diameter, mm (in.)
N = speed in rpm
α = angle of tilt of the regulating wheel, which may be from 0–8°

In-feed centerless grinding is slower than through-feed grinding but is necessary for a workpiece with a shoulder, head, or obstruction that prevents its passing completely through the throat between the wheels. Two examples of in-feed grinding are given in Figure 24-7(B). The work rest and regulating wheel are withdrawn from the grinding wheel to receive a workpiece. They are then moved toward the grinding wheel to grind the work to size. A variation of in-feed grinding utilizes a *cam regulating wheel* that is recessed and relieved over part of its periphery. Work is unloaded, loaded, and fed in as the cutaway side of the wheel passes the work rest (at about 16 rpm). In that way, the work rest and regulating wheel need not be slid to and from the cut; one piece is ground in each revolution of the regulating wheel.

End-feed centerless grinding is for tapered work, as shown in Figure 24-7(C). Either the grinding wheel or regulating wheel or both are trued to a taper. The work is fed lengthwise between the wheels and is ground as it advances until it reaches the end stop.

A medium-size centerless grinder that takes workpieces up to 127 mm (5 in.) in diameter is driven by a 11.2 kW (15 hp) motor and is quoted at $85,000 for manual operation, $95,000 with

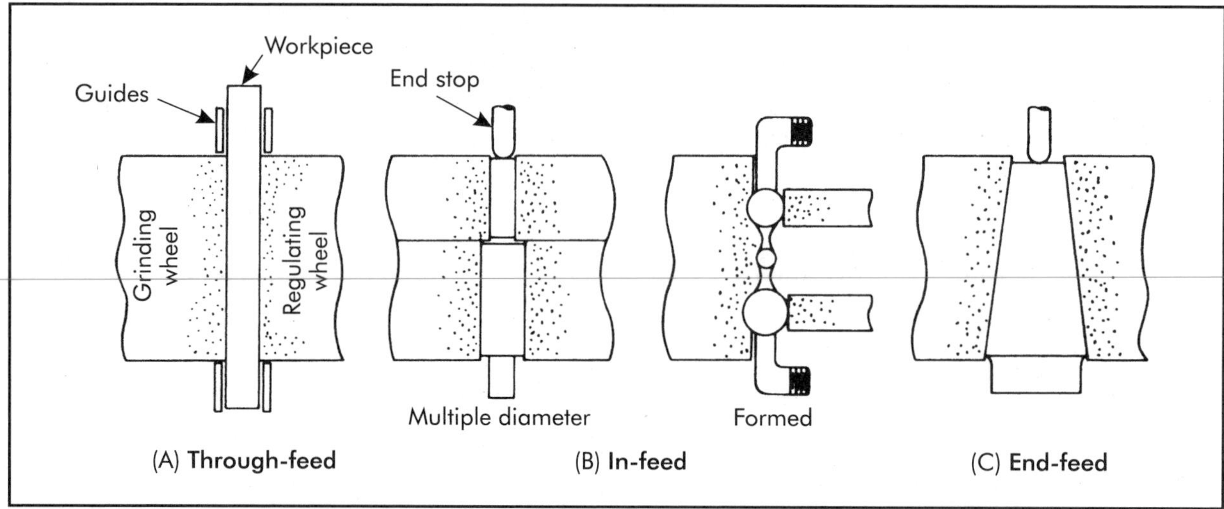

Figure 24-7. Centerless grinding operations.

automatic in-feed, and $350,000 with fully automatic numerical control.

24.1.4 Comparison of Center-type and Centerless Grinders

As a rule, more time is needed to set up a centerless grinder, but the difference is not large for many simple parts. Much can be done to minimize centerless setup time by scheduling similar parts in successive lots over the same machine. Parts with several diameters or curved or tapered profiles usually require special work rests, supports, wheel mounts, truing devices, or other equipment for centerless grinding. Those adjuncts may cost hundreds or thousands of dollars and take considerable setup time. Therefore, such parts are not centerless ground unless produced in large quantities.

Centerless grinding is faster and often cheaper than center-type grinding because:

1. It is almost continuous, especially for through-feed grinding, with a minimum of machine time lost for loading and unloading.
2. The work is fully supported by the work rest blade and regulating wheel and can be subjected to cuts as heavy as it will take without overheating. Plunge cuts can be made over the entire length of a workpiece. Although most centerless grinding wheels are 152–203 mm (6–8 in.) wide, machines are available with wheels up to 1 m (≈40 in.) wide, and power to 75 kW (100 hp), and can plunge grind correspondingly long and slender or frail work (such as thin-walled tubes).
3. With large grinding wheels, wheel wear is relatively small, and a minimum amount of adjustment is necessary to compensate for wear.
4. No axial thrust is present, as it is on work between centers. Long thin pieces are not so likely to be distorted.
5. The action of centerless grinding is such that each workpiece is cleaned up with the removal of the least possible amount of stock. Errors of centering are eliminated.
6. Adjustments for size are made directly on the diameter of the workpiece and that contributes to accurate results. If the regulating wheel and work rest blade are moved 25 mm (.001 in.) toward the grinding wheel, the workpiece diameter is reduced 25 mm (.001 in.).
7. Much of the time, a low order of skill is needed to tend a centerless grinding machine.
8. Center holes are costly if needed for center-type grinding only.

For simple parts, the savings in grinding time for only a few pieces may make up for the longer setup time required for centerless grinding. Centerless grinding has been found profitable in many places for lots less than 100 pieces, sometimes for 12 or less. However, as a rule, center-type grinding is preferable where the work is varied, irregular in shape, or large in size, especially in small quantities.

Equation (19-8) is helpful in showing for what quantities of production a centerless grinder is economical for any particular product. As an example, an armature shaft for an electric motor is to be ground on four diameters with stock removal of 0.25–0.38 mm (.010–.015 in.) on each diameter. The shafts can be finished on a plain cylindrical center-type grinder at the rate of 125 pieces/hr, with 30 min required for setup of each lot. A production rate of 150 pieces/hr can be realized with setup time of 1 hr per lot on the centerless grinder. This machine must have special equipment for the job that costs $2,400, which must be paid for by the first lot. The labor, overhead, and depreciation rate is $18/hr for either machine. The equal cost quantity in this case is

$$Q = \frac{125 \times 150[(1 \times 18) + 2{,}400 - (.5 \times 18) - 0)]}{(150 \times 18) - (125 \times 18)}$$
$$= 100{,}375 \text{ pieces} \qquad (24\text{-}2)$$

The centerless grinder is economical for this job if about 100,000 or more pieces are to be produced. Figure 24-7 indicates a principle of economical grinding for moderate and large quantities. That is to use a wide wheel or several wheels on one mount trued to take care of two or more surfaces at the same time. This is especially applicable to centerless grinding because the work is well supported for multiple cuts. However, the principle is often applied profitably to all other forms of grinding, as suggested by Figure 24-4.

Equation (24-3) provides the basis for ascertaining when it is economical to combine grinding operations. For example, three diameters may be ground one at a time on the stem of a bevel gear pinion with an output of 35 pieces/hr. The combined operation produces 80 pieces/hr but requires a special wheel mount and cam for the truing attachment that costs $1,200 and 1.5 hr more setup time for each lot. The labor rate in the plant is $10.50/hr with a labor dollar overhead rate of 1. A rate of 35% for interest, insurance, taxes, and maintenance is required. The number of pieces in one lot for which the combined operation is justified may be ascertained from

$$N(\tfrac{1}{35} - \tfrac{1}{80})(10.50)(1 + 1)$$
$$= (1{,}200 \times 1.35) + (1.5 \times 21) \qquad (24\text{-}3)$$

Thus $N = 4{,}893$ pieces is the smallest quantity for which it is economical to invest in the special wheel mount. It is presumed that the machine has the power to sustain the faster rate of production.

24.1.5 Internal Grinding

Machines

A *chucking internal grinder* holds the workpiece on a faceplate or in a chuck or fixture and rotates it around the axis of the hole ground, as depicted in Figure 24-8. The revolving grinding wheel is reciprocated lengthwise through the hole and is fed crosswise on a slide to engage the workpiece. The work head, and sometimes the wheel head, can be swiveled to adjust for straight and tapered holes and faces.

A production-type internal grinder with automatic cycling can grind holes up to 100 mm (≈4 in.) in diameter by 200 mm (≈8 in.) long, has a 2.2-kW (3-hp) motor, and costs under $50,000.

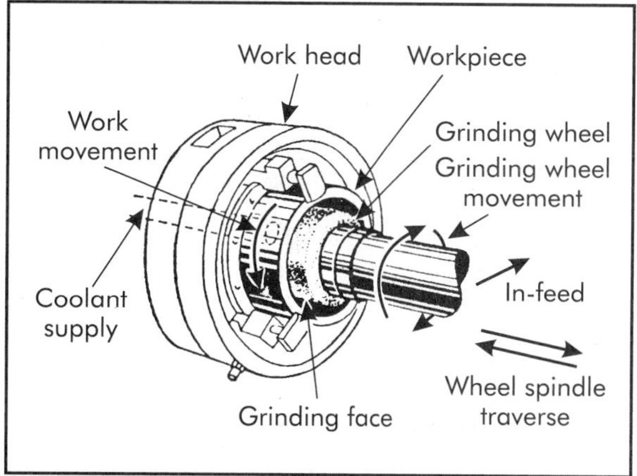

Figure 24-8. Action of a chucking internal grinding machine. (Courtesy Carborundum Abrasives, Inc.)

A universal manual toolroom type grinds holes to 300 mm (≈12 in.) in diameter by 400 mm (≈16 in.) long, has a 3.7-kW (5-hp) motor, and a price of $65,000. An internal grinder of similar size costs over $200,000 with numerical control for fully automatic grinding of round and flat surfaces, steps, profiles, lobes, etc.

A *centerless internal grinder* grinds the bore of a round workpiece concentric with the outside surface. The workpiece is held and revolved amid three rolls in a roll-type centerless internal grinder as shown in Figure 24-9. The large roll is the driver; the rolls provide rigid support for the workpiece. The movements of the grinding wheel are the same as shown in Figure 24-8. On the shoe-type centerless grinder, the workpiece is revolved against two fixed and hard shoes by the action of the grinding wheel and a rotating end-backing plate. The shoes contact more area than the rolls, bridge irregularities, and produce better average concentricity between the hole and outside.

Straight, tapered, continuous, interrupted, open-end, and blind holes and grooves are ground on centerless internal grinders. The wall thickness of a finished piece is quite uniform. The manner of holding the work lends itself well to automatic unloading and loading of pieces.

A planetary internal grinder is designed for parts too large or unwieldy to be conveniently rotated. The workpiece is not revolved. Instead, the grinding wheel is orbited around the axis of the hole to be ground.

A *vertical internal grinder* takes large pieces that can be rotated on a table like on a vertical boring mill.

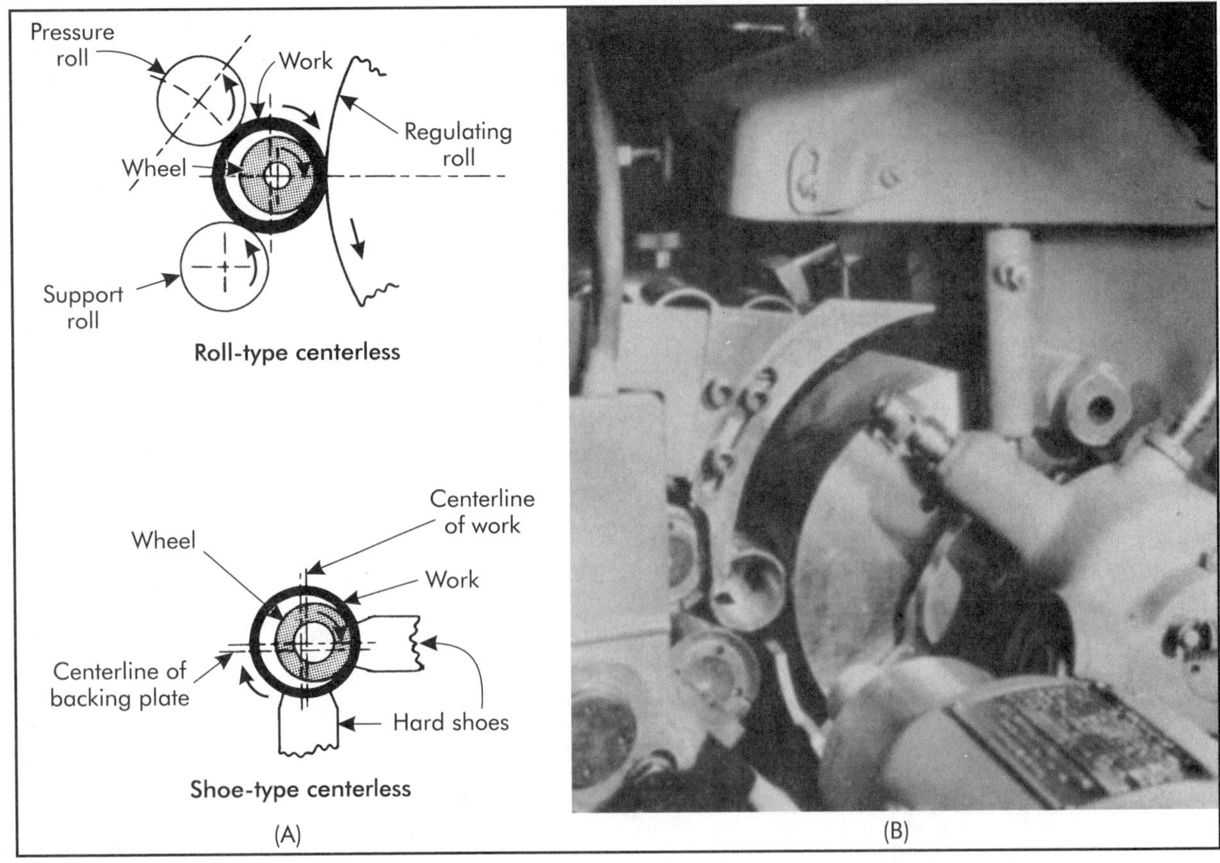

Figure 24-9. (A) Two kinds of internal centerless grinding. (B) View of a roll-type centerless internal grinder. An arm holding a diamond nib for truing the grinding wheel is shown in its retracted position in front of the large roller. The wheel head is in the foreground in the position from which it is advanced to insert the grinding wheel into the hole to be ground. (Courtesy Cincinnati Milacron, Inc.)

Operations

Squareness between a hole and face or shoulder on a workpiece is achieved by grinding both surfaces in the same operation. This may be done with the periphery and end of one wheel, but some machines have two wheel heads on the same cross slide, one with a small wheel for the hole and the other with a large wheel for the face.

In a long hole the wheel is reciprocated (traversed) over the length of the hole to cover all the surface. Even in a short hole the wheel may be reciprocated to distribute wear over its length.

A grinding wheel is commonly fed crosswise at a constant rate to grind a hole to size. Another way is to feed the wheel with a constant force, called *controlled force grinding*. Under a constant force, the grinding action is uniform and better control of size and finish is obtained. Under constant force, the deflection of the grinding wheel spindle is constant and can be compensated for by swiveling the wheel head slightly to produce straight or precisely tapered holes.

Internal grinding can be accomplished on a variety of machines ranging from a simple lathe with a tool-post grinding attachment to a CNC-operated internal grinding machine with multiple internal grinding spindles arranged on a turret head. One lathe tool-post grinder has a high-speed grinding spindle driven by an air motor. The spindle unit is reciprocated with a variable stroke range up to 127 mm (5 in.) to grind deep bores.

For toolroom and medium production work, universal cylindrical grinders are often equipped with either swiveling wheel heads or swing-down units. The swiveling type has a spindle on one end of the wheel head used for outside diameter (OD) grinding with a large diameter grinding wheel. For internal grinding, the wheel head is rotated 180° to bring a small, high-speed spindle into position for grinding internal diameters (ID) with small grinding wheels. The universal wheel head usually includes independent external and internal wheel spindle drives. On a cylindrical grinder equipped with a swing-down unit, the ID spindle unit is mounted on a pivoted bracket above the OD wheel head. The unit is swung down in front of the OD wheel head and clamped in place to perform internal grinding operations. Normally, the ID grinding spindle is driven by a separate electric motor via belt drives.

Dedicated CNC internal grinding machines are available for higher volume production work. Some of these are equipped with four-position turrets and independent drives on each spindle. Indexing of the turret brings any of four tools to the workpiece in a controlled cycle. Since each spindle is independently driven, tools other than grinding wheels may be used at the different stations. Thus, boring bars, internal grinding wheels, and measurement probes can be used in succession to bore, grind, and gage an internal dimension. When equipped with automatic part loading and unloading devices, and automatic tool changing mechanisms, a machine of this type becomes a very versatile production unit or a *machining center.*

24.1.6 Surface Grinders

Surface grinding is primarily concerned with grinding plane or flat surfaces, but also is capable of grinding irregular, curved, tapered, convex, and concave surfaces.

Conventional surface grinders may be divided into two classes. One class has reciprocating tables for work ground along straight lines. This type is particularly suited to pieces that are long or have stepped or curved profiles at right angles to the direction of grinding. The second class covers the machines with rotating worktables for continuous rapid grinding.

Surface grinders may be classified according to whether they have horizontal or vertical grinding wheel spindles. Grinding is normally done on the periphery of the wheel with a horizontal spindle. The area of contact is small, and the speed is uniform over the grinding surface. Small-grain wheels can be used to produce fine finishes. Grinding with a vertical spindle is done on the side of the wheel, which may be solid, sectored, or segmental. The area of contact may be large, and stock can be removed rapidly. A crisscross pattern of grinding scratches is left on the work surface.

The combinations resulting from the wheel and table arrangements just described for surface grinding are shown in Figure 24-10.

Surface grinders with reciprocating tables and horizontal spindles, such as the one depicted in Figure 24-10(A), are popular for toolroom work. Common practice is to reciprocate the workpiece

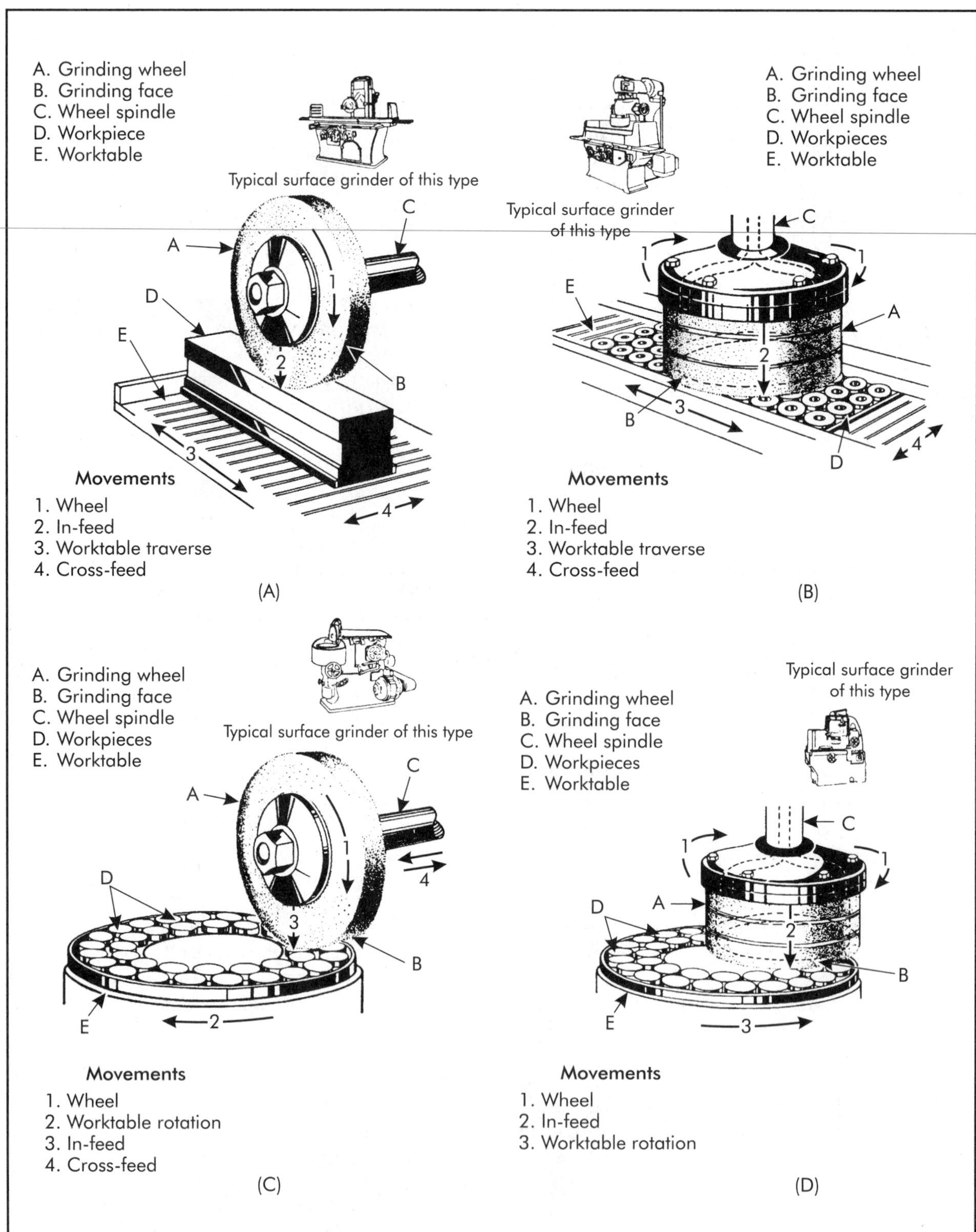

Figure 24-10. Principal kinds of surface grinders: (A) and (B) reciprocating table grinders; (C) and (D) rotary table grinders; (A) and (C) horizontal spindles; (B) and (D) vertical spindles. (Courtesy Carborundum Abrasives, Inc.)

lengthwise at 15–30 m/min (50–100 ft/min) and feed it crosswise one-fourth to one-half the width of the wheel for each stroke. The wheel is moved down into the work incrementally each time the surface is covered until size is reached. For roughing, increments may be 50 μm (.002 in.) or more, and for finishing 25 μm (.001 in.) or less.

Another method, called *creep grinding* or *creep-feed grinding*, is analogous to plunge cutting. Here the wheel is set to cut to full depth and the work is fed under it at a slow rate. This method is not common, but has been found advantageous for high-temperature alloys.

Some surface grinders with reciprocating tables have wheel heads that can be tilted vertically; one variety is called a *combination way and surface grinder*. Examples of the kinds of surfaces that can be ground on such machines with reciprocating tables are presented in Figure 24-11.

The wheel head, rather than the table, is reciprocated on some machines. The table or magnetic chuck is fixed on a long bed. The wheel is mounted with no overhang in a carriage that bridges the table and is traversed on ways on the bed along the sides of the table. The wheel is moved in the carriage for vertical in-feed and horizontal cross-feed.

Surface grinders with vertical spindles and rotary tables, as depicted in Figure 24-10(D), are used for rapid production abrasive machining. The rotary table is usually a magnetic chuck and can hold a large number of small pieces. Larger workpieces also may be ground. Machines of this type are relatively high powered; many have 75 kW (100 hp) and some as much as 373 kW (500 hp) motors. The Williams *continuous through-feed production surface grinder* has a vertical spindle but no moving table. It feeds workpieces under the wheel on an endless belt over a magnetic chuck. The pieces ride against or between side guides that take the grinding forces. Production rates of tens of thousands of pieces per hour are achieved.

The capacity of a surface grinder with a reciprocating table is commonly designated by two numbers specifying the nominal width and length of the working surface of the table; a third figure may be given to specify the largest distance from the top of the table to the face of the wheel. The usual designation for a rotary surface grinder is the diameter of the chuck or table. Specifications for some typical surface grinders are listed in Table 24-2.

24.1.7 CNC Grinding Machines

As with other machine tools, the movement and positioning of the elements of a grinding machine has shifted from manual control to computer numerical control (CNC), particularly for high precision and to facilitate frequent job changes under just-in-time (JIT) manufacturing practices. The shift to automated processes has included grinding since it is, by its very nature, a rather slow process with a relatively low rate of material removal. Traditionally, grinding has been one of the last operations to be performed on a workpiece and, in earlier days, was to be avoided if at all possible because of its high cost. However, new products and the space-age materials used in their construction require grinding because of extremely small tolerance and fine surface finish requirements and the hardness of the material.

Not so long ago, the size tolerances on ground parts were in the 0.010–0.013 mm (.0004–.0005 in.) range. In today's precision manufacturing environment, it is not uncommon to encounter

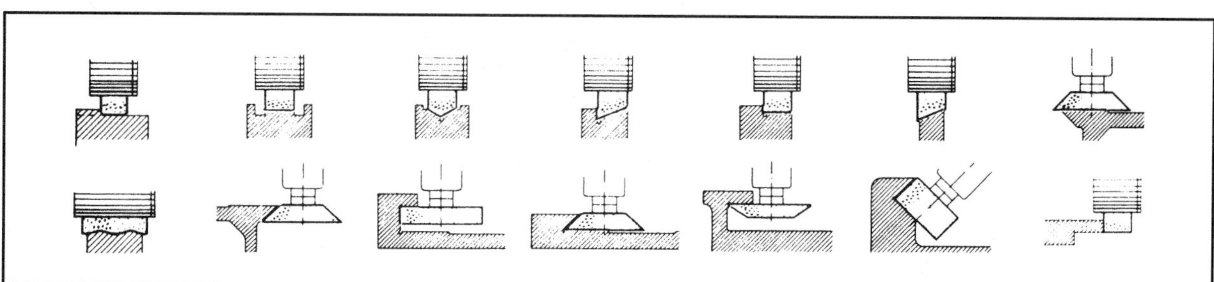

Figure 24-11. Examples of surfaces finished by grinders with reciprocating tables.

Table 24-2. Specifications for several typical surface grinders

Type of table/spindle	Size, mm (in.)	Main Motor, kW (hp)	Weight, Mg (lb)	Cost
Reciprocating horizontal	203 × 635 (8 × 25)	4.1 (5.5)	2.2 (4,840)	$ 15,000
	610 × 3,048 (24 × 120)	149 (200)	18.6 (41,000)	110,000
(fixed table)	152 × 914 (6 × 36)	4.3 (5.8)		25,000
Rotary horizontal	610 (24)	14.9 (20)	5.5 (12,200)	60,000
Rotary vertical	914 (36)	26.1 (35)	6.4 (14,000)	40,000

grinding tolerances of 0.003–0.005 mm (.0001–.0002 in.) or even less. In earlier years, ground surface finishes in the range of 0.40–1.6 μm (16–63 μin.) R_a were more or less the norm. However, today's customers are requiring finer surface finishes within the range of 0.05–0.40 μm (2–16 μin.) R_a for a variety of reasons. These reduced tolerances and finer surface finishes can, of course, be achieved on manually controlled precision grinding machines, but the process is slow and tedious, and the skill requirements are high. The conversion to CNC operation permits a greater degree of consistency (repeatability) in precision grinding, while at the same time reducing setup, cycle, scrap, and wheel dressing time.

Most of the basic types of grinding machines are available with some degree of computer numerical control. These range from rather simple two-axis control on basic cylindrical and surface grinders, to multiple-axis control on certain dedicated types of machines. One type of fully automatic centerless grinder is available with 15 axes of CNC control. The CNC universal cylindrical grinder shown in Figure 24-12 has two-axis control and a swiveling wheel head for either external or internal grinding. The wheel head has a maximum in-feed (X axis) of 203 mm (8 in.) at speeds ranging from 0.05–5,994 mm/min (.002–236 in./min) and in diametral increments as small as 0.0010 mm (.00004 in.). Programmable control of the wheel head along the X axis is accomplished through a servo drive and ball screw with a linear scale feedback.

The CNC surface grinding machine shown in Figure 24-13(A) has programmable control for the table cross-feed (Z axis) and the vertical wheel-feed (Y axis), both in minimum increments of 0.0010 mm (.00004 in.). The machine is also fitted with an optimal table-mounted rotary indexer (Figure 24-13[B]), which provides a third programmable control axis (A axis). The rotary indexer holds and positions workpieces. It can be mounted on the table in a number of different positions to suit the demands of a particular part. Figure 24-13(B) shows the indexer spindle in a horizontal position for grinding a triangular shape on the end of a pressworking punch.

Precision grinding machines with state-of-the-art CNC equipment generally cost 30–40% more than manually controlled machines, but that additional cost is usually offset by the increased production output. In addition, user-friendly software often makes it possible to use operators with lower skill levels on certain types of CNC equipment.

24.1.8 Disk Grinders

Disk grinders are used in production processes on flat surfaces to obtain a high degree of parallelism and close dimensional tolerance between opposite surfaces. For example, the tops and bottoms of automotive cylinder heads need to have surfaces that are flat and parallel, and the distance between those surfaces needs to be held to specified tolerances. Disk grinding is commonly a secondary finishing operation, following rough milling or some other preparatory machining operation. Thus, stock removal by the disk grinding process is usually small, 0.20–0.31 mm (.008–.012 in.), but in some cases, fairly large

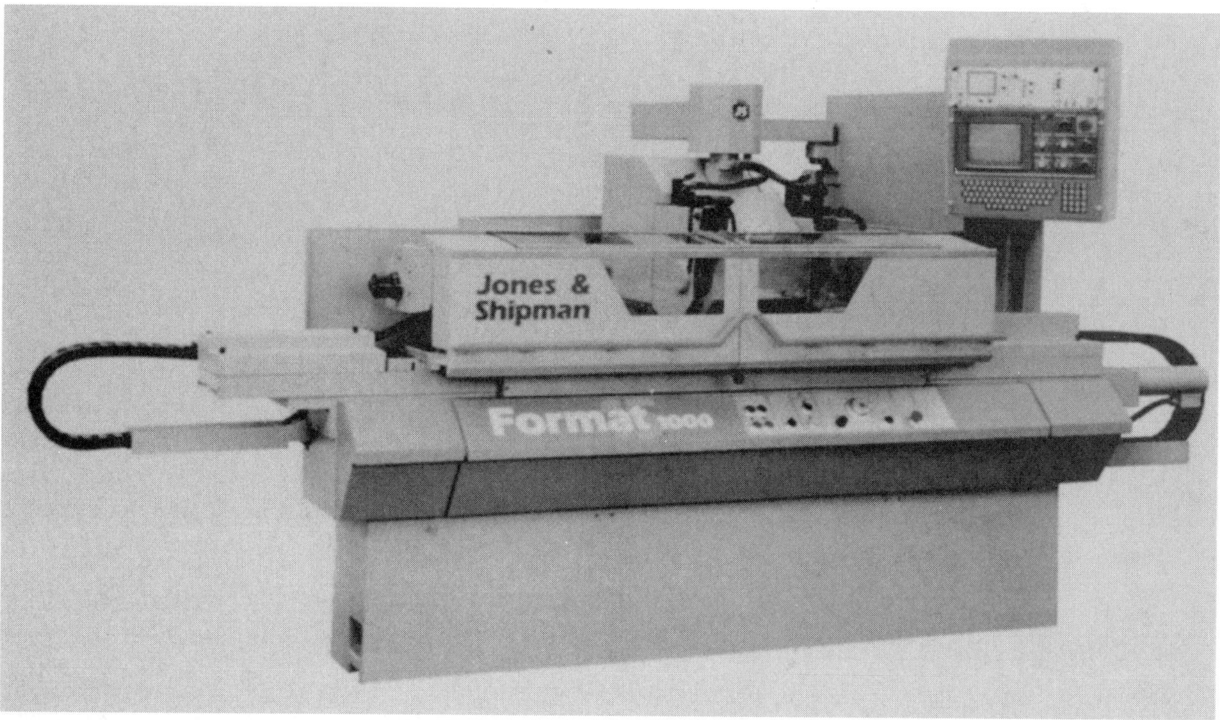

Figure 24-12. CNC universal cylindrical grinder for either toolroom or production environments. (Courtesy Jones & Shipman, Inc.)

amounts up to 3.18 mm (.125 in.) may be removed from large workpieces to obtain the degree of parallelism required. A variety of surface finishes may be achieved, depending on the requirements of the job and the grade of abrasive disk selected.

Disk grinders are available in either horizontal or vertical styles, some with two spindles on which the abrasive disks are mounted. Figure 24-14(A) shows a cutaway view of a vertical single-spindle disk grinder. This machine has a servo-driven worktable that can be used as a rotary or oscillating fixture for grinding one part or several parts consecutively. It will accommodate either conventional or superabrasive grinding disks for grinding parts up to 127 mm (5 in.) in diameter. Typical materials ground are ceramic, glass, aluminum, ferrite, steel, titanium, plastic, and sintered carbide. Figure 24-14(B) illustrates a typical setup for grinding two parts at a time on a six-station indexing table.

Figure 24-15 shows the basic elements of a *horizontal double-disk grinder* with all of the motor and grinding station covers removed and no workholding or feeding mechanisms in place. This machine, with two opposed abrasive disks, is capable of grinding opposite faces of a part simultaneously to achieve a high degree of flatness, parallelism, and close dimensional tolerance between those faces. The slides that carry the spindles for the abrasive wheel heads on this machine are servomotor-driven to provide for adjustment of the spacing between the wheels and to in-feed the heads during the grinding operation. A computerized system controls the position of the wheel heads and provides in-feed movement with accuracies to 0.0003 mm (.00001 in.). The control system also provides automatic wheel dressing and dynamically compensates for wheel wear.

Four methods of holding and feeding parts through the abrasive disks on a horizontal double-disk grinder are illustrated in Figure 24-16. In the rotary method, workpieces are fed between the disks in an arc by means of a servomotor-driven carrier. Upper and lower rails support and guide workpieces through the abrasive disks in the feed-through method. The feeding

Figure 24-13. (A) Three-axis CNC toolroom surface grinder. (B) Table-mounted programmable rotary indexer. (Courtesy Jones & Shipman, Inc.)

motion is accomplished by means of power-driven belts, pick-off chains, rollers, sprockets, magnetic disks, and other devices. The workpiece is held captive in a servomotor-driven, paddle-type fixture in the reciprocating method. In the fourth method, a special type of fixturing is required for some workpieces that do not lend themselves to the other methods.

A new design for the four-cylinder engine connecting rod shown in Figure 24-17 makes it an ideal part to be simultaneously ground on both sides of both ends in a double-disk grinder. This is a good example of a part that has been designed for manufacturability, a process being practiced more and more in the manufacturing industry. The rods are ground in a rotary-type workholding and feeding carrier similar to the one shown in Figure 24-16. In this particular operation, 0.76 mm (.030 in.) of stock is removed from each face with a tolerance between faces of 0.05 mm (.002 in.) and a surface finish of 4 μm (.00016 in.). A production rate of 700 connecting rods/hr is achieved by this process.

24.1.9 Thread Grinders

Threads are produced by a number of methods as described in Chapter 28. They are ground to obtain accuracy and finishes not available in other ways. Tolerances are held for size to ±0.0001 mm/mm (±.000004 in./in.) of pitch diameter and for lead within 7.6 μm in 508 mm (.0003 in. in 20 in.) of length. Hard materials can be threaded more economically by grinding than by other methods. Threads may be cut and then finish ground after heat treatment, or they may be ground from solid stock.

Thread grinding is performed on center-type and centerless machines. They may utilize single-rib-type or multirib-type grinding wheels. The first is a thin wheel with its outer edge trued to the shape of the thread space. The second is a wide wheel with grooves and ridges formed on its periphery.

Center-type thread grinders grind external or internal threads, or both. A typical machine resembles a universal center-type cylindrical grinder but has features to enable it to do its

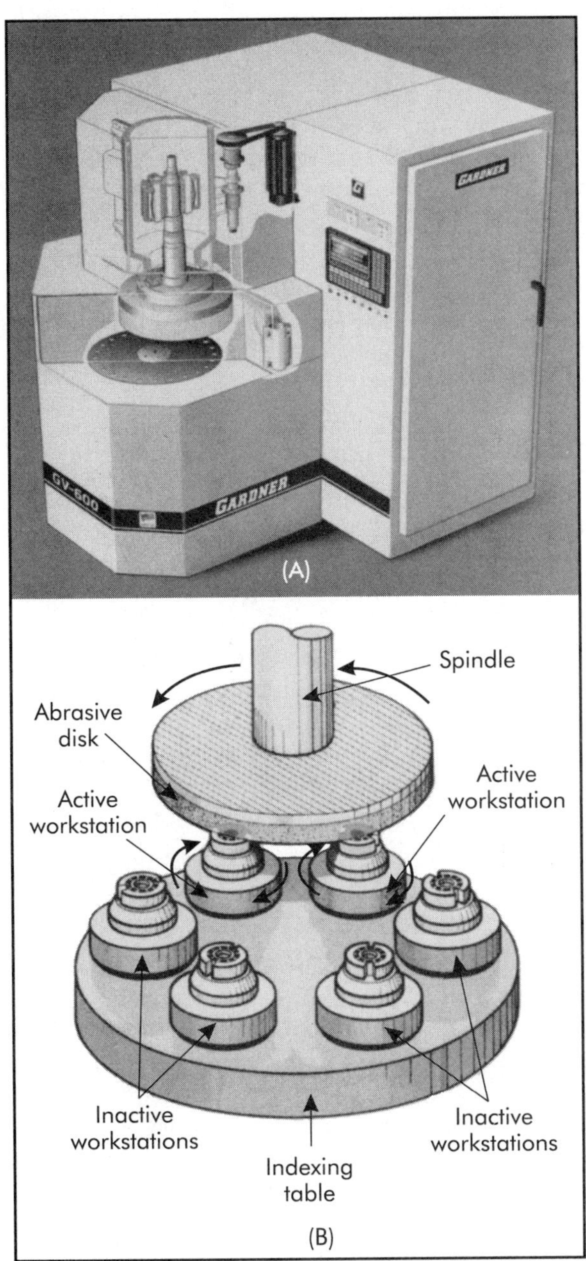

Figure 24-14. (A) Vertical single-spindle disk grinder. (B) Typical disk-grinding setup. (Courtesy Gardner Disk Grinders & Abrasives)

intended work. A master lead screw is geared to the work spindle and causes the table to traverse and the workpiece to turn with the proper lead as it advances. The grinding wheel is tilted to the helix angle of the thread. An attachment trues the wheel to the thread form.

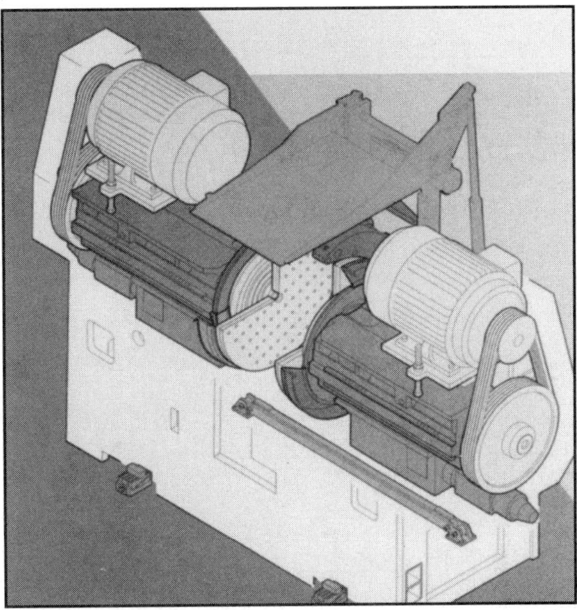

Figure 24-15. Pictorial view of a horizontal double-disk grinder. (Courtesy Gardner Disk Grinders & Abrasives)

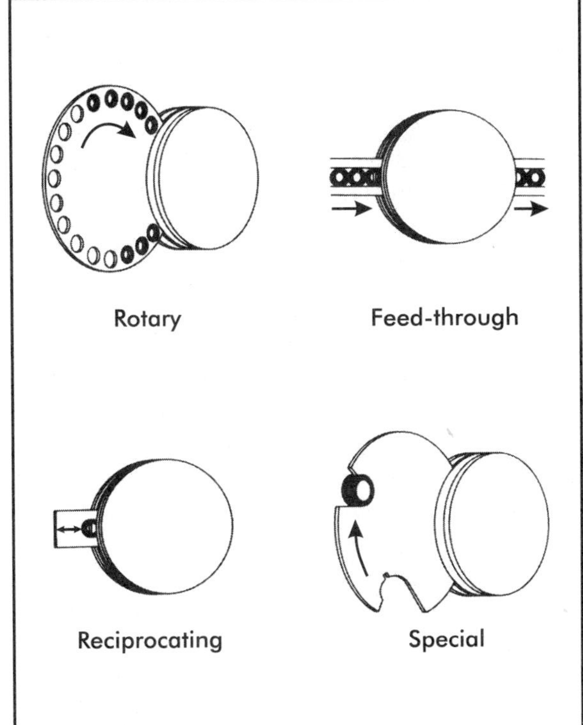

Figure 24-16. Four methods of holding and feeding workpieces through the abrasive disks on a horizontal double-disk grinder. (Courtesy Gardner Disk Grinders & Abrasives)

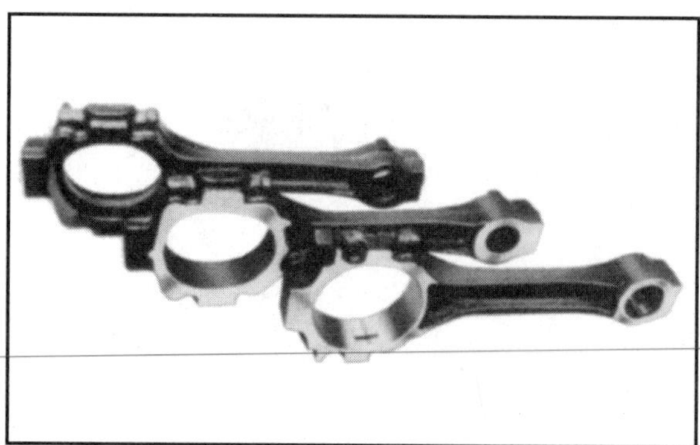

Figure 24-17. Automotive connecting rods shown before and after grinding both sides of both ends on a double-disk grinder. (Courtesy Gardner Disk Grinders & Abrasives)

24.1.10 Tool and Cutter Grinders

Grinders for finishing tools and sharpening cutters are available from simple grinding wheel stands for off-hand work to complex single-purpose machines like automatic face-mill grinders. Probably the most popular and versatile machine for precision sharpening of all kinds of tools is the *universal tool and cutter grinder*. It can sharpen multiple-tooth cutters like reamers, milling cutters, taps, and hobs, as well as single-point tools. It also can do light surface, cylindrical, and internal grinding to finish such items as jig, fixture, die, and gage details.

The main parts of a typical tool and cutter grinder are visible in Figure 24-18(A). An upper table can be swiveled on a lower table, which slides longitudinally on a saddle that provides cross movement. The wheel head can be raised or lowered and swiveled. Many attachments are available, such as radial truing devices, universal tilting work heads of various sizes, indexing devices, combination workholders, and form grinding equipment. Two other types of drill grinders are shown in Figures 24-18(B) and 24-18(C).

Profile or *contour grinders* are capable of reproducing a template form on a flat or round cutter. Some can grind metal surfaces to conform to outlines drawn on paper. An *optical profile grinder* is one on which a view of the zone of contact between wheel and workpiece is highly magnified (10, 20, or 50 times, for example) and cast upon a screen. The form produced by the wheel can be seen as it is ground and compared with a large-scale drawing on the screen of the form desired. Machine movements may be simultaneously controlled by the operator.

A type of tool grinder particularly adaptable to spiral tools, as indicated in Figure 24-19, is the *drill web-thinning grinder*. The work head spindle can be indexed and synchronized with the table movement to grind helices. The major units are fully adjustable to enable much work to be done in one setting that otherwise would require several setups.

24.1.11 Miscellaneous Grinders

Cam and camshaft grinders are essentially modifications of center-type cylindrical grinders, which enable them to finish various forms of round cams, camshafts, and pistons. The headstock and footstock are on a cradle and rock to and from the grinding wheel in response to a master cam that rotates in unison with the workpiece.

Crankshaft or *crank pin grinders* resemble cylindrical center-type grinders but are implemented to grind the offset pins in the throws of crankshafts.

24.1.12 Abrasive Belt Grinders

Continuous-coated abrasive belts of all widths are used for precision and nonprecision grinding of parts of all kinds, mainly to obtain good surface finishes with little stock removal. The same work may be done by grinding wheels, and the abrasive in belt form costs more, but other factors may make the cost of belt grinding less. For example, a belt on a flexible wheel can be readily adapted to an irregular or curved surface; a wheel would have to be trued carefully to the shape at considerable expense. Belts up to 2.5 m (≈8 ft) or so in width are applied over rollers to plunge cut steel slabs, steel and aluminum sheet, plywood panels, and rolls of metal, rubber, etc. This is faster than traversing a grinding wheel back and forth over the surface. Machines are available with abrasive belts instead of grinding wheels and with reciprocating or revolving tables. They do much the same work as surface grinders.

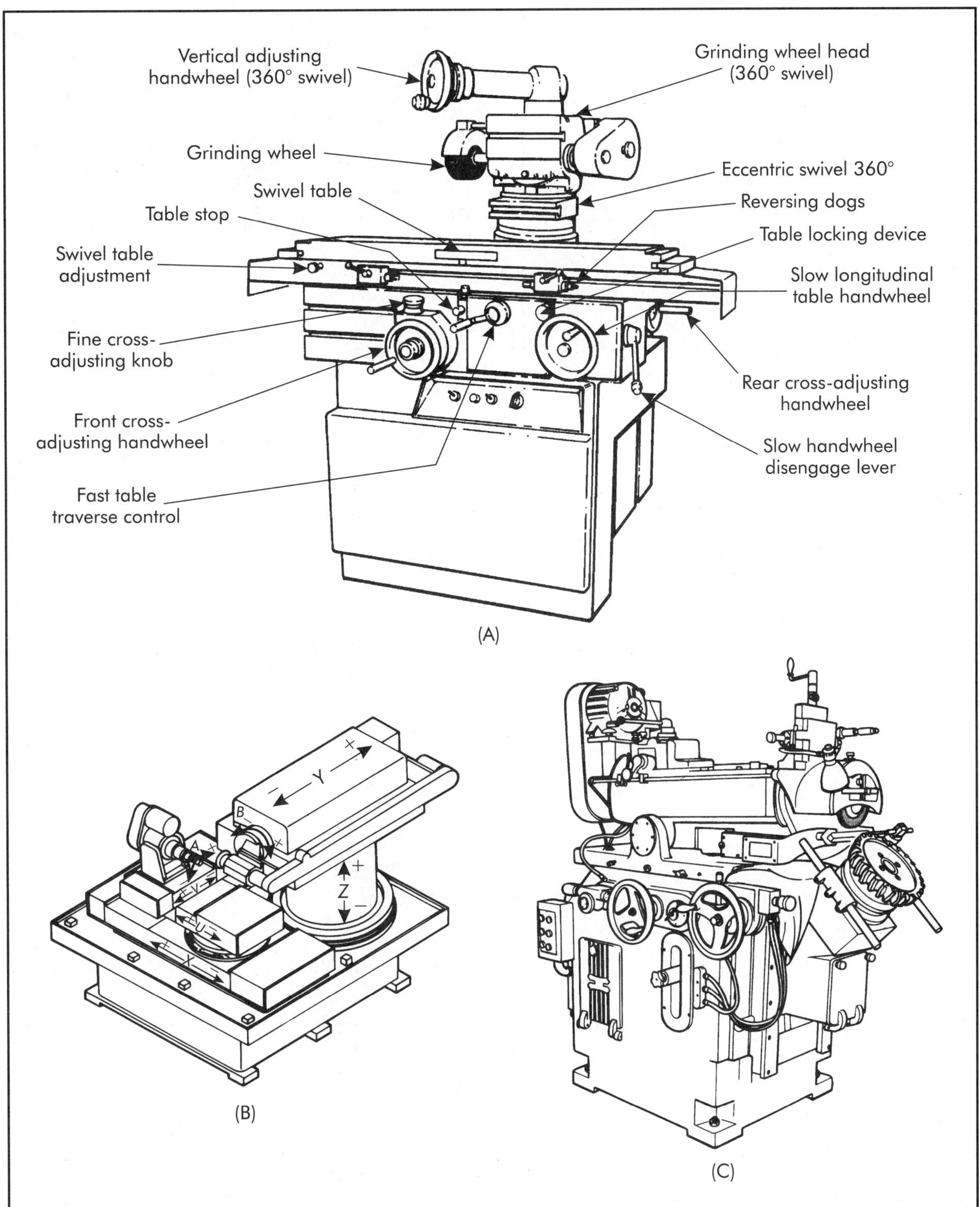

Figure 24-18. (A) An example of a tool and cutter grinder with a wide range of cutter sharpening capabilities; (B) automatic face-mill grinder; (C) eight-axis, CNC end-mill grinder.

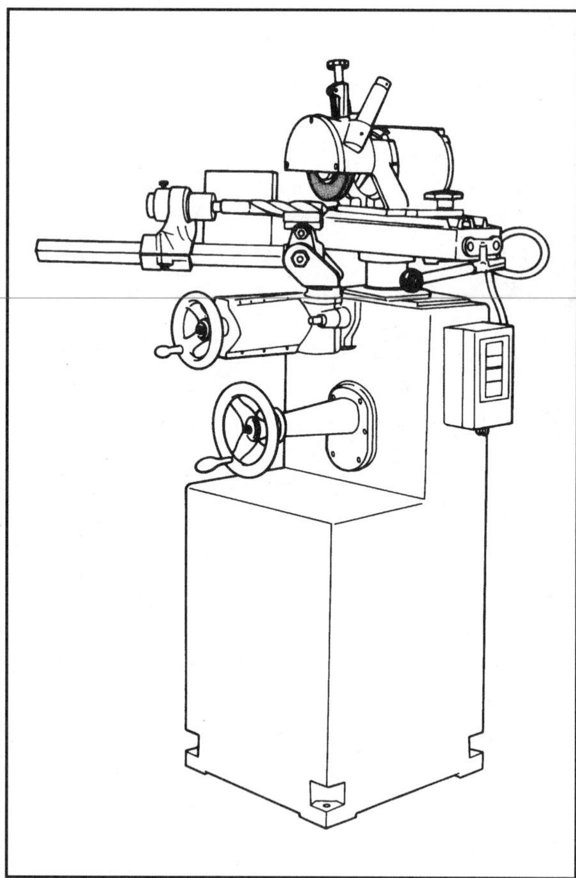

Figure 24-19. Drill web-thinning grinder.

24.2 NONPRECISION GRINDERS

24.2.1 Swing-frame Grinders

A *swing-frame grinder* has a horizontal frame from 1.5–3 m (5–10 ft) long suspended at its center of gravity to enable it to move freely within the area of operation. A configuration of a swing-frame grinder is illustrated in Figure 23-7(D).

24.2.2 Floor-stand and Bench Grinders

A *floor-stand grinder* has a horizontal spindle with wheels usually at both ends and is mounted on a base or pedestal. Figure 24-20 shows examples of a floor-type grinder and a bench-type grinder. A small-size grinder mounted on a bench is called a *bench grinder*. These machines are used for snagging and off-hand grinding cutting tools and miscellaneous parts. Polishing wheels may be run on these grinders.

24.2.3 Portable and Flexible-shaft Grinders

The usual form of portable grinder resembles a portable or electric hand drill with a guard and grinding wheel mounted on the spindle. A similar purpose machine is the flexible-shaft grinder that has the grinding wheel on the end of a long flexible shaft driven by a motor on a relatively stationary stand. Heavy tools are used for roughing and snagging, and lighter ones for burring and die work.

24.2.4 Standard Tool and Cutter Grinders

Figure 24-21 shows other types of standard tool and cutter grinders. A center hole grinder, shown in Figure 24-21(A), is used for generating precision center holes. A helical gear cutter grinder, shown in Figure 24-21(B), is used for sharpening helical gears and has the ability to produce the correct face angle and angle of dish. It is equipped with automatic indexing and automatic stock removal. Figure 24-21(C) is an example of a tap chamfer grinder. Although tap regrinding can be performed on a standard tool and cutter grinder, a specialized tap grinder is much more efficient.

24.3 GRINDING COMPARED WITH OTHER OPERATIONS

Grinding has always been considered mainly a finishing process to be preceded by other methods that remove the bulk of the stock from rough workpieces. For example, assume a shaft made from 25.4-mm (1-in.) diameter barstock must have a 18.99/19.03 mm (.748/.749 in.) diameter with a surface finish of 0.5 μm (20 μin.) for a length of 75 mm (≈3 in.) on one end. Grinding would be the economical way of meeting the specifications, but ordinarily the piece would first be turned to remove most of the stock.

More and more parts are being ground completely from the rough. This is commonly called *abrasive machining*. Abrasive machining is not a replacement for all other methods of metal removal, but it may be superior under certain conditions. Exceedingly hard and some space-age materials are difficult to machine by other

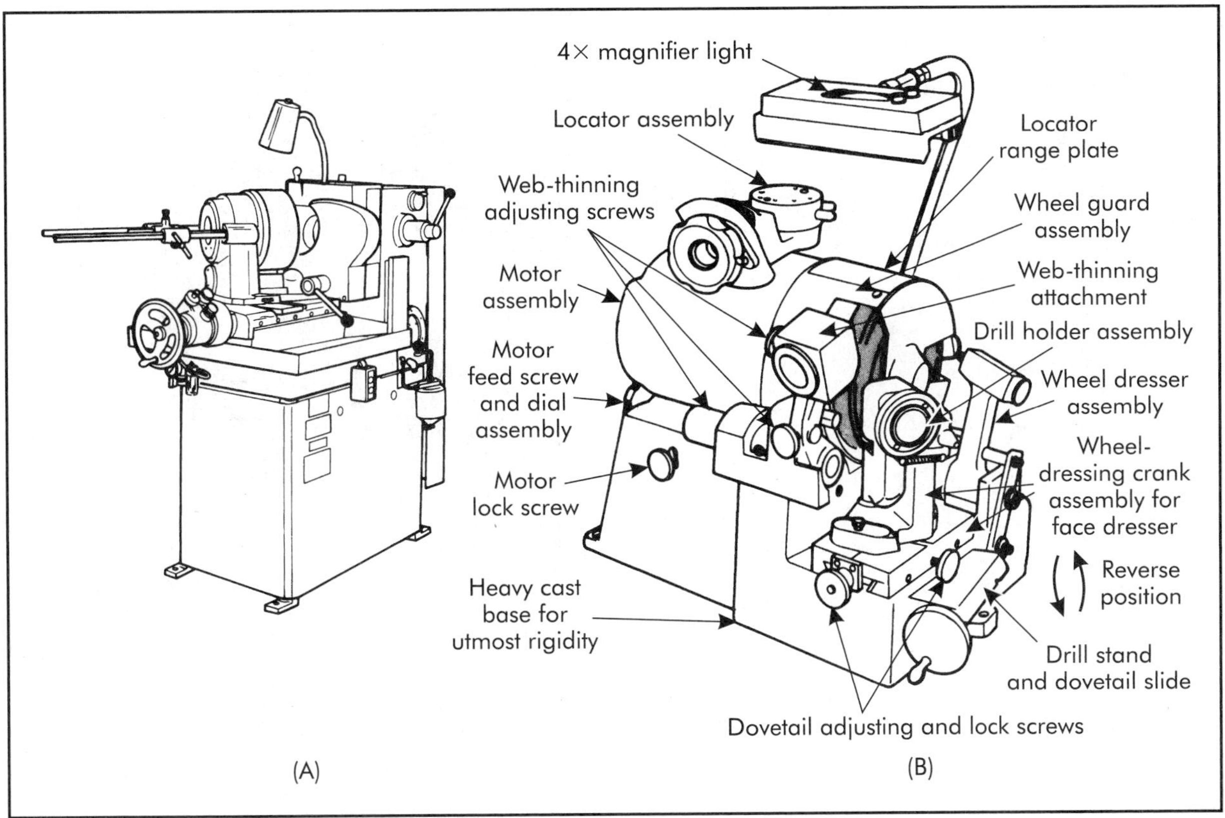

Figure 24-20. (A) Example of a floor-type drill grinder; (B) a bench-type drill grinder.

methods, and for these abrasive machining from the solid may be fastest. Even if grinding is slower than other means of removing a certain quantity of metal, the amount of metal that needs to be removed in grinding really may be much less. Usual practice is to provide thick stock on castings with hard and scaly skins so that cutting tools can get under and not be ruined by the scale. Abrasives are not deterred by the scale. Thus for grinding, 2 mm (≈.063 in.) or less of stock may be quite enough, whereas 3.18–4.78 mm (.125–.188 in.) would be needed otherwise. Furthermore, the finished area may sometimes be less for grinding. For example, a surface may be completely finished except for openings, as in Figure 24-22(A), to minimize interrupted cutting, which is detrimental to cutting tools. A relieved surface, as depicted in Figure 24-22(B), may serve the purpose just as well and can be ground with no damage to the wheel from the interruptions. Stock saved may be of value in itself if the workpiece is made of a costly material.

Grinding forces are substantial, and the maximum rate of grinding depends upon the rigidity of the machine, wheel, and workpiece. This is important in abrasive machining because it must be done fast to be competitive. If the system is too flexible, chatter may occur, the grinding wheel may wear too rapidly, or spark-out time for good finish and size is likely to be too long. Colding reported that when the rigidity of a particular system was increased 10 times, five times as many parts could be ground between wheel dressings. Spark-out time was reduced by a factor of 10–15, which resulted in a decrease of 65–90% in operation cost.

More power is consumed by grinding than by coarser methods, but relatively heavy abrasive machining consumes less unit power than fine grinding, as seen in Figure 23-14. In a typical fine-grinding operation, as much as 12 kW (16

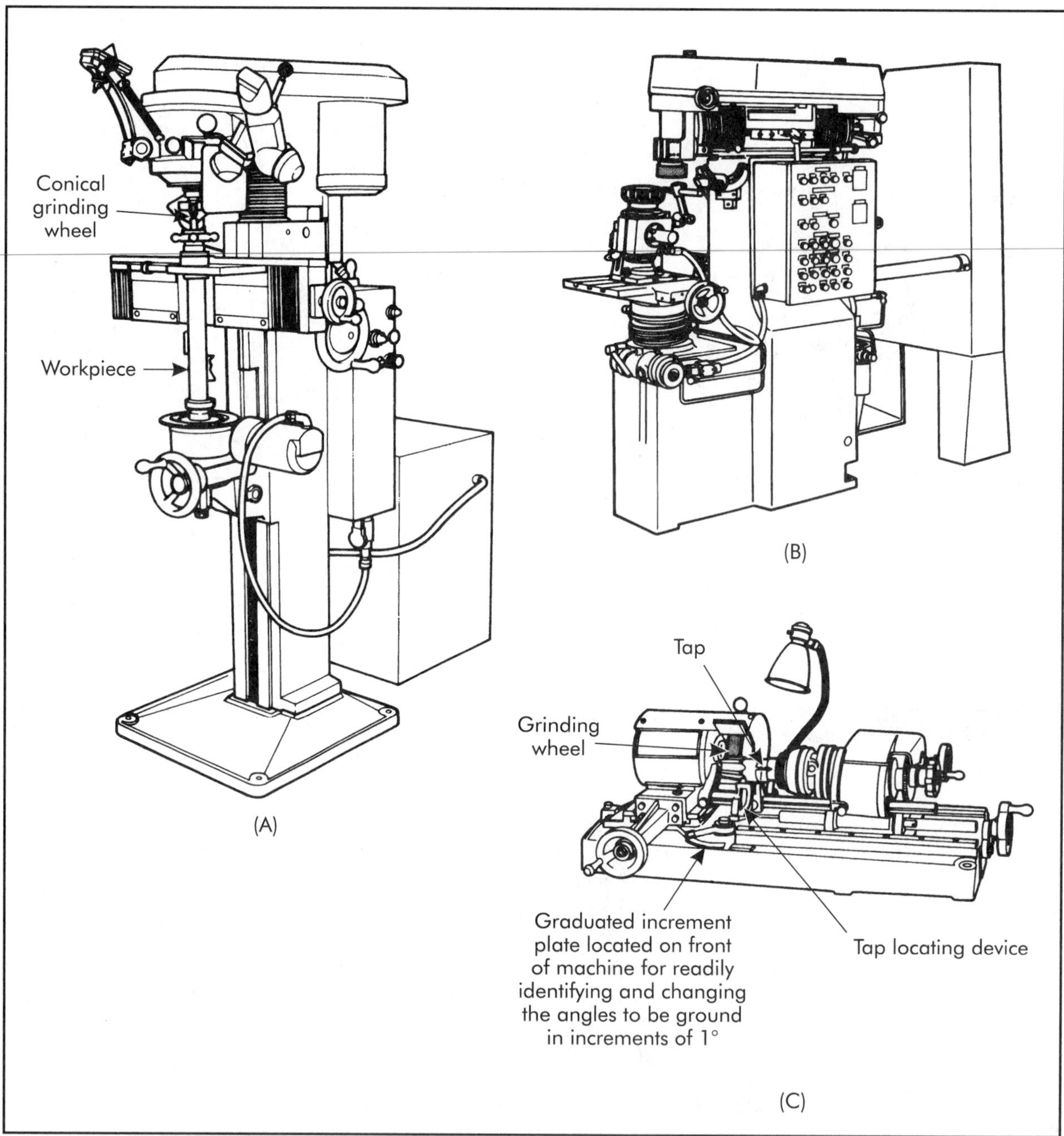

Figure 24-21. Examples of other types of standard tool and cutter grinders: (A) center hole grinder for generating precision center holes; (B) automatic helical gear grinder; and (C) tap chamfer grinder.

hp) may be required to remove about 16 cm^3/min (≈1 $in.^3$/min) of steel. In an ordinary rough-grinding operation, the power consumed may be 7.5 kW (10 hp) for each 16 cm^3/min (≈1 $in.^3$/min) removed. With a rigid and powerful machine and a proper grinding wheel, dressing technique, and operation procedure for abrasive machining, the power expended may well be reduced to 2.2–3 kW (3–4 hp) for each 16 cm^3/min (≈1 $in.^3$/min). This is comparable to .75–2.2 kW (1–3 hp) for rates of 16 cm^3/min (≈1 $in.^3$/min) for turning and milling steel. Powerful and rigid machines de-

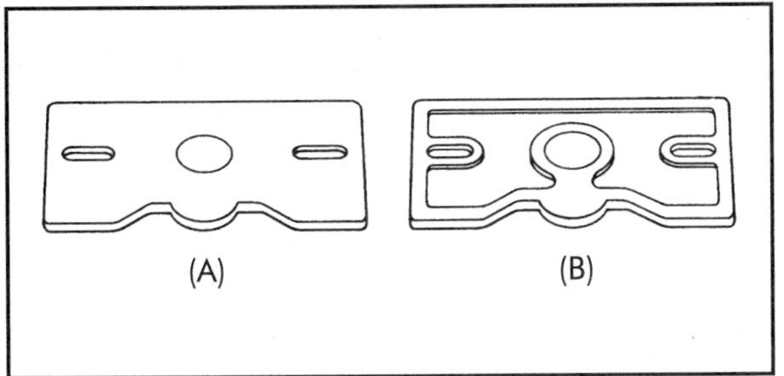

Figure 24-22. Motor base: (A) as originally designed for milling; (B) as redesigned for abrasive machining.

veloped in recent years with capacities of 75–373 kW (100–500 hp) are able to grind at rates competitive with other methods. For one case reported, SAE 4140 steel forgings were milled at the rate of 33 pieces/hr and at a cost of $0.54/piece. On a surface grinder, 58.6 pieces were machined per hour at a cost of $0.11/piece.

New ideas are being explored to develop more efficient machines and methods for abrasive machining. Ways are being found to drive grinding wheels and workpieces at higher speeds and cool them more efficiently. One proposal involves a wheel 3.1–4.6 m (10–15 ft) in diameter on its side revolving at a surface speed of 10,973 m/min (36,000 ft/min) with six, eight, or 10 workstations around its periphery on which pieces are ground all at the same time.

Obvious savings can result from abrasive machining when it means two operations, cutting and grinding, are replaced by a single grinding operation. Even when grinding replaces only a cutting operation, the workpiece may be held more easily on a magnetic chuck alone, with a savings of the cost of a fixture. Small parts are normally ground in a group while held by a magnetic chuck.

In most cases, fine finishes and tolerances less than 25–76 μm (.001–.003 in.) are more easily obtained by grinding than by nonabrasive methods. Even so, the cost of grinding increases for better finishes and smaller tolerances. Some grinding is done to tolerances as small as 0.5 mm (.00002 in.) and surface finishes better than 50 nm (2 μin.) R_a but only at high cost. In many cases, the smallest tolerances and finest finishes can be better obtained by lapping, honing, and superfinishing. As indicated by Figure 16-30, these methods normally cover ranges only exceptionally reached by grinding. They are discussed in Chapter 25.

24.4 QUESTIONS

1. Identify and describe the functions of the primary elements of a plain cylindrical grinder.
2. How is a cylindrical workpiece (barstock) held in position and rotated on a cylindrical grinder?
3. Describe and distinguish between "plunge cut" and "traverse grinding."
4. How is the size of a cylindrical center-type grinder designated?
5. How does a universal cylindrical center-type grinder differ from a plain cylindrical grinder?
6. Show by a sketch the rotational directions of the workpiece and grinding wheel when grinding a cylindrical workpiece between centers on a cylindrical grinder.
7. Identify and give the direction of the feed movements involved in the operation described in question 6.
8. What is a "roll grinder" and what does it do?
9. Describe the action of a centerless grinder. What are its advantages?
10. Describe "through-feed," "in-feed," and "end-feed centerless grinding" and their applications.
11. Name at least six advantages that centerless grinding has over center-type grinding.
12. Show by a sketch the rotational directions of the elements involved in centerless through-feed grinding.
13. Show by a sketch the rotational directions of the elements involved in a chucking internal-grinding operation.

14. Describe an "internal centerless grinder" and state its advantages.
15. What advantage, if any, does the shoe-type internal centerless grinder have over a roll-type grinder?
16. What purpose does a planetary internal grinder serve and how?
17. How does a universal cylindrical grinder accomplish internal grinding?
18. Describe the four principal types of surface grinders.
19. What is "creep-feed surface grinding" and how does it differ from regular surface grinding?
20. What are some of the advantages of a CNC grinder over a manual type?
21. Which major axes are programmable on the CNC universal cylindrical grinder shown in Figure 24-12?
22. Which major axes are programmable on the CNC surface grinder shown in Figure 24-13?
23. What is the primary purpose of disk grinding?

24.5 PROBLEMS

1. The mild steel workpiece shown in Figure 24-23 is to be ground over the full length of the large diameter. The bar will be mounted between centers on a 1.8 × 5.4 m (6 × 18 ft) plain cylindrical grinder driven by a dog on the small diameter. The machine has a 7.5 kW (10 hp) motor and is 80% efficient. The workpiece is to be traversed past a 40-mm (≈1.5-in.) wide grinding wheel at a rate of 15 mm (≈.5 in.) for each revolution. Total stock removal is 0.5 mm (.02 in.) on the diameter, of which 0.08 mm (.003 in.) is to be removed in four finishing passes of the wheel over the workpiece. At the wheel, 7.5 kW (10 hp) is required to grind the material at a rate of about 16 cm^3/min (≈1 in.^{3}/min). Select a work speed and a suitable in-feed per pass for the roughing cuts within practical ranges and the capacity of the machine. How long should the grinding time be for this piece?

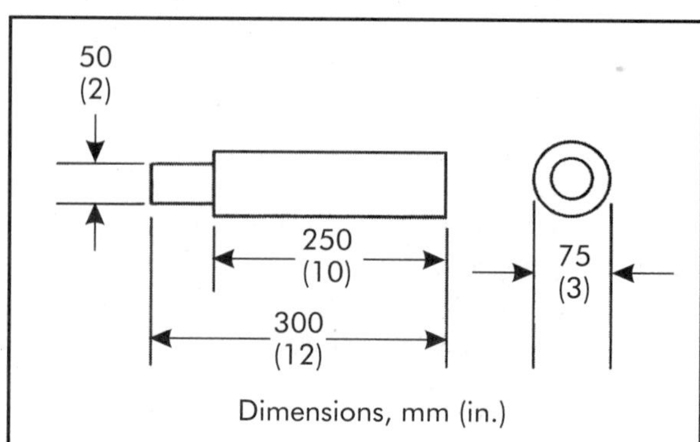

Figure 24-23. Workpiece.

2. A short shaft with three diameters to be ground is produced in 14 lots of 1,000 pieces each during the year. The procedure has been to grind one diameter on all pieces in one lot, then a second, and finally the third. The production rate has been 38 pieces/hr. It is proposed to mount three wheels on the machine to grind the three diameters on one piece at the same time. This is expected to produce 80 pieces/hr, but requires a special wheel mount and 1 hr extra setup time for each lot. The labor and overhead rate is $30/hr and maintenance is 28%/yr. What is the most that can be justified for the cost of the new wheel mount?
3. The workpiece described in Problem 1 is to be ground on a centerless grinder with a 11 kW (15 hp) motor and 80% efficiency. Two passes will be taken with 0.30 mm (.012 in.) of stock removed from the diameter in the first pass, and 0.20 mm (.008 in.) in the second. At the wheel, 7.5 kW (10 hp) is required to grind the material at a rate of about 16 cm^3/min (≈1 in.^{3}/min). Select regulating wheel speeds and tilt angles for best utilization of the machine. How long should the grinding time be per piece?

4. A workpiece has a 25 mm (≈1 in.) diameter for 75 mm (≈3 in.) of length from one end and a 50 mm (≈2 in.) diameter for the remainder of its total length of 180 mm (≈7 in.). A total of 0.40 mm (≈.015 in.) of stock is removed to finish the 50 mm (≈2 in.) diameter. Time required is 3 min per piece on a 1.8 × 3.6 m (6 × 12 ft) plain cylindrical center-type grinder, and 1.75 min on a centerless grinder. Setup time is 20 min/lot on the plain grinder and 40 min on the centerless grinder. The labor, overhead, and depreciation rate is $28/hr on both machines. No special tools are needed. What is the equal cost quantity for this part?
5. The regulating wheel of a centerless grinding machine is turning with a surface speed of 18 m/min (60 ft/min), and its axis is inclined at an angle of 6° with the horizontal. What is the rate of through-feed of the work between the wheels?
6. Three adjacent diameters and shoulders must be ground on a shaft. During the coming year, 100,000 pieces are to be produced. An available grinding machine can be tooled with three wheels in the conventional positions to grind the surfaces. Better action can be obtained if the wheel head can be turned to allow the wheel to make contact with the work at an angle. It is estimated the production rate can be increased form 80 pieces/hr to 100 pieces/hr in that way. However, for that improvement the machine must be rebuilt at a cost of $8,500. The labor and overhead rate in the plant is $30/hr. Interest, insurance, taxes, and maintenance call for a rate of 35%/yr on an investment. Is it worthwhile to rebuild the machine for this job? It is assumed that it will not be impaired for any other work. The rebuilding cost must be recovered within a year.
7. A 15-kW (20-hp) milling machine and a 26-kW (35-hp) surface grinder have about the same costs per hour for labor, most overhead, and depreciation. The milling machine can remove stock about 20% faster than the grinder for a certain class of work. Power costs more on the grinder, but the difference is only around $0.06 for a typical piece. Still, the total grinding cost is less than for milling some workpieces. What factors would you expect to give the grinder the advantage?

24.6 REFERENCES

ANSI B5.53M-1982 (R1987). "Grinding Machines, Cutter and Tool." New York: American National Standards Institute.

ANSI B5.33-1981 (R1987). "External Cylindrical Grinding Machines—Plain." New York: American National Standards Institute.

ANSI B5.42-1981 (R1987). "External Cylindrical Grinding Machines—Universal." New York: American National Standards Institute.

ANSI B5.32-1977 (R1987). "Surface Grinding Machines of the Reciprocating Table Types, Designation and Working Ranges of." New York: American National Standards Institute.

ANSI B32.1-1977 (1987). "Grinding Machines, Surface, Reciprocating Table—Vertical Spindle." New York: American National Standards Institute.

ANSI B11.9-1975 (R1987). "Grinding Machines, Safety Requirements for the Construction, Care and Use of." New York: American National Standards Institute.

ANSI B5.44-1971 (R1986). "Rotary Table Surface Grinding Machines." New York: American National Standards Institute.

ANSI B5.37-1970 (R1987). "External Cylindrical Grinding Machines—Centerless." New York: American National Standards Institute.

Albert, M., ed. 1994. "Precision Grinding: The Last and Best Resort." *Modern Machine Shop*, December.

Grieb, P. 1982. "Putting CBN Wheels to Work." *Manufacturing Engineering*, June.

Lewis, K. B., and Schleicher, W.F. 1976. *The Grinding Wheel, a Textbook of Modern Grinding Practice*, 3rd Ed. Cleveland, OH: The Grinding Wheel Institute.

Mason, F., ed. 1995. "CNC Tool Grinders Get Friendly." *Manufacturing Engineering,* August.

Owen, J. V., ed. 1993. "CNC Makes Grinding Less of a Grind." *Manufacturing Engineering*, May.

Owen, J. V., ed. 1996. "Cost Conscious Grinding." *Manufacturing Engineering,* February.

Society of Manufacturing Engineers (SME). 1995. *Basics of Grinding* video. *Fundamental Manufacturing Processes* series. Dearborn, MI: SME.

Tool and Manufacturing Engineers Handbook, 4th Ed. 1983. Vol. 1. *Machining*. Dearborn, MI: Society of Manufacturing Engineers (SME).

Ultra-finishing Operations

25

The lapping machine, 1928
—Herbert S. Indge, The Norton Company

Heavy cuts in a material leave rough and torn surfaces. The lighter and milder the cut, the better the surface and the smaller the tolerance. Good finishes can be obtained by operating cutting tools at light feeds, but this is slow. As has been explained, grinding is often faster for fine cutting because it removes material by the action of many grains taking small bites. Although grinding may be carried to the extreme to procure fine finishes and a high degree of precision, other abrasive operations that have slower speeds and a milder action usually prove more economical for the best finishes. Such operations are lapping, honing, ultrasonic impact grinding, and superfinishing.

The terms "lapping" and "honing" are often used interchangeably, and it is difficult to make a clear distinction between them. In general, but not always, *lapping* uses a loose abrasive and is applied to external surfaces, while *honing* is done on internal surfaces with bonded abrasives.

When accuracy is not required, some of the costly aspects of grinding and other precision finishing operations may be eliminated by the polishing, buffing, brushing, tumbling, vibratory finishing, and shot-blasting and sandblasting operations. Specialty operations for burr removal include the thermal energy method, and abrasive belt and abrasive flow machining.

25.1 LAPPING

25.1.1 Purpose

Lapping is an abrading process that leaves fine scratches arrayed at random. Its purpose is to improve surface quality by reducing roughness, waviness, and defects to produce accurate as well as smooth surfaces. Lapping pressure is light as compared to grinding, and the work is never overheated.

The lapping process is performed by hand and by machines. Its range of usefulness is large. In some cases, it may be merely an expedient to remove an occasional fault. It is a basic operation in job and tool shops where a typical application is to finish locating and wearing surfaces on precision tools and gages. Gage blocks, the standards of accuracy, are finished regularly by lapping. Machine lapping is common for production. Other typical lapping subjects are surfaces that must be liquid- or gas-tight without gaskets and those from which small errors must be removed, such as gear teeth.

According to Figure 16-30, lapping can be used to produce surface finishes to as low as 0.01 μm (.500 μin.) R_a under special conditions. For average applications, lapped surface finishes range from 0.4 μm (16 μin.) down to 0.05 μm (2 μin.).

25.1.2 Process

Normally, only a small amount of stock is taken off by lapping, up to 1 mm (.04 in.) or so; usually only about 101 µm (.004 in.) or less for roughing, and as little as 2 µm (≈ .0001 in.) for finishing. This is because the fine abrasive works slowly and the surface shape is hard to control if much stock is removed.

Fine loose abrasive mixed with a vehicle, bonded abrasive wheels, or coated abrasives, are used for lapping. Wet lapping with clear or soapy water, oil, or grease may be as much as six times faster than dry lapping.

Most lapping is done by spreading loose abrasive and vehicle on lapping shoes, plates, or quills, called *laps,* which are rubbed against the work. In *flat lapping,* the lap is made of soft close-grained cast iron, commonly with grooves across its face to collect excessive abrasive and dirt. Such a lap soon becomes charged with embedded abrasive particles. When a plate of hardened alloy steel is used, the process is called *free-abrasive machining.* A coarse abrasive can be used and washed away and replaced as needed, because it does not embed in the lap, and the stock removal rate is rapid. Loose abrasive lapping is not often done on soft materials because the abrasive particles become embedded in the workpiece.

In lapping, the work and lap are not rigidly guided with respect to each other, and their relative movements are continually changed. In *equalizing lapping,* the work and lap mutually improve each other's surface as they slide together. This is done in seating mushroom valves, machine lapping gears, and hand lapping plug and ring gages. In *forming lapping,* the work acquires a definite shape from the lap. This is the case for most lapping done with abrasive wheels.

25.1.3 Machines

The basic elements of a *vertical lapping machine* are illustrated in Figure 25-1. In this example, an operator is loading small, flat, rectangular parts in a nested arrangement on the lower lapping wheel. The upper lapping wheel has been tilted back to permit better access to both wheels for loading and unloading parts. The upper wheel is rotated back in place over the lower wheel after the workpieces are

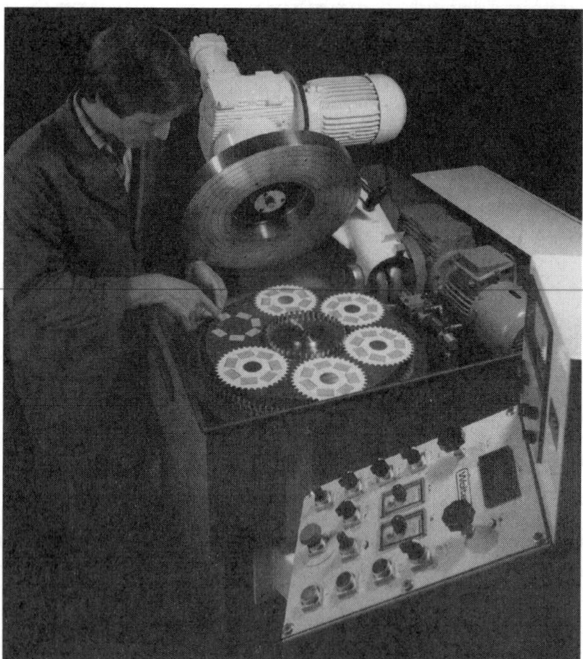

Figure 25-1. Vertical lapping machine with electro-pneumatic control. (Courtesy Peter Wolters of America, Inc.)

positioned in their nesting rings and the two wheels are then rotated in opposite directions to perform a lapping operation. In addition to the rotation of the upper and lower lapping wheel, the nesting rings for the parts are also rotated by means of outer and inner drive pins. This insures uniform lapping action over the surface areas of both sides of the workpiece.

An important feature of the lapping machine shown in Figure 25-1 is the manner in which the lapping load application for the upper wheel is controlled. The necessary lapping force is applied with the aid of electro-pneumatic setting equipment using valves and subsequent adjustable throttles that deliver and exhaust air pressure precisely in accordance with the requirements. This feature makes it possible to achieve a high degree of accuracy and repeatability in the lapping process. For the machine shown, an infinitely variable working force up to 2,500 N (562 lb) is available between the two lapping wheels. The 413-mm (16.26-in.) diameter upper lapping wheel can be rotated at either 54 or 108 rpm, and the 475-mm (18.701-in.) diameter lower wheel can be rotated at 50.1 or 101 rpm. The workpiece drives can be set at either 23.8, 35.8, or 71.6 rpm. Thus, the lapping speeds and pressures for different workpiece materials and thicknesses can be carefully selected and monitored to achieve a high degree of thermal stability of the parts being lapped.

A *centerless lapping machine* is like a centerless grinder (described in Chapter 24) but has extra wide wheels. Lapping wheel speed may be 152–610 m/min (500–2,000 ft/min), and work speed 46–152 m/min (150–500 ft/min). Stock removal is usually 5 μm (.0002 in.) or less, fine surfaces 0.05 μm (2 μin.) R_a (or better) are possible, and close tolerances (1.5 μm [≈ .00005 in.] for size and half as much for straightness) are produced. The machines are designed for continuous production of round parts such as piston pins, bearing races and cups, valve tappets, and shafts. As an example of output, piston pins are produced at the rate of 40 pins/min.

25.2 HONING

25.2.1 Purpose

Honing is an abrading operation mostly for sizing and finishing round holes. To a lesser extent, it is used on external flat and curved surfaces by means of bonded abrasive stones. Because the abrasive is not free to embed in the surface, soft metallic, nonmetallic, as well as hard materials can be honed. Typical applications are the finishing of automobile engine cylinders, bearings, gun barrels, ring gages, piston pins, shafts, and flange faces.

In grinding with the periphery of a rigid wheel, there is near line contact with the work surface, whereas honing has a large area of contact and less pressure. Grinding is done at high speeds; honing at low speeds with a milder action.

As a cutting operation, honing has been used to remove as much as 3.18 mm (≈.125 in.) of stock, but is normally confined to amounts less than 0.25 mm (.010 in.). It normally is preceded by boring or reaming to establish surface shape and position. Size and truth are typically refined to less than 10 μm (≈.0005 in.) by honing. Surfaces can be finished to 0.025 μm (1 μin.) R_a, but 0.2–0.25 μm (8–10 μin.) R_a and more are customary. A surface may be given a crosshatched finish by honing to aid lubrication.

25.2.2 Process

Honing stones are made from standard abrasive and bonding materials as well as from superabrasives, including diamonds. Grain sizes range from 80 grit for roughing, to 320 for finishing materials, to 500 for soft materials. A typical honing toolhead with six stones is depicted in Figure 25-2. With this type of toolhead, the stones are expanded under light pressure while it is working. In operation, the toolhead is mounted on a mandrel that fits into the spindle of a honing machine. Other types of toolheads are available with anywhere from one to several stone positions. On some production types of multiple-spindle honing machines, the toolheads are adjusted to size during setup with each spindle carrying a head of a different size. The honing operation on a part is then performed in sequence with each successive toolhead removing a certain amount of stock to rough and finish a bore to size. For an internal honing operation, the toolhead or workpiece, or both, are rotated and reciprocated in relationship to each other. The two movements are run purposely out of phase to cover all the surface without a regular pattern of scratches. Special honing coolants and oils are available to facilitate the abrasion process and maximize stone life.

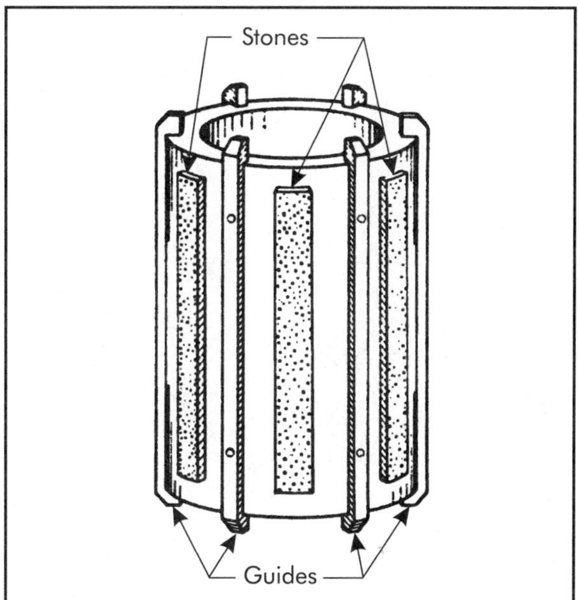

Figure 25-2. Sketch of a honing tool.

25.2.3 Machines

Honing machines may be placed in two general classifications: vertical machines where the hone is attached to and rotated by a vertical spindle, and horizontal machines in which the hone is attached to a horizontal spindle. Generally, the horizontal machines are better suited for long workpieces such as hydraulic cylinders and gun barrels. A variety of machines are available for either manual job-shop type work or semi-automated and automated production honing.

The *single-spindle, horizontal honing machine* (tube hone) shown in Figure 25-3 is capable of precision finishing bore diameters from 25.37–482.60 mm (.999–19 in.) and lengths from 101.6–1,524 mm (4–60 in.). On this machine, tubular workpieces are supported at both ends in a cylindrical housing while the honing toolhead is rotated and reciprocated throughout its length. Variable spindle speeds from 67–336 rpm and honing stroke rates up to 25 m/min (1,000 in./min) are available to accommodate a variety of workpiece materials, abrasive stones, and finishing requirements. The toolhead is expanded by a mechanical linkage during the honing operation to obtain the desired inside diameter on the workpiece. The machine may be manually operated through the keypad at the top of the machine or automatically by means of a programmable logic controller (PLC). A horizontal honing machine of this type and size costs about $55,000.

Vertical honing machines are available with single or multiple spindles, depending on the production rates required. The precision, four-spindle, bore sizing machine shown in Figure 25-4 is equipped with a 12-station turntable that permits parts to be loaded and unloaded while the honing operation is being performed in the four spindle positions. With the four honing spindles and 12 workpiece positions, the machine can be programmed for a variety of cycles and production rates. In one mode of operation, for example, single stroke honing with a diamond hone is used to complete two parts per machine cycle. In single stroke honing, each hone is set to a given size and goes through the bore only once to remove a predetermined amount of stock. Thus, two of the toolheads are set for roughing and two are set for finishing to finish hone two parts per cycle. The machine shown in Figure 25-4 is designed to accommodate parts with bore diameters ranging from 5.92–52.20 mm (.233–2.055 in.) and up to 152 mm (6 in.) in length. Speeds and feeds of the spindles/stroke can be individually controlled to optimize cycle time, tool life, bore geometry, and bore finish. Variable spindle speeds from 0–600 rpm are available along with a variable stroke speed of 12.7–2,540 mm/min (.5–100 in./min). A four-spindle machine

Figure 25-3. Single-spindle, horizontal honing machine for large and long bores. (Courtesy Sunnen Products Company)

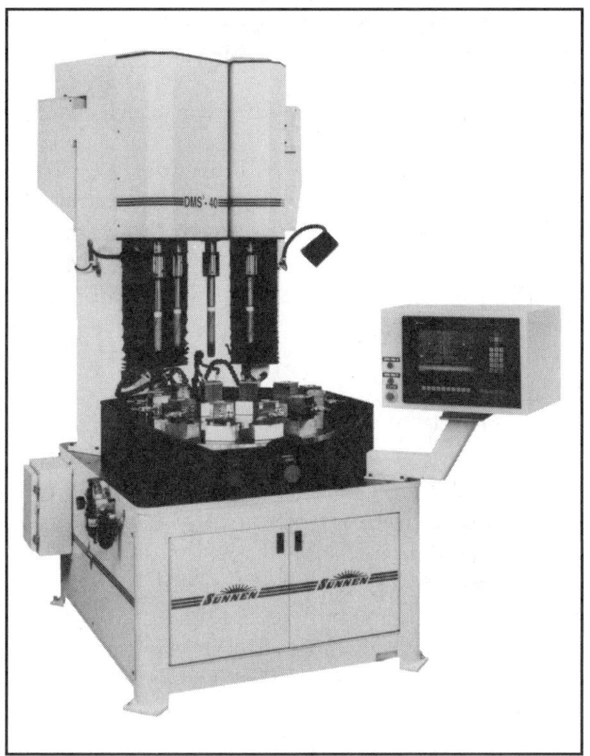

Figure 25-4. Precision vertical four-spindle honing machine with 12-station turntable. (Courtesy Sunnen Products Company)

of this size equipped with a programmable logic controller costs around $100,000.

25.2.4 Fine Grinding

A more recent technology has been developed for the double-side abrasive machining of flat surfaces. This process, referred to as *fine grinding* or *flat honing*, differs from conventional surface grinding in that wheel speeds are relatively slow, about 10 times slower than conventional grinding. Another difference is that the workpieces are not clamped or held firm, but rather are permitted to float and are freely guided between two flat annular-shaped abrasive wheels using workholder guides called carriers. To achieve high stock removal rates as well as a high degree of precision, superabrasive grinding wheels of diamond or cubic boron nitride (CBN) pellets are used. Figure 25-5 shows a pellet grinding wheel used as the lower wheel in a fine grinding machine. Pellet wheels made with diamond or CBN superabrasives provide

Figure 25-5. Pellet grinding wheel used for fine grinding. (Courtesy Peter Wolters of America, Inc.)

the capability of machining the hardest materials at a uniform and consistent cutting rate. Hard steel, sintered iron, crystals, and ceramic materials such as glass, silicon carbide, and aluminum titanium carbide are candidate workpiece materials for this process.

Fine grinding is generally used as a high-precision stock removal process as well as to achieve workpiece flatness, parallelism, and thickness. The finish obtained is semireflective with a precision multidirectional pattern. There are a number of advantages to the fine grinding process, the most significant being that there is considerable reduction in the overall processing time when compared to the several steps required for lapping. In lapping, parts usually require a four-step process after the blanks are cut or formed to the appropriate thickness: (1) pregrinding, (2) cleaning, (3) lapping, and (4) cleaning. For the pellet-wheel fine-grinding process, usually only two operations are required: (1) fine grinding, and (2) cleaning. In addition, the cutting rate is generally much faster than double-side lapping, and close to that achieved in surface grinding. Accuracies for parallelism, flatness, and thickness are much better than for surface grinding and are comparable to double-side lapping.

Figure 25-6 shows some of the results obtained in the application of double-side fine grinding to steel automotive tappet shims. The tappet shim blanks for this operation had a stock allowance of about 0.10 mm (.004 in.) and the stock removal for the fine grinding process was approximately 25 μm/min (.001 in./min). The fine-grinding machine employed in this operation is similar in appearance to a vertical double-side lapping machine with two annular-shaped pellet-grinding wheels mounted on two vertical spindles, one lower and one upper. The 904-mm (35.6-in.) diameter pellet wheels on this machine were able to accommodate 225 tappet shims per load. The machine was equipped with a semi-automatic loading and unloading unit.

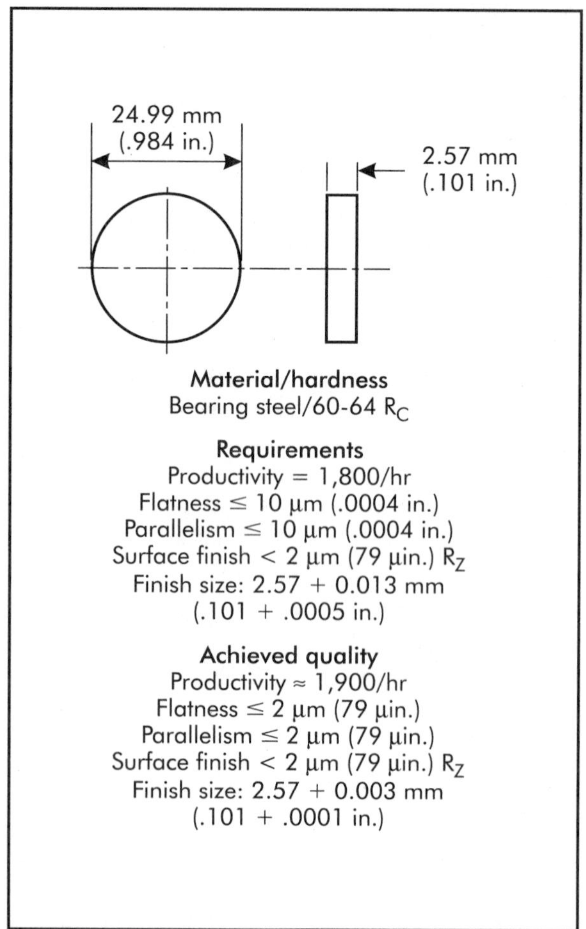

Figure 25-6. Results achieved in the double-side fine grinding of automotive tappet shims with CBN pellet wheels. (Courtesy Peter Wolters of America, Inc.)

25.3 MICROFINISHING OR SUPERFINISHING

25.3.1 Purpose

Microfinishing, also called *superfinishing* or *microstoning,* is generally done by scrubbing with an abrasive stone or stones pressed against a surface as illustrated in Figure 25-7. Microfinishing is generally not used as a dimension-creating operation, although it can correct out-of-roundness as much as 75% and size to less than 30 μm (≈.001 in.). Stock removal is usually in the range of 0.001–0.038 mm (.00005–.0015 in.). Substantial geometrical and dimensional accuracy must be created first, usually by grinding. Thus, microfinishing is generally a second operation used to achieve superior surface finishes as well as eliminate minor surface defects (waviness, chatter, and feed marks) caused by grinding or other manufacturing processes. Another extremely important reason for microfinishing is to remove the amorphous layer of metal that has been previously deformed by machining. This amorphous layer of unstable metal ranges from 0.5–1 μm (20–40 μin.) deep and is subject to early erosion and wear caused by the sliding friction of mating parts. Microfinishing removes this amorphous layer of metal and, in effect, eliminates the so-called "break-in" period for many devices such as automobiles, electric motors, and lawn mowers. Practically perfect surfaces with no apparent scratch pattern may be produced at one extreme by micro-

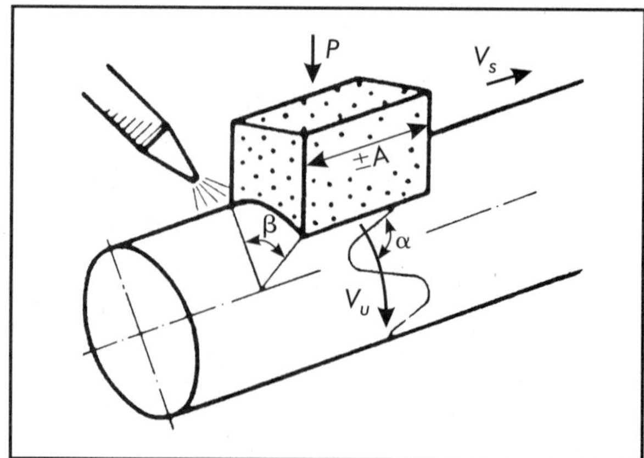

Figure 25-7. Microfinishing with a reciprocating abrasive stone. (Courtesy Supfina Machine Co., Inc.)

finishing. At the other, surfaces with readings of 0.76 μm (30 μin.) R_a and more can be produced with a deliberate crosshatched scratch pattern for lubricating qualities.

25.3.2 Process

For stone microfinishing such as is illustrated in Figure 25-7, a fine-grit abrasive stone with a relatively soft bond is oscillated (A) rapidly while being pneumatically pressed against a cylindrical work surface with a force (P). In addition, the stone is fed linearly along the cylinder at a rate (V_s) while the workpiece rotates to give a surface speed of V_u. The motions are arranged so that an abrasive particle never follows the same path more than once around the workpiece. During the finishing operation, the contact area is flooded with an oil or coolant bath to prevent heat buildup in the workpiece. Although little force is imposed on the abrasive stone, the contact pressure is high at first because the stone touches only a few high spots, and the cutting action is rapid. The stone is able to bridge and equalize a number of surface defects at one time and corrects to the average of the rough profile. When the surface becomes smooth, the pressure decreases, and the stone rides on a film of fluid and ceases to cut. A short surface may be refined to better than 0.1 μm (4 μin.) in less than 1 min. The microfinishing process can be used on a full range of materials, including aluminum, copper, stainless steel, Monel®, hard and soft steel, tungsten carbide, hard chrome, and various metallic and ceramic coated materials. Appropriate types of abrasive stones are available in various grades for microfinishing these materials to the desired specifications.

A more recent variation of the abrasive stone microfinishing process is illustrated in Figure 25-8. This process uses an electrostatically coated abrasive tape between a nonresilient shoe and the workpiece to assure full and consistent workpiece-abrasive contact. New abrasive tape is automatically indexed into place for each part to achieve a uniform size and finish on all parts. In this proprietary process, the two halves of the pivoted tooling are plunge fed into the workpiece while it is being supported and rotated by either a driver/chuck (center type) or a set of driving rolls (centerless). This centerless rotation of the workpiece, combined with the proper tape oscillation and pressure, is able to finish cylindrical parts with an excellent degree of surface texture, roundness, and straightness. The pivoting action of the two halves of the tooling facilitates manual and automatic loading and unloading of parts. One of the obvious advantages of the use of an abrasive tape is that it eliminates the need to compensate for wear on the shoes since they are not in direct contact with the workpiece.

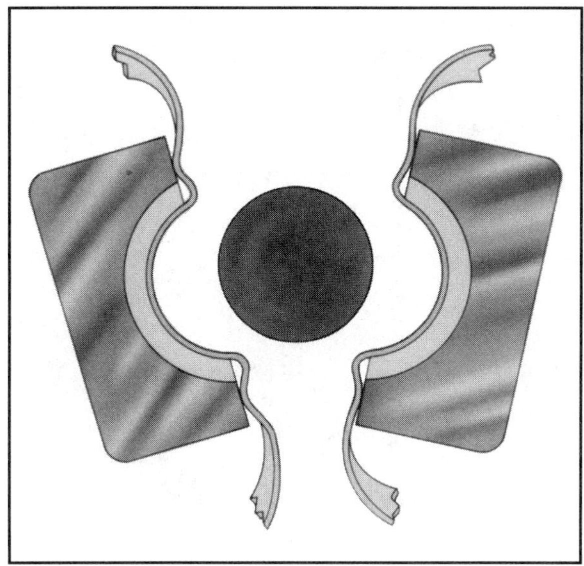

Figure 25-8. The abrasive-tape microfinishing process. (Courtesy Cimtec, Inc.)

25.3.3 Machines and Attachments

A variety of microfinishing equipment and machines are available, ranging from attachments for lathes and other basic machine tools to specialized machines for high-production operations. Figure 25-9 shows a double-stone microfinishing attachment mounted on the cross slide of a lathe. The attachment is being used to finish the surface of a hardened steel roll mounted between centers. A typical set of process parameters for this operation would be as follows:

f_o = frequency of oscillation = 40 Hz
A = amplitude of oscillation = 3 mm (.118 in.)
V_u = workpiece surface speed = 122 m/min (400 ft/min)
f_s = feed rate = 3.18 mm/rev (.125 in./rev)

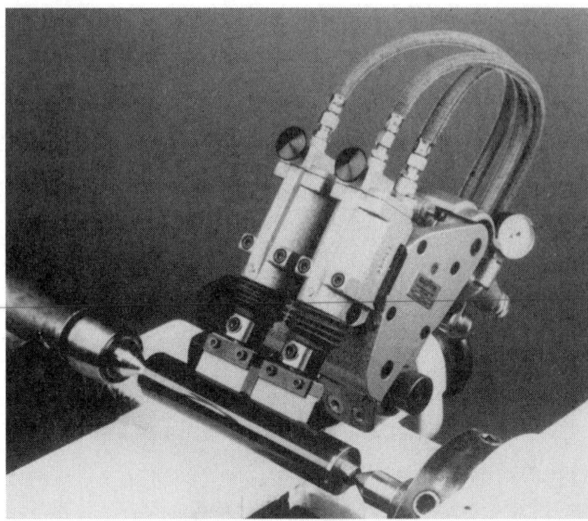

Figure 25-9. Double-stone microfinishing attachment mounted on the cross slide of a lathe. (Courtesy Supfina Machine Company, Inc.)

Figure 25-10. Two-station, centerless abrasive-tape microfinishing machine. (Courtesy Cimtec, Inc.)

As these parameters indicate, the workpiece and abrasive stone speeds in this operation are purposely maintained at a lower rate than would be used in grinding (below sparking limits). This prevents the workpiece from overheating and thereby preserves the metallurgical properties of the surface being finished.

Abrasive-stone microfinishing attachments similar to that shown in Figure 25-9 can be retrofitted to existing lathes, grinding machines, milling machines, and other machines with very little effort, thus eliminating the need to purchase special machines. For high-production purposes, more specialized machines are available for through-feed microfinishing of straight cylindrical parts or plunge-feed microfinishing of several different diameters at the same time on rotating parts such as crankshafts and camshafts.

The *centerless microfinishing* process is represented in the cutaway view of a two-station abrasive-tape microfinishing machine shown in Figure 25-10. This machine employs the abrasive-tape finishing process illustrated in Figure 25-8. The two-station machine can be used to finish two parts at the same time or perform two levels of finishing at separate stations. There are two drive rolls per station that provide support and rotation to the workpiece. The drive rolls are wide enough so that additional tape heads can be placed side-by-side to accommodate several parts per station.

Depending on workpiece requirements, microfinishing may need to be accomplished in two or more steps. For example, a workpiece that requires a near mirror finish may require a coarse grit tape first to remove the amorphous layer of material and, at the same time, correct the workpiece geometry. A second or intermediate step employing a finer grit tape may then be needed to remove marks from the previous stage and further improve upon the geometry. This may be followed by a polishing cycle with a fine-grit tape to achieve a near mirror finish, for example, of 0.5 μm (2 μin.) R_a or less. Cylindrical parts up to 356 mm (14 in.) in length and up to 89 mm (3.5 in.) in diameter can be accommodated by different models of the machine shown in Figure 25-10. With an appropriate abrasive tape and automatic loading and unloading, machines of this type are able to achieve optimum surface finishes on cylindrical parts at rates of 1,000 parts/hr or more.

25.3.4 Ultrasonic Machining

Ultrasonic impact grinding is a means of cutting shapes in hard materials by rapid and forceful agitation of fine abrasive particles in a slurry between the tool and workpiece. Figure 25-11 shows the elements of the operation. The tool is an image of the form to be cut, which may be a hole of almost any shape, a cavity, or figure in relief or intaglio. A sharp, pointed tool may be applied for engraving or die sinking. Vibrations of 15–30 kHz are generated by an electric driver, consisting of an oscillator and high-output amplifier, which supply high-frequency current to a coil around a laminated nickel core. This core has the property of expanding and shrinking under the influence of the alternating current.

The ratio of stock removed to tool wear is about 10:1, and tool life is tolerable. Because it is slow, ultrasonic grinding is practical only for substances harder than 64 R_C and shapes not amenable to regular grinding. In cemented carbide, stock removal is less than 0.5 cm^3/min (.03 $in.^3$/min). Ultrasonic grinding can do much the same work as EDM and ECM described in Chapter 27 (even though slower) and has the advantage that it works on nonmetallic substances while the others cannot. Typical jobs include dies for cemented carbide, slicing semiconductor materials, and engraving delicate patterns in glass. The action does not heat or disturb the material below the workpiece surfaces. Practical finishes obtainable are around 0.25 μm (10 μin.) R_a and tolerances 50 μm (.002 in.) or less.

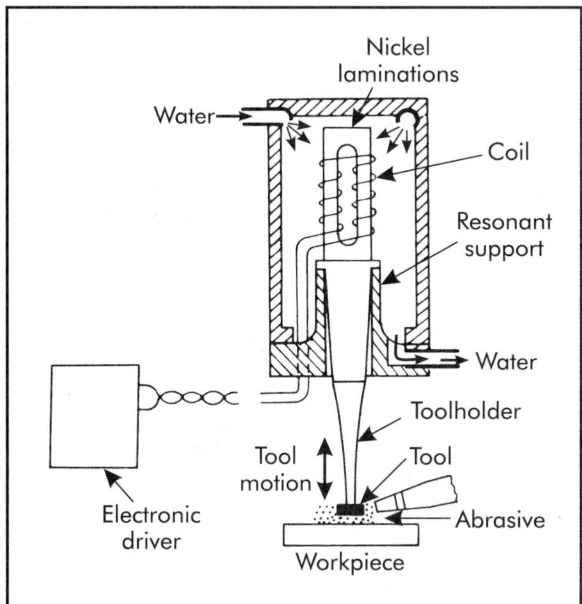

Figure 25-11. Ultrasonic transducer.

Another form of ultrasonic machining applies vibrations to a rotating diamond tool without abrasive slurry. Hard nonmetallic materials can be cut faster in this way than by conventional means.

25.4 BURNISHING AND BEARINGIZING

25.4.1 Roller Burnishing

Roller burnishing cold works metal by forcing hardened, precision-ground rolls to rotate around the surface to be finished. Roller burnishing is used to finish straight and tapered holes, external cylindrical parts, and flat surfaces. Parts made from ductile materials such as steel (alloy and stainless), cast iron, aluminum, copper, brass, bronze, etc., with a hardness less than 38 R_C have all been successfully burnished. The hole burnishing process is accomplished by means of a tool with tapered rolls, which are rotated around and bear on an inversely tapered mandrel as illustrated in Figure 25-12(A). These rolls apply a steady rolling pressure against the work surface causing the high peaks to flow into microscopic cavities and valleys, thus accurately sizing, providing a fine finish, and work hardening the surface. Typically, surface hardness is increased from 5–10% with a penetration of from 0.25–0.76 mm (.010–.030 in.). The improved surface finish combined with a hardened and denser surface resulting from the burnishing process substantially improves the wear and corrosion resistance of a part.

The straight-shank hole-burnishing tool shown in Figure 25-12(B) is adjustable in increments of 0.0025 mm (.00010 in.) from 38.00–39.04 mm (1.496–1.537 in.). A typical range of operating speeds and feeds for this tool while used to burnish a hole in a steel part with a hardness less than 38 R_C would be:

Surface speed = 91–259 m/min
(300–850 ft/min)
Feed rate = 2.29–2.41 mm/rev
(.090–.095 in./rev)

Burnishing does not remove metal, it merely displaces it by working the high peaks of the surface into microscopic valleys. Thus, the amount

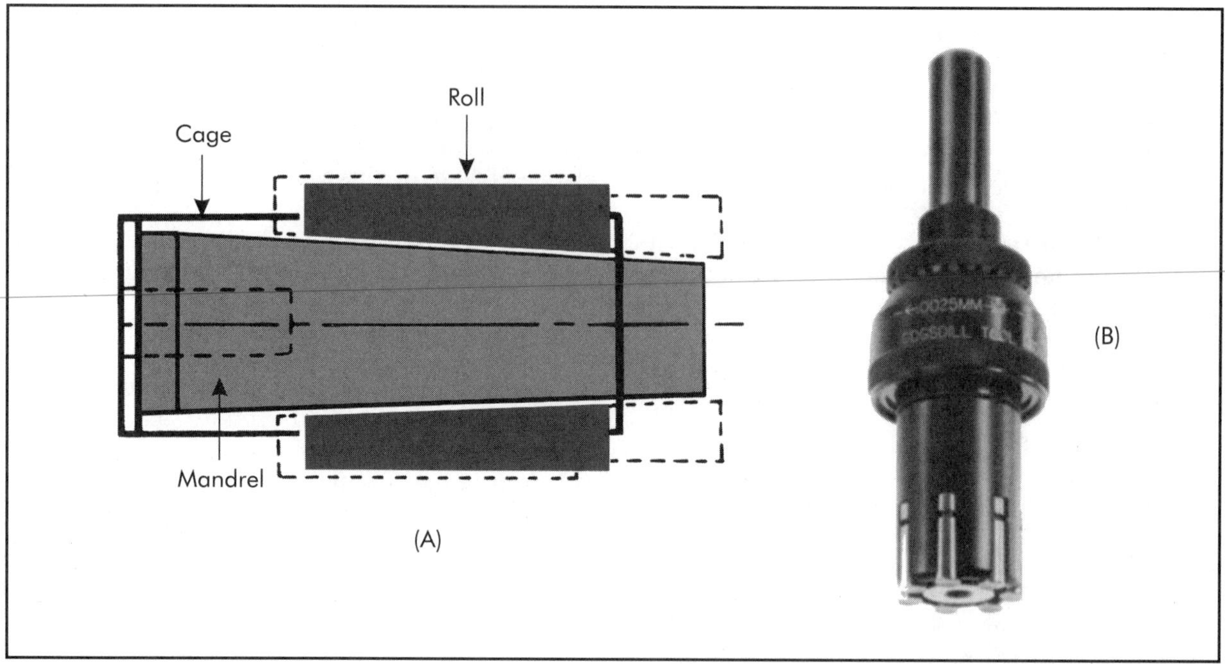

Figure 25-12. (A) The major elements of a roller burnishing tool. (B) Adjustable, straight-shank roller-burnishing tool. (Courtesy Cogsdill Tool Products, Inc.)

of stock that is displaced by burnishing is dependent on the relative roughness of the starting surface. For example, the diameter of a bored hole with a 2–3 μm (80–120 μin.) R_a finish may be enlarged by 0.020–0.030 mm (.0008–.0012 in.) by burnishing. At the other extreme, if that same hole has been honed to from 0.25–0.51 μm (10–20 μin.) R_a, burnishing may enlarge the diameter by only 0.0025–0.0051 mm (.00010–.00020 in.). Burnishing is able to achieve surface finishes in the range of 0.05–0.38 μm (2–15 μin.) R_a, again depending on the condition of the starting surface. A roller burnishing tool with 7 rolls similar to the one shown in Figure 25-12(B) costs about $400. Under normal circumstances, the life of these tools is excellent and they have been known to finish up to 50,000 holes before replacement of inexpensive wear parts is required.

25.4.2 Bearingizing

A *roller bearingizing tool* works by simultaneously rolling and peening the surface. As illustrated in Figure 25-13, precision hardened and ground rolls rotate around a cammed arbor causing the rolls to rise and fall. Thus, when the tool is rotated at high speed, the action of the rolls deliver as many as 200,000 rapid fire blows/min to the work surface. This combination of rolling and peening compacts the peaks and valleys of the work surface into a smooth, hardened, and ultrafine surface finish. Surface hardness may

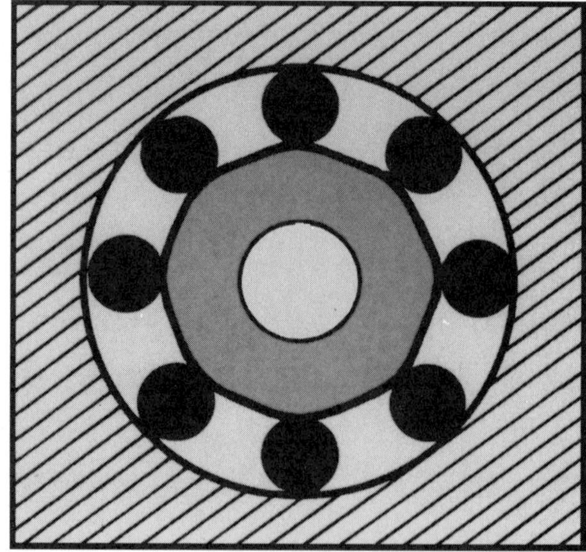

Figure 25-13. The rolls and cammed arbor of a roller bearingizer tool. (Courtesy Cogsdill Tool Products, Inc.)

be increased by this process from 10–30% with a depth of penetration of 0.13–0.38 mm (.005–.015 in.). Surface finishes in the range of 0.05–0.25 μm (2–10 μin.) R_a can be achieved, along with dimensional tolerances as low as ±0.0025 mm (±0.00010 in.). As in burnishing, any ductile or malleable material (laminated, cast, forged, extruded, or sintered) with a hardness less than about 38 R_C can be successfully bearingized. A bearingizing tool of about the same size as the burnishing tool shown in Figure 25-12(B) costs about $600.

25.5 NONPRECISION DEBURRING AND FINISHING PROCESSES

25.5.1 Polishing

Polishing puts a smooth finish on surfaces and may often involve removal of appreciable metal to take out scratches, tool marks, pits, and other defects from rough surfaces. Usually accuracy of size and shape of the finished surface is not important, but sometimes tolerances of 25 μm (.001 in.) or less are held in machine polishing. Polishing wheels, described in Chapter 23, distribute cutting action and conform to curved surfaces on workpieces. The application of abrasives here follows much the same principles as grinding. Commonly several steps are necessary, first to remove the defects and then to put the desired polish on the surface. Much polishing cost can be saved by adequate surface preparation.

The piece may be applied to a wheel by hand for polishing on a floor-stand grinder. More production speed and consistency can be realized on semi-automatic polishing machines when there is enough work to justify the investment. The two general classes of such machines are: (1) those that carry the work in a straight line past one or more wheels, and (2) those that revolve the pieces in contact with the wheels. The time needed by the operator to load a piece must be appreciably less than that required to polish a piece by hand for a machine to be economical. Buffing and power brushing are done in similar ways.

25.5.2 Buffing

Buffing gives a high luster to a surface. It is not intended to remove much metal and generally follows polishing. The work is pressed against cloth or felt wheels or belts on which fine abrasive in a lubricant binder is smeared.

25.5.3 Power Brushing

High-speed revolving brushes are applied to improve surface appearance and remove sharp edges, burrs, fins, and particles. This tends to blend surface defects and irregularities and rounds edges without excessive removal of material. Surfaces may be refined to around 0.1 μm (4 μin.) R_a when desired. Brushing action helps avoid scratches that act as stress raisers.

Common power brushes are wire bristle, hard cord, and tough fiber wheels. They are naturally flexible, able to conform to irregular surfaces, and can get into otherwise hard-to-reach places. Abrasive compounds are often put on brushes. Brushing is done by hand but is readily adaptable to semi-automatic machines, which are fast for production.

25.5.4 Tumbling and Vibratory Finishing

The operation called *tumbling*, *rolling*, or *barrel finishing* consists of loading workpieces in a barrel about 60% full of abrasive grains, sawdust, wood chips, natural or artificial stones, cinders, sand, metal slugs, or other scouring agents, depending on the work and action desired. Water is usually added, often mixed with an acid, a detergent, a rust preventative, or a lubricant. The barrel is closed or tilted and rotated at a slow speed from 1 to more than 10 hr, according to the treatment required. With the right load and speed, the workpieces slide over each other, producing a scouring, trimming, and burnishing action as the barrel turns.

Vibratory finishing does the same work as barrel finishing, but it is done in an open-rubber or plastic-lined tub or trough nearly filled with workpieces and media and vibrated at around 1,000–2,000 Hz with about 3–10 mm (≈.125–.375 in.) amplitude. Various ways of inducing vibrations are employed; a common way is by means of eccentric weights on a revolving shaft. The action makes the entire load rotate slowly in a helical path, but the whole mass is agitated, and scouring, trimming, and burnishing take place throughout the mixture. Therefore, vibratory finishing is much faster (2–10 times and sometimes more) than tumbling in which the action is confined to the pieces in the sliding zone at any one time as depicted in Figure 25-14. Much vibratory finishing is done in batches, but the process is readily adaptable to continuous and automatic flow-through operation because of the steady movement of the mass.

Tumbling gives superior results on a few jobs, such as producing high finishes on some nonferrous metals. It is usually confined to finishing the outside surfaces of pieces. However, vibratory finishing, under proper conditions, also can do well inside of pieces, in recesses, and upon shielded surfaces. Pieces that must be kept apart (such as large or soft ones) may be vibrated in batches in fixtures, whereas they must be tumbled in separate compartments.

Tumbling and vibratory finishing are applied to ferrous and nonferrous metals, plastics, rubber, and wood of small and large sizes. They clean castings, forgings, stampings, and screw machine products; remove burrs, fins, skin, scale, and sharp edges; take off paint and plating; improve surface finish and appearance; and have a tendency to relieve surface strains. Some reduction in size may be experienced but results are uniform in each lot.

Another development of barrel finishing employs two equally loaded barrels on a turret, which revolves in a counterclockwise direction to generate centrifugal forces of 1–25 g. The barrels revolve clockwise at the same time, and the

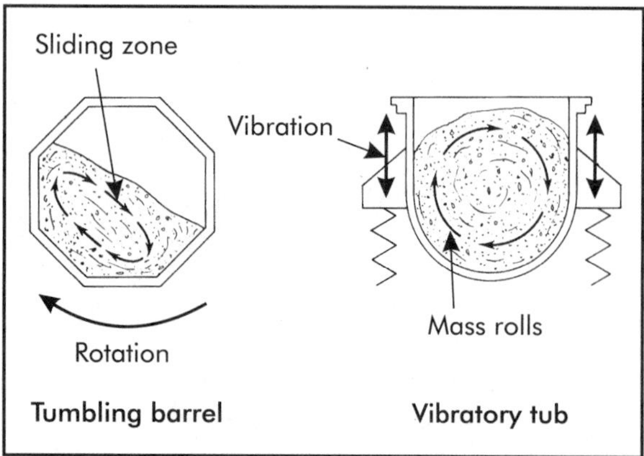

Figure 25-14. Principles of tumbling and vibratory finishing.

parts and media therein convolute under high pressures. In this way, work may be done as much as 50 times faster than simple tumbling, but equipment is expensive and costly development and careful control are necessary for each job. The process is not considered competitive with conventional tumbling or vibratory finishing for most work but has been found superior for certain jobs, such as finishing and deburring precision parts, inducing high surface compressive stresses, and sizing steel balls.

25.5.5 Shot Blasting and Sandblasting

Shot blasting and *sandblasting* are done by throwing particles at high velocity against the work. The particles may be metallic shot or grit; artificial or natural abrasive, including sand; agricultural products such as nut shells; glass beads; or ceramics. The choice of media depends upon what is to be done and the condition of the workpiece. A primary reason for blasting is to clean surfaces. This may mean removing scale, rust, or burnt sand from castings by means of shot or sand, stripping paint by sandblasting objects to be redecorated, cleaning grease or oil from finished parts by means of nut shells, or any number of similar operations. A clean, uniform, and in many cases, final surface finish is obtained by blasting. In addition, shot blasting peens surfaces and leads to the advantages of appreciably increasing fatigue strength and stress corrosion resistance, reducing porosity in nonferrous castings, improving surface wearability as on gear teeth, and improving the oil retentivity of some surfaces.

Four common ways of blasting are by compressed air, centrifugal action, high-pressure water, and a mixture of compressed air and water. Compressed air equipment can be used with any type of abrasive, is easily controlled, simple and relatively inexpensive, and gives ready access to inside surfaces. For centrifugal action, particles are fed to and slung by a rapidly revolving wheel. High flow and a rapid production rate can be obtained in this way, and it is the most popular method. Water with or without a polymer additive and fine particles is applied in jets to produce smooth, satin-like surface finishes and cut thin and thick materials at fast rates. This is called *waterjet machining* (WJM). A fine abrasive propelled by an air jet does what is called *abrasive-jet machining* (AJM). Typical uses include fine deburring, drilling, and trimming of hard, brittle materials, particularly in thin sections, such as silicon wafers. With soft particles, AJM is efficient for delicate cleaning.

Small-lot blasting is commonly done in a cabinet where the operator from the outside manipulates a nozzle through safety gloves and sleeves. In high production, nozzles may be mounted in fixed positions in a large enclosure. Workpieces are carried past the nozzles on rotating tables, conveyors, etc.

A miniaturized version of the sand blasting process is represented by the *microabrasive blasting* process. This is a simple, yet highly effective method of using very small abrasive particles to clean, cut, deburr, or texture a wide variety of parts and surfaces with extreme precision. This is accomplished by uniformly mixing the abrasive with an air stream and then propelling it out of small nozzle tips at high velocity. It may be done manually by holding and guiding the small nozzle like a pencil over the surface to be finished.

Because manual methods are often unsuited for precision work, automated CNC-operated equipment is available for increasing accuracy and productivity. The blasting nozzle on the unit shown in Figure 25-15 can be positioned and moved in two coordinate directions by computer numerical control (CNC). Through the use of CNC, rotating spindles, multiple-axis positioning, slide or shuttle mechanisms, conveyor transports, parts handling, and nozzle manipulation to accommodate a large variety of automated microabrasive applications can be accomplished. Nozzle sizes from 0.38–2.03 mm (.015–.080 in.) are available for the unit shown in Figure 25-15. A variety of abrasive materials ranging from hard silicon carbide to soft, water-soluble bicarbonate of soda are available for different blasting requirements. By selecting the correct type of abrasive, operating air pressure, and appropriate nozzle opening, it is possible to achieve excellent results at several finishing extremes, from deburring hard titanium bone screws to engraving optical filters. A programmable microabrasive blasting unit similar to the one shown in Figure 25-15 costs about $30,000.

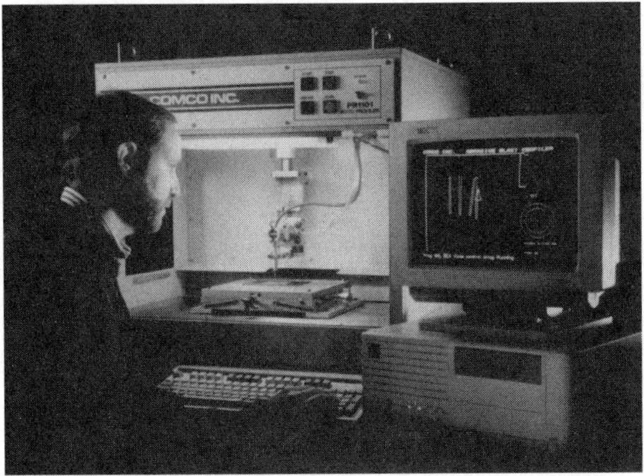

Figure 25-15. CNC-operated microabrasive blasting unit. (Courtesy Comco, Inc.)

25.5.6 Deburring

During most machining processes and press blanking and piercing operations, a small sharp projection of metal, called a *burr*, remains along the edges or around holes in the workpiece. As cutting tools and punches and dies become worn, these burrs become more pronounced and may impair proper fit and function of parts. In addition, they are often unsightly and dangerous. Thus, *burr removal* or *deburring* is often required as a secondary operation for many products. Burrs can be and are removed by many of the processes already described in this chapter and by manual scraping, filing, and trimming. Other methods are also used. One of these is *electrochemical machining* (ECM) described in Chapter 27. Others are described in the following paragraphs.

Abrasive Belt Machining

Abrasive belt machining is commonly used to deburr and finish flat metal parts after they have been blanked and/or pierced. The flat part deburring and finishing machine shown in Figure 25-16(A) employs a composition conveyor belt on which flat parts may be positioned and carried under a moving abrasive belt such as is shown in Figure 25-16(B). A typical operation for this machine would be to deburr and grain-finish ventilated back cover plates for rack-mounted oscilloscope and control circuit cabinets made from 1.14-mm (.045-in.) thick sheet steel. After being pierced on a turret punch press, the cover plates are deburred and given a No. 4 grain finish using a 150-grit SiC abrasive belt. Only a small amount of material is removed from the surface of these plates by the abrasive process and the final part thickness is not important to the function of this part. The grain finish provides an appropriate base for subsequent painting.

The abrasive belt carriers shown in Figure 25-16(A) operate at a constant rpm to provide a belt speed of about 914-m/min (3,000-ft/min). The part conveyor belt speed can be varied from 4.6–13.7 m/min (15–45 ft/min). The distance between the conveyor and abrasive belt on this machine can be adjusted in 0.025 mm (.0010 in.) increments to accommodate flat workpiece thicknesses from 0–203 mm (0–8.00 in.). The 0.940-m (37-in.) abrasive belt width can handle parts up to 0.914-m (36-in.) wide. Abrasive belt machines can be equipped to operate in either a wet or dry mode. The wet machines are more expensive because of the need to maintain corrosion resistance on the machine parts. Some small workpiece shapes and sizes may require nests or carriers on the conveyor belt so that the hold-down rollers in the machine will operate properly. In some cases, the machines will be equipped with vacuum or magnetic beds to hold parts in place for processing.

A variation of the machine shown in Figure 25-16, called a *belt/brush deburring machine* uses a roller brush to round off the edges of flat parts after deburring with an abrasive belt.

Thermal Energy Deburring

Another deburring process, the *thermal energy method* (TEM), burns away burrs. Parts to be deburred are placed in a chamber filled with gasses (that is, hydrogen and oxygen) that are spark ignited. Thin burrs and flash reach high temperature and are burned by excess oxygen; the main body of each part acts as a heat sink and is not affected.

Abrasive Flow Machining

Abrasive flow machining (AFM) or *extrude-honing* is done by forcing an abrasive-laden soft rubber or putty-like compound to flow over surfaces and edges and through holes, slots, and cavities of machined parts. The process can be closely controlled to act uniformly over a part to remove burrs, polish, and even help size surfaces, generate radii, and reach areas not adequately

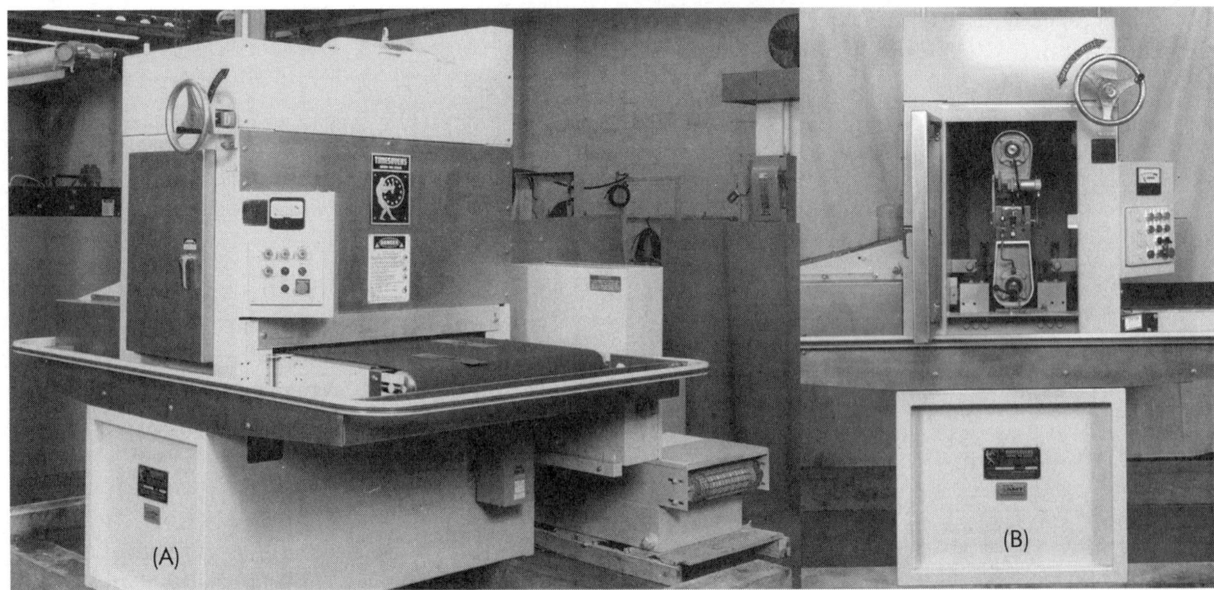

Figure 25-16. (A) Abrasive-belt, flat part deburring and finishing machine. (B) Side view of machine showing abrasive belt drive and carriers. (Courtesy Timesavers, Inc.)

accessible with other methods. The AFM process is capable of generating surface finishes of 0.05 μm (2 μin.) R_a and deburring holes as small as 0.20 mm (.008 in.). Although it is not commonly used to remove large amounts of stock, the AFM process can do a certain amount of sizing within tolerances of ±0.005 mm (±.0002 in.). As with most abrasive processes, stock removal and surface finish are a function of the number of cycles, grit composition and type, and extrusion pressure.

25.6 PROCESS PLANNING

Final finish machining is one of the most labor-intensive and costly aspects of the precision parts manufacturing industry. In some product lines it contributes to up to 15% of the total manufacturing cost. Thus, the process planner must be quite exhaustive in comparing and selecting the appropriate finishing process. Generally, the selection process is based on consideration of the need to satisfy one or more of the following requirements:

1. Generate a smooth, high-quality surface finish to a specified standard.
2. Finish a part to an accurate dimensional size within a small tolerance.
3. Minimize or eliminate geometric inaccuracies resulting from prior operation, that is, inaccuracies in flatness, roundness, straightness, etc.
4. Accomplish one or more of the preceding requirements with appropriate stock removal rates and loading and unloading times to be cost effective.

Usually, by a process of elimination, the process planner is able to select two or three finishing processes capable of satisfying the first three requirements. Then some form of economic comparison is made to select the particular process that is most cost effective.

All of the precision finishing processes are capable of producing fine finishes, some to a lower average threshold than others. Generally, lapping and microfinishing can generate surface finishes in the 0.025–0.051 μm (1–2 μin.) R_a range under average conditions and less under special conditions. However, both of these processes are capable of removing only small amounts of stock efficiently, so care must be taken to create the appropriate geometric and dimensional accuracy in the previous operation. All of the precision finishing processes are capable of correcting some geometric and dimensional inaccuracies, within limits, of course. Some of the processes are limited to certain geometric part configurations. That is, some of them are better suited to the finishing of cylindrical holes, while others are more suitably applied to flat surfaces.

Each of the nonprecision finishing operations has certain areas in which it excels. Polishing may remove large defects and heavy stock around deep scratches and pits at less cost than would be incurred to remove metal to the same depth over the whole piece. Buffing and brushing are quick ways of getting high luster where substantial stock is not needed to be removed. Tumbling and vibratory finishing are economical for cleaning, improvement of appearance, and deburring over the whole piece, especially for large-quantity production. As a rule, blasting is superior for heavy scale and stock removal. Abrasive belt deburring and finishing is used mostly on flat metal stampings. However, these are not the only applications for these processes; their provinces overlap, and often they work best together. For example, a particular part may require polishing to remove surface defects, followed by vibration to form radii on edges and refine the finish.

Processing rates for the various precision and nonprecision finishing processes discussed are best obtained from equipment and abrasive manufacturers.

Quite often, the process planner will perform a simple cost comparison between those processes that are essentially equally capable of achieving the required results for a particular finishing job. As an example of such a comparison, assume a small part can be finished equally well by either tumbling or by polishing and buffing. Time and cost rates for the two processes are as follows:

Polishing and buffing
Labor and overhead rate = \$16.50/hr
Processing time = 3 min/piece

$$\text{Cost} = \frac{3N \times 16.50}{60}$$

where

N = number of pieces

Tumbling
Labor and overhead rate = \$16.50/hr
Loading and unloading time = 12 min/load
Tumbling time = 4 hr/load
Tumbling machine rate = \$6/hr

$$\text{Cost per load} = \frac{12 \times 16.50}{60} + (6 \times 4)$$

The number of pieces for which the cost is the same by either method is N, and

$$\frac{3N \times 16.50}{60} = \frac{12 \times 16.50}{60} + (6 \times 4)$$

Thus, $N = 33$ pieces. For a lot of fewer than 33 pieces, polishing and buffing is the more economical method. It is assumed that at least 33 pieces can be tumbled at once.

25.7 QUESTIONS

1. List the precision finishing processes discussed in this chapter.
2. What is the difference between lapping and honing?
3. Why is a good surface finish important for mating parts where oil seals are used to prevent leakage?
4. Why is it important to remove the amorphous layer of material when finishing a metal part?
5. Describe the difference between lapping and microfinishing.
6. How much stock is generally removed in the lapping process?
7. Give an example of a manufactured part that might be finished by a centerless lapping machine.
8. Why is honing done at lower speeds than grinding?
9. Why are automotive engine cylinders often honed with a crosshatched surface finish?
10. What is the difference between honing and microfinishing?
11. Why are cutting speeds in fine grinding lower than those used in the regular surface grinding process?
12. How does the fine grinding process differ from lapping?
13. What advantage(s) does the abrasive-tape microfinishing process have over the abrasive stone process?
14. Of all of the external surface finishing processes discussed in this chapter, which one has the best potential for correcting geometric inaccuracies? Which has the least?
15. Why does the roller burnishing process change the surface hardness of the part being burnished?
16. What advantage, if any, does bearingizing have over burnishing?
17. Why is it often necessary to remove burrs from machined or stamped parts?

25.8 REFERENCES

ANSI/ASME B46.1-1985. "Surface Texture (Surface Roughness, Waviness, and Lay)." New York: American Society of Mechanical Engineers.

Gillespie, L. K. 1981. "Deburring Technology for Improved Manufacturing." Dearborn, MI: Society of Manufacturing Engineers (SME).

Hettes, F. J. 1991. "Which Brush Should You Choose?" *Manufacturing Engineering,* September.

Koelsch, J. R. 1991. "Taking on Tough Burrs." *Manufacturing Engineering,* October.

Koepfer, G. C. 1994. "Fine Grinding: High Accuracy, High Production." *Modern Machine Shop,* December.

——. 1993. "Microfinishing Comes of Age." *Modern Machine Shop,* December.

Schreiber, R. R. 1992. "NAMRC XX: Challenging the Status Quo." *Manufacturing Engineering,* July.

——. 1994. "The Deburring Process." *Metal Forming,* December.

Other Surface Enhancement Processes

26

Electroplating, 1800—James Wilhelm Ritter, Germany

Manufactured products naturally collect oil, dirt, chips, etc., during processing. This soil must be removed for certain operations, such as inspection, painting, and assembly, and especially for salability. Thus cleaning is an important part of many processes, and the principal ways it is done will be discussed in this chapter.

Coatings are commonly applied to the surfaces of articles for decoration, texture, corrosion resistance, electrical insulation, lubricity, and protection against high temperatures. The common classes are conversion coatings formed by chemical reaction with the surface, organic coatings or paints, metallic platings, and inorganic or vitreous coatings. The chief forms of these and the methods of applying them are described.

26.1 CLEANING

26.1.1 Cleaners

Cleaning is done by mechanical action and by washing with solvents, detergents, and other chemicals. Chunks and layers of sand, dirt, scale, rust, etc., may be removed by sand blasting, tumbling, and other vigorous mechanical methods described in Chapter 25. Such rough cleaning must normally be followed by washing or degreasing prior to painting or plating. Washing and degreasing may be done by *mineral* or *organic solvents* or by *water solutions*. Solvents such as naphtha, kerosene, and chlorinated hydrocarbons (for low fire hazard) dissolve greases, oils, and waxes, but not inorganic dirt. They are relatively expensive but are efficient degreasers and do not harm most surfaces. Water itself is not a good cleaner but is the universal solvent for cleaning agents and is used freely for rinsing, which always must be done after washing for thorough cleanliness. Conditioners like softeners, inhibitors, wetting agents, and detergents make water solutions efficient. The three basic detergents are emulsified solvents and acidic and alkaline detergents. Water *emulsions* of organic solvents combine the advantages of a solvent with those of an agent that disperses soil. They work cold and are economical for light degreasing and precleaning to save time and alleviate contamination in subsequent alkaline cleaning operations. Alkaline solutions, such as caustic soda or trisodium phosphate, often blended with colloidal materials, soap, and other wetting agents, are commonly available in strengths enough to remove any soil, however, they attack some surfaces. They are the most widely used industrial detergents. Mild acids are used for materials attacked by alkalis to remove scale, oxides, and fluxes. Molten salt baths are able to strip tenacious deposits of sand, scale, etc., from forgings and castings.

26.1.2 Methods

The best cleaning method for a product depends on how clean the surfaces must be, the kind of soil, the surface material, the size and shape of the part, and the number of pieces. Some washing is done cold, but much (particularly with alkaline detergents) must be done hot. The easiest way of cleaning and one that can be done with any fluid is *dipping* or *immersion*. This alone may not be enough to dislodge clinging soil, and agitation, barrel tumbling, or rubbing may have to be added. A tank of fluid for washing becomes contaminated in time and thus less effective. Immersion of the work in a series of tanks can help. All methods of cleaning may be done manually or may be highly automated for large-quantity production.

Electrolytic cleaning is a form of immersion. It uses an alkaline solution with the workpiece as the cathode in an electrical circuit. Gas released on the surface helps dislodge foreign substances, and a thin protective layer of tin may be deposited.

Spraying of all kinds of fluids forcefully acts to dislodge solid dirt. The spray may be directed by hand or be part of an automated system as indicated in Figure 26-1.

Vapor degreasing is a method of cleaning with solvents as depicted in Figure 26-1. A nonflammable solvent (for example, trichloroethylene) in the bottom of a tank is vaporized at 38–121° C (100–250° F). Cooling coils near the top condense the vapor and keep it in the tank. The vapor condenses on a relatively cool workpiece; the condensate dissolves grease and runs off, removing dirt and chips. Residue collects at the bottom and is not carried out with the work. Pieces are dry when they come out of the tank. The action is fairly rapid for removal of organic substances. Small light workpieces may become hot quickly and cease to condense vapor before they are cleaned. Some parts may have exceptionally heavy soil to be removed. These may be immersed in warm or violently boiling liquid, sprayed, or dipped in vapor.

Volatile solvents are very toxic and can be dangerous if used or handled improperly. Strict regulations for safety of workers, protection of the environment, and fire prevention have been enacted and must be closely followed by users.

In *electrosonic cleaning*, high-powered and high-frequency sound waves from one or more transducers are focused on a workpiece immersed in a cleaning fluid as indicated in Figure 26-1. This causes the rapid formation and collapse of minute bubbles or cavities in the liquid (called *cavitation*), giving a violent action on exposed surfaces. This is not widely used for removing heavy deposits or jelly-like substances, but is efficient for jarring loose tightly adherent smut and reaching inaccessible recesses. Deburring and rounding of edges can even be done at high power levels.

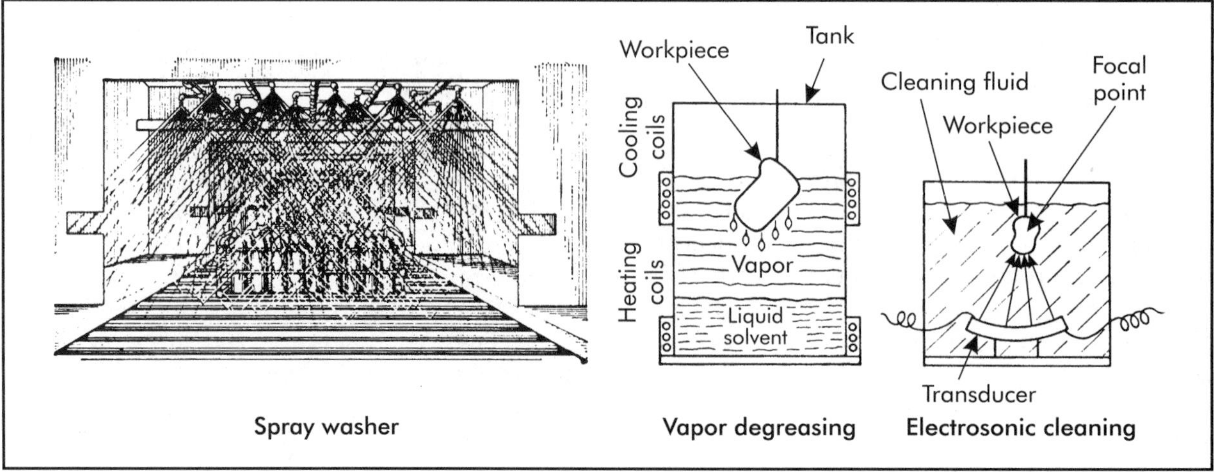

Figure 26-1. Methods of washing and degreasing.

The individual operations just described are commonly combined into processes to suit specific situations. For example, a typical sequence may be: (1) preclean by dipping or spraying with an emulsifiable solvent to remove the bulk of the grease and dirt, (2) rinse by spraying with hot water, (3) dip and clean electrolytically to remove scale and oxide, (4) final hot rinse by dipping, and (5) dry in an air blast.

26.1.3 Pickling and Oxidizing

Pickling is the chemical removal of surface oxides and scale from metals by acid solutions. This is commonly done on rolled shapes, wire, sheets, heat-treated steel parts, wrought and cast aluminum parts, etc. In some applications, such as on aluminum, it is called *oxidizing*.

Common pickling solutions contain sulfuric or hydrochloric acids with water and sometimes inhibitors. Nitric and hydrofluoric acids are used for some applications. A solution may contain 50% acid for cold use but as little as 10% if intended for use at 93° C (200° F). Pickling is usually done by immersion for periods of several minutes or more.

In pickling, the acid cannot get to a surface that is covered with dirt. Thus, parts must be cleaned first. After pickling, the parts must be completely neutralized by an alkaline and then a clear rinse. Any residue of acid will harm paint or other subsequent coating.

26.2 SURFACE COATINGS

26.2.1 Conversion Coatings

Conversion coatings are basically inorganic films formed by chemical reactions with metal surfaces but are often impregnated with organic substances. They are usually much less than 25 μm (.001 in.) thick and normally formed from the original surface, resulting in a tightly bonded coating with no appreciable dimensional change to the part. Common forms are phosphate, chromate, oxide, and anodic coatings.

Phosphate coatings are essentially phosphate salts formed by dipping, spraying, or brushing with acidic solutions of metal phosphates. They are put on iron, steel, and zinc and to a lesser extent on aluminum, cadmium, titanium, and tin to resist corrosion (particularly under paint films), make paint adhere tightly, add lubrication, and resist abrasion.

Chromate coatings are obtained by applying acidic solutions of chromium compounds to steel, zinc, cadmium, aluminum, copper, brass, silver, tin, and magnesium surfaces, occasionally with electrolytic assistance. The thin amorphous film resists abrasion, provides an excellent base for organic coatings, and may be dyed and serve alone as a decorative finish, although it is not lightfast.

Conventional *black oxide coatings* are put on steel by immersion in a boiling solution of sodium hydroxide and mixtures of nitrates and nitrites. Another black oxide process employs steam treatment for drills. Some proprietary processes utilize inorganic and organic substances in baths at room temperature. Various colors of oxide coatings are put on aluminum and other nonferrous metals. Oxide coatings serve as paint bases and final finishes. They are relatively cheap, and when impregnated with oil or wax, furnish good corrosion resistance.

Anodic coatings are applied to aluminum, zinc, beryllium, titanium, and magnesium alloys by electrolytic chemical means. The workpiece is immersed in a solution (commonly sulfuric acid with or without organic additives for aluminum) and connected as the anode in an electrical circuit. Anodic coatings may be from about 2–250 μm (.0001–.010 in.) thick and are good protection against corrosion because they are in reality like, but only more of, the thin film that naturally protects the metal. Dense films over 25 μm (.001 in.) thick are wear resistant. Anodic coatings may be colored in many hues with excellent permanency.

Conversion coatings may be impregnated with fluorocarbon resins, molybdenum disulfide, or graphite by further electrochemical and heat treatment. These are called *synergistic coatings* and can be made to have superior hardness, wear resistance, lubricity, and corrosion resistance. Originally developed for space travel devices, these coatings are being applied to other demanding applications.

26.2.2 Organic Coatings

Organic coatings in the form of thin plastic sheets or strips may be laminated to surfaces but are mostly applied as paints, enamels, or inks. A coating may be put on a finished product or precoated on the stock from which the product is made. Organic coatings can be applied to almost all materials and offer unlimited color and gloss varieties. Thus they provide more decorative possibilities than other coatings. In durability, strength, and corrosion resistance, they generally are superior to conversion coatings but not as effective as some metallic coatings. Usually they cost less than metallic coatings and thus are preferred if they serve satisfactorily.

Paint is the general term for an organic coating and consists of film-forming materials and pigments for coloring, hiding power, and protection. Clear finishes lack pigments. Drying agents also may be added.

Oil paint is a dispersion of metallic pigments, such as white lead, in a vegetable drying oil, such as linseed oil, along with a solvent thinner and perhaps dryers. The thinner evaporates and the oil oxidizes to form the film. Drying time depends on the oil used and the drying agents added, but is relatively long.

A *lacquer* is essentially a solution of plastic resins and plasticizers with or without pigments in a solvent. When the lacquer is applied, the solvent evaporates and leaves a film that can be made attractive by polishing. Lacquer is relatively easy to apply and dries quickly, but the film is not as durable and resistant to some solvents as other coatings.

Varnishes and *enamels* of the older kinds are like oil paints in that they form a film by oxidation of a resin-oil vehicle. Newer synthetic types are based largely or entirely on plastic resins and elastomers and harden by polymerization. Almost all the principal substances described in Chapter 7 are used singly or compounded to obtain various degrees of corrosion, chemical, and environmental resistance, colorability, durability, and other properties. The synthetic resins can be compounded to be equal or superior to conventional enamels for any specific application. Extensive tables that show the relative properties and attributes of the common plastic compounds for coatings are given in reference books and handbooks.

Plastic resin and elastomer coatings come as liquids and powders. Some of the liquids contain a large proportion of organic solvent that evaporates to leave a polymer film. The vapors can cause air and water pollution and require costly equipment to meet environmental standards, so the use of such liquid coatings has greatly curtailed. Other liquid coatings with little or no solvent content have become more popular, although the changeover can be quite expensive because different equipment and techniques are usually required.

The newer liquids are *medium-* and *high-solid paints* and *water-base* or *water-borne paints*. Medium-solid paints have 50–60% solid content (by volume) compared to 30% for conventional paints; high-solid paints are available with much more. As an example of benefits, one user reported solvent emissions with 52% solids almost 60% less than with 30% solids, making it much easier to meet environmental requirements. Energy savings are also a factor. One user gained a 25% savings in the cost of fuel gas for drying with medium-solid paint. Water-base systems are of the water-soluble emulsion and colloidal dispersion types for different kinds of polymers. Water-base paints eliminate fire hazard.

Paints of another class are called two-component systems because they involve the combination of two substances at the time of painting to achieve polymerization of the film. Although these coatings are superior for some applications, in general their capital and operating costs are high and not competitive.

Total elimination of sacrificial solvent has been achieved by *powder coating*. Without solvent, thicker coats can be applied in one pass and painting time is decreased. Some materials are not soluble and can be only used as powders. Powder coating requires heating the work to form a film from the powder.

Specialty finishes include those with metallic pigments, those with oils and dryers that give a crinkled surface, silicone additives for high temperature, and plastisols and organisols. The last listed are abrasion- and chemical-resistant vinyl coatings up to about 3.18 mm (.125 in.) thick. A plastisol contains no solvent, while an organisol does.

Sometimes only one coat of paint is enough but usually two or more layers of different coatings are applied to obtain combinations of desirable properties. An undercoat may be applied to form a good base for a finish coat that provides color, luster, and appearance. Undercoatings include primers that form a bond and inhibit corrosion on the interface and intermediate coats that serve as fillers, smooth surfaces, or sealers. The composition of a paint depends largely upon its intended place in the layers on a surface. Powder coating usually does not need a primer.

Painting Methods

Painting may be done by brush, knife, dip, roller, flow, tumble, silk screen, electrodeposition, fluid bed, or spray methods. Brushing is easily done but is slow, so other methods are used for production. Dipping demands little equipment and can be mechanized easily but requires paint that films out, stirring, and workpieces that can be immersed easily (without pockets).

Flow coating, by pouring paint onto workpieces and recirculating the runoff, is gaining in popularity because it is applicable to many kinds of pieces of small to medium sizes and is fast, thorough, and economical of material.

Paint spraying is the most used application method for industrial painting because it is fast, dependable, versatile, and uniform. It is based on the principle that a liquid stream atomizes when it exceeds a certain speed. The most common system introduces a liquid or powder into a high-velocity stream of compressed air released through a nozzle. In another system, the paint is heated and discharged in a high-velocity stream through a small orifice at pressures up to about 35 MPa (5,000 psi). In still another way, the paint may be slung off the edge of a rapidly revolving disk- or bell-shaped atomizer.

A spray gun or head may be directed by hand, but for continuous production is commonly automated to spray pieces as they pass by on a conveyor. Robots are being utilized more and more for automated paint spraying. Manual touch-up may be necessary at the end of a line, but the cost is normally less than if each piece were completely hand sprayed. An advantage of automated painting is that it can be done at up to 40° C (≈100° F), which is ideal for paint but not for people.

Most industrial painting is done in segregated enclosures or booths that are well ventilated. This promotes cleanliness, which is important for quality control. Enclosure is necessary because of the toxicity and flammability of solvents and to help control air and water pollution. The recovery of excess material is aided by enclosures.

Much paint can be lost when spraying. The loss is reduced by imposing an electrostatic charge on atomized liquid paint or powder. In one *electrostatic painting system* depicted in Figure 26-2(A), a potential of 80,000–150,000 V charges the paint slung from a disk- or bell-shaped atomizer. The droplets are drawn to and deposited uniformly, with almost no loss, on the grounded workpieces of opposite polarity.

When sprayed from an airgun, the particles can be charged inside or outside of the gun as indicated in Figure 26-2(B). The charged particles are propelled to the workpiece of opposite polarity. A "wraparound" effect occurs to deposit the coating material on all exposed surfaces (front and back, inside and outside), although restrictions may cause variations in film thickness. In another system, the particles are discharged into an electrostatic field set up between charged electrodes and the grounded workpiece. The paint particles pick up the electrical charge of the electrode as they pass near it first and are attracted to the work. Spray loss may be kept as low as 5–25% with an electrostatic painting system.

Drips and tears that collect at the bottoms of pieces that have been dipped can be drawn off electrostatically as in Figure 26-2(C). Another way to eliminate excess paint is to whirl the wet pieces in a centrifuge.

Electrodeposition under such names as *electrocoating, electropainting, paint plating,* and *E-coat* is done with a conductive workpiece at one potential dipped into a tank of specially formulated liquid paint of opposite charge (negative in some cases; positive in others). Colloidal particles of paint in solution in the tank are attracted to the work, become soluble on contact, and adhere firmly to the surface. Resistance increases on areas as they are coated, and buildup continues on other areas to deposit a uniform but thin film. Because of resistance, only one coat is feasible, and the process is mostly limited to priming. Prime coats are normally not more than about 20 μm (.0008 in.) thick and show the slightest surface defects. Equipment is costly, and the

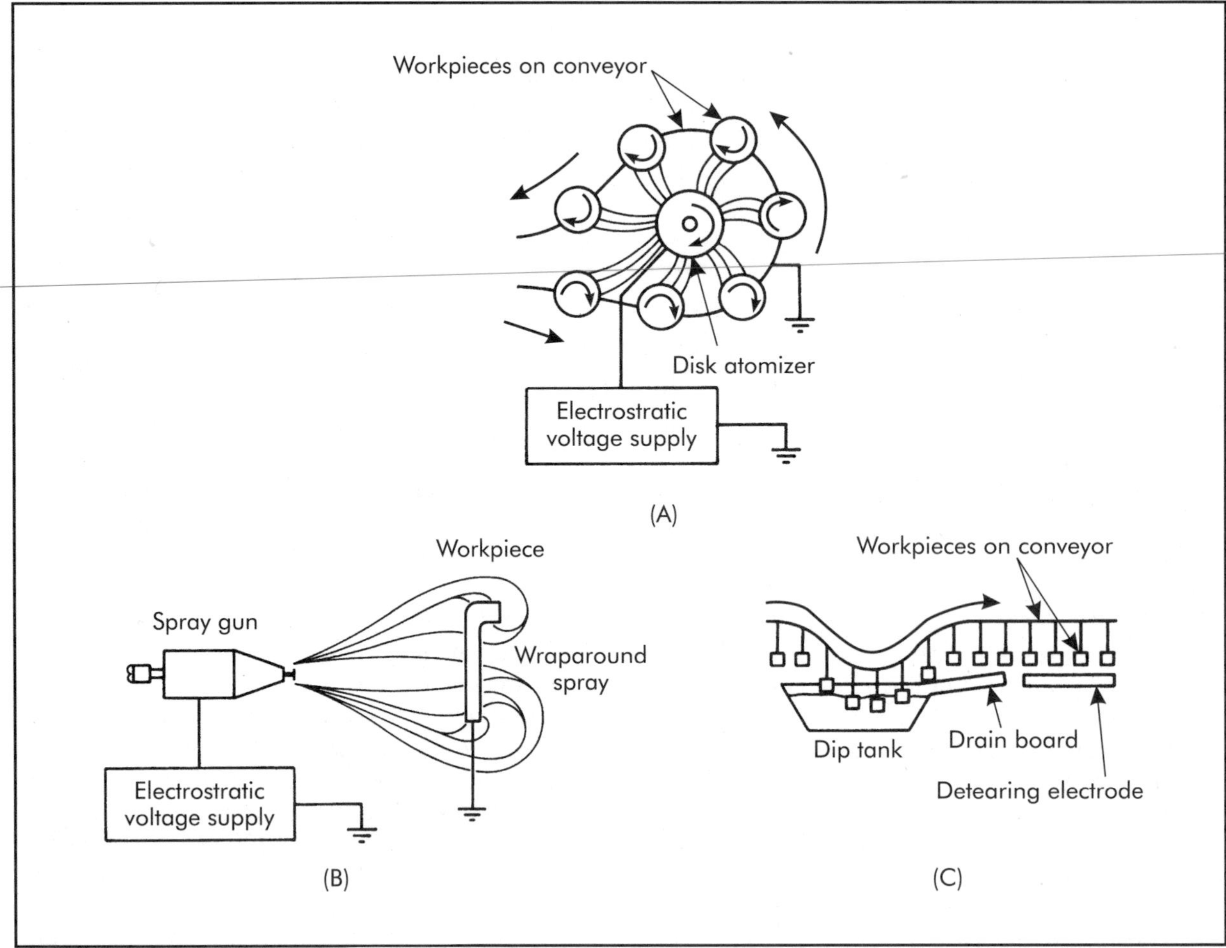

Figure 26-2. Electrostatic painting systems.

process is economical only for large-quantity production.

Fluid bed coating is done by dipping a heated metal piece in an agitated swirling pool of fine powder suspended in a stream of air. The powder particles are a mixture of resin, catalyst, pigment, and stabilizer and melt upon the hot workpiece surface to form a uniform film. The workpiece then is usually oven heated to flow out and cure the coating. In an *electrostatic fluid bed,* a screen called a charging grid polarizes the powder pool, and particles are attracted to the oppositely charged workpiece. Coatings up to 1.5 mm (.06 in.) thick can be obtained in one immersion by powder coating, but any less than 254 μm (.01 in.) are difficult to achieve. Powder coatings are pore-free, smooth, tightly held, and cover edges and corners well. However, a uniform coat is difficult to achieve on a part with thin and thick sections because the heating affect varies for different masses.

Paint may be baked in ovens heated by steam, gas, oil, or electricity. Infrared lamps in banks have become quite popular for baking and drying because the method is fast, clean, and easily changed and adjusted. Attention has been given in recent years to curing organic coatings by radiation, that is, by ultraviolet rays, laser beam, and electron-beam methods. These methods have been reported to save fuel, reduce air pollution, and increase production speeds. For example, one project utilized a specially formulated coating cured by exposure to an electron beam in an inert gas atmosphere for 1–2 sec/piece (as compared to many minutes or even hours in conventional processes). However, the electron beam gun costs several hundred thousand dollars, and the complete production installation over $1 million. Such a process is found economical only for curing 2 million m^2 (≈20 million ft^2) or more per year.

Precoated sheet metal strips with laminated, conversion, galvanized, and painted surfaces in rolls are popular because it is cheaper to coat the stock than the finished pieces. A large variety of coatings are available to withstand almost all fabrication operations and meet end-use requirements.

Painting Costs

The first consideration in the selection of a paint is that it meets service requirements for satisfactory corrosion protection, strength, durability, luster, color, etc. Often several kinds of paint can be found suitable for a particular application. The lowest cost is then the criterion: not the cost per liter (gallon), but the total cost of material, energy, labor, and equipment (including pollution control equipment). An analysis of the actual costs for four common paint materials considered for one plant is given in Table 26-1.

26.2.3 Hot-dip Plating

A low-cost way of putting a protective coating on metal pieces is to dip them into certain molten metals, mainly aluminum, zinc, tin, or lead. The same metals are also electrocoated. Steel mills supply sheets with more than 10 kinds or combinations of dipped or electrocoatings.

Hot-dip galvanizing is done by dipping ferrous parts, such as outside hardware, or passing strips, sheets, etc., continuously through a bath of molten zinc. The work is first cleaned and fluxed in a solution of zinc chloride and hydrochloric acid.

Zinc keeps iron from rusting even if the coating is broken because a galvanic action occurs in the presence of the moist carbon dioxide of the air. Zinc is the more active metal, and the iron does not rust until all the zinc is gone. Most other metals provide a barrier but do not give sacrificial protection. Dipping is an economical way of putting on a heavy and enduring coat of zinc, usually 102–508 μm (.004–.02 in.) thick.

Tin plating or *dipping* is done by immersing cleaned and fluxed steel sheets in a bath of molten tin. They then are passed through rolls in a palm oil bath to remove excess tin. Tinned sheet does not furnish as much outside protection as zinc but is adequate for some uses, such as the insides of food containers, for which it is mostly used. Tin-plated parts solder easily.

Terne plate is steel dipped in an alloy of lead and about 25% tin. It is cheaper than tin plate but has satisfactory corrosion resistance. Lead coatings protect automobile battery boxes and radiator fittings. The lubricating quality of the lead is beneficial on sheets and wire that are going to be drawn.

26.2.4 Electroplating

Electroplating is done on all the common metals and even on many nonmetals (particularly plastics) after their surfaces have been suitably prepared. Plating may be done for protection against corrosion or against wear and abrasion, for appearance, to rework worn parts by increase in size, to make pieces easy to solder, to provide a surface, usually of brass on steel, for bonding rubber, and to stop off areas on steel parts from being carburized during heat treatment. The common (but not the only) plating materials are aluminum, cadmium, zinc, silver, gold, tin, copper, nickel, chromium, and their alloys.

The principle of *electroplating* is illustrated in Figure 26-3. The piece to be plated is immersed in a water solution of salts of the metal to be applied and made the cathode in a direct current circuit. Anodes of the coating metal replenish the solution when the current is flowing, and ions of the metal are attracted to the workpiece to form the coating. Metals that cannot be electrodeposited from water solutions, such as the refractory metals, are deposited out of molten salts at high temperatures. The rate of deposition and the properties of the coating, such as hardness, uniformity, and porosity, depend upon getting a proper balance between the composition of the plating solution, current density, agitation, solution acidity, and temperature. For instance, the higher the current density, the faster the metal is deposited, but a rate above a critical level for a specific solution and temperature results in a rough and spongy plate.

The electroplating process presents several difficulties of concern to the designer. One of

Table 26-1. Finishing costs for common paint materials

Cost Items	Conventional Solvent[g]	Water Borne	High Solids	Polyester Powder[a]
Material				
1. Commercial paint[a], \$/L (\$/gal)	2.51 (9.50)	2.91 (11.00)	4.10 (15.50)	4.41 (2.00)
2. Volume solids (%)	43	40	65	98
3. Reducing agent cost, \$/L (\$/gal)	0.53 (2.00)	0	—	—
4. Mix ratio	5:2	10:3	—	—
5. Mixed coating cost, \$/L (\$/gal)	1.94 (7.36)	2.24 (8.46)	4.10 (15.50)	4.41 (2.00)
6. Volume solids at spray viscosity (%)	31	31	65	98
7. Coverage at 100% efficiency, $m^2/L/\mu m$ ($ft^2/gal/mil$)	310 (498)	310 (498)	650 (1,043)	614 (118)
8. Dry film thickness, μm (mil)	25 (1.0)	25 (1.0)	30 (1.2)	35.6 (1.40)
9. Utilization efficiency, %	60	60	75	99
10. Actual coverage, m^2/L (ft^2/gal)	7.3 (298)	7.3 (298)	16 (652)	17 (83)
11. Applied cost, \$/m^2 (\$/ft^2)	0.2658 (0.0247)	0.3057 (0.0284)	0.2562 (0.0238)	0.2594 (0.0241)
Energy				
12. Annual volume of gas consumed, dam^3 (1,000 ft^3)[b]	187 (6,618)	200 (7,066)	174 (6,150)	101 (3,568)
Total annual cost				
13. Material, \$[c]	295,000	341,000	286,000	289,000
14. Energy (gas), (\$)[d]	19,822	21,200	18,400	10,700
15. Labor and maintenance, \$	131,000	131,000	129,000	74,890
16. Depreciation, \$[e]		9,500	9,000	14,500
17. Total, \$	445,800	502,700	442,400	389,090
18. VOC emissions[f]	C	B	B	A
19. Solid waste disposal	C	C	C	A

[a] Commodity unit for powder is kg (lb) instead of L (gal). Thus powder cost is \$4.41/kg (\$2.00/lb), etc. The specific gravity of polyester powder is 1.6.

[b] Estimated amount of gas for 2,000 hours per year of production time for spray booths and bake ovens. Pretreatment and other requirements assumed the same for all alternatives and not included.

[c] Production of painted surfaces is 1,114,800 m^2/yr (12,000,000 ft^2/yr).

[d] Gas costs \$106/dam^3 (\$3/1,000 ft^3).

[e] This represents a study of the feasibility of replacing a conventional solvent system already in operation. No investment is needed if the present system is continued except for emission control equipment. The investments in the other systems, if selected, are to be depreciated over 10 years on a straight-line basis. Thus the figures shown represent 10% of the investment required for each. Any investments required for hydrocarbon emission control and waste disposal systems are not included. It is expected that the solvent-borne systems will require such equipment to meet federal EPA guidelines, and the water-borne and powder systems may need this equipment to meet regulations in some areas. An indication of the relative merits of the systems with respect to environmental control is given in lines 18 and 19. Actual potential costs must be ascertained and taken into account for a final comparison.

[f] VOC stands for volatile organic compounds in the paint. From the standpoint of the disposal of pollutants, "A" designates least, "B" moderate, and "C" most cost.

[g] The following calculations explain how the analysis was made for the conventional solvent paint.

5. $[(5 \times 2.51) + (2 \times 0.53)]/7 = \$1.94/L$ ($[[(5 \times 9.50) + (2 \times 2.00)]/7 = \$7.36/gal$)
6. $5 \times 0.43/7 = 0.31 = 31\%$
7. $0.31 \times 10^{-3}/10^{-6} = 310\ m^2/L/mm$ ($0.31 \times 0.1337/0.001/12 = 498\ ft^2/gal/mil$)
10. $310 \times 0.6/25.4 = 7.3\ m^2/L$ ($497 \times 0.6 = 298\ ft^2/gal$)
11. $1.93/7.3 = \$0.2644/m^2$ ($7.36/298 = \$0.247/ft^2$)
13. $0.2644 \times 1,114,800 = \$295,000 \approx 0.0247 \times 12,000,000$
14. $187 \times 106 = \$19,822$ ($\approx 6,618 \times 3$)

(Fishkin 1981)

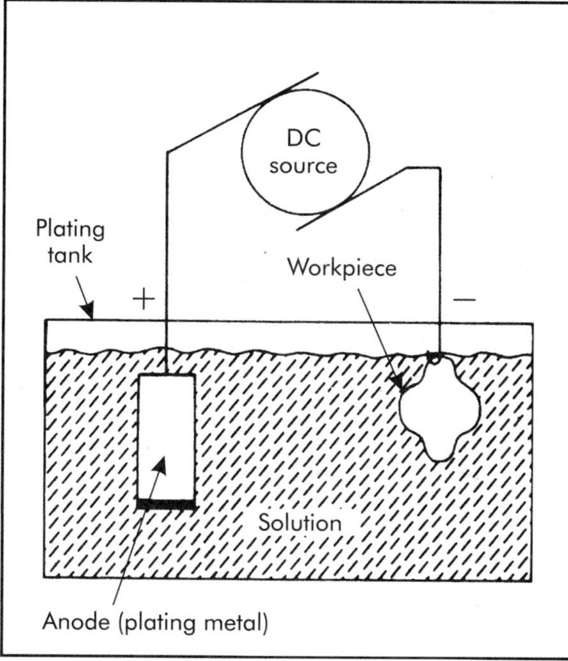

Figure 26-3. Scheme of electroplating.

these is that a plating is not always deposited uniformly. It tends to be thick on projections, thin in recesses, and almost nonexistent in some corners. The designer must avoid design irregularities as much as possible. Much can be done by good planning and control of the operation. Some solutions give better distribution of plating and are said to have better *throwing power* than others. The amount of plate deposited on a surface is related to the distance from an anode. Thus, anodes shaped to match the workpieces and suitably placed can help toward uniform plating.

Plating does not hide defects in the surface of the workpiece, so a surface must be fully finished if it is to be plated for appearance as well as corrosion resistance. Parts, such as automobile bumpers, nickel or chrome plated for durable appearance, are commonly given an initial copper plating of 5–15 μm (.0002–.0006 in.) thickness. This adheres well to and effectively covers the steel, and its surface is easier to buff out than the steel. Decorative chrome plate is ordinarily only around a micrometer or less (a few hundred-thousandths of an inch) thick to maintain brightness over a protective intercoat of nickel. Various combinations of multiple layers are used to achieve best results at lowest costs for various purposes.

Hydrogen released at the cathode causes harmful embrittlement of hardened or cold-worked steel workpieces. The amount of hydrogen can be kept at a minimum by proper control of the operation, and the embrittlement can be alleviated by heating the workpieces immediately after plating.

Equipment

The basic unit for electroplating operations is the tank to hold the solutions. Tanks are constructed of various materials such as lead sheet, rubber, plastics, and tile to resist alkaline and acidic solutions. Large pieces are suspended individually. Small pieces may be mounted on racks. Most are barrel plated or tumbled. A batch of pieces is put in a nonconducting perforated barrel and in touch with suitable contacts. The barrel is lowered into the plating solution and revolved several times each minute. The pieces tumble around and are plated uniformly.

The electroplating process entails cleaning, washing, rinsing, and other treatment in addition to the actual deposition of metal. This means each workpiece or batch must be dipped into and transferred among a number of tanks, whether plated individually, on racks, or in barrels. This is commonly done manually for small lots. Labor is saved and quality is controlled better by automatic and even numerically controlled plating machines for moderate- to large-quantity production. Three general types of machine layouts are illustrated in Figure 26-4. The transfer devices for these machines raise the work from one tank, move it to the next, and then lower it into place at preset intervals. The straight-line and return layouts depicted in Figure 26-4(A) and (B) may be single- or double-line machines, depending upon whether they have one or two rows of work carriers, one above the other. The rotary-type layout (Figure 26-4[C]) takes more floor space but can handle large parts. These machines can be tied in conveniently with conveyor systems. In certain industries, specialized machines have been developed for continuously plating sheet metal, a strip of parts, or wire as it is run through a series of baths.

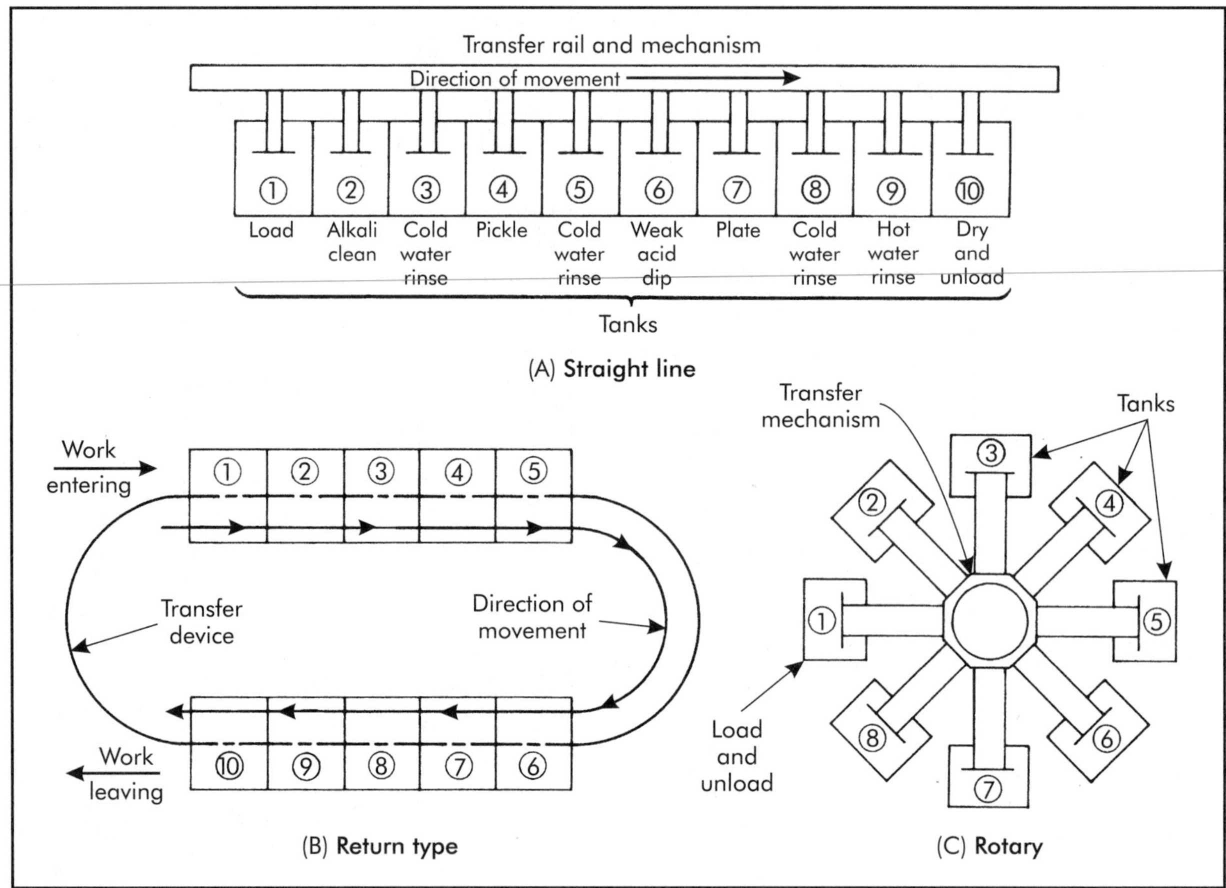

Figure 26-4. Typical plating machine layouts.

High-speed plating techniques have been developed in recent years for large-quantity production. Instead of being immersed in a tank with current passing an average of about 0.5 m (1.5 ft) through relatively still electrolyte as in conventional electroplating, the piece is positioned in a fixture cavity with wall contours matching the part surface, typically with a gap of about 2.5 mm (.1 in.). Electrolyte is pumped through the cavity at around 100 dm^3/min (3.5 ft^3/min). The rapidly moving liquid overcomes ionic depletion in the layer adjacent to the work surface. Thus 20 or more times as much current can be passed effectively as in conventional plating, and metal can be deposited much faster. In one version, called *contour plating,* as much nickel is deposited on automobile bumpers in 1 min as is put on in 1 hr in a tank.

Touch-up, small lot, and selective plating may be done by a brush or swab soaked with plating fluid and connected as the anode. The workpiece is the cathode. This is called *brush plating.*

Used fluids from electroplating tanks can be highly pollutant. In recent years, stringent rules have been set down to control the levels of cyanide and metals in the effluent. A common remedy is to resort to chemical reactions to convert the waste to nontoxic substances. However, in this way valuable materials may be lost, so in some cases evaporation and dialysis are used to concentrate and recover costly substances as well as eliminate pollution.

Costs

Two major direct costs of electroplating are for the metal in the plate and for electricity. These and other costs that depend upon time can be calculated readily from data given in handbooks. For instance, 1 m^2 of zinc plate 25 mm thick weighs 178.5 gram (1 ft^2 of zinc plate .001 in. thick weighs .6 oz).

One faraday of electricity is equal to 96,540 coulombs (A-s) and deposits an equivalent weight in grams of any metal at 100% cathode efficiency. The equivalent weight of zinc is its atomic weight (65.38) divided by its valence (2), or 32.7 grams. Average cathode efficiency for zinc plating is 85%. This means that 85% of the current goes to deposit metal; the remainder into leakage, hydrogen generation, etc. The electricity required to deposit 25 mm on 1 m^2 is

$$(178.5/32.7) \times (96{,}540/3{,}600) \times (1/.85) = 172.2 \text{ Ah} \quad (26\text{-}1)$$

At the usual potential of 6 V, the energy used is

$$172.2 \times 6 = 1.03 \text{ kWh}$$

For 0.001 in. deposited over 1 ft^2, the amount is

$(17/32.7) \times (96{,}540/3{,}600) \times (1/.85)$
$= 16.4$ Ah. The energy then is 16.4×6
$= 98.4$ Wh.

At a permissible current density of 225 A/m^2, the time to deposit a thickness of 25 μm is 172.2/225 ≈ 0.8 hr.

In most cases, electroplates of 25–50 μm (.001–.002 in.) and less thick are adequate for protection and finish. Certain applications call for thicker plates. As much as 0.2 mm (.008 in.) of hard chrome plate may be put on tools and gages for wear resistance and even more where the purpose of the plate is to increase dimensions.

26.2.5 Electroforming

Parts may be formed by electroplating metal on a mandrel, which may be collapsible, coated with a parting agent, dissolvable, or meltable for removal from the product. Copper, nickel, silver, and iron are commonly electroformed.

Electroforming is economical for making parts with special features that are difficult to produce by more common methods. It is advantageous for complex (inside and outside) shapes with thin walls, for laminating metals without heat or pressure, for producing fine and precise details and openings (for example, quite small holes), and for incorporating inserts and flanges (metallic and nonmetallic) difficult to attach otherwise. Walls can be made as thin as a few micrometers to 50 mm but seldom are over 15 mm (≈.0001–2 in. but seldom over ≈0.5 in.). Tolerances have been held to a fraction of the wavelength of light in electroforming diffraction gratings, and those of the order of 25 mm (.001 in.) are commonplace, with surface finishes of 51–203 nm (2–8 μin.) R_a.

26.2.6 Vacuum Deposition

Methods for applying thin coatings of metals and many compounds on metallic and nonmetallic (for example, plastic, glass, and ceramic) materials are of two kinds: *vapor deposition* and *plasma deposition*.

The common vapor deposition process is *evaporated* or *vacuum metallizing* in which aluminum (most often) is flash heated in a high-vacuum chamber and condenses on all directly exposed surfaces of the work. The film is relatively thin, from 25–127 nm (1–5 μin.), and closely reproduces the shape and finish of the original surface, which is commonly precoated with one or two layers of lacquer. A final lacquer or enamel coating may be applied after metallizing to protect and even tint the light metallic film.

Most plasma deposition is called *sputtering*, but there are several forms of the process. Typically, the work is placed near the plating material (called the target) in a chamber that is first evacuated and then given a light backfill of inert gas. A negative potential is applied to the target and the gas is ionized. The gas ions are attracted to the target with sufficient kinetic energy to knock metallic atoms from its surface. These atoms are driven to the work with enough energy to bind them to the surface where they form a film as they collect. If a fraction of reactive gas is admitted to the chamber, metallic compounds can be formed and deposited on the work surface.

In a two-step operation called *ion plating*, the workpiece is reverse-sputtered in an inert-gas atmosphere. This means that the work is given a negative potential and is highly bombarded by gaseous ions to thoroughly clean its surface. Then the plating material in the chamber is heated until atoms evaporate and are ionized and driven to plate the work surface.

Films may be deposited in a vacuum from a few angstroms to about 25 μm (.001 in.) thick to serve as decorative surfaces on items such as costume jewelry, automobile trim, and toys, as reflective and insulating layers on many products, for optical surfaces, and on electronic components. Coatings up to 152 μm (.006 in.) thick have good corrosion and oxidation resistance. Solid film lubricants can be plasma deposited. Vacuum deposition methods can produce thinner films than are usually obtained by electroplating and use less material and energy. These methods do not create pollution problems.

Advantages and Limitations

All the vapor deposition methods can be applied to any nongasing surface, but ion plating is more difficult on nonconducting and heat-sensitive materials. If the part is complex or must be coated on more than one surface, it is usually rotated for complete and uniform coverage. Film adhesion is good on ordinary clean surfaces, and excellent on sputter-cleaned surfaces, a normal part of the ion plating method.

Each vacuum deposition method has certain unique advantages. Vacuum metallizing is best suited to coating with pure metals with moderate melting points. Alloys require special techniques because their constituents evaporate at different temperatures. Vacuum metallizing is fast. A large batch of parts can be treated in a few seconds plus 5–15 min for drawing a vacuum. Continuous coating, such as on long strips, is accomplished by this method. However, equipment cost is high (an installation with a capacity of 10,000 or so parts per hour may cost $100,000 or more) but is justified for large-quantity production. Sputtering is only one twentieth or so as fast as vacuum metallizing but is able to deposit almost any material, for example, refractory metals, stainless steel, Pyrex®, quartz, and Teflon®. Film thickness can be held to ±5% for uniformity and to <13 nm (<.5 μin.) for size tolerance. The workpiece is not heated, and its material integrity is upheld when subjected to sputtering. On the other hand, the workpiece is heated when ion plated, and heat-sensitive materials cannot be so treated. Ion plating is suitable for applying elemental metals to surfaces and has a deposition rate that falls between sputtering and vacuum metallizing.

26.2.7 Other Metal-coating Processes

Metal-coating operations that utilize welding techniques are described in the sections on metal spraying and surfacing, and hard facing in Chapter 14.

Chemical reduction or *electroless plating* is the means of precipitating metal from a chemical solution to form bright films for mirrors and reflectors and as a basis for further electroplating. A variant of the process precipitates a nickel-phosphorus matrix impregnated with blocky-diamond powder to produce a highly wear-resistant surface. In the process called *chemophoresis,* a latex film about 25 μm (.001 in.) thick is chemically deposited on steel and cured by baking.

In *chemical vapor deposition* (CVD) or *gas plating,* a metal halide in an inert-gas or hydrogen stream is passed over a workpiece which is heated to 500–2,000° C (≈900–3,600° F), and deposits the metal or compound on the surface. This process deposits metals and their compounds not readily suited to electroplating or mechanical working, such as the refractory metals, their compounds, and ceramics. It is used for coating, plating, and electroless forming of such products as tubing, crucibles, cutting and forming tools, and electronic components. Common specific usage in industry is the coating of alloy and tool steels, stainless and heat-resisting steels, and cemented carbide with superhard titanium carbide, titanium nitride, chromium carbide, and aluminum oxide singly or in various combinations. The process has good throwing power and can reproduce fine details reliably. The substrate material must be capable of withstanding the process temperatures. CVD has a deposition rate about one-half that of vacuum metallizing and a relatively high cost for materials and equipment.

A *diffusion* or *cementation coating* is a hard and often brittle alloy-rich surface layer formed by heating a piece of metal in intimate contact with another metal in powder, liquid, or gaseous form. The purpose is to obtain corrosion resistance, in some cases against oxidation at high temperatures. Particular processes of this kind are *sherardizing* for zinc on steel, *chromizing* for chromium on steel, *calorizing* for aluminum on steel, *siliconizing* for hard iron surfaces, and *nicrocoating* with a nickel chrome matrix containing other metals or ceramics.

Layers of corrosion-resistant metals may be added to base metals by heating and rolling and thus welding sheets together or applying a powder coating and heating to effect diffusion. This is known as *cladding*. Common examples are aluminum alloys *alclad* with soft aluminum, and steel clad with stainless steel, Monel®, Inconel®, aluminum, or copper in plates, sheets, strips, tubing, and wire.

Mechanical plating is done by depositing powdered zinc, cadmium, tin, or lead on a surface and consolidating it into a continuous uniform coating by the rolling action of glass beads in a vibrating and agitated mass or revolving barrel. Coatings from 2.5–51 µm (.0001–.002 in.) thick are obtainable with adhesion, corrosion resistance, and cost comparable to electroplated coatings. A particular advantage of mechanical plating is that there is no hydrogen absorption and embrittlement that weakens tough steels.

26.2.8 Vitreous Coatings

Vitreous (porcelain or ceramic) enamel is a hard, glass-like, inorganic coating 76–254 µm (.003–.010 in.) thick fused to metal. The main ingredients are a finely ground frit of silicates, feldspar alumina for ceramics, fluxes like borax and soda ash, and various metallic oxides for coloring and other properties. They are applied commonly to certain grades of sheet steel (called enameling irons), but also to copper and bronze, stainless steel, refractory metals, cast iron, and aluminum. One process is to mix the ingredients with clay and water and apply the resulting *slip* by spraying, dipping, flow-coating, or brushing. The coat is fired at 500–900° C (≈900–1,650° F), depending on the substrate. Ceramic coatings are also applied by flame spraying as described in Chapter 13.

Vitreous coatings are smooth, hard, lustrous, and resist high temperatures, but are subject to chipping and cracking. They are used for food preparation and health service equipment, major appliances, containers that resist chemical attack, on surfaces to resist wear and abrasion, and jet engine combustion chambers and exhausts.

26.3 GREEN MANUFACTURING

A manufacturing process that is environmentally sensitive and in which positive steps are taken to reduce or eliminate any negative impact on the environment is often referred to as *green manufacturing*. Today, all manufacturing industries are faced with the challenge of conducting their operations in a way that will not pose a threat to the environment. Because of the myriad of rules and regulations promulgated to protect the environment, many manufacturers find it extremely difficult to achieve a green manufacturing position and still remain economically competitive. Most of these rules and regulations stem from the federally enacted *Clean Air Act* (CAA) of 1955 and the *Clean Water Act* (CWA) of 1948. In addition to the federal regulations, at least 27 states have adopted a variety of measures concerned with the reduction of levels of pollution, waste, and emissions generated by manufacturing and other organizations within those states.

Although compliance with these regulations is often initially viewed by manufacturers as an additional burden to an already extremely competitive and complicated process, they soon come to realize that the development of cleaner processes and the elimination of hazardous wastes makes good business sense. Actually, environmental hazards and wastes represent costs and liability risks that are unnecessary in most cases. Thus, the reduction or elimination of these conditions reduces or eliminates costs and risks and, in reality, improves the competitive position of a business.

26.3.1 Environmental Regulations

Since many finishing operations involve materials that pose some potential hazard to the environment, a brief overview of the major legislation dealing with environmental regulation in manufacturing is presented here. This is not meant to be an exhaustive treatment of this subject and more extensive coverage of these regulations is given in certain references at the end of this chapter.

Clean Air Act (CAA)

The CAA legislation provides for the control of air emissions from manufacturing facilities. Since enactment in 1955, the CAA has been amended four times to broaden its coverage to include, among other things, provisions for national ambient air quality standards, source performance standards, and hazardous air pollution standards. The last amendment in 1990 included provisions to control emissions of 189 hazardous air pollutants. Other provisions of that amendment pertained to the control of ozone-depleting substances. For example, it banned the production of two of the most popular industrial cleaning solvents, CFC-113 and 1,1,1-trichloroethane (methyl) chloroform, as of January 1, 1996. Non-ozone-depleting recycled solvents, such as trichloroethylene, perchloroethylene, and methylene chloride, may still be used as long as emissions are below allowable limits.

Clean Water Act (CWA)

The Clean Water Act, first enacted in 1948 and revised in 1972, regulates, among other things, the discharge of conventional and toxic pollutants into waterways. Effluent discharge limitations for specific pollutants were established for municipalities and industry in the 1972 revision. Thus, industrial effluent discharge is regulated either by a *National Pollutant Discharge Elimination System* (NPDES) permitting process or by standards on the wastewater received from those industries imposed by local municipal treatment facilities. The *Water Quality Act* of 1987 enhanced the scope of the Clean Water Act by imposing controls on nonpoint sources. Under this act, industries must develop and implement plans to reduce the pollutant level of stormwater runoff from their facilities and properties.

Several other very important pieces of legislation have emerged since the CAA and CWA were promulgated. These include the *Resource Conservation and Recovery Act* (RCRA) that regulates the generation, storage, transportation, and treatment or disposal of hazardous wastes, and the *Comprehensive Environmental Response, Compensation, and Liability Act* (CERCLA), also known as *Superfund,* which imposes liabilities for cleanup of hazardous waste sites. In addition, the *Toxic Substance Control Act* (TSCA) requires an evaluation of the possible toxic impact of new substances before they are produced or used in production. Also of concern to manufacturing industries is the *Emergency Planning and Community Right-to-Know Act* (EPCRA) that requires firms to report above-threshold limit release to the environment of certain chemicals and chemical compounds.

Many manufacturers are beginning to realize that the cost of achieving a green manufacturing level are quite high, but the costs associated with noncompliance, neglect, and cleanup are often many times higher. Thus, manufacturers are faced with the necessity of developing and implementing a plan that will, as a baseline, bring them into compliance with applicable environmental regulations. This plan then must become a part of their business plan for achieving success over their competitors.

26.4 QUESTIONS

1. Why is it necessary to clean manufactured parts after processing?
2. What are the principal cleaning fluids and for what is each best suited?
3. Name the three primary types of detergents used for cleaning.
4. Describe the electrolytic cleaning process.
5. Why is spraying more effective than dipping for cleaning?
6. Name one advantage of the vapor degreasing process. What precautions must be taken in vapor degreasing?
7. Describe a typical part cleaning sequence.

8. What is "pickling" and what precautions must be taken with it?
9. What are some of the characteristics of conversion coatings?
10. What is the difference between phosphate and chromate coatings?
11. How do anodic coatings differ from other conversion coatings?
12. What is the typical thickness of an anodic coating?
13. Describe four major types of paint and their uses.
14. Describe the important industrial painting methods.
15. What are the benefits of electrostatic painting and how is it done?
16. Explain the electrodeposition painting process and indicate its primary use.
17. Why and how are ferrous articles galvanized?
18. Describe the electroplating principle and specify the factors that influence its operation.
19. What can be done to help put a uniform plate on an object by the electroplating process?
20. What is the purpose of electroplating different layers of metal on some objects?
21. How can plating cause embrittlement of steel and how can it be avoided?
22. Describe the common methods of plating.
23. Describe the electroforming process and cite its advantages.
24. Explain the difference between the vapor deposition and plasma deposition processes.
25. Describe the chemical vapor deposition process and indicate its use.
26. How are vitreous coatings applied and what are their advantages?
27. What is "green manufacturing" and what are its advantages?

26.5 PROBLEMS

1. Show the calculation for obtaining the applied cost, line 11 from Table 26-1 for:
   (a) conventional solvent paint in $\$/m^2$;
   (b) waterborne paint in $\$/ft^2$;
   (c) high solids paint in $\$/m^2$; and
   (d) polyester powder paint in $\$/ft^2$.
2. A manufacturer is planning to produce an item at the rate of 50,000 pieces/yr. Each piece has an area of 0.33 m^2 (3.600 ft^2) and will be painted with two coats. Paint loss is 15%. A baking enamel that meets requirements costs 42 cents/m^2 (3.9 cents/ft^2) for 25 μm (.001 in.) thickness dry and each coat is 38 μm (.0015 in.) thick. A lacquer that is satisfactory costs 69 cents/m^2 (6.4 cents/ft^2) per 25 μm (.001 in.) thickness dry. A total thickness of 100 μm (≈.004 in.) is needed for the lacquer. If the manufacturer chooses to use the enamel, how much can it afford to spend per year to bake the enamel? It is assumed that the cost of spraying either paint is the same.
3. An automobile bumper has an area of 1.1 m^2 (11.8 ft^2) and is to receive a nickel plate 12.5 μm (≈.0005 in.) thick. Nickel has an atomic weight of 58.69, a valence of 2, and a specific weight of 8.81 gram/cm^3 (5.09 oz/$in.^3$). The cathode efficiency is 95%. The process operates nominally at 6 V. How much nickel and electrical energy are used for each bumper? How long should be the plating time for an allowable current density of 270 A/m^2 (≈25 A/ft^2)?
4. An automobile bumper has an area of 1.1 m^2 (11.8 ft^2) and is to receive a copper plate 12.5 mm (≈.0005 in.) thick. Copper has an atomic weight of 63.57, in this case a valence of 1,

and a specific weight of 8.93 gram/cm^3 (5.16 oz/in.^{3}). The cathode efficiency is 50%. The process operates nominally at 6 V. How much copper and electrical energy are used for each bumper? How long should be the plating time for an allowable current density of 130 A/m^2 ($\approx$12 A/ft^2)?

5. An automobile bumper has an area of 1.1 m^2 (11.8 ft^2) and is to receive a final chromium plate 0.5 μm (.00002 in.) thick. Chromium has an atomic weight of 52.01, a valence of 6, and a specific weight of 6.9 gram/cm^3 (4 oz/in.^{3}). The cathode efficiency is 15%. The process operates nominally at 6 V. How much chromium and electrical energy are used for each bumper? How long should be the plating time for an allowable current density of 1,600 A/m^2 ($\approx$150 A/ft^2)?
6. Instead of plating copper and nickel on a bumper as specified in Problems 3 and 4, the requirements can be met by a single nickel plate 25 μm (.001 in.) thick. In either case, a final chromium plate of 0.5 μm (.00002 in.) must be applied. Copper costs $1.84/kg ($0.835/lb). Nickel costs $5.00/kg ($2.27/lb). Electricity costs 15 cents/kWh. Under what conditions would you recommend the single nickel plate rather than one of copper and one of nickel?
7. An electrostatic spray gun and accessories cost $6,000 but save 30% of the paint used at $1.42/L ($5.38/gal).
   (a) If interest, taxes, etc., are neglected, how much paint must be sprayed to pay for the equipment?
   (b) If interest, insurance, and taxes are 20% per year, how much paint must be sprayed in 1 yr to pay for the equipment?
8. A system is being designed to cadmium-plate parts to a thickness of 12.5 μm ($\approx$.0005 in.) at a rate of 65 m^2/hr (700 ft^2/hr). Cadmium has an atomic weight of 112.41, a valence of 2, and density of 8.65 gram/cm^3 (5 oz/in.^{3}). The system operates with a cathode efficiency of 90%. What must be the minimum capacity rating in kWh of the DC generator at 6 V potential? If the current density must be 215 A/m^2 (20 A/ft^2), how many m^2 (ft^2) of work must the tank be able to accommodate at one time?

26.6 REFERENCES

Burgess, W.A. 1995. *Recognition of Health Hazards in Industry: A Review of Materials and Processes.* New York: John Wiley and Sons.

Cattanach, R.E., Holdreith, J.M., Reincke, D.P., and Sibick, L.K. 1995. *The Handbook of Environmentally Conscious Manufacturing.* Chicago, IL: R.D. Irwin, Inc.

Durney, L., ed. 1984. *Electroplating Engineering Handbook,* 4th Ed. New York: Van Nostrand-Reinhold.

Fishkin, H.W. 1981. "Finishing Systems Economic Overview." *Manufacturing Engineering,* August: 77.

Koelsch, J.R., ed. 1995. "Guilt-free Parts Cleaning." *Manufacturing Engineering,* March.

Metals Handbook, 9th Ed. 1982. Vol. 5. *Surface Cleaning, Finishing, and Coating.* Metals Park, OH: American Society for Metals.

Owen, J.V., ed. 1995. "Managing the Mandates." *Manufacturing Engineering,* March.

Phillipon, E. 1994. "How Clean Do Parts Have to Be?" *Manufacturing Engineering,* October.

Stauffer, R.N. 1981. "Surface Modified Metals." *Manufacturing Engineering,* September.

Tool and Manufacturing Engineers Handbook, 4th Ed. 1985. Vol. 3. *Materials Finishing and Coating.* Dearborn, MI: Society of Manufacturing Engineers (SME).

Vaccari, J.A. 1982. "Changing Paint to Limit Emissions." *American Machinist,* July.

Nontraditional Manufacturing Processes

27

Electro-erosion, 1943—B. Lazerenko, University of Moscow

Most of the processes for molding, forming, joining, and cutting materials have been a traditional part of the manufacturing scenario for many years. These processes, subject to improvements over the years, will probably remain on the scene for many years to come. However, manufacturing is not without technological momentum, bringing about a significant number of innovative techniques and new equipment for use in the highly competitive manufacturing arena. For lack of a better way to characterize them, these new developments have been classified as *nontraditional manufacturing* (NTM) *processes*. Although there are some minor contradictions, these new techniques or processes can be divided into three major groups according to the way they remove or convert metals and other manufacturing materials. These groups include those processes that remove material by:

- chemical reaction;
- electrochemical action; and
- thermal action.

27.1 CHEMICAL MACHINING PROCESSES

The chemical machining processes include those wherein material removal is accomplished by a chemical reaction, sometimes assisted by electrical or thermal energy applications. This group includes chemical milling, photochemical machining, and thermochemical machining.

27.1.1 Chemical Milling

Chemical milling is the name given to a process that removes large amounts of stock by etching selected areas of complex workpieces. It was developed in the aircraft industry as one means of fabricating lightweight parts of large areas and thin sections but has been receiving attention in other industries. One kind of part chemically milled is the wing bulkhead depicted in Figure 27-1. The large cavities are formed by removing metal from a thick slab.

Chemical milling entails four steps: cleaning, masking, etching, and demasking. A neoprene rubber or vinyl plastic mask is sprayed or flowed over all surfaces and baked. It is then slit to a template and stripped from the areas to be etched. Different surfaces can be milled to different depths by removing parts of the mask at different times during the process.

The size of workpiece that can be treated is limited only by the size of the tank into which it is dipped for etching. Caustic soda is the usual etch media for aluminum, and acids for steel, magnesium, and titanium alloys. The action proceeds at the rate of about 25 μm/min (.001 in./min) in the solution if proper concentration is maintained. A

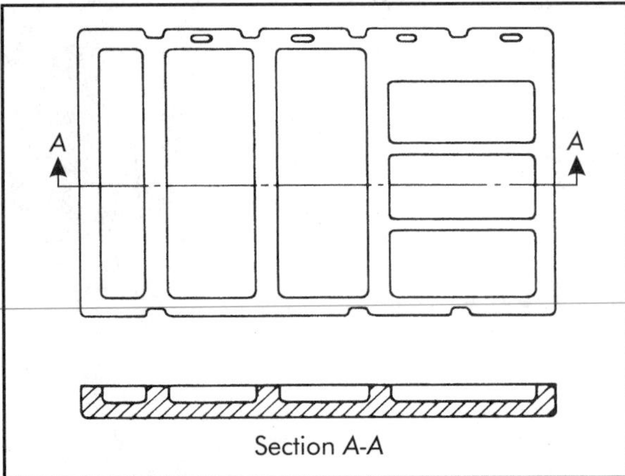

Figure 27-1. Aircraft wing bulkhead chemically milled.

black smut that occurs may be removed by deoxidizing and rinsing. A variation of the process is to withdraw a long piece slowly from the etching tank to make a tapered or conical part.

Tolerances of ±25 mm (±.001 in.) have been reported obtainable in chemical milling but ±1.5 mm (±.060 in.) on length and width and +0.127–0.000 mm (+.005–.000 in.) on depth and thickness are more practical. Surface finish ranges from 0.8–3 μm (≈30–125 μin.) R_a. The deeper the cut, the rougher the finish, but finer-grain material etches more smoothly.

Chemical milling can be done on all kinds of parts: rolled sections, forgings, castings, and preformed pieces. It is not limited by shape, direction of cut, or cutter, and different sizes and shapes of cuts can be made at one time. Both sides of a sheet can be cut at one time to minimize warping and cost. There are no burrs. The evidence points to a higher fatigue strength after chemical milling because of lack of scratches and no residual stresses. Chemical milling is advantageous for complex cuts and parts. However, it is not superior for simple straightforward work because it usually can be done more economically on regular machine tools. Scrap rates are low (less than 3% reported). There are some disadvantages. Only shallow cuts are practical, usually less than 15 mm (≈.5 in.). However, metal removal rates up to about 492 cm^3/min (30 $in.^3$/min) often can be achieved for some materials. Handling and disposal of the chemicals may be a problem and environmental regulations should be reviewed as they relate to this process.

Chemical milling is simple and amenable to automatic operation. Equipment preparation time and costs are low. In one aircraft plant, the usual period of three months to tool an average part for production has been reduced to less than one month where chemical milling is applicable. Masking costs may be high for each piece and chemicals add to the direct costs. Even so, chemical milling is economical in many cases.

27.1.2 Photo-etching

Photo-etching is known by a number of names, including photo-forming, photochemical machining (PCM), micro-milling, and chemical blanking. It utilizes the photoengraving techniques practiced in printing for many years. Parts like those in Figure 27-2 are made from flat 2.5 μm (.0001 in.) foil up to sheets 3 mm (≈.125 in.) thick of some plastics and most metals, including alloys of aluminum, copper, and steel. The sheet is covered with a photosensitive resist on one or both sides and exposed in a camera to an image of the part or parts desired (commonly reduced as much as 20:1 for accuracy). The coating is developed to expose the lines or areas to be eaten away subsequently in an acid bath or high-pressure chemical spray. The resist also may be applied by printing. Pieces up to almost 2 m (≈6 ft) long have been made, but sizes less than about 102 × 152 mm (4 × 6 in.) are most common. Outlines may be laid down to less than 10 μm (.0004 in.), but the edges eaten away are uneven and undercut. Tolerances from ±10% to 50% of stock thickness, depending on material, shape, and size, are practicable.

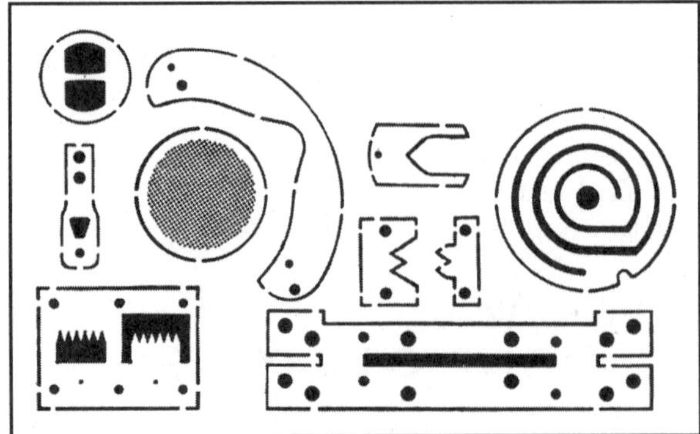

Figure 27-2. Examples of precision parts made from sheet metal by photo-etching. (Courtesy Buckbee-Mears)

The photo-etching process has been used for many years by the microelectronics industry to produce electronic circuit boards and other small, thin parts difficult to form by blanking and piercing.

For most parts, photo-etching is economical for quantities too large to be sheared out individually by hand but not large enough to justify the cost of a die. For instance, a radio manufacturer needed 100 small contactor leaves from sheet brass. They were made 25 to the sheet by photo-etching for a total cost of $75. A die alone could not have been made for that amount. A photo print master usually can be made for one-tenth of the cost of a die. The long lead time to make dies also can be saved. Changes in product design can be made easily because the photo-etching masters can be changed at little cost. Dies may be so expensive for parts that are quite complex or made from thin materials that photo-etching is justified for large quantities.

27.1.3 Thermochemical Machining

Thermochemical machining is used as a secondary operation to remove burrs and fins from parts after they have been either cast or machined. It can be done by either a combustion process or by chemical spraying. In the combustion process, an explosive mixture of hydrogen, oxygen, and natural gas are detonated in a chamber containing the parts. The explosive process produces a thermal shock wave that vaporizes the thin metal burrs and fins without heating up or harming the larger sections of the workpiece where a larger mass is involved. The combustion process is effective in removing burrs and fins from a wide range of materials, particularly those with low thermal conductivity. The chemical spraying process may be employed to remove very fine burrs and fins. However, this process does remove very minute amounts of the exposed surface from the body of the part, which has to be taken into consideration.

27.2 ELECTROCHEMICAL/ ELECTROLYTIC MACHINING PROCESSES

The electrochemical/electrolytic machining processes employ a conductive solution to remove material by setting up an electrolytic action between the tool (cathode) and the workpiece (the anode). These processes include *electrochemical machining, electrochemical grinding,* and *electrogel machining.*

27.2.1 Electrochemical Machining (ECM)

ECM, also called *electrolytic machining,* removes metal essentially by deplating. Its most common application is in the sinking of holes and cavities in the manner depicted in Figure 27-3. Electrolyte is pumped through a gap from less than 25 μm (.001 in.) to over 250 μm (.010 in.) thick between the tool (the cathode) and the workpiece (the anode). The electrical current drains electrons from the workpiece surface and releases ions. In plating the metal, ions are carried to and deposited on the cathode, but in ECM those ions combine with hydroxide ions of the solution and form a sludge that is swept away by the flow of electrolyte. Hydrogen is liberated at the tool and no metal is removed or added there, so the tool does not change shape or size. The electrode (tool) is fed at a constant rate into the workpiece and the hole or cavity is formed to the same shape as the tool. A hump may be left under the opening where the fluid is discharged into the gap. This may have to be removed from a blind hole by a second operation but is ultimately eliminated from a through hole. The electrode is insulated to prevent electrolytic action and metal removal behind the tip and thus keeps the hole straight, as indicated in Figure 27-3. In some cases, the electrode may be left bare and even given a particular profile to form draft or taper in a cavity.

The purposes of the electrolyte are to conduct electricity, chemically combine with the metal ions and remove them, carry away heat, give off gas, and prevent plating at the tool. The most common electrolyte is a saltwater solution. Other substances are added to improve effectiveness with various work materials and in certain operations, and to inhibit corrosion and intergranular attack. The fluid is pumped at the rate of 76–379 L/min (20–100 gal/min). To force it through passageways a fraction of a millimeter (a few thousandths of an inch) thick may take pressures of 1–2 MPa (≈150–300 psi). Such pressures create large forces between the tool and

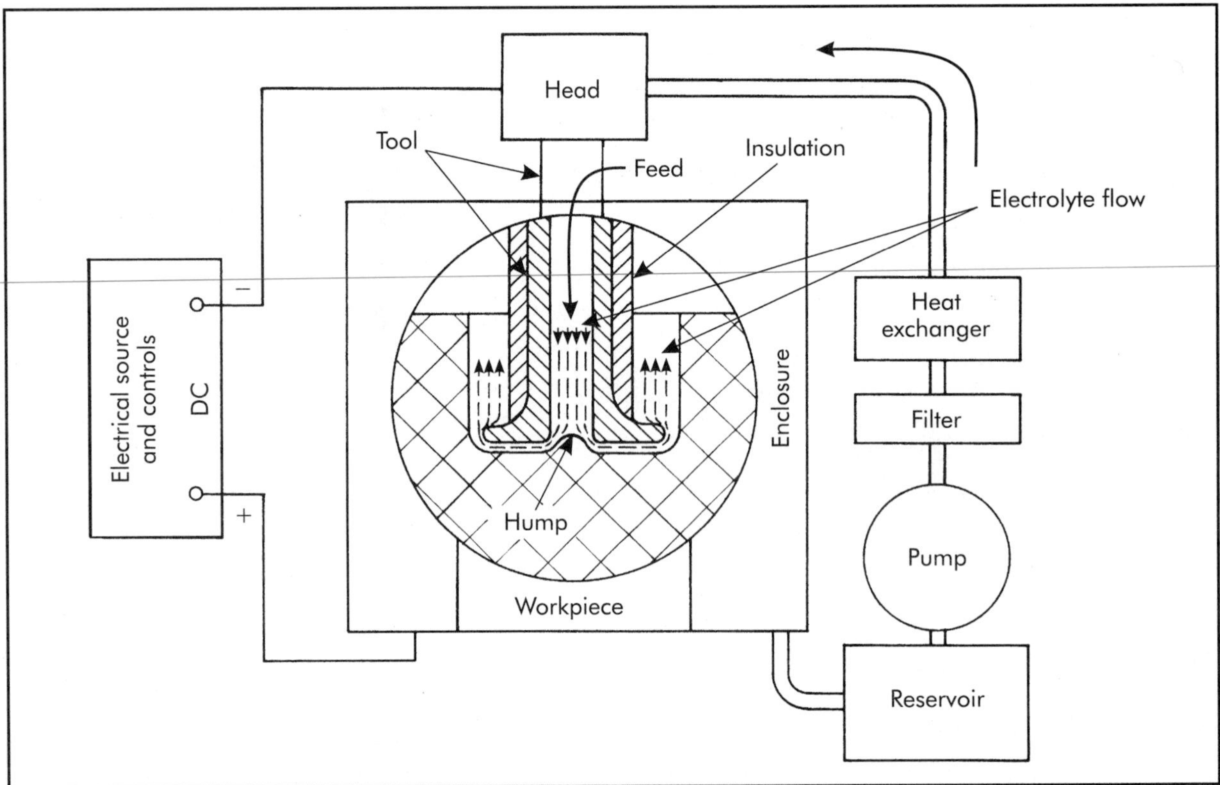

Figure 27-3. Elements of an ECM system.

workpiece, which make sturdy tools and machines necessary. For example, a 150 × 150 mm (≈6 × 6 in.) projected area die cavity with a pressure of 1.5 MPa (217 psi) in the gap may create a force of 34 kN (≈7,600 lb) between the tool and work.

ECM performance can be estimated from Faraday's law: the amount of metal removed is proportional to its equivalent weight and the current. Theoretically, 96,540 coulombs (A-s) takes off one equivalent weight, which is the atomic weight divided by the valence. The rate of metal removal is determined by

$$M = \frac{60IN}{96{,}540nd} \qquad (27\text{-}1)$$

where

M = metal removal rate, cm^3/min
I = current, A
N = atomic weight
n = valence
d = material density, $gram/cm^3$

As an example, the equivalent weight of iron (or steel) is 56/2 and its density is 7.87 gram/cm^3. If the current is 1,000 A, the rate of metal removal is

$$M = \frac{60 \times 1{,}000 \times 56}{96{,}540 \times 2 \times 7.87} = 2.2 \text{ cm}^3/\text{min}$$

$$= .13 \text{ in.}^3/\text{min}$$

Reports indicate efficiency between 70–90%. For most common metals, the attainable rate of metal removal is usually from 1.6–2.3 cm^3/min (.1–.14 $in.^3/min$) for each 1,000 A of current. The lowest number is often used for conservative estimates. Exceptions include tungsten with a valence of 6 and magnesium with an atomic weight of 24.

In ECM, the tool, workpiece, and electrolyte comprise an electrolytic cell governed by Ohm's law, and this determines the rate at which the work can be done. This means that for a given resistance and potential (usually 5–20 V) across the gap,

$$I = E/R \qquad (27\text{-}2)$$

where

I = current, A
E = potential, V
R = resistance, ohms

The resistance across the gap can be expressed as

$$R = \gamma r/A \qquad (27\text{-}3)$$

where

γ = gap thickness, cm
r = resistivity, Ω-cm
A = area of cut, cm^2

These two relationships combine to give

$$I = \frac{EA}{\gamma r} \qquad (27\text{-}4)$$

The rate of metal removal is

$$M = mA$$

where

M = metal removal rate, cm^3/min
m = rate of recession of the work surface being removed, cm/min

From Equation (27-1), $M = CI$, in which the constant factors for a given operation are represented by C. Also, the rate of recession is assumed the same as the advance or feed of the tool, f (cm/min [in./min]); that is, $m = f$. Therefore, $m/C = f/C = I/A$. This relationship is substituted in Equation (27-4) and solved for γ to get

$$\gamma = \frac{EA}{Ir} = \frac{EC}{fr} \qquad (27\text{-}5)$$

This shows that as the feed is increased with all other factors constant, the gap decreases to maintain electrical and electrolytic equilibrium. It must be realized that actually there is not one uniform gap in an ECM operation. There is the frontal gap in the direction of feed, the side gap between the tool and cavity, and the normal gap between other surfaces. These gaps are not usually the same. The model for the ECM process being presented here assumes one uniform gap, which may be thought of as an average gap.

The question as to what is the best feed for the tool in an ECM operation does not have an exact answer. Experience has shown that the faster the feed, the better the accuracy and surface finish obtained and the higher the rate of production. There are two factors that limit the feed in any operation. One is the capacity of the electrolyte; the other is the capacity of the machine. Thus the best feed is the lower of the two feeds set by these limits.

The electrolyte limits the feed in an ECM operation when it is no longer able to conduct the current. Hydrogen is evolved at the cathode and the electrolyte vaporizes in areas where it becomes hot enough to do so. Gases and vapors are not good conductors but normally are swept away, along with the sludge, where electrolyte flow is adequate. Equation (27-5) shows that the gap becomes smaller as the feed is increased. When the gap becomes so small that fluid flow is obstructed, current flow may be impeded by gases or vapors, potential is built up, and an arc ensues with likely catastrophic effects upon the tool and workpiece. The feeds recommended in reference books and handbooks are those found to be on the safe side of electrolytic breakdown. Depending on size, shape, and configuration of the cut and other factors, feeds generally are from 2.5–15.2 mm/min (.10–.60 in./min).

The current flow increases directly with the feed as shown by the relationship $f = CI/A$ in the derivation of Equation (27-5). A machine can be set only to as much feed as the current it can deliver will support, to avoid overload. For example, consider a 50-mm diameter hole to be cut by ECM in hardened steel. Experience indicates that the electrolyte can stand a feed of 10 mm/min in the system being used. To keep within the limit of the electrolyte, the rate of metal removal should be $25\pi/4 = 19.6$ cm^3/min. For a conservative estimating factor of 1.5 cm^3/min/1,000 A for current used, the current required is estimated to be (19.6/1.5)1,000 = 13,000 A. If the machine is rated at 10,000 A, the feed must be reduced to 10/1.3 = 7.7 mm/min to avoid overloading the system. The same considerations apply to estimating the time and energy required for an operation. The feed must first be ascertained as either that limited by electrolyte or that limited by machine capacity, whichever is lower. In the illustration, with a permissible feed of 7.7 mm/min and a hole depth of 2.5 cm, the time for the cut, $t = 2.5 \times 10/7.7 = 3.3$ min. If the voltage is to be set at 15 V, the energy for the cut at an overall efficiency of 75% is $W = [15 \times 10{,}000/(0.75 \times 1{,}000)]\ 3.3/60 = 11$ kWh.

Production of a good workpiece depends largely upon a uniform flow of electrolyte over all areas. If more electrolyte flows in one place, it removes more metal there than in another. Accurate results depend on proper design of the electrode and skillful control of the voltage, current, pressure, and other parameters of the operation. Copper and copper alloy electrodes are the most popular for good conductivity, but other metals are used for specific purposes.

Three common forms of tooling are shown in Figure 27-4. Straight-flow tooling is usually cheapest but does not provide enough fluid control for some jobs. Better surface finish and smaller tolerances are possible with reverse-flow tooling. Mainly because of the need of dams and seals around the work area, reverse-flow tooling is more expensive to design and make. Cross-flow tooling is used on certain kinds of work, such as turbine blades. A fourth class of tooling, called compound or complex tooling, involves a combination of two or three of the basic types. Slots and holes for fluid must be carefully placed in a cathode for effective flow. For the most part, electrode design and fabrication require special skills based on a thorough knowledge of ECM operations. There is no tool wear under normal conditions and tool life is long.

A variation of ECM called *electroshaping* forms external surfaces in the manner depicted in Figure 27-5. A tool ordinarily is made to encompass and machine a whole surface at one time. Another approach is to remove a slug of metal by cutting a groove around it with a thin-wall or bent tube electrode. *Wire cutting* is one name given to such an operation. In this role, ECM may remove stock to reduce weight, produce thin sections (152 μm [.006 in.] is common without distortion), and fabricate external stiffeners.

ECM accuracy depends upon control of several parameters of the operation, such as voltage, feed, fluid temperature, and concentration. Most importantly, these parameters must be kept constant during the operation. However, the accuracy obtainable from ECM is limited. Distinction must be made between the accuracy of the tool frontal-formed surfaces and sidewall surfaces in an ECM cavity. Tolerances of ±7.6 μm (±.0003 in.) have been reported under ideal conditions on tool frontal-formed features, and sidewall tolerances of ±25 μm (±.001 in.) in simple holes. However, ±25 to ±51 μm (±.001 to ±.002 in.) on the surface substantially perpendicular to tool feed and ±127 μm (±.005 in.) on wall dimensions, and up, are more usual tolerances. ECM naturally produces good surface finishes: better than 0.25 μm (10 μin.) R_a with care in many cases, and seldom worse than 2 μm (80 μin.) R_a on frontal surfaces, but commonly as high as 6 μm (250 μin.) R_a on sidewalls. Much that can be done with ECM depends upon product design. Design considerations are illustrated in Figure 27-6.

ECM excels in some respects and is limited in others. It cuts most conductive materials well, no matter how hard they are, and can reproduce almost any shape inside or outside. Materials with many passive inclusions may not be amenable to the process; for example, cast iron is not EC machined well because of the free graphite. A major advantage of ECM is that it does not damage the workpiece surface. The only affect ECM may have upon workpiece properties is that it may remove material previously stressed and thus cause distortion or detract from helpful compressive stresses.

ECM leaves no burrs and in modified forms is used to deburr parts machined by other methods. One form is called *electrochemical deburring* (ECD). Largely because of the lack of burrs, ECM is popular for machining small-diameter and long holes, particularly in clusters. A variation of ECM for rather small holes, such as in turbine blades, is *shaped tube electrolytic machining* (STEM). Sludge formed by the salt electrolyte usual for ECM tends to clog small passages and impede the operation. STEM uses an acidic electrolyte that dissolves the metal, and the ions are easily carried away. For even smaller holes, *electrostream* (ES) *machining* is a similar process with an acidic electrolyte, but requires voltages 10 times or more higher than usual for ECM. It requires exceptionally close control. Hole diameters are reported to be 127–899 μm

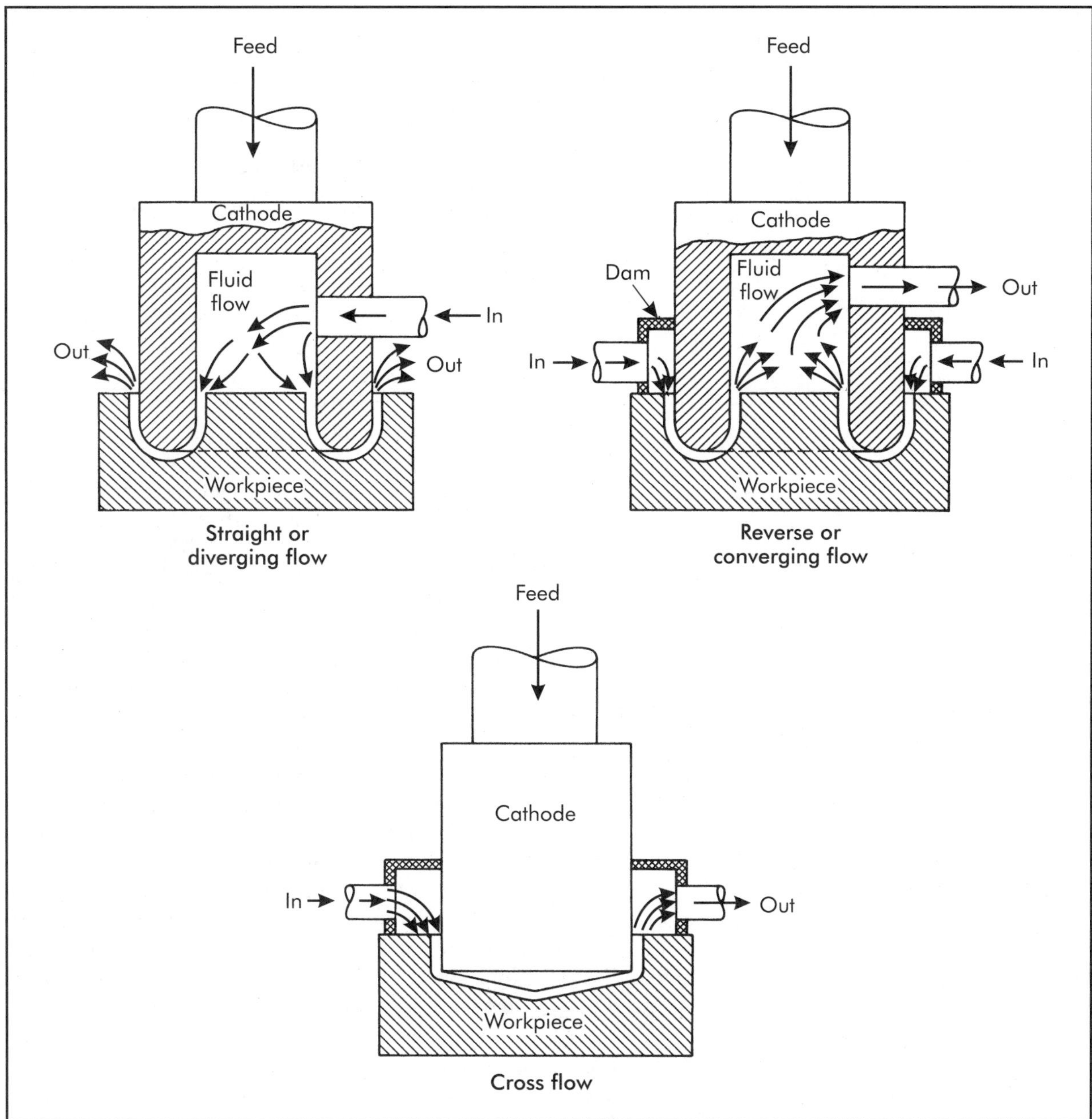

Figure 27-4. Typical ECM tooling.

(.005–.035 in.) with depth-to-diameter ratios up to 40:1.

An ECM machine resembles a vertical press. Smaller machines have C-frames, some large ones two- or four-post frames to carry a vertical ram or quill for feeding the tool. Instead of a bolster or bed as on a press, an ECM machine has a table to carry the workpiece. An important part of the system is the means to deliver and contain the extremely corrosive electrolyte at high pressures. A tight compartment surrounds the work area. The slides and moving parts of the machine must be well protected, so the pump and fittings are of stainless steel. The electrical equipment must supply heavy currents at constant voltage even under varying load, so

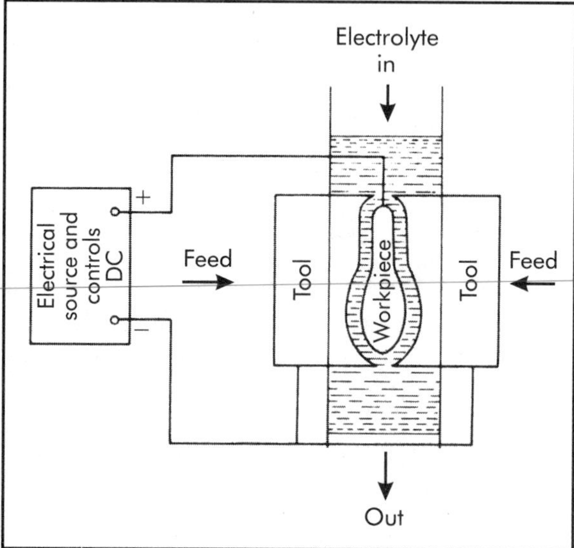

Figure 27-5. Electroshaping system.

controls to prevent short circuits are required. All these factors make an ECM machine relatively expensive.

Comparison of ECM with Other Processes

ECM is not competitive with conventional methods for simple shapes, like round holes, in easily machinable materials. When either the shape or the material is hard to machine, ECM may have the advantage. It has been said that ECM can sink a hole in 450 Bhn steel as fast as a drill, and faster in harder material. When both shape and material are difficult to machine by conventional methods, ECM and electrical discharge machining (EDM) become likely contenders for the job.

EDM can hold tolerances too small for ECM, but in its range ECM is much faster. EDM works regularly in the tolerance range below 51 μm (.002 in.) and ECM above that. On the other hand, what are ordinary surface finishes for ECM are obtainable only at the slowest EDM rates. If the accuracy it can deliver is acceptable, ECM equipment is commonly available for stock removal rates up to about 66 cm^3/min (4 $in.^3$/min), and average performance is around 16 cm^3/min (1 $in.^3$/min). In comparison, maximum EDM rates are less than 16 cm^3/min (1 $in.^3$/min.), and average performance only a small fraction as much. Even so, EDM is usually cheaper for a few pieces because the initial tool cost is high for ECM. A rule of thumb is that ECM is not economical for cavity work unless 20 or more identical pieces are needed.

27.2.2 Electrochemical Grinding

Electrochemical grinding (ECG) or *electrolytic grinding* removes hard conductive materials partly by electrolytic attack, like ECM, and partly by conventional grinding. The proportion varies with the material and operation, but is probably 90% or more by electrolytic action. The abrasive grains mainly scrape away weakened surface residue. With cemented carbide, for instance, the bonding material is dissolved and the more resistant carbide skeleton is torn apart by the abrasive grains. The grinding wheel is the cathode and the workpiece the anode in a circuit like that shown in Figure 27-7. Metal-bonded diamond wheels are most common; aluminum oxide wheels with conductive bonds have been developed and are less costly. The abrasive grains support a gap of about 25 μm (.001 in.) between the workpiece and wheel body. The electrolyte is usually poured on the point of operation and a film is carried into the cut by the abrasive grains. One machine model with a 300-A power supply and rated to remove 0.3 cm^3/min (.02 $in.^3$/min) of stock costs about $20,000.

ECG may be as much as 80% faster than regular grinding but entails extra costs and normally is not economical unless either the workpiece or material, or both, is difficult to grind conventionally. In ECG, grinding force can be kept low, and frail and thin sections EC ground without distortion. Temperature is low and no metallurgical surface changes or residual stresses are induced by electrolytic grinding. Thus the process is faster (several times faster for cemented carbides) and safer for grinding very hard and sensitive materials that crack easily from conventional grinding. The process leaves no burrs. Wheel life is 10 times longer as without electrolysis. Tool cost is low because the wheel does little cutting, and much less wheel truing is required.

To hold tolerances of less than about 25 μm (.001 in.), ECG must be made to do essentially true grinding, and then all the advantages of the process are lost. In practice, the aim is for tolerances of about ±51 μm (±.002 in.) or more. Sur-

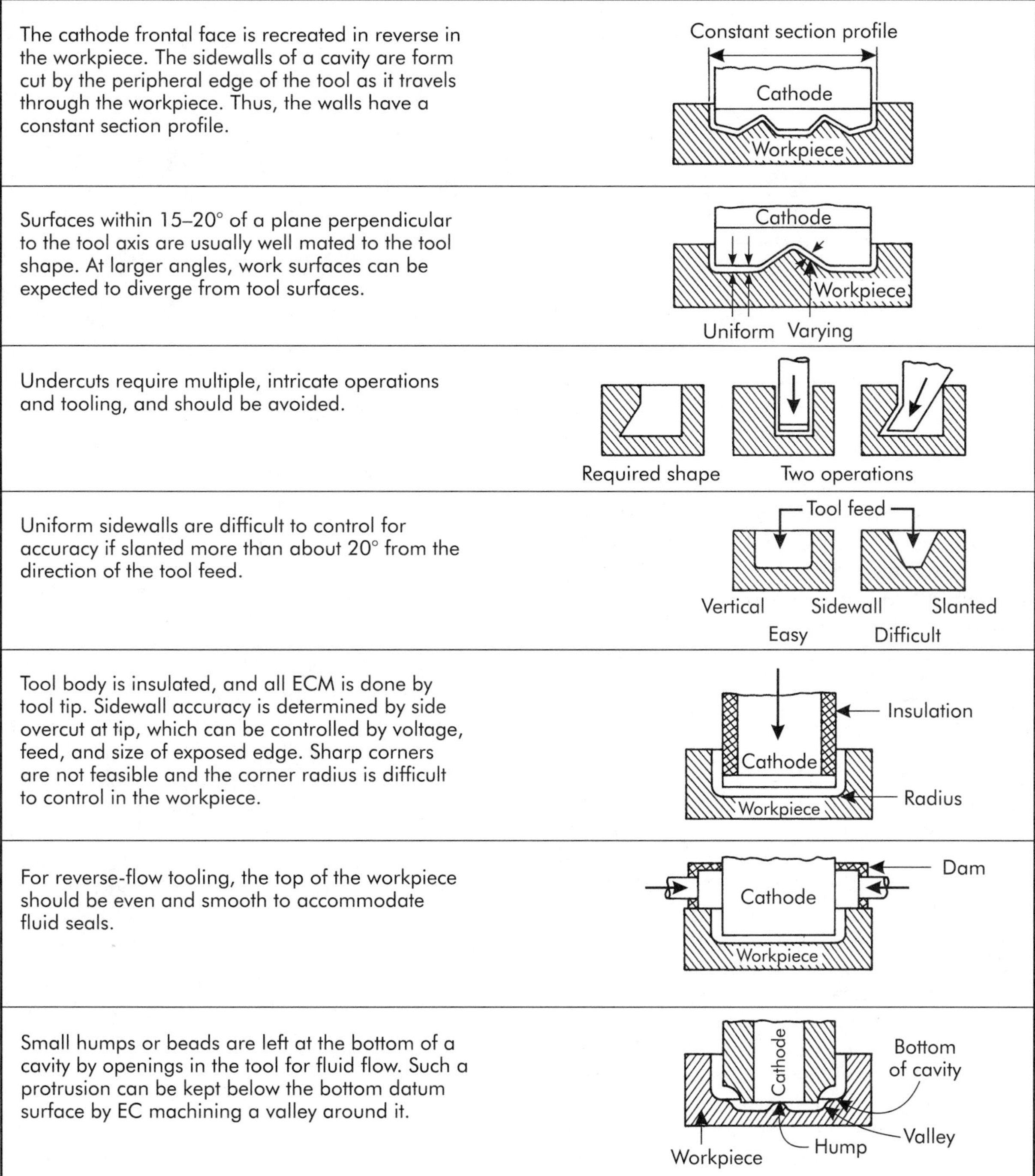

Figure 27-6. Design considerations for ECM.

face finish from ECG is better than any other kind of grinding at the same stock removal rate; a satin finish (with little direction) of 127–254 nm (5–10 μin.) R_a is common.

Electrochemical discharge grinding (ECDG), also called *electrochemical discharge machining* (ECDM), is done with a nonabrasive and easily trued graphite wheel running adjacent to the

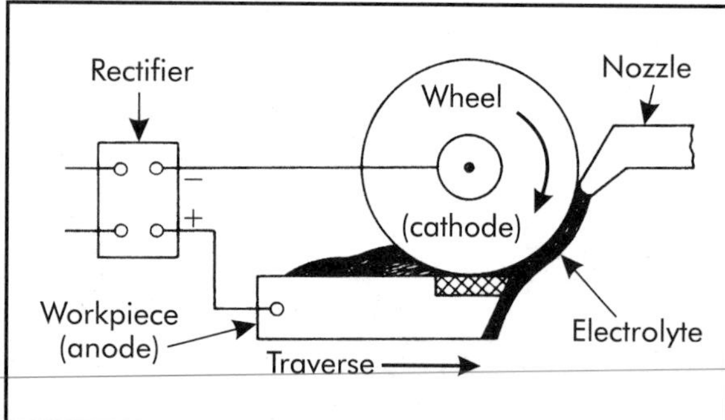

Figure 27-7. Electrolytic grinding system.

workpiece in an electrolyte. The initial action is that of ECM but the residue is removed by electrical discharges (EDM) and not by abrasives as in ECG.

Electrochemical honing (ECH) for straight holes utilizes nonconducting abrasive stones like those shown in Figure 25-2. The tool body is made the cathode and the workpiece the anode in an electrical circuit while a fine abrasive scrubs away film from the workpiece surface left by the electrochemical action. A similar but reversed process, called *hone forming* is described in Chapter 25.

27.2.3 Electrogel Machining (EGM)

EGM is an electrolytic process for removing metal in definite shapes by means of a formed tool made of a stiff gel consisting of cellulose acetate and acid. The molded gel is held against the workpiece as shown in Figure 27-8. As a current (about 8 A at 2.5 V) is passed through the stack, the pattern on the gel is etched in the workpiece at a rate of about 25 μm/min (.001 in./min). Forces are negligible, and there is no effect on the structure of the metal. Thus the mild operation is suited for frail pieces like those made from thin metal honeycomb structures. Steel and nickel alloys, titanium, and high-temperature heat-resistant alloys have been found amenable to the process.

27.3 THERMAL MACHINING PROCESSES

There are essentially four machining technologies that use an extremely high concentration of thermal energy to machine metals and

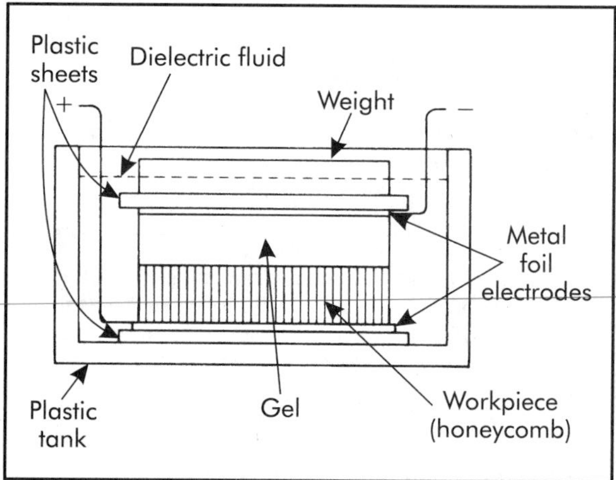

Figure 27-8. Electrogel machining operation.

other engineering materials. These are: (1) electron beam machining (EBM), (2) laser beam machining (LBM), (3) electrical discharge machining (EDM), and (4) plasma beam machining (PBM) (see Chapter 14). These processes have some similar characteristics and some that are dissimilar. For example, the EBM, LBM, and EDM processes remove metal by melting and vaporization, while PBM removes metal by melting and blowing it away. Two of the processes, EDM and PBM, will only machine metals that are conductive. In addition, the machining operations that can be performed by the EBM, LBM, and PBM processes are somewhat limited, for the most part, to cutting (slitting or cutoff) and/or hole cutting (drilling or trepanning).

27.3.1 Electron Beam Machining (EBM)

The *electron beam machining* process is conducted in much the same way as that used for EB welding (Chapter 13). EBM is classified as a micromachining process that employs high-energy electrons, which are focused on a workpiece to melt and vaporize a very small spot of material. The electron beam is sustained in a vacuum (approximately 10^{-4} torr) as illustrated in Figure 13-12, but machining may be performed either inside or outside of the vacuum chamber. The purity of the base material can be maintained if the machining is performed inside the vacuum chamber since no gases or airborne particles can contaminate it. However, it is difficult to accommodate large workpieces in a vacuum chamber. The energy beam can be highly concentrated with power densities over 3 MW/cm^2 (20 MW/in.^{2}) to melt and vaporize ferrous and nonferrous refractory, dissimilar, and even reactive metals.

EBM is particularly applicable to the cutting (piercing) of very small diameter holes in hard materials. Small holes with a 100:1 depth-to-diameter ratio may be pierced to a diameter accuracy of 0.005–0.025 mm (.0002–.0010 in.). The process is adaptable to computer numerical control, which makes it quite useful in the microelectronics industry for a variety of cutting and piercing operations on very hard and thin materials. Care needs to be exercised in the operation of EBM equipment because of X-rays generated by the interaction of the electron beam with the workpiece surface. Thus, appropriate operator training is required.

27.3.2 Laser Beam Machining (LBM)

Although initially used predominately for welding (Chapter 13), lasers are becoming more popular for a variety of machining applications. These include: (1) laser cutting, (2) laser-assisted machining (LAM), (3) laser hole piercing and trepanning, and (4) laser alloying and cladding. The term *laser* is an acronym for "light amplification by stimulated emission of radiation." The term itself refers to a broad spectrum of electromagnetic radiation ranging from ultraviolet (UV) through infrared (IR). This is the range of radiation that is loosely referred to as *light* and it includes wavelengths from 0.1–1,000 μm (3.937–39,370 μin.). Thus, light and radiation mean essentially the same thing in this context, and the laser is basically a light source that is able to capitalize on a number of the remarkable properties of light. One of the very important properties of light is that it can stimulate the emission of light. This property provides the basis for amplification. Building on those properties, the laser is able to focus a highly monochromatic and coherent light source on a very small spot whose diameter is on the order of the wavelength of light.

Two basic types of lasers are available for machining: the continuous power output (CW) type and the pulsed output type. The continuous output types are capable of operating at very low (microwatts) to very high (kilowatts) power output. Pulsed lasers can operate at extremely high (terrawatts) peak power and transmit hundreds of joules (ft-lbf) of energy at each pulse.

Many different types of lasers are available for a variety of commercial applications. Only a few types, however, have been found to be suitable for machining operations. These include the CW or pulsed CO_2 laser, the pulsed ruby laser, the CW or pulsed neodymium yttrium-aluminum-garnet (Nd-YAG) laser, and the pulsed Nd-glass laser. References at the end of this chapter provide sources of information for the operating characteristics of these lasers.

Laser Cutting

Laser cutting operations are commonly done with CW or pulsed CO_2 lasers, or with a pulsed Nd-YAG laser. In most cases, *laser cutting* is gas-assisted with a high-pressure gas used to blow molten metal from the kerf. Oxygen may be used to increase the laser cutting speed when oxidizable materials are being cut. A variety of materials can be cut with a gas-assisted CW CO_2 laser, including most metals, ceramics, composites, glass, paper, plastics, quartz, and wood. For example, sheet steel up to about 5 mm (≈.20 in.) thick is easily cut with a CW CO_2 laser at a power setting of about 500 W (0.7 hp).

The edge quality of most materials cut with a CW CO_2 laser is very good. A certain amount of taper or undercut will be present on both sides of the kerf as illustrated in Figure 27-9. However, kerf undercut can be controlled to a certain extent by the location of the focal point of

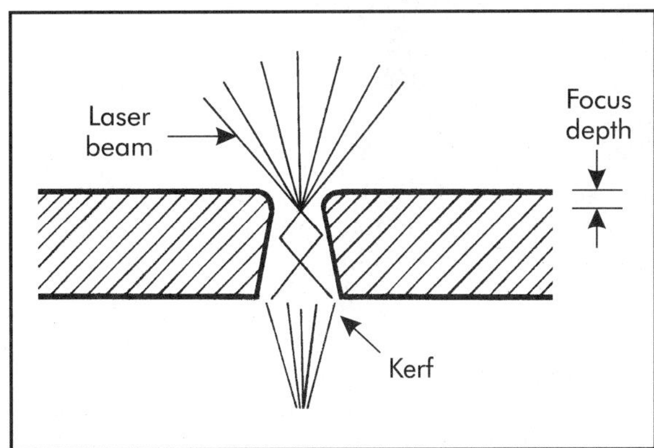

Figure 27-9. Kerf configuration produced by laser cutting.

the laser. Generally, the undercut or taper is decreased as the focal point of the laser is moved deeper into the surface of the workpiece.

In a number of cutting operations, it is necessary to use a pulsed laser because the CW type may not have enough average power to melt and vaporize the material being cut. As an added benefit, the pulsed laser generally minimizes the heat-affected zone. This is particularly important when cutting ceramics and other refractory materials where excess heat along the sides of the kerf would cause microcracking.

Laser cutting of complex shapes from sheet metal with multi-axis CNC laser cutting machines is becoming common practice, often replacing more costly shearing and punching operations. One manufacturer employs a CNC laser-cutting center to cut net-shape circular saw blades at speeds up to 2,540 mm/min (100 in./min). Formerly, a 3.7 m (12 ft) shearing press, three punch presses, two profile milling machines, and additional hand grinding was required to produce the same saw blade. The changeover to CNC laser cutting enabled the saw blade manufacturer to reduce the development cycle on new designs from 4 months to 2 weeks because of the elimination of hard tooling requirements for punch press work and other equipment. In addition, small lot manufacturing became more cost effective.

Laser-assisted Machining (LAM)

Laser-assisted machining employs a laser to preheat and soften the shear area just ahead of the cutting tool as illustrated in Figure 27-10. In this process, the energy from the laser beam is absorbed by the surface of the workpiece and converted mainly to heat. The increase in temperature of the shear area changes the physical properties of the workpiece and effectively reduces the cutting forces required for machining very hard materials. As a result, tool wear is reduced and metal removal rates are increased.

Laser-assisted machining has been used successfully in both turning and milling operations to improve the machinability of superalloys and ceramic materials. Either CO_2 or Nd-YAG lasers may be used for laser-assisted machining, but some experimentation may be required to determine optimum cutting and laser beam parameters for each type. Generally, best results are obtained when the laser beam is focused into an elliptical pattern in front of the tool in such a way as to heat only a vertical cross section of the workpiece slightly in advance of the shear zone.

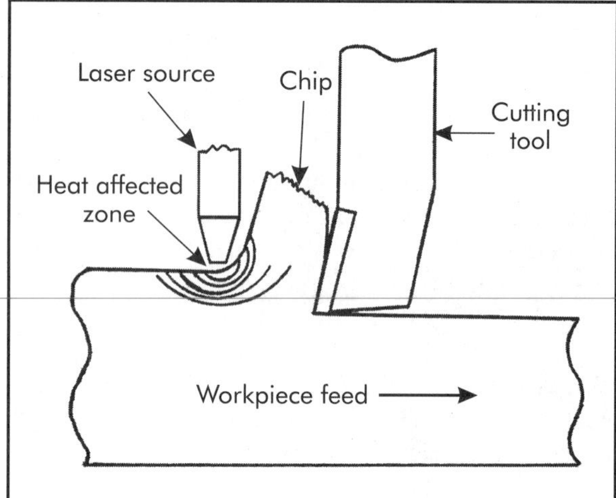

Figure 27-10. Laser-assisted machining arrangement.

Laser Hole Piercing and Trepanning

The laser hole piercing and trepanning operation is frequently used to produce holes in materials of low machinability. Possible applications for this process are practically endless and, compared to conventional drilling, significant cost savings are achieved. The advantages of piercing and trepanning holes with a laser are about the same as for laser cutting or welding. It is a process that can be used to produce holes in practically any material, ranging from very light filter paper to most of the hardest materials, including diamonds. The process is repeatable and there is no tool contact. The major disadvantage to the process is the holes produced do not have precisely straight and smooth walls. Some hole taper is unavoidable because of problems with beam focusing in high aspect ratio holes. However, hole taper can be minimized by repiercing a hole and adjusting the beam focus after the first shot. Usually, some recast materials are pushed to the top of the hole and some dross attaches itself to the underlip. Both of these may have to be removed by secondary operations.

Most hole piercing is done with pulsed CO_2 and Nd-YAG lasers. Trepanning is actually more of a cutting operation and either continuous- or pulsed-power lasers may be used. Generally, rubber and many plastic materials are pierced with pulsed CO_2 lasers, while Nd-YAG lasers are used to pierce most diamonds and other gemstones.

Laser Alloying and Cladding

Laser alloying and *cladding* operations are carried out to modify the surface of a metal to improve its character while retaining certain properties of the substrate material. For either alloying or cladding, the alloying material is applied to the surface and melted with a CO_2 laser beam. In most cases, the alloying material is in powder form or a powder-solvent slurry. For alloying, a significant depth of substrate material is melted and mixed with the melted alloying powder to form a substantial alloyed surface layer after the two materials have solidified. Cladding is done in a similar fashion except the depth of substrate melting is limited to that which will permit the cladding material to bond with the substrate material. In essence, cladding is a *metallurgical weld.*

Generally, both alloying and cladding are done by traversing a defocused or integrated CO_2 laser beam over the surface to be melted. A variety of material may be used in the alloying or cladding processes to improve surface hardness and/or wear resistance. This includes nickel-based chromium carbides and tungsten carbide nickel alloys. A powder form of these materials may be mixed with a solvent or an organic binder and applied to a substrate metal surface prior to laser melting. Care must be exercised in their application to obtain a uniform thickness. Automated spraying of the powdered cladding materials through a small orifice just ahead of the melting laser has been accomplished in a number of situations.

Laser alloying and cladding techniques often produce some slight surface irregularities due to a rippling effect during the melting and subsequent solidification. Thus, some secondary operation finishing may be required to achieve a flat surface.

A photograph of a five-axis CNC laser cladding system is shown in Figure 27-11(A). This machine is designed for the hard-facing of components and resurfacing of worn turbine blades and other turbine components. The unit is fully automated and combines an integrated vision system, a CO_2 laser, and a powdered-metal delivery system. A vision system and CO_2 laser beam is used to obtain proper alignment between the machine and workpiece. Generally, a nickel- or titanium-based Iconel® material is used in resurfacing worn turbine blades. The schematic in Figure 27-11(B) shows the five axes of CNC control. The laser source and powdered-metal delivery nozzle are attached to a vertical (Z axis) slide, while the workpiece is mounted on a stage that can be moved in two linear directions (X and Y) and rotated about two axes (A and C). A laser cladding system of the type shown in Figure 27-11 costs about $500,000.

27.3.3 Electrical Discharge Machining (EDM)

Electrical discharge machining is a means of shaping hard metals and forming deep and complex holes in all kinds of electro-conductive materials. Introduced in Russia in 1943 as an electro-erosion process, it was used to obtain square holes and square corners in a variety of tooling equipment. Introduced later on in the United States, it was used to remove broken drills and taps and was known as a tap buster or disintegrator. At that time, the device used a piece of tubing as an electrode to erode the core from a broken drill or tap so that the remaining metal could be picked out. In essence, the process was used as a sort of drilling machine, except metal removal was accomplished by electrospark or arc erosion rather than with a hard tool.

Over a period of time, the spark erosion process was improved, leading to the development of more sophisticated equipment capable of machining shaped cavities for molds and dies. This led to identification of the process as electrical discharge machining (EDM) and to the identification of the type of equipment used as either a ram or sinker electrical discharge machine.

In 1969, the electrical discharge machining process was adapted to a cutting process through the use of a traveling wire electrode. This permitted

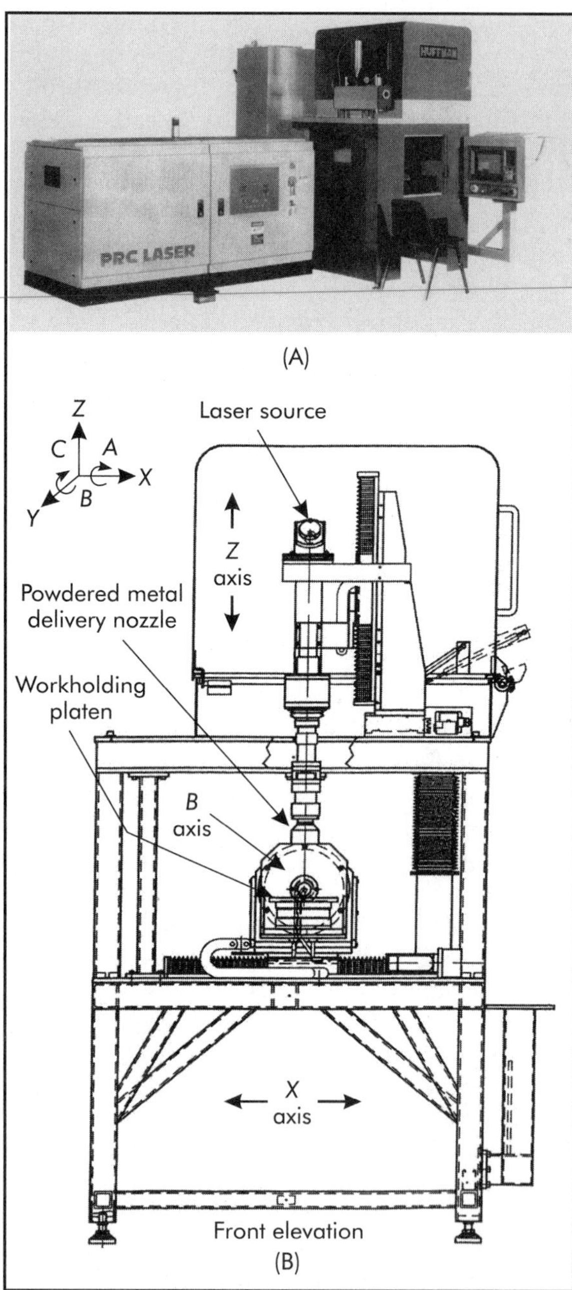

Figure 27-11. (A) Five-axis CNC laser cladding system for resurfacing worn turbine blades. (B) Schematic showing laser source, powdered-metal delivery nozzle, and axes of motion under CNC operation. (Courtesy S. E. Huffman Corporation)

fairly thick sections of hard metals to be cut into intricate shapes for use as punches, dies, and other tooling. Since then, the equipment using the traveling wire electrode has been identified as an *electrical discharge wire-cutting machine* (EDWC).

Process

The electrical discharge process is a thermal process that removes material with heat. Older EDM machines employed resistance-capacitance (RC) or relaxation circuits wherein energy was built up on a capacitor and discharged repeatedly across a gap. Solid-state circuitry such as depicted in Figure 27-12 has proven to be more efficient and faster. The tool (electrode) is brought close to the workpiece surface, 25 μm (.001 in.), and the gap is filled with dielectric fluid. When the transistor bank is triggered by the timing control, the potential polarizes a path over which direct current from the power unit (such as a generator or rectifier) flows as a spark between the closest points of the electrode and workpiece. A minute amount of metal is melted and expelled where the spark strikes the workpiece, leaving a tiny crater, as evidenced by the pocked surface produced and the globular form of the debris. Wear occurs on the electrode. The more energy in a pulse (that is, the more current supplied to the arc), the bigger the chunk torn from the workpiece; too much may create fissures and damage the piece, and is inefficient. A good finish requires light sparking that leaves small craters.

The amount of current depends upon the number of transistors activated in the bank (because each transistor carries only so much current) and the duration of pulse (spark) determined by the setting of the timer to turn the transistors on and off. When one pair of points (closest together) is blasted away by the arc, another pair becomes the closest and the next pulse sends a spark between them, and so on. Thus the arc and material removal move around in the gap between the electrode and workpiece. The reverse polarity with the workpiece negative shown in Figure 27-12 is commonly used for roughing, but straight polarity with the workpiece positive is favored for some applications, especially finishing.

The timer of an EDM system acts to initiate a series of pulses through the course of an operation, regulates the length of each pulse, and the time in between. These intervals and the amount of current are preset for most systems,

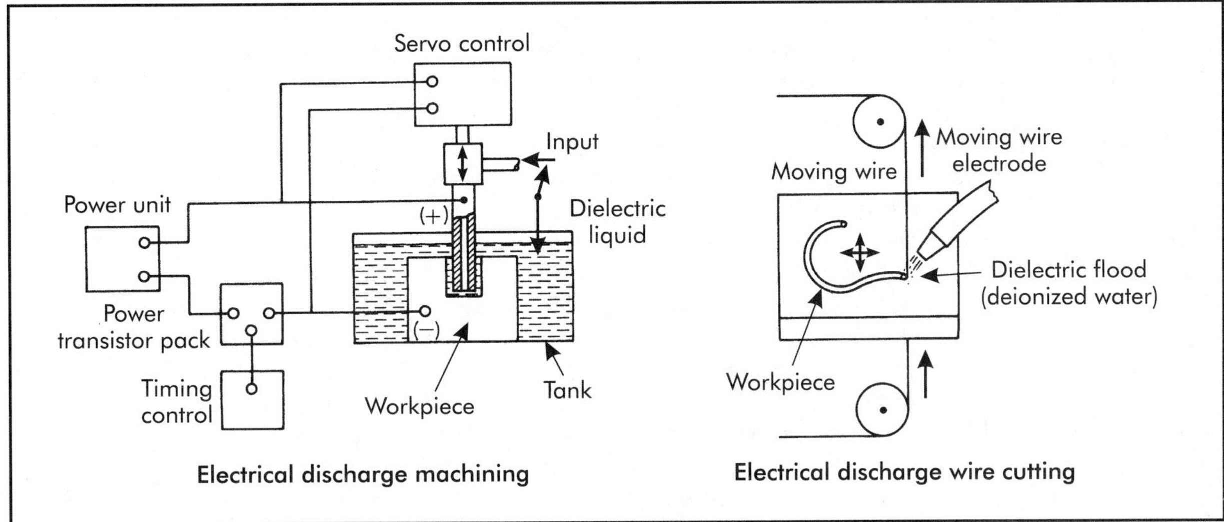

Figure 27-12. Elements of an EDM system.

but in some are varied during the operation to optimize performance through adaptive control circuitry. With time in microseconds, the pulses are very short and occur at high frequencies. In one system, pulse lengths and time between pulses can be set separately in increments from 1–2,999 microseconds; others have coarser controls. Fine settings provide close control for finishing. Some machines have means to group series of pulses into short trains with adjustable intervals between the groups. This gives the controls more time to correct for abnormal conditions such as excessive sparking or short circuiting when they happen.

The gap between the tool and workpiece is maintained by a servo control device governed by the voltage across the gap at the time of spark discharge. In some systems, the tool is given a pulsating motion to avoid dwell of the arc in one spot for too long and to help flush away the fluid. This allows the use of more current and a higher metal removal rate; as much as four times higher is claimed in one case. Instead of being fed straight in, as is common, the tool can be *orbited* on some machines; that is, the tool is rotated about an eccentric axis under servo control to sweep a shape larger than itself, cut up or down tapers, undercuts, or other profiles. On some machines, the tools can be moved in square or rectangular paths or straight lines as they are fed into the work. The workpiece, not the tool, is orbited in some systems. In addition to being able to form cavities of many shapes (some difficult by other means), orbiting helps the EDM action by stirring and flushing the electrolyte and distributing wear on the electrode, enhancing accuracy and finish. As a tool wears, size can be held by increasing the orbit.

The fluid bath around the tool and workpiece performs several functions. As a dielectric, it supports the voltage to assure a high buildup of energy for each discharge. The fluid and the impurities in it supply ions for the path of the arc. The heat of the spark instantaneously vaporizes and decomposes the fluid in its path. The fluid inertia resists rapid expansion and causes high pressure in the discharge column that intensifies the arc, where temperatures are reported in tens of thousands of degrees, and expels the molten metal. The fluid then serves to chill, solidify, flush away the debris, and cool the tool and workpiece. A copious flow of fluid is desirable; common practice is to immerse the tool and workpiece in a bath and pump fluid through holes in the electrode. Light mineral oils, such as kerosene or lubricating oil, are satisfactory fluids for most cases. For particular applications, additives or water compounds have been found helpful. Some impurities are desirable, but filtering is necessary to prevent too much contamination.

Zinc-tin, copper, and tungsten alloys, cemented carbides, aluminum, steel, graphite, and sometimes other materials are used for electrodes to suit various conditions. One may perform better than others with a certain work material. Electrodes may be machined in conventional ways. A number of identical electrodes may be needed to make copies of a forging die, for instance. Zinc-tin alloys are economical for such a purpose because they can be cast and coined easily in a master die. Graphite is mostly used because it is cheap and easily machined, and provides for fast metal removal.

The factors that determine the performance of an EDM operation are the amount of electrical current supplied the proportion of the time during which current flows, and the voltage across the gap at the arc. The time the current is on during any one pulse is usually a preset amount but may be less in some pulses that are delayed, for example, for slow ionization. There is also an idle time set for each cycle. The effective average current is

$$I_e = I_s \ (\Sigma \ t_i/p) \qquad (27\text{-}6)$$

where

I_e = effective average current, A
I_s = amount of current supplied, A
t_i = time the current is on during any one pulse, microsecond
p = substantial time period during which the current flows ($p = \Sigma t_i$), min

The rate of metal removal is large for a large effective current and smaller for a smaller effective current.

For any supplied voltage, there is a maximum gap beyond which the dielectric will not become sufficiently ionized and no discharge will occur. For example, if 100 V is applied to two points closest together but 25 μm (.001 in.) apart, the dielectric ionizes between the points and an arc ensues to transmit all the current that is supplied. (The numbers are likely to be different in another case.) For any smaller gap, the arc voltage is almost proportionately smaller, for example, 40 V for 10 μm (.0004 in.). These numbers are purely hypothetical and what they would be in any particular case depends upon the circumstances.

The operator adjusts the servo-feed device to seek a desired gap potential. In adhering to that potential, the servo maintains a constant gap as it feeds the electrode into the workpiece. Some of the pulses misfire because of impurities in the electrolyte and other factors; occurring more when the gap is large than when it is small. Consequently, for a given input, the effective current increases and so does the rate of metal removal as the gap is held smaller, up to a peak, usually at about 90% of maximum. Any smaller gap appears to impede removal of debris, etc., and performance goes down.

Empirical equations and charts have been derived for various electrode-workpiece material combinations and systems and are available to guide an operator to find optimum settings for the EDM machine. The most advanced systems include programmable controllers (Chapter 31) with memories for storing data on operation parameters. The controller has the ability to apply these data as needed to optimize the action. An average stock removal rate is around 0.8 cm^3/A/hr (.05 in.^{3}/A/hr) of nominal current for roughing. This means a removal rate of 16 cm^3/hr (1 in.^{3}/hr) on a machine with a 20-A capacity. For finishing, the metal removal rate may be one-third to one-half as much.

The electrode is usually held at positive polarity for machining steel, aluminum, cast iron, and some superalloys and wears mostly from loss of ions during the initial stage of each cycle. This is a small part of the action during heavy but not excessive roughing cuts. Some metal removed from the workpiece can be made to adhere to the tool to compensate for wear. Thus, under proper conditions and control in roughing, tool wear can be made so nominal that some machine makers claim practically no wear, but it is seldom possible to achieve this ideal in production. Tool wear is relatively severe for light finishing cuts, but surface finishes are better. An EDM tool wears most at its corners and edges, so these should be rounded. Small corners should be avoided as much as possible in the design of the workpiece.

Size and Accuracy

Size can be held quite closely in EDM, but precision costs more. A hole sunk by conventional EDM has the same shape as the tool; a round tool makes a round hole, etc. An electrode with a sculptured cameo form produces an intaglio of the same form in the workpiece. With numerical control, a simple tool can be moved with respect to the work, or vice versa, to EDM a complex surface. The amount of gap between the tool and workpiece, called the *overcut,* determines the size of the cut. The gap may range from about 5–178 μm (.0002 –.0070 in.) or so, but is usually between 25–76 μm (.001–.003 in.). The higher the rate of metal removal, the larger the overcut must be, because the heavy spark discharges clear away more metal from the sides and space is needed to clear away large chips.

The accuracy obtained in EDM depends largely upon the accuracy of the tool, the wear it gets during operation, and control of the overcut. The more accurate the tool, the more its cost. If tool wear is to be kept minimal in finishing, the stock removal must be kept small. For a constant set of conditions, the overcut can be held uniformly to a tolerance within 5 μm (.0002 in.) around the tool if the overcut is minimal. A common practice is to take roughing and finishing cuts with a series of fresh electrodes, just like several cuts for conventional machining. Typical tolerances held on forging dies are ±76 μm (±.003 in.) with three consecutive electrodes, to ±25 μm (±.001 in.) with seven. In other work, as little as ±2.5 μm (±.0001 in.) tolerance has been reported under carefully controlled conditions.

Surface Finish

Another major consideration in EDM is surface finish, which depends upon how fast the work is done. Low-energy discharges that leave small craters are necessary for fine finishes. The rate of stock removal is then slow. Surfaces have been reported finished to less than 254 nm (10 μin.) R_a at removal rates of 49 mm^3/hr (.003 $in.^3$/hr), but 762 nm (30 μin.) to 4 μm (≈150 μin.) R_a may be expected at ordinary rates. However, an EDM surface has no lay like a mechanically cut surface. It can hold lubricity, and for some purposes is better than a conventional surface with the same or less measured roughness.

Advantages and Limitations

EDM is at an advantage for cutting hard materials, internal shapes (particularly if hard to reach), shapes difficult to generate, and delicate pieces. It can reproduce any shape that can be cut into a tool. Thus, mechanically making a tool and sinking a cavity by EDM may be easier than carving out the cavity. However, it is too slow to compete with conventional machining of common materials, particularly for simple shapes. On the other hand, EDM removes stock almost as fast as grinding from cemented carbide, and even faster than grinding from some of the space-age materials.

Sinking and resinking of forging dies from hardened steel has become almost a monopoly for EDM. One forge shop reports the two halves of a crankshaft forging die took 122 hr by conventional machining but only 68.8 hr by EDM on a 200-A machine. Quantity production applications of EDM are growing, for example, cutting fine and out-of-the-way slots and holes without burrs and to close tolerances on carburetors. With no physical contact and therefore no forces between tool and workpiece, frail pieces such as honeycomb structures can be machined without distortion. An example of a task particularly amenable to EDM is the cutting of 100 holes in molybdenum of 0.20 mm (.008 in.) diameter, 0.25 mm (.010 in.) between centers, with 0.05 mm (.002 in.) wall thickness between holes, and to tolerances in micrometers (ten thousandths of an inch).

Some serious shortcomings of EDM must be recognized and controlled. Surface porosity can result from EDM on sulfurized steel or from sulfurized cutting oils on the surface from previous operations. Of more importance, a carbon-enriched white layer from 2.5 μm (.0001 in.) for light cuts to 127 μm (.005 in.) for heavy cuts is left on hardenable steel by EDM. This is metal that has been melted and resolidified. Beneath that layer there is frequently a second layer of hardened or rehardened steel, usually deeper than the first. Below these two layers is a gradually tempered zone that may range in depth from 50–750 μm (≈.002–030 in.). A severe EDM operation may leave microcracks in the surface. Grinding an EDM surface may cause surface microcracks to develop and propagate under stress. EDM surfaces are quite hard and wear resistant but brittle, and fatigue strength may be cut more than in half. A surface may be improved by light and slow electrical discharge machining after roughing. Stresses may be relieved by annealing. Impaired surfaces may be removed by careful grinding, lapping, or electrochemical milling.

Machines

A ram or *sinker-type electrical discharge machine* looks somewhat like a vertical knee-and-column type milling machine with a ram or quill instead of a cutter spindle and a tank on the table for dielectric fluid. Precision adjustments are provided for the tank and the ram in coordinate directions. On very large machines, the tool may be mounted on a platen on posts. Means are provided to circulate and filter the fluid. Most modern machines have solid-state power supplies and controls. Multiple-tool EDM machines for production have a variable number of separate electrodes, usually on a single machine head with one servo system and connected to a single power supply. Thus, a number of parts can be machined at once for a low unit cost.

A large sinker-type EDM machine with 4-axes of CNC control is shown in Figure 27-13. Often referred to as a *die-sinker,* this type of machine is used in the plastic molding industry to EDM large plastic molds for items such as plastic lawn furniture, lawn tractor hoods, interior dishwasher sections, washing machine tubs, TV housings, recycle bins, and trash containers. The machine shown has a dielectric fluid tank that measures 199.9 × 199.9 × 149.9 cm (78.7 × 78.7 × 59 in.). The machine head can carry electrodes weighing up to 4,990 kg (11,000 lb) and is capable of movement along three axes: X = 134.9 cm (53.1 in.), Y = 140 cm (55.1 in.), and Z = 79.8 cm (31.4 in.). In addition, an integrated gantry head with C-axis movement permits both planetary and contour machining along with standard die-sinking operations.

The mold in the foreground of Figure 27-13 is a plastic injection mold for a TV housing. Generally, several EDM burns are required to complete this type and size of mold. Large graphite electrodes are used to form the large cavity, while smaller rib-type electrodes are employed to complete some of the detail operations.

Figure 27-13. Four-axes CNC sinker-type electrical discharge machine (EDM). (Courtesy Ingersoll GmbH)

Electrical Discharge Grinding (EDG)

A variation of the EDM process, *electrical discharge grinding* (EDG), utilizes a graphite grinding wheel as an electrode for cutting carbide form tools, thin and closely spaced slots, and the like.

Electrical Discharge Wire Cutting (EDWC)

Electrical discharge wire cutting (EDWC) is a variation of EDM done in the manner indicated in Figure 27-12(B). A moving wire between spools and rolls is kept taut by a tensioning device, serves as an electrode, and passes in near contact with the workpiece. Conventionally, the wire is copper of 200, 152, or 102 μm (.008, .006, or .004 in.) diameter or 51 μm (.002 in.) diameter molybdenum. Since the wire is used only once, wear is not a factor. The workpiece is moved in the X and Y directions perpendicular to the wire by numerical control (Chapter 31), and any path can be readily traversed. A third or "steering" axis may be provided to tilt the work with respect to the wire to cut, taper, draft, or relief. The rate of metal removal is governed by the same factors as in conventional EDM, and the rate of travel depends on the thickness of the workpiece. An EDWC machine rating in cm^2/hr ($in.^2/hr$) designates the area of one side of the cut made in a unit time. Thus, with a die block 25 mm (.98 in.) thick, a rate of 15 cm^2/hr (2.33 $in.^2/hr$) is equivalent to feed along the path of cut of 60 mm/hr (2.36 in./hr).

Inside and outside shapes can be cut by EDWC. Common work includes blanking dies, prototype parts, and EDM electrodes. Carbide blanking of progressive dies previously had to be made in accurately ground sections and carefully fitted together. Now they can be produced much more quickly as whole units by EDWC. Tolerances as small as 5 μm (.0002 in.) are commonly held. Average surface finish obtained is estimated to be from 650–1,150 nm (≈25–45 μin.) R_a.

EDWC can do some work better than other methods; tool cost is negligible and the process can be run for long periods without operator attention. Some shops let a machine run on a long cut for a whole weekend with reportedly good results. However, the machines are expensive as they combine EDM with numerical control.

27.4 WATERJET MACHINING (WJM)

The use of a high-pressure jet of water to cut materials is an emerging technology worth consideration in many manufacturing scenarios. Called *waterjet cutting*, *waterjet machining*, or *fluid machining*, this process employs a high-pressure pump or intensifier to pressurize water up to 379,201 kPa (55,000 psi) and force it at two to three times the speed of sound through sapphire nozzles onto extremely small areas.

First used extensively in the cutting of cardboard, printed circuit board, and pressed paper food containers, this technique is now being used to cut a variety of nonmetallic materials, such as acrylics, felt, foam, Mylar®, plastic, polyethylene, polyimide, and rubber. The fluid cutting system provides clean cuts and edges with reduced material waste, increased cutting speeds, and multidirectional cutting. Unlike most of the thermal cutting processes, waterjet machining does not subject the workpiece material to any thermal deformation or mechanical stresses. In addition, the process is dust-free and odorless.

With integrated multi-axis computer control, waterjet cutting can produce intricate shapes from many nonmetallic materials to net or near-net dimensions with minimal kerf width and no burrs. Reasonably simple waterjet cutting systems that can be adapted to an existing *X-Y* table are available for an initial investment of around $60,000. Larger, more sophisticated turnkey systems with a table, catcher tank, gantry, and CNC unit may be obtained for about $200,000. Some systems are designed to accommodate easily interchangeable waterjet or abrasive jet (Chapter 25) cutting heads. This combination provides a great deal of versatility to such a system.

The operating costs of waterjet cutting systems are relatively low when compared to many traditional machining processes. When, however, the cost of waterjet processing operations is higher, the additional cost is quite often offset by improved finished part quality.

27.5 QUESTIONS

1. Name the four steps involved in the chemical milling process.
2. What is the difference between chemical milling and photo etching?
3. Explain the method of removing burrs by the combustion process.
4. Explain why no tool wear occurs in the electrochemical machining (ECM) process.
5. How does feed-rate affect surface finish in ECM?
6. Under what conditions does ECM have an advantage over the conventional methods of machining?

7. How is a grinding wheel made conductive for electrochemical grinding?
8. What types of materials are best suited for electrochemical grinding?
9. Why is electron beam machining (EBM) often referred to as a micromachining process?
10. Why is EBM often used for small hole piercing?
11. What range of electromagnetic radiation is represented by the term "laser"?
12. What kind of lasers are commonly employed in laser cutting operations?
13. How does the location of the focal point of a laser affect taper or undercut in the kerf in a laser cutting operation?
14. How does the pulsed laser impact the heat-affected zone when cutting ceramics or other refractory materials?
15. How does the action of a laser affect the properties of the workpiece during laser-assisted machining?
16. Explain how a laser may be used to trepan a hole.
17. What is the difference between laser alloying and laser cladding?
18. Explain the electrical discharge machining (EDM) process.
19. How is surface finish controlled in the EDM process?
20. Why does electrode wear occur during the EDM process?
21. How is the gap between tool (electrode) and workpiece controlled in the EDM process?
22. Cite at least three advantages of the EDM process.
23. Explain the electrical discharge wire-cutting (EDWC) process.

27.6 PROBLEMS

1. Estimate the time required to sink a simple cavity 7.938 mm (.3125 in.) deep with a cross-sectional area of 32.65 cm^2 (5.060 $in.^2$) in steel using a 20,000-A capacity electrochemical machine.
2. How much average current is required to remove steel by EDM at the rate of 1.92 cm^3/min (.117 $in.^3$/min)?
3. What power input to a 90% efficient EDM circuit is required to remove 4.67 cm^3/min (.285 $in.^3$/min) from steel for an average gap voltage of 50 V?
4. Estimate the average time to sink a cavity with a hexagonal cross section and 10 mm (.4 in.) depth on a 25-A capacity EDM machine. Each side of the hexagon is 15 mm (≈.6 in.).
5. A 25-mm (.9843-in.) square blind hole is to be sunk by ECM in a block of tungsten. The potential across the gap is 15 V. Efficiency is 65%. Feed of 10 mm/min (.394 in./min) is adequate. How much power must be supplied? What is the minimum rating of the machine?
6. An electrochemical machine is to be designed for cavities with cross-sectional areas up to 260 cm^3 (15.86 $in.^3$). Electrolyte will be supplied at 200 L/min (≈50 gal/min) at 1.7 MPa (≈250 psi) maximum pressure. What axial tool force must the machine abe able to withstand?
7. Cemented carbide cutting tools used in a certain plant must be ground on their rake faces to resharpen them. This can be done on an electrochemical grinding machine in the time of 2 min for each tool. The machine and equipment cost $28,000. Depreciation for 5 yr, interest, and taxes amount to $6,000/yr or $30/day. The work also can be done in 4 min for each tool with a diamond wheel on a conventional grinder costing about $8,000, prorated at $10/day. The extra cost of the diamond wheel amounts to $0.04 for each tool ground.

Electricity costs \$0.02 per tool more for the electrochemical method. Labor and related overhead costs amount to \$30/hr for operation of either machine. Other costs are about the same for both machines. How many tools must be ground each day to justify using the electrochemical method?

8. For the situation described in Problem 7, which method should be selected if 12 tools are to be ground each day? If 60 tools are needed every day?
9. In a manufacturing plant it costs \$1.19/kg (\$0.54/lb) to remove metal by chemical milling based on a cost of \$0.79/kg (\$0.36/lb) of etchant. Any costs for masking must be added, including \$1.08/m^2 (\$0.10/ft^2) to apply the mask over the entire exposed surface, the prorated cost of the template, and \$0.13/linear m (\$0.04/linear ft) of cut for slitting and removing the mask. Machine milling costs \$4.41/kg (\$2.00/lb) of metal removed. Setup costs \$18 for each lot. For each piece, let W = weight in kg (lb) of metal to be removed, A = the total surface area in m^2 (ft^2), and L = the length in m (ft) of cuts to be made in the mask. Let T = the cost of the template to be used for all pieces. Set up and solve an expression for N, the number of parts in one lot for which the cost is the same by chemical or machine milling.
10. A part is to be milled under the conditions given in Problem 9. Initially, it is a slab of aluminum 16 mm (.625 in.) thick, 0.8 m (32 in.) wide, and 1.2 m (48 in.) long. Three pockets each 13 mm (.5 in.) deep × 250 mm (≈10 in.) × 760 mm (≈30 in.) are to be cut from one side. If only one part is to be made, how much can be spent on a mask for chemical milling? How much if five parts are to be made?
11. A mask for the part described in Problem 10 costs \$150. What process should be used if only two parts are to be made, both in one lot?

27.7 REFERENCES

Borman, R. 1993. "Speed in Wire EDM—Is It an Illusion or Just Elusive." *Modern Machine Shop,* October: 86–96.

Koelsch, J. R., ed. 1994. "Clearer Thought on EDM Fluids." *Manufacturing Engineering,* June: 59–61.

Matus, T. 1996. "Tech Tips for Plasma Cutting." *Metalforming,* December: 16.

Schreiber, R.R., ed. 1991. "Focus: Lasers." *Manufacturing Engineering,* February: 39.

__________. 1995. "Laser Assist for Machining." *Manufacturing Engineering,* November: 26.

Sebzda, J. 1996. "EDM Tooling—Back to Basics." *EDM Today,* May/June: 48.

__________. 1996. "Enormous EDM Leads to Large Injection Mold Success." *EDM Today,* May/June: 8.

__________. 1995. "EDM Helps Put Precision Diecast Parts at Your Fingertips." *EDM Today,* November/December: 16.

__________. 1994. "Incorporating Plasma System Eliminates Hard Tooling for Indiana Manufacturer." *Modern Machine Shop,* December: 124.

Thread and Gear Manufacturing

28

Precision gear cutting machine, 1855
—Joseph R. Brown

Threads and gears are important and sometimes critical mechanical elements of many manufactured products and the machine tooling. Threads, or screw threads as they are often called, are used in many forms: as fasteners; as a means of transmitting motion such as the lead screw of a lathe; as a means of transmitting force as in clamping; and also for precision measurement. Thus, the form and precision of threads vary in accordance with the functions they are expected to perform.

Gears are used primarily to transmit mechanical motion and/or power in a large variety of manufactured products and in the machines used to produce them. They can be used to convert rotary motion into linear motion or to transmit power between parallel, intersecting, or nonintersecting shafts. As with screw threads, gears vary in form and precision, depending on the requirements of the job.

This chapter discusses some of the most commonly used processes for manufacturing threads and gears.

28.1 SCREW THREADS AND SCREWS

28.1.1 Definitions

A *screw thread* is a ridge of uniform section that lies in a helical or spiral path on the outside or inside of a cylinder or cone. The groove between the ridges is called the *space*. A *straight thread* lies on a cylinder; a *tapered thread* lies on a cone. A thread on an outside surface is an *external thread*. A screw has an external thread. An *internal thread* is found in a nut.

A *right-hand thread* is one that turns clockwise as it moves away from the observer. A *left-hand thread* turns counterclockwise from the same position. A thread is understood to be right-hand unless designated otherwise.

28.1.2 Features

The chief features of an external thread are illustrated in Figure 28-1. Internal threads have corresponding features. They determine the size and shape of a thread.

The *pitch* is the distance parallel to the axis from any point on a screw thread to a corresponding point on the next ridge. The pitch is the reciprocal of the *number of threads* in a unit of length. Thus if a screw has a pitch of 2 mm, it has $\frac{1}{2}$ thread/mm. Another that has 8 threads/in. has a pitch of $\frac{1}{8}$ in.

The *lead* is the distance a screw advances axially in one full turn. The lead is the reciprocal of the *number of turns* required to advance the screw axially a unit of length. Thus a drive screw that requires 100 turns to move forward 1 m (39.37 in.) has a lead of 10 mm (.3937 in.).

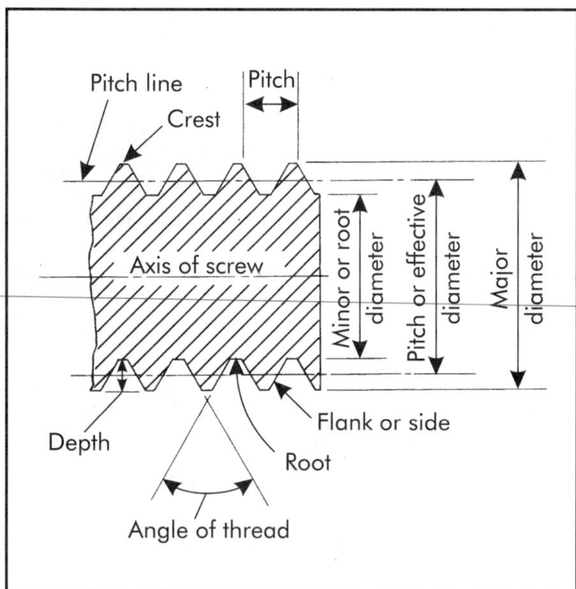

Figure 28-1. Features of an external screw thread.

A *single-thread screw* has only one continuous thread on its surface, as found on most commercial screws, bolts, and nuts. A *multiple-thread screw* has two or more adjacent threads; a *double-thread screw* has two threads; a *triple-thread screw* has three, etc. The lead of a single-thread screw is equal to its pitch; the lead of a double-thread screw is twice the pitch; and the lead of a triple-thread screw is three times the pitch, etc.

The *major diameter* of a straight thread is the diameter of a cylinder on which the crest of an external thread or the root of an internal thread lies. The *minor diameter* applies to the root of an external thread or the crest of an internal thread. The *pitch diameter* of a straight thread is the diameter of a cylinder that cuts the thread where the width of the thread is equal to the width of the space.

28.1.3 Forms

Screw threads are made with a number of cross sections; one list showed 108. Common ones are illustrated in Figure 28-2.

A *sharp V thread* (Figure 28-2[A]) has an angle of 60° and is brought to a point at its crest and root. The thread and space each have the same cross section of an equilateral triangle. This simple thread has little practical use.

Most screws produced in the United States to inch dimensions have the ISO and Unified National Screw Thread Form (Figure 28-3). The crest of the external thread is truncated by $\frac{1}{8}$, and the root by $\frac{1}{6}$, of the depth of a full V thread. Thus the crest has a width F_1 of $\frac{1}{8}$, and the root a width F_2 of $\frac{1}{6}$, of the pitch P. The basic depth of an external thread is $\frac{17}{24}$ of the depth of a V-thread, or depth $D = 0.6134P$. Tolerance is negative on both crest and root diameters. A flat or rounded root is specified for the Unified National (UN) series of threads. The rounded root is partly to allow for wear of cutting tools because they deteriorate faster on the corners. Also, a rounded root increases fatigue strength and resistance to shock loads.

The metric or ISO Thread Form (Basic M Thread) is like the UN Screw Thread Form but with dimensions in millimeters and other differences (Figure 28-2[B]). The crest is truncated by $\frac{1}{8}$, and the root by $\frac{1}{4}$ of a V-thread depth. Thus the crest has a width of F_1 of $\frac{1}{8}$, and the root a width of F_2 of $\frac{1}{4}$ of the pitch P. The basic depth of an external thread is $\frac{5}{8}$ of the depth of a V-thread, or $D = 0.5413P$. The root may be rounded as for UNC threads.

The American National Acme Thread (Figure 28-2[C]) has a 29° angle, which makes it easier to machine with uniform accuracy than a square thread. The basic Acme thread has a depth $D = P/2$, a crest width $F = .3707P$, and a width at the root C from .003–.005 in. less than F. Acme threads can carry heavy loads and are used typically for jackscrews and feed and operating screws for machine tools.

Threads developed on cones or tapers have specific applications, such as for pipe threads and wood screws. The American National Taper Pipe Thread has an angle of 60° between the flanks of the thread. The crests and roots are cut to follow a taper of $\frac{3}{4}$ in./ft.

28.1.4 Standards

At one time, screw threads lacked uniformity in size and shape. Each manufacturer produced screws according to its own system and standards. One product could not be interchanged with another. Standardization of sizes, shapes, and pitches was undertaken to create a condition of order.

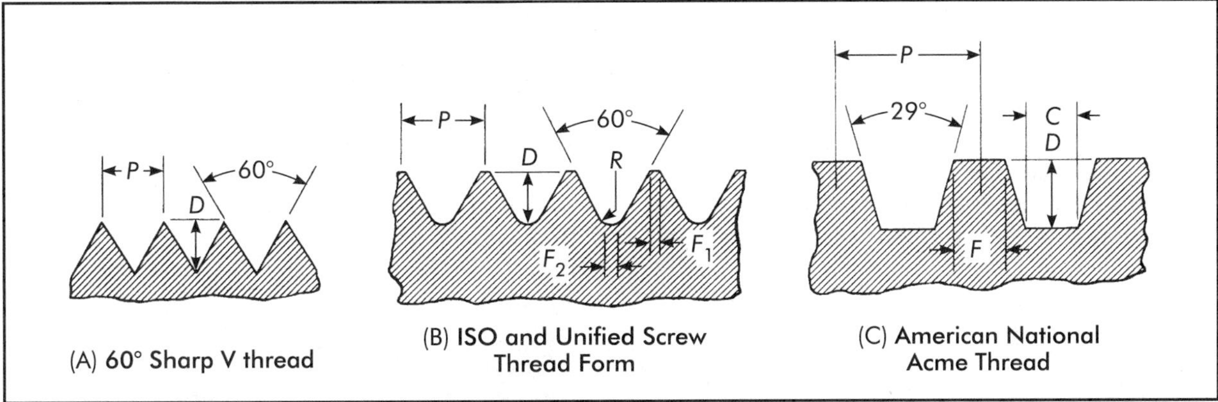

Figure 28-2. Cross sections of several forms of screw threads.

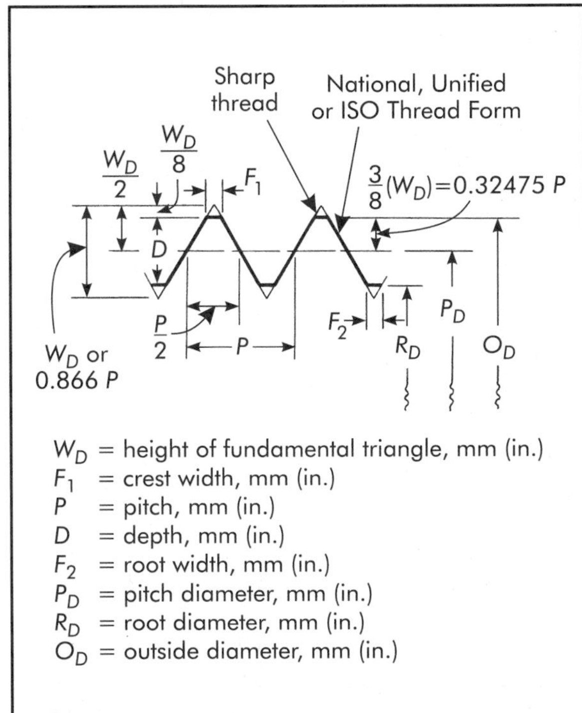

Figure 28-3. Basic profile of metric or ISO M Thread and Unified National Thread Forms.

A number of standards have prevailed over the years. An accord of the United States, Great Britain, and Canada in 1948 set up the Unified Screw Thread Form, which superseded previous standards for most screw threads produced in these countries. The purpose of this uniform thread system was to promote interchangeability of products, particularly war matèriel. The standard designates a coarse thread series by unified national coarse (UNC) and a fine thread series by unified national fine (UNF), and there is a less often used extra-fine thread series called unified national extra-fine (UNEF), among others. Each series specifies the threads per inch (tpi) and basic dimensions and tolerances for certain nominal inch sizes of diameters, like $\frac{1}{4}$, $\frac{5}{16}$, $\frac{3}{8}$, $\frac{7}{16}$, $\frac{1}{2}$, etc. As an example, the basic major diameter of a $\frac{3}{8}$-in. screw is .375 in., but the UNC series specifies 16 tpi and a basic pitch diameter of .3344 in., and the UNF series 24 tpi and a basic pitch diameter of .3479 in. for that size. For each nominal size, the coarse thread series has fewer threads per inch than the fine thread series. In each series, the number of threads per inch decreases as the size increases.

With the adoption of the SI (metric) system by all other countries and the commitment of the United States to convert to that system, metric threads are of increasing importance and are standardized by the American National Standards Institute (ANSI). Screw thread sizes are expressed in metric modules and are not just conversions of older inch standards. ANSI tabulates preferred outside diameters from 1.6–300 mm with standard pitches for each size. Metric threads are designated by the letter "M" followed by the nominal outside diameter in millimeters, then by the symbol "×" and the pitch, for example, "M 20 × 2." This signifies a thread with a 20-mm nominal diameter and a pitch of 2 mm. The standard tables specify only one pitch for each coarse thread diameter; so the diameter alone may suffice after M for a coarse thread. A specification like M 20 × 2 is required for a fine thread. Other symbols are added when needed to designate thread tolerance classes, thread fits, modifications, special features, etc.

Complete specifications for the dimensions, allowances, and tolerances of the various thread forms and sizes are given in standards bulletins and handbooks.

28.1.5 Classes

Screw threads are divided into *classes* to designate the fits between internal and external mating threads. For some applications, a nut may fit loosely on a screw; in other cases, they must fit together snugly. The different fits are obtained by assigning appropriate tolerances on the pitch, major, and minor diameters and allowances to the threads for each class.

The Unified Screw Thread Form Standard recognizes several classes of threads. Those classified as "A" are for screws, "B" for nuts. Classes 1A and 1B are for a loose fit, where quick assembly and rapid production are important and shake or play is not objectionable. Classes 2A and 2B serve for most commercial screws, bolts, and nuts. Classes 3–5 provide for closer fits and for interference fits. Screws from one class may be used with nuts from another class for more fits.

The tolerance class of an external metric thread may be designated by "4g6g." Here, "4g" is the thread tolerance class for pitch diameter, and "6g" is the major diameter. The class of an internal thread might be shown by symbols like "6H." All classes are tabulated in the standards bulletins. A number of classes are available, but there are two preferred classes of fits for general purposes and a third for close fits. The tolerance class may be shown in the thread specification following a dash after the pitch, for example, "M 20 × 2-4g6g."

28.1.6 Measurement

The size of a screw or bolt is designated by its outside diameter. An M 10 ISO metric screw should fit a nut of the same nominal diameter, and a $\frac{1}{2}$-in. UNC screw should fit a nut with the same designation. However, other dimensions also must be measured to assure that the size and form of a screw thread are correct. The major and minor diameters of screw threads are dimensioned to clear the corresponding surfaces of mating threads. Threads make actual contact with each other on their flanks. Measurements must be made in the space against the flanks of a thread to find its true and effective size. The dimensions that need to be measured or gaged directly, indirectly, or compositely, are the outside and root diameters, pitch diameter, thread angle, and pitch or lead.

The outside diameter and pitch specify a thread's size. The other dimensions are calculated from these. Ways of computing the thread depth for the common forms of threads have been explained. The root diameter for a screw is equal to the outside diameter minus twice the thread depth.

The outside diameter of ISO metric and UNC threads is specified over the thread crests and truncated to a width equal to $\frac{1}{8}$ of the pitch. Accordingly,

$$P_D = O_D - 0.6495P \qquad (28\text{-}1)$$

where

P_D = basic pitch diameter, mm (in.)
O_D = outside diameter, mm (in.)
P = pitch, mm (in.)

The basis for this formula is shown in Figure 28-3. As for actual dimensions, the O_D and P_D are given empirical allowances for clearance with mating threads and tolerances for manufacturing. These are specified in the tables of standards bulletins for screw threads.

If the pitch cylinder of a thread were slit along an element parallel to the axis and unrolled into a plane, it would be a strip having a width equal to $\pi \times P_D$. The pitch lines of the thread would lie at an angle across the strip. The angle between one of these lines and normal to the sides of the strip is the lead angle of the thread, as shown in Figure 28-4. This angle is

$$B = \tan^{-1}[L/(\pi \times P_D)] \qquad (28\text{-}2)$$

where

B = lead angle, degrees
L = lead, mm (in.)
P_D = pitch diameter, mm (in.)

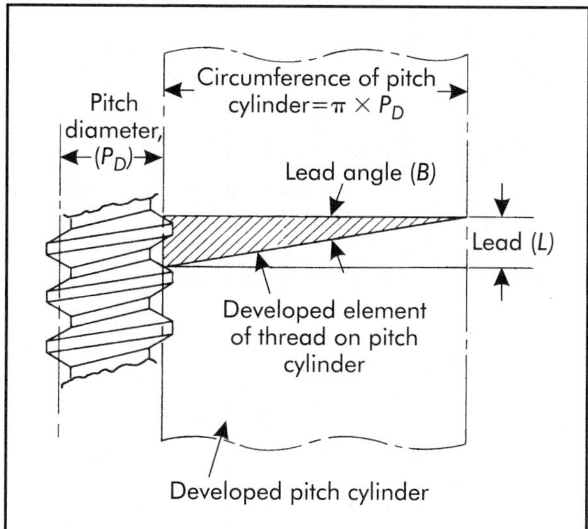

Figure 28-4. Basis for calculation of the lead angle of a screw thread.

The basic dimensions of a $\frac{3}{8}$-16 UNC external thread is calculated as an example.

$$D = .6134 \times \frac{1}{16} = .0383$$

$$R_D = .375 - 2 \times .0383 = .2984$$

$$P_D = .375 - (.6495/16) = .3344$$

So,

$$B = \tan^{-1}[1/(16 \times \pi \times .3344)] = 3°24'$$

where

D = thread depth, in.
R_D = root diameter, in.
P_D = pitch diameter, in.

Screw-thread Micrometer Caliper

A *screw-thread micrometer caliper* is like the standard micrometer caliper described in Chapter 16 but has a spindle with a conical point and an anvil with two V-shaped ridges. It makes contact on the sides of a screw thread and measures the pitch diameter directly. Any one anvil is limited to a small range of pitches. Even with the proper range, the readings are slightly distorted unless the micrometer is set to a standard thread plug and used to measure threads of the same diameter and pitch as the plug.

Measuring with Three Wires

The three-wire method of measuring pitch diameters is more accurate but slower than the use of a screw-thread micrometer caliper. The arrangement of the wires is shown in Figure 28-5. Three wires of any one diameter that would fit within the space might be used, but the preferred diameter or *best wire* (W) for each pitch makes contact on the flanks of the thread at the pitch diameter. A best wire has a diameter equal to $\frac{2}{3}$ of the depth of a full V-thread. That depth is equal to $.866P$. Thus $W = \frac{2}{3} \times .866P = .57733P$.

When a wire of diameter W touches a 60° thread, its center lies at a distance $W/4$ outside the points of contact as indicated in Figure 28-5. Therefore, for a best wire size, the dimension over the wires is $A = P_D + (\frac{3}{2})W$. The effect of the lead angle on the measurement usually is negligible for standard 60° single-thread screws. Sizes of best wires and measurements over wires are given in published tables.

Effect of Pitch, Lead, and Form Errors

The standards do not specify tolerances for thread pitch, lead, or angle. Some error can be expected in these elements, and it detracts from the pitch diameter tolerance, which is specified. For instance, an error of 25 μm (.001 in.) axially in the lead between any two threads in the length of engagement of a screw adds 43 μm (.0017 in.) to the effective pitch diameter in mating with

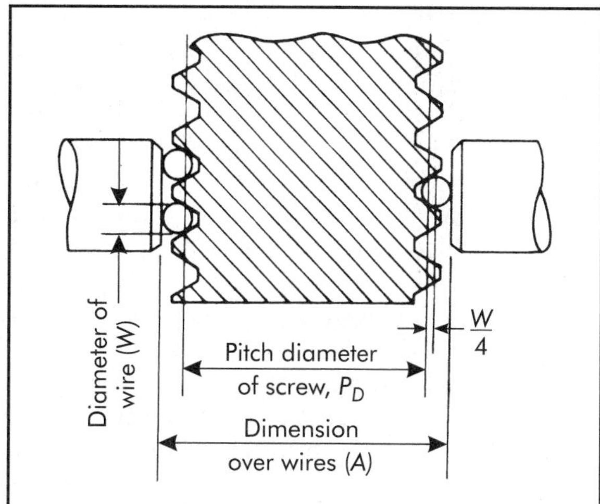

Figure 28-5. Measuring a screw with three wires.

another thread. Thus measuring or gaging the pitch diameter alone across one or two threads is not enough to verify the functional worth of a screw.

The lead of a thread may be checked separately from the pitch diameter. This is done to check precision gages and thread rolls. For production purposes, it is more meaningful to gage the high limit of the pitch diameter of a screw over the normal length of engagement at one time because this indicates how well the screw will fit in a nut. On the other hand, the low limit is gaged over one or a few threads only at one time to minimize the effect that lead and form errors have in making the pitch diameter seem large.

A typical device checks thread lead by comparing the thread on the workpiece with a precise master and showing any difference between the two on a dial indicator.

A set of *screw thread pitch gages* is shown in Figure 16-2 Row 9. Such a gage is held up against a screw to check its pitch approximately.

Optical or projection comparators like the one in Figure 16-21 are widely used for checking thread form and lead.

Thread Gages

Typical plug, ring, and snap gages for threads are described in Chapter 16 in conjunction with Figure 16-13. Such gages are made closely to theoretical size and forms of threads and check how screws and nuts will fit in service. For the most part, each of the thread gages shown in Figure 16-13 are limited to the gaging of a particular type and size of thread. Thus, the fixed-limit progressive-thread snap gage shown in Figure 16-13(D) can only be used to check one particular type and size of thread. Additional gages would need to be purchased if other types and sizes of threads were to be accommodated. For small-lot production of a variety of threads, the thread measuring or gaging system shown in Figure 28-6 would probably be more cost effective.

The thread gaging system shown in Figure 28-6 consists of a gaging unit, signal converter, and printer. The gaging unit has three hardened and ground steel thread rolls that are brought into contact with the threaded part by means of an actuating lever shown at the front.

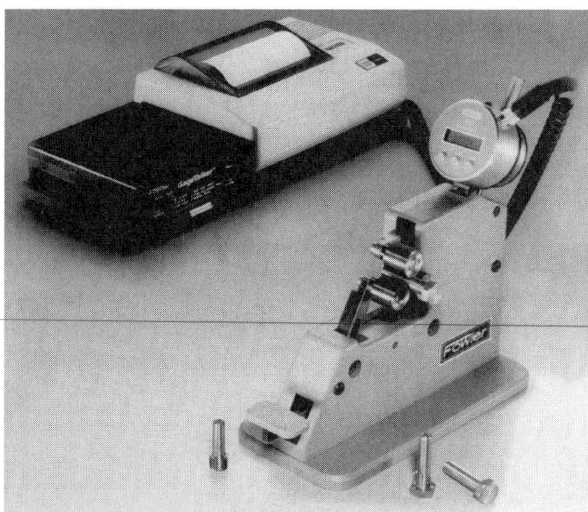

Figure 28-6. Thread measuring or gaging unit with interchangeable thread rolls. (Courtesy Fred V. Fowler Company, Inc.)

The three rolls are removable and may be replaced by a different set of rolls to accommodate different types and sizes of threaded parts. The unit provides measurement of pitch diameter, ease of assembly, and other important thread features. Measurements may be viewed on either a dial or electronic indicator. In addition, digital probes may be used to obtain remote readouts. An electronic indicator is shown on the gage unit in Figure 28-6. A gaging unit and electronic indicator similar to the one shown costs about $1,300. Additional thread roll sets cost around $300.

28.1.7 Making Screw Threads

Screws may be cut or formed. They may be cut with single-point tools on a lathe or with multiple-tooth cutters that include dies, taps, and milling cutters on various types of machines.

Threads are formed on screws and bolts by rolling or pressing on thread rolling machines. Internal and external threads may be rolled in thin sheet metal and tubing, like those on a metal lid for a glass jar or the base of an electric light bulb. They may be ground, as described in Chapter 24, or hobbed like gears, as indicated in the later part of this chapter. Parts with threads are also produced by die casting and plastic molding.

Thread Cutting on a Lathe

Thread cutting on a lathe is called *chasing*. This process is shown in Figure 28-7 and illustrated in Figures 18-1 and 18-3. The single-point tool is ground with its cutting edge of the same shape as the thread space with enough relief to clear the helical sides. Practically every engine and toolroom lathe has an accurate lead screw that drives the carriage, and the tool mounted on it, to chase the thread as the workpiece turns. The number of threads per mm (in.) cut on a lathe depends upon the relative rates of rotation of the work spindle and lead screw, as well as the lead of the lead screw. The necessary rate of rotation of the lead screw for any job is obtained by driving through a selective gear train from the spindle as described in Chapter 18.

A lathe with a sliding gear transmission in the lead screw drive has a plate that tells how to shift the levers to cut any of the threads it can produce. The engineer who designs the lathe must select the gear ratios for each lever position. Because of space limitations and the need to provide capability for cutting many different pitches, this is a problem in machine design and deserves attention. The same problem must be solved on those lathes, usually older ones, that have pick-off change gears between the spindle and lead screw. With *simple gearing*, as shown in Figure 28-8, the rate of rotation of the lead screw depends only upon the stud gear, driven at spindle speed, and the lead screw gear. The idler gear may be any convenient size to connect the others together.

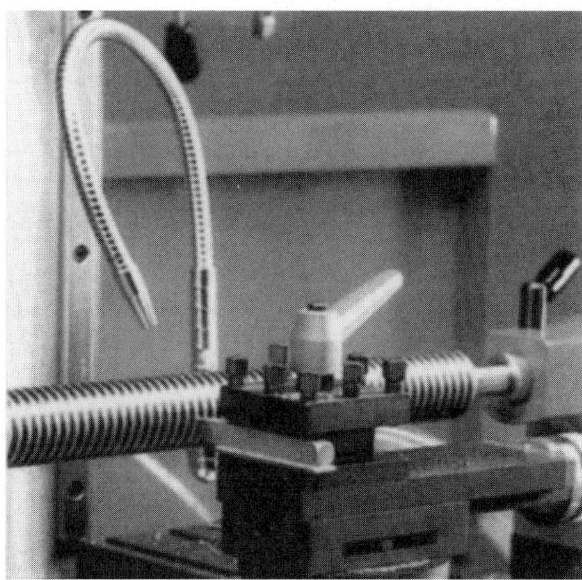

Figure 28-7. Cutting a screw thread on a lathe with a single-point formed tool. (Courtesy EMCO Maier Corporation)

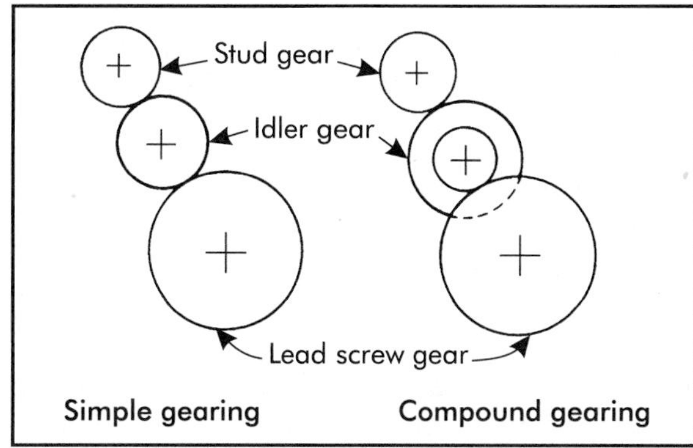

Figure 28-8. Change gears for driving the lead screw on a lathe.

The proper ratio for simple gearing is found from

$$\frac{S_t}{G_t} = \frac{P_t}{L_p} = \frac{L_t}{L_s} \qquad (28\text{-}3)$$

where

S_t = number of teeth on stud gear
G_t = number of teeth on lead screw gear
P_t = pitch of thread to be cut, mm (in.)
L_p = pitch of lead screw, mm (in.)
L_t = lead of thread to be cut, mm (in.)
L_s = lead of lead screw, mm (in.)

The last ratio applies if either thread is a multiple thread. As an example, a thread is to be chased with a pitch of 2.5 mm, and the lead screw has a pitch of 6 mm. Both are single threads. Then

$$\frac{S_t}{G_t} = \frac{2.5}{6} = \frac{10T}{24T} = \frac{15T}{36T} = \frac{20T}{48T}, \text{ etc.}$$

Sometimes simple gear ratios cannot be found to cut a desired thread. Then it is necessary to resort to *compound gearing*. One such arrangement is indicated in Figure 28-8. Two idler or intermediate gears revolve together on one shaft. One is driven by the stud gear, and the other drives the lead screw gear. The product of the ratios of the gear sets must equal the ratio between the pitches or leads of the workpiece and lead screw.

A series of light cuts is taken to chase a thread accurately. The depth of the first cut or two may be as much as 127 μm (.005 in.), but after that the tool is fed to only 51–76 μm (.002–.003 in.) for each cut, and finally 25 μm (.001 in.) or less per pass for the last few finishing cuts. If the cuts are too heavy, the tool may be damaged, the workpiece distorted, or the threads torn. At the end of each pass the tool must be withdrawn and the feed stopped. The tool is then taken back to the beginning and set to depth, and the feed is engaged at the instant that starts the tool on the right path for the next pass. Most lathes have a *threading dial* on the front of the carriage. It is turned by a worm gear meshed with the lead screw and shows the operator when to engage the half nut to start the tool in the right path.

Thread chasing is done at one-third to one-half of the surface speed of turning. This is to conserve the tool and give the operator time to do all that is necessary. The steps to chase a thread have been described in some detail to explain why thread cutting on a lathe is slow, requires skill, and is expensive. Its merit is that it is versatile and calls for little special equipment. External and internal, right- and left-hand, straight and tapered, and practically all sizes and pitches of threads can be chased on screw-cutting engine lathes with regular equipment. Very accurate threads can be produced on a good machine. Other methods are faster but usually require special equipment that is not justified unless a moderate or large quantity of threads of one kind is needed.

As was mentioned, the feed must be stopped and the tool withdrawn at the end of the cut. This is most easily done if a *relief groove* is first cut, as indicated in Figure 28-9. A designer should always specify a groove, if at all possible, at the end of a thread, whether it is to be made on a lathe or in any other way, to provide relief for the tools. It is common practice to make the groove as wide as the pitch of the thread, but it may be wider. If a tool having several teeth is used, such as a die, the groove should be wide enough to take the lead teeth and the first full tooth.

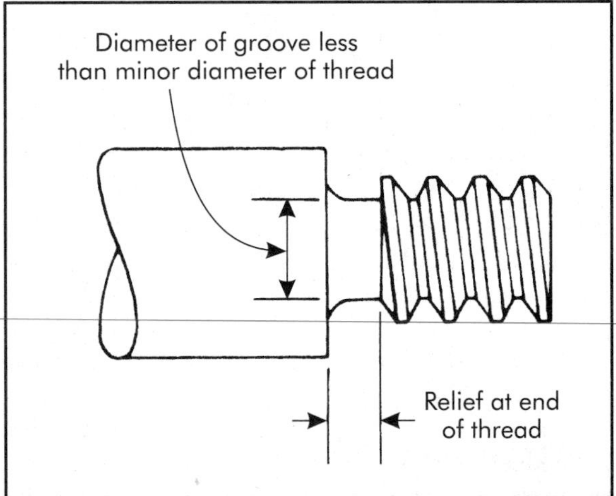

Figure 28-9. Relief needed at the end of a thread.

Thread-cutting Dies

A *threading die* has an internal thread like a nut, but lengthwise grooves in the center hole expose the cutting edges of the thread. The first few threads are on a taper so that the die can be started on a circular workpiece. Dies are made of hardened carbon tool steel or high-speed steel. A set of high-speed steel, solid adjustable, round dies are shown in the top and lower left of the tap and die case in Figure 28-10. Each of the dies have a small slot extending from the outer body diameter into the thread diameter. This permits the die to be adjusted a small amount by means of a set screw to correct for wear. The 21 dies in the case will produce coarse- or fine-pitch threads ranging from $\frac{1}{4}$ in. × 20 threads to 1 in. × 14 threads. Dies of this type are held in a stock, and are usually manipulated manually to cut a thread. The high-speed steel tap and die set shown in Figure 28-10 has three different sizes of stocks to accommodate the different die diameters. It also includes 21 taps and two straight-tap wrenches. The set costs around $1,270.

Spring-adjustable dies and die heads are used on machines, mostly for production. A *spring-adjustable, screw-threading die* resembles and can be sprung like a collet to cut to a desired size.

A *die head* has a body in which four or more serrated blades or chasers are mounted. The

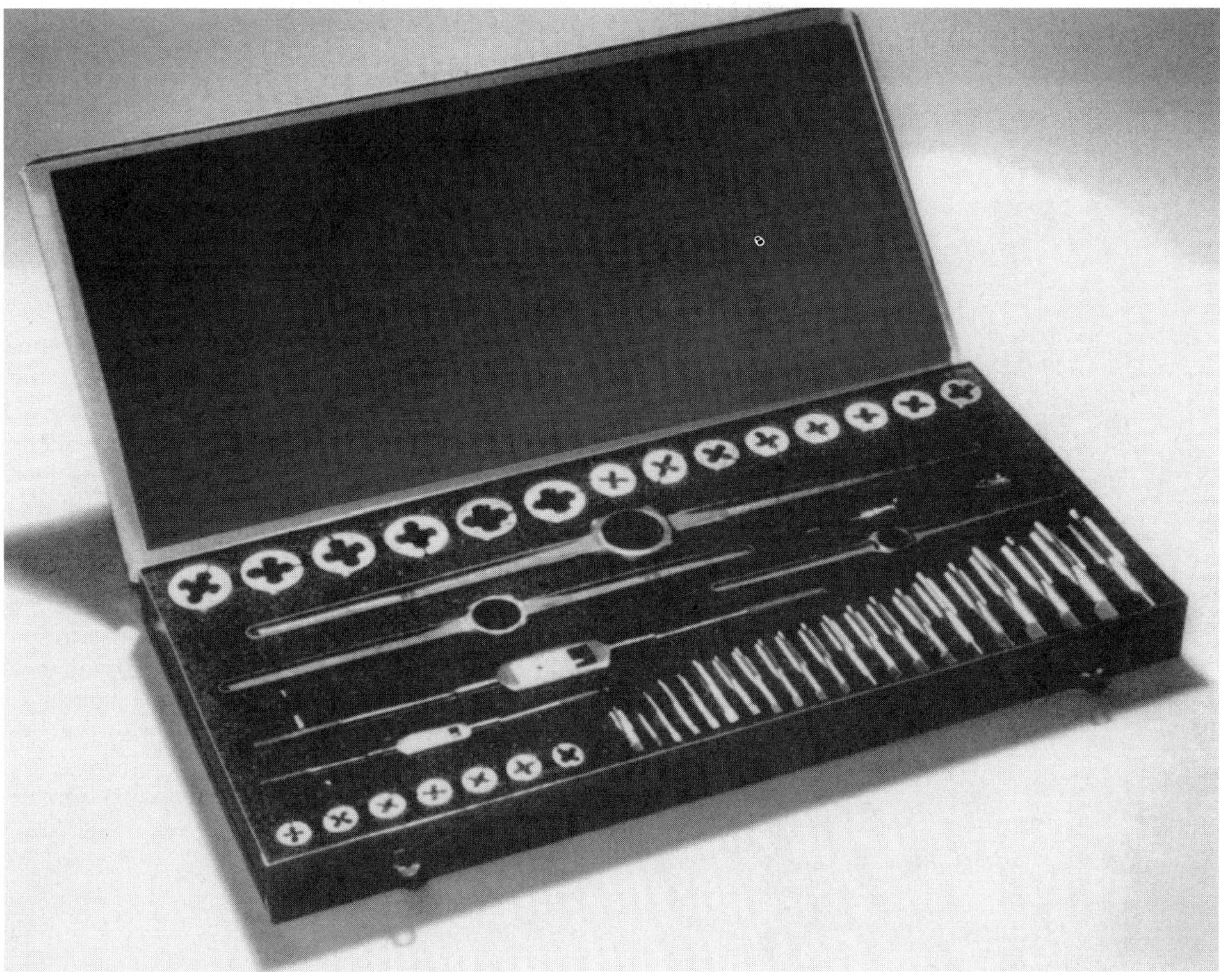

Figure 28-10. High-speed steel tap and die set with die stocks and tap wrenches. (Courtesy Greenfield Industries, Inc.)

blades may be carbon tool steel, high-speed steel, or cemented carbide. They are removed when dull, reground, replaced, and adjusted to cut to a desired size. The chasers are mounted radially in some heads but positioned tangentially to the workpiece in others.

A *self-opening die head* like the one in Figure 28-11 is arranged so that it is tripped, and its chasers snap outward when a thread has been cut to a predetermined length. The workpiece does not then have to be screwed out of the die head. The chasers are returned to cutting position by pulling the handle on the head before another piece is cut. A self-opening die head similar to the one shown in Figure 28-11 for screws .25–1 in. in diameter costs about $830. High-speed steel projection-type chaser sets like the ones shown cost around $90. Chasers are also available for left-hand threads.

Threading Machines

Dies and die heads are commonly used on lathes, turret lathes, automatic lathes and turning machines for cutting threads faster than they can be chased with single-point tools.

Threading machines similar to the one shown in Figure 28-12 are often used for production threading of items such as bolts, studs, automotive parts, and pipes. On this particular machine, the bodies of 1-in. diameter bushings are being die threaded, four at a time, by revolving die heads at two stations. The indexing table has a loading station shown in front and a part ejection station on the right side. The revolving

Figure 28-11. Self-opening die head with extra projection-type chasers. (Courtesy Greenfield Industries, Inc.)

Figure 28-12. Four-station threading machine with four die-head threading spindles, an indexing table, and parts nests. (Courtesy TCE Corporation)

threading dies are fed down over the bushing bodies by lead-screw driven spindles and then the spindles are reversed for withdrawal after threading is complete. A machine of this type costs about $210,000 and can produce threaded bushings at a rate of about 1,800 pieces/hr.

Taps

Holes are usually threaded by taps. A *tap* has a shank and a round body with several radially placed chasers. Taps are made in many sizes and shapes to satisfy a number of purposes. They may be operated by hand or machine. Taps are made to cut all forms of threads. Small taps are solid; large taps may be solid or adjustable. A tap has two or more flutes that may be straight, helical or spiral, or spiral pointed. Those used in production tapping operations may be made of high-speed steel or carbide materials, depending on the severity of the operation. High-speed steel and carbide taps may be coated with abrasion-resistant materials to permit higher cutting speeds and improved performance. For example, a diamond-coated solid carbide tap is available for tapping highly abrasive composite materials.

A tap works under very strenuous conditions because it is buried in metal, where chips are hard to remove, and is fed at an invariable rate. To operate successfully, taps must be supplied with a cutting fluid suitable for the work material. Some taps are constructed to permit the cutting fluid to be conveyed through the tap itself to improve the fluid flow into the cutting zone and assist in the chip removal process.

Hand taps have short shanks with square ends and are made in sets of three for each size. The three are the taper, plug, and bottoming taps, shown in Figure 28-13. As it cuts, a tap must be fed at a constant rate to suit the thread it is cutting. The only way to ease the load of a deep cut on the cutting teeth is to taper the tap so each tooth takes only part of the load.

A taper tap is easier to start in a hole. It will cut a full thread through a hole if advanced far enough, but not to the bottom of a blind hole. In the latter case, a taper tap must be followed by a plug tap and then by a bottoming tap. A high-speed steel tap with a ground thread for a $\frac{1}{2}$-13 UNC costs about $9.00, and a $1\frac{1}{2}$-in. diameter tap costs over $50.00.

Hand taps are driven by machines as well as by hand. In either event, the tap must be started straight and kept aligned with the hole for true threads. The smaller the tap, the more fragile it is. A tap breaks if it meets too much resistance and too much torque is applied. This is a major problem in some operations. Skill is needed to tap small holes by hand and avoid breakage. Torque-limiting tap drivers are commonly used on machines.

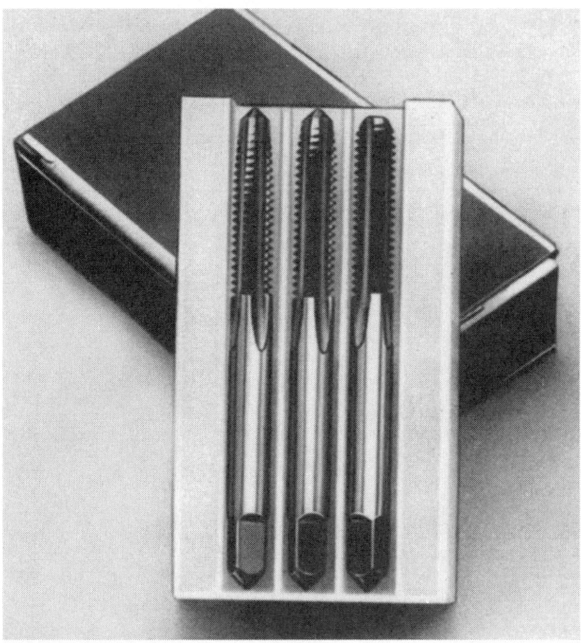

Figure 28-13. Straight-flute hand taps. From left to right: taper, plug, and bottoming taps. (Courtesy Greenfield Industries, Inc.)

Straight-flute taps are easiest to make and sharpen, but spiral or helical flutes help to push chips out of a hole.

Serial hand taps are made in sets of three. Two are undersize for roughing and are used first. A serial tap has one or more rings scribed around its shank to designate its place in the use sequence. Serial taps are used in tough metal, for threads like Acme threads that require a large amount of stock removal, and to provide a light finishing cut for a smooth finish.

Pulley taps and *nut taps* have long shanks to reach into inaccessible places. *Tapper taps* are used for tapping nuts in large quantities on specialized nut-tapping machines. They are made in a number of lengths and shapes for specific applications.

A *collapsible tap* has internal thread chasers in a body; they are withdrawn inward from the thread when it is completed. Then the tap can be pulled out of the hole, and the time to reverse the machine is saved. Some such taps are reset by a hand lever; others by a trip mechanism on the machine.

Pipe taps are tapered and are used to cut internal pipe threads. One style carries a short drill in front of the tap to clean out the hole to be tapped.

Design for Tapping

The larger a hole is drilled for tapping, the easier the thread is to cut. Less thread depth means fewer chips in the hole and less torque with less tap breakage. Theoretically, a full thread is only 5% stronger than a 75% thread and 20% stronger than a 50% thread. Common practice is to use, and the Unified Thread Standard is based on, a 75% thread, but in many cases even less is desirable. Experiments of M. L. Begeman and C. C. Chervenka showed that steel threads from about 6–20 mm ($\approx\frac{1}{4}$ to $\frac{3}{4}$ in.) diameter and not over 40% full depth were as strong as the bolt for an engagement length equal to the diameter. This indicates that in many cases a 50% thread, to allow for tool runout and standard tap drill sizes, is adequate.

Tapping Machines and Attachments

Tapping is done on lathes, turret lathes, automatic lathes, drill presses, and machining centers as a part of the process of machining a threaded workpiece. In addition, special tapping machines are available. These machines often look similar to a drill press and are equipped with tap holders, reversing mechanisms, lead screws, etc., to enhance their tapping ability. They may have one or more spindles.

Self-reversing tapping attachments like the one shown in Figure 28-14 may be inserted into the spindle of either horizontal or vertical computer numerical control (CNC) machining centers. These units are geared to provide increased speeds for the tapping of small holes. For example, the unit shown in Figure 28-14 is capable of tapping $\frac{1}{4}$ in. × 28 threaded holes in aluminum at 3,500 rpm. Other units are available with tapping speeds up to 6,000 rpm. A highly sensitive clutch arrangement within the body of the unit permits tap reversal and withdrawal for either through-hole or bottom tapping. This feature permits the maintenance of a constant tapping speed up to the instant of tap reversal. It also eliminates the need for machine spindle reversal and the resultant wear and tear on spindle drive components.

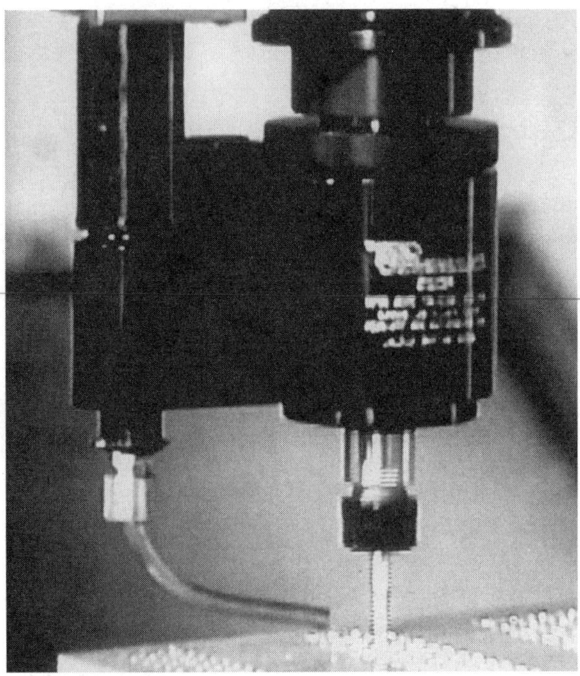

Figure 28-14. Self-reversing tapping attachment. (Courtesy Tapmatic Corporation)

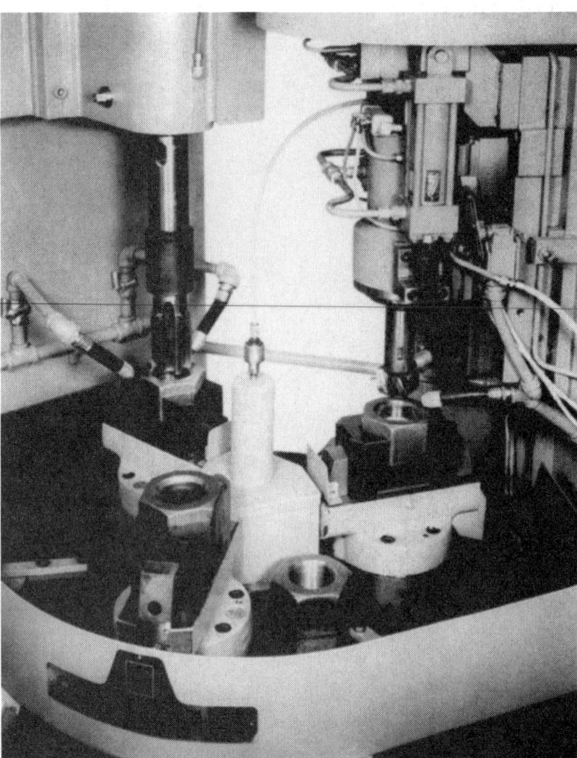

Figure 28-15. Nut-tapping machine with three-station indexing table and two work spindles. (Courtesy TCE Corporation)

A variety of machines are available for tapping operations, including multiple-spindle and gang-type machines. Some of these are general-purpose machines while others are designed for particular tapping operations. Figure 28-15 shows a two-spindle tapping machine tooled to ream and tap threads in 76-mm (3-in.) diameter hex nuts. The indexing table has three stations: (1) loading/unloading in front, (2) reaming for thread inside diameter (ID) dimension on the left, and (3) tapping on the right. The lead-screw driven spindle is reversed and withdrawn at double the tapping rotational rate after threading is complete. Hex nuts are loaded manually into a hydraulically actuated locating/clamping fixture, and a completed nut is unloaded each time the table indexes. A machine of this type and size costs about $185,000, including tooling. With the setup shown in Figure 28-15, about 40 hex nuts/hr can be threaded.

Thread Milling

A conventional thread milling machine is like a lathe with a headstock and tailstock mounted on a bed. A carriage slides between the two on ways and carries a cutter head. The work may be mounted between centers or chucked on the headstock spindle.

Multiple-thread form cutters are used for rapid thread cutting. The cutter does not have a thread or lead. Instead its teeth are arrayed on a series of closed circles. The axes of cutter and work are parallel. The revolving cutter is fed to depth in the work. It is longer than the thread. As the work revolves, the cutter is fed lengthwise to conform to the lead of the thread. The cutter is fed by a lead screw through change gears to the work spindle. At the end of 1.1 revolutions of the work, the cutter is withdrawn from the completed thread. An internal thread may be milled in a similar manner. Milling with multiple-thread form cutters is often as fast as thread cutting with self-opening dies and collapsible taps, and produces more accurate threads and better finishes. It is mostly confined to V-type threads (not necessarily sharp) because too much error is introduced in other threads.

Coarse threads, like those on feed screws and lead screws, are milled with single cutters on a *universal thread miller*. The cutter is tilted to match the lead angle of the thread. As the work revolves, the cutter head and carriage are moved longitudinally on the machine by a lead screw or cam to produce the desired lead. The cutter thus traverses the entire thread, and the length of thread is limited only by the capacity of the machine. A large machine is capable of cutting threads from about 51–305 mm (2–12 in.) diameter and up to 3.66 m (12 ft) long in one setting. For accuracy and finish, thread milling is usually done in two or three passes, the same as other milling, but is still faster than single-point thread chasing for long threads.

Planetary millers do external and internal circular form and thread cutting. They are convenient for cumbersome pieces because the workpiece does not move. Instead, the cutter rotates about its axis and is also made to travel in a circular planetary orbit about the axis of the surface to be cut.

Thread Rolling

Thread rolling is a cold forging rather than a cutting process. It produces external threads by subjecting a blank to pressure between dies in the form of grooved blocks or threaded rolls. The work material is depressed to open the root and raised to fill the crest of the thread. Threads are rolled on machines built especially for this purpose, on lathes and drill presses, or in combination with other operations on turret lathes and automatic bar machines. The principles of operation of the various methods are illustrated in Figure 28-16.

Each type of thread rolling machine operates best in a certain range of sizes, but the ranges of different machines overlap. Where two or three types are acceptable for a certain size, one may be best for one kind of work, another for different work, etc. All types can be arranged with hoppers and automatic loading devices.

The *rotary-planetary die machine* has a continuous action and is fast for commercial screws of less than 13 mm ($\approx\frac{1}{2}$ in.) diameter. It can turn out 6-mm ($\approx\frac{1}{4}$ -in.) diameter screws at 400–1,200/min. The *reciprocating die machine* rolls threads up to 25 mm (≈1 in.) in diameter. Typical rates are 140 pieces/min for 8 mm ($\approx\frac{5}{16}$ in.) diameter and 40–60 pieces/min for 19 mm ($\frac{3}{4}$ in.) diameter threads. It excels for precision threads, wide roots (like those on wood screws), and gimlet point screws. The *cylindrical die machines* (two- or three-die types) are made in several sizes to handle diameters from about 3–150 mm (≈.125–6 in.). They are relatively easy to set up and are preferred for hard materials, tapered and coarse threads, and tube threading. Cylindrical machines can through-feed roll long threads; a $\frac{5}{8}$-11 UNC thread about 1 m (≈3 ft) long is rolled at the rate of 72/min. A basic cylindrical die machine for diameters from 3–50 mm (≈.125–2 in.) and powered by a 5.6 kW (7.5 hp) motor weighs about 1.8 Mg (4,000 lb) and costs about $105,000.

Thread rolling offers a number of advantages. It is capable of producing accurate, uniform, and smooth threads of all classes at high rates of production for all kinds of screws, bolts, and studs. Rolling speeds are equivalent to cutting speeds with high-speed steel. No material is cut away to form the spaces, and up to 27% of material may be saved unless the screw blank is cut from a large piece of stock anyway. The material is cold-worked; its grain pattern continuity is not cut away, and a smooth wearable surface is left on the thread flank. Rolling is superior to cutting for good surface finish. Wear life is lengthened; tensile strength may be increased 10%, and fatigue strength has been found raised as much as 5–10 times for hardened alloy bolts with rolled instead of cut threads. The dies roll rather than rub and retain their original sizes a long time, often for millions of pieces. Because the dies do not lose their true shape over a long time, the work is consistent, and inspection costs are low. Thread rolling is more convenient than die cutting for some setups on automatic bar machines, such as for cutting a thread behind a shoulder.

Thread rolling requires some care and has several limitations. A blank for rolling must have an initial diameter, with close tolerance, less than the pitch diameter of the thread. The rolling operation must be carefully controlled to avoid bending of the piece and seams, slivers, or other imperfections of the thread. A truncation must

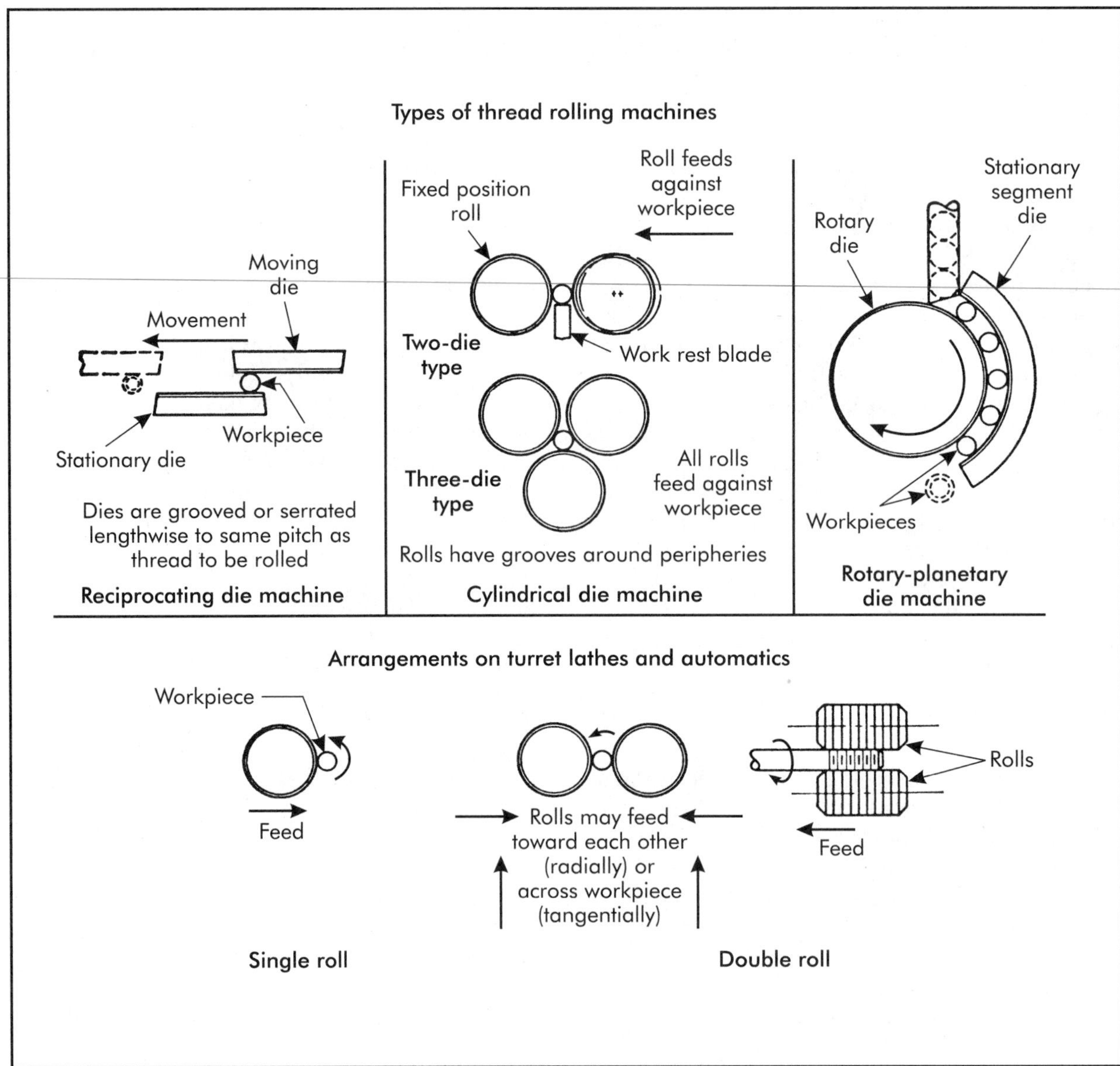

Figure 28-16. Methods of rolling threads.

be expected on the last fraction of a turn at the end of a rolled thread and thus a thread cannot be rolled fully up to a shoulder. Rolling effectiveness is reduced in materials with less than 12% elongation, but heating can help. Rolling ceases to be practical if elongation is less than 5% or hardness is over about 32 R_C. Even above these approximate limits, some materials are difficult to roll. Specifications on the rollability of common materials are given in reference books and handbooks.

Almost all external threads are rolled in large quantities because it is up to 50% faster and cheaper than other methods. One of the advantages is that rolling is amenable to continuous and automatic work handling. Material savings is a big factor. Threads are both rolled and cut in moderate quantities on standard semi-automatic machines, and the total cost is not much different for either method. Commercial, self-opening, thread-rolling heads and rolls for standard sizes of threads are somewhat higher in price than self-opening die heads and chasers. General-purpose threading in small quantities is seldom done by rolling because cutting is more adaptable and versatile for varieties of requirements and circumstances.

Formed Threads in Holes

A *forming* or *fluteless tap* has a tapered end but does not have grooves or flutes like an ordinary tap. It cold forms or swages threads in soft ductile material (even some steels) instead of cutting. The material is compacted and burnished. Grain fibers are distorted but not cut away. The thread is imperfectly formed as a rule but has about the same strength as a cut thread. No chips are made to clog the hole, and lead is easy to control. Torque is high, and a high-speed lubricant is better than a cutting fluid. Speeds up to twice those for cutting have been found practical. Holes must be tap-drilled larger than for conventional tapping. There is a tendency to raise burrs at the end of holes, so countersinking is desirable.

28.2 GEARS

A gear is a machine element that transmits motion in a positive manner through teeth around its periphery. The *spur gear* is the simplest form with teeth parallel to its axis as shown in Figure 28-17. A *rack* is a gear with an infinite radius; it moves in a straight line.

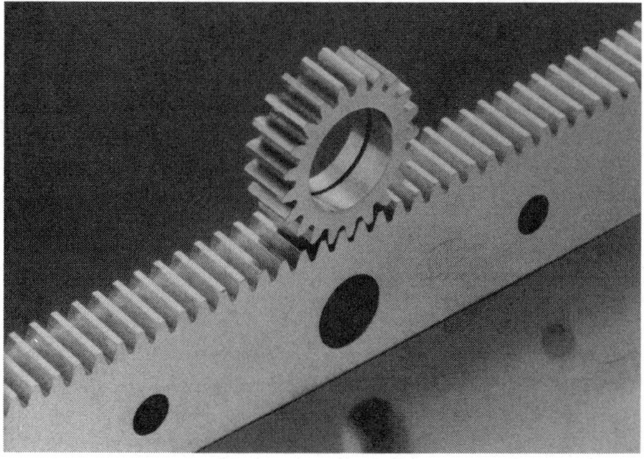

Figure 28-17. Spur gear and rack. (Courtesy Foote-Jones/Illinois Gear Company)

28.2.1 Gear Tooth Curves

The contact surfaces of gear teeth are curved to transmit motion uniformly as they roll and slide together. Most gear teeth have or approximate an involute form because it is simple, easy to reproduce, and allows variations in center distances of mating gears. A gear tooth with the involute curve modified at the extremities is said to have a *composite* form.

An *involute curve* is generated by a point on a straight line rolling on a base circle. The line *AB* in Figure 28-18 is tangent to the base circle, and point *A* traces the involute profile of one side of the gear tooth as the line is rolled on the circle. In the position shown, the angle between the line of action *AB* and the tangent to the pitch circle is called the *pressure angle*. A *pitch circle* is an imaginary ring of the same diameter as a smooth disk that would transmit the same relative motion by friction as the gear does when meshed with another gear. The radius or fillet adds strength to the root of the tooth.

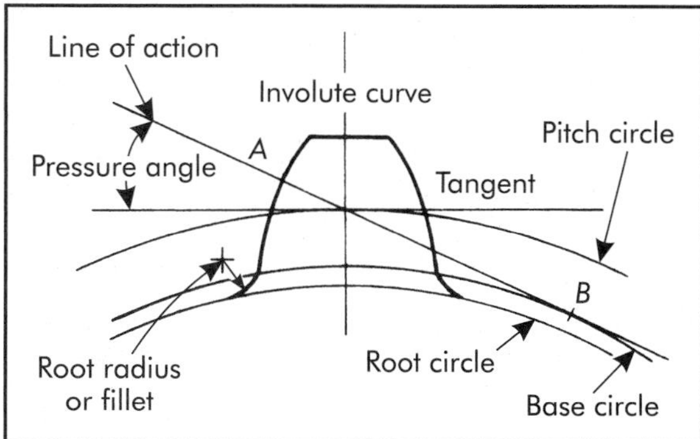

Figure 28-18. Involute gear tooth.

28.2.2 Elements of Gear Teeth

Gear teeth are standardized so they mesh together. Two lengths of gear teeth are common with inch dimensions. One is called the *full-depth tooth* and is longer than comparable sizes of the other, known as the *stub tooth*. Stub teeth are stronger but do not overlap as much as full-depth teeth. Only one length of tooth is common for standard gears with millimeter (metric module) dimensions.

The important dimensions of gear teeth are called elements and are designated in Figure

28-19. In the inch dimension system, the elements are related to a factor called *diametral pitch,*

$$P = N/D \tag{28-4}$$

where

P = diametral pitch
N = number of teeth
D = pitch diameter, (in.)

Any two standard gears of the same diametral pitch and tooth shape will mesh if mounted with the proper distance between their centers. In the millimeter system, the elements are related to a factor called the *module,*

$$m = D/N \tag{28-5}$$

where

m = module
D = pitch diameter, mm
N = number of teeth

Again, gears of the same module and shape mesh together. Care must be taken not to confuse the symbol m with m for meter. P (in.) equals 25.4/m (mm). The module is commonly designated MOD, and the diametral pitch, DP.

Formulas for the main elements of spur gear teeth are given in Table 28-1. They show that the circular pitch and tooth thickness at the pitch line are the same for all gears of the same diametral pitch or the same module. Tooth size (length and thickness) decreases as diametral pitch increases and as module decreases. Helical and bevel gears have much the same basic elements as spur gears plus other elements given in reference books and handbooks.

Although gears in the inch diametral pitch and millimeter module systems have comparable elements, each system has adopted preferred standard values to suit whole number or common fraction tooth and center distance dimensions. Thus standard gears of the two systems are not interchangeable. Generally, tools for one system are not applicable to the other. A set of gears must be designed in one system or the other. Conventional design formulas that determine the diametral pitch or module, number of teeth, and teeth width, as a basis for calculating the elements, may be used for any system with proper conversions. These are discussed in detail in gear design treatises and textbooks.

28.2.3 Types of Gears

Spur gears are the easiest and cheapest to make. They must be mounted on parallel shafts. A *helical gear,* like the one in Figure 28-20, has teeth along helices on a cylinder. The angle between a helix and an element of the pitch cylinder is called the *helix angle.* Helical gears are more expensive than spur gears but are stronger and quieter because the teeth engage gradually and more teeth are in mesh at the same time. They may be mounted on parallel shafts or on nonparallel and nonintersecting shafts.

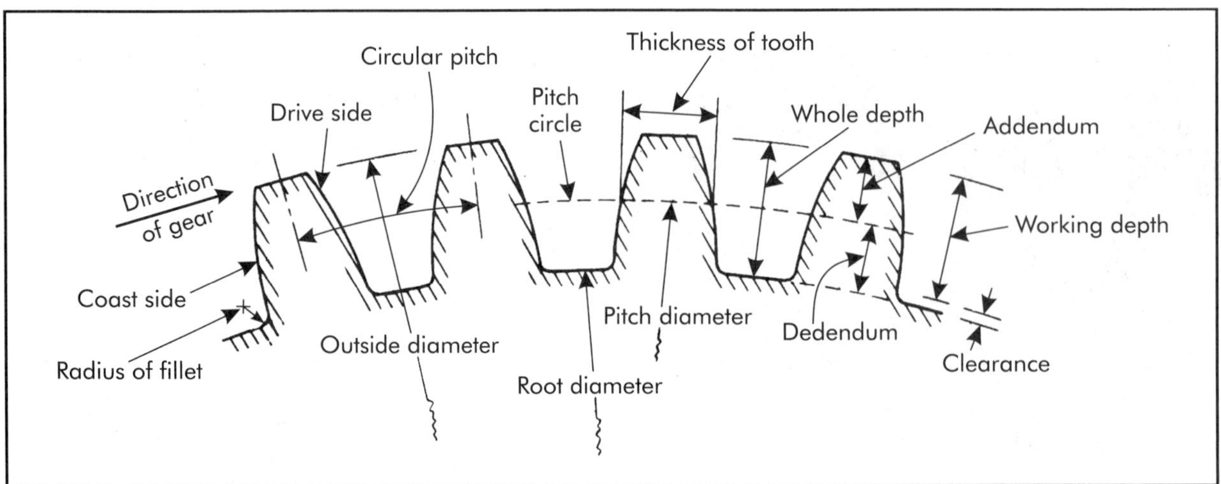

Figure 28-19. Elements of gear teeth.

Table 28-1. Formulas for some major dimensions of spur gears

Element Name	Symbol	Millimeters with 20° Pressure Angle	Inches: Full Depth Teeth with 14½° Pressure Angle	Inches: Stub Teeth with 20° Pressure Angle
Addendum	a	$a = m$	$a = \frac{1}{P}$	$a = \frac{0.8}{P}$
Circular pitch	p	$p = \pi m = \pi \frac{D}{N}$	$p = \frac{\pi}{P} = \frac{\pi D}{N}$	$p = \frac{\pi}{P} = \frac{\pi D}{N}$
Clearance	c	$c = 0.25m$	$c = \frac{.157}{P}$	$c = \frac{0.2}{P}$
Dedendum	b	$b = 1.25m$	$b = a + c = \frac{1.157}{P}$	$b = a + c = \frac{1}{P}$
Number of teeth	N	$N = \frac{D}{m} = \frac{\pi D}{P}$	$N = P \times D = \frac{\pi D}{P}$	$N = P \times D = \frac{\pi D}{P}$
Outside diameter	D_o	$D_o = D + 2m = m(N + 2)$	$D_o = \frac{N+2}{P} = D + 2a$	$D_o = \frac{N+1.6}{P} = D + 2a$
Pitch diameter	D	$D = mN$	$D = \frac{N}{P} = \frac{N \times p}{\pi}$	$D = \frac{N}{P} = \frac{N \times p}{\pi}$
Tooth thickness	t	$t = \frac{\pi}{2} m$	$t = \frac{\pi}{2P} = \frac{1.5708}{P}$	$t = \frac{\pi}{2P} = \frac{1.5708}{P}$
Whole depth	b_t	$b_t = 2.25m$	$b_t = a + b = \frac{2.157}{P}$	$b_t = a + b = \frac{1.8}{P}$

Note: *P* is the symbol for diametral pitch, and *m* for module, in this table.

A helical gear has a decided side thrust that is neutralized in a *herringbone gear*, which has teeth on right- and left-hand helices.

Double-helical or herringbone gears are widely produced by the honing process, normally in large sizes, such as those used for ship propulsion. The two halves of the gear are identical, except opposite in hand. Right-hand and left-hand hobs are used for cutting their respective right-and left-hand helices of the blank. Hob tolerance between the two halves is provided by a narrow gap as shown in Figure 28-21. Overtravel is required to insure complete revolute generation.

A *worm* is like a screw and may have one or more threads, each a tooth. A high ratio can be obtained by engaging a worm with a *worm gear*, as shown in Figure 28-22. Their axes are nonintersecting and usually at right angles. A worm gear with a small helix angle cannot drive the worm, which is an advantage for some applications.

Bevel gears operate on axes that intersect at any required angle but most commonly at right angles. A bevel gear is conical in form. A *straight bevel gear* has straight teeth as shown in Figure 28-23(A). If all the lines along its teeth were extended, they would pass through a common point

Figure 28-20. Helical gear and pinion. (Courtesy Foote-Jones/Illinois Gear Company)

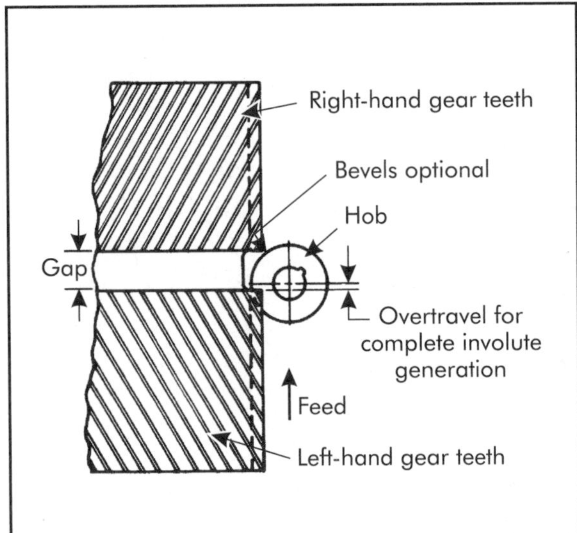

Figure 28-21. Gap requirements for continuous-tooth herringbone gear.

called the *apex*. This apex point coincides with the point of intersection of the axes of mating bevel gears. A pair of bevel gears with equal numbers of teeth and perpendicular axes are called *miter gears*.

A *crown gear* is similar to the straight bevel gear shown in Figure 28-23A, except the crown gear has a flat pitch surface rather than conical.

Figure 28-22. Worm and worm gear. (Courtesy Foote-Jones/Illinois Gear Company)

A *coniskoid gear* is a bevel gear with straight teeth inclined at a spiral angle for longer teeth and more overlap.

The teeth of a *spiral-bevel gear* are curved and oblique. One is shown in Figure 28-23(B). Spiral-bevel gears run smoothly and quietly and are strong because their teeth have what is known as spiral overlap. They are relatively easy to manufacture. A *zerol-bevel gear* has curved teeth, but they lie in the same general direction as straight teeth, as shown in Figure 28-23(C).

Hypoid gears resemble bevel gears but their axes do not intersect, as indicated in Figure 28-23(D). They are quiet and strong. A common application is for automobile rear axle drives.

The gears described so far are the important *external gears*. An *internal gear* is one with teeth inside a cylinder or cone. Internal gears are used for clutches, speed train reducers, and planetary gear trains.

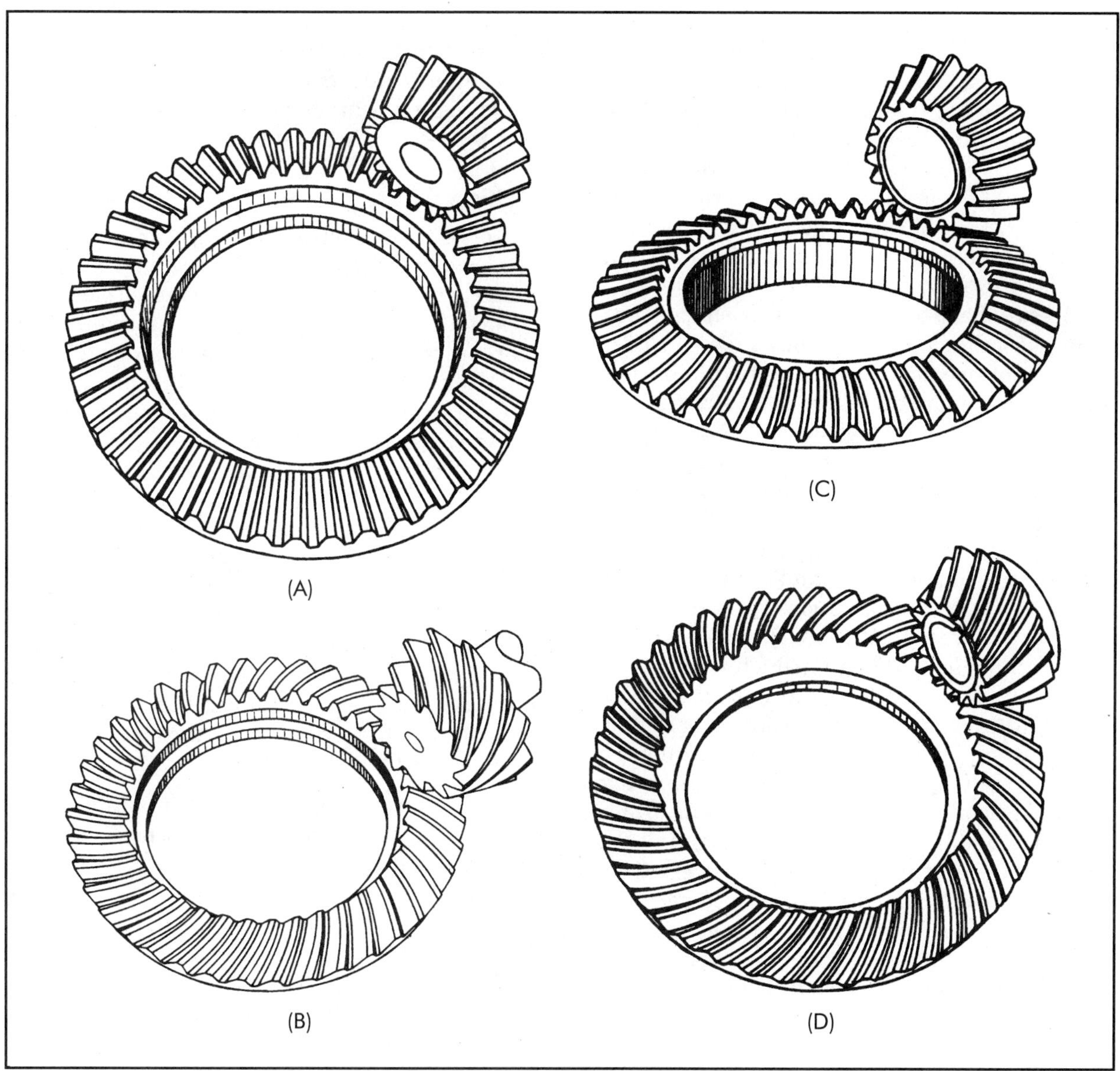

Figure 28-23. (A) Straight-bevel gears; (B) spiral-bevel gears; (C) zerol-bevel gears; and (D) hypoid-bevel gears.

28.2.4 Gear Manufacture

Molding Gears

Gears may be cast in sand or in permanent molds. Gray cast-iron gears are rough, inaccurate, and low in strength, but large sizes can be made at relatively low cost. Hypoid ring gears and pinions rough cast of nodular iron for ease of subsequent machining are found in some automobiles. Many small gears for light service are die cast of zinc, tin, aluminum, and copper alloys with a high degree of accuracy and finish. Gears are sometimes molded of plastic materials where quietness, insulating properties, and only moderate strength for light service are needed. Some gears are pressed and sintered from metallic powders.

Forming Gears

Gears are made by several kinds of hot- and cold-forming operations. Brass or aluminum gear stock is extruded and cut to desired lengths. Some gears are forged hot and finish coined when warm or cold. Even moderate-size steel gears are cold extruded, cold precision forged (Chapter 11), or formed by cold upsetting. One gearmaker reported the cost of a machined gear to be about $60, and that the same could be gear forged for about $20.

All types of gears and splines are rolled from soft steel in ways like those described for threads in this chapter. A form of rolling for finishing gears is called *burnishing*. The work is rolled as it is pressed between or under gear-shaped dies. Rolling is an important and fast process within the limits of size, tooth shape, and workable material for forming gears from solid stock or finishing teeth already rough cut. A leading maker does not recommend rolling gears larger than 102–127 mm (4–5 in.) in diameter, with coarser than about 2 MOD (12 DP) teeth, or with fewer than 16 teeth. It is not considered practicable to roll pressure angles less than 20° or teeth with undercuts. There are exceptions to the rules, and gear teeth coarser than 5 MOD (5 DP), and up to around 305 mm (12 in.) diameter have been reported finish rolled. The gear material should contain little or no sulfur or lead to avoid flaking and should not exceed 28 R_C in hardness.

Good results are obtainable from rolling. The material is work hardened and given a desirable grain flow in the teeth and around the roots, which enhances strength. If residual stresses are excessive, heat treatment may be necessary and cause distortion for which corrections must be made. However, where feasible, rolling produces accurate gears with surface finishes to 102 nm (4 μin.) R_a at rates up to 50 times and more than that of cutting. For example, involute splines on the ends of automobile axle shafts are cold rolled in soft steel at the rate of 6 sec each in quantities of 2.5 million pieces/yr. A major advantage of gear rolling is long die life; more than 1,000,000 pieces are common from one die in a finishing operation, with the last gear before failure identical to the first one checked.

Cold forming is a high-production process and involves risks. Process and tool development is largely an art and may be costly and time consuming. With one machine solely responsible for high production, a breakdown could be a catastrophe. Many gears (for example, starter and transmission pinions) are cold formed or rolled in the automobile industry, but more accurate gears, such as for aircraft, must be machined. One luxury automobile manufacturer rolled transmission gears for several years but gave it up for gear cutting to improve quality.

Cutting Gears

Large gears, such as the ring gear on the base of a revolving crane, are successfully made by flame cutting from thick steel plates. Tolerances reportedly held are ±127 μm (±.005 in.) with plate up to 102 mm (≈4 in.) thick and ±254 μm (±.010 in.) to 203 mm (≈8 in.) thickness.

Gears are cut from cast and forged blanks, barstock, sheet metal, laminated plastic, and molded shapes. Sometimes machining is the most economical method for small quantities. It also can be most economical for large quantities, such as for gears stamped from sheet metal for watches, clocks, toys, and appliances. Generally, gears are machined because it is the only way of getting the degree of accuracy or processing the hard material required by exacting mechanisms like internal combustion engines, aircraft, machine tools, etc. Gear-cutting methods may be divided into three classes as follows:

1. The forming method, which uses a cutter having the same form as the space between the teeth being cut. The cutter may be a single-point tool on a planer or shaper, a rotating cutter on a milling machine, or a broach.
2. The template method, in which a reciprocating cutting tool is guided by a master former or template on a machine called a gear planer. This method is slow and has been displaced for all but very large and coarse pitch gears.
3. The generating method, in which the cutting profile of the tool is like that of a mating gear or rack tooth. The cutter and work roll together as though in mesh to develop the tooth form.

Form Cutting

Spur, helical, worm, and straight bevel gears may be cut on milling machines with standard dividing heads and arbors. Gear cutters are the only tools needed that are not used for other kinds of operations, and their cost is low, from \$25–\$200 each, depending on size. A cutter will cut many gears, if needed. Setup is easy. At one time, most gears were form cut, but generating has proven to be a more efficient method of manufacturing gears. Milling machines are not used in modern gear manufacture except when one or a few gears are made at a time or when more efficient equipment is not available. This is because the cutting of gears on a milling machine is a relatively slow and inaccurate process.

When a gear is cut on a milling machine, it is mounted on a dividing head, usually (but not always) between centers, and the cutter is carried on an arbor. The table is swiveled as illustrated in Figure 21-8 to align the teeth of a helical gear with the cutter; but this is not done for a spur gear. The dividing head is geared to the table for a helical gear (not for a spur gear) so that it revolves as it passes the cutter. One tooth space is cut at a time. After each cut on any gear, the dividing head is manipulated in the usual manner to index the gear to the next space, and so on.

Bevel gears with theoretically correct tooth forms cannot be cut with rotary cutters on a milling machine. Occasionally, a bevel gear may be gashed out on a milling machine in an emergency, but then the teeth must be filed by hand to make them perform satisfactorily.

Gear tooth form cutters. Commercial form-relieved cutters for spur gears with a 14.5° pressure angle and composite form are available in sets. Each set covers one diametral pitch. Each cutter in a set covers a range of certain numbers of teeth and is only an approximation for all but one of the gears in the range. Those made particularly for heavy rough cuts are called *stocking cutters*.

Production form cutting. Gears sometimes are roughed out by form cutting and then finished by generating when manufactured in quantities. Form-cutting machines, commonly called *gashing machines*, are made for the specific purpose of rapidly roughing gears with inexpensive cutters. One type uses a circular form cutter and operates like a milling machine, but has a built-in indexing mechanism and roughs out gears automatically. The operator only has to unload and load the work. Such machines are also commonly used for cutting slots and grooves around parts other than gears. A fully tooled gashing machine for spur and bevel gears up to about 1.5 m (60 in.) in diameter is quoted in excess of \$200,000.

Internal gears and splines with no obstructions may be cut by normal internal broaching. External gears may be cut on surface broaching machines (Chapter 22), a few teeth at a time. A faster method is to push or pull gear blanks (upward to let chips fall) through a *pot broach* or die containing a series of rings of internal teeth (or internal strips of broach teeth) that cut progressively along the length of the broach. A blank enters at the bottom and emerges at the top with a complete set of teeth (spur or helical). Production rates of more than 250 gears/hr are common for both cast iron and steel, and as many as 510/hr for two at a time. Tool life may be well over 1 million gears, with 15,000–25,000 steel gears or 35,000–50,000 cast-iron gears produced per broaching tool grind. This is a large-quantity production process because a pot broach from 750–1,500 mm (≈30–60 in.) long and usable for only one kind of gear may cost up to \$10,000 or more, and a pot broaching machine about \$200,000.

A *shear-speed gear shaper* form cuts internal and external spur gears, splines, clutch teeth, and many special shapes. Single-point tools, one for each tooth space, are arrayed in a circle and point radially. The cutters are mounted around a hollow head for an external gear. The gear blank is mounted on a fixture below and is reciprocated into the head. The tools are fed radially a predetermined amount each stroke and are retracted slightly on the return stroke for clearance. All tooth spaces are cut at the same time.

The shear-speed gear shaper is the fastest gear cutter for many cases and can turn out average gears in 15–50 sec each. A high degree of accuracy is obtained because it depends mostly on the tooling, which is made up especially for each job. The tooling for a job may cost several thousands of dollars, but the cutters can be easily resharpened many times. Because of the tooling cost, the process is economical only for large quantities of parts, but experience has been that tool costs per piece for large quantities are no more and may be less than for other methods.

Shear-speed gear shaping machines are made in several sizes to take gears up to about 508 mm (20 in.) in diameter. A model with a capacity for external gears up to around 178 mm (7 in.) in diameter costs about $150,000.

Gear Generating

Gear hobs. Hobbing is a generating process done with a cutter called a *hob* that revolves and cuts like a milling cutter. Its teeth lie on a helix like a worm. Lengthwise gashes expose the cutting faces that have a contour simulating a rack. The teeth are form-relieved behind the cutting edges. As the hob revolves one turn, the effect is as though the simulated rack were to move lengthwise an amount equal to the lead of the thread on which the teeth lie. When a gear is cut, it is positioned and revolved as if it were in mesh with such a rack. The hob teeth take progressive cuts relative to each tooth space in the manner illustrated in Figure 28-24. A hob cuts all gears having the same tooth form and pitch as the rack it represents regardless of the number of teeth. When a hob becomes dull, all teeth are ground the same amount on the radial cutting face. Because of the way the teeth are relieved, the exposed tooth face after sharpening has substantially the same form as before.

A hob may have one, two, or more threads. When a spur gear with N teeth is being cut, it can

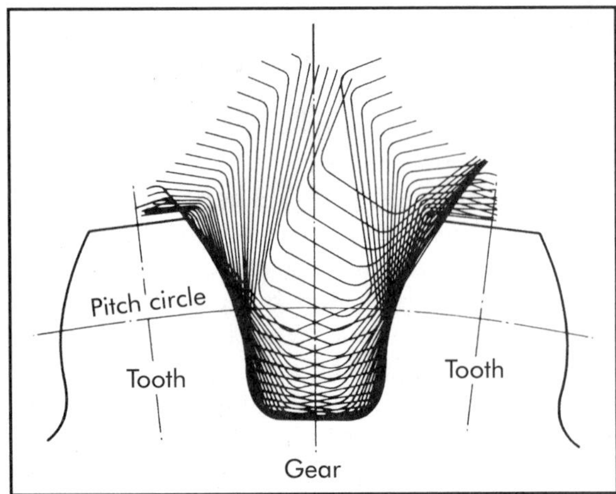

Figure 28-24. Action of a hob as its teeth progress through and cut the tooth space in a gear.

turn only one revolution when a single-thread hob turns N times, once when a double-thread hob turns $N/2$ times, and so on. Thus, a single-thread hob does not produce at as high a rate as a multiple-thread hob turning at the same speed. However, it does cut a more accurate involute because more of its teeth act in each gear tooth space.

Hobs are made of high-speed steel or cemented carbide or with carbide-tipped teeth in several standard classes of accuracy. Hardened but unground hobs are satisfactory for average work, especially for roughing. High-speed steel hobs cost about $125–$1,100 for small to large sizes. Hobs finish-ground all over for accuracy cost from $30–$300 more. Solid cemented carbide hobs may cost 5–15 times more than equivalent high-speed steel hobs. Hobs are made in several styles. The most common is the straight hob.

Hobbing machines. Hobbing machines are among the most popular gear cutters and they are available in many sizes, styles, and grades. Some have horizontal work axes, while others are constructed with vertical axes to provide for increased rigidity and stability. Figure 28-25(A) shows a close-up view of a setup for gang-hobbing four AISI8620 steel helical gears on a horizontal hobbing machine. The action of the hob in cutting teeth on those gears is shown in Figure 28-25(B). The hobbing cutter is set at an angle in a horizontal plane so that its teeth line up with the tooth spaces of the gear. The hob and the gear blanks are rotated at different rates to produce the required number of teeth on the gears. Generally, hob-cutting speeds are about the same as milling cutter speeds for comparable materials and other conditions, and are governed by the same principles. When the hob is engaged with the gear, its teeth wear in a limited zone. When the teeth become dull there after cutting a number of gears, the hob is shifted lengthwise to bring sharp teeth into action. The hob is run in the new position until it is dull again, and the shifting is repeated until all the hob teeth are used. Then the hob is taken off and all the teeth are resharpened.

On the older mechanical-type hobbing machines, the hob spindle and work spindle are connected through index change gears that are selected to make the work gear rotate at the proper speed in relation to the hob. However, most

Figure 28-25. (A) Setup for gang-hobbing four helical gears on a hobbing machine. (B) Hob cutting the teeth on a set of four helical gears. (Courtesy Bourn & Koch Machine Tool Company)

newer hobbing machines constructed in the U.S. are equipped with computer numerically controlled (CNC) servo drives and motors. The CNC operation of hobbing machines provides increased setup flexibility, eliminates the need for change gears, and permits an increase in productivity and quality in gear manufacturing.

Most gear hobbing machines are general-purpose machines, but some are specialized. For large-volume production, hobbing machines are made with several workstations and are often automated. Sizes vary from small machines (often bench-mounted) for instrument gears, to huge ones for gears over 6.1 m (20 ft) in diameter. The size of a gear hobbing machine is usually specified in terms of the maximum outside diameter and face width of a spur gear that can be accommodated on the machine. For example, the machine shown in Figure 28-25 is classified as a model 14-15 machine with a 3.5 normal diametral pitch (NDP) rating, meaning that it will cut gears up to 14 in. (355.6 mm) outside diameter (OD) and a face width of up to 15 in. (381 mm). The helical gears being machined are about 8 in. (203.2 mm) OD × 1 in. (25.4 mm) face width each. The high-speed steel (HSS) hobbing cutter is approximately 5 in. (127 mm) diameter × 5 in. (127 mm) width and costs about $1,000. A CNC hobbing machine of this size would cost about $300,000.

Gear shaping. A gear is shaped by a reciprocating cutter in the form of a single tooth, rack, or pinion. The gear blank and cutter move together as though in mesh, and a series of cuts are taken as depicted in Figure 28-26.

The teeth of the typical pinion-type gear shaper cutter of Figure 28-27 have a true involute form in any section normal to the axis and

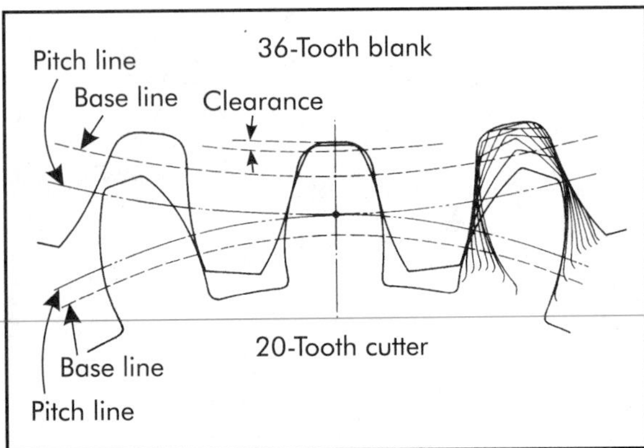

Figure 28-26. Generating action of a rotary-gear shaper cutter. (Courtesy Fellows Gear Shaper Company)

are form-relieved on the sides, outside, and root for clearance. The face of the cutter is dished to provide rake. As this face is ground to resharpen the cutter, the outside diameter becomes smaller, but the teeth retain their involute form and pitch. Cutters for helical gears have helical teeth. A spur gear cutter is capable of generating any gear of the same pitch in the same system, but a helical gear cutter can generate only gears having one pitch and helix angle. Disk-type cutters from stock range in price from about $300–$1,200, depending on size and tooth form.

Gear shapers. The cutter is mounted on the lower end of a reciprocating vertical spindle on most gear shapers, as shown in Figure 28-28, and is revolved as it reciprocates up and down. The spindle is guided by a cam to move the cutter in a helical path to cut helical gears. The reciprocating drive of the cutter spindle may be mechanical or hydraulic. The workpiece is on a vertical spindle below the cutter, and the distance between the spindle axes is set to the required center distance of the cutter and workpiece. The cutter and workpiece are made to revolve in unison as though in mesh. This relationship is arranged by change gears in the drive between the cutter and work spindles on most machines. However, as with hobbing machines, electronic and numerically controlled gear shapers have appeared in recent years, and on such machines the necessary rates and positions for each spindle are specified digitally to an electronic circuitry

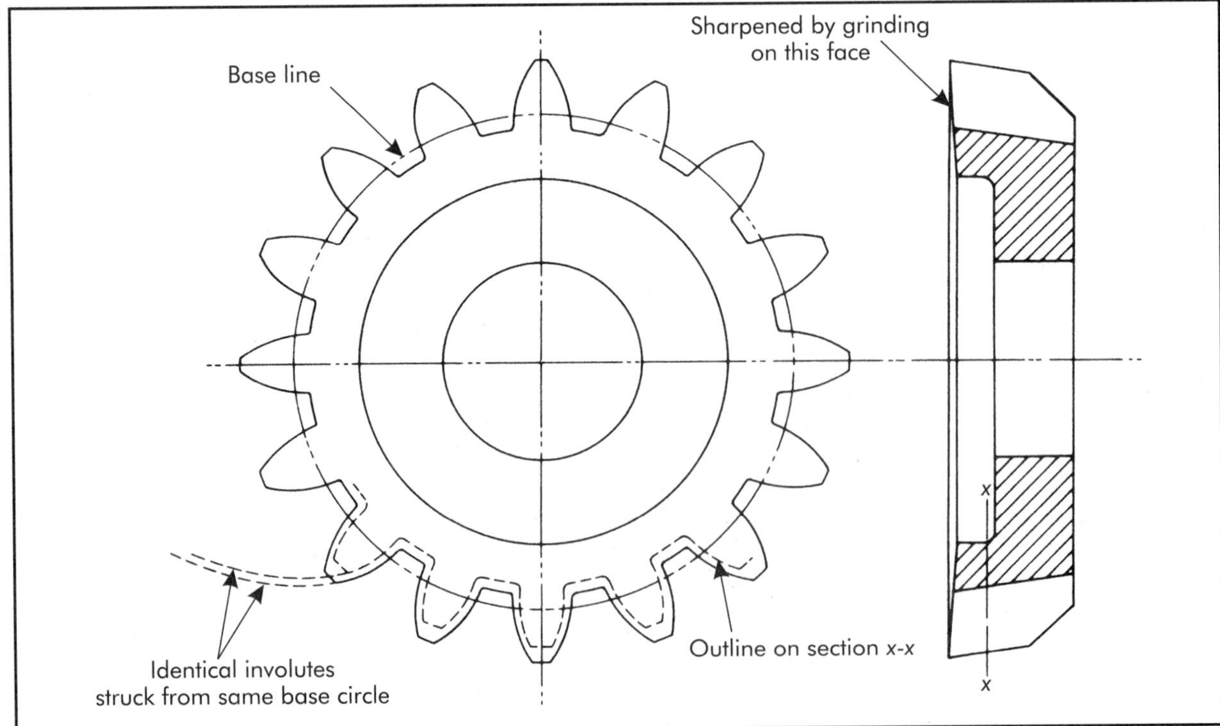

Figure 28-27. Pinion-type gear shaper cutter.

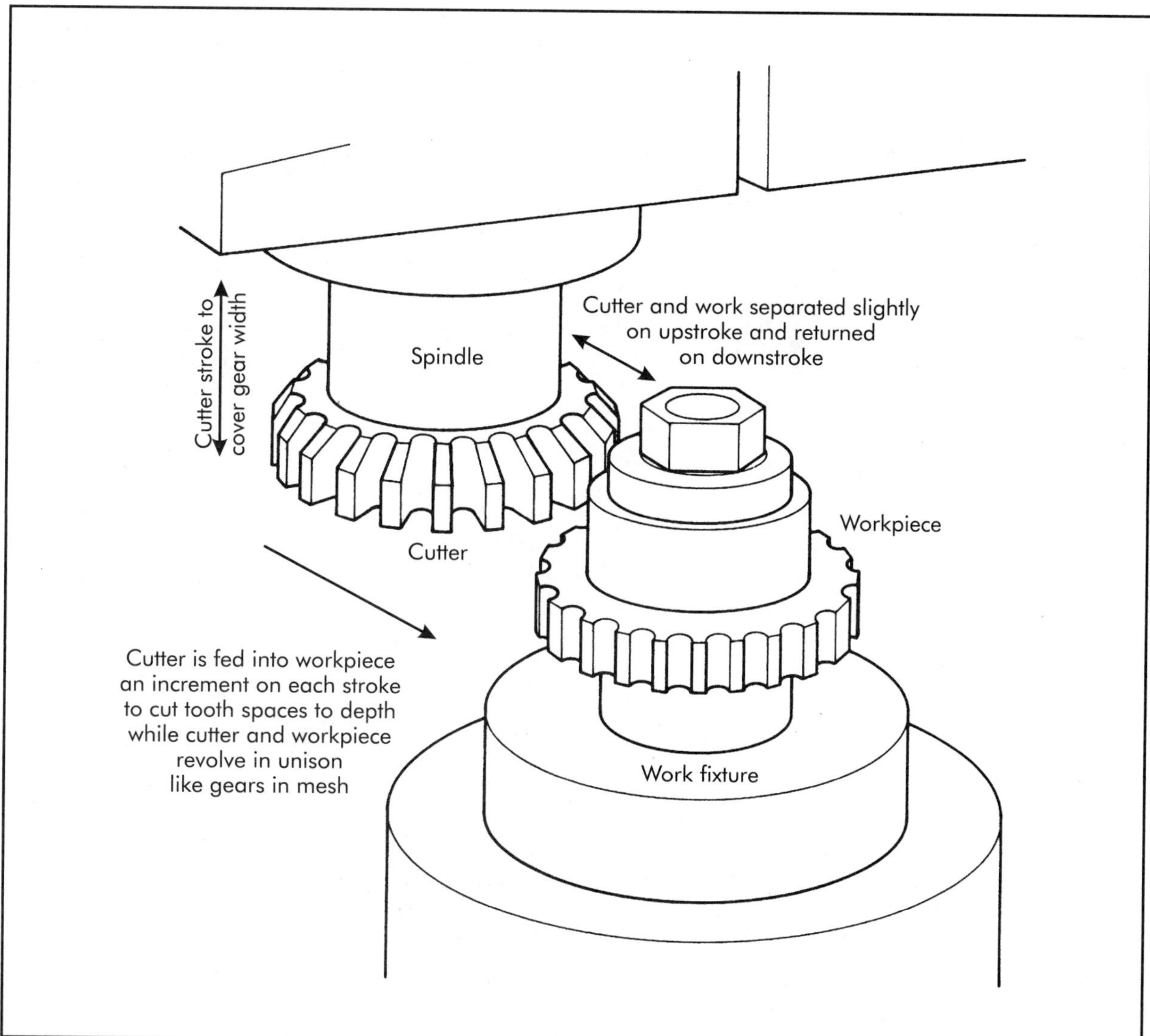

Figure 28-28. Operation zone of a gear shaper.

that keeps the motions going and synchronized through feedback.

The cutter is fed inward to depth while revolving with the workpiece on a gear shaper. One or more cuts are taken in producing a gear, depending upon material, design of the gear and blank, the workholding fixture, and subsequent finishing operations, if any. Sometimes economical results are obtained in production from rough cutting on one machine and finish cutting on another.

A mechanical gear shaper for gears up to about 254 mm (10 in.) in diameter with a 100 mm (≈4 in.) stroke and a 3.7-kW (5-hp) motor costs about $150,000 equipped. A numerically controlled gear shaper with about 450-mm (≈18-in.) diameter capacity has a price in the vicinity of $250,000.

Gear shaping can produce internal gears and splines, gears close to flanges, cluster gears, and continuous herringbone gears (all of which cannot be cut by rotating cutters like hobs), and also racks, cams, and pawls. A shaper in a factory for cutting forms for which it is best suited may well be kept busy in spare time with other gears. The length of stroke of a shaper may limit the width of gear it can cut. However, short approach and overtravel distances make shaping fast for some narrow gears. Tooling costs about the same for gear shaping as for hobbing.

A number of varieties of gear shapers serve various purposes. Some are made with multiple workstations for large-quantity production. Others use rack-type cutters, which are relatively easy to make because the teeth have straight sides. A *rack shaper* is one that cuts racks but uses a pinion-shaped cutter. The machine has a long table that feeds the work past the cutter.

Comparison of Methods

The large majority of all gears are hobbed or shaped. Other methods all have some limitations, such as the range of sizes, kind of work, high tooling and equipment costs, etc. Each excels in certain applications, but none is as universal as hobbing and shaping. More gears are hobbed than are manufactured in any other way. Shaping loses time on the backstroke, while hobbing is more continuous and, in many cases, somewhat faster. For reasons already cited, shaping is more versatile. One authority concedes the advantage to hobbing with respect to tooth spacing and runout accuracy because of continuous indexing. Also, the heat generated in hobbing is dispersed uniformly over the workpiece and cutter. Shaping is considered better for accuracy of tooth profile because of the straight cut it takes along the tooth. Hobbing and shaping are considered about equal for lead accuracy. Many gears semifinished by hobbing and shaping and by other methods are finished by rolling, shaving, and grinding.

Bevel Gear Cutting

Machines for cutting bevel gears may be divided into two classes: (1) for straight teeth, and (2) for curved teeth. The basic machines of universal type are the two-tool straight bevel generator and the spiral bevel and hypoid generator. Others are available for special or supplementary purposes, and the most important of them will be described briefly.

Straight tooth bevel gears. A basic way of cutting straight tooth bevel gears is to employ two cutters reciprocating along the sides of one tooth (or tooth space) at a time in paths that impart the correct taper to the teeth. The cutting action is depicted in Figure 28-29(A). The tools have straight cutting edges on their front ends and simulate the sides of teeth of an imaginary crown gear in mesh with the gear being cut. The cutters are mounted and reciprocated on a cradle that is rocked in time with the roll of the gear to provide the generating action. To begin a cycle, a gear blank is chucked in the work head of the machine, advanced to the cutting position, and then fed directly into the two reciprocating tools to cut a tooth almost to full depth. The workpiece and the cutters are rolled together as though in mesh. The initial roll is to the bottom of the generating cycle to rough shape the tooth. The workpiece is then fed to full depth, and a fast up-roll generates the finished tooth. At the top of the roll, the workpiece is backed away and indexed while the cradle and work head return to the roughing position for the next tooth. The cycle is repeated until all teeth in the gear are cut, and then the machine stops. A machine may be arranged to do only roughing or only finishing.

Two-tool generators are economical for small and moderate quantities of gears. They can cut bevel gears with hubs or flanges that would interfere with rotary cutters. A typical 9-diametral pitch, 30-tooth gear is produced at the rate of 7 gears/hr. A machine for small straight-tooth bevel gears employs two disk-type milling cutters with interlocking teeth on a cradle that rolls with the workpiece to generate the teeth. This method is reported capable of rough and finish cutting 30-tooth gears at a rate of over 12 gears/hr.

Straight-bevel gear generators are available in all sizes for gears from about 14–886 mm (or .75–34.9 in.) in diameter. One with a capacity for gears up to 886 mm (or 34.9 in.) in diameter costs about $390,000 with tooling.

Teeth cut on newer straight-bevel gear generators can be slightly crowned from end-to-end to localize contact and prevent damaging load concentration at the ends of the teeth. These are called *Coniflex*® gears. The same effect is obtained from a difference in curvature of the mating surfaces of bevel gears with curved teeth.

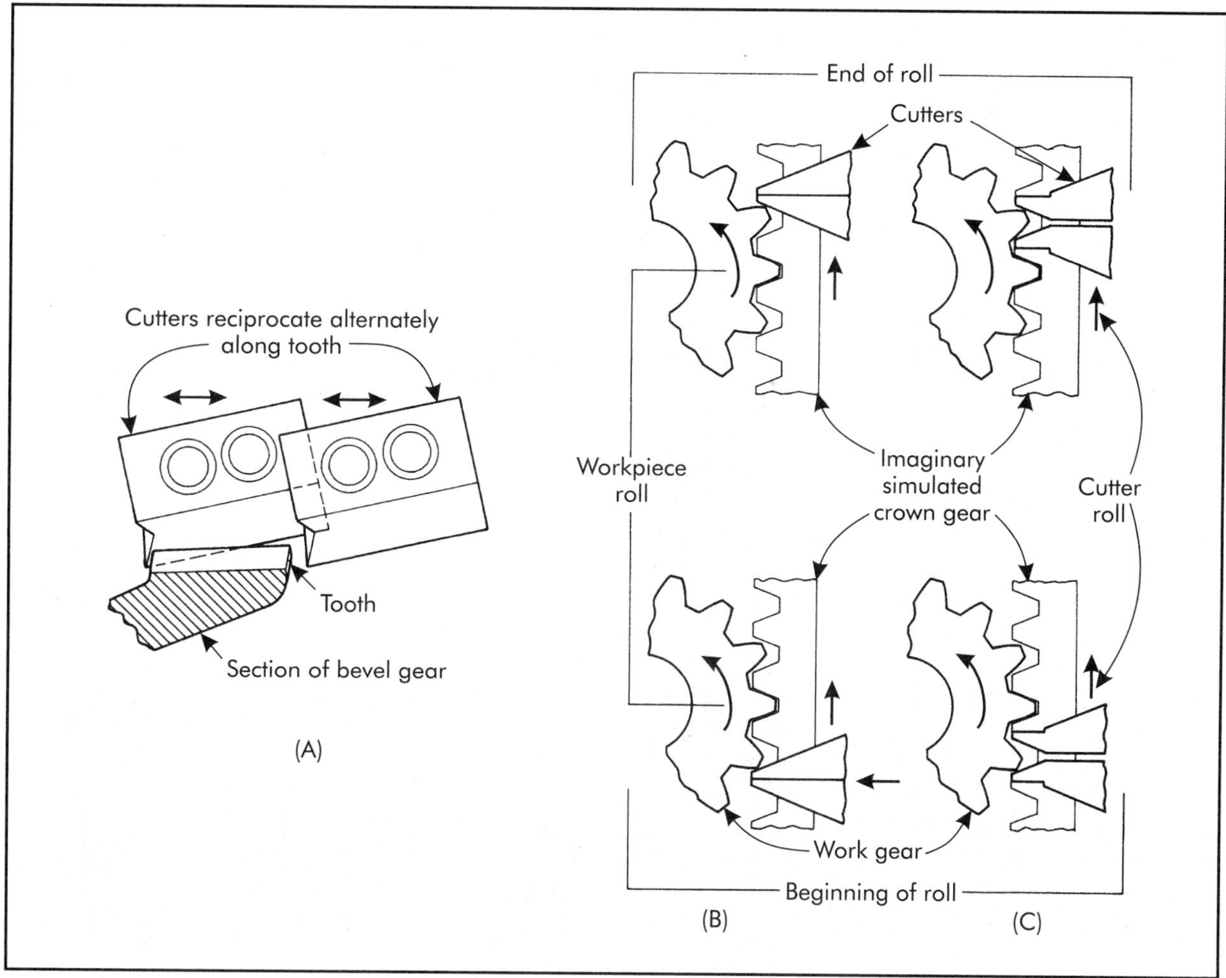

Figure 28-29. Actions of straight-tooth bevel gear generators: (A) reciprocating cutting action along each tooth; (B) relative motions of workpiece and tools on the completing generator; (C) two-tool generator.

Fine-pitch straight-bevel gears are commonly cut from the solid in one operation. Others may be rough cut in one operation and finished in another. This is done for large quantities where machines and tools are available for rapid roughing without generation.

The *Revacycle® process* is the fastest way to produce straight-bevel gears in large quantities. The rotating Revacycle cutter roughs and finishes a tooth space during each revolution it makes. The machine burrs each gear and automatically presents a new one to the cutter. A typical performance is finish cutting to a depth of 1 mm (≈.04 in.) on a 10-tooth gear at a rate of 165 gears/hr. Such a machine for gears up to 150 mm (≈6 in.) in diameter with a face width of 25 mm (≈1 in.) costs about $290,000. Automatic loading and handling devices are often added to enable one operator to take care of a number of machines.

Large straight-bevel and spur gears are cut on the *gear planer*. This is the oldest type of machine capable of cutting bevel gears with teeth tapering in the correct manner. A single-point planing tool is reciprocated across the face of a gear and controlled by a template or former to produce the profile shapes of the teeth.

Curved-tooth bevel gears and hypoid gears. Spiral-bevel and hypoid gears are generated on machines that use a special form of face-milling cutter or face-hobbing cutter. The cutter represents a tooth of a mating gear and is rolled with the gear or pinion being cut to generate the tooth profile. This principle is illustrated in Figure 28-30(A), wherein a face-milling cutter is being used to generate the teeth on a spiral-bevel pinion. A mating spiral-bevel crown gear is shown superimposed between the cutter and pinion to illustrate the action of the cutter in generating the teeth on the pinion.

In the past, conventional bevel gear-generating machines were designed to be mechanically capable of simulating a wide range of gear-generating geometries. These designs incorporated two main components, which provided the appropriate relative motion between the cutter and the workpiece. These were: (1) the *cradle* that held the circular cutter and moved it along a circular path, and (2) a *work head* that positioned and rotated the workpiece in the appropriate manner relative to the rotation of the cutter. To arrange these components for generating a particular spiral-bevel or hypoid gear, setup personnel had to make precise angular and linear adjustments to the components controlling the eight or ten degrees of freedom required on those machines. Thus, considerable time and skill was required to set up the machine for each different type of gear to be machined.

In recent years, machine designers have developed CNC spiral-bevel and hypoid gear-generating machines that require fewer degrees of freedom than the older mechanical models and, in addition, provide more accurate positioning of the cutter and workpiece during the generating process. The schematic in Figure 28-30(B) shows the configuration of a gear generator with three linear and three rotational axes under computer numerical control. With this arrangement, the movements of these six elements can be programmed to follow a series of coordinate positions, which will permit the proper relative movement of the cutter and workpiece. In essence, the generating action is guided by basic machine-setting data translated into six axes positions by a series of equations.

A spiral-tooth bevel gear is being generated by a face-milling cutter on the CNC machine shown in Figure 28-31. This machine will produce face-milled gears up to 445 mm (≈17.5 in.) outside diameter and face-hobbed gears up to 292 mm (11.5 in.). The teeth of the face-milling cutter shown are made of solid carbide with a tin coating to provide relatively long tool life between sharpenings. As illustrated in Figure 28-30, the machine shown in Figure 28-31 is a

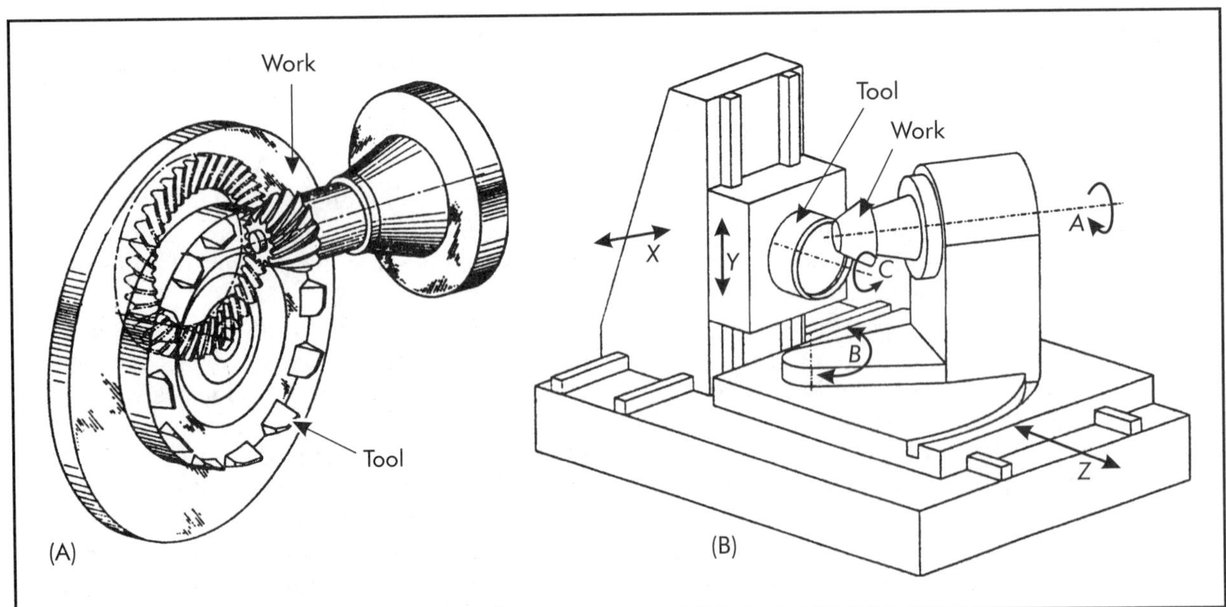

Figure 28-30. (A) A face-milling cutter being used to generate the teeth on a spiral-bevel pinion. (B) Schematic of a CNC spiral-bevel and hypoid gear-generating machine with three linear and three rotational axes of motion. (Courtesy The Gleason Works)

Figure 28-31. Close-up view of a face-milling cutter being used on a CNC gear-generating machine to generate teeth on a spiral-bevel gear. (Courtesy The Gleason Works)

six-axis machine, all under computer control with no mechanical adjustments being required for setup and operation. Complete machine setup, including tooling changes, can usually be accomplished in less than 15 min.

Spiral-bevel and hypoid grinders are available to finish-grind both face-milled and face-hobbed spiral-bevel gears. These machines are used to finish-grind the teeth after the gears have been semifinished and hardened. An abrasive cup wheel with the same effective shape and action as a face-milling cutter is used on these machines.

Gear Tooth Rounding and Chamfering

The ends of the teeth on a gear are commonly rounded or chamfered for strength, easy engagement, etc. Machines are available that use end milling and other types of cutters of suitable shapes for the cuts desired. Typically, the cutter is traversed under cam control around the profile at the end of a tooth as the gear revolves.

Gear Finishing

A gear tooth surface that is hobbed or shaped is composed of tiny flats. Such a surface is satisfactory for some purposes but is not good enough where a high degree of accuracy and stamina are required. The flats can be made very small by cutting at low feeds, but the operation then becomes slow and costly. Often the lowest net cost can be realized by cutting a gear fast with a less accurate and less expensive tool to tolerances of around 25 μm (.001 in.). Then a finishing operation is added to obtain the necessary tooth straightness, size, concentricity, spacing, and involute form with small tolerances. Gears often are heat treated for hardness and strength, which tends to warp them and form scale. Finishing after hardening corrects these deficiencies.

Finishing operations are performed to make gears accurate, quiet, smooth-running, and long-lasting. They include shaving and rolling for soft gears, and grinding and lapping for gears harder than about 40 R_C. Only 100 μm (.0039 in.) or less of stock are left on gear teeth for finishing.

Shaving. A gear is shaved by running it at around 122 m/min (400 ft/min) in mesh with and pressed against a cutter in the form of a gear or rack with gashes or grooves on its tooth faces, as typified in Figure 28-32(A). The edges of the grooves are sharp and actually scrape fine chips from the faces of the teeth of the workpiece. Cutting fluid is applied. Some sliding normally takes place between gear teeth in mesh, and this action is augmented in gear shaving by crossing the axes of the cutter and work gear, generally from 5–15°, and reciprocating the workpiece as it is revolved in mesh with the shaving cutter. The workpiece may be fed radially or tangentially into the cutter. Shaving removes from 25–102 μm (.001–.004 in.) from tooth thickness and can correct from 65–80% of the errors on hobbed or shaped teeth. Shaving can be made to crown gear teeth slightly at their centers to localize tooth contact and keep it away from the ends of the teeth.

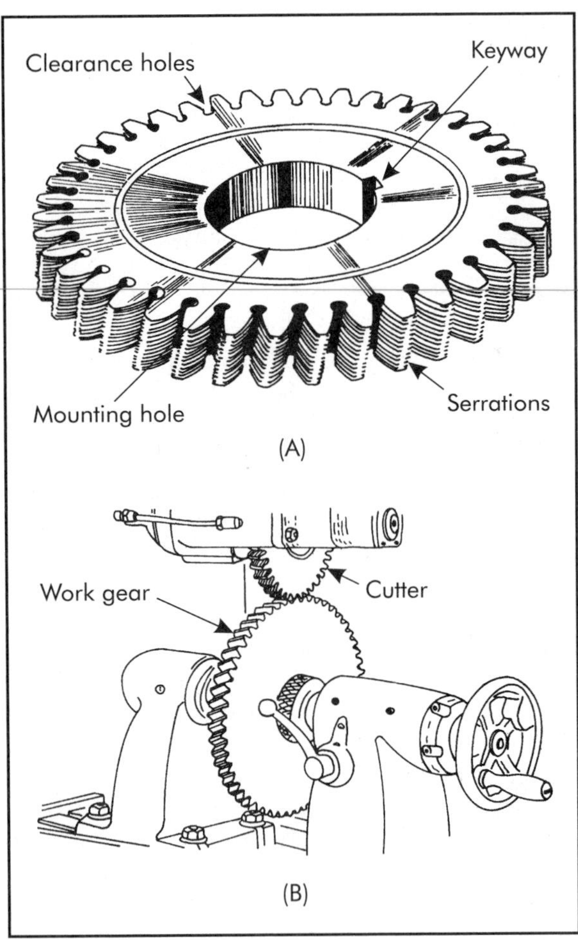

Figure 28-32. (A) A typical rotary gear-shaving cutter; (B) rotary crossed-axes shaving of an external gear.

The gear in Figure 28-32(B) is being shaved by a gear-type shaving cutter on a *rotary shaving machine*. The workpiece is loaded between live centers, raised against the cutter, and reciprocated while being driven in one direction and then the other. Rotary gear-shaving machines are made in many sizes for semi- and fully automatic operations to finish external and internal gears from the smallest to over 4.9 m (16 ft) in diameter. A rotary gear-shaving machine for gears up to 400 mm (≈16 in.) diameter × 150 mm (≈6 in.) wide costs about $170,000.

A shaving cutter in the form of a rack is reciprocated lengthwise at high speed in mesh with a workpiece on a *rack-type shaving machine*. At the same time, the workpiece is reciprocated sideways and fed into the rack. A rack-type shaver produces uniformly precise gears because the rack form is easier to make accurately and has fuller contact with the work than a rotary shaver. The rack-type shaver cannot accommodate cluster or large gears. The cutters are large and expensive but have long lives and a low cost per piece. They are more suitable for long runs. Circular-type tools are better for job lots.

Gear shaving is a low-cost rapid production process. Many gears can be shaved in less than half a minute apiece, some in as short a time as five seconds. Each cutter is suitable only for a single pitch and tooth form, and a demand for many gears is necessary to warrant its cost. Average expected life with four regrinds for a rotary shaving cutter is reported to be 85,000 gears.

Gear tooth grinding. Hardened gear teeth are ground by forming and generating. Three variations of grinding gear teeth with formed wheels are depicted in Figure 28-33(A). The results depend upon the accuracy to which the wheel is trued. Truing is done by guiding the truing diamonds on the machine through a pantograph mechanism from templates typically six times the gear tooth size. A workpiece between centers is reciprocated under the grinding wheel, which is fed down at each stroke until the desired size is reached. Then the workpiece is cleared from the wheel and indexed to the next tooth space. Usual procedure is to rough grind around a gear in this way, and then finish grind with the trued wheel set to depth. Form grinding is done on external and internal spur and helical gears, splines, and similar parts.

When spur or helical gears are ground by generating, the wheel or wheels are trued to simulate a mating rack. On the single-wheel machine in Figure 28-33(B), the workpiece is carried under a reciprocating wheel and rolled as though it were in mesh with the rack. A spur gear is ground with its axis in line with the wheel movement, and a helical gear is swiveled. When one tooth space has been ground, the gear is indexed to grind the next, and so on. Another type of generating gear grinder has two wheels as indicated in Figure 28-33(B). It operates on the same principle as the single-wheel machine, but the wheels are not reciprocated. They are large enough to cover an entire tooth working surface on gears not over about 32 mm (≈1.25 in.) wide.

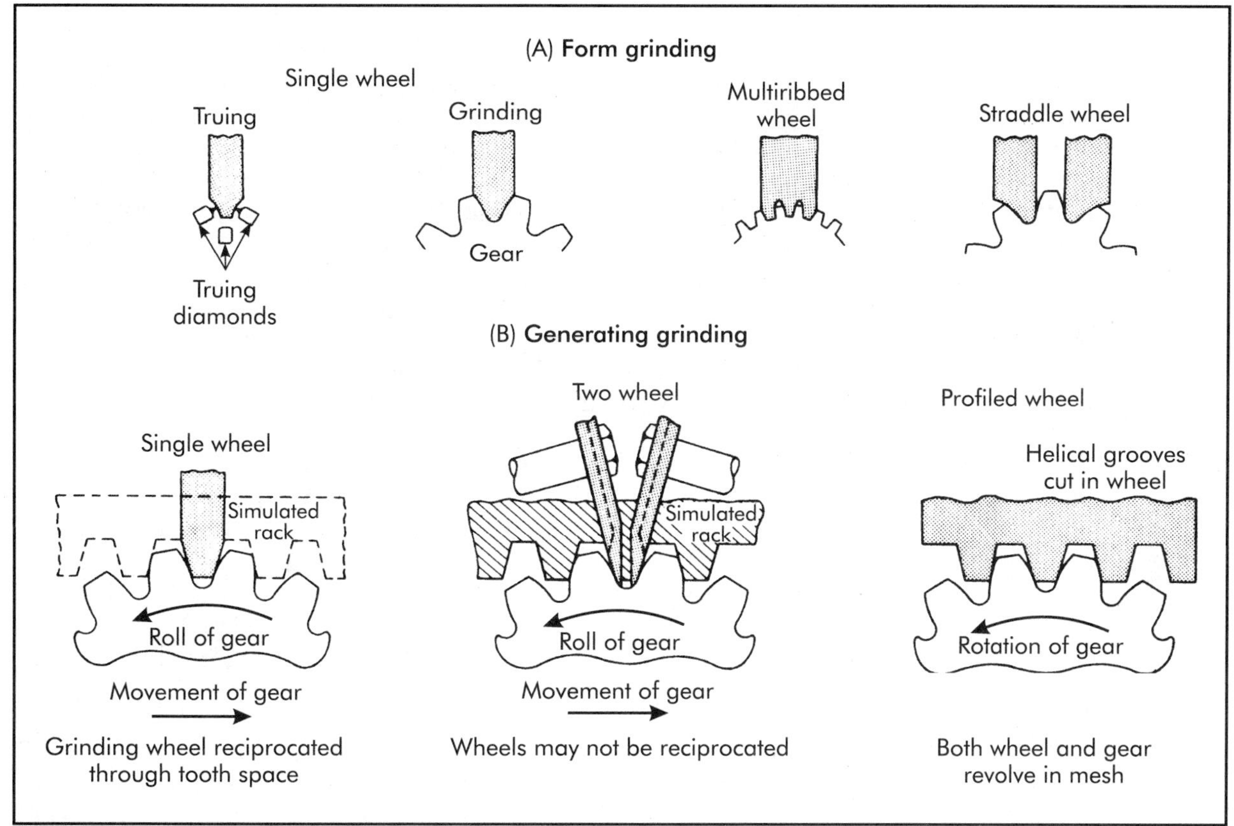

Figure 28-33. Methods of grinding gears.

It is faster than the one-wheel generator for narrow gears.

A third method of generating grinding employs a wide grinding wheel with a helical groove around its periphery. The grooved cross section simulates a basic rack form engaged with the gear as the grinding wheel and workpiece rotate in mesh with an action like that of hobbing. The gear is fed axially to carry the action fully across its teeth. An average of 40–60 gears are reported ground for one wheel dressing. This is the fastest generating method but it requires much time to prepare a wheel for each pitch; an average of 4 hr to true and 20 min to retrue. Tooth-spacing errors can be better controlled by a grooved wheel that covers several teeth at once as compared to grinding one space at a time.

Form grinding has been found faster in some cases, generating in others; but the differences are not large. The area of wheel contact is small for generating, and there is less likelihood of burning and cracking hardened steel, for which care must be exercised in form grinding. The root of a gear space can be finished and blended with the tooth profile better by forming than by generating. Generating leaves small flats on the teeth and subsequent lapping is necessary if they must be removed.

Lapping and honing. Hardened gears are commonly lapped or honed after heat treatment to remove scale, nicks, and burrs, improve surface finish, and correct small errors. Surfaces are finished to 0.5 μm (20 μin.) R_a and better. Occasionally, even ground gears may be given final touches by these methods to eliminate all wheel marks. Essentially, the workpiece is rotated in mesh with a gear form with crossed axes and reciprocated in the manner illustrated for shaving.

Internal and external spur and helical gears are lapped by running the workpiece with a mating gear or with one or more cast-iron toothed laps under a flow of fine abrasive in oil. The work is turned first in one direction and then in the other to lap both sides of the teeth. Several methods are common: among them are *cramp lapping* or forcing the workpiece hard against the lap to speed the action, braking the workpiece against the driving lap, and running several laps at once at different pitch points to reduce tooth spacing errors. A gear-lapping operation time averages from one to several minutes. On average work, a lap can finish several thousand gears and then must be recut.

A flexible gear-lapping tool (called a poli-tool) molded from a rubber-like porous material is run with a lapping compound to put a high polish on gear teeth. Surface finishes as fine as 150 nm (≈5 μin.) R_a have been obtained.

Bevel-type gears are commonly lapped (and kept) in sets by running gear and pinion together with abrasive slurry jet-sprayed into the tooth mesh. Various axial, radial, and twisting motions with braking are added to speed the action and distribute it over the teeth.

Honing is done by helical gear or worm-shaped tools. They may be made of steel, for strength, with abrasive, cemented carbide, or diamond particles embedded in the tooth surfaces. However, most are made of plastic impregnated with abrasive. Their flexibility is advantageous and cost is lower than steel, mainly because each plastic hone can be retrued a number of times. The honing tool is pushed with constant force against the workpiece but must be allowed some float to prevent damage, in contrast to shaving done at a constant center distance between the axes of the workpiece and tool. Commonly, an axial vibratory motion is given to the hone in addition to other movements.

Internal and external spur and helical gears are honed to correct small errors in spacing, lead, or eccentricity, and particularly to make them run quietly. Honing tools cost more, but honing is faster and preferred to lapping for production.

Comparison of finishing operations. Basically, the finishing method for a gear is dictated by the degree of accuracy required and the available equipment. Invariably greater precision demands more costly operations for gears as illustrated in Figure 28-34. In the case of an 8-diametral pitch (DP) gear cited in the example, the most that can be achieved in manufacturing is an involute accuracy of 1.5 μm (.00006 in.) (American Gear Manufacturers Association [AGMA 17]). When limitations of measuring methods are also taken into account, a practical upper limit is 2.8 μm (.00011 in.) (AGMA 15). At the extreme opposite end of the scale, for gears where accuracy is of no concern, there is little cost savings for a tolerance larger than 0.38 mm (.015 in.) (AGMA 3 or 4).

It has been found that most good-quality machines perform well when errors in gear tooth form or spacing do not exceed 5 μm (.0002 in.). Smaller tolerances are required for some products, but cost then rises rapidly. On the other hand, errors of 10–15 μm (.0004–.0006 in.) or larger cause a pronounced inferiority in the running qualities of hardened gears.

As an explanation of the differences in costs for different gear finishing methods, one authority has estimated the time to finish a 10-DP gear with 25 teeth and a 32° left-handed helix angle. The gear was shaved or rolled in as little as 22 sec, ground in 5–15 min (even longer with special care), and honed in 21 sec. Thus shaving, cold rolling, or burnishing before heat treatment and honing afterwards is the cheapest way to finish hard spur and helical gears. This is usual treatment for most automotive, agricultural equipment, aircraft, and machine-tool drive gears. Gears that cannot be shaved or rolled because of shape or small quantities, and spur and helical gears that require the highest degree of accuracy, must be ground after hardening. This is commonly done to make master gears and shaving cutters, for instance. Good-quality straight and spiral-bevel and hypoid gears are lapped or ground for best results.

Care should be taken so that gears are not made too much better than necessary. A machine tool manufacturer found that some gears case hardened to 60 R_C hardness and ground were not appreciably better than others through-hardened to only 38–48 R_C. The softer but tougher and stronger gears could be adequately finished by hobbing alone to tolerances around 10 mm (.0004 in.).

28.2.5 Gear Inspection

The American Gear Manufacturers Association (AGMA) has set up a practical gear-classification system for spur, helical, herringbone, bevel,

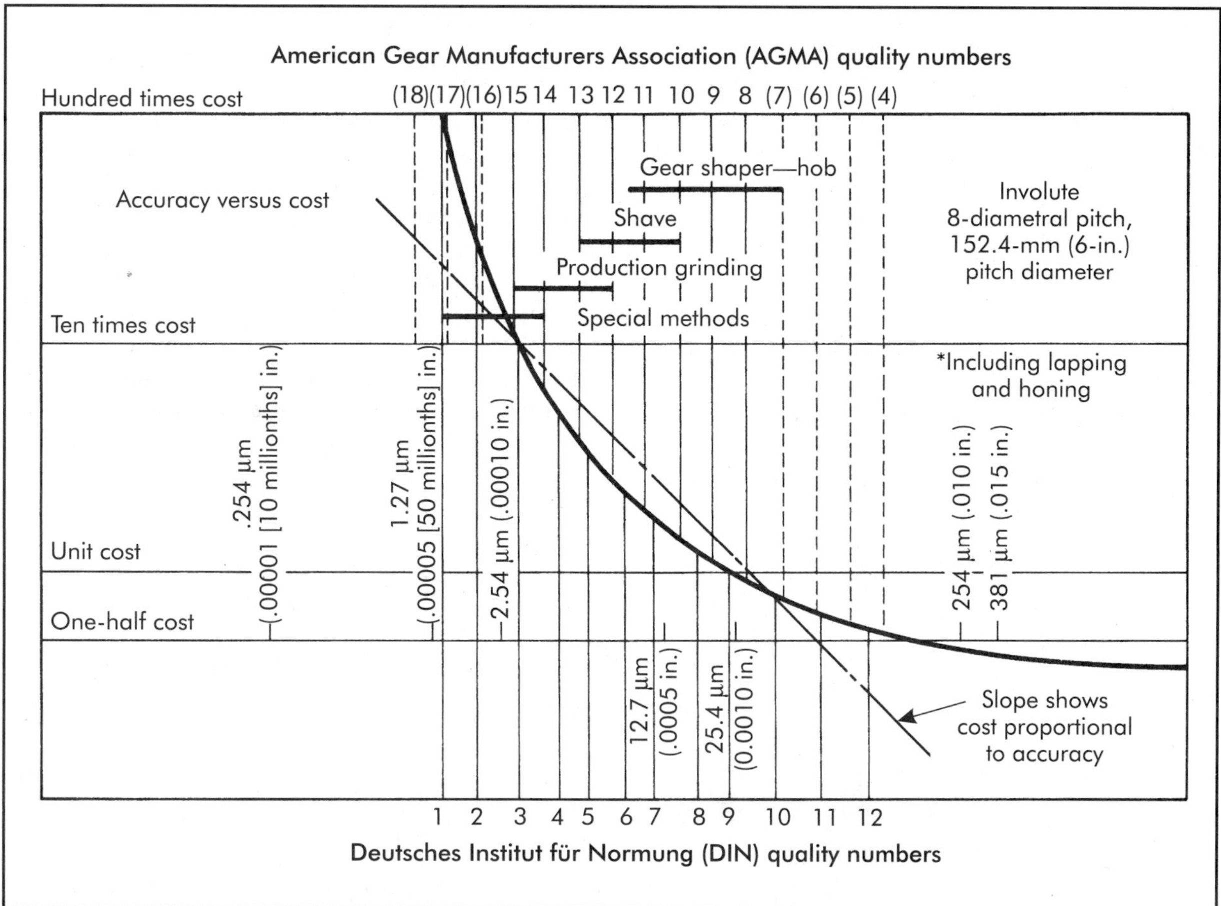

Figure 28-34. Typical comparison of gear quality versus manufacturing cost (Rice 1977).

and hypoid gears based on the broad experience of many manufacturers. Gears are classified in terms of tolerances commonly considered in manufacture and usage, and each class is given a number, for example, AGMA Class No. 8. The larger the class number, the smaller the tolerance. In addition, the system codifies the more usual materials and heat treatments for gears. Thus a gear may be specified by its class number followed by symbols for specific backlash (tooth thinning), material, and heat treatment. Tables giving complete specifications are in the AGMA *Gear Handbook*.

The inspection and measurement of gears involve some techniques not common for other products. Gears are tested for:

- the accuracy of linear dimensions such as outside and root diameters, and tooth thickness and depth;
- tooth profile;
- positions of the teeth as reflected by tooth spacing, runout, radial position, backlash, and helix angle or lead;
- the bearing and finish of the tooth faces; and
- noise.

Gear testing can be divided into two kinds: functional and analytical. Functional checking shows how errors affect the way the gears work together, as when they are rolled together to test for freedom or noise. It is usually the easiest way of testing for acceptable gears. Analytical checking actually measures the important elements. This may be done by taking absolute measurements or by finding how much a gear differs from a master. Analytical checking is usually required to adjust or correct gear cutting or finishing equipment. Usually (but not always), functional checking is done on fine-pitch gears, and analytical checking on coarse-pitch gears. As a rule, a gear is inspected by one method or the other, not both. Examples of these forms of testing are given in the following descriptions of typical gear-checking equipment and methods.

Checking of Gear and Gear Teeth Size

A *gear tooth vernier caliper* such as the one shown in Figure 28-35 measures the thickness of a gear tooth at the pitch line. A gear *tooth comparator* is a similar device that measures the addendum depth to a specified tooth thickness.

Ground rolls of precise diameter to make theoretical contact at the pitch line may be placed in opposite tooth spaces of a gear, and the distance across them measured by a micrometer or comparator. This is like measuring threads across wires. Formulas and tables outlining proper roll sizes and measurements are given in handbooks.

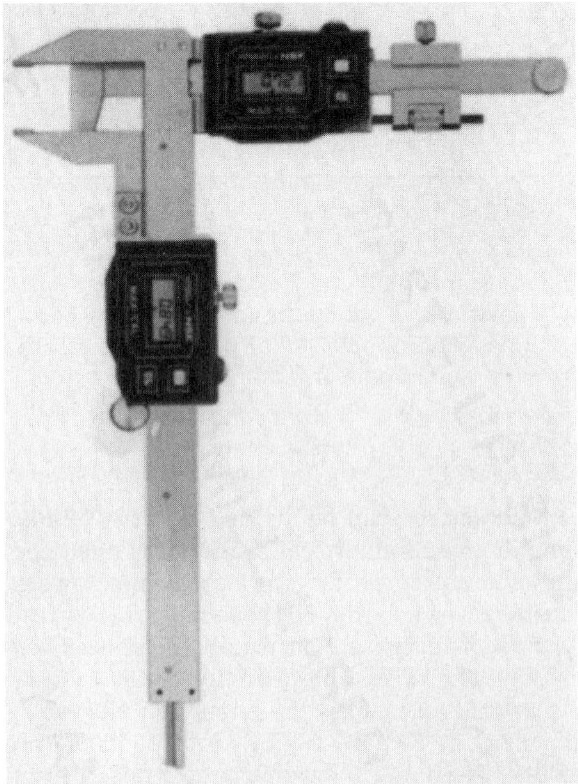

Figure 28-35. Gear tooth caliper for measuring the depth of teeth from the top to the pitch line and thickness of teeth in the pitch line. (Courtesy Fred V. Fowler Company)

Checking Gear Tooth Profile

An optical comparator like the ones in Figures 16-21 and 16-22 are one way of checking the profiles and positions of gear teeth by comparing their enlarged shadows on a screen to a large scale drawing. Fixtures are commonly used to position the gears.

Other schemes involving the use of dial indicators to follow the profile of a gear tooth while being moved along an involute path will show the deviation of the tooth from true involute form. Similar arrangements are made to drive an electrical recorder that traces the deviations on a dial.

Checking Gear Teeth Positions

A typical device to check tooth spacing carries a gear freely between centers. A tapered block is brought in on a slide to locate a tooth space. An indicating finger previously set to a master makes contact with the face of the next tooth. Error in tooth spacing displaces the finger and is indicated on a dial.

The lead or helix angle of a helical gear may be compared with a master cam or the angle to which a sine bar is set (Chapter 16).

To check backlash, a work gear and master gear may be mounted on shafts at fixed center distances. The master gear is held still, and the slight movement or backlash of the engaged gear teeth is measured by an indicator.

Composite errors in tooth spacing, thickness, profile, runout, interference, and eccentricity are reflected in changes either in velocity or center distance when gears are run together. A common way of checking these errors is to run a gear with a very accurate master gear as illustrated in Figure 28-36. The work gear is held against the master gear that is rotated on a fixed center. The movements between the centers of the two gears caused by errors in the work gear are measured by a dial indicator and/or recorded on a moving tape or chart. The movements of the line on the chart indicate errors in the work gear.

Checking Gear Tooth Bearing and Surface Finish

Commonly, bevel and hypoid gears and other gears are inspected by running them together or with masters to determine where the teeth

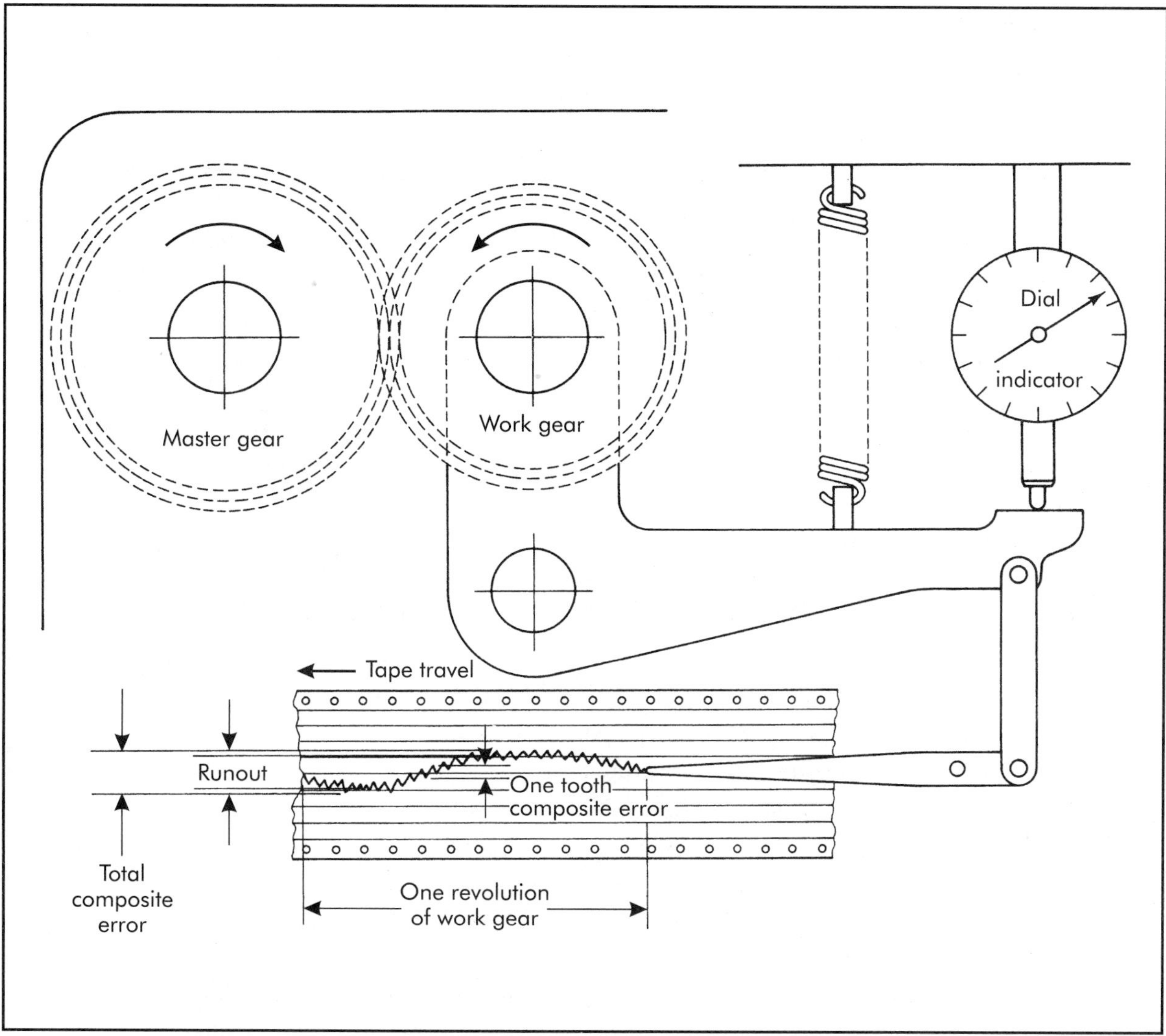

Figure 28-36. Sketch of the way a gear is rolled with a master gear to check for composite errors.

bear. Localized tooth bearing near the center of each face is desirable to avoid concentrating loads at the ends of teeth. The teeth are painted with marking compound that is rubbed away at contact to show bearing areas. Surface finish of tooth faces may be measured in the ways described in Chapter 16.

Checking for Noise

Faulty gears are noisy and can be checked on machines in which they are run in a sound chamber together or with a master. Power is applied to one gear and a brake loads the other. Characteristic sounds have such descriptive names as squeal, whine, growl, knocks, nicks, and marbles; each indicates certain kinds of gear errors to an experienced inspector.

Gear faults also may be detected by measuring frequencies and amplitudes of vibrations (sounds) while the finished product, such as a transmission, is run on a test stand. Careful analysis of the data obtained can point out the errors in gear production that cause the noises and lead to corrections. The matter of noisy gears has become particularly important with government regulation of noise levels from mechanical equipment.

28.3 QUESTIONS

1. Define a "screw thread" and show by means of a sketch the major elements of a straight thread.
2. What is the difference between a straight thread and a tapered thread?
3. Explain the relationship between pitch, lead, and number of threads per unit of length for a single-thread screw and a double-thread screw.
4. Define the "major diameter," minor diameter," and "pitch diameter" of a screw thread.
5. What is the difference between the forms of Unified National and ISO M screw threads?
6. Describe an Acme thread and its uses.
7. What is the purpose of international thread system standards?
8. How is the size of a screw thread specified?
9. What thread characteristic is normally measured by means of a screw-thread micrometer caliper?
10. What thread characteristic is normally measured by means of the three-wire system?
11. How can the lead of a thread be measured or checked?
12. What effect do pitch, lead, and form errors have on a thread?
13. Name at least five methods for cutting or forming external threads.
14. Why is the cutting of a thread on a lathe referred to as "chasing"?
15. What part does the lead screw on a lathe play in cutting a thread?
16. Why should a relief groove be cut at the end of a thread?
17. Why is thread cutting on a lathe slow and costly?
18. Describe the action of a self-opening die head.
19. Describe the shape of the three hand taps needed to tap a blind hole.
20. Why are self-reversing tapping attachments desirable for production tapping operations?
21. How is thread rolling done? What are its advantages and disadvantages?
22. What curve is found on most gear teeth and why is it used?
23. What are the main differences between gears in the diametral pitch system and those in the module system? Are standard gears of the two systems interchangeable?
24. Show by a sketch the major elements of involute gear teeth.
25. What advantages do helical gears have over spur gears?
26. For worm and worm gear assemblies, which is the driving element?
27. What is the primary difference between spiral-bevel and hypoid gears?
28. What are some of the advantages of the rolling process for producing gears?
29. List and describe the three methods that may be used to cut gears.
30. Describe the method used to cut gears on a milling machine.
31. By what processes may internal gears be cut?
32. Describe the generating action of a hobbing cutter in cutting a spur gear.
33. What is the advantage in using a multiple-thread hob for cutting gear teeth?
34. Is it necessary to use different hobbing cutters to produce gears with different numbers of teeth?
35. Describe the action of a gear shaper in cutting a spur gear.

36. Is it necessary to use different gear shaper cutters to produce gears with different numbers of teeth?
37. Describe the action of a two-tool reciprocating generator for cutting straight-tooth bevel gears.
38. Why is it necessary to perform finishing operations on machined gears?
39. Describe the action of a gear-shaving cutter.
40. Describe the two basic methods for grinding gear teeth.

28.4 PROBLEMS

1. What is the nominal thread depth for each of the following threads:
   (a) M 14 (P = 2 mm)?
   (b) M 8 (P = 1 mm)?
   (c) M 20 × 2.5?
   (d) No. 6 (0.138)-32 unified national coarse (UNC)
   (e) $\frac{3}{4}$-10 UNC
   (f) $\frac{5}{8}$-18 unified national fine (UNF)
2. The basic *addendum* of a thread is the radial distance from the outside diameter to the pitch diameter. The basic *dedendum* is the radial distance from the pitch diameter to the root diameter. What are the basic addenda and dedenda of external threads with the specifications listed in Problem 1?
3. Specify the basic outside diameter, root diameter, pitch diameter, and helix angle for each of the external threads listed in Problem 1.
4. Show that the best wire size for a 60° thread has a diameter, $W = \frac{2}{3} \times$ the depth of a full V-thread of the same pitch, and thus that the best wire diameter, $W = .57733\,P$ (see Figure 28-5).
5. Prove that the diameter over best wires on a 60° thread is $A = P_D + (\frac{3}{2})W$ (see Figure 28-5).
6. Find the best wire size and measurement over three wires for a $\frac{3}{4}$-10 UNC thread.
7. What is the error in Problem 6 from neglecting the lead angle?
8. Find the best wire size and diameter over three wires for an M 20 (P = 2.5 mm) thread.
9. Wires of .050 in. diameter are available for checking a $\frac{5}{8}$-11 UNC thread. The dimension measured over these wires is .6315 in. What does this indicate the pitch diameter to be?
10. The lead screw on a lathe has 6 threads/in. Specify the gear ratios required to cut each of the following numbers of threads per in.: (a) 4, (b) 5, (c) 6, (d) 7, (e) 8, (f) 12, (g) 16, (h) 18, (i) 20, (j) 25.
11. It is desired to cut a fine thread of 56 tpi on a lathe with pick-off gears and a lead screw with 4 threads/in. The smallest gear available has 20 teeth, the largest 100 teeth, with all ratios available in between. How should the lathe be geared and with what ratios?
12. What nominal size would you select for a tap drill for a $\frac{5}{8}$-11 UNC thread? Give your reason for your choices if the material is steel.
13. Six high-speed steel taps are to be driven at 18 m/min (≈60 ft/min). They include (2) M 10 (P = 1.5 mm), (2) M 14 (P = 2 mm), and (2) M 20 × 1.5 thread taps. Maximum torque per

tap is estimated to be 16, 25, and 26 Nm (≈140, 220, and 230 lbf-in.), respectively. Machine efficiency is 70%. What size motor should be supplied?

14. Calculate the elements for:

    (a) a 4 diametral-pitch spur gear of 20 teeth with full depth teeth and 14.5° pressure angle;
    (b) a 4 diametral-pitch spur gear of 21 teeth with stub teeth and 20° pressure angle; and
    (c) a 32-tooth spur gear with a module of 4 mm (.16 in.).

15. Compute the important dimensions of a gear with a module of 5 mm (.20 in.), 30 teeth, and a width of 20 mm (.79 in.).

16. Compute the important dimensions for a 10-diametral pitch, 14.5° pressure angle, full-depth tooth spur gear that has 40 teeth and a width of 12.7 mm (.5 in.).

17. Compute the important dimensions for a 10-diametral pitch, 20° pressure angle, stub-tooth spur gear that has 40 teeth and a width of 19 mm (.75 in.).

18. A 30.15-mm (1.187-in.) diameter involute spline in SAE 1040 steel with a 20 R_c hardness can be form-rolled in 18 sec. A set of forming dies costs $2,800 but can turn out 200,000 pieces in its lifetime with no reconditioning necessary. The spline can be hobbed at a rate of 1 min and 36 sec per piece. A hobbing cutter costs $360, must be ground after cutting 200 pieces, and can be resharpened 50 times. The time to sharpen and replace the hob is 20 min. Capital charges are 50% for interest, insurance, taxes, and losses. Labor and overhead are worth a total of $38/hr. If machines are available in the plant for either method, what is the smallest number of pieces in a year's production that justifies the roll-forming method?

19. In the situation described in Problem 18, hobbing machines are available with sufficient available time for the job, but machines must be purchased at a cost of $125,000 each if the parts are to be form-rolled. Each machine can be operated 1,600 hr/year. If 750,000 pieces/yr are required, how long should it take to pay for the machine out of expected savings?

20. A 12-pitch spur gear with 36 teeth and 22 mm (.875 in.) wide can be cut on a gear shaper at the rate of 2 gears in 55 sec. The machine and tooling cost $180,000. On a hobbing machine that costs $56,000, best results are obtained from two cuts for each gear. Six gears are cut at a time, but the total time for each gear is 6 min. One worker operates four hobbing machines or two gear shapers. Tool maintenance and overhead is assumed the same for either machine. Worker time costs $18/hr and 20% of the cost of a machine is charged each year for depreciation, interest, insurance, and taxes. A machine can be operated 2,000 hr/yr.

    (a) Which type of machine should be selected to produce 15,000 gears/yr if the machines are used for other work when not needed for this job?
    (b) Which type of machine should be selected to produce 15,000 gears/yr if the machines have no other use and must be charged all to this job?
    (c) Which type of machine should be selected to produce 100,000 gears/yr even if the machines cannot be used for other jobs?

21. A spur gear with 36 teeth is to be hobbed with a 100-mm (3.94-in.) diameter hob at 30 m/min (≈98.4 ft/min). What should be the work speed in rpm if the hob has a single thread? A double thread?

22. A soft-steel spur gear with 36 teeth, 96.5 mm (3.8 in.) outside diameter, 14.5° pressure angle, and 25.4 mm (1 in.) width is to be cut on a milling machine. What should be:

    (a) the diametral pitch, addendum, tooth thickness, and whole depth?
    (b) the specifications of the cutter?

(c) the number of turns of the dividing head crank to index each tooth?
(d) the cutting time with a feed rate of .003 ipt? Refer to manufacturer's catalog for cutter specifications.

23. A hardened 12-DP gear with 60 teeth and 32 mm (≈1.25 in.) width is shaved before heat treatment. After heat treatment it can be lapped in 3 min. A lap costs $550 and is usable for 2,000 gears. Honing takes 45 sec. A hone costs $1,250 and will finish 1,000 gears before it must be discarded. Production time is worth $36/hr. If satisfactory results can be obtained from either process, for what quantities should each be used for?
24. A gear with 32 teeth and a module of 6 mm (.24 in.) is made of mild steel. Two gears each of 20 mm (.79 in.) wide are put next to each other on a mandrel and cut in one pass by a single-thread hob having an outside diameter of 100 mm (.39 in.) and run at 30 m/min (≈98.4 ft/min). Calculate:
    (a) the rpm of the hob;
    (b) the rpm of the gears;
    (c) the pitch diameter of the hob; and
    (d) the time to cut the gears with a feed of 1.5 mm/rev (.06 in./rev) of the work and a distance for approach and overtravel of 40 mm (≈1.6 in.).

28.5 REFERENCES

ANSI/AGMA 115.01. "Basic Gear Geometry by Alan H. Candee." New York: American National Standards Institute.

ANSI/AGMA 2000-A88. "Gear Classification and Inspection Handbook—Tolerancing and Measuring Methods for Unassembled Spur and Helical Gears (including Metric Equivalents)." New York: American National Standards Institute.

ANSI/AGMA 1012-F90. "Gear Nomenclature: Definitions of Terms with Symbols." New York: American National Standards Institute.

ANSI/ASME B1.1-1989. "Unified Inch Screw Threads (UN and UNR Thread Form)." New York: American National Standards Institute.

ANSI/ASME B1.5-1988. "Acme Screw Threads." New York: American National Standards Institute.

ANSI/ASME B1.8-1988. "Stub Acme Screw Threads." New York: American National Standards Institute.

ANSI/ASME B1.7M-1984. "Screw Threads, Nomenclature, Definitions, and Letter Symbols for." New York: American National Standards Institute.

ANSI/ASME B1.13M-1983 (R1989). "Metric Screw Threads—M Profile." New York: American National Standards Institute.

ANSI B1.18M-1982 (R1987). "Metric Screw Threads for Commercial Mechanical Fasteners—Boundary Profile Defined." New York: American National Standards Institute.

ANSI 18.6.4-1981. "Screws, Tapping and Metallic Drive, Inch Series, Thread Forming and Cutting." New York: American National Standards Institute.

ANSI B 1.9-1973 (R1985). "Buttress Inch Screw Threads." New York: American National Standards Institute.

ANSI B6.1-1968 (R1974). "Coarse-pitch Spur-gear Tooth Forms." New York: American National Standards Institute.

ANSI/SAE MA 1370. "Screw Threads—MJ Profile, Metric, NA." New York: American National Standards Institute.

Koelsch, J.R., ed. 1994. "Hobs in High Gear." *Manufacturing Engineering,* July: 67.

——. 1992. "Turning on Threading & Grooving." *Manufacturing Engineering,* July: 59.

Mason, F., ed. 1996. "Threading Your Way through Thread Gaging." *Manufacturing Engineering,* April: 37.

Rice, Carl S. 1977. "Gear Accuracy versus Cost: Here are the Trade-offs." *Manufacturing Engineering,* April.

Richmind, D. 1992. "Methods of Grinding Spur and Helical Gears." *Manufacturing Engineering,* May: 87.

Society of Manufacturing Engineers (SME). 1996. *Threading Basics* video. *Fundamental Manufacturing Processes* series. Dearborn, MI: SME.

Manufacturing Systems

29

Industrial robot, 1963
—Joseph Engelberger and George Devol

29.1 INTRODUCTION

A basic and important function in human activity is making things. In its broad definition, this is what most people refer to as "manufacturing." *Manufacturing*, also referred to as *production*, is a transformation process that converts input factors into a product of utility and value to humans. Most manufacturing processes can be modeled as shown in Figure 29-1. Factors of input include:

- material—raw material, purchased parts and components, or subassemblies;
- manpower;
- machines—this includes machine tools, computer numerically-controlled machine tools, measuring machines, or computers;
- production tooling—cutting tools, dies, jigs and fixtures, etc.; and
- energy required to run the machines.

The manufacturing process, defined as a transformation process, involves many activities, such as manufacturing planning, manufacturing operations, assembly operations, material handling and storage, quality assurance, and manufacturing control.

29.2 MANUFACTURING SYSTEMS

A *manufacturing system* encompasses a number of activities, each with some input parameters or factors. A system has an output that is the result of converting or processing factors of input into a useful output of some value. By this definition, manufacturing is a system of input factors that primarily include material, manpower, and machines that go through a certain sequence of activities or processes to convert them into tangible items or products. Input factors or resources may be generally classified as human resources, information technology, and physical resources. A manufacturing system is driven by a unique set of mechanisms as shown in Figure 29-2.

Manufacturing systems are characterized by different aspects that include the following factors.

1. The manufacturing system is a unique assembly of machine tools, production workers, production tooling, jigs and fixtures, material handling equipment, and other supplementary devices, supported by a production planning and control system, manufacturing method, technology, and software support systems. These static features may be represented on the layout of the manufacturing plant and facilities.
2. The manufacturing system is defined as a transformation process of different factors of input into finished products. An objective of this transformation process is to maximize productivity and improve efficiency and uti-

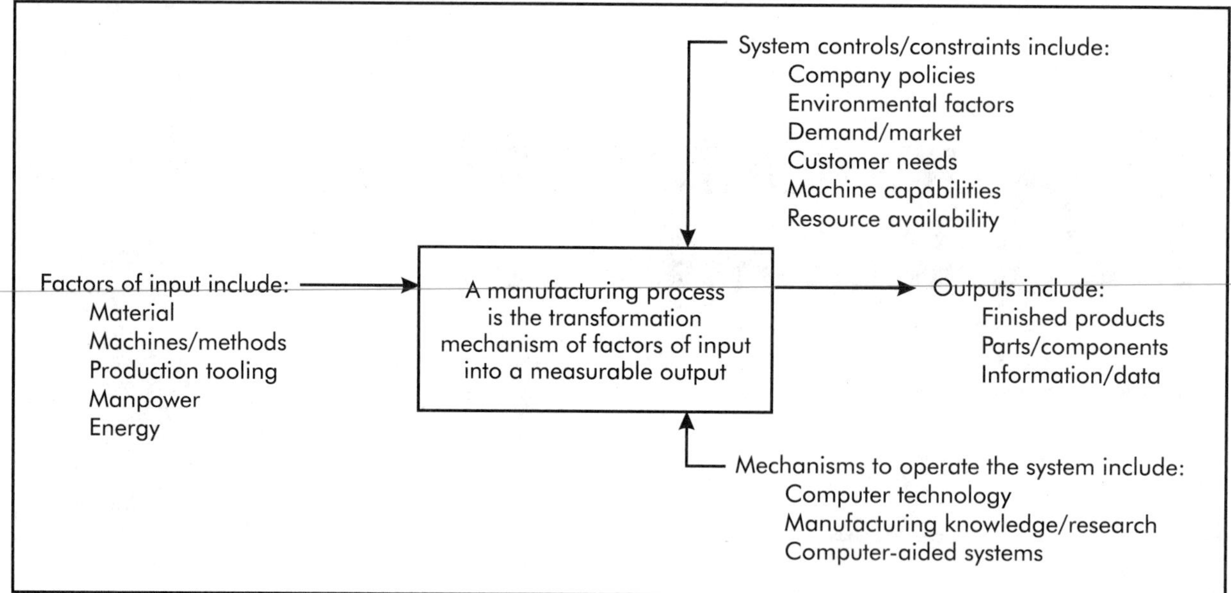

Figure 29-1. Manufacturing process model.

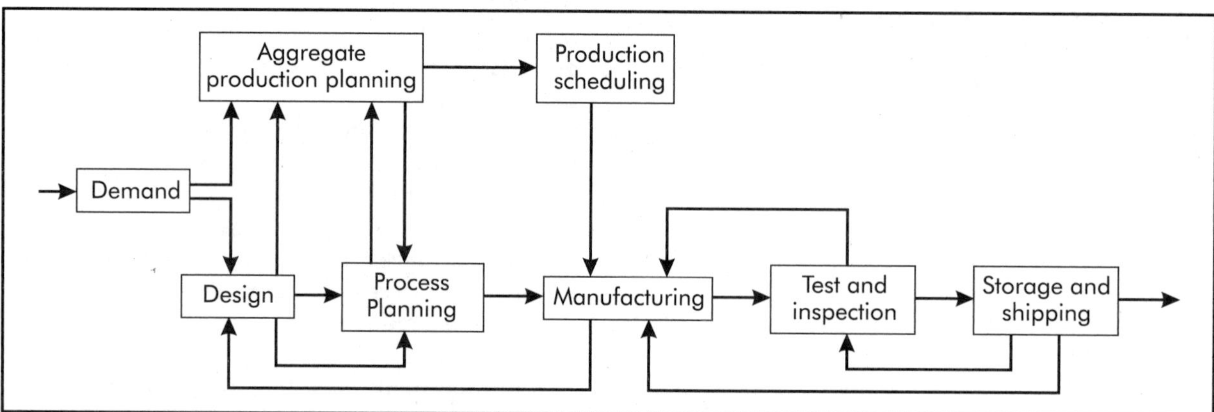

Figure 29-2. Manufacturing system components.

lization of resources. Activities in this transformation or conversion process include product design and development, manufacturing planning and control, manufacturing processes, and product quality assurance.

3. The manufacturing system is considered the operating mechanism of production, commonly known as the manufacturing management system. Activities such as manufacturing resource planning, capacity planning, establishing production objectives, realizing management goals, process planning, aggregate production planning, and other production planning and control activities are included in the manufacturing system. In general, the manufacturing system develops plans and procedures to implement the transformation process of factors of input into finished products to meet production objectives and desired specifications and performance requirements.

29.2.1 Classification

An essential goal of the development of manufacturing systems is to develop and produce parts or components with certain variety, defined quality, and within certain time frames. This goal can be met with different types of manufacturing systems. Many variables define the classification of a manufacturing system. These variables include:

- varieties of parts and components the manufacturing system is capable of producing;
- production rates; and
- level of computer technology employed.

Based on these factors, manufacturing systems can generally be classified as classical or traditional manufacturing systems, such as job shops, flow shops, project shops, and continuous processes. Manufacturing cells and computer-integrated manufacturing (CIM) systems, where computers play a major role in operations, are classified as advanced manufacturing systems.

Most common manufacturing systems are represented in Figure 29-3. These manufacturing systems involve different kinds of machine tools, computer numerically controlled (CNC) machine tools, robots, material handling system(s), special-purpose machines, and transfer lines. The level of automation for these components varies, sometimes significantly, depending on the desired mode of operation.

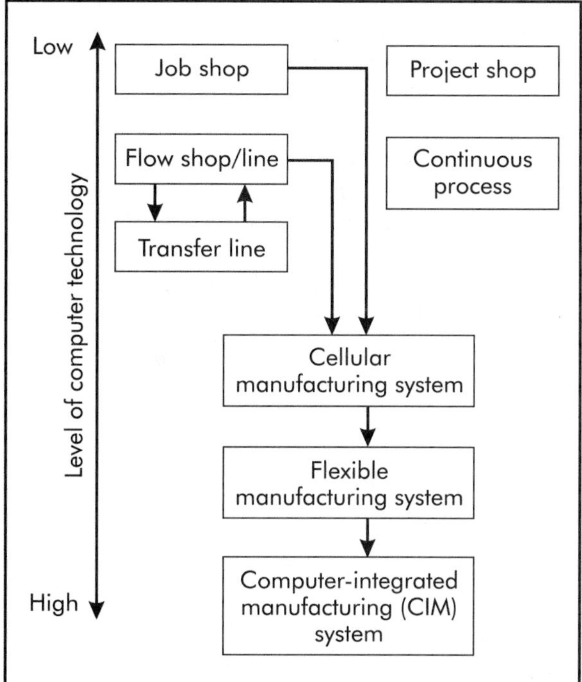

Figure 29-3. Classification of manufacturing systems.

Job Shops

In *job shops*, the production volume is normally low, up to 100 units per year, and product varieties are high, often one-of-a-kind. Job shops are generally capable of producing a large number of different items that require different sequencing and routing through the production equipment. Products made in job shops are normally specialized or customized, such as aircrafts and special tooling and machinery. To deal with wide product variations, labor skills are normally very high and the equipment used is general purpose. Another feature of job shops is the needed flexibility in performing different operations where the layout of equipment and tooling is generally fixed. Larger products with many subassemblies remain at single locations and the completed subassemblies are brought and assembled using different means of material handling and other production tooling equipment.

Because of the unique and different features of each product produced in a job shop, flexibility in product routing through different machines or machine groups is another feature of a job shop environment. Scheduling and production planning problems often arise because of the large number of different products, lower degree of routing and sequencing flexibility, and unpredicted demands for these products. The time each item spends at each machine varies, depending on the product and the operations to be performed. In general, workers move around different machines or stations. Sometimes, in cases of heavy products, such as an airplane or in a ship yard, they move around these products to perform different operations. In some cases, different components, each requiring a different operation sequence, are routed through different machine groups for processing. The layout of machines in this case is referred to as the *process layout*. A schematic representation of a typical job shop environment is shown in Figure 29-4.

Flow Shops

Flow shops or *flow lines* can be distinguished by their affinity to product layout. The most com-

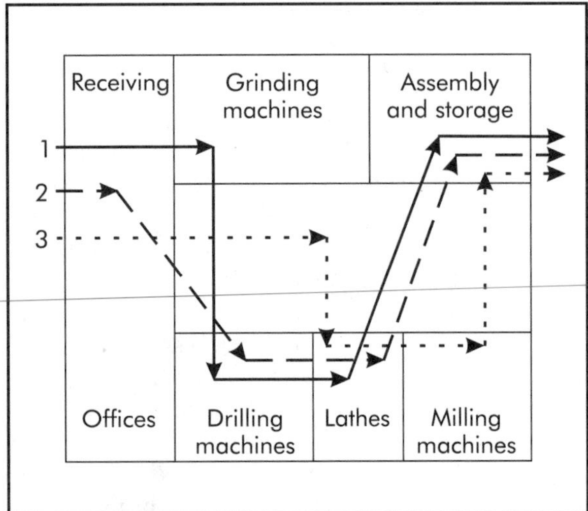

Figure 29-4. A schematic representation of a job shop environment.

mon type of flow-line manufacturing system is the assembly line. In an assembly line, for example, production volumes get very high, resulting in a situation known as mass production. Special-purpose machines and equipment are generally used in flow shops, dedicated specifically to the production of a low variety of products. Initial investment in equipment and hardware, compared to job shops, is very high. On the other hand, the overall workers' skill level is lower than in a job shop. The sequence of operations is fixed; products or items go through the operations one at a time. The time each item spends at each station or machine is fixed. Figure 29-5 shows a schematic representation of a flow shop.

The product sequence of operations dictates the arrangement of production machines and assembly equipment. In general, production and assembly facilities are arranged in line with one machine or workstation of a type. This flow line is dedicated to the production of a batch of a certain product or product group. Setup times are only needed when changing from one product to another. In flow-line manufacturing systems, the workpieces are physically moved through the different machines or workstations on a conveyor or other type of material handling equipment. Examples of applications for high-volume flow-line manufacturing systems include the assembly of television sets and automotive engines.

A transfer line is an example of a flow-line manufacturing system. Transfer machines produce standardized parts in large quantities with minimum worker skill requirement and high productivity. They are probably the most highly automated manufacturing systems and traditionally have been used for the production of single products in very high quantities over long production runs.

Project Shops

Project manufacturing systems are often used where the size or weight of the product justifies performing the different operations in a fixed location. In this fixed-position layout, workers bring the materials, tooling equipment, and machines needed for different operations to the work site. Shipbuilding, construction, and aircraft assembly operations are examples of *project shops*. Manufacturing systems of this type may incorporate job shops or flow lines to make parts or components for the project. Normally, manual or mobile general-purpose machines are used in a project shop. Other characteristics of project shops are the high and specialized workers' skills and large work-in-process inventories. Manufacturing lead times vary, depending on the nature of the project.

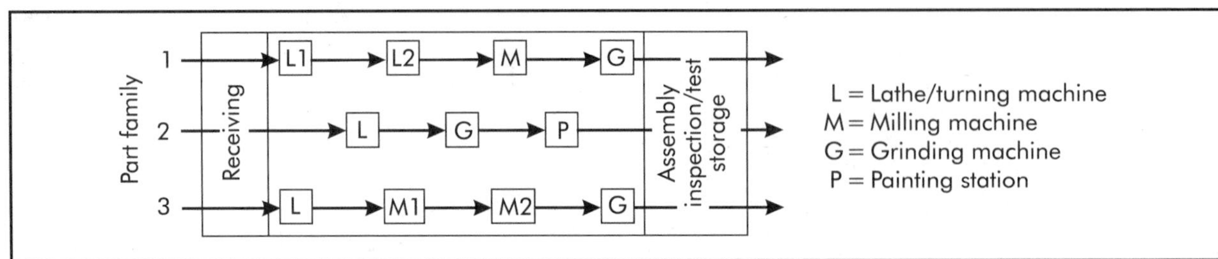

Figure 29-5. A schematic representation of a flow shop environment.

Continuous Processes

Continuous processes are found in oil refineries, food processing plants, or chemical processing operations and involve the continuous physical flow of product. These systems are the least flexible kind of manufacturing system but the most efficient. In continuous processes, work-in-process is the least of all manufacturing systems. These systems are easy to control. Machines that are normally used are specialized and the skill level of workers varies, depending on the types of processes involved and the level of technology employed. Lead times are normally short, but always constant.

Manufacturing Cells

Manufacturing cells or *cellular manufacturing systems* are used to manufacture groups or families of products. Often, U-shaped cells of different manual or numerically controlled (NC) machines comprise the plant layout. A horizontal CNC machining center typically used in flexible manufacturing cells is shown in Figure 29-6.

Figure 29-6. A horizontal CNC machining center typically used in flexible manufacturing cells. (Courtesy Cincinnati Milacron)

Cells may or may not include workers in the system. When workers are involved, the manufacturing cell is referred to as a manned cell. Workers in a manned cell move from one machine to another to service different processes and load and unload parts. In unmanned or unattended cells, robots are often used to perform this function of loading and unloading. A robotic cell is typically formed with a robot servicing three or more computer numerically controlled (CNC) machine tools. An example of a robotic cell is shown in Figure 29-7.

Manufacturing cells may be linked either directly to each other or indirectly through a system of production control, such as kanban, to form a linked-cell manufacturing system (LCMS). Modern trends in cellular manufacturing employ the use of some automated part loading and unloading, material handling, inspection, and storage. Automated guided vehicles (AGV) and automated storage and retrieval systems (AS/RS) are commonly used in cellular manufacturing systems. Cellular manufacturing systems are typically designed for medium production volumes of families of parts with a limited variety of product configurations. Because cellular manufacturing systems are flexible, they are sometimes considered flexible manufacturing systems (FMS).

Cellular manufacturing systems can be easily formed by restructuring job shop layouts. By applying the concept of group technology, which is based on the formation of part families using similarities in design and manufacturing characteristics between different machined parts, cells can be formed to process families of parts and components. The functional layout concept traditionally used in a job shop environment can be eliminated by the formation of cells. Linking cells is possible to operate synchronously with other assembly and subassembly lines. Flow lines, although more efficient than job shops, are not flexible. The inflexibility of flow lines can be eliminated by breaking them into cells.

Computer-integrated Manufacturing (CIM) Systems

Computer-integrated manufacturing (CIM) *systems* employ the computational power of computers, the flexibility of NC machines, and an automated material handling system to manufacture products in a medium range of production volume, product variety, and configurations. The degree of flexibility provided by CIM systems may differ depending on the needs and applications.

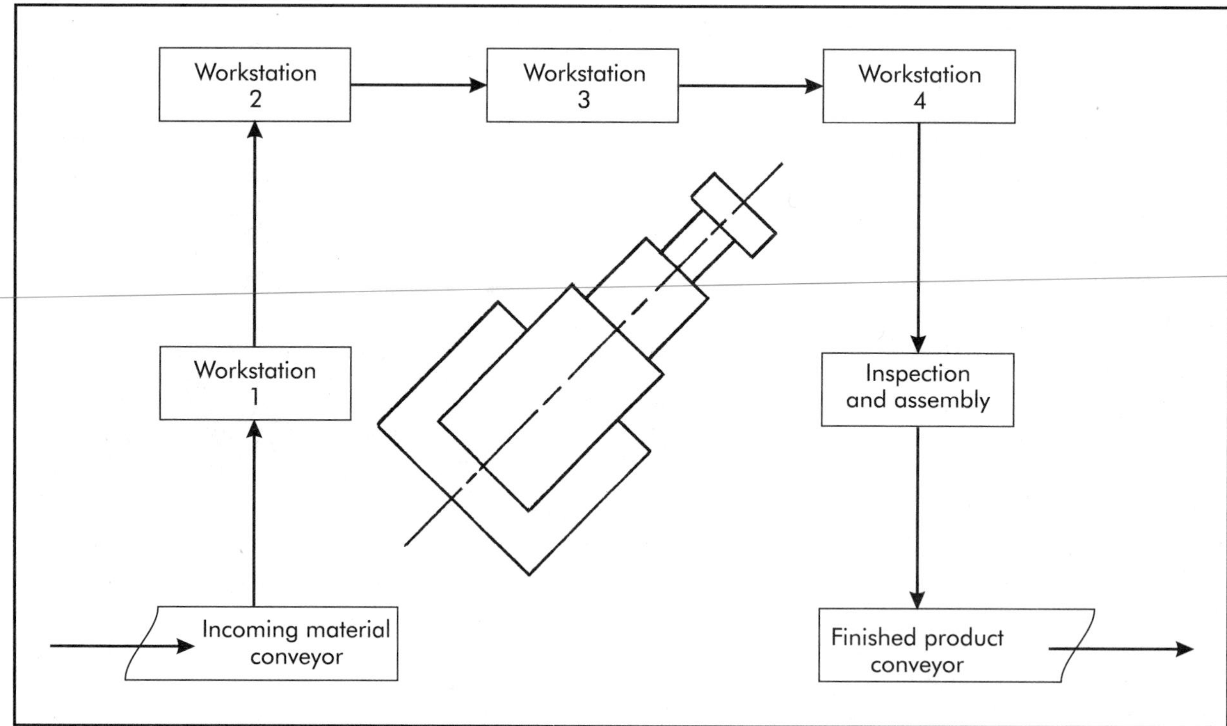

Figure 29-7. A schematic representation of a robotic cell.

Integrated manufacturing systems employ computer technology for all the engineering and business functions of the manufacturing organization. CIM systems encompass all levels of manufacturing activities and manufacturing planning and control, including product design and development, shop floor control, material control, production scheduling, quality assurance, production tooling, analysis and documentation of products and processes, facilities planning and layout, inspection and testing, material handling, material processing, and assembly. Other business functions, such as marketing, financial and accounting activities, and product and customer service also constitute elements of the integrated system.

To facilitate the understanding of CIM, the Computer and Automated Systems Association of the Society of Manufacturing Engineers (CASA/SME) has developed the *CIM wheel*, shown in Figure 29-8. The central core of the wheel, integrated systems architecture, handles common manufacturing data, and information resource management and communications. Various activities of manufacturing have been grouped under three categories. These categories are product/process, manufacturing planning and control, and factory automation. They include the various activities of manufacturing, such as design, material handling, material processing, inspection and test, assembly, and other activities as depicted on the wheel. The outer rim of the wheel represents the necessary support structure.

29.3 CONTEMPORARY MANUFACTURING TECHNOLOGIES

Current advances in manufacturing engineering aim at increased variety, smaller batches, improved quality, increased productivity, and shorter lead times. With the advent of powerful computer technology and new materials and manufacturing processes, new technologies in manufacturing have evolved in recent years in response to ever increasing competition and customer demand for higher quality. Some of these technologies include:

- just-in-time (JIT) manufacturing;
- flexible manufacturing systems (FMS);

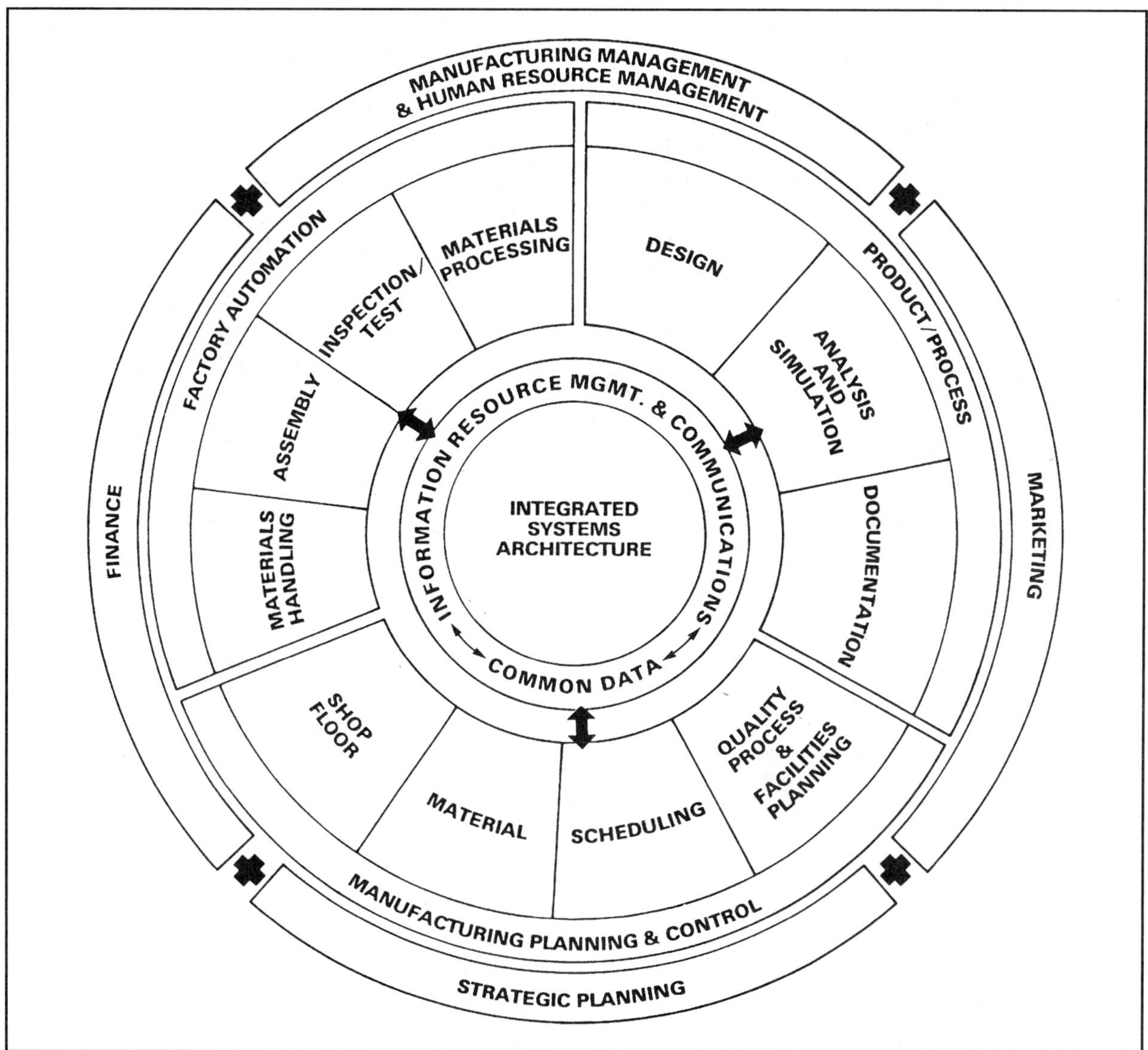

Figure 29-8. The CIM wheel symbolizing the concept of a computer-integrated manufacturing enterprise.

- coordinate measuring machines (CMMs);
- robotics;
- group technology (GT); and
- automation.

29.3.1 Just-in-time (JIT) Manufacturing

Just-in-time (JIT) is a manufacturing concept that aims at minimizing the inventory levels of products. Deliveries from suppliers and movement of products between production stations is carefully monitored so that at each stage the next lot or batch arriving to be processed is received just as the preceding lot has moved to the next stage. Usually, lots or batches arrive in small sizes, hence reducing work-in-process and the number of operators and machines waiting for items to be processed. JIT systems were first developed and implemented at Toyota Motor Company in Japan as an approach to reduce inventory and eliminate waste. In general, these systems are designed to provide a smooth flow of material and items between manufacturing stations with no or minimum stocks. The JIT concept targets reduction of purchased materials and components inventory through in-time delivery and supplier reliability. In the United States, JIT is also known as *zero inventory*.

There are different objectives associated with a JIT manufacturing system that are intended to promote the optimization of the total manufacturing system by developing the policies, procedures, and attitudes required to build a responsive and competitive manufacturing strategy. These objectives are briefly stated as follows:

- optimization of quality and cost;
- ease of manufacturing;
- minimization of the different resources involved in the design and manufacture of parts and products;
- quick responsiveness to customer needs;
- better relationships with suppliers and customers; and
- total commitment to improve the overall system of design and manufacturing.

29.3.2 Flexible Manufacturing Systems (FMS)

A *flexible manufacturing system* (FMS) is capable of producing varieties of parts or products having similar design or manufacturing characteristics, through different routings among different processing or machining stations. A typical FMS consists of a group of machining or processing stations, usually computer numerically controlled (CNC) machine tools served by an automated material handling system and controlled by an integrated control system. Flexible manufacturing systems are based on group technology (GT) concepts where parts or products are classified into families or groups based on design or fabrication similarities.

The number of processing stations in flexible manufacturing systems vary according to the nature and characteristics of the family of products being manufactured. The term *flexible manufacturing cells* (FMC) often arises when the FMS has a limited number of machines, usually three or less.

The basic composition of an FMS may include all or most of the following elements:

- a number of CNC machine tools;
- industrial robots;
- automated material handling systems; and
- automated inspection stations. In most cases, a coordinate measuring machine is used for this purpose.

These elements are arranged in a layout to allow the processing of more than one item at the same time through different routings between different workstations. Flexible manufacturing systems allow the efficient production of medium-size batches of similar products. Research in manufacturing automation has showed that the implementation of flexible manufacturing systems has resulted in an increase in productivity of more than 50% in many cases.

29.3.3 Coordinate Measuring Machines (CMMs)

Over the last decade, *coordinate measuring machines* (CMMs) have become a primary means of dimensional quality control for manufactured parts of complex forms where the volume of production does not warrant the development of functional gaging. The advent of increasingly inexpensive computing power and more fully integrated manufacturing systems will continue to expand the use of these machines into an even larger role in the overall quality assurance of manufactured parts.

A CMM can be most easily classified as a physical representation of a three-dimensional rectilinear coordinate system. CMMs now represent a significant fraction of the measuring equipment used for defining the geometry of workpieces of different shapes. Most dimensional characteristics of many parts can be measured within minutes with these machines. Similar measurements would take hours using older measurement equipment and procedures. Besides flexibility and speed, coordinate measuring machines have several additional advantages.

1. Different features of a part can be measured in one setup. This eliminates errors introduced due to setup changes.
2. CMM measurements are taken from one geometrically fixed measuring system, eliminating the accumulation of errors resulting from using functional gaging and transfer techniques.

3. The use of digital readouts eliminates the necessity for interpretation of readings such as with the dial or vernier-type measuring scales.
4. Minimum operator intervention is required with the use of automatic data recording available on most CMMs.
5. Part alignment and setup procedures are greatly simplified by using software supplied with computer-assisted CMMs. This minimizes the setup time for measurement.

Although coordinate measuring machines can be thought of as a representation of a simple coordinate system for measuring the dimensions of workpieces of different shapes, they are naturally constructed in many different configurations, all of which offer different advantages. CMMs provide a means for locating and recording the coordinate location of points on parts and their relation to one another. They can be classified according to their configurations.

1. A *cantilever CMM* has its probe attached to a vertical machine ram (*Z* axis) moving on a mutually perpendicular overhang beam (*Y* axis) that moves along a mutually perpendicular rail (*X* axis). The cantilever configuration is limited to small and medium-size machines. It provides easy operator access and the possibility of measuring parts longer than the machine table. Figure 29-9(A) shows a cantilever configuration.
2. *Bridge-type CMMs* have a horizontal beam moving along the *X* axis and carrying the carriage to provide the *Y* axis motion as shown in Figures 29-9(B) and 16-23. In the "fixed bridge" configuration, the horizontal beam (bridge structure) is rigidly attached to the machine base and the machine table moves along the *X* axis. A bridge-type coordinate measuring machine provides more rigid construction, which in turn provides better accuracy. The presence of the bridge on the machine table makes it a little more difficult to load large parts. Figure 29-9(C) shows an example of a coordinate measuring machine with a fixed-bridge configuration.
3. In the *column-type CMM,* a moving table and saddle arrangement provides the *X* and *Y* motions and the machine ram (*Z* axis) moves vertically relative to the machine table. A configuration of this machine is shown in Figure 29-9(D).
4. *Horizontal arm CMMs* feature a horizontal probe ram (*Z* axis) moving horizontally relative to a column (*Y* axis), which moves in a mutually perpendicular motion (*X* axis) along the machine base. This configuration provides the possibility for measuring very large parts. Another arrangement of this configuration features a fixed horizontal arm to which the probe is attached and moving vertically (*Y* axis) relative to a column that slides along the machine base in the *X* direction. The machine table moves in a mutually perpendicular motion (*Z* axis) relative to the column. These configurations are shown in Figures 29-9(E) and (F).
5. *Gantry-type CMMs* have a vertical arm (*Z* axis) moving vertically relative to a horizontal beam (*X* axis), which in turn moves along two rails (*Y* axis) mounted on the floor. This configuration provides easy access to the machine and allows the measurement of very large components. Figure 29-9(G) illustrates the gantry-type configuration of coordinate measuring machines.
6. *L-shaped bridge CMMs* have a ram (*Z* axis) moving vertically relative to a carriage (*X* axis), which moves horizontally relative to an L-shaped bridge moving in the *Y*-direction. An L-shaped configuration is shown in Figure 29-9(H).

In addition to classifying coordinate measuring machines according to their physical configuration, they can be classified according to their mode of operation: manually operated, computer-assisted, or direct-computer controlled. In manual machines, the operator moves the probe along the machine's axes to establish and manually record the measurement values provided by digital readouts. In some machines, digital printout devices can be used.

Computer-assisted CMMs can be either manually positioned (free-floating mode) by moving the probe to measurement locations, or manually driven by power-operated motions under the control of the operator. In either case, data

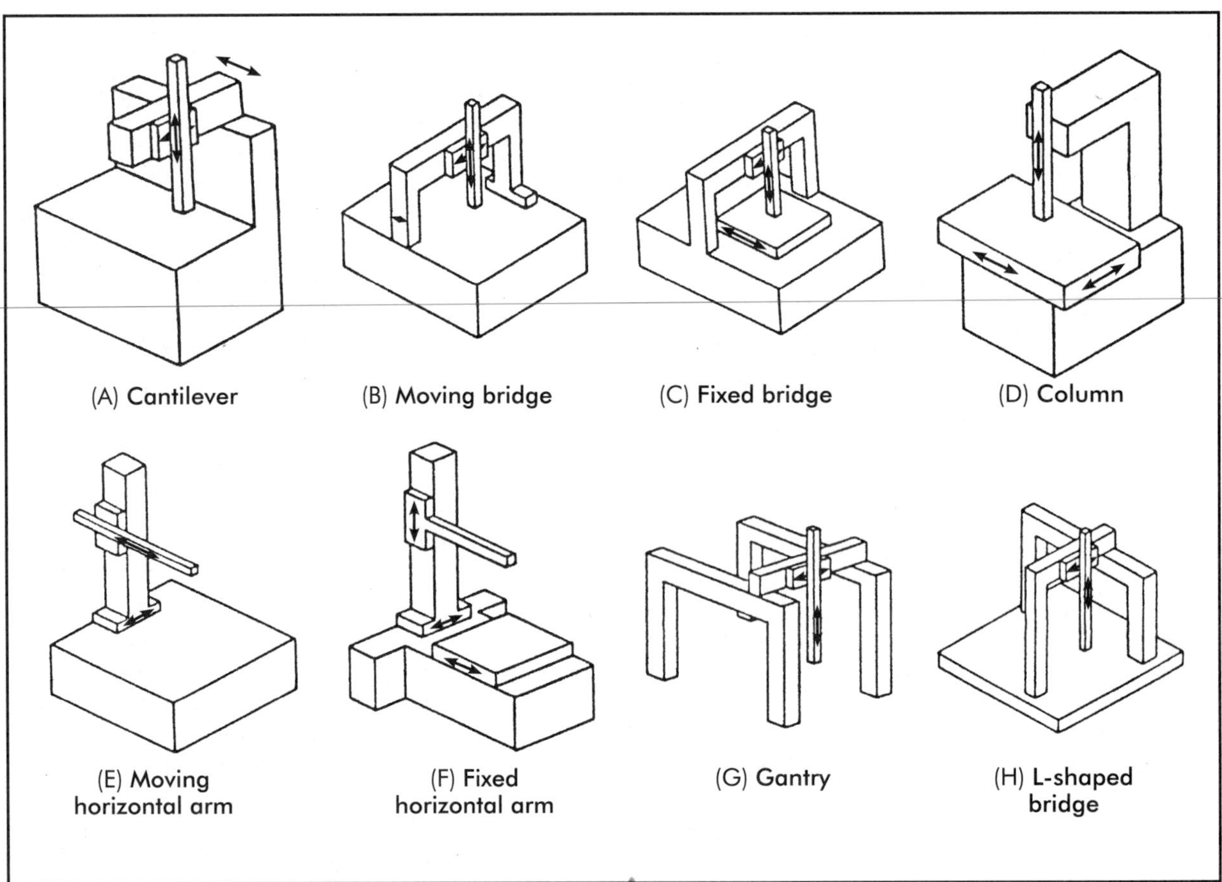

Figure 29-9. Examples of the different configurations of coordinate measuring machines.

processing is accomplished by a computer. Some computer-assisted coordinate measuring machines can perform some or all of the following functions:

- inches/millimeter conversion;
- automatic compensation for misalignment;
- storing of premeasured parameters and measurement sequences;
- data recording;
- a means for disengaging the power drive to allow manual adjustment and manipulation of the machine motions; and
- geometric and analytical evaluations.

Direct computer-controlled CMMs use a computer to control all machine motions and measuring routines and to perform most of the required data processing. These machines are operated in much the same way as CNC machine tools. Both control and measuring cycles are under program control. Off-line programming capability is also available. Additional details on CMMs are given in Chapter 16.

29.3.4 Group Technology (GT)

Group technology (GT) is a manufacturing philosophy that involves the classification of items or parts according to similarities in their design features and manufacturing characteristics and grouping them into part families. This concept smoothes the flow of material within the manufacturing system and increases productivity. Group technology may be defined as a philosophy that identifies and realizes the underlying similarities of items and their associated processes used for manufacture. The success of group technology is reliant upon the existence of a data base that describes and identifies the different parts, processes, and their associated parameters and characteristics. In group technology, parts or products are classified into families using different approaches. One approach uses visual inspection of all the parts produced. Then experience and expert judgment are used to classify these parts into families. Another approach involves the use of information gathered from route sheets to classify parts. This approach is known as *production flow analysis*. A third approach for the identification of similarities between parts is to classify and code parts by means of an appropriate coding scheme. Figure 29-10 describes an example showing the basic structure of a classification and coding system. Many references describe other schemes for parts classification and coding. A typical classification and coding scheme may include the following major design and manufacturing attributes:

- design-related attributes—basic dimensions, external shape, internal shape, material type and composition, accuracy requirements, functional characteristics; and
- manufacturing-related attributes—major manufacturing process(es), tolerances, surface finish, length-to-diameter ratio, and production rate.

A well-designed classification and coding system provides several advantages. Benefits and advantages of a well-designed classification and coding system are cited in references given at the end of this chapter. These benefits include:

- engineering design—design standardization and rationalization, design retrieval, reduced number of similar parts and components and number of drawings, elimination of part and component details;
- manufacturing—reduction of the number of setups and associated cost and time, improved layout and space utilization, reduced material handling time and equipment, improved utilization of machine tool requirements, improved capacity planning among machine groups, improved resource allocation, increased use of flexible manufacturing cells and universal machines and equipment, improved production planning activities, decreased manufacturing costs, improved quality, reduced throughput time;

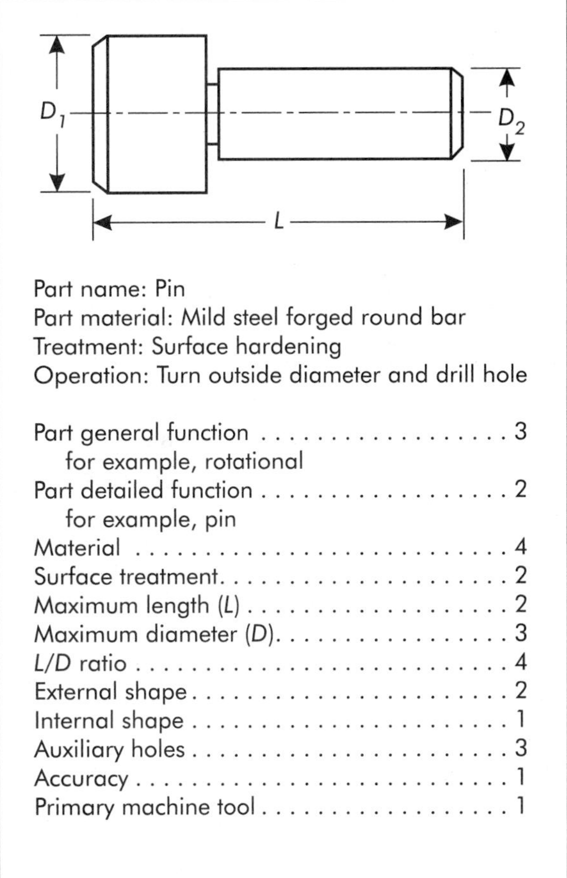

Figure 29-10. An example of a coded part using a classification and coding system.

- manufacturing engineering—reduced number of process plans, reduced process planning effort and time, reduced number of NC programs and programming time, improved flexibility and reusability of process plans, reduced number of production tooling, jigs, and fixtures, standardization of part routings among machine groups, reduction of production tooling design and product analysis;
- production control—reduced work-in-process inventory, improved predictability of production costs, reduced part routing and movement among machine groups, easier identification of problems and bottlenecks, improved production scheduling and equipment monitoring, improved capacity planning effort and time, improved supplier relationships; and

- quality control—improved overall part and product quality and control, improved design quality, ease of built-in quality into products and defect identification, and reduced sampling and inspection time.

29.3.5 Robotics

The word *robot* originates from a Czech word. Carl Capek, a Czech playwright, wrote in 1921 the drama R.U.R. (*Rosum's Universal Robots*), in which he introduced the word "robot" or "forced labor." Today, robots have captured the imagination of many professions.

New Webster's Dictionary defines a robot as "a mechanical device designed to do the work or part of the work of one or more human beings." The Robot Institute of America (RIA) defines the robot as "a programmable, multifunctional manipulator designed to move material, parts, tools, or specialized devices through variable programmed motions for the performance of a variety of tasks."

Applications

Applications of robots in many manufacturing and service industries have made significant contributions to the productivity of the manufacturing industry. Robots may be used for loading and unloading of parts and components between different machines or machining stations, performing jobs in many hostile or hazardous environments, performing repetitive and tedious jobs more efficiently than human workers, and for many finishing jobs and inspection operations.

When choosing a robot for a certain task or operation, it is necessary to consider the following factors:

- arm configuration;
- arm reach;
- load-carrying capacity;
- type of robot drive;
- application capabilities;
- end effector interchangeability;
- desired accuracy and resolution;
- control method;
- programming capabilities; and
- other physical features.

Types

The nature of robot movements required defines its configuration and the number of movements defines its degrees of freedom. For example, the robot shown in Figure 29-11 has the following degrees of freedom: (1) column rotation or swivel, (2) vertical arm travel along the column, (3) arm reach, (4) wrist pitch or wrist bend, and (5) wrist roll.

Additional degrees of freedom may be added to the robot, depending on the desired motion sequence, robot application, and type of robot drive. For example, the robot shown in Figure 29-11 may have a sixth degree of freedom added by providing a rotation about the axis perpendicular to the axes of wrist roll and wrist bend rotation.

Figure 29-11. Direct drive robot with five degrees of freedom operating on a candy packaging line. (Courtesy Adept Technology, Inc.)

Robots may generally be classified according to their physical configuration.

1. *Cartesian robots* may be gantry-mounted or mounted on a track on the floor. A robot of this configuration has a rigid structure and three perpendicular axes in the *X*, *Y*, and *Z* directions as shown in Figure 29-12(A). As in other configurations, the gripper or end effector has three rotations, normally known as rotational or angular degrees of freedom. They are known as the pitch, roll, and yaw.
2. *Cylindrical robots* employ cylindrical coordinates as shown in Figure 29-12(B), where the configuration consists of a vertical column to provide rotation or swivel about the base and an arm assembly to vertically move relative to the column. The arm also moves in and out to provide a third degree of freedom. The end of the robot arm normally provides two additional degrees of freedom, wrist bend and wrist roll.
3. *Spherical robots* employ polar coordinates. The configuration has a sliding bar providing reach by sliding relative to the arm. The arm provides rotation about a vertical and a horizontal axes. Figure 29-12(C) shows a robot with a spherical configuration.
4. *Articulated robots* employ revolute coordinates and work like a human arm. They swivel about the base and rotate about the shoulder and elbow joints. In most cases, the gripper or the robot end effector provide three angular rotations: pitch, roll, and yaw. Figure 29-12(D) shows an articulated robot.

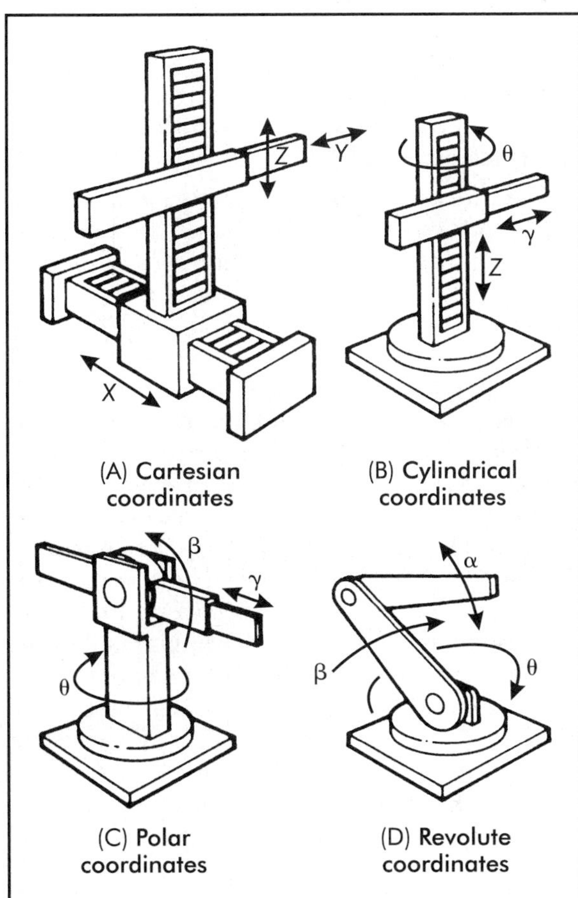

Figure 29-12. Different configurations of industrial robots.

A special case of articulated or jointed arm robots is the selective compliance arm for robotic assembly (SCARA) robot, which allows the rotation of joints in a horizontal plane. An example of a SCARA robot is shown in Figure 29-11. SCARA robots have been specifically designed for electronic part assembly, which requires motions in a horizontal plane to locate components in space and insert such components along a vertical axis.

Control

Robots perform certain movements to accomplish tasks. To do this, the desired movement of each joint must be controlled so that the robot performs a desired sequence-of-motion cycle. The movements of joints are affected by the type of robot drive governing the motion control of each joint. The type of motion control may be classified as follows.

1. In *limited sequence control*, robot capability is limited to pick-and-place tasks. Robots having this type of motion control are often referred to as pick-and-place robots. Limited sequence control is the simplest control system. Typically, robots having this kind of control are capable of picking up small loads and moving them rapidly from one location to another. Their movements

are limited in number, and their control systems are rudimentary and may consist of a series of switches or valves. Length of movement may be determined by mechanical stops. Programming capabilities for these robots are limited. Programming may be done by placing tabs or pins in a plugboard, making pneumatic connections as needed, or push-button setting of binary controls. Reprogramming is slow and is not expected to be done. Robots of this type of control often have pneumatic drives. Precise positioning cannot be accomplished because of the lack of a servo control for robot joints.

2. Playback robots having a *point-to-point* (PTP) *control* system have the capability of moving to and from a number of points but not necessarily in the same sequence. PTP control provides additional capabilities, versatility, and reprogramming capability afforded by a position servo at each joint. This feature adds precision to point location. Programming is done by teaching, where the operator initially leads the robot through its routine. Locations and motion sequence are programmed into memory and subsequently played back during operation.
3. A robot having *continuous-path control* has the same capabilities as the playback robot with point-to-point control. These robots, however, have much more memory storage and a processor capable of coordinating the movements of the joints and interpolating a given path between end points. When programmed, the robot is capable of moving precisely between two different points along a certain path. These robots are used in sophisticated industrial applications such as spray painting and arc welding. Servo control is used to maintain continuous control over the position and speed of the manipulator.

29.3.6 Automation

Automation involves the use of computers and other automated machinery for the execution of tasks that a human laborer normally performs. Automated machinery may range from simple sensing devices to robots and other sophisticated equipment. Automation of operations may range from the automation of a single operation to the automation of an entire factory.

There are many different reasons to automate. Increased productivity is normally the major reason for many companies desiring a competitive advantage. Automation also offers low operational variability. Variability is directly related to quality and productivity. Other reasons to automate include the presence of a hazardous working environment, the presence of hostile environments, and the high cost of human labor. The decision regarding automation is often associated with some economic and social considerations.

Major Alternatives

In recent years, the manufacturing field has witnessed the development of major automation alternatives. Some of these include:

- computer-aided manufacturing (CAM);
- numerically controlled (NC) machines;
- robots;
- flexible manufacturing systems (FMS); and
- computer-integrated manufacturing (CIM) systems.

Computer-aided manufacturing (CAM). CAM refers to the use of computers in the different functions of production planning and control. CAM includes the use of NC machines, robots, and other automated systems for the manufacture of products. It also includes computer-aided process planning (CAPP), group technology (GT), production scheduling, and manufacturing flow analysis. In *computer-aided process planning,* computers are used to generate process plans for the manufacture of different products.

Numerically controlled (NC) machines. NC machines are programmed to execute operational sequences on parts or products. Individual machines may have their own computers for that purpose and they are normally known as *computer numerically controlled* (CNC) machines. In other cases, many machines may share the same computer and are normally referred to as *direct numerically controlled machines.*

Robots. Robots are automated equipment that, through programming, may execute different tasks that are normally handled by a human operator. In manufacturing, robots are used to handle a number of tasks that include assembly, welding, painting, loading and unloading, inspection and testing, and finishing operations.

Flexible manufacturing systems. A flexible manufacturing system may simply be defined as a group of computer numerically controlled machine tools, robots, and an automated material handling system used for the manufacture of a number of similar products or components using different routings among the machines. Flexible manufacturing systems have proven to increase manufacturing productivity by at least 50%.

Computer-integrated manufacturing systems. Computer-integrated manufacturing encompasses many manufacturing functions linked through an integrated computer network. These manufacturing or manufacturing-related functions include production planning and control, shop floor control, quality control, computer-aided manufacturing, computer-aided design, purchasing, marketing, and other functions. The objective of a CIM system is to allow changes in product design, reduce costs, and optimize production requirements.

29.4 EMERGING TECHNOLOGIES

Customer satisfaction and the pursuit of perfection in the design and manufacture of products in a shorter time are among many driving forces for emerging manufacturing technologies. Manufacturing engineering research and development efforts have resulted in the birth of techniques and philosophies that are targeting improved quality of manufacturing products, wider range of customer acceptance, shorter lead times, and improved manufacturing systems productivity. Among such technologies are total quality management, concurrent engineering, and rapid prototyping and manufacturing. Each of these technologies are thoroughly detailed in many of the chapter references.

29.4.1 Total Quality Management (TQM)

Total quality management (TQM) is an approach for continuously improving the quality of products through the participation of all units in the organization. TQM is a philosophy rooted in Dr. W. Edwards Deming's 14 points for management essential to establish a TQM environment. These points are extensively discussed and presented in many references. The following comments highlight Deming's philosophy to improve manufacturing productivity.

1. A strategy must be adopted for the improvement of the product or service. This strategy must be customer-oriented and translate customer needs into the products and processes that address these needs effectively. The strategy should be built on statistical thinking for process control and is oriented toward a continuous (never-ending) improvement of the process as diagramatically shown in Figure 29-13.
2. For quality assurance, 100% inspection is not an acceptable way to improve quality levels. Dependence on mass inspection can be ceased by adopting process control models directed toward eliminating causes of out-of-control situations.
3. Companies should develop and forge long-term relationships with suppliers and provide assistance to them for training and implementation of statistical methods. The development of supplier certification programs often provides an incentive for those suppliers to develop and maintain process control-oriented quality programs.
4. Training and continuous education are essential ingredients for the process of continuous improvement.
5. Leadership and trust must be instituted within all departments and barriers should be eliminated between departments.

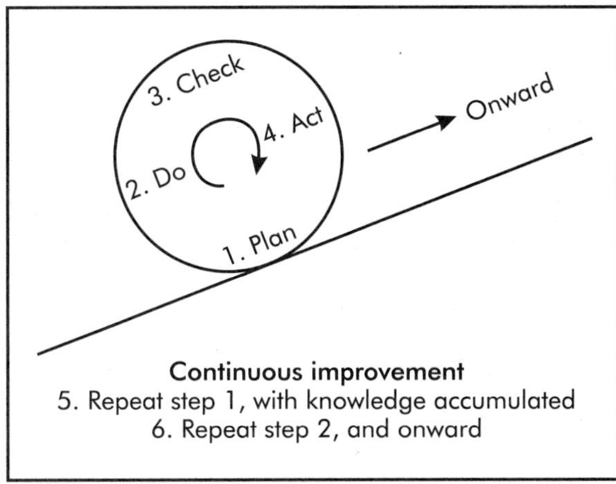

Figure 29-13. The never-ending improvement process.

To accomplish the transformation to total quality management, thinking about customer satisfaction must be a team oriented goal. Figure 29-13 shows the plan-do-check-act (PDCA) cycle, also known as the Shewhart cycle, used to accomplish the transformation.

Total quality management is an organizational, process-oriented philosophy based on three concepts: customer satisfaction, implementation of statistical tools, and continuous improvement. The customer is not just a person in a market outside the organization. A customer is also the next person working on the same product or service.

Statistical Process Control (SPC)

Statistical methods are tools based on simple concepts to help measure and improve quality. *Statistical process control* (SPC) is a set of tools used to improve the quality of products and services through the implementation of simple statistical concepts. SPC is at the heart of successful implementation of total quality management.

The main purpose of traditional quality control is to prevent the production of products that do not meet certain acceptance criteria. This could be accomplished by performing inspection on products that, in many cases, have already been produced. Action could then be taken by rejecting those products or putting them in the scrap pile. Some products would go to be reworked, which is costly and time-consuming. In many cases, rework is more expensive than producing the product in the first place. This situation often results in the following:

- decreased productivity;
- customer dissatisfaction;
- loss of competitive position; and
- higher cost.

To avoid such results, quality must be built into the product and the processes. SPC involves the integration of quality control in each stage of producing the product. In fact, SPC is a powerful collection of tools that implement the concept of prevention as a shift from the traditional quality by inspection/correction.

The main objective of any SPC study is to reduce variation. Any process can simply be considered as a transformation mechanism of different input factors into a product or service. Since inputs exhibit variation, the result is a combined effect of all variations. This, in turn, is translated into the product. The purpose of SPC is to isolate the natural variation in the process from other sources of variation that can be traced or whose causes may be identified.

There are two different kinds of variation that affect the quality characteristics of products. *Common causes of variation* are those variations that are inherent in the process. They are inevitable and can be represented to the best of our knowledge by a normal distribution. Common causes are also called *chance causes* of variation. A stable process exhibits only common causes of variation. The behavior of a stable process is predictable or consistent, and the process is said to be *in statistical control. Special causes of variation*, also called *assignable causes*, are not part of the process. They can be traced, identified, and eliminated. Control charts are designed to hunt for them as part of SPC efforts to improve the process. A process with the presence of special or assignable causes of variation is unpredictable or inconsistent, and the process is said to be *out of statistical control*.

Among the many tools for quality improvement, also known as TQM tools, the following are the most commonly used (see Figure 29-14):

- histograms;
- cause-and-effect diagrams;
- Pareto diagrams;
- control charts;
- scatter or correlation diagrams;
- run charts; and
- process flow diagrams.

Histograms. A *histogram* is a visual display that shows the variability of a process or other quality characteristic. It can be used to illustrate the specification limits on a product in relation to the natural limits for the process used to produce it. Histograms also can be used to identify possible causes of a problem experienced by the process. They are powerful analytical tools for understanding the process.

Pareto charts. A *Pareto chart* focuses the attention on those few problems that cause the most trouble in a process. With Pareto charts, facts about the greatest improvement potential can be easily identified.

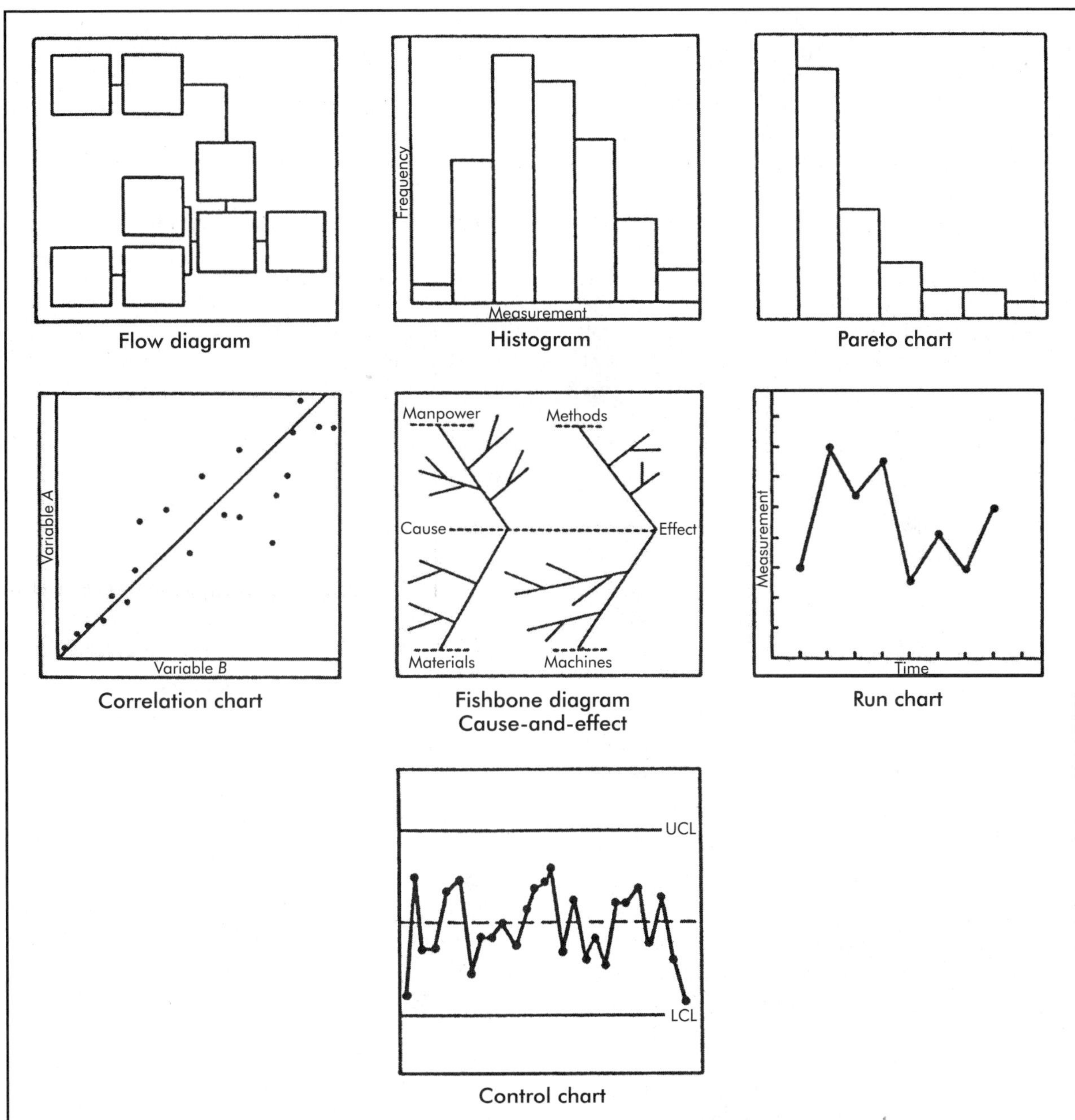

Figure 29-14. Basic tools of statistical process control.

Cause and effect diagrams. Also called *Ishikawa diagrams* or *fishbone diagrams*, *cause-and-effect diagrams* provide a visual representation of the factors that most likely contribute to an observed problem or an effect on the process. The relationships between such factors can be clearly seen, problems can be identified, and the root causes corrected.

Scatter diagrams. Also called *correlation charts*, *scatter diagrams* show the graphical representation of the relationship between two variables. In statistical process control, scatter diagrams are normally used to explore the relationships between process variables and may lead to identifying possible ways to increase process performance.

Control charts. As a graphical representation of process performance, *control charts* provide a powerful analytical tool for monitoring process variability and other changes in process mean or variability deterioration. Control charts were basically developed to hunt for special causes of variation.

Run charts. Depicting process behavior against time, *run charts* are important tools for investigating changes in a process over time. Any changes in process stability or instability can be judged from a run chart.

Process flow diagrams. As graphical representations of the process, *process flow diagrams* show the sequence of different operations that make up a process. Flow diagrams are important tools for documenting processes and communicating information. They also can be used to identify bottlenecks in a process sequence, points of rework or other phenomena in a process, or to define points where data or information about process performance needs to be collected.

29.4.2 Rapid Prototyping and Manufacturing

Rapid prototyping and manufacturing (RP&M) is a new technology that is capable of producing physical objects using graphical data generated by computers. RP&M was initially termed *stereolithography* or three-dimensional printing. Figure 29-15 illustrates the process of stereolithography. The use of rapid prototyping and manufacturing technology can be a crucial component of part quality improvement.

Rapid prototyping is the process of rapid creation of a physical solid model (prototype) for an object from its design data without the use of tools or traditional manufacturing processes. It involves the physical generation of a 3D computer-based model of an object from its CAD drawing. Rapid prototyping starts with a CAD file and ends with the physical model. Computers play a vital role in the successful accomplishment of the rapid prototyping process.

There are several technologies for rapid prototyping. These are stereolithography, selective laser sintering, solid ground curing, fused deposition modeling, and laminated object manufacturing. These techniques are discussed in detail in many books and other publications, some of which are listed at the end of this chapter.

One of the most usable techniques for rapid prototyping is *stereolithography*, which involves directing a laser beam of appropriate power and wavelength onto a surface of a liquid resin, forming patterns of solidified layers. The beam movement is guided by a computer that gets its movement instructions from specially designed software. Through multiple-layer solidification, the object is built over a span of time, the amount of which depends on the resin used, the power of the laser, the exposure time, and the configuration of the object being built. The process is not simple. This is due to the variety of materials being used and the complexity of interaction between various elements of the system, namely the object and its geometry, the software, computer control, the laser, and the material.

The following steps are based upon recent advances in fabricating functional parts through the use of rapid prototyping. They have been demonstrated to achieve improved part quality.

1. Design the prototype using CAD.
2. Build the prototype with the appropriate rapid prototyping technique.
3. Inspect the part for errors.
4. Correct any errors using CAD.
5. Verify the corrected part.
6. Iterate to improve the design.
7. Optimize the design using and testing different design variations.
8. Fabricate a functional test model utilizing the developed prototypes.
9. Perform functional testing on the functional test model.
10. Once achieving satisfactory results, proceed to manufacture.

The different operational steps in the rapid prototyping process may be summarized as follows.

1. The first step is the creation of a CAD model. It involves the generation of a three-dimensional computer-aided design model of the object. The best representation comes from CAD systems that utilize 3-D solid modeling.
2. The CAD file goes to a rapid prototyping translator, where the boundary surfaces of

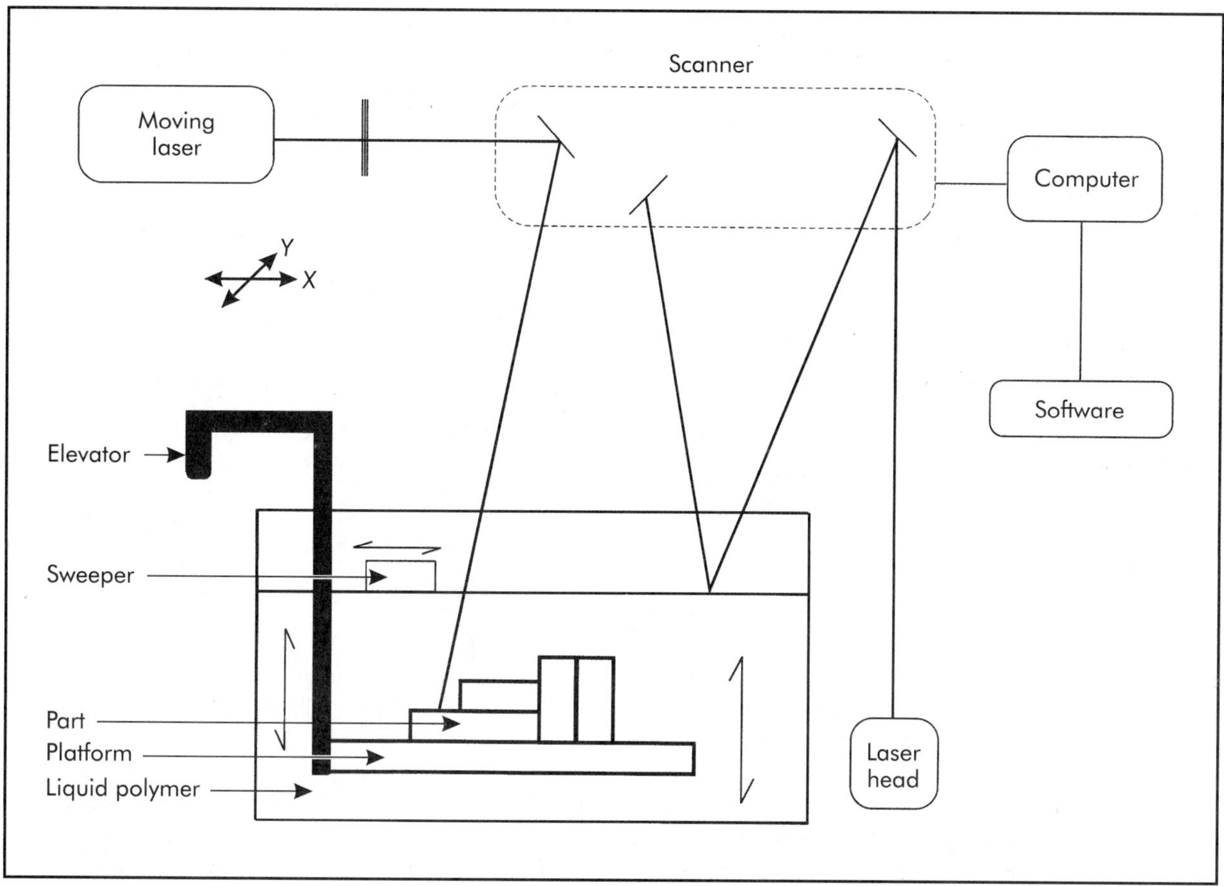

Figure 29-15. Stereolithography process.

the object are represented as numerous tiny triangles. The translator program scans the data file and defines the object in triangular format.

3. Next, supports are generated in a separate CAD file. Supports are used for different reasons: (1) to ensure that the recoater blade will not strike the platform upon which the part is being built, (2) to ensure that any small distortions of the platform will not lead to problems during part building, and (3) to provide a simple means of removing the part from the platform upon its completion.
4. Both the part and the supports are "sliced." The part is mathematically sectioned by the computer into a series of parallel horizontal planes. Also during this step, the layer thickness, intended building style, cure depths, desired hatch spacing, line width compensation value, and shrinkage compensation factor(s) are selected. The object and support programs and files are merged.
5. At the preparation step, certain operational parameters are selected, such as the number of recoater blade sweeps per layer, the sweep period, and the desired "z-wait." *Z-wait* is the amount of time, in seconds, that the system is instructed to pause after recoating. The purpose of this intentional pause is to allow any resin surface nonconformities to undergo fluid dynamic relaxation. The output of this step is the selection of the relevant parameters. Default parametric values may be used.
6. Resin polymerization begins and a physical, three-dimensional object is created through a sequence of steps and parameter extractions.
7. Once the last layer (the top layer) is completed, the control program elevates the

created object above the liquid resin and drains excess or trapped resin into the container. Cleaning solvent is used to wipe away any excess resin, and to clean the object and the platform on which the object is built.

8. The curing process is done by laser during object preparation. Using ultraviolet radiation in a special postcuring device, the object is fully polymerized. The mechanical strength of the object (prototype) is improved by allowing other operations to be performed, depending on the objective of the prototype.

29.4.3 Concurrent Engineering (CE)

Concurrent engineering (CE) is an environment of necessity for companies wanting to compete successfully in the global market. Product development cycle times that are not competitive often result in greater losses than those found with either development or manufacturing cost overruns. Significant reduction in the total cycle time can benefit companies in many ways. Companies can be more responsive to customer needs and more capable of improving time-to-market. Reducing total cycle time can accelerate improvements in cost, quality, yields, productivity, and effectiveness. Today's customers demand responsiveness and quality. This is an investment that pays for itself.

CE is a systematic approach to the integrated, concurrent design of products and their related processes, including manufacture and support. This approach is intended to cause the developers, from the outset, to consider all elements of the product life from concept through disposal, including quality, cost, design, schedule, and user requirements. Factors that influence a customer's purchase decisions, such as delivery, cost, and quality, normally change in four dimensions: organization, organizational infrastructure, requirements, and product development, with quality as a prerequisite. These dimensions also define the concurrent engineering environment. In addition, a company must transform the "five forces of change" into well-managed resources for product development. These forces of change are: technology, tools, tasks, talent, and time. By understanding the need to balance these dimensions, a concurrent engineering approach may be applied. Figure 29-16 depicts concurrent engineering as a process driven by resources, controlled by the four key dimensions, and operated by the five forces of change.

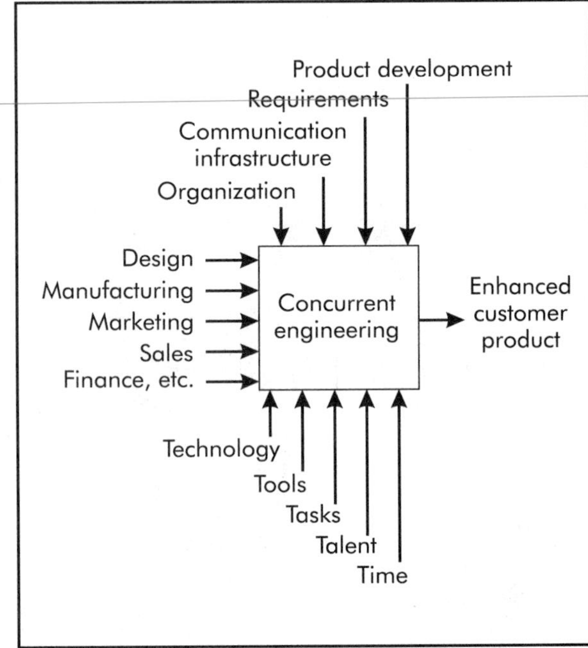

Figure 29-16. The process of concurrent engineering.

29.5 COMPONENTS OF AN INTEGRATED MANUFACTURING SYSTEM

The integration of manufacturing systems is the process of having all elements of the system work in such a way that the objective of the system is realized. *Integrated manufacturing systems* (IMS) include all activities involved in the planning, control, production, assembly, and marketing of products. To close the loop of a fully integrated manufacturing system, other business and management functions, such as finance, manufacturing management, human resource management, and strategic planning, should be integrated into the system.

29.5.1 Manufacturing and Assembly Cells

The concept of *cellular manufacturing* is increasingly emerging in manufacturing. Research efforts have revealed that current trends in manufacturing are targeted toward the manufacture of products in small-to-medium lot sizes. This is realized by the creation of manufacturing and assembly cells. Manufacturing cells are the most flexible type of computer-integrated manufacturing systems. A flexible manufacturing system (FMS) may be considered a manufacturing cell. However, it provides more flexibility and normally works on a higher level of computer and manufacturing control. Cells may be formed by a single stand-alone NC machine tool, a single machining center, or an integrated multimachine cell or machining centers. A concept illustration of an automated four-station manufacturing cell with a closed-loop pallet conveyor is shown in Figure 29-17. The arrangements shown in Figure 29-18 show one-, two-, and three-station freestanding manual load systems.

Manufacturing and assembly cells provide many benefits, such as:

- better utilization of machine tools and other production equipment;
- reduced work-in-process (WIP) inventory;
- efficient use of available floor space;
- reduced setup and manufacturing lead times; and
- better response to customer requests.

Unlike flexible manufacturing systems, manufacturing cells lack the presence of central computer control. They are generally controlled by cell controllers or by their own independent but interfaced machine controllers, such as the vertical machining center shown in Figure 29-19. Cells are typically constrained by the number of cutting tools and tool pockets available in the tool changers of single or multimachine cells as illustrated in Figure 29-20. This limits the part spectrum that can be run through the cell without stopping the machine and manually exchanging tools to accommodate the configurations of different parts. In Figure 29-21, a setup opera-

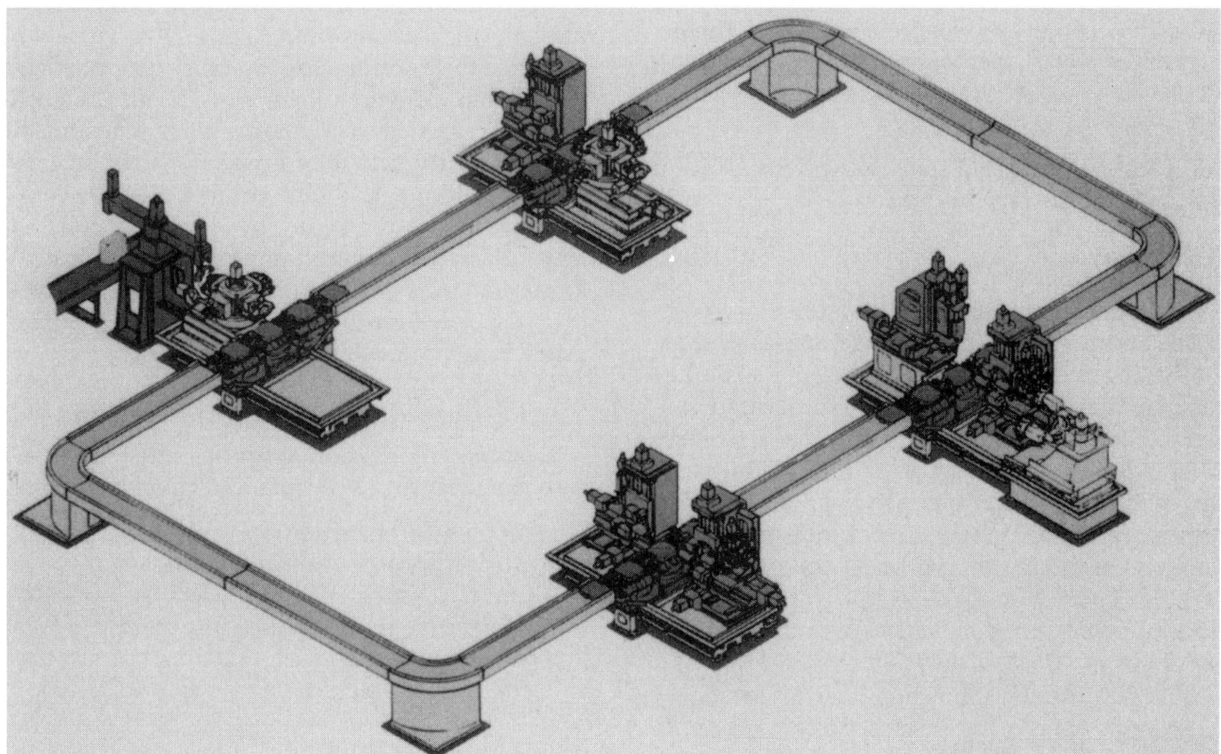

Figure 29-17. Illustration of an automated four-station system with a closed-loop pallet conveyor. (Courtesy Kingsbury Corporation)

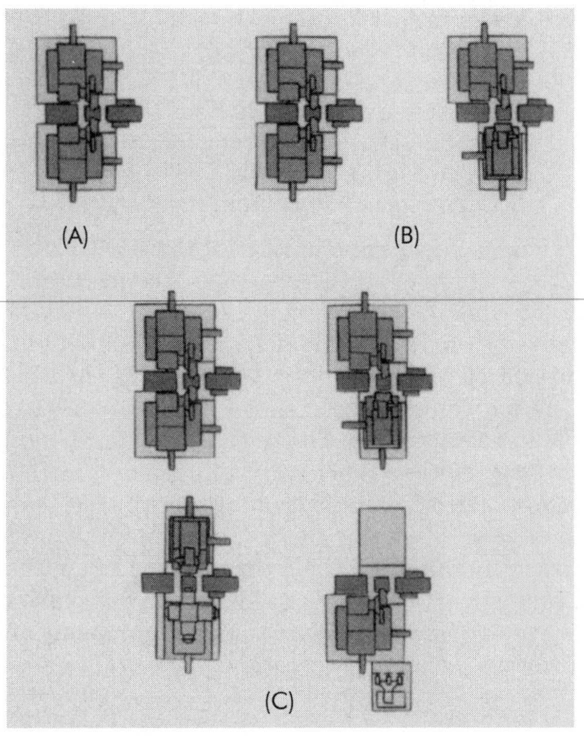

Figure 29-18. Illustrations of single and multimachine systems: (A) one-station freestanding manual load, (B) two-station freestanding manual load, and (C) three-station freestanding manual load. (Courtesy Kingsbury Corporation)

tor adds a tool to the tool changer drum on a vertical spindle machining center.

29.5.2 Cell Linkages

Integrated manufacturing systems are composed of linked cellular manufacturing systems. Designing cellular manufacturing cells is a key task in the implementation of integrated systems. Several techniques can be used to form cells. Digital simulation is gaining wide popularity for the design and analysis of manufacturing systems. Another technique uses mini versions of machines and small robots to emulate the operation of the system, aiding in the development of system software and the analysis of system productivity. The installation of a full-scale system on the shop floor follows.

A more efficient approach to convert functional layouts and job shop systems into group layouts or cellular manufacturing systems is *group technology*. The concept of group technology is based on utilizing the similarities between different parts and components. By grouping similar parts into families of parts, a number of processes can be identified to form a manufacturing cell. Such conversion affects different production planning and control functions, product and tool design, production planning and control activities, quality assurance, marketing and other business, and human resources.

Production flow analysis (PFA) is another approach used to analyze the flow of material in the plant and to provide the foundation for the implementation of cellular manufacturing. PFA uses the existing information on route sheets to determine machine and labor requirements. Based on the output of this analytical process, cells can be formed. Classification and coding schemes also may be used to sort items and form families accordingly. Once a family is created, a cell can be designed to produce the family.

Once cells are formed, they are linked together to form an integrated system of manufacturing. Cells may be linked directly, that is, the output of a certain cell represents an input to another, or by using kanban. An illustration of how kanban works is shown in Figure 29-22. Kanban provides instructions for the routing of parts in the system, delivers information about the need for different parts, and connects the output of one cell to the input of another. Carts or containers holding equal amounts of parts or components move between the different cells in loops. Each cart or container has one withdrawal kanban to take parts from a certain cell and one product-order kanban to obtain parts for the other connecting cell.

29.5.3 Systems Integration Elements

Systems integration activities must incorporate the following concepts and considerations:

- setup minimization;
- work-in-process minimization;
- quality assurance integration;
- preventive maintenance integration;
- automation; and
- computer interface.

Setup Minimization

The main objective of reducing setup times is to improve productivity by increasing the fre-

Figure 29-19. Cells may be controlled by cell controllers or by their own independent but interfaced machine controllers. (Courtesy Giddings and Lewis Corporation)

Figure 29-20. The number of available tool pockets on each machine limits the cell capability. (Courtesy Giddings and Lewis Corporation)

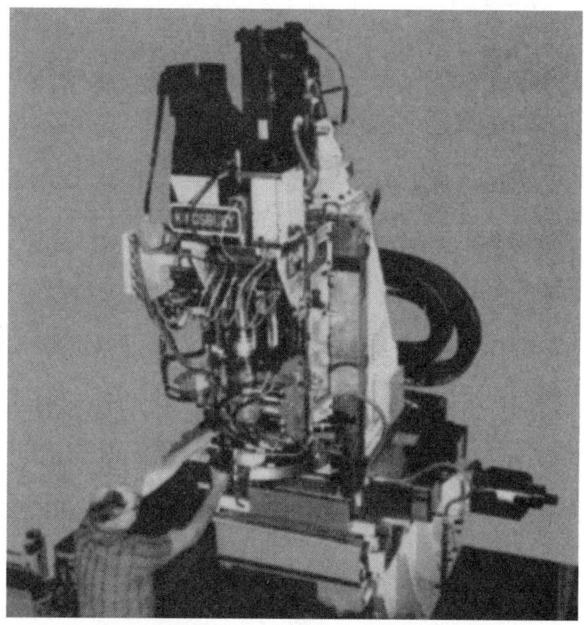

Figure 29-21. A setup operator adds a tool to the tool changer drum on a vertical spindle machining center. (Courtesy Kingsbury Corporation)

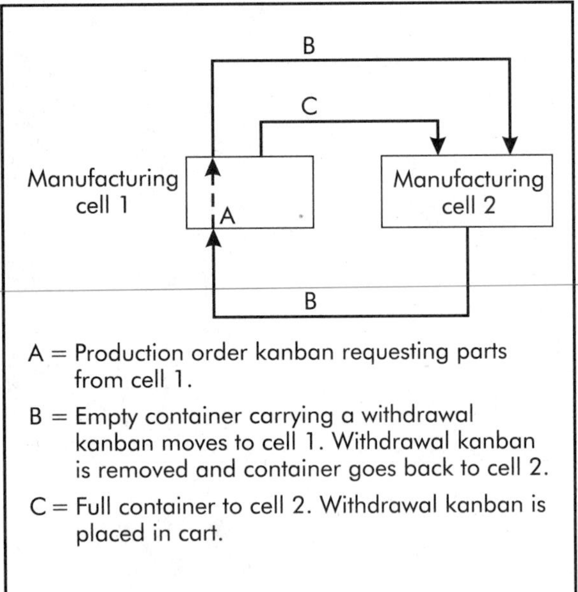

Figure 29-22. A kanban system linking different cells.

quency at which the production lot is produced. Reduction or minimization of setup times may be achieved by using different techniques:

- the use of adapters in jigs allows one jig to accommodate different parts, which eliminates the need for setup each time a different part is to be loaded onto the machine;
- training workers to practice rapid setup techniques; and
- implementation of the single-minute exchange of dies (SMED) system.

Work-in-process (WIP) Minimization

In integrated manufacturing and assembly systems, the level of work-in-process inventory must be minimized to facilitate the flow of parts or components between different cells. Kanban systems can help solve WIP inventory problems by providing adequate information about the level of inventory at each cell. WIP inventory reflects several facts about cell performance. It reflects the quality level of the cell and provides information about cell efficiency, machine breakdowns, setup times, shortage of parts, and human resources allocated to the cell. In many cases, WIP inventory levels in integrated manufacturing systems can be significantly reduced and controlled.

Quality Assurance Integration

A current trend in automated manufacturing is the integration of quality assurance activities into the manufacturing system. Such integration reduces the chance of producing nonconforming parts and improving the productivity of the system. There are several quality assurance techniques. Statistical process control (SPC) methods are effective tools and constitute a fundamental element in the integrated quality assurance system. Many quality control techniques and philosophies have been developed to help improve quality productivity, such as quality circles, zero defects, total quality management (TQM), and others.

Integrated quality systems are emerging and constitute a major component of integrated manufacturing systems. Perhaps the most suitable candidate to realize the integration of quality assurance functions in automation is the coordinate measuring machine (CMM), discussed earlier. The effective use of computers for CMM applications is a principal feature differentiating available coordinate measuring machines' systems. A computer-controlled coordinate measuring machine is shown in Figure 29-23. The value of a measurement system depends to a large extent on the sophistication and ease of use of the associated software and its functional capabilities. The functional capabilities of a CMM depend on the number and types of applications programs available. The following is a listing of many of the features of systems software available for CMMs:

- ability to print out instructions, measurement sequence, zero reference, etc.;
- automatic compensation for misalignment of the workpiece with the machine axes;
- conversion between Cartesian and polar coordinates;
- tolerance calculations to reveal out-of-tolerance conditions;
- ability to define geometric elements such as points, lines, circles, planes, cylinders, spheres, and cones, and their intersections;
- automatic redefinition of coordinate systems or machine axes, printout, or origin and inspection planes;
- inspection of special shapes or contours such as gears and cams;

- multiple-point hole checking using least squares techniques for determining best fit center, mean diameter, roundness, and concentricity;
- evaluation of geometric tolerance conditions by defining types of form and positional relationship such as roundness, flatness, straightness, parallelism, or squareness;
- hole diameter and location checking, considering minimum and maximum material conditions; and
- other software features for statistical analysis, which include graphical data display, histograms, integration of areas under a curve, contour plotting, and automatic part or lot acceptance or rejection based on statistical evaluation.

Figure 29-23. Example of a computer-controlled coordinate measuring machine with a touch-trigger probe. (Courtesy Carl Zeiss, Inc.)

Another example of a coordinate measuring machine used for industrial applications has a moving bridge configuration and can measure up to 100 coordinate points/min. This machine provides an excellent capability for video camera applications. Software options are available to provide additional capabilities.

Preventive Maintenance Integration

Machine tools and other equipment in integrated manufacturing systems need maintenance. Preventive maintenance is a needed function that should be integrated into the manufacturing system. Ways to allow preventive maintenance, while maintaining system performance, should be studied. Parallel processing of parts, multiple copies of production equipment, reallocation of human resources, and investigation of different ways to improve reliability are possible alternatives.

Automation

Employing automation is a faster way to solve problems associated with quality, reliability, scheduling, and other production planning and control functions. Automation reduces the probability of errors and eliminates the presence of human mistakes. The conversion of manned cells to unmanned or unattended cells can solve many problems commonly associated with manned manufacturing systems.

Computer Interface

The computerization of cells is a natural ingredient in integrated manufacturing systems. It provides for the exchange of information between different components and cells of the system. Since many of the activities in integrated manufacturing systems use common data bases, computer interfacing is becoming a significant achievement. All components of the integrated manufacturing systems are computer-assisted functions. Functions such as computer-aided design (CAD), computer-aided manufacturing (CAM), computer-aided process planning (CAPP), computer-aided inspection (CAI), and other computer-based production planning and control systems need an interface to facilitate their operation and improve the overall system performance.

29.5.4 The New Manufacturing Enterprise Wheel

The *CIM wheel* of the Computer and Automated Systems Association of the Society of Manufacturing Engineers (CASA/SME) symbolizes a computer-integrated manufacturing enterprise. The *CIM wheel*, as presented earlier this chapter, represents the concept of enterprise-wide teaming. This concept demonstrates that manufacturing has entered the information age, where computers have emerged to manage the manufacturing enterprise.

Looking beyond the automation and integration inside the enterprise, CASA/SME has developed the new *manufacturing enterprise wheel*. It focuses on the customer, includes people and teaming, considers the virtual enterprise, and depicts manufacturing enterprise integration. The new manufacturing enterprise wheel is shown in Figure 29-24. It describes six fundamental elements for competitive manufacturing.

1. The customer is the central focus. The key to success is a better understanding of the market and evolving customer needs. Marketing, design, manufacturing, and support functions must be aligned to meet customer demands.
2. The role of people and teamwork in the organization include the means of organizing, hiring, training, motivating, measuring, and communicating to ensure teamwork and cooperation. This side of the enterprise is captured in ideas such as self-directed teams, teams of teams, the learning organization, leadership, metrics, rewards, quality circles, and corporate culture.
3. The revolutionary impact of shared knowledge and systems to support people and processes include both manual and computer tools to aid research, analysis, innovation, documentation, decision-making, and control of every process in the enterprise.
4. Key processes from product definition through manufacturing and customer support fall within three main categories: product/process definition; manufacturing; and customer support. A further breakdown of these categories reveals the 15 key processes that complete the product life cycle.
5. Enterprise resources (inputs) and responsibilities (outputs) include capital, people, materials, management, information, technology, and suppliers. Reciprocal responsibilities include employee, investor, and community relations, as well as regulatory, ethical, and environmental obligations. In the new manufacturing enterprise, administrative functions are a thin layer around the periphery. They bring new resources into the enterprise and sustain key processes.
6. The manufacturing infrastructure includes customers, suppliers, competitors, prospective workers, distributors, natural resources, financial markets, communities, governments, and educational and research institutions. Success in a manufacturing enterprise depends on customers, competitors, suppliers, and other factors of the environment.

29.6 QUESTIONS

1. Define "manufacturing." Explain the different inputs and interactions in a typical manufacturing process.
2. What are the different aspects that characterize manufacturing systems? Briefly discuss each aspect.
3. What are the different variables involved in the classification of manufacturing systems?
4. What are the different types of manufacturing systems? What is the common element in all such systems?
5. Provide a comparative study for a job shop, flow shop, project shop, and continuous process as manufacturing systems, considering the following factors:

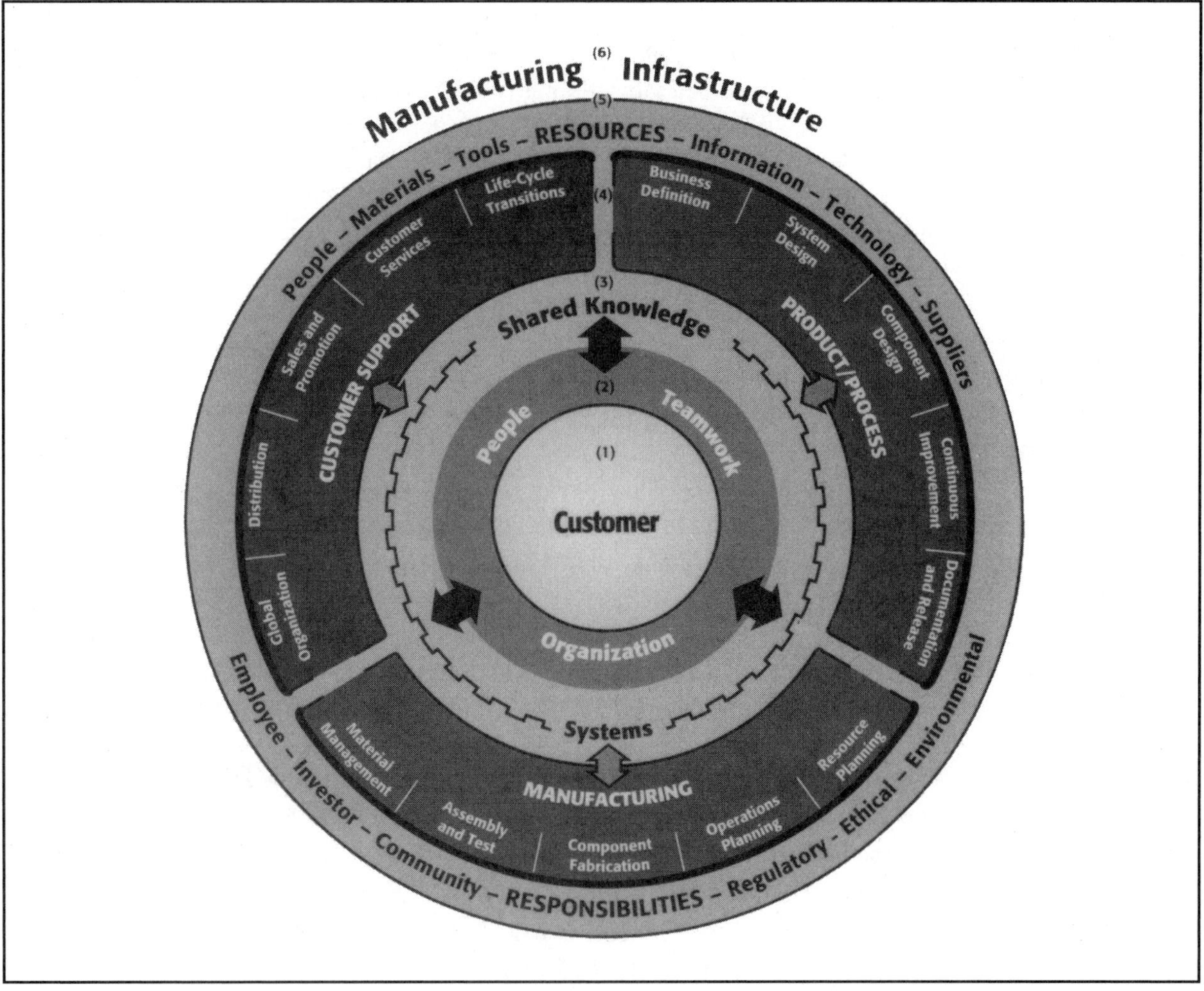

Figure 29-24. The manufacturing enterprise wheel *developed by the Computer and Automated Systems Association of the Society of Manufacturing Engineers (CASA/SME).*

- types of machine tools used;
- setup times;
- workers' skill level;
- work-in-process (WIP) inventory;
- lot sizes; and
- manufacturing lead time.

6. How can group technology (GT) be used to design a cellular manufacturing system by converting a job shop functional layout?
7. What are the different objectives of a JIT manufacturing system?
8. What is a "flexible manufacturing system"? What does the word "flexible" mean in the context of manufacturing systems?
9. Discuss the different features of coordinate measuring machines.
10. List the different functions that may be performed using coordinate measuring machines.
11. What differentiates a flexible manufacturing system from a transfer line?

12. What are the main areas in which industrial robots may be used?
13. What are the basic components of a robot?
14. How many degrees of freedom does a jointed-arm robot have? If the robot is mounted on an *X-Y* table, how many additonal degrees of freedom are added?
15. Define the following terms: "group technology"; "classification and coding"; "cellular manufacturing systems"; and "rapid prototyping and manufacturing."
16. Explain the benefits of a well-designed classification and coding system as related to product design and manufacturing.
17. What features does a coordinate measuring machine have that make it more adaptable to automated manufacturing?
18. What are the four most common robot configurations?
19. What are the different design attributes a typical classification and coding system may include?
20. How can classification and coding systems be used to form manufacturing cells?
21. When choosing a robot for a certain operation, what factors should be considered?
22. Robots may be classified according to their mode of motion control. Provide a brief discussion for each of these modes.
23. What is "automation"? How has it been utilized in manufacturing systems engineering?
24. Define and explain the following terms. Give examples of how each term is used in manufacturing systems: "computer numerically controlled machines"; "computer-aided process planning"; and "CAD/CAM."
25. What does each of the following acronyms mean: SPC; FMS; CIM; and CMM.
26. Define "total quality management (TQM)." What are its main tenets?
27. How can TQM be used to improve manufacturing productivity?
28. Why are most manufacturing companies trying to adopt TQM?
29. What are the basic tools of statistical process control? Explain briefly the use of each tool as it is used to improve manufacturing productivity.
30. Rapid prototyping and manufacturing is an emerging new technology in manufacturing. Schematically illustrate how it works using diagrams and flow charts.
31. Using rapid prototyping techniques, list the different steps involved to achieve improved part quality.
32. What is "concurrent engineering"? Why is it widely adopted in manufacturing organizations?
33. Explain the different benefits manufacturing and assembly cells can provide.
34. How are cells linked in computer-integrated manufacturing systems?
35. What is "PFA"? How can it be used to form manufacturing cells?
36. Briefly explain the different elements of an integrated manufacturing system.
37. Briefly explain the efforts of SME to help manufacturing organizations better understand computer-integrated manufacturing.

29.7 REFERENCES

Clausing, D. 1992. "Concurrent Engineering." *Design Productivity International Conference Proceedings.* Honolulu, HI: Design Productivity International, February 6–9: 35-44.

Dallas, D. B. 1980. "The Advent of the Automated Factory." *Manufacturing Engineering,* November.

Elshennawy, A. K., and Ham, I. "Performance Improvement of Coordinate Measuring Machines by Error Compensation." *Journal of Manufacturing Systems,* Vol. 9: 151-158. Dearborn, MI: Society of Manufacturing Engineers (SME).

Elshennawy, A. K., Ham, I., and Cohen, P. 1988. "Evaluating the Performance of Coordinate Measuring Machines." *Quality Progress,* January: 59-65.

Groover, M. K. 1987. *Automation, Production Systems, and Computer-integrated Manufacturing.* Englewood Cliffs, NJ: Prentice-Hall.

Hamilton, R. 1992. "A Process Called Stereolithography." *Outside Business Magazine.*

Kendrick, T. 1995. "SPC on the Line." *Quality,* January: 35-39.

Rapid Prototyping Association/SME. 1997. "Rapid Prototyping Original Equipment Manufacturers' Update Report." Reports from the *Rapid Prototyping and Manufacturing '97 Conference.* Dearborn, MI: Society of Manufacturing Engineers, April 22–24.

Rembold, U., Nnaji, B. O., and Storr, A. 1993. *Computer-integrated Manufacturing and Engineering.* Wokingham, England: Addison-Wesley.

Rosenblatt, A. and Watson, G. F. 1991. "Concurrent Engineering." *IEEE Spectrum,* July: 22-37.

Vasilash, G. S. 1982. "The Road to the Automated Factory: 1970-1981." *Manufacturing Engineering,* January.

Wohlers, T. 1997. "Rapid Prototyping: State of the Industry. *Worldwide Progress Report,* April.

Wolak, J. 1994. "The New SPC Net Effect." *Quality,* August: 23-27.

Flexible Program Automation

30

Flexible manufacturing systems, 1969
—David Williamson, Great Britain

30.1 CLASSES OF AUTOMATION

Automation is the technology of using computerized or automatic equipment and devices, at different levels, to perform, execute, and control different functions of manufacturing. Automation was presented as a contemporary manufacturing technology in Chapter 29. Automation in manufacturing has emerged into different areas, such as manufacturing processes, material handling, inspection and measuring machines, assembly, and quality assurance. When a company decides to partially or fully automate its production and manufacturing lines, such a decision has one or all of the following objectives:

1. Reducing human involvement and thus decreasing the possibility of human errors. This eventually raises the level of quality and increases the productivity gains of the company.
2. Better planning and control of the manufacturing processes and thus decreased manufacturing costs.
3. Increased process performance resulting in better quality of products.
4. More efficient operations with computer control.
5. Improved machine reliability by employing computerized preventive maintenance programs.
6. Better utilization of machines and other equipment, which improves productivity and reduces bottlenecks in production activities. Setup time is decreased.
7. Increased efficiency and improved safety levels through the use of automated machines and equipment.
8. Minimization of waste, thus decreasing manufacturing costs.

Automation is a word that has many meanings in industry today. The term was coined shortly after World War II at the Ford Motor Company to describe the automatic handling of materials and parts between processes and operations. In effect, automation represents the combination of automatic operations into integrated groups. This is said to be *fixed program automation* and exists in molding lines in foundries, punch press lines, and die casting, as well as in assembly and inspection operations. It is far advanced in the process industries, such as the petroleum and chemical industries, where production rates are high. The processes are simple and easy to control, and the product is easy to move.

Flexible program automation is another class of automation aimed at the output of a variety of parts or assemblies with different configurations, each produced in small or medium lot sizes. This concept of flexibility in routing the different parts among machines is carried out in *flexible manufacturing systems,* described in this chapter and in Chapter 29. The same concept is also carried out by numerical control applications in manufacturing as described in Chapter 31. In this chapter, flexible program automation will be emphasized as it represents the most versatile concept in manufacturing systems.

30.1.1 Fixed Program Automation

Fixed program automation, also called *hard* or *fixed-position automation,* performs a certain operation or series of operations on a particular part or family of similar parts. These parts are mostly standardized products, such as engine blocks, gears, and pistons. This class of automation is difficult to adapt to new parts and is not flexible to accommodate major changes in part design and configuration. The operations and the transfer of parts from station to station are automatic. Two forms of fixed automation are the synchronous and nonsynchronous systems depicted in Figure 30-1.

A *synchronous system*, also called a *dependent system*, may have a series of machine operation units in line, in a transfer machine, or around an indexing table on a rotary or circular index machine. When the work is completed at all stations, all parts in the circuit are moved simultaneously to succeeding stations and another cycle is started. The time for a cycle depends upon the slowest operation, and the highest efficiency is obtained when all operation times are made the same and minimized. For tool changes and breakdowns, all stations are shut down, and efficiency decreases as the number of stations increases.

A *nonsynchronous system*, also called an *independent unit system* or *power-free system*, provides for a bank of parts to supply each machine, which permits the system to operate at its fastest rate. Any machine in the system may be shut down temporarily without stopping the others. Such a system has a high overall efficiency and can be optimized by using the best combination of machines. For *straight-line flow,* the banks of parts between machines may be contained in hoppers, magazines, or elevators. *Parallel flow* is an arrangement of conveyor-stored banks between machines. The conveyors may be stacked one above the other in the least possible space adjacent to the line.

Rotary Index Systems

In a rotary index machine, the workpieces are carried on a circular table with a vertical axis. This type of system normally takes less floor space and is cheaper to construct than an equivalent in-line system, but it is limited to a few stations. A system can be expected to cost several hundred thousand dollars. Typically, on machines with more stations or for larger workpieces on larger-diameter circles, the workpieces are carried on pallets that are moved from station to station on a circular track. Another less common design has a small table, called a trunion, with a horizontal axis. The toolheads are arrayed on both sides, and the trunion indexes the work in a vertical plane from station to station. The rotary index machine shown in Figure 30-2 has four spindles with tool changers and multiple chuckings of parts. Under the control of one operator, 99 operations are performed on seven faces of two different parts with a production rate of 35 parts/hr. Figure 30-3 shows a rotary machine with six stations and two vertical and two horizontal spindles with 30 tool storage modules.

Transfer Lines

A common type of fixed program transfer line is comprised of a transfer machine, such as the one shown in Figure 30-4. This 14-station inline machine has free transfer and produces 250 parts/hr. It has been widely adopted for medium- and large-volume production. A five-station machine, equipped with three stations having tool changers with added diverse milling and contouring capabilities is depicted in Figure 30-5. The stations can back up one another and continue production if one is down, thus improving up time. The straight-line work flow conforms naturally to continuous production and fits readily into desirable plant layouts. Considerable flexibility is permitted in planning for the number, kinds, and positions of stations with floor space and cost being the only limiting factors. Workpieces can be reoriented readily at stations in the line, and the least restriction is imposed upon approach of the tools to the part

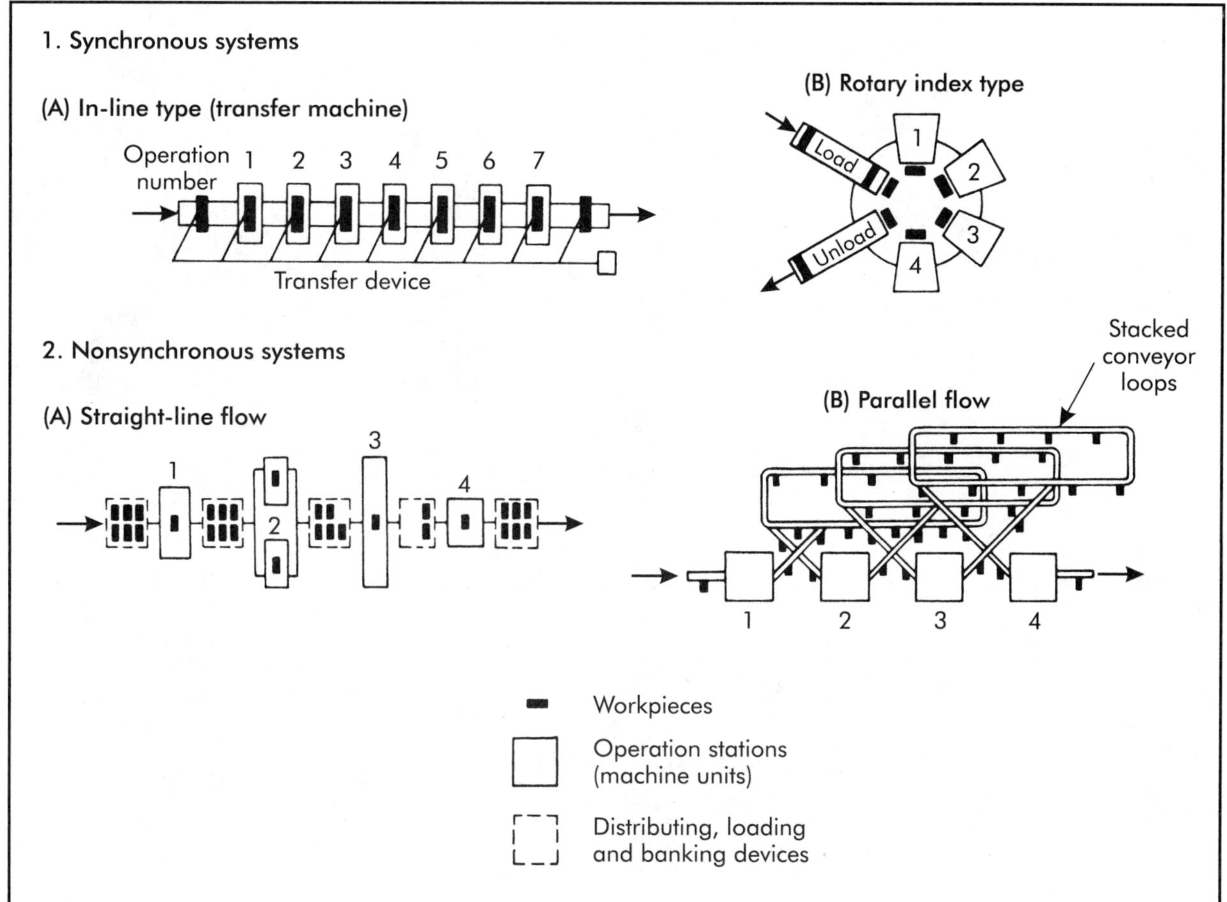

Figure 30-1. Diagrams of fixed-program systems of automation.

from any direction. Vibration, heat, and strain at any one station can be isolated from the others.

In addition to the unfavorable factors already cited for synchronous systems, transfer machines have some of their own. Some parts can be handled only on pallets. In addition, parts moved from station to station need to be relocated and reclamped at each station. Usually, gaging and automatic tool compensation require extra stations, and that adds to cost. When a transfer machine is built to make one part, and after a period of time the part becomes obsolete, the machine must be scrapped or renovated. An auto manufacturer reports that on average the cost to change over and retool for another part is about 90% of the original cost.

A transfer machine or line is basically an integrated material-handling device and is applicable to all kinds of operations in addition to machining. Transfer lines are widely used for such operations as welding, molding, assembly, and inspection. Transfer machines are found in a number of forms. One type for large workpieces has toolheads moved sequentially to and from each workpiece while it is located and clamped in one position. One workpiece may be set up in its fixture while another is being machined.

A common trend in transfer machine design is to make the systems more versatile so they can run varieties of parts and/or smaller batches economically. This may be done, for instance, by use of interchangeable toolheads for different parts.

The biggest savings from a transfer machine is in labor cost. The transfer machine shown in Figure 30-4 needs only one operator. In contrast, a conventional line to do the same work would employ five operators. Other savings may result

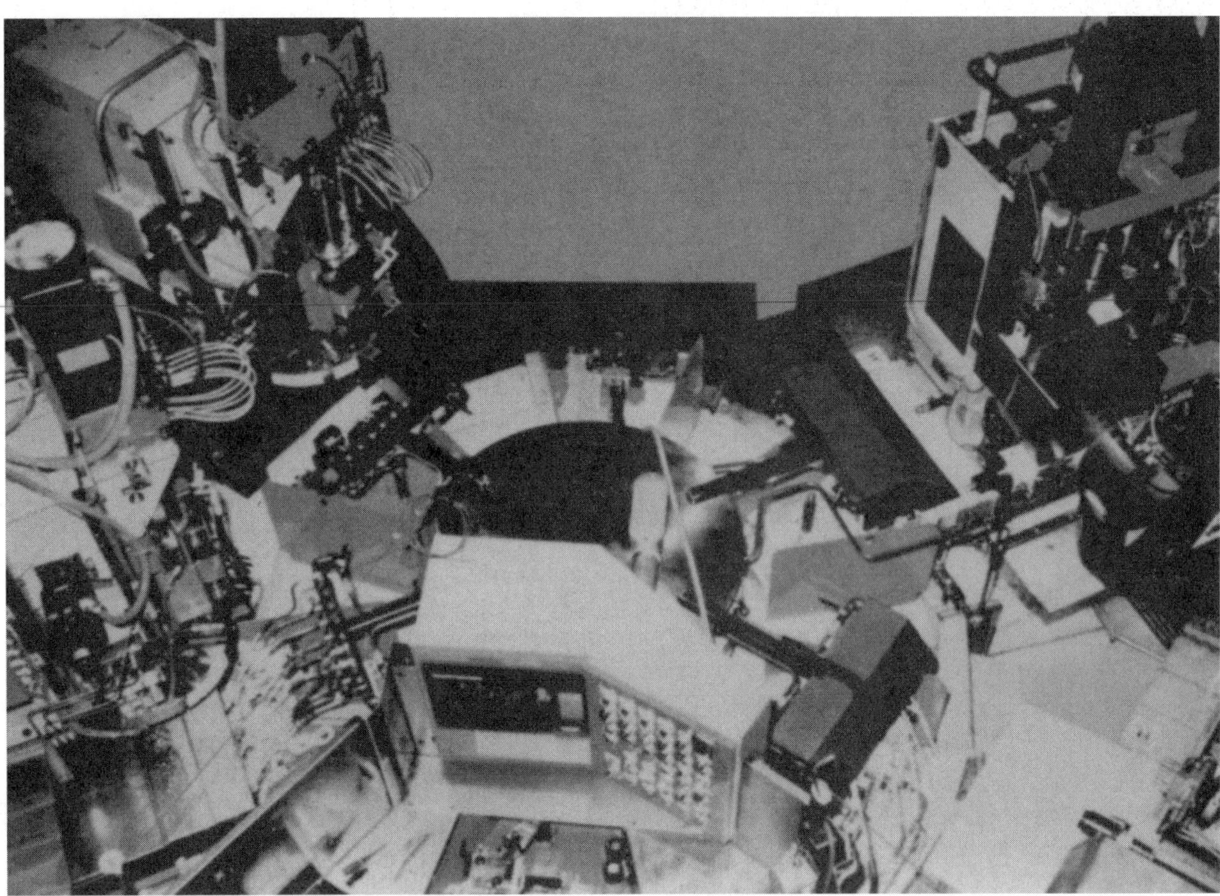

Figure 30-2. Rotary index machine. (Courtesy Kingsbury Corporation)

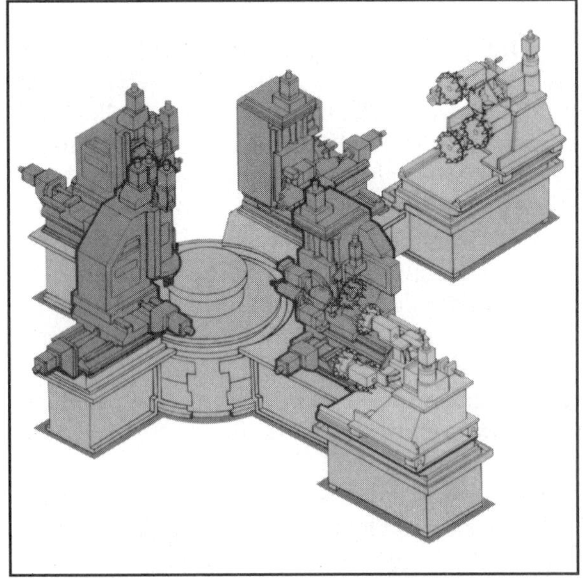

Figure 30-3. Six-station rotary index machine. (Courtesy Kingsbury Corporation)

Figure 30-4. In-line transfer machine. (Courtesy Kingsbury Corporation)

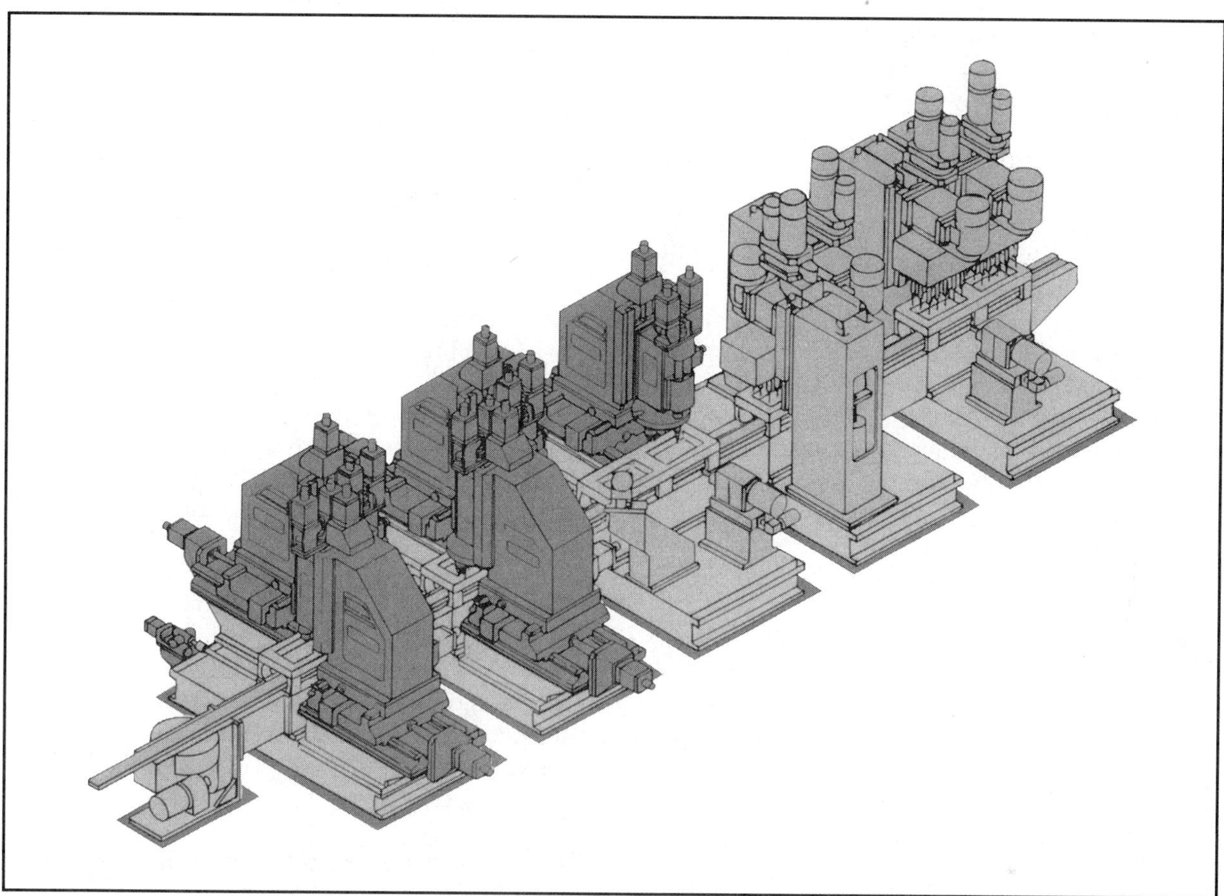

Figure 30-5. Five-station in-line transfer machine. (Courtesy Kingsbury Corporation)

from better quality, less scrap, less floor space used, higher productivity, and on-machine inspection.

The largest cost in transfer machine operation is the relatively high initial investment required. A major disadvantage is the high cost of design changes to the part or parts before they can be produced. To contain costs, the part or parts to be made must not be changed often and they must be initially designed to be made in the easiest way with all the needed surfaces for locating and handling. The machine must be designed to be as flexible as possible, should changes need to be made. Certain features may be considered for efficiency. Safety of equipment and personnel should be of major concern. This includes foolproofing the machine with interlocking to prevent wrecks and personnel injuries, allowing for preinspection of rough workpieces, and providing a means to stop the machine when it misses. The machine must perform everything that is done by an operator. This includes cleaning and disposing of chips, rough checking and orienting of workpieces, and watching for trouble. Tool problems must be solved before a transfer machine is designed to assure as much dependability as possible. To minimize interruptions, provisions should be made to preset sharp tools before they are put on the machine. Tools may be mounted in quick-change holders. Tool tracks make the preset tools readily available when needed. Signals can alert the attendants to the time to change tools. Schedules may be set up for changing the tools in batches.

Nonsynchronous Systems

Some nonsynchronous fixed program systems, such as the one shown in Figure 30-1, cost about 20% more than the total for the machines equipped with mostly standard items. Because the machines are relatively independent, few interlocking controls are needed, and the units do not have to be precisely aligned with each other. Thus design and construction are simplified. The standard units are easy to service and repair parts may be readily available.

Other advantages of the independent unit systems are that parts can be positioned readily in the best way for each operation, there is more room for chip removal, and loading times are at a minimum since parts are waiting at each station. If automatic controls fail, standard machines can be arranged to operate manually. A good application of independent stations is in assembly lines where some operations may be manual and not easy to pace with other automatic operations. Scrap control may be provided by including inspection units in the connecting devices between machines.

The disadvantage of independent systems, however, is the size; an independent unit system is not compact and larger floor space is needed. These systems are not suitable for larger part sizes and more complex shapes.

Automatic Tool Compensation

As a cutting tool wears, the size that it cuts changes. In manual operations, the operator checks the part from time to time and corrects the setting or changes the tool as needed. In many semi- or fully automatic operations, the tools are run for a predetermined period wherein they normally do not go out of limits and after which they are preset or changed.

Automated systems are equipped with automatic tool compensation. A sensory device measures the size after an operation. If a limit is exceeded, it sends back a signal to adjust the tool to compensate for errors. An example of automatic tool compensation for a chamfering and turning operation is shown in Figure 30-6. The cutting tool can be moved to the workpiece by a solenoid-actuated cam a total distance of 50 mm (.002 in.) in steps of 5 mm (.0002 in.). The increments are small so that adjustment from one limit back into the tolerance zone in one step does not push the dimension outside the opposite limit. The tool also can be indexed around its axis in 100 steps to bring fresh bits of cutting edge into action. When the gage head senses a workpiece close to or over the high limit, it closes contacts to transmit an impulse to a storage relay system. If the next piece is well within size, the relay system returns to its starting position; if it is near but not over the high limit, the relay system energizes the solenoid that sets the tool in a notch. If the second piece is oversize, a chipped tool is indicated, the machine is shut off, and a signal is turned on to show trouble.

The workpiece in Figure 30-6 is shown gaged on the machine. Quite often, pieces are gaged as they come off the machine. When gaging indicates out-of-limit performance, tool compensation can be made for the next piece, and so on.

For automatic tool compensation to be effective, the machine and tooling must be rugged and stable. Many other factors besides tool wear affect the results in an operation, and they must be minimized. If the machine were quite erratic, the regulating device would continually activate, causing overcompensation, and soon compounding the errors. Some consider it desirable that the machine be able to run at least 40 pieces without adjustment and still hold within one-half of the tolerance.

Control of Automatic Operations

One kind of manufacturing operation consists of a sequence of steps, each of which is started when the one before it is finished. Each step is timed by how long it takes to do its task and not by an external clock. This is illustrated by the basic face milling operation of Figure 30-7. For simplicity, only the motions are considered; other functions, such as motor and coolant controls, are not included in this discussion. Each workpiece is located and clamped on a fixture platen before it comes to this operation. As one platen is transferred out of the station, a new one is moved in, located, and clamped in position. A limit switch (LS1) is closed when the platen is properly located and firmly clamped in place. Then the milling head with a revolving face mill is fed from left to right across the surface to be machined. After the cutter has cleared the workpiece at the end of the stroke, a limit switch (LS2) is tripped. The slide on which the milling head rides is retracted to back the cutter from the work, and the head is returned at rapid rate from right to left. When the cutter is out of the way,

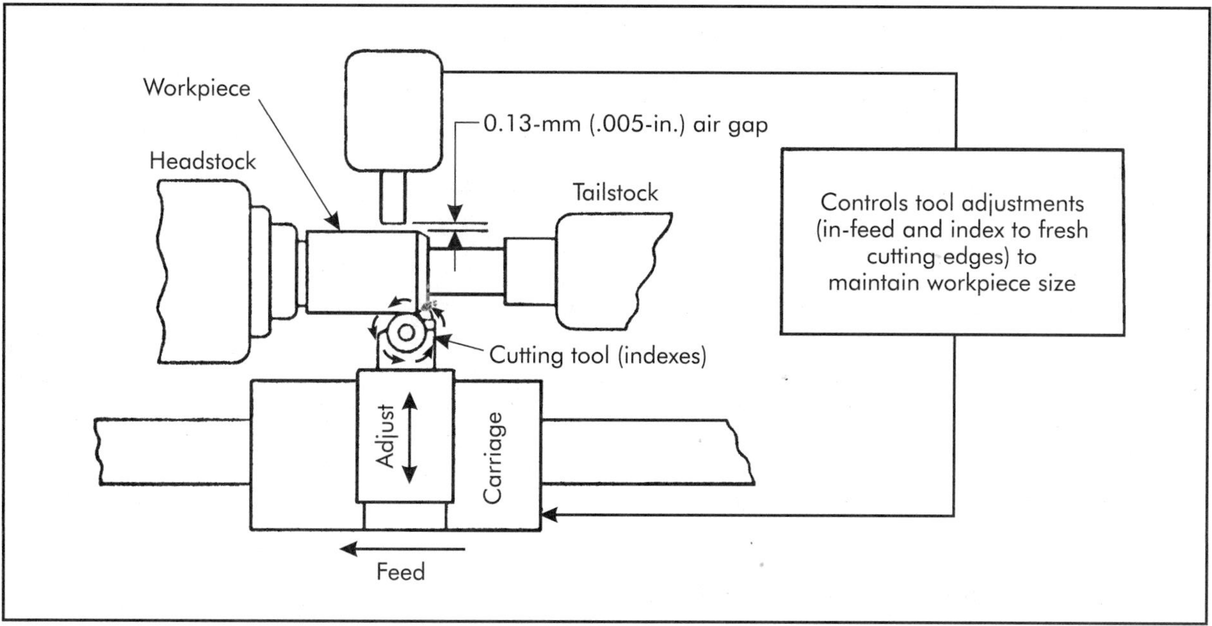

Figure 30-6. Automatic tool compensation system.

the platen is unclamped and raised from its locators and then transferred out of the station. Each sensor (limit switch) that sends a signal to the controller is designated by an input symbol in the tabulation of inputs and outputs from the controller. Each step is started by an output from the controller, which energizes an activator such as a solenoid valve or motor control. Some control systems utilize air or hydraulic media, but most are electrical.

The logic of an operation like that illustrated in Figure 30-7 is commonly expressed by either a *ladder diagram* or a set of *Boolean equations.* Both are shown for this case. In explanation, the first line of the Boolean equations states that when the sensor (LS1) shows that the platen has been unclamped and raised from the locators and the milling cutter is fully retracted, the finished piece may be transferred out of the station (on its platen), and a new piece should be brought into position and its platen located and clamped. The second line specifies conditions that must exist to warrant advancing the cutter to its starting position, and so on. The top rung of the ladder diagram corresponds to the first Boolean equation, the second rung to the second equation, and so on. The uprights on the side of the ladder represent two electrical lines with a voltage (V) between them. It may be 110 V alternating current (AC) or direct current (DC), but for safety is often less. When the switches on a rung are all closed, the voltage acts on a circuit that energizes necessary drivers to perform the step in the operation.

The output device (CR) of a controller may be an electrical-mechanical relay or equivalent solid-state device. These can be mounted on a panel and wired to the sensors to conform to the ladder diagram. This is called a hard-wired system and is common for small systems. One authority recommends hard-wiring up to six relays. For large and complex systems, the preference is for a programmable controller.

Another kind of manufacturing operation is one that is timed by an external clock. The time that a workpiece is undergoing an operation is determined by a programmed external clock that governs the transfer carriers. Some processes contain both timed and untimed operations. For example, an automatic die-casting machine starts a cycle with closing and locking the die. The next step of injecting a shot of metal is not started until sensors signal that the die is securely closed. Then when the die is filled, it is kept closed for a predetermined time to allow the metal to freeze. The next step is to free and

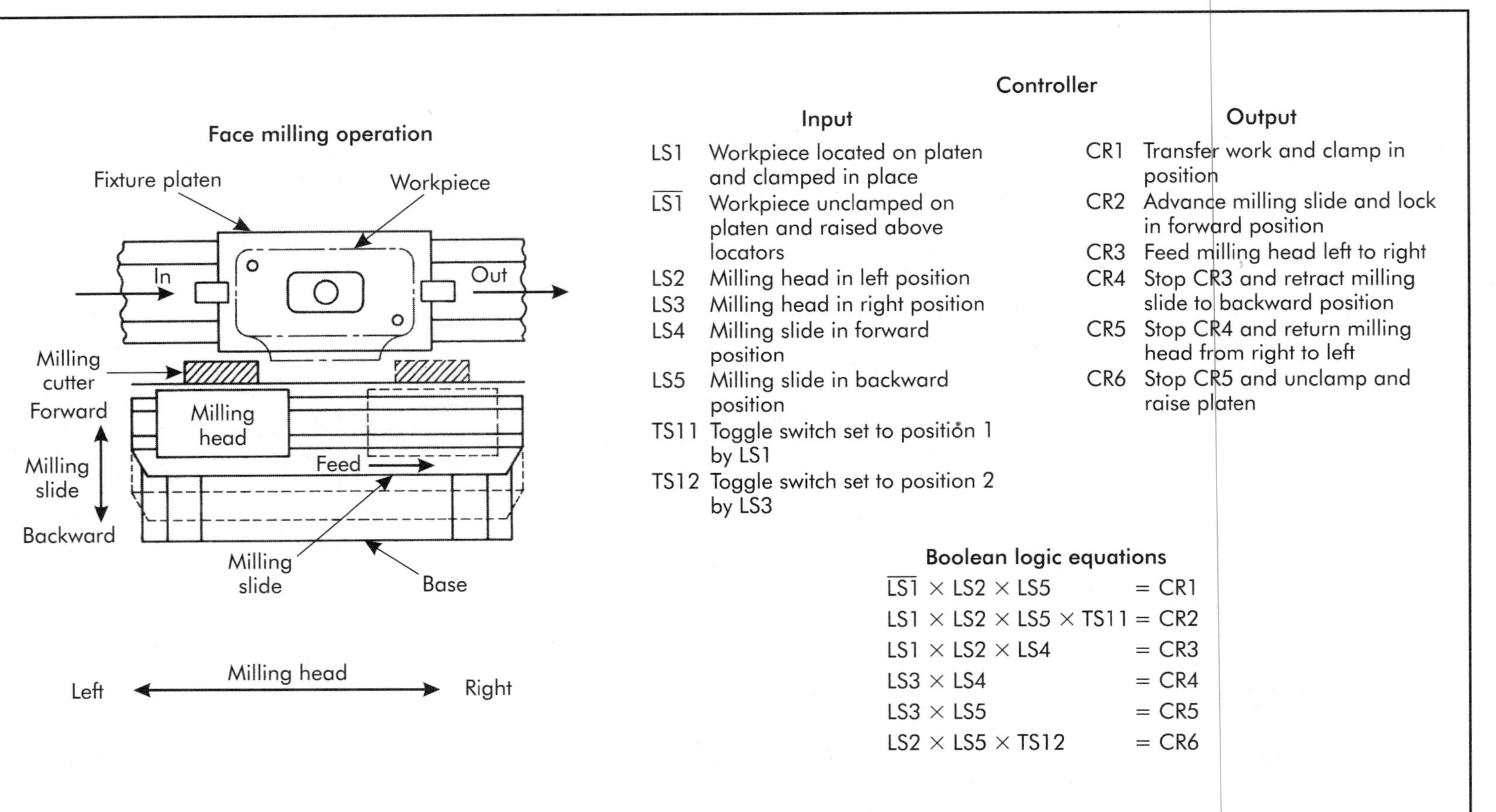

Figure 30-7. A simple manufacturing operation, its logic, and the elements of a programmable controller.

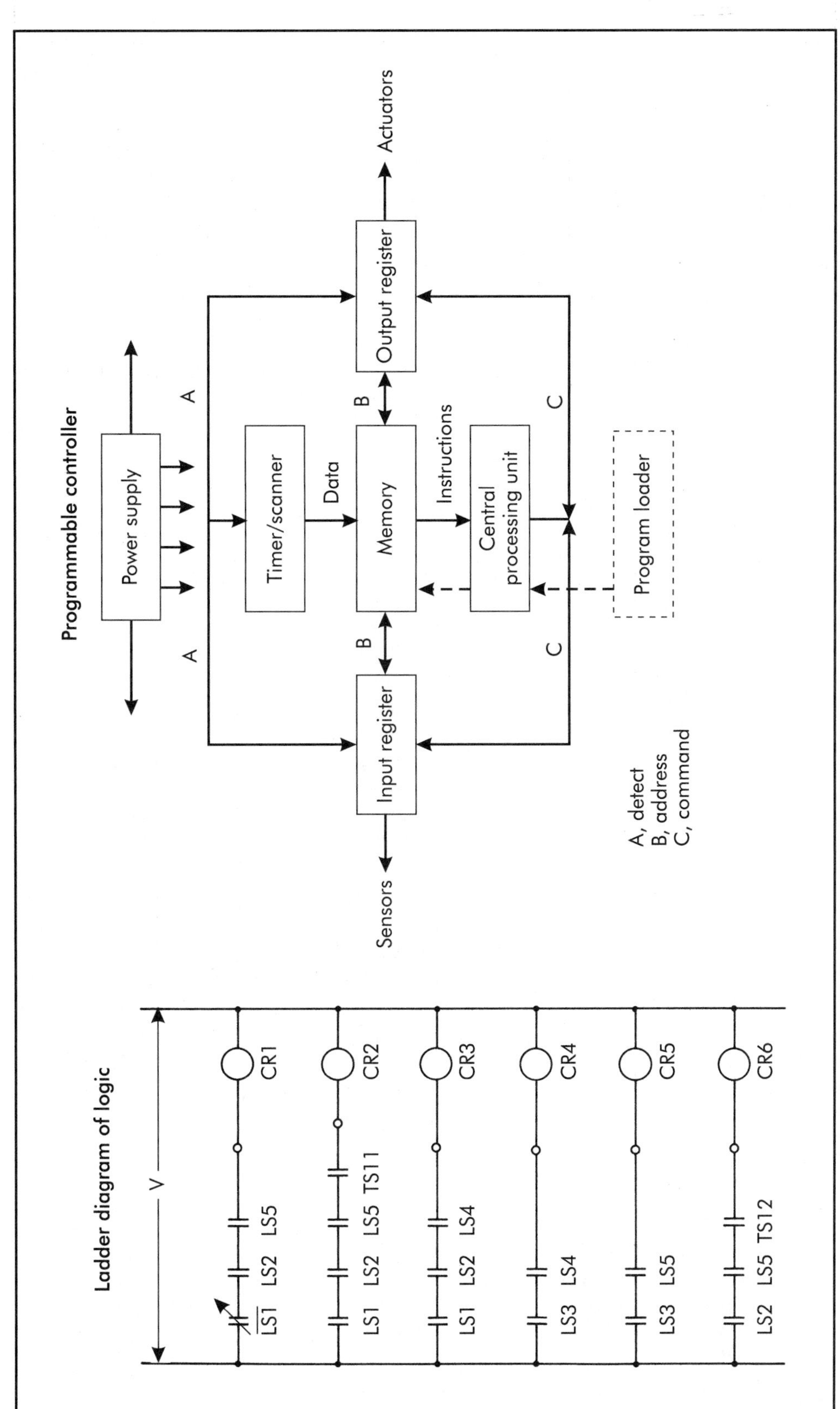

Figure 30-7 (continued).

eject the casting; a sensor signals when the workpiece is clear of the die and a new cycle may be started. Processes such as these may be controlled by a commercial timer with relays hardwired into the machine. A common device is a rotating drum indexed by a timer or by signals from sensors. Pegs are inserted into rings of holes around the drum to trip switches at selected intervals. Drum controllers are limited in scope; programmable controllers are preferred for large and complex systems.

Programmable Logic Controllers

A *programmable controller*, also known as a *programmable logic controller* (PLC), is a general-purpose control system that accepts input from such sources as push buttons, limit switches, and temperature, pressure, and flow sensors. It is capable of generating outputs to such devices as load relays, solenoid valves, motor starters, stepping motors, and even servo drives. There are many makes and they differ in detail but are based on the same principles.

A diagram of a basic programmable controller is shown in Figure 30-7. The program for a process is stored in the solid-state memory. The simplest is a read-only memory (ROM), not changeable in some systems but changeable in others by an auxiliary program or memory loader. As a rule, a program is not changed (except for corrections) while the controller is dedicated to a particular process. Input/output registers serve to convert signal levels from outside to inside (and vice versa) and to isolate the controller from outside transients. The scanner checks the input and output registers continuously and informs the memory and CPU of the status of input and output. It also serves to synchronize the activities of the controller. The central processing unit (CPU) or logic section makes all decisions based on the condition of inputs and outputs to conform to the program in memory and commands the output to change. All of this is executed repeatedly in milliseconds, which is much faster than most activities under control.

Most PLCs have added features beyond the essential. One survey showed 95% of them with timing and counting functions. Many have read/write memories (RWM) that can be changed at any time by push button, keyboard, or computer loading. RWMs are more volatile than ROMs. Cathode ray tube (CRT) display is available to show what is stored in the memory and the status of the operation. Memory capacity ranges in different makes from less than 1,000 to tens of thousands of words of information. Almost all PLCs accept logic or ladder formats for programming the memory, but a few utilize computer-like word programs. A sizable proportion of PLCs have computer capabilities. A survey showed about one-third with arithmetic, comparison, and decision functions, and over one-fourth with data handling capabilities. Such PLCs can communicate and interrelate with computers and other PLCs, and can coordinate and monitor automated material handling and multilateral transfer lines. Fault diagnosis accessories are available. Prices range from under $500 for a basic controller for a dozen or so functions to tens of thousands of dollars for some able to control thousands of functions and provide full auxiliary services.

Programmable logic controllers lie between computers and hard-wired relays and control synchronous and nonsychronous systems in manufacturing for all kinds of processes. Without the power of computers, PLCs are much cheaper than computers, and where more than a few functions are controlled, PLCs are more compact and cheaper than hard-wired systems. The main advantage of PLCs with respect to hard-wired systems is that the PLC logic can be changed much more easily, quickly, and cheaply. This may be helpful to test a system, improve a system, or change a system to accommodate changes in part design. A PLC need not become obsolete; if a machine is no longer needed, its PLC may be easily reprogrammed for another part. Common extras with PLCs, but not hard-wired controllers, include scheduled tool change alarms, system diagnostics, real-time cycle analysis, machine utilization reports, and production control summaries. Solid-state devices on PLCs without moving parts are more reliable than electromechanical relays on which contacts notoriously wear and corrode. And, PLCs are not as sensitive as computers and can be designed to survive better in harsh and dirty industrial environments.

It is important to recognize the differences between PLCs and numerical control (NC) systems. The purposes of the two systems are different. A PLC is dedicated, at least for a significant period of time, to one process or operation and works to one program in that time. With a limited number of variables, little feedback is required. On the other hand, the task of an NC system is to accept and carry out ever-changing instructions. Thus the NC system is normally more complex and costly.

The PLC operates only in on-off modes; it accepts signals and gives commands only for on and off. The NC system operates by numbers, although the numbers may be expressed in binary form. Thus NC can recognize and work to any given dimension in its range, whereas a PLC cannot. The NC can do what the PLC can, but costs much more.

30.2 MANNED CELL PARTIAL AUTOMATION

Classical manufacturing systems, such as job shops, employ general-purpose machines, adding some flexibility to the different jobs and operations that typically require different setups and process sequences. Functional or process layouts are used and setups vary, depending on the nature of the operations. Highly skilled workers are a major component of such systems.

Converting job shops into cells requires restructuring and regrouping different machines to manufacture part families by applying group technology concepts (discussed in Chapter 29). When cells are formed in this way, they are known to be manned cells, in which human workers get involved in many activities, including loading and unloading of parts; moving parts, tooling, and materials to and from different machines; providing adjustments and setups for different operations; and inspecting parts for dimensional accuracy. Cells may be linked together at some assembly or subassembly stations, or by a kanban system. Flow shops also may be restructured to form cells.

Restructuring and regrouping machines within a job shop or flow shop to form cells requires adjustments to some operating and support functions in the plant, including machines. This may necessitate the use of some level of automation. Long setup times, work positioning and clamping, machine operation, and material handling are just a few candidate activities for potential automation. The objective is to provide more flexibility to the system. The change to manned cells greatly increases worker productivity. Workers in manufacturing cells can service more than one machine or process, and may perform inspection and other duties as needed.

Cells can make families of parts with design similarities and utilize flexible workholding devices, also called universal fixtures, and tool changers to respond rapidly to changes in part design. A dependable material handling system is required to allow smooth movements of parts and components throughout the plant. Better machine control must be maintained to assure the quality of manufactured parts.

The following sections discuss the impact that partially automating manned cells will have on different functions in the plant. These affected functions include material handling, work positioning and clamping, inspection, machine operation, and cell control.

30.2.1 Material Handling

Material handling refers to the movement of parts, tooling and fixtures, materials, components, and other equipment in a manufacturing plant. Figure 30-2 shows the manufacturing system as a series of work flow activities. In a typical manufacturing operation, workpieces are moved between stations, loaded and unloaded on different machines, and moved for inspection and shipping. Tools, dies, molds, jigs and fixtures, and various other equipment are also moved between different machines and workstations in the plant. The movement of these workpieces, tools, and equipment is performed either manually or by some mechanical means. In transfer and flow lines, this is done by special-purpose manipulators, transfer mechanisms, conveyors, and positioners, each adapted to a particular purpose. When parts or the work to be done are varied but the volume is large, manual handling is utilized. In recent years, more efficient methods of material handling have been adapted to accommodate the high and ever-increasing level of automation in manufacturing operations.

Effective manufacturing planning and control systems should have material handling as an integral part of the plant operations. The selection of a suitable material handling system for a particular manufacturing operation depends on several factors, such as the:

- distance of movement of workpieces among machines and different workstations,
- location of machines, as well as the position and orientation of workpieces on these machines,
- size, shape, and weight of workpieces,
- type of plant layout and the route that the material handling system would take,
- level of automation and degree of integration of the material handling system with other system components, and
- amount of hazard and other environmental factors involved, such as foundry operations, and forging.

Another major factor influencing the selection of a material handling system is its compatibility with the rest of the machines and other components of the manufacturing system, such as the computer control mode, and other fixturing and support equipment. The degree of flexibility that the manufacturing system has also provides certain constraints to the selection of material handling equipment.

Among the many different types of material handling systems used in manufacturing, the following is a partial list of the most common material handling equipment:

- roller conveyors,
- rail-guided transporters,
- automated guided vehicles (AGVs),
- elevated tracks,
- forklift trucks,
- robots, and
- other mechanical, electrical, or hydraulic devices.

Automated guided vehicles are used extensively in flexible manufacturing systems and other automated systems to move parts between machines and workstations. They are powered by rechargeable batteries and use precision stop mechanisms for the positioning of loads in different locations. Common among the different types of AGVs are wire-guided vehicles and light-guided vehicles. AGVs may apply several load transfer techniques, providing versatility in their material handling tasks. These techniques include pallet truck, unit load, or fork trucks and are illustrated in Figure 30-8. AGVs may operate by picking up pallets of finished parts from one station and moving them to some designated location, such as a CNC machine tool, machining center, cleaning and deburring station, inspection station, or any other station requiring the presence and service of automated guided vehicles. After the operation is finished, the AGV moves the pallet to a different station.

Some advanced material handling systems provide more than moving parts and components. An automated storage and retrieval system (AS/RS) is a computer-controlled material handling system that can be used in automated or partially-automated cells, where human intervention is minimized or greatly reduced. An AS/RS is capable of moving material from and into the storage area and providing accurate information about inventory planning and control functions.

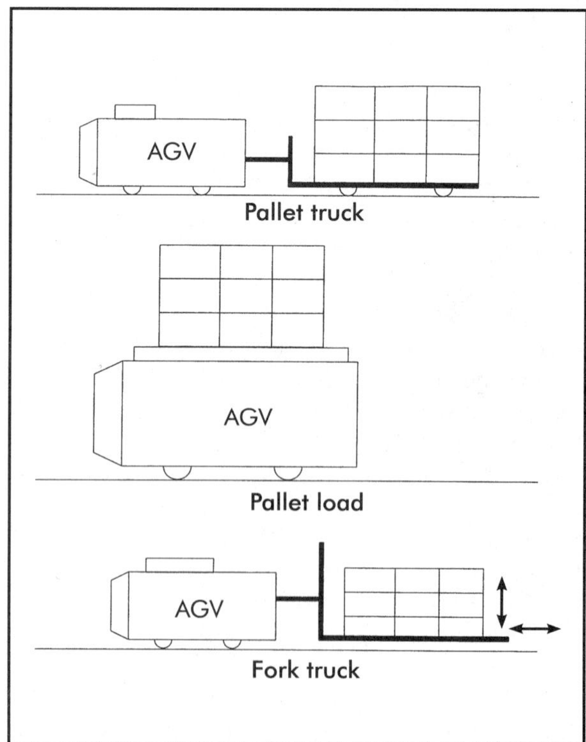

Figure 30-8. Load transfer techniques with AGVs.

30.2.2 Work Positioning and Clamping

In conventional machine tools, work is positioned and clamped on the machines manually using chucks, collets, mandrels, and various types of fixtures and clamping devices. In automated systems, workholding devices are designed to operate at different levels of automation. Power chucks, for example, are driven by mechanical, hydraulic, or electrical means. In automated manufacturing, work positioning and clamping devices are designed with some flexibility that makes them adaptable to automation. The term *flexible fixtures* is often used in association with flexible automation. These are devices that are capable of accommodating a range of part configurations and sizes without the need to change the setup, adjust the fixture, or for human intervention. These flexible fixtures often use sensing devices to compensate for excessive or small clamping forces as needed. Clamping forces must be precisely estimated and applied to flexible fixtures. Flexible workholding and positioning devices must position and clamp the workpiece precisely and accurately to maintain the level of quality expected from automated systems.

Unless properly designed, a flexible fixture may create problems during the manufacturing operation. In manual operations, the operator normally keeps the surface of the workpiece clean and free of any loose chips. The presence of loose chips between the workpiece and the fixture in unmanned systems can be a serious problem and may result in out-of-tolerance workpieces.

There are some considerations for the design of work positioning and clamping devices. The devices must hold the workpiece precisely to prevent any translational or rotational movement of the workpiece, clear the tool path to avoid collision, and not obstruct the loading or unloading of workpieces. Clamping forces must be maintained all the time the machine is in operation.

Single-minute exchange of dies (SMED) is another concept that makes use of the similarity between parts and the processes used to produce them to significantly reduce the setup time to less than 10 min. This is particularly important in automated manufacturing systems.

Pallets may be used for loading and unloading of parts on NC machines and machining centers. Examples of pallet applications in manufacturing are presented in Chapter 31. Pallets are used to manually or automatically load and unload parts, and to position and orient part surfaces on different sides of the machine.

30.2.3 Inspection

Inspection is one of the oldest functions associated with manufacturing operations. It has three different phases of inspection to assure the quality of parts and their conformance to specifications.

1. Incoming material inspection, also called *preprocess inspection*, in which raw material and components are checked for conformance to certain specifications that are set by design.
2. *In-process inspection*, during which tools and techniques are implemented to assure the process performs in a state of control and parts are produced within the required specification limits. In many cases, statistical process control (SPC) techniques are used for this purpose.
3. Finished product inspection, also called *post-process inspection*, in which the part is inspected for certification and conformance to customer requirements.

In each of these phases, the amount of inspection, that is the number of units inspected within the lot, varies. It may be either a *100% inspection* (also called screening inspection), or *sampling inspection*. In 100% inspection, all parts are inspected for conformance to specifications. Sampling inspection relies on the inspection of only a sample of the whole batch or lot of production. Each has its own advantages and disadvantages. For example, 100% inspection is not suited for those parts that must be subjected to destructive testing procedures and is not the best inspection technique when the cost of inspection and testing is very high or the inspection time is very long. Sampling inspection runs the risk of rejecting a good lot based on the inspection of a nonrepresentative sample, or accepting a bad lot based on the inspection of an accidentally good sample.

Statistical process control (SPC) methods are normally associated with inspection programs. SPC is a very powerful collection of tools and methods used to improve process performance. SPC is discussed in several sections throughout this book.

Inspection may be performed using contact or noncontact inspection methods. Contact inspection methods involve the use of a mechanical device, usually a stylus or probe, which makes contact with the part being measured or inspected. This operation may be accomplished manually or automatically. Manual or computer-driven coordinate measuring machines may be used to perform this function using contact probes, also called hard probes.

Noncontact inspection methods use sensors to measure desired part features. Some of the advantages that noncontact inspection methods provide are: 1) they do not require special positioning of the part, and 2) they provide relatively shorter inspection times. Familiar technologies in noncontact inspection include the use of lasers, machine vision, and sensors. Some coordinate measuring machines have applications associated with machine vision and noncontact inspection techniques.

30.2.4 Machine Operation

The capability of partially automated manufacturing cells is largely dependent on the different machines in the cell, their operating features, and mode of operation. A cell producing a part family or a large number of similar parts needs to have machines capable of performing a large number of processes on the same part. Different cells may produce different part families or two parts of the same family. In either case, a worker will tend to both cells, servicing machines, providing necessary tooling, and performing different adjustments and setups.

In some cases, a machining center may be used as the main component of the cell. Since the machining center has tool storage capability, the role of workers is minimized and limited to loading and unloading parts. Some CNC machines and machining centers use universal or flexible fixtures. Pallets may be used to transfer fixtures between different machines. Workers in these cells are mainly responsible for setting up and securing fixtures and monitoring part loading and unloading.

Machines that compose a partially automated cell have computer-assisted programming capability. Other features that these machines have determine the degree of system flexibility. Machine operation is largely dependent on the following features:

- number of axes on each machine,
- machine performance measures—repeatability and positional accuracy,
- chip removal system,
- spindle horsepower, and
- other features, such as the coolant system, tool storage capacity, mode of operation, machine table dimensions, and programming capability.

30.2.5 Digital Numerical Control

One of the most noted applications of automation in manufacturing is numerical control (NC) and its progressive technologies of computer numerical control (CNC) and direct (or distributed) numerical control (DNC), industrial robots, and computer-integrated manufacturing systems (CIMS). These technologies are presented in Chapter 29 and other sections throughout this book. Numerical control is discussed in Chapter 31.

Numerically-controlled machines may be linked together by a host computer to form a DNC system, in which the program of instruction for the NC machines is directly downloaded to each NC machine, or through a hierarchical chain as shown in Figure 30-9. The communication and dialogue between the DNC computer and the NC machines takes place in real time and no NC tapes or tape readers are involved. As DNC systems operate under computer control, they are most suitable for unmanned cell automation. Their application in manned systems is possible and can provide better control of machine operation. Productivity and quality are immediate outputs realizing higher gains.

30.2.6 Fuzzy Control Logic

Fuzzy logic is an emerging processing technique that makes machines think like humans. These fuzzy control systems attempt to control the manu-

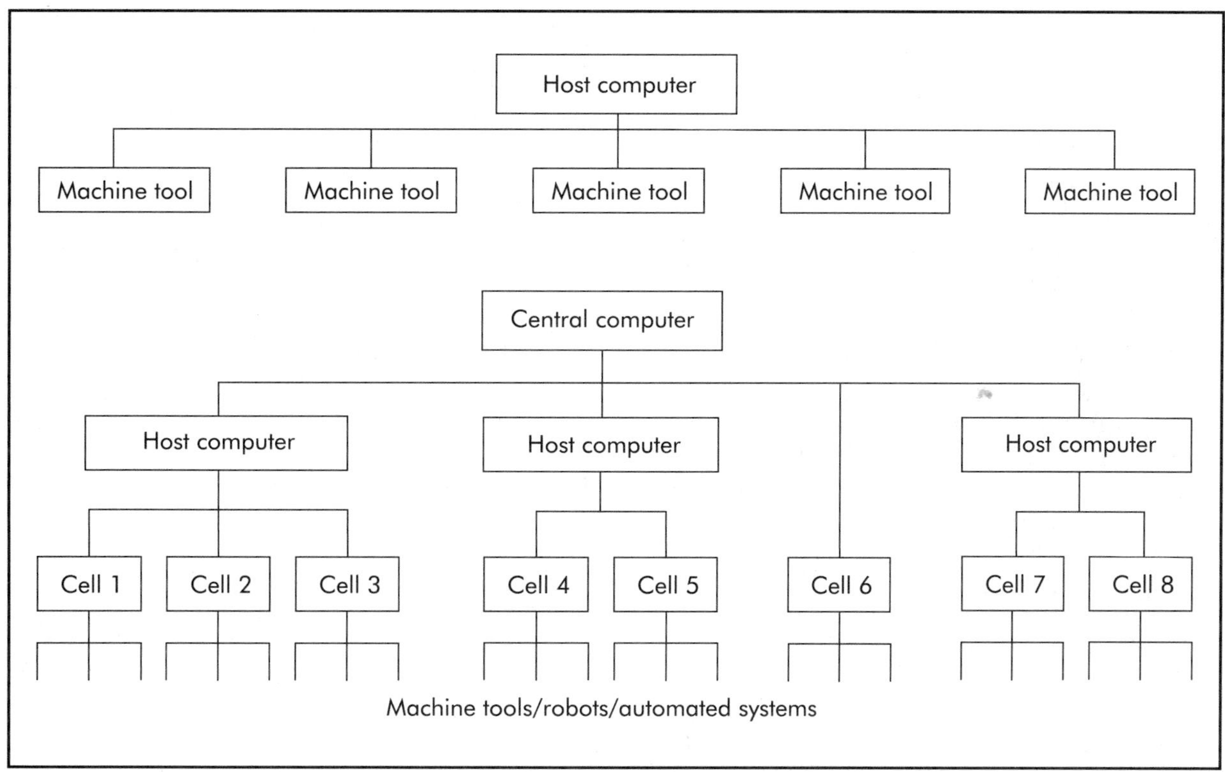

Figure 30-9. Digital numerical control systems.

facturing processes according to a predefined set of rules based on the laws of fuzzy logic.

Control systems need to be able to respond to the changing parameters of the process on time. Even in newer adaptive control systems, the presence of an operator standing by is necessary. With fuzzy logic, the control system monitors and adjusts the process constantly without continuous human intervention.

Fuzzy logic uses algorithms and mathematical models to describe the system and the relationships between its parameters. These relationships are described using rules based on the best available experience of skilled operators and are programmed into the controller to cover all possibilities. The rules act as the system brain that thinks and responds in a very similar way to humans. An example of a simple fuzzy system is illustrated in Figure 30-10. A fuzzy logic control system has many benefits:

1. It eliminates the difficulties encountered with mathematical modeling by removing the excess precision inherent in the process.
2. It provides convenience and a user-friendly programming interface.
3. It permits machine improvement by integrating practical rules originating from daily practice into the system. A rule, in a fuzzy logic sense, is simply a collection of the best experiences available from the best

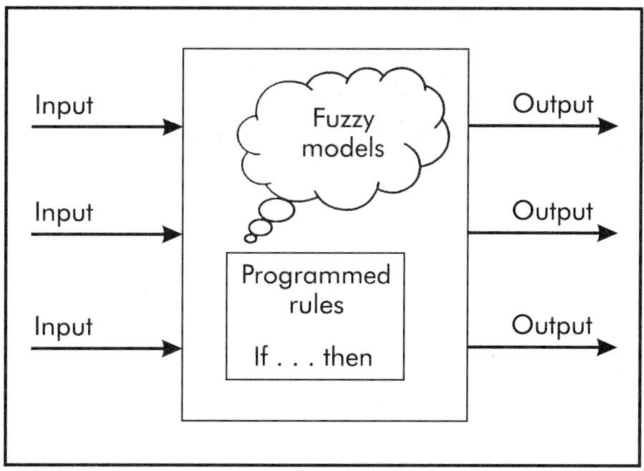

Figure 30-10. An illustration of a simple fuzzy control system.

operators translated into system-readable commands.
4. It requires less computer memory. Rules are put into the system in plain English.
5. Rules are often independent. If one rule fails, others may work and thus system efficiency is increased.
6. Once rules are preprogrammed into the control system, there is no need for skilled operators, which decreases the cost and improves productivity.

Calculations using fuzzy logic can be controlled in less time with better results. A comparison of machining speeds on an EDM machine at the Industrial Electronics and System Development Lab at Mitsubishi Electric Corporation in Japan is shown in Figure 30-11. Figure 30-12 presents another example of the savings realized by fuzzy logic control systems in mold cavity production at Mitsubishi where it resulted in about 15 (or over 22%) less hours than if the cavity was produced by conventional methods.

30.3 UNMANNED CELL AUTOMATION

Unmanned cells can be viewed as a cooperating collection of machines, assisted by an automated measuring system and an automated loading/unloading device, such as a robot. These machines and devices are controlled by a computer or a cell controller.

The concept of cellular manufacturing, whether using manned or unmanned cells, has evolved with the introduction of group technology (GT) into manufacturing systems. Each cell is designed to produce a part family that requires processing using the same machines, tooling, jigs and fixtures, and similar operations.

Unmanned cells use pallets that secure parts while they are being machined. The pallets can be automatically removed or replaced from the machine. Parts, in turn, are secured to the pallets, not the machine table. To increase flexibility, some cells have machining centers with pallet changers, a tool storage magazine, and automatic tool monitoring. This flexibility can be further increased by utilizing robots with sensing devices attached to the gripper. The robot can be programmed to pick a variety of parts, load parts into the NC machine, and unload parts and move them to the next workstation.

As demonstrated in Figure 30-13, manufacturing cells work in the medium range of product variety and production volume. They provide improved productivity and a higher degree of flexibility than flow lines, transfer lines, or job shops.

30.3.1 Machining Centers

The design of NC and computer-controlled machines has rapidly evolved into what are called *machining centers*. These machines have the versatility and flexibility that make them a primary choice in machine tool selection. Machining centers operate under numerical control and programming to control their operations; they are basically a computer-controlled machine capable of simultaneously performing a variety of operations on different surfaces and in different directions on a workpiece. Since the development of machining centers is intimately related to nu-

Electrode: 10 mm (.39 in.) in diameter | Workpiece: tool steel
Peak current: 10 amps | On time: 64 μ/sec (microseconds)

Constant settings (no adaptive control)	80 min	Machining unstable, test stopped	Depth: 9 mm (.35 in.)
Constant settings (adaptive control)	140 min	Machining stable but slow	Depth: 15 mm (.59 in.)
Skilled operator present	95 min	Machining stable and fast	Depth: 15 mm (.59 in.)
Fuzzy logic (no operator present)	90 min	Machining stable and fast	Depth: 15 mm (.59 in.)

Figure 30-11. System performance with and without fuzzy logic control (Wilkins 1995).

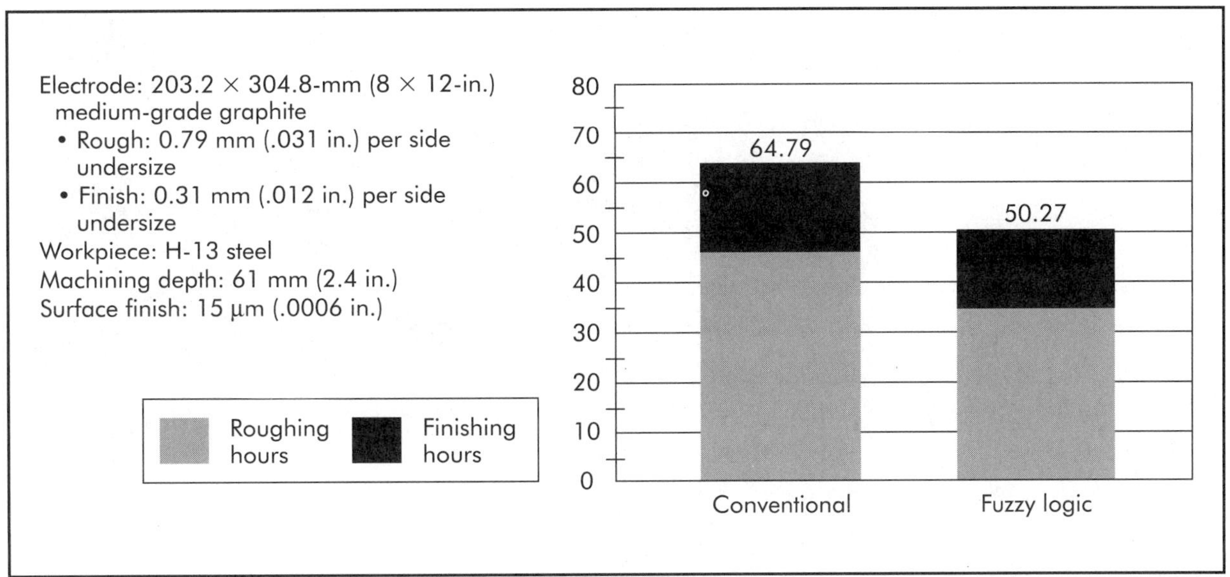

Figure 30-12. Fuzzy logic at work for mold cavity (Wilkins 1995).

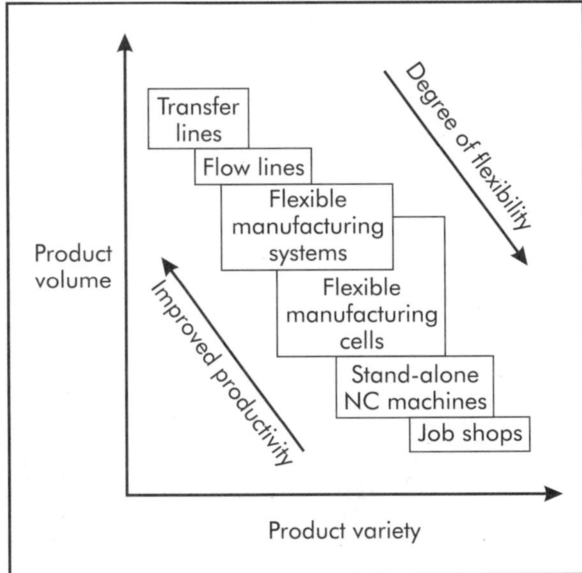

Figure 30-13. Product variety versus production volume for different manufacturing systems.

merical control, more discussion on the operation and components of machining centers is presented in Chapter 31.

Machining centers are equipped with automatic tool changers with which cutting tools may be automatically retrieved from tool storage, a magazine, or tool chain to perform the required operations on workpieces. For more flexibility and versatility, a part-checking station may be available in some machining centers, where parts are checked for accuracy and conformance to specifications automatically using touch probes. Workpieces in a machining center may be placed on pallets that provide orientation in different directions to allow the machining of different surfaces of the workpiece.

There are two main designs of machining centers: horizontal spindle and vertical spindle. A universal machining center, another design of machining center, has the capability of machining many surfaces of the workpiece in different orientations. Horizontal spindle machining centers, also called horizontal machining centers, perform machining on large workpieces on several surfaces. An example of a horizontal machining center is shown in Figure 30-14. Another class of horizontal machining centers is *turning centers*, which are computer-controlled lathes with many added features.

Vertical spindle machining centers, also called vertical machining centers, are used primarily to perform machining on flat workpieces with long holes and cavities. An example of a vertical machining center performing a drilling operation on a workpiece is shown in Figure 30-15.

Machining centers come in different sizes and with varying features with an average cost ranging from $75,000 to over $1 million. They are

Figure 30-14. Horizontal machining center. (Courtesy Giddings and Lewis)

highly automated machines with the capability of handling and machining varieties of parts with a high degree of accuracy and precision. The multi-axes feature of a machining center adds versatility to the machine and the capability of machining different workpieces or workpiece surfaces simultaneously. The added feature of a gaging and part-checking station on some machining centers contributes to their popularity as being highly versatile and flexible machines in automated manufacturing systems.

30.3.2 Decouplers

Decouplers are used in manufacturing systems to provide certain control functions for the process. They are mostly applicable in partially automated systems, where operator intervention is still required. Decouplers are normally placed between the machines, machine operations, or processes to force the system to allow the performance of a certain operation. They provide different functions, such as flexibility and quality control. A decoupler may be used to stop the machine to allow the inspection of a part for flaws, to delay the process until the part is cured, to delay the process to allow a certain action on the part, or to serve as an in-process check point. For example, in a forging operation, a process delay decoupler stops the part's handling at a certain point and delays its movement to the next workstation to allow the part to cool down for a certain period beyond the cycle time of the manufacturing system. Decouplers represent an important component of partially automated systems.

30.3.3 Robots

A large part of manufacturing involves the movement of parts, tooling, material, and other equipment. This is done in transfer lines and flow lines by special-purpose manipulators, conveyors, and positioners, each performing a

Figure 30-15. Vertical machining center. (Courtesy Giddings and Lewis)

certain function. Robots have been primarily used in manufacturing for loading and unloading workpieces and manipulating tools. More discussion on robot applications in manufacturing systems is presented in Chapter 29. Robots are presented here as a component of unmanned manufacturing cells and as a part of flexible program automation.

In manufacturing, a robot has been defined as a programmable device capable of performing complex actions in a wide variety of operations. It is a manipulator that normally can be programmed to do various repetitious actions without human intervention. The robot is directed by an information processor following instructions that are programmed in its memory. Robots are powered by air, hydraulic, electrical, or mechanical means, often by a combination of these in one unit. Some robots are mounted on wheels and can even be made to follow a course laid out by a wire in the floor.

Robots may be classified according to their physical configuration as: (a) Cartesian or rectilinear robots, (b) spherical or polar robots, (c) cylindrical robots, and (d) revolute or jointed-arm robots. See Figure 30-16. Each of these classifications has its own applications based on its suitability to the nature of jobs and compatibility with other components. Robots also may be classified according to their mode of motion control as: (a) pick-and-place, (b) point-to-point, or (c) continuous path. These robot classifications are discussed further in Chapter 29.

Pick-and-place robots are the simplest. They are capable of picking up small loads from one location and moving them to some other location. Their movements are limited in number and their control systems are rudimentary and may consist of a series of switches or valves tripped by dogs. Length of movement may be determined by fixed stops. They may be programmed to perform a certain sequence of movements and reprogrammed to perform a different sequence as necessary. A point-to-point transfer robot is capable of moving between a number of points and not necessarily in the same sequence. Versatility and fast reprogrammability aid in precise point location. This is imparted by a position servo at each joint. A continuous-path robot has a much larger memory and is capable of coordinating joint movements and interpolating a given path between end points.

Most robots go through a programmed routine on the assumption that the environment or conditions of the operation remain the same. Typical jobs are depicted in Figure 30-16. At the basic level, the robot takes a hot billet out of a furnace and places it in a die on a forging press. Then it transfers the forging to a trimming press and the scrap to a bin. In the top view, a robot carries a gun on a yoke to rivet or spot weld assemblies as they pass on a conveyor. Through a camera or some other means, the robot tracks the pieces and works on them as they pass.

To take on more difficult operations, a new generation of robots was developed with sensory capabilities enabling them to adapt to changing conditions in an operation. Another direction for expansion to increase robot capabilities is to make robots able to discriminate among different kinds of parts. One system can discriminate

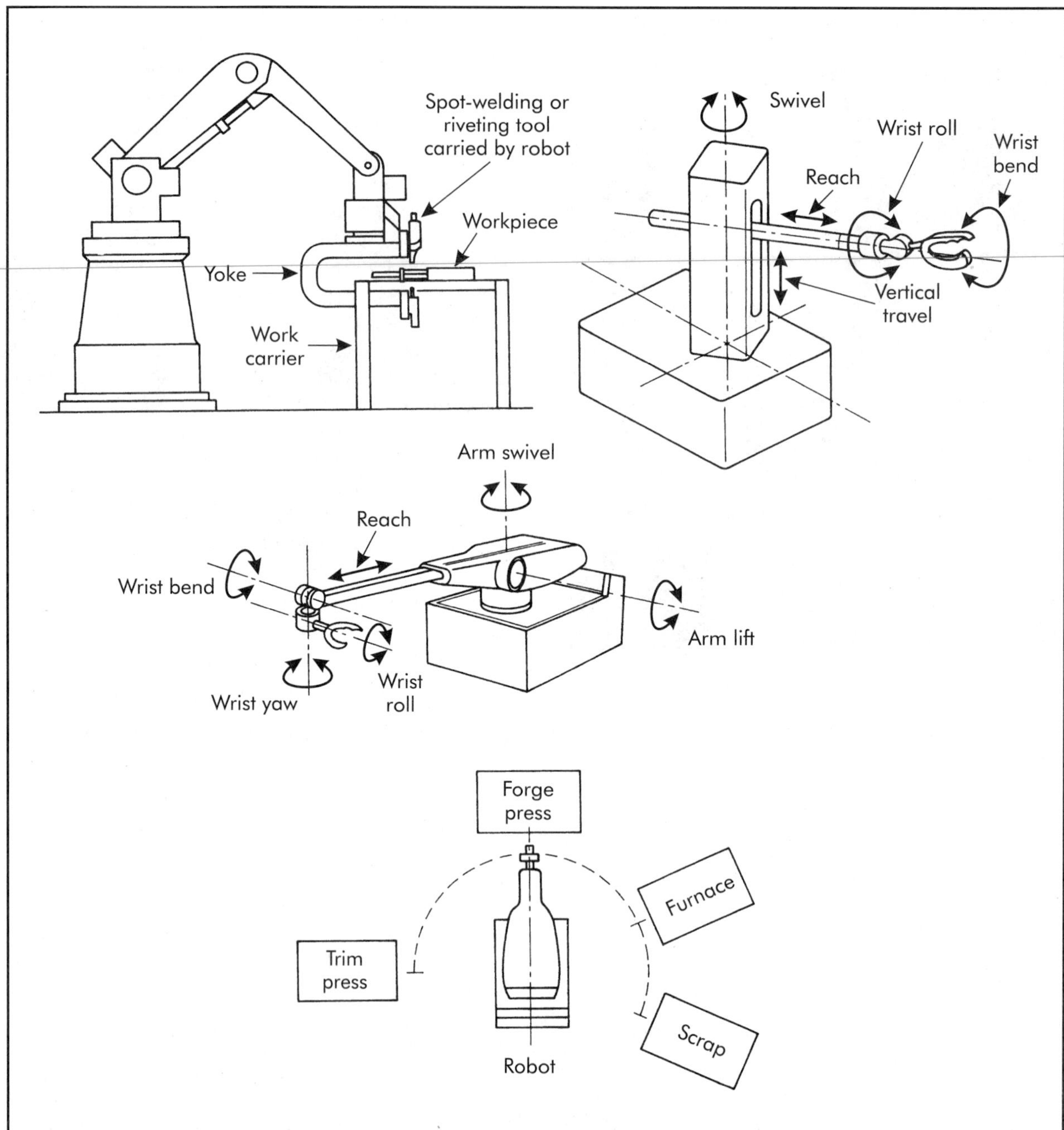

Figure 30-16. Three models of robots.

among several different parts passing on a conveyor, pick them off one by one, and deposit each at the station where they belong. These are called adaptive or visionary robots. Touch-sensitive fingers have been developed with an artificial skin made up of a grid of small electrode points. Feedback from the fingers enables a robot with programmed computer control to feel for spots and obtain a secure grip on quite irregular workpieces. In another direction, robots now can perform assembly operations. For example, one is reported able to select and pick up an assortment of parts and arrange them into such assemblies as radio speakers or electric motors.

Human beings can do the same work as robots, often more economically, but there are conditions under which a robot is superior. Robots can work tirelessly in dirty, hot, tiring, monotonous, unhealthy, and unsafe environments. They can keep up a steady pace and make less scrap. Robots can lift heavy loads. They labor in foundries, die casting, plastic molding, forging, welding, machining, assembling, and many other operations. Other applications for robots include material handling, loading and unloading, and transferring workpieces through the manufacturing plant. They perform operations, such as spot welding of automobile and truck components, arc welding, and arc cutting. Robots can do other jobs like spray painting, automated assembly, inspection, and gaging.

30.3.4 Automated Inspection

An objective of an automated inspection system is to assure conformance of parts to a desired set of specifications. Other objectives may include providing a means for corrective measures on the process and dealing with nonconforming parts. Statistical process control (SPC) methods are widely used as part of inspection programs.

Automated inspection may generally be classified as on-line inspection and off-line inspection.

On-line inspection is generally performed at the machine or workstation. Some machining centers have the capability of performing this function in an independent gaging station within the machining center work area. This is done by electronic probes, also called touch trigger probes, or other electronic devices. The machine spindle, under the control of a computer, searches for the appropriate probe or device and performs measurements automatically. On-line inspection is an effective technique in automated manufacturing; it calls for an immediate corrective action to be taken on the process immediately after a faulty operation or a nonconforming part is detected. Therefore, on-line inspection is a cost inspection method that provides a means for improving quality and lowering the cost of inspection.

A major disadvantage of on-line inspection is the longer machine cycle time required because inspection takes place on the machine. This could be avoided by having a separate station for inspection away from the main work area. This feature is available on some larger machining centers.

Off-line inspection uses separate machines or devices to perform inspection. Coordinate measuring machines (CMMs), discussed in Chapter 29, are a prime candidate to perform off-line inspection. Different configurations of CMMs are available to suit different part configurations and other inspection requirements and procedures. CMMs provide more functions than just inspecting the part for conformance to specifications; they provide a means for process control. In these cases, CMMs are referred to as coordinate measuring systems.

30.4 COMPUTER INTEGRATION

The advent of computers and computer technology and their emergence in manufacturing have resulted in the development of many technologies that have helped the manufacturing industry realize dramatic increases in productivity and profitability. These technologies are listed according to each area of manufacturing-related activity:

- Product Design and Development—computer-aided design (CAD); computer-aided engineering (CAE); group technology (GT); rapid prototyping (RP).
- Manufacturing Engineering—computer-aided process planning (CAPP); group technology (GT); computerized bill of materials.
- Production Planning—material requirement planning (MRP); manufacturing resource planning (MRPII); computerized bill of materials.
- Manufacturing Processes—numerical control (NC); computer numerical control (CNC); direct numerical control (DNC); robotics; statistical process control (SPC); flexible manufacturing systems (FMS); manufacturing cells; automated assembly; shop floor control; fuzzy logic control systems; just-in-time (JIT).

- Inspection and Testing—computer-aided inspection (CAI); computer-aided testing (CAT); automatic inspection systems; coordinate measuring machines (CMMs).
- Material Handling—automated guided vehicles; automated storage and retrieval systems (AS/RS).

Common in all phases of manufacturing are other technologies involving computer integration, such as concurrent engineering (CE), computer-integrated manufacturing (CIM), and total quality management (TQM).

30.4.1 Computer-aided Design

Computer-aided design (CAD) is a major element of automated manufacturing systems. It involves the use of computers to generate and manipulate designs and product models that can be used to manufacture parts and products. The evolution and use of computers and their high computing ability has led to shorter product development times. Development of a product starts with an idea, which is then translated into parameters and designed as illustrated in Figure 30-17.

A CAD system consists of two major components in addition to the user or designer: computer with input/output devices and CAD system software.

The CAD software may require a mainframe computer, a minicomputer, or a microcomputer, depending on the sophistication and capabilities of the CAD package. A CAD package provides the functionality to the system. It should have the ability to draw and redraw and to perform calculations, such as areas, volumes, moments of inertia, and other functions needed for the design. Other desirable features of a CAD package are user-friendly dialoguing and communication, two- and three-dimensional drawing capabilities, 3-D solid modeling generation, dimensioning and tolerancing, labeling, and editing capabilities.

Input/output devices are generally used as a medium to transfer information to and from a computer. Many input devices may be used with CAD software. The standard default input device is the keyboard used to transmit input data to the system. Mouses, joysticks, and track balls also may be used to manipulate the cursor movements. Other CAD systems may use a light pen for interactive graphics, a digitizer that translates drawings into digital signals and sends

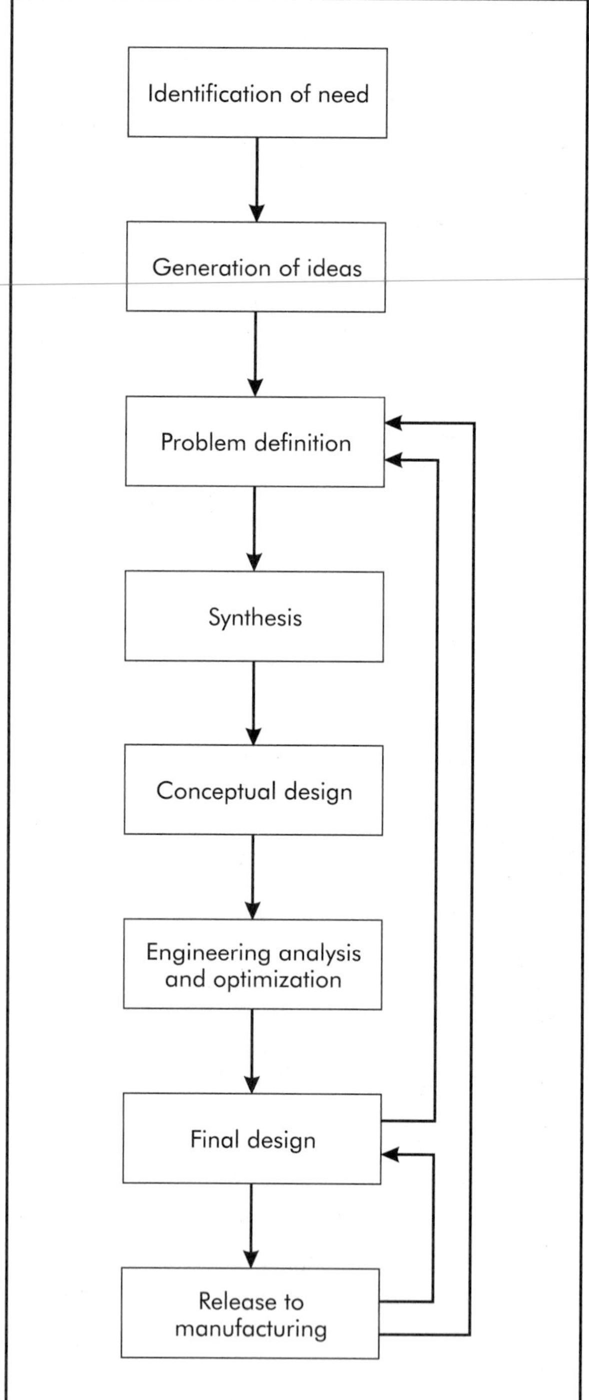

Figure 30-17. Steps in the design process.

these signals to a compute, or a graphics scanner that scans a drawing and converts it into a CAD system readable format.

The cathode-ray tube (CRT) is the standard output device for a CAD system. CRTs provide temporary output. Permanent output or hard copies of the design may be generated using peripheral devices, such as plotters or graphics printers. One of the fastest computer output devices available is the video display unit (VDU), which provides a temporary output that can be instantly updated by a computer.

Improved functionality now allows CAD to reproduce an actual prototype of the product using computer representation of graphics. This is called rapid prototyping and is discussed in Chapter 29. It uses a computer-guided laser to develop a three-dimensional prototype of the product. The technique is based on using the data generated by CAD drawings to produce plastic prototypes. The parts or products are made by mathematically slicing CAD drawings or designs into very thin cross sections throughout the part. A laser beam traces each layer in a container of treated chemicals that solidify as they are irradiated.

Computer-aided engineering (CAE) is a term that is often used in conjunction with CAD. It involves the integration of product performance and structural characteristics, computers, and information technology and management into an integral concept for product design and development. CAE is aimed at the integration and automation of different engineering functions in the product development process, including design, testing, data management, process planning, tool design, numerical control, manufacturing engineering, reliability, and quality control.

30.4.2 Computer-aided Process Planning

The sequence of operations that a part goes through during its manufacture is called a *process plan*. Process plans may be developed manually using route sheets that identify operations, machine tools used for each operation, and cutting or operating conditions, such as feed, speed, and depth of cut. Other information about cutting tools, jigs and fixtures, and other equipment also may be listed on route sheets.

The emergence of computers in manufacturing has made it possible to automatically generate process plans. This is called *computer-aided process planning* (CAPP). There are two types of CAPP systems: *retrieval computer-aided process planning* and *generative computer-aided process planning*.

In retrieval (also called variant) CAPP systems, parts are grouped into families utilizing group technology (GT) concepts. Group technology is a manufacturing philosophy and is discussed in Chapter 29. Standard plans for each part family are generated and stored in the computer. When a plan needs to be developed for a new part, the computer searches for similar process plans that may then be modified as needed. The steps involved in variant process planning are illustrated in Figure 30-18. Variant process planning starts with a search for an existing part family after a code (using some classification and coding scheme) is entered into the computer. The computer then displays the operation sequence or route sheet for the part. If the search for a part family fails, a standard plan is displayed and then edited and modified according to the part configuration.

Unlike variant process planning, generative process planning systems create process plans for each new part using manufacturing information available in a data base. The creation of generative plans is done automatically using knowledge about manufacturing process engineering that is encoded into a software system, including tool selection, machine tool to be used, and cutting parameters.

Generative CAPP systems are still limited in their capabilities to deal with complex parts and processes. They are most compatible with automated and computer-integrated manufacturing systems, where new parts can be easily planned and process plans may be rapidly developed.

30.4.3 Computer-aided Manufacturing

Computer-aided manufacturing (CAM) refers to the use of computers and computing power to assist in all functions involved in the manufacture of a product, including process planning,

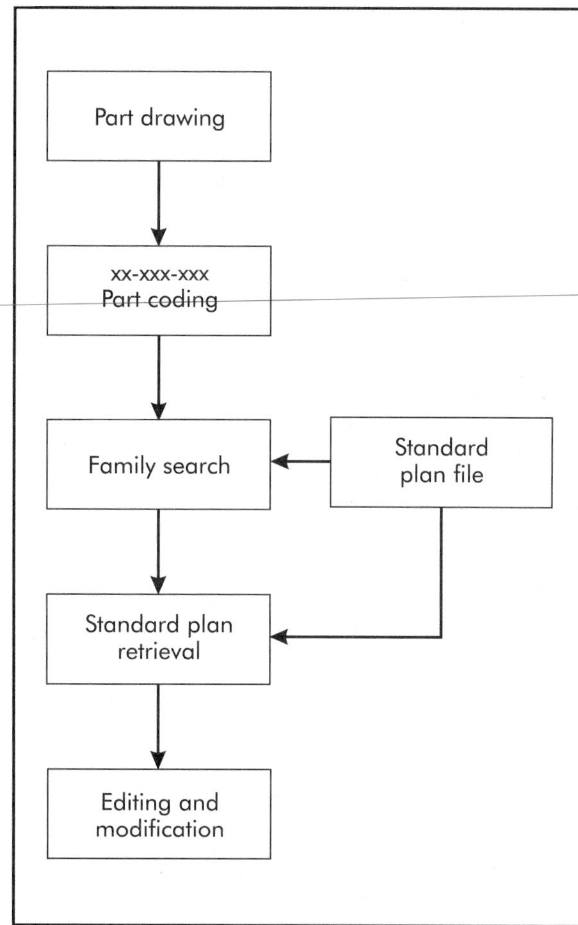

Figure 30-18. Variant process planning.

production planning and control, material processing, and quality control. CAM encompasses many technologies, such as computer-aided part programming, computer numerical control (CNC), direct numerical control (DNC), adaptive control, automated material handling systems, and automated assembly and inspection systems.

Because of potential benefits, computer-aided design (CAD) and computer-aided manufacturing are combined to provide CAD/CAM systems. These systems provide flexibility and allow sharing of data bases between design and manufacturing. CAD/CAM systems also help reduce costs by eliminating the effort of transferring data from design to manufacturing. Productivity of manufacturing systems utilizing CAD/CAM software has improved as the need for prototype development is eliminated.

30.4.4 Computer Integration of an Unmanned Cell

Current research efforts in manufacturing automation have resulted in a significant increase in productivity, in addition to other advantages, such as reduced lead time, reduced setup times, minimum work-in-process (WIP) inventory, higher levels of quality, and better utilization of resources.

Manufacturing cells are the most common types of unmanned manufacturing. Unlike manned cell operation, unmanned cell operation is operated under computer control. Unmanned cells may be composed of any combination of the following:

- NC or CNC machine tools,
- an automated materials handling system,
- robots, and
- an automated inspection system.

The nature of operations performed and the configuration of parts produced define the system structure, its mode of operation, and the number and type of machines and other systems used. In many unmanned cells, the need for a material handling system may be eliminated, as part movements may be performed by a robot, as in robotic cells.

Computer control of unmanned cells is hierarchical, as shown in Figure 30-19.

30.5 ECONOMIC JUSTIFICATION OF AN AUTOMATED SYSTEM

When an appreciable quantity of pieces is to be produced, the question usually is not whether

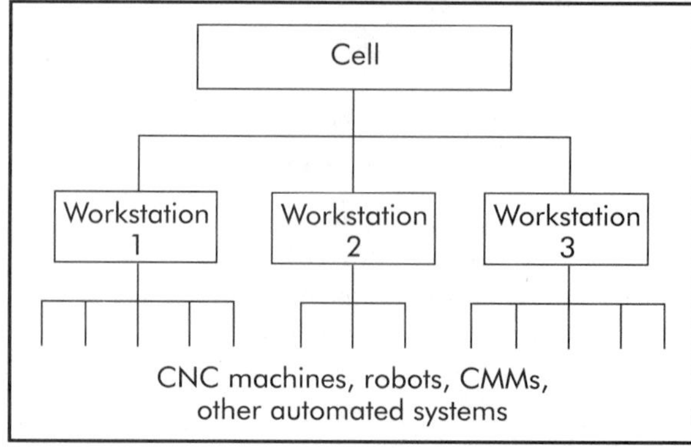

Figure 30-19. Unmanned cell control structure.

to automate but rather how much automation is justified. An illustration is furnished by a case for blanking and then notching electric motor laminations. Seventeen different round blanks are needed in lots of 1,300–8,100 pieces, a total of 70,000 blanks a week. There may be anywhere from 18–58 notches required around the outside of a blank, in 50 different shapes and sizes.

One blanking press and four notching presses are capable of meeting production requirements. Several arrangements are possible, and all have been studied, but for simplicity only the two most feasible will be explained here and are indicated in Figure 30-20. A detailed study shows that the following annual savings may be expected from the fully automated system as compared to no automation:

Operator time at blanking press (760 hr at $18.00/hr)	$ 13,680
Overhead saved at blanking press	2,430
Operator time at notching presses (8,320 hr at $18.00/hr)	149,760
Overhead saved at notching presses	26,600
Total annual savings from full automation	$192,470

All four notching presses must process the same blanks being turned out by the blanking press at any one time. This necessitates three additional sets or 150 notching dies. The costs of equipment for the fully automated system are:

150 duplicate notching dies at $2,500 each	$375,000
Automatic distribution mechanism	150,000
Automatic press loading, feeding, and stacking devices	298,000
Total cost of equipment for full automation	$823,000

A partially automated system, also depicted in Figure 30-20, dispenses with the automatic distribution mechanism and requires handling and storing the blanks in lots by means of pallets and forklift trucks. With this system, the blanking press can be run at its fastest speed because it is not held back by the notching presses when they are turning out laminations with large numbers of notches. Although handling costs are more, the annual savings for partial automation is not much less than for full automation and is estimated to be $176,360. The

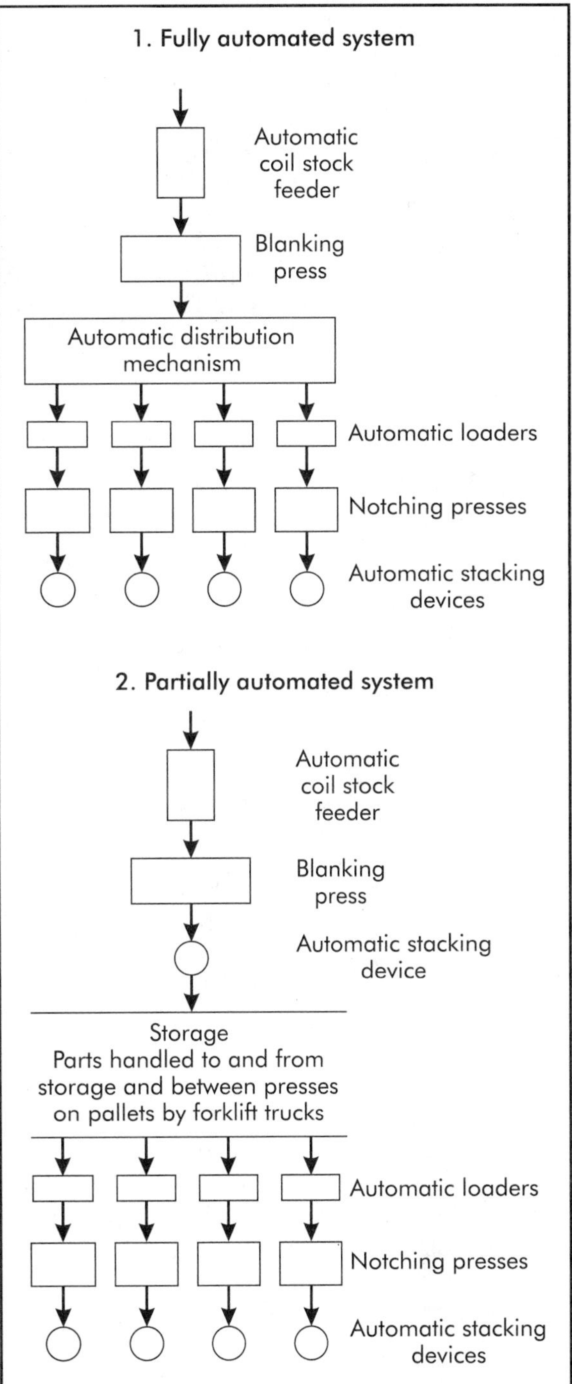

Figure 30-20. Two arrangements for producing motor laminations.

equipment costs are much less. The notching press can be run independently, and the 150 duplicate notching dies are not needed. Neither is the distribution mechanism required. Thus the total cost of equipment for partial automation is only $298,000 for automatic press loading, feeding, and stacking devices. The return on the investment (total savings ÷ equipment cost × 100 for each alternative) is 23% for full automation, but a handsome 59% for partial automation.

The complete manufacturing effort may be divided into handling of materials, fabrication of pieces, finishing, inspection, assembly, and packaging. The completely automated plant through all these functions is an ideal, but substantial savings may be realized by applying the principles of automation to each step of production as justified. The rule is that the areas that promise to be most profitable should be attacked first, and each area should be exploited to the extent that the return justifies the investment.

30.6 QUESTIONS

1. What does "automation" mean?
2. What are the different classes of automation? Briefly discuss each.
3. What are the advantages and disadvantages of transfer machines?
4. What are the advantages and disadvantages of nonsynchronous systems?
5. Briefly describe how an automatic tool compensation system works.
6. Describe a typical programmable logic controller.
7. What is a "robot"? Describe a typical robot used in manufacturing.
8. What are the different factors influencing the selection of a material handling system?
9. Describe the load transfer techniques normally used with AGVs.
10. What is the difference between computer-aided design and computer-aided engineering?
11. How is a variant computer-aided process planning system generated?
12. Briefly explain how a fuzzy control system works.
13. What are the major advantages of using fuzzy logic control systems in manufacturing?
14. How would you make an analysis to determine whether to automate an operation?

30.7 PROBLEM

A rotary index machine has three stations. Workpieces (one at a time) are unclamped, unloaded, loaded, and clamped by an operator in a jig at station one. At the same time, a six-spindle drill head at station two descends and six holes are drilled in a piece. Also at the same time, three holes are reamed in a piece by a three-spindle drill head at station three. When both heads are finished and raised out of the way and the operator signifies that he or she is clear, each piece is indexed to the next station. Indexing time is .02 min. It is estimated to take .20 min to unload and load a workpiece, .25 min to drill the six holes simultaneously, and .15 min to ream three holes at the same time. The drilling and reaming heads are raised so that the tools clear each workpiece by at least 75 mm (≈3 in.). They descend at a rapid rate until the tools are within 5 mm (≈.2 in.) of the pieces and then continue at feed rates.

(a) Write a set of Boolean equations to express the logic of this operation.
(b) Construct a ladder diagram to express the logic of this operation.
(c) Estimate the number of pieces that can be machined in 50 min of continuous operation.

30.8 REFERENCES

Askin, R. G. and Standridge, C. R. 1993. *Modeling and Analysis of Manufacturing Systems.* New York: John Wiley and Sons.

Astrom, K. J., and Wittenmark, B. 1989. *Adaptive Control.* Reading, MA: Addison-Wesley.

Groover, M. K. 1987. *Automation, Production Systems, and Computer-integrated Manufacturing.* Englewood Cliffs, NJ: Prentice-Hall.

Kusiak, A. 1990. *Intelligent Manufacturing Systems.* Englewood Cliffs, NJ: Prentice-Hall.

Lee, M. H. 1989. *Intelligent Robotics.* New York: Halsted Press.

Luggen, W. W. 1991. *Flexible Manufacturing Cells and Systems.* Englewood Cliffs, NJ: Prentice-Hall.

Maleki, R. A. 1991. *Flexible Manufacturing Systems: The Technology and Management.* Englewood Cliffs, NJ: Prentice-Hall.

Rembold, U., Nnaji, B. O., and Storr, A. 1993. *Computer-integrated Manufacturing and Engineering.* Wokingham, England: Addison-Wesley.

Wilkins, C. 1995. "Fuzzy Logic: The New Brain in EDM Machines." *EDM Today,* January/February: 12–22.

Numerical Control

31

CNC milling machine, 1953
—Massachusetts Institute of Technology, Cambridge, MA

31.1 INTRODUCTION

Numerical control (NC) is a form of programmable automation in which the mechanical movements of machine parts are controlled by a program that contains alphanumeric codes. These coded data control the amount and sequence of the motion of the cutting tool relative to the part being processed. This chapter provides an overview of numerical control and programming. Manual and computer-assisted programming also will be presented. Metal cutting operations are emphasized in this chapter, but it should be noted that the concepts are valid for other applications.

Numerical control has been applied to all kinds of metal cutting machine tools and to many other areas, such as arc welding and flame cutting. NC machines are designed to meet each need most efficiently and economically and, therefore, many types and sizes of NC machines exist. All are based on certain underlying principles.

NC started out with punched-tape input, and has been called *tape control,* but other media are also used for input. Punched cards and tapes have been applied to direct machines for over 150 years: in Jacquard looms and cloth-cutting machines and on the player pianos of a bygone era. The first NC machine, a milling machine, was developed and exhibited in 1953 at the Massachusetts Institute of Technology under the sponsorship of the U.S. Air Force. In the period since then, NC machine tools have come to comprise an appreciable portion of all machine tools made and sold in the world.

NC has taken a prominent and secure place in industry, but so far has not been able to replace all other methods. Along with its merits, NC has certain disadvantages, and these will be pointed out later in this chapter.

31.2 ELEMENTS OF NUMERICAL CONTROL

In a numerical control (NC) system, programs containing coded data direct the machine tool and control its movements to perform specified functions. In general, there are three basic functional and operational elements in a numerical control system:

1. Data input. Numerical information to process a certain part is read and stored in a tape reader or computer memory.
2. Data processing. Programs are read into the *machine control unit* (MCU), also called control or controller.
3. Data output. After the numerical information is read into the MCU, it is translated into commands to the servomotor. The ser-

vomotor moves the table on which the part is clamped in a specified position. The table movement may be accomplished through linear or rotary movements, or by other means, such as lead screws, gear-and-pinion systems, stepping motors, or other devices.

In NC systems, the data required to produce a part are stored on a punched tape or other medium called the *part program*. A part programmer prepares the part program for an NC machine tool. Part programs include information about part dimensions and geometric features, along with the different motion instructions required to machine the part.

NC machines have different axes of motion; each has its own driving device. *Coordinate systems* are used to define the direction of motion of different machine axes. A coordinate system using absolute or incremental positioning concepts is normally used. Part programs are read and decoded by the MCU. The MCU operates the drives and causing movements of the different machine axes to perform the desired action on the part.

Numerical control impacts manufacturing operations and the development of integrated manufacturing systems. The use of NC machines has several advantages over other conventional methods:

1. Improved flexibility of operation. A new program need only be developed to produce parts of complex shapes and configurations.
2. Accuracy and repeatability are improved and maintained through a full range of cutting conditions.
3. Shorter setup and machining times.
4. Shorter manufacturing lead times. Programs may be developed rapidly and stored programs may be easily modified to accommodate similar part configurations.
5. Improved productivity and part quality. Operator's influence is minimized and the need for highly skilled operators is reduced.
6. Machine adjustments are easier in NC systems using computers and digital readouts.

NC systems also impact other production planning and control activities: design changes are easily made, tooling costs are reduced, and inventory is reduced. However, there are a few disadvantages or limitations to the use of NC systems that include:

1. The need for a highly skilled maintenance crew as special and preventive maintenance procedures are required.
2. A relatively high initial investment.
3. The need for trained NC programmers and the cost of computing time.

Despite all of these limitations, the use of NC systems can be economically justified considering the many other benefits they provide.

31.2.1 NC Operation

NC Input

Instructions to an NC system consist of a stream of numbers assembled into a program. Typical programs are displayed later in this chapter. Programs are made up and fed into the unit that controls an NC machine in a number of ways, as illustrated in Figure 31-1. It is necessary to show a number of different paths for the flow of information in NC programming, as in Figure 31-1, because there are many systems and different procedures. Specifications for the product to be made generally are conveyed by an engineering part drawing or print. A programmer lists the steps and operations, as well as the dimensions and conditions for each step to produce the part on a programming process sheet. The instructions are then entered from a keyboard to a device that prepares tapes, cassettes, disks, or cards, or directly into a general-purpose computer or the control of the machine. Much of the drudgery of repeating steps or making routine calculations may be done by a computer.

In some advanced plants, particularly in the aerospace industry, product specifications are computer-generated by the designers and transmitted directly to a computer that prepares the NC program. Drawings and program sheets for reference may be made up on automatic drafting machines instructed by the computers. At the other extreme are some NC systems wherein programming is done at the machine by the operator. The instructions for each step are entered by push buttons on a panel at the workstation.

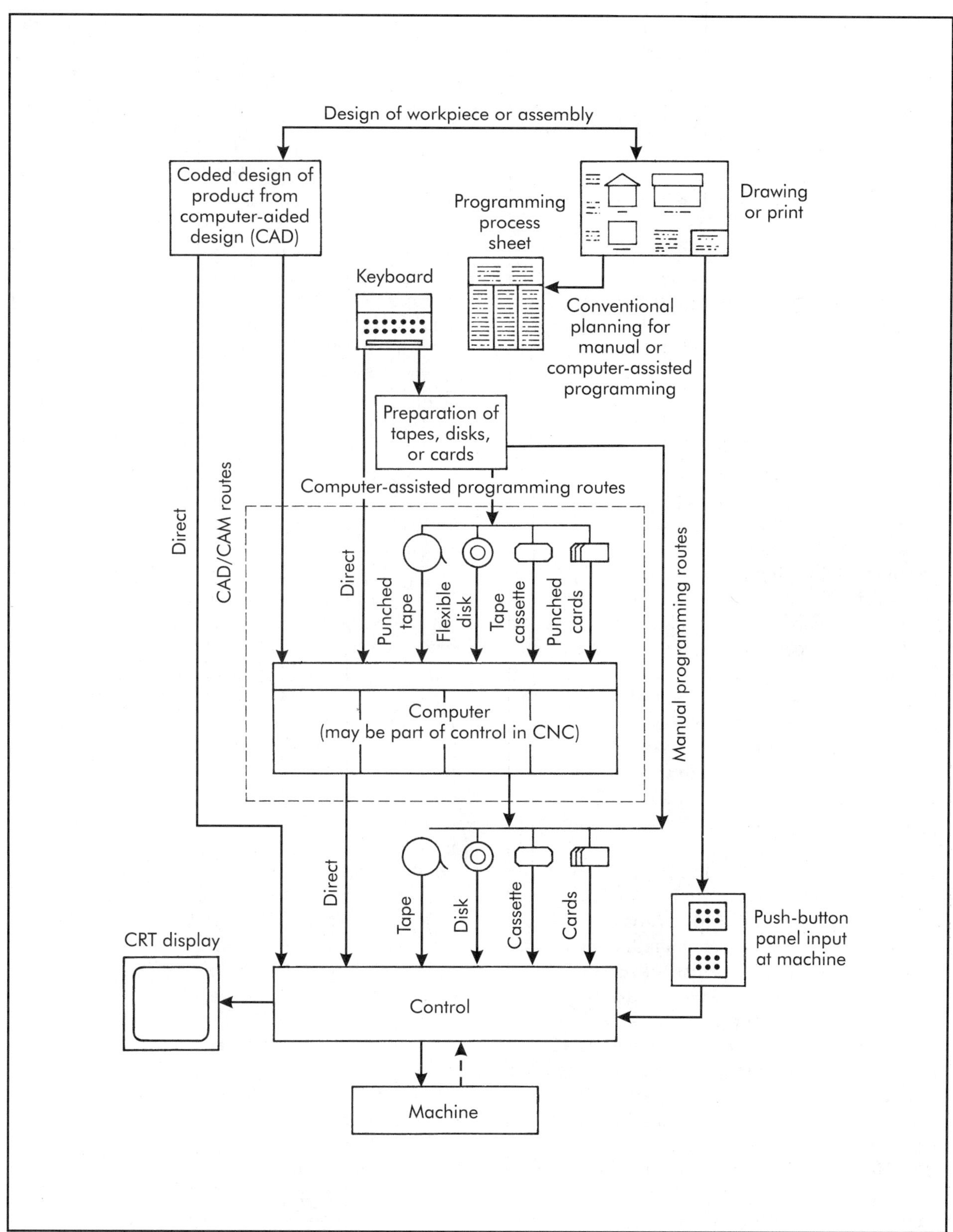

Figure 31-1. Procedures for NC programming.

This is done one step at a time in turn. A system of this sort, sometimes called *digital control* or *manual data input* (MDI), does not need a reader for tape or other medium and can cost 20% less than full NC. Most NC systems lie between these two extremes.

The route that a program takes to the control depends upon the kind of program written, the equipment available, and the input accepted by the control. The program that goes to the control must be quite detailed, as a rule, and such a program may be written by the programmer for a simple and short operation and thus bypass the computer. However, this is inefficient for long programs, where the programmer writes a "broad brush" program that is fed into a computer directly or through tape, cards, etc. The computer is programmed to produce detailed instructions for the control. From the programmer or from the computer, programs are conveyed to the control in most systems by punched tape, flexible magnetic disk, or magnetic tape in cassettes.

Some systems have an external computer connected directly to the NC control. An NC machine is relatively slow because it must pace the process, so one computer can service a number of machines through their controls. This arrangement of a host computer and its NC satellites is called *direct numerical control* (DNC) or *distributed numerical control.* An application of DNC is in the flexible manufacturing system as explained in Chapter 30. DNC systems range from small ones where one computer essentially supplies an expanded memory for a few machines, to systems with over 100 machines directed and served by a hierarchy of computers. The external computers may not only prepare and store NC programs but also perform management reporting and production control functions, thus augmenting computer-aided manufacturing (CAM). DNC is efficient while working, but computer breakdown is a problem if it controls many machines. This problem is avoided by tape or disk or a small computer backup at each machine. DNC calls for a sizable investment, so large numbers of DNC systems have not been built. However, it is a step toward the ultimate automatic factory and is growing.

Some controls offer output for a cathode ray tube (CRT) or other display. This may show the commands being executed on the machine at the moment or other information desired by the operator. Often, two-way communication is provided through a display and keyboard. Thus programs may be changed or optimized at will or troubles in the program or machine diagnosed and corrected through interaction. An NC control normally has a push button station through which the operator may exert control. This station may range from only a quick stop in some systems to many more options in others.

An NC punched tape is 1-in. wide, perforated tape with eight columns. The Electronic Institute of America (EIA) standard is the usual NC input. A newer code is the American Standard Code for Information Exchange (ASCIE), which is the same as the International Organization for Standardization (ISO) code. It is the recognized standard in many areas, for instance for long-distance transmission of data. Many newer controls can take input in both codes.

The column of small holes along the EIA tape is to engage a drive sprocket and is not part of the coding. Each row across the tape represents a *character* (number, letter, or symbol). A number is written in *binary coded decimal form* by the first four positions of each row.

Binary code is the natural numbering system for digital computers and NC machines because it requires only two conditions to represent a number, such as a switch on or off in a circuit or a hole punched or unpunched in a tape at a particular position. A punched hole stands for one; an unpunched hole for zero. These are called *bits,* an abbreviation for binary digits.

A group of characters constituting a normal unit of information makes up a *word,* such as an *X*-axis dimension. A group of words considered to be a unit is called a *block,* and it is separated from other such units by a hole punched in column eight and called the "end of block" character. A block, in effect, designates the disposition of all the controllable elements of an NC machine at each step of an operation. A block is essentially a line of the program.

The physical arrangement of the words in a block is called the *format*, and three standard forms are recognized. In *fixed-sequential format*, each word has a definite length and location in any one system. Words must be presented in a specific order, and all possible words preceding the last desired word must be present in the block. Each word in *tab-sequential format* is preceded by a tab character formed by holes punched in positions two through six in a row. Words must be in a specific order but may vary in length. In *word-address format,* each word in a block starts with one or more characters that identify the meaning of the word. Each format in the order just given requires fewer characters than the one preceding it for an average program, but more complex circuitry in the translator.

The *one character per row code* is standard for position and straight-cut control systems; it and a *one-word-per-track code* are both acceptable for continuous path systems. The latter code is also binary with one word along each column or track. The words for any one block are read in parallel along the tape, and end of block (EOB) characters separate the blocks.

NC Controls

Machine components actuated to do a job normally cannot respond to numbers that people work with and put into NC controls. The numbers must be converted into timely signals of sufficient magnitude to activate the machine components. Thus a control must perform certain functions to convert the numbers into adequate signals; all controls do not need to perform all the functions. The functions performed by controls are:

1. To read and verify each block.
2. To identify each word and assign it to a register or memory address for temporary storage. Common procedure is for a block of information to be read and put into storage while the preceding block is being processed. The numbers going in or being processed may be displayed on a CRT screen.
3. To convert each word into a signal. The commands for coolant, starting a spindle, etc., may initiate turning on or off an electric current to energize or de-energize a control valve, motor starter, etc. The position words, such as for the *X, Y,* or *Z* axis, are commonly converted into a series of pulses. This is called a *digital signal.* In many systems, each pulse represents a movement increment of 2.5 μm (.0001 in.).
4. To arrange the signals to carry out the commands for feed and speed rates. The feed rate for an increment of movement along an axis may be expressed by the rate at which pulses are emitted. A speed command for spindle speeds in steps may be translated into electric currents to the gear shifters in the spindle drive.
5. To interpolate from the point established by one block to the next. If a particular path must be traversed between two points, points in between can be calculated on a computer during programming so that movements between the interior points will not deviate from the prescribed curve by more than the tolerance allowed. This is not often necessary because many newer controls are able to synchronize machine movements to produce slopes and common curves like circular and parabolic arcs. Some controls can interpolate over almost any path.
6. To coordinate the feed rates of the various axes if contouring is to be done.
7. To convert the digital signals into *analog signals* required to actuate many systems. An analog signal is a physical magnitude, such as the voltage of a direct current or a phase shift angle of an alternating current.
8. To amplify the signals into electric currents, pressures, forces, etc., required to drive the machine components.
9. To compare a response with each signal if the system employs feedback (most do). If there is a discrepancy between command and response, another signal must be given to correct the error.

Older controls and many still made and in use today are composed of a collection of electronic units, each with circuits to perform these functions as required in the particular system. These are called *hard-wired controls.* As a next step, *computer numerical control* (CNC) developed. Controls of this kind are like general-purpose computers, and software directs the control. A CNC machine has a central processing unit (CPU) that performs arithmetic and logic processes to direct the functions of the control. The CNC machine also has a memory that stores an executive program (software), which directs the CPU to carry out the control functions. The memory also stores the part programs, wholly or partially. Some CNC machines can build up, refine, and do editing of part programs as would otherwise be done by external computers. A CNC system also has input and output (IO) registers. CNC technology was built on a macroelectronic basic, that is, around a *minicomputer.* In recent years there has been a trend to large-scale integration (LSI) of the circuitry. Such controls with one or a few small chips as the CPU are called *microprocessors* and are based on *microcomputers*. Microprocessors have become cheaper and more reliable and take less space than minicomputers. In some cases, a minicomputer may be faster, more flexible, expandable, and able to handle more peripheral accessories.

Software gives CNC a number of advantages. With a hard-wired control, the tape must be rerun each time the operation is repeated. This is not necessary with the part program stored in the CNC system memory. Also, the program in the CNC memory can be updated, revised, and edited more easily. In most cases, even the executive program can be revised. Subroutines may be incorporated in the executive program to provide added features. Canned subroutines can carry out whole drilling, tapping, boring, and other machining cycles from a single block of tape command; family of parts programming from only one tape; built-in trouble diagnosis; and many other tasks.

31.2.2 The Coordinate System

Positions in numerical control are defined with the Cartesian coordinate system, such as illustrated in Figure 31-2. An axis of an NC machine is considered a direction of possible programmed machine movement. On a three-axis milling machine, for example, the *X* axis is the direction of the table travel, the *Y* axis is the direction of the saddle travel, and the *Z* axis is the direction of spindle travel. The *Z* axis is always perpendicular to both the *X* and *Y* axes. As a general rule, a positive movement of the *Z* axis moves the tool away from the workpiece. In some NC systems, three rotational axes may be specified. These rotational axes, shown in Figure 31-2 as *A*, *B*, and *C*, are used to rotate the workpiece at a certain angle for machining different surfaces or to orient the spindle at a certain angle relative to the workpiece. Each point in the coordinate system is defined by an *X*, *Y*, and *Z* value in the form.

The number of directions of simultaneous movement provides another classification for NC machines. In this sense, a numerically controlled machine may be classified as a two-, three-, four-, or five-axis machine. In a five-axis milling machine, the *X* axis is the direction of the table travel, the *Y* axis is the direction of the saddle travel, and the *Z* axis is the direction of spindle travel. The fourth and fifth axis define rotations of the table, the spindle, or both.

In general, the definitions of machine tool axes follow certain guidelines for different machine tool and workpiece configurations (see Figure 31-3).

Part-rotating machines, such as lathes, have their *Z* axis parallel to the spindle. A positive *Z* motion moves the cutting tool away from the part being machined. The X axis is the direction of tool movement and a positive *X* motion moves the cutting tool away from the part.

Tool-rotating machines, such as drilling, boring, or milling machines, have their *Z* axis parallel to the spindle and a positive *Z* motion moves the cutting tool away from the part. For horizontal milling machines, where the tool is rotating on a horizontal spindle, the positive *X* axis is pointing to the right, as shown in Figure 31-3(C). The *X* axis on a vertical milling machine or on a drill is parallel to the table.

31.2.3 Positioning

There are two methods to define positions in the coordinate system: absolute positioning and incremental positioning. The difference between these two methods is illustrated in Figure 31-4. In *absolute positioning*, programmed locations are referenced to a fixed zero reference point. The machine control unit (MCU) uses this point to position the machine at the desired or programmed location. In *incremental positioning*, also called *relative positioning*, movements are based on the change between two successive locations. The next position is defined relative to the present location.

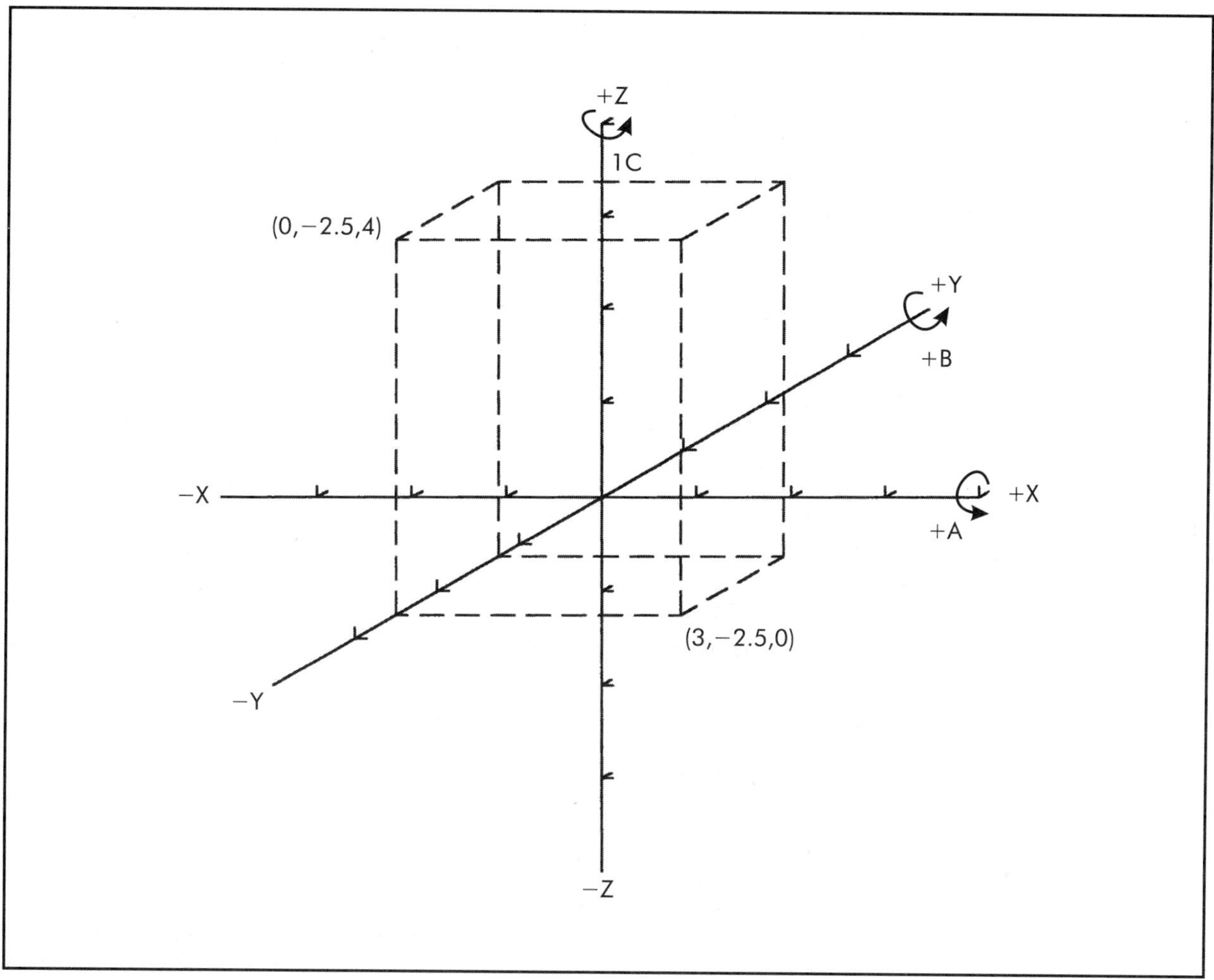

Figure 31-2. Coordinate system in numerical control.

In Figure 31-4, the tool is presently located at point *A* (2, 2). Movements to points *B*, *C*, and *D* are required. In absolute positioning, point *B* is defined as (9, 2). This means that point *B* is specified by $X = 9$, $Y = 2$. In incremental positioning, point *B* is defined as (7, 0) . This means that an incremental movement of $X = 7$ and $Y = 0$ has been accomplished. Points *C* and *D* are defined in a similar way.

Both absolute and incremental positioning systems can be used in point-to-point or continuous path (or contouring) programming, discussed later in this chapter. A drawback in using incremental positioning is the accumulation of errors. In Figure 31-4, for example, if positioning error occurs at point *A*, points *B*, *C*, and *D* will carry the accumulated error. In absolute programming, only point *A* will be in error since points are specified from a fixed reference.

31.2.4 Closed- and Open-loop Systems

NC systems may be classified as *open-loop systems* and *closed-loop systems.* The type of control system determines the accuracy of the NC machine. In open-loop systems, normally the machine table does not provide feedback about its position status. In other words, open-loop systems do not provide positioning feedback to the machine control unit (MCU). Open-loop systems normally use stepping motors. Closed-loop systems use various feedback devices, such as transducers and sensors for positioning error detection or correction.

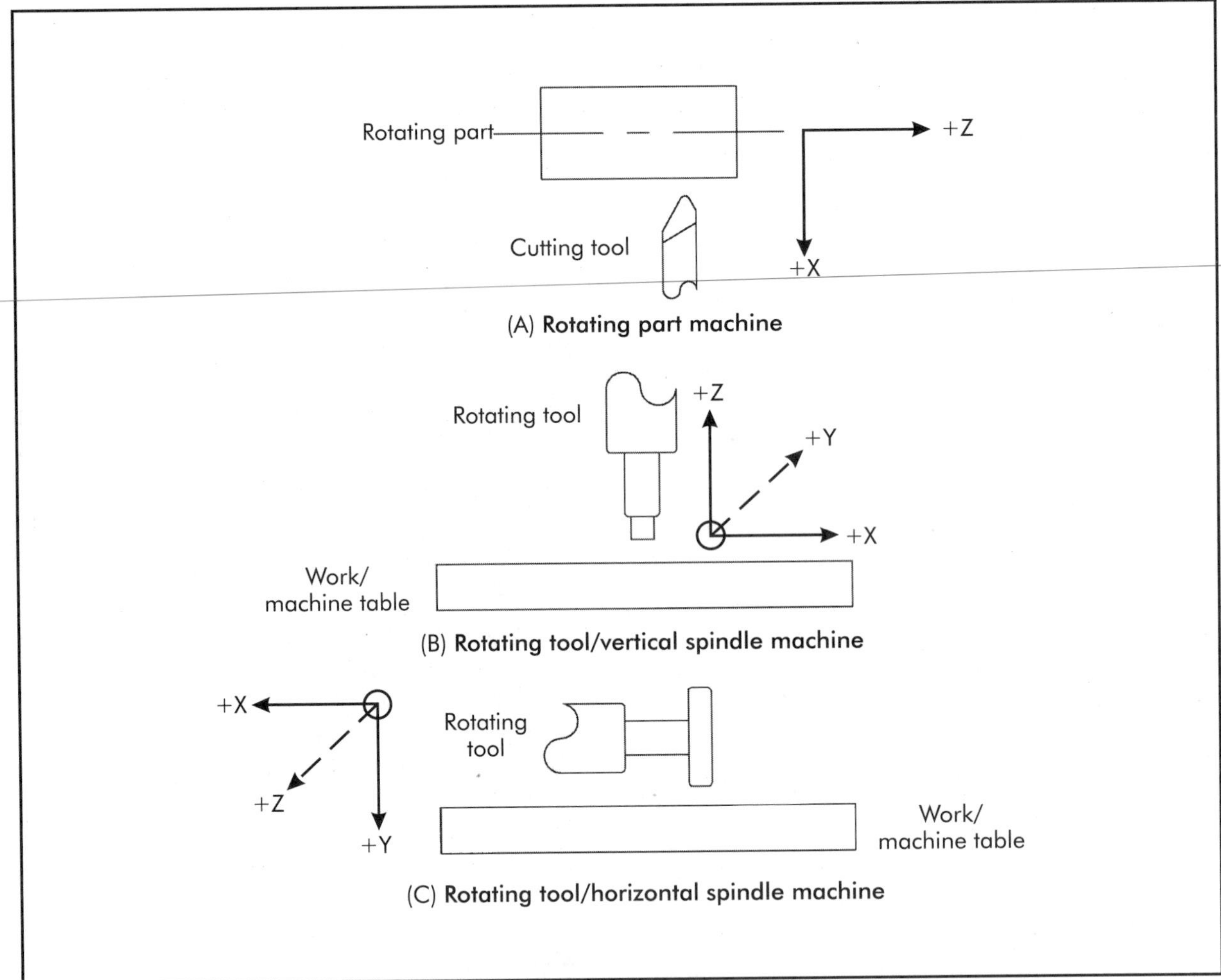

Figure 31-3. An illustration of coordinate systems.

An NC machine may be controlled through an *open-loop* or a *closed-loop circuit* as depicted in Figure 31-5. The open-loop system is the simplest and cheapest but does not assure accuracy. A signal that is an order for a certain action (such as to turn on the coolant, start the spindle, or move the table to a certain position) is issued by the control unit. This travels through the drive mechanism that imposes the action upon the controlled member of the machine.

The basic drive mechanism for a table or other machine member in an open-loop system is a *stepping motor*, also called a *digital* or *pulse motor*. It has, for example, 49 poles around the armature inside of 50 poles around the stator. Two poles are always aligned. Poles are energized to align the next two poles and turn the armature through a definite small angle each time a pulse of electricity is received. The motor turns the lead screw, which moves the table a corresponding distance. If, for example, a pulse stands for a 0.001 mm (.00004 in.) movement, and a movement of 0.05 mm (.002 in.) is desired, the control unit issues 50 pulses; the motor turns through 50 steps and turns the lead screw, and the table (or other member) is driven the 0.05 mm (.002 in.) required. The rate at which the pulses are issued determines the rate of feed. The power of a stepping motor is limited, and if the resistance to movement is large, the motor may stall and miss steps. There is no feedback to report the omission and accuracy

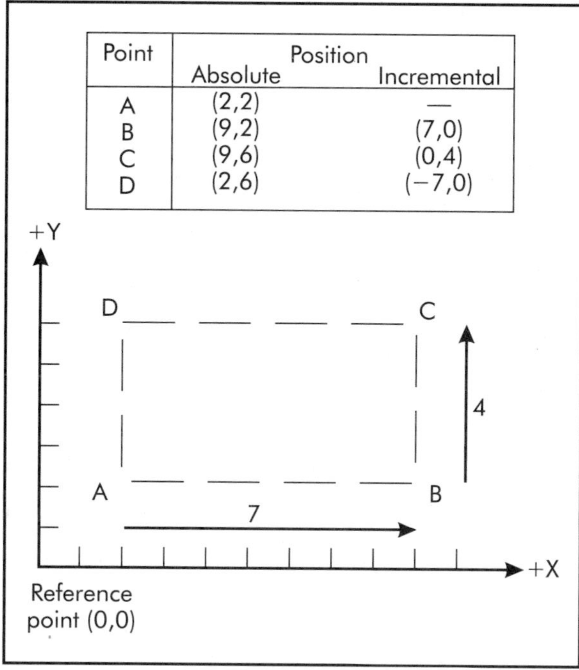

Point	Position Absolute	Incremental
A	(2,2)	—
B	(9,2)	(7,0)
C	(9,6)	(0,4)
D	(2,6)	(−7,0)

Figure 31-4. Absolute and incremental positioning.

is lost. Direct drive with a stepping motor is confined to light service at fractional horsepower. For more power and heavy service, a stepping motor is matched with a hydraulic motor through a servo-valve for table drives up to 7.5 kW (10 hp) and more.

It is common in closed-loop systems for the pulsed command signal that moves the table (or other member) of the machine to be converted to a steady analog signal. That signal turns on the power to the drive mechanism. Hydraulic and AC but mostly DC electric motors are all used for NC machine drives. The intensity of the signal is determined by the feed velocity ordered. For a higher feed, pulses are issued at a faster rate and/or the analog signal is larger, which turns on more power to the motor to drive the table faster. A tachometer in the motor drive sends back a signal that is compared to the feed rate order to assure the required feed rate is being delivered and to make corrections if it is not. Commonly, a resolver is driven by the lead screw and returns a signal showing how far the table moves. More accuracy for a longer period of time is obtained with the feedback device attached directly to the table, but such devices are more costly. The feedback signal is compared to the order. Where the feedback equals the order and signifies the required movement has been completed, the power is shut off and the movement stops. Since they are controlled through separate channels, all movements may act at the same time. Indeed, it is necessary that they act simultaneously and be synchronized accurately for contouring.

31.2.5 Adaptive Control

An added feature of some NC systems is *adaptive control,* which is a means of continuously monitoring the critical parameters of an operation to maintain optimum conditions. Adaptive control has been applied largely but not solely to machining operations in manufacturing. One purpose is to detect faults, such as worn or broken cutting tools or hard spots in material, and correct for them. Another purpose is to periodically adjust speeds, feeds, and other variables for optimum performance. Sensors constantly measure critical parameters in an operation and report to a computer programmed to compare the current status to an ideal. If a difference persists for a predetermined period of time, the computer orders corrections to be made.

For adaptive control of a metal-cutting operation, the common arrangement is to measure deflection, speed, and power at the machine tool spindle. Deflection is measured by strain gages or magnetic transducers placed around the spindle. Deflections can be translated into forces on the cutter by performing tests on the machine. The torque of the spindle may be calculated from power and speed. Other conditions, such as vibrations and work surface reflection, have been monitored but not in many systems. An external computer may be used for adaptive control with a hard-wired control, or the computer may be part of a CNC control.

Relationships among parameters such as speed, feed, depth of cut, type of cutter, material, forces, and power are determined for each operation to be performed on the machine under adaptive control, like the relationships discussed in Chapter 17 but in more detail. From these models and the capacities of the machine, working limits are set for the variables. This

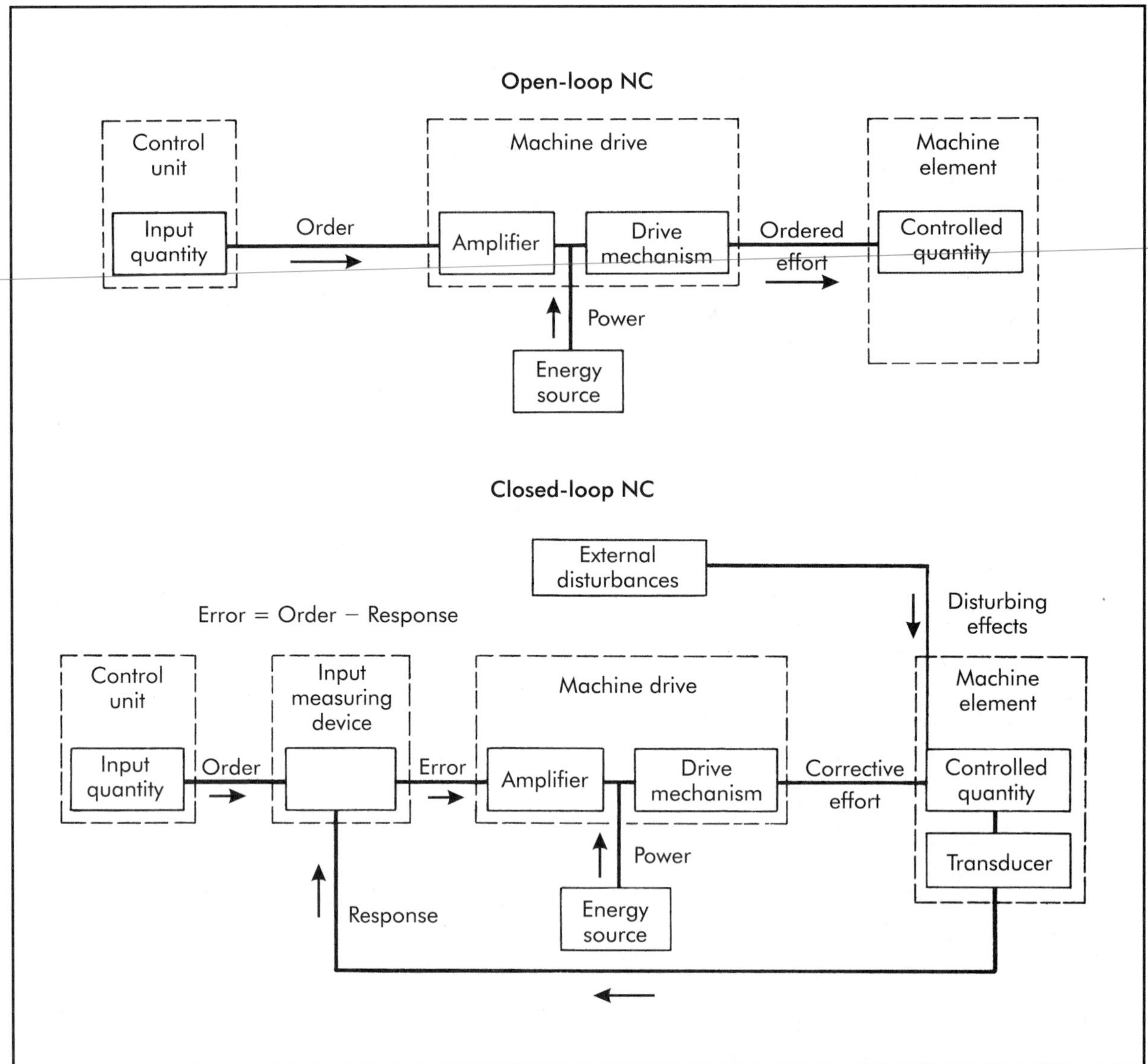

Figure 31-5. Schematics of open- and closed-loop NC systems.

information is stored in the computer and called out by the adaptive control program as needed. Thus, if no forces or power are detected during a cut, a broken tool is indicated, and the computer may order the tool replaced if the machine has a tool changer. Otherwise, it may stop the machine and signal the operator to change the tool. If deflections and torque of the spindle indicate cutting forces larger than the tool can stand, the computer orders an override of the originally programmed feed rate to reduce it to a tolerable level. As a cutter becomes dull, forces and power increase. The computer may be programmed to calculate the expected tool life under prevailing conditions and adjust the feed or speed to obtain optimum tool life. When the forces and power rise to the point where the tool is so dull that it must be removed for economical sharpening or to prevent catastrophic breakdown, the computer calls for a change. Where tool life or strength are not the limiting factors, the computer may regulate the feed to utilize the full capacity of the machine and prevent any harmful overload. Feed rate may be automatically raised in gaps between full cuts or when cutting resistance decreases.

In another area, acoustic-type adaptive control is used for production spot welding. Sounds that occur during deformation, melting, solidification, cracking, and other changes are measured. Sound intensity, current flow, and electrode forces are all taken into account to control the operation for optimum weld quality, electrode life, and productivity for the material being welded. One system, called an expulsion limit controller (ELC), turns the current off when expulsion or splatter occurs to prevent over-welding and obtain maximum nugget size.

31.3 NUMERICAL CONTROL (NC) SYSTEMS

A numerical control system consists of the machine control unit (MCU), also called the *control,* and the machine tool. The MCU stores the part program and converts it into actions to be carried out by the machine tool to process the part. The MCU may contain a tape reader if programs are loaded from punched tape.

Three classes of NC systems are commonly recognized. One is called *point-to-point* or *positioning NC*, where a cutting tool and workpiece are positioned with respect to each other before a cut is taken, as exemplified by an NC drill press. The path between points is of little concern and is not particularly controlled. Another class, known as a *straight-cut NC system*, involves movements between points like point-to-point NC but along straight or curved paths determined by the machine ways or slides. An NC turret lathe is an example. A third class is *continuous-path NC*, or more formally *continuous tool-path control,* which does contouring or profiling of lines, curves, and surfaces of all shapes. One way of achieving continuous path NC is to move the workpiece or cutter from point to point along a straight line or curve (a parabolic path between points is used on some machines) with many points spaced closely enough together so that the composite path of the cut approximates a desired curve within the required limits. Another way is to drive machine members, for example, a saddle and the table on it, along coordinate axes at varying velocities controlled so that the resultant motion is along a particular line or curve.

31.3.1 NC Machines

Machines may be classified according to the number of axes of numerically controlled movements with respect to Cartesian *X-Y-Z* coordinates. There may be other movements not numerically controlled. A two-axis machine has the table move longwise and crosswise in a horizontal plane; a three-axis machine has an additional vertical movement of the spindle, for example. Four-, five-, or six-axis machines provide additional linear or rotary movements.

There are many configurations of advanced manufacturing systems in which CNC machines constitute a major component. An example of a CNC machine is shown in Figure 31-6. One concept of forming machining systems is to form a freestanding machining station that has its own CNC control system. To form a multiple-station configuration, a number of stations and a parts

Figure 31-6. CNC vertical milling machine. (Courtesy Clausing Industrial)

transfer system are added. Such a system can operate in a synchronous or nonsynchronous mode. This system may need other modules to increase its productivity, such as a multiple-spindle head changer, palletized part handling or a pallet shuttle, and tool storage modules.

The horizontal spindle shown in Figure 31-7 has three-axis motion capability and a sprocket-type 10 tool changer. The "flip spindle" and follow-along tool sprocket arrangement make it possible for 1.7 second changes. Three more sprockets are carried on an optional storage module, giving a total of 400-tool capacity. Referring to Figure 31-7, the two-spindle arrangement (1) "flips" 180° alternately. While one spindle is working, the sprocket moves toward the rear spindle, takes out a tool, rotates, inserts a new tool, and retracts. The sprocket changer and spindle unit (3) move to the transfer point where a sprocket is removed, rotated, and another 10-tool sprocket is transferred to the spindle unit. Tool change time from one sprocket to another is less than 22 seconds.

A toolhead indexer and head changer is another add-on feature that enhances the capabilities of CNC machines. The head indexer/head changer arrangement shown in Figure 31-8 features quick head change capability. This arrangement features a face tooth coupling that locks the index table in position. The no-lift rotary motion of the table maintains a tight seal at all times.

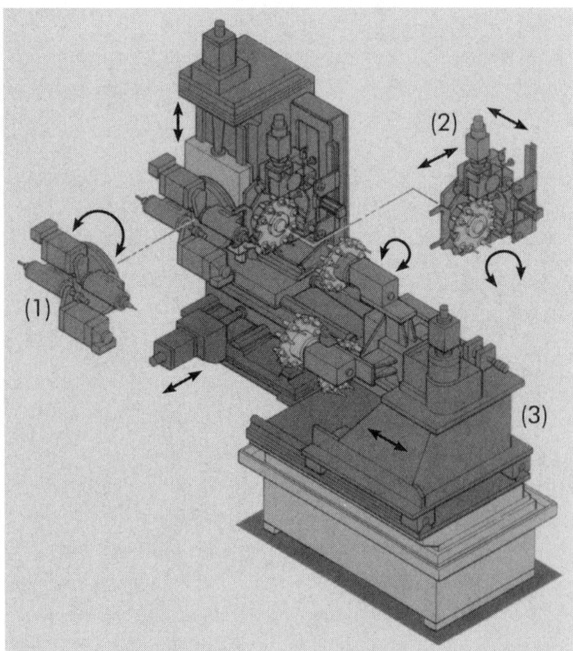

Figure 31-7. A horizontal spindle with a tool changer arrangement. (Courtesy Kingsbury Corporation)

31.3.2 Machining Centers

The principle of numerical control was first applied to conventional types of machine tools. Most NC machines may still be classified as NC lathes, NC milling machines, and so on. Further developments have led to NC systems of more versatility, each capable of totally machining a wide variety of parts. These are called *machining centers*. Machining centers have more flexibility than other machine tools. They are computer-controlled machine tools capable of performing a variety of machining operations on different surfaces and in different directions on a part. An example of a vertical machining center is shown in Figure 31-9. The machining center is capable of milling, drilling, boring, reaming, and tapping in one setup, on different surfaces of the part. Machining centers also come with horizontal spindles, known as horizontal machining centers, such as the one shown in Figure 31-10.

The capability of the machining center to perform different operations on different surfaces

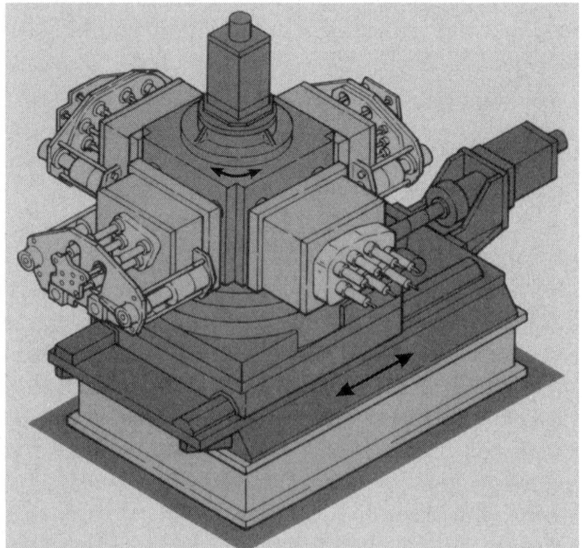

Figure 31-8. Head indexer/head changer arrangement. (Courtesy Kingsbury Corporation)

Figure 31-9. A vertical machining center with cutaway view of vertical spindle and tool storage/transfer unit. (Courtesy Cincinnati Milacron)

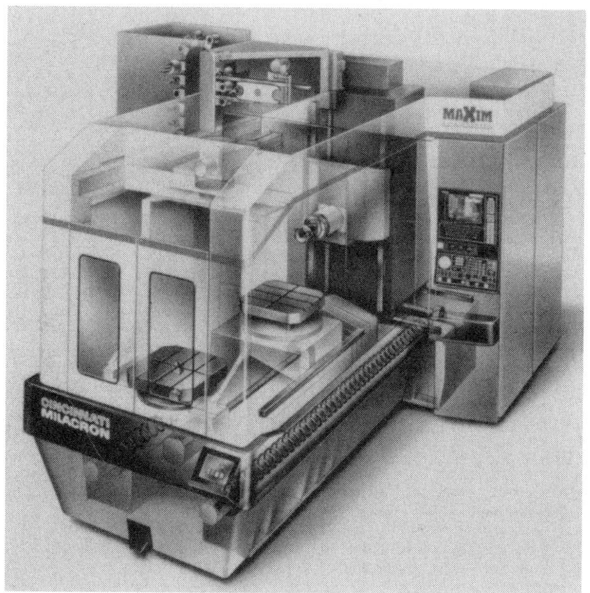

Figure 31-10. A horizontal machining center with cutaway view of horizontal spindle, worktable, and tool storage/transfer unit. (Courtesy Cincinnati Milacron)

of a workpiece is achieved by using pallet shuttles. A freestanding pallet shuttle for manual loading/unloading from one side of the machine is shown in Figure 31-11.

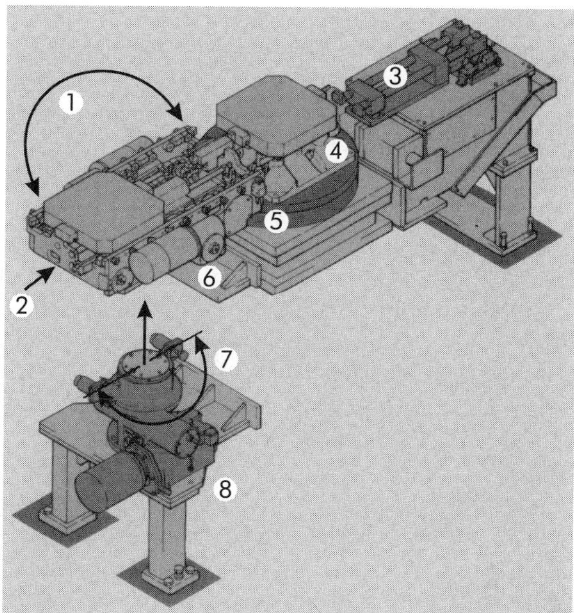

Figure 31-11. A freestanding pallet with manual load/unload from one side of the machine. (Courtesy Kingsbury Corporation)

31.3.3 NC Accuracy

The smallest increment of motion that can be specified by a signal in the system is the *resolution* of the NC system. Obviously, for a series of pulses, the value of one pulse is the resolution. Typical on high-quality NC systems is a resolution of 2.5 μm (.0001 in.), but some have more and others less. *Accuracy* of the machine is a different matter because it depends on mechanical performance. It signifies how closely a machine member can be moved to any specified point. The advertised accuracy for any axis of a typical machining center is ±13 μm within 300 mm (±.0005 in. within 11.8 in.) and ±20 μm (±.0008 in.) over full range. The accuracy of most NC machines lies in the range of ±5 to 25 μm (±.0002 to .001 in.). An accurate part has no errors if machined with all specified tolerances. Modern NC or CNC machines are capable of maintaining a tolerance of 0.254 μm (±.00001 in.) or better. Of equal importance is the *repeatability* or *precision* of an NC machine; that is, how closely the table (or other member) can be returned to an initial position repeatedly. The repeatability of a machining center is ±8 μm (±.0003 in.).

The ultimate accuracy of a machining operation is dependent on what the tools can do. For example, an NC jig boring machine was found accurate to 5 μm (.0002 in.) in table positioning, but holes drilled on the machine were out of true position by as much as 81 μm (.0032 in.) because of the tendency of drills to wander. An inexpensive NC drill press with table positioning accuracy estimated to be 25–50 μm (.001–.002 in.) consistently drilled holes to within 97 μm (.0038 in.) of specified location. Experience has indicated that setup and tooling errors are commonly appreciably more than laxity of the machines.

An NC machine must have design features not always required for conventional machine tools. A skilled operator may be depended upon to compensate for the deficiencies of a machine tool, but all requisites must be programmed for or built into an NC machine. Much that a skilled operator would do can be planned for in an NC operation, such as taking several boring cuts where hole location is important. An NC machine must be rigid, tight, accurate, and true because the operator cannot nurse it along as he or she might a manual machine. The drive system must be stiff to minimize windup and enhance stability of the servo drive. Low inertia is required for rapid response to changing command signals. Friction forces must be low and uniform for stability, rapid response, and positioning accuracy. This is achieved through the use of rolling contact bearings and hydrostatic bearings. Even so, NC machines are available at various levels of quality, and economy dictates that a machine be no more expensive than needed for the purposes intended.

Reliability is defined as the ability of the machine to maintain consistent function under specified machining conditions for a certain period of time without failures. Reliability is associated with other terms, such as maintainability, availability, and maintenance. Preventive maintenance normally increases machine reliability and elongates the mean time between failure (MTBF). MTBF is the average length of time the machine is operating without failures or breakdowns. If properly maintained, NC machines should have a long MTBF, low maintenance cost, and high production rates.

Several factors affect the accuracy of NC and CNC machine tools:

- machine parameters, including machine resolution, accuracy, and repeatability; drive system capabilities; number of machine axes; machine size and dimensions; and machine control capabilities;
- machine errors, including geometric and kinematic errors, dynamic errors, spindle rotational errors and runout, and static errors and loading deformation;
- thermally-induced errors due to thermal expansion and different thermal coefficients of machine elements affecting machine accuracy;
- positioning and dimensional errors of machine axes, and
- other errors due to tool deformation and clamping, cutting process parameters, friction and bearing effects, and workholding methods.

31.3.4 NC Adjuncts

Adaptive control is a desirable feature of NC systems. Tests have shown production increases of 23–90% with an average of 37% from a fully utilized adaptive control system in metal cutting. Commercial adaptive control systems are available at various levels of sophistication. Most of them are confined to reacting to tool failures and hazardous operating conditions. To do more is much more expensive. Even at the lowest level, adaptive control is desirable for unmanned machining centers.

Verification and *correction* of NC programs is necessary, particularly when working with expensive materials, tools, and preprocessed pieces. As a minimum, an NC program should be examined carefully by a second person, just like an engineering drawing or project of any kind. Trial runs are commonly made on wood, aluminum, or foamed plastics. This is not really economical because it takes time on the machine tool and may upset production schedules for other parts. Because of corrections that must be made, the time to tryout a program is usually longer than to make a part afterward, and time on an NC machine may cost $20–120/hr. It is reported that in one aerospace plant over a 12-month period, 3,777 programs were verified by trial at a cost of over $1 million.

An NC program can be run on a computer with a plotter if the job is a point-to-point or plane-contouring operation, but this is not feasible for three-dimensional cutting. Even so, an ordinary plot reveals only gross errors and may give some indication of the possibility of collisions, but it does not prove how well small tolerances are held. Groups in the aerospace industry are developing a system for verification of two- and three-dimensional NC programs by computer simulation. Specifications for the work, fixtures, and cutting tools together with the NC program are fed in as data. The verification program graphically simulates (with dynamic animation) the machining process on a CRT screen. It also displays relationships between the cutter, workpiece, holding fixture, and machine tool. The cutter path is dynamically compared to models of the raw stock, part geometry, and holding fixture to verify part configuration and detect collisions, interferences, and unmachined conditions. Surfaces with excess stock or that are cut out of tolerance are shown in color.

Many newer NC systems have diagnostic capabilities to find the causes of problems and display instructions on a CRT screen to correct the faults. Diagnostics check the mechanical as well as the electrical parts of the system. Two kinds of diagnostic systems are in general use. One is *resident diagnostics*, which is programmed in the CNC control. The other is known as a *diagnostic communication system*, and its main program is contained in an outside computer, typically at the plant of the manufacturer of the NC machine, and is reached by telephone lines. Since experience has indicated that about 95% of conventional troubleshooting time is spent in finding the cause of the trouble, a diagnostic system saves considerable machine downtime. Less skill is also needed. In addition to diagnosing breakdown emergencies, a diagnostic system can indicate when routine preventive maintenance is needed by detecting failing components before they collapse.

31.3.5 NC Measuring and Inspection Machines

Some continuous-path control is found on measuring machines but is questionable because of the dynamic effects that occur during machine movements. As a result, continuous-path measurements are made mostly along straight lines and circles. More precise results are obtainable from point-to-point measuring systems, where the apparatus and workpiece come to rest when readings are taken.

Recent advances in NC measuring machines have achieved a factor of 10 increase in machine performance. Such achievements were made possible by studying the different error sources affecting machine performance and developing algorithms to automatically compensate for their effects on machine positioning and accuracy.

An NC measuring system measures a piece by means of a probe that is displaced and which emits a signal upon contact. There are two types of systems as indicated in Figure 31-12. In *null probing*, the signal from the probe is fed back to the control unit of the machine to position the workpiece to put the probe in a neutral position. The dimension to which the slide carrying the workpiece has moved from the reference zero position is then displayed on a screen or printed record. In *deviation probing*, the workpiece is moved the amount of the specified dimension by the NC system as it would be in machining. The displacement of the probe when the final position is reached is shown on a scale, screen, or record and represents the difference between the actual and required dimensions. Most measuring systems with two or three axes employ

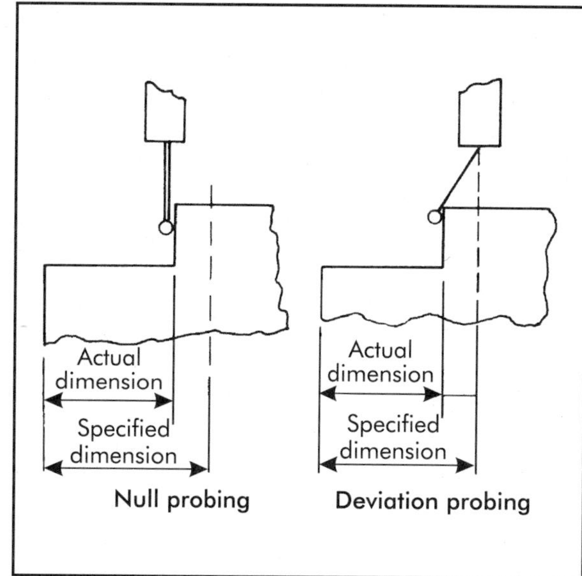

Figure 31-12. Two forms of NC measuring systems.

both null and deviation probing so the user has either available as desired.

Further discussion of numerically controlled measuring machines can be found in Chapter 29.

31.4 PROGRAMMING FOR NUMERICAL CONTROL

The detailed instructions for an NC operation may be listed by hand, but this would become a long and tedious task, particularly for continuous-path and involved point-to-point operations. This can be done more quickly in many cases by digital computers. Compiler programs are available that enable computers to take simple statements of specifications and convert them into detailed instructions for NC machines. Compilers differ and may be classified as to whether the input must be in a restricted numerical form, in an English-like language describing the tool action, or in words merely defining the conditions and areas for machining. A computer is able to translate input specifications into instructions because it can make decisions to pick the proper course of action for a situation and follow a routine to delineate the steps to be taken. Continuing development of NC is toward incorporating it into a total system of information processing through the stages of design and manufacture of products.

31.4.1 Manual Programming

One way of programming an NC machine manually, as depicted in Figure 31-1, is to write a programming process sheet or planning sheet from the part drawing or print. This process sheet contains all the instructions needed by the control for the operation and can be recorded on tape or disk to be read into the control as needed, or entered directly into the memory of the control, depending on the system. Or, an operator may compose one block at a time and enter it into the control through a push button panel or keyboard. The process sheet also commonly contains tool specifications and instructions to the operator.

Each line of coordinate data and machine function instruction is a block and is punched in tape between end-of-block symbols. The tape is all that is needed to actuate the control unit of the NC machine. After setup, the machine operator has little control over the operation; the machine only does precisely what it is told, and therefore the programmer must understand, anticipate, and specify explicitly what must be done.

In a manual data input (MDI) system, programming may be done directly from the part print (without a process sheet) by an operator using an input pad or keyboard into the NC control. Modern processors for such a system offer such auxiliary aids as canned cycles and linear, circular, parabolic, and even helical interpolation to simplify programming. MDI is practical mostly for simple jobs on less complex machines and for small lots that are not to be run again. The reason is that the machine is idle (an average estimated time of 30 min per job) while the operator is programming. A few systems allow programming while work is in progress, but that has not been widely adopted because operator attention is needed for machining. Thus jobs that require moderately long programs (over about 20 lines) and program editing are more efficiently programmed off the machine so that costly NC time is minimized. MDI does have the advantages of requiring little or no tape-preparation equipment and only simple organization and few steps for least chance of error.

All methods of NC programming must involve more than just listing steps in operations; planning starts in product design with the selection of parts most suitable for NC fabrication. For instance, NC machines commonly operate on an *absolute system* in which all dimensions are referred to a zero point; optimum results are achieved if the workpiece is dimensioned in the same way. As each part is programmed, it must be analyzed to ascertain the best orientation for it on the machine, the most efficient sequence of operations (such as mill, drill, rough bore, and finish bore for accurate hole location), the proper cutters to use, and the necessary fixtures or other holding means for security and reliability. For a point-to-point positioning operation, the programmer normally takes three steps in resolving an operation. They are to: (1) select the best zero point, (2) ascertain the order of positioning the coordinate points, and (3) determine what cuts are to be made at each point. Once these items have been determined, the programming is merely a matter of expressing the parameters in the format required for the machine.

A point-to-point positioning program has been presented as an example of manual programming, but continuous-path contouring or profiling may be programmed in substantially the same way, only the task is likely to be longer and require more calculations. The instructions for profiling solely along straight lines and circles may have to include only the first and last points of each line or curve and the slope of a line or radius of a circle if the NC machine control is capable of interpolating between the ends, as many are. Still the points where lines and curves intersect must be calculated. Desk calculators may be used for simple cases and elaborate systems using tables are available for the ease of manual programming.

A manually prepared part program uses several codes for designating different parameters or functions for the machine. The Electronic Institute of America (EIA) standard provides a line format for point-to-point machining jobs, such as drilling operations. Different codes, such as the following, are used in manual programming:

- *N code* is used for the sequence number that identifies each block within an NC program.
- *G code* is used as a preparatory function to tell the MCU or the control that a certain control function is requested. For example, G01 requests a linear interpolation mode, G03 requests a counterclockwise (CCW) circular interpolation, and so on.
- *X, Y,* and *Z* give the coordinate location of the tool. Machines with more than three axes require additional words, such as *A* or *B*.
- *F code* specifies the feed rate (in. per min) in a machining operation.
- *S code* specifies the cutting speed (revolutions per min [rpm]) in a machining operation.
- *T code* specifies the tool to be used in a machining operation and is normally used on machines that are equipped with an automatic tool changer or head indexer/tool changer, such as the one shown in Figure 31-8.
- *M code* is used to designate a specific operation mode of an NC machine. For example, M03 requests spindle clockwise (CW), M08 requests coolant on, M09 coolant off, and so on.

A typical program is demonstrated in Figure 31-13 for performing the drilling on the part shown in Figure 31-14. In this example, G00 indicates that a point-to-point system is to be used for indicating position, G01 is for linear interpolation, and M02 indicates the end of program and completion of the workpiece.

31.4.2 Computer-assisted NC Programming

The need for a computer (either as part of the control or outside) is obvious for finding the coordinates of points on and at intersections of slopes, curves, and surfaces. Manual programming of even simple point-to-point operations is tedious and time consuming when done for a continuous stream of jobs, and computer assistance yields fewer mistakes and lower costs. Figure 31-15 shows the difference between manual and computer-assisted part programming for NC systems.

The computer is directed by a master program, called a *processor, compiler*, or *interpreter,* which tells the computer how to process

```
N010G00G90X-2.000Y0.000F10.0S800T1M03
N020X2.000Y2.000Z0.500
N030G01Z-1.000
N040G00Z0.500
N050X8.000
N060G01Z-1.000
N070G00Z0.500
N080Y4.000
N090G01Z-1.000
N100G00Z0.500
N110X2.000
N120G01Z-1.000
N130G00Z0.500
N140X-2.000Y0.000M02
```

Figure 31-13. A typical program for drilling.

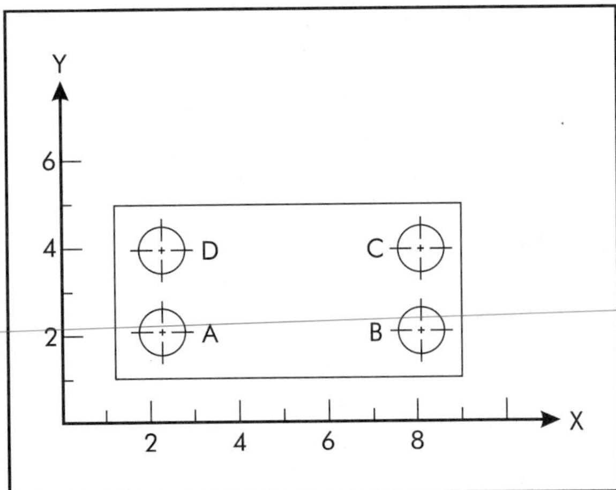

Figure 31-14. An example of a part to be drilled.

the input statements to derive the necessary instructions for the NC machine.

A processor for NC programming performs several definite functions and commonly has sections to carry out each function. One section is the *translator* that converts the input into the computer code so it can be processed into the series of basic commands required for the NC control. A *calculator section* performs the necessary computations to establish all points, tool center offsets, feed rates, etc. An *editorial and diagnostic section* detects errors and organizes the program. The result of this stage is what is called the *centerline* or *cutter location* (CL) file or tape. Overall, an *executive control section* supervises and coordinates all the other sections in the processor.

CL instructions must be converted to suit each NC machine control because almost every model of NC machine is different in some respects from all others. Thus, most controls require a final computer program called a *post-processor* to convert the CL data to suit the particular type of control. A processor may be dedicated to a particular machine, with its output suitable to the specific controller, so a separate post-processor is not needed. This is called a *one-pass system*. Another may have a separate post-processor for each of a number of NC machines and is called a *two-pass system*. Some older machine controls have been retrofitted, and new machines are being made with controls that accept a standardized single form of input called *CL exchange data*. This is derived from the CL file by a post-processor called a *CL exchange converter.* All such controls are compatible and different post-processors are not needed.

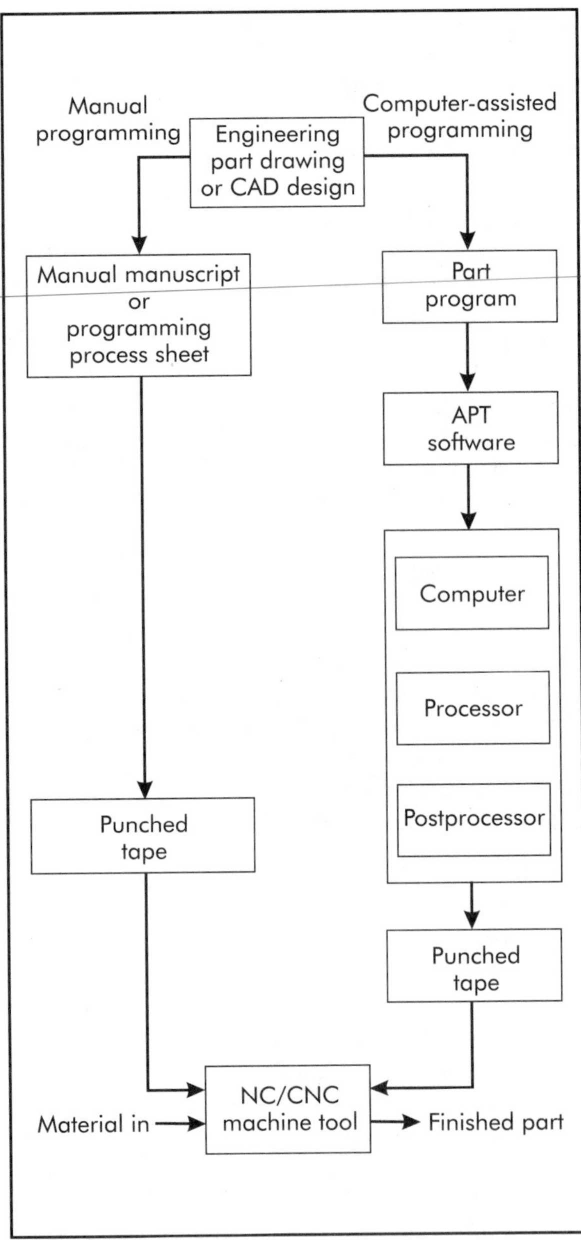

Figure 31-15. Manual versus computer-assisted part programming.

Most NC-language processors serve for a particular kind of operation, such as milling, turning, point-to-point drilling, or electrodischarge wire cutting. Whatever their purposes, NC processors may be divided into three groups in accordance with their format and complexity. Further explanation of these groups will be given in the following sections together with some examples. There are many program languages. The examples given are merely typical and do not cover the details of all varieties.

Computer-assisted part programming offers many advantages over manual programming.

1. Programming time is significantly reduced due to the capability of the programming system to accept a large amount of data on machining parameters, machine tool and process variables, tooling information, and other auxiliary functions related to the machine tool. Calculations required to execute the program and to define surfaces and other part features are performed by the software and result in reduced effort and programming time.
2. The possibility of human errors is reduced.
3. It offers ease of changing and editing programs to accommodate different part families and part configurations.
4. There is lower cost due to the reduced programming time.

Group I Programs

A *group I program* accepts the definition of the patterns of points, arrays, lines, curves, surfaces, etc. The input is largely, and in some cases entirely, numerical and in a fixed format. The kind of work to be done may be specified by a code, and the processor fills in the details of execution. Quite often the order of the instructions determines the final sequence of steps in an operation.

A computer is able to translate the specifications for a job into the detailed instructions for an NC machine because it can make prearranged decisions and follow prescribed routines. From the basic capability of adding positive or negative numbers, the computer can perform all forms of arithmetic. A flow chart is a conventional means of showing what takes place in a program to direct a computer to do a certain job. Symbols designate functions: a diamond depicts a point in the program where a decision must be made to proceed along one of two or more possible paths, a rectangle where general information processing, such as calculating, is done. There are many other symbols. The computer proceeds from step to step in the directions designated by the arrows, choosing its paths according to the decisions it must make, which are dependent upon the conditions encountered.

Most group I programs use separate routines from storage for each array of holes or type of curve (circle, parabola, etc.) and for each different use of the same curve. Consequently, considerable storage space is taken up in the computer by the program. This, along with the rigid format of input, limits the number of situations that can be handled.

Some group I programs are tutorial; a series of statements or questions are displayed one by one on a CRT tube, and the programmer must reply to each by keying in the requested data. An excerpt from such a program is presented in Figure 31-16. It covers the milling of a pocket in a workpiece by the zigzag movements of an end mill as depicted on the left in the illustration. To the right are shown some typical lines displayed on the CRT screen and the data furnished by the programmer. The latter appears on the screen as soon as entered in response to each query. When the information is complete, the computer calculates the points for the cutter locations and renders the instructions to the control, which may be punched in tape, recorded on disk or tape, or stored in the memory of a CNC.

Group II Programs

The input of a *group II program* is in the form of an easy-to-learn, English-like, word description of tool positions or paths. The format is not rigidly fixed, and punctuation rather than position distinguishes and separates entries. The product is defined in terms of lines and circles and other geometric shapes. Extensive use is made of symbolic notations. Practically no manual precalculation is required. For instance, the coordinates of line intersections and curves do not have to be specified. All this makes it possible to write specifications for many different jobs. However, it adds to the size and complexity of the compiler program.

The quasi-English languages of NC computer programs possess restricted vocabularies and formats. Recognized words consist of definitions

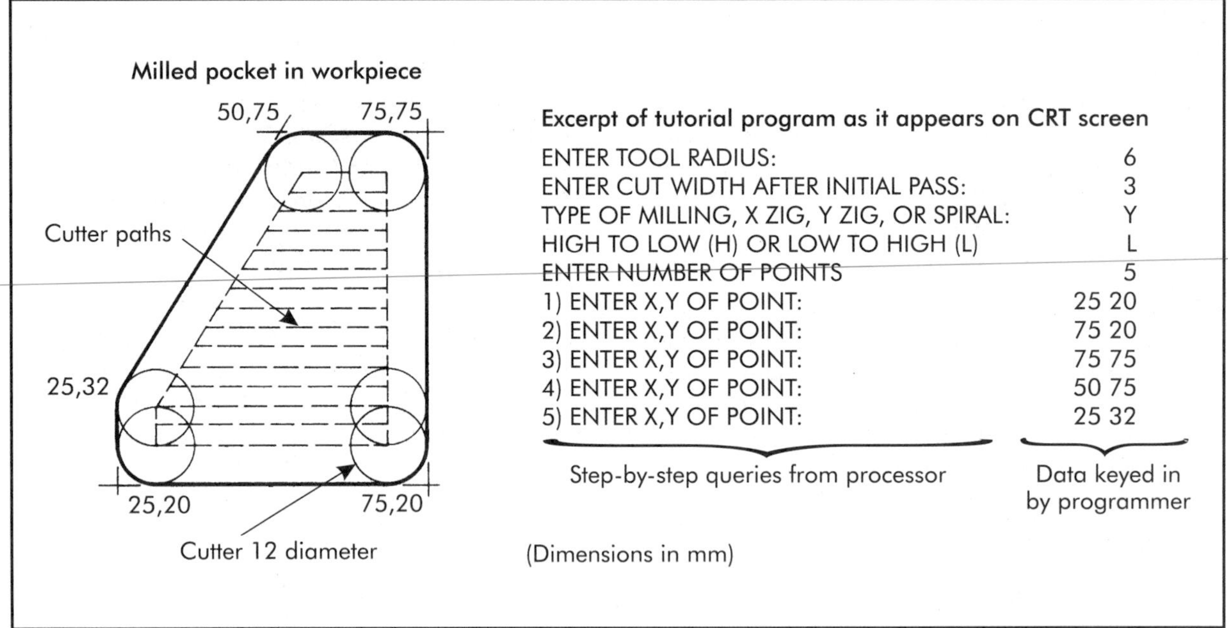

Figure 31-16. Example of a tutorial program for NC.

like POINT, LINE, CIRCLE; action words like GO, STOP, CUT, GOBACK; modifiers like RIGHT, LEFT, TO, PAST; and punctuation marks. An abridged dictionary of terms based on actual languages is given in Table 31-1. As a rule, additional words or symbols may be defined and used throughout the program. A typical format is described in the next section.

The routines of group II programs are more generalized than those of group I. For instance, one procedure may calculate the points along any curve instead of a routine for each specific curve. This eliminates the need to store a large number of routines in the memory of the computer.

APT. One of the best known group II working programs, automated programmed tools (APT), originated in the 1950s at MIT. It was first developed for milling and was used largely in the aerospace industry. Later additions were made for other kinds of operations. Developments have continued. In 1961, version APT III was produced. In 1971, the development of APT was taken over by an organization known as Computer-Aided Manufacturing International (CAMI). APT no longer represents the forefront of computerized manufacturing technology but has been developed since then as one of the important components of expanding management and manufacturing technology. In 1977, the American National Standards Institute (ANSI) published the standards bulletin entitled *Programming Language APT, ANSI, X3.37* to establish the form for and the interpretation of programs expressed in the APT language. The purpose was to promote the interchangeability of processing all APT systems on a wide variety of computers. To do this, the standard established the syntax of the APT language, the rules governing the interpretation of the syntax, and the format of the tool positions in the output data file. However, the standard does not prescribe how APT programs are processed or implemented.

According to the ANSI standard, the APT language is composed of a hierarchy of five proper subsets. Each subset has the capability of all lower-level subsets. The simplest (subset 1) contains no geometric definitions; these are introduced in subset 2. Subsets 1 and 2 are suitable for point-to-point operations. Contouring in a plane begins with subset 3, three-dimensional contouring with subset 4, and multiaxis contouring with subset 5. With higher levels more and more comprehensive definitions and symbols are

Table 31-1. Abridged dictionary of an NC language

Definitions

Points: SYMBOL = POINT/X, Y
Lines: SYMBOL = LINE/X_1, Y_1, X_2, Y_2
SYMBOL = LINE/X, Y, ATANGL, A
Circles: SYMBOL = CIRCLE/CENTER, X, Y, RADIUS, R
XSMALL, YSMALL = On the side of the small values of X or Y
XLARGE, YLARGE = On the side of the large values of X or Y
TOLER/U = The maximum deviation permitted from a curve by approximation
TLRAD/U = Tool radius (in.)

Motions and Modifiers

TLLFT = Place the center of the tool on the left side of the line
TLRGT = Place the center of the tool on the right side of the line
TLON = Place the center of the tool on the line
GOLFT = Go left on the next line or curve
GORGT = Go right on the next line or curve
GOFWD = Go forward on the next line or curve
GOBACK = Go backward on the next line or curve
GO/TO = Move the edge of the tool to the next line or curve
GO/ON = The tool center is to be set on the next point, line, or curve
GO/PAST = The trailing edge of the tool is to be tangent to the next line or curve
INDIR, VECTOR/X, Y = Move the tool in the direction of a vector with the specified components
GO DLTA/U, V = Move the tool by an amount $\Delta X = U$, $\Delta Y = V$
FROM/X, Y or FROM/SYMBOL = The first motion statement of the program
TANTO = Directs the tool to move until it is tangent to a line or curve when used as a modifier
FAR = Directs the tool to be moved until it comes to the second intersection with the following curve
END = This part is done
FINI = All parts are done

Punctuation

/ Forward slash separates names or actions from modifiers
, Comma always used between every two words or symbols not separated by other punctuation
= Equal to or stands for
( ) Parentheses enclose a function
+ Means add, but is not needed to signify a positive number
– Means subtract and must always be placed before a negative number
All numbers must be expressed to the first decimal place like 3.0 or 4.2.

*The letters X, Y, U, and V stand above in places of numbers.

available, and lengthy routines (such as do-loops) may be incorporated. Subsets 4 and 5 may be group III programs.

An APT system consists of an APT language and an APT processor, which is available for a fee to those companies who do not have their own system. There is no one monolithic system. The early APT developed into a ponderous system, mainly for the aircraft industry, which required monstrous computers. It was beyond the needs and capabilities of most potential users, so consequently smaller and simpler versions were created. This is reflected in the present-day hierarchy set down in the ANSI standard previously described. Systems in the realms of the simpler subsets are commonly devoted to work on certain families of machine tools and can be processed on small- to moderate-size computers. Well known and widely used of these are ADAPT and AUTOSPOT. A number of proprietary APT-based systems are commercially available. Prominent among these are COMPACT, SPLIT, UCC-APT, and UNI-APT. It is reported that some 15 different versions of APT exist in the United States and probably as many more overseas. Of course, the APT family is not alone; it is estimated that there are over 60 part programming languages.

A sample program for plane contouring is displayed in Figure 31-17 to give a basic view of the APT language. The program consists of a series of statements that include: (1) geometric definitions (PT1 = POINT/5, 4), (2) parameter definitions (FEDRAT/80), and (3) motion instructions (GO/TO, BASE). Items (1) and (3) may be combined, like FROM/(SETPT = POINT/0, 0). All programming is in terms of a right-hand rectangular (*X*, *Y*, and usually *Z*) coordinate system. The cutter moves with respect to points and surfaces fixed in position. Many alternate ways are available for defining points, lines, circles, ellipses, parabolas, hyperbolas, and vectors. For instance, a circle may be specified by its center and radius, center and a tangent line, center and point on the circumference, three points on its circumference, center and a tangent circle, or radius and intersecting tangent lines. A mean curve may even be faired through a given set of points. In all, there are over 1,000 rules in the APT language for defining words, syntax, statements, etc. Over 75 postprocessors have been written to adapt APT programs to different NC machines.

Curves and other circular arcs can be produced to a specified tolerance by using interpolation. A series of straight line segments is generated with the desired tolerance specifications using INTOL and OUTTOL statements as shown in Figure 31-18.

Points may be defined in different ways, such as:

- By using rectangular coordinates:

symbol = POINT/X, Y, Z

- By intersection of two lines:

symbol = POINT/INTOF, line 1, line 2

- By intersection of a line and a circle:

XLARGE
XSMALL
symbol = POINT/YLARGE, INTOF, line, circle
YSMALL

- By intersection of two circles:

XLARGE
XSMALL
symbol = POINT/YLARGE, INTOF, circle 1, circle 2
YSMALL

- As the center of a circle:

symbol = POINT/CENTER, circle

Examples of point definitions are presented in Figures 31-19 through 31-23.

A line may be defined in several ways, such as:

- By two points:

symbol = LINE/x1, y1, z1, x2, y2, z2

- Through a point and tangent to a circle:

symbol = LINE/point, RIGHT, TANTO, circle
symbol = LINE/point, LEFT, TANTO, circle

- Tangent to two circles:

symbol = LINE/RIGHT, TANTO, circle1, RIGHT, TANTO, circle2
symbol = LINE/RIGHT, TANTO, circle1, LEFT, TANTO, circle2
symbol = LINE/LEFT, TANTO, circle1, RIGHT, TANTO, circle2
symbol = LINE/LEFT, TANTO, circle1, LEFT, TANTO, circle2

- Through a point and at angle with the X- or Y-axis:

symbol = LINE/point, ATANGL, degrees, XAXIS
symbol = LINE/point, ATANGL, degrees, YAXIS

- Through a point and parallel to a line:

symbol = LINE/point, PARLEL, line

- Through a point and perpendicular to a line:

symbol = LINE/point, PERPTO, line

- Parallel to a line at some distance:

XLARGE
symbol = LINE/PARLEL, line, XSMALL, distance
YLARGE
YSMALL

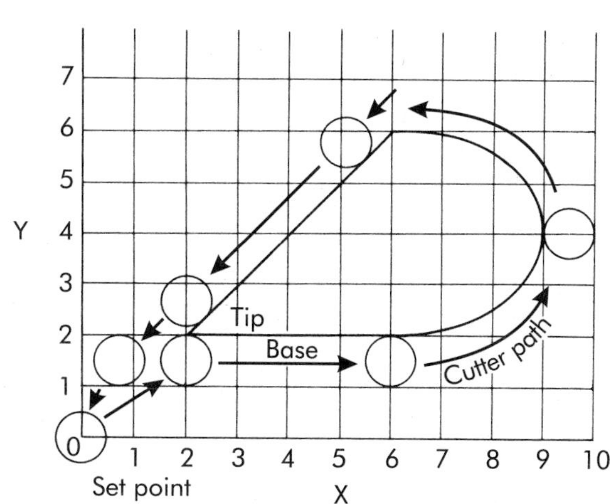

Input program	Explanation
CUTTER/1 $$ FLAT END	Use a 1-in. diameter cutter with a flat end.
INTOL/.005	All cuts must be within .005 in. of specifications.
FEDRAT/80	The initial rate of feed (rapid traverse) is at 80 in./min.
HEAD/1, MOD/1	Operate tool in mode No.1 in head No. 1.
SPINDL/2800	Run spindle at 2,800 rpm.
COOLNT/FLOOD	Apply flood of coolant.
PT1 = POINT/5,4	Defines a reference point, called PT1, at coordinates (5,4).
FROM/(SETPT = POINT/0,0)	The tool is to start from the point called SETPT, which is defined as having coordinates (0,0).
INDIRP/(TIP = POINT/2,2)	Move the tool toward the point called TIP, which is a point with coordinates (2,2).
BASE = LINE/TIP, AT ANGL, 0	A line called BASE is defined as passing through the point TIP at a zero angle with the horizontal.
FEDRAT/5	The rate of feed is 5 in./min.
GOTO, BASE	Move cutter to touch the line called BASE.
TLRGT, GORGT/BASE	Go right along the line called BASE with the tool on the right.
GOFWD/(ELLIPS/CENTER, PT1, 3,2,0)	Go forward along an ellipse with focus at PT1, semimajor axis 3 in., semiminor axis 2 in., and major axis at 0° with horizontal.
GOLFT/(LINE/6,6,2,2), PAST, BASE	Continue left along a line joining points (6,6) and (2,2) and past the line BASE.
FEDRAT/80	Return at rapid traverse rate.
GOTO/SETPT	Go straight to the point SETPT and stop.
COOLNT/OFF	Turn off the coolant.
SPINDL/OFF	Turn off the spindle.
END	This is the end of this part of the program.
FINI	This is the end of the whole program.

Figure 31-17. APT program for a hypothetical operation.

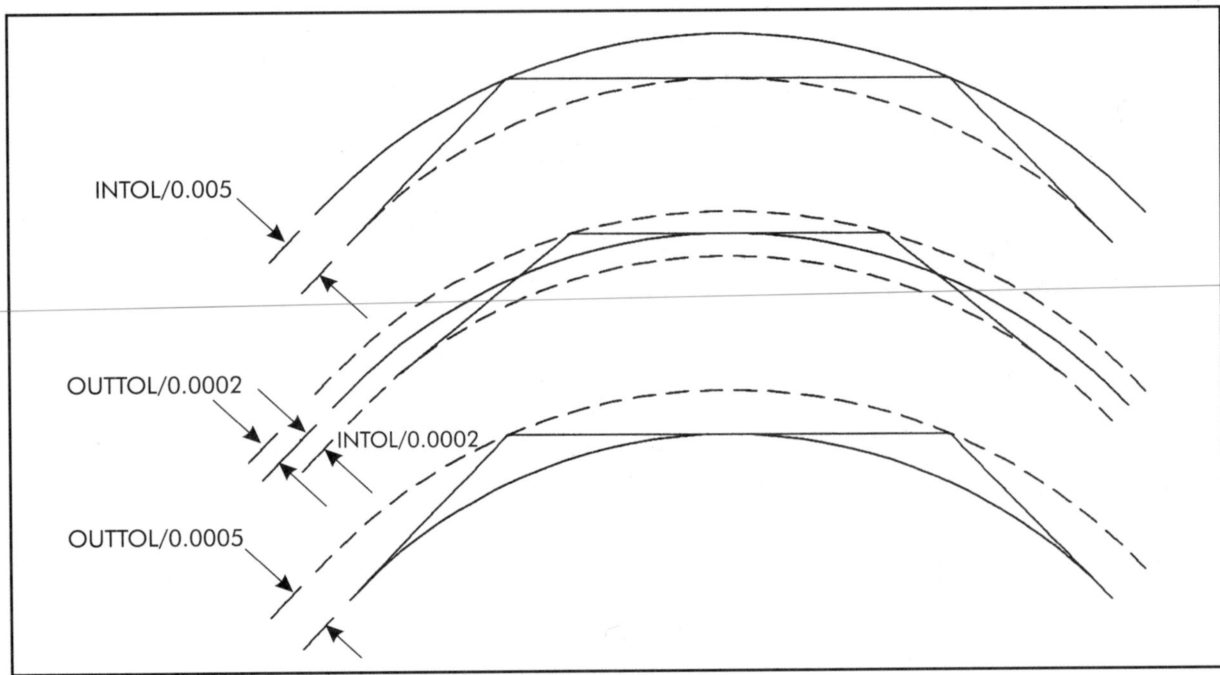

Figure 31-18. Examples of INTOL and OUTTOL statements.

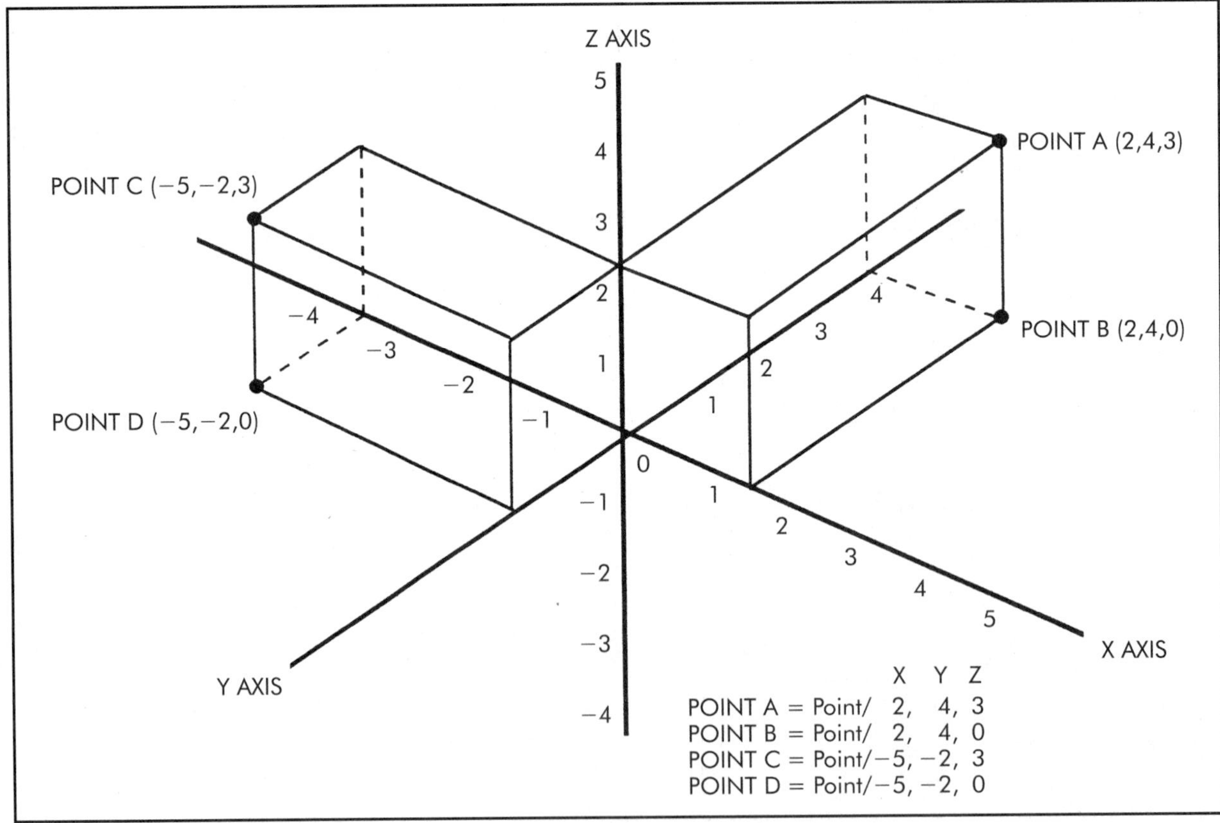

Figure 31-19. Examples of points defined by Cartesian coordinates.

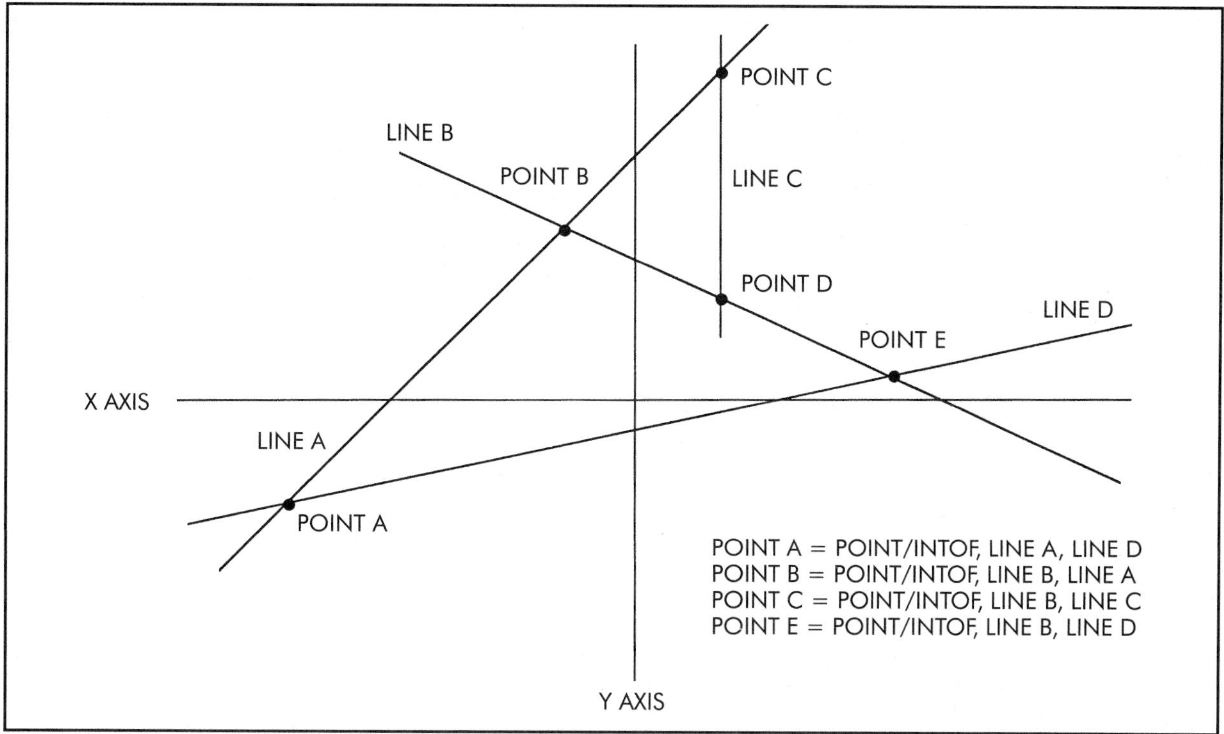

Figure 31-20. Examples of points defined by the intersection of two lines.

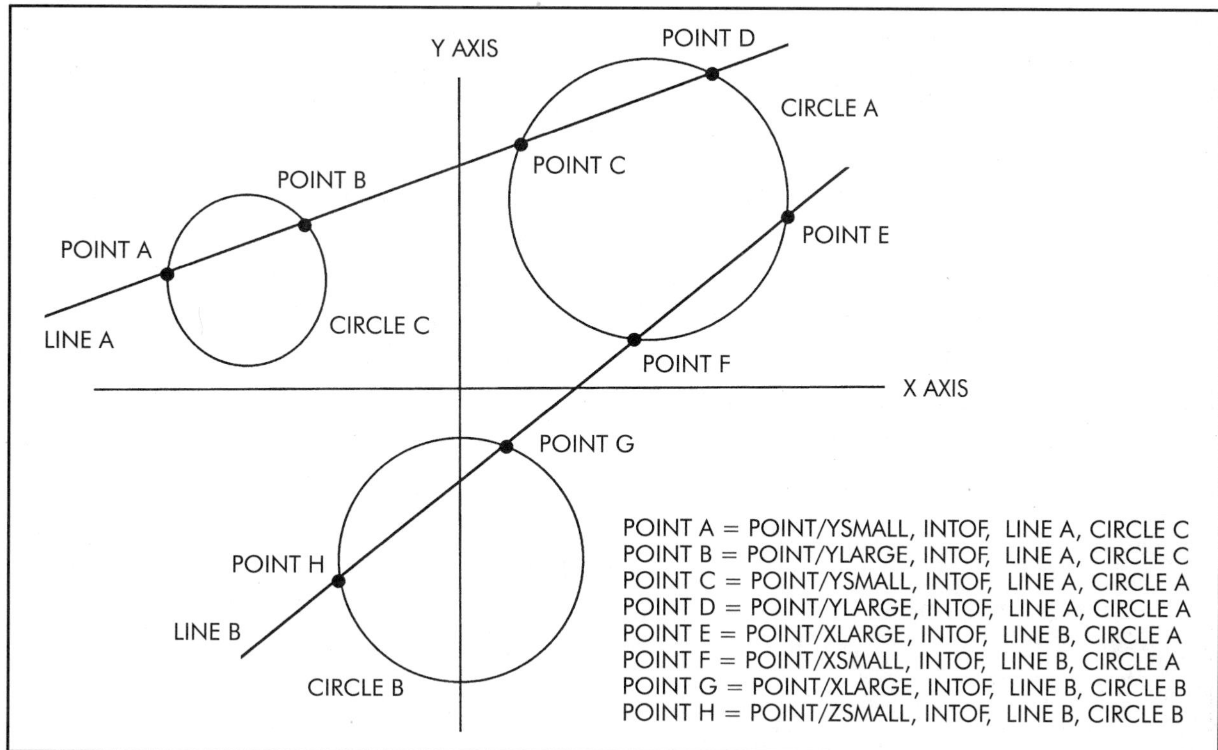

Figure 31-21. Examples of points defined by the intersection of a line and a circle.

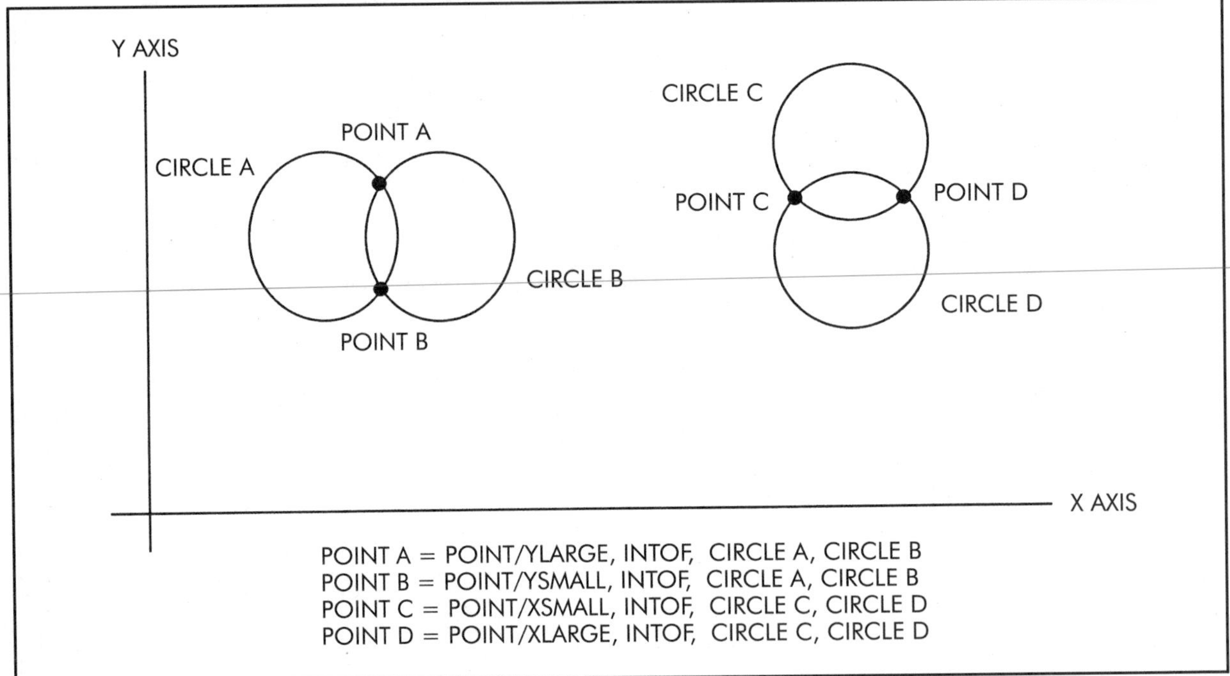

Figure 31-22. Examples of points defined by the intersection of two circles.

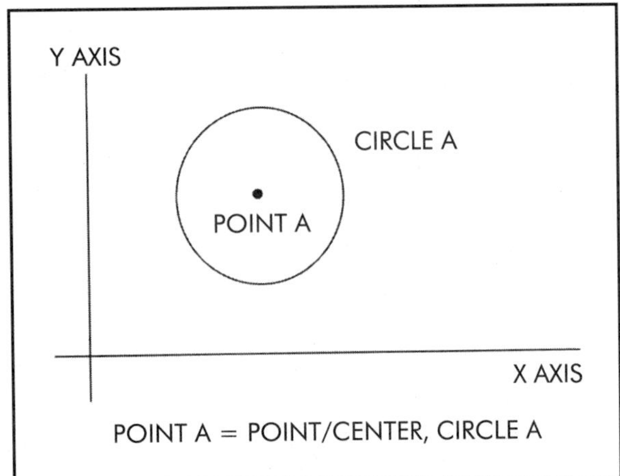

Figure 31-23. A point may be defined as the center of a circle.

Examples to illustrate these definitions are given in Figures 31-24 through 31-29.

A circle may be defined in different ways, such as:

- By the coordinates of its center and radius:

symbol = CIRCLE/x, y, z, radius
symbol = CIRCLE/x, y, radius
symbol = CIRCLE/CENTER, point, RADIUS, radius

- By its center and tangent to a given circle. There are two possibilities in this case. The modifiers LARGE and SMALL indicate if the circle is to be chosen with the largest or the smallest radius.

symbol = CIRCLE/CENTER, point, LARGE, TANTO, circle
symbol = CIRCLE/CENTER, point, SMALL, TANTO, circle

- By its radius and tangent to two intersecting lines:

symbol = CIRCLE/XLARGE, line1, XLARGE, line2, RADIUS, radius
symbol = CIRCLE/XSMALL, line1, XSMALL, line2, RADIUS, radius
symbol = CIRCLE/YLARGE, line1, YLARGE, line2, RADIUS, radius
symbol = CIRCLE/YSMALL, line1, YSMALL, line2, RADIUS, radius

- By its center and a point on its circumference:

symbol = CIRCLE/CENTER, center, point

- By three points on its circumference:

symbol = CIRCLE/point1, point2, point3

- By its center and tangent to a line:

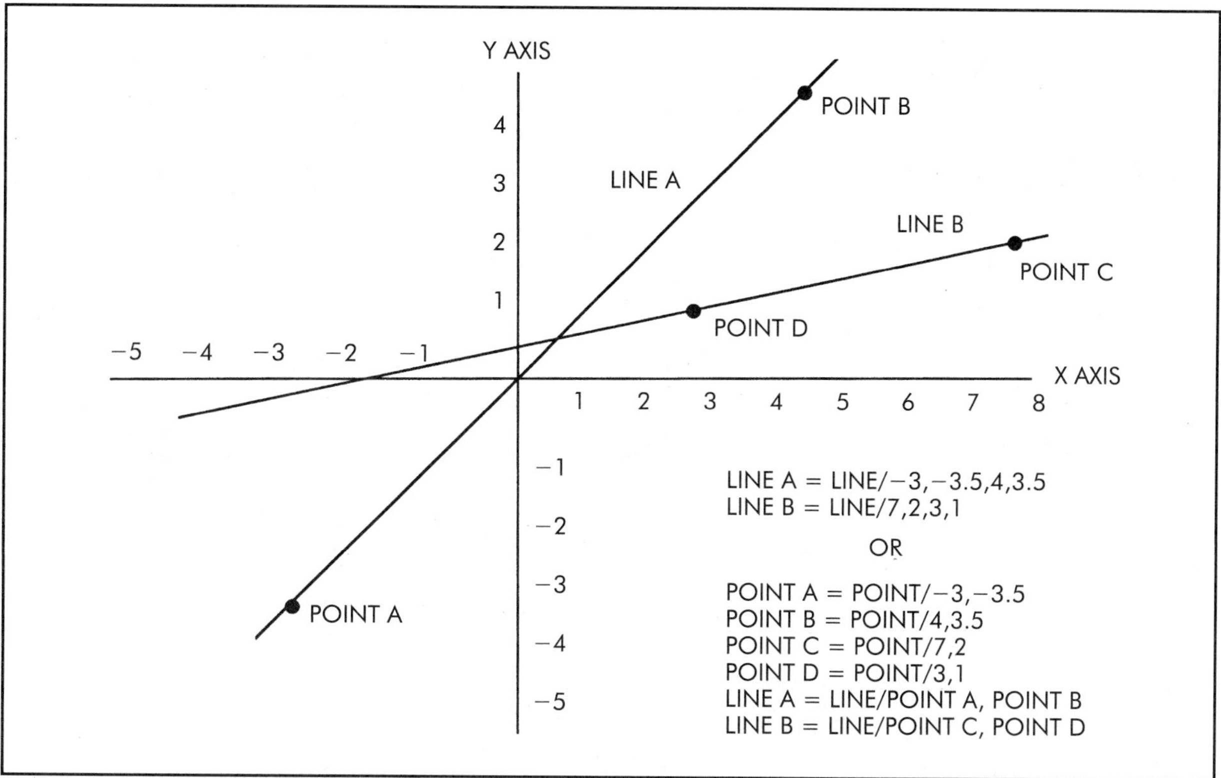

Figure 31-24. Examples of lines defined by coordinates of two points or passing through two points.

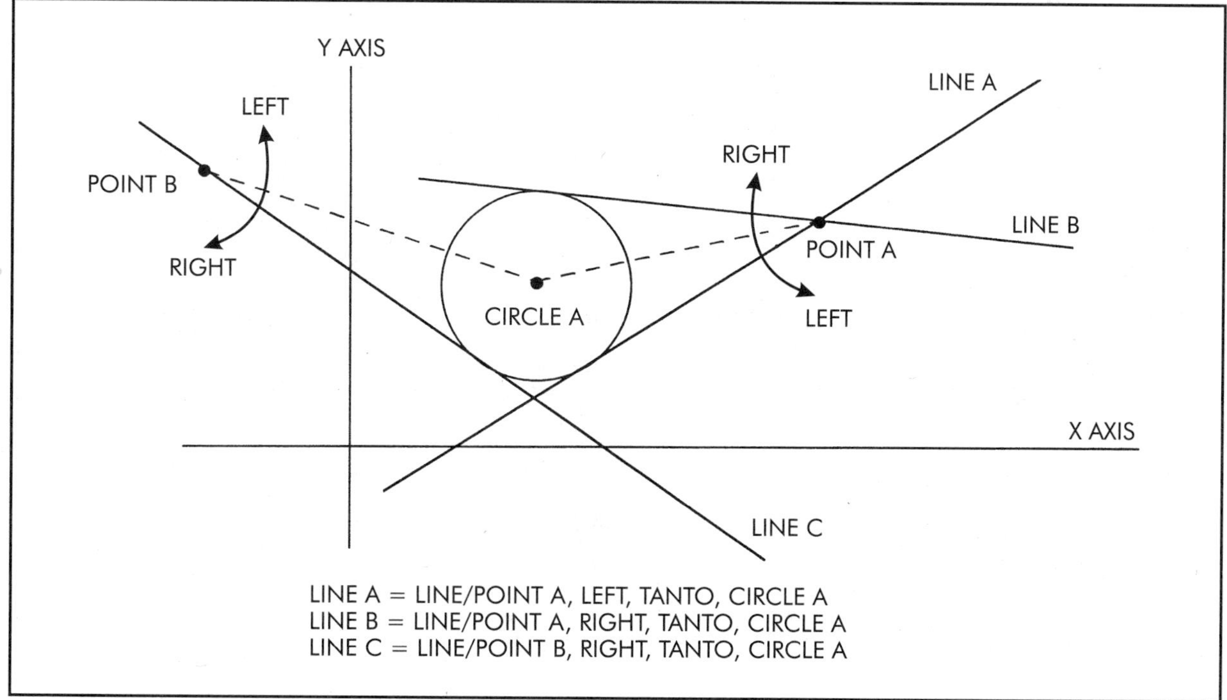

Figure 31-25. Examples of lines defined by a point and tangent to a circle.

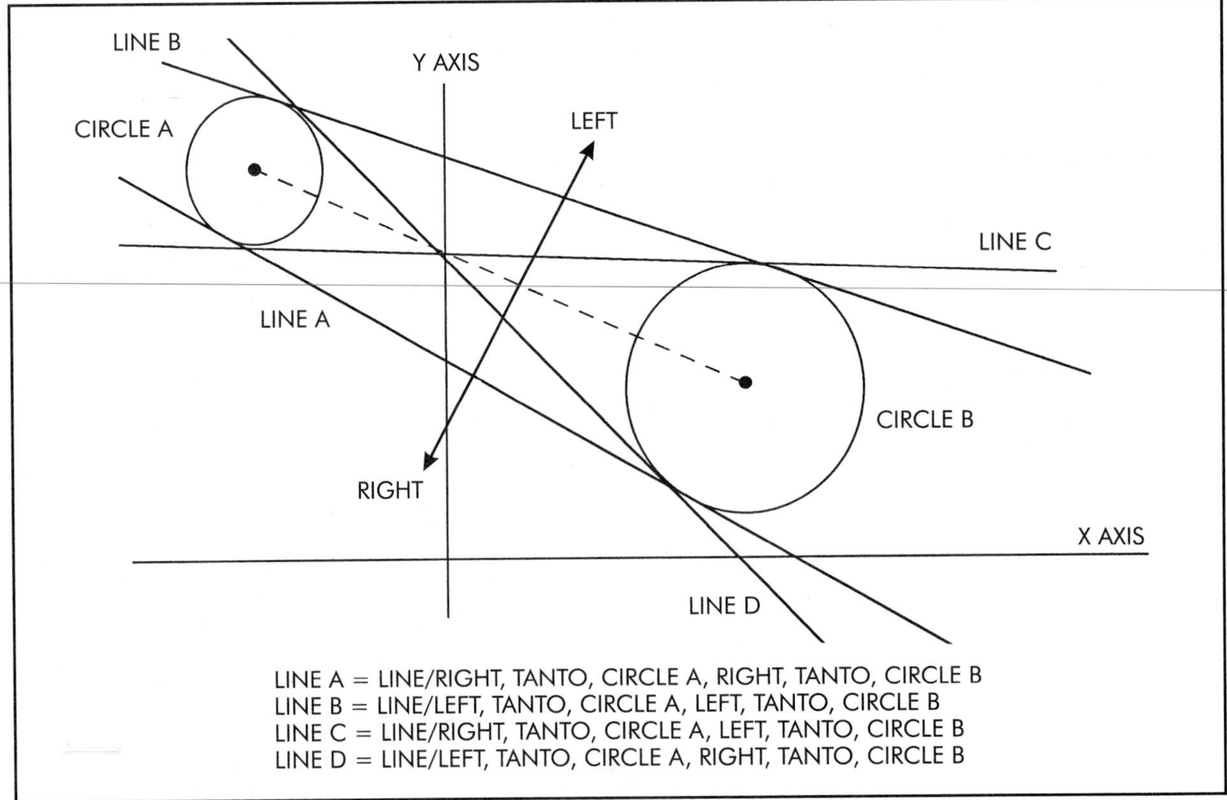

Figure 31-26. Examples of lines defined as tangent to two circles.

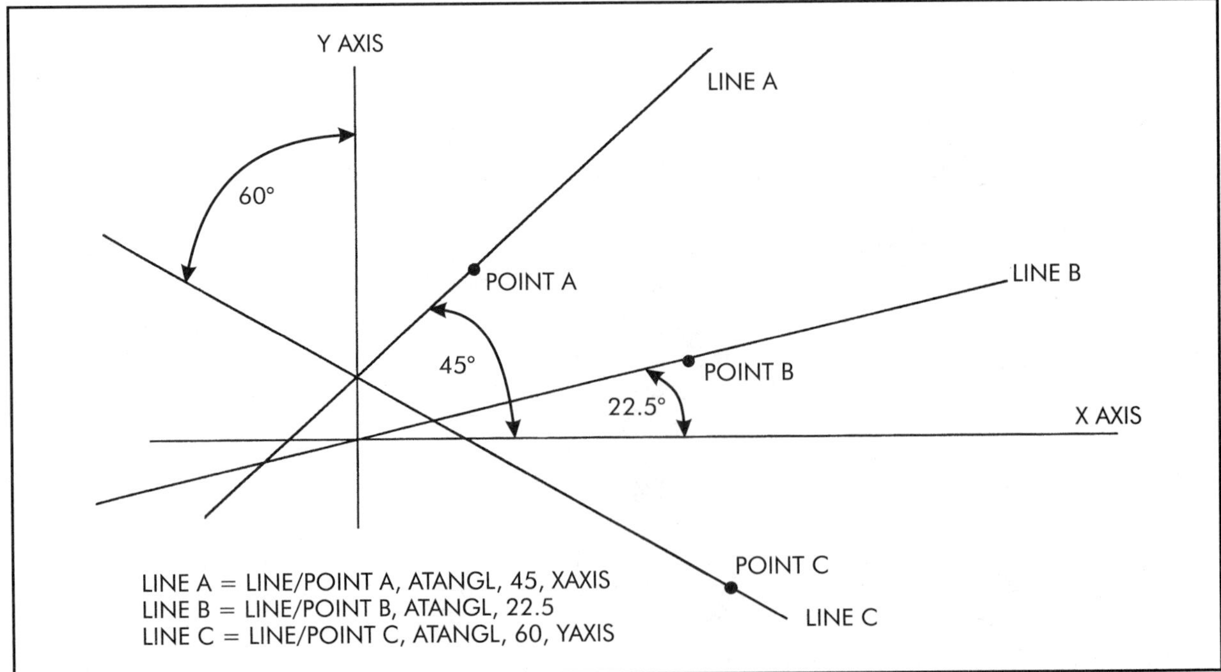

Figure 31-27. Examples of lines defined by a point and at an angle to the X or Y axis.

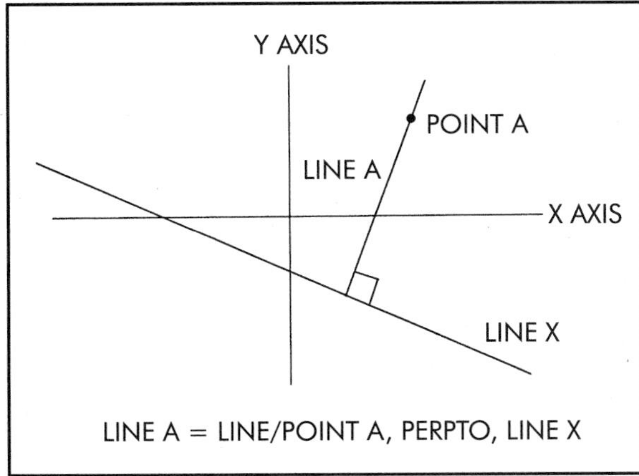

Figure 31-28. Examples of lines defined by a point and perpendicular to a line.

symbol = CIRCLE/CENTER, point, TANTO, line

Circle definitions are illustrated in Figures 31-30 through 31-35.

APT requires that three surfaces be defined to control the movement of the cutter, as follows:

1. *Part surface* is the surface on which the end or bottom of the cutting tool is riding.
2. *Drive surface* is the surface against which the edge of the cutting tool is moving. This is in fact the actual part surface being cut.
3. *Check surface* is the one at which the actual cutting tool movement is to stop.

In addition to geometric statements, APT provides different motion commands that can be used for point-to-point, straight-cut, and contouring programming. It also provides statements for identifying cutting parameters and auxiliary definitions. Examples of such statements and commands are defined in Table 31-2. The following figures illustrate these commands. Figure 31-36 shows a simple part to be machined on a CNC machining center. A possible APT program listing for this part is shown in Figure 31-37. No actual cutting will be performed since the *Z* coordinate is set to zero. To actually cut and machine the part to the desired depth of .25 in., the following two surfaces need to be defined:

PLN1=PLANE/PT1, PT2, PT3. This plane is defined by three points and is actually the part surface.

PLN2=PLANE/PARLEL, PLN1, ZSMALL, 0.25. This plane is set as being .25 in. below PLN1.

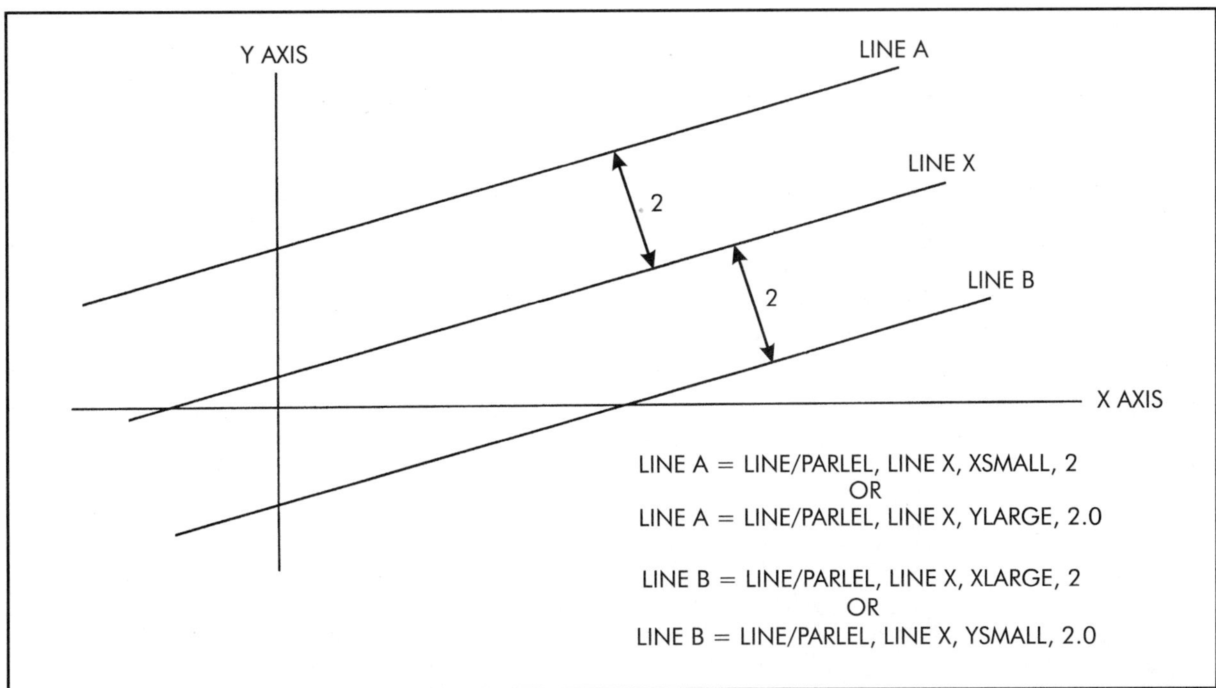

Figure 31-29. Examples of lines defined as parallel to another line at some distance.

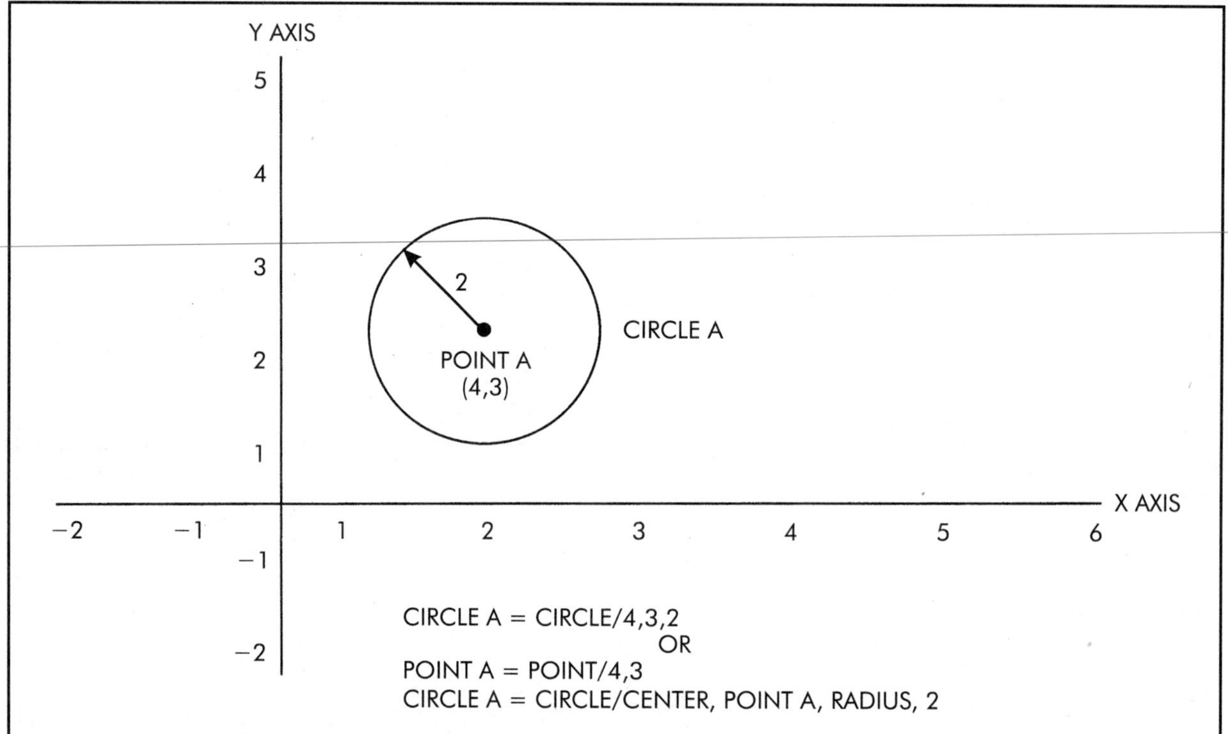

Figure 31-30. Example of a circle defined by center coordinates and radius or by its center point and radius.

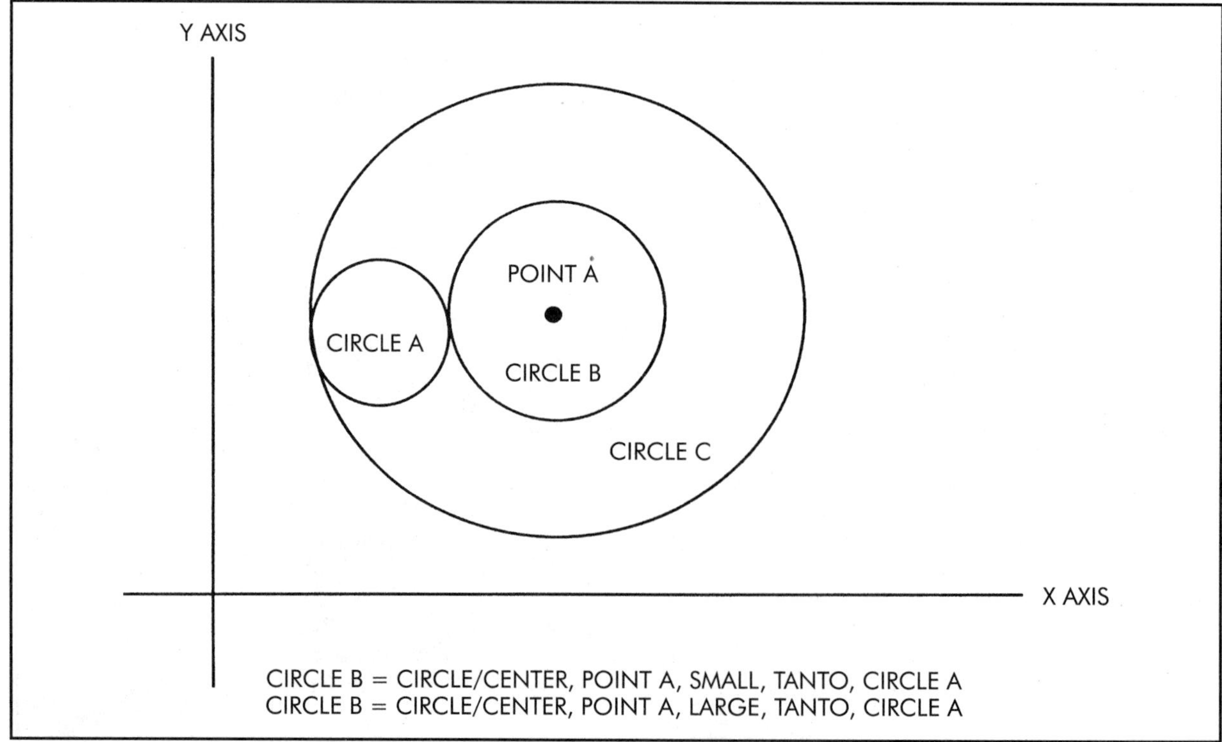

Figure 31-31. Example of a circle defined by its center point and tangent to another circle.

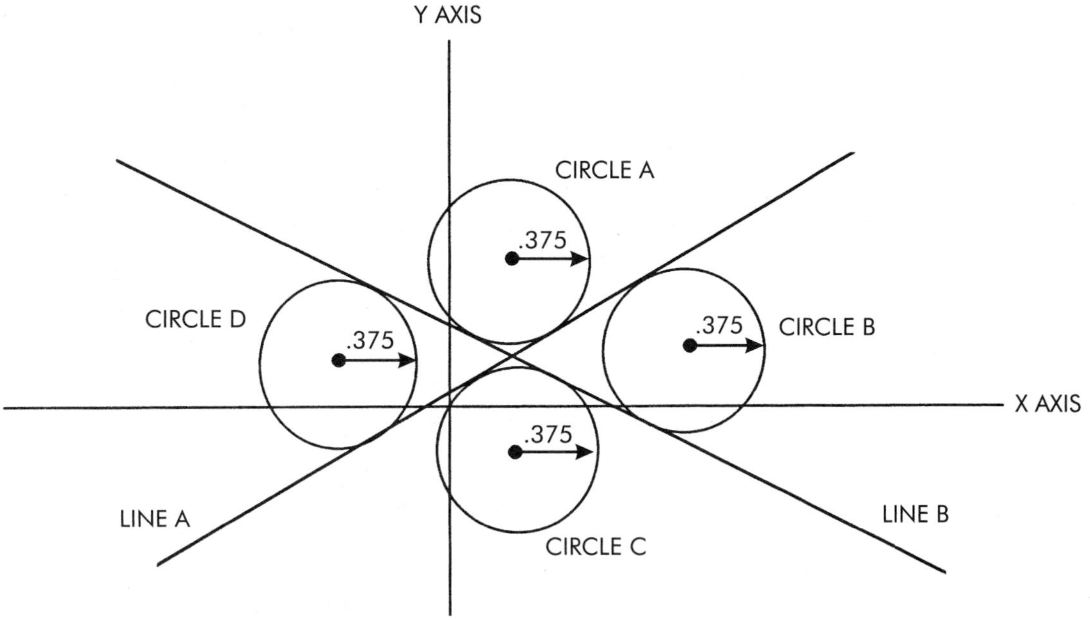

CIRCLE A = CIRCLE/YLARGE, LINE B, YLARGE, LINE A, RADIUS, 0.375
CIRCLE B = CIRCLE/YSMALL, LINE A, XLARGE, LINE B, RADIUS, 0.375
CIRCLE C = CIRCLE/YSMALL, LINE B, YSMALL, LINE A, RADIUS, 0.375
CIRCLE D = CIRCLE/XSMALL, LINE B, YLARGE, LINE A, RADIUS, 0.375

OR

CIRCLE A = CIRCLE/XLARGE, LINE B, XSMALL, LINE A, RADIUS, 0.375
CIRCLE B = CIRCLE/XLARGE, LINE A, YLARGE, LINE B, RADIUS, 0.375
CIRCLE C = CIRCLE/XSMALL, LINE B, XSMALL, LINE A, RADIUS, 0.375
CIRCLE D = CIRCLE/YSMALL, LINE B, XSMALL, LINE A, RADIUS, 0.375

Figure 31-32. Examples of circles defined as tangent to two lines and a radius.

CIRCLE A = CIRCLE/CENTER, POINT B, POINT C

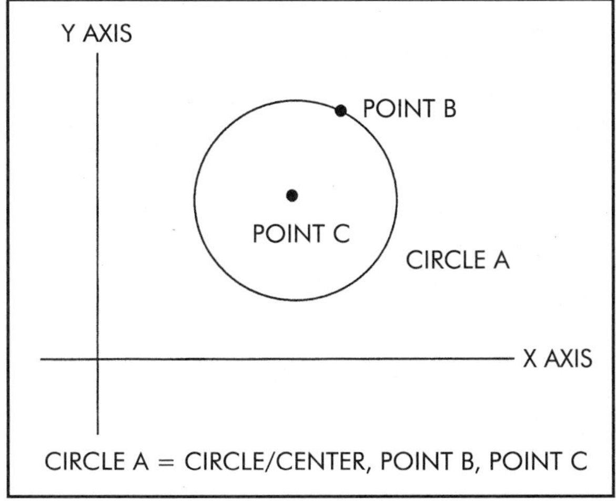

Figure 31-33. Example of a circle defined by its center and a point on the circumference.

CIRCLE A = CIRCLE/POINT A, POINT B, POINT C

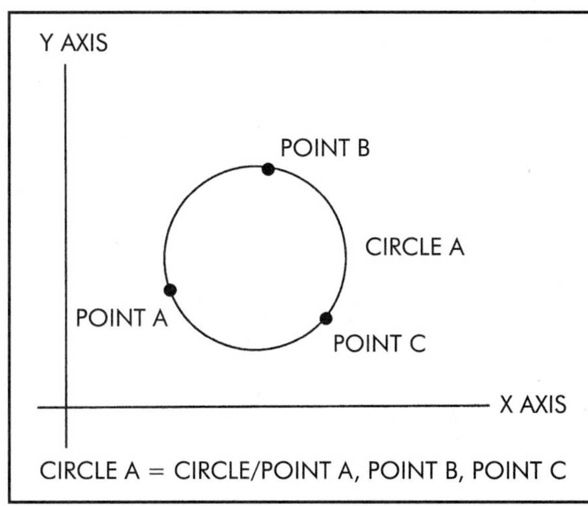

Figure 31-34. A circle may be defined by three points on the circumference.

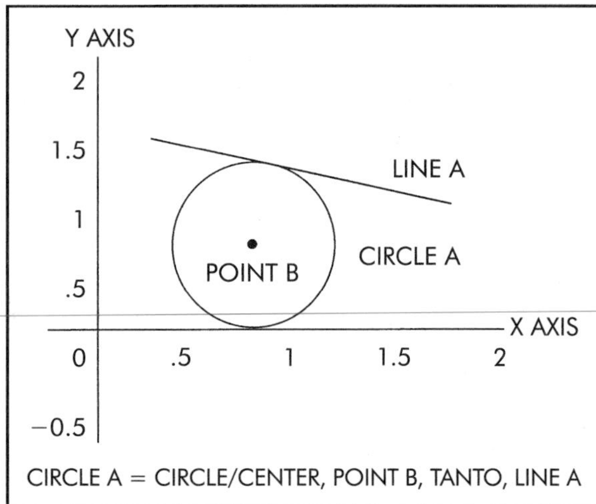

Figure 31-35. A circle may be defined by its center and tangent to a line.

The program will be essentially the same as the one shown in Figure 31-37, except that *Z* coordinates will be defined at .25 in. and the tool is to be moved on PLN2. The program listing will look like the one shown in Figure 31-38.

Group III Programs

In contrast to the group I and group II programs where tool paths must be specified, a *group III program* accepts a description of a shape to be machined and automatically generates all the tool paths to do the job most efficiently. The language is abridged English with symbolic notations and considerable flexibility. Programs in this class are rather long and logically complex and can be processed only on the largest computers.

A typical group III program recognizes and performs the calculations for any surface that can be expressed by an equation in three-dimensional analytic geometry. Common surfaces may be described in a number of ways and may be divided into parts for convenience. In addition, the program provides for a filleted surface to make a smooth transition between two intersecting analytical surfaces.

An example of a simple group III program is given in Figure 31-39. The first set of statements defines the symbols arbitrarily assigned to pertinent surfaces in the operation: TOP for the upper surface of the workpiece, CLR for a plane above which the cutter may be moved without interference, and WORK for the surface to be machined. The next set of instructions sets forth the conditions of the operation: the idle tool position, the path over which the cutter moves to cutting position to clear the work, feed rate, radius of the ball end of the cutter, tolerated error

Table 31-2. Examples of APT motion commands

Command	Definition
FROM/point or point location	Start from current location
GOTO/point or point location	Move to designated location
GODLTA/incremental coordinated	Perform incremental positioning
GO/TO	Move to the next drive surface
GO/PAST	Move past the next drive surface
GO/ON	Move on next drive surface
GOLFT/	Move left along the drive surface
GORIGHT/	Move right along the drive surface
GODOWN/	Move down along the drive surface
GOUP/	Move up along the drive surface
GOFWD/	Move forward along the drive surface
GOBACK/	Move backward along the drive surface
MACHIN/	Specifies the machine to be used
COOLNT/	Turns coolant ON or OFF
FEDRAT/	Feed rate (in. per min)
SPINDLE/	Spindle speed (rpm)
TURRET/	Specific tool to be called
FINI	End of program
END	Forces the machine to stop

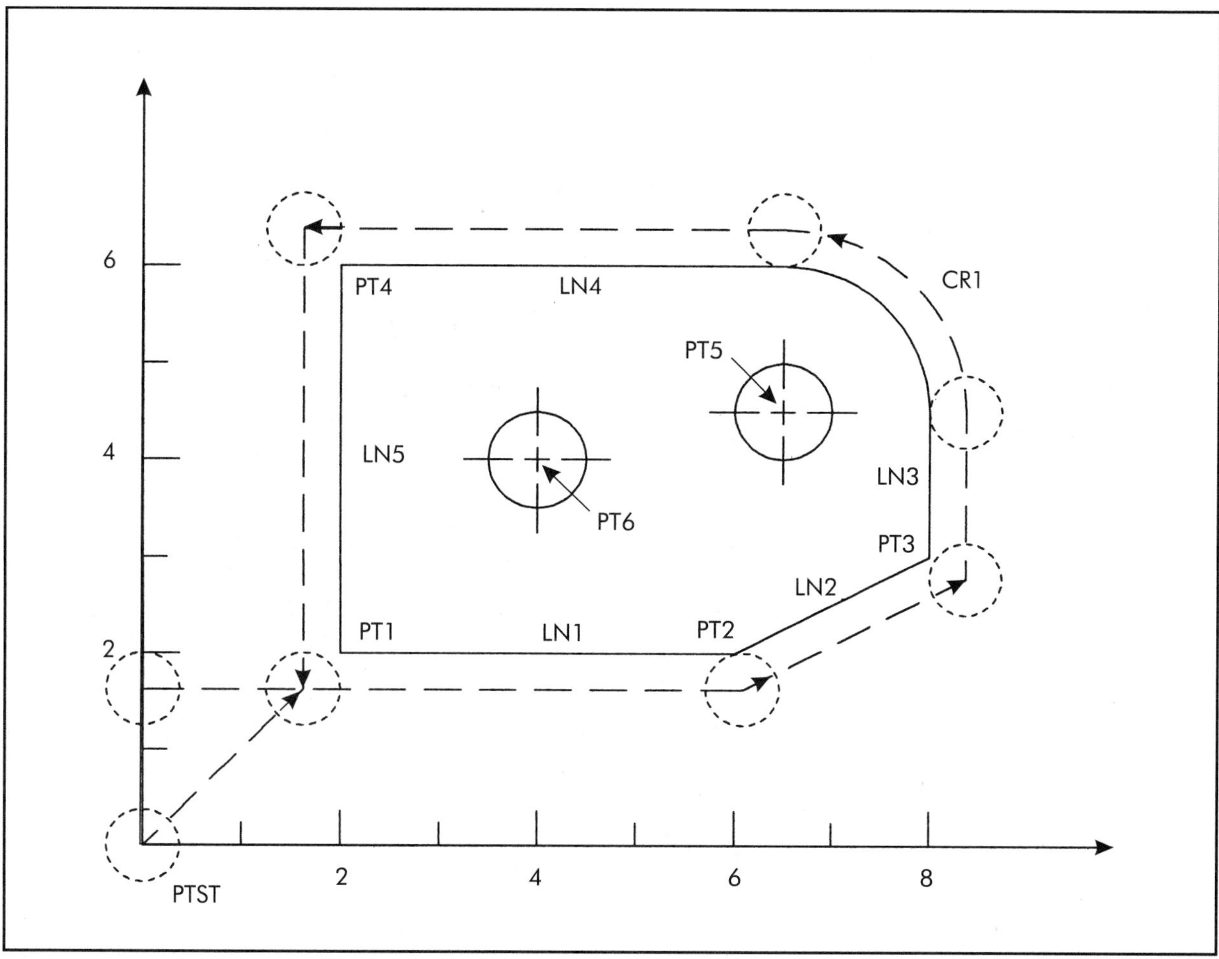

Figure 31-36. Example of a simple part.

of the machined surface, and coolant flowing. In the third set, REALM is an area to be machined in one pass of the tool, in this case one surface. The cutter is directed to start outside of the area at point (75, 0, 25) and proceed in the $+Y$ direction. The region to be machined lies to the left. The external boundary of the cut is the intersection of WORK and TOP, and the cutter path is to be along a spiral. The last set of statements disposes of the tool and closes the operation.

Advanced Programs

Computer graphics systems are tutorial in nature for interactive NC programming between programmer and computer, commonly some versions of APT. With a typical system, the programmer picks part dimensions and angles off the part drawing in the order dictated by instructions that appear automatically on a CRT screen. These are entered into the system through a keyboard. With some systems, the programmer enters the proper commands (for example, TLLFT, GOTO, etc.) to apply to the data; with others, commands are issued automatically to suit the input data. The tool path is generated automatically in a simulated three-dimensional perspective (in some systems on a second CRT screen) to provide a visual check of the program created. Programming may be done for turning, milling, punching, and die-making operations. One computer graphics system is offered for less than $20,000.

A *preprocessor* is a program that prepares input statements for an NC processor from a series of point coordinates. For instance, a program is to be prepared to machine a mold to

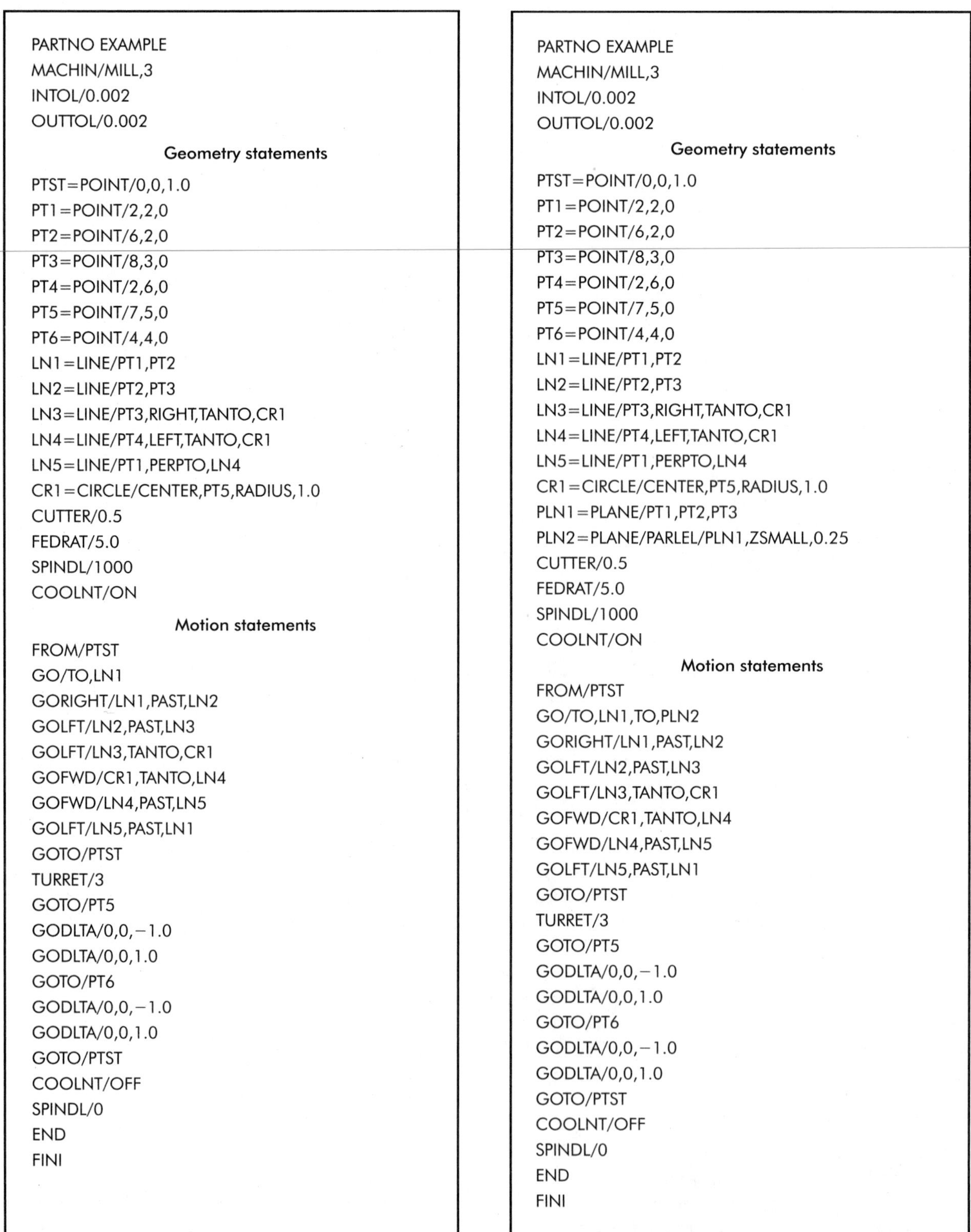

```
PARTNO EXAMPLE
MACHIN/MILL,3
INTOL/0.002
OUTTOL/0.002
            Geometry statements
PTST=POINT/0,0,1.0
PT1=POINT/2,2,0
PT2=POINT/6,2,0
PT3=POINT/8,3,0
PT4=POINT/2,6,0
PT5=POINT/7,5,0
PT6=POINT/4,4,0
LN1=LINE/PT1,PT2
LN2=LINE/PT2,PT3
LN3=LINE/PT3,RIGHT,TANTO,CR1
LN4=LINE/PT4,LEFT,TANTO,CR1
LN5=LINE/PT1,PERPTO,LN4
CR1=CIRCLE/CENTER,PT5,RADIUS,1.0
CUTTER/0.5
FEDRAT/5.0
SPINDL/1000
COOLNT/ON
            Motion statements
FROM/PTST
GO/TO,LN1
GORIGHT/LN1,PAST,LN2
GOLFT/LN2,PAST,LN3
GOLFT/LN3,TANTO,CR1
GOFWD/CR1,TANTO,LN4
GOFWD/LN4,PAST,LN5
GOLFT/LN5,PAST,LN1
GOTO/PTST
TURRET/3
GOTO/PT5
GODLTA/0,0,-1.0
GODLTA/0,0,1.0
GOTO/PT6
GODLTA/0,0,-1.0
GODLTA/0,0,1.0
GOTO/PTST
COOLNT/OFF
SPINDL/0
END
FINI
```

Figure 31-37. An APT program for the part shown in Figure 31-36.

```
PARTNO EXAMPLE
MACHIN/MILL,3
INTOL/0.002
OUTTOL/0.002
            Geometry statements
PTST=POINT/0,0,1.0
PT1=POINT/2,2,0
PT2=POINT/6,2,0
PT3=POINT/8,3,0
PT4=POINT/2,6,0
PT5=POINT/7,5,0
PT6=POINT/4,4,0
LN1=LINE/PT1,PT2
LN2=LINE/PT2,PT3
LN3=LINE/PT3,RIGHT,TANTO,CR1
LN4=LINE/PT4,LEFT,TANTO,CR1
LN5=LINE/PT1,PERPTO,LN4
CR1=CIRCLE/CENTER,PT5,RADIUS,1.0
PLN1=PLANE/PT1,PT2,PT3
PLN2=PLANE/PARLEL/PLN1,ZSMALL,0.25
CUTTER/0.5
FEDRAT/5.0
SPINDL/1000
COOLNT/ON
            Motion statements
FROM/PTST
GO/TO,LN1,TO,PLN2
GORIGHT/LN1,PAST,LN2
GOLFT/LN2,PAST,LN3
GOLFT/LN3,TANTO,CR1
GOFWD/CR1,TANTO,LN4
GOFWD/LN4,PAST,LN5
GOLFT/LN5,PAST,LN1
GOTO/PTST
TURRET/3
GOTO/PT5
GODLTA/0,0,-1.0
GODLTA/0,0,1.0
GOTO/PT6
GODLTA/0,0,-1.0
GODLTA/0,0,1.0
GOTO/PTST
COOLNT/OFF
SPINDL/0
END
FINI
```

Figure 31-38. A modified APT program for the part shown in Figure 31-36, which allows actual cutting.

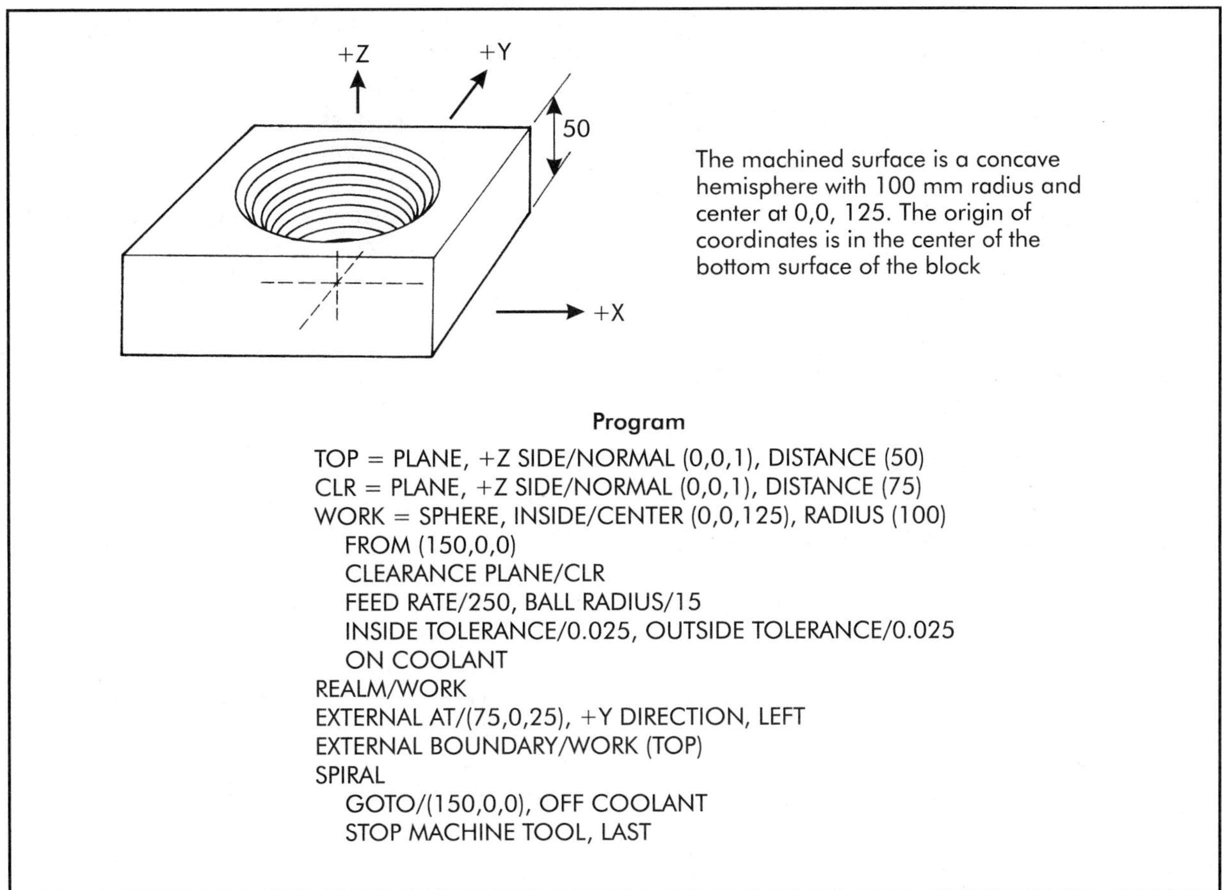

Figure 31-39. Sample of a simple Group III program.

duplicate one already made by hand. The mold cavity is a surface of revolution, and a coordinate measuring machine is utilized to read out the coordinates of points along the profile. In another case, the designer of an object may define a profile by specifying the coordinates of the points establishing a series of lines, arcs, or curves. In any case, the preprocessor takes the statements of the geometric and parameter definitions and motion instructions for a selected operation to enable a processor (for example, an APT processor) to prepare the commands for the NC machine.

A worthwhile goal for NC programming, as well as for all computer programming, is to achieve voice input and thus eliminate the slower and tedious routine of typing or punching cards or tapes. *Voice programming* for NC (VNC) is available but not widely used. It is most readily applicable to programming an English-like programming language with a limited vocabulary, such as a simple version of APT. In a typical system, the programmer delivers the words through a microphone into a translator that compiles them into a program executed in the usual way by a processor for the language used. The translator can be trained to recognize the voice of an operator simply by his or her repeating each word in the vocabulary 5–10 times. The parameters associated with the operator's voice are then stored in memory for later matching and identification. As each line is dictated, it is written on a CRT display so it can be checked for errors.

31.5 SELECTION OF A PROGRAMMING METHOD

The first choice should be between manual and computer-assisted programming. Authorities recommend that a shop new to NC should start with manual programming for at least 3–6 months. This provides closer hands-on experience with the NC system in the initial learning period and the opportunity to find out what manual programming costs really are in the shop. A survey of experts has shown a profile for manual programming for up to four tapes a month: average preparation time was not in excess of 15 min per block, and no more than 2 hr average debugging time was required per tape.

Computer-assisted programming is justified, of course, when its cost is lower compared to manual methods. True computer-assisted programming costs may not be easy to find for a plant without previous experience; costs from other plants or from system vendors may not be applicable. Factors to consider when selecting a particular language system are: (1) the ability of the language to handle part complexity (now and in the future) without being too elaborate; (2) the dependability of the system supplier; (3) hardware requirements, including computer, terminals, peripheral equipment, etc.; (4) availability and cost of programmer training; (5) availability of technical assistance and support; and (6) actual unit programming costs.

There are some other general considerations for the selection of a programming method, such as:

- type and capabilities of computing equipment and machines,
- cost allocation and time involved in programming,
- complexity and dimensions of the part, and
- capabilities and skill level of manufacturing personnel and workers.

It is also important that NC programmers and operators be familiar with the manufacturing processes being used and knowledgeable about part materials, processing parameters, and other machining conditions. Program verification is also important before actual processing and production begins. It may be done using simulation of the process or by making a part out of less expensive materials, such as wood or aluminum.

31.6 NC IN THE TOTAL MANUFACTURING SYSTEM

Figure 31-1 illustrates the conventional way of programming NC, manually or with computer assistance, from a description of a workpiece on a drawing. This has been practiced because as the NC system has been developed, it has been attached to the existing industrial system of product design and manufacture. The ideal is to process information automatically from original design to fabrication. With such a refined system, only the functional specifications and economic conditions of a part (such as a drive housing to support certain loads, occupy certain space, and to be of minimum weight and cost) would be given by the designer. The system would automatically determine all the necessary dimensions, tolerances, and other details of the required part and proceed to generate instructions to make the product in the best way and provide a drawing if desired.

Progress toward the ideal in manufacturing systems is being made. The first step is to compile the product specifications in numerical form. This can come naturally because designers are more frequently turning to computers to aid them in their work with the use of tools such as computer-aided design (CAD). Some parts may be represented by mathematical models. An example is a functional cam, for which equations relating to follower displacement are all that are necessary. Many aerospace shapes may be defined mathematically. One system enables a designer to make a sketch with a light pen on a CRT screen. The lines may be altered, erased, corrected, rotated, etc., under the direction of the designer. The computer also can be called upon for calculations and file management. When the design is finished, its dimensions and other specifications are stored in an orderly program. Some shapes, like those of automobile bodies, are designed for appearance rather than mathematical considerations. Models of clay or other substances may be made for such products, and their dimensions determined at necessary points by photographic or mechanical methods. However obtained, the design data base can be stored and transmitted on tapes, disks, etc., for further use. Input from a CAD system for NC programming is shown in Figure 31-1.

Design data are processed in a number of ways, more or less automatically, to plan production and program for NC. An interactive graphic on a CRT screen is one form of CAD. This interface may be employed with additions for NC programming. The programmer can trace cutter paths on the screen with a light pen and designate specific points for action. He or she can type on a keyboard the commands in APT or another language, and the computer fills in the dimensions and compiles the program. Such interactive systems used in the aircraft industry are reported much faster than entirely manual programming. Computer-aided NC programming may be used to make tools, as well as parts. In the automobile industry, NC is not employed as a rule for actual production, but is widely utilized to make the dies for shaping sheet metal and forgings.

The configuration of a workpiece is conventionally depicted by a drawing, which may be the basis for programming the NC operation to make a part as illustrated in Figure 31-1. Drawings can be made quickly from a computerized data base on an NC drafting machine. Authorities have estimated that NC drafting is on an average 10 times faster than manual. A typical automatic drafting machine has a pen or scriber on a head that slides along a bridge. The bridge is supported at each end on rails and moves along a table. Coordinate movements are made to conform to the data input to the NC control.

The initial step in manufacturing, once the design specifications have been finished, is to plan the processes. The process specifications include a list of operations, including machines and tools required to make a part efficiently. These specifications, together with the design specifications, are passed on for material procurement, cost estimating, production scheduling and control, inspection, etc. At one time, all these functions were performed manually, but in recent years some or all have been automated using *computer-aided manufacturing* (CAM) systems. It is reported that there are over 2,500 stand-alone CAD/CAM installations worldwide. The initial step toward a full CAM system is to utilize *computer-aided process planning* (CAPP).

NC programming is at least a sequel to process planning; in the ideal it is an integral part. CAPP and automatic NC programming start with a technique called *group technology* (GT) in which the parts made in a plant are coded and segregated into groups or families. All the parts in one family have common features or processing. Each new part is assigned to its proper family. For example, a family might be all shafts from 25.4–76.2 mm (1–3 in.) in diameter, 152.4–203.2 mm (6–8 in.) in length, and with no more than six diameters. An optimum routing is prepared specifying the operations, machines, and tools for any part in a family. The routing includes operations for permissible deviations such as threaded ends, cross holes, material differences, etc., in a family of shafts. These extra steps are eliminated to route any shaft that does not require them. An optimum program is written for each NC operation on the routing. The routings and NC programs are stored in the memory of the computer without dimension and other specifics for any particular part. When a part is to be processed, the CAPP program identifies its family from its coding, retrieves the proper routing and NC programs, and inserts into them the necessary dimensions ascertained from the design data base. Detailed cutter paths may be calculated, if required, to conform to workpiece dimensions. In the end, the computer issues routings for production control and other functions, and NC programs on tape, disk, or directly from NC control storage.

31.7 ECONOMICS OF NUMERICAL CONTROL

The big drawback of NC is its initial cost. NC machines cost from around 1.5–5 times as much as conventional machines of like size, depending upon the capacity of the control and accessories. Maintenance of the NC equipment requires a high order of skill and trained personnel, although this is being alleviated by diagnostic systems and replaceable circuit boards for controls. Programmers are needed for most NC machines.

NC is not the best method for all jobs; it is more economical for certain kinds of work and quantities as indicated by Figure 31-40. In many cases, one or a few simple pieces (for example, to turn a diameter or mill a face) can be machined by a skilled operator in less time at a lower cost per hour on general-purpose conventional machines than would be required to program and run the job on an NC machine. On the other hand, with more intricate parts even one piece may be made more economically by NC. Once

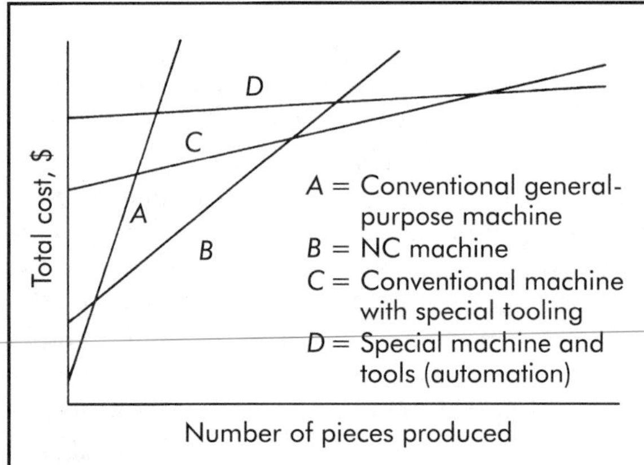

Figure 31-40. Comparison of the cost of NC with other methods.

the programming has been done, the time is often less per piece on the NC machine, which is then more economical for more than a few pieces.

Figure 31-41 indicates average savings and productivity gains from NC as compared to conventional machine tools. These results were found in a study at the University of Michigan of 356 companies employing 4,648 NC machines. Productivity gains are more impressive than time savings. As an example, assume that the time to produce a part on an NC machine is 40% of the time required on a conventional machine; that is a savings of 60% in time. The conventional machine takes 1 hr to make one part and the NC machine can make one part in 24 min or 2.5 parts/hr, an increase in productivity of 150%.

There are a number of reasons why NC machines can often work faster than manually operated machines and thus save time as indicated in Figure 31-41. Idle time is a minimum (the machine does not demand a coffee break), fatigue is nonexistent, human mistakes are avoided, all planning must be done beforehand, rejects are fewer, and less time is needed for checking and inspection. A study by one manufacturer showed NC machines cutting 80% of the time as against less than 25% of the time for conventional machine tools. Tool life may be better because speeds and feeds are set as planned and continuously maintained. Much setup can be done off the machine and by programming.

NC systems offer savings other than those reflected by reductions in operating time. Scrap and rework were reported reduced over 25%, material handling decreased 20–50%, and inspection cost reduced by 30–40%. Since operator errors were better controlled by NC, parts were

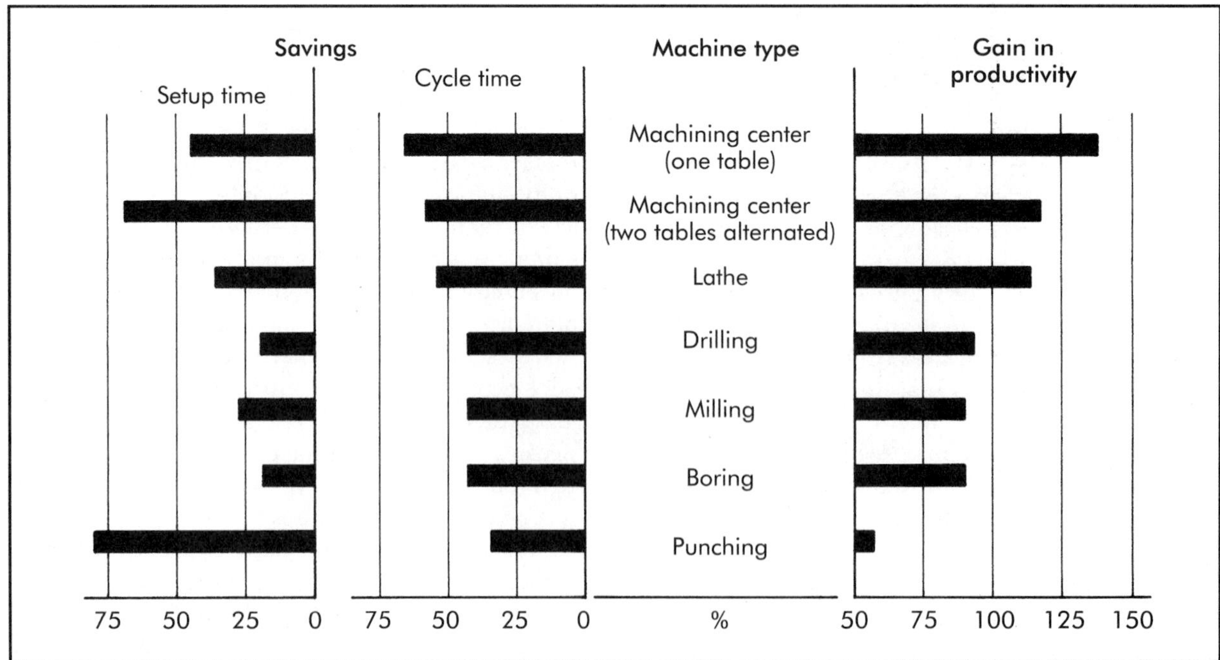

Figure 31-41. Average savings and productivity gains from NC machines.

produced more accurately, and assembly costs were reported reduced by 10–20%. An NC machining center may be able to do in one operation as much as several different conventional machines, thus saving considerable floor space. Multiple control of machine tools and even remote control for dangerous material is practical for NC. Tooling in the form of tapes or disks is easily stored and preserved.

Special tooling, such as fixtures, jigs, or templates, is commonly added to conventional machines for moderate to large quantities of pieces. Preparation or lead time is commonly long for special tools. A job often can be started much sooner with NC, which may not require as much or any special tooling. The tooling costs, tool storage charges, setup costs, and the costs of making changes are also much less for NC. In one case, the time to prepare tools was reported to be 240 hr for a conventional machine, but only 60 hr (including 22 hr of programming) for an NC machine. Even if the NC cycle time per piece is more after the job has been set up, and it is not always so, a moderate to large number of pieces must be run to make up for the difference in preparation cost, as indicated by Figure 31-40. With NC flexibility, setup cost is often less and smaller lot sizes are economical and, less floor space is needed for materials in process and storage.

As indicated on the right of Figure 31-40, NC cannot compete with fixed-program special-purpose machines and tools to produce large quantities of pieces. For instance, automatic bar machines controlled by cams are simple, direct, and fast for turning; and transfer machines are specialized in making certain products very efficiently. NC machines cannot do the same jobs at as low a cost when the quantities required are large. On the other hand, NC may supplement special machines in some aspects. For instance, NC may be part of a flexible machining line that machines large quantities of different kinds of pieces but not a large quantity of any one part at one time.

31.8 QUESTIONS

1. What is "numerical control"? What are its principal features?
2. Draw a diagram of the different ways of preparing input for an NC machine. Briefly describe them.
3. Differentiate between CNC and DNC.
4. Specify and explain the functions that an NC control may perform.
5. What is a five-axis NC machine?
6. What is meant by point-to-point, straight-cut, and continuous-path motion control in NC machines?
7. In NC coding, what is meant by "bit," "word," and "block"?
8. Explain the binary-coded decimal form of coding.
9. Describe the fixed-sequential, tab-sequential, and word-address formats.
10. What is the difference between an open-loop and closed-loop NC system?
11. How does a typical closed-loop system operate?
12. What is a "machining center"? To what factors does it owe its efficiency?
13. In NC systems, differentiate between machine resolution, machine repeatability, and the accuracy achieved in an operation.
14. What is "adaptive control"? How is it commonly used in NC systems?
15. List the advantages of CNC over conventional NC systems.
16. What are the major factors affecting the accuracy of CNC machines? Briefly discuss each.
17. Explain the difference between manual and computer-assisted part programming.
18. What is the name of a processor for computer-assisted NC systems?
19. What is a "post-processor"? Why is it used?

20. What are the advantages of computer-assisted NC programming?
21. What is meant by the statement "NC programming is at least a sequel to process planning"?
22. In describing the performance of a CNC machine, what is meant by machine geometric errors, kinematic errors, and thermal distortion?

31.9 PROBLEMS

1. Rough and finish operations on a workpiece include boring a hole and profiling the front and two sides. For a conventional milling machine, 210 hr is required to design and make a jig, fixture, and template at a toolroom rate of $16/hr. Cycle time is 5.5 hr/piece at a rate of $30/hr. For a numerically controlled machine, 71 hr is required in the toolroom for a fixture; 46 hr at $20/hr for the drawing, process sheet, and punched tape; and 1 hr at $50/hr on an electronic computer. The cycle time is 3 hr/piece at $65/hr. For how many pieces should the numerically controlled system be used?
2. A job can be set up on a standard jig boring machine in 1 hr. Each piece takes 6 hr for boring. If the job is done on a tape-controlled jig boring machine, 2 hr is needed for programming in the office. Then each piece can be bored in 5 hr. The shop rate is $24/hr on the standard jig borer and $28/hr on the numerically controlled jig borer. Office work costs $18/hr. When is the numerically controlled machine justified?
3. The following four machine tools are available.
   (1) An engine lathe at $16/hr. Setup time is 15 min for any job on this machine. To adjust for a rough cut on any diameter takes .2 min, and for a finish cut .78 min on this machine, but not on others. Adjustments must be made for each piece.
   (2) A tracer lathe at $16/hr for which a template must be made for each job at a cost of $50 plus $30 for each diameter turned. Setup time is 20 min.
   (3) An NC lathe at $20/hr on which the setup time is 10 min for any job. Programming for any job requires 30 min plus 15 min for each diameter at a rate of $12/hr.
   (4) An automatic lathe at $18/hr. Setup time is 30 min plus 15 min per diameter. Tools and cams cost $200 plus $50 for each diameter turned. All the diameters on one piece can be turned at once, so the time per piece is the time for the longest cut. However, all pieces in a lot must be rough cut and then finish cut, so extra time of .50 min to unload and load each piece is required.

   A workpiece 50 mm (≈2 in.) in diameter and 75 mm (≈3 in.) long is to be turned at 92 m/min (≈300 ft/min) at a feed of 0.50 mm/rev (.020 in./rev) rough and 0.25 m/rev (.010 in./rev) finish. A 60-min productive hour is assumed.

   (a) For what quantities of pieces would each machine be economical?
   (b) Which machine should be selected for one lot of (1) 1 piece? (2) 25 pieces? (3) 50 pieces? (4) 500 pieces? (5) 1,500 pieces? (6) 2,500 pieces?
4. The machines described in Problem 3 are available for turning a workpiece with the following diameters and cutting times:

Diameter (mm [in.])	Length (mm [in.])	Cutting time (min)
50 (≈2.0)	75 (≈3.0)	.78
57 (≈2.25)	13 (≈.5)	.15
80 (3.15)	50 (≈2.0)	.82
110 (≈4.30)	38 (≈1.5)	.88
120 (≈4.72)	20 (≈.8)	.49

The cutting time is the total for the rough and finishing cuts and includes time for the approach and overtravel. A 60-min productive hour is assumed. Only one setup is required.

(a) For what quantities should each machine be selected?
(b) Which machine should be selected for one lot of (1) 1 piece? (2) 10 pieces? (3) 25 pieces? (4) 100 pieces? (5) 1,000 pieces? (6) 5,000 pieces?

5. A cast-iron compressor body requires 152 operations consisting of 50 holes drilled, 12 bored, 38 co-sunk, 36 tapped, and 16 surfaces milled. These operations can be done on NC machining centers in 1 hr/piece at a rate of $40/hr. Programming and setup costs are negligible. Conventional machines may be tooled for the job at a cost of $300,000. The annual charge for the tools is $130,000, and the rate on the machines is $24/hr, including labor and overhead. It is estimated that about 1 hr will be required per piece by this method. A CNC transfer machine to produce 188 pieces/hr will cost $1,800,000 with an annual charge of $800,000. The labor and overhead rate is $16/hr. For what quantities is each method suitable?
6. Write an APT program for the part shown in Figure 31-42. Cutter diameter is .5 in., suggested spindle speed is 800 rpm, and feed rate is 8 in./min.
7. Write an APT program for the part shown in Figure 31-43. Use the same conditions described in Problem 6.
8. Write the geometry, motion, and other auxiliary APT statements for the part shown in Figure 31-44. Assume the following conditions: the cutting tool is presently at (−1, −1, 0.2); the *Z*-coordinate value of 0 is to be .2 in. above the part surface; spindle speed is 1,200 rpm; feed rate is 8 in./min; coolant is to be ON during actual machining; the cutting tool is in turret holder 5; and the tool diameter is .5 in.
9. Write the APT geometry and motion statements for the parts shown in Figure 31-45.
10. Make a flowchart to show how a routine may enable a computer to calculate the coordinates of holes on a circle.
11. Simulate the program of Figure 31-39 to construct a program for machining a partial cylindrical groove in the top of a block 4 in. × 6 in. × 2 in. high. The axis of the cylinder is parallel to and 5 in. above the bottom of the block in the *Z*-*Y* plane depicted in Figure 31-39. Its radius is 4 in. Specify the cutter path to be ZIGZAG.

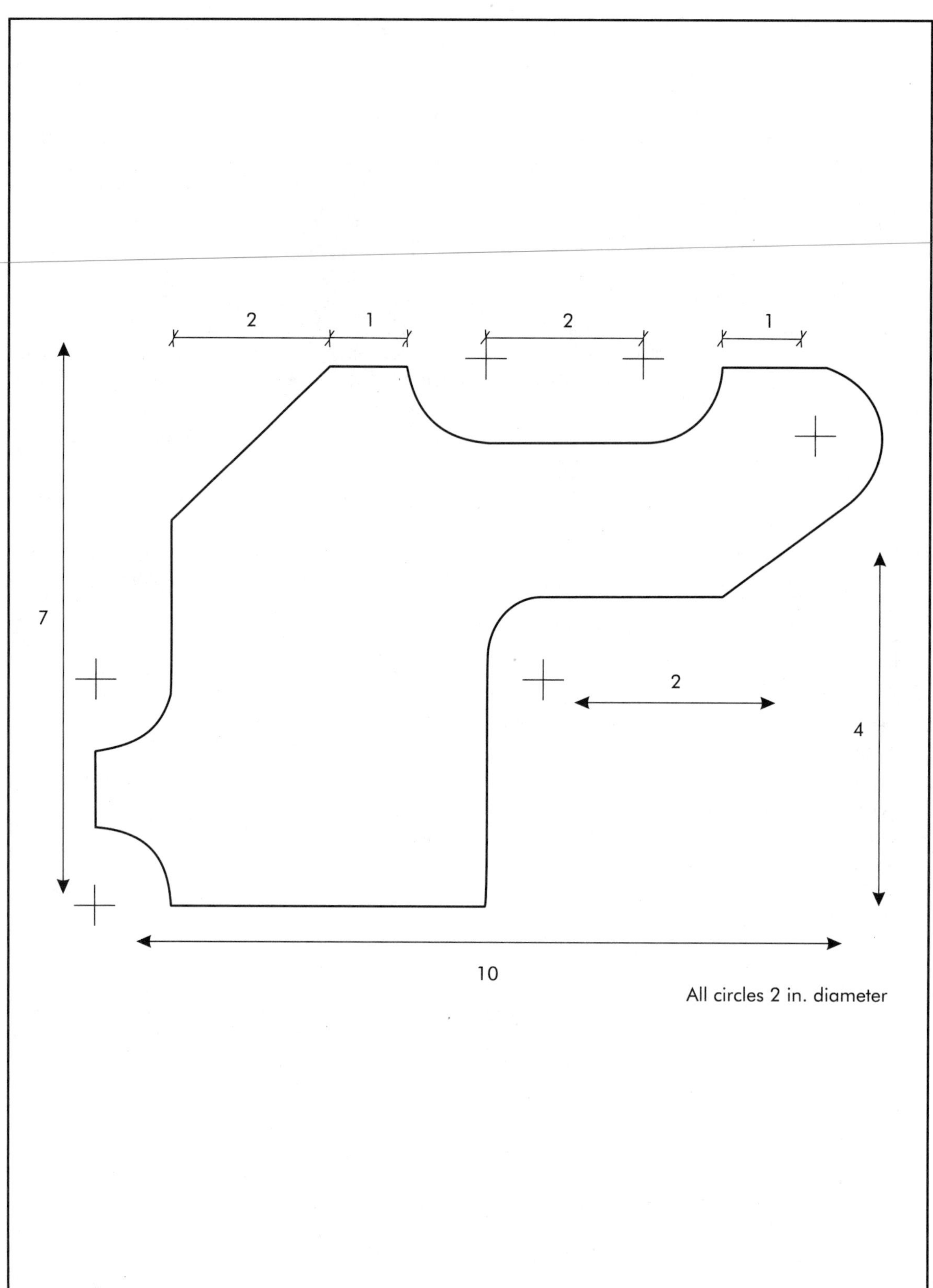

Figure 31-42. Part for Problem 6.

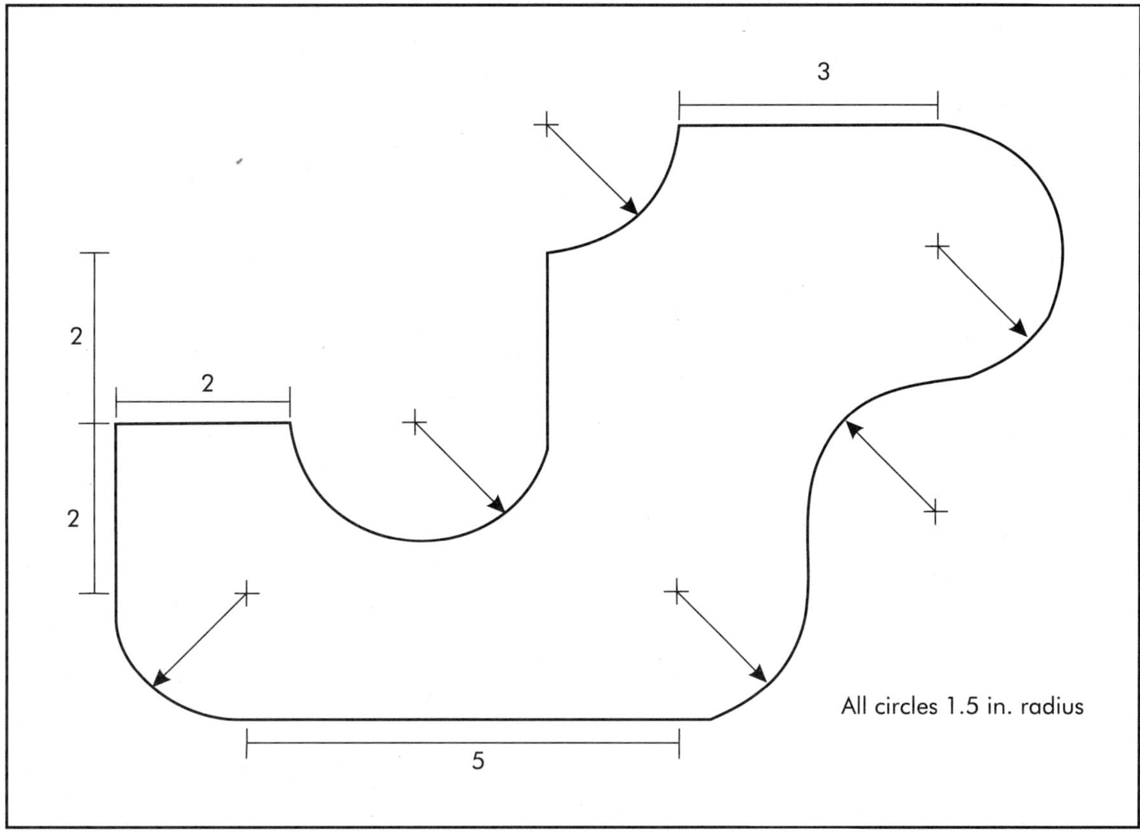

Figure 31-43. Part for Problem 7.

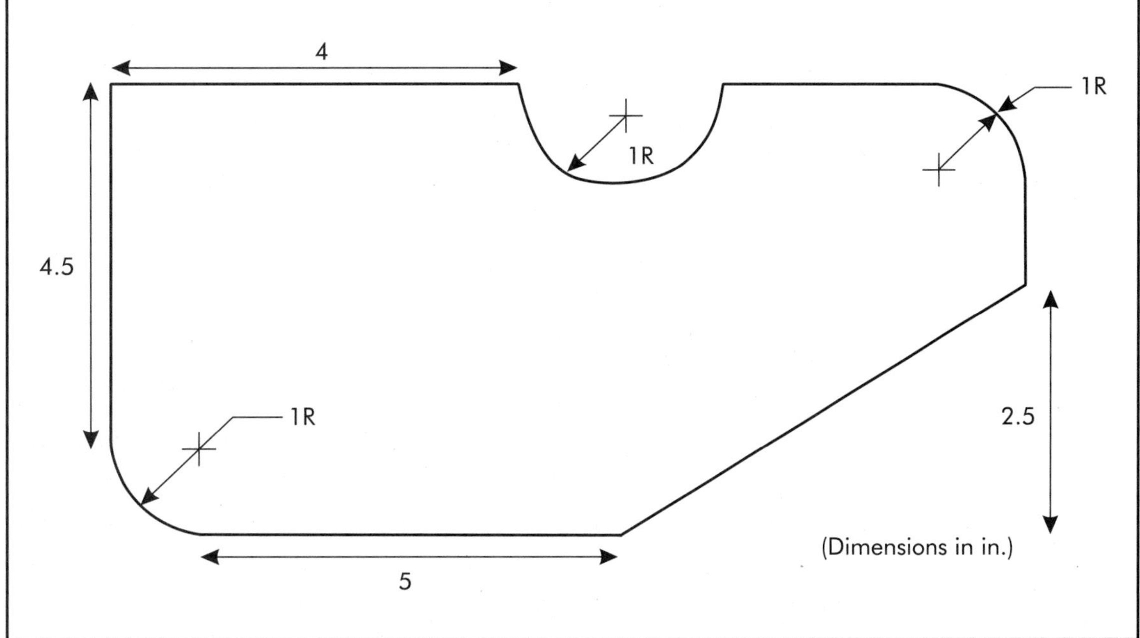

Figure 31-44. Part for Problem 8.

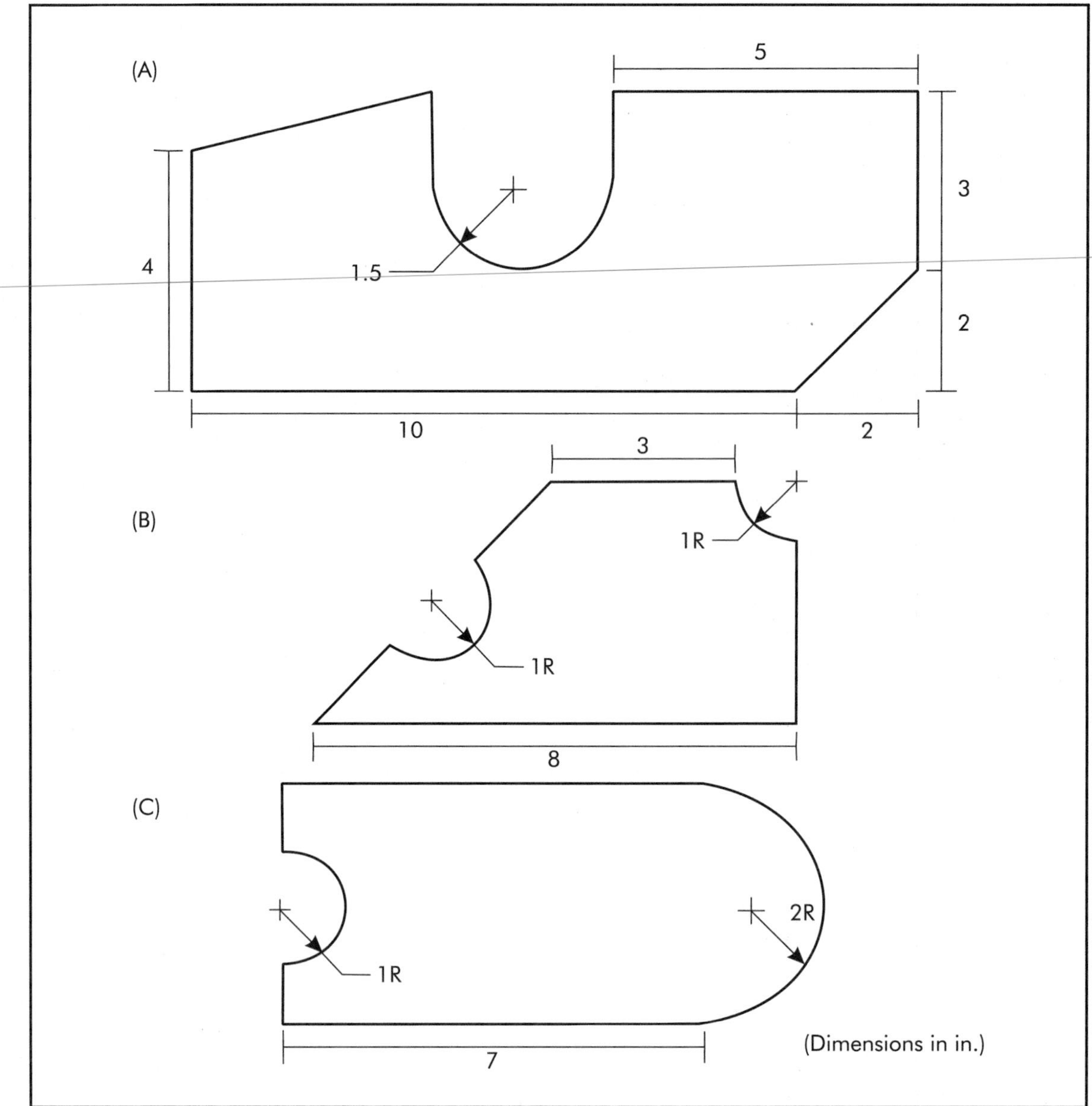

Figure 31-45. Parts for Problem 9.

31.10 REFERENCES

Aronson, R. B. 1994. "Machine Tools: Part 7, Machine Tools of the Future." *Manufacturing Engineering*, July.

Childs, J. J. 1982. *Principles of Numerical Control*, 3rd ed. New York: Industrial Press.

Groover, M. K. 1987. *Automation, Production Systems, and Computer-integrated Manufacturing*. Englewood Cliffs, NJ: Prentice-Hall.

Horath, L. 1993. *Computer Numerical Control of Machines*. New York: Macmillan Publishing Company.

Koren, Y. 1983. *Computer Control of Manufacturing Systems*. New York: McGraw-Hill.

Kral, I.H. 1987. *Numerical Control Programming in APT*. Englewood Cliffs, NJ: Prentice-Hall.

Luggen, W. W. 1984. *Fundamentals of Numerical Control*. Albany, NY: Delmar Publishers.

——. 1991. *Flexible Manufacturing Cells and Systems.* Englewood Cliffs, NJ: Prentice-Hall.

Pollack, H. W. 1990. *Computer Numerical Control.* Englewood Cliffs, NJ: Prentice-Hall.

Rembold, U., Nnaji, B. O., and Storr, A. 1993. *Computer-integrated Manufacturing and Engineering.* Wokingham, England: Addison-Wesley.

"System/370 Automatically Programmed Tools—Advanced Contouring Numerical Control Processor." 1986. *Programming Reference Manual*, 4th ed. White Plains, NY: IBM Corporation.

Index

A

B

C

D

E

F

G

H

J

K

L

M

N

O

P

Q

R

S

T

U

V

W

X

Y

Z